INTELLIGENT SYSTEMS AND AUTOMATION

To learn more about AIP Conference Proceedings,
including the Conference Proceedings Series, please visit the webpage
http://proceedings.aip.org/proceedings

INTELLIGENT SYSTEMS AND AUTOMATION

1st Mediterranean Conference on Intelligent Systems and Automation (CISA '08)

Annaba, Algeria   30 June – 2 July 2008

EDITORS

Hichem Arioui
*University of Evry Val d'Essonne
Evry, France*

Rochdi Merzouki
*Ecole Polytechnique de Lille
Villeneuve d'Ascq, France*

Hadj A. Abbassi
*Badji Mokhtar University
Annaba, Algeria*

All papers have been peer-reviewed.

SPONSORING ORGANIZATIONS
University of Evry, France
National Agency of Research in Universities, Algeria
University of Annaba, Algeria
National Scientific and Research Centre, France
Foreign France Ministry

Melville, New York, 2008
AIP CONFERENCE PROCEEDINGS ■ 1019

Editors

Hichem Arioui
IBISC CNRS 40
Rue du Pelvoux, CE1455
Courcouronnes, 91020
Evry cedex
France
E-mail: hichem.arioui@ibisc.fr

Rochdi Merzouki
LAGIS UMR CNRS 8146
Avenue Paul Langevin –
59655 Villeneuve d'Ascq
France
E-mail: Rochdi.Merzouki@polytech-lille.fr

Hadj A. Abbassi
Automatic and Signals Laboratory
Badji Mokhtar University
Annaba, Algeria
E-mail: Abbassi@univ-annaba.org

Authorization to photocopy items for internal or personal use, beyond the free copying permitted under the 1978 U.S. Copyright Law (see statement below), is granted by the American Institute of Physics for users registered with the Copyright Clearance Center (CCC) Transactional Reporting Service, provided that the base fee of $23.00 per copy is paid directly to CCC, 222 Rosewood Drive, Danvers, MA 01923, USA. For those organizations that have been granted a photocopy license by CCC, a separate system of payment has been arranged. The fee code for users of the Transactional Reporting Services is: 978-0-7354-0540-0/08/$23.00

© 2008 American Institute of Physics

Permission is granted to quote from the AIP Conference Proceedings with the customary acknowledgment of the source. Republication of an article or portions thereof (e.g., extensive excerpts, figures, tables, etc.) in original form or in translation, as well as other types of reuse (e.g., in course packs) require formal permission from AIP and may be subject to fees. As a courtesy, the author of the original proceedings article should be informed of any request for republication/reuse. Permission may be obtained online using Rightslink. Locate the article online at http://proceedings.aip.org, then simply click on the Rightslink icon/"Permission for Reuse" link found in the article abstract. You may also address requests to: AIP Office of Rights and Permissions, Suite 1NO1, 2 Huntington Quadrangle, Melville, NY 11747-4502, USA; Fax: 516-576-2450; Tel.: 516-576-2268; E-mail: rights@aip.org.

L.C. Catalog Card No. 2008927816

CD-ROM ISBN 978-0-7354-0541-7
BOOK ISBN 978-0-7354-0540-0
ISSN 0094-243X

Printed in the United States of America

CONTENTS

Preface xi
Committees xiii
CISA'08 Sponsors and Supports xv

PARALLEL SESSION CONTRIBUTION

SESSION A1: COMPUTER VISION

The Real Time Correction of Stereoscopic Images: From the Serial to a Parallel Treatment 3
    Z. Irki, M. Devy, K. Achour, and M. S. Azzaz

A Three-dimensional Computerized Facial Reconstruction Using Non-linear Registration of a Reference Head 9
    A. Kermi, M. T. Laskri, and I. Bloch

Image Segmentation with Spiking Neuron Network 15
    B. Meftah, A. Benyettou, O. Lezoray, and M. Debakla

Hybrid Feature Extraction-based Approach for Facial Parts Representation and Recognition 20
    C. Rouabhia and H. Tebbikh

SESSION A2: INTELLIGENT TRANSPORTATION SYSTEM

Robust Second Order Sliding Mode Observer for the Estimation of the Vehicle States 27
    A. Chaibet, L. Nouveliere, S. Hima, and S. Mammar

Modeling of Steer-by-Wire System Used in New Braking Handwheel Concept 33
    K. Messaoudène, N. Ait Oufroukh, and S. Mammar

Sensor Position Identification and Vehicle State Estimation Using the Extended Kalman Filter 41
    H. Slimi, C. Sentouh, S. Mammar, and L. Nouveliere

Two Wheeled Vehicle Dynamics Synthesis for Real-time Applications 47
    S. Hima and H. Arioui

Intelligent Transportation Systems: Automated Guided Vehicle Systems in Changing Logistics Environments 53
    H. L. Schulze, S. Behling, and S. Buhrs

SESSION B1: ROBUST CONTROL

An Initial Corrector Using Different Approaches for Youla Parametrization via LMI Optimization 61
    S. Ziani and S. Filali

Guaranteed Trajectory Tracking of a Small-Size Autonomous Helicopter in a Smooth Uncertain Environment 67
    T. Cheviron, A. Chriette, and F. Plestan

Robust Control of an Uncertain Physical Process 74
    B. Benyahia, A. Choukchou-Braham, and B. Cherki

Stabilization of an Under-actuated Mechanical System by Sliding Mode Control 80
    A. Choukchou-Braham, C. Bensalah, and B. Cherki

The Observer Adaptive Backstepping Control for a Simple Pendulum 85
    M. Mokhtari, N. Golea, and S. Berrahal

SESSION B2: INTELLIGENT CONTROL

Usage of the HMM-based Speech Synthesis for Intelligent Arabic Voice 93
    T. S. Fares, A. H. Khalil, A. El-Fatah, and A. Hegazy

Hybrid Takagi-Sugeno Fuzzy FED PID Control of Nonlinear Systems 99
    B. Hamed and A. El Khateb

Adaptive Fuzzy Control of a Class of SISO Nonaffine Nonlinear Systems 103
    S. Doudou and F. Khaber
New Stability Conditions of Takagi-Sugeno Fuzzy Systems via LMI 109
    F. Bourahala and F. Khaber
Backstepping Control Augmented by Neural Networks for Robot Manipulators 115
    M. Belkheiri and F. Boudjema
Direct Torque Control System for Permanent Magnet Synchronous Machine with Fuzzy Speed PI Regulator 120
    K. Nabti, K. Abed, and H. Benalla

SESSION C1: MODELING AND IDENTIFICATION

Optimal Reduced-order Approximation of Fractional Dynamical Systems 127
    R. Mansouri, M. Bettayeb, and S. Djennoune
Parameters Identification for Motorcycle Simulator's Platform Characterization 133
    L. Nehaoua and H. Arioui
Temperature Effects on Chemical Reactor 139
    M. Azzouzi
Design of Full-band and Low-pass FIR Differentiators: A Comparative Study 143
    C. Mekhnache, Y. Ferdi, and A. Taleb-Ahmed

SESSION C2: VIRTUAL REALITY AND INTERFACES

Design of a 3D Navigation Technique Supporting VR Interaction 149
    P. Boudoin, S. Otmane, and M. Mallem
Force Feedback Control of Robotic Forceps for Minimally Invasive Surgery 154
    C. Ishii and Y. Kamei
Using Virtual Reality to Dynamically Setting an Electrical Wheelchair 156
    S. Dir, O. Habert, and A. Pruski
A Planning Architecture for Mobile Robotics 162
    J. Guitton, J.-L. Farges, and R. Chatila
Control by Sliding Mode of a Trajectory Follow-up for a Mobile Robot 168
    S. Berrahal, D. Ameddah, and M. Mokhtari
Modeling and Analyzing Mixed Reality Applications Using Timed Automata 173
    J.-Y. Didier, B. Djafri, and H. Klaudel

SESSION D1: PREDICTIVE CONTROL

Predictive Direct Torque Control for Induction Motor Drive 181
    A. Benzaioua, M. Ouhrouche, and A. Merabet
Parameter Tuning of Fractional $PI^\lambda D^\mu$ Controllers with Integral Performance Criterion 186
    K. Bettou and A. Charef
Preparation Model Based Control System for Hot Steel Strip Rolling Mill Stands 191
    S. E. Bouazza, H. A. Abbassi, and A. K. Moussaoui
Adaptive Fuzzy Hysteresis Band Current Controller for Four-Wire Shunt Active Filter 197
    F. Hamoudi, A. Chaghi, H. Amimeur, and E. Merabet
Model Predictive Control of the Permanent Magnet Synchronous Motor in State Space with Input Constraints 203
    Lh. Arab, A. Belemhedi, M. A. Ahmed, and N. Habani

SESSION D2: INTELLIGENT CONTROL

FPGA Implementation of Multilayer Perceptron for Modeling of Photovoltaic Panel 211
    H. Mekki, A. Mellit, H. Salhi, and K. Belhout

Hybrid Approach to Reinforcement Learning 216
    B. Boulebtateche, M. Fezari, and M. Boughazi

Neural and Fuzzy Adaptive Control of Induction Motor Drives 221
    Y. Bensalem, L. Sbita, and M. N. Abdelkrim

Neural Network Identification for a C5 Parallel Robot 228
    M. E. Daachi, B. Achili, B. Daachi, and D. Chikouche

Extended Kalman Filter Based Neural Networks Controller for Hot Strip Rolling Mill 234
    A. K. Moussaoui, H. A. Abbassi, and S. Bouazza

SESSION E1: COMPUTER VISION

A Direct Frequency-based Phase Algorithm for Motion Estimation 243
    M. Boughazi, B. Boulebtateche, M. Fezari, L. Zouaoui, and N. Bonnet

MRI Images Compression Using Curvelets Transforms 249
    M. Beladgham, I. B. Hacene, A. Taleb-Ahmed, and M. Khélif

Recognition of Handwritten Arabic Words Using a Neuro-Fuzzy Network 254
    A. Boukharouba and A. Bennia

Indexing Color Images Using Color Band Moments and a Relevance Feedback 260
    M. Ayad, A. Bessaid, H. Bechar, and A. Taleb-Ahmed

SESSION E2: INTELLIGENT AND FLEXIBLE MANUFACTURING

On-line Scheduling of Automatics and Flexible Manufacturing System Using SARSA Technique 269
    N. Aissani and B. Beldjilali

Intelligent Production Monitoring and Control Based on Three Main Modules for Automated Manufacturing Cells in the Automotive Industry 275
    U. Berger, R. Kretzschmann, and A. V. Vargas

Synthesis of a Discrete Controller Based on Grafcet 281
    H. Hamdi, H. Alla, and I. El Ani

Decentralized Method for Complex Task Allocation in Massive MAS 287
    Z. Brahmi and M. M. Gammoudi

SESSION E3: PRODUCT DESIGN AND ROBOTICS

Antenna Automation for NOAA Satellite Images Reception 297
    W. L. Rahal, N. Benabadji, and A. H. Belbachir

On the Partial Attitude Control of Axi-Symmetric Rigid Spacecraft 302
    C. Jammazi

VLSI Cells Placement Using the Neural Networks 308
    H. Azizi, L. Zouaoui, and S. Mokhnache

3D Curves with a Prescribed Curvature and Torsion for a Flying Robot 313
    Y. Bestaoui

Fuzzy Visual Path Following by a Mobile Robot 319
    A. Hamissi and A. Bazoula

Integration of Additional Constraints to Inverse the Differential Kinematic Model for a Nonholonomic and Redundant Mobile Manipulator 324
    I. Akli and N. Achour

SESSION F1: MOTION AND ROBOST CONTROLS

Numerical Scheme for Viability Computation Using Randomized Technique with Linear Programming 333
    B. Djeridane

Nonlinear Control of the Doubly Fed Induction Motor with Copper Losses Minimization for Electrical Vehicle 339
    S. Drid, M.-S. Nait-Said, M. Tadjine, and A. Makouf

Sub-Optimal Motion Planner of Wheeled Mobile Manipulators with Under-Actuated Platform 346
    M. Haddad, T. Chettibi, H. E. Lehtihet, W. Khalil, and F. Boyer

Minimum Time Trajectory Planning for a Micro Quadrotor Aerial Robot 353
    Y. Bouktir and T. Chettibi

SESSION F2: ADAPTIVE CONTROL AND FAULT DETECTION

Comparative Study of Adaptive Type-1 and Type-2 Fuzzy Controls for Nonlinear Systems under Uncertainty 363
    S. Mokaddem and F. Khaber

Control of Perturbed Systems in the Frequency Domain 369
    K. Ghedjati and M. Abdelaziz

FOCOVE: Formal Concurrency Verification Environment for Complex Systems 375
    D. E. Saïdouni, A. Benamira, N. Belala, and F. Arfi

Data Reconciliation and Gross Error Detection: A Filtered Measurement Test 381
    Y. Himour

Intelligent Diagnosis of Degradation State under Corrosion 383
    D. Isoc, A. Ignat-Coman, and A. Joldiş

SESSION F3: VISUAL SERVOING AND MODELING

Kinematic Visual Servo Controls of an X4-Flyer: Practical Study 391
    O. Bourquardez, N. Guenard, T. Hamel, F. Chaumette, R. Mahony, and L. Eck

Modelling and Control of Flexible Airship 397
    S. Bennaceur, A. Abichou, and N. Azouz

A Technique Combining Optimal Filtering and Periodicity 408
    S. Zenati and A. Boukrouche

Telerobotics Using a Gestural Servoing Interface 414
    H. Abdelmoumene and N. E. Berrached

Driving Forces Study in the Ultrasonic Motor 420
    M. Djaghloul, Z. Boumous, and S. Belkhiat

A Behavior-Based Visual Servoing Control Law 427
    M. Marey and F. Chaumette

Stochastic Wireless Channel Modeling, Estimation and Identification from Measurements 433
    M. M. Olama, Y. Li, S. M. Djouadi, and C. D. Charalambous

A Hybrid Approach for MILP Partitioning Problem 439
    R. Boudour, D. Farfar, and M. T. Kimour

POSTER SESSION
(Organizer: J-Y Didier)

Classical Control System Design: A Non-graphical Method for Finding the Exact System Parameters 447
    M. T. Hussein

Application of GMMs to Speech Recognition Using Very Short Time Series 450
    S. Friha and N. Mansouri

Particle Swarm Optimization for Image Deblurring 454
    A. Toumi, A. Taleb-Ahmed, K. Benmahammed, and N. Rechid

Improved Topology Control Algorithm for MANETs 460
    H. Hamad

Training RBF Networks Using DDA Algorithm Combined with Genetic Algorithm 466
    G. Khensous, B. Messabih, and N. Benamrane

A Sliding Mode Controller Using Nonlinear Sliding Surface Improved with Fuzzy Logic: Application to the Coupled Tanks System 470
    A. Boubakir, F. Boudjema, and C. Boubakir

Presentation of a GIS for Road Network 476
    B. Benarbia and N. Berrached

Solving Capelin Time Series Ecosystem Problem Using Hybrid ANN-GAs Model and Multiple Linear Regression Model 482
    K. M. Eghnam and A. F. Sheta

Automatic Generation of Observers for the Dala Robot with TTG 487
    S. Bensalem, M. Bozga, M. Gallien, F. F. Ingrand, M. Krichen, and S. Tripakis

3D Molecular Modeling Systems: State of Art 493
    M. Essabbah, S. Otmane, and M. Mallem

A Strategy for Unicycle's Formation Control Based on Invariance Principle 498
    M. A. El Kamel, L. Beji, and A. Abichou

Development of Algorithms for the Classification of the Benign and Malignant Tumors 503
    L. Zouaoui, H. Azizi, M. Boughazi, and H. Akdag

Improvement of the Performances of the Genetic Algorithms by Using an Adaptive Search Space Reduction and the Transformtion 507
    L. Yousfi and N. Mansouri

Indirect Adaptive Fuzzy Power System Stabilizer 512
    K. Saoudi, Z. Bouchama, M. N. Harmas, and K. Zehar

Improvement of Arab Digits Recognition Rate Based in the Parameters Choice 516
    C. Hadri, M. Boughazi, and M. Fezari

Fuzzy Neural Order Robust of the Non-Linear Systems 520
    F. Madour and K. Benmahammed

A Verbal Guidance System for Severe Disabled People 526
    A. Redjati and M. Bousbia-Salah

Fabrication of Wall Shear Stress Sensor for Micro Flow Measurement 531
    M. Laghrouche, A. Adane, J. Boussey, S. Ameur, D. Meunier, and M. Tahanout

Proposal and Study for an Architecture Hardware/Software for the Implementation of the Standard H264 536
    K. Messaoudi, S. Toumi, E. Bourennane, and M. Boutalbi

Estimation of Contact Forces of a Four-wheel Steering Electric Vehicle by Differential Sliding Mode Observer 541
    K. Bouibed, A. Aitouche, A. Rabhi, and M. Bayart

Management Traceability Information System for the Food Supply Chain 547
    S. Bendriss, A. Benabdelhafid, and J. Boukachour

E-Maintenance Scenarios Based on Augmented Reality Software Architecture 553
    S. Benbelkacem, N. Zenati-Henda, and M. Belhocine

Numerical Approximation of Null Controllability for 1-D Linear Parabolic-Hyperbolic Equations 559
    A. Salem

Interacting with a Virtual Tool on a Real Object 563
    B. Bayart, J.-Y. Didier, and A. Kheddar

A Robust and Non-blind Watermarking Scheme for Gray Scale Images Based on the Discrete Wavelet Transform Domain 565
    A. Bakhouche and N. Doghmane

PLENARY SESSION PRESENTATIONS
(Abstracts only)

Higher Order Sliding Modes Observation, Identification, and Fault Detection 573
    L. Fridman

Integrated Design of Mechatronic Systems 574
    G. Dauphin-Tanguy

Recent Results in Visual Servoing 575
    F. Chaumette

Design of Supervision Systems: Theory and Practice 576
    B. O. Bouamama

Applications in Robotics and Controls 577
    K. Youcef-Toumi

Some Challenging Issues in Humanoid Robotics 578
    A. Kheddar

Computer Assisted Surgery and Current Trends in Orthopaedics Research and Total Joint Replacements 579
    F. Amirouche

Author Index 581

Preface

Welcome to the first edition of CISA, Mediterranean Conference on Intelligent Systems and Automation, 2008, held in Annaba, Algeria. The organizers of this event are the IBISC, CNRS and University of Evry (France) joint laboratory, and LASA laboratory of Autoamtic and signals of Badji Mokhtar University (Algeria).

This conference will give to researchers from Mediterranean region and also from all over the world, the opportunity to discuss on the new challenges of the scientific research in the fields of robotics and automation. This can be realised by increasing the multilateral collaborations, as well as to share opinions on the new directions and the actions to be led to encourage the research area in countries of the south Mediterranean region.

During this edition, 231 full papers were submitted to the conference and 108 of them were selected for presentation and publication. This is a significant participation and acceptance rates compared to the traditional event on the same topics. With not less 18 represented nationalities, we have covered the major scientific countries in the world. We hope that this first edition will be durable for the future, by the participation of more Asian and American researchers in the conference. Thus, one of the most important features of the conference will be reached, namely organization of meetings and exchanges of experience in different areas between experimented scientists and students from all over the world.

We thank our sponsors: Badji Mokhtar University and its President; Evry Val d'Essonne and its President, National Centre of Scientific Research (CNRS), French Embassy at Algiers, for their generous support.

We would like to express our best regards to all the authors and members of the program and local organizing committees that has made this conference a real success.

We especially thank Professors. Farid Amirouche, François Chaumette, Genevieve Dauphin-Tanguy, Leonid Fridman, Abderrahmane Kheddar, Belkacem Ould Bouamama and Kamal Youcef-Toumi for their fruitful presentations, encouragement, support and guidance during the preparation of the special supplement.

Hichem ARIOUI	Rochdi Merzouki	Hadj Ahmed Abbassi
Associate Professor	Associate Professor	Full Professor
IBISC-CNRS FRE 3190	LAGIS-CNRS UMR 8146	Head LASA Laboratory
Evry, France	Villeneuve d'Ascq, France	Annaba, Algeria

CISA'08 Organizing and Scientific Committees

General chair
Hadj Ahmed Abbassi, Algeria,
Hichem Arioui, France.

Honorary Chair
Mohamed Tayeb Laskri, Algeria,
Kamal-Youcef Toumi, United States.

Program committee
Program chair
Mohamed Mourad Lafifi, Algeria,
Rochdi Merzouki, France.

Awards
Nasreddine Debbache, Algeria,
Etienne Colle, France.

Publication
Brahim Boulebtateche, Algeria,
Salim Hima, France,
Adel Merabet, Canada
Messaoud Ramdani, Algeria.

Posters and demo
Naima Ait Oufroukh, France,
Mounir Bousbia Salah, Algeria.

Co-general chair
Mohamed Larbi Saidi, Algeria,
Nicolas Séguy, France.

Organizing committee
Organizing chair
Djemil Messadeg, Algeria.

Web and registration
Jean-Yves Didier, France,
Mohamed Fezari, Algeria.

Finance
Sophie Abriet, CNRS, France
Salah Kermiche, Algeria.

Publicity
Ali Amouri, France,
Salah Bensaoula, Algeria.

Local arrangements and social events
Rabah Lakel, Algeria,
Lamri Nehaoua, France,
Djeghaba Messaoud, Algeria.
Mohamed Boughazi, Algeria,

International Scientific committee

Fakhr-eddine Ababsa, France,
Bessam AbdulRazak, United States,
Azgal Abichou, Tunisia,
Omar Ait-Aider, France,
Mouldi Bedda, Kingdom of Saudi Arabia,
Lotfi Beji, France,
Khier Benmahammed, Algeria,
Yasmina Bestaoui, France,
Maamar Bettayeb, United Arab Emirates,
Mohamed Seghir Boucherit, Algeria,
François Chaumette, France,
Yacine Chitour, France,
Abdelhamid Chriette, France,
Boubaker Daachi, France,
Nourdinne Doghmane, Algeria,
Stéphane Espié, France,
Zhi-qiang Feng, France.
Paolo Fiorini, Italy,
Leonid Fridman, Mexico,
Tarek Hamel, France,

Pierre Joli, France,
Abderrahmane Kheddar, Japan,
Jean Lerbet, France,
Prabhat Kumar Mahanti, Canada,
Robert Mahony, Australia,
Said Mammar, Netherlands,
Mourad Nouicer, Algeria,
Lydie Nouvellière, France,
Yuri Orlov, Mexico,
Samir Otmane, France,
Mohand Ouhrouche, Canada,
Belkacem Ould-Bouamama, France,
Shahram Payandeh, Canada,
Edwige Pissaloux, France,
Hocine Rezine, Algeria,
Zaki Sari, Algeria,
Philippe Soueres, France,
Olaf Stursberg, Germany,
Abdelmalik Taleb-Ahmed, France,
Costas Tzafestas, Greece,
Mimoun Zelmat, Algeria,

CISA'08 Sponsors and Supports

 University of Evry Val d'Essonne, France

 National Agency of Research in Universities (ANDRU), Algeria

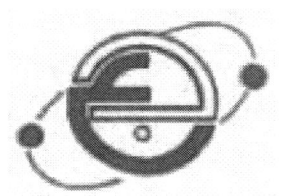 University of Annaba, Algeria

 National Scientific and Research Centre (CNRS), France

 Foreign France Ministry

PARALLEL SESSION CONTRIBUTION
SESSION A1: COMPUTER VISION

The Real Time Correction of Stereoscopic Images: From the Serial to a Parallel Treatment

Zohir IRKI*, Michel DEVY**, Karim ACHOUR***, Mohamed Salah AZZAZ*

*Ecole Militaire Polytechnique,
BP17Bordj ElBahri, Alger, Algérie*
**LAAS: 07Avenue du Colonel Roch, Toulouse, France*
***CDTA: Baba H'sen, Alger, Algérie*

zohir_irki@yahoo.fr

Abstract: The correction of the stereoscopic images is a task which consists in replacing acquired images by other images having the same properties but which are simpler to use in the other stages of stereovision. The use of the pre-calculated tables, built during an off line calibration step, made it possible to carry out the off line stereoscopic images rectification. An improvement of the built tables made it possible to carry out the real time rectification.
In this paper, we describe an improvement of the real time correction approach so it can be exploited for a possible implementation on an FPGA component. This improvement holds in account the real time aspect of the correction and the available resources that can offer the FPGA Type Stratix 1S40F780C5.
Keywords: Radial Distortions, Corrected Image, Distorted Image, Correction's Tables, Real Time.

1. INTRODUCTION

The stereoscopic images are images perceived of the same scene but acquired from different points of view. Often, because of a bad camera's positioning and/or of some technical defects related to their design, the acquired images need to be improved before sending them to later treatment stages.

Among these improvements, the more important one is called the rectification, i.e. the correction which aims at replacing the two original images by two other ones, giving same photometric information on the perceived scene, but having properties adapted for the two principal stages of the stereovision: (1) the 3D point reconstruction requires that the two cameras respect perfectly the classical pinhole model, i.e. optical rays are not disturbed by optical distortions, and (2) the matching between corresponding pixels (pixels, projections of the same 3D point on the left and right images).is made simpler and faster if it is guaranteed that a left pixel (u,v) can be only matched with a right one on the same line, so with coordinates (u, v'). The difference (v-v'), called disparity, is in inverse ratio of the depth of the projected 3D point.

This work deals with an improvement of the real time correction technique of stereoscopic images proposed in [4] and developed starting from an off line rectification approach requiring memorizing parts of original images. The purpose of this improvement is the correct management of the memory resources so that the proposed method can be implemented on an FPGA component Type Stratix 1S40F780C5.

This paper will be started by presenting the correction of stereoscopic images. Then a description of the real time correction approach is given before passing to the improvements made to this approach in order to adapt it to a possible implementation. By the end, some simulation results will be presented.

2. PRINCIPLE OF THE STEREOSCOPIC IMAGES CORRECTION

The correction of stereoscopic images aims at replacing the two original images by two other images giving same photometric information on the perceived scene, but having properties adapted for the two principal stages of stereovision:

- The 3D point reconstruction starting from the image coordinates of two pixels (u_1,v_1) and (u_2,v_2) taken in the left and the right image;
- The mapping between these pixels.

Correction can be subdivided in two tasks: the alignment of captured images and the correction of the distortions in images.The alignment of images consists in carrying out changes on the original images so that a point, of the scene, perceived in the two images has the same line coordinate. This task can be done easily by manipulating the matrices which represent the system of cameras.

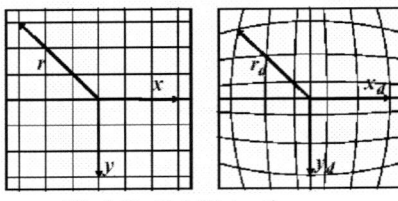

Fig.1 Radial Distortions

The complex task in correction is the correction of the radial distortions. So the problem of the real time correction can be summarized to correct the radial distortions in original images and this in the real time.

The radial distortions come from the defects of curve on the lenses of the camera's objective. These distortions appear in the image plan like a translation along the ray joining the projection of the optical center (not far from the center of the image) and the considered pixel.

On the image plan, in metric coordinates, we distinguish:
- The ideal coordinates (x, y), which correspond to the perfect projection;
- The Distorted coordinates (x_d, y_d).

The relation between these coordinates is given by the system of equations:

$$\begin{cases} x_d = x.(1 + k_1 r^2 + k_2 r^4 + k_3 r^6) \\ y_d = y.(1 + k_1 r^2 + k_2 r^4 + k_3 r^6) \end{cases} \quad (1)$$

Where:
- k_i : Are the coefficients of distortion on the metric coordinates. These coefficients are given following the calibration of the stereoscopic bench.
- $r^2 = x^2 + y^2$: Is the distance to the principal point.

The off line correction [1] consists in calculating, for each pixel (u,v) of the corrected image, the position of the corresponding distorted pixel (u_d, v_d) of the original image. Then to allot the value of brightness's function (grey scale) of the distorted pixel to the corrected one.

This operation consists of the sweeping of the corrected image and to resolve the system of equations refer in (1) in order to determine the distorted position corresponding to the corrected position. By considering that the size of the corrected image is equal to the size of the distorted one; two cases can be distinguished for a pixel of the corrected image:
1. The corresponding position in the distorted image is out the limits of the image. This is due to the fact that one of the coordinates pixel is either negative or higher than the maximum value. This case arises especially for edge's pixels. In this case, the grey scale value of the corrected pixel is taken null.
2. The corresponding position in the distorted image is limited by the edges of the distorted image. In this case, the grey scale of the calculated position is allotted to the corrected pixel.

The calculated corresponding position can take an integer value as it can take a real one. However on an image, the function of brightness is known in integer pixels. This fact carries out us to deduce the value of brightness's function in real positions starting from the values known in the integer positions. This value can be:
- The value of the nearest integer pixel [2]. In this case, we speak about approximation;
- Calculated starting from the four nearest integer values. In this case, we use a bilinear interpolation [3].

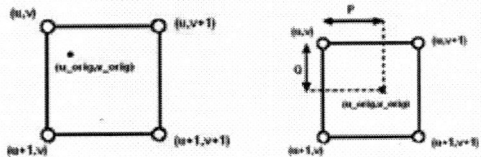

Fig.2. Approximation and Interpolation

In this technique, to each corrected pixel **only one** distorted pixel corresponds. As the treatment will be done on a video signal (flow of images) and for reasons of implementation on a programmable component [5], the memorizing of the couples (*corrected pixel, distorted pixel*) will make it possible to build a correction's table and to avoid calculation with each new image.

According to the method to assign the grey scale value to a corrected pixel, the elements of this table can take several representations. The simplest is the representation which associates to any corrected pixel (u,v), the nearest integer distorted pixel (u_d, v_d) to its correspondent.

Another representation consists in representing, in addition to the correspondent, the weights of the interpolation.

The built table will be known as ***indirect*** correction's table because it makes it possible to build the rectified image starting from a sweeping in the indirect direction (from the rectified image towards the distorted image). The use of this table requires memorizing certain lines of the distorted image. The number of lines which must be memorized depends on the nature of the distortion. Thus, this method is qualified as an off line method [3].

3. THE REAL TIME CORRECTION

In opposite to the off line method, where to each corrected pixel can correspond one distorted pixel in maximum, the real time approach supposes that a distorted pixel can correspond to several corrected pixels. This assumption is due to the approximation of the system's solution towards the nearest integer pixel. So, it's necessary to determine for each distorted the set of the corrected pixels to which it corresponds.

Also, when memorizing data during the off line method is made by using an indirect table (in reality, this table is composed by two tables having the same nature, the first used to memorise lines coordinates, the second for columns coordinates), the memorizing data during real time approach can be done, as we will see, by using two tables of different nature. The tables used for the memorizing of the relation (***distorted pixel, corrected pixels***) will be known as **direct Tables**. To distinguish between these tables, we will call the first Table of the ***address***. The second will be called table of ***coordinates***. The size of the address's table is equal to the number of the pixels of the distorted image. The size of the table of coordinates is initially unknown; it will be deduced after having constructed the address's table.

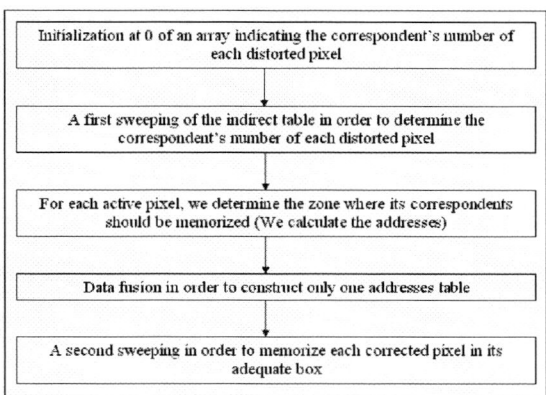

Fig.3. Steps of the Real Time Rectification

In what follows, we recall how to build these tables of correction.

After the initialization at 0 of an array of counters associated to the pixels of the distorted image, we use the built indirect tables to determine for each distorted pixel the number of its corresponding corrected pixels. This is done by incrementing the counter associated to a distorted pixel (u_i, v_i) each time where this pixel is met on the indirect tables. It's clearly understood that in this approach, a first sweeping on the indirect tables is essential. At the end of this sweeping, for each distorted pixel, the number of the corresponding corrected pixels will be known. Thus, a distorted pixel can:
1. Have one or several correspondents;
2. Have no correspondents.

In the first case, the distorted pixel will be called an **Active** pixel because it will contribute to calculate the grey scale of a set of corrected pixels. In the other case, the pixel will be called a **Passive** pixel. So, the first stage of the real time approach will make it possible to classify the distorted pixels in two categories: Active and Passive.

For an active pixel, the number of the correspondents is a unique factor, so it's characteristic. But, since we must know all these correspondents, this factor remains insufficient for the completely characterise a distorted pixel. For that, it is necessary important to know the emplacement of these pixels in the table of correspondents.

When we memorize the correspondents, a zone (in the table of the coordinates) is reserved for each active pixel. The size of this zone is equal to the number of correspondents of the considered distorted pixel. In this zone, correspondents will be placed one after other. A zone is distinguished by the address of its first element. Knowing that active pixels are numbered according to the order of their appearance, and then the first zone, distinguished by the address 0, will be reserved to the first active pixel. To the ith active pixel, we reserve the zone distinguished by the address:

$$Ad_i = \sum_{j=1}^{i-1} N_j - 1 \qquad (2)$$

Where N_j represents the number of correspondents of the jth active pixel.

So knowing the number of the correspondents of an active pixel as well as the address where its first correspondent is placed, we will be able to know all its other correspondents. Thus, a distorted pixel is characterized by a number and an address if it is active.

For not being obliged to distinguish between an active pixel and a passive one, a fusion between the number of correspondents and the address will be made. This fusion will make it possible to generate the final table of the addresses in which one information is composed by two informations. The purpose of this fusion is to optimize the space of memorizing.

During programming tests, it was noted that the maximal number of correspondents for an active pixel is 7. This observation and the fact of coding the number of correspondents using 4 bits push us to make fusion as follows:

$$F_{caracteristic} = N_{correspondent} + 16 * Ad_{correspondent} \qquad (3)$$

Where:
- $F_{cacarcteristic}$: The characteristic factor of a distorted pixel;
- $N_{correspondent}$: The number of correspondents of a pixel;
- $Ad_{correspondent}$: The address of memorizing of the first correspondent.

At the end of this stage, we will be able to calculate the size of the table of correspondents. In reality, the table of correspondents is destined to memorize all the correspondents of the active pixels. So the size of this table is given by:

$$T = \sum_{i=1}^{NB_{pa}} N_i \qquad (4)$$

Where:
- NB_{pa}: The number of the active pixels;
- N_i: The number of correspondents of the ith active pixel.

After having calculated the size of the table of correspondents, it is necessary to clarify the elements of this table. It should be noted that the table of correspondents has the same nature that the indirect table, it is composed of two tables.

The task of clarifying elements is carried out by a second

sweeping of the indirect table. For this, we must initialise at 0 an array of counters. The size of this array is equal to the number of the active pixels. So for each active pixel we associate an element of this array. This array will be used to determine the position where to memorize a corrected pixel correspondent of an active pixel.

During the second sweeping of the indirect table, we note that the corrected pixel (U_c, V_c) corresponds to the distorted pixel (U_d, V_d). The values U_c, V_c will be memorized in the position indicated by the address:

$$Ad = Ad(U_d, V_d) + Cp(U_d, V_d) \quad (5)$$

Where:
- $Ad(U_d, V_d)$: The address where is memorized the first correspondent of the distorted pixel (U_d, V_d), this address is known because it has been calculated at the time of the preceding stage;
- $Cp(U_d, V_d)$: The current value of the position corresponding to the distorted pixel (U_d, V_d) in the array of counters.

After memorizing of the corrected pixel, we must update the value of $Cp(U_d, V_d)$ such as:

$$Cp(U_d, V_d) = Cp(U_d, V_d) + 1 \quad (6)$$

This update will make it possible to memorize the next correspondent of the distorted pixel (U_d, V_d) in the adequate position.

At the end of the second sweeping, all the corrected pixels will be memorized in their adequate positions related to their distorted correspondent pixels. Here, all the necessary tables for the real time correction will be built. The number of these tables is 2:
1. Table of the addresses intended to memorize the number of the correspondents and the address where is memorized the first correspondent of a distorted pixel.
2. Table of correspondents intended to memorize the correspondents only for the active pixels.

The following figure shows the result of the application of the real time approach.

original image　　　　　　　　corrected image

Fig.4. Correction of the Radial Distortions

The application of the method appears like a suppression of a zone from the image. This is due to the fact that some distorted pixels do not have correspondents in the corrected image. The distortion is definitely visible on the straight lines present in the original image. We notice that the curve of these straight lines is disappeared after the correction.

4. FROM THE SERIAL TO THE PARALLEL TREATMENT

In what follow, we will evaluate the memory space required to memorize data on a component and this for the treatment of an image having a size of 572x768. For that, we will make assumptions to raise the memory size. These assumptions are [3]:
- The size of the corrected image is equal to the size of the original image;
- Each pixel of the corrected image has at least one correspondent in the original image.

These conditions force the tables of correspondents to have the same size that the corrected image. The size of the table of the addresses is of course equal to the size of the original image.

The memory space required would be thus the space occupied by the table of the addresses plus that occupied by the two tables of the correspondents, or
- 572*768*32=14057472 bits for the table of the addresses;
- 572*768*16*2=14057472 bits for the two tables of the correspondents.

The total required memory is more than **28 Mbits.**
For a stereoscopic application, the Look up Tables (LUT) of the left image are not the same of the right image. This fact complicates the requirement in memory. The required memory size has to be doubled.

The component with which we suppose to work is a component FPGA Altera in which the memory resources are approximately 3 Mbits. For the current application the needed space is approximately more than **56 Mbits**. Thus, we can't use this component for implementing the real time approach. The necessary space is very large compared to that offered by the component. To make the tables exploitable, their sizes should be reduced so that we make them memorable in a memory capacity lower than that offered by the component. For each image we must allocate to the maximum a space of **1, 5 Mbits**.

An analysis of the tables of the correspondents shows that the absolute value of the maximum difference between a coordinate of a distorted pixel and the same coordinate of one of its correspondents in the corrected image is 57. This value can be coded on 7 bits. The difference can be positive as can be negative. Therefore, if a bit is reserved to code the sign of the difference, the displacement of a pixel on a given axis could be coded on 8 bits. This analysis thus enables us to not represent the correspondents in the form of co-ordinates but in the form of displacements of distorted pixel. A displacement according to the axis u and another according to the axis v.

While adopting this representation, a correspondent will take only 16 bits in the memory. Thus for each correspondent, there will be a profit of 16 bits. And consequently, the tables of the correspondents are reduced to only one table that it's called ***table of displacements.***

Another very important remark will contribute to reduce the required memory. This remark is related to the size of the images with which the tables of correction will be built (the tables of correction represent all the tables that we must build without reference of the type or size).Indeed, larger is the image, larger are the tables. And since the construction of the tables is an off-line operation, then it is possible to build reduced tables by using reduced images. Some constructions were made and showed that the reduction of the original image intended to build the tables involves a reduction of the required memory intended to contain the tables of correction, but the reduction is not proportional.

The following table illustrates the required memory according to the factors of reduction. The size of the original image is 572*768.

Table 1. The required memory according to the factors of reduction [4].

Factor of reduction	1	4	16
Table of adresses	14,057472	3,514368	0,872448
Table of displacements	7,028736	1,757184	0,436224
Total (Mbits)	21,086208	5,271552	1,308672

This table shows that to work within the limits of the resources offered by the component, it is necessary to build the tables by using an image 16 times smaller than the original image. This reduced image is obtained by the reduction of the original numbers of the lines and columns by a factor of 4.

The manner with which the data are arranged in the memory does not allow the real time treatment because it is a serial treatment which requires the simultaneous reading from several memory boxes of the same physical memory at the same time. However, only one memory box can be read to each clock cycle. This is due to the fact that we can access to the memory only once at each cycle. On the other hand, we can read several different memories at the same time. Thus, to solve this problem, of a practical nature, it is necessary to use a parallel architecture. The parallel treatment consists in not sorting the correspondent of a distorted pixel in a zone starting from a known address but in memory boxes having the same address in different physical memories.

The use of a parallel architecture will increase the required memory because it is necessary to reserve a sufficient number of memories in order to memorize all the correspondents. The number of these memories is not other than the maximum number of correspondents for a distorted pixel. The following table gives the distribution of the pixels of a distorted reduced image (size 143*192) according to their numbers of correspondents.

Table 2. Distribution of the pixels according to the number of correspondents.

0	1	2	3	4	5	6	7
10811	8751	5981	1111	794	6	2	0

From this table we can deduce:

- The required number of memories is 6 ;
- The number of the active pixels is 16645, thus the latest box will have 16644 as address. This value can be coded using 15 bits.

In this case, the table of addresses will be intended to memorize only the address where the correspondents of a distorted pixel are stored. For an active pixel having a number of correspondents lower than 6, the last correspondent have to be duplicated so that the number of the correspondents reaches 6. The purpose of this duplication is to standardize the treatment. Under these conditions, the required memory for only one image is 2.147 Mbits what exceeds the allowed resources limited to 1.5 Mbits [3].

The image used for the development of the method is not that used in the PICAS$O project [3]. It is an image of size higher than that of the images which are provided by the stereoscopic bench of the project. The project's images are of size equal to 492*656. Thus, the size of the reduced image used for the construction of the tables is 123*164 (20172 pixels).

reduced image   **reduced corrected image**

Fig.5. Image acquired by the bench PICAS$O

By using this image for the construction of the tables, there will be the following table giving the distribution of the pixels distorted according to the number of the correspondents:

Table 3. Distribution of the pixels according to the number of correspondents. (Project PICAS$O image)

0	1	2	3	4	5
5661	10003	3712	603	193	0

The noticed differences are due to the parameters of the used cameras, in particular the coefficients of distortion.

From this table we can deduce:

- The required number of memories is 4 ;
- The number of the active pixels is14511, thus the latest box will have 14510 as address. This value can be coded using 14 bits.

The required memory will be **1,332144** Mbits. This value comprises a certain useless part due to the duplication of the correspondents of the same pixel in the same address. In our case, the useless memory size is equal to **0,608576 Mbits**. It is a value equal to **45.68%** of the total memory requires. The useless memory is distributed on the four memories used but at different rates. The following table gives the number of boxes of 16 bits occupied in each memory as well as the rate of the useful memory.

Table 4. Useful Memory

	Useful boxes	Rate of useful boxes
Memory 1	14511	100%
Memory 2	4508	31.06%
Memory 3	796	5.48%
Memory 4	193	1.33%

To reduce the useless memory, we limit the number of correspondents of a distorted pixel. The number of the used physical memories will be the imposed maximum number of correspondents. The following table shows the profit in memory (in Mbits) according to the imposed maximum number.

Table 5. Profit in Memory according to the imposed number of correspondents

Imposed Number of Correspondents	Profit in Memory (Mbits)	Required Memory (Mbits)
1	0.754572	0.577572
2	0.522396	0.809748
3	0.290220	1.041924

The limitation of the number of correspondents will have an effect on the quality of the corrected image. This effect is represented on the following figure.

Fig.6. Effect of the limitation of the number of correspondents

In fact, the quality of the corrected image is better as more the imposed number of correspondents is large. For a limitation at 3, we notice that there are some black spots (holes) distributed especially at the ends of the image. The following table shows the number of the holes for each limitation as well as the rate compared to the whole of the corrected image.

Table (6): Number of holes according to the limitation of the number of correspondents

Number of correspondents	Number of holes	Rate (%)
1	4508	22.35
2	796	3.95
3	193	0.96

The correction of distortions is associated to another program which makes the correction of a stereoscopic bench before calculating the disparity image. Consequently, the limitation of the number of correspondents will act on the quality of the disparity image [5].

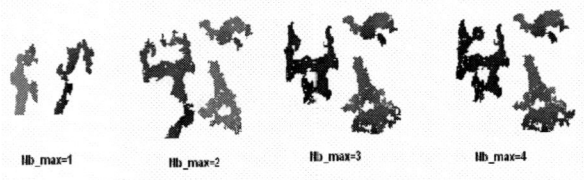

Fig.7. Limitation influence on the disparity image

It is noticed that when the maximum number of the correspondents is fixed at 3, the obtained disparity image is very close to the image that it is obtained when no hole is tolerated (all the correspondents are considered). It is thus preferable to impose this choice in order to optimize the required memory. But the useless memory remains important. It is equal to **36, 42%** of the required memory.

5. CONCLUSION

In this work, we showed the feasibility of the real time correction of the stereoscopic images by using pre calculated tables starting from taken images and the parameters of the cameras used for the capture of these images.

From the off line technique used by researchers in the LAAS/CNRS, we reversed the relation which connect the rectified images to the original distorted stereoscopic images. The new relation is obtained without having the need to solve a complex system of equations of order 7 and this by using only the results of the off line method [4].

The use of the pre calculated tables will make it possible to implement the method on a programmable component like FPGA. In this context, some works have been done with the aim to implement the off line method [5].

The next step of this work is the validation of the method by implementing it on a programmable component (FPGA). This after the definition of an architecture able to especially support the management of the access to the various used memories in reading mode or in writing one.

REFERENCES

[1] M.Devy: ''Présentation de la stéréovision passive''; Rapport interne LAAS 2005.
[2] V.Lemonde : ''Stéréovision Embarquée sur Véhicule: de l'Auto-Calibrage à la Détection d'Obstacles'' ; Thèse de Doctorat, INSA Toulouse, Novembre 2005.
[3] Z.Irki : '' Rectification Temps Réel des Images Stéréoscopique'', Rapport de Stage, LAAS 2006 ;
[4] Z.Irki, M.Devy, JL.Boizard, P.Fillatreau: ''An approach for the Real Time Correction of Stereoscopic Images''; ECMS2007, Liberec, Czech Republic
[5] A.Naoulou: ''FPGA Based Architecture for Real Time Computation of the Census Transform and Correlation in Various Stereovision Contexts'', LAAS report December 2004;

A Three-Dimensional Computerized Facial Reconstruction Using Non-Linear Registration of A Reference Head

A. Kermi[*,**], M.T. Laskri[*], I. Bloch[**]

*University Badji Mokhtar of Annaba, Department of Computer Sciences, LRI GRIA - BP.12, 23000, Annaba, Algeria
(E-mail:{adel.kermi, laskri}@univ-annaba.org)
**ENST (TELECOM ParisTech), Dép. TSI, CNRS UMR 5141 LTCI – 46, Rue Barrault 75634, Paris, Cedex 13, France
(E-mail:{kermi, bloch}@enst.fr)

Abstract: This paper presents a 3D computerized facial reconstruction method based on recent techniques of medical imaging. It is based on a reference 3D image from which skull and skin are extracted, and an image of the skull of the unknown head for which the skin should be reconstructed. The facial reconstruction is obtained by deforming the 3D image of the reference skull to the one of the unknown skull. This transformation is based on a non-linear registration method guided by B-splines Free-Form Deformations (FFD). Then, the same deformation is applied to transform the reference skin to obtain a new skin that we consider near the unknown skin. This method has been evaluated on sets of skull/skin data, extracted from 3D-MRI of individual heads having similar anthropological features.

Keywords: 3D computerized facial reconstruction, skull, 3D-MRI, non-linear registration, FFD.

1. INTRODUCTION

Facial reconstruction has emerged as an increasingly important tool in several scientific areas, especially in forensic sciences and archaeology. In both areas, the basis of all work is a skull of a dead person which is often the only source of information available about the unknown physiognomy of the individual [1-6]. In those situations, it is essential to estimate the face of an unknown individual from the shape of his/her cranial skeleton, for judicial or historical aims to help the skulls identification and people search.

Different facial reconstruction techniques have been developed and described in the literature and are currently used in practice; they are two-dimensional (2D), three-dimensional (3D), manual or computerized. A general review of these various methods can be found in reference books of forensic anthropology [3,4] and more recently in [5-9]. The 3D computerized facial reconstruction techniques can be divided into three main categories. The first category concerns morphometry based techniques. These methods imitate closely 3D manual techniques using a limited number of landmarks representing soft tissues thicknesses. In general, these methods consist in placing a generic model of head surface around a polygonal model of unknown skull. Then, the model of the head surface is adjusted or interpolated through the virtual landmarks previously placed on the skull surface therefore allowing to obtain a candidate face for unknown skull. That is done by applying some 3D transformations such as B-splines functions [1], or radial basis functions [11]. Other more recent methods, classified into the second category, use volumetric deformation and registration based techniques. These techniques are very present in many medical image processing applications to map one dataset onto another based on a certain correspondence between these different datasets. Thus, it is also prominent to use them for facial reconstruction as well. Generally, these techniques are based on deformation approaches of a reference head to unknown skull based on crest lines [12], disc fields [7,13], feature points [14], Euclidean distance maps [15], or semi-landmarks [16]. The reference head is often selected based on similarities in morphological characteristics (i.e. sex, age and ethnic group). The 3D facial reconstruction method, presented in this paper, belongs to this category. The third category relates to the techniques based on the morphology and the anatomy of the human head. The morphology of the face is obtained by including the muscles and the fat, before ending the reconstruction, by putting on the skin layer. Wilhelms and Van Gelder [17] presented computer-graphics algorithms to model the bones, muscles and underlying skin. The approach of Kähler et al. [10] consists in fitting a model of virtual head based on anatomy, including the skin and the muscles, on the surface of unknown skull by using statistical soft-tissue thicknesses data and the correspondences between some skull and skin landmarks.

This paper is organized as follows. In Section 2, we describe our facial reconstruction method. After evaluating the method in Section 3, we propose some ways to be explored in order to improve the quality of the reconstruction in Section 4. This paper extends the works presented in [26] and [27]. Therefore, we present more results.

2. PROPOSED FACIAL RECONCTRUCTION METHOD

2.1 Method description

The goal of our work is to experiment and validate a facial reconstruction method. The work assumption and the principle are similar to those of [7] and [12]. The work assumption leans on the morphological homology which is expressed as follows: if two skulls have similar shapes presenting the same morphological features (age, sex and race), the corresponding faces should have some main characteristics in common.

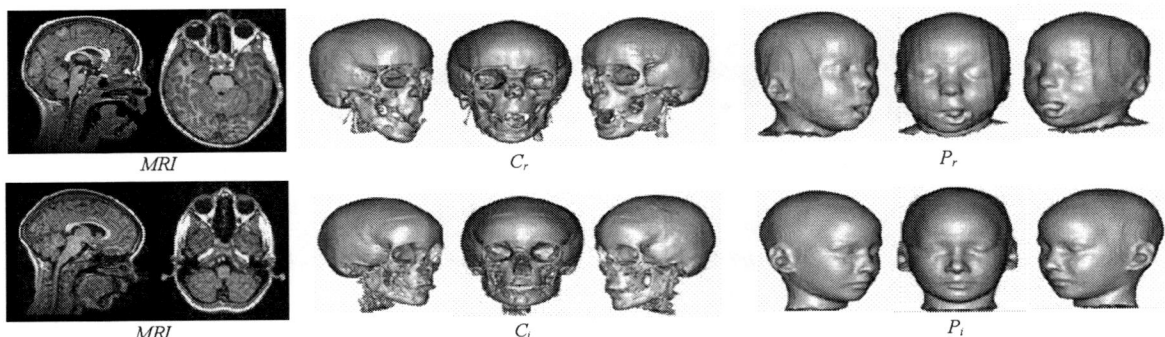

Fig. 1. Results of the segmentation and 3D visualization using Anatomist (http://brainvisa.info). Sagittal and axial slices extracted from The 3D-MRI of the head (left), surface of the skull (middle) and surface of skin (right).

We propose a 3D facial reconstruction method that utilizes recent techniques of medical imaging. It is based on a global volumetric deformation approach of the 3D image of a reference skull towards the one of an unknown skull. This transformation is based on a non-linear registration method guided by B-splines Free-Form Deformations. Then, the same transformation is spatially extrapolated and is applied to deform the reference skin to obtain a new skin which is considered as being close to the unknown skin. Therefore, we seek to find the unknown skin only from the unknown skull and from one reference skull/skin set.

2.2 Collection of data and 3D representation

In contrast to the computerized facial reconstruction methods described in the literature, we use data extracted from three-dimensional magnetic resonance images (3D-MRI) of individual heads rather than x-ray computer tomography images (CT). In our study, thirteen sets of skulls and skins have been segmented from 3D-MRI of individual heads using mathematical morphology and topological constraints, as developed in [18,19]. Thus, we obtain the sets of voxels corresponding to the skulls and skins, necessary for our method. For 3D visualization, we apply the "Marching Cubes" algorithm [20] on the voxels data in order to obtain triangulated surfaces of skull and skin. Fig. 1 shows an example of the segmentation results. The first set (in the top) represents the reference head including a reference skin P_r and a reference skull C_r. The second set (in the bottom), consisting of a skin P_i and a skull C_i, is to be considered as the head of the unknown person whose skin is to be reconstructed. The skin P_i will be used as test in order to validate the quality of the reconstruction.

2.3 Facial reconstruction process

The facial reconstruction is obtained by initially calculating a volumetric transformation of the reference skull image C_r to the skull image C_i (i.e. $C_i = T(C_r)$). We then apply the same transformation T to the reference skin image P_r in order to obtain an approximation of the unknown skin P^* (i.e. $T(P_r)=P^*$). Fig. 2 describes the process of the facial reconstruction method. The evaluation of the method will be done by comparing P^* with P_i on data where C_i and P_i are both known. From an image processing point of view, the problem is essentially expressed as a registration problem between two images, with the objective to find a transformation allowing to modify the shape of the skull C_r contained in an image (known as source) so that it has the same shape as the skull C_i contained in the other image (known as target). This transformation requires a non-linear registration. This is due to the fact that the skulls to be registered have some complex shapes that can strongly vary from one individual to another, in a non-linear manner.

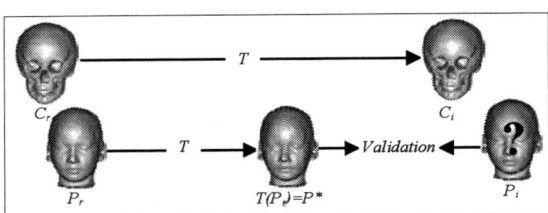

Fig. 2. The facial reconstruction process.

2.3.1 Deformation model: Free-Form Deformations

To calculate the transformation T, we employ a non-linear registration guided by a B-splines Free-Form Deformations model (FFD). The choice of this technique over other non-linear transformations models, existing in the image processing literature, is due on the one hand to the complexity of the skull shapes to register and on the other hand to the constraints imposed by our application: a sufficiently high number of degrees of freedom for more local deformations and the non utilization of anthropological landmarks to help the registration process.

The FFD, introduced by Sederberg et al. [21], is a parametric model which provides a flexible non-linear transformation due to the fact that no assumptions are made on the images or shapes to register. This model has been successfully used in various medical imaging applications, such as mammogram registration [22], brain registration [23], cardiac segmentation [24], or the registration of CT images with positron emission tomography images (PET) in the thoracic and abdominal regions [25].

In the proposed technique, the registration algorithm is divided into two stages: a global transformation stage T_{global}, providing a good initial approximation for the transformation T and a non-linear transformation stage T_{local}, allowing for more local deformations:

$$T(x,y,z) = T_{global}(x,y,z) + T_{local}(x,y,z) \quad (1)$$

The goal of the global transformation is to provide an initialization to the non-linear registration as close as possible to the desired final result. This transformation is affine. It is composed of a rotation, a translation and a scaling transformation.

While the affine registration computes only a global motion between two images, a non-linear registration models a local motion using FFD [21]. In this technique, deformations of the volume (the 3D image of the skull C_r in our application) are achieved by the optimization of an underlying mesh of control points associated to B-splines functions. The control point displacements are then interpolated to obtain a smooth and continuous C^2 transformation. A B-splines based FFD can be written as a 3D tensor product of one-dimensional cubic B-splines, producing a transformation separately for each axis. Let Φ denote an uniformly spaced grid of $n_x \times n_y \times n_z$ control points $\phi_{i,j,k}$ with a spacing of δ, where $-1 \leq i < n_x - 1, -1 \leq j < n_y - 1, -1 \leq k < n_z - 1$.

Then, the non-linear transformation for each image point x, y, z is computed as:

$$T_{local}(x,y,z) = \sum_{l=0}^{3}\sum_{m=0}^{3}\sum_{n=0}^{3} B_l(u)B_m(v)B_n(w)\phi_{i+l,j+m,k+n} \quad (2)$$

where $i = \lfloor x/\delta \rfloor - 1, j = \lfloor y/\delta \rfloor - 1, et\ k = \lfloor z/\delta \rfloor - 1$, denote the index of the control point cell containing (x,y,z), and $u = x/\delta - (i+1), v = y/\delta - (j+1), w = z/\delta - (k+1)$ are the relative positions of (x,y,z) in the three dimensions. The B-splines functions B_l are given by:

$$\begin{aligned}B_0(u) &= (1-u)^3/6 \\ B_1(u) &= (3u^3 - 6u^2 + 4)/6 \\ B_2(u) &= (-3u^3 + 3u^2 + 3u + 1)/6 \\ B_3(u) &= u^3/6\end{aligned} \quad (3)$$

Registration is obtained by minimizing a cost function, as proposed in [25], which uses at the same time a dissimilarity criterion C_{dis} between the A and B images and a regularization criterion C_{reg} of the desired transformation:

$$C = C_{dis}(A, T(B)) + C_{reg}(T) \quad (4)$$

The FFD method requires the optimization of the control points of the grid to minimize a given dissimilarity criterion. The choice of this criterion is straightforward in our case, as we are working with binary images (sets of voxels) of skulls having been extracted from the 3D-MRI of individuals' heads. Therefore, we used the criterion of Root Mean Square (RMS) as dissimilarity measure to determine the optimal parameters of the deformation. This metric computes the difference between the intensities of the N voxels X_i of the image A and the N transformed voxels $B(T(X_i))$ for a transformation T of the N voxels X_i of the image B corresponding to the voxels X_i of the image A (Equation (5)). In our case, A and B are binary images. This means that, $A(X_i)$ and $B(X_i)$ take only two values 0 or 255.

$$C_{dis}(A, T(B)) = \sqrt{\frac{1}{N}\sum_{i}^{N}(A(X_i) - B(T(X_i)))^2} \quad (5)$$

To force the regularity of the desired transformation, we introduce a regularization term C_{reg} composed of a local spring force which consists in pulling each node towards the centroid of its neighboring nodes to avoid the intersection between the nodes. The regularization term is given by:

$$C_{reg}(T) = \lambda \left| C_{x_c, y_c, z_c} - \phi_{i,j,k} \right| \quad (6)$$

where λ is a constant of regularization, and C_{x_c, y_c, z_c} denotes the position of the centroid of each neighboring node of the control node $\phi_{i,j,k}$. The regularization term was introduced into the optimization procedure to avoid undesirable deformations (strong and different deformations) and to preserve the topology of the structure to be readjusted. The parameter λ allows the control of regularization. Its value has been experimentally determined. With a grid of 10 control points per dimension, we found that a value of $\lambda = 0.5$ provides a good compromise for the two terms of the cost function.

2.3.2 Optimization procedure

The optimization procedure varies the displacements of the control points $\phi_{i,j,k}$ in such a way that the dissimilarity between the images A and B is minimized. The optimization of the transformation parameters is based on an iterative gradient descent technique over the entire grid of control points. At each iteration m, we compute a local gradient estimation of the cost function for each control point by finite differences method. Then the transformation parameters are iteratively changed along this gradient until such an update does not yield an improvement of the cost function measure:

$$\phi_{i,j,k}^{m} = \phi_{i,j,k}^{m-1} + \mu \frac{\nabla C}{\|\nabla C\|} \quad (7)$$

where μ is the optimization step size, $\nabla C = \partial C / \partial \phi_{i,j,k}^{m-1}$ is the gradient estimation, that must be re-calculated after each iteration of the procedure, and $\|\nabla C\|$ is the gradient norm of the cost function.

For every control point $\phi_{i,j,k}^{m}$, displacements of one step μ are tried along each axis in positive and negative directions, and the cost function is evaluated. If one of these displacements provides better dissimilarity measures (i.e. better registration results), the control point displacement is accepted and updated. This procedure is repeated until no substantial gradient can be determined or the maximal number of iterations is reached. This procedure is embedded in a multi-step framework (the initial optimization step μ is divided by 2 at each step), in order to cope with more global deformations at the beginning and more local ones at the end.

3. RESULTS

We tested our method on thirteen sets of skull/skin data, segmented from thirteen 3D-MRI of individual heads: nine MRI of children heads (*A, B, C, D, E, F, G, H,* and *I*), and four MRI of adults heads (*J, K, L,* and *M*). For each facial reconstruction, we first select manually two skull/skin couples having approximately the same morphological characteristics. Thus, we choose to compute the transformation only on the nearest skull of the unknown skull for which the skin should be reconstructed. For each skull C_i, we thus select the nearest skull from our database and we then apply the transformation, computed between these skulls, on the reference skin P_r (known) associated to the nearest skull C_r.

In FFD-based registration techniques, several parameters must be specifically tuned, mostly the number of control points of the grid that determine its resolution and in consequence the degree of local deformations of the transformation. Another aspect to be taken into account is the computational cost of the registration algorithm because a higher number of control points involves a higher quantity of parameters to optimize. However, convergence time notably increases with a more densely populated grid. During our tests, we have found a good trade-off between these two aspects with a grid of 10 control points per dimension. The other parameters of the method have been determined experimentally. The total number of iterations has been fixed to 50. The number of steps has been tuned to 4, and the initial optimization step μ is fixed to 8. Thus, μ decreases from 8 to 0.5 through the optimization procedure.

We illustrate the results of the facial reconstruction with four sets of skull/skin data: the two sets of Fig. 1 segmented from two MRI of children heads C and D, and two sets extracted from two MRI of adults heads J and K. The first two MRI have a size of 256 x 256 pixels in the xy plane and 132 slices, with dimensions of voxels of approximately 1.0 x 1.0 x 1.3 mm^3, and the other two sets have a size of 256 x 256 pixels in the xy plane and 183 slices, with dimensions of voxels of approximately 0.9 x 0.9 x 0.9 mm^3.

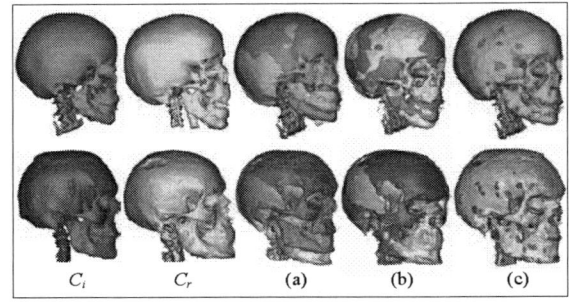

Fig. 3. Two examples of registration results between two children skulls C and D (top) and between two adults' skulls J and K (bottom). Deformation of the skull C_r (green) to the skull C_i (red): (a) before registration, (b) after T_{global} and (c) after T_{local}.

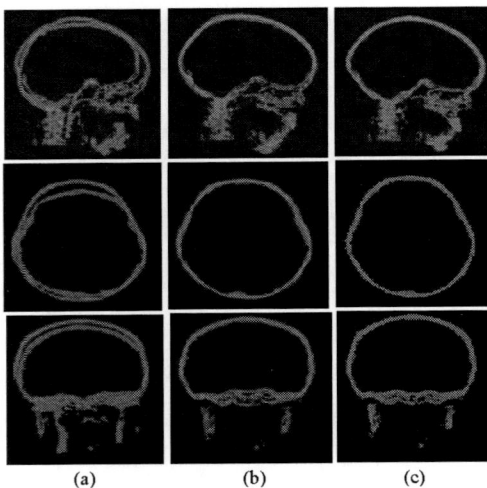

Fig. 4. From top to bottom: the same sagittal, axial, and coronal slices of registration results between the two children skulls C and D. Transformation of C_r (green) to C_i (red): (a) before registration, (b) after T_{global} and (c) after T_{local}.

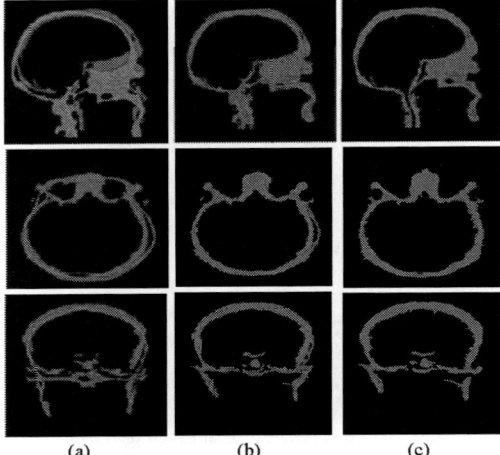

Fig. 5. From top to bottom: the same sagittal, axial, and coronal slices of registration results between the two adult skulls J and K. Transformation of C_r (green) to C_i (red): (a) before registration, (b) after T_{global} and (c) after T_{local}.

Fig. 3 presents two examples of registration between two children skulls C and D (in the top) and between two adults skulls J and K (in the bottom). This figure shows the results obtained by the registration algorithm through its two stages of global and local transformations of C_r in green color to C_i in red color. For better visualizing the registration effect, this figure represents the superposition of surfaces of the two skulls (a) before registration, (b) after T_{global} transformation and finally (c) after the non-linear transformation T_{local}. Sagittal, axial and coronal slices of the same results are represented in Fig. 4 and in Fig. 5. We show in Fig. 6 and Fig. 7 sagittal, axial and coronal slices of the results obtained by the application of the same transformation T, computed between skulls, on the reference P_r.

In general, the validation of the facial reconstruction results is a complex problem because of the important subjectivity which can be involved in the identification and recognition steps. In our case, we have the real skin P_i which will be used as test in order to validate the quality of the reconstruction. The visual inspection is the most obvious facial reconstruction evaluation method, but it remains qualitative and insufficient. For the two preceding facial reconstruction examples, Fig.8 and Fig.9 show the comparison between the skin P^* reconstructed from the unknown skull C_i by the algorithm (on the right), and the real skin P_i (in the middle). In general, we can observe a good correspondence of the upper part of the skins; however some differences can be perceived near the chin, mainly due to the problem of opening degrees of the mandible and the biomechanical characteristics of facial tissues.

A more quantitative facial reconstruction evaluation is provided by the histograms of distances differences between the skins P_i and P^*, and between unknown skull C_i and skull deformed after the global and local transformations. Fig. 10 presents the quantitative evaluation for the facial reconstruction of the unknown child skull $\mathcal{D}$. This figure shows clearly the improvements of registration, since both histograms (a) and (b) are increasingly concentrated towards low values.

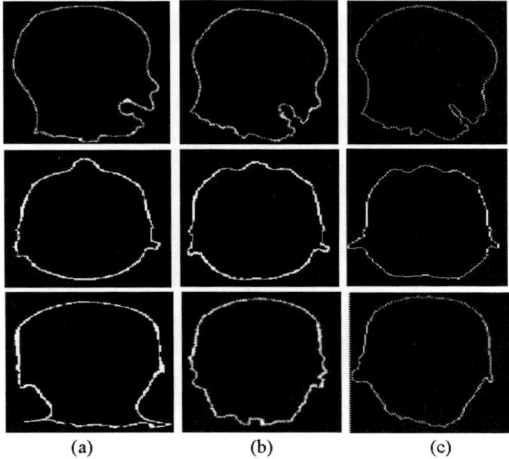

Fig. 6. From top to bottom: the same sagittal, axial, and coronal slices of results of the application of the same transformation, computed between the two children skulls C and $\mathcal{D}$, on the reference skin P_r. (a) before registration, (b) after T_{global} and (c) after T_{local}.

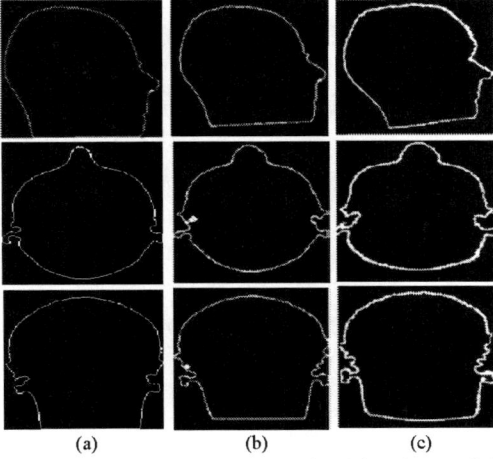

Fig. 7. From top to bottom: the same sagittal, axial, and coronal slices of results of the application of the same transformation, computed between the two adults skulls $\mathcal{J}$ and $\mathcal{K}$ on the reference skin P_r. (a) before registration, (b) after T_{global} and (c) after T_{local}.

During our tests, a methodology similar to the leave-one-out approach is used. Each skull/skin set of the database is in turn considered as unknown and is used as unknown skull and skin. We present in Table 1 the quantitative evaluation of results obtained by the registration through its two stages for the thirteen cases. This table provides the selection of the nearest skull/skin couple of unknown skull C_i, the distance between skulls before registration, the distance between skulls after T_{global}, the distance between skulls after T_{local}, and the distance between P^* and P_i.

The mean distance of the thirteen facial reconstructions is 3.1 mm. The best facial reconstruction is realized for the subject $\mathcal{F}$ with a distance of 2.0 mm. The worst case is realized for the subject C with a distance of 4.5 mm.

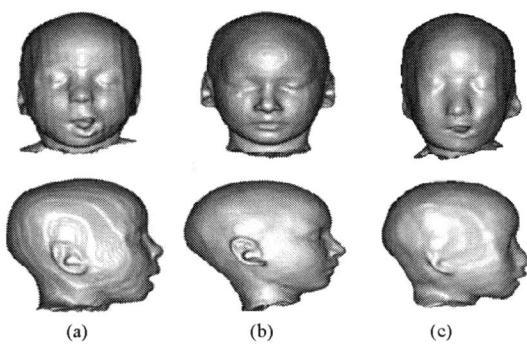

Fig. 8. Frontal and lateral views of facial reconstruction results of child case $\mathcal{D}$. (a) surface of reference skin P_r, (b) surface of real skin P_i and (c) surface of reconstructed skin P^*.

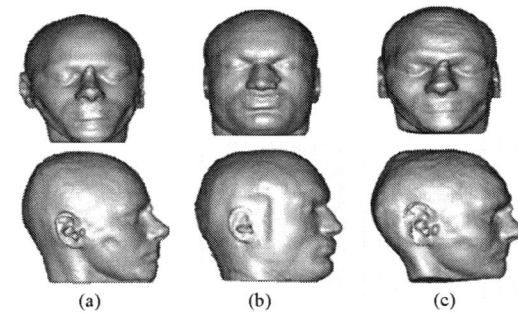

Fig. 9. Frontal and lateral views of facial reconstruction results of adult case $\mathcal{K}$ (a) surface of reference skin P_r, (b) surface of real skin P_i and (c) surface of reconstructed skin P^*.

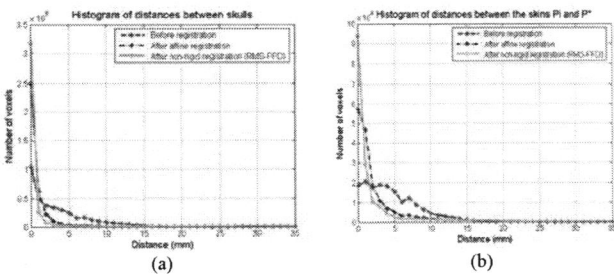

Fig. 10. Histograms of distances (child case $\mathcal{D}$): (a) between C_i and transformed skull and (b) between P_i et P^*.

In the case of non-linear registration between skulls, the mean distance is 0.4 mm and the distance varies between 0.2 mm (for the subject $\mathcal{M}$) and 0.6 mm (for the subject $\mathcal{H}$). In the case of worst facial reconstruction realized for the subject C, the quality of registration between skulls is good (0.4 mm). In the case of best reconstruction realized for the subject $\mathcal{F}$, the quality of registration between skulls is also good (0.3 mm). These evaluations lead us to say that there are few relationships between the registration quality between skulls and the quality of facial reconstruction, even though the well registered subjects are rather better reconstructed and the badly registered subjects are poorly estimated.

In Fig. 9, the error of facial reconstruction for the subject $\mathcal{K}$ is larger than the error indicated in Table 1. This is due to the average effect of errors computation. In most regions of the skin, the errors are equal to zero or approximately zero (the number of voxels having the zero distance is important as seen on the distance histograms), and in few regions the error is large. A refinement using muscle thickness is still necessary in particular in the lower half of the head.

Table 1. Mean distances in millimeters (mm) for the thirteen cases at the different registration stages.

unknown skull C_i	A	B	C	D	E	F	G	H	I	J	K	L	M	Mean
Reference subject C_r/P_r	B	A	F	C	G	D	F	E	H	L	J	K	J	
Distance between C_i and C_r	5.3	5.3	4.7	3.2	6.7	4.7	6.1	5.6	6.0	4.0	2.8	6.7	7.4	5.3
Distance between skulls after T_{global}	0.6	0.8	1.0	0.9	0.6	0.6	0.9	0.8	0.9	0.6	0.6	0.8	0.5	0.7
Distance between skulls after T_{local}	0.5	0.5	0.4	0.5	0.3	0.3	0.4	0.6	0.3	0.3	0.3	0.3	0.2	0.4
Distance between P^* and P_i	2.4	2.4	4.5	2.9	2.3	2.0	2.8	4.1	3.5	3.4	3.3	3.9	2.2	3.1

4. CONCLUSION AND PERSPECTIVES

In contrast to the most facial reconstruction methods described in the literature, our method does not use the knowledge of mean values of facial soft tissues thickness at specific anatomical landmarks. It is based on a global volumetric deformation approach of a reference face (or more exactly a skin) to obtain a face for the unknown skull. This deformation is based on a non-linear registration technique guided by FFD. By this method, we are able to obtain a facial reconstruction relatively close to the real skin. One advantage of the method is its simplicity. Moreover, it allows the generation of different results from the same unknown skull. This is done by using other reference heads or by modifying some transformation parameters such as the number of control points in the grid.

The difference between the skin that we calculate and the real skin is explained by the fact that we do not take into account, during the transformation, the problem of the mandible degree of opening, the soft tissues thicknesses at characteristic points and the biomechanical characteristics of facial tissues.

Current results are promising and we believe that they could be greatly improved. An interesting perspective consists in developing a deformable model in order to refine the surface of the reconstructed skin under certain constraints. These constraints are on the one hand a distance criterion (soft tissues thicknesses at some fixed anthropological landmarks) and on the other hand a regularity criterion according to the shape of reference skin.

REFERENCES

[1] K.M. ARCHER, *Craniofacial Reconstruction Using Hierarchical B-spline Interpolation*, Master's thesis of Applied Science, University of British Columbia, August 25, 1997.

[2] G. ATTARDI, M. BETRO, M. FORTE, R. GORI, A. GUIDAZZOLI, S. IMBODEN, and F. MALLEGNI, *3D facial reconstruction and visualization of ancient Egyptian mummies using spiral CT data*, ACM SIGGRAPH'99, Sketches and applications, Los Angeles, USA, pp. 223-239, August 1999.

[3] M.Y. ISCAN, Craniofacial Image Analysis and Reconstruction, In Iscan, Mehmet Yasar and Helmer, Richard P. (Ed.), *Forensic Analysis of the Skull*, 1 (New York: Wiley-Liss, 1993, 1-10).

[4] W.M. KROGRAM and M.Y. ISCAN, *The Human Skeleton in Forensic Medicine*, In Charles C. Thomas, Publisher, Springfield, Illinois, 1986.

[5] K.T. TAYLOR, *Forensic Art and Illustration*, (CRC Press, New York, 2001).

[6] A.J. TYRRELL, M.P. EVISON, A.T. CHAMBERLAIN, and M.A. GREEN, Forensic Three-dimensional Facial Reconstruction: Historical review and Contemporary developments, *Journal of Forensic Science*, Vol. 42, n. 4, pp. 653-661, 1997.

[7] L.A. NELSON and S.D. MICHAEL, The application of volume deformation to three-dimensional facial reconstruction: A comparison with previous techniques, *Forensic Science International*, Vol. 94, n. 3, pp. 167-181, 1998.

[8] W.A. AULSEBROOK, M.Y. ISCAN, J.H. SLABBERT, and P. BECKER, Superimposition and reconstruction in forensic facial identification: A survey, *Forensic Science International*, Vol. 75, n. 2-3, pp. 101-120, October 1995.

[9] J.G. CLEMENT and M.K. MARKS, *Computer-graphic Facial reconstruction*, (Elsevier Ltd, 10 Academic Press, USA, 2005).

[10] K. KÄHLER, J. HABER, and H.P. SEIDEL, Reanimating the Dead: Reconstruction of Expressive Faces from Skull Data, *ACM Transaction On graphics (SIGGRAPH Conference Proceedings)*, ACM, Vol. 22, n. 3, pp. 554–561, July 2003.

[11] P. VANEZIS, M. VANEZIS, G. MC COMBE, and T. NIBLETT, Facial reconstruction using 3-D computer graphics, *Forensic Science International*, Vol. 108, n. 2, pp. 81-95, 2000.

[12] G. QUATREHOMME, S. COTIN, G. SUBSOL, H. DELINGETTE, Y. GARIDEL, G. GREVIN, M. FIDRICH, P. BAILLET, and A. OLLIER, A Fully Three-Dimensional Method for Facial Reconstruction Based on Deformable Models, *Journal of Forensic Sciences*, Vol. 42, n. 4, pp. 649-652, 1997.

[13] S.D. MICHAEL and M. CHEN, *The 3D reconstruction of facial features using volume distortion*, Proceedings of the 14th Annual Conference of Eurographics, London, UK, pp. 297-305, 1998.

[14] M.W. JONES, Facial reconstruction using volumetric data, In T. Ertl, B. Girod, G. Greiner, H. Niehann, H.P. Seidel (Eds.), Proceedings of the Vision, Modeling, and Visualization, Stuttgart, Germany, IOS Press, Amsterdam, pp.135-150, 2001.

[15] D. VANDERMEULEN, P. CLAES, P. SUETENS, S. DE GREEF, and G. WILLEMS, *Volumetric deformable face models for cranio-facial reconstructions*, Proceedings of ISPA'05, Zagreb, Croatia, pp. 353-358, 2005.

[16] M. BERAR, M. DESVIGNES, G. BAILLY, and Y. PAYAN, 3D Semi-Landmarks-Based Statistical Face Reconstruction, *Journal of Computing and Information Technology*, Vol. 14, n. 1, pp. 31-43, 2006.

[17] J. WELHELMS and A. VAN GELDER, Anatomically based modeling, *Computer Graphics*, Annual Conference Series, Vol. 31, pp. 173-180, August 1997.

[18] J. BURGUET, N. GADI, and I. BLOCH, *Realistic models of children heads from 3D-MRI segmentation and tetrahedral mesh construction*, Proceedings of the 2nd International Symposium on 3D Data Processing, Visualization, and Transmission, pp. 631-638, 2004.

[19] P. DOKLADAL, I. BLOCH, M. COUPRIE, D. RUIJTERS, R. URTASUN, and L. GARNERO, Topologically Controlled Segmentation of 3D Magnetic Resonance Images of the Head by using Morphological Operators, *Pattern Recognition*, Vol. 36, n. 10, pp. 2463-2478, October 2003.

[20] W.E. LORENSEN, and H.E. CLINE, Marching Cubes: A High Resolution 3D Surface Construction Algorithm, *Computer Graphics*, Vol. 21, n. 3, pp. 163-169, July 1987.

[21] T. SEDERBERG and S. PARRY, Free from deformation of solid geometric models, *SIGGRAPH'86*, Dallas, USA, Vol. 20, pp. 151-160, August 1986.

[22] D. RUECKERT, I. SONADA, C. HAYES, D.L.G. HILL, M.O. LEACH, and D.J. HAWKES, Nonrigid Registration Using Free-Form Deformations: Application to Breast MR Images, *IEEE Transactions on Medical Imaging*, Vol. 18, n. 8, pp. 712-721, August 1999.

[23] T. HARTKENS, D.L.G. HILL, A.D. CASTELLANO-SMITH, D.J. HAWKES, C.R. MAURER JR., A.J. MARTIN, W.A. HALL, H. LIU, and C.L. TRUWIT, *Using points and surfaces to improve voxel-based non-rigid registration*, MICCAI'02, pp. 565-572, 2002.

[24] J.M.P. LOTJONEN, Segmentation of MR images using deformable models: Application to cardiac images, *International Journal of Bioelectromagnetism*, Vol. 3, n. 2, pp. 37-45, 2001.

[25] O. CAMARA, G. DELSO, O. COLLIOT, A. MORENO-INGELMO, and I. BLOCH, Explicit Incorporation of Prior Anatomical Information Into a Nonrigid Registration of Thoracic and Abdominal CT and 18-FDG Whole-Body Emission PET Images, *IEEE Transactions on Medical Imaging*, Vol. 26, n. 2, pp. 164-178, February 2007.

[26] A. KERMI, I. BLOCH, and M. T. LASKRI. Une approche combinant recalage non rigide et modèle déformable pour la reconstruction faciale tridimensionnelle. In *proceedings of the 2nd Algero-French Conference on Medical Imaging (JETIM'2006)*, pp. 91-96, Algiers, Algeria, November 2006.

[27] A. KERMI, I. BLOCH, and M. T. LASKRI. A Non-Linear Registration Method Guided by B-Splines Free-Form Deformations for Three-Dimensional Facial Reconstruction. *International Review on Computers and Software*, Vol. 2, n. 6, pp. 209-219, November 2007.

Image Segmentation with Spiking Neuron Network

B. Meftah*. A. Benyettou**. O. Lezoray***. M. Debakla*

*Equipe EDTEC (LRSBG), Centre Universitaire Mustapha Stambouli, Mascara, Algérie
** Laboratoire SIMPA, BP 1505, Université Mohamed Boudiaf (USTO), Oran, Algérie
***Université de Caen, GREYC UMR CNRS 6072, 6 Bd. Maréchal Juin, F-14050, Caen, France

Abstract: The process of segmenting images is one of the most critical ones in automatic image analysis whose goal can be regarded as to find what objects are presented in images. Artificial neural networks have been well developed. First two generations of neural networks have a lot of successful applications. Spiking Neuron Networks (SNNs) are often referred to as the 3rd generation of neural networks which have potential to solve problems related to biological stimuli. They derive their strength and interest from an accurate modeling of synaptic interactions between neurons, taking into account the time of spike emission. SNNs overcome the computational power of neural networks made of threshold or sigmoidal units. Based on dynamic event-driven processing, they open up new horizons for developing models with an exponential capacity of memorizing and a strong ability to fast adaptation. Moreover, SNNs add a new dimension, the temporal axis, to the representation capacity and the processing abilities of neuralnetworks. In this paper, we present how SNN can be applied with efficacy in image segmentation.

Keywords: Classification, Clustering, Learning, Segmentation, Spiking neuron network.

1. INTRODUCTION

Image segmentation consists in subdividing an image into its constituent parts and extracting these parts of interest. A large number of segmentation algorithms have been developed since the middle of 1960's [1], and this number continually increases from year to year in a fast rate.
Simple and popular methods are threshold-based and process histogram characteristics of the pixel intensities of the image. Of course, thresholding has many limitations: the transition between objects and background has to be distinct and the result does not guarantee closed object contours, often requiring substantial post-processing. Region-based methods have also been developed; they exploit similarity in intensity, gradient, or variance of neighboring pixels. Watersheds methods can be included in this category. The problem with these methods is that they do not employ any shape information of the image, which can be useful in the presence of noise. Meanwhile, artificial neural networks are already becoming a fairly renowned technique within computer science. Since 1997, Maass[2],[3] has quoted that computation and learning has to proceed quite differently in SNNs. He proposes to classify neural networks as follows:

- 1st generation: Networks based on McCulloch and Pitts' neurons as computational units, i.e. threshold gates, with only digital outputs (e.g.perceptrons, Hopfield network, Boltzmann machine, multilayer perceptrons with threshold units).
- 2nd generation : Networks based on computational units that apply an activation function with a continuous set of possible output values, such as sigmoid or polynomial or exponential functions (e.g. MLP, RBF networks). The real valued outputs of such networks can be interpreted as firing rates of natural neurons.
- 3^{rd}generation of neural network models: Networks which employ spiking neurons as computational units, taking into account the precise firing times of neurons for information coding.

The use of spiking neurons promises high relevance for biological systems and, furthermore, might be more flexible for computer vision applications [4]. Many of the existing segmentation techniques, such as supervised clustering use a lot of parameters which are difficult to tune to obtain segmentation where the image has been partitioned into homogeneously colored regions. In this paper, a spiking neural network approach is used to segment images with unsupervised learning.
The paper is organized as follows: in the first Section, related works present in the literature of spiking neural network (SNNs). The second Section is the central part of the paper and is devoted to the description of the SNN segmentation method and its main features. The results and discussions of the experimental activity are reported in the third Section. Last Section concludes.

2. SPIKING NEURAL NETWORK

2.1 Biological background

Neurons are remarkable among the cells of the body in their ability to propagate signals rapidly over large distances. They do this by generating characteristic electrical pulses called action potentials, or more simply spikes that can travel down nerve fibers.

Neurons are highly specialized for generating electrical signals in response to chemical and other inputs, and transmitting them to other cells. Some important morphological specializations are the dendrites that receive inputs from other neurons and the axon that carries the neuronal output to other cells. The elaborate branching structure of the dendrite tree allows a neuron to receive inputs from many other neurons through synaptic connections [6].

The membrane potential $U_j(t)$ of a postsynaptic neuron N_j varies continuously through time (cf. Figure 1). Each action potential, or spike, emitted by a presynaptic neuron connected to generates a weighted PostSynaptic Potential (PSP) which is function of time.

If the W_{ij} synaptic weight is excitatory, the EPSP is positive: Sharp increasing of the potential $U_j(t)$ and then smoothly decreasing back to null influence. If W_{ij} is inhibitory then the IPSP is negative: Sharp decreasing $U_j(t)$ and then smoothly increasing. At each time, the value of $U_j(t)$ results from the addition of the still active PSPs variations. Whenever the potential $U_j(t)$ reaches the threshold value ϑ of N_j, the neuron fires or emits a spike, that corresponds to a sudden and very high increase of $U_j(t)$, followed by a strong depreciation and a smooth return to the resting potential U_0 [7].

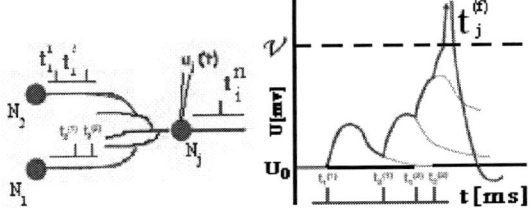

Fig. 1. Emission of spike

2.2 Models of spiking neurons

Since the works of Santiago Ramon y Cajal and Camillo Golgi, a vast number of theoretical neuron models have been created, with a modern phase beginning with the work of Hodgkin and Huxley [8].

We divide the spiking neuron models into three main classes, namely threshold-fire, conductance based and compartmental models. Because of the nature of this paper we will only cover the class of threshold-fire and specially spike response model (SRM).

The SRM as defined by Gerstner [9] is simple to understand and to implement. The model expresses the membrane potential u at time t as an integral over the past, including a model of refractoriness.

Let $F_i = \{t_i^f; 1 \leq f \leq n\}$ denote the set of all firing times of neuron N_i and
$\Gamma_j = \{i;\ N_i\ is\ presynaptic\ to\ N_j\}$ denote the set of all presynaptic neuron to N_j.

The state $U_j(t)$ of neuron N_j at time t is given by:

$$u_j^{(t)} = \sum_{t_j^{(f)} \in F_j} \eta_j\left(t - t_j^{(f)}\right) + \sum_{i \in \Gamma_j} \sum_{t_i^{(f)} \in F_i} w_{ij} \epsilon_{ij}\left(t - t_i^{(f)} - d^k\right) \quad (1)$$

η_j models the potential reset after a spike emission, w_{ij} describes the response to presynaptic spikes. For the kernel functions, a choice of usual expressions is given by:

$$\eta_j(s) = -\vartheta e^{\left(-\frac{s}{\tau}\right)} H(s) \quad (2)$$

where H is the Heaviside function, ϑ is the threshold and τ a time constant defining the decay of the PSP. The function $\epsilon(t)$ is an α -function as:

$$\epsilon(t) = \frac{t}{\tau} e^{\left(1 - \frac{t}{\tau}\right)} \ for\ t > 0$$
$$else\ \epsilon(t) = 0 \quad (3)$$

3. SEGMENTATION USING SPIKING NEURAL NETWORK

However, before building a SNN, we have to explore three important issues: network architecture, information encoding and learning method. After that we will use the SNN to segment images.

3.1 Network architecture

The network architecture consists of a fully connected feedforward network of spiking neurons with connections implemented as multiple delayed synaptic terminals (Fig 2).

The network consists of an input layer, a hidden layer, and an output layer. The first layer is composed of three inputs neurons (RGB values) of pixel. Each node in the hidden layer has a localized activation $\Phi^n = \Phi(\|X - C_n\|, \sigma_n)$ where $\Phi^n(.)$ is a radial basis function (RBF) localized around C_n with the degree of localization parameterized by σ_n.

Choosing $\Phi(Z, \sigma) = e^{-\frac{z^2}{2\sigma^2}}$ gives the Gaussian RBF. This layer transforms real values to temporal values. The activations of all hidden nodes are weighted and sent to the output layer. Instead of a single synapse, with its specific delay and weight, this synapse model consists of many sub-synapses, each one with its own weight and delay d^k.

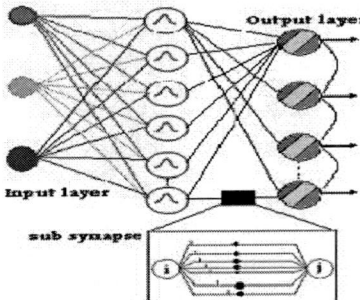

Fig. 2. Network architecture

3.2 Information encoding

The first question that arises when dealing with spiking neurons is how neurons encode information in their spike trains, since we are especially interested in a method to translate an analog value into spikes. We distinguish essentially three different approaches [9] in a very rough categorization:
Rate coding: the information is encoded in the firing rate of the neurons.
Temporal coding: the information is encoded by the timing of the spikes.
Population coding: information is encoded by the activity of different pools (populations) of neurons, where a neuron may participate of several pools.
We have used the temporal encoding proposed by Bohte et al. in [10]. By this method, the input variables are encoded with graded and overlapping activation functions, modeled as local receptive fields.
Each neuron of entry is modeled by a local receiving field. A receiving field is a Gaussian function. Each receiving field i have a center C_i given by the equation (4) and a width σ_i given by the equation (5) such as: m is number of receptive fields in each population and $\gamma = 1.5$.

$$c_i = \frac{i - 1.5}{m - 2} \quad (4)$$

$$\sigma_i = \frac{1}{\gamma(m - 2)} \quad (5)$$

3.3 Learning method

The approach presented here implements the Hebbian reinforcement learning method through a winner-takes-all algorithm [11]. For unsupervised learning, a Winner-Take-All learning rule modifies the weights between the input neurons and the neuron first to fire in the output layer using a time-variant of Hebbian learning: if the start of a PSP at a synapse slightly precedes a spike in the output neuron, the weight of this synapse is increased, as it had significant influence on the spike-time via a relatively large contribution to the membrane potential. Earlier and later synapses are decreased in weight, reflecting their lesser impact on the output neuron's spike time. For a weight with delay d^k from neuron i to neuron j we use:

$$\Delta w_{ij} = \eta L(\Delta t_{ij}) \quad (6)$$

And

$$L(\Delta t) = (1 + \beta)e^{\frac{(\Delta t - \alpha)^2}{2(k-1)}} - \beta \quad (7)$$

with $\kappa = 1 - \frac{\vartheta^2}{2\ln\frac{\beta}{1+\beta}}$

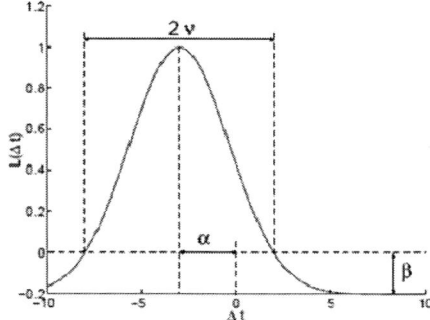

Fig. 3. Gaussian learning function

The learning window is defined by the following parameters:
- v : this parameter, determines the width of the learning window where it crosses the zero line and affects the range of Δt_{ij}, inside which the weights are increased.
- Inside the neighborhood the weights are increased, otherwise they are decreased.
- β: this parameter determines the amount by which the weights will be reduced and corresponds to the part of the curve laying outside the neighborhood and bellow the zero line.
- α: because of the time constant τ of the EPSP, a neuron i firing exactly with j does not contribute to the firing of j, so the learning window must be shifted slightly to consider this time interval and to avoid reinforcing synapses that do not stimulate j.

4. EXPERIMENTAL RESULTS AND DISCUSSION

We have chosen an image from Berkeley Segmentation Dataset and Benchmark [12] defined in pixel grid of 250x250 pixels (Fig 4).

Fig. 4. Original image

To show the influence of the number of neurons at exit on the number of areas of the segmented image, we had fixed the number of sub-synapses at 14 between two neurons, the step of training to 0.35, the choice of the base of training is random starting from the image source of 5% and numbers of receiving fields with 18 (6 for each value of intensity) and we varied the number of classes at exit. The images obtained are shown in Figure 5:

Fig. 5. Segmented image with 5 and 10 classes

To show the influence of the number of sub-synapses on the number of areas of the segmented image we had fixed the number of area at exit at 10, the step of training to 0.35, the choice of the base of training is random starting from the image source of 5% and numbers of receiving fields with 18 (6 for each value of intensity) and we varied the number of subsynapses. The images obtained are shown in Figure 6:

Fig. 6. Segmented image with 4 and 14 sub-synapses

To show the influence of the number of receptive fields on the number of classes of the segmented image we had fixed the number of area at exit at 10, the step of training to 0.35, the choice of the base of training is random starting from the image source of 5%, the number of sub-synapses at 14 and we varied numbers of receiving fields. The images obtained are shown in Figure 7:

Fig. 7. Segmented image with 4 and 6 receptive fields for each value of intensity.

To show the influence of the percentage of simple training on the number of classes of the segmented image we had fixed the number of area at exit at 10, the step of training to 0.35, the number of sub-synapses at 14 and numbers of receiving fields with 18 (6 for each value of intensity) and we varied the number of percentage of simple training. The images obtained are shown in Figure 8:

Fig. 8. Segmented image with 5% and 20 % of simple training.

To see if segmentation is close to the original image, an error metric is needed. The error between the original image and the quantized image is generally used. For this evaluation we had used the Peak Signal Noise Ratio (PSNR), the Mean Square Error (MSE), the Mean Absolute Error (MAE) and Normalized Color Difference (NCD) are therefore considered to evaluate the segmentation.
Table 1 summarizes the evaluation obtained for segmented images.

Table1. Segmentation evaluation

	MSE	PSNR	MAE	NCD
Result of fig.5 (5 classes)	819.241	63.105	20.199	0.159
Result of fig.5 (10 classes)	400.23	73.44	15.058	0.122
Result of fig.6 (4 sub-synapses)	1993.18	50.278	31.943	0.240
Result of fig.7 (4 receptive fields)	1071.88	59.227	24.706	0.195
Result of fig.8 (20% of simple training)	341.96	75.710	14.13	0.116

5. CONCLUSION

In this paper we applied spiking neural networks to image segmentation. At first, the network is build, a subset of the image pixel is taken to be learned by the network and finally the SNN processes the rest of the image to have as a result an important number of cluster (classes) quantized the image.

REFERENCES

[1] Y.J. Zhang. An Overview of Image and Video segmentation in the Last 40 Years. *Proceedings of the Sixth International Symposium on Signal Processing and Its Applications*, 148-151, 2001.

[2] W. Maass. Networks of spiking neurons: The third generation of neural network models. *Neural Networks*, 10:1659- 1671, 1997.

[3] W. Maass. On the relevance of time in neural computation and learning. *Theoretical Computer Science*, 261:157-178, 2001.

[4] S. J. Thorpe, A. Delorme and R.VanRullen. Spike-based strategies for rapid processing. *Neural Networks*, 14(6-7), 715-726, 2001.

[5] P.Dayan, L.F.Abbott. Computational and mathematical of modeling neural systems. *Theoretical neuroscience*, MIT Press 2004.

[6] H.Paugam-Moisy. Spiking neuron networks a survey. IDIAP-RR 06-11, 2006.

[7] A.L.Hodgkin, A. F Huxley. A quantitative description of ion currents and its applications to conduction and excitation in nerve membranes. *Journal of Physiology*, 117:500- 544, 1952.

[8] W.Gerstner, W.M.Kistler. Spiking Neuron Models. *The Cambridge University Press*, Cambridge, 1st edition, 2002.

[9] S.M.Bohte, H. La Poutre, J.N.Kok. Unsupervised Clustering with Spiking Neurons by Sparse Temporal Coding and Multi-Layer RBF Networks. *IEEE transactions on neural networks*, 13(2), Mar 2002.

[10] A. P. Braga, T. B. Ludemir, and C.P.L.F. André de Carvalho. Artificial Neural Networks - Theory and Applications. *LTC Editora, Rio de Janeiro*, 1st edition, 2000.

[11] M. Oster and S.C. Liu. A winner-take-all spiking network with spiking inputs. In Proceedings of the 11th IEEE International Conference on Electronics, *Circuits and Systems*, 11: 203-206, 2004.

[12] D. Martin, C. Fowlkes, D. Tal and J. Malik.A Database of Human Segmented Natural Images and its Application to Evaluating Segmentation Algorithms and Measuring Ecological Statistics. *Proc. 8th Int'l Conf. Computer Vision.* 2:416—423,2001.

Hybrid Feature Extraction-based Approach for Facial Parts Representation and Recognition

C. Rouabhia*. H. Tebbikh*

*Laboratory of Automatic and Informatics of Guelma -LAIG-
Université 8 mai 45, BP 401 - 24000 - Guelma, Algérie
(e-mail: c_rouabhia@yahoo.fr, tebbikh@yahoo.com)

Abstract: Face recognition is a specialized image processing which has attracted a considerable attention in computer vision. In this article, we develop a new facial recognition system from video sequences images dedicated to person identification whose face is partly occulted. This system is based on a hybrid image feature extraction technique called ACPDL2D (Rouabhia et al. 2007), it combines two-dimensional principal component analysis and two-dimensional linear discriminant analysis with neural network. We performed the feature extraction task on the eyes and the nose images separately then a Multi-Layers Perceptron classifier is used. Compared to the whole face, the results of simulation are in favor of the facial parts in terms of memory capacity and recognition (99.41% for the eyes part, 98.16 % for the nose part and 97.25 % for the whole face).

Keywords: Hybrid feature extraction, Facial parts, Connectionist approach.

1. INTRODUCTION

Human face is a particular visual stimulus because faces are rich in social information, very similar in structure (the eyes on dimensioned, the nose in the medium and the mouth in the lower part of the face,...) with minor differences from person to person: the eyes can be round or thin, the nose can be pointed or be blunted, the mouth can be thin or small and the skin's colour can be clear or dark,... The emotions influences, also, on these elements, for example an astonished person can have open mouth and make large eyes and an annoyed person can have a serious glance and the mouth closed,... Though we are good at recognizing persons despite all this differences, machine recognition of faces is very difficult because of variations due to head poses, lighting, facial expression, partial occlusion, and size of the face image. To perform face recognition by computer, the system involves several steps:

- Face segmentation (detection, localization) from a simple or complex background aims to determine the position of the face in the image,
- Feature extraction aims to reduce dimension of the face image and
- Face identification and/or authentication reports the person identity of unknown individual.

Psychophysics and neuroscience studies in face recognition suggest the possibility of global descriptions serving as front end for finer, feature-based perception (Zhao et al. 2003). If dominant features are present, holistic descriptions may not be used, so this paper aims to study the discriminatory power of the eyes and the nose parts separately for face recognition compared to the whole face. So, we develop neural system for person identification based on the shape of the facial elements and a hybrid feature extraction in 2D space called ACPDL2D which fusions two-dimensional principal component analysis (2DPCA) and two-dimensional linear discriminant analysis (2DLDA) proposed, respectively, in (Yang et al. 2004) and (Visani et al. 2004). In the 2nd part of this paper, we will introduce the reader to connectionist approaches applied in the field of face recognition. The 3rd section details the principle and all the steps of the proposed system: acquisition of the face database, facial elements localization, principle of the proposed hybrid method and its algorithm. Experimental results in face recognition, comparison as well as the conclusion will make the object of 4th and the 5th section respectively.

2. CONNECTIONIST APPROACHES

The artificial neural networks have a number of properties that make them highly attractive in many fields; parallel structure, learning, noise and fault tolerance, auto-adaptability, universality and good generalization ability.

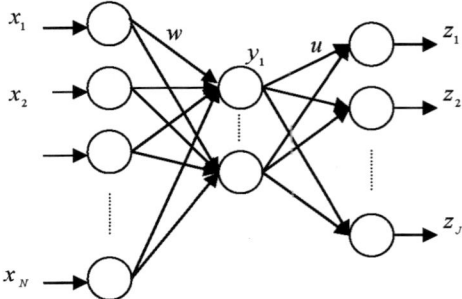

Fig. 1. Architecture of a multi-layers perceptron

One of the first artificial neural networks used for face recognition was the WISARD (Stonham, 1984) containing a separate network for each person. It has been applied for recognizing both identity and expression. A database of 16 subjects was used and full image (153×214) templates were input to a self-adapting single layer network. In (Fleming and Cottrell 1990), the authors have used nonlinear units to train a network via back-propagation. They studied the performances of the network using a full face template (64*64) for classification. In (Lin et al. 1997), a fully automatic system called Probabilistic Decision-Based Neural Network (PDBNN) has been proposed. It had three different modular to face detection, eye localization and face recognition and could achieve up to 96% correct recognition. In this paper, we are interested in the Multi-Layers Perceptron (MLP) which has been used for several tasks such as feature extraction, verification, detection and face recognition.

Algorithm 1: Back-propagation

Inputs :
  Q : number of patterns contained in the training set,
  N : number of nodes in input layer,
  J : number of nodes in output layer,
  d : desired responses,
  η : learning rate,
  Stopping criterion.

Outputs:
  Convergence error,
  Classification rate.

- Random initialisation of weights and biases of the input layer: U et b^o
- Random initialisation of weights and biases of the hidden layer: W et b^H

for $q = 1$ until Q

(A) ⎧
1. compute the responses of hidden and output layers:
$$Y = f(W^T . X_q^T + b^H) \text{ and } Z = g(Y^T . U + b^o)$$
2. compute the quadratic error: $E_q = d - Z$
3. update the weights and biases of the output layer:
$$U = U + \eta.Y.(E_q .* Z .* (1 - Z))$$
$$b^o = b^o + \eta.(E_q .* Z .* (1 - Z))$$
4. update the weights and biases of the hidden layer:
$$W = W + \eta.X^T.(E_q .* Z .* (1 - Z)U^T) .* Y .* (1 - Y)$$
$$b^H = b^H + \eta.(E_q .* Z .* (1 - Z)U^T) .* Y .* (1 - Y)$$

end

5. compute the mean square error:
$$EQM = \frac{1}{Q} \sum_{q=1}^{Q} \sum_{j=1}^{J} (d_j^{(q)} - z_j^{(q)})^2$$
6. If the stopping criterion is satisfied go to step **7**, **else** repeat **(A)**
7. **End**

Where $.*$ denotes the element-wise product

MLP is the most popular feed-forward neural networks consisting of a set of sensory units that constitute the input layers, one or more hidden layers and an output layer of computation nodes (see figure 1). Neurons in each layer are connected to all the neurons in the previous layer. The input signal propagates through the network in forward direction, from left to right and on layer-by-layer. The well known Back-propagation algorithm is often used for training the MLP. It is a gradient descent technique that consists to minimize the difference between the actual output vector of the network and the desired output vector. The speed of convergence, the generalization, and the training success are very depending on a proper choice of learning parameter, initial weights and the network topology (number of neurons in hidden layer) (for more details about MLP, see Sim. 1998). So, when applying MLPs to classification tasks, it is almost always rewarding to try more than one network topology. The algorithm 1 details the back-propagation algorithm.

3. THE PROPOSED FACIAL PARTS-BASED SYSTEM

This section is dedicated to describe the facial parts system; motivation, principle and algorithm of the proposed method. The need to recognize people whose facial elements are partially occulted either by the wearing of glasses, a muffler, or even by a beard and/or a moustache, etc., the discriminatory information of these elements and the performances of ACPDL2D method for feature extraction and dimensional reduction (Rouabhia et al. 2007) motivate us to propose the facial parts-based neural network system as shown in figure 2.

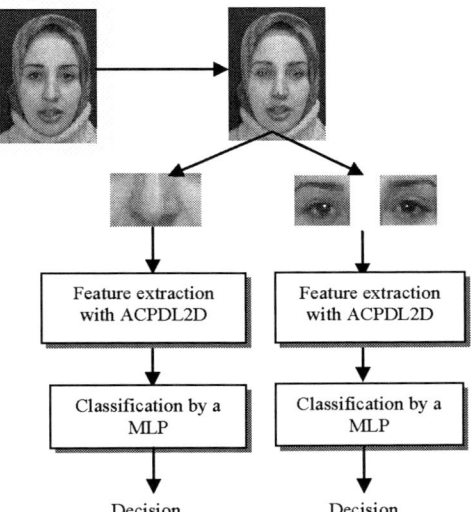

Fig. 2. The proposed facial parts-based system

This system contains different steps:

- Face database acquisition,
- Localization of the eyebrows, the eyes, the nose and facial parts extraction,
- Feature extraction using a hybrid approach ACPDL2D,

- Classification by a multi-layers perceptron trained using the popular back-propagation algorithm (see algorithm 1).

3.1 Video sequences face database

The face database we used for our applications was acquired during 2003/2004 in our laboratory (Laboratory of Automatic and Informatics of Guelma). It is composed of 400 frontal face images, a sequence of ten video images for each 40 persons (20 females and 20 males). The illumination was controlled, the background was black and each person is asked to look into the camera and pronounce a sentence. The images were acquired at a resolution of 177×144 pixels. More details about the face database are in (Boualleg, 2004).

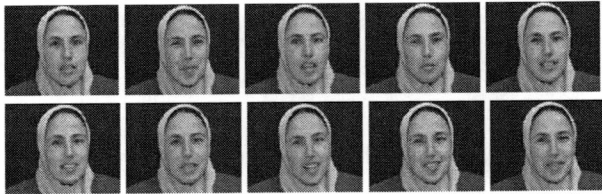

Fig. 3. Video sequence of ten images of one person in the face database

3.2 Facial features localization

To localize the eyes and the nose, we have adopted firstly the technique proposed in (Leroy et al. 1997), but we have encountered difficulties for the nose localization from the original images in particular with the bearded men, the women wearing a veil and near frontally face. To improve the localization, we have combined the technique of (Leroy et al. 1997) with the partition and the geometry of the face as shown in figure 4.

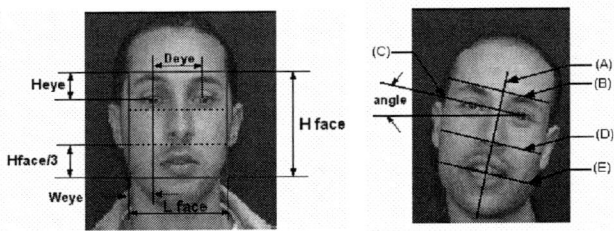

Fig. 4. Partition and geometry of the face

The anthropometric proportions (Wong et al. 2001) shown in figure 4 (at left) are based on the distance between the localized centers of iris:

$$H_{face} = 1.8 \cdot D_{eye} \quad (1)$$

$$L_{face} = 2 W_{eye} + D_{eye} \quad (2)$$

$$W_{eye} = 0.225 \cdot H_{face} \quad (3)$$

$$H_{eye} = \frac{1}{5} H_{face} \quad (4)$$

The orientation of the face is estimated by the angle of inclination of the eyes (figure 4, at right). The nose is situated in the second partition of the face and it is the point of intersection of the lines (A) and (D). To locate the eyebrows, we calculate the image gradient and search the maximum value along the vertical line passing by the iris located before. Figure 5 shows some examples of facial element localization.

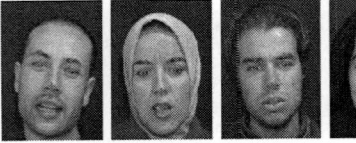

Fig. 5. Some examples of facial elements localization

3.3 Facial parts used for face recognition

Rather the whole face, the idea is to represent each person using only two eyes parts or a nose part as shown in figure 6. Each eye rectangle contains the eyebrow and the eye and the nose part image contains only the nose without the moustache to ensure a correct recognition based only on the shape of this facial element because men can leave or remove their moustache.

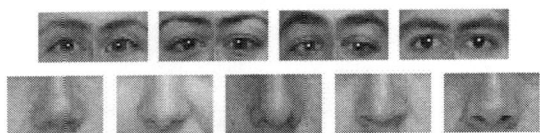

Fig. 6. Eyes and nose images used for recognition

3.4 Hybrid feature extraction: principle and algorithm

The ACPDL2D is a two-dimensional hybrid approach proposed for feature extraction and dimensional reduction. It is based on the fusion of 2DPCA (Yang et al. 2004) and 2DocLDA (Visani et al. 2004). The principal is to perform hybrid bilateral projection in 2D space. Firstly, the image facial part is projected using 2DPCA (left multiplication) then 2DocLDA (right multiplication) is used for the second feature extraction. This new representation, gives matrices of reduces size, called feature matrices or feature images which will use for the classification.

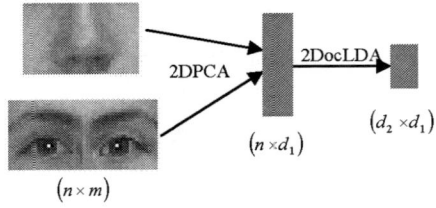

Fig. 7. Dimensional reduction of the facial parts image using ACPDL2D method

Dimensional reduction is due to the redundancies reduction among both columns and rows of the images. Figure 7 shows the ACPDL2D projection scheme. Both 2DPCA and 2DLDA are a unilateral-projection based scheme which removes redundancies among row or columns. Let X an $(n \times m)$

image, 2DPCA and 2DocLDA returns respectively an $n \times d_1$ and an $d_2 \times m$ whereas ACPDL2D returns an $d_2 \times d_1$ feature matrice where d_1 and d_2 are much smaller than the rows n and the columns m of images.

a. Training phase

The training set is composed of M images of facial part. The training phase contains two steps: i) image feature extraction using ACPDL2D method and ii) MLP training. This phase returns feature matrix for each facial part, weights and the optimal topology of MLP. The resulting feature matrices must be transformed into vectors which are the input of MLP. The architecture of the network is as follows:
- an input layer: the number of neurons is equal to the dimension of the feature vector, ($N = d_2 \times d_1$),
- a hidden layer: the number of neurons is determined empirically and
- an output layer that comprises 40 neurons; one neuron for each class.

Algorithm 2: ACPDL2D
Input: M Facial parts images
Output: Feature matrices and rate recognition
---

1. Compute the global mean: $\bar{X} = \frac{1}{M} \sum_{j=1}^{M} X_j$ where M is the total number of facial parts images in the training set.
2. Compute the image covariance matrix:
$$G_t = \frac{1}{M} \sum_{j=1}^{M} (X_j - \bar{X})^T (X_j - \bar{X})$$
3. Form a matrix $R = [R_1 R_2 \cdots R_{d_1}]$ which the columns are the first d_1 eigenvectors of G_t in decreasing order of eigenvalues,
4. Feature extraction: $Y_j = \hat{X}_j . R$ pour $j = 1, \cdots, M$
5. Facial parts reconstruction: $X_j = Y_j . R^T + \bar{X}$
6. Form a new database « Ω » containing of the feature matrices $Y_j (n \times d_1)$
7. Compute the class mean of the database Ω:
$$\bar{Y}_c = \frac{1}{n_c} \sum_{i=1}^{n_c} Y_i \quad \text{pour} \quad c = 1, \cdots, C$$
8. Compute the global mean of Ω: $\bar{Y} = \frac{1}{M} \sum_{j=1}^{M} Y_j$
9. Compute the within and between class covariance:
$$\Sigma_w = \sum_{c=1}^{C} \sum_{i=1}^{n_c} (Y_i - \bar{Y}_c)(Y_i - \bar{Y}_c)^T \text{ and } \Sigma_b = \sum_{c=1}^{C} n_c (\bar{Y}_c - \bar{Y})(\bar{Y}_c - \bar{Y})^T$$
10. Form a matrix $L = [L_1 L_2 \cdots L_{d_2}]$ which the columns are the first d_2 eigenvectors of $(\Sigma_w^{-1} \Sigma_b)$ in decreasing order of eigenvalues,
11. Feature extraction: $Z_j = L^T . Y_j$ for $j = 1, \cdots, M$
12. Transform the feature matrices $Z_j (d_2 \times d_1)$ into vectors
13. Training the MLP using algorithm 1.

b. Test phase

Each facial part from the test set is projected using ACPDL2D, then classified using the saved weights. The algorithm 2 details the proposed neural ACPDL2D method.

4. Experimental results

To perform comparison between the two facial parts, we have used two databases. The first one is composed of 400 eyes images of an (21×46) and the second one contains 400 nose images of an (21×23). The evaluation of the results obtained is carried out by randomly partitioning each database in a training set and a test sets of 200 images each one. This operation is repeated six times and the final recognition rate is the average of the all. The classification tests are performed several times according to the parameters of the PMC before preserving an optimal topology. The training process was terminated when the MSE reached 0.01.

Table 1 demonstrates experiments conducted with different number of eigenvectors on both eyes and nose images.

As shown in (Kang et al. 2004), the energy of an image is concentred on its first small eigenvectors. So, after a series of tests designed in the first projection stage using 2DPCA, we have preserved 8 eigenvectors for eyes images and 10 eigenvectors for nose images but in the second stage using 2DocLDA, the eigenvectors number is varied. With 8 eigenvectors, the eyes images achieve the highest recognition rate (99.41 %); the final feature image is a (8×8) square matrix. The nose part achieved the highest recognition rate (98.16 %) with 10 eigenvectors.

Table 1. Recognition rate

Eyes-ACPDL2D						
Size	2×8	4×8	6×8	8×8	10×8	12×8
Rate (%)	98.41	99.08	99.25	**99.41**	99.41	99.33
Nose-ACPDL2D						
Size	2×10	4×10	6×10	8×10	10×10	12×10
Rate (%)	97.25	**98.16**	97.66	97.75	97.41	97.41

4.1 Comparison of the facial parts with the whole face

The system performances are also compared to the whole face. We have segmented the face to eliminate the background, hair and veils. In figure 8, two examples of faces segmented are demonstrated. Each image is (73×56) pixels.

Fig. 8. Examples of the whole faces in the training set

Three methods have been compared; 2DPCA, 2DLDA and ACPDL2D. In all the case, MLP was used for classification. Because both 2DPCA and 2DLDA depend on the proper choice of the number of eigenvectors, we have plot in figure 9 and figure 10 respectively the magnitude of the eigenvalues in decreasing order and design a series of tests using different eigenvectors (Rouabhia et al. 2007). We have preserved 8 eigenvectors using 2DPCA, 6 eigenvectors with 2DocLDA and the feature matrix obtained from ACPDL2D is a (8×8).

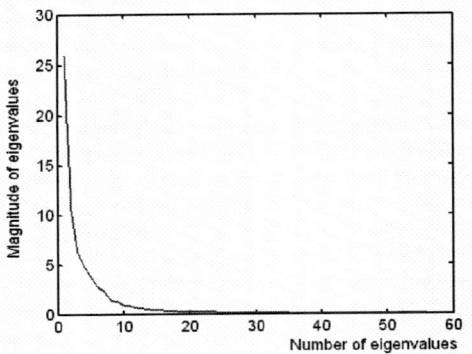

Fig. 9. Magnitude of the eigenvalues in decreasing order obtained with 2DPCA

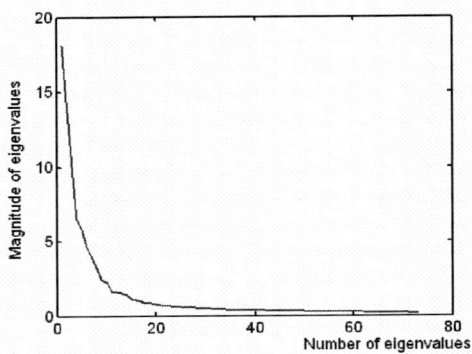

Fig. 10. Magnitude of the eigenvalues in decreasing order obtained with 2DocLDA

Table 2 reports the best rate recognition obtained with the whole face compared to the geometric approach Neuro-ACP proposed in (Boualleg et al. 2004). Through experiments, we can conclude that facial parts are more efficient than the whole face and geometric distances. Eight distances have been calculated between the eyes, the nose and the mouth.

Table 2. Comparison of the performances (Rouabhia et al. 2007)

	Neuro-ADL2DoC	Neuro-ACPDL2D	Neuro-ACP2D	Neuro-ACP
Rate (%)	97.25	97.00	96.50	94.50
Size ($d_2 \times d_1$)	6 × 56	8 × 8	73 × 8	---

5. CONCLUSIONS

In this paper, we are interested in human face recognition. We have proposed a neural system based on the facial parts from video sequences. Our system takes advantages of image compression using a bilateral hybrid projection method called ACPDL2D, generalization ability of the MLP trained using the back-propagation algorithm and the discrimination of the facial elements. Also, it is insensitive to background, hairstyles and overcomes the problem of the big size of the face image. This is due to firstly, the 2D spatial information is kept, secondly, the eyes and the nose parts are much smaller than the whole face image and, thirdly, redundancies are removed among both rows and columns using ACPDL2D method. Through comparison experiments done between the eyes, the nose and the whole face on the video sequences database of 40 individuals, we conclude that the eyes are the most discriminating (99.41 %) followed by the nose (98.16 %) and the whole face (97.25 %). Our neural facial parts based system could be dedicated to person identification from frontally face images and whose faces are partly occulted.

REFERENCES

Boualleg, A/H. (2004). La reconnaissance automatique des visages. *Mémoire de magister*, université 8 mai 45 de Guelma.

Fleming, M. and G. Cottrell (1990). Categorization of faces using unsupervised feature extraction. *Proceeding of IJCNN*, Vol.2.

Leroy, B., A.Chouakria, L.Herlin and E.Diday (1996). Approche géométrique et classification pour la reconnaissance de visage. *INRIA*, France.

Lin, S.H., S.Y.Kung and L.J.Lin (1997). Face recognition/detection by probabilistic decision-based neural network. *IEEE Trans. Neural Networks*, vol.8, pp.114-132.

Rouabhia, Ch., A/H.Boualleg and H.Tebbikh (2007). Approche bidimensionnelle hybride Neuro-ACPDL2D pour la reconnaissance automatique de visages. *In Proceedings of the 4th International Conference: Science of Electronic, Technologies of Information and Telecommunications*: (SETIT'2007), March 25-29, 2007, Tunisia

Sima, J. (1998). Introduction to neural networks. *Technical report*, No. V-755.

Stonham, T.J. (1984). Practical face recognition and verification with WIZARD. *Aspects to Face Processing*, pp.426-441.

Visani, M., C.Garcia, J.M.Jolion (2004). Two Dimensional-Oriented Discriminant Analysis for Face Recognition. *In Proc. of the Int. Conf. On Computer Vision and Graphics ICCVG'04* à paraître dans la série Computational Imaging and Vision, Varsovie, Pologne.

Wong, K-W., K-M. Lam, W-C. Siu (2001). An efficient algorithm for human face detection and facial feature extraction under different conditions, *Centre for Multimedia Signal Processing, Department of Electronic and Information Engineering, The Hong Kong Polytechnic University, Hung Hom, Hong Kong, Pattern Recognition* 34.

Yang, J., D.Zhang, A.F. Frangi and J-Y.Yang (2004). Two dimensional PCA: a new approach to appearance-based face representation and recognition. *IEEE Transaction on Pattern Analysis and Machine Intelligence*, Vol.26, No.1.

Zhao, W., R.Chellappa, P.J.Philips and A.Rosenfeld (2003). Face Rrecognition: A literature Survey. *ACM Computing Surveys*, Vol.35, No.4, pp.399-458.

SESSION A2: INTELLIGENT TRANSPORTATION SYSTEM

Robust Second Order Sliding mode Observer for the Estimation of the Vehicle States

A. Chaibet* L. Nouveliere** S. Hima*** S. Mammar****

*Université d'Évry val d'Essonne, France. IBISC/CNRS-FRE 3190, 40 rue du Pelvoux CE1455, 91025, Evry, Cedex, France (e-mail: achaibet@estaca.fr).
**Université d'Évry val d'Essonne, France. IBISC/CNRS-FRE 3190, 40 rue du Pelvoux CE1455, 91025, Evry, Cedex, France (e-mail: nouveliere@iup.univ-evry.fr).
***Université d'Évry val d'Essonne, France. IBISC/CNRS-FRE 3190, 40 rue du Pelvoux CE1455, 91025, Evry, Cedex, France (e-mail: salim.hima@iup.univ-evry.fr).
**** M. Said Mammar with INRETS/LCPC - LIVIC Laboratoire sur les Interactions Véhicule-Infrastructure-Conducteur. 14, route de la Minière, Bât 824, 78000, Versailles, France (e-mail: said.mammar@inrets.fr)

Abstract: This paper is dedicated to the observation of non measurable variables for automotive systems. A non linear observer, based on a sliding mode approach, is presented for the estimation of the dynamic states of the vehicle. The considered technique is applied to the estimation problem for an automated vehicle following. Both the simulation and the experimental results are addressed to demonstrate the effectiveness of the sliding mode observer for different maneuvers, in terms of performances and robustness.

Keywords: Non linear observer, Sliding mode, robustness, driver assistance.

NOMENCLATURE

v_x/v_y	Longitudinal/lateral vehicle speeds
T_c	Equivalent drive and brake torque
T_{rr}	Rolling resistance torque
C_x/C_y	Longitudinal/lateral aerodynamic drag coefficients
m	Vehicle mass
I_{eff}/I_z	Effective longitudinal inertia/ inertia moment about the yaw axis through the vehicle center of gravity
F_{yf}/F_{yr}	Cornering forces at the front tires/rear tires
l_f/l_r	Distances of the front and rear tires from vehicle's center of gravity
α_f/α_r	Slip angle of the front and rear tires
c_f/c_r	Cornering stiffness of the front and rear tires
δ_f	Steering angle

1. INTRODUCTION

Regarding the automated vehicle concept, the vehicle must be able to achieve several autonomous functions. Among these functions, there are : the heading variation, the lane change maneuver and the vehicle following control in order to maintain a safety distance [1]. To realize them, the vehicle must be equipped with proprioceptive and exteroceptive sensors that cannot measure all the vehicle states. To measure the whole vehicle states, the dynamic is not feasible for different reasons as economic and technical feasibility. In order to overcame these constraints, a solution is to use of the observer concept.

One can find in the literature several kinds of observers for vehicle state variables. For example, in [5], [4] a linear observer is presented (Proportional Integral observer) and a Kalman filter for the lateral control of a vehicle. [9] developed a non linear observer for the estimation of the lateral and longitudinal velocities of automotive vehicles. We opt to a sliding mode observer based on the algorithm of an optimal control for the nonlinear system. The estimation of the vehicle state variables retains a complexity of the model and strong non linearities. Moreover, the vehicle is subjected to many disturbances like the state of the roadway : dry, softened, snow-covered, icy.... The sliding mode observer is known to be a robust technique that is appropriate for the estimated uncertain systems. High robustness is maintained against various kinds of uncertainties such as external disturbances and measurement error. This non-linear observer will be used to estimate the leader and follower vehicle state variables. The presented paper is organized as follows. In Section 2 a vehicle modelling and positioning is addressed. In section 3, the description of the observer design is given. Section 4 deals with the validation of the proposed observer by some simulation examples of scenarios of driving. Then the analysis of the robustness of the observer is evaluated with respect to these various uncertainties. Finally section 5 is devoted to the obtained experimental results and their analysis. Conclusion and perspectives of this work are presented in section 6.

2. VEHICLE MODELING AND POSITIONING

In this section a simplified model of the vehicle dynamic often used for control design is presented, under the assumptions that there is no longitudinal slip between the tire and the road and that the pitch, the roll and the vertical dynamics are neglected.

For more clearness, the indices (l,s,r) will respectively represent the leader, the follower and the relative variables. The overall dynamic can be described by three equations: the first and the second correspond respectively to the longitudinal and lateral translation dynamics and the last one describes the yaw motion [3, 1]

$$\begin{cases} \dot{v}_{x_s} = \dfrac{T_c - T_{rr}}{I_{eff}} - \dfrac{C_x v_{x_s}^2}{m} + v_{y_s}\dot{\psi}_s \\ \dot{v}_{y_s} = \dfrac{F_{y_f} + F_{y_r} - C_y v_{y_s}^2}{m} - v_{x_s}\dot{\psi}_s \\ \ddot{\psi}_s = \dfrac{1}{I_z}\left[l_f F_{y_f} - l_r F_{y_r}\right] \end{cases}$$

The expression of the cornering tire forces are given as follows

$$\begin{cases} F_{y_f} = 2c_f \alpha_f \\ F_{y_r} = 2c_r \alpha_r \end{cases}$$

where

$$\begin{cases} \alpha_f = \delta_f - \dfrac{v_{y_s} + l_f \dot{\psi}_s}{v_{x_s}} \\ \alpha_r = -\dfrac{v_{y_s} - l_r \dot{\psi}_s}{v_{x_s}} \end{cases}$$

The model can be rewritten in a canonical form

$$\dot{v}_{x_s} = f_0 + g_0 T_c \quad (1)$$
$$\dot{v}_{y_s} = f_1 + g_1 \delta_f \quad (2)$$
$$\ddot{\psi}_s = f_2 + g_2 \delta_f \quad (3)$$

$$\begin{cases} f_0 = \dfrac{-T_{rr}}{I_{eff}} - \dfrac{C_x v_{x_s}^2}{m} + v_{y_s}\dot{\psi}_s \\ g_0 = \dfrac{1}{I_{eff}} \\ f_1 = \dfrac{-2c_f}{m}\dfrac{v_{y_s} + l_f \dot{\psi}_s}{v_{x_s}} - \dfrac{2c_r}{m}\dfrac{v_{y_s} - l_r \dot{\psi}_s}{v_{x_s}} - \dfrac{C_y v_{y_s}^2}{m} - v_{x_s}\dot{\psi}_s \\ g_1 = \dfrac{2c_f}{m} \\ f_2 = \dfrac{-l_f}{I_z} 2c_f \dfrac{v_{y_s} + l_f \dot{\psi}_s}{v_{x_s}} + \dfrac{l_r}{I_z} 2c_r \left(\dfrac{v_{y_s} - l_r \dot{\psi}_s}{v_{x_s}}\right) \\ g_2 = \dfrac{l_r}{I_z} 2c_f \end{cases} \quad (4)$$

The aim of this paper is to estimate the vehicle states in the case of a vehicle following. So we must give the mathematical formulation of the longitudinal (inter distance) and lateral (lateral displacement) dynamic errors [2]

$$\begin{cases} \dot{d}_{xr} = -v_{x_l} + d_{yr}\dot{\psi}_l + v_{x_s}\cos\psi_r + (v_{y_s} + \dot{\psi}_s l_f)\sin\psi_r \\ \dot{d}_{yr} = -v_{y_l} + (l_r - d_{xr})\dot{\psi}_l - v_{x_s}\sin\psi_r + \\ (v_{y_s} + \dot{\psi}_s l_f)\cos\psi_r \end{cases} \quad (5)$$

The derivative of (5) gives

$$\begin{cases} \ddot{d}_{xr} = a_0 + b_0 T_c + c_0 \delta_f \\ \ddot{d}_{yr} = a_1 + b_1 T_c + c_1 \delta_f \end{cases} \quad (6)$$

where

$$\begin{cases} a_0 = -\dot{v}_{x_l} + d_{yr}\ddot{\psi}_l + \dot{d}_{yr}\dot{\psi}_l - f_0 \cos\psi_r - v_{x_s}\sin\psi_r \dot{\psi}_r \\ \qquad + (v_{y_s} + l_f \dot{\psi}_s)\cos\psi_r \dot{\psi}_r + (f_1 + f_2 l_f)\sin\psi_r \\ b_0 = g_0 \cos\psi_r \\ c_0 = (g_1 + g_2 l_f)\sin\psi_r \\ a_1 = -\dot{d}_{xr}\dot{\psi}_l + d_{xr}\ddot{\psi}_l - (\dot{v}_{y_l} - \ddot{\psi}_l l_r) - f_0 \sin\psi_r \\ \qquad - (v_{y_s} + l_f \dot{\psi}_s)\sin\psi_r \dot{\psi}_r + (f_1 + f_2 l_f - v_{x_s}\dot{\psi}_r)\cos\psi_r \\ b_1 = -g_0 \sin\psi_r \\ c_1 = (g_1 + g_2 l_f)\cos\psi_r \end{cases} \quad (7)$$

with

$$\psi_r = \psi_l - \psi_s \quad (8)$$

3. SYNTHESIS PROCEDURE AND OBSERVER DESIGN

Because of the strong non linearities of the system and the parameters uncertainties in the modelling (speed, adhesion,...), we adopt a sliding mode observer scheme based on a suboptimal control algorithm. This one allows the finite time stabilization of uncertain and nonlinear systems with incomplete states measurement [7, 8]. Let us give the different measurable and non measurable variables in the following table, where the last column designates if the considered state is measurable (m) or not measurable (nm).

v_{xl}/v_{xs}	Longitudinal speed of leader vehicle/ follower	m/m
ψ_l/ψ_s	Yaw angle of leader/follower vehicle	nm/m
$\dot{\psi}_l/\dot{\psi}_s$	Yaw rate of leader/follower vehicle	nm/m
$\ddot{\psi}_l$	Yaw acceleration of leader vehicle	nm
$\dot{v}_{yl}$	Lateral acceleration of leader vehicle	nm
d_{xr}/d_{yr}	Longitudinal distance/ lateral displacement	m/m
$\dot{d}_{xr}/\dot{d}_{yr}$	Variation of longitudinal distance/ Variation of lateral displacement	nm/nm
$\dot{\psi}_r$	Variation of relative heading	nm

Table 1: Measurable and not measurable variables

3.1 Estimation of the variation of the relative yaw angle

It is assumed that the relative heading ψ_r for both vehicles is measured by a stereoscopic video system. The aim is to estimate the unknown variable $\dot{\psi}_r$ from ψ_r. The relative yaw dynamic can be expressed as follows

$$\begin{cases} \dot{x}_1 = x_2 \\ \dot{x}_2 = f_2 + g_2 \delta_f \end{cases} \quad (9)$$

with $x_1 = \psi_r$ and $x_2 = \dot{\psi}_r$. x_2 is not a measurable variable and f_2 and g_2 are smooth functions, assumed as uncertain.
Firstly, the observer equations are given by

$$\begin{cases} \dot{\hat{\psi}}_r = \hat{v}_{lacet\_r} \\ \dot{\hat{v}}_{lacet\_r} = \eta_r(t) \end{cases}$$

where $\eta_r(t): \mathscr{R}^+ \to \mathscr{R}$ is an auxiliary input to be appropriately selected to guarantee the observer convergence in finite time. $\dot{\hat{\psi}}_r = \hat{v}_{lacet\_r}$ is the estimate of the relative yaw angle variation.
Now, the second step consists in defining the observation errors

$$e_{\psi_r} = \psi_r - \hat{\psi}_r \quad (10a)$$
$$e_{v_{lacet\_r}} = v_{lacet\_r} - \hat{v}_{lacet\_r} \quad (10b)$$

From the (10a) and (10b), one can deduce the equations of the dynamic error

$$\begin{cases} \dot{e}_{\psi_r} = e_{v_{lacet\_r}} \\ \dot{e}_{v_{lacet\_r}} = \ddot{\psi}_r - \eta_r(t) \end{cases} \quad (11)$$

with

$$\left| \ddot{\psi}_l - f_2 - g_2 \delta_f \right| \leq M \quad (12)$$

where M is a positive constant. Furthermore, $\ddot{\psi}_s$ satisfies

$$-\ddot{\psi}_{smax} \leq \ddot{\psi}_s = f_2 + g_2 \delta_f \leq \ddot{\psi}_{smax} \quad (13)$$

Now, the problem is to find η_r such that the errors (10a) and (10b) are steered to zero in finite time.

Observer algorithm for the variation of the relative yaw angle [8, 7]

i) Set $\gamma^* \in (0,1]$

ii) Set $e_{\psi_r max} = e_{\psi_r}(0)$

Repeat, for $t > 0$, the following steps

iii) If $[e_{\psi_r}(t) - \frac{1}{2} e_{\psi_r max}][e_{\psi_r max} - e_{\psi_r}(t)] > 0$ then

$\gamma = \gamma^*$ else set $\gamma = 1$

iv) If $e_{\psi_r}(t)$ is an extremal value then set $e_{\psi_r max} = e_{\psi_r}(t)$

v) Apply the control law $\eta_r(t)$:

$$\eta_r(t) = -\gamma U_{Max} \text{sign}(e_{\psi_r}(t) - \frac{1}{2} e_{\psi_r max})$$

Until the end of the control time interval.

with $U_{Max} > \max(\frac{M}{\gamma^*}; \frac{4M}{3-\gamma^*})$

Remark 1. The value of M is obtained empirically by using the bounded states of vehicle characteristics.

3.2 Estimation of the yaw rate of the leader

The yaw rate $\dot{\psi}_l$ of the leader is not measurable. But we know that the yaw rate of the follower is measured by a gyroscope. Therefore, we can use the estimate of the variation of the relative yaw angle, previously developed, in order to deduce the variation of the yaw angle for the leader

$$\dot{\psi}_r = \dot{\psi}_l - \dot{\psi}_s \quad (14)$$

$$\hat{\dot{\psi}}_l = \hat{\dot{\psi}}_r + \dot{\psi}_s \quad (15)$$

3.3 Estimation of variation of the inter distance

It is assumed that the inter distance is obtained by a LiDAR. In order to estimate the variation of the longitudinal distance $\dot{d}_{xr}$, we should express the dynamic of this variable. Consider the system

$$\begin{cases} \dot{x}_1 = x_2 \\ \dot{x}_2 = a_0 + b_0 T_c + c_0 \delta_f \end{cases} \quad (16)$$

where $x_1 = \dot{d}_{xr}$ and $\dot{x}_2 = \ddot{d}_{xr}$. The second equation is similar to the equation (6), where T_c et δ_f are the control inputs and a_0, b_0, c_0 are smooth functions.

Considering the following system

$$\begin{cases} \dot{\hat{d}}_{xr} = \hat{v}_{xr} \\ \dot{\hat{v}}_{xr} = \eta_{v_{xr}}(t) \end{cases}$$

where $\hat{d}_{xr}$ corresponds to the estimate of the variation of the longitudinal distance, $\eta_{v_{xr}}(t)$ is an auxiliary input which ensures the convergence on finite time of the observer.

Defining the following observer errors

$$e_{d_{xr}} = d_{xr} - \hat{d}_{xr}$$
$$e_{v_{xr}} = \dot{d}_{xr} - \hat{\dot{d}}_{xr}$$

the equations of the dynamic error can be written as

$$\begin{cases} \dot{e}_{d_{xr}} = e_{v_{xr}} \\ \dot{e}_{v_{xr}} = a_0 + b_0 T_c(t) + c_0 \delta_f(t) - \eta_{v_{xr}}(t) \end{cases}$$

The aim is to find out the positive constant M_1 to satisfy the constraint below

$$|a_0 + b_0 T_c + c_0 \delta_f| \leq M_1 \quad (17)$$

in order to steer $\hat{d}_{xr}$ to d_{xr} in finite time. The bounding of the functions a_0, b_0, c_0 ensures the existence of M_1.

Remark 2. The following table summarizes the different variables which intervene in the convergence of the algorithm

$\gamma^* \in (0,1]$	$\gamma^* = 0.75$
$x_{Max} = x_1(0)$	$e_{d_{xr} max} = e_{d_{xr}}(0) = 10m$
$x_1(t)$	$e_{d_{xr}}(t)$
U_{Max}	$U_{Max} > \max(\frac{M_1}{\gamma^*}; \frac{4M_1}{3-\gamma^*}) = 15$

Table 2 : Variables for the estimation of the interdistance

3.4 Estimation of the variation of the lateral displacement

It is assumed that the lateral displacement is given by a video sensor. Once again, we follow the same way as previously. The different steps for the estimation of the variation of the lateral displacement is summarized in the following table :

$\gamma^* \in (0,1]$	$\gamma^* = 0.75$
$x_{Max} = x_1(0)$	$e_{d_{yr} max} = e_{d_{yr}}(0) = 0.05$
$x_1(t)$	$e_{d_{yr}}(t)$
U_{Max}	$U_{Max} > \max(\frac{M_2}{\gamma^*}; \frac{4M_2}{3-\gamma^*}) = 5$

Table 3 : Variables for the estimation of the lateral displacement

3.5 Estimation of the lateral acceleration of the vehicle leader

It is assumed that the lateral velocity of the follower is measured and the relative velocity is obtained by a LiDAR or by a video sensor. Therefore it is possible to obtain the lateral velocity of the leader. In such a case, we know this measure and one seeks to estimate the lateral acceleration of the leader. The different steps can be summarized with the observer equations

$$\begin{cases} \dot{\hat{v}}_{yl} = \hat{a}_{yl} \\ \dot{\hat{a}}_{yl} = \eta_{a_{yl}}(t) \end{cases}$$

where $\hat{v}_{yl}$ is the estimate of the acceleration and $\eta_{a_{yl}}(t)$ is the auxiliary input to be determined. The observer errors are :

$$e_{v_{yl}} = v_{yl} - \hat{v}_{yl} \quad (18a)$$
$$e_{a_{yl}} = a_{yl} - \hat{a}_{yl} \quad (18b)$$

and the dynamic error equations are

$$\begin{cases} \dot{e}_{v_{yl}} = e_{a_{yl}} \\ \dot{e}_{a_{yl}} = F(v_{yl}, a_{yl}) - \eta_{a_{yl}}(t) \end{cases}$$

The input $\eta_{a_{yl}}(t)$ and the gain M_3 are chosen such as

$$|F(v_{yl}, a_{yl})| \leq M_3$$

The different numerical values of the algorithm are addressed in the following table :

$\gamma^* \in (0,1]$	$\gamma^* = 0.75$
$x_{Max} = x_1(0)$	$e_{v_{yl}\max} = e_{v_{yl}}(0) = 0$
$x_1(t)$	$e_{v_{yl}}(t)$
U_{Max}	$V_{Max} > \max(\frac{M_3}{\gamma^*}, \frac{4M_3}{3-\gamma^*}) = 2$

Table 4 : Variables for the estimation of the vehicle leader acceleration

4. SIMULATION RESULTS

In this section, several simulations have been carried out in order to show the effectiveness of the proposed algorithm for a car-following problem.

4.1 Car following on a circular road

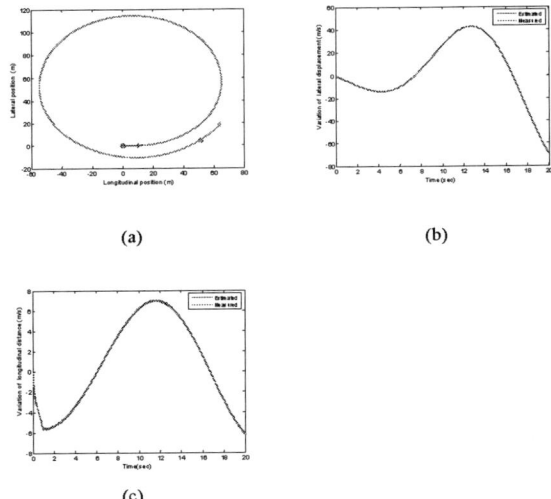

Fig. 1. Circular road (a): lateral and longitudinal positions of the leader (*,-.) and the follower (o,-), (b): variation of the lateral displacement, (c) : variation of the longitudinal distance.

One considers that both vehicles are initially located on the same lane distant from $10m$. They are evolving on a circular trajectory with the same longitudinal speed of $16m/s$. To carry out this trajectory, one applies the same steering angle for both vehicles (2.9 deg). The evolution of both vehicles in the (x,y) plane can be shown on figure 1a). According to the

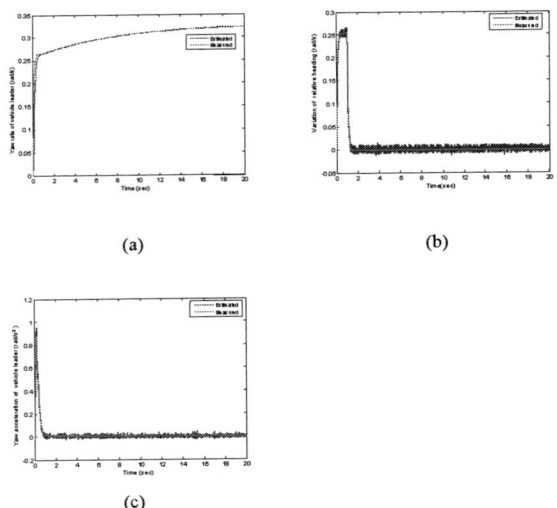

Fig. 2. Circular road (a): yaw rate of the leader vehicle,(b): variation of relative yaw angle, (c) : lateral acceleration of the leader vehicle

figures 1 and 2, one can see the juxtaposition of the real state variables with the estimated ones obtained by the observer. The convergence in finite time of the observers is satisfied. It can be also noted that the steady state error is near zero.

4.2 Lane change maneuver

This scenario is summarized as follows : initially, both vehicles are located at the same right lane. They evolve at the same velocity of 16 m/s, with an initial inter distance of $10m$. This speed is maintained roughly constant by applying a constant composite torque. The leader initiates a lane change maneuver and one can admit that the follower generates an identical steering angle profile. It can be noticed that this control is carried out in opened loop. The objective here is only to highlight the performances of the observer. The figure 3a) shows the profile of both vehicles in the (x,y) plane. On all the figures, the measured signals (in red) and their estimates (in blue) are simultaneously represented. One can notice that

- The estimated state variables converge quickly towards the real state variables.
- The obtained performances are good as well in dynamic as in static.
- The observation errors are steered to zero in finite time

4.3 Robustness of the observer

This part deals with the evaluation of the robustness of the proposed observer against the degradation of the road adhesion. This degradation results in a reduction of the cornering stiffnesses c_f and c_r from 20% and 70% of their nominal values. These reductions correspond respectively to a wet and very slipping road. The results appear on figures 5 and 6. The plot of the obtained signals with the observer are almost identical to those of the real signals. These results thus show the robustness of the observers with respect to the parametric uncertainties (variation of the adhesion) for both cases of wet and very slipping road.

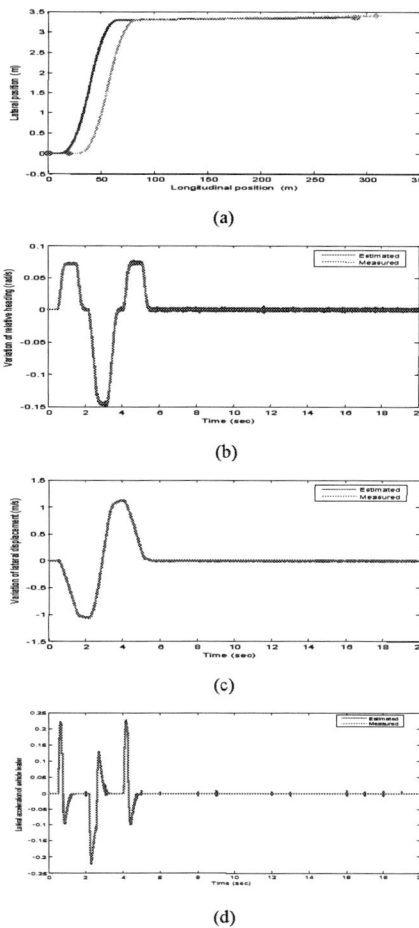

Fig. 3. lane change maneuver. (a): lane change maneuver lateral and longitudinal positions of the leader (*,-.) and the follower (o,-), (b): variation of relative yaw angle, (c): variation of the lateral displacement, (d): lateral acceleration of the vehicle leader.

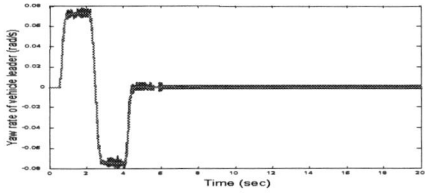

Fig. 4. lane change maneuver : yaw rate of the leader vehicle

5. EXPERIMENTAL RESULTS

The performances of the proposed observer applied with real data collected during experiments have been designed in this section. Measures have been carried out with the prototype vehicle VIPER [1] on the test track of the LIVIC laboratory in Versailles-Satory. This vehicle has been instrumented with sev-

[1] http://www.inrets.fr/ur/livic/

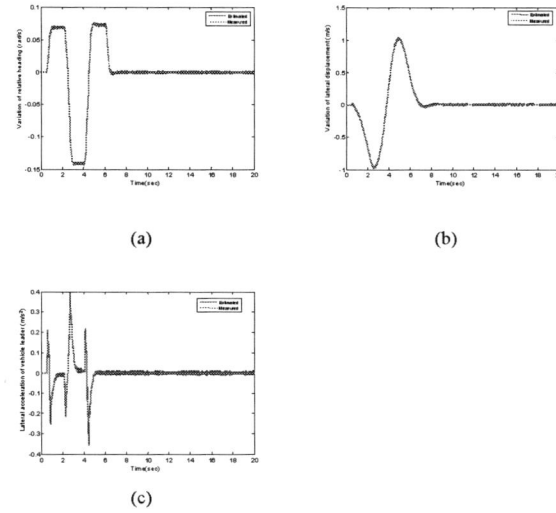

Fig. 5. case of wet road (a): variation of the relative yaw angle, (b): variation of the lateral displacement, (c): lateral acceleration of leader vehicle

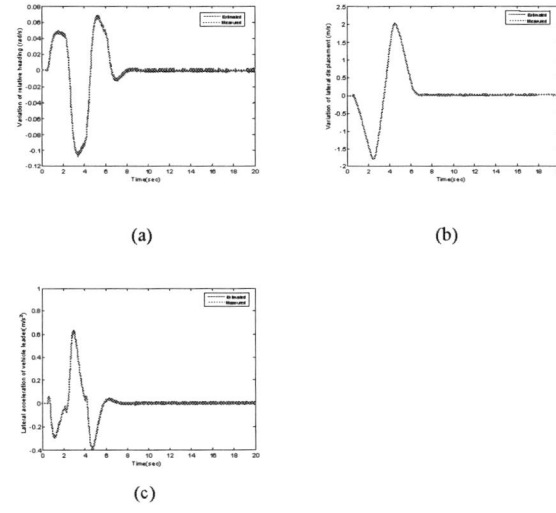

Fig. 6. case of slippery road (a): variation of the relative yaw angle, (b): variation of the lateral displacement, (c): lateral acceleration of leader vehicle

eral sensors (gyroscope, INS sensors, topometer and video sensor). The vehicle trajectory is raised by a RTK-GPS (the plot of the test track is given on the figure 7a)). The figure 7b) gives an outline of the measured and the estimated lateral displacement. We can note that both plots are great identical which shows the convergence of the observer in a finite time with a negligible steady state error. According to the measurement of the lateral displacement, the variation of the lateral displacement is estimated and presented on the figure 8a). One can notice that the estimated curve follows the measured one. In order to illustrate the performances of our observer approach, we can compare the longitudinal velocity given by the topometer with that obtained by the observer (see figure 8b). Once again, a fast

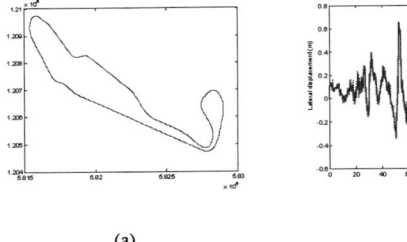

Fig. 7. Experimental results for path following. (a): test tracks, (b): lateral displacement

convergence of the estimated velocity towards the measured speed is ensured. Thereafter, we have estimated the longitudinal acceleration using the measured longitudinal speed. It appears on the figure 8c)). Both responses of the acceleration measured by the INS sensor and that rebuilt by the observer are similar. In the same way than previously, we carried out the estimate of the yaw angle, the measured yaw rate is measured by a gyroscope sensor (see figure 8d) and e)). According to these figures the dynamic and static performances are very satisfactory.

6. CONCLUSION

The main interest of this paper is to propose a new approach of the second order sliding mode observer to estimate vehicle state variables in a problem of car-following and path-following. Simulations have been carried out to illustrate the ability of this approach to give well performances of the states estimation. Afterwards, an evaluation of the robustness of this observer versus uncertainties on the model parameters has also been verified. Experimental tests confirm the adopted approach.

REFERENCES

[1] A. Chaibet, L. Nouveliére, S. Mammar M. Netto, Backstepping Control Synthesis For Both Longitudinal And Lateral Automated Low Speed Vehicle, IEEE, Intelligent Vehicles, Las vegas, 2005.

[2] A. chaibet, Contrôle Latéral et Longitudinal pour le Suivi de Véhicule, PhD thesis, IBISC-LSC, Université D'Evry Val D'Essonne, 2006.

[3] D. Swaroop, Seok Min Yoon, The design of a controller for vehicle following in an emergency lane change maneuver, California PATH Working Paper UCB-ITS-PWP-99-3, 1999.

[4] D. Koenig, S. Mammar, ., "Design of Proportional-Integral Observer for Unknown Input Descriptor Systems", *IEEE Transactions on Automatic Control*, vol 47, No 12, 2002.

[5] S. Mammar, D. Koenig, "Reduced order unknown input kalman filter: application for vehicle lateral control", *American Control Conference*, 2003

[6] S. Mammar, D. Koenig, Vehicle Handling improvement by Active Steering, Vehicle System Dynamics, vol 38, No3, 211-242, 2002.

[7] G. Bartolini, A. Ferrara E. Usai, Applications of a Sub Optimal Discontinuous Control Algorithm for Uncertain Second Order Systems, International Journal of Robust and Nonlinear Control, 7, 4, 297-310, 1997

[8] A. Ferrara P. Pisu, Minimum Sensor Second-Order Sliding Mode Longitudinal Control of Passenger Vehicles, IEEE

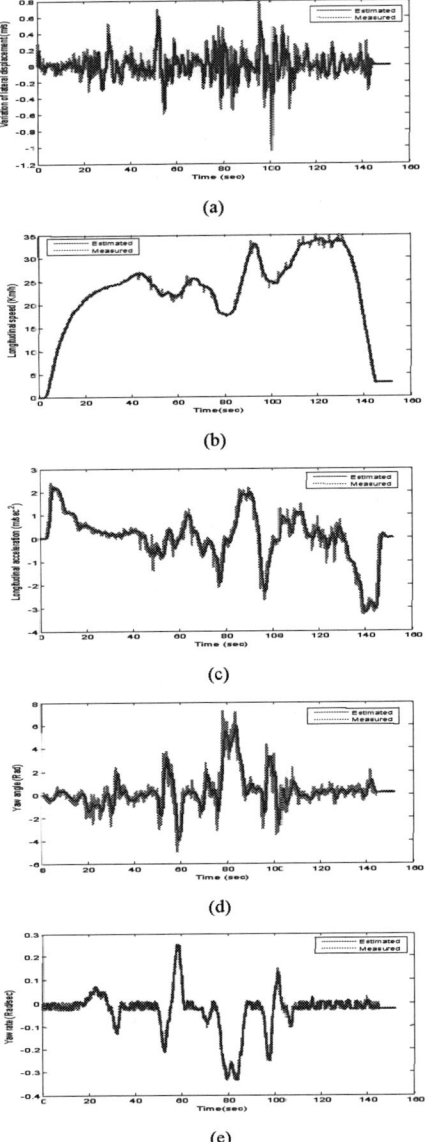

Fig. 8. Experimental results for path following (a): variation of the lateral displacement, (b): longitudinal velocity, (c): longitudinal acceleration, (d): yaw angle, (e): yaw rate

Transactions on Intelligent Transportation Systems, 5,1, March 2004.

[9] L. Imsland T. A. Johansen, T. I. Fossen, H.F Grip, J.C. Kalkkuhl, A. Suissa, Vehicle velocity estimation using nonlinear observers, *Automatica*, June 2006.

Modeling of Steer-by-Wire System Used in New Braking Handwheel Concept

K. Messaoudène*, N. Ait Oufroukh*, S. Mammar*,**

*Informatics, Integrative Biology and Complex Systems, 40 Rue de Pelvoux, 91020 Evry Cedex, France
(e-mail: {Kamel.messaoudene, Naima.aitoufroukh}@ibisc.fr).
** INRETS/LCPC - LIVIC Vehicle-Infrastructure-Driver Interactions Laboratory,
14 route de la Minière, Bât 824, 78000, Versailles, France (e-mail: mammar@inrets.fr)

Abstract: The handwheel is one of the primary control mechanisms of automobile thus interaction between the handwheel and the driver is critical to safety. The driver applies forces that direct the vehicle while the handwheel communicates feedback information to the driver of the forces experience by the car within its environment. The handwheel also provides a predictable mechanical feel to the driver to allow smooth and safe control. Many researchers tried to reproduce this feeling by creating steer-by-wire systems. This paper explores this new concept of handwheel and it describes the modeling steps of the components including the restitution mechanism for force feedback and its various links with the vehicle lateral dynamics and the pneumatic contacts. The aim is to explore the possibility to combine a braking device within the steer-by-wire system in order to provide a more suitable and ergonomic device to the driver.

Keywords: Handwheel, Brake, Steer-by-wire, Modeling, Vehicle's dynamics, Pneumatic contact

1. INTRODUCTION

The X-by-wire technology, which is widely spread in aeronautic application is now entering ground vehicles. It aims to eliminate the mechanical links between control devices (handwheel, pedals, etc) and actuators (engine, brakes, steer) and replace them by the electronic links, associating sensors and the electric engines (Normand, 2005). Specialists often present this technology as the next big revolution in the automotive industry. Car manufactures are developing this concept since now many years ago and it is really close to the market. Some comfort oriented applications of x-by-wire are already available such as in functions concerning window and seats which are controlled electrically. It is now quiet ready to spread to vehicle control functions such steer, brake, shift and suspension.

The purpose of this paper is to explore the feasibility of an integrated braking device to the handwheel which can make easier vehicle control. First of all, the concept of proposed braking handwheel device is described. The advantages provided by this new solution and its operating modes are presented. In section 3, the x-by-wire technology and its particular application to steer are explored. The usefulness of this technology for the proposed device realisation are presented.. The modeling of as steer-by-wire system is carried out in section 4. The different synthesis, which would allow restoring a handwheel haptic feedback, is studied by means of vehicle lateral dynamics and tires contact model.

The estimation of the vehicle lateral velocity, which is necessary for the haptic device, is obtained using a two-wheels vehicle model (bicycle) combined with a kalman observer. In section 5, simulation results will an overview of different driver torque inputs and the resulting handwheel dynamics, vehicle dynamics and. The calculation of the aligning moment from which the motor torque, which restores handwheel force feedback is also presented Conclusion and future works are given in the last section.

2. OPERATING MODE OF BRAKING HANDWHEEL

The device proposed in this paper aims in particular driving assistance for paraplegic peoples but can be used by all the drivers. It consists in a new automotive brake/accelerate device coupled to a steer-by-wire system (Figure 1). The originality of the proposed device is in the braking tool which is integrated to the handwheel. Brake is obtained by exerting a pressure force on the handwheel. The acceleration is simply obtained through an already available solution which consists of a circle behind the handwheel. This device allows the driver to brake and accelerate while still keeping the hands on the handwheel in all driving situations. This also offers the advantage of a significant optimization in braking time and a more flexible and comfortable driving. In fact recent studies on the driver's behaviour have shown that in case of abrupt brake, the driver tightens the hands more strongly on the handwheel (PSA, 2005), thus, the proposed braking device will take advantage of this driver reaction and will provide a kind of instinctive braking.

For safety reasons and in order to respect the ergonomic aspect of this device, the braking movement of the handwheel is released only by slackening the accelerating circle. In addition this functioning mode makes the device in operation mode of the accelerator and brake pedals on a normal vehicle. Thus the translation movement of the handwheel is inhibited when the accelerator is actuated. This inhibition is released as soon as the accelerating circle is slackened (Messaoudène et al., 2007, a).

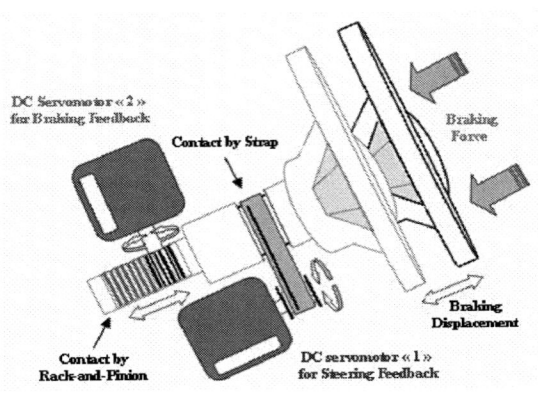

Fig. 1. Braking handwheel concept

The proposed system can be designed as passive or active. However an active type offers the possibility of a progressive brake which can also make the manoeuvre more intuitive. This solution will permit the management of the in-depth handwheel displacements and the control of feedback braking torque. It can also open many others prospects in the active braking assistance field. Servomotors which are shown Figure 1 are used for the implementation.

The realization of such a device requires in-depth studies in several domains with both theoretical and practical aspect. In this context, some theoretical studies have been achieved; they are related to the operation mode of the braking device on handwheel particularly the modelling of the dynamic behavior of the bust in the driving situation, and modeling of vehicle's dynamic (Messaoudène *et al.*, 2007, b). In the following, another topic is described; it consists in the modeling of a steer-by-wire system.

3. ARCHITECTURE OF STEER-BY-WIRE SYSTEM

The main advantage of steer-by-wire systems is the elimination of the steering column. A European Legislation now allows this elimination (Sécurité routière, 2004). Concerning the driver's security, this new alternative permits a considerable reduction of risks in case of collision, the entry of the direction column inside the cockpit being often synonymous of serious injuries for the driver. Besides, the disappearance of this particularly heavy and cumbersome organ simplifies vehicle cabin design and entails a reduction of the consumption of fuel. This would represent a non-negligible factor toward the challenge of environment (Wilwert, 2005).

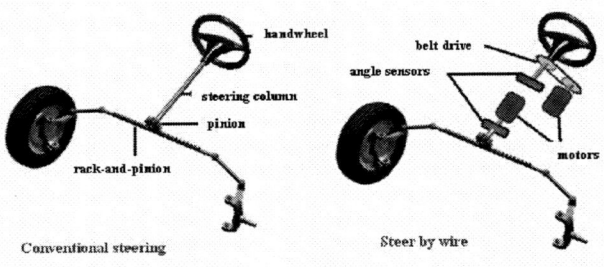

Fig. 2. Steer-by-wire concept

In a conventionally steered vehicle, the handwheel is mechanically connected to the wheels, and the driver controls the steering system of the vehicle through this mechanical linkage. In a steer-by-wire vehicle, this mechanical link has been removed, and some sort of actuator controls the wheels in response to commands from the handwheel (Fig. 2). Steer-by-wire vehicles require also an artificial force feedback to replicate the forces generated by the tires in a conventional steering system. In addition, steer-by-wire systems open up new possibilities for vehicle control (Coe, 2004). For example, the variable steering ratio of the steer-by-wire system brings a remarkably increased comfort to the driver. This function enables the steering ratio between the handwheel and the wheels to adapt according to the driving conditions. In parking and urban driving, this ratio should be smaller in order to reduce the amplitude of the handwheel rotation (Wilwert, et al., 2005).

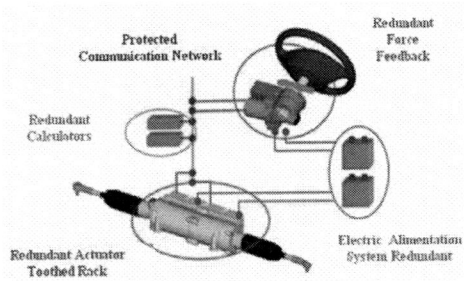

Fig. 3. Components of steer-by-wire system

The steer-by-wire system still has a rack-and-pinion manipulated not with the steering column, but by double electric engines. The handwheel is equipped of course of an angular position sensor, and also of an electric engine in order to provide to the driver a corresponding effort to the one normally generated by wheels. The driver has thus a feel of the road, even though it is artificial (Yih, 2005) and synthesized from measured and estimated variables. The steering being a very sensible function of safety, all system elements are generally in redundancy: calculators, batteries, electric circuits, actuator of rack-and-pinion, motor for steering feedback (Fig. 3).

4. MODELING OF STEER-BY-WIRE SYSTEM

The restitution of handwheel force feedback is a paramount step. Indeed, during steering the driver applies a torque on the handwheel in order to give to the vehicle the desired trajectory. In counterpart, the efforts due to the tire/road contact and the efforts related to the dynamics of the vehicle go up through the steering column until the handwheel. These efforts feel on the handwheel is obtained thanks to the steering column which allows these torque exchanges. These exchanges are of great importance for safety as they provide to the driver several information on the road adhesion limits for example. Being given that the steer-by-wire system does not have a steering column, one have to produce artificially the handwheel force feedback (Ueda, 2002).

Works on handwheel force feedback have focused on transmitting in a comfortable way to the driver the mechanical aligning moment resulting from the steering geometry. This is largely due to the fact that the only available data for torque synthesis is the aligning moment. Many researchers have also investigated nonlinear modelling of steering feel (Segawa, 2002), as well as means to ensure that the road wheels track the handwheel adequately. Other works have developed force feedback for alternative steering devices such as a joystick (Tokunaga, 2002).

The motor torque τ_{motor}, which ensures torque feedback, is a combination of several torques, which are as follows (Switkes, 2004):

- Damping torque τ_{damp}
- Inertia torque $\tau_{inertia}$
- Aligning moment τ_{align}

4.1 Handwheel dynamics

The goal of the handwheel feedback is to reproduce most accurately the same feel as when driving a conventional vehicle. For that, the handwheel dynamics must be in its general form as the one of a conventional steering system. For such a system, the dynamics are of the form of a differential equation of second order. By taking into account several elements of which J_{eq} the equivalent inertia of the steering system, b_{eq} the coefficient of dynamic frictions (or coefficient of viscous frictions) and θ_h the handwheel angle:

$$J_{eq}\ddot{\theta}_h(t) + b_{eq}\dot{\theta}_h(t) = \tau_{driver} - \tau_{motor} \quad (1)$$
$$\text{with } \tau_{motor} = \tau_{damp} + \tau_{inertia} + \tau_{align}$$

The general diagram of the steer-by-wire concept is presented in Figure 4. It has as input the torque τ_{driver} that the driver exerts on the handwheel. Equation (1) determines the handwheel angle θ_h and steering angle θ_s, where K_s is the steering ratio, such that $\theta_s = K_s \theta_h$.

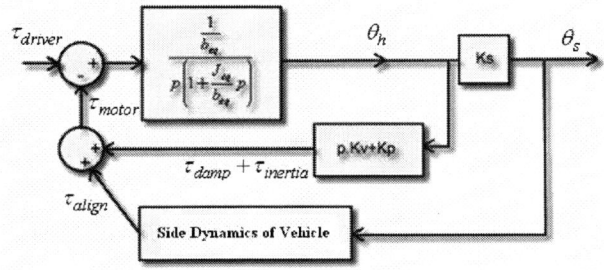

Fig. 4. Block diagram for handwheel force feedback

Firstly, the damping and inertia torques are determined in order to overcome the absence of the mechanical linkage. The expression of the sum of these two torques is chosen as follows:

$$\tau_{damp} + \tau_{inertia} = (K_v p + K_p)\theta_h \quad (2)$$

With K_v and K_p are coefficients fixed to have more or less significant handwheel feedback.

4.2 Calculation of aligning moment

The aligning moment is the origin of the major feedback on the handwheel of conventional vehicles. This moment tends also to bring back the wheels in the vehicle longitudinal axis. The aligning moment is caused primarily by the lateral dynamic's behavior of the vehicle (Mohellebi, 2005). Therefore, it is significant to carry out a study on this aligning moment, which would be also the principal component of the motor torque (connected to the handwheel). The purpose here is to determine an aligning moment from a vehicle model which presents the side slip angle as a state variable. According to the tire dynamics, the aligning moment is given by the equation below:

$$\tau_{align} = F_\alpha (c+a)\cos\phi_c \quad (3)$$

The parameter "c" represents the intersection of the axis of pivot with the roadway and the point of contact of the wheel with the roadway, this distance is called the geometrical trail. When the vehicle moves, a pneumatic trail "a" is added to a geometrical trail. Pneumatic trail is defined as being the distance between the point of application of the side force and the point of the reaction force developed by the tire. The angle created by the axis of pivot and the vertical axis is noted ϕ_c and is called trail angle (William, 1995). The geometrical trail is obtained from the following relation:

$$c = r\tan(\phi_c), \text{ with "} r \text{" a ray of the wheel}$$

All these variables are depicted in Figure 5.

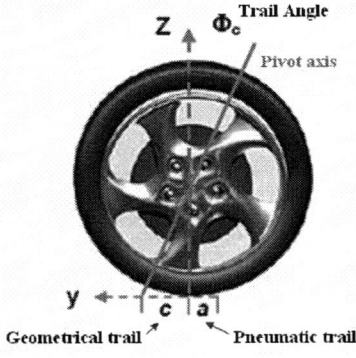

Fig. 5. Geometry of the front tire of vehicle

It is now necessary to express the lateral force F_α, in order to be able to obtain the aligning moment. For this, a study on the behavior of tires dynamics is conducted in the section below.

4.3 The lateral dynamics of the tires

A tire is designed to develop a sufficiently significant side force so that the vehicle is not deviated of its trajectory during a curve. This side force depends on several elements which are mainly the state the tire and the type of road on which the vehicle rolls. Without constraints, a wheel is in equilibrium position; it rolls according to its longitudinal axis. if a side force is exerted, the direction of the movement of the wheel deviates of its longitudinal axis and causes a deformation at the tire/road contact (Mendoza, 2000). This deformation generates an angle α called side slip angle, which give raise to a lateral force F_α shown in Figure 6. This force is mainly a function of the slip angle and the static load on the tire noted N_f. It consists to the normal force which is exerted on the tire (depends on the mass of the vehicle). The contact surface sheared by the side force is moving in the opposite vehicle's trajectory. This displacement creates a pneumatic trail.

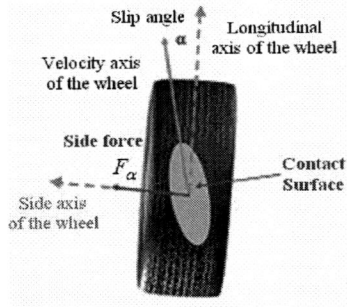

Fig. 6. Side force on the tire

The general form of the lateral force is shown in Figure 7. It is characterized by three zones, which are the linear zone, the transitory zone and the saturation zone (Lechner, 2001).

- **Linear Zone:** The variation of the side force is linear according to the slip angle, the contact is comparable thus to a pure stiffness.

- **Transitory zone:** The variation of the side force in this case is nonlinear, the side force is evolving more slowly with the increase in the slip angle up to a maximum value of the slip angle, and this value depends on the adherence properties of tire.

- **Saturation zone:** The increase of the slip angle does not involve any more increase of side force; it represents the limit of the tire adherence. In this case, the tires cannot provide additional adhesion force.

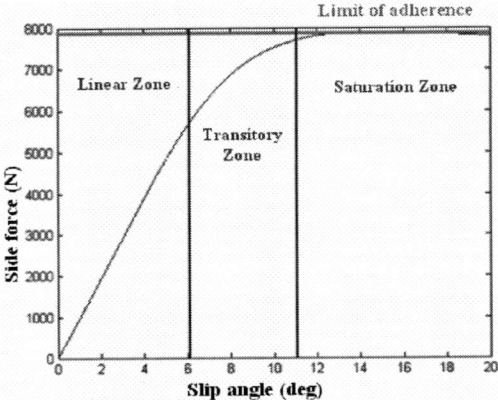

Fig. 7. Characteristic curve of side force

Various models of side force of the tires are existing, the most known being the Pacejka model "the magic formula" (Pacejka, 1997). It is an empirical model based on experimental measures allowing to fix the model parameters. Another empirical model with simpler parameters setting is also available (Marchant, 1995). The obtained curve is less precise but is generally sufficient for the realization of a simulation for a handwheel force feedback. Using this model, the tire side force takes the following:

$$F(\alpha) = \mu(\alpha) N_f \qquad (4)$$

$\mu(\alpha)$ is the adhesion coefficient of the tires. Mainly, this coefficient is a function of slip angle of wheels, it approximated by the following equation:

$$\mu(\alpha) = 2 C_p \alpha_p \frac{\alpha}{\alpha_p^2 + \alpha^2} \qquad (5)$$

C_p is a coefficient which characterizes the maximum adhesion of the tires for each type of road surfacing, for example (for the asphalt C_p =0.8, for the wet asphalt C_p =0.2). Product $C_p N_f$ represents the maximum side force that the tire supports before this one reaches the limits of adherences (saturation zone), α_p is slip angle corresponding to the maximum value of the slip force.

4.4 Two-wheels vehicle model (bicycle)

The dynamic model of the vehicle establishes relation between lateral and yaw motion and tire/road contact forces. For this study, a two-wheels vehicle model with linear forces is. This model is shown on Figure 8 (Imine, 2003).

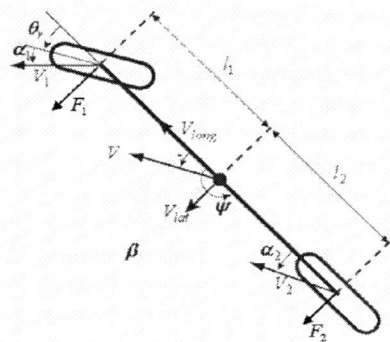

Fig. 8. Bicycle model of vehicle

The components of interest for the computation of the aligning moment are primarily the slip angle at the vehicle centre of gravity (function of the side velocity) and the yaw rate. In order to write the translation and rotation motion equations, it is necessary to establish the sum of the side forces and the sum of the moments around the yaw axis. The sum of the lateral forces is given by:

$$\sum F_y = (F_1 + F_2) \quad (6)$$

The side forces of the front and rear tires are noted respectively F_1 and F_2, they depend on the front slip angles α_1 and rear slip angle α_2 respectively. The slip angles depend on the yaw angle and slip angle in the centre of gravity of the vehicle. Because when a vehicle moves on a rectilinear trajectory, the longitudinal velocity vector of the vehicle V_{long} is parallel to the axis of this trajectory. However, as soon as the vehicle is in situation of curve trajectory, a lateral velocity V_{lat} appears and deflects the velocity of the vehicle displacement of an angle β, being the slip angle at the centre of gravity. This angle is obtained as follows:

$$\beta = \arctan\left(\frac{V_{lat}}{V_{long}}\right) \quad (7)$$

For normal driving situations (low lateral velocity), this angle is generally small and linear approximation can be made:

$$\beta = \left(\frac{V_{lat}}{V_{long}}\right)$$

It is important to specify that the side velocity of the front wheels is not completely the same as one of the vehicle. Because of the yaw movement, the front wheels would have a velocity equal at the lateral velocity of the vehicle to which the component $l_1\dot{\psi}$ is added. Thus, the slip angle of the front wheels is defined as being the difference between the steering angle and the angle created between the velocity vector of the wheel and the vehicle longitudinal. Then one can give the expression of the slip angle of the front and rear wheels by the following equations:

$$\begin{cases} \alpha_1 = \theta_r - \beta - \dfrac{l_1\dot{\psi}}{V_{long}} \\ \alpha_2 = -\beta + \dfrac{l_2\dot{\psi}}{V_{long}} \end{cases} \quad (8)$$

The side forces contribute also to the yaw movement of the vehicle. Indeed, the side forces exert on the tire a rotation around its centre of gravity. The resultant yaw moment is given by:

$$\sum \tau_\psi = F_1 l_1 - F_2 l_2 \quad (9)$$

Assuming cornering stiffness C_1 for the front wheel and C_2 for the rear one, the model equations are given by:

$$\begin{cases} \left(\dfrac{-C_1-C_2}{mV_{long}}\right)\beta + \left(-1+\dfrac{C_2 l_2 - C_1 l_1}{m(V_{long})^2}\right)\dot{\psi} + \left(\dfrac{C_1}{m}\right)\theta_r - \dot{\beta} = 0 \\ \left(\dfrac{C_2 l_2 - C_1 l_1}{I_z}\right)\beta + \left(\dfrac{-C_1 l_1^2 - C_2 l_2^2}{I_z V_{long}}\right)\dot{\psi} + \dfrac{C_1 l_1}{I_z}\theta_r - \ddot{\psi} = 0 \end{cases} \quad (10)$$

These equations take into account several variables, which are listed in Table 1:

Table 1. Parameters used in modeling

l_1 / l_2	Distance from the centre of gravity to the front / rear axis [m]
C_1 / C_2	Cornering stiffness of the front / rear tires [N.rad^{-1}]
m	Total mass of vehicle [kg]
I_z	Inertia moment of vehicle in Z axis [kg.m²]
θ_r	Steering angle of the front wheels [rad]
β	Slip angle in the centre of gravity [rad]
$\dot{\psi}$	Yaw rate [m.s^{-1}]
V_{long}	longitudinal velocity [m.s^{-1}]

According to the differential equations, one can write the following second order state-space equation:

$$\begin{bmatrix} \dot{\beta} \\ \ddot{\psi} \end{bmatrix} = \begin{bmatrix} \dfrac{-C_1-C_2}{mV_{long}} & -1+\dfrac{C_2 l_2 - C_1 l_1}{m(V_{long})^2} \\ \dfrac{C_2 l_2 - C_1 l_1}{I_z} & \dfrac{-C_1 l_1^2 - C_2 l_2^2}{I_z V_{long}} \end{bmatrix}\begin{bmatrix} \beta \\ \dot{\psi} \end{bmatrix} + \begin{bmatrix} \dfrac{C_1}{m} \\ \dfrac{C_1 l_1}{I_z} \end{bmatrix}\theta_r \quad (11)$$

The model is simulated with the following numerical values: V_{long} =16 m/s, $C_1 = C_2 = 50000$ N/rad, $l_1 = 1.0065$ m, $l_2 = 1.4625$ m, $I_z = 2454$ kg.m², $m = 1500$ kg.

A steering angle profile, representing a slalom maneuver is used. The curves of the slip angle and yaw rate, obtained under Simulink are shown on the Figure 9.

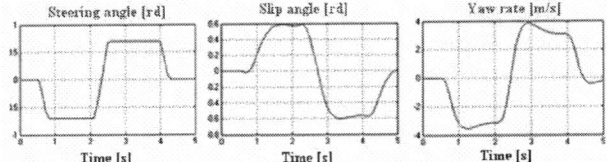

Fig. 9. Curves of two-wheels vehicle's model

4.5 Lateral velocity estimation by Kalman filter

The Kalman filter consists to determining the estimators of system's variables when the environment presents random disturbances. In this part, we deal with the stochastic aspect (noise) of the observer's concept. The goal is to determine an optimal system which minimizes the error variance between the value of real variable and its estimate.

A. *Discrete Kalman filter*

The Kalman filter makes it possible to solve the problems of prediction and estimate in a recursive way in the time. Being given a discrete stochastic linear system whose dynamic evolution is modelled by using state equation and observation or measurement equation:

$$\begin{cases} x_{k+1} = A_k . x_k + B_k . u_k + w_k \\ y_k = C_k . x_k + v_k \end{cases} \quad (12)$$

Where k represents the successive moments of time, x_k is system state, y_k is the output (measurement or observation), u_k the unquestionable input. w_k is the input noise, v_k is the measurement noise. A_k, B_k, C_k are the matrices of the state equation. The state noises and measurement noises are supposed to be white, centred and whose variance expression is as follows:

$$\begin{cases} W_k = E(w_k . w_k^T) \\ G_k = E(v_k . v_k^T) \end{cases} \quad (13)$$

Then, we solve simultaneously a prediction and estimate problems. The recursivity is decomposed into these three following steps:

- Prediction of x_{k+1} by using $y_0, y_1, y_2, \ldots, y_k$. Value will be noted $\hat{x}_{k+1/k}$ (predicted value of x_{k+1} knowing the observations claiming until k.
- New measurement y_{k+1}.
- Estimate of x_{k+1} by using $\hat{x}_{k+1/k}$ and y_{k+1}. Estimated will be noted $\hat{x}_{k+1/k+1}$.

B. *Kalman filter algorithm*

According to Kalman, the difficulty of filtering for the dynamic system is to find best estimate $\hat{x}_k$ of state x_k at the instant k, by employing the observations carried out until the discrete instant $k-1$, according to the criterion of the minimum conditional variance. The Kalman algorithm is articulated in the following way:

- Initialization:

$$\begin{aligned} X_{0/0} &= E[X_0] \\ \Sigma_{0/0} &= \Lambda_0 \end{aligned} \quad (14)$$

- Recurrence with step (k+1), we determine in the order:

$$\begin{aligned} \Sigma_{k+1/k} &= A_k \Sigma_{k/k} A_k^T + \Gamma_k \\ H_{k+1} &= \Sigma_{k+1/k} . C_{k+1}^T (\Omega_{k+1} + C_{k+1} . \Sigma_{k+1/k} . C_{k+1}^T)^{-1} \\ X_{k+1/k} &= A_k X_{k/k} + B_k U_k \\ X_{k+1/k+1} &= \Sigma_{k+1/k} + H_{k+1} . (Y_{k+1} - C_{k+1} . X_{k+1/k}) \\ \Sigma_{k+1/k+1} &= [I - H_{k+1} C_{k+1}] \Sigma_{k+1/k} \end{aligned} \quad (15)$$

Matrix H_k is called "gain of Kalman" or "correction matrix". The construction of this algorithm permits to achieve two goals:

- Linear filter minimizing a priori variance of estimation error.
- Filter maximizing the posterior probability of the estimated parameters. That is applicable only on the assumption of Gaussian noises.

Knowing that Kalman filter gives importance to the model or to the measurement according to the importance of the error variance of the measurement noise compared to the state variance.

C. *Application of the Kalman algorithm in our case*

A Kalman filter is used for the estimation of the lateral velocity of the vehicle which is not directly measurable by means of conventional sensors. In fact the COREVIT sensor price is of about 20 thousands euros It is necessary for us to estimate this lateral velocity using only the yaw rate by exploiting measurement of the steering angle (Figure 10).

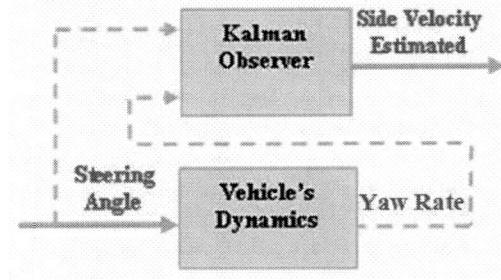

Fig. 10. Block diagram of Kalman observer

The proposed Kalman filter as presented before is applied on the discrete time version of the vehicle model which is of the form $X_{k+1} = A_k X_k + B_k U_k$. According to first order discretization method, the matrices A_k and B_k are given by:

$$\begin{cases} A_k = I + A.\Delta T \\ B_k = B.\Delta T \end{cases} \quad (16)$$

With I the identity matrix of the same dimension as A and ΔT is sampling period. The measurement and state noises were added to this system. Figures 11 and 12 illustrate the estimate of side velocity for respectively zero and nonzero initial values.

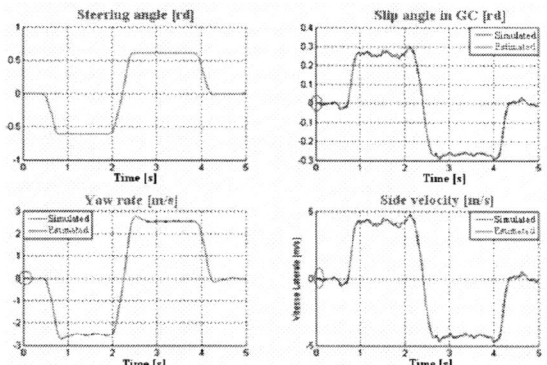

Fig. 11. Estimation with zero initial conditions

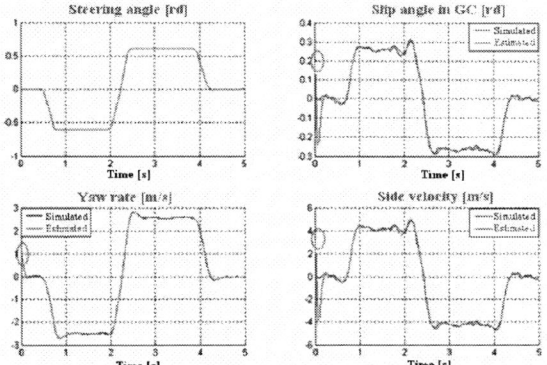

Fig. 12. Estimation with nozero initial conditions

These results show clearly the Kalman filter is able to estimate the lateral velocity and thus the side slip angle.

5. SIMULATION RESULTS

Having all the components, the whole system is modelled in Simulink. It presents several blocks consisting in: driver torque profile, dynamics of the handwheel, dynamics of the vehicle, computation of the aligning moment and the motor torque, which provides handwheel force feedback.

The driver torque used in this simulation is a signal of 5s time duration, varying from -5Nm to 5Nm. It corresponds to the driver input on the handwheel when negotiating a slalom maneuver (see Figure 13).

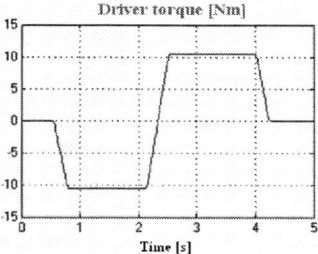

Fig. 13. Driver torque profile

The dynamics of the handwheel are represented by the differential equation of the second order, which was given previously. The damping torque and inertia torque whose curves are illustrated in Figure 14. This figure also shows the evolution of the handwheel angle resulting from the combination of the motor torque and driver torque.

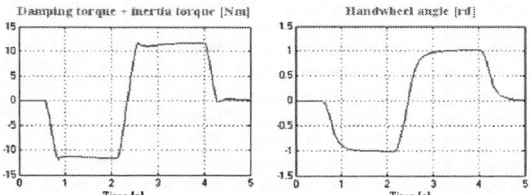

Fig. 14. Curves obtained from the handwheel dynamics

The vehicle dynamics enable us to have access to the yaw rate and the slip angle at the centre of gravity of the vehicle. These variables allow the calculation of the aligning moment. The obtained curves are shown in Figure 15.

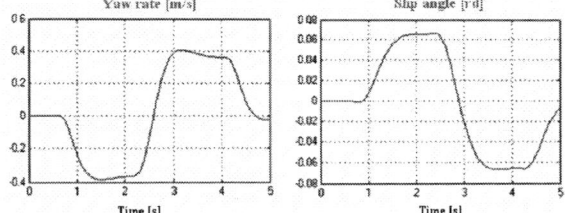

Fig. 15. Curves obtained for bicycle vehicle's model

The aligning moment is obtained from the vehicle state, the dynamics of the handwheel and some other parameters related to the tire pneumatic characteristics. Figure 16 illustrates the curve of the aligning moment at the handwheel during driving in the chicane at a constant velocity of 13 m/s equivalent to 47 km/h.

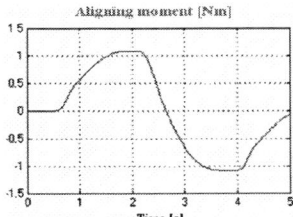

Fig. 16. Aligning moment curve

The goal of this modeling is to reproduce on a steer-by-wire system a reaction forces feel on traditional steering system, in particular the haptic feedback on handwheel. This artificially force feedback is restored to the handwheel of steer-by-wire device by the motor which is located near to the handwheel. The curve of this torque is presented below on Figure 17.

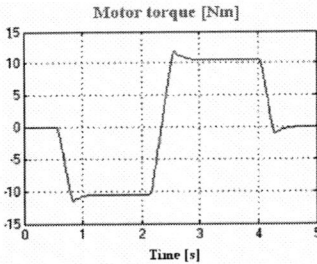

Fig. 17. Motor torque curve

6. CONCLUSIONS

New solution of braking device on handwheel was described in the first part of this paper. We showed its operation mode and the various advantages brought by this new device in terms of safety, comfort and ergonomics. This device is integrated in a steer-by-wire system, for that, we devoted the remainder of this paper to the modeling of this new steering technology. A study of the dynamics of the vehicle was necessary and more particularly the lateral dynamics, in order to be able to have all the elements necessary to the modeling of steer-by-wire device of the vehicle. A key element is the restitution to the driver of the aligning moment. This is achieved by the use of a vehicle lateral model coupled with a Kalman filter in order to estimate the side slip angle which is not directly measurable. Simulation results showed that the Kalman filter provide a good estimation of the state variables.

The modelling of the steer-by-wire system was carried out under Simulink, and the results which rise from these simulations have conclusive behaviour. The goal of this modelling part is to show that it is possible to restore accurately the aligning moment as with a conventional steering system. The model described in this paper will be validated by experimental tests in future works and can be used for a variety of force feedback schemes, like control schemes made possible by steer-by-wire. With effective handwheel force feedback and advanced steer-by-wire algorithms, the cars of the future will be higher performance, safer, accessible for all and more fun to drive.

REFERENCES

[1] Coe, A. P, Switkes and J.C. Gerdes. (2004). Using Mems Accelerometers to Improve Automobile Handwheel State Estimation for Force Feedback. Paper, IMECE 2004-62183.

[2] Imine, H. Observation d'états d'un véhicule pour l'estimation du profil dans les traces de roulement. Ph.D Thesis. University of Versailles, France, 2003.

[3] Lechner, D. Analyse du comportement dynamique des véhicules routiers légers : Développement d'une méthodologie appliquée à la sécurité primaire. Ph.D Thesis. Centrale School of Lyon-France, 2001.

[4] Marchant, S. Réalisation d'un simulateur automobile. Ph.D Thesis. University of Toulouse- France, CNRS LAAS, 1995.

[5] Mendoza, R.A. Sur la modélisation et la commande des véhicules automobiles. Ph.D Thesis. University of Grenoble, 2000.

[6] Messaoudène, K., N. Ait Oufroukh and S. Mammar. (2007). Driving Assistance for Paraplegic People " Brake on Handwheel Coupled to the Steering-by-wire System". IFAC'07 Toulouse-FRANCE. 2-3-4 September 2007.

[7] Messaoudène, K., N. Ait Oufroukh and S. Mammar. (2007). Modeling of the Bust Dynamics in the Driving Situation for a New Braking Device on Handwheel. LT'07 Sousse - TUNISIE. 17-20 Novembre 2007.

[8] Mohellebi, H. Conception et réalisation de systèmes de restitution de mouvement et de retour haptique pour un simulateur de conduite à faible coût dédié à l'étude comportementale du conducteur. Ph.D Thesis. University of Evry- France, 2005.

[9] Normand, M. (2005). X-by-wire Prototype. Le Monde, 11 September 2005.

[10] Pacejka H. B and I. J. Besselink. Magic formula tyre Model with transient properties. Vehicle System Dynamics, 1997.

[11] PSA. (2005). By-wire C5 Citroën Demonstrator. Press Release, 24 June 2005.

[12] Segawa, M., Kimura, S., Kada, T., and Nakano, S. 2002. A study of reactive torque control for steer by wire system. AVEC'02.

[13] Sécurité Routière. (2004). Réglementation et Législation dans l'automobile. France.

[14] Switkes, P,. A. Coe and J.C. Gerdes. (2004). Handwheel Force Feedback for Lanekeeping Assistance: Combined Dynamics and Stability. Paper, AVEC'04.

[15] Tokunaga, H., Misaji, K., Takimoto, S., Shimizu, Y., and Shibata, K. 2002. Steer feel evaluation 'method based on analytical method of equivalent linear system using the restoring force model of power function type'. AVEC'02, pp. 849–854.

[16] Ueda, E., Inoue, E., Saki, Y., Hasegawa, M., Takai, H., and Kimoto, S. 2002. The development of detailed steering model for on-center handling simulation. AVEC'02, pp. 657–662.

[17] William.F and L. Milliken. Race car vehicle dynamics, 1995.

[18] Wilwert, C. (2005). Influence des Fautes Transitoires et des Performances Temps Réel sur la Sûreté des Systèmes X-by-Wire. Thesis. Polytechnic National institute of Lorraine, France, 24 March 2005.

[19] Wilwert, C., N. Navet, Y.Q. Song and F. Simonot-Lio. (2005). Design of Automotive X-by-Wire Systems. Paper.

[20] Yih, P. (2005). Steer-by-wire: Implications for vehicle handing and safety. Stanford university. USA.

Sensor Position Identification and Vehicle State Estimation Using The Extended Kalman Filter.

H. Slimi*. C. Sentouh*. S. Mammar*. L. Nouveliere*

*Informatics, Integrative Biology and Complex Systems,
40 Rue de Pelvoux, 91020 Evry Cedex, France
{ Hamid.Slimi, Chouki.Sentouh, said.mammar, nouveliere }@iup.univ-evry.fr.

Abstract: In this paper, an extended Kalman filter (EKF) based identification method for longitudinal and lateral sensor's positions is proposed. The previous works which have shown the effectiveness of Kalman observer for estimating states and parameters of the vehicle have motivated the presented one. This method is very effective because it provides to know at the same time an estimate of the sensor location and the vehicle state components. It uses a lateral model of the vehicle with three degrees of freedom (lateral translation, yaw rate and roll rate).

Keywords: Identification, Estimation, Kalman Observer, Bicycle Model, Sensor.

1. INTRODUCTION

Road safety becomes a crucial point in vehicle design and imposes increasingly systems integration of driving aid. During the last decade, important work have been carried out in this context and today's cars are equipped with a wide range of physical and intelligent sensors, which make real time measurements to continuously inform the user about the functioning or a malfunctioning of his vehicle. However, the accuracy of the data provided by the various sensors depends greatly on the vehicle's center of gravity location that constantly changes during vehicle movement (braking, acceleration or deceleration).

The main objective of this paper is to develop an integrated identification and estimation method which allows both vehicle state estimation, horizontal and vertical sensor position identification. It uses an *extended Kalman filter*.

The simulation results show the effectiveness of this method face to this hard.

2. MODELING

Several road vehicle models have been developed in the literature [1] [3] [5]. Their main objective is the analysis vehicle dynamics evolution and new concepts and devices that improve road safety. The choice of model is closely related to the expected goals; in fact the vehicle is a three-dimensional complex mechanical system, which presents strongly nonlinear features. Model simplifications have thus to be conducted accordingly.

2.1. VEHICLE MODEL

A. Model equations

For the purpose of this paper, the vehicle dynamics are represented here by a single track, or bicycle model which includes the roll motion [4]. To achieve this simplification, it is assumed that the slip angles of the inner and outer wheels are the same. **Fig 1.1** shows a schematic diagram of the vehicle model. δ represents the steering angle, v is the vehicle velocity while $r \equiv \dot{\psi}$ is the yaw rate and β is the sideslip angle at the center of the vehicle frame. F_{yf} and F_{yr} are the front and rear lateral tire forces and α_f, α_r are respectively the front and rear tire slip angles.

Lateral forces exhibit several operational conditions that make them nonlinear function of the tires side slip angles. For non excessive values of these variables, a linear tire model can be retained, F_{yf} and F_{yr} become:

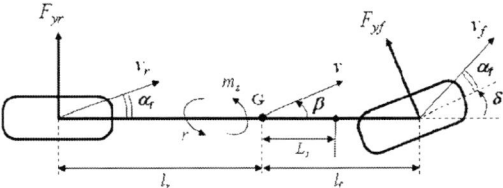

Fig.1. Bicycle vehicle model.

$$\begin{cases} F_{yf} = C_{sf}\alpha_f = C_{sf}(\delta - \beta - \dfrac{l_f \dot{\psi}}{v}) \\ F_{yr} = C_{sr}\alpha_r = -C_{sr}(\beta - \dfrac{l_r \dot{\psi}}{v}) \end{cases} \quad (1)$$

Where C_{sf} and C_{sr} are the total front and rear cornering stiffness

The equations of the lateral, the yaw and roll motions of the vehicle can be then linearized and lead to the following system equations:

$$\begin{cases} \dot{x}_1 = [-\dfrac{I_e C_0}{I_x m v}]\beta + [-1 - \dfrac{I_e C_1}{I_x m v^2}]r + [\dfrac{h_r(mgh_r - k_r)}{I_x v}]\varphi + (\dfrac{h_r l_r}{I_x v})\dot{\varphi} + (\dfrac{I_e C_{sf}}{I_x m v})\delta \\ \dot{x}_2 = (\dfrac{C_1}{I_z})\beta + (\dfrac{C_2}{I_z v})r + (\dfrac{l_f C_{sf}}{I_z})\delta \\ \dot{x}_3 = \dot{\varphi} \\ \dot{x}_4 = (\dfrac{C_0 h_r}{I_x})\beta + (\dfrac{C_1 h_r}{I_x v})r + (\dfrac{mgh_r - k_r}{I_x})\varphi + (\dfrac{b_r}{I_x})\dot{\varphi} + (\dfrac{C_{sf}}{I_x})\delta \end{cases} \quad (2)$$

$$\overline{\overline{A}} = \begin{bmatrix} -\dfrac{I_e C_0}{I_x M\,v} & -1 - \dfrac{I_{eq} C_1}{I_x m v^2} & \dfrac{h(mgh_r - k_r)}{I_x v_x} & -\dfrac{h_r}{I_x v_x} \\ -\dfrac{C_1}{I_z} & -\dfrac{C_2}{I_x v} & 0 & 0 \\ 0 & 0 & 0 & 1 \\ -\dfrac{C_0 h_r}{I_x} & -\dfrac{C_1 h_r}{I_x v} & \dfrac{mgh_r - k_r}{I_x} & -\dfrac{b_r}{I_x} \end{bmatrix}$$

$$\overline{\overline{B}} = \begin{bmatrix} \dfrac{I_e C_{sf}}{I_x m\,v} \\ \dfrac{l_f C_{sf}}{I_z} \\ 0 \\ \dfrac{C_{sf}}{I_x} \end{bmatrix},$$

Where, I_z is the moment of inertia of the vehicle about its yaw axis, m is the vehicle mass. l_f and l_r are the distances of the front and rear axles from the G. I_x is the moment of inertia about the roll axis, k_r is the roll stiffness and b_r is the roll damping coefficient. h_r is the height of G from the roll center. The others model data and numerical values are shown in Table 1.

$$\begin{cases} C_0 = Cs_f + Cs_r \\ C_1 = l_f Cs_f - l_r Cs_r \\ C_2 = l_f^2 Cs_f + l_r^2 Cs_r \\ I_e = I_x + m h_r^2 \end{cases} \quad (3)$$

These linear model equations can be written in the state space form, defining a four components state vector:

$$\overline{\dot{x}} = \overline{\overline{A}}\,\overline{x} + \overline{\overline{B}}\,\delta \quad (4)$$

Where

$\overline{x} = \begin{bmatrix} \beta & r & \varphi & \dot{\varphi} \end{bmatrix}^T$ the state vector.

Table.1. Nomenclature of the Vehicle Model.

Csf	Stiffness coefficient of the front tire	N.rad^{-1}
Csr	Stiffness coefficient of the rear tire	N.rad^{-1}
V	Vehicle speed	m.s^{-1}
m	Vehicle mass	kg
I_z	Moment of inertia about the yaw axis	kg.m^2
I_x	Moment of inertia about the roll axis	kg.m^2
k_r	Roll stiffness coefficient	N.m.rad^{-1}
b_r	Roll damping coefficient	N.m.s.rad^{-1}

B. Problem formulation

Suppose now that a sensor is located at a horizontal distance L_x and a vertical distance L_z from the center of gravity G, the height of G is itself denoted by h_r. This height is also *subject to variation* we should be replacing and by their expressions, which are now variables because G varies. This is depicted in figures 1 and 2.

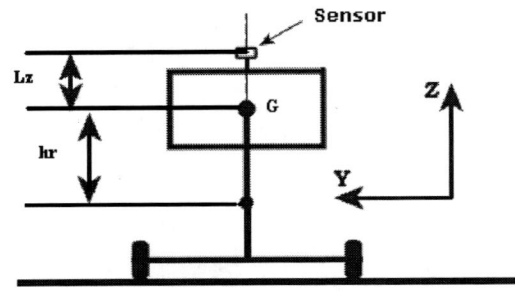

Fig.2. Vehicle CG and sensor positions

The following equations can be written:

- **For the longitudinal axis**

$$\begin{cases} l_f + l_r = L \\ l_f = L_x + d \\ L_r = L - l_f = L - L_x - d \end{cases} \quad (5)$$

- **For the vertical axis**

$$H = h_r + L_z \Rightarrow h_r = H - L_z \quad (6)$$

The distances d and H are supposed to be known, they represent respectively the distance between the sensor and the front axle and the total height of the sensor with respect to the unsparing mass frame.

Replacing l_f, l_r and h_r by their expressions and assuming L_x and L_z as constants, one can get the following augmented system:

$$\begin{cases} \dot{x}_1 = [-\frac{I_e C_0}{I_z m v}]\beta + [-1 - \frac{I_e C_1}{I_z m v^2}]r + [\frac{(H-L_z)(mg(H-L_z)-k_r)}{I_z v}]\varphi + (-\frac{(H-L_z)l_r}{I_z v})\dot{\varphi} + (\frac{I_e C_{tf}}{I_z m v})\delta \\ \dot{x}_2 = (-\frac{C_1}{I_z})\beta + (-\frac{C_2}{I_z v})r + (\frac{(L_x+d)C_{tf}}{I_z})\delta \\ \dot{x}_3 = \dot{\varphi} \\ \dot{x}_4 = (-\frac{C_0(H-L_z)}{I_x})\beta + (-\frac{C_1(H-L_z)}{I_x v})r + (\frac{mg(H-L_z)-k_r}{I_x})\varphi + (-\frac{b_r}{I_x})\dot{\varphi} + (\frac{C_{tf}}{I_x})\delta \\ \dot{x}_5 = 0 \\ \dot{x}_6 = 0 \end{cases} \quad (7)$$

With

$$\begin{cases} C_0 = Cs_f + Cs_r \\ C_1 = (L_x + d)Cs_f - (L - L_x - d)Cs_r \\ C_2 = (L_x + d)^2 Cs_f + (L - L_x - d)^2 Cs_r \\ I_e = I_x + m(H - L_z)^2 \end{cases} \quad (8)$$

$x = [\beta \quad r \quad \varphi \quad \dot{\varphi} \quad L_x \quad L_z]$ is the new vector state. This augmented system is highly non-linear.

C. Matrix measure

The lateral speed provided by the sensor is given by the following formula:

$$V_{cap} = v + L_x . r \quad (9)$$

Therefore the first component of the measurement vector is given by:

$$y_1 = x_1 + (\frac{L_x}{v}).x_2 + (\frac{\beta}{v}).L_x \quad (10)$$

The final measurement matrix has thus the following form

$$C = \begin{bmatrix} 1 & \frac{x_5}{v} & 0 & 0 & \frac{x_1+x_2}{v} & 0 \\ 0 & 1 & 0 & 0 & 0 & 0 \\ 0 & 0 & 1 & 0 & 0 & 0 \\ 0 & 0 & 0 & 1 & 0 & 0 \end{bmatrix}$$

In following section an *Extended Kalman filter (EKF)* based estimator will be developed in order to estimate the dynamic states of the vehicle and to identify the positions (longitudinal and vertical) of the sensors.

First of all, some generalities of *(EKF)* are provided.

3. EXTENDED KALMAN FILTER (EKF)

The *extended Kalman filter (EKF)* applies for non-linear system with the principle linearization of the process and the observations around the current estimate of the state $\hat{x}$. Consider the non-linear discrete-time system with its measurement vector z given by:

$$\begin{cases} x_{k+1} = f(x_k, u_k, w_k) & x \in \Re^n \\ z_k = h(x_k, v_k) & z \in \Re^m \end{cases} \quad (11)$$

Where w_k and v_k represent the process noise and the measurement noise respectively.

The extended Kalman Filter algorithm is built around three successive steps [2]; the first step is to linearize the equation of evolution, around the state estimated $\hat{x}$. The second step is the prediction of state and the uncertainty on the prediction, by calculating the covariance P_{k+1}^-

$$\begin{cases} \hat{x}_{k+1}^- = f(\hat{x}_k, u_k, 0) \\ P_{k+1}^- = A_k P_k A_k^T + W_k Q_k W_k^T \end{cases} \quad (12)$$

Where

P_{k+1}^- : The covariance;

Q_k is the process noise covariance at step k

W_k : the Jacobian matrix of partial derivatives of $f(x_k, u_k, w_k)$ with respect to w_k

The final step applies a correction term when a new observation becomes available. An estimate of the state is computed from that predicted and the difference between the predicted measure and the available:

$$\begin{cases} \hat{x}_{k+1} = \hat{x}_{k+1}^- + K_{k+1}(z_{k+1} - h(\hat{x}_{k+1}^-)) \\ P_{k+1} = (I - K_{k+1} H_k) P_{k+1}^- \end{cases} \quad (13)$$

Where:

K_k represents the Filter gain which minimizes the variance of the estimation error, it is given by:

$$K_{k+1} = P_{k+1}^- H_k^T (H_k P_{K+1}^- H_k^T + V_k R_k V_k^T)$$ (14)

Where:

R_k is the measurement noise covariance equation at step k, V_k: the Jacobian matrix of partial derivatives of $h(x_k, v_k)$ with respect to v_k and H_k: the Jacobian matrix of partial derivatives of $h(x_k, v_k)$ with respect to x_k.

In what follows, the *extended Kalman filter* is applied to the augmented non-linear vehicle model

4. SIMULATION RESULTS

The sampling period is set at 0.01 second.
We will give more confidence to the model in developing the *Kalman* filter, taking:

$$Q_k = 0.8 * I_6$$
$$R_k = 0.5 * I_4,$$

In order to illustrate the proposed identification and estimation procedure, different simulation tests are conducted on two scheduled scenarios. The longitudinal speed is fixed to 23m/s along all the simulations.

Scenario.1.

In this first case the control input of the vehicle (δ) is chosen with the profile shown in **Fig.3**.

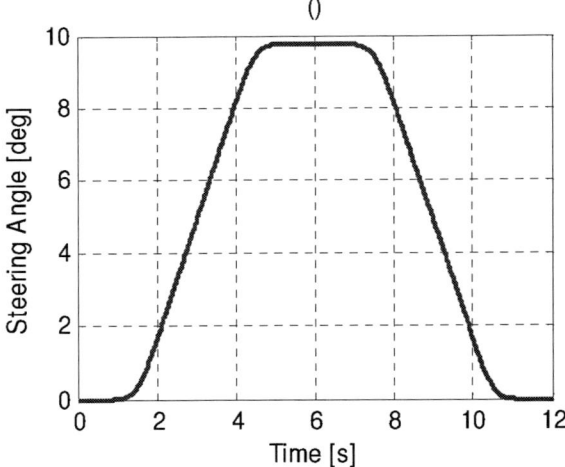

Fig.3. Steering Angle.

The **Fig.4.** gives a representation of states in dotted line and their estimates in dashed lines. It shows a good EKF performing.

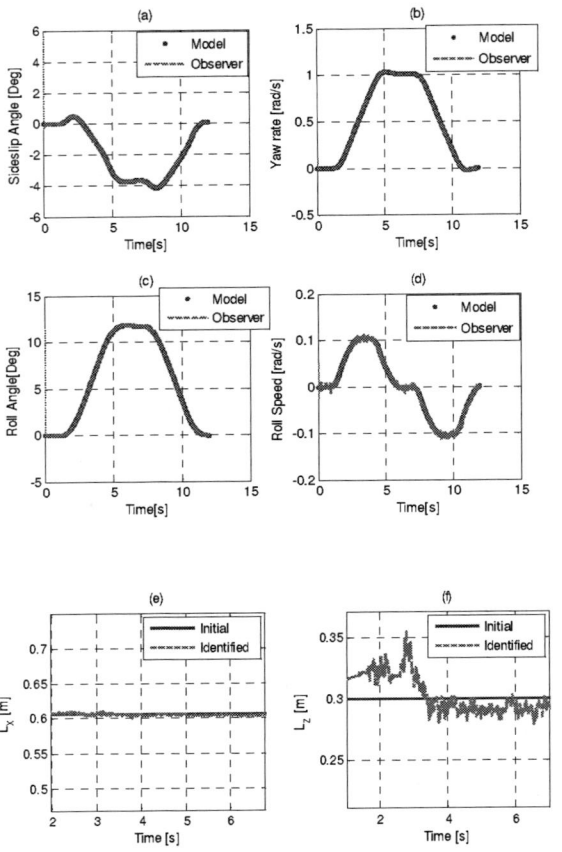

Fig.4. sideslip angle, yaw rate, roll angle, roll speed, L_x and L_z estimation.

Fig.4. (a) represents sideslip angle given by the *EKF*. The estimate follows faithfully the sideslip angle.

Fig.4. (b) represents the yaw rate simulated and estimated, as in the precedent case, the estimator rebuilt yaw rate with good accuracy.

Fig.4. (c), one can observe the quality of the estimate of the roll angle which coincides perfectly with that of the simulated model.

Fig.4. (d) provides an estimate as a result of the simulation of roll rate one can also conclude to a good estimate of the state.

The **Fig.4. (e)** gives the results for the estimation of the horizontal distance between the sensor center of gravity compared with the constant value. The result shows that the distance ranges (small variations) around this initial value.

Finally **Fig.4. (f)** provides an estimate of the vertical distance of the sensor relative to the center of gravity, as in the preceding case, the estimated varies around the truth constant value.

Scenario. 2.

In the second case the steering wheel (δ) is given in **Fig 5.** It simulates a lane change maneuver.

As in the scenario 1, the simulation results for this scenario show the good estimate of dynamic states and the identification of vertical and longitudinal sensor's position using the *extended Kalman filter*.

These results are shown in **Fig.7.** and **8**

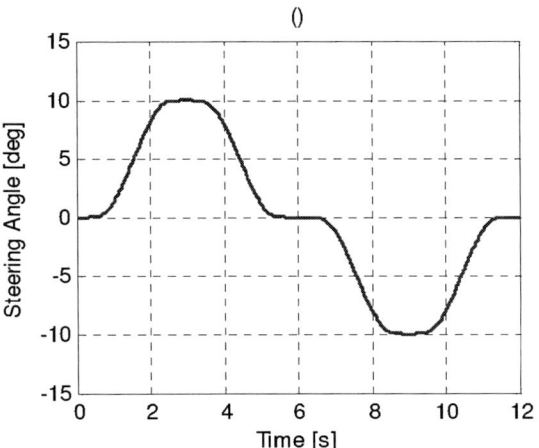

Fig.5. Steering Angle.

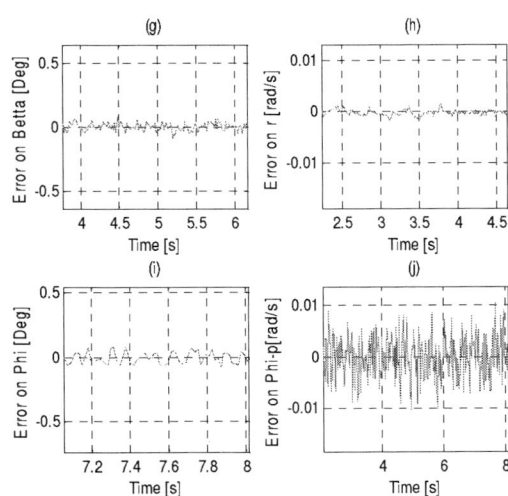

Fig.7. Error of estimation on the scenario.1.

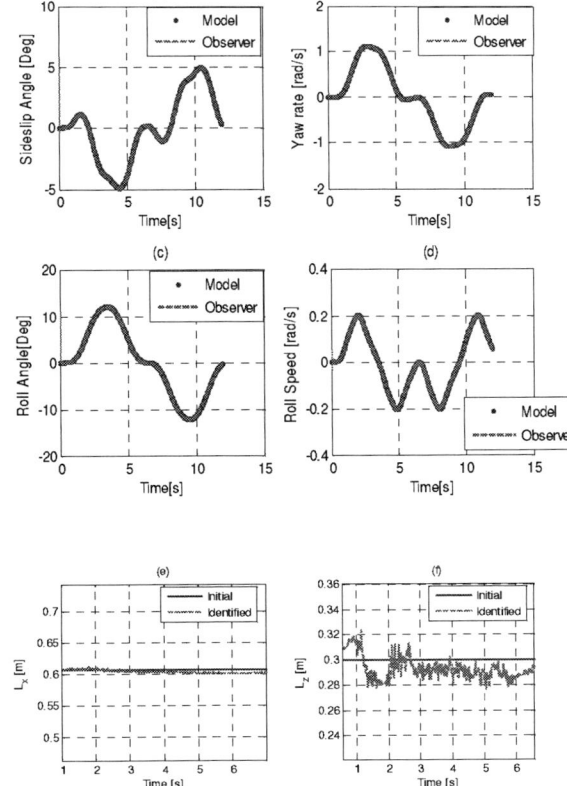

Fig.6. sideslip angle, yaw rate, roll angle, roll speed, L_x and L_z estimation.

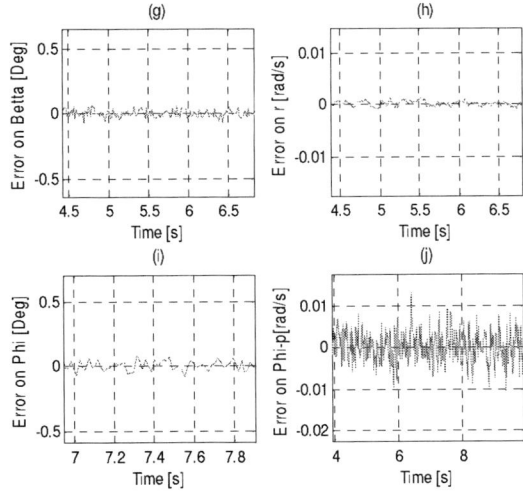

Fig.8. Error of estimation on the scenario.2.

5. CONCLUSIONS

In this paper a method for simultaneous identification of sensors position and estimating states of a ground vehicle has been presented. The non-linearities introduced in the bicycle model suggested to use the *extended Kalman filter* which is particularly suitable.

In light of the quality of simulation results obtained for the estimation of the state and the horizontal and vertical sensor's position, therefore, the variation in the center of gravity, one can conclude that the *EKF* works well, it not only allows to estimate the states, but also offers an opportunity to identify the exact position of sensors at any time with respect to the center of gravity.

REFERENCES

[1] J. Ryu, E. and J.C. Gerdes, Estimation of Vehicle Roll and Road Bank Angle, American Control Conference, Boston, MA, 2004.

[2] G. Welch, G. Bishop, An Introduction to the Kalman Filter, Department of Computer Science University of North Carolina at Chapel Hill Chapel Hill, NC 27599-3175, Avril 2004.

[3] H. B. Pacejka, E. Bakker and L. Lidner, A new tire model with an application in vehicle dynamics studies. SAE paper, 1989.

[4] C. R. Carlson and J. C. Gerdes, "Optimal rollover prevention with steer by wire and differential braking", in ASME Dynamic Systems and Control Division (Publication) DSC, vol. 72, no. 1, Washington, D.C., 2003, pp. 345.354.

[5] J. Stephant, A. Charara, Virtual Sensor : Application to Vehicle Sideslip Angle and Transversal Forces, IEEE Transactions on industrial electronics, Vol.51, No.2, April 2004.

[6] M. SATRIA, M. C. BEST, State estimation of vehicle handling dynamics using non-linear robust extended adaptive Kalman filter, Vehicle System Dynamics, Vol.41, 2004, pp. 103-112.

[7] C. Sentouh, S. Glaser S. Mammar and Y. Bestaoui Estimation des paramètres d'un modèle dynamique de véhicule par filtrage de Kalman étendu, Conférence Internationale Francophone d'Automatique, France, 2006.

[8] M. Labarrère, J. P. Krief et B. Gimonet, Le filtrage et ses applications, Cépaduès Editions.

[9] H. Kwakernaak and R. Sivan, Wiley inter, Linear optimal control systems - science, John Wiley and Sons, 1972.

[10] S.J. Julier and J.K. Uhlmann. A new extension of the kalman filter and nonlinear systems. In Proceedings of the SPIE Aerosense International Symposium on Aerospace/Defense Sensing, Simulation and Controls, Orlando, Florida, September 2001.

[11] B. Mourllion, D. Gruyer, A. Lambert, and S. Glaser. Kalman filters predictive steps comparison for vehicle localization. In IEEE/RSJ International Conference on Intelligent Robots and Systems, pages 1251–1257, Edmonton, Alberta, Canada, August 2005.

[12] R.E Kalman. A new approach to linear filtering and prediction system. Transactions of the ASME-Journal of Basic Engineering, 82(D) :35–45, 1960.

[13] K. Ito and K. Xiong. Gaussian filters for nonlinear filtering problems. In IEEE Transaction on Automatic Control, volume 45 of 5, May 2000.

Two Wheeled Vehicle Dynamics Synthesis for Real-Time Applications

Salim Hima[*] and Hichem Arioui[*]

[*] *Informatics, Integrative Biology and Complex Systems CNRS-FRE 3190, 40 Rue de Pelvoux, 91020 Evry Cedex, France.*
e-mail: {Salim.Hima, Hichem.Arioui}@ibisc.univ-evry.fr

Abstract: In this paper, we present a modeling technique for deriving the motorcycles equation of motion. The proposed technique is based on the recursive Newton-Euler approach and adapted to tree structure with floating base multibody systems. The derived model presents a low number of arithmetic operations, and hence, suitable for implementation into a two wheeled vehicles real-time applications such as driving simulators. The synthesized model takes in consideration the main wrenches that affect the behavior of motorcycle such as: pneumatic, aerodynamic, suspensions, contact constraints and control inputs.

Keywords: Motorcycle driving simulator, motorcycle dynamics, Recursive Newton-Euler algorithm.

1. INTRODUCTION

IN the two last decades, the safety of ground vehicles has known a significant progress. Due to the reduction of the electronics and the informatics equipments, several intelligent systems have been developed to optimize the performance of the ground vehicles. In general, the development of such systems is based on the vehicle behavioral model that allows the development and testing them without need for real prototypes. In such applications, real-time constraint is not crucial. However, this constraint become important when the application is centered around a human operator such as driving simulators.

In fact, driving simulators have initially proposed in aeronautic field to train a novice pilots, enhance the performance of experimental ones and also testing a new systems and designs in a safe environment and without need of a real aircraft. Few years later, this technology has been extended to four wheeled vehicles field and recently to two wheeled vehicles field.

The idea behind this technology is to immerse the rider in a virtual traffic environment by properly exciting his perception organs with a convenient cues. In literature, we distinguish two type of driving simulators: a fixed base and motion base simulators. The former can merely provide the rider with a visual and auditive cues, while the second add an appropriate acceleration cue component by moving the simulator motion-base, on which the rider is rising, within its limited workspace.

One of the central components of driving simulator is a vehicle's behavioral model. When the driver acts on the vehicle control inputs, these actions are sensed and transmitted to the vehicle behavioral model bloc. This bloc computes the state and accelerations of the vehicle and transmit them to different simulator components (visual, auditive, and motion-base) in order o update their states. It is important from perception point of view to reduce the delay time between driver actions instant and the updating the whole simulation component states instant.

In the framework of the SIMACOM project, a new motorcycle motion-base simulator has been designed, see Nehaoua [2007]. The proposed design is customized to reproduce the relevant motorcycle situations in real driving. In this paper, we are interested to derive the motorcycle behavioral model with a low arithmetic operation in order to reduce the computation time.

In literature, several model are proposed in the scope of control purposes. These models are partial and focused on a local behaviors and neglect some element such as suspension or road irregularity. However, driving simulator need for a broad model in order to cover a very large spectrum of rider's actions.

Multibody mechanics offers a convenient framework to derive a motorcycle behavioral model. In Cossalter [2002] authors have adopted Lagrange formalism to derive their motorcycle motion model. A direct application of this approach, leads to unattractive performances in term of number of operations and implementation facilities. An algorithmic formulation alternative of this formalism, that presents the advantage to be easily debugged and easy to implement, has been presented in Hima [2007].

Exploiting the recursive scheme in computing some kinematic quantities, very efficient algorithms have been proposed to solve the inverse dynamics of open chain multibody systems, see Featherstone [2000], Khalil [1987], Hollerbach [1980]. So, the direct dynamics can be derived from the inverse dynamics using the articulated body algorithm (ABA), Featherstone [2000]. This algorithm is known to be fast, accurate and stable Khalil [1999].

In this paper, we extend ABA algorithm to derive a fast direct model of motorcycles suitable for real-time application such as two wheeled interactive simulator. In the second section of this paper, a brief presentation of our simulator is given. The third Section is devoted to present the derivation steps of motorcycle forward dynamics. In the fourth section, simulation results is illustrated. Finally, conclusion remarks and future works are outlined.

---

[*] This work was supported by the French National Research Agency (ANR) in the Framework of SIMCOM project.

2. SIMACOM PROJECT PRESENTATION

The primary goal of SIMACOM project is to propose a low cost replicable motorcycle simulator. It's dedicated to driving schools for training a novice riders and perform the experienced riders aptitudes face to life-threatening situations in a safe environment. Mainly, the proposed simulator architecture is designed around five subsystems: a visual subsystem, acoustic subsystem, motion-base platform, motorcycle model and haptic subsystems, see Fig 1. The visual subsystem allows the immersion of the rider in a virtual traffic environment, managed by ARCHISIM software, and provides him with velocity cue.

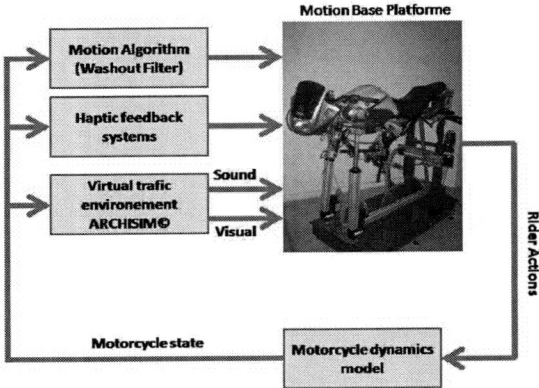

Fig. 1. SIMACOM project simulator architecture

In Panerai [2001] the authors have stated that, an acoustic cues deprivation leads to a systematic increase tendency of the speed by drivers. To prevent this problem, and making the simulation more realistic, an acoustic subsystem is placed to reproduce sounds resulting from the juxtaposition of many sources where the main ones come from the engine and tires. In addition, inertial cues are generated by stimulating the rider vestibular organ by moving him with a proper motions. For this aim, a motion-base platform, which consists of a real motorcycle structure hanged on a parallel robot, is designed to reproduce the main motorcycle situations such as bending in braking situation, leaning in cornering situation and also the rear tire skidding situation. That's why, two prismatic actuator are placed in the front and connected to saddle body by way of spherical joints. So, the pitch and roll of the motion-base are directly controlled by changing the legs lengths using two DC motors. In addition, illusion of rear tire skidding is created by placing a sliding system at the rear of the motion-base to perform a horizontal displacement, see Fig 2.

Changes in previous described subsystems states, are made in response to the rider actions captured by placing a sensors in the motorcycle control organs (throttle angle, braking and clutch leavers). Once rider's action are sensed, they are sent to motorcycle dynamics model to compute motorcycle's motion (configuration, velocities and accelerations). These informations are then provided to visual and acoustic subsystem to update the virtual environment and the adequate sounds. In the same time, motion informations are also provided to washout filter in order to transform it into motions compatible with motion-base workspace limitations.

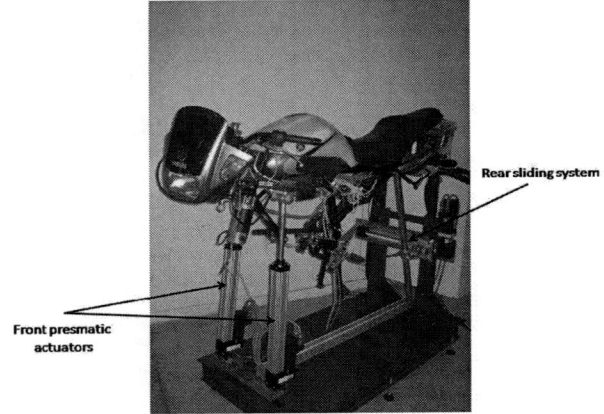

Fig. 2. Motion-base platform prototype

When riding a motorcycle, the ground-tire interaction forces and the gyroscopic precession force of the front wheel are transmitted to handlebar and so felt by the rider. Furthermore, tension forces exerted on the rider arms in acceleration and deceleration phases, caused by the rider bust inclination, are present. In order to reproduce these forces and make the simulator more efficient, two haptic feedback subsystems are integrated to our simulator (Fig 3). The first subsystem consists of a DC motor connected to the steering axis by way of pulley-belt system. It permits the application of a torque on the steering axis to give the rider sensation of the tire-ground and gyroscopic forces. This torque is computed from the motorcycle dynamics model. Moreover, the second subsystem consists of a mechanism that endow the handlebar with a longitudinal translation motion controlled by a brushless actuator, see Fig 3. In fact, in acceleration phase, the handlebar translates far from the rider and hence, pulls the rider arms to create illusion of the bust inclination away from the handlebar. Otherwise, in deceleration phase, handlebar translates toward the rider and so, push the rider arms to create illusion of the bust inclination toward handlebar.

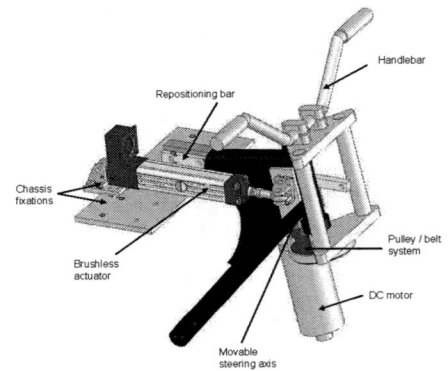

Fig. 3. Haptic feedback systems in the handlebar

For further description of this simulator and kinematic study of the motion-base, readers can consult Nehaoua [2007].

3. MOTORCYCLE MODELLING

Motorcycle direct dynamics can be derived using mechanics of multibody systems theory. Mainly, the motorcycle is composed by the saddle body, the front upper part (handlebar and upper part of the front suspension), the front lower part of the front suspension, the swinging arm and finally the front and rear tires. All these parts are connected between them with a simple joints. the Handlebar and the swinging arm are attached to saddle body by a simple rotoid joints. The font lower part is linked to the front upper part by prismatic joint. In addition, the rear suspension is connected to saddle with a rotoid joint from one side and to the swinging arm with a rotoid joint also from the other side. This creates a closed kinematic loop, and so, leads to a modelling difficulties. Finally, rear and front wheels are respectively connected to the other tips of swinging arm and front lower part as illustrated in Fig 4.

The motion of the motorcycle is referenced to an inertial frame $\mathcal{R}_0(O_0, x_0, y_0, z_0)$. Moreover, to each body B_i an orthonormed frame is rigidly attached. In order to completely define the configuration of the motorcycle, a set of 11 DOFs have been considered. A full characterization of motorcycle's situation can be done by the position and orientation of the saddle $(x, y, z, \phi, \theta, \psi)$, the steering angle or handlebar orientation, δ, the elongation of the front suspension, L, the rotation angle of the swinging arm with respect to the saddle, θ_s (elongation of the rear suspension), and finally, the rotation angles of the front and rear tires, θ_f and θ_r.

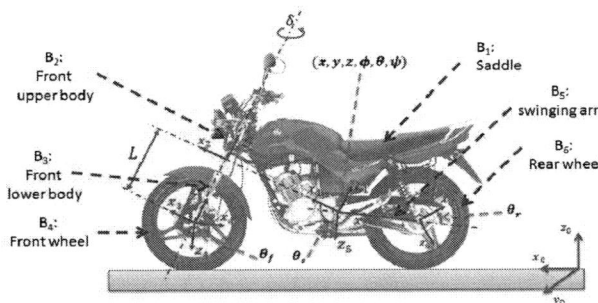

Fig. 4. Motorcycle configuration parametrization

3.1 Direct Dynamics

Application of the classical formalisms from mechanics of multiboby systems can leads to an inefficient dynamic models in term of the number of arithmetic operations. However, recursive techniques present another alternative to reduce greatly the computational burden and that should allow a real-time computation of the direct dynamics of the motorcycle. In Khalil [1999] and Hollerbach [1980], Recursive Newton-Euler Algorithm, RENA, have been shown to be the fast one. Originally, this technique is developed to solve the inverse dynamic model of the open chain manipulator with a fixed base for control purposes. By projecting the dynamics of each body in its attached frame, the acceleration of the joint variable can be easily derived without need to compute the total inertia matrix of the whole system and invert it. In addition, this technique, named Articulated Body Algorithm (ABA), is more numerically stable then inertia matrix inversion method, see Khalil [1999].

To illustrate the ABA, consider the multibody system sketched in Fig 5. Let $\mathbb{V}_i = (v_i^T, \omega_i^T)^T$ be the twist vector of B_i where v_i is its linear velocity and ω_i is its angular velocity. $\mathbf{k}_i$ is the ith joint axis, $\mathbb{F}^{e_i} = (F^{e_i T}, M^{e_i T})^T$ is the external wrench acting on B_i, $\mathbf{L}_i$ the position vector of $(i+1)^{th}$ joint origin with respect to i^{th} joint origin, $\mathbf{S}_i$ is the position vector of the B_i center of gravity with respect to the ith joint origin, $^i\mathbf{R}_{i-1}$ the transformation matrix from B_{i-1} to B_i, $\mathbb{F}_i = (F_i^T, M_i^T)^T$ is the link wrench vector transmitted to B_i from B_{i-1}, m_i and $\mathbf{J}_i$ are, respectively, the mass and the inertia tensor of the body B_i, q_i is i^{th} joint variable and τ_i be the actuator force/torque acting on the joint i. In the sequel of this paper, the expression $^i\mathbf{X}_j$ means that the vector $\mathbf{X}_j$ is projected into the frame associated to B_i.

Let σ_i be a binary variable defining the joint type and given by:

$$\sigma_i = \begin{cases} 1 & \text{if joint i is a prismatic} \\ 0 & \text{if joint i is a revolute} \end{cases} \quad (1)$$

Let $^i\mathbb{T}_{i-1}$ be a 6×6 matrix defined as:

$$^i\mathbb{T}_{i-1} = \begin{pmatrix} ^i\mathbf{R}_{i-1} & -^i\mathbf{R}_{i-1}{}^{i-1}\tilde{\mathbf{L}}_{i-1} \\ 0_{3\times 3} & ^i\mathbf{R}_{i-1} \end{pmatrix} \quad (2)$$

where $\tilde{}$ is the skew-symmetric matrix operator for a vector $\mathbf{v} = (x, y, z)^T$ defined by:

$$\tilde{\mathbf{v}} = \mathbf{v}\times = \begin{pmatrix} 0 & -z & y \\ z & 0 & -x \\ -y & x & 0 \end{pmatrix} \quad (3)$$

Let $^i\mathbb{J}_i$ be the global inertia matrix of B_i given by:

$$^i\mathbb{J}_i = \begin{pmatrix} m_i\mathbf{I}_{3\times 3} & -m_i{}^i\tilde{\mathbf{S}}_i \\ m_i{}^i\tilde{\mathbf{S}}_i & \mathbf{J}_i \end{pmatrix} \quad (4)$$

and finally let $^i\mathbf{a}_i = [\sigma_i \mathbf{k}_i^T \quad \bar{\sigma}_i \mathbf{k}_i^T]^T$.

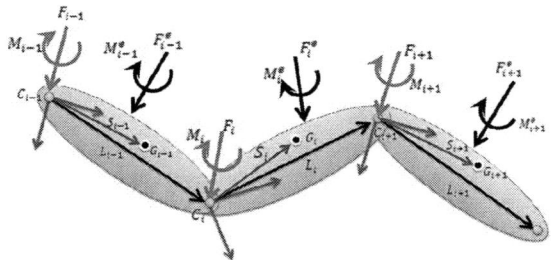

Fig. 5. Open chain multibody system

The ABA is based on three steps. In the first step (forward recursion), velocities, Coriolis, centrifuge and external forces are updated for each body. The second step (backward recursion), the total forces and inertia matrices felt by each joint are computed. in final step (forward recursion), the joints accelerations and dynamic torsors are computed for each body. Setting $\omega_0 = \dot{\omega}_0 = 0$ and $\mathbb{V}_0 = (-\mathbf{g}, 0_{1\times 3})^T$, where $\mathbf{g}$ is the gravity vector. So, the ABA is given by:

First Recursion
for $i = 1 \to n$ **do**

$$^{i}\mathbb{V}_{i-1} = {}^{i}\mathbb{T}_{i-1}{}^{i-1}\mathbb{V}_{i-1} + \dot{q}_i{}^i\mathbf{a}_i$$

$$^{i}\gamma_i = \begin{bmatrix} {}^{i}\mathbf{R}_{i-1}\left({}^{i-1}\omega_{i-1} \times \left({}^{i-1}\omega_{i-1} \times {}^{i-1}\mathbf{L}_{i-1}\right)\right) \\ +2\sigma_i\left({}^{i}\omega_{i-1} \times \dot{q}_i{}^i\mathbf{k}_i\right) \\ \bar{\sigma}_i\left({}^{i}\omega_{i-1} \times \dot{q}_i{}^i\mathbf{k}_i\right) \end{bmatrix}$$

$$^{i}\beta_i = {}^{i}F_i^e - \begin{bmatrix} {}^{i}\omega_i \times \left({}^{i}\omega_i \times m_i{}^i\mathbf{S}_i\right) \\ {}^{i}\omega_i \times \left({}^{i}\mathbf{J}_i{}^i\omega_i\right) \end{bmatrix}$$

end

Second recursion

$${}^{n}\beta_n^* = {}^{n}\beta_n$$
$${}^{n}\mathbb{J}_n^* = {}^{n}\mathbb{J}_n$$

for $i = n \to 1$ **do**

$$^{i-1}\beta_{i-1}^* = {}^{i-1}\beta_{i-1} - {}^{i}\mathbb{T}_{i-1}^T \left[-{}^{i}\beta_i^* \right.$$
$$\left. + {}^{i}\mathbb{J}_i^{\star i}\gamma_i + \frac{{}^{i}\mathbb{J}_i^{*i}\mathbf{a}_i \left\{ {}^{i}\mathbf{a}_i^T \left({}^{i}\mathbb{J}_i^{*i}\gamma_i + {}^{i}\beta_i^*\right) + \tau_i \right\}}{{}^{i}\mathbf{a}_i^{T i}\mathbb{J}_i^{*i}\mathbf{a}_i} \right]$$

$$^{i-1}\mathbb{J}_{i-1}^* = {}^{i-1}\mathbb{J}_{i-1} + {}^{i}\mathbb{T}_{i-1}^T \left[{}^{i}\mathbb{J}_i^* - \frac{{}^{i}\mathbb{J}_i^{*i}\mathbf{a}_i{}^i\mathbf{a}_i^{T i}\mathbb{J}_i^*}{{}^{i}\mathbf{a}_i^{T i}\mathbb{J}_i^{*i}\mathbf{a}_i} \right]$$

end

Third recursion
for $i = 1 \to n$ **do**

$$^{i}\dot{\mathbb{V}}_{i-1} = {}^{i}\mathbb{T}_{i-1}{}^{i-1}\dot{\mathbb{V}}_{i-1}$$
$$\ddot{q}_i = \mathbf{H}_i^{-1} \left[-{}^{i}\mathbf{a}_i^{T i}\mathbb{J}_i^* \left({}^{i}\dot{\mathbb{V}}_{i-1} + {}^{i}\gamma_i\right) + \tau_i + {}^{i}\mathbf{a}_i^{T i}\beta_i^* \right]$$
$${}^{i}\mathbb{F}_i = \begin{bmatrix} {}^{i}\mathbf{F}_i \\ {}^{i}\mathbf{M}_i \end{bmatrix} = {}^{i}\mathbb{K}_i{}^i\dot{\mathbb{V}}_{i-1} + {}^{i}\alpha_i$$
$${}^{i}\dot{\mathbb{V}}_i = {}^{i}\dot{\mathbb{V}}_{i-1} + {}^{i}\mathbf{a}_i\ddot{q}_i + {}^{i}\gamma_i$$

end

Algorithm 1. Recursive Newton-Euler algorithm for direct dynamics

Removing the rear suspension and replace it with two forces acting on the saddle and swinging arm at rear suspension fixation points, the motorcycle can be considered as a tree structure system with floating base. Hence, algorithm 1 is used to compute the accelerations of relative configurations $(\delta, L, \theta_f, \theta_s, \theta_r)$, and forces acting on their corresponding articulations for the open chains (B_2, B_3, B_4) and (B_5, B_6). The floating base (saddle) accelerations are often obtained by considering 6 virtual bodies linked between them by a single saddle's degree of freedoms. This leads to consider a serial chain with 6 bodies, and hence, apply RENA algorithm again. In our approach, and to reduce the number of arithmetic operations, we have isolated the saddle body only, and replace the front and the rear chains by their transmitted dynamical wrenches $(\mathbb{F}_2, \mathbb{F}_5)$ at the 2nd and 5th joints origins. In this case, we can consider the saddle as a spatial rigid body subjected to external forces. So, the saddle's accelerations are then given by:

$$\begin{aligned} {}^{1}\dot{\mathbb{V}}_1 &= {}^{1}\mathbb{J}_1^{-1} \left(\mathbf{C}(\eta, {}^{1}\mathbb{V}_1) + \mathbf{G}(\eta) + {}^{1}\mathbb{F}_2 + {}^{1}\mathbb{F}_5 + {}^{1}\mathbb{F}_{rs} \right) \\ \dot{\eta} &= \mathbb{R}^{1}\mathbb{V}_1 \end{aligned} \quad (5)$$

where $\eta = (x, y, z, \phi, \theta, \psi)$ is the configuration vector of the saddle. ${}^{1}\mathbb{V}_1$ is the vector of the linear and angular velocities of saddle body with respect to its frame. ${}^{1}\mathbb{J}_1$ is the inertia matrix of the saddle body expressed in its frame. $\mathbf{C}$ and $\mathbf{G}$ are, respectively, the Centrifuge/Coriolis and weight wrenches. $\mathbb{F}_{rs}$ is rear suspension force acting on the saddle. $\mathbb{R}$ is the transformation matrix form B_1 frame to inertial frame $\mathcal{R}_0$ and can be given by:

$$\mathbb{R} = \begin{pmatrix} c_\psi c_\theta & -s_\psi c_\phi + c_\psi s_\theta s_\phi & s_\psi s_\phi + c_\psi c_\phi s_\theta \\ s_\psi c_\theta & c_\psi c_\phi + s_\phi s_\theta s_\psi & -c_\psi s_\phi + s_\theta s_\psi c_\phi \\ -s_\theta & c_\theta s_\phi & c_\theta c_\phi \end{pmatrix} \quad (6)$$

where $c_x = \cos(x)$ and $s_x = \sin(x)$.

3.2 External forces

In this section, we summarize the principal forces and moments affecting the motorcycle behavior.

Suspension forces In order to absorb the shocks caused by the road irregularities and bumps, the motorcycle is equipped with a rear and front suspension systems. In this paper a linearly dependence of the forces generated by the suspensions on the elongations and their rates of change is considered:

$$\mathbf{F}_i = K_i (L_i - L_{i0}) + C_i \dot{L}_i \quad (7)$$

where $i \in \{rs, fs\}$ designates the rear or front suspension, L_{i0} is its free load elongation, K_i is the suspension stiffness and C_i is the damper coefficient. According to Fig 6, the suspension length can be expressed as:

$$\begin{aligned} L_{fs} &= \sqrt{\left({}^{2}P_{fs_u} - {}^{2}P_{fs_l}\right)^T \left({}^{2}P_{fs_u} - {}^{2}P_{fs_l}\right)} \\ L_{rs} &= \sqrt{\left({}^{1}P_{rs_u} - {}^{1}P_{rs_l}\right)^T \left({}^{1}P_{rs_u} - {}^{1}P_{rs_l}\right)} \end{aligned} \quad (8)$$

where P_{i_u} and P_{i_l} denote respectively the position of the upper and lower extremities of suspension i. Hence, the rate of change of suspensions lengths are given by:

$$\begin{aligned} \dot{L}_{fs} &= \frac{1}{L_{fs}} \left({}^{2}\dot{P}_{fs_u} - {}^{2}\dot{P}_{fs_l}\right)^T \left({}^{2}P_{fs_u} - {}^{2}P_{fs_l}\right) \\ \dot{L}_{rs} &= \frac{1}{L_{rs}} \left({}^{1}\dot{P}_{rs_u} - {}^{1}\dot{P}_{rs_l}\right)^T \left({}^{1}P_{rs_u} - {}^{1}P_{rs_l}\right) \end{aligned} \quad (9)$$

Hence, the suspense wrenches are given by:

$$\begin{aligned} {}^{1}\mathbb{F}_{rs_u} &= \left[\frac{\left({}^{1}P_{rs_u} - {}^{1}P_{rs_l}\right)^T}{L_{rs}} \mathbf{F}_{rs}, {}^{1}P_{rs_u} \times \frac{\left({}^{1}P_{rs_u} - {}^{1}P_{rs_l}\right)^T}{L_{rs}} \mathbf{F}_{rs} \right]^T \\ {}^{5}\mathbb{F}_{rs_l} &= \left[\frac{\left({}^{5}P_{rs_l} - {}^{5}P_{rs_u}\right)^T}{L_{rs}} \mathbf{F}_{rs}, {}^{5}P_{rs_l} \times \frac{\left({}^{5}P_{rs_l} - {}^{5}P_{rs_u}\right)^T}{L_{rs}} \mathbf{F}_{rs} \right]^T \\ {}^{2}\mathbb{F}_{fs_u} &= \left[\frac{\left({}^{2}P_{fs_u} - {}^{2}P_{fs_l}\right)^T}{L_{fs}} \mathbf{F}_{fs}, {}^{2}P_{fs_u} \times \frac{\left({}^{2}P_{fs_u} - {}^{2}P_{fs_l}\right)^T}{L_{fs}} \mathbf{F}_{fs} \right]^T \\ {}^{3}\mathbb{F}_{fs_l} &= \left[\frac{\left({}^{3}P_{fs_l} - {}^{3}P_{fs_u}\right)^T}{L_{fs}} \mathbf{F}_{fs}, {}^{3}P_{fs_l} \times \frac{\left({}^{3}P_{fs_l} - {}^{3}P_{fs_u}\right)^T}{L_{fs}} \mathbf{F}_{fs} \right]^T \end{aligned}$$
$$(10)$$

Fig. 6. Suspension systems wrenches

Tire/road contact wrench Behavior of the ground vehicles depends significantly on the nature of the interaction between the tire and the road. Indeed, in the last two decades, and due to the importance of this phenomena in the security of ground vehicles, several efforts are done to establish mathematical models characterizing the friction forces and moments of the tire/road contact. The well known and widely used one is Pacejka Magic formula. This model captures in steady-state motion, the tire/road forces and moments, in algebraic equations form, with respect to kinematic quantities, longitudinal slip κ and lateral slip β, and the load. The steady-state behavior restriction still valid in the scope that, the effect of the tire dynamics is generally small compared to the effect of the complete motorcycle dynamics.

Initially, Pacejka model has been developed for four wheels vehicles. In this case, wheels stay approximately vertical and consequently, the camber angle can be neglected. Other varieties of this model have been proposed and adapted to motorcycle tires by including the camber angle. In fact, this angle could achieves $50°$ for racing motorcycles when cornering and hence, could'nt be ignored, Lot [2004].
Among a six components of the tire/road contact wrench, we have considered the principal ones: longitudinal and lateral forces and the aligning moment, Sharp [2004], Pacejka [1997].

Aerodynamic wrench When a body moves in a fluid, this leads to the apparition of aerodynamic forces and moments proportional to the relative motion of the body with respect to the fluid, applied to the center of pressure. The most important components are the drag and the lift. In our developed motorcycle model, we have consider these two forces. The drag force opposes the motion of the motorcycle, and hence limit the maximum achievable velocity and also acceleration performance. In otherwise, the lift force can have an undesirable effects on the motorcycle performances. Indeed, it reduces the load on the wheels and, thus, decreases the tire adherence. In general, the drag and the lift forces are expressed in a linear form with respect to the square of the motorcycle linear velocity relative to air:

$$F_D = \frac{1}{2}C_D \rho S \|V_a\|^2$$
$$F_L = \frac{1}{2}C_L \rho S \|V_a\|^2 \quad (11)$$

where C_D and C_l are dimensionless aerodynamic coefficient, ρ is the air density. $V_a = V - V_\infty$ is motorcycle relative velocity

where V is the motorcycle velocity and V_∞ is the air velocity. S is the frontal motorcycle area exposed to the air flow.

Contact constraints forces Keeping the tires in contact with the road imposes two constraints. In order to satisfy these constraints we have placed two spring-dapper systems between the road and the tires contact points. This formulation is more advantageous than using Lagrangian multipliers which need to include constraints stabilizers to prevent constraints divergence and also can capture carcasses elasticity features.

Input controls The last considered external forces acting on the motorcycle behavior coming from the rider actions through the engine torque, brakes torques and finally steering torque. We have modeled these actions by a pure torque and we don't integrate their dynamic models, which will be considered in future works.

4. SIMULATION RESULTS

In the lack of a wide controller to handle the whole motorcycle behavior, we have synthesized a basic autopilot based on a linear PID controller for motorcycle velocity regulation in street forward motion. In this simulation we have modeled the engine by a first order filter to take into account the delay of the engine reaction. The delivered torque by the engine is then bounded by the maximum allowable torque. The resultant torque is applied at the rear tire axis. To test the longitudinal behavior of the developed motorcycle, we have considered a scenario with three phases, acceleration phase up to $15m/s$, a second acceleration phase up to $30m/s$, and finally a deceleration phase down to $20m/s$, see Fig 7. In this figure, the steady error between the desired and the motorcycle errors in the three phases are different. It is most important when the motorcycle travels with a high speed. The origin of this effect is related to the aerodynamic drag force which is proportional to the square of the motorcycle speed. In Fig 8, requested autopilot control torque and effective control torque delivered by the engine at the rear tire axis are sketched. The effect of the engine delay and torque limitation can be observed on the convergence time of the speed profile. Fig 9 illustrates the effective motorcycle driving forces generated by the rear and the front tires pneumatic models.

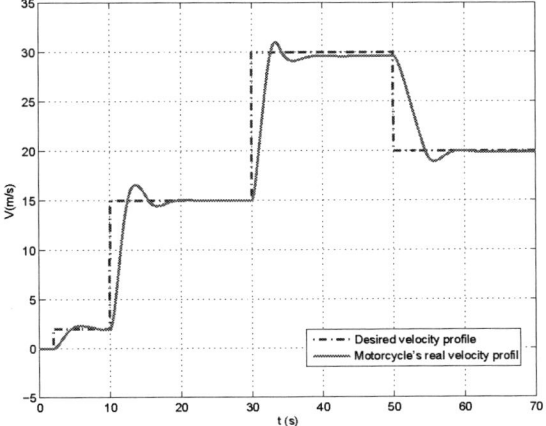

Fig. 7. Motorcycle velocity regulation

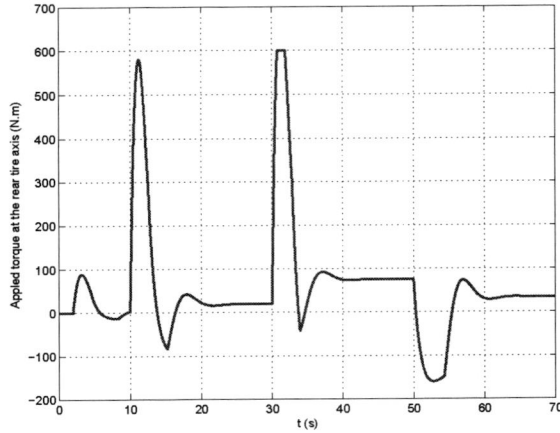

Fig. 8. Rear tire control torque

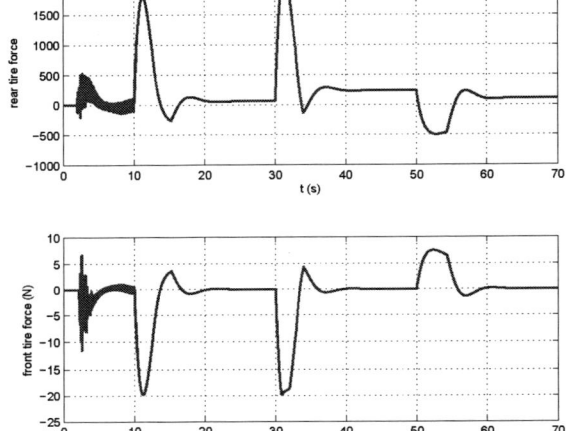

Fig. 9. Rear and front tire/road contact forces

5. CONCLUSION AND FUTURE WORKS

In this paper we have proposed a motorcycle dynamics model with a low arithmetic operations number. This model is based on a Newton-Euler recursive algorithm adapted to serial mutlibody chain with floating base. A fast model is needed in some application such as interactive driving simulators, to reduce the delay time between the rider action and the reaction of the motion base platform in order to enhance the rider immersion in the simulation environment.

For the future works, a more global steering controller will be synthesized to evaluate a very complex maneuvers of the synthesized model such as cornering. Besides, this model can be refined by considering the tire/road contact point migration in the tire circumference when cornering, motorcycle engine dynamics model, brakes dynamics model, and finally, a qualitative evaluation will be done by implementing the synthesized model into our simulator.

REFERENCES

L. Nehaoua, S. Hima, H. Arioui, N. Seguy and S. Espié Design and Modeling of a New Motorcycle Riding Simulator. *American Control Conference*, pages 176-181, 2007.

W. Khalil and J. F. Kleinfinger Minimum operations and minimum parameters of the dynamic models of tree structure robots. *IEEE Journal on Robotics and Automation*, Vol RA-3, NO. 6, 1987.

W. Khalil and E. Dombre. *Modélisation, identification et commande des robots*. Hermes science publications, Paris, 2nd edition,1999.

J. M. Hollerbach *A recursive lagrangian formulation of manipulator dynamics and a comparative study of dynamics formulation complexity*. IEEE Transaction on systems, man and cybernetics, vol smc-10, No 11, pages 730-736, 1980.

F. Panerai et al *Speed and safety distance control in truck driving: comparison of simulation and real-world environment*. Proceedings of driving simulation conference, pages 91-107, 2001.

A. Champion, Ming-Yu Zhang, J. M. Auberlet and S. Espie, *Behavioral simulation of a high-density traffic network involving an adaptive ramp metering system*. IEEE International Conference on Systems, Man and Cybernetics, vol 5, 2002.

V. Cossalter and R. Lot, *A Motorcycle Multi-Body Model for Real Time Simulations Based on the Natural Coordinates Approach*. Vehicle System Dynamics, vol 37, No. 6, pages 423-447, 2002.

S. Hima, L. Nehaoua, N. Séguy and H. Arioui, *Motorcycle Dynamic Model Synthesis for Two Wheeled Driving Simulator*. To appear in International IEEE conference on intelligent transportation systems, 2007.

R. Featherstone and D. Orin, *Robot dynamics: equations and algorithms*. IEEE International Conference on Robotics and Automation, Vol. 1, pages 826-834, 2000.

A. A. Shabana *Dynamics of multibody systems*. John wiley & sons,1989.

R. Lot, *A Motorcycle Tire Model for Dynamic Simulations: Theoretical and Experimental Aspects*. Meccanica, an International Journal of Theoretical and Applied Mechanics, Vol. 39, pages 207-220, 2004.

R. S. Sharp and S. Evangelou and D. J. Limebeer, *Advances in Modelling of Motorcycle Dynamics*. Multibody System Dynamics, Vol. 12, pages 251-283, 2004.

H. B. Pacejka and I. J. M. Besselink, *Magic Formulation Tyre Model with Transient Properties*. Vehicle System Dynamics, Vol. supplement 27, pages 234-249, 1997.

Intelligent Transportation Systems:
Automated Guided Vehicle Systems in Changing Logistics Environments

Prof. Dr.-Ing. habil. L. Schulze*. Dipl.-Ing. S. Behling*. Dipl.-Ing. S. Buhrs*.

*Department Planning and Controlling of Warehouse and Transport Systems (PSLT),
Leibniz Universität Hannover, Callinstr. 36, 30167 Hannover, Germany (e-mail: mail@pslt.uni-hannover.de).

Abstract: The usage of Automated Guided Vehicle Systems (AGVS) is growing. This has not always been the case in the past. A new record of the sells numbers is the result of inventive developments, new applications and modern thinking. One market that AGVS were not able to thoroughly conquer yet were rapidly changing logistics environments. The advantages in recurrent transportation with AGVS used to be hindered by the needs of flexibility. When nowadays managers talk about Flexible Manufacturing Systems (FMS) there is no reason not to consider AGVS. Fixed guidelines, permanent transfer stations and static routes are no necessity for most AGVS producers. Flexible Manufacturing Systems can raise profitability with AGVS. When robots start saving billions in production costs, the next step at same plants are automated materials handling systems. Today, there are hundreds of instances of computer-controlled systems designed to handle and transport materials, many of which have replaced conventional human-driven platform trucks. Reduced costs due to damages and failures, tracking and tracing as well as improved production scheduling on top of fewer personnel needs are only some of the advantages.

Keywords: Automated Guided Vehicle Systems, Worldwide Developments, Applications

1. INTRODUCTION

The main task of the material flow is to supply and to dispose operational units, like machines, plants and work areas, in time. A failure of the material flow can result in material holdup or a material undersupply. That clarifies the significance of an efficient material flow for the utilization of these operational units.

The material flow is an in-plant "service", which creates no direct value. However, it guarantees that the value-added processes in the enterprises can run accurate. It is often regarded as an "unimportant" process, whose impact on the profitableness is underestimated. Especially the quality of the material flow depends heavily on the commitment of the employees, because of the man based operations. An alternative is the automation of the material flow [5].

In this context the fundamental advantage of Automated Guided Vehicle Systems with their high flexibility and efficiency is highlighted. For years Automated Guided Vehicle Systems proved that they are suited for a wide range of tasks in the area of material flow. This applies equally to transport indoor in buildings and outdoor in the grounds of a company. A substantial advantage is that the vehicle units are planable and calculable according to predefined strategies. Automation also makes a substantial contribution to the traceability of goods which is demanded worldwide.

The AGVs have a high individuality. They are usually developed and constructed for the demands of a special application and are therefore unique. Due to this individuality a special problem is the wide variety of maintenance and part logistics. Research and development lead to new approaches to this question. Examples are the modularization and the standardization of AGVS. Further points are the reduction of complexity of the modules and the establishment of compatibility between various AGVS-producers. Another main direction of the development is the subsequent automation of standard fork lift trucks.

Applications of AGVS exist in all fields of industry and trade. Typical examples are the use in assembly lines, warehouses, order picking systems, production plants and hospitals.

In the long run automation of processes must be regarded as an aspect of ensuring industrial locations as well. The opinion that automation is a "job killer", yields to the insight that the development of automated transport systems and the associated processes safeguard qualified jobs [7].

Fig. 1. AGV at the Container Terminal Altenwerder, Germany [Source: Gottwald, Germany]

The past years are characterized by significant technological advancements. They contributed to increased attractiveness of AGVS for the users and essentially concern modularity, standardization, energy concepts, navigation systems, automation of series vehicles and safety systems [10].

2. TECHNOLOGICAL DEVELOPMENTS

Based on the AGVS statistic Europe, which is created and administered by the Department of Planning and Controlling of Warehouse and Transport Systems (PSLT), the main application of AGVS and the sectors in which they are used are analyzed. With this database developments and current trends in the AGVS sector are identified.

2.1 Sensor and Safety Systems

A new promising feature of modern systems is a collision avoidance system. The idea is to equip the AGVS, forklift trucks and other free or guided vehicles with a system to determine each vehicle's position and instantaneously broadcast it to other vehicles in order to forewarn drivers or the AGVS navigation to avoid collisions at an early time. This will especially reduce the number of accidents caused by collisions between Automated Guided Vehicles and manual trucks. The avoidance system can also be used to improve AGVS route planning.

Laser range scanners have long been used in the AGVS market. For safety equipment and laser navigation this technology has proved an indispensible potential. It is only a question of time when AGVS will be delivered with 3D laser technology as navigational or safety system. Further efforts will also make the use of wall-mounted reflectors unneeded. This will make AGVS even more precise and accurate at lower installation costs beyond the limitations of the current triangulation laser navigation method, released more than ten years ago.

2.2 Modularity and Standardization

Modularity is a modern strategic production method that AGVS suppliers have recently taken into account. More than anything this movement was driven by the needs of shorter delivery times. The standardization of AGVS leads to two important advantages. On the one hand the commitment on lifetime to one AGVS-manufacturer is abandoned, since the users can contact competitors in the context of modernizations or extensions. On the other hand production costs of the AGVS-manufacturers can be lowered by the standardization [6].

Cost reduction in the stages of development and realization of AGVS is a must for all manufacturers. Point to start from is that each manufacturer develops the vehicles on a base of a component system to reduce the variety of parts. Similar components can be used in different vehicle types. The module principle simplifies the logistic of replacement parts, increases the availability of parts and reduces the capital lockup of the manufacturer.

2.3 Contactless Transfer of Energy

A technical innovation is the inductive power transfer, which is increasingly used in AGVS. For this technology a wire is placed in the floor. By energizing this wire a magnetic field is generated above the floor. At the vehicle's lower surface a pickup containing an inductor is attached. The induced voltage is used for the vehicles power supply.

There are two basic principles of the inductive power transfer. In one case the vehicle is supplied continuously with energy by inductive power transfer. The primary conductor must be installed on the entire driving course. Thus the vehicle does not need batteries [11].

In the second case the vehicle has a battery on board and can thus compensate an interruption of the inductive power supply. The battery of the vehicle could be charged inductively at one point, at several points or on a defined course section.

For AGVS the inductive power transfer opened new operational areas, which were reserved so far for other conveyors. However, for complex driving courses that are branched out strongly or on floors, which contain too much metal or cannot be mill cut, the battery will remain the preferred energy concept.

A target of the development is to realize the power transfer as well as the data communication and the inductive guidance over the same wire. Thus the expenditure for the floor installation will be reduced. Alternatively an antenna at the vehicle can be used for data communication, guidance and positioning.

The inductive power transfer does not require rails for guidance or for power transfer. The danger of small parts e.g. screws or pins fall into the rails or slots is eliminated. This is particular important in assembly areas, for example in the automotive industry.

Today the inductive power transfer represents an alternative to the conventional battery loading station. Of the AGVS put into operation in 2006 about 8 % of all AGVS were supplied with energy inductively.

2.4 Navigation and Communication

The transportation task of AGVS requires efficient and intelligent routing on top of collision avoidance. Many producers can compute these aspects in real-time guaranteeing a proper responding time and behavior. Additionally it can handle priorities or a given time schedule.

Dynamic routing for AGVS is the consecutive calculation of the best path considering a certain time-frame with respect to the changing interdependencies. Mathematical models and modern computers make it possible to accomplish this in less than a second per request. With this strategy the flexibility of a productions system can be maintained and improved with the use of AGVS. On the other hand conventional transportation systems do not provide this behavior.

Guidance systems without a guideline provide high routing flexibility for the AGVS. New and also proven systems are the distance and angle measurement with laser navigation [1]. Another flexible system realizes guidance by transponders, which allow a point-to-point-ride.

A tendency similar to the navigation systems can be recognized in communication systems. Conditional on the guideline-free navigation, currently radio data transmission is applied in more than 80 % of the AGVS as communication technology. The data communication by infrared or inductive technology becomes less important.

2.5 Automation of Series Vehicles

The automation of series-production vehicles represents a further line of development. That concerns equally industrial trucks and motor trucks [12]. The series vehicles are automated by the AGVS-manufacturers. Industrial trucks are preferred in particular if both manual and automatic handling has to be realized. To increase the flexibility the AGV can be equipped with more than one load handling attachment.

**Fig. 2. Automated motor truck
[Source: Götting KG, Germany]**

The automation of motor trucks has an increasing impact on the market of shuttle transportation between different buildings of one plant [4]. Naturally the truck can be used during the daytime shift manually to increase the driving speed and in automatic mode during the rest of the time.

3. OPERATING COSTS AS SCALE OF ECONOMY

Contrary to man-operated industrial trucks the operating costs of AGVS are only marginally affected by the development of the labor costs. From this it results that relating to the labor costs a high calculative planning reliability can be achieved in the long-term. This is a general advantage of all automated material flow systems. On the assumption that the labor costs will rise even more strongly in the future than in the past, AGVS will increase above average in comparison to personnel intensive material flow systems.

The development of the labor costs, as for example a start-up financing for the creation of a job or a shortening of subsidies, may not remain unconsidered. Each of these factors can either promote or restrain the development of the AGVS-market.

The investment in a plant with an AGVS is usually higher than for a plant with man-operated industrial trucks. That has consequences on both the cost-accounting interest and the height of the depreciation. For AGVS higher cost-accounting interests result. The height of the interest rate has to be oriented at the development on the capital market. If the interest rate decreases, the profitableness of AGVS is affected positively.

The depreciation has to be regarded under two criteria, namely according to tax law and cost accounting criteria. It has to be an aim of the plant operators to estimate the economic lifetime of the plant as short as possible. Therewith it should be reached that the depreciation of the fixed capital can be made valid for taxation as promptly and completely as possible.

The labor costs, the interest trend at the capital market and the amortization period belong to the substantial economic factors, which determine the development of the AGVS-market. The development of these factors cannot be affected by the AGVS-manufacturers, the factors affect the market from the outside.

The amortization period defined by the technical lifetime is applied to the cost comparison method, thus for the system decision. A long technical lifetime affects the system comparison positively. The technical lifetime is specified internally considering the tasks and the operating conditions.

For the success of the European AGVS-manufacturers on non-European markets the rates of exchange are relevant. With a low US-Dollar price per Euro the European AGVS-manufacturers can make attractive offers for the international market. In the year 2006 about 23 % of the AGVS by European manufacturers were installed outside of Europe.

4. WORLDWIDE INSTALLATION OF AGVS

Applications can be found throughout all industrial branches, from the automotive, printing and pharmaceutical sectors over metal and food processing to aerospace and port facilities. The increasing interest in AGVS is reflected in the sales figures which reached a new peak in 2006 with a volume of 200 Mio. EUR according to a yearly survey among European AGVS producers carried out by the PSLT.

The current developments promise that automated transport systems will have a high value for the operating company in the future as well. A set of additional conditions must be fulfilled in order to maintain the latest positive trend regarding the start-up of AGVS.

The trends of the markets and therefore the development of the AGVS-manufacturers are also of particular importance for the investment decisions of the users. It must be ensured that the acquired technology is future-oriented and that the manufacturer is present at the market segment of AGVS in the long term. The selected AGVS-manufacturer should be available for service and support of the system as well as for spare part logistics for a long time [8].

In comparison to the year 2000 about a quarter of the AGVS-manufacturers are on the one hand "new" vendors. On the other hand the "old" vendors offer new and different achievement profiles today. Both aspects point out the dynamic on the vendor side, which offers with more than twenty five European AGVS-manufacturers a large variety.

A substantial indicator for the market tendency of AGVS is the annual number of AGVS put into operation. The key number for the European manufacturers is issued by the PSLT of the Leibniz Universität Hannover, Germany, is based on the information of the AGVS-manufacturers.

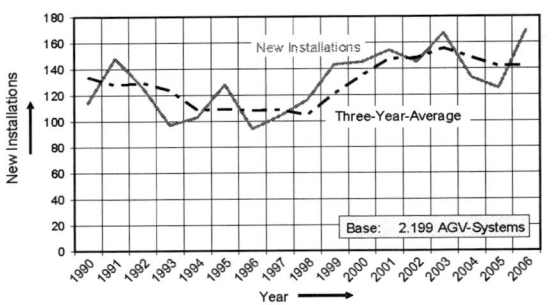

Fig. 3. Vehicles in openings of European AGVS-manufacturers

The number of AGVS put into operation world-wide by European AGVS-manufacturers sums up to over 3.300 new systems with about 27.500 Automated Guided Vehicles in total.

After an intermediate flattening related to the number of the Automated Guided Vehicles and AGVS put into operation a significant growth can be registered from 1999 until 2006. Thus in the three-annual average the level rose in the meantime over 140 new systems per year. In 2006 a new peak with 169 AGVS was reached.

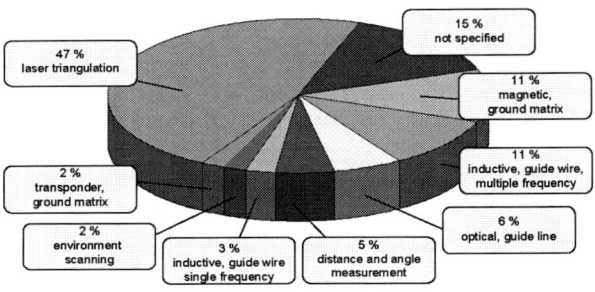

Fig. 4. Guiding technologies in AGVS put into operation by European AGVS manufacturers 2006

A similar process is ascertained for the number of Automated Guided Vehicles put into operation. It also is a trend that the average system size measured in vehicles per system rises. The average number of vehicles per system amounts now again over six vehicles. The complex systems increase again and the requirements on planning, engineering, project management, realization and putting into operation rise. This trend is also pointed out by the fact that the average equipment price allocated on the vehicles increased.

In 2006 a considerable number of about 6 % among all AGVS in Europe was realized within the outdoor sector. These applications of AGVS within port facilities for the containerization or as shuttle transportation on the work area already proved their fitness to practice and are to be regarded as concrete delivery offers of the manufacturers [13].

It is obvious that the development of the AGVS-market depends directly on the strategies of the users. An example for this is the choice of location of manufacturing plants, which is of utmost economic importance for the companies. Geographical and political frontiers play almost no more roles in this matter today. The aims are to produce as economically as possible and to be as close as possible to the sales market. As a consequence the requirements of internationalization and globalization approach to the AGVS-manufacturers. They must be able to install and to support AGVS world-wide.

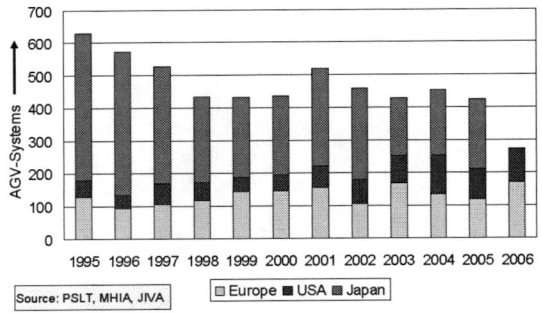

Fig. 5. Worldwide openings of AGVS of European, American and Japanese manufacturers

In these relations some characteristic numbers clarify the variability of the requirements for the operators in Europe, Japan and the USA. Comparing to Japanese and American Automated Guided Vehicle Systems, European systems have the largest average number of vehicles per system. Japanese systems have the smallest number. Most AGVS are operating in Japan, but most of these systems are very "simple". This is also reflected in the installed navigation systems. In Europe laser navigation was realized in about 47 % of the systems in the year 2006 whereas in Japan the magnetic procedures predominate.

Until July of 2005 there are altogether less than 60 AGVS operating in China with less than 400 vehicles. The application in the tobacco industry is the widest where 20 enterprises have used AGVS so far. Most of them are using laser guidance technology.

Because of the popularization and the superiority in practice of the AGV more and more companies and institutions in China start to research on AGV technology [3]. Selected points in research on AGVs are the order management, route design and communication [2].

5. APPLICATIONS

AGVS can be found in virtually any area of industrial production, trade and service. The main application areas are production, connection of different work areas, order picking, warehousing and assembly. AGVS are especially employed in cases of consistent material flow connections. On the other hand trends like laser guided vehicles and flexible software lead to advanced application areas.

The realization of the material flow processes in the warehousing and order picking sector is characterized by high volume of traffic from defined sources to defined drains. Thus it is a standard application area of AGVS. The specialty of AGVS employed in this area is a high loading capacity. Regularly the units loaded by these AGVs are standardized pallets, therefore the vehicles are equipped with standard loading devices. Due to the high amount of goods to transport, these systems often consist of more than 100 vehicles. This demands a sophisticated central controlling unit and optimizing approaches for routing and path-finding.

Fig. 6. AGVS for assembly line [Source: TMS, Germany]

Another area with a high application rate of AGVS are the assembly lines. In this sector the load is inhomogeneous and changing. Therefore the loading devices must be fitted to the specific application. The AGV often not only transports the load from one assembly station to the next but represents an assembly station on its own. In this case the AGV can be considered as a mobile workbench. Another assembly application is the pick-up AGV which has the work piece mounted on it and virtually represents a conveyor for both the worker and the work piece.

In contrast to the expanded, complex systems with a high investment, small and comparatively low priced systems can be found in many environments. Systems with only one or very few vehicles can be realized without any central controlling unit. Few vehicles with robust techniques can be applied in any industrial environment. These systems stand out by low investments and simple maintaining.

New developments at the PSLT are techniques that facilitate AGVS to follow a designated person. The basic requirement for human-robot interaction is the system's ability to distinguish between a person and its surroundings and moreover to identify that person. This can be achieved by using sensors such as digital cameras or laser range scanners. The main focus is centered on computer vision and its practical integration into an AGV-System. Experimental results show good performances of the system. Logistic operations like order picking would immensely profit from such a function. An interesting perspective is a consignment store with employees focusing on picking while trolleys are following automatically. When they are loaded the trolleys will carry their load away. Empty replacements are provided by the central control in time.

Fig. 7. "AnGiV" at the PSLT´s AGV Symposium 2006

In 2006 the system was presented at the AGV Symposium that is conducted by the PSLT at the Leibniz Universität Hannover every two years.

Many more new applications for AGVS can be found in the logistics sector. An example is the implementation of an AGV to a stationary pallet stretch wrapping device. Thereby manufacturers can eliminate the use of pallet conveyors, reduce labor costs, increase plant floor safety, and eliminate product and conveyor damage caused by human errors.

6. CONCLUSIONS

Significant technological advancements contributed to increase the attractiveness of Automated Guided Vehicle Systems for the users. They essentially concern the modularity, the standardization, the navigation system, the energy concept, the automation of series vehicles and the safety system. All in all it is to be noted that many new impressive reference systems and the technology innovations of the last years opened new and demanding applications for Automated Guided Vehicle Systems.

For the manufacturers of AGVS the internationalization and globalization represents a new challenge, which offers large chances for the future. Therefore it is necessary to acquire a large scaled knowledge of the different national operator requirements and to create transparent basic conditions for the line operation. An emphasis represents surely the global support of the AGVS.

In addition, laws and rules, which regulate the use of Automated Guided Vehicles, can have a substantial influence on the market trend. The dense administration structures particularly in Germany and generally in Europe are repressive to the application of AGVS. A fact that has been preset by law and has ever been neglected by the AGVS producers is the driving speed of the vehicles. The future will show if the rising demands of the applications will affect a breakthrough on this topic.

The transportation of pallets from a certain position to another is the classical market for AGVS which will still exist in the future as it does today. In addition various mobile robot applications like autonomous mining robots or automatic vehicles for agriculture will be of significance for the AGVS branch. The modularization will help bringing different initial projects together to the benefit of all.

A very special topic for AGVS will be China's market in the future. The first systems were already taken into operation. Great efforts are undertaken by European vendors to install reference assets. In this connection long-term export possibilities for AGVS-manufacturers are of particular interest. On the other hand China itself is currently developing AGVS for their own market [14].

For almost any internal material flow task an AGVS can be made up that meets the demands. This ranges from the simplest systems with one vehicle to huge, complex systems with over 100 vehicles. Since the flexibility of AGVS will steadily rise they offer possibilities to companies for their position in the global competition. It is necessary to use these chances for the reduction of costs and improvement of quality.

REFERENCES

[1] Gremm, F., "Laser steuert FTS zielgenau - FTS sichert 24-h-Betrieb", in Schweizer Maschinenmarkt, vol. 10, no. 1/2, pp. 22-24, 2002 (in German).

[2] Guo F., Yuan X., Yu D., Yuan K., "Research on Key Technology of AGVs", Journal of University of Science and Technology Beijing, 1999 (in Chinese).

[3] Jin Y., Xiang G., "Application of AGVs in Modern Logistics", Authoritative Forum, 2005 (in Chinese).

[4] Mäkelä, H., Numers, T., "Development of a navigation and control system for an autonomous outdoor vehicle in steel plant", in Control Engineering Practice, vol. 9, no. 5, pp. 573-583, 2000.

[5] Schulze, L., Lucas, M., "Logistics and Automation State of the Art and Trends in Europe", presented at Greater China Supply Chain and Logistics Forum and Academic Conference 2005, 30.10.2005, Nanjing, China

[6] Schulze, L., Lucas, M., Baumann, "Automated Guided Vehicle Systems - Trends in Technology and Application", in Proceedings of IMECE2005, ASME International Mechanical Engineering Congress and Exposition, Florida, USA, 2005.

[7] Schulze, L., "Tendenzen der Fahrerlosen Transportsysteme", in Logistik für Unternehmen, vol. 17, no. 4-5, pp. 40-43, 2003 (in German).

[8] Schulze, L., Lucas, M., Runge, J., "Mehr Planungssicherheit für Hersteller und Betreiber: Zukunftsprognosen für die FTS-Technik?", in Logistikwelt 2003, vol. 10, no. 1, pp. 66-70, 2003 (in German).

[9] Schulze, L., "Technologie und Anwendung - Fahrerlose Transportsysteme für die Praxis", Proceedings FTS-Fachtagung 2004, Universität Hannover, 16.06.2004 (in German).

[10] Schulze, L., Runge, J., "Europa-Marktübersicht FTS 2003, Neuaufstellung bei den Herstellern - Innovative Technologien für Betreiber", in Logistikwelt 2003, vol. 10, no. 1, pp. 72-88, 2003 (in German).

[11] Schulze, L., "Innovation und Modernisierung - Fahrerlose Transportsysteme zum Kundennutzen", Proceedings FTS-Fachtagung 2002, Universität Hannover, 19.09.2002 (in German).

[12] Schulze, L., Frenzel, M., Runge, J., "Eine Vision wird Wirklichkeit - Fahrerlose Nutzfahrzeuge - heute im Werksgelände, morgen auf der Straße", in Logistikwelt 2000, vol. 7, no. 1, pp. 58-61, 2000 (in German).

[13] Schulze, L., Runge, J., "Europa Marktübersicht FTS-Anlagen - Daten und Fakten für die Investitionsentscheidung", in Logistikwelt 2000, vol. 7, no. 1, pp. 66-84, 2000 (in German).

[14] Zhang Z., "Review of AGVs Technology", Logistics Technology and Application, 2005 (in Chinese).

SESSION BI: ROBUST CONTROL

An initial corrector using different approaches for Youla parametrization via LMI optimization

S. ZIANI* S. FILALI**

Laboratory of automatic and robotic, Mentouri University, Constantine,
zianide_s@yahoo.fr
**CESAME, Université catholique de Louvain, Avenue G. Lemaitre 4, B-1348, Louvain-la-Neuve, Belgium*
safilali@hotmail.com

Abstract: This paper presents an approach by multiobjective optimization of the output feedback design in discrete time. Our goal is to find a controller stabilizing the system with temporal or frequential specifications. This is achieved using Youla parametrization based on an initial corrector chosen, in combination with different Lyapunov functions; via LMI optimization. A comparison of different approaches is made in different cases, ie when the initial corrector is a LQG, or H2, or H∞, another goal of this work is also reducing the LMI conservatism.

Keywords: Youla Parameter, LMI Optimization, LQG, H2/H∞ control, Multiobjective control

1. INTRODUCTION

In recent years linear matrix inequalities (LMI's) have emerged as a powerful tool to approach control problems that appear hard if not impossible to solve in an analytic fashion. Although the history of LMI's goes back to the fourties with a major emphasis of their role in control in the sixties (Kalman, Yakubovich, Popov, Willems), only recently powerful numerical interior point techniques have been developed to solve LMI's in a practically efficient manner (Nesterov, Nemirovskii 1994). A LMI is a constraint of affining on the design variables, the characteristics of attenuation such as the placement of poles, robust stability, execution LQG, or of RMS, gains which can be expressed like LMIs. These characteristics define a multiobjective problem that can be solved numerically via convex optimization under LMI constraints [1]. The LMI optimization treats a problem with contradictory objectives; our objective is to found an optimal solution who is a compromise between all the defined objectives. The method of synthesis presented rests on the Youla parametrization [2,3]. Indeed, we use the fundamental properties of the Q-Parameter to present a methodology to obtain a representation of the inter-connected systems G, J and Q of closed loop $F_L(P,K)$ (linear Fractional Transformation or LFT Lower). We consider the LQG (H2 and H∞ respectively) controller as an initial corrector for the Youla parametrization, Where the Q-Youla gives access to all the correctors K who stabilize the closed loop via the parameter Q. Hence it exist a corrector satisfying the specifications, we can find it by convex optimization [4]. These parameterizations transform the initial problem into a convex LMI problem. This formalism is particularly adapted to the multi-criteria design because it possible to juxtapose the criteria without losing convexity [1, 5, 6]. In this work we can follow these three steps to treat a problem of control by LMI convex optimization. The first one is the formulation of the initial problem to an optimization problem. The second is how to get a convex formulation, and the last one is the construct of the command law. Each stage of this process modifies the initial problem, and so induced a difference between the practical solution (found) and the theoretical optimal solution. Then the notion of the conservatism (Complexity / Calculability) [3] of the problem became another problem. The principle of multiobjective [7] is to satisfy several criteria simultaneously. A comparison between different approaches is done when the initial corrector is LQG, H2, H∞.

2. DISCRETE TIME LFT SYSTEM

The system P is stabilized by a controller K, satisfying the conditions of Youla parametrization defined by the interconnected systems P, J and Q, figure 1 and 2.

The closed loop system is: $T = P*K = P*J*Q = G*Q$ (1)

* : is the LFT "Linear Fractional Transformation Lower".

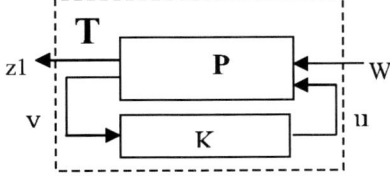

Fig.1. Closed loop system

$$P = \begin{bmatrix} A & B_1 & B_2 \\ C_1 & D_{11} & D_{12} \\ C_2 & D_{21} & D_{22} \end{bmatrix} \quad (2)$$

$$K = \begin{bmatrix} A_K & B_K \\ C_K & D_K \end{bmatrix} \quad (3)$$

$$G = \begin{bmatrix} G_{11} & G_{12} \\ G_{21} & G_{22} \end{bmatrix} \quad (4) \quad \text{or} \quad G = \begin{bmatrix} A_G & B_G \\ C_G & D_G \end{bmatrix} \quad (5)$$

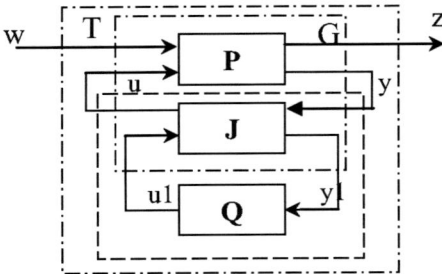

Fig.2. General form of the Youla parameterization

Q: is a transfer stable
J: is representing the inter-connected system that the controller estimator/ states feedback structure [2, 3, 8]:

$$J = \begin{bmatrix} K & 0 & \tilde{V}_0^{-1} \\ V_0^{-1} & -V_0^{-1}N \end{bmatrix} \quad (6)$$

Ko is the initial corrector characterized by two gains K_f and K_c (observer gain and states feedback).

3. LINEARIZATION

The state representation of closed loop system is given by:

$$T = \begin{bmatrix} A_{cl} & B_{cl} \\ C_{cl} & D_{cl} \end{bmatrix} = \begin{bmatrix} A+B_2D_KC_2 & B_2C_K & B_1+B_2D_KD_{21} \\ B_KC_2 & A_K & B_KD_{21} \\ C_1+D_{12}D_KC_2 & D_{12}C_K & D_{11}+D_{12}D_KD_{21} \end{bmatrix} \quad (7)$$

Introducing the static Q-parameter, with the fundamental properties of Youla parametrization [2, 7, 8] and system J we get T as :

$$T = \begin{bmatrix} A-B_2.K_C & B_2.K_C+B_2QC_2 & B_1+B_2.Q.D_{21} \\ 0 & A-K_f.C_2 & B_1-K_f.D_{21} \\ C_1-D_{12}.K_C & D_{12}.K_C+D_{12}.Q.C_2 & D_{11}+D_{12}.Q.D_{21} \end{bmatrix} \quad (8)$$

The goal is to separate the matrix system from the corrector, we get two new matrices:

$$T = \begin{bmatrix} B_2 \\ 0_{nax1} \\ D_{12} \end{bmatrix} . Q . \begin{bmatrix} 0_{1xna} & C_2 & D_{21} \end{bmatrix} + \\
\begin{bmatrix} A-B_2.K_c & B_2.K_c & B_1 \\ 0_{naxna} & A-K_f.C_2 & B_1-K_f.D_{21} \\ C_1-D_{12}.K_c & D_{12}.K_c & D_{11} \end{bmatrix} \quad (9)$$

Then $T = P*J*Q = T_1.Q.T_2 + T_0$ (10)

$$T_1 = \begin{bmatrix} B_2 \\ 0_{nax1} \\ D_{12} \end{bmatrix} \quad (11) \quad T_0 = \begin{bmatrix} A-B_2K_c & B_2K_c & B_1 \\ 0_{naxna} & A-K_fC_2 & B_1-K_fD_{21} \\ C_1-D_{12}K_c & D_{12}K_c & D_{11} \end{bmatrix} \quad (13)$$

$$T_2 = \begin{bmatrix} 0_{1xna} & C_2 & D_{21} \end{bmatrix} \quad (12)$$

Usually whatever J the function T is affine in Q.

Investigate the case where J is represented by the observer (Eq.6) In same way for To, we separate matrices systems from K_c and K_f we yields:

$$T_0 = T_3.K_c.T_4 + T_5.K_f.T_6 + T_7 \quad (14)$$

$$T_3 = \begin{bmatrix} -B_2 \\ 0_{nax1} \\ -D_{12} \end{bmatrix} \quad (15) \quad T_5 = \begin{bmatrix} 0_{naxna} \\ I_{naxna} \\ 0_{1xna} \end{bmatrix} \quad (16)$$

$$T_4 = \begin{bmatrix} I_{naxna} & -I_{naxna} & 0_{nax2} \end{bmatrix} \quad (17)$$

$$T_7 = \begin{bmatrix} A & 0_{naxna} & B_1 \\ 0_{naxna} & A & B_1 \\ C_1 & 0_{1xna} & D_{11} \end{bmatrix} \quad (18)$$

$$T_6 = \begin{bmatrix} 0_{1xna} & -C_2 & -D_{21} \end{bmatrix} \quad (19)$$

The function T depends linearly of initial corrector Ko (K_c, K_f) and from Youla parameter Q all other matrices have the data system.

$$T = T_1.Q.T_2 + T_3.K_c.T_4 + T_5.K_f.T_6 + T_7 \quad (20)$$

4. PROBLEM FORMULATION

Our problem is

$$\min_K \|T\|_\infty \text{ (Respectively } \min_K \|T\|_2 \text{)} \quad (21)$$

"Which is the minimization of the maximum value of the ratio between the energy of the output signal y(t) and energy input signal w(t); it is clear that the minimization of this norm is necessary to maximize the rejection of disturbance and tracking".

Then:

$$\min_K \|T\|_\infty = \min_K \|T_1.Q.T_2 + T_3.K_c.T_4 + T_5.K_f.T_6 + T_7\|$$

$$\min_K \|T\|_\infty \leq \min_K \|T_1.Q.T_2\|_\infty + \min_K \|T_3.K_c.T_4\|_\infty + \min_K \|T_5.K_f.T_6\| + \min_K \|T_7\| \quad (22)$$

Equation (22) shows that the infinite norm of the closed loop system is depend on the Youla parameter Q, and the initial corrector Ko.

$$T = f(Q, K_c, K_f) \quad (23)$$

Our objective is to optimize K which means optimizing the two systems J and Q. We fix J and optimize Q we will study the choice of interconnected system J on the responses and performances improvement in closed loop, for the following three cases of K_o who are $K_{o-H\infty}, K_{o-H2}, K_{o-LQG}$.

The initial problem is mathematically difficult to solve, we need to make these two changes

4.1. Definition: change on the complexity of the problem:

The initial problem is

$$\min_K \|T\|_\infty \Leftrightarrow \min_{J,Q} \|T\|_\infty \quad (24)$$

With the observer J we can have $\min_{K_c,K_f,Q} \|T\|_\infty$ (25)

By using $\min_Q \|T\|_\infty$ we can fixed K_c and K_f (26)

4.2. Definition: change in computability of the problem:

Using the LMI tools, the problem turns into an optimization LMI problem

$$\min_Q (\gamma_1, \gamma_2)$$

Such as
$$\begin{cases} \|T\|_\infty < \gamma_1 \\ \|T\|_2 < \gamma_2 \\ \gamma_1 > 0, \ \gamma_2 > 0 \end{cases} \quad (27)$$

To reduce the conservatism of the LMI problem, from the complexity and calculability, we can take into consideration the following remarks:

• Optimize γ such as $\gamma = \tau\gamma_1 + (1-\tau)\gamma_2^2$ / $\tau = 0.8 \in [0, 1]$ without losing the convexity of problem LMI, decrease the number of objective also decrease the computing time.

• Choose the Toeplitz structure of Lyapunov functions for reducing the number of the decisions variables and the memory capacity.

The final problem becomes:

$$\min_Q(\gamma)$$

Such as $\gamma > 0$, $\|T\|_\infty < \gamma$, $\|T\|_2 < \tau\gamma$ (28)

5. ALGORITHM

In the following algorithm we search K such as the system $F_L(P, K)$ checks :

$$\begin{cases} \|T\|_\infty < \gamma_1 \\ \|T\|_2 < \gamma_2 \\ F_L(P,K)) \text{ is } \alpha\text{-stable} \end{cases} \quad (29)$$

5.1. Step 1: Introduction of the weight functions:

To achieve the desired objectives, the augmented system is:

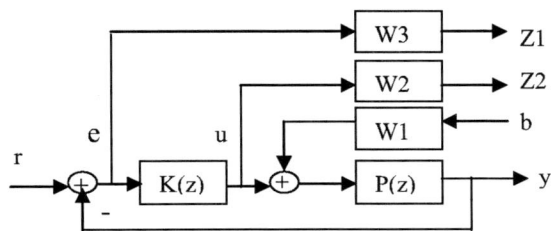

Fig.3. Block diagram of the augmented system

Define $(W1, W2, W3)$, using $H\infty$ formulation problem [9], and by using small gain theorem

$$\|W3S\|_\infty < \gamma \Leftrightarrow \forall \omega \in R \quad |S(j\omega)| < \frac{\gamma}{|W3(j\omega)|} \quad (30)$$

$$\|W2KS\|_\infty < \gamma \Leftrightarrow \forall \omega \in R \quad |K(j\omega)S(j\omega)| < \frac{\gamma}{|W2(j\omega)|} \quad (31)$$

$S=(1+KP)^{-1}$ is the sensitivity (32)
KS: is the sensitivity in disturbance. (33)
$W3$ is selected for rejection of the disturbance in low frequency and to reduce the static error.

$$W3(s) = \frac{s/M + wc}{s + wc \cdot eo} \quad (34)$$

M: peak of S; eo: the offset; wc: the bandwidth

$$W2(s) = \frac{s + wc2}{s + wc2 \cdot c} \quad (35)$$

For a good control law, we selected c, constant $\gg 1$ (36)

5.2. Step 2: Compute J as:

• $J_{LQG} = J(K_{o\_LQG})$

The inter-connected J_{LQG} system has the following form[3, 4]

$$J_{LQG} = \left[\begin{array}{c|cc} A - B_2 F_{LQG} - L_{LQG} C_2 & L_{LQG} & B_2 \\ \hline -F_{LQG} & 0 & I_{nu} \\ -C_2 & I_{ny} & 0 \end{array}\right] \quad (37)$$

• $J_{H2} = J(K_{o\_H2})$

Using the two dual algebric Riccati equations [8, 10]:

$$A^T X_2 + X_2 A - X_2 B_2 B_2^T X_2 + C_1^T C_1 = 0 \quad (38)$$

$$A Y_2 + Y_2 A^T - Y_2 C_2^T C_2 Y_2 + B_1 B_1^T = 0 \quad (39)$$

The evolution matrices $(A-B_2F_2)$ and $(A-L_2C_2)$ (40) are stable and compute by:

$$F_{H2} = -(B_2^T X_2 + D_{12}^T C_1) \ (41) \text{ and } L_{H2} = -(Y_2 C_2^T + B_1 D_{21}^T) \ (42)$$

The H2 corrector is defined by :

$$J_{H2} = \left[\begin{array}{c|cc} A - B_2 F_{H2} - L_{H2} C_2 & L_{H2} & B_2 \\ \hline -F_{H2} & 0 & I_{nu} \\ -C_2 & I_{ny} & 0 \end{array}\right] \quad (43)$$

• $J_{H\infty} = J(K_{o\_H\infty})$

Using the dual algebric Riccati equations [8, 10]:

$$X_\infty A + A^T X_\infty + X_\infty(\gamma^{-2} B_1 B_1^T - B_2 B_2^T) X_\infty + C_1^T C_1 = 0 \quad (44)$$

$$Y_\infty A^T + A Y_\infty + Y_\infty(\gamma^{-2} C_1^T C_1 - C_2^T C_2) Y_\infty + B_1 B_1^T = 0 \quad (45)$$

The two gains $F\infty$ and $L\infty$ is given by:

$$F_{H\infty} = -(B_2^T X_\infty + D_{12}^T C_1) \quad (46) \quad L_{H\infty} = -Z_\infty H_{\inf} \quad (47)$$

or $H_{\inf} = -(Y_\infty C_2^T + B_1 D_{21}^T)$, $Z_\infty = -(I - \gamma^{-2} Y_\infty X_\infty)^{-1}$ (48)

$$J_{H\infty} = \left[\begin{array}{c|cc} A - B_2 F_{H\infty} - L_{H\infty} C_2 & L_{H\infty} & B_2 \\ \hline -F_{H\infty} & 0 & I_{nu} \\ -C_2 & I_{ny} & 0 \end{array}\right] \quad (49)$$

5.3. Step 3: Formulations of the LMI inequalities

• $\|T\|_\infty < \gamma_1$ is characterized by the inequality[11]: $\exists X_1 = X_1^T > 0$

$$\begin{bmatrix} -X_1^{-1} & A_{cl} & B_{1,cl} & 0 \\ A_{cl}^T & -X_1 & 0 & C_{1,cl}^T \\ B_{1,cl}^T & 0 & -\gamma_1 I & D_{1,cl}^T \\ 0 & C_{1,cl} & D_{1,cl} & -\gamma_1 I \end{bmatrix} < 0 \quad (50)$$

• $\|T\|_2 < \gamma_2$ is characterized by the inequality [12]:

$\exists\ X_2 = X_2^T > 0$, and $\exists\ Y = Y^T > 0$

$$\begin{bmatrix} -X_2^{-1} & B_{2,cl} & 0 \\ B_{2,cl}^T & -Y & D_{2,cl}^T \\ 0 & D_{2,cl} & -I \end{bmatrix} < 0 \quad \begin{bmatrix} -X_2^{-1} & A_{cl} & 0 \\ A_{cl}^T & -X_2 & C_{2,cl}^T \\ 0 & C_{2,cl} & -I \end{bmatrix} < 0$$

$$\text{trace}(Y) < \gamma_2^2 \tag{51}$$

- α-stability for P*K is characterized by the inequality [3]:

$$\exists\ X_3 = X_3^T > 0$$

$$\begin{bmatrix} \alpha^2 X_3 & A_{cl}^T \\ A_{cl} & X_3^{-1} \end{bmatrix} > 0 \tag{52}$$

X_1, X_2, et X_3 are the desired Lyapunov functions; the (A_{cl}, B_{cl}, C_{cl}, D_{cl}) represent the system matrix of the closed loop. And α is the α-stability value ($\alpha \in [0, 1]$).

5.4. Step 4: Introduction of the Q-parameter

5.4. Step 5: Linearization of the matrix inequalities

Apply the results of step3 (§5.3.) to the closed loop system T for the linearization of the optimization variables (X_i, Q). Let X_i be the Lyapunov function associated with objective i {i=1...3 ⇔ H2 ; H∞ ; α-stability} X_i is partitioned as G (A_G):

$$A_G = \begin{bmatrix} A_1 & A_3 \\ 0 & A_2 \end{bmatrix} \quad X_i = \begin{bmatrix} W_i & Z_i \\ Z_i^T & Y_i \end{bmatrix}$$
According $\tag{53}$

By using the corollary S-procedure lemma [3]:

$$\begin{bmatrix} W_i & Z_i \\ Z_i^T & Y_i \end{bmatrix} \mapsto \begin{bmatrix} R_i & S_i \\ S_i^T & T_i \end{bmatrix} = \begin{bmatrix} W_i^{-1} & -W_i^{-1} Z_i \\ -Z_i^T W_i^{-1} & Y_i - Z_i^T W_i^{-1} Z_i \end{bmatrix} \tag{54}$$

And the congruence Lemma with the matrix [3]:

$$M_i = \begin{bmatrix} R_i & 0 \\ S_i^T & I \end{bmatrix} \tag{55}$$

While applying to the system T these changes one will have:

- *Compute H∞:*

By using the congruence lemma to the matrix:
$$\Pi_1 = \text{diag}(M_1\ M_1\ I\ I) \tag{56}$$

We obtain from (Eq.50):

$$\begin{bmatrix} -R_1 & 0 & A_1R_1A_1S_1-S_1A_2+A_3+B_uQC_{y\wedge} & B_{1,1}+B_u+B_uQD_{y\wedge1}-S_1B_{1,2} & 0 \\ 0 & -T_1 & 0 & T_1A_2 & T_1B_{1,2} & 0 \\ * & * & -R_1 & 0 & 0 & R_1^T C_{1,1}^T \\ * & * & 0 & -T_1 & 0 & C_{1,2}^T + C_y^T Q^T D_{1u\wedge}^T + S_1^T C_{1,1}^T \\ * & * & * & * & -\gamma_1 I & D_{11}^T + D_{y\wedge1}^T Q^T D_{1u\wedge}^T \\ * & * & * & * & * & -\gamma_1 I \end{bmatrix} < 0$$
$$\tag{57}$$

- *Compute H2:*

By using the congruence lemma with the two matrix:
$$\Pi_{21} = \text{diag}(M_2\ M_2\ I) \quad (58) \qquad \Pi_{22} = \text{diag}(M_2\ I\ I) \quad (59)$$

We obtain from (Eq.51)

$$\begin{bmatrix} -R_2 & 0 & A_1R_2A_1S_2-S_2A_2+A_3+B_uQC_{y\wedge} & B_{1,1}+B_u+B_uQD_{y\wedge1}-S_2B_{1,2} & 0 \\ 0 & -T_2 & 0 & T_2A_2 & T_2B_{1,2} & 0 \\ * & * & -R_2 & 0 & 0 & R_2^T C_{2,1}^T \\ * & * & 0 & -T_2 & 0 & C_{2,2}^T + C_y^T Q^T D_{2u\wedge}^T + S_1^T C_{2,1}^T \\ * & * & * & * & & -I \end{bmatrix} < 0$$

$$\begin{bmatrix} -R_2 & 0 & B_{2,1}+B_u\wedge QD_{y\wedge2}-S_2B_{2,2} & 0 \\ 0 & -T_2 & & 0 \\ * & * & -Y & D_{22}^T + D_{y\wedge2}^T Q^T D_{2u\wedge}^T \\ * & * & 0 & -I \end{bmatrix} < 0$$

$$\text{trace}(Y) < \gamma_2^2 \tag{60}$$

- *Compute for α-stability:*

Using the congruence lemma to the matrix:
$$\Pi_3 = \text{diag}(I\ M_3) \tag{61}$$

We obtain from (Eq.52)

$$\begin{bmatrix} \alpha^2 R_3 & * & * & * \\ 0 & \alpha^2 T_3 & * & * \\ A_1 R_3 & A_1 S_3 - S_3 A_2 + B_2 QC_2 + A_3 & R_3 & * \\ 0 & T_3 A_2 & 0 & T_3 \end{bmatrix} > 0 \tag{62}$$

The inequalities (57), (60), (62) are linear on the decisions variables R_1, S_1, T_1, Q, γ_1 for the H∞ problem; R_2, S_2, T_2, Q, γ_2^2 for the H2 problem and R_3, S_3, T_3, Q for the α-stability problem. These three problems are coupled by the static feedback-output Q and use independent Lyapunov functions. (As mentioned in the Abstract)

5.6. Step 6: LMI problem formulation

Find the solution of the LMI optimization problem [13]

$$\min_Q (\gamma_i)$$
$$\begin{cases} \|T_i\|_\infty < \gamma_i \\ \|T_i\|_2 < \tau.\gamma_i \\ P*K \text{ is α-stable} \end{cases} / i = 1, 2, 3 \tag{63}$$

5.7. Step 7: Construct of the inter-connected systems and compute the command law.

6. APPLICATION

We consider the system P defined by [14]:

$$P(z) = \frac{0.2879z^2 + 0.03516z - 0.2217}{z^3 - 2.158z^2 + 1.874z - 0.6908} \tag{64}$$

Under the specifications:
1. Settling time tr < 4s
2. Bandwidth for S wc=2rd/sec and for the KS wc2=2rd/sec.
3. Peak of $|S|$ < 8 db and peak of $|KS|$ < 8 db
   With parameters:
   α-stability value =0.88, w1 = 0.2,

$$w2 = 0.20 \frac{z-1}{z-0.007454} \ ; \quad w3 = 0.50 \frac{z-0.4804}{z-1} \tag{65}$$

Table 1. Parameters of comparison

Parameters	K_{o_LQG}	K_{o_H2}	$K_{o\_H\infty}$
Number of matrix inequalities LMI	6	6	6
Number of decision variables	194	194	194
Number of objectives	1	1	1
Lyapunov F^{cts}	Different	Different	Different
Lyapunov F^{cts} Structure	Toeplitz	Toeplitz	Toeplitz
Optimality tolerance	0.0001	0.0001	0.0001
α-stable value	0.88	0.88	0.88
Settling time t_r(sec)	2.6	≈ 1.5	≈ 3
Overshoot D (db)	1.13	1.28	0
Peak of the $|S|$ (db)	6.71	57.6	10.01
Peak of the $|KS|$(db)	6.80	41.3	7.93
Q_{opt}	0.9684	3.9881	0.6171
γ_{opt}	1.5936	4.5580	11.9075
$\|T\|_\infty$	1.1268	1.6916	1.3225
$\|T\|_2$	0.4996	0.9888	0.4250
Time computing LMI(Approximate)	3.2350	11.7960	11.9530

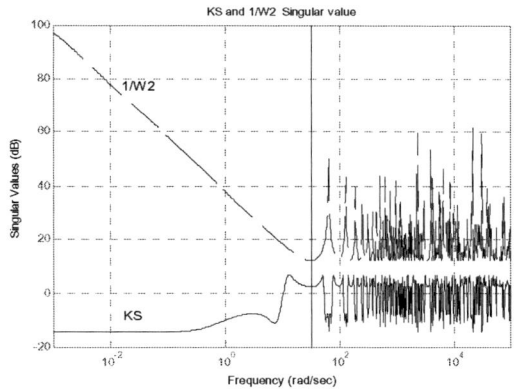

Fig.4. Sensitivity KS (K_{o_LQG}) and 1/W2

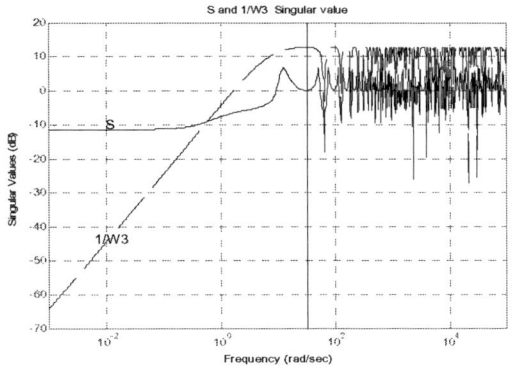

Fig.5. Sensitivity S (K_{o_LQG}) and 1/W3

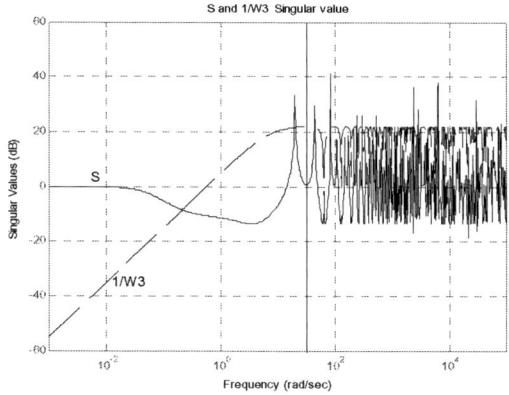

Fig.6. Sensitivity S (K_{o_H2}) and 1/W3

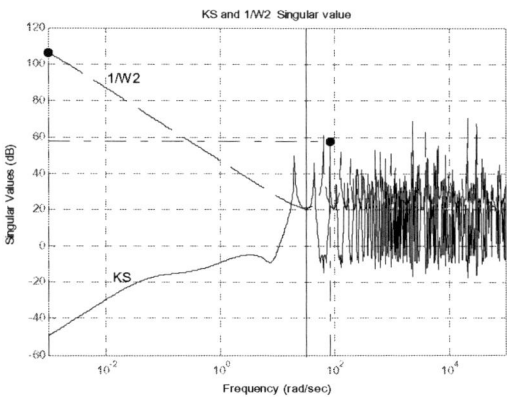

Fig.7. Sensitivity KS (K_{o_H2}) and 1/W2

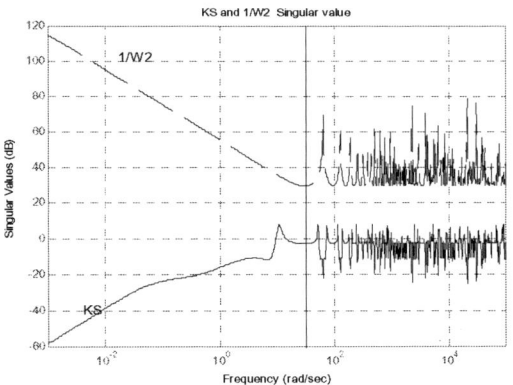

Fig.8. Sensitivity KS ($K_{o\_H\infty}$) and 1/W2

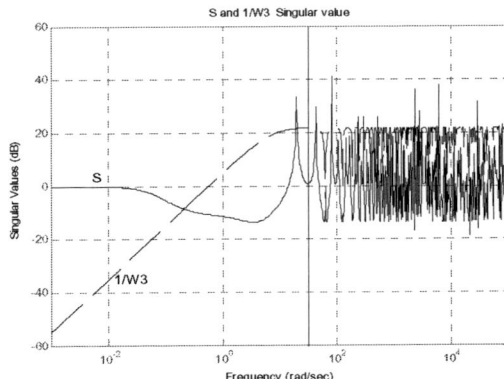

Fig.9. Sensitivity S ($K_{o\_H\infty}$) and 1/W3

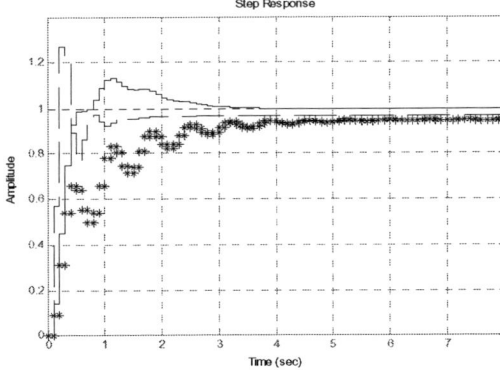

Fig.10. System response (- Ko_{LQG}, - - Ko_{H2}, ** $Ko\_{H\infty}$)

7. CONCLUSION

In this paper, we have introduced a methodology based on the properties of the Youla parameterization, which makes possible the inter-connected systems representation of the closed loop. This parameterization ensures the convexity of the problem.

According to this work of comparison, it is concluded that the selected initial controller in Youla parameterization is essential and a determining factor for the performances of the system. The initial controller defines also completely the LMI optimization. The only problem is the difficulty to design the three initial controllers (Ko_LQG, Ko_H∞, Ko_H2) that satisfy the required specifications, satisfying the conditions of the Riccati equations.

REFERENCES

[1]. A. Molina-Cristobal, I.A. Grifin. P.J. Fleming and D.H. Owens (2006), *Linear matrix inequalities and evolutionary optimization in multiobjective control*, IJSS, **V.37**, N°8.20.

[2]. P. Rodríguez, D. Dumur, (2002), *Robustification of GPC controlled system by convex optimization of the Youla parameter*, Proceedings IEEE Conference on Control Applications, Glasgow.

[3]. B. Clément, G. Duc, (Juin.2000), *Multi-objective control via Youla parametrization and LMI optimization: application to a flexible arm*, IFAC Symposium robust control and design, Prague.

[4]. B. Clément, G. Duc, (2001), *Synthèse multicritère utilisant la parametrisation de Youla et l'optimisation convexe*, Hermès, Paris.

[5]. Genci Capi, Masao Yokota, (Dec.2006), *Optimal multi-criteria Humanoid robot gait synthesis an evolutionary approach*, IJCIC, **V.2**, N°6,

[6]. Chung Seop Jeong, Edwin E. yaz, (Aug.2006), *Nonlinear observer design with general criteria*, IJCIC, **V.2**, N°4.

[7]. C.W. Scherer, P. Gahinet, M. Chilali, (1997), *Multiobjective output feedback control via LMI optimization*, IEEE Trans. on automatic control, **V. 42**.

[8]. J. Maciejowski, (1989), *Multivariable feedback design*, Addisson-Wesley, Wokingham.

[9]. G. Duc, S. Font, (1999), *Commande H∞ et μ analyse*, Hermès, Paris.

[10]. J. Doyle, K. Glover, P.P. Khergonekar, B.A Francis, (1989), *State-space solution to standard H2 and H∞ control problems*, IEEE Trans. A.C., **V.34**.

[11]. S. Boyd, L. El Ghaoui, E. Feron, V. Balaskrishnan, (Jun.1994), *Linear Matrix inequalities in systems and control theory*, SIAM Publications.

[12]. J.P. Folcher, L. El Ghaoui, (Aug.1994), *State feedback design via LMI : Application to a benchmark problem*, IEEE conference on control application.

[13]. Pascal Gahinet, Arkadi Nemirovski, Alan J. Laub, Mahmoud Chilali; (2001) *LMI Control Toolbox*.

[14]. Richard C.Dorf, R. H. Bishop, (1995), *Modern control systems*, 7th edition, Addison-Wesley publishing company, England, PP 434.

Guaranteed trajectory tracking of a small-size autonomous helicopter in a smooth uncertain environment

T. Cheviron, * A. Chriette and ** F. Plestan **

LRBA, DGA, Vernon, France (e-mail: Thibault.Cheviron@irccyn.ec-nantes.fr).
**IRCCyN, UMR CNRS 6597, Ecole Centrale de Nantes, Nantes, France (e-mail: chriette,plestan@irccyn.ec-nantes.fr)*

Abstract: This paper deals with the problem of disturbance (wind gusts) reconstruction acting on a scale autonomous helicopter in 3D flight. A general class of disturbance is addressed, namely any uniform time-varying tridimensional wind gust occurring in the vicinity of the UAV. Using an unknown input observer, it is shown that the disturbance can be accurately reconstructed online. Singular perturbation assumption is made to separate translational and rotational dynamics. Robust backstepping techniques are then used to stabilize each subsystems by taking into account the disturbance estimation procedure. Using a Lyapunov analysis, it is proven that the closed loop system coupling two non linear controllers with two nonlinear observers converges to an asymptotical set of which radius may be tuned as small as desired by increasing some gains. Simulation results validate the work.

Keywords: Autonomous helicopter, robust nonlinear control, unknown input observer, backstepping, perturbation compensation, trajectory tracking.

1. INTRODUCTION

With a range of applications in both civilian and military scenarios, the development of automated aerial robots has now become a very challenging field of robotics research, particulary towards the development of autonomous scale model helicopter due to their high payload to power ratio and its VTOL (Vertical Take Off and Landing) capacity [1, 11, 13, 20]. Such vehicles have strong commercial potential in constrained environment in remote surveillance applications such as monitoring traffic congestion, regular inspection power cables, to name but a few of the possibilities. Autonomous navigation ability of such robotic vehicles states numerous problems in sensing and control. For instance, a key challenge is development of controllers able to track trajectories (*i.e.* straight line, circle, hovering flight) with a given precision imposed by the mission (*i.e.* infrastruture inspection, surveillance of a predefined area of interest) in spite of wind. Many previous works focus on (linear, nonlinear, robust, ...) control, including a particular attention on the analysis of the stability [6], but very few works have been made on the influence of wind gusts acting on the flying system, whereas it is a crucial problem for outdoor applications, especially in constrained environment. In [?], two controllers (nonlinear and H_∞) are designed for a nonlinear reduced-order model of a 3DOF helicopter, but this work is limited due to the considered nature of wind gusts and the considered model. In [23], a control strategy stabilizes the position of the flying vehicle in wind gusts environment in spite of unknown aerodynamic efforts, this work being based on robust backstepping approach and estimation of the unknown aerodynamic efforts assumed to be constant with adpatative techniques.
The results displayed in the present paper are original for several reasons. In order to develop such flight control systems for maneuverable autonomous miniature helicopter, dynamic models which are accurate for their flight envelope are needed. However, in order to design nonlinear control law, minimal complexity models are preferred. Then, a 6DOF model (detailed in [3]) is proposed focusing on the key effects in the dynamics of small-size helicopter [6, 20] and nonlinear full-scale rotorcraft models [24, 22].
Thereafter, the authors have still worked on the estimation of time-varying disturbances. The basic idea consists in considering the disturbance observability definition in order to write the unknown input as a function of measurement vector, input and their time derivatives. Many approaches have been developed in order to solve robust differentiation problem in a noisy context : finite-time differentiator based on Δ-modulation and higher order sliding mode techniques [16], algorithms inspired by signal processing tools as regularization [7]. Two methods (high gain differentiator, higher order sliding mode differentiator) based on the explicit differentiation of measurements have been evaluated in [3] and [4] respectively. Despite the simplicity of these methods, their high sensitivity with respect to noise reduce their applicability on real experimental set-up. In the current paper, an original approach is proposed. The objective is to design an high gain unknown input observer observer [17] which estimates the perturbation (wind gusts) viewed as an *unknown input*. The controller takes then advantage of the pyramidal structure of the system to design a nonlinear hierarchical backstepping controller [23] using the estimation of the perturbation. The main interest of a such structure is that there is *a priori* no assumption on the perturbations dynamics apart from boundness. Singular perturbation assumption leads to consider only slow time varying perturbation for the translational dynamics; however, bounded but dynamical perturbation

may be compensated for the rotational dynamics.

The difficulty is to ensure the stability for the connected systems, *i.e.* the stability of the system with the controller receiving state and unknown input estimation from the observer. For linear systems with linear observers, the proof is trivial under the well-known superposition principle. This latter property is lost in nonlinear systems. Then, the third original point of the current paper, which is a hard task, consists in providing a formal proof of the stability of observer based controller via Lyapunov stability analysis. This work leads to bring into relief an asymptotical residual set of which radius may be tuned as small as desired by increasing some gains. Consequently, under some condition about the tuning of control and observer gains, it is proven that the tracking trajectory error may be increased by a constant defined as small as the mission required in spite of any bounded perturbation. This result was studied in [5] using viability theory but computational difficulty prevented from constructive tuning conditions because of the great dimension of the problem.

The paper is organised as follows. Section 2 displays the model of the 6DOF helicopter. In Section 3, the observability of the system with regard to the perturbation is analyzed and a high gain unknown input observer is designed. Section 4 gives the control strategy based on a hierarchical backstepping approach. In Section 5, the stability and the robustness of the closed loop are discussed. Simulations applied to a Vario Benzin-Acrobatic model helicopter are given in Section 6.

2. MODEL OF THE SMALL-SIZE HELICOPTER

This section displays the 6DOF dynamic model of miniature helicopter taking the case of the Vario Benzin-Acrobatic. This one is viewed as a rigid body, in a general 3D flight mode. Once the nonlinear model is obtained, it is broken down in three connected subsystems relying on translational dynamics, rotational dynamics and flapping angle dynamics respectively.

2.1 Dynamics of reduced scale helicopter [3]

Let $\mathcal{I} = \{E_x, E_y, E_z\}$ denote a right-hand inertial frame stationary with respect to the Earth. Let $\mathcal{B} = \{E_1^b, E_2^b, E_3^b\}$ be a (right-hand) body fixed frame for the helicopter. Let $\xi = (x, y, z)^T$, v, R and $\Omega = (p, q, r)^T$ be respectively the linear position and linear velocity, attitude and angular velocity of the helicopter such that $\xi \in \mathcal{I}$, $v \in \mathcal{I}$, $R \in SO(3)$ and $\Omega \in \mathcal{B}$. Let m denote the mass of the helicopter and let $\mathbf{I}$ denote the constant inertia matrix around the center of mass expressed in $\mathcal{A}$ and assumed to be diagonal. Following previous developments on dynamics of reduced scale autonomous helicopter in quasi-stationary flying conditions [6] and in a larger flight envelope [1][3], assume that the motion of a miniature helicopter is the motion of a rigid object subjected to forces and torques due to the action of the rotor blades (see Figure 1). As affine function of main and tail rotor collective angle (ie: θ_m, θ_t), main and tail rotor lift (ie: T_m, T_t) control helicopter altitude and yaw dynamics respectively. Besides, longitudinal and lateral cyclic (ie: A_1, B_1) angles control the main rotor flapping dynamics and so helicopter roll and pitch dynamics. When this dynamics tends to equilibrium, the total lift of the main rotor T_m is tilted in comparison to the motor shaft [1][6]

$$T_m = |T_m| e_m = |T_m| \begin{bmatrix} a_{1s} \\ -b_{1s} \\ -1 \end{bmatrix} \quad (1)$$

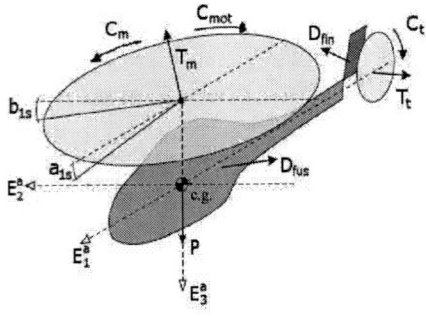

Fig. 1. Reference frames, forces and torques.

where a_{1s} and b_{1s} are the longitudinal and lateral flapping angles assumed to be small. Then, equations of motion for a helicopter are given by the Newton-Euler equations [9] augmented by flapping angles dynamics of a teetering rotor [3]. However well known coupling between the mechanism force and moment generation appears in the dynamics which produces small "parasitic" forces named *small body forces* introducing zero-dynamics into the system [11, 13, 6]. In order to overcome the effect of the small-body forces, a simpler approach consists in designing a control law which neglect this term, and then in relying on controller robustness features to overcome any perturbation [11, 6]. Once the zero dynamics does not appear anymore, the passivity feature of dynamics model (See [3]) can be written as two connected subsystems. This idea is motivated by the higher priority of attitude stabilization than position control. As in [1][6], quasy steady state asumption of flapping angle is made (ie: $\dot{a}_{1s} \approx 0$, $\dot{b}_{1s} \approx 0$). One gets

$$\begin{aligned}(\Sigma_1) & \begin{bmatrix} \dot{\xi} \\ m\dot{v} \end{bmatrix} = \begin{bmatrix} v \\ -un + mgE_3 + \Psi_F \end{bmatrix} \\ (\Sigma_2) & \begin{bmatrix} \dot{R} \\ \mathbf{I}\dot{\Omega} \end{bmatrix} = \begin{bmatrix} R sk(\Omega) \\ -sk(\Omega)\mathbf{I}\Omega + K\Gamma + \Psi_C \end{bmatrix} \end{aligned} \quad (2)$$

with $n = RE_3$ and $E_3 := [0, 0, 1]^T$,

$$\begin{aligned} \Psi_F &= f_1(D_{fus}, D_{fin}) + F_{ext}, \\ \Psi_C &= f_2(D_{fus}, D_{fin}, C_m, C_{mot}, C_t) + M_{ext}, \end{aligned} \quad (3)$$

$$u = |T_m| \quad \Gamma = [-a_{1s}|T_m| \; b_{1s}|T_m| \; |T_t|]^T \quad (4)$$

$sk(\Omega)$ is the skew antisymmetric matrix induced by Ω,

$$K = \begin{bmatrix} 0 & -l_m^3 & -l_t^3 \\ l_m^3 & 0 & 0 \\ 0 & 0 & l_t^1 \end{bmatrix}.$$

$l_i = \sum_{k=1}^{3} l_i^k E_k^a$ (i stands for m, t) represents the distance between the center of gravity and the application point of the force[1]. Ψ_F and Ψ_C stand for unknown inputs gathering unmodelled physical terms (*i.e.* drag $D_{fus,fin}$, air resistance torque of rotors $C_{m,t}$ C_{mot}, engine torque C_{mot}) and aerodynamic effects of wind gust (*i.e.* F_{ext}, $M_{ext} = sk(\kappa)F_{ext}$ where κ is the lever arm of aerodynamics efforts induced by wind gusts).

3. UNKNOWN INPUT RECONSTRUCTION

In order to improve the performances of the controller, the perturbations and unknown terms, Ψ_F and Ψ_C, are reconstructed.

---
[1] Assume $l_m^1 = l_m^2 = 0$.

After the analysis of the observability of the unknown input, the method used for the estimation of the unknown input is presented. It is based on the synthesis of a high gain state observer [8].

3.1 Observability analysis

Let Ψ_F and Ψ_C denote all disturbances acting on the translational and rotational dynamics respectively. The objective is to determine the necessary measurements which provide an analytical expression of Ψ_F and Ψ_C. From system (2), one gets

$$\begin{aligned} \Psi_F &= m\dot{v} - mgE_3 + uRE_3 \\ \Psi_C &= \mathbf{I}\dot{\Omega} + \Omega \wedge \mathbf{I}\Omega - K\Gamma - uK_0 \end{aligned} \quad (5)$$

From (5), it is obvious that the disturbances are observable in the sense of [21] and are evaluated from measurements of only ξ, v, R and Ω. By a practical point of view, ξ, η, R, v and Ω are generally estimated by complementary filters or Extended Kalman Filters ensuring multisensor data fusion (inertial measurement unit, GPS, laser telemeters, $\cdots$).

3.2 Unknown input Observer

Consider nonlinear system (2). Unknown inputs Ψ_F and Ψ_C are added to both nominal dynamics (Σ_1) and (Σ_2). Then, in a sake of clarity and without loss of generality for the current application, consider the following class of nonlinear system with unknown input

$$\begin{aligned} \dot{\lambda} &= f(\lambda, U) + \Psi \\ y &= \lambda \end{aligned} \quad (6)$$

with $\lambda \in \mathcal{X} \subset \mathbb{R}^n$ the state variable, where $\mathcal{X}$ is a compact subset of $\mathbb{R}^n$ and $\lambda = v$ (resp. Ω). $U \in \mathbb{R}^m$ is the control input such that $U := un$ (resp. Γ), and $\Psi \in \mathbb{R}^n$ the unknown input such that $\Psi := \Psi_F$ (resp. Ψ_C). Extend the state vector as $X = \begin{bmatrix} X^{1T} & X^{2T} \end{bmatrix}^T := \begin{bmatrix} \lambda^T & \Psi^T \end{bmatrix}^T$: one gets

$$\begin{aligned} \dot{X} &= AX + \bar{f}(X^1, U) + \bar{\epsilon} \\ y &= CX \end{aligned} \quad (7)$$

with $\bar{\epsilon} = \begin{bmatrix} 0_{n\times 1}^T & \epsilon_{n\times 1}^T \end{bmatrix}^T$, ϵ being a vector whose each component ϵ_i is an unknown bounded real-valued function,

$$\bar{f}(X^1, U) = \begin{bmatrix} f(X^1, U) \\ 0_{n\times 1} \end{bmatrix} \quad A = \begin{bmatrix} 0_{n\times n} & Id_{n\times n} \\ 0_{n\times n} & 0_{n\times n} \end{bmatrix}$$

and $C = [Id_{n\times n}, 0_{n\times n}]$.

A1. The function $\bar{f}$ is globally Lipschitz with respect to X^1 uniformly in U.
A2. The function $\epsilon(t)$ is bounded.

Let Δ_θ be the block diagonal matrix defined as (with $I_{n\times n}$ the $n\times n$-dimensional identity matrix and $\theta > 0$ a real number)

$$\Delta_\theta = \text{diag}\left[I_{n\times n}, \frac{1}{\theta} I_{n\times n}\right] \quad (8)$$

Let S be the unique solution of the algebraic equation

$$S + A^T S + SA - C^T C = 0 \quad (9)$$

Theorem 1. ([8]). Under assumptions A1-A2, the dynamic system

$$\dot{\hat{X}} = A\hat{X} + \bar{f}(\hat{X}^1, U) + \theta \Delta_\theta^{-1} S^{-1} C^T (y - C\hat{X}) \quad (10)$$

where A, C, S, θ and Δ_θ defined previously, is such that the estimation error $e(t) := X(t) - \hat{X}(t)$ satisfies

$$\| e(t) \| \leq \theta \sigma(S) e^{-\frac{\theta - c_1}{2}t} \| e(0) \| + \frac{c_2 \beta_\epsilon}{\theta - c_1} \quad (11)$$

with $\sigma(S) = \sqrt{\lambda_{max}(S)/\lambda_{min}(S)}$, λ_{max} (resp. λ_{min}) the largest (resp. smallest) eigenvalue of S, $c_1 = 2\sigma(S)^2 \zeta$, $c_2 = 2\sigma(S)\sqrt{\lambda_{max}(S)}$ and the real ζ such that

$$\| \bar{f}(\hat{X}^1, U) - \bar{f}(X^1, U) \| \leq \zeta \| e \|$$

∎

Viewed the definition of X^1, it means that the unknown input Ψ is estimated through nonlinear observer (10). Note that the asymptotic estimation error can be made as small as desired by choosing values of θ large enough. An accommodation between precision and filtration is then ensured by the choice of θ.

4. CONTROL DESIGN

The problem considered here is a smooth path tracking. Consider the desired trajectory $\xi^d = [x^d, y^d, z^d]^T$: the main objective consists in controlling the yaw velocity and in tracking this trajectory in spite of wind gusts [6] [10]. Given the structure of systems (2), the disturbance decoupling problem has no solution [12]: it is not possible to *exactly* reject perturbations. In the sequel, an original approach is proposed to solve the problem of disturbance compensation by using perturbation reconstruction with a hierarchical backstepping control [15]. The control strategy takes benefit from the cascade structure (2) due to a realistic time scale separation assumption. To each subsystem is affected a controller which is designed such its corresponding output is supposed constant for the "next" controllers in the structure. This fact is ensured by tuning, for a given controller, smaller gains than ones for next controllers in the cascade.

Each controller is based on backstepping approach due to the passivity feature of the dynamics [23]. This approach consists in designing, *step-by-step*, Lyapunov's candidate functions: as a matter of fact, the system is written in subsystems, from which, for each one, a Lyapunov candidate function is designed in order to ensure its stability.

4.1 Position control

Consider system (Σ_1): the control law must define the magnitude and the orientation of the main rotor lift. Let $\delta_1 = \xi - \xi^d$ denote the tracking trajectory error and define an intermediate state $\sigma = K_1 \delta_1 + m\delta_2$ with $K_1 > 0$ and $\delta_2 = v - v^d = \dot{\delta}_1$. If $\sigma = 0$, it should be noted that ξ converges exponentially toward ξ^d (ie: $\dot{\delta}_1 = -\frac{K_1}{m}\delta_1$). Taking the time derivative of the storage function $\S_\sigma = \frac{1}{2} \| \sigma \|_2^2$, one gets:

$$\begin{aligned} \dot{\S}_\sigma &= -K_2 \| \sigma \|_2^2 + \sigma^T (\alpha_1^d - un) + \sigma^T \Delta \Psi_F \\ \alpha_1^d &= mgE_3 + sat(\hat{\Psi}_F) + K_1 \delta_2 + K_2 \sigma \\ \Delta \Psi_F &= \Psi_F - sat(\hat{\Psi}_F) \end{aligned} \quad (12)$$

where $K_2 > 0$, and $sat(\hat{\Psi}_F)$ is the saturated estimation of disturbance vector Ψ_F given by (10). Backstepping algorithm stops at this step because control input un of translational dynamics appears. Assuming $\Delta \Psi_F = 0$, $un = \alpha_1^d$ ensures the strict decreasing of S_σ toward zeros. The analytical expression of the main rotor lift is then deduced:

$$u = \| mgE_3 + sat(\hat{\Psi}_F) + K_2\sigma + K_1\delta_2 \|_2 \quad (13)$$

The link between closed loop linear and attitude dynamics is performed by the calculation of a desired attitude R^d which must be tracked asymptotically by the attitude controller (ie: $R \to R^d$). The key point of this so-called Attitude Set Problem is that the desired orientation R^d is fully defined by the vectorial constraint on $n^d = R^d E_3 = \frac{\alpha_1^d}{u}$ combined with the desired yaw.

4.2 Attitude control

Now consider the system (Σ_2) which must track the previously calculated desired orientation R^d. Recalling singular perturbation hypothesis, R^d must be considered constant in the following design. Define the candidate Lyapunov storage function measuring the similarity between R and R^d thanks to Frobenius norm:

$$S_3 = \frac{1}{4} \| I_{3\times 3} - \tilde{R} \|_F^2 = \frac{1}{2} tr(I_{3\times 3} - \tilde{R}) \quad (14)$$

Using the fact that the trace of a symmetric and antisymmetric matrix product is equal to zero and that $tr(AB) = -2vex(A)^T vex(B)$ [2], if A and B are both antisymmetrical matrix, S_3 time derivative calculation yields to:

$$\dot{S}_3 = \frac{1}{2} tr(sk(\Omega)\tilde{R}) = -K_3 \| vex(\pi_a(\tilde{R})) \|_2^2 + vex(\pi_a(\tilde{R}))^T \delta_4 \quad (15)$$

with $K_3 > 0$, $\delta_4 = K_3 vex(\pi_a(\tilde{R})) - \Omega$. Consider an other storage function S_4 ensuring the convergence of δ_4 to zero:

$$S_4 = S_3 + \frac{1}{2}\delta_4^T \mathbf{I}\delta_4 \quad (16)$$

Calculating the time derivative of S_4,

$$\begin{aligned}\dot{S}_4 &= -K_3 \| vex(\pi_a(\tilde{R})) \|_2^2 - K_4 \| \delta_4 \|^2 + \\ & \quad \delta_4^T(K\Gamma - \alpha_2^d) + \delta_4^T \Delta\Psi_C \\ \alpha_2^d &= sk(\Omega)\mathbf{I}\Omega - sat(\hat{\Psi}_C) - K_3\mathbf{I}vex(\pi_a(sk(\Omega)\tilde{R})) + \\ & \quad (1 + K_3K_4)vex(\pi_a(\tilde{R})) - K_4\Omega \\ \Delta\epsilon_2 &= \Psi_C - sat(\hat{\Psi}_C)\end{aligned} \quad (17)$$

where $K_4 > 0$, and $sat(\hat{\Psi}_C)$ is the saturated estimation of disturbance vector Ψ_C given by (10). As control input Γ appears, the backstepping process ends, the analytical expression of the control torque is then deduced:

$$\Gamma = K^{-1}\Big(sk(\Omega)\mathbf{I}\Omega - K_3\mathbf{I}vex(\pi_a(sk(\Omega)\tilde{R})) - sat(\hat{\Psi}_C) + (1 + K_3K_4)vex(\pi_a(\tilde{R})) - K_4\Omega\Big) \quad (18)$$

From the definition of $\Gamma = [-a_{1s}u, b_{1s}u, |T_t|]^T$, quasi steady state hypothesis for flapping dynamics and aerodynamics model (**??**), expressions of collective angles (for main and

---
[2] for all vector $x \in \mathbb{R}^3$, $vex(sk(x)) = x$ and for all matrix A, $\pi_a(A) = \frac{A - A^T}{2}$

tail rotor) and lateral and longitudinal cyclic angle can be determined.

5. STABILITY AND ROBUSTNESS OF THE CONNECTED SYSTEM

When connecting the two subsystems, disturbances appear in the translational and rotational dynamics due to, on the one hand, the non immediate convergence of Re_3 to $R^d e_3$, on the other hand, the unknown input reconstruction error. Considering that trajectory tracking error is bounded by an a priori constant (ie: $\| \delta_1 \|_\infty < \epsilon$, $\epsilon > 0$) and unknown inputs are uniformly continuous and bounded. Under these assumptions, this section is dedicated to prove the practical convergence of the system to a compact domain of which radius may be set by control and observer gains.

5.1 Step 1:

Let $S_1 = \frac{1}{2} \| \delta_1 \|_2^2$ denote a candidate Lyapunov function of the system : $\dot{\delta}_1 = -\frac{K_1}{m}\delta_1 + \frac{1}{m}\sigma$. Calculating its time derivative, it yields to [3]:

$$\begin{aligned}\dot{S}_1 &= \delta_1^T\left(-\frac{K_1}{m}\delta_1 + \frac{1}{m}\sigma\right) \\ &\leq -W_1(\| \delta_1 \|_2) + \frac{\mu^2}{m} \| \sigma \|_2^2\end{aligned} \quad (19)$$

With $W_1(\| \delta_1 \|_2) = \frac{1}{m}\left(K_1 - \frac{1}{\mu^2}\right) \| \delta_1 \|_2^2$, $0 < \mu$. Considering σ as a disturbance for the dynamics of δ_1 and assuming that $\| \sigma \|_2 < \epsilon_\sigma < +\infty$, $\frac{1-\zeta}{2} \| \delta_1 \|_2^2 \leq S_1 \leq \frac{1+\zeta}{2} \| \delta_1 \|_2^2$, $0 < \zeta << 1$ and $\frac{1}{m}\left(K_1 - \frac{1}{\mu}\right) \| \delta_1 \|_2^2 \leq W_1(\| \delta_1 \|_2)$, the Nonlinear Damping Lemma [15] may be applied:

- δ_1 converges asymptotically to the residual set:

$$\mathcal{R}_{\delta_1} = \{\delta_1 : \| \delta_1 \|_2 \leq \kappa \bar{\delta}_1\} \quad (20)$$

With $\kappa = \sqrt{\frac{1+\zeta}{1-\zeta}}$ and $\bar{\delta}_1 = \frac{\mu\epsilon_\sigma}{\sqrt{K_1 - \frac{1}{\mu}}}$.

- δ_1 is uniformly bounded:

$$\| \delta_1 \|_\infty \leq \max\{\kappa\bar{\delta}_1, \kappa \| \delta_1(0) \|_2\} \quad (21)$$

To satisfy $\| \delta_1 \|_\infty < \epsilon$, one deduces:

$$\begin{aligned}\| \delta_1(0) \|_2 &< \frac{\epsilon}{\kappa} \\ \epsilon_\sigma &< \sqrt{K_1 - 1}\epsilon, \ K_1 > 1\end{aligned} \quad (22)$$

5.2 Step 2:

Let now analyze the dynamics of σ. Combining main rotor thrust expression (13) with (12) and the error term to stabilize orientation dynamics, it follows:

$$\dot{S}_\sigma = -K_2 \| \sigma \|_2^2 + \sigma^T(\Delta\Psi_F - uR(I_{3\times 3} - \tilde{R})e_3) \quad (23)$$

First, let note that the main rotor thrust is bounded:

$$0 \leq u \leq \left(\frac{K_1}{m} + K_2\right)\epsilon_\sigma + M < +\infty \quad (24)$$

---
[3] $\forall \eta > 0, \forall X, Y \in \Re^n, X^T Y \leq \| X \|_2^2 + \frac{1}{\eta} \| Y \|_2^2$

where $M = mg + sat(\hat{\Psi}_F) + \frac{K_1^2}{m}\epsilon$. Using Rodrigue's formula [4], the error term coupling translational and rotational dynamics may be increased by:

$$\parallel uR(I_{3\times 3} - \tilde{R})e_3 \parallel_2 \leq 2\sqrt{2}|\sin\frac{\tilde{\theta}}{2}| \left((\frac{K_1}{m} + K_2)\epsilon_\sigma + M \right) \quad (25)$$

Increasing the reconstruction error of Ψ_F by $\Delta\Psi_{Fmax}$ which is developed in the unknown input observer stability proof (11), it yields to:

$$\dot{S}_\sigma \leq -a_1 S_\sigma + a_2 \sqrt{S_\sigma} \quad (26)$$

where $a_1 = 2(K_2 - 2\sqrt{2}|\sin\frac{\tilde{\theta}}{2}|(\frac{K_1}{m}+K_2))$ and $\sqrt{2}(2\sqrt{2}|\sin\frac{\tilde{\theta}}{2}|M + \Delta\Psi_{Fmax})$. As this system cannot diverge in finite time [?] assuming $a_1 > 0$, the previous differential inequality may be integrated [8]:

$$\forall t \geq 0, \parallel \sigma(t) \parallel_2 \leq e^{-\frac{a_1}{2}t} \parallel \sigma(0) \parallel_2 + \frac{\sqrt{2}a_2}{a_1} \quad (27)$$

To satisfy $\parallel \sigma \parallel_2 < \epsilon_\sigma$, one deduces:

$$\begin{array}{ll} \parallel \sigma(0) \parallel_2 & < \mu_1\sqrt{K_1 - 1}\epsilon \\ \Delta\Psi_{max} & < \mu_2 K_2\sqrt{K_1 - 1}\epsilon \\ |\tilde{\theta}| & < \tilde{\theta}_{min} \end{array} \quad (28)$$

with $\tilde{\theta}_{min} = \mu_3 \frac{K_2\sqrt{K_1-1}}{M}\epsilon$, $0 < \mu_i << 1$, $i = \{1,2,3\}$. As a consequence, the attitude controller must be always very accurate, which is coherent with the singular perturbation assumption.

5.3 Step 3:

Let now analyze the convergence of $\tilde{R}$ and $\tilde{\Omega}$ toward $I_{3\times 3}$ and zero respectively via the candidate Lyapunov function (16) of rotational dynamics. Writing the time derivative of S_4 with Rodrigue's formulation, one gets:

$$\begin{aligned} \dot{S}_4 &= -4K_3\cos^2\frac{\tilde{\theta}}{2} - K_4 \parallel \tilde{\Omega} \parallel_2^2 + \tilde{\Omega}^T \Delta\Psi_C \\ &\leq -W_4(\parallel X_R \parallel_2) + \mu^2 \parallel \Delta\Psi_C \parallel_2^2 \end{aligned} \quad (29)$$

where $W_4(\parallel X_R \parallel_2) = X_R^T \Sigma_{R,\mu} X_R$, $X_R = \left[\sin\frac{\tilde{\theta}}{2}, \tilde{\Omega}^T\right]$ and $\Sigma_{R,\mu} = diag\left(4K_3\cos^2\frac{\tilde{\theta}}{2}, K_4 - \frac{1}{\mu^2}\right)$. It is obvious that three strictly positive constants a_{min}, a_{max}, b_{min} may be defined satisfying $a_{min} \parallel X_R \parallel_2^2 \leq S_4 \leq a_{max} \parallel X_R \parallel_2^2$ and $b_{min} \parallel X_R \parallel_2^2 \leq W_2(\parallel X_R \parallel_2)$. Increasing the reconstruction error of Ψ_C by $\Delta\Psi_C$ (cf. unknown input observer stability proof (11)). Nonlinear Damping Lemma [15] may be applied as in Step 1:

- X_R converges asymptotically to the residual set:

$$\mathcal{R}_{X_R} = \{X_R : \parallel X_R \parallel_2 \leq \bar{\kappa}\frac{\mu}{b_{min}}\Delta\Psi_{Cmax}\} \quad (30)$$

with $\bar{\kappa} = \sqrt{\frac{a_{max}}{a_{min}}}$.

- X_R is uniformly bounded:

$$\parallel X_R \parallel_\infty \leq \max\{\frac{\bar{\kappa}\mu}{b_{min}}\Delta\Psi_{Cmax}, \bar{\kappa} \parallel X_R(0) \parallel_2\} \quad (31)$$

---
[4] $\tilde{R} = I_{3\times 3} + \sin\tilde{\theta}sk(\tilde{a}) + (1 - \cos\tilde{\theta})sk^2(a)$ with $\tilde{\theta} \in [0, 2\pi]$ and $a \in \mathbb{R}^3$, $\parallel \tilde{a} \parallel_2 = 1$

To satisfy $|\tilde{\theta}| < \tilde{\theta}_{min}$, we may impose that $\parallel X_R \parallel_\infty < \tilde{\theta}_{min}$ and one deduces:

$$\begin{array}{ll} \parallel X_R(0) \parallel_2 & < \frac{\tilde{\theta}_{min}}{\bar{\kappa}} \\ \Delta\Psi_{Cmax} & < \frac{b_{min}}{\mu\bar{\kappa}}\tilde{\theta}_{min} \end{array} \quad (32)$$

5.4 Synthesis

From the previous analysis, the fulfilment of the following conditions guarantees that $\forall t \geq 0, \parallel \xi - \xi^d \parallel_\infty < \epsilon$:

$$\begin{array}{ll} \parallel \delta_1(0) \parallel_2 & < \frac{\epsilon}{\kappa} \\ \parallel \sigma(0) \parallel_2 & < \mu_1\sqrt{K_1 - 1}\epsilon \quad , K_1 > 1 \\ \parallel X_R(0) \parallel_2 & < \bar{\mu}_3 \frac{K_2\sqrt{K_1-1}}{M}\epsilon \quad , 0 < \bar{\mu}_3 << 1 \\ \Delta\Psi_{Fmax} & < \mu_2 K_2\sqrt{K_1 - 1}\epsilon \quad , 0 < \mu_2 << 1 \\ \Delta\Psi_{Cmax} & < \mu_4 \frac{K_2\sqrt{K_1-1}}{M}\epsilon \quad , 0 < \mu_4 << 1 \end{array} \quad (33)$$

It results that the asymptotic tracking precision ϵ may be chosen as small as desired by tuning high control and observer gains. However, due to the picking phenomenon occurring in the transient phase of high gain unknown input observer (ie: $\forall t \in [0, \tau]$) [?], another tracking precision ϵ_0 must be defined such that $\forall t \in [0, \tau], \parallel \xi - \xi^d \parallel_\infty < \epsilon_0$ and $\forall t > \tau, \parallel \xi - \xi^d \parallel_\infty < \epsilon$.

6. SIMULATIONS

This section displays results of simulations for both backstepping controller and perturbation rejection. The parameters used for dynamic model (??) are based on the Vario Benzin-Acrobatic 23cc with

$$\mathbf{I} = \begin{bmatrix} 0.4 & 0 & 0 \\ 0 & 0.56 & 0 \\ 0 & 0 & 0.22 \end{bmatrix}. \quad (34)$$

The aim is to fly in straight line from $\xi(0) = [0,0,0]^T$ to $\xi^d = [1,1,-1]^T$. The magnitude of the initial force input is chosen to be $u(0) = mg$; in other words, the helicopter is initially in hover flight. Furthermore, wind gust is viewed as a slowly time varying tridimensional impact

$$\Psi_F = \begin{bmatrix} 5(1 + e^{-(\frac{t-15}{5})^2}) + 2\sin(2\pi 0.1 t) \\ 4(1 + e^{-(\frac{t-15.5}{4})^2}) + 2\sin(2\pi 0.05 t) \\ -0.5(1 + e^{-(t-16.8)^2}) \end{bmatrix} \quad (35)$$

The lever arm of aerodynamic efforts induced by wind gusts is assumed to depend on the attitude and set to

$$\Psi_C = sk(\begin{bmatrix} -0.15\sin(pitch)\sin(yaw) \\ 0 \\ -0.15\cos(roll)\sin(pitch) \end{bmatrix})\Psi_F$$

The simulations have been made with a 4th order Runge Kunta integration scheme with a sample frequency equal to 50Hz (as the experimental set-up under construction). Control gains are tuned as $K_1 = 1, K_2 = 1.5, K_3 = 9, K_4 = 10$, while gains of unknown input observer are tuned as $\theta_{\Psi_F} = 5$ and $\theta_{\Psi_C} = 20$. The efficiency of the disturbance compensation method is illustrated by Figures 2: this figure displays the cartesian tracking error in three cases: the first one (reference one) supposes no wind gust, the second one with not estimated wind gust, and the third one with wind gusts and their estimation. Simulations clearly show that the performances of control are improved with the estimation of perturbations. Figure 3 shows helicopter attitude to bring into relief the efficiency of the attitude controller. Both figures 4-5 display the estimation of perturbation, which shows the efficiency of the perturbation reconstruction. Note also that backstepping controllers with disturbance reconstruction compensate average 80% of aerodynamic effect of a wind gust in comparison with a classical backstepping control law as proposed in [6] or [10].

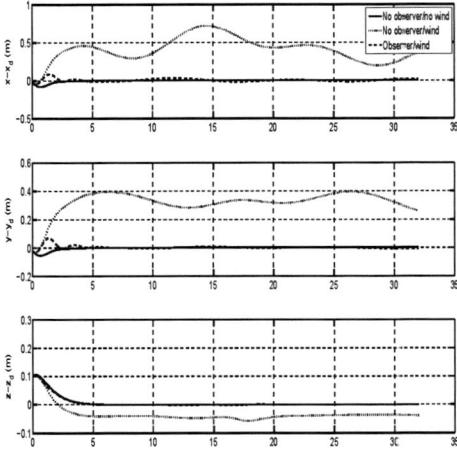

Fig. 2. **Without noise.** Cartesian positions ($x\ y, z$) tracking errors versus time (sec.). **Solid line**: No observer/no wind gusts, **Dotted line**: No observer/presence of wind gusts, **Dashed line**: Observer/presence of wind gusts.

7. CONCLUSION

The paper has proposed an original approach, based on hierarchical backstepping control law and perturbation reconstruction to track a 3D trajectory with a 6 DOF nonlinear dynamics of miniature helicopter in presence of non constant wind gusts. In this way, a closed loop estimation scheme based on high gain unknown input observer has been chosen to reconstruct wind gusts. The assumption of two connected subsystems with different time-constants leads to simple controllers. The stability and the robustness of the closed loop coupling nonlinear control law with high gain nonlinear obsrver is proven using a Lyapunov analysis. Guaranteed trajectory tracking in an uncertain environment (ie: disturbances are assumed smooth and bounded) is then theorically performed with an a priori precision. Finally, realistic simulation results display the interest of the approach. Future works will consist in evaluating the discussed method on an experimental set-up (ie: a Vario Benzin Acrobatic).

8. ACKNOWLEDGEMENTS

The authors thank Tarek Hamel, CNRS-I3S Sophia antipolis, for his useful advice. This work is in line with a Ph.D. thesis supported by DGA and in collaboration with IRCCyN.

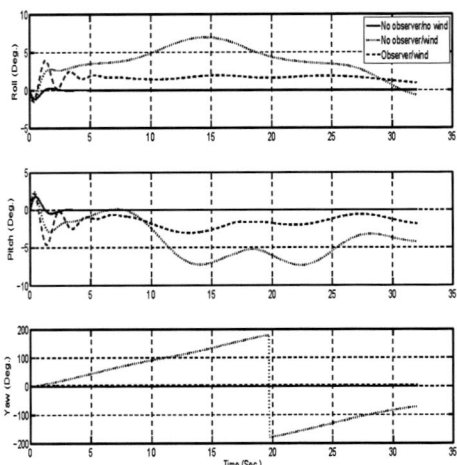

Fig. 3. **Without noise.** Helicopter attitude angles versus time (sec.). **Solid line**: No observer/no wind gusts, **Dotted line**: No observer/presence of wind gusts, **Dashed line**: Observer/presence of wind gusts.

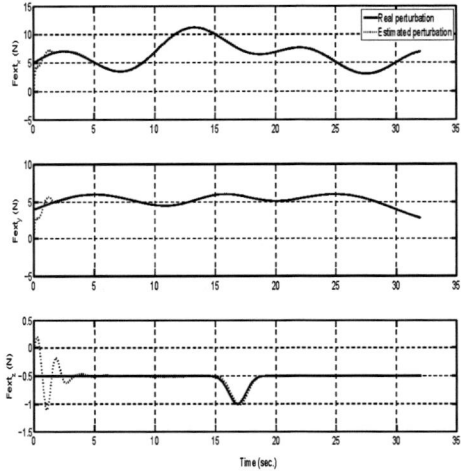

Fig. 4. **Without noise.** Reconstruction of perturbation F_{ext} (Real (solid line) and estimated (dotted line) values) versus time (sec.).

REFERENCES

[1] J.C. Avila Vilchis, B. Brogliato, A. Dzul, and R. Lozano, "Nonlinear modelling and control of helicopters", *Automatica*, Vol.39, pp.1583-1596, 2003.

[2] S. Bouabdallah, and R. Siegwart, "Backstepping and sliding-mode techniques applied to an indoor micro quadrator", *Proc. IEEE International Conference on Robotics and Automatic ICRA'05*, Barcelona, Spain, 2005.

[3] T. Cheviron, A. Chriette, and F. Plestan, "Robust Control of a scale autonomous helicopter", *Proc. AIAA Guidance,*

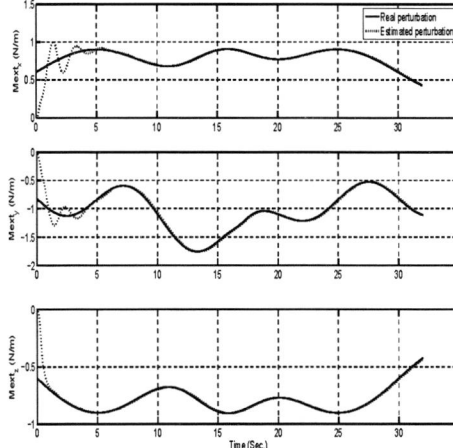

Fig. 5. **Without noise.** Reconstruction of perturbation M_{ext} (Real (solid line) and estimated (dotted line) values) versus time (sec.).

Navigation and Control Conference, Keystone, Colorado, 2006.

[4] T. Cheviron, F. Plestan, and H. Chriette, "Wind gusts compensation acting on an autonomous helicopter using variable structure differentiation scheme", *Proc. 8th International IFAC Symposium on Robot Control SYROCO*, Bologna, Italy, 2006.

[5] T. Cheviron, A. Chriette, and F. Plestan, "Guaranteed robust guidance and control of an autonomous scaled helicopter in presence of wind gusts", *Proc. European Micro Air Vehicle Conference and Flight Competition EMAV 2006*, Braunschweig, Germany, 2006.

[6] A. Chriette, T. Hamel, and R. Mahony, "Zero dynamics analysis for ibvs control of under-actuated rigid body dynamics", *Proc. International IFAC Symposium on Robot Control SYROCO*, Wroclaw, Poland, 2002.

[7] S. Diop, J.-W. Grizzle, and S. Ibrir, "On regularized numerical observers", *Proc. IEEE Conference on Decision and Control CDC*, Phoenix, Arizona, 1999.

[8] M. Farza, M. M'Saad, and L. Rossignol, "Observer design for a class of MIMO nonlinear systems", *Automatica*, Vol.40, 135-143, 2004.

[9] R. Goldstein, "Classical mechanics - 2nd Edition", Addison-Wesley, USA, 1980.

[10] T. Hamel, R. Mahony, and A. Chriette, "Adaptative estimation of aerodynamic forces for hover control of a reduced scale autonomous helicopter", *Proc. International IFAC Symposium on Robot Control SYROCO*, Vienna, Austria, 2000.

[11] J. Hauser, S. Sastry, and G. Meyer, "Nonlinear control design for slightly non minimum phase systems: Application to v/stol aircraft", *Automatica*, Vol. 28, pp.651-670, 1992.

[12] A. Isidori, "Nonlinear control systems - 2nd Edition", Spinger-Verlag, Berlin, Germany, 1989.

[13] T.J. Koo, and S. Sastry, "Output tracking control design of a helicopter model based on approximate linearization", *Proc. IEEE Conference on Decision and Control CDC*, Tampa, Florida, 1998.

[14] P.V. Kokotovic, "A Control Engineer's Introduction to Singular Perturbation", Stanford University, Stanford, CA, pp. 1-12.

[15] M. Krstic, I. Kanellakopoulos, and P. Kokotovic, "Nonlinear and adaptative control design", John Wiley & Sons, Chichester, England, 1995.

[16] A. Levant, "Higher-order sliding modes, differentiation and output-feedback control", *International Journal of Control*, Vol.76, pp. 924-941, 2003.

[17] F.L. Liu, M. Farza, and M. M'Saad, "Observateur à entrées inconnues pour une classe de systèmes non linéaires" [in French], CIFA, Bordeaux, 2006.

[18] R. Mahony, T. Hamel, and A. Dzul, "Hover control via approximate Lyapunov control for a model helicopter", *Proc. IEEE Conference on Decision and Control CDC*, Phoenix, Arizona, 1999.

[19] R. Mahony, and T.Hamel, "Robust trajectory tracking for a scale model autonomous helicopter", *International Journal of Nonlinear and Robust Control*, Vol.14, No.12, pp.1035-1059, 2004.

[20] B.F. Mettler, "Identification modeling and characteristics of miniature rotorcraft", Kluwer Academic Publisher, 2003.

[21] J. Moreno, "Unknown input observers for SISO nonlinear systems", *Proc. IEEE Conference of Decision and Control CDC*, Sydney, Australia, 2000.

[22] G.D. Padfield, "Helicopters Flight Dynamics: The Theory and Application of Flying Qualities and simulation Modeling", Blackwell Science LTD, 1996.

[23] J.-M. Pflimlin, P. Souères, and T. Hamel, "Hovering flight stabilization in wind gusts for ducted fan UAV", *Proc. IEEE Conference on Decision and Control CDC*, Atlantis, Paradise Island, The Bahamas, 2004.

[24] R.W. Prouty, "Helicopter, Performance, Stability, and Control", Krieger Publishing Company, 1986.

[25] H. Sira-Ramirez, R. Castro-Linares, and E. Licéaga-Castro, "Regulation of the longitudinal dynamics of an helicopter system: a liovillian systems approach", *Proc. American Control Conference ACC*, San-Diego, California, 1999.

Robust control of an uncertain physical process

B. Benyahia * A. Choukchou-Braham ** B. Cherki ***

Université Aboubekr Belkaid BP 19 - 13000 - Tlemcen Algérie
(e-mail: benyahiab@yahoo.fr)
** *(e-mail: Amel_choukchou@yahoo.fr)*
*** *(e-mail: b_cherki@yahoo.fr)*

Abstract: This paper presents the synthesis of a multivariable robust control, for a regulation of a semi-industrial physical process. This control known as H_∞ by Loopshaping is calculated using the nominal model, whose uncertainties are included by taking -a priori- the maximum of guarantee. The synthesized control laws are tested on the real system, and the experimental results obtained are presented and discussed.

Keywords: H_∞ by loopshaping, nominal model, physical process, robust control, uncertainties.

1. INTRODUCTION

The automatic systems operate in a real environment, which is often variable because of the noises and disturbances [5]. The identified models are uncertain "*nominal*", they are obtained after many simplifications, among which negligence of some dynamics. Consequently, the control laws calculated by using these models often proved their limits in industrial environment. In such cases, we search to design a robust controllers, who should adapt to the real environment, and to ensure the correct operation of the process for all his uncertain models. In fact, the problems of the robustness consist in trying to take the maximum of guarantees -a priori- so that the control law synthesized on a model works indeed on the physical system.

In this paper, we present a robust control strategy based on the H_∞ by Loopshaping approach for a multivariable regulation in real time, of air flow and temperature output of a semi-industrial process.

The paper is structured as follows. In section 2, we present the physical process and the used acquisition and control cards. Section 3 is devoted to the process identification. The essential of the H_∞ by Loopshping theory is the subject of section 4. Section 5 deals with the synthesis of an H_∞-Loopshaping robust multivariable controller. Finally, in section 6 we give the experimental results obtained with some discussions.

2. MATERIALS AND METHODS

2.1 Regulation and testing ground

The used process is a multivariable regulation bench on air and water represented by Fig. 1. It allows to carry out i) mono-loops regulations such as: flow and pressure regulation on air and water, temperature control on air and level regulation on water. ii) a multivariable regulation such as: flow/level regulation on water and flow/temperature regulation on air which we chose for our application.

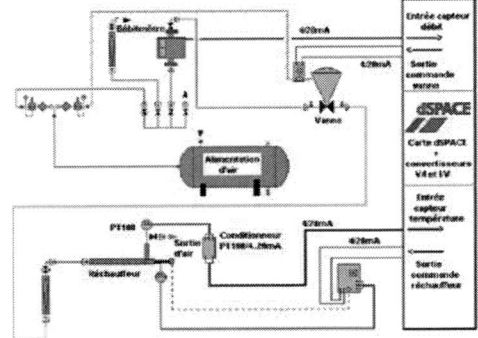

Fig. 1. Wiring diagram of the flow/temperature system

This global system is composed of two sub-systems: the flow sub-system and the temperature sub-system.

The flow system alone is a canalization portion equipped with a pneumatic valve and a flow sensor, in which the air circulates to the pressure of 3 bars. The disturbances are carried out by a manual tap at the output. The flow sensor delivers a signal of 4..20 mA to the regulator, this one corrects the consign/measure difference by an action on the pneumatic valve increasing or thus decreasing the flow.

The temperature system alone is a heat exchanger in which the air heated by a resistance circulates. The electric energy is transformed into thermal energy by resistance and is partially transmitted to the air. The PT100 with its conditioner provides to the regulator a current of 4..20 mA proportional to measurement and thereafter a control will be given to heating resistance so that the output is near to the consign in spite of the disturbances.

Starting from the preceding sub-systems, we can carry out the wiring of our multivariable global system (see Fig. 1), including the dSPACE card DS1102 on which we program the synthesized control law.

2.2 dSPACE card and conversion cards

dSPACE card: It is one of the products of the German company dSPACE, it is a very powerful card designed for the development of the digital computers at very high speed and to make real time simulations in all the fields: robotics and industrial processes, vehicles controls and the trajectories, space and aeronautical, signals treatment and telecommunication, electric and servo-hydraulic actuators, ...

This card is based on the TMS230C31 DSP (*Digital Signal Processor*) of the Texas Instrument technology, moreover it has a TMS320P14 controller for the numerical inputs/outputs. As it is shown by Fig. 2, the DS1102 supplemented by other peripherals such as A/D and D/A converters, numerical I/O subsystem and the incremental sensors interface, becomes a whole system in only one card offering solutions to a broad set of the numerical control tasks [6].

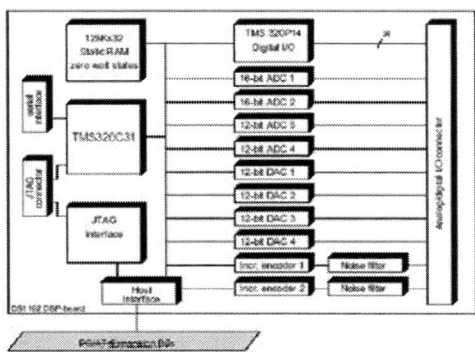

Fig. 2. Architecture of the dspace DS1102 card

Conversion cards: The sensors and controls signals used in the regulation pilot (process) are of 4..20 mA, standard scale largely used in industrial medium. The inputs and outputs of the dSPACE card are voltages of 0..10 V, now, the problem is that the process only understands 4..20 mA currents, whereas the card only exploits 0..10 V voltages.

To cure this problem, we carried out conversion cards: current/voltage (4..20 mA/0..10 V) and voltage/current (0..10 V/4..20 mA). These cards should be powerful, stable and in particular precise, because information (or control) received (or delivered) by these cards and which is used in the control program will give evidently erroneous results, affecting the correct operation of the process (instability and imprecision). The conversion card current/voltage is followed by an anti-aliasing filter.

3. PROCESS IDENTIFICATION

The functional diagram of the multivariable global system is represented by Fig. 3, it is about a system having two inputs (control temperature and control flow) and two outputs (temperature and flow). The identification problem is to find the models of the 4 blocks: $G_{11}(s)$ between the temperature input and output, $G_{22}(s)$ between the flow input and output, $G_{12}(s)$ between the flow input and the temperature output, $G_{21}(s)$ between the temperature input and the flow output.

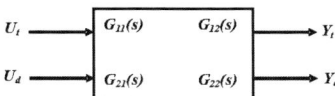

Fig. 3. The functional diagram of the process

The identification methods used are classical methods, which give simple models '*nominal*'. The use of these models thereafter allows to test the robustness of the synthesized controller.

The real step pesponses of the blocks $G_{11}(s)$, $G_{12}(s)$ and $G_{22}(s)$ are filtered, acquired and recorded by the dSPACE card. They are represented by Fig.4.

a. Temperature response (input and output are temperature)

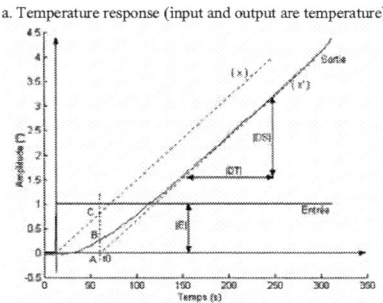

b. Flow response (input and output are flow)

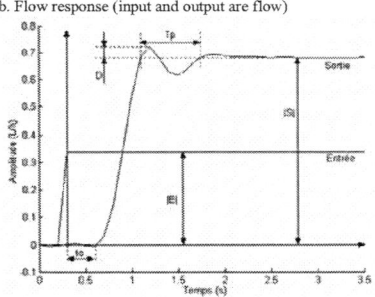

c. Temperature response (flow input and temperature output)

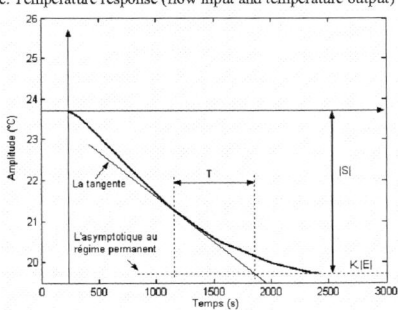

Fig. 4. Real step responses of the systems G_{11}, G_{22}, G_{12}

We remark that the G_{11} system is of the integrating type, the transfer function calculated (*strej'c method*) for this system is:

$$G_{11}(s) = \frac{0.0168}{s(1+44s)} \quad (1)$$

For the G_{22} system, it is about a second order with delay, its transfer function found (*second order method*) is [7]:

$$G_{22}(s) = \frac{40.3e^{-0.3s}}{s^2 + 6.012s + 20.15} \quad (2)$$

The system represented by the first block of coupling G_{12}, which represents the flow influence on the temperature behaves like a first order. The final model obtained (*first order method*) is [7]:

$$G_{12}(s) = \frac{-0.2}{1 + 1179.2s} \quad (3)$$

The second block of coupling G_{21}, which represents the temperature influence on the flow is [7]:

$$G_{21}(s) = 0 \quad (4)$$

This is obtained after having carried out the experiment which showed us that a change of input temperature (step) does not affect the flow output.

Finally the transfer matrix of the global process is:

$$G(s) = \begin{bmatrix} \dfrac{0.0168}{s(1+44s)} & \dfrac{-0.2}{1+1179.2s} \\ 0 & \dfrac{40.3e^{-0.3s}}{s^2 + 6.012s + 20.15} \end{bmatrix} \quad (5)$$

We use the G_{22} model without delay in order to test thereafter the robustness of the synthesized control law.

4. H_∞ BY LOOPSHAPING PROBLEM

The diagram block of the H_∞ by loopshaping problem is given by Fig. 5.

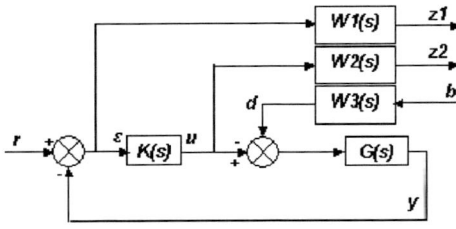

Fig. 5. Diagram bloc of H_∞ by Loopshaping problem

This general diagram can practically represent any system, in which the signal r is the reference and b the disturbance. Vector z contains the signals to be minimized: $z1$ corresponds to the error signals ε, which one wishes to maintain to 0, weighted by the $W1(s)$ function; $z2$ corresponds to the control u, which one wishes to save, weighted by the function $W2(s)$ [2].

We put this diagram in the standard form of the H_∞ problem, which is indicated by Fig. 6.

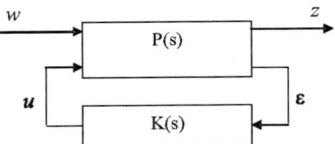

Fig. 6. Standard form of H_∞ problem

In this structure, w is the input vector, which regroups the external disturbances, the references and the noises, y there are available measurements to the $K(s)$ controller, u and z respectively represent the controls and the signals to be minimized [1]. $P(s)$ is the system known as '*increased*'.

The relations between (w, u) and (z, y) are given by:

$$\begin{bmatrix} Z(s) \\ Y(s) \end{bmatrix} = P(s) \begin{bmatrix} W(s) \\ U(s) \end{bmatrix} = \begin{bmatrix} P_{11}(s) & P_{12}(s) \\ P_{21}(s) & P_{22}(s) \end{bmatrix} \begin{bmatrix} W(s) \\ U(s) \end{bmatrix} \quad (6)$$

The transfer function between $Z(s)$ and $W(s)$ is given by the following linear fractional transformation (LFT):

$$F_l(P, K) = P_{11}(s) + P_{12}(s)K(s)(I - P_{22}(s)K(s))^{-1}P_{21}(s) \quad (7)$$

The H_∞- Loopshaping problem consists to stabilize internally the $P(s)$ system and to ensure $\|F_l(P, K)\|_\infty < 1$. It is a disturbance rejection problem, i.e. to minimize the w effect on the system behavior.

From Fig.5, we can calculate $F_l(P, K)$:

$$F_l(P, K) = \begin{pmatrix} W_1 S & W_1 W_3 SG \\ W_2 KS & W_2 W_3 KSG \end{pmatrix} \quad (8)$$

Where S is the sensitivity function, G is the initial system and W_i are weighting functions chosen starting from the specifications in the following way:

$W_1(s)$ is a weighting imposed on the error signal ε, it is selected so that the Bode diagram of $\frac{1}{|W_1|}$ present a sufficiently small gain in low frequency, which allows to cancel the disturbances and consequently to reduce the error ε. Generally, in high frequencies $\frac{1}{|W_1|}$ approaches to 1, after it cuts the 0 dB axis to the desired band-width. The $W_2(s)$ choice imposed on the control u must be carried out so that $\frac{1}{|W_2|}$ has sufficient gain in low frequencies, (what allows a control u at the system in the band-width), and is attenuated in high frequencies in order not to amplify the noises. $W_3(s)$ can be chosen constant and adjusted by taking care that the function S follows to the more meadows the model $\frac{1}{|W_1|}$ [1] [2].

The resolution of the H_∞ problem uses then a 4-blocks criterion:

$$\|F_l(P, K)\|_\infty = \left\| \begin{pmatrix} W_1 S & W_1 W_3 SG \\ W_2 KS & W_2 W_3 KSG \end{pmatrix} \right\|_\infty < 1 \quad (9)$$

According to the H_∞ norm properties, we have:

$$\|W_1 S\|_\infty < 1 \Longrightarrow \|S\|_\infty < \frac{1}{|W_1|}$$

$$\|W_2 KS\|_\infty < 1 \Longrightarrow \|KS\|_\infty < \frac{1}{|W_2|}$$

$$\|W_1 W_3 SG\|_\infty < 1 \Longrightarrow \|SG\|_\infty < \frac{1}{|W_1 W_3|}$$

$$\|W_2 W_3 KSG\|_\infty < 1 \Longrightarrow \|KSG\|_\infty < \frac{1}{|W_2 W_3|}$$

Now, we give the resolution method of the H_∞ problem using the Riccati equations and the state space techniques. For that, it is necessary to calculate the increased system $P(s)$ by writing z_1, z_2 and ε according to r, b and u (see Fig. 5), which gives:

$$P(s) = \begin{bmatrix} W_1 & W_1 W_3 G & -W_1 G \\ 0 & 0 & W_2 \\ 1 & W_3 G & -G \end{bmatrix} \quad (10)$$

Thereafter, we must find a state space realization and description of the $P(s)$ system:

$$\begin{bmatrix} \dot{x}(t) \\ z(t) \\ y(t) \end{bmatrix} = \begin{bmatrix} A & B_1 & B_2 \\ C_1 & D_{11} & D_{12} \\ C_2 & D_{21} & D_{22} \end{bmatrix} \begin{bmatrix} x(t) \\ w(t) \\ u(t) \end{bmatrix} \quad (11)$$

With

$x \in \Re^n$, $w \in \Re^{n_w}$, $u \in \Re^{n_u}$, $z \in \Re^{n_z}$, $y \in \Re^{n_y}$.

To solve the problem, one must satisfy the following assumptions, they allow to ensure that one seeks to solve a problem known as 'well posed' [1] [8]:

H1- (A, B_2) is stabilisable, and (C_2, A) is detectable.

H2- $Rang(D_{12}) = n_u$ and $Rang(D_{21}) = n_y$. These two rows must be fulls.

H3- $\forall \omega \in \Re$, $Rang \begin{pmatrix} A - j\omega I_n & B_2 \\ C_1 & D_{12} \end{pmatrix} = n + n_u$.

$\forall \omega \in \Re$, $Rang \begin{pmatrix} A - j\omega I_n & B_1 \\ C_2 & D_{21} \end{pmatrix} = n + n_y$.

H4- $D_{11} = 0$, $D_{12}^T (C_1 \; D_{12}) = (0 \; I_{n_u})$.

$D_{22} = 0$, $\begin{pmatrix} B_1 \\ D_{21} \end{pmatrix} D_{21}^T = \begin{pmatrix} 0 \\ I_{n_y} \end{pmatrix}$.

The solvability of the standard problem is tested using the following theorem.

Theorem 1 [1] Under the assumptions H1-H5, the H_∞ standard problem has a solution if and only if the three following conditions are fulfilled:

i- there is an solution $X_\infty \geq 0$ of the Riccati algebraic equation:

$$A^T X_\infty + X_\infty A + C_1^T C_1 + X_\infty (\gamma^{-2} B_1 B_1^T - B_2 B_2^T) X_\infty = 0 \quad (12)$$

such as:

$$Re(\lambda_i [A + X_\infty (\gamma^{-2} B_1 B_1^T - B_2 B_2^T) X_\infty]) < 0, \; \forall \, i \quad (13)$$

ii- there is an solution $Y_\infty \geq 0$ of the Riccati algebraic equation:

$$A Y_\infty + Y_\infty A^T + B_1 B_1^T + Y_\infty (\gamma^{-2} C_1^T C_1 - C_2^T C_2) Y_\infty = 0 \quad (14)$$

such as:

$$Re(\lambda_i [A + Y_\infty (\gamma^{-2} C_1^T C_1 - C_2^T C_2)]) < 0, \; \forall \, i \quad (15)$$

iii- The maximum eigenvalue of $X_\infty Y_\infty$, must verify the following condition:

$$|VP_{\max}(X_\infty Y_\infty)| < \gamma^2 \quad (16)$$

γ being given, in the case of the loopshaping H_∞ problem $\gamma = 1$

The following theorem gives the elaborate corrector parameters for the H_∞ problem.

Theorem 2 [2] let us suppose that H1-H5 are checked, and the conditions of theorem 1 are respected, then the controller is given by:

$$K(s) = C_c (sI - A_c)^{-1} B_c \quad (17)$$

With :

$$A_c = A + (\gamma^{-2} B_1 B_1^T - B_2 B_2^T) X_\infty$$
$$-(I - \gamma^{-2} Y_\infty X_\infty)^{-1} Y_\infty C_2^T C_2$$
$$B_c = (I - \gamma^{-2} Y_\infty X_\infty)^{-1} Y_\infty C_2^T$$
$$C_c = -B_2^T X_\infty$$

Stabilizes internment the system and satisfies $\|F_l(P, K)\|_\infty < \gamma$.

5. CONTROL LAW SYNTHESIS

A simplified diagram of the multivariable system is given by Fig. 7.

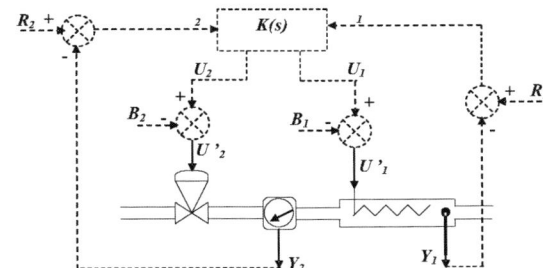

Fig. 7. Simplified diagram of the multivariable system

Let us put the system in its standard form shown by Fig. 5 and 6, let us search a state space description of $P(s)$ by finding for each transfer matrix G, $W1$, $W2$ and $W3$ a state space representation.

$$G : \begin{pmatrix} \dot{x} = Ax + B(u - b) \\ y = Cx \end{pmatrix}$$

$$W_1 : \begin{pmatrix} \dot{x}_1 = A_1 x_1 + B_1 (r - y) \\ z_1 = C_1 x_1 + D_1 (r - y) \end{pmatrix}$$

$$W_2 : \begin{pmatrix} \dot{x}_2 = A_2 x_2 + B_2 u \\ z_2 = C_2 x_2 + D_2 u \end{pmatrix}$$

$$W_3 : \begin{pmatrix} \dot{x}_3 = A_3 x_3 + B_3 d \\ b = C_3 x_3 + D_3 d \end{pmatrix}$$

After some calculation, we obtain:

$$P(s) = \begin{bmatrix} A & 0_{5x2} & 0_{5x2} & -BC_3 & 0_{5x2} & -BD_3 & B \\ -B_1 C & A_1 & 0_{2x2} & B_1 DC_3 & B_1 & B1DD3 & -B_1 D \\ 0_{2x5} & 0_{2x2} & A_2 & 0_{2x2} & 0_{2x2} & 0_{2x2} & B_2 \\ 0_{2x5} & 0_{2x2} & 0_{2x2} & A_3 & 0_{2x2} & B_3 & 0_{2x2} \\ -D_1 C & C_1 & 0_{2x2} & D_1 DC_3 & D_1 & D_1 DD_3 & -D_1 D \\ 0_{2x5} & 0_{2x2} & C_2 & 0_{2x2} & 0_{2x2} & 0_{2x2} & D_2 \\ -C & 0_{2x2} & 0_{2x2} & DC_3 & I_{2x2} & DD_3 & -D \end{bmatrix} \quad (18)$$

We fix the specifications according to: errors static due to the disturbances lower or equal to 10%, sufficient stability, speed of answer and in particular robustness with respect to the model uncertainties.

We announce that the choice of various weightings Wi is delicate in the multivariable case, because they are matrices. Nevertheless, an initial choice can be carried out in the following way ([3]):

- $W_1(s) = diag\{W_{1ii}(s)\}$, with:

$$W_{1ii}(s) = \frac{\frac{s}{M_i} + \omega_{Bi}}{s + \omega_{Bi} A_i}; \quad A_i << 1$$

ω_{Bi} represented the frequency cut-off of ith system.
One can choose W_{1ii} in the following way: to calculate a corrector for each sub-system (flow and temperature) by other synthesis methods (classical methodl: PID, poles placement,...). To trace the amplitudes of the diagonal elements of resulting $S = (I + GK)^{-1}$ according to the frequencies and to choose: $W_{1ii}(s) = \frac{1}{|S_{ii}(j\omega)|}$.

- A reasonable initial choice of the matrix $W2(s)$ is $W_2(s) = I$, thereafter, one will be able to introduce changes on $W_2(s)$: $W_2(s) = K_i I$ in low frequencies, and/or to impose attenuations in high frequencies [3].
- $W_3(s) = diag\{W_{3ii}(s)\}$ with $|W_{3ii}(j\omega)| \ll 1$ in low frequencies and wide in high frequencies [3].

The weightings matrices chosen finally are:

$$W_1(s) = \begin{bmatrix} \frac{\frac{s}{1.2} + 0.01}{s + 0.0005} & 0 \\ 0 & \frac{s + 0.4}{1.1s + 0.00011} \end{bmatrix}$$

$$W_2(s) = \begin{bmatrix} 0.1 & 0 \\ 0 & 0.5 \end{bmatrix} \quad (19)$$

$$W_3(s) = \begin{bmatrix} 0.1 & 0 \\ 0 & 0.1 \end{bmatrix}$$

The calculation of the H_∞ multivariable controller, as well as the optimal value of γ is made under MATLAB, by using the function '*hinfsyn*'.

For a γ_{opt} value of 0.9295, the obtained discrete multivariable corrector has as a transfer matrix:

$$K(s) = \begin{bmatrix} K_{11}(s) & K_{12}(s) \\ K_{21}(s) & K_{22}(s) \end{bmatrix} \quad (20)$$

The resulting correctors have high orders (order of 7), practically, this is not wished. We applied to these correctors an order reduction method known as *balanced reduction*, which consists in eliminating the corrector modes which would be little observables and/or little commandables ([3] [4]). We obtain after some calculation the following reduced discritized correctors:

$$K_{11red}(z) = \frac{1.672z^2 - 3.341z + 1.668}{z^3 - 2.578z^2 + 2.162z - 0.5838} \quad (21)$$

$$K_{12red}(z) = \frac{0.0289z^3 - 0.08429z^2 + 0.08189z - 0.0265}{z^4 - 3.153z^3 + 3.475z^2 - 1.491z + 0.1692}$$

$$K_{21red}(z) = \frac{0.001094z^2 - 0.00157z + 0.0004769}{z^3 - z^2 - 0.0001671z - 0.0002077}$$

$$K_{22red}(z) = \frac{0.1042z^2 - 0.1072z + 0.02356}{z^3 - 1.013z^2 + 0.013z - 0.0002193}$$

6. EXPERIMENTAL RESULTS

The practical implementation of the multivaraible reduced corrector on the real system gave satisfactory results. On the Fig. 8, we present the real step response of the temperature system. It is fast, the output joins the input (35 $C°$) in response time of 3.5 min, such it is stable and the error static is about 0.1 at $0.2°C$ ($\approx 4\%$).

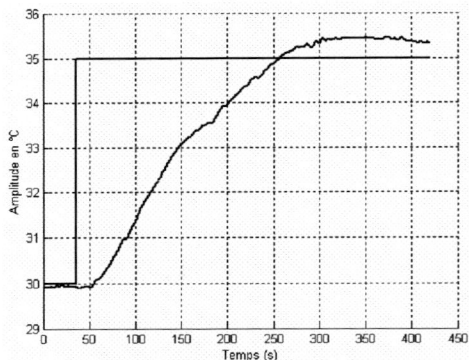

Fig. 8. Step reponse of temperature system

The Fig. 9 shows the disturbance rejection (which is a flow step of 500 L/h at 1000 L/h). We note that this rejection is done successfully by the corrector, the output stabilizes finally around 35 $C°$, but in a slow time of 6 min. We think that this is due to the choice of W_{222} which is insufficient, it will still be necessary to increase its value (to amplify the control).

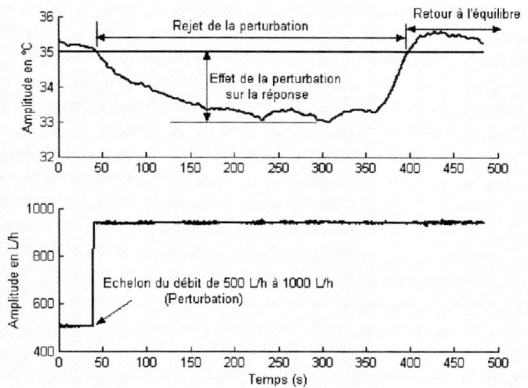

Fig. 9. Disturbance rejection in the temperature system

Concerning the real step response of the flow system, which is shown by the Fig. 10, we notice that it is stable, precise (in all cases, the error is lower or equal 10%) and present a rise time of about 8 sec, while the response time to 1% is of about 12 sec.

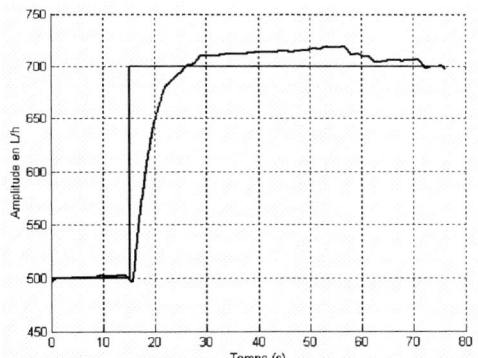

Fig. 10. Step reponse of the flow system

The corrector also reject completely the disturbance applied to the system (it is a closing of a manual tap placed at the system output during 3 or 4 sec, and its complete or partial reopening). The system finds quickly its equilibrium around 700 L/h, and the error static is negligible (see Fig. 11).

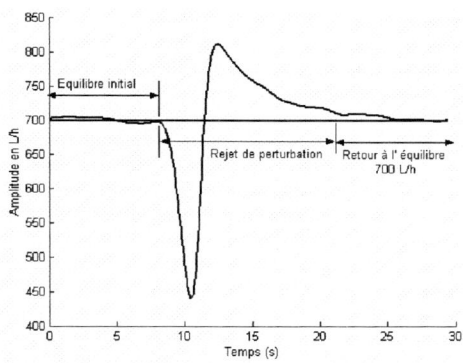

Fig. 11. Disturbance rejection in the temperature system

We can conclude that the H_∞ synthesized corrector better operated his role in the system control, in particular in the disturbances rejection. However, this corrector is calculated on the basis of nominal model, in more the model of the flow system was taken without delay and although this model has a hysteresis on its flow sensor (the origin of many practical problems). The robustness with respect to model uncertainties was quite assured.

7. CONCLUSION

A robust control approach of a multivariable physical process, based on the H_∞-Loopshaping technique was presented in this paper. The control law is calculated on the basis of system nominal mode, this model is uncertain and is much simplified: delay negligence, linearizations effects around the operation point.

The experimental results obtained show that this H_∞-Loopshaping approach appears a suitable and robust control strategy with respect to model uncertainties.

REFERENCES

[1] D. GILLES and F. STÉPHANE. Commande H_∞ et μ-analyse. GERMES, Paris, 1999.
[2] A. DANIEL, C. CHRISTELLE, A. PIERRE, G. MICHEL and F. GILLES. Robustesse et commande optimale. CÉPADUÈS, Toulouse, 1999.
[3] S. SIGURD and P. LAN. Multivariable feedback control. WILEY, England, 2003.
[4] K. ZHOU with K. GLOVER and J. DOYLE. Robuste and optimal control. PRINTICE HALL, New Jersey, 1996.
[5] C. BEN M. H_∞ Control and its application. SPRINGER, London, 1998.
[6] DS1102 User's Guide. Floating-Point controller board, DS1102. dSPACE GmbH, germany, 2001.
[7] P. DE LARMINAT and Y. TOMAS. Automatique des systèmes linéaires T2 : Identification. FLAMMARION SCIENCES, Paris, 1977.
[8] A. OUIS. Sur la fragilité des synthèses optimales. Magister en Electronique, Université de Tlemcen, 2001.

Stabilization of an under-actuated mechanical system by sliding mode control

A. Choukchou-Braham [*] C. Bensalah [**] B. Cherki [***]

[*] *Université Aboubekr Belkaid BP 19 - 13000 - Tlemcen Algérie*
(e-mail: Amel_choukchou@yahoo.fr)
[**] *(e-mail: choukri79@yahoo.fr)*
[***] *(e-mail: b_cherki@yahoo.fr)*

Abstract: The objective of this work is to stabilize an under-actuated mechanical system Ball and Beam by applying a sliding mode control. The model obtainted by Lagrange formalism, is strongly nonlinear, a classical linearization by feedback can't be applied due to the defect of the relative degree. We propose first to apply an approximate linearization by feedback associated with robust relative degree that linearize the Ball and beam, next, we stabilize the linearized system by a sliding mode control law.

Keywords: Robust control, under-actuated system, Ball and Beam system, robust relative degree, approximate linearization, sliding control.

1. INTRODUCTION

As there is no general procedure to deal with nonlinear systems usually, we linearize locally the system so that the linear procedures of analysis and synthesis can be applied. We thus speak about the linearization of a system around its equilibrium point. The disadvantage of this approach is that the domain of validity of the control laws thus determined is reduced. To increase this domain, researchers thought about new method of linearization.

The first technique was the exact linearization by feedback and diffeomorphism due to Hunt[8] and Isidori[6], this technique takes into account all the dynamics and nonlinearities of the system, it allows a perfect control and stabilization. But this technique requires somes assumptions that are often not verified. Specially this procedure can't be applied for under-actuated mechanical systems. Formally, under-actuated systems are systems that have fewer actuators then configuration variables, this restriction on control authority makes the control design for these systems rather complicated. The second technique of linearization is the approximate linearization by feedback appeared thereafter. It consists of linearizing a system at a certain order and neglecting certain nonlinear dynamics of higher order to deal with systems which don't verify the conditions of the first technique.

However, it is well known that applying exact or approximate linearization leads to less robust systems. On the other hand, if after a linearization, a robust control approach such as H_∞ or sliding mode is used to design control laws, then the lack of robustness is compensated.

During the last decades, sliding mode control became very popular for stabilization and for trajectories tracking of either linear or nonlinear systems such as: mechanical systems (inverteted pendulum), electromechanics systems (motors)..., considering its simplicity of implementation and also its robustness. The idea is to build a control law which brings the trajectories toward a certain surface called sliding surface. In the linear case, this technique of control is similar to a poles placement approach [5].

We propose in this work to stabilize an under actuated mechanical system, Ball and Beam, by the means of a sliding mode control. For a fixed gain of control, the basin of attraction of the system depends on the sign function used.

This paper contains the following sections: First, we give some preliminary assumptions and definitions. In section 3, we present the approximate linearization by feedback approach. Section 4 will be devoted to the design of the control law by sliding mode based on the approximate linearized model. In the last section, we illustrate the effectiveness of this technique on the Ball and Beam system by giving a comparison on the impact of the choice of the sign function on the domain of convergence. We finish by a conclusion and some prospects.

2. THEORETICAL FRAMEWORK

2.1 Definitions and notations

In this section, we give some definitions and assumptions in order to clarify different notations and problems presented thereafter.

Consider the nonlinear system affine in control:

$$\begin{cases} \dot{x}(t) = f(x(t)) + g(x(t))u \\ y(t) = h(x(t)) \end{cases} \quad (1)$$

with $x \in \mathbb{R}^n$ the state vector, $u \in \mathbb{R}$, the input control and $y \in \mathbb{R}$ the output of the system. The functions $f(x)$ and $g(x)$ are supposed smooth, $h(x)$ is a diffentiable function and $f(0) = h(0) = 0$, that is the origin is the equilibrium point of the system (1).

Notation: We note $\dot{y}(t)$, the derivative of h w.r.t time i.e:

$$\dot{y}(t) = \frac{\partial h}{\partial x}\dot{x}(t) = dh_{x(t)} \cdot (f(x(t)) + g(x(t))u) \quad (2)$$

and $L_f h(x)$, the Lie derivative of $h(x)$ along the vector field f: $L_f h(x) = dh_x \cdot f(x)$ with $L_f^i h(x) = L_f \cdot L_f^{i-1} h(x)$.

definition 1 The relative degree associated with the system (1) at $x = x_0$, is given by the integer γ such that :

- $L_g L_f^i h(x)\big|_{x=x_0} = 0 \quad \text{for} \quad i \leq \gamma - 2$
- $L_g L_f^{\gamma-1} h(x)\big|_{x=x_0} \neq 0$ (3)

The relative degree of a nonlinear system represents in a general way the number of diffentiations of the output needed for one of the inputs to appear explicitely.

Theorem 1 [6] The nonlinear system (1) is exactly linearizable by feedback around the origin, if there is a regular output $y = h(x)$ such that the relative degree is exactly equal to the dimension of the state space of the system ($\gamma = n$). □

The model obtained after linearization is controllable. If $\gamma < n$, we seek with a robust relative degree which will be equal to n after a modification in the vector field f or g. Under these conditions, the system can be approximatly linearized at the origin and thus becomes locally controllable.

3. APPROXIMATE LINEARIZATION BY FEEDBACK

Let us suppose that the relative degree of (1) is less than the dimension of the system. Hence, according to theorem 1 (1) can't be exactly linearized by feedback. In the vector fields f and g, some terms constitutes an obstruction to the linearization, that means, the relative degree in presence of these terms is less than the dimension of the space. So, if we neglect some of these terms we can achieve a full relative degree called robust relative degree, hence leading to an exact linearization of the new system [10]. By this way, (1) can be transformed in a system which can be exactly linearized using the change of coordinates given by the following diffeomorphism :

$$\varphi(x) = \begin{bmatrix} \varphi_1(x) \\ \varphi_2(x) \\ \vdots \\ \varphi_n(x) \end{bmatrix} = \begin{bmatrix} z_1 \\ z_2 \\ \vdots \\ z_n \end{bmatrix} = \begin{bmatrix} h(x) \\ L_{\tilde{f}} h(x) \\ \vdots \\ L_{\tilde{f}}^{n-1} h(x) \end{bmatrix} \quad (4)$$

where $\tilde{f}$ and $\tilde{g}$ are obtained from f and g after eliminating of some bad terms. The application $\varphi : x \longmapsto z$ is called: robust coordinates transformation. In the new coordinates, system (1) becomes as follow :

$$\begin{cases} \dot{z}_1 = z_2 \\ \dot{z}_2 = z_3 \\ \vdots \\ \dot{z}_{n-1} = z_n \\ \dot{z}_n = L_{\tilde{f}}^n h(\varphi^{-1}(z)) + L_{\tilde{g}} L_{\tilde{f}}^{n-1} h(\varphi^{-1}(z)) u \end{cases} \quad (5)$$

we note by φ^{-1} the inverse of the application φ sufficently defined in the neighborhoods of x_0. Suppose that the function $L_{\tilde{g}} L_{\tilde{f}}^{n-1} h(\varphi^{-1}(z)) \neq 0$, then, it is always possible to find a new control v such that :

$$u = \frac{1}{L_{\tilde{g}} L_{\tilde{f}}^{n-1} h(\varphi^{-1}(z))} (-L_{\tilde{f}}^n h(\varphi^{-1}(z)) + v) \quad (6)$$

The result system is written in the following form

$$\begin{cases} \dot{z}_1 = z_2 \\ \dot{z}_2 = z_3 \\ \vdots \\ \dot{z}_{n-1} = z_n \\ \dot{z}_n = v \end{cases} \quad (7)$$

which is linear and locally controllable.

4. SLIDING CONTROL OF AN APPROXIMATE LINEARIZED SYSTEM

In this section, we present a method for stabilization an approximate linearized system by feedback, using sliding control methodology. This technique of control forces the states trajectories of the system to converge toward a surface S in finite time, and to remain there. This surface ($S \subset \mathbb{R}^n$), referred to as a sliding surface , cut the space phase in two parts, a positive part and a negative one. A control law, which alternates between Umax and Umin forming a hysteresis, forces the trajectories to evolve around this surface.

The sliding surface S is defined as follows:

$$S = \{ z | \ s(z) = 0 \} \quad (8)$$

All the problems of this technique of control lie in the choice of this sliding surface.

Theorem 2: [1] Consider the system (5) with the sliding surface

$$S = \{ z \in \mathbb{R}^n | \ z_n + a_{n-1} z_{n-1} + \cdots + a_2 z_2 + a_1 z_1 = 0 \} \quad (9)$$

such that the polynome :

$$P(s) = s^{n-1} + a_{n-1} s^{n-2} + \cdots + a_1 \quad (10)$$

is Hurwitz. Let D a compact and convexe vicinity of $z = 0$, then, there is a control u given by :

$$u_{eq} = -k \, sign(L_{\tilde{g}} L_{\tilde{f}}^{n-1} h(\varphi^{-1}(z))) \times sign(s(z)) \quad (11)$$

with:

$$k_1 \left| L_{\tilde{g}} L_{\tilde{f}}^{n-1} h(\varphi^{-1}(z)) \right| \geq \left| \sum_{i=1}^{n-1} a_i z_{i+1} + L_{\tilde{f}}^n h(\varphi^{-1}(z)) \right|, \text{ and }$$

$k > k_1 > 0$ for $z \in D$, such that the system (5) is asymptoticaly stable for $z(0) \in D_0$ and for $u = u_{eq}$ where $D_0 \subset D$ is an open vicinity of $z = 0$.

proof: Recall that :
$$s(z) = z_n + a_{n-1} z_{n-1} + \cdots + a_2 z_2 + a_1 z_1$$

then,
$$\begin{aligned} \dot{s}(z) &= \dot{z}_n + a_{n-1} \dot{z}_{n-1} + \cdots + a_2 \dot{z}_2 + a_1 \dot{z}_1 \\ &= L_{\tilde{f}}^n h(\varphi^{-1}(z)) + L_{\tilde{g}} L_{\tilde{f}}^{n-1} h(\varphi^{-1}(z)) u + \\ &\quad a_{n-1} z_n + \cdots + a_2 z_3 + a_1 z_2 \\ &= \alpha_s(z) + \beta_s(z) u \end{aligned}$$

where

$\alpha_s(z) = \sum_{i=1}^{n-1} a_i z_{i+1} + L_f^n h(\varphi^{-1}(z))$

and

$\beta_s(z) = L_{\tilde{g}} L_f^{n-1} h(\varphi^{-1}(z))$.

Suppose that: $\left|\dfrac{\alpha_s(z)}{\beta_s(z)}\right| \leq k_1$ with $k_1 > 0$. To design the control that leads to an asymptotic stability around $z = 0$, we use the Lyapunov theory. The idea is to choose $s(z)$ as a Lyapunov function: $V(z) = \dfrac{1}{2} s(z)^2$, next, we must find a control that make $V(z)$ a decreasing function ie $\dot{V}(z) < 0$. The derivative of this function is:

$$\begin{aligned}\dot{V}(z) &= s(z).\dot{s}(z) \\ &= s(z).(\alpha_s(z) + \beta_s(z).u) \\ &= s(z).\alpha_s(z) + s(z).\beta_s(z).u \\ &\leq k_1.|s(z)|.|\beta_s(z)| + s(z).\beta_s(z).u\end{aligned}$$

Let: $u_{eq} = -k \times sign(\beta_s(z)) \times sign(s(z))$ with $k = k_1 + k_0$ ($k_0 > 0$). Substitute u_{eq} in the last equation:

$$\begin{aligned}\dot{V}(z) &\leq k_1 |s(z)|.|\beta_s(z)| \\ &\quad + s(z)\beta_s(z)(-k.sign(\beta_s(z)).sign(s(z))) \\ &\leq k_1 |s(z)|.|\beta_s(z)| \\ &\quad - k.s(z).sign(s(z)).\beta_s(z).sign(\beta_s(z)) \\ &\leq k_1 |s(z)|.|\beta_s(z)| - k |s(z)|.|\beta_s(z)| \\ &\leq k_1 |s(z)|.|\beta_s(z)| - (k_1 + k_0).|s(z)|.|\beta_s(z)| \\ &< -k_0.|s(z)|.|\beta_s(z)|\end{aligned}$$

For $u = u_{eq}$, the Lyapunov function is decreasing. Hence, the system is asymptoticaly stable at the vicinity of $z = 0$. □

5. STABILIZATION OF THE UNDER-ACTUATED BALL AND BEAM SYSTEM BY SILDING CONTROL

5.1 Modeling

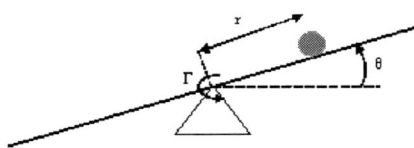

Fig. 1. Ball and Beam system

The Ball and Beam system(BBS) depicted in figure (fig 1), is an under-actuated system. In fact, the number of variables to control (ball position r and beam angle θ) is greater then the number of input control (torque applied to the beam center Γ). The application of the Lagrange formalism to this system, enables us to have the system of equations according to:

$$\begin{cases} m\ddot{r} + mg\sin(\theta) - r\dot{\theta}^2 = 0 \\ (m.r^2 + I)\ddot{\theta} + 2mr\dot{r}\dot{\theta} + mg\cos(\theta) = \Gamma \end{cases} \quad (12)$$

where r is the ball position, θ the beam angle, Γ the torque applied to the beam, (m, I) respectively the mass, the inertia of the ball and g the gravity constant.

Applying the change of control:

$$u = \frac{1}{I + mr^2} mr(-2\dot{r}\dot{\theta} - g\cos(\theta) + \Gamma) \quad (13)$$

and noting by $x_1 = r, x_2 = \dot{x}_1, x_3 = \theta$ and $x_4 = \dot{x}_3$, we obtain the new dynamic:

$$\begin{cases} \dot{x}_1 = x_2 \\ \dot{x}_2 = \mathcal{B}(x_1 x_4 + g\sin(x_3)) \\ \dot{x}_3 = x_4 \\ \dot{x}_4 = u \end{cases} \quad (14)$$

with $\mathcal{B} = 0.72$ and $g = 9.8$

5.2 Synthesis of the control law by sliding mode

In order to apply the sliding control (11), to the BBS we need to express the system (14) in the form (5). For that, we need first, to calculate the relative degree of the BBS. It is easy to show that:

$$\begin{aligned} L_g h &= L_g L_f h = 0 \\ L_g L_f^2 h &= 2\mathcal{B} x_1 x_4 \end{aligned}$$

Note that $L_g L_f^i h(0) = 0$ for $i = \{0, 1\}$, but $L_g L_f^2 h(0) \neq 0$, hence $\gamma = 3$, that is the relative degree of the BBS is less then the dimension of the state space so that the BBS can't be exactly linearized by feedback. However, the linearization technique of Kokotovic [10] for example, which is based on the elimination of certain terms in the vector field g allows to define a robust relative degree that allows linearization, but this linearization is only approximate (because of the elimination of certain dynamics). This procedure applied to the BBS is given by:

$$\begin{aligned} h(x) &= & x_1 & = z_1 \\ \dot{z}_1 &= & x_2 & = z_2 \\ \dot{z}_2 &= & \mathcal{B}(x_1 x_4^2 - g\sin(x_3)) & = z_3 \\ \dot{z}_3 &= & \underbrace{\mathcal{B}(-x_4 \cos(x_3) + x_2 x_4^2)}_{z_4} + 2\mathcal{B} \underbrace{x_1 x_4}_{\psi_3(x)} \cdot u \end{aligned} \quad (15)$$

$$\dot{z}_4 = \mathcal{B}(gx_4 \cos(x_3) + x_4^4 x_2 - gx_4^3 \sin(x_3)) \\ + \mathcal{B}(2x_4 x_2 - g\sin(x_3)) \cdot u$$

Since the term $\psi_3(x) = x_1 x_4 = r\dot{\theta}$ does not allow us to define the relative degree, simply, we will eliminate it, so that the robust relative degree will be equal to 4. Note that only the vector field g is modified, hence, in what follow, g will be replaced by $\tilde{g}$.

According to the transformation (4) (which is identical to the transformation (15)), the system (14) takes the following form:

$$\begin{cases} \dot{z}_1 = z_2 = x_2 \\ \dot{z}_2 = z_3 = -\mathcal{B}g\sin(x_3) + \mathcal{B}x_1 x_4^2 \\ \dot{z}_3 = z_4 = -\mathcal{B}gx_4 \cos(x_3) + \mathcal{B}x_2 x_4^2 \\ \dot{z}_4 = L_f^4 h(\varphi^{-1}(z)) + L_{\tilde{g}} L_f^3 h(\varphi^{-1}(z)) u \end{cases} \quad (16)$$

According to theorem 2, the sliding surface coefficients $a_{i, i=\{1,2,3\}}$ given by (9) must satisfy the condition on the stability of the polynome (10). Arbitrarly we choose, the following poles: -1, -2, -3 which leads to the polynome $s^3 + 6s^2 + 11s + 6$ consequently $a_1 = 6, a_2 = 11$ and $a_3 = 6$. In this case, the control law u_{eq} is written as :

$u_{eq} = -k \times sign(\mathcal{B}g\cos(x_3) + 2\mathcal{B}x_2x_4) \times$

$sign(\mathcal{B}(-gx_4\cos(x_3) + x_2x_4^2 - a_3g\sin(x_3) + a_3x_1x_4^2) + a_2x_2 + a_1x_1)$

5.3 Simulations results

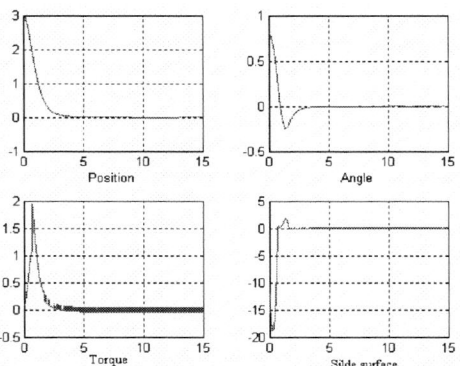

Fig. 3. States trajectories and control input of the BBS for the initial condition $x_1(0) = 3, x_3(0) = 0.8$ and $k = 2$

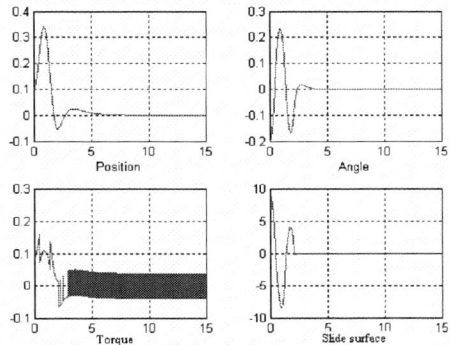

Fig. 4. States trajectories and control input of the BBS for the initial condition $x_1(0) = 0.1, x_3(0) = -0.2$ and $k = 2$

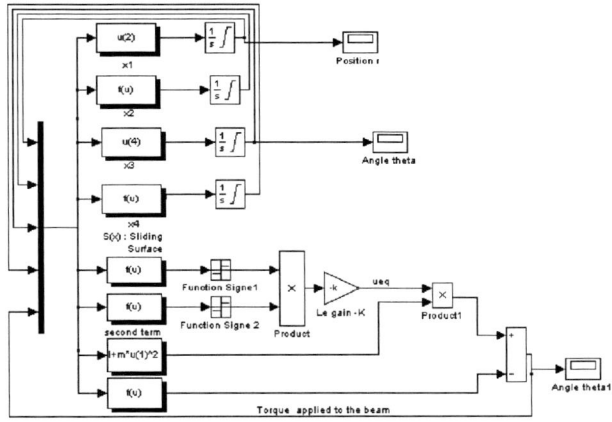

Fig. 2. BBs simulation schema with sliding control

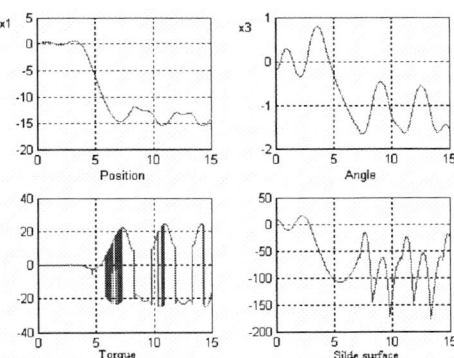

Fig. 5. States trajectories and control input of the BBS for the initial condition $x_1(0) = 0.2, x_3(0) = -0.2$ and $k = 2$

In order to observe the performance of the proposed control law based on sliding mode, we performed simulations on MATLAB using SIMULINK (fig 2). Moreover, these simulations allow us to define the domain D_0 for which the system (12) with the real control Γ is asymptoticaly stable. In a first time, we apply the sign function defined by :

$$sign(x) = \begin{cases} 1 \text{ si } x > 0 \\ 0 \text{ si } x = 0 \\ -1 \text{ si } x < 0 \end{cases} \quad (17)$$

The results presented in figures fig. 3 and fig. 4 show that the BBS is asymptoticaly stable for initial conditions $x_1(0)$ and $x_3(0)$ choosed close to the origin . However, the performances of the system decay when we choose greater initial conditions in order to increase the basin of attraction D_0 (fig. 5).

Remark For all figures position is mesured in meter, angle in radian and torque in Newtom.meter.

In the other hand, if we suppose that the BBS is controlled by DC motor for example, and according to the form of the torque input Γ (figures fig. 3, fig. 4), the motor can be damaged because of the Chattring phenomenon . This effect can be anticipated if we use a more smooth sign function, for example the use of the function **sign(x)=arctan(x)**, decreases oscillations and make broader the domain of convergence (figures fig. 6, fig. 7, fig. 8).

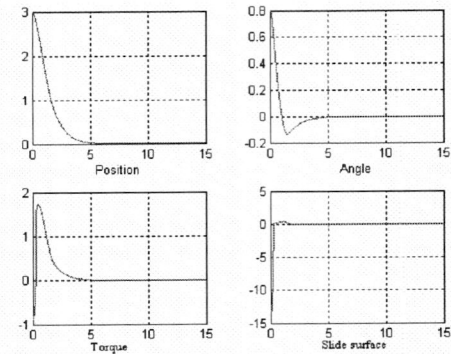

Fig. 6. States trajectories and control input of the BBS for the initial condition $x_1(0) = 3, x_3(0) = 0.8, k = 2$ and cotangent function

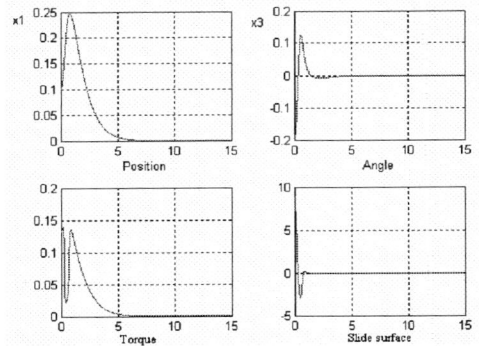

Fig. 7. States trajectories and control input of the BBS for the initial condition $x_1(0) = 0.1, x_3(0) = -0.2$. $k = 2$ and cotangent function

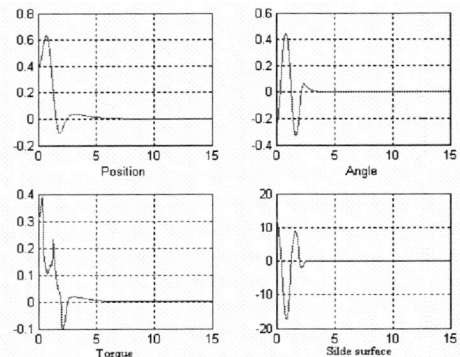

Fig. 8. States trajectories and control input of the BBS for the initial condition $x_1(0) = 0.4, x_3(0) = -0.25, k = 2$ and cotangent function

6. CONCLUSION

In this paper, we have applied a strategy of linearization for a system that initially could'nt be linearized by feedback since it belongs to the class of under-actuated mechanical systems. This approach consists on eliminating some terms in some dynamic which dones'nt allow the definition of the relative degree. The elimination of these terms allows the definition of the robust relative degree leading to an approximated linearized model.

To compensate the weak robustness due to this procedure of linearization, we choose to design the control law that stabilize the system through a robust control approach namely the sliding control mode. For this, first we designed a sliding surface using the coordinates obtained from the approximate linearization, next, we designed the control law that stabilize the trajectories around this surface.

The control scheme has been tested on the Ball and Beam system and good performance has been obtained. Moreover, we showed that the basin of attraction can be modified by choosing a more smooth sign function.

The proposed control strategy is applicable to a wider class of under-actuated mechanical systems as the inverted pendulum. However, we think that the elimination of terms influences directly the basin of attraction of the stabilization of the system. In fact, this basin could be broader if we take in account the eliminated term in the conception of the control.

REFERENCES

[1] D. A. Voytsekhovsky, R. M.Hirschorn. *Stabilization of Single-Input Nonlinear Systems Using Higher Order Compensating Sliding Mode Control*. 44th IEEE Conference on Decision and Control, and the European Control conference 2005.

[2] B.C. Chang, H. Kwtny, S. Hu. An Application of Robust Feedback Linearization to a Ball and Beam Control System. IEEE. 0-7803-4104-X. 1998.

[3] R. Olfati-Saber, A. Megretski. *Controller Design for the Beam and Ball System*. Proc. of the 37th conf. on Decision and Control, pp.4555-4560, Tampa, FL, Dec. 1998.

[4] H. Khalil. *Nonlinear Systems*. Third edition. Prentice Hall, 2002.

[5] C. Edwards, S. K. Spurgeon. *Sliding Mode Control*. Taylor and Francis, 1998.

[6] A. Isidori. *Nonlinear Control Systems*. Springer-Verlag, London, 1995.

[7] J. E. Slotine. *Applied Nonlinear Control*. Prentice-Hall, 1991.

[8] L.R Hunt, R. Su and G. Meyer, *Global Transformations of Nonlinear Systems*. IEEE. Trans.Autom, Contr. Vol Ac-28,No 1,pp.24-31, 1983.

[9] L.R Hunt, J. Turi. *A new Algorithm For Constructing Approximate Transformations For Nonlinear Sytems*. IEEE. Trans.Autom, Contr. Vol 38, No 10, pp1553-1556, 1993.

[10] J. Hauser, S. Sastry, P. Kokotovič. *Nonlinear control via approximate input-output linearization : the ball and beam example*. Proc. 28th IEEE CDC, Tampa, Florida, pp.1987-1989. 1989.

[11] J. Hauser. A. Banaszuk, *Approximate linearization via around trajectory : Application to trajectory planning*. Proc 28th IEEE CDC, San Diego, California USA, pp. 7-11,1997.

The Observer Adaptive backstepping Control for a Simple Pendulum

M. Mokhtari*. N. Golea**. S. Berrahal***

* *Department of electronics, Batna University, 05000 Batna, Algeria,*
(e-mail : messaoud.mokhtari@yahoo.fr).
** *EE Institute, Oum El-Bouaghi University, 04000 Oum El-Bouaghi, Algeria*
(e-mail : n.golea@lycos.com)
*** *Department of electronics, Batna University, 05000 Batna, Algeria,*
(e-mail : razik61@yahoo.fr).

Abstract: In this paper, the adaptive backstepping control with an observer is presented for the class of nonlinear systems. The problem of observability has a practical importance, because certain intern variables are some times inaccessible to measurement or "expensive" to measure. Generally the physical utilization of a sensor is impossible for reasons of implementation or cost, which do not allow the measurement of all the states. We will see how we can start from measurements of inputs and outputs of the process, to reconstruct (we also say to observe), the complex state vector. The subsystem, which carries out this reconstructing, is called the observer. To clarify the approach further, an example of simple pendulum is studied and the results of simulation clearly show the power of this approach.

Keywords: Nonlinear systems, Adaptive, Control, Lyapunov, backstepping, Observer

1. INTRODUCTION

In recent years, there was much progress concerning the control of the nonlinear systems. Adaptive backstepping control is considered as one of the principal results in this field of research [1][2][3]. Its development is based on the Lyapunov function which will guarantee the stability of the system to be controlled. The design methods yields a force control law with a parameter adaptation and update control law. In this paper, we will shed the light on the technique of control backstepping with observer and we will deduce its efficacy.

2. THE OBSERVER ADAPTIVE BACKSTEPPING CONTROL

The observer has like inputs the inputs and the outputs of the real process and like output the estimated value (rebuilt) of the state of this process (figure 1).

Thus, the problem consists in reconstructing this observer. For a given process, a system is defined by its equations of state, which is the output that gives an estimate of the real state of the process. This estimate comprises an error which must tend towards zero. When this property is satisfied, the observer is known as asymptotic.

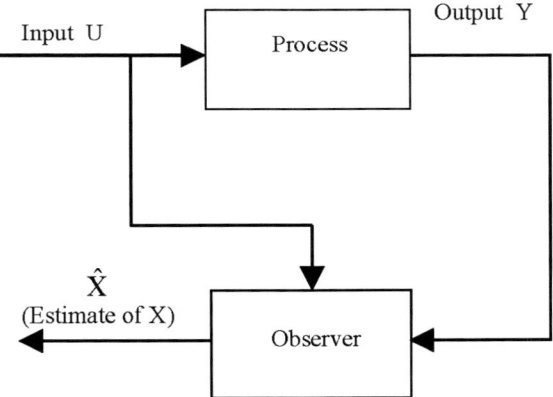

Fig. 1. General diagram of the observer

To achieve the aim that has already laid down in this part, we must adopt all the assumptions in order to introduce the observer. We will treat an example according to the usual steps of the adaptive backstepping control.

The first principle consists in exposing two diagrams that may clear up the difference between the non adaptive control (figure 2) and the adaptive control (figure 3) with observer.

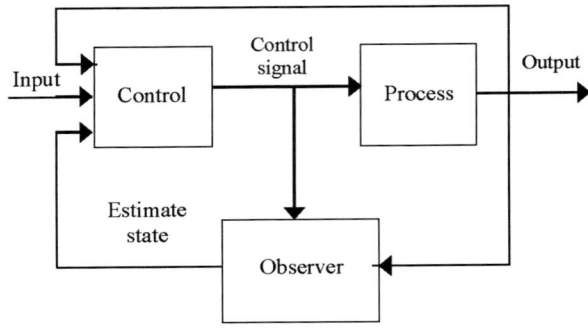

Fig. 2. General diagram of the no adaptive control with the observer

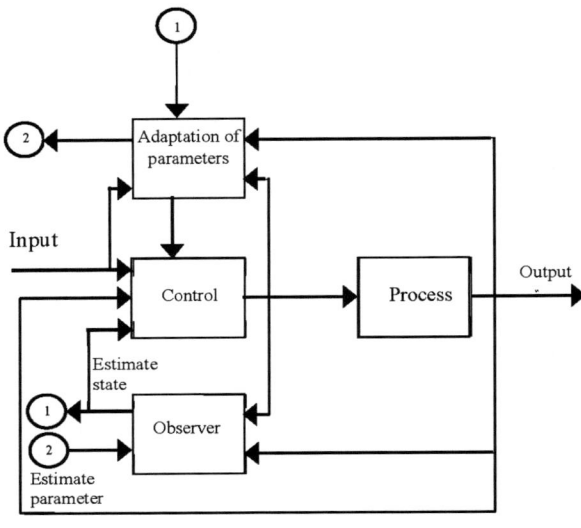

Fig. 3. General diagram of the adaptive control with the observer

The conditions and assumptions needed in order to synthesize an observer-based adaptive backstepping controller are [1]:

1. Full-state measurement is not available,
2. An output function y=h(x) must be defined, and
3. All system nonlinearity must be a function of output only.

These additional conditions lead to the form:

$$\dot{x}_1 = x_2 + \varphi_1(y)^T . \theta$$
$$\dot{x}_2 = x_3 + \varphi_2(y)^T . \theta$$
$$\vdots \qquad (1)$$
$$\dot{x}_{n-1} = x_n + \varphi_{n-1}(y)^T . \theta$$
$$\dot{x}_n = \beta(y).u + \varphi_n(y)^T . \theta$$
$$y = x_1$$

Where each $\varphi_i : R \to R^P$ is a vector of nonlinear functions of output, and $\theta \in R^P$ is a vector of constant coefficients which scales the nonlinearity φ_i.

In order to design an observer, equation (1) is rewritten as a sum of:

- The known linear part,
- The unknown nonlinear part, and
- The control function.

These give us:

$$\dot{x} = \underbrace{A.x}_{\text{Linear part}} + \underbrace{\varphi^T(y).\theta}_{\text{Nonlinear part}} + \underbrace{B.g(y).u}_{\text{Control}} \qquad (2)$$

where:

$$A = \begin{bmatrix} 0 & 1 & 0 & 0 & 0 & 0 \\ 0 & 0 & 1 & 0 & 0 & 0 \\ 0 & 0 & \ldots & \ldots & 0 & 0 \\ 0 & 0 & 0 & 0 & 1 & 0 \\ 0 & 0 & 0 & 0 & \ldots & 1 \\ 0 & 0 & 0 & 0 & 0 & 0 \end{bmatrix};$$

$$\varphi(y) = \begin{bmatrix} \varphi_1^T(y) & \varphi_2^T(y) & \ldots & \varphi_i^T(y) & \ldots & \varphi_n^T(y) \end{bmatrix};$$
$$B^T = \begin{bmatrix} 0 & 0 & \ldots & 0 & \ldots & 1 \end{bmatrix};$$
$$x = \begin{bmatrix} x_1 & x_2 & \ldots & x_i & \ldots & x_n \end{bmatrix}^T;$$
$$\theta = \begin{bmatrix} \theta_1 & \theta_2 & \ldots & \theta_i & \ldots & \theta_p \end{bmatrix}^T$$

3. BACKSTEPPING CONTROLLER DESIGN FOR SIMPLE PENDULUM USING AN OBSERVER

3.1 Model of a simple pendulum

It is about an unstable system described perfectly by a nonlinear model obtained by applying the laws of physics. The simple pendulum is a traditional example [4].

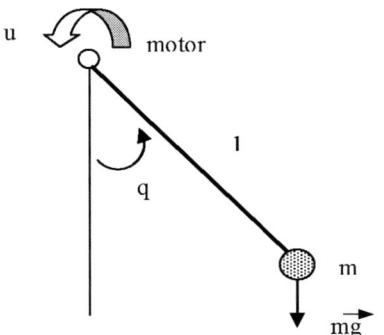

Fig. 4. General diagram for simple pendulum

The interest of design is to stabilize the pendulum in its position of unstable balance vertical. This system is treated like the model ($q, \dot{q}$), where q is the rod angle from the vertical position as shows in the figure 4.

The Lagrangian is given by:

$$L = E_C - E_P$$
$$E_C = \frac{1}{2} m.l^2.\dot{q}^2 \qquad (3)$$
$$E_P = m.g.l.(1 - \cos q)$$

what implies:

$$L = \frac{1}{2} m.l^2.\dot{q}^2 - m.g.l.(1 - \cos q) \qquad (4)$$

The differential equations are:

$$\frac{\partial L}{\partial \dot{q}} = m.l^2.\dot{q}$$
$$\frac{d}{dt}\left(\frac{\partial L}{\partial \dot{q}}\right) = m.l^2.\ddot{q} \qquad (5)$$
$$\frac{\partial L}{\partial q} = -m.g.l.\sin q$$

According to the expression of Lagrange, the equation of the system will be expressed by:

$$m.l^2.\ddot{q} + m.g.l.\sin q = u \qquad (6)$$

Where u is the controller applied to the pendulum rod, m is the mass located at the end of the rod, g is the gravitational constant and l is the half length of the rod.

3.2 Backstepping controller design assuming that the nonlinear functions can be evaluated

➤ *Model*
To simplify the state presentation and to apply the algorithm of backstepping, the equation 6 can be written as

$$\ddot{q} = -\frac{g}{l}.\sin q + \frac{1}{m.l^2}.u \qquad (7)$$

While choosing the following state variables:

$x_1 = q$: represent the angular position,
$x_2 = \dot{q}$: represent the angular velocity.

Then, the general model that we will use for this application is as it follows:

$$\dot{x}_1 = x_2$$
$$\dot{x}_2 = -\frac{g}{l}.\sin x_1 + \frac{1}{m.l^2}.u \qquad (8)$$
$$y = x_1$$

We define the parameters as it follows:

$$\theta_l = -\frac{g}{l} \;;\; \theta_u = \frac{1}{m.l^2}$$

These transformation lead to the form:

$$\dot{x}_1 = x_2$$
$$\dot{x}_2 = \theta_l.\varphi(x_1) + \theta_u.u \qquad (9)$$
$$y = x_1$$

with the nonlinear function $\varphi(x_1) = \sin x_1$

➤ *Observer*
In what follows, it is considered that only the position x_1 is measurable and it is supposed that the speed is constant.

The observer is defined as:

$$\dot{\hat{x}} = \zeta(t) + \lambda(t).\theta_l + \upsilon(t).\theta_u \qquad (10)$$

Where: $\zeta \in R^2$, $\lambda \in R^2$ and $\upsilon \in R^{2 \times M}$. The terms θ_l and θ_u, used in the relation (10), are the "true" unknown parameters, thus, we can not implement. In addition, the filter banks ζ, λ and υ must be individually implemented as it follows:

$$\dot{\zeta}(t) = A.\zeta - K.\zeta_1 + K.y \qquad (11)$$

$$\dot{\lambda} = A.\lambda - K.\lambda_1 + \begin{bmatrix} 0 \\ \varphi(y) \end{bmatrix} \qquad (12)$$

$$\dot{\upsilon} = A.\upsilon - K.\upsilon_1 + \begin{bmatrix} 0 \\ u \end{bmatrix}, \qquad (13)$$

where: $A = \begin{bmatrix} 0 & 1 \\ 0 & 0 \end{bmatrix}$ and $K = \begin{bmatrix} k_1 \\ k_2 \end{bmatrix}$

Given (10), (11), (12) and (13), the observer error dynamics are described by the equation:

$$\begin{aligned}\dot{\varepsilon} &= \dot{x} - \dot{\hat{x}} \\ &= \dot{x} - \left(A.(\zeta + \lambda.\theta_1 + v.\theta_u) + K.(y - (\zeta_1 + \lambda_1.\theta_1 + v_1.\theta_u)) \right. \\ &\quad \left. + \begin{bmatrix} 0 \\ \varphi(y) \end{bmatrix}.\theta_1 + \begin{bmatrix} 0 \\ u \end{bmatrix}.\theta_u \right) \\ &= A.\varepsilon - K.\varepsilon_1 \end{aligned} \quad (14)$$

where: $A = \begin{bmatrix} 0 & 1 \\ 0 & 0 \end{bmatrix}$, $\varepsilon = \begin{bmatrix} \varepsilon_1 \\ \varepsilon_2 \end{bmatrix}$ et $K = \begin{bmatrix} k_1 \\ k_2 \end{bmatrix}$

which then leads to:

$$\dot{\varepsilon} = \begin{bmatrix} \varepsilon_2 - k_1.\varepsilon_1 \\ -k_2\varepsilon_1 \end{bmatrix} = A_0.\varepsilon \quad (15)$$

where $A_0 = \begin{bmatrix} -k_1 & 1 \\ -k_2 & 0 \end{bmatrix}$, and K is chosen such that A_0 is Hurwitz (the solutions of equation $s^2 + k_1.s + k_2 = 0$ with negative real parts).

> Step 1

We adopt the following coordinate transformation:

$$z_1 = y - y_r \quad (16)$$

$$z_2 = v_2.\hat{\theta}_u - \dot{y}_r - \alpha_1 \quad (17)$$

Where α_1 is the virtual control as yet undefined.

Knowing that the first step consists in identifying the virtual control, we choose a $P \in R^{2x2}$ such that $P > 0$ and $P^T = P$ where $P.A_0 + A_0^T.P = -I$. The first Lyapunov function is chosen as:

$$V_1 = \frac{1}{2}z_1^2 + \frac{1}{2.g_1}\tilde{\theta}_1^2 + \frac{1}{2.g_u}\tilde{\theta}_u^T.\tilde{\theta}_u + \frac{1}{d_1}\varepsilon^T.P.\varepsilon \quad (18)$$

Its derivative can be written:

$$\begin{aligned}\dot{V}_1 &= z_1.\dot{z}_1 + \tilde{\theta}_1\left(-\frac{1}{g_1}\dot{\hat{\theta}}_1\right) + \tilde{\theta}_u^T\left(-\frac{1}{g_u}\dot{\hat{\theta}}_u\right) - \frac{1}{d_1}\varepsilon^T.\varepsilon \\ &= z_1.(\dot{y} - \dot{y}_r) + \tilde{\theta}_1\left(-\frac{1}{g_1}\dot{\hat{\theta}}_1\right) + \tilde{\theta}_u^T\left(-\frac{1}{g_u}\dot{\hat{\theta}}_u\right) - \frac{1}{d_1}\varepsilon^T.\varepsilon \end{aligned} \quad (19)$$

Let us note that:

$$\dot{y} = \dot{x}_1 = x_2 = \hat{x}_2 + \varepsilon_2 = \zeta_2(t) + \lambda_2(t).\theta_1 + v_2(t).\theta_u + \varepsilon_2 \quad (20)$$

We will have then:

$$\begin{aligned}\dot{V}_1 &= z_1.(z_2 + \alpha_1 + \zeta_2 + \lambda_2.\hat{\theta}_1) + z_1.\varepsilon_2 + \tilde{\theta}_1\left(z_1.\lambda_2 - \frac{1}{g_1}\dot{\hat{\theta}}_1\right) \\ &\quad + \tilde{\theta}_u^T\left(z_1.v_2^T - \frac{1}{g_u}\dot{\hat{\theta}}_u\right) - \frac{1}{d_1}\varepsilon^T.\varepsilon \end{aligned} \quad (21)$$

Choosing the virtual control to add a stability term, we cancel all the known terms except the z_2 term and we dampen out the unknown observer error:

$$\alpha_1 = -c_1 z_1 - d_1 z_1 - (\zeta_2 + \lambda_2.\hat{\theta}_1) \quad (22)$$

This leads to the Lyapunov function derivative:

$$\begin{aligned}\dot{V}_1 &= -c_1 z_1^2 + z_1.z_2 - d_1\left(z_1 - \frac{1}{2.d_1}\varepsilon_2\right)^2 + \frac{1}{4.d_1}\varepsilon_2^2 \\ &\quad - \frac{1}{d_1}\varepsilon^T.\varepsilon + \tilde{\theta}_1\left(z_1.\lambda_2 - \frac{1}{g_1}\dot{\hat{\theta}}_1\right) + \tilde{\theta}_u^T\left(z_1.v_2^T - \frac{1}{g_u}\dot{\hat{\theta}}_u\right) \\ &\leq -c_1 z_1^2 + z_1.z_2 - \frac{3}{4.d_1}\varepsilon^T.\varepsilon + \\ &\quad \tilde{\theta}_1\left(z_1.\lambda_2 - \frac{1}{g_1}\dot{\hat{\theta}}_1\right) + \tilde{\theta}_u^T\left(z_1.v_2^T - \frac{1}{g_u}\dot{\hat{\theta}}_u\right) \end{aligned} \quad (23)$$

> Step 2

In the second step, the first Lyapunov function is augmented with a z_2 term along with an additional observer error term, which gives us:

$$V_2 = V_1 + \frac{1}{2}z_2^2 + \frac{1}{d_2}\varepsilon^T.P.\varepsilon \quad (24)$$

Its derivative is written like this way:

$$\dot{V}_2 = \dot{V}_1 + z_2.\dot{z}_2 - \frac{1}{d_2}\varepsilon^T.\varepsilon \quad (25)$$

Differentiating V_2, we get:

$$\begin{aligned}\dot{V}_2 &\leq -c_1 z_1^2 + z_2(z_1 + \dot{z}_2) - \frac{3}{4.d_1}\varepsilon^T.\varepsilon - \frac{1}{d_2}\varepsilon^T.\varepsilon \\ &\quad + \tilde{\theta}_1\left(z_1.\lambda_2 - \frac{1}{g_1}\dot{\hat{\theta}}_1\right) + \tilde{\theta}_u^T\left(z_1.v_2^T - \frac{1}{g_u}\dot{\hat{\theta}}_u\right) \end{aligned} \quad (26)$$

For compactness of notation, we define $c_1^* = c_1 + d_1$ and $u = \alpha_2/\hat{\theta}_u$, then, we can develop the term $(z_1 + \dot{z}_2)$ by using the following expression:

$$\begin{aligned}(z_1 + \dot{z}_2) &= \alpha_2 + z_1 - k_2.(\zeta_1 + \lambda_1\hat{\theta}_1 + v_1.\hat{\theta}_u) \\ &\quad + c_1^*.(\zeta_2 + \lambda_2\hat{\theta}_1 + v_2.\hat{\theta}_u) + c_1^*.(\lambda_2\tilde{\theta}_1 + v_2.\tilde{\theta}_u) \\ &\quad - c_1^*.\dot{y}_r + k_2 y - \ddot{y}_r + c_1^*.\varepsilon_2 + \lambda_2\dot{\hat{\theta}}_1 + v_2.\dot{\hat{\theta}}_u + \varphi(y).\hat{\theta}_1 \end{aligned} \quad (27)$$

By defining the control u, such that:

$$\begin{aligned}\alpha_2 &= -c_2.z_2 - d_2.(c_1^*)^2.z_2 - \{z_1 - k_2.(\zeta_1 + \lambda_1\hat{\theta}_1 + v_1.\hat{\theta}_u) \\ \alpha_2 &= +c_1^*.(\zeta_2 + \lambda_2\hat{\theta}_1 + v_2.\hat{\theta}_u) \\ &\quad - c_1^*.\dot{y}_r + k_2 y - \ddot{y}_r + \varphi(y).\hat{\theta}_1 + \lambda_2.g_1.\tau_1 + v_2.g_u.\tau_u\} \end{aligned} \quad (28)$$

Where τ_l and τ_u will be defined later, we obtain:

$$(z_1 + \dot{z}_2) = -c_2.z_2 - d_2.(c_1^*)^2.z_2 + c_1^*.(\lambda_2\tilde{\theta}_l + v_2.\tilde{\theta}_u + \varepsilon_2) + \lambda_2\dot{\hat{\theta}}_l + v_2.\dot{\hat{\theta}}_u - (\lambda_2.g_l.\tau_l + v_2.g_u.\tau_u)$$

(29)

and expression (26) will have the following structure:

$$\dot{V}_2 \leq -c_1 z_1^2 - c_2 z_2^2 - \frac{3}{4.d_1}\varepsilon^T.\varepsilon - \frac{3}{4.d_2}\varepsilon^T.\varepsilon - \lambda_2.z_2 g_l\left(\tau_l - \frac{1}{g_l}\dot{\hat{\theta}}_l\right)$$
$$+ \tilde{\theta}_l\left(\tau_l - \frac{1}{g_l}\dot{\hat{\theta}}_l\right) - v_2.z_2 g_u\left(\tau_u - \frac{1}{g_u}\dot{\hat{\theta}}_u\right) + \tilde{\theta}_u^T\left(\tau_u - \frac{1}{g_u}\dot{\hat{\theta}}_u\right)$$
$$\leq -c_1 z_1^2 - c_2 z_2^2 - \frac{3}{4.d_1}\varepsilon^T.\varepsilon - \frac{3}{4.d_2}\varepsilon^T.\varepsilon$$
$$+ \left(-\lambda_2.z_2 g_l + \tilde{\theta}_l\right)\left(\tau_l - \frac{1}{g_l}\dot{\hat{\theta}}_l\right) + \left(-v_2.z_2 g_u + \tilde{\theta}_u^T\right)\left(\tau_u - \frac{1}{g_u}\dot{\hat{\theta}}_u\right)$$

(30)

where:

$$\tau_l = (c_1^* z_2 + z_1)\lambda_2$$
$$\tau_u = (c_1^* z_2 + z_1)v_2^T$$

(31)

Restating the definition of the control, and defining the update laws:

$$u = \frac{\alpha_2}{\hat{\theta}_u} = \frac{1}{\hat{\theta}_u}\left[-c_2.z_2 - d_2.(c_1^*)^2.z_2\right.$$
$$-\{z_1 - k_2.(\zeta_1 + \lambda_1\hat{\theta}_l + v_1.\hat{\theta}_u) + c_1^*.(\zeta_2 + \lambda_2\hat{\theta}_l + v_2.\hat{\theta}_u)$$
$$\left. - c_1^*.\dot{y}_r + k_2 y - \ddot{y}_r + \varphi(y).\hat{\theta}_l + \lambda_2.g_l.\tau_l + v_2.g_u.\tau_u\}\right]$$

(32)

$$\dot{\hat{\theta}}_l = g_l\tau_l = g_l.(c_1^* z_2 + z_1)\lambda_2$$
$$\dot{\hat{\theta}}_u = g_u\tau_u = g_u.(c_1^* z_2 + z_1)v_2^T$$

(33)

The final derivative of Lyapunov function is given by:

$$\dot{V}_2 \leq -\sum_{j=1}^{2} c_j z_j^2 - \sum_{i=1}^{2}\frac{3}{4.d_i}\varepsilon^T.\varepsilon$$

(34)

Based on the Lyapunov function $V = V_2$, we have shown that $\dot{V} < 0$, $\forall (z, \varepsilon) \neq 0$, implying asymptotic stability of the system and the observer.

4. SIMULATION RESULTS

4.1 Regulation: $yr = 1$

In this observer-based experiment, the reference of real rod angle is equal to 0.7 rad. The selected parameters of the pendulum are: $m=10$ kg, $l=1$m and $g=10$; $\theta_l = -10$; $\theta_u = 10$; $\hat{\theta}_{l0} = -9,3$; $\hat{\theta}_{u0} = 9,5$.

Table 1. Design Parameters for Observer-Based Controller

Description	value
Lyapunov gain	$c_3 = c_4 = 9$
Observer gain	$k_1 = 30$; $k_2 = 300$
Damping gain	$d_1 = d_2 = 1$
Spline parameter adaptation gain	$g_l = 700$
Load parameter adaptation gain	$g_u = 11$

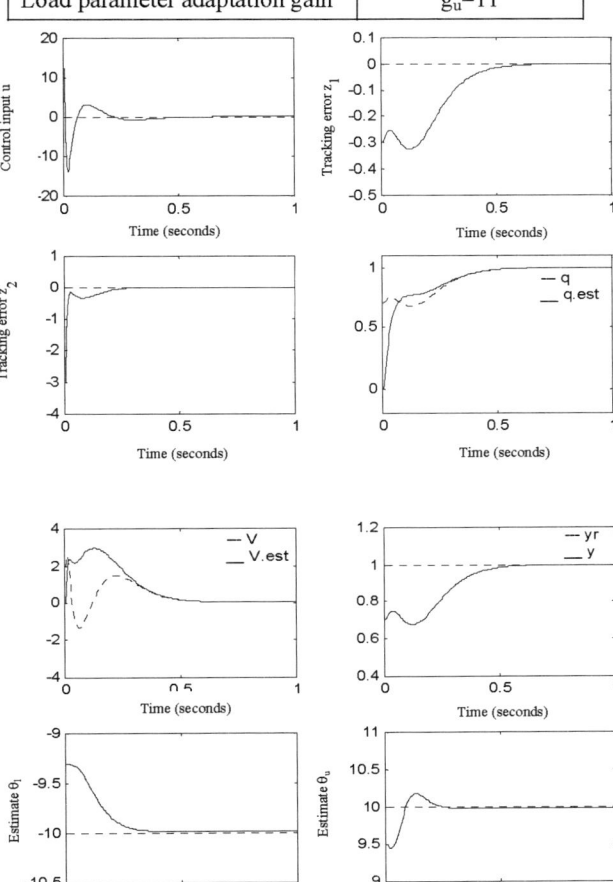

Fig. 5. Results of the observer adaptive backstepping control for a simple pendulum "constant input"

It is clearly noticed that the error tends towards zero and the state x_1 follows perfectly the trajectory of reference y_r. This means that this technique of control gives a good response by a good choice of the constants c3 and c4.

4.2 Pursuit : $yr = sin(2\pi t)$

In this case, we will maintain the same parameters and "the true" values and estimated values.

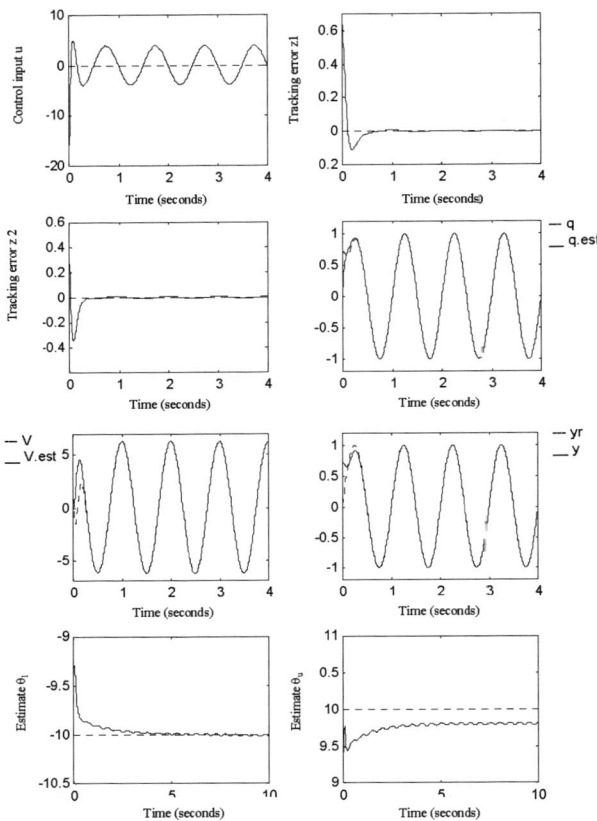

Fig. 6. Results of the observer adaptive backstepping control for a simple pendulum "variable input"

5. CONCLUSIONS

In this paper, observer-based adaptive backstepping is synthesized, using the simple pendulum as an example.

The adaptive backstepping control is studied by introducing the observer approach. The application of this technique for the nonlinear systems gives good results. The convergence and the total stability of the controlled system well express the advantage of this strategy of control.

The backstepping control with an observer was also studied to make the classical backstepping control more robust by solving the problem of measurement states.

In this case, the technique of control presented rests on the use of an observer by supposing that the states of the system are not all measurable. To carry out the objectives of continuation and regulation, a choice of an observer was used. But, it is valid only for the systems which can be represented in triangular form. This observer makes it possible to carry out the continuation as well as the regulation in a perfect way.

6. REFERENCES

[1] R. Milman, *Adaptive Backstepping Control of the Variable Reluctance Motor*, thesis, Department of Electrical and Computer Engineering, University of Toronto, 1997.

[2] B. Yaon, *Adaptive Robust Control of Nonlinear Systems with Application to Control of Mechanical Systems*, thesis, University of California, New York, 1996.

[3] M. Kristić, I. Kanellakopoulos, and P.V. Kokotović, *Nonlinear and Adaptive Control Design*, John Willey and Sons, Inc., New York, NY, 1995.

[4] A. Benaskeur and A. Desbiens, "Application of backstepping to the stabilization of the Inverted Pendulum," *Electrical and Computer Engineering IEEE, Canadian Conference*, vol. 1, May 24-28, 1998, pp. 113-116., 1998.

SESSION B2: INTELLIGENT CONTROL

Usage of the HMM-Based Speech Synthesis for Intelligent Arabic Voice

Tamer S. Fares[1], Awad H. Khalil[2] and Abd El-Fatah A. Hegazy[3]

1 Department of Computer Science, Modern Academy In Maadi, Cairo, Egypt
tamerfares@hotmail.com
2 Department of Computer Science & Engineering, The American University in Cairo, Egypt
akhalil@aucegypt.edu
3 Department of Computer Science, Arab Academy for Science, Technology & Maritime Transport in Cairo, Egypt
hegazy@aast.edu

Abstract: The HMM as a suitable model for time sequence modeling is used for estimation of speech synthesis parameters, A speech parameter sequence is generated from HMMs themselves whose observation vectors consists of spectral parameter vector and its dynamic feature vectors. HMMs generate cepstral coefficients and pitch parameter which are then fed to speech synthesis filter named Mel Log Spectral Approximation (MLSA), this paper explains how this approach can be applied to the Arabic language to produce intelligent Arabic speech synthesis using the HMM-Based Speech Synthesis and the influence of using of the dynamic features and the increasing of the number of mixture components on the quality enhancement of the Arabic speech synthesized.

1- INTRODUCTION

Speech synthesis is the process of converting a text into speech. The Simplest way to produce synthetic speech is to play long prerecorded samples of words or sentences. Diphone concatenation is a kind of synthesis for the Arabic language that has proved its effectiveness [1]

The HMM-based approach of speech synthesis consists of two phases as shown in Fig (1); the training stage and the synthesis stage. First in the training phase, Mel-cepstral coefficients are obtained from speech database by mel-cepstral analysis [3].Dynamic features (i.e. delta and delta-delta) and mel-cepstral coefficients are calculated then the Arabic phoneme HMMs are trained using these features.

In the synthesis phase, an arbitrary Arabic text to be synthesized is transformed into Arabic phoneme sequence. According to this phoneme sequence, a sentence HMM which represents the whole text to be synthesized is constructed by concatenating phoneme HMMs.From the sentence HMM, a speech parameter sequence is generated using the algorithm for speech parameter generation from the HMM. By using the MLSA filter [4] [5], speech is synthesized from the generated mel-cepstral coefficients.

The paper organization is composed of six sections. The next sections discuss the Arabic language and its phonemes characterization. Section 3 explains all the procedures carried out by the synthesizer engine from the database training to the synthesis of a given utterance. Section 4 describes the contextual clustering and how we use the Arabic phonemes to construct decision tree.

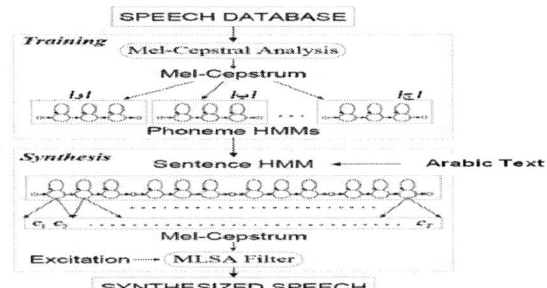

Fig (1) Functional Scheme of an HMM-Based speech synthesis Approach

Section 5, two subjective tests are presented: the effect of using the dynamic features and increasing the number of mixture components in the enhancement of the quality of the Arabic voice, finally last section summarizes the result.

2- THE ARABIC LANGUAGE [17]

The Arabic language is the liturgical language of many Arabic peoples. Arabic either refers to the standard Arabic or to the many dialectal variations of Arabic. The Arabic language consists of 28 consonant letters. There are 25 consonants, 3 long vowels (الألف-الياء -الواو). There are 3 short vowels (الفتحة-الضمة-الكسرة)called diacritization marks are written above or below the consonant that precedes them, and two can be semi-vowels (الياء-الواو) [1].

2.1 The Arabic Sounds

The Arabian phonemes sound is determined by the International Phonetic Alphabet (IPA) revised to year

2005 for Arabic consonants. Table 1 shows the classification of the Arabic phonemes consonant

Table (1) International Phonetic Alphabet

		Bilabial	Labiodental	Dental	AlveoDental	Alveopalatal	Palatal	Velar	Lab-Velar	Uvular	Pharyngeal	Glottal
Nasal	Voiced	م			ن							
	Unvoiced											
Plosive	Voiced	ب			د							
	Unvoiced				ت			ك		ق		ء
Emphatic Stop	Voiced				ض							
	Unvoiced				ط							
Fricative	Voiced			ذ	ز					غ	ع	
	Unvoiced		ف	ث	س	ش				خ	ح	ه
Emphatic Fricative	Voiced				ظ							
	Unvoiced				ص							
Africate	Voiced					ج						
	Unvoiced											
Glide	Voiced						ي		و			
	Unvoiced											
Lateral	Voiced				ل							
	Unvoiced											
Trill	Voiced				ر							
	Unvoiced											

2.2 Transcribing Arabic Phonemes

Transliterating Arabic Sounds into English writing system has always been problematic. This stems from the fact that Arabic has nine consonant sounds that are not found in English. The International phonetics Association (IPA) has a comprehensive transcription scheme that explains the way on how to transcribe every sound in the modern language. This paper proposes a new transcription for the Arabic phonemes that try to reflect the real pronunciation of all the Arabic phonemes sounds. Table 2 show the proposed transcription scheme for Arabic phonemes consonant. The Diacritization mark SHADA ˝ over any letter mean the pronunciation repetition of this letter. The SHADA may be accompanied by the TANWIN.The TANWIN mark can be TANWIN FATH, TANWIN DAM or TANWIN KASR [1], this is pronounced as it is a 2 successive phonemes (diphone).The TANWIN FATH (FATHA followed by NOUN), the TANWIN DAM (DAMA Followed by NOUN) and for the TANWIN KASR (KASRA followed by NOUN).Table 3 shows the proposed transcription scheme for different TANWIN marks

Table (2) Proposed Transcription Scheme for Consonant and Vowels

Phonemes In Arabic	Proposed Scheme	Consonant	
		ع	o
		غ	gh
ب	b	ف	f
ت	t	ق	q
ث	th	ك	k
ج	g	ل	l
ح	ha	م	m
خ	kh	ن	n
د	d	ه	h
ذ	dh	ء	ca
ر	r	**Long Vowels**	
ز	z	ا	aa
س	s	و	w
ش	sh	ي	y
ص	sa	**Short Vowels**	
ض	da	الفتحة	a
ط	ta	الضمة	u
ظ	za	الكسرة	i

Table (3) Proposed Transcription Scheme for TANWIN

TANWIN	Proposed Scheme
ً	an
ٌ	on
ٍ	in

3- ENGINE DESCRIPTION

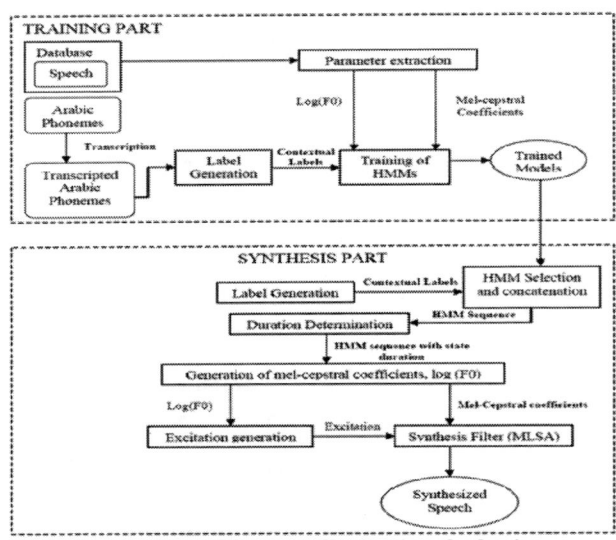

Fig (2): Block Diagram illustrating the basic procedures conducted by the speech synthesis engine

3.1 Training Part

The Training part in the speech synthesis is similar to that used in speech recognition. The main difference is both spectrum and excitation parameters are extracted from a speech database and modeled by context-dependent

HMMs which are represented in the cepstral coefficients and their dynamic features and in the log F0 and its dynamic features respectively [7].Fig (2) describes the engine for synthesis Arabic voice

3.1.1 Speech Parameter Extraction

a. Fundamental Frequency

A sequence of fundamental frequency logarithms $\{\log(F0^1), \ldots, \log(F0^N)\}$, including voicing decision information (If F0=0 the frame is considered unvoiced), where N is the total number of frames of all the utterances from the training database, is extracted in a short time basis.

b. Mel-Cepstral Coefficients

Sequence of mel-cepstral coefficient vectors which represent speech envelope spectra [22], $\{c^1, \ldots, c^N\}$, is obtained. Each mel-cepstral coefficient vector $c^i=[c^i_0 \ldots c^i_M]^T$, (where i indicated the frame number),is derived through an M-th order mel-cepstral analysis, taking into account the already extracted sequence of log(F0) in order to remove signal periodicity. The procedures of spectral analysis smoothed by fundamental frequencies are performed as described in [21].

3.1.2 Label Generation

For each speech utterance a monophone sequences was defined which are tied together to form a fullcontext which describe the corresponding transcription label for each speech utterance. Table 4 & 5 describe the label format and give example for the label generation of Arabic Sentences respectively.

Table (4) Contextual Format

m1	Phone before previous phone
m2	Previous phone
m3	Current phone
m4	Next phone
m5	Phone after next phone

Table (5) Example of Monophone and Fullcontext Generation

Example	Monophone Sequence	FullContext
مشهورة	m	x ^ x - m + a = sh
	a	x ^ m - a + sh = h
	sh	m ^ a - sh + h = w
	h	a ^ sh - h + w = d
	w	sh ^ h - w + d = on
	d	h ^ w - d + on = x
	on	w ^ d - on + x = x

Last tables present the contextual label format. This format studies the contextual factors. These factors might include context dependent terms, such as preceding/succeeding phone, syllable, and phrase. For each phone from the speech database there is a corresponding contextual label which includes all its related features (i.e. nature of each phoneme, the output pronunciation of each phoneme (Nasal, Larynx, trill), and the stress related factor) .Thus, there is a wide range of different contextual labels which can result during the training stage of the synthesizer. It is impossible to capture all these effects in the contextual factors, their possible combinations together for studying will increase exponentially. Section 4 describes how this problem can be solved [2].

3.1.3 HMM Training [11]

Each HMM corresponds to a no-skip S-State left-to-right model, with S=5.Each output observation vector o^i for the i-th frame consists of 4 streams, $o^i = [o_1^{iT} \cdots o_4^{iT}]^T$ as illustrated in Fig (3), where :

- Stream 1 (o^i_1) : vector composed of mel-cepstral coefficients $\{c^i_0, \ldots, c^i_M\}$, their corresponding delta $\{\Delta c^i_0, \ldots, \Delta c^i_M\}$, and delta-delta $\{\Delta^2 c^i_0, \ldots, \Delta^2 c^i_M\}$.
- Stream 2,3,4 (o^i_2, o^i_3, o^i_4) : composed respectively of fundamental frequency logarithm, $\log(F0^i)$, its corresponding delta, $\Delta\log(F0^i)$, and delta-delta, $\Delta^2\log(F0^i)$

The observation vector o^i is output by an HMM state s according to a probability distribution given by

$$\beta_s(o^i) = \prod_{j=1}^{4} \left[\sum_{l=1}^{R_j} \omega_{sjl} N(o^i_j; \mu_{sjl}, \Sigma_{sjl}) \right]^{\gamma_j} \quad (1)$$

Where $N(.;\mu,\Sigma)$ means Gaussian distribution with mean μ and variance Σ, ω_{sjl} is the weight for l-th mixture component of the j-th stream vector o^i_j output by state s, and γ_j is the output distribution weight for the j-th stream , with R_j being the corresponding number of mixture components. The first stream vector, $o^i_1 = [c^{iT} \Delta c^{iT} \Delta^2 c^{iT}]^T$ is modeled by single-mixture continuous Gaussian distributions where its dimensionality is $3(M+1)$.For the second, third and fourth scalar streams, $o^i_2 = \log(F0^i)$, $o^i_3 = \Delta\log(F0^i)$ and $o^i_4 = \Delta^2\log(F0^i)$, the output probability is modeled by multi-space Gaussian distributions with two mixture components [10].

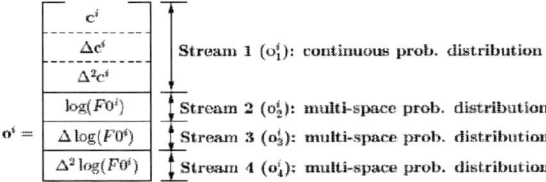

Fig (3): Structure of each feature vector o^i [11]

For each HMM k, the durations of the S states are considered as vectors, $d^k = [d_1^k ... d_S^k]^T$, where d_s^k represents the duration of the s-th state. Further, each of the duration vectors, $\{d^1 ... d^K\}$, where K is the total number of HMMs representing the database, is modeled by an S-dimensional single-mixture Gaussian distribution. The output probabilities of the state duration vectors are thus re-estimated by Baum-Welch iterations in the same way as the output of the speech parameters [19].

3.2 Synthesis Part [11]

In the synthesis part, the utterance to be synthesized is converted to a context-dependent label sequence and the utterance HMM is constructing by concatenating the context-dependent HMMs according to this label sequence. Diagram that explains the synthesis phase is shown in Fig 4.

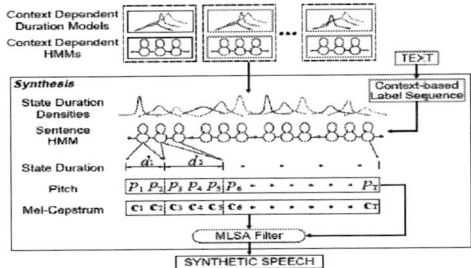

Fig (4): Synthesis Part [6]

3.2.1 Label Generation and HMM Selection / Concatenation

The Synthesis procedure starts with the conversion of a given sentences into contextual labels, which are eventually used to select corresponding leaves from each one of the $2S+1$ decision-trees generated by the context-clustering procedure in the training stage. In the end of this step, 3 logical HMM sequences, whose states correspond to the selected leaves, are derived for each of the following parameters: (1) mel-cepstral coefficients; (2) logarithms of fundamental frequencies; and (3) state durations. The sequences for item (1) and (2) consist of HMMs with S states, whereas the sequence for state durations comprises single-state HMMs.

3.2.2 Parameter Determination

The three above mentioned HMM sequences are then used to derive mel-cepstral coefficients and log($F0$). The whole procedure is conducted as follows. Initially, the duration vectors $\{d^1, ..., d^K\}$, where K is the number of HMMs in each sequence, are determined from the K S-dimensional Gaussian distributions, defining the state sequence $s = \{s_1, ..., s_L\}$, with L being the number of frames of the utterance to be synthesized and s_i the HMM state wherein the i-th frame belong to [12]. After that, mel-cepstral coefficient vectors, $\{c_1, ..., c_L\}$ and logarithms of the fundamental frequencies $\{\log(F0^1), ..., \log(F0^L)\}$ are determined from each corresponding HMM sequence in a way to maximize their output probability given s, taking into account the delta and delta-delta components, according to the algorithm described in [9].

3.2.3 Excitation Construction and Filtering

The last step of the synthesis process is divided into two parts. In the first one an excitation signal is derived from the sequences of generated fundamental frequency logarithms, $\{\log(F0^1), ..., \log(F0^L)\}$ using the same approach described in the high-quality vocoding method of [21], which is based on mixed excitation construction according to frequency subband strengths. In the second part, speech waveform is generated with the utilization of the Mel Log Spectrum Approximation (MLSA) filter [7], whose corresponding coefficients are derived from the sequence of generated mel-cepstral coefficients, $\{c_1, ..., c_L\}$.

4- CONTEXTUAL CLUSTERING

To overcome the problem that the paper mentioned in the Section 3.1.2, a decision-tree based context clustering technique [13][14] is applied to distributions for spectrum, F_0, and state duration in the same manner as HMM-based speech recognition. The decision-tree based context clustering algorithm have been extended for MSD-HMMs in [15]. Since each of spectrum, F_0 and duration has its own influential contextual factors, they are clustered independently (Fig 5). State durations of each HMM are modeled by a n-dimensional Gaussian, and context-dependent n-dimentional Gaussians are clustered by a decision tree[8]. Note that spectrum and F_0 part of the state output vector are modeled by multivariate Gaussian distributions and multi-space probability distributions, respectively [2].

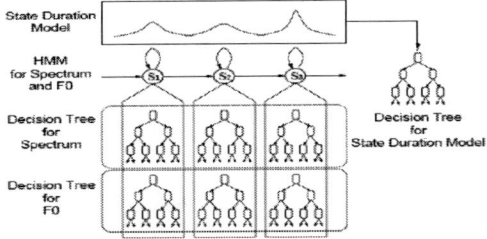

Fig (5): Decision Trees for context clustering [2][6]

The determination of the questions for context-clustering represents an important issue in order to achieve synthesized Arabic speech with good quality.

Questions based on phonetic/phonemic characteristics: For the phones, the questions are based on phonetic/phonemic characteristics of Arabic Language.

- Consonant (ء-ب-ت-ث-ج-ح-خ-د-ذ-ر-ز-س-ش-ص-ض-ط-ظ-ع-غ-ف-ق-ك-ل-م-ن-ه-)
- Vowels (ي- و- ا- الفتحة – الضمة -الكسرة)
- Short Vowels (الفتحة – الضمة - الكسرة)
- Voiced Consonant (ن- م- د- ب- ض- ز- ذ- ص – ظ – ج- ل- ر)

Consonant: Stop, fricative, bilabial, labiodentals, dental, alveolar, palatal, velar, unvoiced, voiced and nasal. According to this classification, some examples of questions are listed below:

- Is the current phone a voiced fricative?
- Is the next phone a unvoiced fricative?
- Is preceding phone a nasal?

5- EXPERIMENTATION

Arabic independent sentences in [16] are used because contain rich and balanced words. Database consists of 367 sentences; 2 to 9 words per sentence. The statistical results show that the database contains 1835 words. The sentences were recorded by a fluent Arabic speaker and read with the complete diacritization. Speech signals were sampled at a rate of 16 kHz and windowed by a 25ms Blackman window with a 5ms shift. The Speech samples used in the test were synthesized using spectral sequences generated without dynamic features from models trained using dynamic and static features.

Speech samples used in the test were synthesized using spectral sequences generated without dynamic features from models trained using dynamic and static features. Assuming that both state and mixture sequence is observable (state 1 in [9]), Fig 6 & Fig 8 shows generated spectra for a Arabic sentence fragment " هَذَ ا المَعهَدُ " from the sentence " هَذَا المَعهَدُ مَشهُودٌ لَهُ بِالنَّجَاح " without and with the dynamic features respectively.

Fig 7 compares two spectra obtained from single-mixture HMMs and 16-mixture HMMs, without dynamic features.

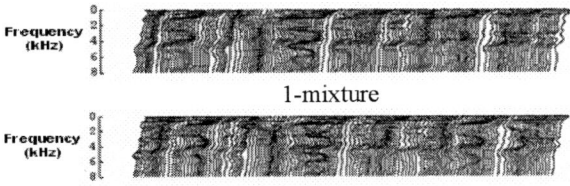

1-mixture

16-mixture

Fig (6) Generated spectra for a sentence fragment " هَذَ ا المَعهَدُ " without the dynamic Features

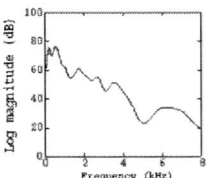 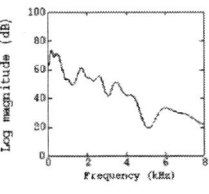

1-mix   16-mix

Fig (7) Generated spectra for a sentence fragment " هَذَ ا المَعهَدُ " without the dynamic Features

Fig 9 compares two spectra obtained from single-mixture HMMs and 16-mixture HMMs, with the using of dynamic features, for the same temporal position of the sentence fragment.

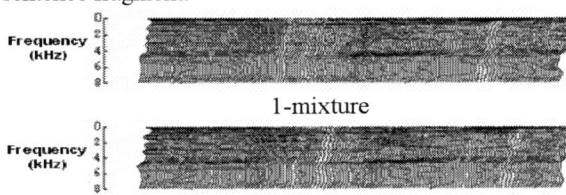

1-mixture

16-mixture

Fig (8): Generated spectra for a sentence fragment " هَذَ ا المَعهَدُ " with the dynamic features

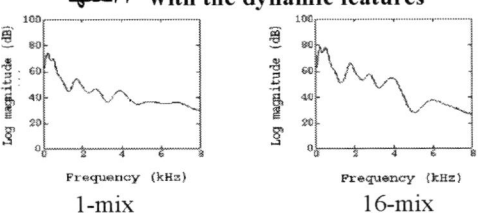

1-mix   16-mix

Fig (9): Spectra obtained from 1-mixture HMMs and 16-mixture HMMs with dynamic features

In the parameter generation algorithm [9], without dynamic features, a speech parameter sequence that maximize $P(O|Q, \lambda, T)$ becomes a sequence of mean vectors. As a result, discontinuities occur in the generated spectral sequence at transitions of states (Fig 6). It is seen from Fig 8 & Fig 9 that with increasing mixtures and the usage of the dynamic features, the formant structure of the generated spectra get clearer because the formant structure corresponding to each mean vector $\mu_{q,i}$ might be vague since $\mu_{q,i}$ is the average of different speech spectra. From informal listening of the synthetic speech, it has been observed that the quality of the Arabic synthetic speech is considerably improved by increasing mixtures and the usage of dynamic features.

6- CONCLUSION

This paper has described the HMM-based speech synthesis system, and how can be applied to the Arabic language to synthesis intelligent Arabic voice. The system is based on a technique wherein the speech waveform is generated from parameters directly derived from HMMs. The main advantages of this approach, when compared with the other techniques, corresponds to the possibility of obtaining synthesized Arabic speech with good quality and the modification of voice characteristics styles, even from relatively small speech databases. Experimental results have shown that we can reproduce clear formant structure from multi-mixture HMMs as compared with that produced from single-mixture HMMs and demonstrate the effectiveness of the use of dynamic features to produce Arabic intelligent sounding speech [17].

7- ACKNOWLEDGEMENT

The authors would like to thank Dr Heiga Zen (Dept of Comp. Science and Engineering Nagoya Institute of Technology, Japan) for his support which resulted in a significant improvement of this paper.

8- REFERENCES

[1] T. S. Fares, A.F. Hegzay, A. H. Khalil "Investigating An Arabic Text To Speech System Based On Diphone Concatenation".IJICIS,Vol 7, No. 1,pp.49-69 January 2007

[2] Keiichi Takuda , Heiga Zen , Alan W.Black AN HMM-BASED SPEECH SYNTHESIS SYSTEM APPLIED TO ENGLISH . IEEE TTS Workshop 2002. Santa Monica. California, USA.

[3] K.Takuda, T. Kobayashi, T.Fukada , H. Saito and S.Imai, "Spectral estimation of speech based on mel-cepstral representation," IEICE Trans. A, vol. J74-A, no.8 , pp.1240-1248, Aug. 1991 (in Japanese).

[4] S. Imai, K. Sumita, and C. Furuichi, "Mel log spectrum approximation (MLSA) filter for speech synthesis," IECE Trans. A, vol.J66-A, no.2, pp.122–129, Feb. 1983 (in Japanese).

[5] T. Fukada, K. Tokuda, T. Kobayashi, and S. Imai, "An adaptive algorithm for mel-cepstral analysis of speech," Proc. ICASSP-92, pp.137–140, Mar. 1992

[6] [24] T. Yoshimura, K. Tokuda, T. Masuko, T. Kobayashi, and T. Kitamura, "Simultaneous modeling of spectrum, pitch and duration in HMM-based speech synthesis," IEICE Trans. D-II, vol.J83-D-II, no.11, pp.2099–2107,Nov. 2000 (in Japanese).

[7] T. Fukada, K. Tokuda, Kobayashi T., and S. Imai, "An adaptive algorithm for mel-cepstral analysis of speech," in ICASSP, 1992, pp. 137–140.

[8] S.J. Young, and P. C. Woodland, "State Clustering in Hidden Markov model-based continuous speech recognition ," Computer Speech and Language, vol.5, no.3, pp.369-383, 1994

[9] K. Tokuda, T. Yoshimura, T. Masuko, T. Kobayashi, and T.Kitamura,"Speech parameter generation algorithms for HMM-based speech synthesis," in ICASSP, 2000, pp. 1315–1318.

[10] K. Tokuda, T. Masuko, N. Miyazaki and T. Kobayashi, "Hidden Markov Models Based on Multi-Space Probability Distribution for Pitch Pattern Modeling," Proc. of ICASSP, 1999.

[11]R.Maia, H.Zen, K.Tokuda, T.Kitamura, F.G.V Resende Jr. "AN HMM-BASED BRAZALIAN PORTUGUESE SPEECH SYNTHESIZER AND ITS CHARACTERITICS" Revista da Sociedade Brasileira de Telecomunicacoes, Vol 21,No.2,pp.58-71 August 2006

[12] T. Yoshimura, K. Tokuda, T. Masuko, T. Kobayashi and T. Kitamura, "Duration Modeling in HMM-based Speech Synthesis System," Proc. of ICSLP, vol.2, pp.29–32, 1998.

[13] J. J. Odell, "The Use of Context in Large Vocabulary Speech Recognition," PhD dissertation, Cambridge University, 1995.

[14] K. Shinoda and T. Watanabe, "MDL-based context-dependent subword modeling for speech recognition," J. Acoust. Soc. Jpn.(E), vol.21, no.2, pp.79–86, 2000.

[15] T. Yoshimura, "Simultaneous modeling of phonetic and prosodic parameters, and characteristic conversion for HMM-based Text-To-Speech systems" PhD dissertation, Nagoya Institute of Technology,2002.

[16] Alghamdi, Mansour, Abdulaziz Alhumayid and Muneer ad-Dusooqee (2003) Arabic Sound Database: Sentences, Computer and Electronics Research Institute (HK-28), King Abdulaziz City for Science and Technology, Riyadh (in Arabic).

[17] T.S.Fares, A. H. Khalil, A.F.Hegazy "AN HMM-BASED SPEECH SYNTHESIS SYSTEM APPLIED TO ARABIC" IJICIS,Vol 8, July 2007

[18] R. Maia, H. Zen, K. Tokuda, T. Kitamura, and F. G. Resende, "Towards the development of a Brazilian Portuguese text-to-speech system based on HMM," in Proc. of the European Conf. on Speech Communication and Technology (EUROSPEECH),2003.

[19] T. Yoshimura, K. Tokuda, T. Masuko, T. Kobayashi, T. Kitamura,"Mixed-excitation for HMM-based speech synthesis," in Proc. of the European Conf. on Speech Communication and Technology (EUROSPEECH), 2001.

[20] H. Zen and T. Toda, "An overview of Nitech HMM-based speech synthesis for Blizzard Challenge 2005," in Proc. Of the European Conf. on Speech Communication and Technology (EUROSPEECH), 2005.

[21] H.Kawahara,I.Masuda-Katsuse, &A. de Cheveign´e, "Restructuring speech representations using a pitch-adaptive time frequency smoothing and an instantaneous-frequency-based F0 extraction, vol. 27, no. 3-4, pp. 187–207, Apr.1999.

[22] T. Fukada, K. Tokuda, T. Kobayashi, and S. Imai, "An adaptive algorithm for mel-cepstral analysis of speech," in Proc. of the IEEE Int. Conf. on Acoustics, Speech, and Signal Processing (ICASSP), 1992.

Hybrid Takagi-Sugeno Fuzzy FED PID Control of Nonlinear Systems

BASIL HAMED and AHMAD EL KHATEB

Electrical and Computer Engineering Department, Islamic University of Gaza
P.O.BOX 108, Gaza, Palestine
(Emails: bhamed@iugaza.edu; akhateb@iugaza.edu)

Abstract: The new method of proportional–integral-derivative (PID) controller is proposed in this paper for a hybrid fuzzy PID controller for nonlinear system. The important feature of the proposed approach is that it combines the fuzzy gain scheduling method and a fuzzy fed PID controller to solve the nonlinear control problem. The resultant fuzzy rule base of the proposed controller contains one part. This single part of the rules uses the Takagi-Sugeno method for solving the nonlinear problem. The simulation results of a nonlinear system show that the performance of a fed PID Hybrid Takagi-Sugeno fuzzy controller is better than that of the conventional fuzzy PID controller or Hybrid Mamdani fuzzy FED PID controller.

Keywords: Fuzzy gain scheduling, fuzzy fed proportional–integral-derivative (PID), Takagi-Sugeno, nonlinear system.

I. INTRODUCTION

The PID control is widely used in industrial applications although it is a simple control method. Stability of PID controller can be guaranteed theoretically, and zero steady-state tracking error can be achieved for linear plant in steady-state phase. Computer simulations of PID control algorithm have revealed that the tracking error is quite often oscillatory, however, with large amplitudes during the transient phase.

To improve the performance of the PID controllers, several strategies have been proposed, such as adaptive and supervising techniques.

Fuzzy control methodology is considered as an effective method to deal with disturbances and uncertainties in terms of ignorance and ambiguity. Fuzzy PID controller combining fuzzy technology with traditional PID control algorithm has become the most effective domain in artificial intelligence control [1],[15],

The most common problem resulted early depending on the complexity of FLC is the tuning problem. It is hard to design and tune FLCs manually for the most machine problems especially used in industries like nonlinear systems. For alleviation of difficulties in constructing the fuzzy rule base, there is the conventional nonlinear design method [2] which was inherited in the fuzzy control area, such as fuzzy sliding control, fuzzy scheduling [3],[8], and adaptive fuzzy control [4],[14]. The error signal for most control systems is available to the controller if the reference input is continuous. The analytical calculations present two-inputs FLC employing proportional error signal and velocity error signal. PID controller is the most common controller used in industries, most of development of fuzzy controllers revolve around fuzzy PID controllers to insure the existence of conventional controllers in the overall control structure, simply called Hybrid Fuzzy Controllers [5],[6].

The key idea of the proposed method is as follows: First, the fuzzy gain scheduling method is applied to linearize the nonlinear system at frozen times. A fuzzy fed PID controller is designed for each frozen system by replacing the conventional PID controller by an incremental FLC, the integral part of the PID controller is fed by a deferential feedback gain, this fed PID controller is the new method used in this paper and it gives the best results any way. Second, fuzzification of the reference input is performed for the system, while the control space of error signals is linearly partitioned after normalization. Third, the fuzzy rule base is constructed in a recursive way to obtain better nonlinear control as well as to guarantee closed-loop stability of the frozen system.

It should be emphasized that because the proposed approach utilizes some modern control theorems, tuning the hybrid fuzzy controller is much easier than tuning a conventional fuzzy logic controller.

In Section II, the gain scheduling method is introduced as an effective nonlinear control method for nonlinear systems. In Section III, a novel fuzzy fed PID controller is proposed. We show that recursive design of the fuzzy rule base can guarantee stability of local closed-loop systems. In Section IV, control of a pole-balancing robot illustrates how the proposed design method can be easily applied to a nonlinear robotics system. Concluding remarks are given in the last section.

II. NONLINEAR CONTROL PROBLEM

Generally, most of robotics systems are nonlinear systems. One common task in robotics system control is to demand the robot or parts of the body to follow a given reference trajectory [12]. Tracking control of system dynamics may change significantly. Hence, instead of trying to model the system, a more feasible solution is to schedule the gains at each operating point. Since human expert can describes the system in a natural language better than mathematical equations, fuzzy control is also commonly used in nonlinear control of robotics systems [9],[13].

A. Gain Scheduling Method

Nonlinear systems can be generally expressed by the following nonlinear autonomous system equation:

$$\dot{x} = f(x) + g(x)u \qquad (1)$$

Where $x=[x_1,x_2,..x_m]^T$ is the state vector, $u=[u_1,u_2,..u_m]^T$ is the control input vector, f(x) and g(x) are vector functions of states.

Assume $x^d(t) \in R^{n \times 1}$ is the given reference trajectory whose corresponding reference input is $u^d(t)$

$$\dot{x}^d = f(x^d) + g(x^d)u^d \quad (2)$$

Taking Lyapunov linearization around the operating points (x^d, u^d), we have

$$\dot{x} = \dot{x}^d + A(x^d)(x - x^d) + B(x^d)(u - u^d) \quad (3)$$

Where

$$A(x^d) = \left.\frac{df}{dx}\right|_{x=x^d}$$

$$B(x^d) = g(x^d) \quad (4)$$

Let $e = x - x^d, \dot{e} = \dot{x} - \dot{x}^d$ and (5)

System (3) is equivalent to

$$\dot{e} = A^d e + B^d u^e \quad (6)$$

where A^d and B^d are assumed to can be transformed into diagonal CCF, which is valid for many robotics systems. Because the reference trajectory x^d is a function of time, the nonlinear system (1) can be linearized at frozen time τ so that the tracking problem of the nonlinear system is transformed into a stabilization problem of the linear system (6) in the error state space. The equilibrium points are shifted from the reference trajectory points $x^d(\tau)$ to the origin. However, the aforementioned conventional gain-scheduling method employs linearization between two consecutive operating points. If the system states vary significantly along the time axis, global stability will be a problem. An alternative solution is to utilize fuzzy rules containing expert knowledge to perform smoother interpolation of all the operating points in the control envelope [2],[11].

B. Fuzzy Gain Scheduling

At some frozen times τ_i the corresponding control input can be approximated by (2), which is $x^d(\tau_i)$ or x_i shortly. In partitioning the state space, this x_i will be the centers of membership functions (MFs), LX_i [16]. The nonlinear system given by (1) can, therefore, be transformed into several local linearized systems

$$R^i : IF\ x^d\ is\ LX^i, THEN\ \dot{e} = Ae + Bu^e \quad (7)$$

where A_i and B_i are system state matrices corresponding to x_i.

The control law to be designed is

$$R^i : IF\ x^d\ is\ LX^i, THEN\ u = u^d + u^e \quad (8)$$

where u^d is the control input corresponding to the reference input x^d and u^e is the control input derived from error inputs.

The conventional approach of using the gain scheduling method is to design a linear controller for each local system in (7). The main advantage of this approach is that the powerful linear control theory may be applied. However, some simple nonlinear controllers like fuzzy PID controllers could be a better choice for handling the system nonlinearities. Then, the fuzzy PID controllers for local systems may be embedded in the global fuzzy gain scheduling rules to improve the integrity of the design, Moreover, the fuzzy fed PID controller will give the optimal solution more than any previous controller.

III. HYBRID FUZZY CONTROLLER DESIGN

In this section, a fuzzy fed PID controller is proposed for enhanced control of the local linearized systems. By employing recursive feedback and appropriate tuning of conventional derivative gain, the fuzzy fed PID controller guarantees sector conditions of the output [10],[13]. Local stability analysis also explores the relationship between the conventional derivative gain and the fuzzy gain. Although the proposed controller is developed as a hybrid fuzzy fed PID controller, the overall structure shows its potential to be a new form of stand alone fed FLC as depicted in Fig.1.

A fuzzy PID controller with fuzzy is proposed by constructing from simple linear fuzzy rules in an incremental way. But in this section, a new type of fuzzy PID controller is proposed based on fuzzy fed PID control structure using the Takagi-Sugeno (T-S).

The fuzzy fed PID controller is constructed in an incremental way by employing both error signals and recursive feedback signals as inputs to fed PID the main idea is found in the integral side where the integral side when fed by a deferential feedback gives us a null overshoot and steady state error, the enhancement is very significant using Fuzzy fed PID controller. The most widely adopted conventional PID controller structure used in industrial applications is the following structure [7]:

$$u_{PID}(t) = K_P^C e_v(t) + K_I^C e_p(t) + K_D^C e_a(t) \quad (9)$$

where K_P, K_I, and K_D are the conventional proportional, integral, and derivative gains, respectively, and $u_{PID}(t)$ is the controller output and $e_v(t)$ is the velocity error signal, $e_p(t) = \int e_v(t)$ is the proportional error signal and $e_a(t) = de_v(t)/dt$ is the acceleration error signal.

The items in (9) form the PID control and can be replaced by the following linear fuzzy rules:

$$R^j : IF\ e_p\ is\ LE_p^j\ AND\ e_v..,THEN\ u_{PID}\ is\ DU_{PID}^j \quad (10)$$

Where LE_p^j and LE_v^j are the linguistic values of error signals of the jth fuzzy rule and DU_{PID}^j is the desired function value of the output $u_{PID}(t)$

The first look of fed PID gives the following equation:

$$u_{PID}(t) = K_P^C e_v(t) + (0.5)K_I^C e_p(t) + K_D^C e_a(t) \quad (11)$$

But the real output is differ when the fed PID controller is used where the fed PID controller has overshoot and steady state error less than the conventional PID controller.

Note that the output feedback from the integrator is taken from the output of the defuzzification process which gives the best results showing in the illustrative example.

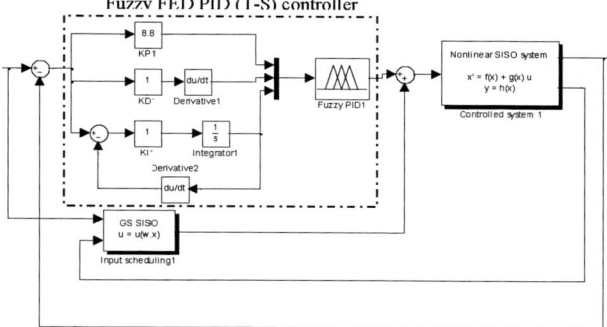

Fig. 1: Overall Control Structure

IV. ILLUSTRATIVE EXAMPLE

In the example, the proposed controller is used to Takagi-Sugeno fuzzy control with an inverted pendulum robot, that robot is used in the most of our applications because of nonlinearity problem and marginally stability. The dynamic equation of the inverted pendulum robot is given by

$$\ddot{\theta} = \frac{(m_p + m_c)g\sin\theta - m_p \dot{\theta}^2 l \sin\theta - F\cos\theta}{(m_p + m_c)l(4/3 - m_p \cos^2\theta)}$$

Where θ is the angle between the pendulum and the vertical, the angular velocity is expressed by $\dot{\theta}$, the force which acts on the cart is F, the gravity acceleration g is 9.8m/sec2, m_c and m_p are the mass of cart and the mass of pole respectively, and l is the half length of the pendulum. The system equation is written as follow:

$$\dot{x} = f(x) + g(x)u$$

Where

$$f(x) = \begin{bmatrix} \dot{\theta} \\ \dfrac{(m_p + m_c)g\sin\theta - m_p \dot{\theta}^2 l \sin\theta \cos\theta}{(m_p + m_c)l(4/3 - m_p \cos^2\theta)} \end{bmatrix}$$

$$g(x) = \begin{bmatrix} 0 \\ \dfrac{-F\cos\theta}{(m_p + m_c)l(4/3 - m_p \cos^2\theta)} \end{bmatrix}$$

The last two equations are used for simulation without a previous technique of linearization because of two methods are used, the first one is the gain scheduling method which divides the system into small areas to relent using of iterations, the second method is the fuzzy PID controller which use the linguistic formulas and it by default make a linearization of the nonlinear system. The addition of the two methods is called hybrid fuzzy PID controller. Beside the point that the amount of masses and measurements of the pendulum, the most point to be focused is the fed Fuzzy PID controller is make lower overshoot and minimum steady state error., this technique always make the best results shown in Fig 3, the fuzzy rules of the fed PID controller using Takagi-Sugeno shown bellow is better than the results of Hybrid fuzzy FED PID controller [17]:

For the fuzzy proportional integrator differentiator:
1. If (I1 is -ve) and (I2 is -ve) and (I3 is -ve) then (O/P1 is F1)
2. If (I1 is -ve) and (I2 is -ve) and (I3 is zero) then (O/P1 is F1)
3. If (I1 is -ve) and (I2 is -ve) and (I3 is +ve) then (O/P1 is F1)
4. If (I1 is -ve) and (I2 is zero) and (I3 is -ve) then (O/P1 is F1)
5. If (I1 is -ve) and (I2 is zero) and (I3 is zero) then (O/P1 is F2)
6. If (I1 is -ve) and (I2 is zero) and (I3 is +ve) then (O/P1 is F2)
7. If (I1 is -ve) and (I2 is +ve) and (I3 is -ve) then (O/P1 is F2)
8. If (I1 is -ve) and (I2 is +ve) and (I3 is zero) then (O/P1 is F3)
9. If (I1 is -ve) and (I2 is +ve) and (I3 is +ve) then (O/P1 is F3)
10. If (I1 is zero) and (I2 is -ve) and (I3 is -ve) then (O/P1 is F1)
11. If (I1 is zero) and (I2 is -ve) and (I3 is zero) then (O/P1 is F2)
12. If (I1 is zero) and (I2 is -ve) and (I3 is +ve) then (O/P1 is F2)
13. If (I1 is zero) and (I2 is zero) and (I3 is -ve) then (O/P1 is F2)
14. If (I1 is zero) and (I2 is zero) and (I3 is zero) then (O/P1 is F2)
15. If (I1 is zero) and (I2 is zero) and (I3 is +ve) then (O/P1 is F2)
16. If (I1 is zero) and (I2 is +ve) and (I3 is -ve) then (O/P1 is F2)
17. If (I1 is zero) and (I2 is +ve) and (I3 is zero) then (O/P1 is F2)
18. If (I1 is zero) and (I2 is +ve) and (I3 is +ve) then (O/P1 is F3)
19. If (I1 is +ve) and (I2 is -ve) and (I3 is -ve) then (O/P1 is F1)
20. If (I1 is +ve) and (I2 is -ve) and (I3 is zero) then (O/P1 is F2)
21. If (I1 is +ve) and (I2 is -ve) and (I3 is +ve) then (O/P1 is F2)
22. If (I1 is +ve) and (I2 is zero) and (I3 is -ve) then (O/P1 is F2)
23. If (I1 is +ve) and (I2 is zero) and (I3 is zero) then (O/P1 is F2)
24. If (I1 is +ve) and (I2 is zero) and (I3 is +ve) then (O/P1 is F3)
25. If (I1 is +ve) and (I2 is +ve) and (I3 is -ve) then (O/P1 is F3)
26. If (I1 is +ve) and (I2 is +ve) and (I3 is zero) then (O/P1 is F3)
27. If (I1 is +ve) and (I2 is +ve) and (I3 is +ve) then (O/P1 is F3)

The output (O/P) constants are -0.5, 0, and 0.5 for the three functions respectively.

Fig 2 illustrates the membership functions of the inputs (I's) to the controller desired:

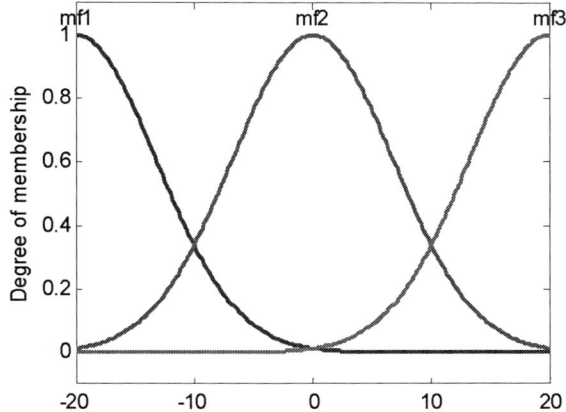

Fig.2: membership functions of the inputs to the controller

Fig 3 illustrates the step response of hybrid fuzzy FED PID controller versus conventional PID controller using mamdani technique:

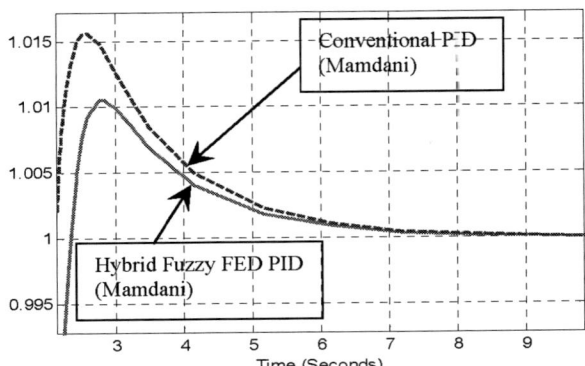

Fig.3: Stabilization control of the PID versus FED PID (Mamdani)

Fig 4 illustrates the step response of hybrid fuzzy FED PID controller (Takagi-Sugeno) versus conventional PID controller:

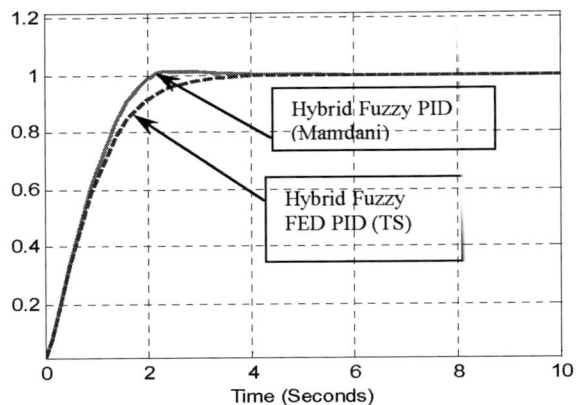

Fig.4: Stabilization control of the PID versus FED PID (T-S)

Fig 5 illustrates the step response of hybrid fuzzy FED PID controller (Takagi-Sugeno) versus hybrid fuzzy FED PID controller (Mamdani):

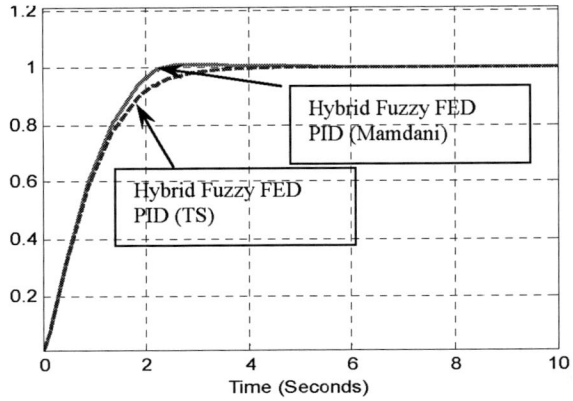

Fig.5: Stabilization control of the FED PID (M) versus FED PID (T-S)

V. CONCLUSION

In this paper, the new approach of control design of a hybrid fuzzy FED PID controller is proposed using the T-S method. The proposed design method focuses on constructing the fuzzy rule base. The proposed controller demonstrates excellent control performance for nonlinear robot which depends on the hybridizing of the gain scheduling method and fed PID T-S controller which gives the best control specifications towards the conventional PID, fuzzy PID and hybrid fuzzy PID . The proposed problem is considered one of the hottest and useful topics in the area of fuzzy control field related with robotics systems.

REFERENCES

[1] E H Mamdani. *Application of Fuzzy Algorithms for the Control of a Dynamic Plant*. Proc. IEE 1974.12

[2] H. K. Khalil, *Nonlinear Systems*. New York: Macmillan, 1992.

[3] G. Chen, T. Pham, *Introduction to fuzzy sets, fuzzy logic, and fuzzy control systems*. New York: CRC, 2001.

[4] L.X. Wang, *A course in Fuzzy systems and controls*. Upper Saddle River, NJ: Prentice-Hall, 1997.

[5] W. Li, *"Design of a hybrid fuzzy logic proportional plus conventional integral-derivative controller,"* IEEE Trans. Fuzzy Syst., vol. 6, pp. 449–463, Nov. 1998.

[6] M.J. Er and Y.L. Sun, *"Hybrid fuzzy proportional-integral plus conventional derivative control of linear and nonlinear systems,"* IEEE Trans. Ind. Electron., vol. 48, pp. 1109–1117, Dec. 2001.

[7] K. K. Tan, Q. G.Wang, and C. C. Hang, *Advances in PID Control*. New York: Springer-Verlag, 1999.

[8] Z. Y. Zhao, M. Tomizuka, and S. Isaka, *"Fuzzy gain scheduling of PID controllers,"* IEEE Trans. Syst. Man Cybern., vol. 23, pp. 1392–1398, Oct. 1993.

[9] K. Kiimbla, M. Jamshidi, *"Hybrid Fuzzy Control Schemes for Robotic Systems,"* IEEE Trans. 1995.

[10] Ch. Clifton, A. Homaifar and M. Bikdash, *"Design of generalized sugeno controllers by approximating hybrid fuzzy PID controller,"* IEEE Trans. 1996.

[11] O. Castillo, P. Melin, *"A General Method for Automated Simulation of Non-Linear Dynamical Systems using a New Fuzzy-Fractal-Genetic Approach,"* IEEE Trans. 1999.

[12] H. B. Kazemian, *"The SOF-PID Controller for the Control of a MIMO Robot Arm,"* IEEE Trans. Fuzzy Syst.., vol. 10, pp. 523-532, Aug 2002

[13] Y. Lei Sun, M. Joo Er, *"Hybrid Fuzzy Control of Robotics Systems,"* IEEE Trans. Fuzzy Syst., vol. 12, pp. 755-765, Dec 2004.

[14] L. X. Wang, *Adaptive Fuzzy Systems and Control: Design and Stability Analysis*. Upper Saddle River, NJ: Prentice-Hall, 1994.

[15] L. A. Zadeh, *"Fuzzy logic = computing with words,"* IEEE Trans. Fuzzy Syst., vol. 4, pp. 103–111, 1996.

[16] K. J. Astrom, B. Wittenmark, *Adaptive Control*. USA: Adison-Wesley, 1989.

[17] B. Hamed, A. EL Khateb, *"Hybrid Fuzzy FED PID Control of Nonlinear Systems"*.

Adaptive fuzzy control of a class of SISO nonaffine nonlinear Systems

S. Doudou, F. Khaber

Laboratoire QUERE, Département d' Electrotechnique, Université Ferhat Abbas 19000 - Sétif- Algérie
dousof2003@yahoo.fr, Farid.Khaber@ieee.org

Abstract: The aim of this paper is to control a nonaffine nonlinear system single input single output (SISO). Based on the implicit function theory, the existence of an unknown ideal controller is demonstrated. A fuzzy system is used to approximate this controller and its parameters are updated according to gradient descend method. The closed-loop control structure stability is guaranteed using Lyapunov analysis. An illustrative simulation example is given to demonstrate the feasibility of the proposed method.

Key Words: Fuzzy Systems, Fuzzy control, Adaptive control, Gradient descend method, nonaffine nonlinear Systems

1. INTRODUCTION

Recently, the control of complex nonlinear dynamic systems has become a topic of considerable importance of research. Several adaptive techniques have been developed to control the nonlinear systems. Conceptually, there are two distinct approaches that have been formulated in the design of an adaptive fuzzy controller: direct and indirect.

The most of works in the fuzzy control literature are dedicated to the control problem of the affine in the control nonlinear systems, i.e, systems characterized by inputs appearing linearly in the system state equation. Results are available for nonaffine nonlinear systems in which the control input appears in a nonlinear form [1,2,4,5,6,7,11,12]. The parameters adaptation laws used in the above indirect [1,5,11] and direct plans [2,4,6,7,12] are designed, based on a Lyapunov approach.

In this work we introduce a direct adaptive fuzzy control approach for a class of SISO nonaffine nonlinear systems. The basic idea is to use a fuzzy system to construct an unknown ideal controller that can accomplish the control objectives. The implicit function theory is used to show the existence of this unknown controller. The parameters fuzzy adaptive law are designed, based on the gradient descend method. The total stability of closed-loop system is studied by using a Lyapunov approach.

This paper is organized as follows: a class of SISO nonaffine nonlinear systems and the control objectives are described in section 2. Section 3 presents a brief description of the used fuzzy systems. The proposed direct adaptive control is presented in section 4 with its adaptive law and the total stability analysis of the system. In section 5, the proposed control is used to control a simple nonaffine nonlinear system. Section 6 concludes this work.

2. PROBLEM FORMULATION

Consider the nonaffine nonlinear system single input single output (SISO) represented by the following normal form

$$\begin{cases} \dot{x}_i = x_{i+1}, & i = 1,...,n-1. \\ \dot{x}_n = f(x,u) \\ y = x_1. \end{cases} \quad (1)$$

where $x = [x_1,...x_n]^T \in \Re^n$, is the state vector of the system in the normal form which is assumed available for measurement, $u \in \Re$ is the control input, $y \in \Re$ is the system output and $f(x,u)$ is an unknown smooth nonlinear function. The control objective is to design an adaptive fuzzy controller for system (1) such that the system output $y(t)$ follows a desired trajectory $y_d(t)$ while all signals in the closed-loop system are bounded.

In this paper, we will make the following assumptions regarding the system (1) and the desired trajectory $y_d(t)$.

Assumption 1: The function $f_u(x,u) = \partial f(x,u)/\partial u$ is nonzero and bounded for all $(x,u) \in \Omega_x \times \Re$. This implies that $f_u(x,u)$ is strictly either positive or negative for all $(x,u) \in \Omega_x \times \Re$, without loss of generality, it is assumed that it exists a constant c such that $f_u(x,u) \geq c \geq 0$ for all $(x,u) \in \Omega_x \times \Re$. Note that $f_u(x,u)$ can be assumed strictly negative and the controller can be similarly derived.

Assumption 2: The desired trajectory $y_d(t)$ and its time derivatives $y_d^i(t), i=1,...,n$ are smooth and bounded.

Define the tracking error vector $e \in \Re^n$ as

$$e = [e_1, e_2,...,e_{n-1}]^T \quad (2)$$

where $e_1 = y_d - y$. Then, from (1), we get

$$e_n = y_d^n - f(x,u) \quad (3)$$

which can be written in matrix form

$$\dot{e} = A_0 e + b\{y_d^n - f(x,u)\} \quad (4)$$

where $A_0 = \begin{bmatrix} 0 & 1 & 0 & \cdots & 0 \\ 0 & 0 & 1 & \cdots & 0 \\ \vdots & \vdots & \vdots & \ddots & 0 \\ 0 & 0 & 0 & 0 & 1 \\ 0 & 0 & 0 & 0 & 0 \end{bmatrix} \in \Re^{n \times n}$ and $b = \begin{bmatrix} 0 \\ 0 \\ \vdots \\ 0 \\ 1 \end{bmatrix} \in \Re^n$

Let $K = [k_0, k_1, \ldots, k_{n-1}]^T$ be a positive constant vector selected such that the matrix $A_c = A_0 - bK^T$ is stable. Thus, for any given positive definite symmetric matrix Q, there exists a unique positive definite symmetric matrix solution P to the following Lyapunov algebraic equation:

$$A_c^T P + P A_c = -Q \quad (5)$$

Let a signal v be defined as

$$v = y_d^n + K^T e + \beta \tanh(b^T Pe/\varepsilon) \quad (6)$$

where $\tanh(.)$ is the hyperbolic tangent function, β is a (large) positive constant, and ε a (small) positive constant.

Remark 1. The motivation of using the term $\beta \tanh(b^T Pe/\varepsilon)$ in the signal v, is to ensure some robustness, against modeling error, for the adaptive fuzzy controller which will be proposed later. The term $\beta \tanh(b^T Pe/\varepsilon)$ is a smooth approximation of the discontinuous term $\beta \text{sign}(b^T Pe/\varepsilon)$ usually used in robust control. So, β is selected larger than the magnitude of the uncertainty and it will affect the convergence rate of the tracking error, and ε is chosen very small to best approximate the sign function and it will affect the size of the residual set to which the tracking error will converge.

The sign function sign(.) is not used here to avoid problems associated with it as chattering.

By adding and subtracting the term, $K^T e + \beta \tanh(b^T Pe/\varepsilon)$ to the right-hand side of (4), we obtain

$$\dot{e} = A_c e - b\beta \tanh(b^T Pe/\varepsilon) - b\{f(x,u) - v\} \quad (7)$$

From Assumption 1 and the fact that the signal v, defined in (6), does not explicitly depend upon the control input u, the partial derivative of $f(x,u) - v$ with respect to the input u satisfies

$$\frac{\partial (f(x,u) - v)}{\partial u} = \frac{\partial f(x,u)}{\partial u} > 0 \quad (8)$$

thus, based on the implicit function theorem [2, 6], we know that the nonlinear algebraic equation $f(x,u) - v = 0$ is locally solvable for the input u for each (x,v). So, there exists some ideal controller $u^*(x,v)$ satisfying the following equality for all $(x,v) \in \Omega_x \times \Re$:

$$f(x, u^*(x,v)) - v = 0 \quad (9)$$

Therefore, if the control input u is chosen as the ideal control law, i.e., $u = u^*$ the closed-loop error dynamic (7) is reduced to

$$\dot{e} = A_c e - b\beta \tanh(b^T Pe/\varepsilon) \quad (10)$$

Let us consider the following positive function:

$$V = e^T Pe \quad (11)$$

Using (10) and (5), the time derivative of (11) becomes

$$\dot{V} = -e^T Q e - 2\beta b^T Pe \tanh(b^T Pe/\varepsilon) \quad (12)$$

Since the term $b^T Pe \tanh(b^T Pe/\varepsilon)$ is always positive, we conclude that $\dot{V}$ is a negative semi-definite function and that the tracking error $e(t)$ and its derivatives $e_{(i)}(t), i = 1,\ldots,n-1$, go to zero as t goes to ∞.

However, the implicit function theory only guarantees the existence of the ideal controller $u^*(x,v)$ for system (1), and does not prescribe a technique for constructing it even if the dynamics of the system are well known.

3. DESCRIPTION OF THE USED FUZZY SYSTEMS

In this section we use a Takagi-Suegeno fuzzy system which performs a mapping from an input vector $z \in \Omega_z \subset \Re^m$ to a scalar output variable $y_f \in \Re$, where $z = [z_1,\ldots,z_m]^T$, $\Omega_z = \Omega_{z1} \times \ldots \times \Omega_{zm}$ and $\Omega_{zi} \subset \Re$

If we define M_i fuzzy sets F_i^j, $j=1,\ldots,M_i$, for each input z_i, then the fuzzy system will be characterized by a set of if–then rules of the form [10]

$$R^k : \text{if } z_1 \text{ is } G_1^k \text{ and}\ldots\text{and } z_m \text{ is } G_m^k \text{ Then} \\ y_f \text{ is } y_f^k (k = 1,\ldots,N) \quad (13)$$

where $G_i^k \in \{F_i^1,\ldots,F_i^{Mi}\}, i=1,\ldots,m$, y_f^k is the crisp output of the kth rule, and N is the total number of rules. By using the singleton fuzzifier, product inference engine, and center-average defuzzifier, the final output of the fuzzy system is given as follows [10]:

$$y_f(z) = \frac{\sum_{k=1}^{N} \mu_k(z) y_f^k}{\sum_{k=1}^{N} \mu_k(z)} \quad (14)$$

where $\mu_k(z) = \prod_{i=1}^{m} \mu_{G_i^k}(z_i)$, with $\mu_{G_i^k} \in \{\mu_{F_i^1},\ldots,\mu_{F_i^{Mi}}\}$, and $\mu_{F_i^j}(x_i)$ is the membership function of the fuzzy set F_i^j.

By introducing the concept of fuzzy basic functions [10], the output given by (14) can be rewritten in the following compact form:

$$y_f(z) = w^T(z)\theta \quad (15)$$

where $\theta = [y_f^1,\ldots,y_f^N]^T$ is a vector grouping all consequent parameters, and $w(z) = [w_1(z),\ldots,w_N(z)]^T$ is the set of fuzzy basic functions defined as

$$w_k(z) = \frac{\mu_k(z)}{\sum_{j=1}^{N}\mu_j(z)}, \quad k=1,\ldots N.$$

The fuzzy system (15) is assumed to be well defined so that $\sum_{j=1}^{N}\mu_j(z) \neq 0$ for all $z \in \Omega_z$.

4. ADAPTIVE FUZZY CONTROLLER DESIGN

In this section, we propose to use a fuzzy system to adaptively build this unknown ideal implicit controller in section 2.

4.1. Control law

To develop the control law, we assume that the fuzzy system described in section 3 can approximate the unknown implicit ideal controller $u^*(x,v)$ as follows:

$$u^*(z) = w^T(z)\theta^* + \delta(z) \quad (16)$$

where $z = [x^T, v]^T$, $\delta(z)$ is the fuzzy approximation error, θ^* is an ideal parameter vector which minimizes the function $|\delta(z)|$ over an operating compact set Ω_z, and $w(z)$ is a fuzzy basic functions vector assumed suitably specified by the designer. In this paper, we assume that the used fuzzy system does not violate the universal approximation property on the compact set Ω_z, which is assumed large enough so that the variable z remains inside it under closed-loop control. So it is reasonable to assume that the fuzzy approximation error is bounded for all $z \in \Omega_z$, i.e., $|\delta(z)| \leq \bar{\delta}$ where $\bar{\delta}$ is a positive constant. Since the ideal parameter vector θ^* is unknown, so it should be estimated by a suitable adaptation law. Let θ be an estimate of the ideal vector θ^* and define the control law as the adaptive fuzzy approximation of the ideal controller (16), i.e., the control law for system (1) is chosen as

$$u(z) = w^T(z)\theta \quad (17)$$

4.2. Parameter Adaptation Law

Our goal in this subsection is to design an adaptive algorithm for the parameter vector θ such that the fuzzy controller (17) approximates the unknown controller (16), i.e., the adaptive algorithm should be designed to make the error between u and u^* as small as possible. Furthermore, the adaptive law should guarantee the boundedness of the estimated parameters. To this end, let us define the error between the controllers u and u^* as

$$e_u = u^* - u \quad (18)$$

Using (16) and (17), (18) becomes

$$e_u = u^* - w^T(z)\theta = w^T(z)\tilde{\theta} + \delta(z) \quad (19)$$

where $\tilde{\theta} = \theta^* - \theta$ is the parameter estimation error vector. By invoking the mean value theorem, there exists a constant λ with $0 < \lambda < 1$, such that the nonlinear function $f(x,u)$ can be expressed around u^* as

$$f(x,u) = f(x,u^*) + f_{u\lambda}(u - u^*) \quad (20)$$

where $f_{u\lambda} = \partial f(x,u)/\partial u|_{u=u_\lambda}$ avec $u_\lambda = \lambda u + (1-\lambda)u^*$.

By substituting (20) into the error equation (7), we get

$$\dot{e} = A_c e - b\beta \tanh(b^T P e/\varepsilon) \\ - b\{f_{u\lambda}(u-u^*) + f(x,u^*) - v\} \quad (21)$$

Using (8), (20) becomes

$$\dot{e} = A_c e - b\beta \tanh(b^T P e/\varepsilon) - b f_{u\lambda}(u-u^*) \quad (22)$$

which can be rewritten in the following form:

$$e_{(n)} + K^T e + \beta \tanh(b^T P e/\varepsilon) = f_{u\lambda}(u^* - u) = f_{u\lambda} e_u \quad (23)$$

We notice here that u^* is an unknown quantity, so the signal e_u defined in (18) is not available. Equation (23) will be used to overcome this difficulty. Indeed, from (23), we see that even if the signal e_u is not available for measurement, the quantity $f_{u\lambda} e_u$ is measurable. This fact will be exploited in the design of the parameters adaptive law.

Now, consider a quadratic cost function that measures the discrepancy between the implicit controller and the fuzzy controller, defined as

$$J(\theta) = \frac{1}{2}e_u^2 = \frac{1}{2}(u^* - u)^2 = \frac{1}{2}(u^* - w^T(z)\theta)^2 \quad (24)$$

The gradient descent method is used here to minimize the cost function (24). Hence, by applying the gradient descent method [3], we obtain as an adaptive law for the parameters θ, the following first-order differential equation:

$$\dot{\theta} = -\eta(t)\nabla_\theta J(\theta) \quad (25)$$

where $\eta(t)$ is a positive time-varying parameter. From (24), the gradient of $J(\theta)$ with respect to θ is

$$\frac{\partial J(\theta)}{\partial \theta} = -w(z)e_u.$$

Therefore, the gradient descent algorithm becomes

$$\dot{\theta} = \eta(t)w(z)e_u \quad (26)$$

The adaptive law (26) cannot be implemented since the signal e_u is not available. In order to make (26) computable, from (23), we select the design parameter $\eta(t)$ as $\eta(t) = \eta_0 f_{u\lambda}$ where η_0 is a positive constant. Thus, (26) becomes

$$\dot{\theta} = \eta_0 w(z)\{f_{u\lambda} e_u\} \quad (27)$$

Using (23), we get:

$$\dot{\theta} = \eta_0 w(z)\{e_{(n)} + K^T e + \beta \tanh(b^T P e/\varepsilon)\} \quad (28)$$

As shown by [3], the adaptive law (28) cannot guarantee the boundedness of the parameters $\tilde{\theta}$ in the presence of the approximation error, which is unavoidable in such adaptive schemes. So, to improve the robustness of adaptive law (27) in the presence of the approximation error, we modify it by introducing a σ - modification term as follows [3]:

$$\dot{\theta} = \eta_0 w(z)\{e_{(n)} + K^T e + \beta \tanh(b^T Pe/\varepsilon)\} - \eta_0 \sigma \theta \quad (29)$$

where σ is a small positive constant.

Let us consider the following positive function:

$$V_\theta = \frac{1}{2\eta_0}\tilde{\theta}^T\tilde{\theta} \quad (30)$$

Using (23) and (29), the time derivative of (30) can be written as

$$\dot{V}_\theta = -\tilde{\theta}^T(w(z)f_{u\lambda}e_u - \sigma\theta) \quad (31)$$

With (19), (31) becomes

$$\dot{V}_\theta = -f_{u\lambda}e_u^2 + f_{u\lambda}\delta(z)e_u + \sigma\tilde{\theta}^T\theta \quad (32)$$

Using the 2 inequalities

$$\sigma\tilde{\theta}^T\theta = -\frac{\sigma}{2}\|\tilde{\theta}\|^2 - \frac{\sigma}{2}\|\theta\|^2 + \frac{\sigma}{2}\|\tilde{\theta}+\theta\|^2 \leq -\frac{\sigma}{2}\|\tilde{\theta}\|^2 + \frac{\sigma}{2}\|\theta^*\|^2$$

$$-e_u^2 + \delta(z)e_u = -\frac{1}{2}e_u^2 + \frac{1}{2}\delta^2(z) - \frac{1}{2}(e_u - \delta(z))^2$$

$$\leq -\frac{1}{2}e_u^2 + \frac{1}{2}\delta^2(z)$$

Equation (32) can be bounded as

$$\dot{V}_\theta \leq -\frac{1}{2}f_{u\lambda}e_u^2 + \frac{1}{2}f_{u\lambda}\delta^2(z) - \frac{\sigma}{2}\|\tilde{\theta}\|^2 + \frac{\sigma}{2}\|\theta^*\|^2 \quad (33)$$

Since the parameters θ^* are constants, and the functions $\delta(z)$ and $f_{u\lambda}$ are assumed bounded in this paper, so we can define a positive constant bound ψ as

$$\psi = \sup_t(\frac{1}{2}f_{u\lambda}\delta^2(z)) + \frac{\sigma}{2}\|\theta^*\|^2$$

Then, (33) can be simplified to

$$\dot{V}_\theta \leq -\alpha V_\theta + \psi - \frac{1}{2}f_{u\lambda}e_u^2 \leq -\alpha V_\theta + \psi \quad (34)$$

where V_θ is given by (31) and $\alpha = \sigma\eta_0$.

Now we can prove the following theorem, which shows the boundedness of the parameters error vector $\tilde{\theta}$.

Theorem 1. If assumption 1 is satisfied and the approximation error $\delta(z)$ in (19) is bounded then, the adaptive law (29) guarantees that:

- The parameters error vector $\tilde{\theta}$ is bounded and converges to the residual set: $\Omega_\theta = \left\{\tilde{\theta} \mid \|\tilde{\theta}\|^2 \leq 2\eta_0\psi/\alpha\right\}$.

- The estimation error e_u is bounded and it is $(\bar{\delta}^2 + \sigma)$ - small in the mean square sense.

Proof. Equation (34) implies that for $V_\theta \geq \psi/\alpha$, $\dot{V}_\theta \leq 0$ and, therefore, V_θ et $\tilde{\theta}$ are bounded which, together with $\delta(z) \in L_\infty$ and $w(z) \in L_\infty$, implies that $e_u \in L_\infty$.

By integrating (34), we can establish that

$$\|\tilde{\theta}(t)\|^2 \leq \|\tilde{\theta}(0)\|^2 \exp(-\alpha t) + 2\eta_0\psi/\alpha \quad (35)$$

which implies that $\tilde{\theta}$ converges to the residual set

$$\Omega_\theta = \left\{\tilde{\theta} \mid \|\tilde{\theta}\|^2 \leq 2\eta_0\psi/\alpha\right\} \quad (36)$$

Because $V_\theta \in L_\infty$, $0 < f_{u\lambda} < \infty$ and $|\delta(z)| \leq \bar{\delta}$, by integrating (33), we can obtain:

$$\int_t^{t+T} e_u^2 d\tau \leq (\bar{\delta}^2 + \frac{\sigma}{c_0}\|\theta^*\|^2)T + \frac{2}{c_0}(V_\theta(t) - V_\theta(t+T))$$
$$\leq c_1(\bar{\delta}^2 + \sigma)T + c_2 \quad (37)$$

with $c_0 = \inf_t(f_{u\lambda})$, $c_1 = \max(1, \|\theta^*\|^2/c_0)$ and $c_2 = 2\sup_t((V_\theta(t) - V_\theta(t+T))/c_0)$. Expression (38) implies that e_u is $(\bar{\delta}^2 + \sigma)$ - small in the mean square sense [3].

Remark 2. Because of the system uncertainty, the signal $e_{(n)}$ in the adaptive law (29) is not available for measurement. In this paper, $e_{(n)}$ is approximated by

$$e_{(n)}(t) \approx \frac{e_{(n-1)}(t) - e_{(n-1)}(t - \Delta(t))}{\Delta(t)},$$ where $\Delta(t)$ is a small positive constant. Notice that because of the integral structure of the adaptive law (29), the only place where the signal $e_{(n)}$ intervenes, the effect of noise measurement will be considerably reduced.

4.3. Tracking error convergence

To analyze the tracking error convergence, let us consider the following Lyapunov function:

$$V_e = e^T Pe \quad (38)$$

Differentiating (38) with respect to time and using (5) and (22), we obtain

$$\dot{V}_e = -e^T Qe - 2b^T Pe\beta \tanh(b^T Pe/\varepsilon) + 2b^T Pe f_{u\lambda}(u^* - u) \quad (39)$$

with (19), (39) becomes

$$\dot{V}_e = -e^T Qe - 2b^T Pe\beta \tanh(b^T Pe/\varepsilon) + 2b^T Pe f_{u\lambda}(w^T(z)\tilde{\theta} + \delta(z)) \quad (40)$$

From (35), we get

$$\|\tilde{\theta}(t)\| \leq \|\tilde{\theta}(0)\|\exp(-0.5\alpha t) + \sqrt{2\eta_0\psi/\alpha} \quad (41)$$

Using (42) and the fact that $w(z), \delta(z)$ and $f_{u\lambda}$ are bounded, we can write

$$\left| f_{u\lambda} \times (w^T(z)\tilde{\theta} + \delta(z)) \right| \leq \psi_0 \exp(-0.5\alpha t) + \psi_1 \quad (42)$$

where ψ_0 and ψ_1 are some finite positive constants.

with (42), (40) can be bounded as

$$\dot{V}_e \leq -e^T Q e - 2b^T P e \beta \tanh(b^T P e / \varepsilon) + 2|b^T P e|(\psi_0 \exp(-0.5\alpha t) + \psi_1) \quad (43)$$

By assuming that the design parameter β is chosen large enough such that $\beta \geq \psi_1$, and using the following inequality $-x \tanh(x/\varepsilon) + |x| \leq \kappa\varepsilon$ [8], with $\kappa = 0.2785$, so Eq. (43) can be reduced to

$$\dot{V}_e \leq -e^T Q e + 2|b^T P e|\psi_0 \exp(-0.5\alpha t) + 2\psi_1 \kappa\varepsilon \quad (44)$$

Using the inequality,

$$2|b^T P e|\psi_0 \exp(-0.5\alpha t) \leq 0.5\|e\|^2 + 2\|b^T P\|^2 \psi_0^2 \exp(-\alpha t), \quad (45)$$

becomes

$$\dot{V}_e \leq -(\lambda_{\min}(Q) - 0.5)\|e\|^2 + 2\|b^T P\|^2 \psi_0^2 \exp(-\alpha t) + 2\psi_1 \kappa\varepsilon \quad (46)$$

where $\lambda_{\min}(Q)$ denotes the minimum eigenvalue of the matrix Q and it is assumed chosen such that $\lambda_{\min}(Q) > 0.5$, Eq. (46) can be rewritten as

$$\dot{V}_e \leq -\alpha_e V_e + 2\|b^T P\|^2 \psi_0^2 \exp(-\alpha t) + 2\psi_1 \kappa\varepsilon \quad (47)$$

where $\alpha_e = (\lambda_{\min}(Q) - 0.5)/\lambda_{\max}(P)$ with $\lambda_{\max}(P)$ is the maximum eigenvalue of the matrix P.

Now we can prove the following theorem, which shows the convergence of the tracking error to a neighborhood of the origin.

Theorem 2. Consider the system (1). Suppose that Assumptions 1–2 are satisfied and the fuzzy approximation error in (19) is bounded. Then the control law defined by (17) with adaptation law given by (29) guarantees the boundedness of the signals x and u, and the convergence of the tracking error to the residual set:

$$\Omega e = \left\{ e \mid \|e\| \leq \sqrt{2\psi_1 \kappa\varepsilon / (\lambda_{\min}(P)\alpha_e)} \right\}.$$

Proof. Equation. (47) implies that for $V_e \geq (2\|b^T P\|^2 \psi_0^2 \exp(-\alpha t) + 2\psi_1 \kappa\varepsilon)/\alpha_e$, $\dot{V}_e \leq 0$, therefore, the tracking error vector e is bounded which, together with the boundedness of the desired trajectory and its derivatives, implies that the state vector x is bounded.

Moreover, since the term $2\|b^T P\|^2 \psi_0^2 \exp(-\alpha t)$ in the right-hand side of (47), goes to zero when $t \to \infty$, we can conclude that the function V_e will be asymptotically bounded as

$$V_e \leq 2\psi_1 \kappa\varepsilon / \alpha_e \quad (48)$$

and therefore the tracking error will converge asymptotically to the residual set

$$\Omega e = \left\{ e \mid \|e\| \leq \sqrt{2\psi_1 \kappa\varepsilon / (\lambda_{\min}(P)\alpha_e)} \right\} \quad (49)$$

Since we have established in Theorem 1 that $\tilde{\theta} \in L_\infty$, so we have $\theta \in L_\infty$ which, together with $w(z) \in L_\infty$ implies that $u \in L_\infty$.

5. SIMULATION RESULTS

In this section, to illustrate the validity of the proposed adaptive fuzzy controller, a SISO nonaffine nonlinear system is simulated. This system is described by the following differential equations [7]:

$$\begin{cases} \dot{x}_1 = x_2 \\ \dot{x}_2 = x_1^2 + 0.15u^3 + 0.1(1 + x_2^2)u + \sin(0.1u) + d(t) \end{cases} \quad (50)$$

where $d(t) = 0.5\sin(10t)$ is an external disturbance included in order to test the robustness of the adaptive controller against external disturbances.

The control objective is to force the system output $y(t) = x_1(t)$ to track the desired trajectory $y_d(t) = \sin(t) + \cos(0.5t)$.

Within this simulation, the unknown ideal implicit controller is approximated by a fuzzy system in the form of (15). The input variables of the fuzzy system are chosen as $z = [z_1, z_2, z_3]^T$,

where, $z_1 = x_1$, $z_2 = x_2$ and

$$z_3 = \ddot{y}_d + K^T E + \beta \tanh(b^T P E / \varepsilon)$$

For each variable z_i, $i = 1,2,3$, we define three Gaussian membership functions as

$$\mu_{F_i^1}(z_i) = \exp(-0.5(z_i + 2)^2), \mu_{F_i^2}(z_i) = \exp(-0.5(z_i)^2) \text{ and}$$

$$\mu_{F_i^3}(z_i) = \exp(-0.5(z_i - 2)^2).$$

The initial conditions of system are $x(0) = [0.6, 0.5]^T$, and the initial values of the estimates parameters $\theta(0)$ are set equal to zero. The design parameters used in this simulation are selected as follows.

$$\eta_0 = 5, \beta = 20, \varepsilon = 0.05, \sigma = 0.02, K = [1,2]^T, Q = \text{diag}[10,10]$$

$P = \begin{bmatrix} 15 & 5 \\ 5 & 5 \end{bmatrix}$. A Gaussian white noise with mean zero and variance 0.02 is added to the measurements x_1 and x_2.

The simulation results for the state variables x_1 and x_2 are shown in Fig.1 and Fig. 2. It can be seen that actual trajectories converge rapidly to the desired ones. The control input signal shown in Fig. 3 is bounded.

These simulation results demonstrate the tracking capability of the proposed controller, its robustness against modeling errors and disturbances and its effectiveness for control tracking of uncertain nonaffine nonlinear systems.

6. CONCLUSION

In this work, we have presented a stable direct adaptive control using a fuzzy system to control a nonaffine nonlinear system. The proposed ideal implicit controller and its adjustable parameters are updated by using the gradient descend method. The closed-loop system stability is guarantee by a Lypaunov analysis. The simulation results show the efficiency of the approach proposed for control of nonaffine nonlinear systems. The future work will focus on the exploitation of this approach to control MIMO nonaffine nonlinear systems.

REFERENCES

[1] R. Boukezzoula, S. Galichet and L. Foulloy (2003), Fuzzy adaptive control for non-affine systems, *IEEE Internat. Conf. Fuzzy Systems*, pp. 543–548.

[2] S.S. Ge and J. Zhang (2003), Neural-network control of nonaffine nonlinear system with zero dynamics by state and output feedback, *IEEE Trans. Neural Networks,* **14 (4)** pp. 900–918.

[3] P. Ioannou and J. Sun (1996), *Robust Adaptive Control*, Prentice-Hall, Englewood Cliffs, New Jersey.

[4] S. Labiod and T.M. Guerra (2007), Adaptative fuzzy control of a class of SISO nonaffine nonlinear systems, *Fuzzy Systems and Systems*, **158 (10)**, pp. 1126–1137.

[5] Y.J. Lui and W. Wang (2007), Adaptative fuzzy control for a class of uncertain nonaffine nonlinear systems, *Information Sciences*, **177 (18)**, pp. 3901–3917.

[6] J.-H. Park and S.-H. Kim (2004), Direct adaptive output-feedback fuzzy controller for nonaffine nonlinear system, *Control Theory Appl,* **151(1)**, pp. 65–72.

[7] J.-H. Park, G.-T. Park, S.-H. Kim and C.-J. Moon (2005), Direct adaptive self-structuring fuzzy controller for nonaffine nonlinear system, *Fuzzy Sets and Systems,* **153 (3)**, pp. 429–445.

[8] M.M. Polycarpou, and P.A. Ioannou (1996), A robust adaptive nonlinear control design, *Automatica,* **32 (3)**, pp. 423–427.

[9] Y. Tang, N. Zhang, and Y. Li (1999), Stable fuzzy adaptive control for a class of nonlinear systems, *Fuzzy sets and Systems,* **104** pp. 279–288.

[10] L.X. Wang (1994), *Adaptive Fuzzy Systems and Control*, Prentice-Hall, Englewood Cliffs, New Jersey.

[11] P.-S. Yoon. J-H. Park. G.-T. Park (2001), Adaptative fuzzy control of nonaffine nonlinear systems using Takagi-Sugeno fuzzy models, in: proc. *IEEE Internat. Fuzzy Systems,* pp. 642-645

[12] T. Zhang, S.S. Ge and C.C. Hang (1998), Direct adaptive control of non-affine nonlinear system using multilayer neural networks, *ACC*, pp. 515–519.

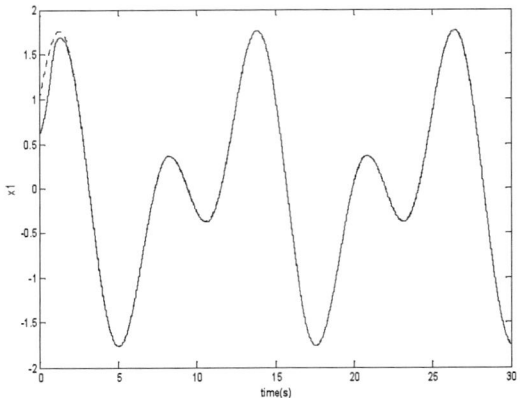

Fig. 1. Response of the variable x_1: actual (solid line); desired (dotted line).

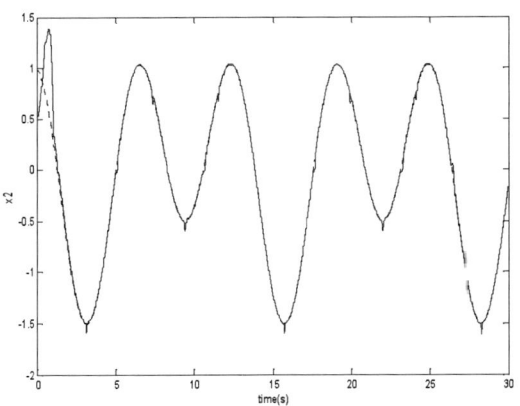

Fig. 2. Response of the variable x_2: actual (solid line); desired (dotted line).

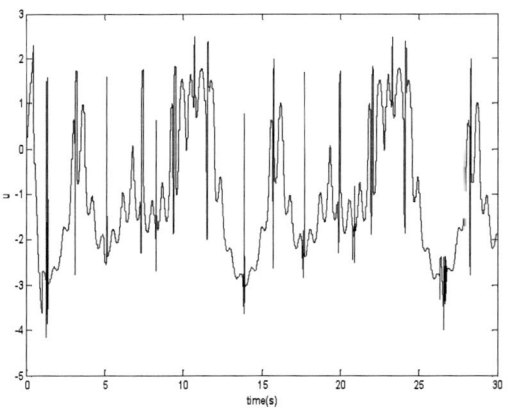

Fig. 3. Control input signal

New Stability Conditions of Takagi-Sugeno Fuzzy Systems via LMI

F. Bourahala, F. Khaber

QUERE Laboratory, electrotechnic Department, Ferhat Abbas University, Setif (19000)
(bourahala1981@yahoo.fr, farid.Khabe@ieee.org)

Abstract: This paper presents new stability conditions for the continuous Takagi-Sugeno (T-S) fuzzy systems by using the Linear Matrix Inequality (LMI) approach. These new conditions are applied to design problems of fuzzy regulator. First, Takagi-Sugeno fuzzy models and some stability results are recalled. To design fuzzy control systems, nonlinear systems are represented by Takagi-Sugeno fuzzy models. The concept of parallel distributed compensation (PDC) is employed to design fuzzy controllers from the Takagi-Sugeno fuzzy models. New stability conditions are obtained by relaxing the classical stability conditions. The stability conditions (classic and relaxed) of the closed-loop system are expressed in terms of LMI. Design examples for nonlinear systems demonstrate the utility of the relaxed stability conditions and the LMI procedures.

Keywords: Takagi Sugeno fuzzy model, relaxed conditions, Linear Matrix Inequality, Fuzzy control, PDC concept, Nonlinear system.

1. INTRODUCTION

Since last years, many works have been focused on the stability and the stabilization of closed-loop fuzzy systems including a fuzzy controller. Specially, the approach using T-S fuzzy models [10], many researches have focused on this model-based approach for controlling nonlinear systems. The issue of stability and stabilization of T-S fuzzy control systems has been considered extensively in nonlinear stability frameworks [7], [9]. However, the present results are only sufficient and require conservative assumptions. For example, Tanaka and Sugeno presented sufficient conditions for the stability of T-S models [6] using a quadratic Lyapunov approach. The stability depends on the existence of a common positive definite matrix guarantying the stability of all local subsystems. These stability conditions may be expressed in linear matrix inequalities (LMIs) form [11]. The obtaining of a solution is then facilitated by using numerical toolboxes for solving such problems. Recently LMI constraint has been added to compute a decay rate and using the concept of parallel distributed compensation (PDC) [4]. However, if the number of submodels is large, it might be difficult to find a common matrix. Moreover, these constraints are often very conservative and it's well known that, in a lot of cases, a common positive definite matrix does not exist, whereas the system is stable. To overcome this limitation, some works have been developed in order to establish new stability conditions by relaxing some of the previous constraints, frameworks [9], [3] and [1].

This paper presents new stability conditions based on Lyapunov stability theorem for the continuous T-S fuzzy systems by LMI approach. A LMI formulation and it's potential of resolution are used in order to relax the classical stability conditions [9], [3]. First, we recall T-S fuzzy models and their stability results. To design fuzzy regulators, nonlinear systems are represented by T-S fuzzy models. The concept of parallel distributed compensation is used to design fuzzy regulators from the T-S fuzzy models. We define new stability conditions, which improve the conservativity of the previous results. LMI-based design procedures for fuzzy regulators are constructed using the PDC and the relaxed stability conditions. Other LMIs with respect to decay rate are also derived. Design examples for nonlinear systems demonstrate the utility of the relaxed stability conditions and the LMI procedures. The organization of this paper is as follows:

Section 2 presents T–S fuzzy model and its stability issues. Section 3 presents new stability conditions by relaxing the previous obtained stability conditions. Section 4 presents the LMI satisfying fuzzy regulators. Two numerical examples are given in section 5 to illustrate the utility of the relaxed stability conditions. Finally, a conclusion is given in section 6.

2. TAKAGI–SUGENO FUZZY MODEL AND STABILITY

A. Takagi–Sugeno Fuzzy Model

The fuzzy model proposed by Takagi and Sugeno [1] is described by fuzzy IF-THEN rules which represent local linear input-output relations of a nonlinear system. The ith rule of the T–S fuzzy model is of the following form.

Plant Rule i

IF $z_1(t)$ is M_{i1} and....and $z_p(t)$ is M_{ip}

THEN $\begin{cases} \dot{x}(t) = A_i x(t) + B_i u(t) \\ y(t) = C_i x(t) \end{cases}$ $i = 1, 2, ..., r$ (1)

where $M_{ij}(j = 1,..,r)$ is the fuzzy set, r is the number of IF-THEN rules, $x(t) \in R^n$ is the state vector, $u(t) \in R^m$ is the input vector, $y(t) \in R^q$ is the output vector, $A_i \in R^{n.n}$, $B_i \in R^{n.m}$ and $C_i \in R^{q.n}$. $z(t) = [z_1(t) \; z_2(t) \ldots z_p(t)] \in R^p$ a premise vector depending linearly or not on $x(t)$. Each linear consequent equation represented by $A_i x(t) + B_i u(t)$, is called "subsystem."

$M_{ij}(z_j(t))$ is the grade of membership of $z_j(t)$ in the M_{ij} where

$$w_i(z(t)) = \prod_{j=1}^{p} M_{ij}(z_j(t)), i = 1, 2, \ldots, r \quad (2)$$

with $w_i(z(t)) \geq 0$

Because the membership functions take the values in [0 1].

and $h_i(z(t)) = \dfrac{w_i(z(t))}{\sum_{i=1}^{r} w_i(z(t))}$ with $\sum_{i=1}^{r} h_i(z(t)) = 1$, $\forall t$

$h_i(z(t))$ is a fuzzy basis function (FBF) satisfying :

$$\begin{cases} \sum_{i=1}^{r} h_i(z(t)) = 1 \\ 0 \leq h_i(z(t)) \leq 1, \; \forall i : 1, \ldots, r \end{cases}$$

The total model is obtained by aggregation of the r local models. It is assumed in this paper that the fuzzy basis function $h_i(z(t))$ depends only on the measurable state variables of the system [1].

$$\begin{cases} \dot{x}(t) = \sum_{i=1}^{r} w_i(z(t))\{A_i x(t) + B_i u(t)\} \Big/ \sum_{i=1}^{r} w_i(z(t)) \\ y(t) = \sum_{i=1}^{r} w_i(z(t)) C_i x(t) \Big/ \sum_{i=1}^{r} w_i(z(t)) \end{cases} \quad (3)$$

These equations can be rewritten of the following way:

$$\begin{cases} \dot{x}(t) = \sum_{i=1}^{r} h_i(z(t))\{A_i x(t) + B_i u(t)\} \\ y(t) = \sum_{i=1}^{r} h_i(z(t)) C_i x(t) \end{cases} \quad (4)$$

B. Quadratic stability of the T-S fuzzy model

The following theorem of stability presents the sufficient conditions of quadratic stability using the approach of Lyapunov in the case of the continuous T-S fuzzy systems of (4).

Theorem1 [6]: The equilibrium of the continuous fuzzy control system of (4) is asymptotically stable in the large if there exist a common positive definite matrix P such that

$$A_i^T P + PA_i < 0 \quad i = 1, 2, \ldots, r \quad (5)$$

The condition (5) can be only sufficient if no characteristic of the functions $h_i(z(t))$ is taken in account.

3. NEW STABILITY CONDITIONS

A. Fuzzy regulator design via parallel distributed compensation

Fuzzy regulator is required to satisfy an asymptotic convergence of the state i.e. $x(t) \to 0$ when $t \to \infty$.

This requirement implies stabilization of control systems. The PDC concept [4] is used to elaborate a linear law of control of a nonlinear model. The concept PDC is used to calculate a linear law of control for each subsystem. The idea is to design a regulator by feedback laws for each subsystem. The global control law is obtained by interpolation of the local linear control laws.

Regulator Rule i

IF $z_1(t)$ is M_{i1} and....and $z_p(t)$ is M_{ip}

THEN $u(t) = -F_i x(t)$, $\quad i = 1, 2, \ldots, r.$

The global control law is:

$$u(t) = -\dfrac{\sum_{i=1}^{r} w_i(z(t)) F_i x(t)}{\sum_{i=1}^{r} w_i(z(t))} = -\sum_{i=1}^{r} h_i(z(t)) F_i x(t) \quad (6)$$

The design of the fuzzy regulator is to determine the local feedback gains F_i in the consequent parts.

By substituting (6) into (4), we obtain:

$$\dot{x}(t) = \sum_{i=1}^{r} \sum_{j=1}^{r} h_i(z(t)) h_j(z(t)) \{A_i - B_i F_j\} x(t) \quad (7)$$

Equation (7) can be rewritten as:

$$\dot{x}(t) = \sum_{i=1}^{r} h_i(z(t)) h_i(z(t)) G_{ii} x(t) + 2 \sum_{i<j}^{r} h_i(z(t)) h_j(z(t)) \left\{ \dfrac{G_{ij} + G_{ji}}{2} \right\} x(t) \quad (8)$$

Where $G_{ij} = A_i - B_i F_j$.

By applying the stability conditions for the open-loop system (theorem 1) to (8), we can derive stability conditions for the closed-loop System.

Theorm2: [8] The equilibrium of the continuous fuzzy control system described by (8), is globally asymptotically stable if there exist a common positive definite matrix P such that

$$G_{ii}^T P + P G_{ii} < 0 \quad (9)$$

$$\left(\dfrac{G_{ij} + G_{ji}}{2} \right)^T P + P \left(\dfrac{G_{ij} + G_{ji}}{2} \right) \leq 0, \quad i < j \quad (10)$$

for all i and j, excepting the pairs (i, j) such that

$h_i(z(t))h_j(z(t)) = 0, \forall t$.

To check stability of the fuzzy control system, it has long been considered difficult to find a common positive definite matrix $P > 0$ satisfying the conditions of theorems 1 and 2. The problem to find the common matrix $P > 0$ can be solved numerically i.e. the stability conditions of theorems 1 and 2 can be expressed in LMIs. For example, to check the stability conditions of theorem 2, we need to find $P > 0$ satisfying the following matrix inequalities

$$\begin{cases} P > 0 \\ G_{ii}^T P + PG_{ii} < 0 \quad i = 1,...r \\ \left(\dfrac{G_{ij} + G_{ji}}{2}\right)^T P + P\left(\dfrac{G_{ij} + G_{ji}}{2}\right) \leq 0, \quad i < j \end{cases}$$

These inequalities can be put in the form of LMI (with change of the variables).

The determination of the matrix P poses a convex feasibility problem. Numerically, this feasibility problem can be solved very efficiently by means of the most powerful tools available to date in the mathematical programming literature. For instance, recently developed interior-point method [2] is extremely efficient in practice.

B. Relaxed stability conditions for fuzzy model

If the number of IF-THEN rules r is large, it might be difficult to find a common $P > 0$. The stability conditions of (9) and (10) are conservative because they ask the stability of all models (dominant and crusaders).

The following theorem allows to reduce this conservatism while taking into account the interactions between the subsystems (characterized by the number r of active subsystems every instant). The conditions obtained impose only the stability of the dominant models. The idea is to relax the stability conditions of (9) and (10).

Theorem 3: [3] The equilibrium of the fuzzy control system described by (8) is globally asymptotically stable if there exist a common positive definite matrix P and a common positive semidefinite matrix Q such that

$$G_{ii}^T P + PG_{ii} + (s-1)Q < 0 \quad (11)$$

$$\left(\dfrac{G_{ij} + G_{ji}}{2}\right)^T P + P\left(\dfrac{G_{ij} + G_{ji}}{2}\right) - Q \leq 0, \quad i < j \quad (12)$$

for all i and j, excepting the pairs (i, j) such that $h_i(z(t))h_j(z(t)) = 0, \forall t$, where $1 < s \leq r$ is the maximum of the number of the fuzzy subsystems that are fired at an instant.

Proof: Consider a candidate Lyapunov function $V(x(t)) = x^T P x(t)$, where $P > 0$.

$$\dot{V}(x(t)) = \sum_{i=1}^{r}\sum_{j=1}^{r} h_i(z(t))h_j(z(t)) x^T(t)$$
$$\left[(A_i - B_i F_j)^T P(A_i - B_i F_j)\right] x(t)$$

$$= \sum_{i=1}^{r} h_i^2(z(t)) x^T \left[G_{ii}^T P + PG_{ii}\right] x(t) + \sum_{i<j}^{r} 2h_i(z(t))h_j(z(t)) x^T(t)$$
$$\cdot \left[\left(\dfrac{G_{ij} + G_{ji}}{2}\right)^T P + P\left(\dfrac{G_{ij} + G_{ji}}{2}\right)\right] x(t)$$

where $G_{ij} = A_i - B_i F_j$

knowing that

$$\sum_{i=1}^{r} h_i^2(z(t)) - \dfrac{1}{r-1}\sum_{i<j}^{r} 2h_i(z(t))h_j(z(t)) \geq 0$$

with $\sum_{i=1}^{r} h_i(z(t)) = 1, h_i(z(t)) \geq 0$

Checked by

$$\sum_{i=1}^{r} h_i^2(z(t)) - \dfrac{1}{r-1}\sum_{i<j}^{r} 2h_i(z(t))h_j(z(t))$$
$$= \dfrac{1}{r-1}\sum_{i<j}^{r} \{h_i(z(t)) - h_j(z(t))\}^2 \geq 0.$$

From the condition (12), we have

$$\dot{V}(x(t)) \leq \sum_{i=1}^{r} h_i^2(z(t)) x^T \left[G_{ii}^T P + PG_{ii}\right] x(t)$$
$$+ \sum_{i<j}^{r} 2h_i(z(t))h_j(z(t)) x^T(t) Q x(t)$$
$$\leq \sum_{i=1}^{r} h_i^2(z(t)) x^T \left[G_{ii}^T P + PG_{ii}\right] x(t)$$
$$+ (s-1)\sum_{i=1}^{r} h_i^2(z(t)) x^T Q x(t)$$
$$= \sum_{i=1}^{r} h_i^2(z(t)) x^T \left[G_{ii}^T P + PG_{ii} + (s-1)Q\right] x(t)$$

Remark: It is assumed in the derivations of theorems 1 and 3 that the weight $w_i(z(t))$ of each rule in the fuzzy controller is equal to that of each rule in the fuzzy model for all t.

Example: This example illustrates the effectiveness of the new stability conditions. Considering a continuous fuzzy plant composed of the following two rules ($r = s = 2$)

$$\begin{cases} R_1 : IF \ x_1 \ is \ M_{11} \quad THEN \ \dot{x}(t) = A_1 x + B_1 u \\ R_2 : IF \ x_1 \ is \ M_{12} \quad THEN \ \dot{x}(t) = A_2 x + B_2 u \end{cases}$$

where

$$A_1 = \begin{bmatrix} 2 & -10 \\ 1 & 0 \end{bmatrix}, B_1 = \begin{bmatrix} 1 \\ 0 \end{bmatrix}, A_2 = \begin{bmatrix} 15 & -10 \\ a & 0 \end{bmatrix}, B_2 = \begin{bmatrix} b \\ 0 \end{bmatrix}$$

The local feedback gains F_1 and F_2 are determined by selecting [-1 -2] as the eigenvalues of the subsystems in the

PDC concept. Fig. 1 and 2 shows the feasible areas satisfying the conditions of theorems 2 and 3 for the variables a and b, respectively.

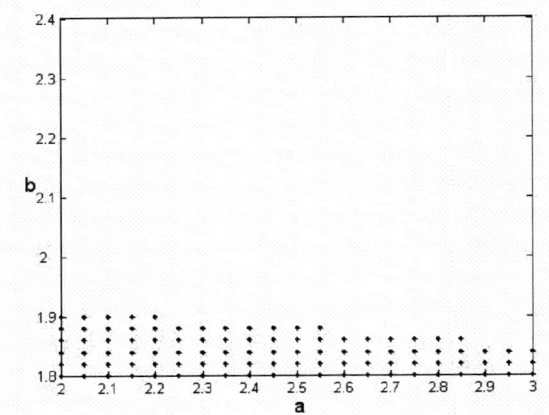

Fig. 1. Regions feasible of stability based on the theorem 2.

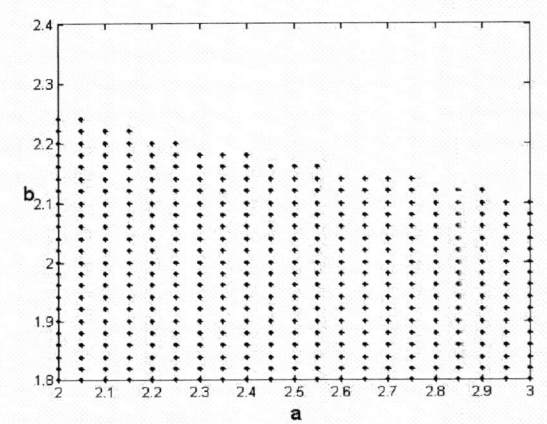

Fig. 2. Regions feasible of stability based on the theorem 3.

It can be seen in previous figures that the theorem 2 demonstrates the most conservative results and theorem 3 shows the most relaxed results, (the theorem 3 provides more relaxed results than theorem 2), i.e. the possibility to have a practically stable system is large with relaxation (important number of pairs (a,b)).

4. AN LMI APPROACH TO FUZZY CONTROLLER

Let us recall that the common P problem for fuzzy controller design can be numerically solved using LMI-based technique.
The LMI-based designs allow us to realize a total and systematic design satisfying not only stability but also decay rate, etc. The LMIs for stability and decay rate will be derived in this section [5].

A. Stability

From the relaxed stability conditions of theorem 3, the design problem to determine the common matrix P and the feedback gains F_i can be defined under the form of LMIs as follows:

Find $X > 0, Y \geq 0$, and M_i satisfying:

$$\begin{cases} -XA_i^T - A_iX + M_i^T B_i^T + B_i M_i - (s-1)Y > 0 \quad i = 1,...,r \\ 2Y - XA_i^T - A_iX - XA_j^T - A_jX \\ \quad + M_j^T B_i^T + B_i M_j + M_i^T B_j^T + B_j M_i \geq 0 \quad i < j \end{cases} \quad (13)$$

where

$$X = P^{-1}, M_i = F_i X, Y = XQX \quad (14)$$

The above conditions are LMIs with respect to variables X, Y and M_i. We can find a positive definite matrix X, a positive semi-definite matrix Y and M_i satisfying the LMIs or determination that no such X, Y and M_i exist. This is a convex feasibility problem.

From (14) the feedback gains F_i, a common matrix P and a common Q can be obtained as

$$P = X^{-1}, \quad F_i = M_i X^{-1}, \quad Q = PYP \quad (15)$$

B. Decay Rate

The degree of stability (decay rate) of the system (8) is the biggest value α such as:

$$\exists P > 0 : \frac{dV(x(t))}{x(t)} + 2\alpha V(x(t)) \leq 0 \quad (16)$$

where $\alpha > 0$ represents the decay rate of the trajectories of the system (8), (maximal value). Let's note that to maximize the decay rate is a generalized eigenvalue minimization problem GEVP in P and α.
The formulation of this problem is as follows:

For the theorem 2
Maximize α
X, M_i
Subject to $X > 0$

$$\begin{cases} -XA_i^T - A_iX + M_i^T B_i^T + B_i M_i - 2\alpha X > 0 \quad i=1,..,r \\ -XA_i^T - A_iX - XA_j^T - A_jX + M_j^T B_i^T \\ \quad + B_i M_j + M_i^T B_j^T + B_j M_i - 4\alpha X \geq 0 \quad i < j \end{cases} \quad (17)$$

For the theorem 3
Maximize α
X, Y, M_i
Subject to $X > 0, \quad Y \geq 0$

$$\begin{cases} -XA_i^T - A_iX + M_i^T B_i^T + B_i M_i - (s-1)Y - 2\alpha X > 0 \quad i=1,..,r \\ 2Y - XA_i^T - A_iX - XA_j^T - A_jX \\ \quad + M_j^T B_i^T + B_i M_j + M_i^T B_j^T + B_j M_i - 4\alpha X \geq 0 \quad i < j \end{cases} \quad (18)$$

where

$X = P^{-1}, M_i = F_i X, Y = XQX$

Remark: A fuzzy regulator that satisfies the LMI conditions of (17) and (18) is a stable fuzzy controller.

5. SIMULATION.

Let us consider a nonlinear system which is a DC motor controlling an inverted pendulum, this system can be represented by the T-S fuzzy model with two sub systems [12]. The fuzzy rules of this system are described as follows:

Plant Rule 1

$$IF\ x_1(t)\ is\ F_1\ THEN \begin{cases} \dot{x}(t) = A_1 x(t) + B_1 u(t) \\ y(t) = C_1 x(t) \end{cases}$$

Plant Rule 2:

$$IF\ x_1(t)\ is\ F_2\ THEN \begin{cases} \dot{x}(t) = A_2 x(t) + B_2 u(t) \\ y(t) = C_2 x(t) \end{cases}$$

where $x(t) = [x_1(t)\ x_2(t)\ x_3(t)]^T$, $x_1(t)$ is the angle of the pendulum, $x_2(t) = \dot{x}_1(t)$ is the angular speed and $x_3(t)$ is the current of the motor.

$$A_1 = \begin{bmatrix} 0 & 1 & 0 \\ 9.8 & 0 & 1 \\ 0 & -10 & -10 \end{bmatrix}, B_1 = \begin{bmatrix} 0 \\ 0 \\ 10 \end{bmatrix}, C_1 = [1\ 0\ 0]$$

$$A_2 = \begin{bmatrix} 0 & 1 & 0 \\ 0 & 0 & 1 \\ 0 & -10 & -10 \end{bmatrix}, B_2 = \begin{bmatrix} 0 \\ 0 \\ 20 \end{bmatrix}, C_2 = [1\ 0\ 0]$$

$$x(0) = [0\ 10\ 0]$$

with

$$F_1(x_1(t)) = \begin{cases} \frac{\sin(x_1(t))}{x_1(t)}, & x_1(t) \neq 0 \\ 1, & x_1(t) = 0 \end{cases}$$

$$F_2(x_1(t)) = 1 - F_1(x_1(t))$$

$F_1(x_1(t))$ and $F_2(x_1(t))$ are fuzzy sets of the state variable $x_1(t)$. This fuzzy model exactly represents the dynamics of the nonlinear system in $-\pi \leq x_1(t) \leq \pi$

To illustrate the utility of the relaxed stability conditions and the LMI-based design procedures, two examples are proposed. In the first example, we design a stable fuzzy regulator by considering the decay rate without relaxation, and in the second example, we design a stable fuzzy regulator by considering the decay rate with relaxation

Example 1: Decay Rate (without relaxation): this problem is defined as follows:

Maximize α
 $X, Y, M_1, ..., M_r$

Subject to $X > 0$, $Y \geq 0$ and (17)

The feedback gains F_i and a common matrix P can be obtained as:

$$F_1 = [214.8714\ 57.9745\ 3.7930]$$
$$F_2 = [83.0103\ 22.8201\ 1.4541]$$
$$P = 10^{+3} \times \begin{bmatrix} 1.8165 & 0.4539 & 0.0297 \\ 0.4539 & 0.1190 & 0.0082 \\ 0.297 & 0.0082 & 0.0007 \end{bmatrix} > 0$$

The dotted line in Fig. 3 shows the responses of $y(t) = x_1(t)$ and the corresponding control input $u(t)$ is represented in Fig.4 (The dotted line)

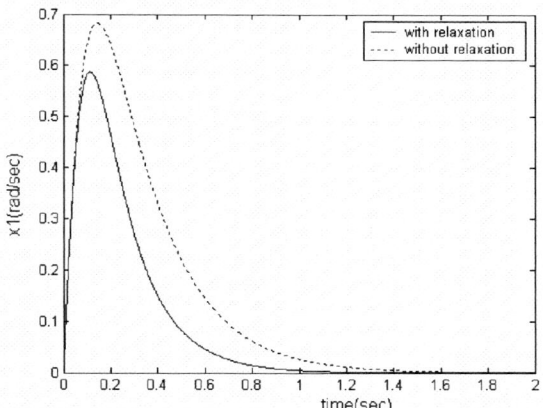

Fig.3. The Responses of the pendulum (without and with relaxation)

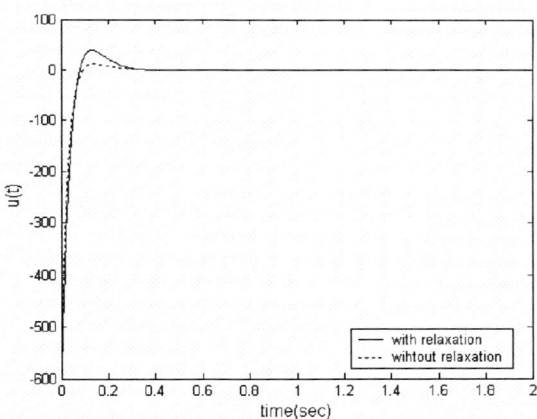

Fig.4. Control signals of the inverted pendulum (without and with relaxation)

The following example presents the new stability conditions of theorem 3.

Example 2: Decay Rate (with relaxation): Let us consider the total stability conditions described into (11) and (12), this problem is defined as follows:

Maximize α
 $X, Y, M_1, ..., M_r$

Subject to $X > 0$, $Y \geq 0$ and (18)

The feedback gains F_i, and the common matrix P and Q can be obtained as:

$$F_1 = [282.3128 \quad 62.4176 \quad 3.2238]$$
$$F_2 = [110.4644 \quad 24.9381 \quad 1.2716]$$
$$P = \begin{bmatrix} 105.1085 & 20.4393 & 1.0529 \\ 20.4393 & 4.2999 & 0.2368 \\ 1.0529 & 0.2368 & 0.0157 \end{bmatrix} > 0$$
$$Q = \begin{bmatrix} 1.4320 & 0.2998 & 0.0163 \\ 0.2998 & 0.0632 & 0.0034 \\ 0.0163 & 0.0034 & 0.0002 \end{bmatrix} \geq 0$$

The continuous line in fig.3, shows the response of $y(t) = x_1(t)$ and the corresponding control input $u(t)$ is represented in Fig.4 (The continuous line)

Simulation results

The stable fuzzy controller design satisfying the decay rate is feasible. It can be seen that the designed fuzzy controller stabilizes the inverted pendulum system, i.e. $x_1 \to 0$ when $t \to \infty$.

It is found in this figure that the speed of response of the decay rate fuzzy controller with relaxation (satisfying the conditions in theorem 3) is better than that of the fuzzy controller without relaxation (satisfying the conditions in theorem 2) i.e. the proposed fuzzy controller with relaxation provides the fastest settling time. However, it offers the largest magnitude of control signal.

In general, the proposed decay rate fuzzy controller (with relaxation) offers better performance than that fuzzy controller (without relaxation) in terms of maximum magnitude of control signal and settling time.

6. CONCLUSION

This paper has presented new stability conditions for the continuous T-S fuzzy systems by using the LMI approach. They have been applied to design problems of fuzzy regulators. First, T-S fuzzy models and some stability results have been recalled. To design fuzzy regulators, nonlinear systems have been represented by T-S fuzzy models. The concept of parallel distributed compensation has been employed to design fuzzy regulators from the T-S fuzzy models. New stability conditions have been obtained by relaxing the classical stability conditions. To relax these stability conditions, it was supposed that the maximum of the number of the fuzzy subsystems that are fired at an instant is lower than the total number of rules. Convex optimization techniques involving LMIs have been utilized to find a common Lyapunov function and stable feedback gains using the PDC concept and the relaxed stability conditions. Other LMI's with respect to decay rate have been also derived and utilized in the design procedures. Some examples for nonlinear systems and simulation results have demonstrated the widening of the feasibility area by relaxed stability conditions, and in addition allowed the synthesis of the feedback controllers stabilizing closed-loop system without or with relaxation.

REFERENCES

[1] C. M.Marcelo, A. Edvaldo and R. G. Avellar (October 203). On Relaxed LMI-Based Designs for Fuzzy Regulators and Fuzzy Observers. *IEEE Trans. on fuzzy systems*, **vol. 11, no. 5**, pp. 613-623.

[2] C. Scherer et S. Weiland (October 2000). *Linear Matrix Inequality in Control*, version 3.

[3] E. Kim (October 2000). New approaches to relaxed quadratic stability condition of fuzzy control systems. *IEEE Trans. on fuzzy systems*, **vol. 8, no. 5**, pp522-533.

[4] H. Wang, K. Tanaka and M. Griffin (1995). Parallel distributed compensation of nonlinear systems by Takagi and Sugeno's fuzzy model. in Proc. *FUZZ-IEEE, Yokohama, Japan*, pp. 531–538.

[5] H. Wang, K. Tanaka, and M. Griffin (1996). An approach to fuzzy control of nonlinear systems: Stability and design issues. *IEEE Trans. on fuzzy systems*, **vol. 4**, pp. 14–23.

[6] K.Tanaka et M. Sugeno (1992). Stability analyses and design of fuzzy control systems, *fuzzy sets and systems*. **vol.45, no. 2**, pp.135-15.

[7] K. Tanaka and M. Sano (1994). On the concept of fuzzy regulators and fuzzy observers. in *Proc.3rd IEEE Int. Conf. Fuzzy Syst., Orlando, FL*, **vol. 2**, pp. 767–772.

[8] K.Tanaka and M.Sano (May 1994). A robust stabilization problem of fuzzy control systems and its applications to baking up control of a truck-trailer. *IEEE Trans. on Fuzzy Systems*, **vol. 2**, pp. 119.134.

[9] K. Tanaka and T. Ikida, et H. O .Wang (1998). Fuzzy regulators and fuzzy observers: Relaxed stability conditions and LMI- based designs. *IEEE Trans. on Fuzzy Systems*, **vol.6, no.2**, pp. 250-264.

[10] M. Takagi and M.Sugeno (1985). Fuzy identification of systemes and its applications to modiling and control. *IEEE Trans. on systems Man and Cybernetics*, **vol.15**, pp.116-132.

[11] S. Boyd and al. (1994). *Linear Matrix Inequalities in Systems and Control Theory*. in Philadelphia, PA: SIAM.

[12] S. Kawamoto and al (1996). Nonlinear control and rigorous stability analysis based on fuzzy system for inverted pendulum. in Proc. *FUZZ-IEEE*, **vol. 2**, pp. 1427–1432

Backstepping Control Augmented by Neural Networks For Robot Manipulators

Mohammed Belkheiri * Farès Boudjema **

* Université Amar Telidji BP G37 - 03000 - Laghouat Algérie
(e-mail: mbelkhiri@yahoo.com).
** Ecole Nationale Polytechnique BP 162 - 16200 - Elharrach, Alger, Algérie
(e-mail:fboudjema@yahoo.fr)

Abstract:
A new control approach is proposed to address the tracking problem of robot manipulators. In this approach, one relies first on a partially known model to the system to be controlled using a backstepping control strategy. The obtained controller is then augmented by an online neural network that serves as an approximator for the neglected dynamics and modeling errors. The proposed approach is systematic, and exploits the known nonlinear dynamics to derive the stepwise virtual stabilizing control laws. At the final step, an augmented Lyapunov function is introduced to derive the adaptation laws of the network weights. The effectiveness of the proposed controller is demonstrated through computer simulation on PUMA 560 robot.

Keywords: Robot Manipulators, Backstepping control, Neural Networks, Lyapunov.

1. INTRODUCTION

Many papers have been published in the area of adaptive control for nonlinear systems. Initially, most of the results presented required full state feedback. For example, Slotine and Li [1] assume full state feedback and use Lyapunov stability analysis to design a globally stable adaptive controller. More recently, output feedback control for nonlinear systems has been explored. Khalil and Esfandiari [2] use output feedback control for a constant, unknown parameter system with no zero dynamics. They prove semiglobal stabilization using saturation and an observer, which estimates the states. Khalil [3] further analyzes the technique of using an observer. He proves via asymptotic analysis that if the speed of the high-gain observer is fast enough, then the adaptive output feedback controller achieves the same performance as the full state feedback controller. Jankovic [4] uses a high-gain observer with saturation for semi-global stabilization to the origin. He also states that if the regressor vector is persistently excited, exponential tracking of the reference signal can be achieved. Similar ideas were then used for model reference adaptive control algorithms. Schwartz [5] presents an adaptive output feedback controller that belongs to the class of model reference adaptive control. This algorithm uses a linear observer to estimate the states for a robotics case. Further, he proposes an algorithm [6] that avoids the use of observers all together. This paper proposes a new direct adaptive controller that augments an existing PD controller by an adaptive neural networks that compensates for unknown terms. At present, robotic manipulators are commonly used in manufacturing industry. In many cases, their end-effectors are required to move from one place to another and follow some desired trajectories. A great number of researches have been done for the trajectory tracking control of robot manipulators [7, 8, 9]. Most of the existing robot manipulator systems use only classical PID controllers which are easily implemented.

The objective of this work is to augment the existing PD controller by an adaptive element to improve the tracking performance. The rest of paper is organized as follows: Section 2 is devoted for presenting Robot manipulator Modeling, then in Section 3 we will present the Backstepping control for robots. In section 4 we will augment the backstepping regulator by an adaptive neural network to robustify it. Simulation results are presented in Section 5. Finally we end by conclusion.

2. SYSTEM MODEL AND PROPERTIES

In general, the dynamic model of an n-link rigid-body robot manipulator can be written in the following matrix equation [7, 8]:

$$M(q)\ddot{q} + C(q,\dot{q})\dot{q} + G(q) = \tau \quad (1)$$

where $q \in \Re^n$ is the angular position vector, $M(q) \in \Re^{n \times n}$ is the positive definite symmetric inertia matrix, $C(q,\dot{q}) \in \Re^{n \times n}$ is the Coriolis-Centrifugal matrix, $G(q)$ is the gravity vector, and $\tau \in \Re^n$ is the control input vector. Equation (1) can be rewritten in the state space model

$$\begin{aligned}\dot{x} &= A(x) + B(x)\tau \\ y &= Cx\end{aligned} \quad (2)$$

where $x = [x_1 \ x_2]^T = [q \ \dot{q}]^T$

$$A(x) = \begin{bmatrix} x_2 \\ -M^{-1}(x)[C(x)\dot{q} + G(x)] \end{bmatrix}$$

$$B(x) = \begin{bmatrix} 0 \\ M^{-1}(x) \end{bmatrix}, \ C(x) = [I \ 0]$$

In this system, it is assumed that the measurement of the position vector q and the velocity vector $\dot{q}$ is available for feedback.

Some important properties of the system model are given as follows:

Property 1
$$m_m\|x\|^2 \leq x^T M(q)x \leq m_M\|x\|^2, \forall x \in \Re^n,$$
where m_m and m_M are known positive constants.

Property 2
$$\|Cq,\dot{q}\| \leq c_M\|\dot{q}\|,$$
where c_M is a positive constant.

Property 3
$$C(q,y)x = C(q,x)y, \forall x,y \in \Re^n.$$

Property 4

$\dot{M}(q) - 2C(q,\dot{q})$ is skew matrix, that is for any $x \in \Re^n$ and $x \neq 0$, we have
$$x^T\left[\dot{M}(q) - 2C(q,\dot{q})\right]x = 0$$

3. BACKSTEPPING CONTROLLER

Defining the link position tracking error e as
$$e = q - q_d \quad (3)$$
where q_d is the desired position trajectory. The control objective is to solve the trajectory tracking problem. The path generation method is not discussed in this note.

Now, we use backstepping schemes to design the nonlinear controller with the velocity observer.

Defining $z_1 = e$ to be the regulated variable in the backstepping design procedure, then we compute the derivative of z_1 as
$$\dot{z}_1 = \dot{e} = \dot{q} - \dot{q}_d = e_v \quad (4)$$

Using the virtual control e_v, we choose the following stabilizing function
$$\alpha_1 = -K_p z_1 \quad (5)$$
where K_p is a diagonal proportional gain matrix with positive elements $k_i > 0$.

Defining the state error vector
$$z_2 = e_v - \alpha_1 = \dot{q} - \dot{q}_d + K_p z_1$$

The augmented error dynamics take the following form:
$$\begin{cases} \dot{z}_1 = -K_p z_1 + z_2 \\ \dot{z}_2 = \ddot{q} - \ddot{q}_d + K_p(-K_p z_1 + z_2) \end{cases} \quad (6)$$

According to equations (2,6) we can write:
$$\begin{cases} \dot{z}_1 = -K_p z_1 + z_2 \\ \dot{z}_2 = K_p z_2 + M^{-1}[\tau - C\dot{q} - G] - \ddot{q}_d + -K_p^2 z_1 \end{cases} \quad (7)$$

The Backstepping control method suggests the following control signal:
$$\tau = \hat{M}(q)\ddot{q}_d + \hat{C}(q,\dot{q})\dot{q}_r + \hat{G}(q) \\ -K_d(\dot{q} - \dot{q}_d) - K_p z_1 + \tau_{ad}, \quad (8)$$

where the control K_p, K_d are the positive definite diagonal gain matrices analogous to the well known PD regulator; and $\hat{M}, \hat{C}$, and $\hat{G}$ are the best available approximations of the real M, C, and G respectively. The above control law ensures the stability of the system and the convergence of the position error if the adaptive part τ_{ad} cancels the effect of the modeling error $\triangle$ (see [10]).

4. NEURAL NETWORK AUGMENTATION

Usually in the literature, neural networks are used in function approximation in the modeling phase[12], however we will use them in controller design given some conditions are fulfilled. In this note, the error term $\triangle$ will be approximated using a simple feed-forward neural network of two layers as shown on fig. 1.

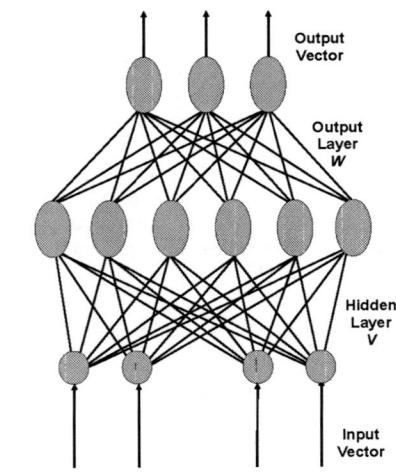

Fig. 1. A Neural Network with two layers

The approximated error signal can be expressed as:
$$\triangle = W^T \Phi(V,\mu) + \varepsilon(\mu) \; \forall \mu \in D \quad (9)$$
where

$V \in D_V \subset \Re^{N_1}$ is the hidden layer weight vector
$W \in D_W \subset \Re^{N_2}$ is the output layer weight vector
Φ is a set of basis functions
μ is the network input vector, and
ε is the neural network reconstruction error

Assuming that the approximation reconstruction error is bounded on some domain D by $\|\varepsilon(\mu)\| \leq \epsilon^*$, $\forall \mu \in D$.

The weight vectors are assumed to be bounded $\|W\| \leq W^*$ and their adjustment can be done online. However the whole controller implementation will be difficult as the number of the parameters increases for complex problems. So the first layer parameters can be adjusted offline in such a way $\Phi(V,\mu)$ is a basis [12] whereas the adaptation law for the second layer is included in the controller design

The adaptive control signal u_{ad} is designed to cancel the unknown nonlinear terms $\triangle$.
$$\tau_{ad} = \hat{W}^T \Phi(\mathbf{V},\mu) \quad (10)$$

where $\hat{W}$ are the estimates of W weights, the initial values of these estimates $\mathbf{W}_0$ can simply be set to zero, while μ is an implementable input vector to the NN defined as:

$\mu = \begin{bmatrix} z^T & \bar{y}_d^T & \bar{u}_d^T & 1 \end{bmatrix}$ with $\bar{y}_d$, and $\bar{u}_d$ are vectors of delayed values of the output and control signals respectively [13].

4.1 Hidden layer weight offline selection

As stated earlier, we have many choices in designing the hidden layer weights depending on the used type of neural network. In this note we will show a design alternative based on Radial Basis Function Neural Networks (RBFNN), which are highly recommended for function approximation due to their simple training. An RBFNN is composed of two layers; a hidden layer contains a set of neurons with their associated centers, the output of each neuron gives the distance between the input vector μ and its center ν_i. The output ψ of the network is a linear combination of the outputs of the hidden layer as.

$$\psi(\mu_j) = \sum_{i=1}^{N_1} w_i \varphi_{ij} \qquad (11)$$

where $\varphi_{ij} = exp\left(\frac{\|\mu_j - \nu_i\|}{\rho}\right)$ is a Gaussian activation function of the distance between the input μ_j and the ith center ν_i.

Defining $W = [\,w_1 \cdots w_{N1}\,]$, $V = [\,\nu_1 \cdots \nu_{N1}\,]$. The design of an RBF network to approximate a given function consists of selecting V in such a way that we construct a set of basis functions [14] and W is adjusted online using an adaptive law to be illustrated in the next subsection. Therefore we have an identification problem based on model (15) which includes the selection of $N1$ model terms $V = [\,\nu_1 \cdots \nu_{N1}\,]$ from a full model set of $M > N1$ terms $U = [\,\mu_1 \cdots \mu_M\,]$ (typically hundreds or even thousands of terms) while μ_i is defined earlier as a tapped delay line of possible values of the input/output and the estimate of z.

We construct the regression matrix Φ_L corresponding to the set of the input vectors U_L a subset of the starting set of centers U.

$$\Phi_L = \begin{bmatrix} \varphi_{11} & \cdots & \varphi_{1L} \\ \vdots & \ddots & \vdots \\ \varphi_{L1} & \cdots & \varphi_{LL} \end{bmatrix}. \qquad (12)$$

Notice that this matrix is symmetric and all the diagonal elements are ones.

It has been shown that the orthogonal algorithm can be employed in selecting the optimal model structure V and to estimate the parameters simultaneously [14]. The orthogonal term selection is formulated using the error reduction ratio vector $ERR_L = [\,err_i \cdots err_L\,]$ defined by:

$$ERR_L = \frac{W^{\mathrm{T}} \Phi_L^2 W}{\Psi^{\mathrm{T}} \Psi} \qquad (13)$$

To find $N1$ optimal model terms V a stepwise approach is applied to the full model set U. At each step, the model term with the maximum err_j value from all of the model terms excluding the previously selected terms is chosen. The selection is terminated at the $N1^{th}$ step where a desired tolerated error tol is reached.

$$1 - \sum_{k=1}^{N_1} err_k < tol \qquad (14)$$

4.2 Neural network weights adaptation

Now, let us consider the stability analysis of the resulting closed-loop system from the following Lyapunov function candidate:

$$V_a(E) = \frac{1}{2} z_1^T K_1 z_1 + \frac{1}{2} z_2^T M(q) z_2 + \frac{1}{2} \tilde{W}^{\mathrm{T}} F^{-1} \tilde{W} \qquad (15)$$

where $F > 0$ is an adaptation gain.

Let $\tilde{W} = \hat{W} - W$, the generalized error vector $E = \begin{bmatrix} z_1 & z_2 & \tilde{W}^{\mathrm{T}} \end{bmatrix}^{\mathrm{T}}$ Differentiating V_a with respect to E gives

$$\dot{V}_a = z_1^T K_1 \dot{z}_1 + \frac{1}{2} z_2^T \dot{M}(q) z_2 + z_2^T M(q) \dot{z}_2 + \frac{1}{2} \tilde{W}^{\mathrm{T}} F^{-1} \dot{\tilde{W}} \qquad (16)$$

From (6)-(7), the derivative of (16) is computed as

$$\dot{V}_a = z_1^T K_1 [-K_p z_1 + z_2] + \frac{1}{2} z_2^T \dot{M}(q) z_2 + \frac{1}{2} \tilde{W}^{\mathrm{T}} F^{-1} \dot{\tilde{W}} \\ + z_2^T M(q) \left[K_p z_2 + M^{-1}(\tau - C\dot{q} - G) - \ddot{q}_d + -K_p^2 z_1 \right] \qquad (17)$$

$$\dot{V}_a = z_1^T K_1 [-K_p z_1 + z_2] + \frac{1}{2} z_2^T \dot{M}(q) z_2 + \tilde{W}^{\mathrm{T}} F^{-1} \dot{\tilde{W}} \\ + z_2^T \left[\tau - C\dot{q} - G - M(q)\left(\ddot{q}_d + K_p^2 z_1 - K_p z_2\right) \right] \qquad (18)$$

Using property (4) of the skew matrix:

$$\dot{V}_a = -z_1^T K_p K_1 z_1 + z_1^T K_p z_2 + \tilde{W}^{\mathrm{T}} F^{-1} \dot{\tilde{W}} \\ + z_2^T \left[C(q,\dot{q}) z_2 + \tau - C\dot{q} - G - M(q)\left(\ddot{q}_d + K_p^2 z_1 - K_p z_2\right) \right] \qquad (19)$$

Substituting for τ by its value from (8)

$$\dot{V}_a = -z_1^T K_p K_1 z_1 + z_1^T K_p z_2 + \tilde{W}^{\mathrm{T}} F^{-1} \dot{\tilde{W}} \\ + z_2^T [C(q,\dot{q}) z_2 + \hat{M}(q)\ddot{q}_d + \hat{C}(q,\dot{q})\dot{q}_r + \hat{G}(q) \\ - K_d(\dot{q} - \dot{q}_d) - K_p z_1 + \tau_{ad} - C\dot{q} - G \\ - M(q)\left(\ddot{q}_d + K_p^2 z_1 - K_p z_2\right)] \qquad (20)$$

$$\dot{V}_a = -z_1^T K_p K_1 z_1 - z_2^T (M(q) K_p - K_d) z_2 + \tilde{W}^{\mathrm{T}} F^{-1} \dot{\tilde{W}} \\ - z_2^T M(q) K_p^2 z_1 + z_2^T (\tau_{ad} - \Delta) \qquad (21)$$

Using (9),(10) yields :

$$\dot{V}_a = -z_1^T K_p K_1 z_1 - z_2^T (M(q) K_p - K_d) z_2 + \tilde{W}^{\mathrm{T}} F^{-1} \dot{\tilde{W}} \\ - z_2^T M(q) K_p^2 z_1 + z_2^T (\hat{W}^{\mathrm{T}} \Phi(\mathbf{V}, \mu) - W^{\mathrm{T}} \Phi(V,\mu) - \varepsilon(\mu)) \qquad (22)$$

Using the fact that $\tilde{W} = \hat{W} - W$ Equation (22) can be written as

$$\dot{V}_a = -z_1^T K_p K_1 z_1 - z_2^T (M(q) K_p - K_d) z_2 + \tilde{W}^{\mathrm{T}} F^{-1} \dot{\tilde{W}} \\ - z_2^T M(q) K_p^2 z_1 + z_2^T (\tilde{W}^{\mathrm{T}} \Phi(\mathbf{V}, \mu) - \delta) \qquad (23)$$

Hence, we can derive this adaptive law for the network parameters W which will make the derivative of the augmented Lyapunov function to be negative definite in some domain given some conditions are fulfilled.

$$\dot{\hat{W}} = -F \left[\Phi(\mathrm{V},\mu) z_2 + 2G(\hat{W} - W_0) \right] \qquad (24)$$

Substituting (24) in (23) for the adaptation law yields

$$\dot{V}_a = -z_1^T K_p K_1 z_1 - z_2^T (M(q) K_p - K_d) z_2 \\ - 2\tilde{W}^{\mathrm{T}} G(\hat{W} - W_0) - z_2^T M(q) K_p^2 z_1 - z_2^T \delta \qquad (25)$$

Using the fact that the neural networks weight vector is bounded, and applying Property 1 and 2 in Section 2, the derivative of the augmented Lyapunov candidate can be upper bounded as

$$\dot{V}_a = -k_1 k_p \|z_1\|^2 - (k_d - m_M k_p)\|z_2\|^2 + m_M k_p^2 \|z_1\|\|z_2\| \\ - G\|\tilde{W}\|^2 - G\|\hat{W} - W_0\|^2 + G\|W - W_0\|^2 + \|z_2\|\epsilon^* \qquad (26)$$

Putting $\underline{N} = -k_1 k_p \|z_1\|^2 + G\|\hat{W} - W_0\|^2$, and

$c_2 = (k_d - m_M k_p)$ yields

$$\dot{V}_a \leq -\underline{N} - c_2\|z_2\|^2 + \|z_2\|\epsilon^* \\ -G\|\tilde{W}\|^2 + G\|W - W_0\|^2 \quad (27)$$

Completing the squares using $\epsilon^*\|z_2\| = -\frac{1}{2}(\epsilon^* - \|z_2\|)^2 + \frac{1}{2}\epsilon^{*2} + \frac{1}{2}\|z_2\|^2$, we get

$$\dot{V}_a \leq -\underline{N} - c_2\|z_2\|^2 - G\|\tilde{W}\|^2 + G\|W - W_0\|^2 \\ -\frac{1}{2}(\epsilon^* - \|z_2\|)^2 + \frac{1}{2}\epsilon^{*2} + \frac{1}{2}\|z_2\|^2 \quad (28)$$

Further, it can be written as

$$\dot{V}_a \leq -\underline{N} - (c_2 - \frac{1}{2})\|z_2\|^2 - G\|\tilde{W}\|^2 - \frac{1}{2}(\epsilon^* - \|z_2\|)^2 \\ +G\|W - W_0\|^2 + \frac{1}{2}\epsilon^{*2} \quad (29)$$

The following conditions

$$\|\tilde{z}_2\| > \sqrt{\frac{2\epsilon^{*2} + 2G\|W - W_0\|^2}{2c_2 - 1}} \\ \|\tilde{W}\|^2 > \sqrt{\frac{\epsilon^{*2} + G\|W - W_0\|^2}{G}} \quad (30)$$

with $c_2 > \frac{1}{2}$, and $G > 0$ ensures that $\dot{V}_a \leq -\underline{N}$. Which is negative definite in some domain defined by conditions in (30).

From (14) and (26), it implies that the closed-loop system is exponential stable according to Lyapunov stability theorem. The resulting controller is obtained in a constructive manner for a robotic system and it can be implemented by augmenting the PD regulator available in most industrial systems, and we have a freedom in selecting the linear controller gains to achieve a desirable performance.

In other worlds, not only the stability of the system can be guaranteed, but also the trajectory tracking error will converge to zero finally.

5. SIMULATION RESULTS

In the simulations, a PUMA robot manipulator is considered with the D-H parameters summarized in Table 1

Table 1. DH Parameters for PUMA 560

Link i	α_i	a_i	θ_i	d_i
1	$\pi/2$	0	0	0
2	0	0.4318	0	0
3	$-\pi/2$	0.0203	0	0.1505
4	$\pi/2$	0	0	0.4318
5	$-\pi/2$	0	0	0
6	0	0	0	0

In the computer simulation, We have considered only the first three joints for simplifications, the aim of each joint to follow a given reference trajectory generated in operational space. We have first started by generating control torques without Neural Networks figure (2). Then We have designed a Neural Network for one link only, the tracking error is reduced as shown in figure (3). The tracking performance is improved by adding Neural Networks to compensate for uncertainties in each case. Figure (4) highlights the simulation results for the Augmented controller for the two joints 1 and 2. Since the tracking for joint 3 is not perfect, it affects joint 1. Finally Figure (5) presents the perfect tracking where the joints of the robot follow the prescribed path perfectly. The Networks weights history is highlighted in fig. 6 and the control effort in fig. 7.

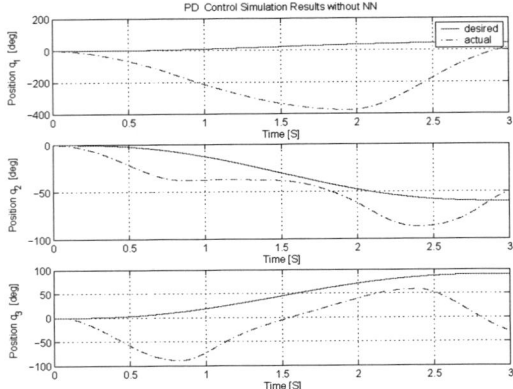

Fig. 2. Simulation results without Neural Network

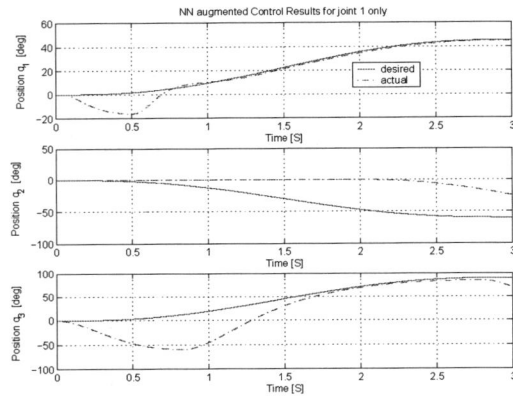

Fig. 3. Simulation results with Neural Network applied to joint 1

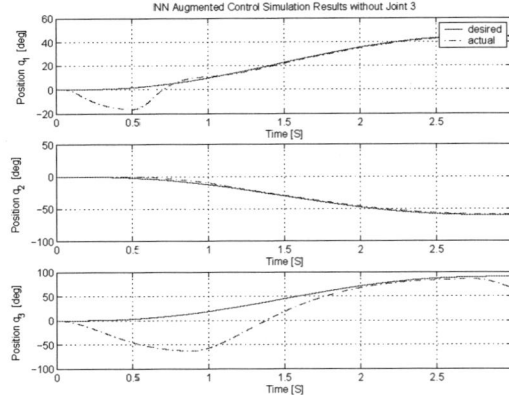

Fig. 4. Simulation results with Neural Network for joint 1 & 2

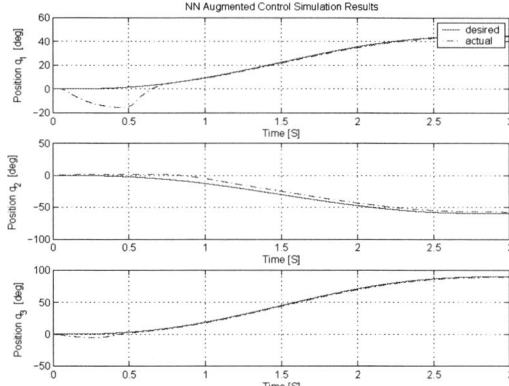

Fig. 5. Control with Neural Networks Results

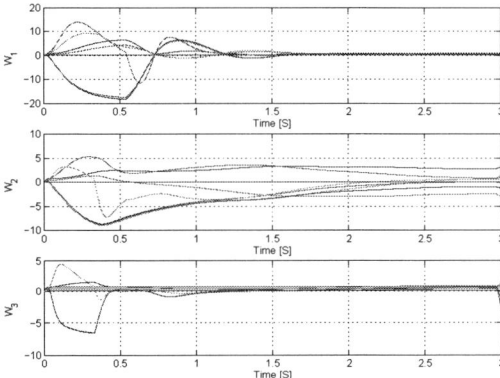

Fig. 6. Neural Networks parameters history

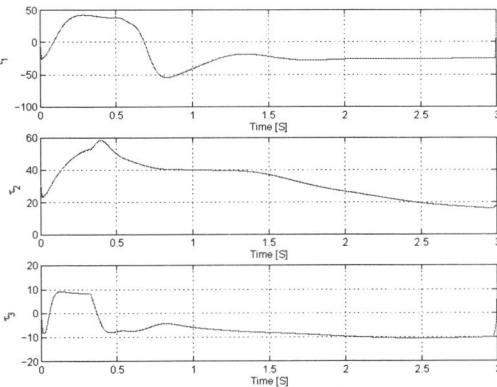

Fig. 7. Control Effort

6. CONCLUSION

In this paper, a nonlinear backstepping design scheme has been developed for the control of a robot manipulator augmented by a neural network to achieve control objectives in the presence of system uncertainties. The resulting closed-loop system is exponential stable to trajectory tracking error according to Lyapunov stability theorem. With this design approach, the simulation results of puma robot manipulator seemed to be excellent for the tracking in the operation space. For further research, we will try to estimate the velocity vector that is needed for feedback and we have to ensure that the tracking and observation errors will converge to zero.

REFERENCES

[1] E.Slotine and W.Li, On the Adaptive Control of Robot Manipulators, *Int.J. of Robotics Res* Vol. 6, No. 3, pages 49-59, 1987.

[2] H. Khalil and F. Esfandiari, Semiglobal Stabilization of a Class of nonlinear Systems Using Output Feedback, *IEEE Trans. Automatic Control.* Vol. 38, No. 9, pages 1412-1415, 1993.

[3] H. Khalil, Adaptive Output Feedback Control of nonlinear Systems Represented by Input-Output Models, *IEEE Trans. Automatic Control.* Vol. 41, No. 2, pages 177-188, 1996.

[4] M. Jankovic, Adaptive Output Feedback Control of nonlinear Feedback Linearizable Systems, *Int. J. Adaptive Control and Signal Processing.* Vol. 10, pages 1-18, 1996.

[5] H. Schwartz, Model Reference Adaptive Control For Robotic Manipulators without velocity measurements, *Int. J. Adaptive Control and Signal Processing.* Vol. 8, pages 279-285, 1994.

[6] H. Schwartz, An MRAC Output Feedback Controller for Robot Manipulators, *Mediterranean Conference on Control and Automation*, pages 2293-2301, 1999.

[7] J.J. Craig, Adaptive Control of Mechanical Manipulators, *Addison-Wesley*, 1988.

[8] S. Nisosia and P. Tomei, Robot control by using joint position measurements, *IEEE Transactions on Automatic Control*, Vol.35, No.9, pages 1058-1061, 1990.

[9] H. Berghuis and T. Nijmeijer, Tracking control of robots using only position measurement, *Proceedings of the 30th Conference on Decislon and Control*, pages 1039-1040, Brighton, England, Dec 1991.

[10] M. Belkheiri and F. Boudjema, Adaptive Neural Network controller for a class of uncertain nonlinear systems, *Transactions on Systems, Signals and Devices*, Vol 2. N. 1 pages 1-18, Shaker Verlag, Germany, August 2007.

[11] N. Hovakimyan, A. J. Calise and N. Kim. Adaptive output feedback control of a class of multi-input multi-output systems using neural networks. *Int. Journal of Control*, vol. 77, no. 15, 2004.

[12] W. Yu and X. Li, Some new results on system identification with dynamical neural networks, *IEEE Trans. on Neural Networks*, vol. 12 no. 2, march 2001.

[13] F. Nardi. Neural Network based Adaptive Algorithms for Nonlinear Control. *PhD thesis*, The Academic Faculty of The School of Aerospace Engineering, Georgia Institute of Technology, November 2000.

[14] R. R. Selmic and F. L. Lewis. Identification of nonlinear systems using RBF neural networks: application to multimodel failure detection. *XVI intern. conf. on Material Flow, Mach. And Dev. in Industry*, University of Belgrade, December 2000.

Direct Torque Control System for Permanent Magnet Synchronous Machine with Fuzzy Speed Pi Regulator

K. Nabti*. K. Abed. H. Benalla

*Electrotechnic's Laboratory of Constantine, Faculty of Engineering Sciences
Mentouri University, Campus Zerzara, Constantine, ALGERIA (e-mail: Idor2006@yahoo.fr).

Abstract: The Permanent Magnet Synchronous Machine (PMSM) speed regulation with a conventional PI regulator reduces the speed control precision, increase the torque fluctuation, and consequentially low performances of the whole system. With utilisation of fuzzy logic method, this paper presents the self adaptation of conventional PI regulator parameters Kp and Ki (proportional and integral coefficients respectively), using to regulate the speed in Direct Torque Control strategy (DTC). The ripples of both torque and flux are reduced remarkable, small overshooting and good dynamic of the speed and torque. Simulation results verify the proposed method validity.

Keywords: PMSM, Direct Torque Control, Fuzzy PI regulator,

1. INTRODUCTION

The PMSM control difficulty resides in the coupling of control variables such as flux and electromagnetic torque. Two principal strategies were developed almost at the same time in two different research centers, Direct Torque Control strategy was first introduced by Takahashi. I, in 1986 [5]. And Depenbrock. M, developed a similar idea in 1988 under the name of Direct Self Control [2]. The DTC is one of the recent researched control schemes which are based on the decoupled control of stator flux and torque providing a quick and robust response with a simple implementation in AC drives. DTC has the advantages of simplicity, good dynamic performance, and insensitive to motor parameters except the stator resistance. In DTC strategy the speed sensor is not essential for the flux and torque estimation. Basically Direct Torque Control employs two hysteresis controllers to regulate the stator flux and torque, which results in approximate decoupling between the flux and the torque control. The key issue of DTC design is how to choose a suitable stator voltage vector to keep the stator flux and torque in their hysteresis band.

The conventional DTC disadvantages are: the high torque ripples and the slow transient response to the step changes in torque during start-up. Several techniques have been developed to improve the torque performance [1], [3], [4], [6], [7]. The paper [6] realizes Fuzzy-PI speed regulation for induction machine; we utilize this idea for PMSM in this paper. The fuzzy control is basically nonlinear and adoptive in nature, giving robust performances in the face of parameter variations and load disturbance effects. The regulator inputs are speed error and its change. Self-adaptation PI regulator, based on conventional PI regulator, consist to adjust dynamically the conventional PI regulator parameters. Simulation results showing that the method improves the Direct Torque Control system performances.

2. DIRECT TORQUE CONTROL PRINCIPLES

2.1. Flux and torque estimation and control

In stationary reference frame, the machine stator voltage space vector is represented as follows:

$$V_s = R_s \cdot i_s + \frac{d\psi_s}{dt} \quad (1)$$

$$V_s = V_{s\alpha} + jV_{s\beta} = \sqrt{\frac{2}{3}}\left[V_{aN} + V_{bN} \cdot e^{\left(j\frac{2\pi}{3}\right)} + V_{cN} \cdot e^{\left(j\frac{4\pi}{3}\right)}\right] \quad (2)$$

Where: R_s, i_s, ψ_s stator resistance, current and flux respectively,

V_{aN}, V_{bN}, V_{cN} the three phase voltage inverter outputs given as follows:

$$V_{aN} = V_{sa} = \frac{U_c}{3}(2 \cdot C_1 - C_2 - C_3)$$

$$V_{bN} = V_{sb} = \frac{U_c}{3}(2 \cdot C_2 - C_1 - C_3) \quad (3)$$

$$V_{cN} = V_{sc} = \frac{U_c}{3}(2 \cdot C_3 - C_2 - C_1)$$

U_c is the inverter DC supply voltage, C1, C2, C3 are the switching table outputs, and they are relevant to the switching strategy.

Frome (1) we can estimate the stator flux as follow:

$$\psi_s = \psi_{s0} + \int (V_s - R_s \cdot i_s) \quad (4)$$

$$\|\psi_s\| = \sqrt{\psi_{s\alpha}^2 + \psi_{s\beta}^2} \quad (5)$$

ψ_s is the stator flux vector, and ψ_{s0} is its initial value.

For simplicity, it is assumed the stator voltage drop $R_s \cdot i_s$ is small and neglected, the stator flux variation can be expressed as: $\Delta\psi_s \approx V_s \cdot \Delta t$

The selected voltage vector

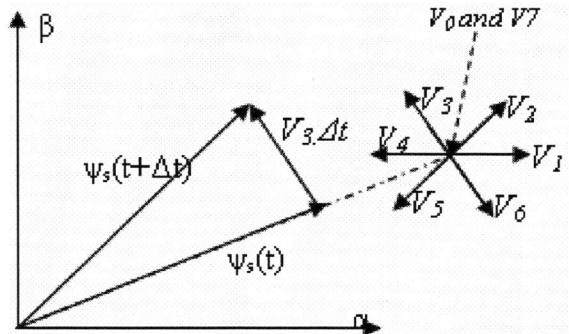

Fig. 1. Stator flux variation in stationary (α, β) frame

The machine's electromagnetic torque (*Tem*) can be evaluated as follow:

$$Tem = P(\psi_{s\alpha} \cdot i_{s\beta} - \psi_{s\beta} \cdot i_{s\alpha}) \quad (6)$$

In the DTC technique, the inverter switches are controlled using the flux and torque errors, and the position of the stator flux within the six-region control of the motor. The flux and torque errors are evaluated as follows:

$$\Delta\psi = \psi_{sref} - \psi_s \quad (7)$$

$$\Delta T = T_{ref} - T_{em} \quad (8)$$

$$\theta = tg^{-1}\left(\frac{\psi_\beta}{\psi_\alpha}\right) \quad (9)$$

Where θ is the angle between stator flux vector and α axis, $\psi_{s\alpha}, \psi_{s\beta}$ are the stator flux components in *(αβ)* reference frame.

In order to determine the inverter switching pattern using flux and torque errors, for the flux and torque control two hysteresis controllers are employed. The inverter is switched based on these errors and position of the stator flux within the six-region control, in such a way that the inverter output voltage vector minimizes the flux and torque errors and determines the out-put flux rotation direction of these controllers.

2.2. Switching table:

Table.1. DTC switching table

Flux	Couple	1	2	3	4	5	6
$K_\psi=1$	$K_{Cem}=1$	V_2	V_3	V_4	V_5	V_6	V_1
	$K_{Cem}=0$	V_7	V_0	V_7	V_0	V_7	V_0
	$K_{Cem}=-1$	V_6	V_1	V_2	V_3	V_4	V_5
$K_\psi=0$	$K_{Cem}=1$	V_3	V_4	V_5	V_6	V_1	V_2
	$K_{Cem}=0$	V_0	V_7	V_0	V_7	V_0	V_7
	$K_{Cem}=-1$	V_5	V_6	V_1	V_2	V_3	V_4

3. CONTROL STRUCTURE

Figure (2) illustrates the PMSM drive scheme considered in this investigation. The drive consists of a Fuzzy PI speed controller, flux and torque controllers, space flux position, and PMSM. The rotor speed w is compared with the reference speed w_{ref}. The resulting error and its rate of change are processed in the fuzzy PI speed controller for each sampling interval. The output of this is considered to be the reference torque T_{ref}.

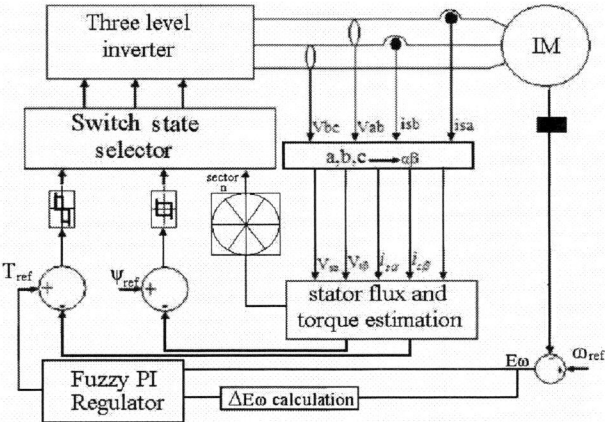

Fig.2 - A Direct Torque Control Diagram

4. FUZZY CONTROLLER

Speed error "$E\omega$," and it rate of change $\Delta E\omega$ are used as inputs to the fuzzy controller. Proportional coefficient Kp and integral coefficient Ki are the outputs of the fuzzy controller. The number of fuzzy segments is chosen to have maximum control with a minimum number of rules. Triangle membership functions have been used. The fuzzy membership functions of input and output variables are shown in (Fig. 3) [6].

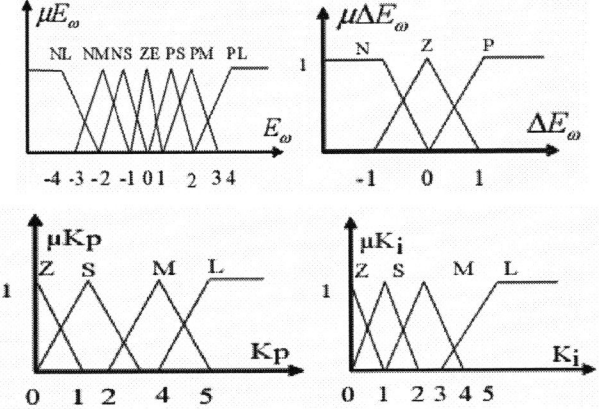

Fig. 3 - The fuzzy membership functions of input and output variables.

From the experience of simulation and experiment, the range of coefficients kp and kI are [1, 5] and [0.005, 0.02], respectively [6]. The speed error universe of discourse is divided into seven fuzzy sets. {NL (negative large), NM (negative middle), NS (negative small), ZE (zero), PS (positive small), PM (positive middle), PL (positive large)}. rate of change $\Delta E\omega$ includes 3 fuzzy subsets, it is not necessary to subdivide it, because it's changes quickly in DTC. Out put membership KP and KI, both contain four fuzzy subsets as shown in (fig. 3).

There are total of 21 rules as listed in table 1. Each control rule can be described using the inputs variables speed error "$E\omega$", it rate of change $\Delta E\omega$, and output variables (controller parameters ki and kp). The ith rule R_i can be expressed as:

R_i: If $E\omega$ is A_i, $\Delta E\omega$ is B_i then KP is V_i, KI is W_i.

Where A_i, B_i, V_i and W_i denote the fuzzy subsets.

Table. 2- Fuzzy Rules Base

Kp, Ki		\multicolumn{7}{c}{$E\omega$}
		NL	NM	NS	ZE	PS	PM	PL
$\Delta E\omega$	N	L, Z	M, S	S, M	M, L	S, M	M, S	L, Z
	Z	L, Z	M, S	L, M	Z, L	L, M	M, S	L, Z
	P	L, Z	M, M	L, L	Z, L	L, L	M, M	L, Z

The inference method used in this paper is Mamdani's procedure (inference) based on min-max decision. The firing strength (applied fuzzy operators) αi, for i^{th} rules is given by:

$$\alpha_i = \min(\mu_{A_i}(E\omega), \mu_{B_i}(\Delta E\omega)) \qquad (10)$$

By fuzzy reasoning, Mamdani's minimum procedure gives:

$$\mu'_{V_i}(KP) = \min(\alpha_i, \mu_{V_i}(KP))$$
$$\mu'_{W_i}(KI) = \min(\alpha_i, \mu_{W_i}(KI)) \qquad (11)$$

Where μ_A, μ_B, KP_V and KI are membership functions of sets A, B V and W of the variables $E\omega$, $\Delta E\omega$, Kp and KI, respectively.

Thus, the membership functions μ_v and μ_w of the outputs KP and KI are given by:

$$\mu_V(KP) = \max_{i=1}^{21}(\mu'_{V_i}(KP))$$
$$\mu_{W_i}(KI) = \max_{i=1}^{21}(\mu'_{W_i}(KI)) \qquad (12)$$

The Maximum criterion method is used for defuzzification. The final single-valued output is obtained by this method.

5. SIMULATION RESULTS

To verify the technique proposed, digital simulations based on MATLAB/SIMULINK have been implemented. The PMSM used for the simulations has the following parameters [1]: U = 240V, f = 50Hz, Rs = 1.4Ω; Ld = 5.6*10⁻³ H; Lq = 5.8*10⁻³ H; Jf = 0.00176Kg/m²; fv = 0.00038818Nm/rad/s; ffe = 0.1546V/rad/s; p = 6, and conventional PI regulator ki = 0.441; kp = 0.0553. For proposed DTC and conventional DTC the dynamic responses of speed, flux, and torque for the starting process with 7Nm load torque applied at 0.7s.

We Inverse the speed at t=1s.

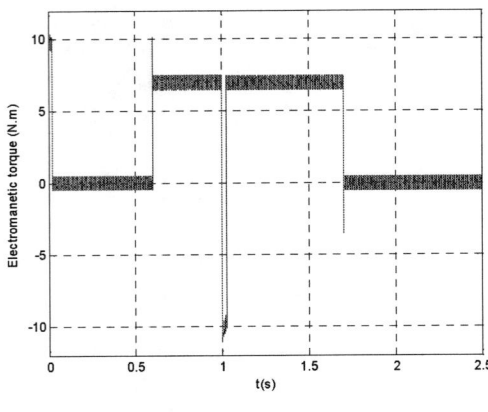

(a1)

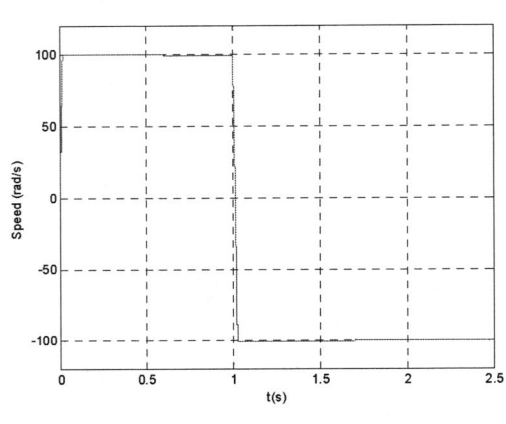

(b1)

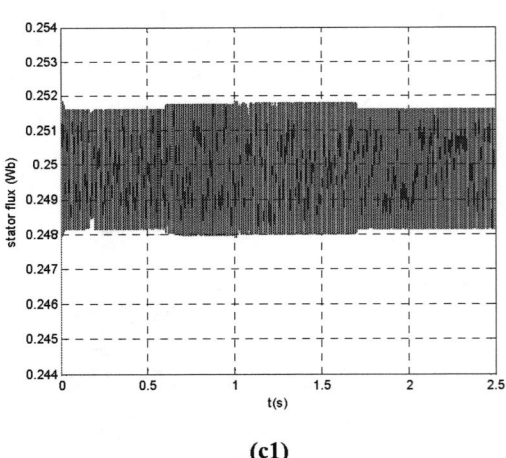

(c1)

Fig.4-(a1) torque response, (b1) speed response, (c1) flux response, for DTC with fuzzy PI regulation.

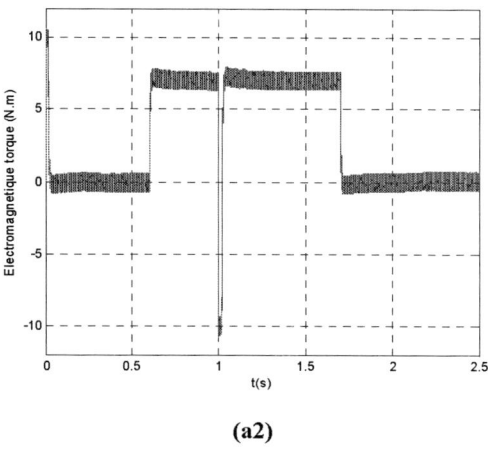

(a2)

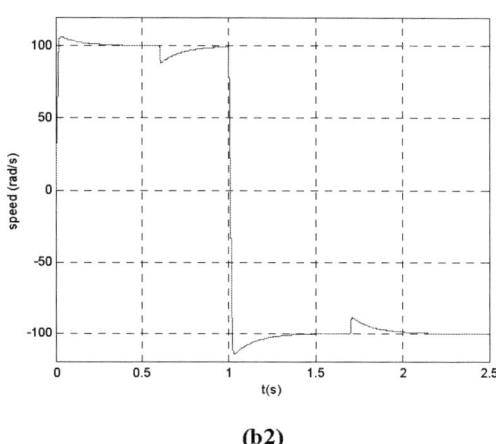

(b2)

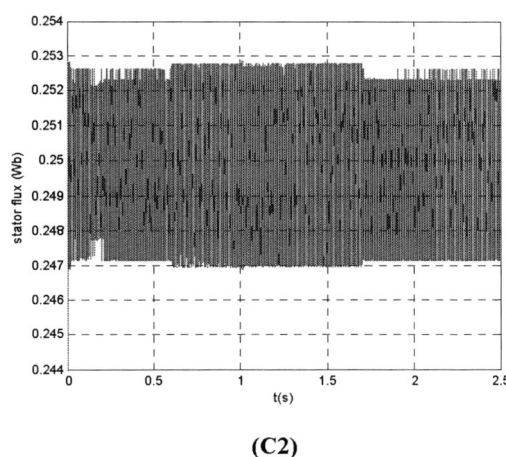

(C2)

Fig.5-(a1) torque response, (b1) speed response, (c1) flux response, for DTC with conventional PI regulator.

We compare between speed response for DTC with conventional regulator and fuzzy PI speed regulator.

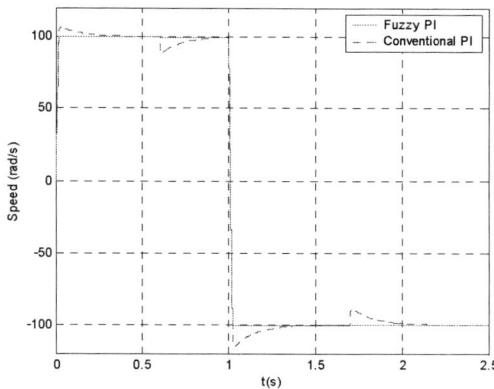

Fig.6. Speed responses for DTC conventional PI regulator and DTC fuzzy PI regulator.

The simulation results show that flux and torque responses are very rapid for two DTC methods. By proposed DTC technique, the ripple of torque in steady state is reduced remarkably compared with conventional DTC. A ripple of flux is decreased also.

In fuzzy PI regulation good dynamic responses of torque with neglected influence on speed which restored its reference quickly.

6. CONCLUSION

In this paper, a fuzzy logic direct torque control scheme using fuzzy PI regulator technique is presented. Using fuzzy logic technique, the kp and ki can be obtained dynamically which give a fast speed response. The simulation results suggest that FLDTC can achieve precise control of the stator flux and torque .Compared to conventional DTC, presented method the steady performances of ripples of both torque and flux are considerably improved.

7. REFERENCES

[1] Del Toro. X, Calls. S, Jayne. M. G, Witting. P. A, Arias. A, Romeral. J.L. 4-7 May 2004. *Direct torque control of an induction motor using a three-level inverter and fuzzy logic.* On Vol. 2, Page(s):923 - 927 vol. 2. Industrial Electronics, 2004 IEEE International Symposium.

[2] Depenbrock. M. 3(4), 1988. *Direct self-control (dsc) of inverter-fed induction machine.* 420-429. IEEE Transactions on Power Electronics.

[3] Jawad. F, Sharifian. M.B.B. 20 August 2001. *Comparison of different switching patterns in direct torque control technique of induction motors.* 60 (2001), 63–75.ELSEVIER, Electric Power Systems Research.

[4] Longji. Z, Wang. R. 14-16 Aug. 2004. *A novel direct torque control system based on space vector pwm.* Page(s) 755 - 760 Vol.2. Power Electronics and

Motion Control Conference, IPEMC 2004. The 4th International.

[5] Takahashi. I, Noguchi. T. 22(5), 1986. *A new quick-response and high-efficiency control strategy of an induction machine*. 820-827. IEEE Transactions on Industry Applications.

[6] Yang. J, Huang. J. 18-21 August 2005. *Direct torque control system for induction motors with fuzzy speed pi regulator*. 568-573. Proceedings of the Fourth International Conference on Machine Learning and Cybernetics, Guangzhou.

[7] Zelechowski. M, Kazmierkowski. M. P, Elaabjerg. F. 20-23 June 2005. *Controller design for direct torque controlled space vector modulated (dtc-svm) induction motor drives*. On Vol. 3, Page(s) 951 – 956. Industrial Electronics ISIE 2005. Proceedings of the IEEE International Symposium.

SESSION C1: MODELING AND IDENTIFICATION

Optimal Reduced-order Approximation of Fractional Dynamical Systems

R. Mansouri * M. Bettayeb ** S. Djennoune *

*Laboratoire de Conception et Conduite des Systèmes de Production
(L2CSP), Mouloud Mammeri University of Tizi Ouzou (15000), Algeria.
({Rachid_mansouri_ummto, s_djennoune}@yahoo.fr).
**Department of Electrical & Computer Engineering University of Sharjah,
United Arab Emirates. (maamar@sharjah.ac.ae)

Abstract: This paper deals with the fractional approximation of systems by integer reduced models. This approximation does not assume any restriction since the considered system can be commensurate or non commensurate, scalar or multivariable. Two model reduction schemes, namely, balanced truncation and singular perturbation balanced truncation are applied. To show the performance of this reduction, the reduced model obtained is compared to the integer model having the same order obtained by direct approximation. Several fractional models, are then considered. The obtained results are encouraging and attractive especially for the purpose of simulation and control of multivariable fractional systems.

Keywords: Fractional systems, Model approximation, State space representation, Integral Representation, Multivariable model, Model reduction, Hankel singular values

1. INTRODUCTION

The importance of the non integer differentiation (integration) also called fractional derivative and integral is now well established since it's used in many fields of science and technology. The complexity of the theory and the lack of adequate mathematical tools, to analyze and simulate fractional systems, are the principal reasons of its marginalisation in the past. Non integer system simulation, known to be with long memory, was one of the first problems to solve for obvious reasons [1, 2]. This difficulty is especially due to the global character of the non integer differentiation and integration operator. Indeed, the non integer system output at a given moment depends on all its past, knowledge which is difficult to take into account [3]. Several solutions are then proposed.

The first method is analytical; it uses Mittag-Leffler function [4] to determine the expression of the output system temporal response. The second method, uses discrete models and can be obtained in two different ways. A direct method that is based on Grünwald-Letnikov definition [3], indirect method, which consists in discretizing the differentiation operator [5, 6]. The third method consists in replacing the fractional differentiation operator by an integer transfer function which approximates it in a given frequency range [7, 8, 9]. This last method is that which is generally used.

Nevertheless, it gives a high order integer model, which is turn high order control laws expensive to realize. One can use a less accuracy approximation in order to obtain a relatively reduced-order model. However, for a given low order, the approximation error may not be acceptable. We show in this paper that this solution is not attractive in particular in the multivariable non integer systems case. A more adequate solution is then proposed: the use of reduction-order model techniques. This consists in approximating the non integer model by a large scale model in a very broad frequency range and using a high number of cells. The model thus obtained is certainly of high order but is more accurate. With the reduced-order methods, using the singular values, we can then obtain a low order model while keeping its dominant dynamic characteristics. The interest of these reduced-order techniques is the characterization of the H_∞ norm bounds of the approximation error, according to the distribution of singular values of the system, thus making it possible to impose a priori the approximation error.

To show the performance of this method, the reduced-order model obtained is compared with the model, having the same order, obtained by a direct approximation. Three numerical examples are then presented to illustrate the results.

2. NON INTEGER ORDER SYSTEMS

2.1 Definition of the non integer derivative

Let $f(t)$ be a causal function ($f(t) = 0$ for $t < 0$). The non integer Caputo's derivative, more adapted to represent physical systems, denoted $D^\alpha f(t)$, is given by: [10, 11].

$$D^\alpha f(t) = \frac{1}{\Gamma(r-\alpha)} \int_{t_0}^{t} \frac{f^{(r)}(\tau)}{(t-\tau)^{\alpha-r+1}} d\tau \quad (1)$$

$r > 0$ is an integer number such that $(r - 1 < \alpha < r)$, and Γ is the Euler Gamma function, defined by:

$$\Gamma(\lambda) = \int_{0}^{\infty} v^{\lambda-1} e^{-v} dv \quad (2)$$

2.2 State Space Representation

The fractional state space representation is defined, as in the integer case by two equations: [3, 12].
- A state equation where each state $x_i(t)$ is differentiated to a

* This paper is financially supported by L2CSP Laboratory

fractional order α_i, is called generalized state space model.
- An output equation as in the integer case.

$$\begin{cases} D^{(\alpha)}(x) = Ax + Bu \\ y = Cx + Du \end{cases} \quad (3)$$

where:

$$D^{(\alpha)}(x) = \begin{bmatrix} D^{\alpha_1} x_1, & D^{\alpha_2} x_2 & \ldots & D^{\alpha_r} x_n \end{bmatrix}^T$$

$x \in \Re^n, u \in \Re^m, y \in \Re^p, A \in \Re^{n \times n}, B \in \Re^{n \times m}, C \in \Re^{p \times n}, D \in \Re^{p \times m}$.

In the commensurate case all states $x_i(t)$ are differentiated to the same fractional order α, the state space model becomes:

$$\begin{cases} D^{\alpha} x = Ax + Bu \\ y = Cx + Du \end{cases} \quad (4)$$

where:

$$D^{\alpha}(x) = D^{\alpha} \begin{bmatrix} x_1, & x_2 & \ldots & x_n \end{bmatrix}^T$$

The properties of the fractional state space model are different from those of the integer case. Nevertheless, the controllability and observability properties are equivalent. [3, 13]. The transfer function representation of the fractional system (3) is done as in the integer case, it's given by:

$$G_{frac}(s) = C \left[\left(s^{(\alpha)} I_d - A \right)^{-1} \right] B + D \quad (5)$$

where: I_d is the identity matrix, and

$$s^{(\alpha)} I_d = diag \begin{bmatrix} s^{\alpha_1}, & s^{\alpha_2} & \ldots & s^{\alpha_n} \end{bmatrix} \quad (6)$$

2.3 State Space Representation Using Integral Operator

Usually, the state space representation of an integer linear time invariant system is:

$$\begin{cases} \dot{x} = Ax + Bu \\ y = Cx + Du \end{cases} \quad (7)$$

it can be simulated using the block diagram of Figure (1) when the state vector is the integrator output $x(t)$.

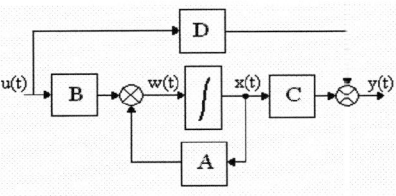

Figure 1. State representation system block diagram

In Equation (7), the state variables are the integral operator outputs; in our case we choose the integral operator input as state space vector and write the corresponding state space representation using the integral function instead of the derivative one. This yields to:

$$\begin{cases} w = A \int_{t_0}^{t} w(\tau) d\tau + Bu \\ y = C \int_{t_0}^{t} w(\tau) d\tau + Du \end{cases} \quad (8)$$

(8) is the integral representation of (7). The corresponding transfer function model is then given by:

$$G(s) = C \left[\frac{1}{s} I_d \right] \left[\left(I_d - A \left[\frac{1}{s} I_d \right] \right)^{-1} \right] B + D \quad (9)$$

The factorization of the Laplace variable s yields the classical formula of $G(s)$.

$$G(s) = C \left(s I_d - A \right)^{-1} B + D \quad (10)$$

The block diagram of the non-integer order system (3), is the same as given by Figure (1) in which the integer integrator is replaced by the generalized fractional integrator $I^{(\alpha)}(x)$:

$$I^{(\alpha)}(x) = \begin{bmatrix} I^{\alpha_1} x_1 & I^{\alpha_2} x_2 & \ldots & I^{\alpha_n} x_n \end{bmatrix} \quad (11)$$

In this case, Equation (8) can be expressed by:

$$\begin{cases} w = A I^{(\alpha)}(w) + Bu \\ y = C I^{(\alpha)}(w) + Du \end{cases} \quad (12)$$

The corresponding fractional transfer function model is:

$$G_{frac}(s) = C \left[\frac{1}{s^{(\alpha)}} I_d \right] \left[\left(I_d - A \left[\frac{1}{s^{(\alpha)}} I_d \right] \right)^{-1} \right] B + D (13)$$

where:

$$\frac{1}{s^{(\alpha)}} I_d = diag \begin{bmatrix} \frac{1}{s^{\alpha_1}}, & \frac{1}{s^{\alpha_2}} & \ldots & \frac{1}{s^{\alpha_n}} \end{bmatrix} \quad (14)$$

This new state representation, using the integral function, constitutes the key development of the approximation model presented in this paper.

3. FRACTIONAL SYSTEM APPROXIMATION

3.1 Fractional Integral Operator Approximation

Non integer systems simulation and realization require the approximation of an infinite dimensional fractional integration operator by a finite dimension rational model. Its poles and zeros are recursively distributed on a limited frequency range. To obtain a good accuracy in a bandwidth $[\omega_{min}, \omega_{max}]$, in particular in high frequencies, the poles and the zeros are distributed in a broader frequency range $[\omega_A, \omega_B]$ [7] such as:

$$\omega_A = \frac{\omega_{min}}{\sigma} \quad \text{and} \quad \omega_B = \sigma \, \omega_{max} \quad (15)$$

σ being an adjustment parameter chosen arbitrarily. In [7], author proposes to take $\sigma = 10$. The rational transfer function, denoted $\Im_{\alpha}(s)$, which approximates the fractional integral will have the particular form:

$$I^{\alpha} \approx I^{\alpha}_{[\omega_{min}, \omega_{max}]} = \Im_{\alpha}(s) = \frac{1}{s} D^{(1-\alpha)}(s) \quad (16)$$

$D^{(1-\alpha)}(s)$ being the $(1-\alpha)^{th}$ fractional derivative approximated by:

$$D^{(1-\alpha)}(s) = G_{\alpha} \frac{\prod_{i=-N}^{N} \left(1 + \frac{s}{\omega_{z,i}} \right)}{\prod_{i=-N}^{N} \left(1 + \frac{s}{\omega_{p,i}} \right)} \quad (17)$$

G_{α} is an adjustment parameter such that the fractional integrator $(1/s^{\alpha})$ and the integer transfer function $\Im_{\alpha}(s)$ have the same gain for $\omega_u = 1rd/s$. $\omega_{z,i}$ and $\omega_{p,i}$ are the zeros and poles of $\Im_\alpha(s)$, respectively. N is the number of cells. The transitional frequencies $\omega_{p,i}$ and $\omega_{z,i}$ are determined by the following recursive relations [7].

$$\begin{cases} \omega_{z,-N} = \omega_A \sqrt{\delta} \\ \omega_{p,i} = \eta \omega_{z,i} & i = -N, ..., N \\ \omega_{z,i+1} = \delta \omega_{p,i} & i = -N, ..., N-1 \end{cases} \quad (18)$$

with:

$$\delta = \left(\frac{\omega_B}{\omega_A}\right)^{\frac{\alpha}{2N+1}}, \quad \eta = \left(\frac{\omega_B}{\omega_A}\right)^{\frac{1-\alpha}{2N+1}}, \quad G_\alpha = (\omega_B)^\alpha \quad (19)$$

$\Im_\alpha(s)$ is strictly proper; it presents a fractional order integrator behaviour inside the frequency range $[\omega_{min}, \omega_{max}]$ and acts as a conventional integrator outside it. The state space model corresponding to $\Im_\alpha(s)$ is given by:

$$\begin{cases} \dot{z} = A_\alpha z + B_\alpha f \\ I^\alpha f = C_\alpha z \end{cases} \quad (20)$$

The input is the function $f(t)$ to be integrated and the output is the approximation of it's fractional α order integral. $z(t)$ is the $(2N+1)$ dimensional state vector. A_α, B_α and C_α are matrices of appropriate dimensions depending on the parameters of the fractional integrator approximation. A non integer order system is approximated by substituting, in the integral representation (18), the integration operator by its approximation $\Im^\alpha(s)$. In the following, a generalization of this method is summarized when the system is given by its state space representation (3). This approximation does not assume any restriction since the considered systems can be commensurate or not, scalar or multivariable. Moreover, the integer model parameters depend explicitly on the non integer system and the parameters related to the fractional integration operation (20). It can thus be very useful not to simulate the fractional system only, but to study its dynamic characteristics [14]. Another approximation using the derivative state space representation can be found in [15].

Theorem 1. [14]. Given $(A_{\alpha_i} B_{\alpha_i} C_{\alpha_i})$ the state space model approximating the non integer integrator $(1/s^{\alpha_i})$ in the frequency range $[\omega_{min}, \omega_{max}]$, then the integer model which approximate the fractional system (3) in the same frequency range is given by:

$$Sys_{int} : \begin{cases} \dot{Z} = A_G\, x + B_G\, u \\ y_a = C_G\, x + D_G\, u \end{cases} \quad (21)$$

y_a being the output vector approximation, and

$$\begin{cases} A_G = A_I + (B_I A C_I) \\ B_G = B_I B \\ C_G = C C_I \\ D_G = D \end{cases} \quad (22)$$

$A_G \in \Re^{(2N+2).n \times (2N+2).n}$, $B_G \in \Re^{(2N+2).n \times m}$, $C_G \in \Re^{p \times (2N+2).n}$, $D_G \in \Re^{p \times m}$. A_I, B_I, C_I are given by:

$$\begin{cases} A_I = Block-diagonal[A_{\alpha_1} \quad A_{\alpha_2} \quad ... \quad A_{\alpha_n}] \\ B_I = Block-diagonal[B_{\alpha_1} \quad B_{\alpha_2} \quad ... \quad B_{\alpha_n}] \\ C_I = Block-diagonal[C_{\alpha_1} \quad C_{\alpha_2} \quad ... \quad C_{\alpha_n}] \end{cases} \quad (23)$$

The fractional integration of each state variable w_i of the state space model using integral function (8) is approximated by the model (20). The integration of the state vector w of (8) is then approximated by setting n models (20) in parallel. The proof of this theorem is detailed in [14]

4. SVD BASED MODEL REDUCTION METHODS

Since the eighties, powerful singular values based model reduction schemes have been developed and successfully applied to many practical problems. The main advantage of these methods is without any doubt the characterization of the H_∞ norm bounds of the approximation error, which is very important for robust closed loop control system design [16, 17, 18, 19, 20].
In the following, two model reduction schemes are summarized: The balanced truncation using a direct truncation of the states associated with the small singular values of the system [20], and the singular perturbation balanced truncation which, contrary to the first technique, does not neglect the states associated to the small singular values but only dynamics of the fast modes [19].

Given a linear time invariant multivariable system Σ with a minimal realization:

$$G = \left[\begin{array}{c|c} A & B \\ \hline C & D \end{array}\right] = C\left(sI - A\right)^{-1} B + D \quad (24)$$

The corresponding state space model is:

$$\begin{cases} \dot{x} = A\, x + B\, u \\ y = C\, x + D\, u \end{cases} \quad (25)$$

We represente a full nth-order model by: $G_n = \left[\begin{array}{c|c} A_n & B_n \\ \hline C_n & D_n \end{array}\right]$ where $A_n \in \Re^{n \times n}$, $B_n \in \Re^{n \times m}$, $C_n \in \Re^{p \times n}$, $D_n \in \Re^{p \times m}$, and let $G_r = \left[\begin{array}{c|c} A_r & B_r \\ \hline C_r & D_r \end{array}\right]$ be the rth-order reduced model where $A_r \in \Re^{r \times r}$, $B_r \in \Re^{r \times m}$, $C_r \in \Re^{p \times r}$ and $D_r = D_n$.

4.1 Hankel singular values

Let $\mathcal{P}$ and $\mathcal{Q}$ the unique Hermitian positive definite solutions to the Lyapunov equations:

$$A\mathcal{P} + \mathcal{P}A^T + BB^T = 0\,; \quad A^T\mathcal{Q} + \mathcal{Q}A + C^TC = 0 \quad (26)$$

The Hankel singular values of G, noted $\sigma_i(G)$, such that $\sigma_1 > \sigma_2 > \cdots > \sigma_n > 0$ are the eigenvalues square roots of $\mathcal{P}\mathcal{Q}$.

$$\sigma_i(G) = \sqrt{\lambda_i\left(\mathcal{P}\mathcal{Q}\right)} \quad (27)$$

4.2 Balanced system representation

A state space realization is called balanced if the controllability and observability gramians $\mathcal{P}$ and $\mathcal{Q}$ are equal to the matrix Σ given by:

$$\Sigma = diag\left(\sigma_1, \sigma_2, \cdots \sigma_n\right) \quad (28)$$

A linear transformation which gives this balanced state space representation is a matrix T solution to:

$$\mathcal{P}\mathcal{Q} = T\, \Sigma^2\, T^{-1} \quad (29)$$

4.3 Balanced truncation method

Partition the balanced realization associated to the original system (24), as:

$$G(s) = \left[\begin{array}{c|c} A_b & B_b \\ \hline C_b & D_b \end{array}\right] = \left[\begin{array}{cc|c} A_{b\,11} & A_{b\,12} & B_{b\,1} \\ A_{b\,21} & A_{b\,22} & B_{b\,2} \\ \hline C_{b\,1} & C_{b\,2} & D_b \end{array}\right] \quad (30)$$

where

$$A_b = T^{-1}AT, \ B_b = T^{-1}B, \ C_b = CT \text{ and } D_b = D \quad (31)$$

with the balanced gramian Σ partitioned conformably as:

$$\Sigma = \left[\begin{array}{c|c} \Sigma_1 & 0 \\ \hline 0 & \Sigma_2 \end{array}\right] \quad (32)$$

such that, Σ_1 and Σ_2 have no common diagonal entries and Σ_1 contains the largest singular values and Σ_2 contains the smallest. Balanced truncation scheme consists in carrying out a truncation of $n - r$ state variables x_2 associated to Σ_2. The rth-order reduced model thus obtained is given by:

$$G_{rb} = \left[\begin{array}{c|c} A_{b\,11} & B_{b\,1} \\ \hline C_{b\,1} & D_b \end{array}\right] \quad (33)$$

The reduced-order model G_{rb} has the following properties:

- G_{rb} is stable and the realization (33) is a minimal balanced realization.
- The H_∞ norm bounds of the approximation error is:

$$\|G - G_{rb}\|_\infty = 2 \sum_{j=r+1}^{n} \sigma_j \quad (34)$$

In general, the balanced truncation approximation results in reduced-order models with good transient properties but not necessarily good steady-state properties.

4.4 Singular perturbation balanced truncation method

From the balanced realization (30), instead of eliminating the state variables x_2 associated to Σ_2, we only neglect their dynamic, ($\dot{x}_2 = 0$). The corresponding reduced-order model is given by:

$$G_{psb} = \left[\begin{array}{c|c} A_{b\,11} - A_{b\,12} A_{b\,22}^{-1} A_{b\,21} & B_{b\,1} - A_{b\,12} A_{b\,22}^{-1} B_{b\,2} \\ \hline C_{b\,1} - C_{b\,2} A_{b\,22}^{-1} A_{b\,21} & D - C_{b\,2} A_{b\,22}^{-1} B_{b\,2} \end{array}\right] \quad (35)$$

G_{psb} has the following properties:

- G_{psb} is stable and is a minimal balanced realization with balanced gramian Σ_1.
- The H_∞ norm bounds of the approximation error is also:

$$\|G - G_{psb}\|_\infty = 2 \sum_{j=r+1}^{n} \sigma_j \quad (36)$$

- The steady-state error is zero, $\left(G_{psb}(0) = G(0)\right)$.

5. SIMULATION RESULTS

In this section we apply the two model reduction schemes summarized in Section 4 to three numerical examples. To obtain a good accuracy approximation, the fractional integration operator is initially approximated in a broader frequency range, and using a high number of cells. the integration operator is then replaced, in the non integer system, by its approximation. An integer large order model is then obtained and reduced using the two reduced model techniques. On the other hand, the fractional system is directly approximated with a model having the same order than the reduced-order model. The integration operator is then approximated in the same frequency range, but with a reduced number of cells.

5.1 Fractional integration operator approximation

The integration operator $s^{-0.65}$ is approximated in the frequency range $[10^{-4}, 10^{+4}]$ using $CRONE$ method [7] with $N = 20$. A $41th$-order model is obtained. This one is then reduced to $5th$-order. On the other hand, $s^{-0.65}$ is directly approximated with a $5th$-order model. It is approximated in the same frequency range, but with $N = 2$ cells. Figure (2) shows the first 10 Normalized Hankel Singular values, Figure (3) shows the Bode diagram and Figure (4) the step response.

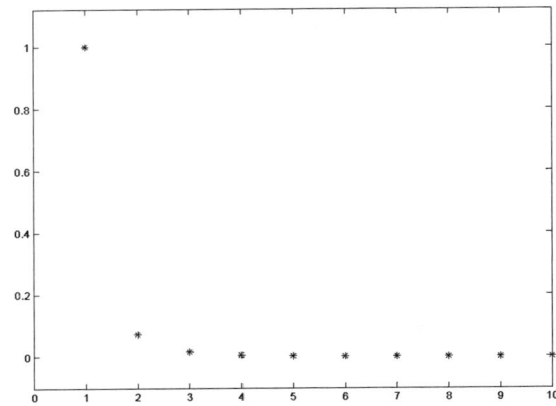

Figure 2. Normalized Hankel Singular values

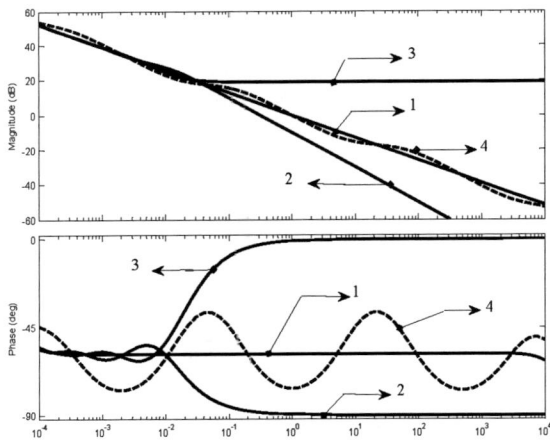

Figure 3. Bode diagram of the different integer approximations.
1-High order model, 2-Balanced truncation, 3-Singular perturbation balanced truncation, 4-Direct approximation.

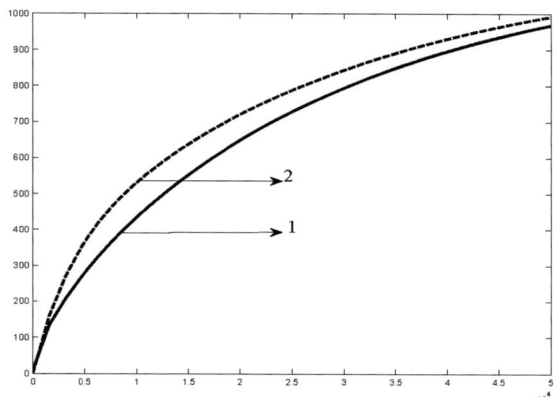

Figure 4. Step responses of different integer approximations.
1-High order model, Balanced truncation, and Singular perturbation balanced truncation, 2-Direct approximation.

The Bode diagram (gain diagram in particular) shows that direct approximation with a reduced number of cells (curve 4) seems to give the best results. Nevertheless, step response shows that reduced-order models (curve 1) obtained using the reduced order techniques are better. Curve 1 represents the step response of the original 51-order and its reduced-order models. Curve 2 represents the response of the $5th$-order model obtained by a direct approximation.

5.2 Fractional system approximation

The reduced-order integer model which approximates the non integer order system is very attractive for the large order system case. To show the interest of the reduced-order method, consider the explicit fractional system given by the transfer function model (37).

$$G_{frac}(s) = \frac{164}{s^{0.8} - 20\, s^{0.4} + 164} \frac{800}{s^{1.2} + 40\, s^{0.6} + 800} \quad (37)$$

The corresponding state space model is.

$$Sys_{frac} : \begin{cases} D^{(\alpha)}(x) = Ax + Bu \\ y = Cx + Du \end{cases} \quad (38)$$

where:

$$A = \begin{bmatrix} 0 & 1 & 0 & 0 \\ -164 & 20 & 0 & 0 \\ 0 & 0 & 0 & 1 \\ 800 & 0 & -800 & -40 \end{bmatrix}, B = \begin{bmatrix} 0 \\ 164 \\ 0 \\ 0 \end{bmatrix},$$

$$C = [\,0\ 0\ 1\ 0\,],\ D = [0]$$

$$\alpha = [\,0.4\ 0.4\ 0.6\ 0.6\,]$$

System (38) is approximated using the reduced-order methods presented in Section 3, where the fractional integration operator is approximated in the frequency range $[10^{-4}, 10^{+4}]$ with $N = 10$. A $84th$-order model is then obtained and reduced to $10th$-order. To approximate model (38) with approximately an $10th$-order model, we must use only $N = 1$ cell. This is impossible because the approximated integer model is unstable. To obtain an stable model we must use $N = 3$ cells. The approximated model is an $28th$-order. The results obtained are illustrated in figures (5), (6) and (7).

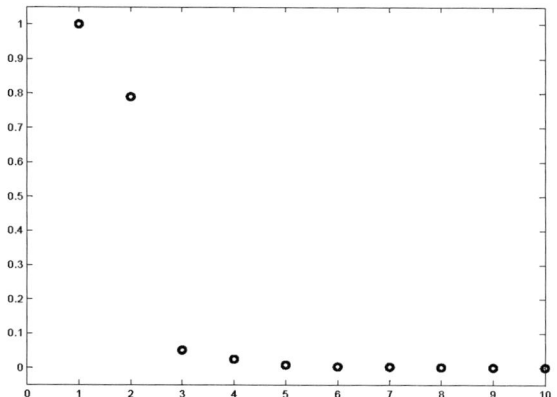

Figure 5. Normalized Hankel Singular values

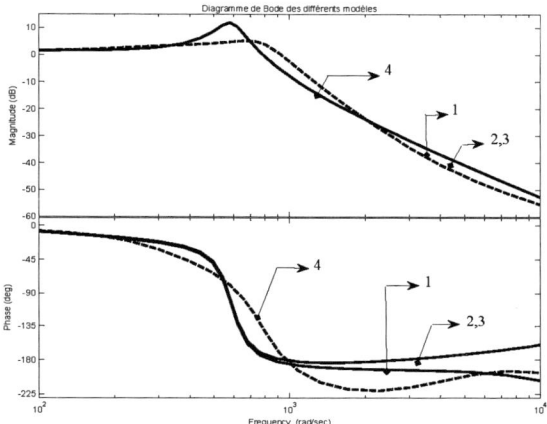

Figure 6. Bode diagram of the different integer approximations.
1-High order model, 2-Balanced truncation, 3-Singular perturbation balanced truncation, 4-Direct approximation.

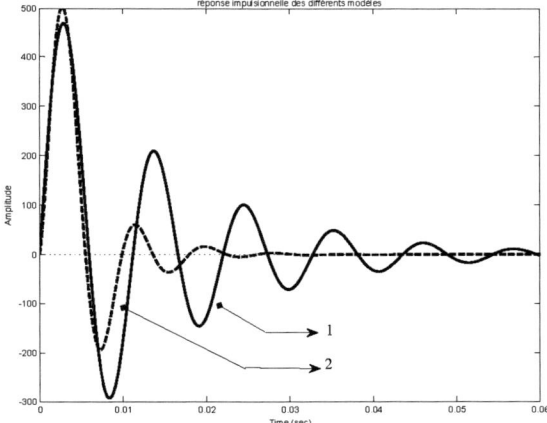

Figure 7. Impulse responses of different integer approximations. 1-High order model, Balanced truncation, and Singular perturbation balanced truncation, 2-Direct approximation.

In this case, the Bode diagram shows that balanced truncation, and singular perturbation balanced truncation methods (curve 2, 3) agree very well with the full-order model whereas the method using a direct approximation (curve 4) is the least accurate. These results are confirmed also in terms of impulse response. Moreover, this shows that the reduced-order models can be reduced again. This is not the case when a direct approximation is used.

All these results are summarized in Table (1) which represents the normalized error H_∞-norm defined by:

$$\varepsilon = \frac{\|G_n - G_r\|_\infty}{\|G_n\|_\infty} \qquad (39)$$

	reduced-order model	Direct approximation
Fractional integrator	$1.5\,10^{-3}$	$1.02\,10^{-1}$
explicit model $r = 10$	$4.7\,10^{-2}$	$6.3\,10^{-1}$
explicit model $r = 28$	$9.3\,10^{-4}$	$6.3\,10^{-1}$

Table 1. Normalized H_∞ norm errors

These results show that the reduction methods give approximation errors smaller than the direct approximation.

6. CONCLUSION

In this paper we presented an approximation method of non integer system by a reducer-order integer model using two model reduction schemes; balanced truncation and singular perturbation balanced truncation. Two numerical examples are then considered. The results show that the reduced-order model obtained using these methods is better than that obtained when the non integer model is approximated using a reduced number of cells. This method is very attractive, not only for simulation of non integer systems, but especially for the non integer controllers realization. This suggests a systematical way for non integer order controllers design, such that $PI^\lambda D^\mu$ or $CRONE$ controllers, the third generation in particular.

REFERENCES

[1] S. Manabe. The non-integer integral and its application to Control systems. *English Translation Journal of Japan*, Vol. 3-4, pages 83-87, 1961.

[2] M. Ichise, Y. Nagayanagi and T. Kojima. An analog simulation of non integer order transfer functions for analysis of electrode processes. *Journal of Electroanalytical Chemistry and Interfacial Electrochemistry*, Vol. 33, pages 253-265, 1971.

[3] A. Oustaloup. La Dérivation non entière : Théorie, synthèse et application. Editions Hermès, Paris, 1995.

[4] I. Podlubny. Fractional differential equations. Academic Press, San Diego, 1999.

[5] Y. Q. Chen and K. L. Moore. Discretization schemes for fractional-order differentiators and integrators. *IEEE Transactions on Circuits and Systems-I: Fundamental Theory and Applications*, Vol. 49, n 3, pages 363-367, 2002.

[6] B. M. Vinagre, I. Podlubny, A. Hernandez and V. Feliu. Some approximations of fractional order operators used in control theory and applications. *Fractional Calculus & applied Analysis*, Vol. 3, n 3, pages 231-248, 2000.

[7] A. Oustaloup. La commande CRONE: du scalaire au multivariable, 2nd Edition. Hermès, Paris, 1999.

[8] A. Charef, H.H. Sun, Y.Y. Tsao and B. Onaral. Fractal system as represented by singularity function. *IEEE Transactions on Automatic Control*, Vol:37, pages 1465-1470, 1999.

[9] A. Charef. Modeling and analog realization of the fundamental linear fractional order differential equation. *Nonlinear Dynamics*, Vol. 46, pages 195-210, 2006.

[10] R. L. Magin. Fractional calculus in bioengineering. Begell House Punlishers, Inc., Connecticut, 2006.

[11] S.G. Samko, A.A. Kilbas and O.I. Marichev. Fractional integrals and derivatives. Gordon and Breach Science Publishers, 1993.

[12] D. Matignon. Représentations en variables d'état de modèles de guides d'ondes avec dérivation fractionnaire. Ph.D. Thesis, Université de Paris-Sud, Orsay, 1998

[13] M. Bettayeb and S. Djennoune. A note on the controllability and the observability of fractional dynamical systems. In: Proceedings of the 2nd IFAC Workshop on Fractional Differentiation and its Applications, july 19-21 Porto, 2006.

[14] R. Mansouri, M. Bettayeb and S. Djennoune. Multivariable fractional order system approximation using integral representation. *submitted to Mathematical and Computer Modelling February 2008*.

[15] R. Mansouri M. Bettayeb and S. Djennoune. Multivariable fractional order system approximation using derivative representation. *International Journal of Applied Mathematics*, Vol. 20, n 7, pages 983-1003, 2007.

[16] M. Bettayeb, L. M. Silverman and M. G. Safonov. Optimal approximation of continuous-time systems. *19th IEEE Conference on Decision and Control*, Albuquerque New Mexico, 21-24 December, pages 10-12, 1980.

[17] K. Glover. All optimal Hankel-norm approximations of linear multivariable systems and their H_∞-error bounds. *International Journal of Control*, Vol. 39, pages 1115-1193, 1984.

[18] D. Kavranoglu and M. Bettayeb. Characterization of the solution to the optimal H_∞ model reduction problem. *Systems and Control Letters*, Vol. 20, pages 99-107, 1993.

[19] . Y. Liu, and B. D. O. Anderson. Singular perturbation approximation of balanced systems. *International Journal of Control*, Vol. 50, pages 1379-1405, 1989.

[20] B. C. Moore. Principal component analysis in linear systems: Controllability, Observability and model reduction. *IEEE Transactions on Automatic Control*. Vol. AC-26, pages 17-31, 1981.

Parameters Identification for Motorcycle Simulator's Platform Characterization ⋆

L. Nehaoua * H. Arioui **

Institut National de Recherche sur les Transports et leur Sécurité, INRETS, 2 Av Gl Malleret-Joinville, 94114, Arceuil, France (e-mail: nehaoua@ inrets.fr).
**IBISC LSC-LAMI CNRS-FRE 2873, 40 rue Pelvoux 91020, Evry, France(e-mail: Hichem.Arioui@iup.univ-evry.fr).*

Abstract: This paper presents the dynamics modeling and parameters identification of a motorcycle simulator's platform. This model begins with some suppositions which consider that the leg dynamics can be neglected with respect to the mobile platform one. The objectif is to synthesis a simplified control scheme, adapted to driving simulation application, minimising dealys and without loss of tracking performance.
Electronic system of platform actuation is described. It's based on a CAN BUS communication which offers a large transmission robustness and error handling. Despite some disadvanteges, we adapted a control solution which overcome these inconvenients and preserve the quality of tracking trajectory.
A bref description of the simulator's platform is given and results are shown and justified according to our specifications.

1. INTRODUCTION

Human applications are becoming more widespread in different areas of scientific research at the international level. This is for several reasons, in order to prevent hazardous driving situations, learning novice drivers and detect behavioral anomalies (somnolence, drags,...etc). On the other hand, the context of road safety is an international problem and constitutes a daily struggle for different governments to reduce the number of accidents through prevention, as well as testing new systems of security and assistance.

Driving simulators were extensively used in the aeronautical and automotive areas. They present an inexpensive tools for training future bikers and develop new technological features. The main theories used in this area have been developed for flight simulation. However, the adaptation of these technologies to other simulators (cars and motorcycles) is possible by taking in care some characteristics of land vehicles. Indeed, their dynamics is more constrained and human - machine interaction is more rich. This situation is much more complicated for two-wheeled vehicles, minimization of risk and the lack of visibility leads to fatal consequences, while knowing that the power-mass ratio is more important.

Very little two-wheeled vehicles simulators are constructed comparing to those of cars. Main works were acheived by Honda, a leading manufacturer in the field of two-wheeled vehicles (Miyamaru et al. [2002]Chiyoda et al. [2000]). Honda has developed several prototypes from 1988, one of them has been heavily marketed, supported by a Japanese law demanding a few hours of riding simulation before awarding the motorcycle licence. PERCRO, a laboratory based in Italy, in the context of the MORIS project, have constructed a motorbike simulator based on a classical 6 degrees of freedom (DOF) Stewart platform (Ferrazzin et al. [2003]). Finally, a 5DOF simulator is designed in the Department of Mechanical Engineering at Padova University, espacially dedicated to the competition motorcycle (Cossalter et al. [2004]).

The main difficulty in developing a driving simulator is the choice of the mechanical platform engineering. The position of the different axes of rotation should be properly investigated in order to avoid false cues or unwanted motion. Actuation system is one of more important feature because it defines the simulator's bandwidth and hence, the motion frequency components that it can reproduce. The choice of electronic acquisition must be taken from a wide range of commercial or locally built solutions. Analog flow control offers best performances in spite of great vulnerability against noise measurements and sensors multiplication. Today, servocontrollers are more powerful and present digital solutions which are robust, requiring fewer sensors. The choice of a communication that facilitate the error management increase greatly the safety of subjects and thus satisfy one of the first requirements of the driving simulation.

2. PRELIMINARIES

The choice of the simulator architecture is guided by the necessary needs to have a sufficient perception during the riding simulation. Our goal is to reproduce the important inertial effects perceived for the application needs but not all the motorcycle movements. So, the real amplitude of the various DOF was not a dominating object during the design phase.

Based on these considerations, the number of degrees of freedom for our architecture is determined. We aim to develop a mechanical platform for training and behavioral study of two-wheeled vehicles users in a normal urban traffic and in situations of risk. Therefore, motorcycle dynamics is analyzed (Cossalter [2002]) in order to establish a precise specifications for the development of a first prototype (Figure 1). After several investigations, three movements were privileged:

⋆ This research was supported by funds from the NATIONAL RESEARCH AGENCY in SIMACOM project.

- Roll: is the most important mouvment for the reproduction of turns, slalom and lane change. Unlike car driving, the rider make an effort on handlebars to tilt his motorcycle. This inclination generates a gravity force that compensates the centrifugal one and thus help to keep the rider balance in turn. Other riding styles are also possible (riding with upper body inclination), these will not be taken into account in our first prototype.
- Pitch: this rotation allows to reproduce longitudinal accelerations present during acceleration and braking phases. This effect is vital for more realism of the riding simulation. Unlike speed which can be reproduced with the simple visual projection, acceleration can't be felt without body excitation via a mechanical motion. Special techniques called motion cueing algorithms can transform the actual motorcycle motion in an achievable movement by the simulator, which can create an acceleration illusion without causing sensory conflict (Nehaoua et al. [2007], Nehaoua et al. [2006]).
- Yaw for slippage of the rear wheels in accident situations.

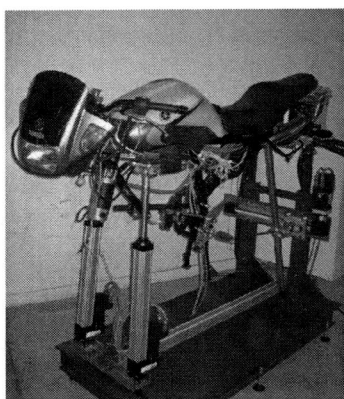

Fig. 1. Constructed riding motorcycle simulator

3. INVERSE KINEMATICS FORMULAION

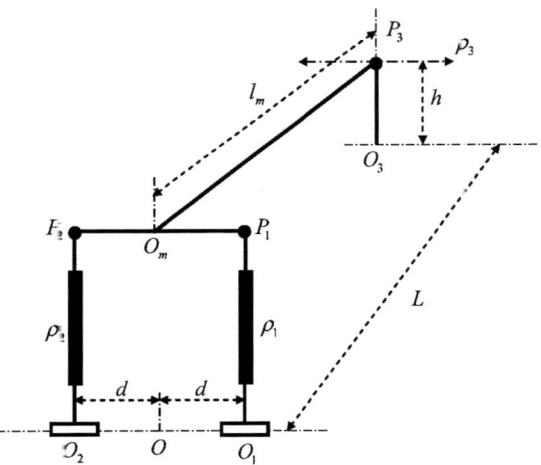

Fig. 2. Kinematics scheme of the simulator's platform

Let $\Re(O, i, j, k)$ be the fixed reference and $\Re_m(O_m, i_m, j_m, k_m)$ the platform mobile reference. O_1, O_2 and O_3 are respectively the attachment points of the two legs and the rear slide with the simulator's base. P_1, P_2 and P_3 are respectively the attachment points of the two legs and the rear slide with the upper mobile platform. The configuration of the reference frame $\Re_m$ is characterized by the position (x_m, y_m, z_m) of its origin point and the three Euler orientation angles (ψ, θ, φ), corresponding respectively to the yaw, pitch and roll angles. We take the Z-Y-X convention to compute the rotation matrix, as following:

$$R = R_\psi R_\theta R_\varphi = \begin{pmatrix} r_{11} & r_{12} & r_{13} \\ r_{21} & r_{22} & r_{23} \\ r_{31} & r_{32} & r_{33} \end{pmatrix} \quad (1)$$

or in a detailed form, where $c \equiv \cos$ and $s \equiv \sin$:

$$\mathbf{R} = \begin{pmatrix} c\theta c\psi & s\varphi s\theta c\psi - c\varphi s\psi & c\varphi s\theta c\psi + s\varphi s\psi \\ c\theta s\psi & s\varphi s\theta s\psi + c\varphi c\psi & c\varphi s\theta s\psi - s\varphi c\psi \\ -s\theta & s\varphi c\theta & c\varphi c\theta \end{pmatrix} \quad (2)$$

The vector $\mathbf{OP_3}$ is given in the fixed base reference by $\mathbf{OP_3}^O = (-L, \rho_3, h)^T$. Using the transformation matrix R the same vector can be written as following:

$$\mathbf{OP_3}^O = \mathbf{OO_m}^O + R\mathbf{O_m P_3}^m \quad (3)$$

where, $\mathbf{OO_m}^O = (x_m, y_m, z_m)^T$ and $\mathbf{O_m P_3}^m = (-l_m, 0, 0)^T$. Replacing the different vectors components in equation (3) we can deduce the coordinates of the mobile reference origin O_m and the rear slide displacement ρ_3 such as:

$$\begin{cases} x_m = -L + l_m r_{11} \\ y_m = 0 \\ z_m = h - l_m r_{31} \\ \rho_3 = -l_m r_{21} \end{cases} \quad (4)$$

where L, h and l_m are geometric constants (figure 7).

Now, the leg vector equation for $i = 1, 2$ is:

$$\mathbf{O_i P_i}^O = \mathbf{O_i O}^O + \mathbf{OO_m}^O + R\mathbf{O_m P_i}^m \quad (5)$$

where, $\mathbf{O_m P_i}^m = (0, (-1)^{i+1}l, 0)^T$, $\mathbf{O_i O}^O = (-1)^i d\mathbf{j}$ and d is the coordinate of the two cylindrical joints O_1 and O_2. l is the constant length between points P_1 and P_2. Replacing in equation (5) we can deduce the components of vectors $\mathbf{O_i P_i}^O = \rho_i u_i$ as following:

$$\rho_i \mathbf{u_i} = \begin{pmatrix} -L + l_m r_{11} + (-1)^{i+1} l r_{12} \\ 0 \\ h + l_m r_{31} + (-1)^{i+1} l r_{32} \end{pmatrix} \quad (6)$$

and

$$\dot{d} = l r_{22} \quad (7)$$

where, $\mathbf{u_i}$ is the unit vector along the leg axis and $\rho_i^2 = \mathbf{O_i P_i}^T \mathbf{O_i P_i}$ are the legs lengths.

To determine the inverse Jacobian matrix, we note the Euler angle rates vector by $\dot{\mathbf{q_r}} = (\dot{\psi}, \dot{\theta}, \dot{\varphi})^T$. The velocity of the leg elongation is given by:

$$\dot{\rho_i} = \mathbf{O_i \dot{P}_i}^T \mathbf{u_i} \quad (8)$$

Deriving equation (5) and replacing in (8) we find that:

$$\dot{\rho_i} = (-1)^i \mathbf{u_i}^T \dot{d}\mathbf{j} + \mathbf{u_i}^T . \mathbf{O\dot{O}_m} + (\Omega \times \mathbf{O_m P_i})^T \mathbf{u_i} \quad (9)$$

where $\Omega = \gamma \dot{\mathbf{q_r}}$ is the mobile platform angular velocity expressed in the fixed reference frame, and γ is the matrix transformation between the angular velocities and Euler rates. Equation (9) can be written using the property of the mixed vector product $(\mathbf{u} \times \mathbf{v}).\mathbf{w} = (\mathbf{w} \times \mathbf{u}).\mathbf{v}$ as following:

$$\dot{\rho}_i = (-1)^i \mathbf{u_i}^T \dot{d}\mathbf{j} + \mathbf{u_i}^T.\mathbf{O}\dot{\mathbf{O}}_\mathbf{m} + (\mathbf{u_i} \times \mathbf{P_i O_m})^T \Omega \quad (10)$$

For the rear slide velocity, we derive the equation of $\rho_3 = \mathbf{j}^T \mathbf{O}\dot{\mathbf{P}}_3$. Rearranging, we obtain:

$$\dot{\rho}_3 = \mathbf{j}^T \mathbf{O}\dot{\mathbf{O}}_\mathbf{m} + (\mathbf{j} \times \mathbf{P_3 O_m})^T \Omega \quad (11)$$

From equations (10) and (11) and knowing that $\mathbf{u_i}^T \mathbf{j} = 0$, we can deduce that:

$$\rho = J_{-1} W \quad (12)$$

where, $\rho = (\rho_1, \rho_2, \rho_3)^T$, $W = (\mathbf{O}\dot{\mathbf{O}}_\mathbf{m}, \dot{\mathbf{q}}_\mathbf{r})^T$ and the inverse Jacobian is:

$$J_{-1} = \begin{bmatrix} \mathbf{u_1}^T & (\mathbf{u_1} \times \mathbf{P_1 O_m})^T \\ \mathbf{u_2}^T & (\mathbf{u_2} \times \mathbf{P_2 O_m})^T \\ \mathbf{j}^T & (\mathbf{j} \times \mathbf{P_3 O_m})^T \end{bmatrix} \quad (13)$$

The platform is designed to perform three rotations of $\pm 15°$, The inverse Jacobian matrix is always invertible within this domain, so there is no singularity of the platform within its workspace.

4. DYNAMICS OF THE PLATFORM

In this section, a simple dynamics formulation of the simulator's platform will be demonstrated. The main objective is to propose a control scheme adapted for our riding application and, to characterize the platform capabilities. For this, we neglect at first time the contribution of legs dynamics and we focus on the upper platform motion. Applying Newton-Euler equations on the mobile platform gives (Dasgupta and Mruthyunjaya [1998]):

$$\begin{aligned} m_p \mathbf{g} + \mathbf{F_1} + \mathbf{F_2} + \mathbf{F_3} &= m_p \mathbf{O}\ddot{\mathbf{G}}_p \\ m_p \mathbf{O_m G_p} \times \mathbf{g} + \mathbf{O_m P_1} \times \mathbf{F_1} + \mathbf{O_m P_2} \times \mathbf{F_2} + \\ \mathbf{O_m P_3} \times \mathbf{F_3} &= m_p \mathbf{O_m G_p} \times \mathbf{O}\ddot{\mathbf{G}}_p + \mathbf{I_p}\dot{\Omega} + \Omega \times \mathbf{I_p}\Omega \end{aligned} \quad (14)$$

where $\mathbf{F_i}$, $i = 1..3$ are actuation forces of the front two legs and rear slide. m_p, $\mathbf{I_p}$ platform masse and inertia matrix. $\mathbf{O_m G_p}$ position of the mobile platform center of gravity G_p with respect to point O_m. All vectors and matrices are expressed in the global reference frame (O, i, j, k). Ω is the rotational velocity yet expressed in the inverse kinematics section. Combining the two equations into one algebraic formulation:

$$\mathbf{J}_{-1}^T \mathbf{F} = m_p \begin{bmatrix} \mathbf{I_3} \\ \widetilde{\mathbf{O_m G_p}} \end{bmatrix} (\mathbf{O}\ddot{\mathbf{G}}_p - \mathbf{g}) + \begin{bmatrix} \mathbf{0}_{3 \times 1} \\ \mathbf{I_p}\dot{\Omega} + \Omega \times \mathbf{I_p}\Omega \end{bmatrix} \quad (15)$$

with $\mathbf{I_3}$ is 3×3 identity matrix, $\mathbf{F} = \begin{bmatrix} \mathbf{F_1} & \mathbf{F_2} & \mathbf{F_3} \end{bmatrix}^T$, $\mathbf{0}_{3 \times 1} = \begin{bmatrix} 0 & 0 & 0 \end{bmatrix}^T$ and $\mathbf{J}_{-1}$ is the inverse jacobian matrix. $\mathbf{O}\ddot{\mathbf{G}}_p$ is the acceleration of the platform center of gravity with respect to global frame given by:

$$\mathbf{O}\ddot{\mathbf{G}}_p = \mathbf{O}\ddot{\mathbf{O}}_\mathbf{m} + \dot{\Omega} \times \mathbf{O_m G_p} + \Omega \times (\Omega \times \mathbf{O_m G_p}) \quad (16)$$

We Note $W = (\mathbf{O}\dot{\mathbf{O}}_\mathbf{m}, \Omega)^T$ the platform wrench. Then, $\mathbf{O}\ddot{\mathbf{G}}_p$ can be written in more convenient equation as:

$$\mathbf{O}\ddot{\mathbf{G}}_p = \begin{bmatrix} \mathbf{I_3} & -\widetilde{\mathbf{O_m G_p}} \end{bmatrix} \dot{W} + \widetilde{\Omega}^2 \mathbf{O_m G_p} \quad (17)$$

where $\widetilde{\Omega}$ is the skew matrix of vector Ω given by:

$$\widetilde{\Omega} = \begin{pmatrix} 0 & -\Omega_3 & \Omega_2 \\ \Omega_3 & 0 & -\Omega_1 \\ -\Omega_2 & \Omega_1 & 0 \end{pmatrix} \quad (18)$$

Replacing in equation (17) into (15) and with some algebraic manipulations we deduce the simplified dynamic model of the simulator's platform as:

$$\mathbf{M}\dot{W} + \mathbf{C} + \mathbf{G} = \mathbf{J}_{-1}^T \mathbf{F} \quad (19)$$

where, $\mathbf{M}$ is mass matrix, $\mathbf{C}$ is a nonlinear vector function of the angular velocity and $\mathbf{G}$ is the gravity term given as following:

$$\mathbf{M} = \begin{bmatrix} m_p \mathbf{I_3} & -m_p \widetilde{\mathbf{O_m G_p}} \\ m_p \widetilde{\mathbf{O_m G_p}} & \mathbf{I_p} - m_p \widetilde{\mathbf{O_m G_p}}^2 \end{bmatrix} \quad (20)$$

$$\mathbf{C} = \begin{bmatrix} m_p \widetilde{\Omega}^2 \mathbf{O_m G_p} \\ \widetilde{\Omega} \mathbf{I_p} \Omega + m_p \widetilde{\mathbf{O_m G_p}} \widetilde{\Omega}^2 \mathbf{O_m G_p} \end{bmatrix} \quad (21)$$

$$\mathbf{G} = -m_p \begin{bmatrix} \mathbf{I_3} \\ \widetilde{\mathbf{O_m G_p}} \end{bmatrix} \mathbf{g} \quad (22)$$

At this point, this model is convenient for describing the dynamics of fully actuated 6DOF platform. However, our architecture is a 3DOF where the three rotations and the three translation are dependent. Then, we must include three algebraic constraint equations. A simple formulation of Lagrange multiplyers is added to the model formula as:

$$\mathbf{M}\dot{W} + \mathbf{C} + \mathbf{G} + \Phi_q^T \lambda = \mathbf{J}_{-1}^T \mathbf{F} \quad (23)$$

with, Φ_q is the jacobian of the constraint matrix $\Phi(q,t)$ and λ are the Lagrange multiplyers. Due to the symmetrical representation of the mechanical platform, the algebraic constraints can be deduced from the coordinates expression of the vector $\mathbf{OO_m} = (x_m, y_m, z_m)^T$, so:

$$\Phi(q,t) = \begin{cases} x_m + L - l_3 c\theta c\psi - h_3(c\varphi s\theta c\psi + s\varphi s\psi) = 0 \\ y_m = 0 \\ z_m - h + l_3 s\theta - h_3 c\varphi c\theta = 0 \end{cases} \quad (24)$$

By derivation with respect to time variable, it results that $\dot{\Phi}(q,t) = \Phi_q W$, where:

$$\Phi_q = \begin{bmatrix} 1 & 0 & 0 & q_1 & q_2 & q_3 \\ 0 & 1 & 0 & 0 & 0 & 0 \\ 0 & 0 & 1 & 0 & q_4 & q_5 \end{bmatrix} \quad (25)$$

and:

$$\begin{cases} q_1 = l_3 c\theta s\psi + h_3(c\varphi s\theta s\psi - s\varphi c\psi) \\ q_2 = l_3 s\theta c\psi - h_3 c\varphi c\theta c\psi \\ q_3 = h_3 s\varphi s\theta c\psi - h_3 c\varphi s\psi \\ q_4 = l_3 c\theta + h_3 c\varphi s\theta \\ q_5 = h_3 s\varphi c\theta \end{cases} \quad (26)$$

Knowing that $\mathbf{O}\dot{\mathbf{O}}_\mathbf{m} = \Gamma \dot{\mathbf{q}}$, $\Omega = \gamma \dot{\mathbf{q}}$, then $W = \mathbf{A}\dot{\mathbf{q}}$ and $\dot{W} = \mathbf{A}\ddot{\mathbf{q}} + \dot{\mathbf{A}}\dot{\mathbf{q}}$, where $\mathbf{A} = \begin{bmatrix} \Gamma & \gamma \end{bmatrix}^T$, we can deduce a reduced representation of the dynamics model such:

$$\mathbf{M}'\ddot{\mathbf{q}} + \mathbf{C}' + \mathbf{G} = \mathbf{J}_{-1}^T \mathbf{F} - \Phi_q^T \lambda \quad (27)$$

where $\mathbf{M}' = \mathbf{MA}$, $\mathbf{C}' = \mathbf{M}\dot{\mathbf{A}}\dot{\mathbf{q}} + \mathbf{C}$ and $\mathbf{q} = (\psi, \theta, \varphi)$.

5. IDENTIFICATION

In this section, we expose an identification procedure to estimate the mass and inertia parameters. For this, the dynamics model must be written in linear form with respect to different parameters to be estimated. A simple manner to achieve this, is to express the dynamics model in the local frame of the upper platform, where the inertia matrix is constant. In the present work, we will continu with the previous formulation in the global frame, because we intend in the next works to extend this procedure to identify the whole platform parameters using a more complex model, which include the legs dynamics. From equation (15):

$$\mathbf{J}_{-1}^T\mathbf{F} - \Phi_q^T \lambda = m_p \begin{bmatrix} \mathbf{I}_3 \\ \widetilde{\mathbf{O}_m\mathbf{G}_p} \end{bmatrix}(\ddot{\mathbf{OG}}_p - \mathbf{g}) + \begin{bmatrix} 0_{3\times 1} \\ \mathbf{I_p}\dot{\Omega} + \Omega \times \mathbf{I_p}\Omega \end{bmatrix} \quad (28)$$

Matrix inertia in the global frame is function of the platform configuration such, $\mathbf{I_p} = \mathbf{R}\mathbf{I_0}\mathbf{R}^T$, where $\mathbf{I_0}$ is the constant inertia matrix of the platform about its center of gravity, supposed diagonal in a first approximation. So:

$$\mathbf{I_p} = \begin{bmatrix} r_{11} & r_{12} & r_{13} \\ r_{21} & r_{22} & r_{23} \\ r_{31} & r_{32} & r_{33} \end{bmatrix} \begin{bmatrix} I_1 & 0 & 0 \\ 0 & I_2 & 0 \\ 0 & 0 & I_3 \end{bmatrix} \begin{bmatrix} r_{11} & r_{21} & r_{31} \\ r_{12} & r_{22} & r_{32} \\ r_{13} & r_{23} & r_{33} \end{bmatrix} \quad (29)$$

Simplifying, we can deduce a linear formulation of $\mathbf{I_p}$ such that $\mathbf{I_p} = \mathbf{R_1}I_1 + \mathbf{R_2}I_2 + \mathbf{R_3}I_3$, where:

$$\mathbf{R_1} = \begin{bmatrix} r_{11}^2 & r_{11}r_{12} & r_{11}r_{13} \\ r_{11}r_{21} & r_{21}^2 & r_{21}r_{31} \\ r_{11}r_{31} & r_{21}r_{31} & r_{31}^2 \end{bmatrix} \quad (30)$$

$$\mathbf{R_2} = \begin{bmatrix} r_{12}^2 & r_{12}r_{22} & r_{12}r_{32} \\ r_{12}r_{22} & r_{22}^2 & r_{22}r_{32} \\ r_{12}r_{32} & r_{22}r_{32} & r_{32}^2 \end{bmatrix} \quad (31)$$

$$\mathbf{R_3} = \begin{bmatrix} r_{13}^2 & r_{13}r_{23} & r_{13}r_{33} \\ r_{13}r_{23} & r_{23}^2 & r_{23}r_{33} \\ r_{13}r_{33} & r_{23}r_{33} & r_{33}^2 \end{bmatrix} \quad (32)$$

Replacing this in the last term of equation (28) yeilds:

$$\begin{bmatrix} 0_{3\times 1} \\ \mathbf{I_p}\dot{\Omega} + \Omega \times \mathbf{I_p}\Omega \end{bmatrix} = \mathbf{R_1}''I_1 + \mathbf{R_2}''I_2 + \mathbf{R_3}''I_3 \quad (33)$$

where, $\mathbf{R_i}'' = \begin{bmatrix} 0_{3\times 1} & \mathbf{R_i}' \end{bmatrix}^T$ and $\mathbf{R_i}' = \mathbf{R_i}\dot{\Omega} + \widetilde{\Omega}\mathbf{R_i}\Omega$

Finally, the dynamics model can be written in a linear formulation of the different parameters as following:

$$\Phi_p \mathbf{p} = \mathbf{J}_{-1}^T \mathbf{F} \quad (34)$$

with $\Phi_p = \begin{bmatrix} \mathbf{M_g} & \mathbf{R_1}'' & \mathbf{R_2}'' & \mathbf{R_3}'' & \Phi_q \end{bmatrix}$, $\mathbf{p} = \begin{bmatrix} m_p & I_1 & I_2 & I_3 & \lambda \end{bmatrix}$ and $\mathbf{M_g} = \begin{bmatrix} \mathbf{I}_3 & \widetilde{\mathbf{O}_m\mathbf{G}_p} \end{bmatrix}^T (\ddot{\mathbf{OG}}_p - \mathbf{g})$.

Next, several methods were developed in litterature for the parametric identification. We choose the adaptive gradient method due to its simplicity in an off-line or an online implementation. Then, if $\tau = \mathbf{J}_{-1}^T \mathbf{F}$ this method consist at optimizing a quadratic cost function $\mathbf{J} = 1/2(\tau_{ref} - \tau)^2$, where τ_{ref} relates to measured actuation torques. Adaptation law is expressed as following:

$$\dot{\mathbf{p}} = -\mathbf{K}\frac{\partial \mathbf{J}}{\partial \mathbf{p}} \quad (35)$$

where $\mathbf{K}$ is adaptation matrix coefficient, ajusted to ensure rapid convergence and are also lied to the different excitation trajectories (slow or rapid reference trajectory). Finally, the different paramterts are obtained by integrating the following equation:

$$\dot{\mathbf{p}} = \mathbf{K}\Phi_p^T(\tau_{ref} - \Phi_p \mathbf{p}) \quad (36)$$

6. EXPERIMENTATION

6.1 Actuation system

Simulator's platform is electrically actuated. The front two legs are lead-secrew type driven by a Parker Hannifin SMBAB6045 brushless servomotor and a Lust CDD3000 servocontroller. The rear slide is actuated by an SMBAB82300 brushless servomotor coupled with Tecnoingranaggi MP060.1.10 reductor. The different servocontrollers offer a various Preset solutions to drive the platform (in position, velocity or torque scheme) and are equipped with CAN (Controller Area Network) modules for digital transmission. This constitutes a robust solution against noise and offers a big simplicity for the task management and errors handling. More, with CAN modules we can acquire position, velocity, actual torque and phase current without installing more sensors. This is a very flexible solution but it prensents three major disadvantages:

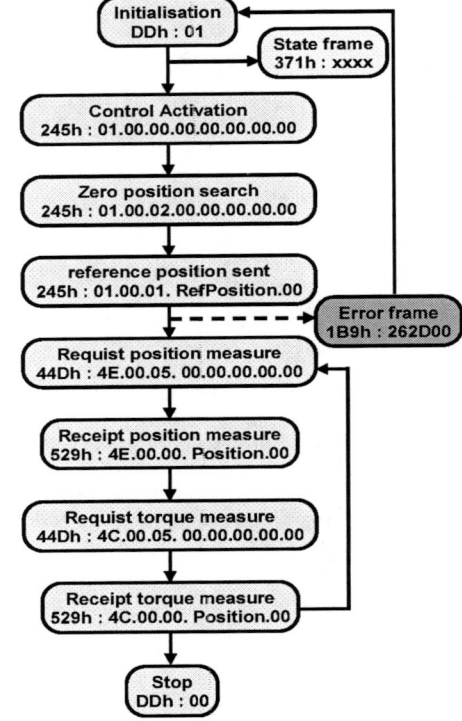

Fig. 3. CAN frames organization: for exemple initialisation is done by a frame with identifier 0DD (in hexa). An error frame resulting from course limitation is sent with an identifier 1B9

(1) all parameters (position, velocity, actual torque and phase current) are codded on the same CAN frame and hence, a parallel acquisition is not possible (constructor limitation)

(2) parameter channel is treated with low priority and every request and response CAN frame is fixed at 10ms (to acquire just one measure of position and torque, we must send two request frames and receipt two responses frame which present 20ms, figure 3)

(3) Comparing to analog flow control, this solution present some delays related to the computing time of the servocontroller and the transmission delays.

For our riding simulator application, such delays may creat a serious problems and contribute to the generation of the simulator's sickness. To overcome this limitations, and knowing that in driving simulation, delays minisation is more important than an accurate trajectory tracking, we have optimized the inner servocontroller control scheme without using an external control loop.

6.2 Control

We aim to explore the capabilities of two control methods, the first based on torque computing scheme, which need a good parameters identification. The second controller based on the optimisation of the inner control loop of the servocontroller to satisfy prior specifications (acceptable tracking error and sufficient stability margins).

Torque computing method allows the compensation of gravity and nonlinear terms with a feedforward term, as:

$$\tau = \tau_{ff} + \tau_{fb} = (\mathbf{M}'\ddot{\mathbf{q}}_d + \mathbf{C}' + \mathbf{G}) + \mathbf{M}'(\mathbf{K_p}\mathbf{e} + \mathbf{K_d}\dot{\mathbf{e}}) \quad (37)$$

where $\mathbf{e} = \mathbf{q}_d - \mathbf{q}$ is the tracknig error. $\mathbf{K_p}$, $\mathbf{K_d}$ are gain matrices to be adjusted. Remplacing torque τ in the equation of the dynamics model gives:

$$\mathbf{M}'\ddot{\mathbf{q}} + \mathbf{C}' + \mathbf{G} = (\mathbf{M}'\ddot{\mathbf{q}}_d + \mathbf{C}' + \mathbf{G}) + \mathbf{M}'(\mathbf{K_p}\mathbf{e} + \mathbf{K_d}\dot{\mathbf{e}}) \quad (38)$$

Then:

$$\ddot{\mathbf{e}} + \mathbf{K_p}\mathbf{e} + \mathbf{K_d}\dot{\mathbf{e}} = 0 \quad (39)$$

which means that $\mathbf{K_p}$, $\mathbf{K_d}$ can be adjusted by a classical control method such pole placement.

For the second method, each servocontroller contains three inner control loops, a Proportionnel position loop and two proportionnal-Integral loops for speed and current control. A pre-control solution is also implemented to compensate for external load inertia and friction by a prediction of articulation velocity and acceleration (figure 4).

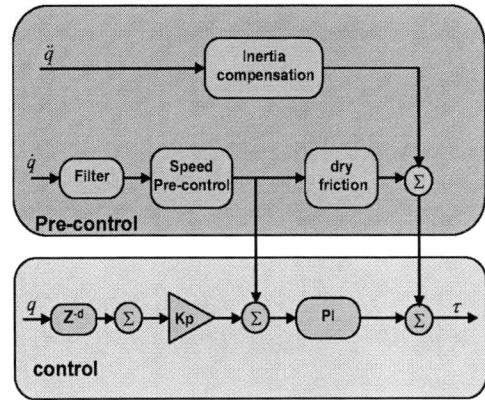

Fig. 4. Servocontroller inner control loops

Figure 5 presents the measured legs displacement for a step reference of $1cm$. It is shown that the performance of the computed torque method is better than of that obtained by the servocontroller loops. However, for our application, this is sufficient with a great advantage of reducing communications delays and time computing. An additional effort may be made to better optimize the Proportionnel and PI parameters of servo inner loops control. Noting that, in this experimentation, the pre-control part is disabled and this, carrefully used, may greatly increase control performances.

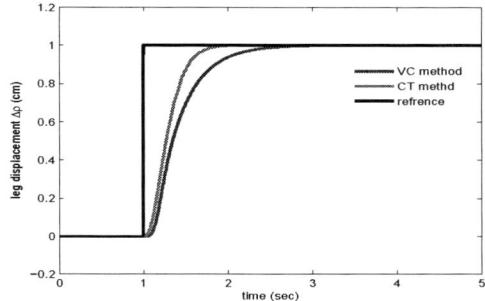

Fig. 5. Measured legs displacement for a step reference. (VC):using servocontroller control loops. (CT): using computing torque method

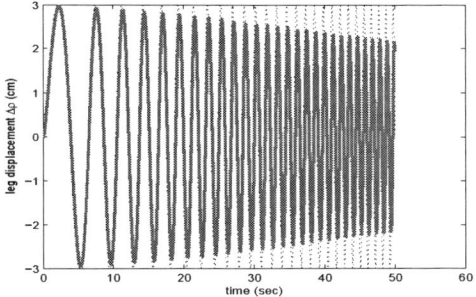

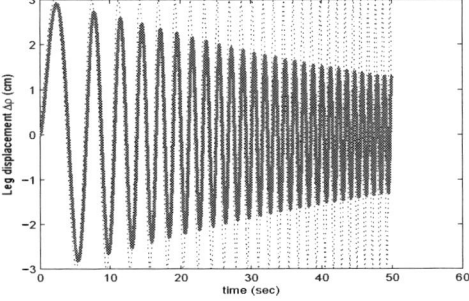

Fig. 6. Measured legs displacement for a chirp sine reference. (a): using computing torque method. (b):using servocontroller control loops.

Figure 6 shows the measured legs displacement for a chirp sine reference displacement with a variable frequency. Computed torque method allows more bandwith than the second method, this represents an important feature in driving simulation. Transitory components of motorcycle motion are difficult to reproduce with our simulator platform. These components are in the range of 3-5Hz, and are mainly used to reproduce an illusion of acceleration and braking. For this, psychophysical experimentation will be conducted to validate the simulator applicability in a motorcycle riding view point, and hence, propose different solutions for the next prototype.

6.3 Identification

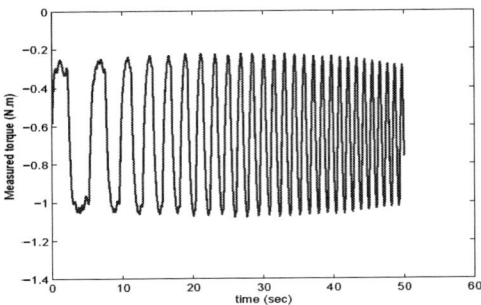

Fig. 7. Mesured leg torque

Figure 7 shows the measured torque corresponding to the chirp sine reference displacement of figure 6. Velocity and acceleration are obtained by numerical derivation. All variables are saved for un an off-line estimation procedure. Each DOF is separaly actuated in order to excit matrix inertia element by element. Adaptation gains are tuned by trail-error to have a fast convergence and according to the amplitude and frequency of excitation trajectory. Finally, identified parameters are presented in figure 8.

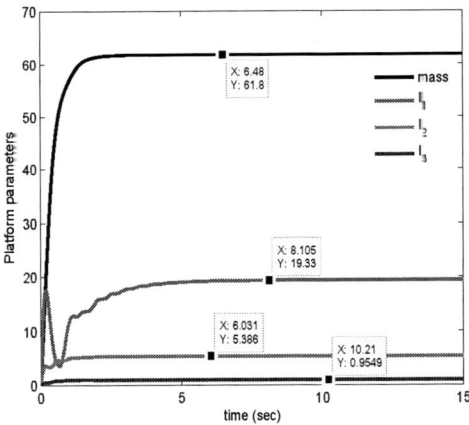

Fig. 8. Identified platform parameters

7. CONCLUSION

In the first part of this paper, the important points for the development of a riding simulator are exposed. We have justified the choice of the platform architecture and the actuation system to drive the different servomotors. Next, The inverse kinematics of the platform was presented allowing to transform the motion cueing algorithm trajectories into actuator inputs. Also, a detailed dynamics modeling of the simulator's platform is developed, based on some simplification. Algebraic constraints equations are included and a reduced representation of the dynamics model is demonstrated. An identification method is detailed which allows to estimate the platform parameters and to get a linear form of the dynamics model without re-writting it in a local reference frame.

Experiments were conducted to the present platform. A robust and powerfull electronic solution is described and a comparison between two control strategies is done. The main problems are high-lighted and adapted solutions are proposed to minimise delays and overcome the different limitations. Results are very sufficient for our riding simulation application.

Future work will be focus on the inclusion of the dynamics model into servocontroller parameters optimisation. Indeed, Proportionnel and PI control loops are optimized for single axis regulation. Pre-control coefficients will be adapted to have more performance. Psychphysical experimentations will be conducted to validate these approaches in riding motorcycle aspect.

REFERENCES

S. Chiyoda, K. Yoshimoto, D. Kawasaki, Y. Murakami, and T. Sugimoto. Development of a motorcycle simulator using parallel manipulator and head mounted display. In *Driving Simulation Conference(DSC00)*, Paris,France, 2000.

V. Cossalter. *Motorcycle Dynamics*. Race Dynamics Inc, ISBN: 0-9720514-0-6, Milwaukee, USA, 2002.

V. Cossalter, A. Doria, and R. Lot. Development and validation of a motorcycle riding simulator. In *World Automotive Congress(FISITA2004)*, Barcelona, Spain, May 2004.

B. Dasgupta and T.S. Mruthyunjaya. Closed-form dynamic equations of the general stewart platform through the newton-euler approach. *Mech. Mach.Theory*, 33:993–1012, 1998.

D. Ferrazzin, F. Barnagli, C.A. Avizzano, G.Di Pietro, and M. Bergamasco. Designing new commercial motorcycles through a highly reconfigurable virtual reality-based simulator. *Journal of Advanced Robotics*, 17(4):293–318, 2003.

Y. Miyamaru, G. Yamasaki, and K. Aoki. Development of a motorcycle riding simulator. *Society of Automotive Engineers of Japan*, 23:121–126, 2002.

L. Nehaoua, H. Arioui, H. Mohellebi, and S. Espié. Restitution Movement for a Low Cost Driving Simulator. In *Proceedings of the 2006 American Control Conference (ACC06)*, pages 2599–2604, Minneapolis, Minnesota, June 2006.

L. Nehaoua, S. Hima, H. Arioui, N. Seguy, and S. Espie. Design and modeling of a new motorcycle riding simulator. *American Control Conference (ACC07)*, pages 171–181, 2007.

Temperature effects on chemical reactor

M. Azzouzi

University "Politehnica" of Bucharest, Faculty of Automatic Control and Computer Science, Str 13 Septembrie, 06003 - Bucharest, Romania. University of Djelfa, Institute of Electronics, BP 3117- 17000- Djelfa Algeria (e-mail: enailia@yahoo.fr)

Abstract: In this paper we had to study some characteristics of the chemical reactors, from which we can understand the reactor operation in different circumstances; from these and the most important factor that has a great effect on the reactor operation is the temperature, it is a mathematical processing of a chemical problem that was already studied, but it may be developed by introducing new strategies of control; in our case we deal with the analysis of a liquid-gas reactor which can make the flotation of the benzene to produce the ethylene; this type of reactors can be used in vast domains of the chemical industry, especially in refinery plants where we find the oil separation and its extractions whether they are gases or liquids which become necessary for industrial technology, especially in our century.

Keywords: Chemical reactor, Temperature, Automatic control, Closed loop.

1. INTRODUCTION

Automatic control, although its widespread application began in 1930's, was first introduced several centuries ago, it saw the beginning in chemical industry in oil refineries, a field which tended to develop separately and to become known as process control. So it is a vast area that had been developed to minimize all the error effects on the analyzed systems and to understand its operation. The fact that helps to invent a lot of methods, by which we can have the possibility to control the behaviour of any plant regardless of its application purpose [2] [4].

Thermo chemistry was one of the important fields that exploited the control strategies in chemical reactions because a lot of factors need to be controlled during the chemical transfer; it has to do with the heat effects that accompany reactions and leads to a discussion of the nature of the equilibrium state [6]. Then in this case we need to satisfy the process stability under some conditions.

A chemical plant represents a complex arrangement of different units (reactors; separation units such as distillation, absorption, extraction, chromatography and filtration; heat exchanges; pumps; compressors; tanks ;...). These units must be either maintained close to their steady states for continuous operation or for the optimal trajectories flow in continuous operation [7].

The task of a control system is to ensure the stability of the process, to minimize the influence of disturbances and perturbation, and to optimize the overall performance, these objectives are achieved by maintaining some variables (temperature, pressure, concentration, position, speed, quality,...) close to their desired values or using set point which can be fixed or time dependent.

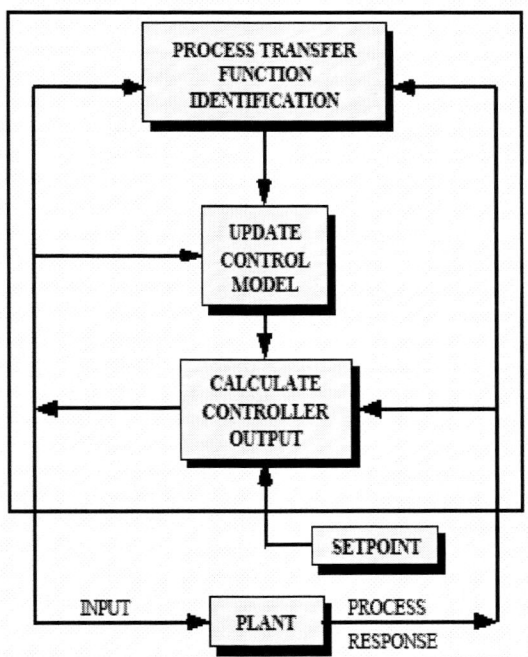

Fig (1): The closed loop advanced control system [3]

A Closed-Loop system utilizes feedback to measure the actual system operating parameter being controlled such as temperature, pressure, flow, level, or speed. This feedback signal is sent back to the controller where it is compared with the desired system set point. The controller develops an error signal that initiates corrective action and drives the final output device to the desired value.

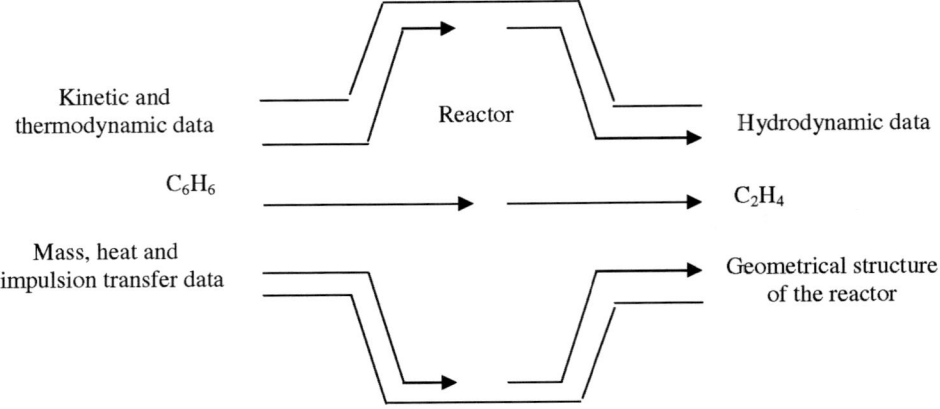

Fig (2): Principal factors that determine the reactor

2. CHEMICAL REACTOR

The chemical reactors have long been the system in which the method of analysis of dynamical systems have found ready applications and to which they have brought much insight. Some of the results of stability analysis, singularity theory, bifurcation, and period in forcing are surveyed.

The reactors engineering processes the calculation methods of the chemical transformation of the matter, where are placed some chemical reactives. Generally the reactor doesn't have great weighting in chemical plants investment, however these characteristics contain the amount plants (reactive, matter preparation, temperature degrees and pressure values and in downstream (reaction products separation).

When we want to design a process control system, we must first study the process and determine its characteristics, the process variables are classified in the chemical reactor process like the operative parameters temperature, flow, pressure, concentration,..., and the output that is presented by results in the product nature, the rate of the reactive conversion, the product distribution and the yielding, not to forget the states which are kinetic and thermodynamic data, mass, heat and impulsion transfer data, hydrodynamic data, the geometrical structure of the reactor [2].

The form of the reactor into which reactants flow continuously, from which interacted feed and products emerge and in which the contents are mixed has long been the work-horse of both practical and theoretic reactor analysis.

Our treatment of reactor theory thus has exclusively with no isothermal systems. In practice, however, the majority of them are being a topic more suitable for discussion when considering particular reactor types, which is exactly where we are now.

3. MATHEMATICAL MODEL

In our case we have a lot of unknown variables that must be shown or specified to have the longest interval of reactor stability operation, it is a difficult to give the exact values of these factors, but we have some intervals which must be followed [5].

The mathematical model of reactor can be built in the general form of mathematical equation to describe the most essential properties of the real object and reflect them in mathematical from where we can understand the reactor operation [1]. The simplest example of no isothermal reaction is the reversible exothermal reaction $A \rightarrow B$ [2].

Most reaction rates are sensitive to temperature, and most laboratory studies regard temperature as an important means of improving reaction yield or selectivity. Our treatment has so far ignored this point, the reactors have been isothermal, and the operating temperature, as reflected by the rate constant, has been arbitrary assigned.

Temperature effects should be considered even for no isothermal reactors, since the operating temperature must be specified as a part of the design. For no isothermal reactors, where the temperature varies from point within the reactor, the temperature dependence directly enters the design calculation [8].

Important reactions are accompanied by significant heat effects which must be recognized and incorporated into the analysis or design of the reactor. The problem of what to do about reaction systems that are both no isothermal was put off.

Then as being a topic more suitable for discussion when considering particular reactor types, which is exactly where we are now.

The rate constant for elementary reactions is almost always expressed by Arrhenius relation which is:

$$k = e^{-\Delta F_0 / RT} \qquad (1)$$

$$\frac{\partial}{\partial T}\left(\frac{\Delta F_0}{T}\right)_{P=ct} = -\frac{\Delta H_0}{T^2} \qquad (2)$$

Where:

ΔH_0 in the reaction heat.

ΔF_0 Free energy of reaction.

T is the reactor temperature.

R gases constant.

By combining these to expressions, we can find:

$$\frac{d \ln k}{dT} = \frac{\Delta H_0}{RT^2} \quad (3)$$

With the integration of $\Delta H_0 = ct$ we find:

$$k = Ae^{-\Delta H_0/RT} = k_0 e^{(\Delta H_0/RT_0 - \Delta H_0/RT)} \quad (4)$$

where: k_0 is the k value for $T = T_0$.

On simple reaction $A \longrightarrow B$, we know that $k = p_B/p_A$ and if we consider initially A pure in total pressure π, the conversion of A in equilibrium will be:

$$f_{Ae} = \frac{p_B}{\pi} = \frac{p_B}{p_A + p_B} = \frac{k}{1+k} \quad (5)$$

We note that $Y = RT/\Delta H_0$ and we find:

$$f_{Ae} = \frac{1}{1 + \left(\frac{1}{A}\right) e^{\frac{1}{Y}}} \quad (6)$$

4. MATLAB SIMULATION

However the reaction type - exothermal with extraction of temperature or endothermic with a consummation of temperatures, there are great effects of temperature values on the reactor operation, and our goal is to arrive to the desired temperatures degree which assures the best results like the product concentrations, quality and variety and with an economic consumption of energy to effectuate the reaction, the variation of this temperature value can influence the conversion of A.

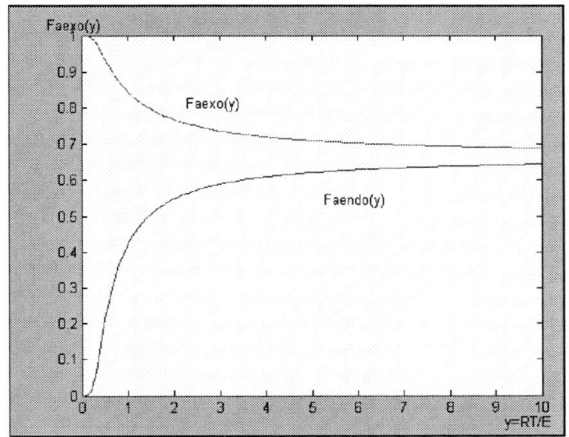

Fig (3): Temperature effect on conversion at equilibrium for endothermic and exothermic reactions

In the simple reaction $A \longrightarrow B$, we have:

$$f_A = \frac{kt_R}{1 + kt_R} \quad (7)$$

The stationing time t_R also has a great effect on this reactor operation, we can remark below, the exothermic reactions according to it.

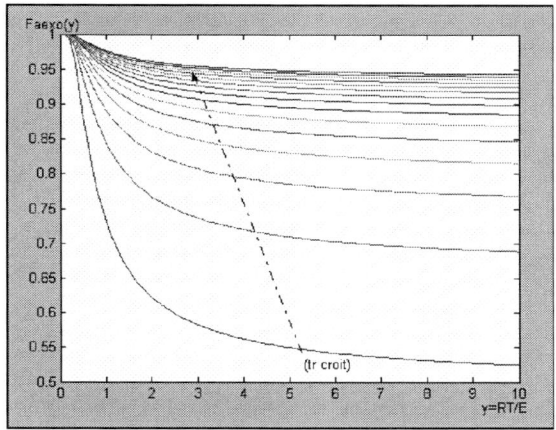

Fig (4): Temperature and stationing time effects on the conversion en equilibrium (exothermic irreversible reaction)

By using the equation (6), we find:

$$f_A = \frac{1}{1 + (1/k_0 t_R) e^{1/z}} \quad (8)$$

Where: $z = RT/E$

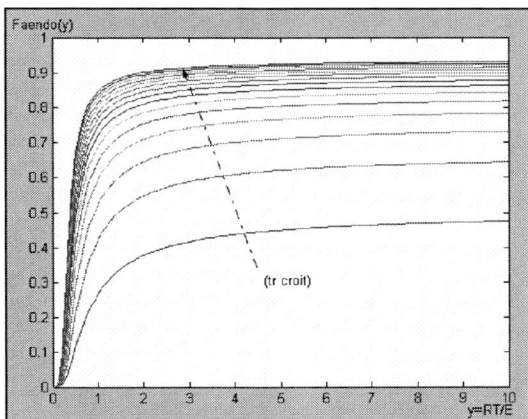

Fig (5): Temperature stationing time effects on the conversion in equilibrium (endothermic irreversible reaction)

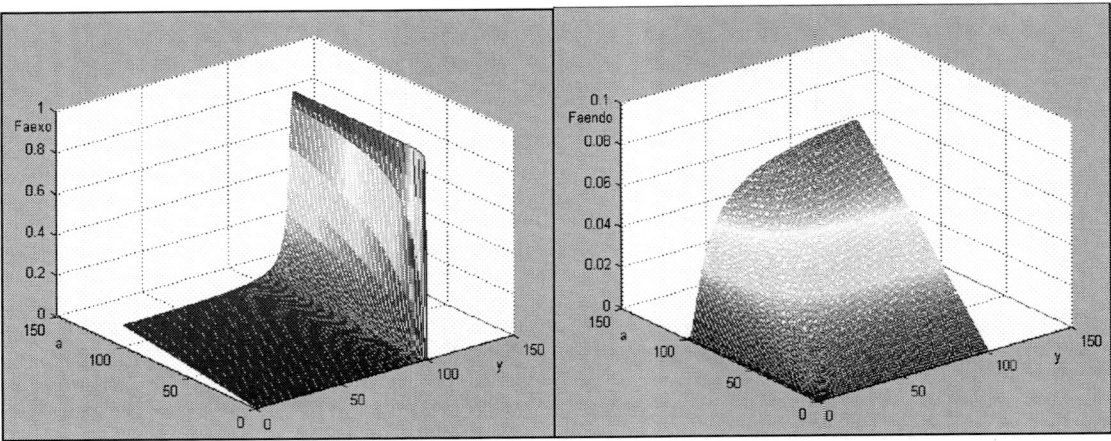

Fig (6): Temperature and stationing time effects on the conversion in equilibrium for exothermic reaction and respectively endothermic reaction

5. CONCLUSIONS

In isothermal or no isothermal reactors, the temperature has a great effect on the reactor operation, and then on its production, so in chemical plants we must control all the operation coefficients, the pressure, the flow, the quantity of the desired products or their concentrations, and the temperature [8].

In the following figures we demonstrate the effect of the temperature and the stationing time in the same time. From which we can understand that the exponential dependence of the rate constants on the temperature means that the reaction rate can change rapidly as the temperature changes. The larger the value of activation energy, the greater will be the change in the reaction rate for a given change in temperature. This work let us conclude that:

- The temperature has a great effect on the temperature operation;
- We can keep the stability of this reactor by optimizing the temperature value that has good performances;
- Many chemical problems were caused by some errors in their control and coefficients; it arrived to explosions and the death of humans, so the control of these plants offers human and material safety.

REFERENCES

[1] Mtros. Y. S (1985). *Unsteady process in catalytic reactors*, edition ELSEVIER, 364p, USSR.

[2] Minneapoli. A (1985). The mathematical background of chemical reactor analysis II the stirred tank, Rutherford, Minnesota *Reacting flows: combustion and chemical reactors, part(1)*, editor ITHALCA, 496p, New York.

[3] Advanced control of batch reactor temperature, M. Huzmezan, B. Gough, S. Kovac, L. Le, G. Roberts, American Automatic Control Council, pp1156-1161, New York 2002, USA.

[4] M.Azzouzi, F. Halal (2007), Comparative analysis of PD classical control and PD fuzzy control in two liquid level tanks, The 4th Conference on Management and Control of Production and Logistics, IFAC MCPL 2007, V. 2, pp487-491, Sibiu, ROMANIA.

[5] But. J (1980), *Reaction kinetics and reactor design*, printice Hale, New jersey, 429p, USA.

[6] Aris. R (1969) *Elementary Chemical Reactor Analysis*, 352p, UK.

[7] Corriou, J. P (2004), Process control: theory and applications, prentice Hall, 752p, New York.

[8] Nauman. E. B (2001), Handbook of Chemical Reactor Design, Optimization, and Scale up, 590p, Kindle edition, New York, USA.

[9] Hayes. R. E (2001). Introduction to Chemical Reactor Analysis, CRC First edition, 416p, New York. USA.

Design of Full-Band and Low-Pass FIR Differentiators: A Comparative Study

C. Mekhnache[*], Y. Ferdi[*], and A. Taleb-Ahmed[**]

[*]LRES Laboratory, Department of Electrical Engineering,
University of Skikda, BP. 26, Route d'El-Hadaiek, 21000, Skikda, Algeria
(e-mails: cmokhnache@yahoo.fr; yferdi@yahoo.fr)
[**]LAMIH UMR CNRS 8530 Laboratory, University of Valenciennes
and Hainaut Cambresis, Le Mont Houy, 59313 Valenciennes, France
(e-mail: Abdelmalik.Taleb-Ahmed@univ-valenciennes.fr)

Abstract: Digital differentiators are useful in many fields of sciences and engineering. They can be designed using two approaches, namely, FIR filters design and IIR filters design. This paper is concerned by the first one in which great interest in the design of digital differentiators has encouraged the development of various design methods. The widely used methods for FIR differentiators are those based on criteria L_1, L_2, L_∞ and that based on Taylor series. A comparison between these methods is carried out in terms of approximation accuracy and computational complexity. Numeric examples are presented to illustrate the performance of each method. It was found that the design results obtained by least squares method for fullband and low-pass differentiators are better than the other ones.

Keywords: FIR Differentiators, optimal methods, least squares method, eigenfilter algorithm, Taylor series, Remez exchange algorithm

1. INTRODUCTION

Digital differentiators are used to perform numerical computation of derivatives of discrete-time signals. They can be utilized in many applications including control systems (Franklin et al., 1990; Su et al., 2005; Khan. et al. 2007), biomedical engineering (Laguna, et al., 1990), mechanical vibrations (Pintelon, et al., 1990), radar and sonar (Skolnik, 1980). Since fullband differentiators cause amplification of high frequency noise, it is desirable in some applications to use low-pass differentiator in order to reduce the effect of this undesirable phenomenon. Both full-band and low-pass differentiators can be designed using FIR filters design techniques. These includes Taylor series based method (Khan, et al., 1999), the minimax approach based on the Remez exchange algorithm (Antoniou, 2003), least squares method (Mollova, et al., 2001), eigenfilter method (Pei. et al., 1996), method based on L_1 criteria (Chen. et al., 1995). Since there exist little work regarding comparison between these methods, the purpose of this paper is to implement these techniques and to compare their performance in terms of time complexity and accuracy, and thus to help the designer selecting the appropriate design techniques. This paper is organized as follows. In section 2, FIR differentiators design is given. In section 3, we present a brief of used techniques for FIR differentiator approximation in this contribution. Design examples and their magnitude frequency responses and error curves are shown in section 4. Conclusions are provided in section 5.

2. FIR DIFFERENTIATORS DESIGN

The ideal analogue differentiator is a linear time-invariant system described by the transfer function:

$$H_a(s) = s \quad (1)$$

Its frequency response is given by

$$H_a(\omega) = j\omega = \omega e^{j\pi/2}, \ -\infty < \omega < \infty \quad (2)$$

The magnitude and phase responses are represented in Fig.1 in the frequency range $-50 \le \omega \le 50$. Since the maximum frequency which may be present in discrete-time signals is $\omega_s/2$, where $\omega_s = 1/T_s$ is the sampling frequency, digital differentiator must have the frequency response

$$H_d(\omega) = j\omega, \ -\omega_p \le \omega \le \omega_p \quad (3)$$

with $\omega_p = \omega_s/2$ for digital full-band differentiator and $\omega_p = \alpha\omega_s/2$ ($0 < \alpha \le 1$) for digital low-pass differentiator. The frequency ω_p represents the passband edge frequency. The problem of designing FIR differentiator is to find the coefficients of the impulse response $h(n)$, $n=0,1,...N-1$ such that the digital transfer function

$$H(z) = \sum_{n=0}^{N-1} h(n) z^{-n} \quad (4)$$

best approximates the transfer function (3) in some sense. The transfer function (4) has a delay while the ideal one in (3) has not. This problem is solved by considering the approximation of the transfer function

$$G_d(\omega) = H_d(\omega) e^{-j\tau\omega} \quad (5)$$

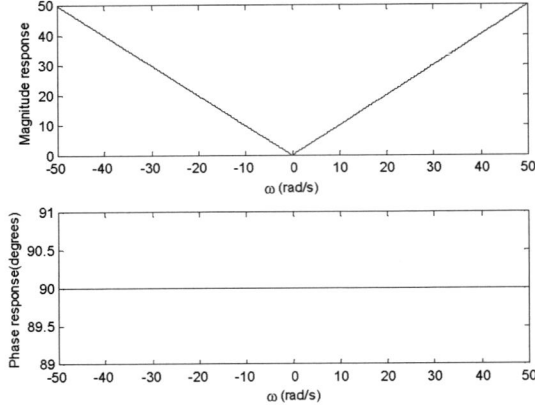

Fig. 1. Frequency response of the ideal analogue differentiator

instead of (3). In this paper five methods were implemented in MATLAB on Pentium IV (1.8 GHz, 128 Mb RAM), to determine the FIR filter coefficients (4) to approximate (5). A brief review of each method is given in the following section.

3. TECHNIQUES FOR FIR DIFFERENTIATOR APPROXIMATION

The techniques for FIR differentiator approximation can be grouped into five groups:

3.1 Least squares approach (Mollova, et al., 2001):

This method involves formulating an error function based on the absolute mean-square difference between the amplitude responses of the practical $M(\omega)$ and the ideal differentiators $D(\omega)$ as a quadratic function. The filter coefficients are obtained by solving a system of linear equations.

$$E_{mse} = [1/(\omega_s/2)]\int_0^{\omega_p}(D(\omega)-M(\omega))^2 d\omega \quad (6)$$

3.2 Minimax approach (Antoniou, 2003):

It is based on minimizing the maximum error between the frequency response of the ideal differentiator and that of the designed one:

$$\min_{b} imiser\{\max_{\omega}|E(\omega)|\} \quad (7)$$

where b is vector of differentiator coefficients.

3.3 Eigen filter approach (Pei. et al., 1996):

This method, as the least squares one, is optimal in L_2 sense. It is based on minimizing a quadratic measure of the error between the desired amplitude response and the actual amplitude response of the designed filter. In this method the desired amplitude response is equal to the amplitude response of the designed filter at any arbitrary reference frequency.

The filter coefficients are obtained by computing the eigenvector that corresponds to the smallest eigenvalue of a real symmetric and positive-definite matrix.

3.4 Method based on L_1 criteria (Chen. et al., 1995):

It is based on minimizing the absolute error function:

$$j = \sum_{i=1}^{L} W_{L1}(\omega_i)|E(\omega_i)| \quad (8)$$

where $W_{L1}(\omega_i)$ is the error weighting function, $i=1,2,...L$, and L is the number of frequency points.
First the design problem is reformulated as a linear programming problem in the frequency domain, then a method based on modification of Karmarkar's algorithm is used to solve this problem.

3.5 Taylor series (Khan, et al., 1999):

Taylor series expansion of input signal $x(t)$ at $t=t_0$ can be written as:

$$x(t) = x(t_0) + \sum_{m=1}^{\infty}[(t-t_0)^m]/(m!)]x^{(m)}(t_0) \quad (9)$$

where $x^{(m)}(t_0)$ denotes the mth derivative of $x(t)$ at $t=t_0$. If we discretize and truncate the series after $2N$ terms, a set of $2N$ equations will be obtained which can be written as:

$$x_l - x_0 = \sum_{m=1}^{2N}[(lT_s)^m]/(m!)]x_0^{(m)} \quad (10)$$
$$l = (2n-1)/2, -N \leq n \leq N$$

where x_l denotes the value of $x(t)$ at $t=lT_s$.
By solving this equation system the first derivative of input signal $x(t)$ can be found.

4. DESIGN PERFORMANCE EVALUATION

For FIR differentiator approximation, the following specification parameters have to be considered: the filter coefficients number N, the sampling frequency $F_s=\omega_s/2\pi$, the passband edge frequency $f_p=\omega_p/2\pi$, the desired value of the error E_d between the magnitude of the frequency response of the ideal differentiator and the magnitude of the approximated one. On the other hand, each design method has its specific parameters. For eigenfilter method: reference frequency ω_r. In the method based on L_1 criteria, the constraints is imposed at a fixed frequency point ω_0, other parameters which have to be considered in this method are the number of iterations (max-iter) and error weighting function $W_{L1}(\omega)$. The performance assessment of the design methods were carried out by evaluating the computation time, the peak error in the passband $(max|D(\omega)-M(\omega)|, 0\leq\omega\leq\omega_p)$, the frequency interval where the error between the magnitude of the frequency response of the ideal differentiator and the magnitude of the approximated one is less than a certain value. Two examples are provided to demonstrate the effectiveness of methods.

4.1 Example 1: Low-pass differentiator

In this example we want to obtain a low-pass differentiator, with length $N=31$, $F_s=1$, the number of frequency points L is $8N$, $E_d \leq 3\times10^{-3}$ and $f_p=0.45$, $(\alpha=0.9)$. For this design example we use optimal methods based on L_1, L_2, L_∞. For eigenfilter method ω_r is 0.25. For the method based on L_1 criteria $\omega_0=0$, max-iter is 13 and $W_{L1}(\omega)$ is set to 1.

The magnitude responses of the ideal and the approximated differentiators for each method are illustrated in Fig.2.

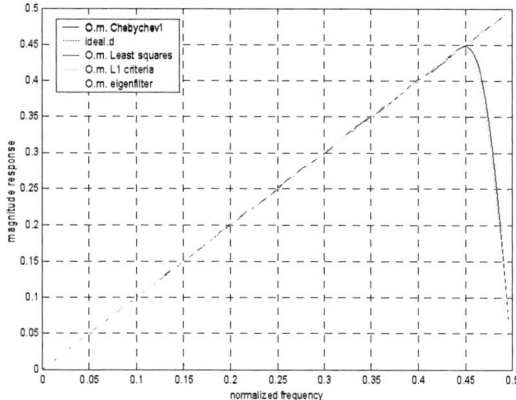

Fig. 2. The magnitude responses of the ideal and low-pass digital differentiators

It can be seen from this figure that the magnitude frequency responses of the designed differentiators obtained by previous method are very close to the ideal one except in a small region near $f=0.5$.

The error curves between the ideal differentiator and the designed one $E(\omega)=D(\omega)-M(\omega)$ are presented in Fig. 3.

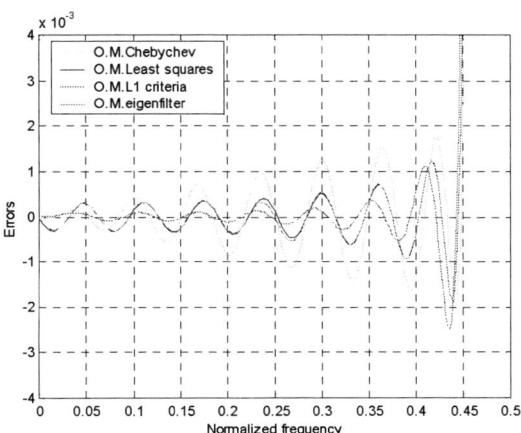

Fig. 3. Curves between the ideal differentiator and designed ones.

Based on this figure and using tic-toc command in MATLAB 5.3 for obtaining the computation time, the results are reported in table 1.

Table 1. Comparison of the optimal methods (in sense: L_1, L_2, L_∞)

	Computation time (seconds)	Frequency interval $(E_d \leq 3\times 10^3)$	Peak error
Optimal method in L_∞ sense	0.16	0-0.4476	0.0017
least squares method	0.05	0-0.4476	0.0021
Optimal method in L_1 sense	103.09	0-0.4435	0.0052
eigenfilter method	0.05	0-0.4476	0.0021

Methods based on L_2 criteria (eigenfilter method, least squares one) present the lower computation time, the highest one is for method based on L_1 criteria. Notice that the peak error of the latter is higher than the other ones. It can be seen that the results obtained by least squares method are the same results of eigenfilter method, but this method does not take the ideal response into account but rather the frequency response of the practical filter at an arbitrary frequency. In fact, the filter that is designed depends on the reference frequency. Then the least squares method is better than the other design methods.

4.2 Example 2: fullband differentiator

This example is concerned with full band differentiator, with length $N=42$, $F_s=1$, $L=8N$, $E_d \leq 25\times10^{-4}$. For this design example we use optimal methods based on L_1, L_2, L_∞ and the method based on Taylor series. For eigenfilter, ω_r is 0.25. For the method based on L_1 criteria, $\omega_0=0$, max-iter is 13 and $W_{L1}(\omega)$ is set to 1.

The ideal and designed amplitude responses are shown in Fig. 4.

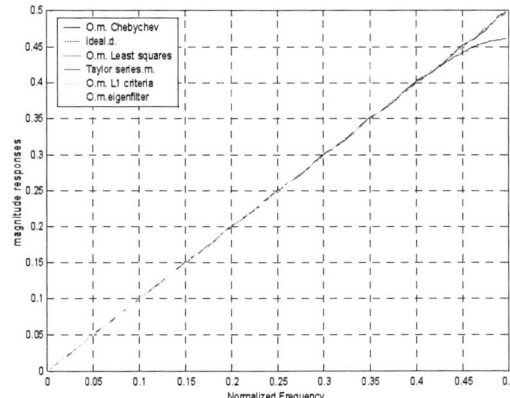

Fig. 4. Amplitude responses of the ideal and fullband differentiators

For further illustration we present curves for the designed differentiators in Fig.5.

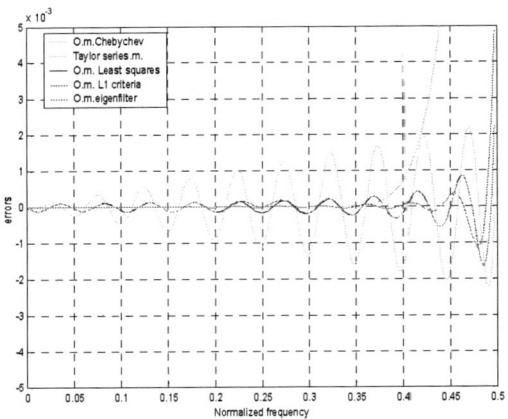

Fig. 5. Curves for designed differentiators

The results concerning the peak error in the passband, computation time and frequency interval where the error is less than 25×10^{-4} are listed in table 2.

Table 2. Comparison of the optimal methods (in sense: L_1, L_2, L_∞) and that based on Taylor series

	Computation time (seconds)	Frequency interval ($E_d \leq 25\times10^{-4}$)	Peak error
Optimal method in L_∞ sense	0.11	0-0.4970	0.0023
Taylor series based method	0.05	0-0.4226	0.0364
least squares method	0.05	0-0.4970	0.0022
Optimal method in L_1 sense	317.25	0-0.4911	0.0049
Eigenfilter method	0.05	0-0.4970	0.0022

It can be seen from this table that the method based on Taylor series, eigenfilter method and least squares method are much faster than the other ones, particularly that based on L_1 criteria. The largest frequency interval where the error is less than 25×10^{-4} is obtained using methods based on L_∞, L_2 criteria. For these methods the peak errors in the passband are very small. We conclude that least squares method performs better than the other ones.

5. CONCLUSIONS

In this paper, we have compared the existing methods commonly used to design fullband and low-pass digital differentiators. Two numerical examples have been presented to demonstrate the effectiveness and accuracy of these methods. Designed low-pass differentiators are found to be very close to ideal one over the entire frequency range except in a narrow frequency band near the half of the Nyquist frequency edge. The first comparative study shows that the method based on L_2 criteria is faster than the others. It has the largest frequency interval for a fixed value of error, and the smallest peak error. The drawback of eigenfilter method is the using of the amplitude response of the designed filter at any arbitrary reference frequency instead of the amplitude response of the ideal differentiator. Then we have concluded that the least squares method is the best one. The obtained full band differentiators are very close to ideal one over the entire frequency range. We have attained the same conclusion from the second illustrative example: the least squares method is effective because of the broad band where the frequency response of designed differentiator by this approach is very close to ideal one and its lower time computation and peak error in the passband.

REFERENCES

[1] Antoniou. A., (2003). *Efficient Remez algorithms for the designing of nonrecursive filters*. p. 1-39. University of Victoria.

[2] Chen. C.K., Lee Ju-Hong. (1995). *Design of higher order digital differentiator using L_1 error criteria*. IEEE Trans on circuits and systems-II: Analog and digital signal process. **Vol. 42 N°. 4**, p. 287-291.

[3] Franklin. G.F., Powell. J.D., Workman M.L. (1990). *Digital control of dynamics systems*. 3, second ed. Addison wesley, Reading, MA.

[4] Khan. I.R, Ohba. R. (1999). *New design of full-band differentiators based on Taylor series*, IEE Proc. Vis. Image Signal Process, **Vol. 146 N°. 4**, p.185-189.

[5] Khan. I.R., Masahiro O (2007). *Finite-Impulse-Response Digital Differentiators for Midband Frequencies Based on Maximal Linearity Constraints*, IEEE trans. Circuits and systems-ii: express briefs, **Vol. 54 N°. 3**, p. 242-246.

[6] Laguna. P, Thakor N.V., Caminal P., Jane R. (1990). *Lowpass differentiator for biological signal with known spectra; application to ECG signal processing*, IEEE Trans. Biomed Eng. **Vol. 37 N°. 4**, p. 420-425.

[7] Mollova G., R.Unbehauen,(2001). *Analytical design of higher order differentiators using least-square technique*, Electronics letter. **Vol. 37 N°. 17**, p. 999-1000

[8] Pei. S.C., Jong-Ju .Shyu, (1996). *Eigen filter design of higher-order digital differentiator*, IEEE Trans. Acoustics, Speech, and Signal processing. **Vol. 37 N°. 4**, p. 505-511.

[9] Pintelon R., Shoukens J., (1990). *Real-time integration and differentiation of analog signals by means of digital filtering*, IEEE Trans. Instrum.Measur. **Vol. 39 N°.6**, P. 923-927.

[10] Skolnik. M. I. (1980). *Introduction to radar systems*. second ed, McGraw-Hill. New york.

[11] Su. Y.X, Dong Sun, B.Y. Duan (2005). *Design of an enhanced nonlinear PID controller*, Mechatronics, **Vol. 15 N°. 8**, p.1005-1024.

SESSION C2: VIRTUAL REALITY AND INTERFACES

Design of a 3D Navigation Technique Supporting VR Interaction

Pierre Boudoin*   Samir Otmane*   Malik Mallem*

*Informatics, Integrative Biology and Complex Systems,
40 Rue de Pelvoux, 91020 Evry Cedex, France
(e-mail: pierre.boudoin@ibisc.fr, samir.otmane@ibisc.fr, malik.mallem@ibisc.fr).

Abstract: Multimodality is a powerful paradigm to increase the realness and the easiness of the interaction in Virtual Environments (VEs). In particular, the search for new metaphors and techniques for 3D interaction adapted to the navigation task is an important stage for the realization of future 3D interaction systems that support multimodality, in order to increase efficiency and usability.

In this paper we propose a new multimodal 3D interaction model called Fly Over. This model is especially devoted to the navigation task. We present a qualitative comparison between Fly Over and a classical navigation technique called gaze-directed steering. The results from preliminary evaluation on the IBISC semi-immersive Virtual Reality/Augmented Realty EVR@ platform show that Fly Over is a user friendly and efficient navigation technique.

Keywords: 3D Interaction Techniques, Multimodality, Navigation Task, Virtual Environment

1. INTRODUCTION

Multimodality is a powerful paradigm to increase the realness and the easiness of the interaction in Virtual Environments (VEs). In particular, the search for new metaphors and techniques for 3D interaction adapted to the navigation task is an important stage for the realization of future 3D interaction systems that support multimodality, in order to increase efficiency and usability.

In this paper we propose a new multimodal 3D interaction model called Fly Over. This model is especially devoted to the navigation task. We present a qualitative comparison between *Fly Over* and a classical navigation technique called *gaze-directed steering*. The results from preliminary evaluation on the IBISC semi-immersive Virtual Reality/Augmented Realty EVR@ platform show that Fly Over is a user friendly and efficient navigation technique.

2. RELATED WORK

The navigation task is probably the most utilized task in VEs [Bowman *and al.*, 2005]. The aim of this task is to give the user the feeling he is moving naturally and easily in a VE, whereas avoiding sickness feelings.

A lot of navigation techniques have already been developed. However, they are highly dependent from hardware interface. Indeed:

• Gaze-directed steering technique [Mine, 1995] needs user's head tracking;

• Pointing technique [Bowman *and al.*, 1997] needs user's hand tracking;

• Map-based travel technique [Bowman *and al.*, 1998] needs a 2D display and a pointer;

• Grabbing the air technique [Mapes, *and al., 1995*] needs pinch-gloves.

These techniques are efficient for an isolated navigation task. But if we consider a global action in a VE (including navigating, selecting or manipulating objects), different devices may be needed and switches between tasks and devices may be difficult to handle for the user.

Hence, multimodal framework is needed. We have studied different multimodal frameworks, especially devoted to devices management. Two drew our attention:

The Sylvia Irawati's team [Irawati *and al.*, 2006] proposed a complete framework. The most interesting part is their using of objects ontology to make the interaction more natural and user-friendly. Moreover the object ontology they've proposed supports constraints definitions. It could be interesting to be inspired by some parts of their framework.

The Ed Kaiser's team [Kaiser *and al.*, 2003] worked on mutual disambiguation. Their work testified an interesting approach to manage multimodality with the use of what they called Multimodal Integrator. The aim of this integrator is to find the best multimodal interpretation with the preliminary rated inputs. The principle is to unify inputs data in:

Amalgamating redundant or complementary data via a logical test set;

• Taking care about the spatiotemporal aspect of data;

• Taking care about data's hierarchy.

3. THE FLY OVER MODEL

We propose a new 3D interaction model - called Fly Over - based on the following four constraints:

• To be compatible with all common 2D, 3D or 6D devices (mouse, hand/head/finger tracking, force feedback) that could

return a 2D/3D position/orientation of the user or an object he manipulates;
• To maintain the same logic of use for all devices, even if the employed technologies are very different from each other;
-to be natural;
-to be associated to a short training duration.

3.1. Generic model specification

The Fly Over model may be depicted as a blob, which is composed of interaction areas, modeled as concentric spheres (see figure 1). Each interaction area is a subspace of the whole task space. For example, figure 1 depicts a 6D task space in VE parted into two 3D subspaces.

The action of the user on VE may be summarized, at each time: t, by a vector P (6 components for a 6D task). Each interaction area Zi is associated to specific sub-vector Pi of P.

As stated by constraint 1, the user's action on the device he utilizes may control a pointer (modeled as a tiny sphere in figure 1) in VE. The presence of the pointer into area Zi is translated into a modification of the sub-vector Pi.

The crossing from one interaction area to another allows the user to modify consecutively all the components of vector P.

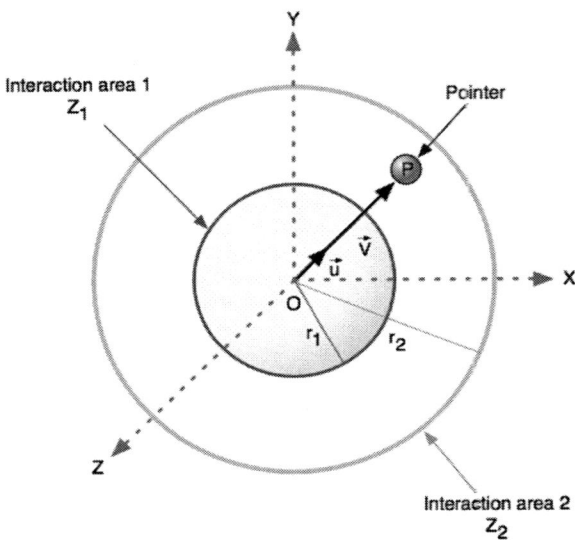

Fig. 1. Fly Over blob with two 3D interaction areas Z1 and Z2 designed for a 6D task in VE.

In order to fulfill the last two conditions (natural technique and fast learning), we decided to use the simple virtual hand technique to handle the virtual pointer move in VE.

3.2. Generic model parameters

We can access to five parameters via the generic model. These parameters are computed with the only knowledge of position of the effector controlled by the user.

-First parameter: Position vector of pointer
$$\vec{V} = \overrightarrow{OP}$$
This parameter indicates the position of the effector relative to the blob. It may be interpreted as a position, a translation vector, a direction and so defining an orientation

-Second parameter: presence in the interaction area 1
$$\begin{cases} P \in Z_1 & Si \quad \|\vec{V}\| < r_1 \\ P \notin Z_1 & Si \quad \|\vec{V}\| > r_1 \end{cases}$$

-Third parameter: Presence in the interaction area 2
$$\begin{cases} P \in Z_2 & Si \quad r_1 < \|\vec{V}\| < r_2 \\ P \notin Z_2 & Si \quad \|\vec{V}\| > r_2 \ ou \ \|\vec{V}\| < r_1 \end{cases}$$

-Fourth parameter: Global intensity
$$I = \|\vec{V}\|$$

-Fifth parameter: Intensity in the 3D interaction area 2
$$I_2 = \|\vec{V}\| - r_1$$

3.3. Fly Over for navigation task: Fly Over – N

The generic Fly Over model was firstly designed for navigation tasks. Indeed, we have previously noticed that:
• Managing simultaneously translation and rotation in VE on a (semi-)immersive VR platform may cause nausea for the user;
• The users naturally choose their orientation in order to have the aimed object in front of them, and then translate to the object.

So, these observations were compatible with the fact that, within the Fly Over model, it is possible to decouple the 6D navigation task into two 3D subspaces: a subspace dedicated to the position of the user in VE and another one dedicated to the orientation of the user in VE. This leads to the Fly Over – N model depicted in figure 2.

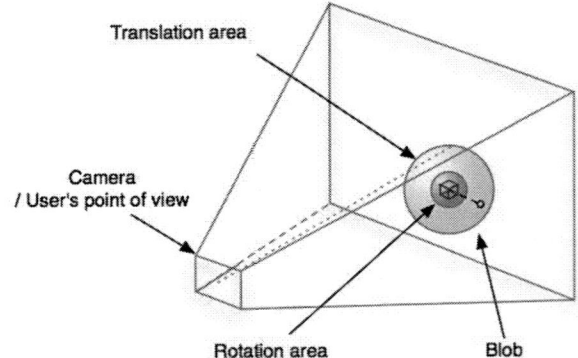

Fig. 2. Fly Over – N model. The 6D navigation space is parted into two 3D subspaces: Z1 dedicated to rotations and Z2 dedicated to translations.

The position of the pointer in the rotation area leads to a rotation of the user in VE with magnitude I whereas the position of the pointer in the translation area leads to a translation of the user in VE with magnitude I2.

The generic algorithm described in section 3.2 becomes:

```
START:
IF  (  ‖V⃗‖ < r₁  )  THEN
   Orientation( V⃗ ) ;
ELSE
IF  (  r₁ < ‖V⃗‖ < r₂  )  THEN
   //Computing the translation vector ;
   T⃗= Normalize( V⃗ ) * I₂ ;
   Translation( T⃗ ) ;
END
```

Where:

- Orientation($\vec{V}$): execute the necessary rotations in order to direct the camera in the direction given by $\vec{V}$.

- Translation($\vec{V}$): execute the translation given $\vec{V}$.

- Normalize($\vec{V}$): normalize the vector $\vec{V}$ in order to transform it in unit vector.3.3. Visual assistances going with Fly Over - N

In order to help users when they navigate in VE, visual assistances are displayed. The blob is displayed in a translucent way and is placed in front of the virtual camera, which represents the point of view of user (see figure 2).

The first interaction area will be blue tinted and the second won't be tinted and vice versa, depending on the presence of the pointer into these areas (see figure 3). A wire-frame cube is displayed to symbolize the effective orientation of the user in VE.

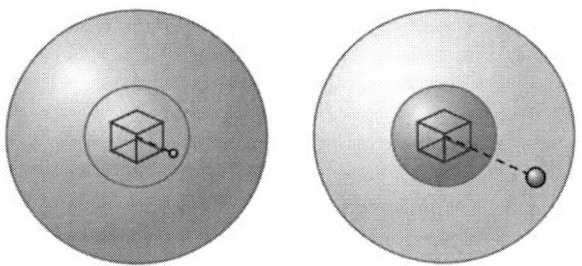

Fig. 3. Visual assistances in the Fly Over blob.

4. HARDWARE AND SOFTWARE IMPLEMENTATION

Our experiments has been performed on the IBISC Lab. semi-immersive multimodal EVR@ platform (see figure 4), which permits to follow the gestures of the user's hand and finger positions (wireless Flystick 1 coupled to two ARTTrack1 infrared cameras, wireless 5DT data gloves Ultra 14) and has a 6D force feedback device (SPIDAR-G [Sato, 2006]).

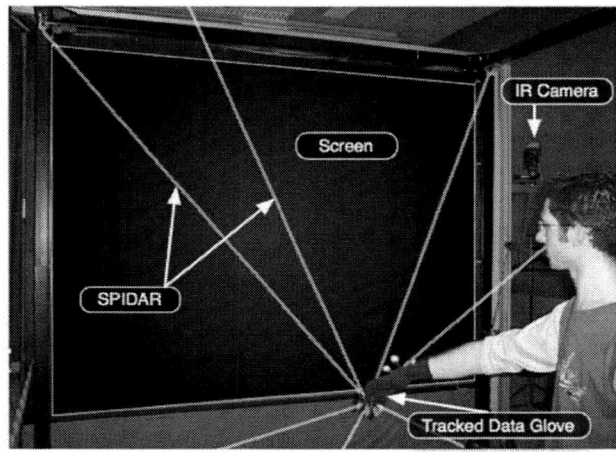

Fig. 4. The EVR@ platform. We can see Data Gloves, Optical Tracking System and SPIDAR in use.

Each device is associated with a specific server. We utilized the VRPN library [Taylor II *and al.*, 2001] to implement the gathering of all our data from the different servers and Virtools™ to make the interactive virtual environments needed in our experiments.

5. PRELIMINARY EVALUATION

In order to realize a preliminary evaluation, we compared Fly Over – N (FO-N) to gaze-directed steering (GDS). A total of 12 young students, including 10 males and 2 females participated to the experiment. 2 of them considered themselves as experts in using VE systems whereas 4 considered themselves as intermediate and 6 as beginners. However, none of them have already utilized FO–N nor GDS.

VE was the representation of a part of the IBISC Lab. For FO–N, the device used to navigate was a Flystick. Figure 5 shows the experimental setting.

The participants were asked to follow 3 times as precisely as possible a trajectory in VE depicted with a thin red line, going from point A to point B (see figure 6). Duration of the experiment was not considered.

The target trajectory was built to be sinuous. The main question was: is it easier to follow the target trajectory with the help of FO-N than with GDS?

Data showed that the use of FO-N gives smoother trajectories than GDS (figure 7). However, there exists a bias for FO-N: users are doing trajectories that are near from the target trajectory but not centered on it.

Participants were given qualitative questionnaires after the experiment: Q1-Did you find easy to learn the FO-N/GDS?, Q2-Did you found easy to navigate with FO-N/GDS?, Q3-Did you found easy to follow the target trajectory?, Q4-Did you feel sickness? The possible answers were Agree, Neutral and Disagree.

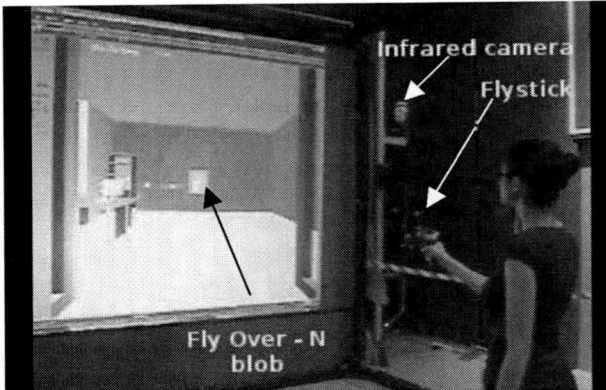

Fig. 5. Experimental setting with the use of FO–N on the IBISC semi-immersive EVR@ platform. Users navigate by moving a Flystick in their hand, which position is computed by two infrared cameras.

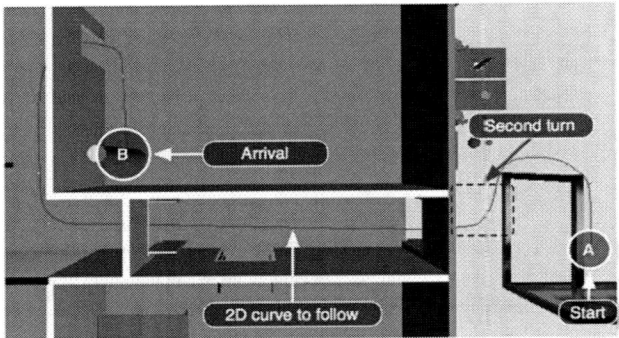

Fig. 6. Course to follow with target trajectory.

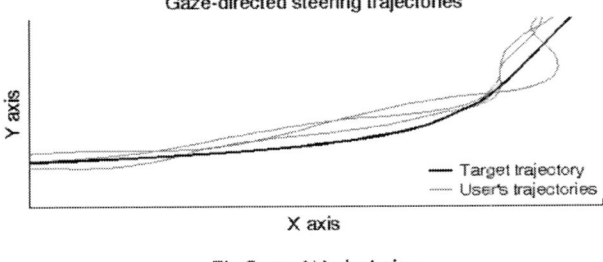

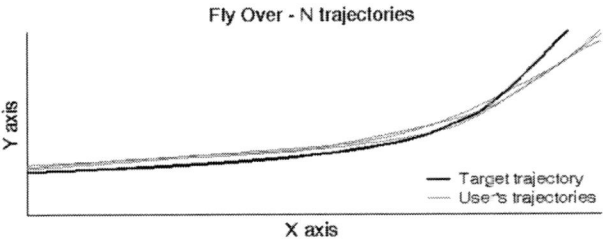

Fig. 7. Trajectories comparison for one user in the second turn of GDS and FO- N techniques.

The results show that FO-N was preferred and seems to be easier and more usable than GDS. Indeed, for Q1, we got a 9A-3N-0D for FO-N and a 8A-3N-1D for GDS. For Q2, we got a 8A-3N-0D for FO-N whereas GDS obtained a 2A-1N-9D which tend to show that users were not easy with GDSI. For Q3, we got a 5A-4N-3D for FO-N whereas GDS obtained a 3A-3N-6D. Finally, for Q3 we obtained a 0A-1N-11D for FO-N whereas GDS got a 3A-1N-8D, which means that all the users felt comfortable with FO-N whereas a minority of them felt sickness with GDS.

6. CONCLUSION AND PROSPECTS

In this paper, we propose a new multimodal 3D interaction model, called Fly Over. This model is generic and is based on two main ideas. First, all basic 3D interaction tasks may be turned into a simple pointing task. Second, the 6D space of the user (3D position and 3D orientation) may be seen as a set of hyperspaces in which a separate pointing task may be applied. Due to these ideas, Fly Over may be utilized the same way with various 2D, 3D or 6D devices.

The generic model has been applied to a 2D navigation task and has been compared to the gaze-steering technique. Preliminary qualitative results obtained on the IBISC semi-immersive Virtual Reality EVR@ platform shows that Fly Over generates smoother trajectories and is well accepted by the users.

Ongoing work is concerning the evaluation of Fly Over for a real 3D navigation task in submarine environments (French ANR Digital Ocean project). We predict that the splitting between 3D orientation and 3D position will have a benefic effect on the easiness of the 3D navigation task.

As the Fly Over model is generic, future work will be held on the application on manipulation and system command tasks. Our goal will be to show if our technique leads to a continuity feeling between tasks when switching from a device to another, and if the total training time is lessen, as we suppose to be.

ACKNOWLEDGMENTS

This work is supported by the VarSCW (Virtual and augmented reality Supported Collaborative Work) project. We wish to thank "Le Conseil Général De l' Essonne", "Le C.N.R.S." and the IBISC laboratory for funding this project

REFERENCES

[Bowman and al., 1997] D.A. Bowman, D. Koller, L. Hodges (1997). Travel in Immersive Virtual Environments: an Evaluation of Viewpoint Motion Control Techniques. In: *Proceedings of the Virtual Reality Annual International Symposium*, pp. 45-52.

[Bowman and al., 1998] D.A. Bowman, J. Wineman, L. Hodges, D. Allison (1998). Designing Animal Habitats Within an Immersive VE. *IEEE Computer Graphics & Applications*, **18**(5), pp. 9-13.

[Bowman and al., 2005] D.A. Bowman, E. Kruijff, J.J. LaViola, I. Poupyrev (2005). *3D user interfaces: Theory and Practice*, pp. 1-26, 87-287. Addison- Wesley.

[Irawati *and al.*, 2006]   S. Irawati, D. Calderòn, H. Ko (2006). Spatial Ontology for Semantic Integration in 3D Multimodal Interaction Framework. In: *VRCIA*, pp. 129-135. Hong Kong.

[Kaiser *and al.*, 2003]   E. Kaiser, A. Olwal, D. McGee, H. Benko, A. Corradini, X. Li, P. Cohen and S. Feiner S (2003). Mutual Disambiguation of 3D Multimodal Interaction. In: *ICMI-PUI '03*.

[Mapes *and al.*, 1995]   D. Mapes, J. Moshell (2005). A Two-Handed Interface for Object Manipulation in Virtual Environments. In: *Presence: Teleoperators and Virtual Environments*, **4**(4), pp. 403-416.

[Mine, 1995]   M. Mine (1995). *Virtual Environment Interaction Techniques (Technical Report TR95-018)*. UNC Chapel Hill CS Dept.

[Sato, 2006]   M. Sato (2006). A String-based Haptic Interface "SPIDAR. In: *ISUVR2006*, **191**.

[Taylor II *and al.*, 2001]   R.M. Taylor II, T.C. Hudson, A. Seeger, H. Weber, J. Juliano, A.T. Helser (2001). VRPN: A Device-Independent, Network-Transparent VR Peripheral System. In: *ACM Symposium on Virtual Reality Software and Technology*, pp. 56-61.

Force Feedback Control of Robotic Forceps for Minimally Invasive Surgery

Chiharu Ishii* and Yusuke Kamei**

*School of Global Engineering, Kogakuin University,
139 Inume-cho, Hachioji-shi, Tokyo 193-0802, Japan (e-mail: c-ishii@cc.kogakuin.ac.jp).
**Graduate School of Kogakuin University, Tokyo, Japan (e-mail: am07020@ns.kogakuin.ac.jp)

Abstract: Recently, the robotic surgical support systems are in clinical use for minimally invasive surgery. For improvement in operativity and safety of minimally invasive surgery, the development of haptic forceps manipulator is in demand to help surgeon's immersion and dexterity. We have developed a multi-DOF robotic forceps manipulator using a novel omni-directional bending mechanism, so far. In this paper, in order to control the developed robotic forceps as a slave manipulator, joy-stick type master manipulator with force feedback mechanism for remote control is designed and built, and force feedback bilateral control system was constructed for grasping and bending motions of the robotic forceps. Experimental works were carried out and experimental results showed the effectiveness of the proposed control system.

Keywords: Force feedback, Robotic forceps, Minimally Invasive Surgery

1. INTRODUCTION

Minimally Invasive Surgery has an excellent characteristic that can reduce a burden of patients. However, it causes difficult operation for surgeons due to the inflexibility of surgical instruments and small work space. Recently, the robotic surgical support systems such as 'da VINCI' [1] are in clinical use. In such robotic surgical systems, particularly, the development of multi-DOF robotic forceps manipulators so as to realize complex human finger action in laparoscopic surgery is one of the most important subjects. We have developed a multi-DOF robotic forceps manipulator using a novel omni-directional bending mechanism with screw drive mechanism that we call double-screw-drive (DSD) mechanism [2]. DSD mechanism has two linkages for bending motion, which are consisted of screwed universal joint. One side of the screwed universal joint is left-handed screw and the other side is right-handed screw. The omni-directional bending motion is achieved by rotating these two linkages.

For further improvement in operativity and safety of minimally invasive surgery, the development of haptic forceps manipulator, in which operation forces are measured by force sensor and haptic feedback is provided to help surgeon's immersion and dexterity, is in demand. In this paper, in order to control the developed DSD robotic forceps as a slave manipulator, joy-stick type master manipulator with force feedback mechanism for remote control is designed and built. Based on the force reflecting servo type bilateral control law, position and force feedback control law was constructed for master-slave control system of bending and grasping motions, and experimental works were executed.

The part of this study is supported by the Grants-In-Aid for Scientific Research from the Ministry of Education, Culture, Sports, Science and Technology in Japan.

2. TELEOPERATION SYSTEM

2.1 Double-Screw-Drive Robotic Forceps

Configuration of bending part of the developed DSD robotic forceps is shown in Fig. 1.

Fig. 1. DSD robotic forceps

DSD mechanism has three linkages. Two are used to achieve omni-directional bending motion, and one is used to rotate a gripper. The opening and closing motions of the gripper is achieved by wire-driven. The main specifications of the DSD robotic forceps are described as follows.

1. Diameter of the forceps is 10[mm], which enables to insert the forceps into a trocar.
2. Bending range is ± 90[degrees] in horizontal and vertical direction, respectively.
3. The gripper can rotate arbitrarily.
4. In the grasping motion, only one side of the jaw can move. The other side of the jaw is fixed.

2.2 Manipulator for Remote Control

In a laparoscopic surgery, multi-DOF robotic forceps manipulators are operated by remote control. In order to control the DSD robotic forceps as teleoperation system, joy-stick type manipulator for remote

Fig. 2. Teleoperation system

control was designed and built by reconstruction of a ready-made joy-stick and forceps. The teleoperation system is shown in Fig. 2.

2.3 Force Feedback Mechanism

Force feedback mechanisms for grasping force and bending force are illustrated in Fig. 3 and Fig. 4, respectively.

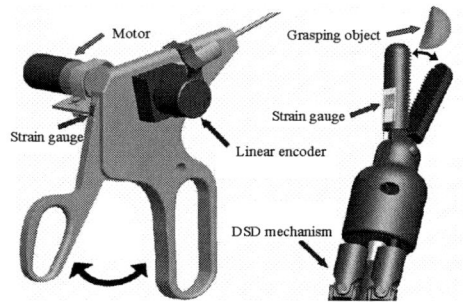

Fig. 3. Mechanism of grasping force feedback

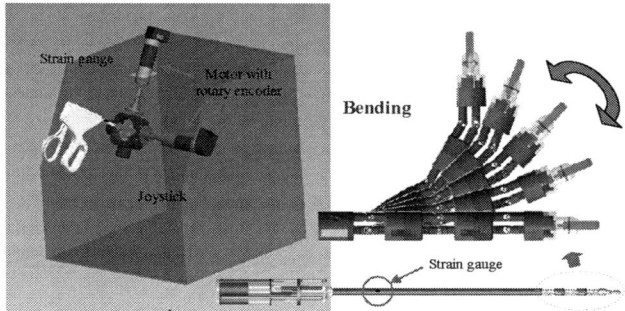

Fig. 4. Mechanism of bending force feedback

In both cases, operation force is detected by strain gauge, and variation of position is measured by encoder.

3. BILATERAL CONTROL SYSTEM

For the grasping and bending motions, the force reflecting servo type bilateral control law was constructed, respectively. The control system is shown in Fig. 5. Then, the control inputs are given as follows.

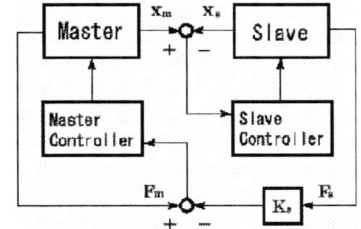

Fig. 5. Bilateral control system

$$\tau_m = K_f(F_m - K_s F_s), \quad (1)$$
$$\tau_s = K_p(X_m - X_s) \quad (2)$$

where, subscripts m and s denote master and slave respectively, X_m and X_s represent the positions, F_m and F_s stand for the forces, K_f and K_p are feedback gains of force and position, τ_m and τ_s are torque control inputs to the motors, and K_s is the amplification ratio coefficient of the slave force.

4. EXPERIMENTS

Experimental works were carried out along the following conditions.

(a) Grasping experiment: From 7.5[s] to 12.5[s], the gripper grasped the sponge.

(b) Bending experiment: A 200[g] weights pet bottle filled with water was hung up on a tip of the forceps, and lift up and down were repeated. The bending motion was restricted to the vertical direction only.

Experimental results are shown in Fig. 6 and Fig. 7. In Fig. 6, master force was amplified ten times so that surgeon can easily feel the reaction force.

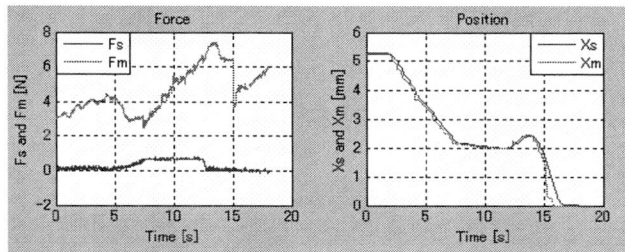

Fig. 6. Experimental results for grasping (Ks=10)

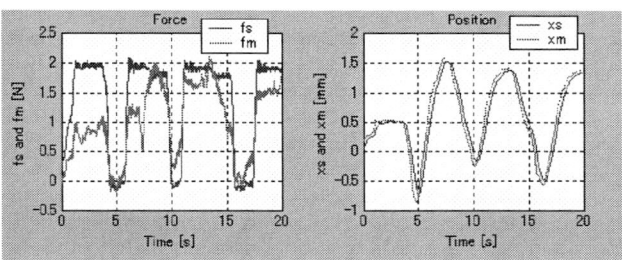

Fig. 7. Experimental results for bending (Ks=1)

In both cases, tracking performance of the position is well, and operator felt the reaction force properly.

5. CONCLUSIONS

In this paper, force feedback bilateral control system was proposed to the grasping and bending motions of robotic forceps for minimally invasive surgery. Experimental results showed the effectiveness of the control system. Improvement of tracking performance of the force, and in order to enhance the safety, exchange of the force sensor from strain gauge to optical fiber are left as future works.

REFERENCES

[1] http://www.intuitivesurgical.com
[2] Ishii, C. and Kobayashi, K. (2007). Development of a New Bending Mechanism and Its Application to Robotic Forceps Manipulator. *Proc. of 2007 IEEE Int. Conf. on Robotics and Automation*, pp.238-243.

Using Virtual Reality to Dynamically Setting an Electrical Wheelchair

S. DIR, O. HABERT, A. PRUSKI

Laboratory of Automation and Cooperative Systems (LASC),
7 rue Marconi, 57070 METZ Technopole, FRANCE
E-mail: dir@univ-metz.fr, habert@univ-metz.fr ,pruski@univ-metz.fr

Abstract: This work uses virtual reality to find or refine in a recurring way the best adequacy between a person with physically disability and his electrical wheelchair. A system architecture based on *"Experiment → Analyze and decision-making → Modification of the wheelchair"* cycles is proposed. This architecture uses a decision-making module based on a fuzzy inference system which has to be parameterized so that the system converges quickly towards the optimal solution. The first challenge consists in computing criteria which must represent as well as possible particular situations that the user meets during each navigation experiment. The second challenge consists in transforming these criteria into relevant modifications about the active or non active functionalities or into adjustment of intrinsic setting of the wheelchair. These modifications must remain most stable as possible during the successive experiments. Objectives are to find the best wheelchair to give a beginning of mobility to a given person with physically disability.

Keywords: Artificial intelligence, Assistive tools, handicap, decision making, virtual reality

1. INTRODUCTION AND CONTEXT

A technical assistance is often a possible solution to offset a handicap. The mobility loss of a person can be partly offset by an electrical or non-electrical wheelchair. This one is controlled by using the onboard-person residual mobility functions in conjunction with numerous low and high level assistive technologies. An evaluation of any assistive system for people with disability is essential in order to determine the real needs. The feedback coming from tests in real situation ensure a source of very profitable information for the developments and search to be carried out. We could note that required work and concentration during experiments can be tiring for numerous people with disability. Moreover, building contexts to study particular situations (environments setting, danger situation ...) can prove to be complex.

Because of the diversity of the users, the numerous contexts which can be used and the technical evolutions of the wheelchairs, the experimental evaluation way which consists in putting in situation people in real built environments has reached its limits. New tools must be used to give much more flexibility during the experimental steps. Virtual reality brings these possibilities. For more than 20 years, the use of virtual reality has been essential in many fields. It allows experiments free from any danger and out of all material constraints. The flexibility brought by the virtual reality makes possible to define infinite experimental situations. According to the problem to be solved and its context, the proportion of virtual in the real world is more or less significant. The total and active immersion in a virtual world is often used for control training of more or less complex machines. It is only since ten years that people who work in the field of the handicap use virtual reality-based-simulations. Thus some projects in this field were born [1], [2], [3], [4], [5], [6], [7], [8], although it remains a marginal if it is compared to the whole of the simulators applications. Among these projects, one distinguishes mainly the following problems:

- The study of the accessibility of the wheelchairs to the indoor/outdoor public or private infrastructures
- Assistance of wheelchair choice
- The help in the development of new advanced functionalities
- The evaluation of the couple man/machine which is based on navigations in environments to determine the best wheelchair.

Our work is within the framework of this latter problematic.

The phase of prescription of a wheelchair is carried out by a team which includes a doctor prescriber, an occupational therapist and the user itself. Goals of this team are to globally evaluate the driving abilities of the onboard person which depend on its disabilities. Because of a too significant handicap which induces security problems, part of the people which disability never reaches or fails the phase of wheelchair prescription. They are estimated unfit to drive a wheelchair and it is impossible to evaluate the help that assistive tools could bring them. A statistical study [12] conducted in 200 clinics in the USA, showed that 9 to 10% of people who receive training in the use of electric wheelchairs are unable to use them for security reasons. The setting in situation in virtual environments brings a solution for the security aspects and allows a broad range of parameter setting while minimizing material constraints.

2. SYSTEM ARCHITECTURE

2.1 The experimental cycle

The system shown in figure 1 is the association of a simulator, a configurator and an evaluator. The experiment begins by specifying the initial conditions relating to the user, the indoor environment and the wheelchair. The simulator puts the onboard person in situation of navigation in the virtual world. The observations data make possible to calculate relevant criteria which allow a decision-making system to stop the experiment or to propose a wheelchair modification. Any proposed modifications call high-level assistive navigation tools. These tools already exist or are under development in the laboratory working in handicap fields. The configurator will be able to also propose modifications about some intrinsic parameters of the wheelchair (linear and angular speed, acceleration or deceleration, kinematics, dimension...). This cycle continues as long as the system estimates that it is necessary to improve the equipment of the wheelchair.

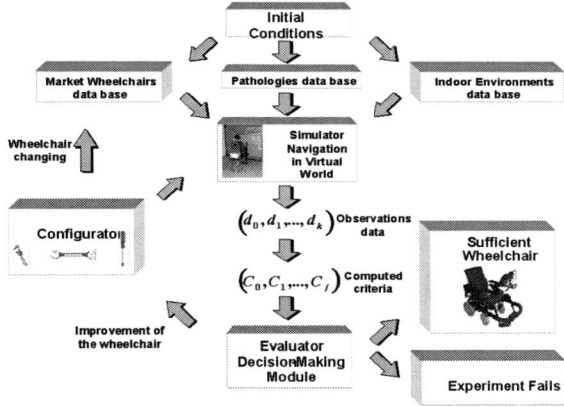

Figure 1: The system architecture

2.2 The Initial Conditions

These initial conditions take into account the opinion of the experts (the prescriber, the occupational therapist and possibly an engineer) in the best way of carrying out the cycle of experimentation. Initial conditions concern:

- The user's Pathologies. These indications allow to specify the residual functional capacities of the user and to deduce the choice of the wheelchair. It will target the basic system command of the wheelchair. The experts could be able to also decide to establish criteria about the mobility level to achieve. The stop conditions according to the physical and psychological state of the patient can also be established.

- The basic wheelchair whose characteristics are chosen by the experts. Kinematics, anthropomorphical constraints, an adapted basic control system of the wheelchair (joystick, interface ...)

- Indoor environment to carry out the virtual reality experiments. This environment will gather whole of the difficulties to be submitted to the user. This environment will be selected according to the kind of environment in which the user of the wheelchair will be brought to daily navigate.

2.3 The simulator

The use of a virtual reality based simulator is a solution which solves at least partly security and reliability problems. In this work a museum model of environment has been designed by using 3D modeling software. It is possible to easily integrate this model in any 3D creation application environment development. We have used Virtools® which enable to add animations and behaviors to these 3D environments to design realistic and interactive virtual reality applications (see figure 2).

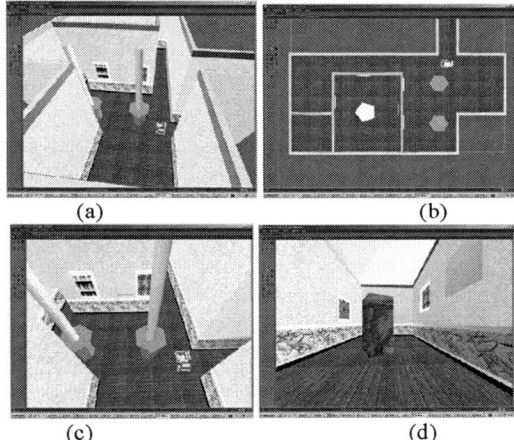

Figure 2: 3D environments examples: (a), (b), (c) are views with an external camera (d) a view given by an onboard camera

The simulator allows putting in situation the person by using several immersion methods. The first method consists in using a head-mounted displays associated with a 3D movement sensor. The immersion seems to be less constraining (weight of the head-mounted displays, field of vision and problem of nausea or claustrophobia phenomenon) by projecting the environment on a wide flat-screen or curved-screen.

A system of rollers equipped with position encoders allows carrying out the test with any kind of wheelchair. The user controls his wheelchair as in a real environment. The data coming from the position encoders allow computing and representing the wheelchair in the virtual scene. The user has the choice between two kinds of immersion, the head mounted device one and the projection of the scene on a wide flat-screen. The onboard visualization mode allows the user to see the environment and part of the virtual wheelchair.

According to the degree of equipment of the wheelchair to be tested or the new functionality to be evaluated, exteroceptive sensors can be modeled in the virtual scene (ultrasound or infrared sensors, camera ...). The incoming sensory data are used by the control algorithms of the wheelchair. Assistances already available are: direction following, wall following, free space following, obstacle avoidance, doorway crossing, pre-stored or real time computed trajectories following ...

Other more complex assistances are under study and the simulator is a fundamental tool to develop them. More details about the simulator and the VAHM project can be found in the following publications [7], [8], [9].

2.4 The Evaluator or decision-making system

The simulator is the central module of the dynamic system of configuration. Criteria are computed from observation of each experiment as a whole. The i criteria of the vector $(C_0, C_1 ..., C_i)$ in evaluation step are the inputs of the decision-making module. The n outputs $(S_0, S_1, ..., S_n)$ represent the n functionalities which it is possible to install or to configure. Term "functionality" is taken in the broadest sense and represents at the same time the high-level functionalities and the intrinsic setting of the wheelchair parameters. From these criteria, it is possible to decide:

- If the goal to find a wheelchair which allow restoring a certain level of mobility to the tested person has been reached. To do this, we will take into account some initial conditions concerning the mobility level to be given to the onboard person.
- If it is necessary to stop because they is no convergence towards an acceptable solution. We will also use criteria linked to the initial conditions (maximum experiments that the person can support) and to integrate later data resulting from physiological measurements.
- If a next experiment has to be started. Mission of the evaluator is to propose some modifications of the wheelchair in terms of high-level functionalities and/or the setting of intrinsic parameters.

2.5 The configurator

Goal of this module is to pass on the suggested changes coming from the evaluator to the wheelchair. The different functionalities are defined like agents which are in an active or inactive state. The intrinsic parameters are data which can be modified at any time. Among the functionalities already virtualized in a platform coming from an other work [4],[5],[6], we can mention obstacle avoiding, wall following, back Tracking, wall following, free-space following, direction Following, path planning (need the use of a model of the environment) .. Other assistive tools as Man-Machine interfaces, particular constrained controls could be considered. Concerning the intrinsic parameters, acceleration, deceleration, minimum/maximum linear and angular speed, kinematics (holonomic or non holonomic) and distance between the wheels can be modified. It is also possible that the configurator would completely change the wheelchair proposed while the definition of the initial conditions

3. DECISION MAKING TOOL

3.1 The fuzzy inference system

The simulator is partly based on a fuzzy inference system used as a decision-making tool. Such a system is well-suited when the modeling of the phenomena is incomplete or when a human being is involved in this modeling. Indeed, it is a question in our work to aggregate various data categories and to use them in order to highlight behaviors or difficult situations meet by the user. The analysis must lead in modifying the wheelchair in an optimal way. The used FISPRO software [10] allows both the design and the possibilities of optimizing the fuzzy inference system. To validate the initial phase of this work, inputs and outputs have been limited. Using a fuzzy inference system will gradually enable us to raise the complexity of the problem (by adding other observation data to compute other criteria) while optimizing it (adjustment of the membership functions). The fuzzy system uses the fuzzy sets and the membership functions to describe the values of the vector of the uncertain criteria. The base of rules will be given by the experts who have to integrate any functionality which correspond to an analyzed situation.

3.2 The computed criteria

The simulation step consists in navigating in a virtual world. We have chosen the visit of a museum where, while the navigation, the user will meet a wide number of particular situation as walls following, obstacle avoidance in more or less encumbered environment, passage door ... The simulation observation data (successive position of the wheelchair, sensory data if available, collision detection and semantic information about surrounding objects of each wheelchair position ...) are stored in a file. These situations are characterized by different temporal or quantitative greatness which concern the entire experiment or concern the collisions aspects. It is also necessary to gather all the information about the local or global obstruction of the environment. The following criterion has been chosen.

- **Time Out (TO):** The maximum time the experiment lasts beyond it must be imperatively stopped. Its value is fixed by the medical experts.
- **Collision Number Beginning (CNB):** A counter is incremented each time a beginning collision is detected.
- **Average Length of Collisions (ALC):** Using the previous counter, it is possible to compute the duration of all collisions which occur during the simulation. ALC is sum of all collision divide by the number of collision beginning
- **Stop Number (SN):** A counter increases each time

the speed is 0 or less than V_{min}.

- **Number of Blocking Situations (NBS):** A counter increases each time a stop duration is more than T_{bl} seconds.
- **Oscillation (OS):** A number which is computed only in walls area. (Semantic data about the kind of the observed obstacle is used). It's a number that depends on the number of direction changing and on the magnitude of each oscillation.
- **Right Obstruction (RO):** Average distance given by the wheelchair right sensors. RO is the sum of each sensory data for a given position. All sums are averaged considering the whole of the points.
- **Left Obstruction (LO):** Average distance given by the wheelchair left sensors.
- **Front Obstruction (FO):** Average distance given by the wheelchair front sensors.

In this work, the fuzzy logic has been chosen as a decision-making tool. Criteria resulting from the simulation represent the inputs of de decision-making module. Output of the evaluator is a vector of functionalities. According to the weight of each member of the vector, the associated functionality will be or will not be installed on the wheelchair. A basic threshold is used.

The inputs are represented by their membership functions to describe the values of the vector of the uncertain criteria. An example concerning the number of collisions is represented figure 3.

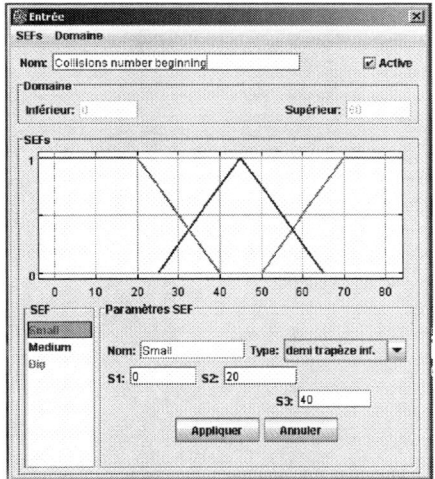

Figure 3: Membership function of CNB criteria

Our proposition is that any output is represented by a number between 0 and 100. Following fuzzy sub-sets will be associated:
{
    VLoC (*Very Low Conceivable*),
    LoC (*Low Conceivable*),
    MeC (Medium *Conceivable*),
    HiC (High *Conceivable*),
    VHiC (*Very High Conceivable*)
}

An example about these sub-sets about obstacle avoidance is shown in figure 4. Figure 5 shows both the criteria and the functionalities to be defined in FISPRO. The implementation of the base of the fuzzy rules in FISPRO is made from an analysis that we carried out. This step will be detailed in the following section.

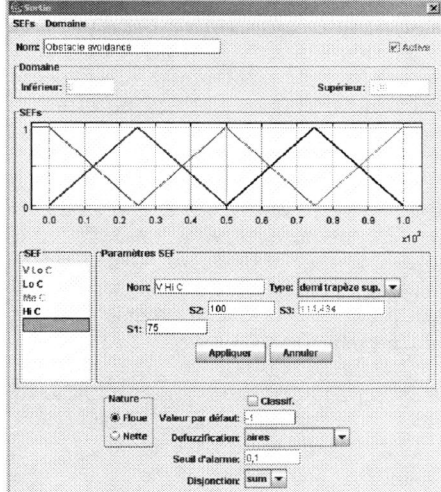

Figure 4: Membership function of obstacle avoidance

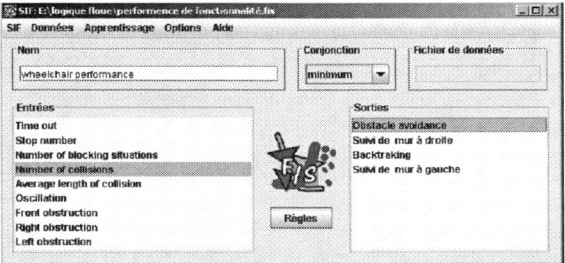

Figure 5: Implementation of inputs and outputs in FISPRO

4. SITUATION ANALYSIS

4.1 The used Platform

A multi agent based control coming from a previous work is used. It is ensuring the best choice of the control in a given context according to user's preferences. In fact, this control tries to merge the man and the machine, in order to make a symbiosis to create the machine easily acceptable and the control more efficient. The system is based on an architecture where independent behaviors, based on sensors referenced controls, are in concurrency. The very first control is given by the user thanks to the interface. This control consists in a direction control order. According to the pathology and the used interface, this information can be inaccurate or slow to obtain, and the system has to take these phenomenon into account. More details can be found in [11].

4.2 Situation Analysis

The criteria analysis gives information about the problem met by the onboard person while the navigation.

For each particular situation highlighted by one or more criteria, one or more functionalities can be proposed to resolve this situation.

Table 1 recapitulates the analysis of various situations that have been retained in the first step of our work. Columns represent the criteria and lines the different potential functionalities. Any line is an analysis of a particular situation met by the person while he's navigating. All considered situations are not exhaustive but are the result of a brainstorming allowing a first design of the fuzzy inference system. When a criterion is relevant for a given functionality, the corresponding cell is noted "X", and "-" if not. Each line is then analyzed to build fuzzy rules tables. For this, the criteria are analyzed two by two in order to affect a fuzzy subset to each combination. At this level, role of the expert is predominant. Table 2a) and 2b) gives an example in the case of the fuzzy rules of the obstacle avoidance. The first two lines reflect an analysis of two distinct situations for which the functionality of obstacle avoidance can be considered. The obstacle avoidance is an interesting functionality when the analysis of an experiment shows some collision beginnings which last a certain amount of time. Depending on the number of those beginnings of collision and the average length of these collisions, the expert can complete table 2a. The expert fills the cells of the table by considering 3 subsets: Small, Medium and Big. Figure 6 illustrates the reasoning of the first row of table 1 and explains the reasoning of the expert to fill the table 2a). When the situation is impossible, the corresponding rule is disabled.

The following analysis is held:

- If the number of collisions beginnings is high and the average length of the collisions is small then the user often hits the obstacles and this, without leading blocking situations. In such a case, activation of obstacle avoidance functionality is very relevant.

- If the number of collision beginnings is high and the average length of the collisions is high, they is no explanation to give and solution to bring because the experiment is time-limited. Such a situation must not lead to strengthening the activation of the obstacle avoidance functionality.

- If the number of collision beginnings is low and the average length of the collisions is high, using obstacle avoidance could be relevant but there is no certainty that this is the best solution

- If the number of collision beginnings is low and the average length of the collisions is low, the user doesn't need the obstacle collision functionality.

Table 2b corresponds to the fuzzy rules of the second line of table 1. Figure 7 shows another example of reasoning concerning the OS criterion and the left or right wall following functionalities.

Criteria / Functionalities	TO	SN	NBS	CNB	ALC	OSC	FO	RO	LO
Obstacle Avoidance (OA)	-	-	-	X	X	-	-	-	-
Obstacle Avoidance (OA)	-	X	-	-	-	-	X	-	-
Right Wall Following (RWF)	-	-	-	-	-	X	-	X	-
Left Wall Following (LWF)	-	-	-	-	-	X	-	-	X
Backtracking (BAC)	X	-	X	-	-	-	-	-	-
Backtracking (BAC)	-	-	-	X	X	-	-	-	-

Table 1: recapitulate the analysis of different situations ("X" means that criterion is taken into account to compute fuzzy rules of corresponding functionality, and "-" the opposite)

ALC \ CNB	Small	Medium	Big
Small	V Lo C	V Lo C	Me C
Medium	Hi C	Me C	Me C
Big	V Hi C	Hi C	Rules is disabled

(a)

FO \ SN	Small	Medium	Big
Small	V Lo C	V Lo C	Lo C
Medium	Hi C	Me C	Me C
Big	V Lo C	Hi C	V Hi C

(b)

Table 2a and 2b: fuzzy rules tables of obstacle avoidance

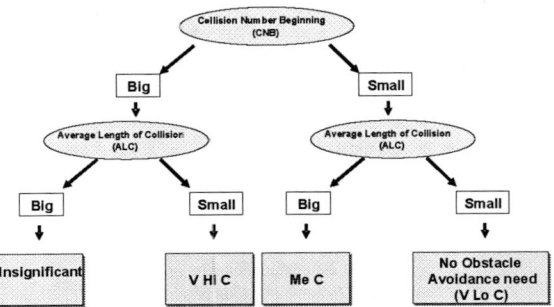

Figure 6: Obstacle avoidance analysis according CNB and ALC criteria

5. CONCLUSION AND STATE OF THE WORK

Virtual environment has been built. It is easily modifiable and interchangeable depending on the experiments to be performed. The module which collects the observation data has been developed and can be improved at any time by adding new data. Another module retrieves these data, computes the criteria and saves the results in a file compatible with the FISPRO tool. The latter is very useful to set and simulate the parameters of the fuzzy inference system. It is necessary to use such a tool to keep the consistency of the decision making system when the complexity increases. The next challenge is to improve the system by adding new criteria and new rules to allow the decision making system to give relevant decision about the functionalities to activate or not. The conduct of the experiment over several cycles leads to more precisely analyze the system to ensure its stability and convergence towards optimal and relevant solutions.

REFERENCES

[1] C.H. Tsai, S.W. Yang and Y.-Y. Fan (2006). Using virtual reality biofeedback to improve manipulative skill of manual wheelchair users. *Journal of Biomechanics*, 39, Supplement 1, S170.

[2] P.M.Grant, C.S.Harrison, B.A.Conway (2004) Wheelchair Simulation. *Cambridge Workshop on Universal Access and Assistive Technology (CWUAAT)*, Cambridge.

[3] Hasdai, Aya, Jessel, S. Adam, Weiss, L. Patrice (1998). Use of a computer simulator for training children with disabilities in the operation of a powered wheelchair. *American Journal of Occupational Therapy / AJOT*, 3 52, 215-20.

[4] Harrisson, A. Derwent, G. Enticknap, A. Rose F.D. Attree E.A., (2000). Application of Virtual Reality Technology to the Assessment and Training of Powered Wheelchairs. *Proceedings of the 3rd International Conference On Disability, Virtual Reality & Associated Technologies* 15-21, Alghero, Italy.

[5] Stott, I. Sanders, The Use Of Virtual Reality To Train Powered Wheelchair Users And Test New Wheelchair (2000). *Systems International Journal of Rehabilitation Research*, 4 23, 321-26.

[6] Majid Majdolashrafi, Majid Nili Ahmadabadi, A. Ghazavi (2002). A Desktop Virtual Environment to Train Motorized Wheelchair Driving. *IEEE International Conference on Systems, Man, and Cybernetics Hammamet*, Tunisia.

[7] G. Bourhis, O. Horn, O. Habert, A. Pruski, (1997). The Autonomous Mobile Robot SENARIO: A sensor Aided Intelligent Navigation System for Powered Wheelchairs. *IEEE Robotics & Automation Magazine*, 4, 60-70.

[8] G. Bourhis, O. Horn, O .Habert, A. Pruski (2001). An Autonomous Vehicle for People with Motor Disabilities. *IEEE Robotics & Automation Magazine*, 8, 20 -28.

[9] G. Bourhis, Y. Agostini (1998). The VAHM Robotized Wheelchair: System Architecture and Human-Machine Interaction. *Journal of Intelligent and Robotic Systems*, 22 n°1, 39-50.

[10] Fuzzy Inference System professional, INRA http://www.inra.fr/bia/M/fispro

[11] A. Pruski, M. Ennaji, Y. Morère (2002). VAHM : A User adapted intelligent wheelchair. *IEEE Conference on Control Application*, Galsgow, Scotland, UK.

[12] L. Fehr, WE. Langbein, SB. Skaar (2000). Adequacy of power wheelchair control interfaces for persons with severe disabilities: A clinical survey. *J Rehabil ResDevel*, 37(3), 353-60.

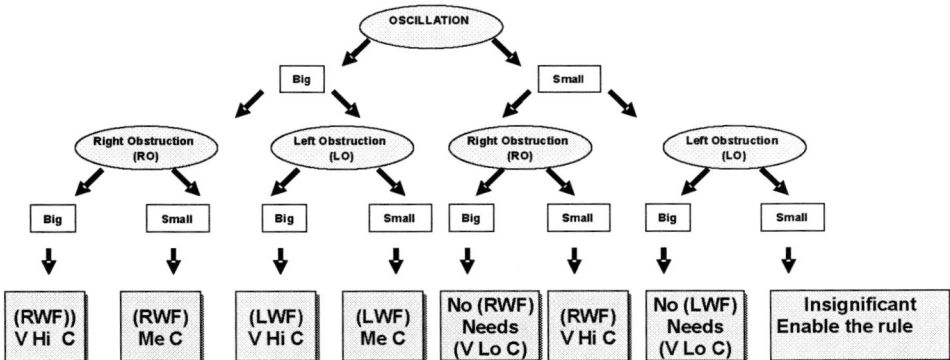

Figure 7: Left and right wall following analyze according (OS), (RO) and (LO) criteria

A planning architecture for mobile robotics

Julien Guitton * Jean-Loup Farges * Raja Chatila **

* ONERA - DCSD, 2 av. Edouard Belin, 31055 Toulouse Cedex 04, France
(e-mail: julien.guitton@onera.fr, jean-loup.farges@onera.fr).
** LAAS - CNRS, 7 av. du Colonel Roche, 31077 Toulouse Cedex 04, France
(e-mail: raja@laas.fr).

Abstract: Mobile robots such as explorer rovers need task and path planning abilities in order to fulfill their assigned missions: path planning to plan their movements and task planning to plan their actions. The coupling between these two kinds of planning presents open issues such as the description of the environment and the consideration of geometric constraints that must be verified in order to act and move during an action. This paper addresses these issues by proposing an architecture in which a hierarchical task planner sends requests to a path planner in order to check the feasibility of actions. Requirements allowing the path planner to produce an answer are presented as well as the description of planning operators. Finally, we specify the mechanism and the communication language by which the task planner produces requests and takes into account answers.

Keywords: Agent reasoning, autonomous mobile robot, task planning, path planning, interaction language.

1. INTRODUCTION

Autonomy of mobile robots is caracterized by the ability to act and move in their environment without human intervention. For applications with complex environment and goals, the agent's autonomy implies symbolic and geometric reasoning. For instance, in order to accomplish its assigned mission, an autonomous planetary rover needs to plan all the actions it has to execute (*e.g.* make a sample of rock, take a picture) and to find a path between the different action areas.

This kind of problems was mainly addressed as orienteering problems [1] or in a multi-robot context as task allocation and path planning problems [2]. In the last case, a sequence of actions was usually pre-calculated and the goal turned out to be an assignment of each action to the different robots. In [3], a generic mission planner was developed. This planner tries to address mission problems by coupling a general purpose grammar interpreter with dynamic planning techniques. The grammar allows to define a set of tasks to be achieved. However, in this work, geometric constraints concerning the accomplishment of a task are not taken into account.

There is a need to investigate more specifically on the choice of actions (based on a symbolic description) that a robot will have to perform in order to fulfill its mission and on the inherent link between these actions and the movements needed to accomplish them.

A general purpose planner is not able to solve this kind of problems due to the difficulty associated with efficient and accurate representation of the environnement geometric concepts in its first order language. The use of a domain-dependant planner (*i.e.* a specific planner) is not recommended because it addresses only one specific kind of problem. In other words, such a specialized planner is not reusable for a slightly different problem (*e.g.* modification of the robot properties). Therefore, there is a need to have different planners interacting together. For instance, a task planner for the selection of the symbolic actions and a path planner to define the movements of the robot.

A naive approach is to define a hierarchy of planners. Once we have obtained a symbolic plan of actions, a specialized planner can calculate the movements of the agent. This approach raises the following questions [4]: (1) What kind of information should the path planner feed back to the task planner ? (2) How can the task planner use this information to generate a new plan ? In addition, with this method the sequence of actions must be fully calculated before realizing that it is not feasible in terms of movements. Another approach is to use a path planner during the action planning. This idea of relying on specific planners during planning [5] was used for the application with a mobile robot in [6]. In this work, the link between the generic planner and an itinerary planner was done using special symbolic attributes. However, the specialized planner builds its work on an accessibility graph and does not integrate reasoning about the geometric constraints of the problem. Moreover, the issue of dependency between subproblems of movement is not addressed. The other approaches dealing with the link between task planning and geometric reasoning are mainly approaches for one kind of specific problems [7]. In [8], an integrated planner is presented, in which the definition of states of the world contains symbolic and geometric informations. These informations are defined by the use of a set of types and predicates that allow to establish a link between symbolic planning and manipulation planning.

This paper outlines some proposals allowing to bring together task planning and path planning in order, for a mobile robot, to accomplish its assigned mission. We take as an example a data acquisition mission (*e.g.* taking pictures and collecting samples) for a mobile robot. In the first part, we present the existing relationship between symbolic reasoning and geometric reasoning in the context of mobile robotics. Next, we propose and detail an architecture allowing a cohabitation between this two types of reasoning. We also introduce the concepts we wish to implement within this architecture, and explain how to implement them at the different levels of the architecture. Finally, based on a example of data aquisition mission for a mobile robot, we show the interactions between the two planners leading to the construction of the final plan.

2. LINK BETWEEN TASK PLANNING AND PATH PLANNING

In this part, we propose to study the link between planning the actions of an agent and planning its displacements through an example of data aquisition mission for a mobile exploration robot. This example will be used throughout this paper.

2.1 Case study

An exploration robot must do some sampling of rock and take pictures of specified areas of the environment. This example is taken from the International Planning Competition (IPC) of 2002 [9]. We add the following assumptions:

- the environment is not provided as a set of predicates intended for the task planner but with topological and geometric data which could be used only by a specialized planner like a path planner;
- the actions can specify a set of geometric constraints in their preconditions. For instance, a picture must be taken at 20 meters of the target according to a given angle ;
- the actions can be durative and require movements. For instance, to film an area of the environment during 10 seconds while maintaining the robot speed and heading.

We do not make any assumption about the environment internal representation of the specialized planner. The figure 1 represents a map of the environment in which some possible paths have been calculated. These paths have been constructed with the use of a visibility graph.

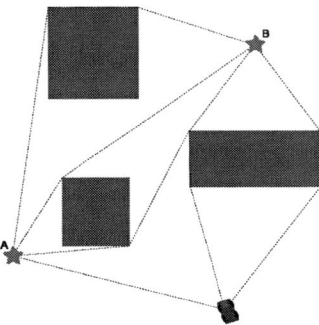

Fig. 1. Example of an environment map representing some possible paths for the mobile robot

The objective is that the robot fulfills its goals which are to sample rock and take pictures. These goals are previously defined but possibly without particular order. The robot's abilities and the geometric constraints provided for each type of actions must be respected.

In the example represented in the figure 2, a robot r has to film the objective A (represented by the star) during 10 seconds. The filming must be done in the bottom right corner, *i.e.* with an angle range between 270 and 360 degrees from the objective.

The robot must position itself perpendiculary to the objective and maintain its heading during the action. Its speed must be constant during the travel. All of these constraints are expressed in the planning operator `film_objective`.

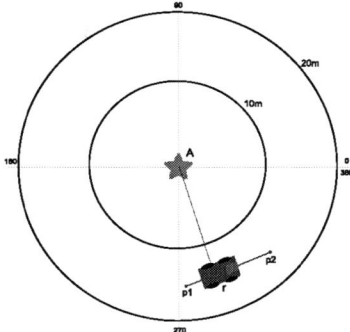

Fig. 2. The robot must film the objective A

2.2 Problems

With this example, three subproblems concerning the movements of the robot can be identified:

- At which position must the robot execute the action ?
- What is its movement during the action ?
- How should it move between the different actions ?

In addition to these three questions, a general problem can be identified : How to schedule the achievement of the actions according to the courses of the robot in the environment ?

Indeed, dealing with the movements locally, *i.e.* between each action, can result in a suboptimal solution with regard to the total distance traveled by the robot.

2.3 Proposals

The type of problem presented below requires the use of a task planner and a path planner. The task planner aims at defining a sequence of actions to carry out the objectives of the mission. The path planner is used to define the movements of the robot. We propose an architecture in which task planning and path planning are interleaved.

The main proposal is as follow : the movements are not symbolic actions (*i.e.* managed by the task planner) but are *geometric preconditions* to symbolic actions. For instance, in order to sample some rock the robot must be able to move to the relevant area. Depending on the type of robot, these preconditions may include only static geometric constraints or kinematic constraints. For instance, the turning radius may be a function of the speed of the robot.

The exchanges between the two planners are carried out by a set of *requests* sent from the task planner to the path planner. These requests specify the geometric and kinematic constraints which must be respected in order to plan an action.

In addition, the path planner can provide some *advices* to the task planner. These advices aim at re-scheduling the planned actions according to the specialized planner feedbacks. They aim at solving the global optimality problem concerning the agent movements by proposing alternatives to the current plan.

3. PLANNING ARCHITECTURE

This part details the planning architecture of the robot deliberative module allowing to connect a general purpose planner and one or several specialized planners dedicated to the different mobile robots. For now, the architecture is composed of a task planner and only one specialized path planner. The two planners are connected through a communication interface (figure 3).

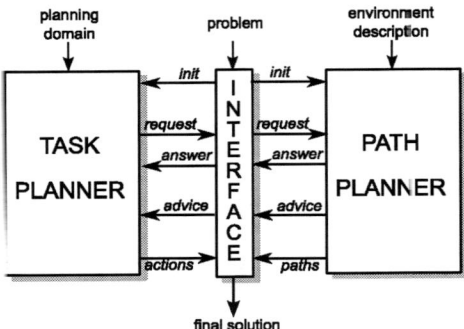

Fig. 3. Planning architecture

At the initialization phase, a human operator provides the informations needed by the architecture in order to solve the problem. These informations are the symbolic description of the possible actions the robot can do, the environment map and description, and the problem to solve. At the end of the processing, the interface sends back the solution plan to the operator. This solution plan contains the symbolic actions and the movement actions calculated by the two planners.

3.1 The task planner

The aim of the task planner is, starting from an initial state representing the state of the environment (in a symbolic formalism) and the state of the robot at the beginning of the mission, to reach a final state in which all the goals of the mission will be achieved. The achievable actions are expressed by a set of planning operators. The sequence of necessary actions to the achievement of an objective can be defined as a *recipe*. These recipes are expressed so as to be able to start them at any level during the execution, according to the current environment and robot state.

Thus, we propose a task planning algorithm based on hierarchical planning techniques. The task planner is similar to SHOP2 [10]. Hierarchical planning techniques allow to express recipes thanks to the use of *methods*. The aim of such an algorithm is to decompose a high-level task in a sequence of subtasks until obtaining primitive tasks (*i.e.* non decomposable tasks). These primitive tasks correspond to the tasks achievable directly.

Our hierarchical planning algorithm allows to express a set of partially ordered tasks as in Nau's work [11]. In the example below, the three tasks are independent and can be achieved in any order (thanks to the use of the keyword unordered).

```
(:unordered
  (film_objective rover1 locationA)
  (film_objective rover1 locationB)
  (sample_rock rover1 locationA)
)
```

A problem for the task planner is expressed in the form of a planning domain and a planning problem. The domain is a set of operators (primitives tasks) and methods (high-level tasks). A problem is composed of the initial state which describes the initial symbolic state of the world and the goals in the form of a sequence of high-level tasks to decompose.

3.2 The path planner

The path planner must be able to provide a route between two points or two geographical areas specified by a set of geometric constraints. Various internal representations of the environment can be chosen : cells decomposition, use of a grid, representation with a visibility graph... However, this choice can affect the quality of the final solution in terms of movements of the robot. Moreover, it must respect the kinematic constraints of the robot (*e.g.* its speed).

The goal of the path planner is to answer to the requests of the general purpose planner by indicating, on the one hand, if a possible path between the specified points exists and, on the other hand, if the robot can be moved during the action according to the specified constraints.

It must also be able to maintain an historic of previous movements in order to suggest to the task planner some optimizations as, for example, the inversion of two or more tasks in order to decrease the total traveled distance of the robot. Moreover, the search for a path can produce more results than the path itself. These results and the already computed paths can be re-used to optimize the entire path.

3.3 Interface between the two planners

The interface between the general purpose planner and the specialized one allows a communication between them. They are not directly connected to permit the integration of a specialized planner corresponding to the kind of problem which will be treated. For instance, in the above example, a path planner can be used. Whereas, in a manipulation problem in which an articulated arm has to grasp objects, a motion planner will be more suitable.

This interface will also allow to deal with problems in a multi-robot context. In this kind of problem, a general purpose planner will be linked with several specialized planners in charge of the different robots.

The first purpose of the interface is to initialize the two planners. It receives data from the the mission management system of the robot and sends to the task planner the planning domain description, and to the path planner the environment description.

After the initialization phase, the interface aims at controlling the exchanges between the two planners. It transfers requests from the task planner to the corresponding specialized planner (in the case of a multi-agent architecture) and, according to the information provided by the path planner, it can propose some advices to the task planner.

4. IMPLEMENTATION

We propose to delegate the actions of movement to the specialized planner. This one is called by the task planner through a set of requests.

Our proposals can be implemented at different levels of the architecture: at the symbolic level, *i.e.*, in the general purpose planner, by adding geometric preconditions to the existing symbolic preconditions ; at the communication level, by using a communication protocol between the different modules of the architecture.

4.1 At the symbolic level

To be executed, an action must respect some preconditions. These preconditions are expressed in the operators and correspond to the state in which the world must be so that the action is achievable. In the same way, we propose to add *attitude preconditions* which allow to specify the attitude (position, orientation, ...) the robot must have in order to perform the action. We also propose to add *behaviour preconditions* which are durative conditions and define the behaviour of the robot in terms of movements during the action. These attitude and behaviour preconditions allow to express a set of geometric and kinematic constraints which will be provided to the specialized planner. There is a continuity between the attitude preconditions and the behaviour preconditions, *i.e.* the robot properties defined in the attitude preconditions (*e.g.* r.speed and r.heading) remain available for the behaviour preconditions as long as they were not modified.

```
(Operator (film_objective ?r ?o)
  ;; preconditions
  ((rover ?r) (objective ?o)
   (camera ?c) (has_camera ?r)
   (is_calibrated ?c)
  ) ;; attitude preconditions
  ((agent ?r) (object ?o)
   (distance(r.pos, o.pos)>=10)
   (distance(r.pos, o.pos)<=20)
   (abs_angle(r.pos, o.pos)>=270)
   (abs_angle(r.pos, o.pos)<=360)
   (rel_angle(r.pos, o.pos, r.heading)=90)
   (r.speed = r.speed_max)
  ) ;; behaviour preconditions
  ((agent ?r) (object ?o)
   (duration 10)
   (constant(r.heading))
   (constant(r.speed))
  ) ;; effects
  ((has_film ?r ?o) )
)
```

During the planning process and following the test of action preconditions, the task planner sends a request to the path planner in order to know if the robot can be correctly positioned. If the answer is positive and if the action requires some movements during its achievement, the task planner sends a new request to the path planner concerning these movements. If all the preconditions are satisfied, the task planner can apply the effects of the action to the current state of the world (figure 4).

4.2 Communication language

From the attitude and behaviour preconditions, the specialized planner must find a path allowing the positioning of the robot for the achievement of the action and then move it during the achievement itself. If there is no solution, *i.e.* no possible position or movement, it sends back a failure message. In this case, the task planner must find another action to plan.

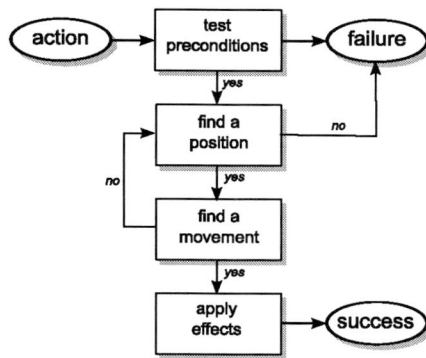

Fig. 4. Decomposition of an action precondition test

This exchange between the planners is done through communication primitives, *i.e.* with the use of a common vocabulary. This language allows to identify *concepts* (e.g. agent, object), *global properties* (e.g. duration), *concept properties* (e.g. r.heading, o.position) and *rules* (e.g. =, constant) contained in the requests.

Concepts allow to identify the agents and physical objects of the environment, which will be handled by the planners. With each concept we associate a set of properties called the *concept properties*. These properties indicate the agent and object properties handled. *Global properties* specify properties of a movement subproblem as, for instance, the maximal duration of a robot move. *Rules* allow to define the geometric constraints of the agents compared to the objects of the environment, as well as the kinematic constraints of the robot. They are applied to the concepts properties and global properties.

The *concept properties* are specific to each agent. But it is possible to define *ontologies of properties*. We may have one ontologie for each type of mobile robot. A specific agent is able to interpret only a fixed number of properties concerning its own abilities. For instance, a planetary rover will not be able to interpret the *vertical speed property*. However, this concept property may be essential for an unmanned aerial vehicle. The designer is in charge of defining a planning domain containing only properties interpretable by the specialized planner.

These different elements of the communication language must be encapsulated in messages which will be exchanged between the planners. The messages are expressed in the form of requests and advices.

4.3 Requests and queries

We can distinguish two types of requests: *planning requests* and *system requests*. Planning requests allow the general purpose planner to ask the specialized planner to find movements satisfying the given geometric constraints.

Definition 1. (Planning request).
A planning request R is defined as a tuple
$< Type(R), Agent(R), Id_{action}(R), \mathcal{C}(R) >$

- $Type(R)$ is the type of the request. For example, *attitude* to define an attitude precondition request, or *behaviour* for a behaviour precondition request ;
- $Agent(R)$ allows to know which agent is concerned by the request in the case of a multi-agent problem ;

- $Id_{action}(R)$ is an action Id allowing to maintain a coherence between the task planner and the path planner resolution ;
- $C(R)$ is the set of geometric constraints to be respected during the robot movements.

An example of planning request is :
`request(attitude, robot1, #1, {distance(r.pos, o.pos)>=10},...})`

System requests is used by the interface in order to give some instructions to the different planners. These instructions are used to initialize, to guide and to finalize the plan construction.

Queries are messages sent by the task planner to the path planner in order to ask for an *advice*. They are used when the task planner is in front of a choice, as for example, *do something at point A or do something at point B*. The answer of the specialized planner is called an *heuristic advice*.

For example, in order to choose between doing an action at point A or at point B, according to the current robot position, the query sent by the task planner will be :
`query(min_distance, robot1, {A,B})`

4.4 Advices

Advices are messages sent by the path planner to the task planner. We can distinguish three types of advices : *heuristic advices* which are answers to the high-level planner queries, *optimization advices* which expressed some proposals of the specialized planner in order to optimize the final plan, and *repair advices* which are used to propose movement alternatives allowing to avoid a deep backtrack of the task planner.

The heuristic and optimization advices can be seen as optional requests. If they are not taken into account, the planners can still construct a valid plan but this one will be less optimized.

Definition 2. (Advice).
An advice A is defined as a tuple
$< Type(A), Content(A), Agent(A), set\{Id_{action}(A)\} >$

- $Type(A)$ is the type of the advice : heuristic, optimization, repair ;
- $Content(A)$ is the content of the advice ;
- $Agent(A)$ allows to know which agent is concerned by the advice in the case of a multi-agent problem ;
- $set\{Id_{action}(A)\}$ is a set of Ids for actions concerned by this advice.

An example of *optimization advice* for the task planner is:
`advice(optimization, reverse, robot1, {#2,#3})`

In this example, the path planner asks the general purpose planner if it can reverse the realization order of task #1 and task #2 (because task #1 and #3 must be realized on the same objective).

4.5 Example of interactions between the planners

In the previously seen example, the robot must take pictures of objectives A (action #1) and B (action #2), and sample rock at point A (action #3).

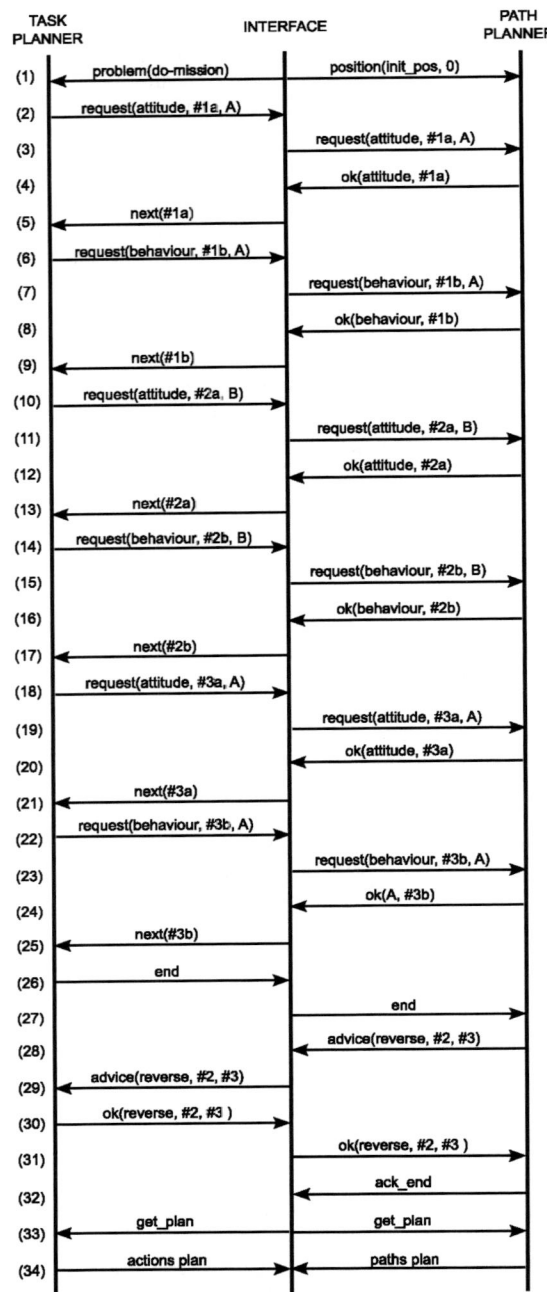

Fig. 5. Sequence diagram of the exchanged messages during the construction of the plan (using a simplified formalism).

The sequence diagram (figure 5) represents the sequence of exchanged messages between the different modules during the resolution of the example problem.

After the initialization phase (1), the task planner tries to plan the first action (*film the objective A*). It sends a request (2) to the path planner in order to test the attitude preconditions of this action. This request contains the set of constraints expressed in the planning operator `film_objective`. The path planner replies with a positive answer (4), *i.e.*, required movements are possible in order to position the robot for starting the action achievement. So, the task planner sends a second request which aims at verifying the behaviour preconditions (6).

In the same way, the two planners exchange messages in order to plan the action `film_objective` at the point B (10-17), and the action `sample_rock` at the point A (18-25).

At this point, the task planner sends a end-message to the interface (34), *i.e.* there is no more task to achieve. The general purpose planner has successfully decomposed its high-level tasks into primitive actions which can be achieved by the robot. The interface informs the specialized path planner of the completion of the mission planning process.

As the specialized planner has calculated all the needed movements, it can now try to optimize the entire robot trajectory. Indeed, the robot visits the point A twice. This is due to the local search of movements without having a global vue of the planned trajectory. So, the path planner proposes an optimization advice to the task planner (28). This advice proposal is to reverse the achievement order of actions #2 and #3. Indeed, since the point A was already visited, the inversion of these two tasks will allow to optimize the solution in term of traveled distance. The sequence of the three actions is expressed as an unordered sequence, *i.e.*, these actions can be planned in any order. The task planner reply with a positive answer to this adviceIndeed, changing the order of these action achievements does not impact the rest of the symbolic plan (30).

The final plan can be constructed and sent back to the mission management system (33). This construction is done by assembling the plan of actions produced by the symbolic reasoner and the plan of movements produced by the geometric reasoner. The correspondence between symbolic actions and movements is done with the use of action Id which are present in the exchanged requests.

Compared to a traditional hierarchical planning architecture in which a symbolic plan is found before trying to satisfy the entire set of geometric constraints, the proposed architecture is able to react immediately to a failure of some geometric preconditions.

5. CONCLUSION AND FUTURE WORKS

We have proposed an architecture allowing to deal with robotics problems in which a mobile robot has to perform some tasks in the environnement, and illustrated it through an example of data acquisition mission for a mobile robot. This architecture is based on a strong coupling between a symbolic and a geometric reasoning. The symbolic reasoning is made by a hierarchical task planner allowing to express the tasks to perform as recipes (through the use of methods). The geometric aspects of the problem are handled by a specialized planner.

The functioning of our architecture is as follow : the task planner calls the path planner by sending requests expressing the necessary conditions for the agent to move in the environment according to the constraints. These constraints are formulated using specific preconditions of the planning operators : the attitude preconditions and the behaviour preconditions.

The communication between the two planners is done through an interface with the use of a common vocabulary. The exchanged messages are expressed in the form of requests containing the necessary geometric constraints to be respected in order to achieve the actions. The other type of exchange concerns the sending of advices allowing, on the one hand, to guide the symbolic plan construction based on information given by the path planner and, on the other hand, to re-schedule the already-planned actions in order to optimize the mission.

Thereafter, we propose to extend our architecture in a multi-agent context in which several mobile robots will have to cooperate to accomplish the mission. However, since the notion of optimization for a multi-robot mission plan is more complex than for a single robot, we will have to study the efficientness of our architecture in this case. We will especially study over-subscribed problems in which the planning architecture will have to choose objectives in a huge set of possible actions. We will also study the architecture's ability to react in the case of a necessary replanning due to events occuring during the execution as well as how the already computed plans can be re-used to efficiently repair the current failure. Finally we will implement our propositions within the framework of search-and-rescue missions for a team of unmanned aerial vehicles.

REFERENCES

[1] D.E. Smith. Choosing objectives in over-subscription planning. In *International Conference on Automated Planning and Scheduling*, 2004.

[2] G. Rabideau, T. Estlin, S. Chien, and A. Barrett. A comparison of coordinated planning methods for cooperating rovers. In *AIAA Space Technology Conference and Exposition*, pages 133–140, 1999.

[3] B.L. Brumitt and A. Stenz. GRAMMPS: A generalized mission planner for multiple mobile robots in unstructured environments. In *IEEE International Conference on Robotics and Automation*, volume 2, pages 1564–1571, May 1998.

[4] J.-C. Latombe. *Robot Motion Planning*. Kluwer Academic Publishers, 1991.

[5] S. Kambhampati, M.R. Cutkosky, J.M. Tenenbaum, and S.H. Lee. Integrating general purpose planners and specialized reasoners: case study of a hybrid planning architecture. In *IEEE transactions on Systems, Man and Cybernetics*, pages 1503–1518, 1993.

[6] B. Lamare and M. Ghallab. Integrating a temporal planner with a path planner for a mobile robot. In *AIPS Workshop Integrating Planning, Scheduling and Execution*, pages 144–151, 1998.

[7] F. Zacharias, C. Borst, and G. Hirzinger. Bridging the gap between task planning and path planning. In *IEEE International Conference on Intelligent Robot and Systems*, pages 4490–4495, 2006.

[8] S. Cambon, F. Gravot, and R. Alami. aSyMov: Towards more realistic robot plans. In *International Conference on Automated Planning and Scheduling*, 2004.

[9] International Planning Competition (IPC 3). Hosted at the *Artificial Intelligence Planning and Scheduling conference*, 2002. *http://planning.cis.strath.ac.uk/competition/*.

[10] D. Nau, T.-C. Au, O. Ilghami amd U. Kuter, J. W. Murdock, D. Wu, and F. Yaman. SHOP2 : An HTN planning system. *Artificial Intelligence Research*, 20, 380–404, 2003.

[11] D. Nau, H. Munoz-Avila, Y. Cao, A. Lotem, and S. Mitchell. Total-order planning with partially ordered subtasks. In *International Joint Conference on Artificial Intelligence*, 2001.

Control by sliding mode of a trajectory follow-up for a mobile robot

*S.Berrahal** *D.Ameddah*** *M.Mokhtari****

*Electronics, Laboratory of study advanced,
Univresité Hadji Lakhdar Batna 05000Algerie(e-mail:razik61@yahoo.fr)
**Université Hadji Lakhdar - 05000 - Batna Algérie
e-mail:d_jimparis@Lycos.com
*** Université Hadji Lakhdar - 05000 - Batna Algérie
e-mail:mokh_mw@yahoo.fr

Abstract: this paper proposes a sliding mode control method for wheeled mobile robot. we develop the constraint and dynamic equations and we explain the state system. To control this robot, we use a variable structure control to track desired trajectory. we adjust this controller to implement it in the view to delete the chattering. Different simulations are performed to show the efficiency of this controller.

Keywords: Sliding mode, mobile robot, dynamic modelling, trajectory tracking.

1. INTRODUCTION

As our system is non-linear, so a lot of research was presented on the control of a mobile robot nonholonome, as the command by returning state linéarisant [Slotine 91], [93 Fossard] system to command a predefined trajectory example with a "PID" controller, unfortunately, the results are not acceptable with technology, so we chose the command by sliding mode, which is part of the class of variable command structure, it was proposed and developed by several researchers among them [Utkin 77], [93 Hung Shim] and [95]. This article examines a particular a follow-up of a right trajectory (right) of a mobile robot nonholonome, we adopted two surfaces of slips to this robot which owns at the the rear two driving wheels that assure a follow-up acceptable of this right by this robot. Along with our simulations-Matlab Simulink, we noticed our system almost stable, and its performance is obtained, but it exists a small inconvenience that is the "chattering" that can be minimized with the proper choice of parameters. The advantage of using sliding mode control include fast response, good transient performance and robustness with regard to parameter variations. Researchers in this domain proposed a sliding mode control law for stabilizing a nonholonomic system, also this command is robust when approached dynamic modelling is used.

2. DYNAMIC MODELLING OF THE MOBILE ROBOT

The used mobile robot is a parallelepiped platform with wheels that can move thanks to the two leading driving wheels placed to the rear, the two wheels before are some mad wheels, their role is to maintain the equilibrium of the flat form. (to see fig 1).

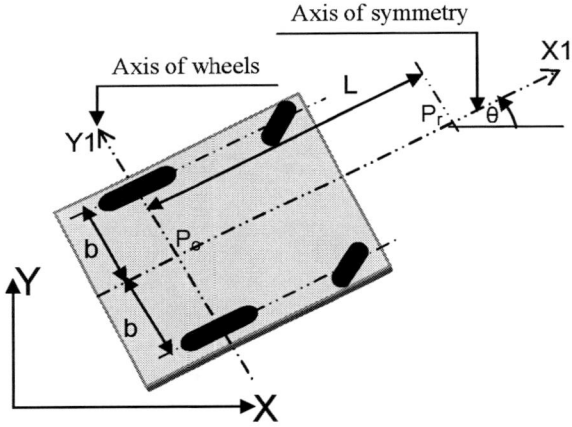

Fig. 1. Coordinated of the mobile robot

(X, Y): System of coordinates in the absolute benchmarkR
(X1, Y1) : System of coordinates in the benchmark bound to the R1 robot.
P_0: Center of gravity of the robot with its coordinates (x_G, y_G).
b: Distance between each of the driving wheels and the axis of symmetry of the robot.
r: ray of each wheel.
c: constant equal to r/2b.
m_c: masse of the mobile robot without wheels and without rotor of the motor.
m_r: mass of every driving wheel including the rotor of the motor.
I_c : Moment of inertia of the mobile robot.
I_m : Moment of inertia of every wheel around its diameter.
I_r : Moment of inertia of every rear wheel around its axis.

L: distance between the center of gravity Po and the order Pr situated on the axis X1.

ϕ_l : the angular position of the left rear driving wheel.

ϕ_r : the angular position of the right rear driving wheel.

θ : the angle between the two coordinate systems.

3. MODEL CONSTRUCTION

In order to establish the dynamic model of the robot, we choose the vector of generalized coordinates according to:

$$q = \begin{bmatrix} x_G & y_G & \phi_l & \phi_r \end{bmatrix}^T$$

The mobile robot is submitted to constraints that are:

- The impossibility to move in the lateral direction.
$$\dot{y}_G \cos\theta - \dot{x}_G \sin\theta = 0 \quad (1)$$
- The constraint of rolling without slip for every wheel.
$$\dot{x}_G \cos\theta + \dot{y}_G \sin\theta - b\dot{\theta} - r\dot{\phi}_r = 0 \quad (2)$$
$$\dot{x}_G \cos\theta + \dot{y}_G \sin\theta + b\dot{\theta} - r\dot{\phi}_l = 0 \quad (3)$$

Let's make the subtraction between (2) and (3) one will have:

$$\dot{\theta} = \frac{r}{2b}(\dot{\phi}_r - \dot{\phi}_l)$$

While integrating this equation, and considering the null initial conditions. One gets the following holonome constraint:

$$\theta = c(\phi_r - \phi_l) \quad (4)$$

With the help of the previous constraint equations we determine the matrix of constraint as:

$$A(q).\dot{q} = 0 \quad (5)$$

with:

$$A(q) = \begin{bmatrix} -\sin\theta & \cos\theta & c.Lr & -c.Lr \\ \cos\theta & \sin\theta & -\dfrac{r}{2} & -\dfrac{r}{2} \end{bmatrix} \quad (6)$$

The formalism of Lagrange is based on the calculation of energies brought into play the system. We then calculate the first energy implicated by the movement of the robot, to know the kinetic energy.

We get the following expression then:

$$L = \tfrac{1}{2} m(\dot{x}_G^2 + \dot{y}_G^2) - 2 m_{roue} L_r \dot{\theta}(-\dot{x}_G \sin\theta + \dot{y}_G \cos\theta) + \tfrac{1}{2} I . c^2 \dot{\theta}^2 + \tfrac{1}{2} I_{roue} (\dot{\phi}_l^2 + \dot{\phi}_r^2)$$

with $\begin{cases} m = m_c + 2m_r \\ I = I_c + 2I_m + 2m_r(b^2 + L_r^2) \end{cases}$ (7)

The second energy implicated is the potential energy U, as the mobile robot only moves on the horizontal plan then the expression of U(q)=0. We can now calculate the Lagrangien (L = T - U = T) corresponding merely to the expression of the kinetic energy, according to the formalism of Lagrange, we write the following relation:

$$\frac{d}{dt}\left(\frac{\partial L}{\partial \dot{q}_i}\right) - \frac{\partial L}{\partial q_i} = \tau' - A^T \lambda \quad (8)$$

Where τ' the generalized outside efforts is vector, this vector amounts to the vector of torque applied by actuators. λ is the vector of multipliers of Lagrange owed to the kinematics constraints of the system, we can write the system then in the form:

$$M(q)\ddot{q} + V(q,\dot{q}) = E(q)\tau - A^T(q)\lambda. \quad (9)$$

with $\tau' = E(q)\tau$

$$M(q) = \begin{bmatrix} m & 0 & 2m_r c.Lr\sin\theta & -2m_r c.Lr\sin\theta \\ 0 & m & -2m_r c.Lr\cos\theta & 2m_r c.Lr\cos\theta \\ 2m_r c.Lr\sin\theta & -2m_r c.Lr\cos\theta & I_r + I c^2 & -I c^2 \\ -2m_r c.Lr\sin\theta & 2m_r c.Lr\cos\theta & -I c^2 & I_r + I c^2 \end{bmatrix}$$

$$V(q) = \begin{bmatrix} 2m_r Lr.\dot{\theta}^2 \cos\theta \\ 2m_r Lr.\dot{\theta}^2 \sin\theta \\ 0 \\ 0 \end{bmatrix} \quad E(q) = \begin{bmatrix} 0 & 0 \\ 0 & 0 \\ 1 & 0 \\ 0 & 1 \end{bmatrix}$$

$$\lambda = \begin{bmatrix} \lambda_1 \\ \lambda_2 \end{bmatrix} \quad \tau = \begin{bmatrix} \tau_1 \\ \tau_2 \end{bmatrix} \quad (10)$$

In order to eliminate multipliers of Lagrange λ, do we use the angular speeds of each of the wheels to determine the matrix $S(q)$ as:

$$\begin{cases} A(q)S(q) = [0] \\ \dot{q} = S(q)\eta \end{cases} \text{with } \eta = \begin{bmatrix} \dot{\phi}_l \\ \dot{\phi}_r \end{bmatrix} \quad (11)$$

And on the other hand we have:

$$\ddot{q} = S(q).\dot{\eta} + \dot{S}(q).\eta \quad (12)$$

The resolution of this system permits us to clarify the matrix $S(q)$:

$$S(q) = \begin{bmatrix} c.Lr\sin\theta + \frac{r}{2}\cos\theta & -c.Lr\sin\theta + \frac{r}{2}\cos\theta \\ -c.Lr\cos\theta + \frac{r}{2}\sin\theta & c.Lr\cos\theta + \frac{r}{2}\sin\theta \\ 1 & 0 \\ 0 & 1 \end{bmatrix}$$

While multiplying the equation by $S^T(q)$ and let's note that:

$S^T(q)A^T(q) = [0]$ et $S^T(q)E(q) = I_{2\times 2}$ and we also take $V(q,\dot{q}) = 0$ [2], we will have:

$$S^T(q)M(q)\ddot{q} = S^T(q)E(q)\tau = \tau \quad (12')$$

Let's replace $\ddot{q}$ by its expression in (12') we will have:

$$S^T(q)M(q)S(q).\dot{\eta} + S^T(q)M(q)\dot{S}(q).\eta = \tau$$

Let's choose the following variable of state:

$$x = \begin{bmatrix} q \\ \eta \end{bmatrix}$$

The equation (9) can be replaced under the form of state follow:

$$\dot{x} = f(x) + g(x)\tau \quad (13)$$

$$\dot{x} = \begin{bmatrix} \dot{q} \\ \dot{\eta} \end{bmatrix} = \begin{bmatrix} S(q)\eta \\ f_1 \end{bmatrix} + \begin{bmatrix} [0] \\ (S^T(q)M(q)S(q))^{-1} \end{bmatrix} \tau \quad (14)$$

with $f_1 = (S^T(q)M(q)S(q))^{-1}(-S^T(q)M(q)\dot{S}(q))\eta$

The system is put in the form of state representation, and for more clarity let's put a new variable of entry u as:

$$\tau = (S^T(q)M(q)S(q))(u - f_1)$$

We can put the system then in the form:

$$\dot{x} = \begin{bmatrix} S(q)\eta \\ [0] \end{bmatrix} + \begin{bmatrix} [0] \\ I \end{bmatrix} u \quad (15)$$

with $f(x) = \begin{bmatrix} S\eta \\ [0] \end{bmatrix}$ and $g(x) = \begin{bmatrix} [0] \\ I \end{bmatrix}$

I is the matrix of rank identity2.

4. CONTROL BY SLIDING MODE

The position of the robot is chosen like vector of exit of the system since it corresponds to the goal to reach that we are ourselves stationary, this vector is defined by:

$$Z = \begin{bmatrix} z_x \\ z_y \end{bmatrix} = \begin{bmatrix} x_G + L\cos\theta \\ y_G + L\sin\theta \end{bmatrix}$$

This vector corresponds to a point on the axis of abscissas of the reference mark binds to the X1(Pr robot), the vector speed associate to this point is:

$$\dot{Z} = \begin{bmatrix} \dot{x}_G - L\dot{\theta}\sin\theta \\ \dot{y}_G + L\dot{\theta}\cos\theta \end{bmatrix}$$

we have the system of the expression (15):

$$\dot{x} = \begin{bmatrix} S(q)\eta \\ [0] \end{bmatrix} + \begin{bmatrix} [0] \\ I \end{bmatrix} u \quad (16)$$

This system has the particularity to possess some features we permit to simplify its representation of state by partitioninit in two under vectors x_1 and x_2 where the dimension of x_2 is equal to the dimension of u, either of dimension m=2 the dimension of x_1 is therefore n-m =4. we can then express the system (16) under the following form:

$$\dot{x} = f(x) + g(x)u$$

$$\begin{cases} \dot{x}_1 = f_1(x) + g_1(x)u \\ \dot{x}_2 = f_2(x) + g_2(x)u \end{cases} \text{ where } \begin{cases} f_1(x) = S(q)\eta \\ g_1(x) = [0] \\ f_2(x) = [0] \\ g_2(x) = I \end{cases} \quad (17)$$

This permits us to express the system under its reduced form:

$$\begin{cases} \dot{x}_1 = f_1(x) \\ \dot{x}_2 = g_2(x)u \end{cases}$$

We now consider $s(z)$ the vector defining the functions of commutation permitting to define surfaces of slip of the system of m dimension. By definition, these functions have for expression:

$$s(z) = \begin{bmatrix} s_1(z_x) \\ s_2(z_y) \end{bmatrix} \text{ where } \begin{cases} S_1 = \dot{\tilde{z}}_x + \lambda_x \tilde{z}_x \\ S_2 = \dot{\tilde{z}}_y + \lambda_y \tilde{z}_y \end{cases}$$

with $\dot{\tilde{z}}_i = \dot{z}_i - \dot{z}_{id}$ $i = x, y$
$\tilde{z}_i = z_i - z_{id}$

On the other hand we have:

$$z = \begin{bmatrix} z_x \\ z_y \end{bmatrix} = \begin{bmatrix} x_G + L\cos\theta \\ y_G + L\sin\theta \end{bmatrix} \text{ and } \dot{z} = \begin{bmatrix} \dot{z}_x \\ \dot{z}_y \end{bmatrix} = \begin{bmatrix} \dot{x}_G - L\dot\theta\sin\theta \\ \dot{y}_G + L\dot\theta\cos\theta \end{bmatrix}$$

z_{id} and $\dot{z}_{id}$ are the respective coordinates of the vector position and speed wanted. λ_x and λ_y are coefficients of surfaces. These coefficients permit to determine the time of response of the system. In order to determine the dynamics of the system, we use the method told "the reaching law approach» [2], its permits to specify the dynamics of approach of the system directly as:

$$\dot{s} = -Qsign(s) - Kh(s)$$

With Q and K two diagonals matrixes with positive elements:

$$sign(s) = \begin{bmatrix} sign(s_x), sign(s_y) \end{bmatrix}^T$$

$$h(s) = \begin{bmatrix} h(s_x), h(s_y) \end{bmatrix}^T$$

The function $h(s)$ is a function depending on the surface of slip. We now determine the law of order.

$$\dot{s} = \frac{ds}{dt} = \frac{\partial s}{\partial x}\dot{x} = \frac{\partial s}{\partial x}(f(x) + g(x)u) = -Qsign(s) - Kh(s) \quad (18)$$

In short we will have:

$$u = -\left(\frac{\partial s}{\partial x_2} g_2(x)\right)^{-1} \left(Qsign(s) + Kh(s) + \frac{\partial s}{\partial x_1} f_1(x)\right) \quad (19)$$

5. SIMULATIONS

At the time of this simulation, the robot must follow a straight trajectory of reference. The initial position of the robot in the absolute benchmark is $(x, y, \theta) = (0, 0, 0)$, the final position to reach is $(x_f, y_f, \theta_f) = (2, 0, \pi/6)$.

Gains used at the time of this simulation are

Q=20m/s²; K=5Hz; λ=0.5 Hz.

Parameters of the robot are:

**m_c=99Kg;   m_r=0.5Kg;   b=0.5148m;   r=0.0228m;
L=1.05m;   Lr=0.3m;   Ir=8.26*10^-3   Kg.m²;
Ic=14.4*10^-3 Kg. m².**

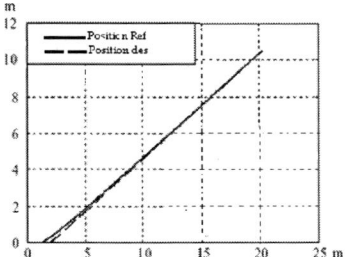

Fig.2. Simulation of right follow-up

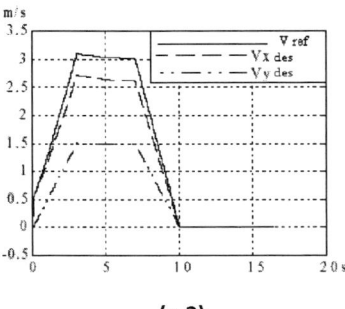

(a.3)

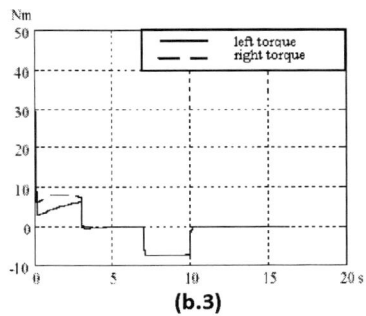

(b.3)

Fig.3. Simulation of right follow-up, speeds (a), torques (b)

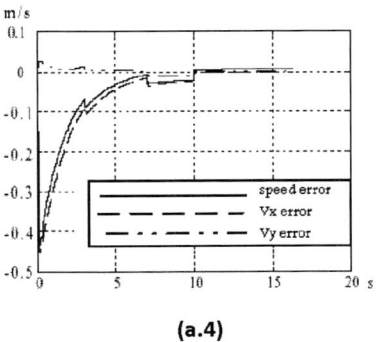

(a.4)

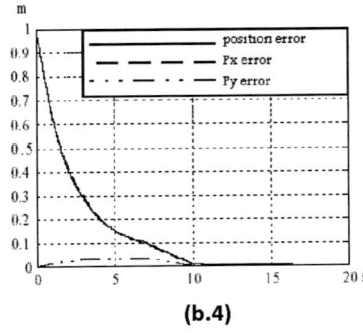

(b.4)

Fig 4. Simulation of right follow-up, speeds error (a), position error (b)

6. INTERPRETATIONS

We notice that the follow-up in position (figure.2) is perfected, so the presence of a disconnecting on speed error during the phase of deceleration (figure.3.a), the robot is less sensitive to variations of speed (it reacts less quickly), however position error become very small at the end of 10 seconds, we also distinguish jumps of torque generate by this profile (right) (figure.3.b), it doesn't influence on actuators, one also notices that speed error and position error tend to zero (figures.4a and 4b).

7. CONCLUSION

In this article did we proposed an order of follow-up of trajectory of a robot mobile nonholonome by sliding mode, at the time of simulations we obtained good results, so we showed the importance of values of λ which gives the dynamics of our system while keeping the physical sizes, then we deduct that the control by sliding mode that is robust in relation to other types of control.

8. BIBLIOGRAPHY

[1] H.S.Shim, J.H.Kim, K.Koh ''Variable structure control of nonholonomic mobile robots''. IEEE conf. Robot, Automat, May 1995. pp 1694-1699.
[2] P.Ruaux, G.Bourdon, S.Delaplace "Dynamic control of wheeled mobile robot using sliding mode" Romancy1996 Udine, Italie, pp 205-112, 1996.
[3] J.M.Yang, I.H.Choi, J.H.Kim " Sliding mode motion of non holonomic mobile robots ",IEEE control system, vol19, N°19. April 1999. pp 15-23.
[4] Hung JY." Variable structure control: A Survey". IEEE Transactions on Industrial Electronics1993; 40(1): 2-22.
[5] A .Ishigame, T.Furukawa "Sliding mode controller design based on fuzzy inference for nonlinear systems". IEEE transactions on industrial Electronics, vol. 40, N°1, February 1993.
[6] A.J. Fossard, « Commande à structure variable, systèmes de régulation, Collaboration automatique » Editions Masson, 1993.

Modeling and analyzing mixed reality applications using timed automata

Jean-Yves Didier, Bachir Djafri, Hanna Klaudel

IBISC Laboratory, CNRS FRE 2873, Université d'Evry, France
`{jean-yves.didier,bachir.djafri,hanna.klaudel}@ibisc.fr`

Abstract: We propose a compositional modeling framework for Mixed Reality (MR) software architectures in order to express, simulate and validate formally the real-time properties of such systems. Our approach is first based on a functional decomposition of such systems into generic components. The obtained elements as well as their typical interactions give rise to generic representations in terms of timed automata. A whole application is then obtained as a composition of such defined components. The approach is illustrated on a case study modeled by timed automata synchronizing through channels and including a large number of time constraints. This system has been simulated in UPPAAL and checked against basic behavioral properties.

Keywords: Mixed reality systems modeling, real-time, timed automata, formal analysis, simulation, model-checking.

1. INTRODUCTION

Mixed Reality (MR) hardware systems tend to be more and more complex. This is partly due to new ways of interacting with such systems and the wide range of available human computer hardware interfaces.

For example, in classical immersive virtual reality applications, users can choose between data-gloves, a wide range of options between motion trackers (Welch and Foxlin [2002]), camera, force-feedback devices like haptic arms to interact with the system. In mobile augmented reality, outdoor systems will cope with a lot of different sensors in order to recover the position and orientation of the user in its own workspace. Such result is achieved usually by combining inertial sensors, cameras and D-GPS or GPS (Feiner et al. [1997], Roberts et al. [2003]).

MR developers have to interface such heterogeneous devices with their own MR applications. However, the challenge is not only restricted to sensor's heterogeneity, but it also relies on variety and novelty of algorithms and techniques developed in parallel with new hardware. Moreover, those new fields of research tend to enrich the feeling of presence of the user in VR environment and also to fuse VR with reality. One lead is to propose several kinds of interactions modalities to the end user by combining visual, aural and haptic feedbacks (Bayart and Kheddar [2006]), each of these sensory modalities giving rise to its own rendering loop. Indeed, sensors as well as rendering loops have their own time constraints, different from each other.

Nowadays, the current process for developing MR applications relies mostly on fast response and high hardware performances to cope with real-time constraints. However, for some applications (for example, teleoperation or haptic ones) the respect of time constraints may be critical. Therefore, it may be worth to validate the application, before testing it on actual hardware, by modeling it and applying formal method techniques to prove its robustness in terms of the absence of deadlocks and temporal integrity. The benefits may be twofold: it may avoid unnecessary cost related to a possible deterioration of hardware, and in the case of design errors, it allows to identify their source and to be corrected and validated again.

In this paper we are interested in specification and validation of heterogeneous, reconfigurable, open but not hot pluggable MR systems. We propose a compositional modeling framework for software architectures in order to express, simulate and validate formally their real-time properties.

Our intention here is not proposing a new formalism, but taking advantage of known real-time specification and verification techniques in the design and the programming of MR systems. We chose for this purpose to use a prominent model of timed automata (Alur and Dill [1994] and its associated tool UPPAAL Larsen et al. [1997]). UPPAAL allows simulating systems detecting deadlocks and verifying, through model-checking, various reachability properties. Typically, it can answer the designer questions which may look like "starting from its initial state, can the system reach a given state in a given delay?".

Our approach is first based on a functional decomposition of such systems into generic components. The obtained elements as well as their typical interactions give rise to generic representations in terms of timed automata and the whole application is then obtained as a composition of such defined components.

2. RELATED WORK

Facing the challenge of heterogeneity of sensors and algorithms has been one of the motivations leading to develop modular software architectures and frameworks for MR. During these past years, almost thirty different projects of frameworks have been developed for MR (Endres et al. [2005]).

Amongst the most remarkable ones, we can mention the StudierStube led by the Technical Universities of Vienna and Graz (Austria) (Fuhrmann [1999]). This project is based on the OpenInventor API and uses the concept of distributed scenegraphs. One of the sub-projects of StudierStube mainly focuses on sensor configuration issues (Reitmayr and Schmalstieg [2001]) and on data processing aspects after being acquired by sensors. This is performed by using an object-oriented approach combined with software engineering practices, like configuration files written in XML.

The DWARF project (for Distributed Wearable Augmented Reality Framework) relies on distributed services. Each tracker becomes a service broadcasting data to other services (that could be filters, rendering loops,...) using an extension of CORBA (Bauer et al. [2001]).

ImageTclAR (Owen et al. [2003]) aims to provide a rapid prototyping environment to test and design MR applications. People can use proposed components or develop their own ones in C++ whereas the whole logical glue between components is written using Tcl interpreted scripts.

Tinmith (Piekarski and Thomas [2003]) is an API for developing mobile AR systems. It uses an object store, which is based on Unix file hierarchy, and allows applications to register callbacks on these objects. Once these objects change (for example, when one sensor acquires a new data set) an event is sent to trigger these callbacks.

The AMIRE project (Haller et al. [2003]) emphasizes component based development. This project embeds a graphical tool to connect and configure components. Data concerning the configuration of the application are stored in an XML file.

Finally, the ARCS (for Augmented Reality Component System) project (Didier et al. [2006]) is also a component-based framework with graphical tools to help to design applications. It focuses on the component life span in running MR applications and on the reconfigurability of data flow between components at runtime.

These projects are mostly emphasizing modular development or even component based engineering to deal with the heterogeneity challenge. Of course, our approach is related to the above, but we are especially interested in real-time aspects of creating MR applications. Usually, the real-time in the MR field of research may be understood in different ways and seen as a real world time in simulation or as low latency of man-machine interactions, or even as a preservation of imposed time constraints. In this paper, we aim at addressing the last point of view which focuses on operational deadlines, from events to system response.

Real-time systems may be specified using numerous dedicated methods and formalisms. Most of them are graphical semi-formal notations allowing a state machine representation of the behavior of the system. Among the most popular formalisms, we may quote Statecharts (Harel [1987] or UML/RT Douglass [1997]). Such visual representations do not enable to verify the properties of systems and it is necessary to associate a formal semantics to them, based in general on automata, process algebras (Harel et al. [1987]), Petri nets (Reisig [1985] or temporal logics Manna and Pnueli [1992]). We chose to use timed automata (Alur and Dill [1994]), which have the advantage to be relatively simple to manipulate and possess adequate expressivity in order to model time constrained concurrent systems. Moreover, there exists for this model powerful implemented tools (e.g., UPPAAL (Larsen et al. [1997])) allowing model-checking and simulation.

3. TIMED AUTOMATA

A timed automaton (see figure 1) is a finite state automaton provided with a continuous time representation through real-valued variables, called *clocks*, allowing to express time constraints. Generally, a timed automaton is represented by an oriented graph, where the nodes correspond to locations in which the system may be and the arcs correspond to the transitions between these locations. The time constraints are expressed through *clock constraints* and may be attached to locations as well as to transitions. A clock constraint is a conjunction of atomic constraints which compares the value of a clock x, belonging to a finite set of clocks, to a rational constant c. Each timed automaton has a finite number of locations, one of them being tagged as *initial*. In each location, the time progression is expressed by a uniform growth of the clock values. In that way, in a state at each instant, the value of the clock x corresponds to time passed since the last reset of x. A clock constraint, called an *invariant*, may be associated to each location and has to be satisfied in order for the system to be allowed to stay in this location. Transitions between locations are instantaneous and conditioned by clock constraints, called *guards*, and may also reset some clocks. They may also carry labels allowing synchronization.

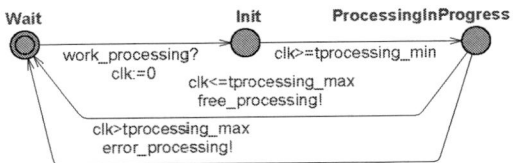

Fig. 1. Example of a timed automaton modeling the processing of a task, where clk is a clock. After the reception of a signal *work_processing*!, the automaton spends at least *tprocessing_min* time in the location *Init*. Then, it sends the signal *free_processing*! if the processing time does not exceed *tprocessing_max*, otherwise, it emits *error_processing*!.

In UPPAAL (Larsen et al. [1997]), which is used in our modeling, a timed automaton is a finite structure handling, in addition to a finite set of clocks evolving synchronously with time, a finite set of integer-valued and Boolean variables. A model is composed of a set of timed automata, which communicate using binary synchronization through transition labels and a syntax of emission/reception. By convention, a label k! indicates the emission of a signal on a channel k. It is supposed to be synchronized with the signal of reception, represented by a complementary label k?. Absence of synchronization labels indicates an internal action of the automaton. The execution of the model starts in the initial state (corresponding to the initial location of each automaton with all variable values set to zero), and is a succession of reachable states. The state change may occur for three reasons:

- by time progression corresponding to d time units in the locations of the components, provided that all the location invariants are satisfied. In the new state, the clock values are increased by d and the integer variables do not change;
- by a synchronization if two complementary actions in two distinct components are possible, and if the corresponding guards are satisfied. In the new state, the corresponding locations are changed and the values of clocks and of integer

variables are modified according to the reset and update indications;
- by an internal action if such an action of a component is possible, it may be executed independently of the other components: the location and the variables of the component are modified as above.

Another peculiarity of UPPAAL, useful in expressing a kind of synchronicity of moves, is the notion of "committed" locations, labeled in the figures by a special label C; see, for instance, the location *ActiveInProgress* in the first automaton of figure 3(a). In such a location, delaying is not permitted. This implies an immediate move of the concerned component. Thus, two consecutive transitions sharing a committed state are executed without any intermediate delay.

The behavior of a complex system may be represented by a single timed automaton being a product of a number of other timed automata. The set of locations of this resulting automaton is the Cartesian product of locations of the component automata, the set of clocks is the union of clocks, and similarly for the labels. Each invariant in the resulting automaton is the conjunction of the invariants of the locations of the component automata, and the arcs correspond to the synchronization guided by the labels of the corresponding arcs.

4. A SOFTWARE ARCHITECTURE FOR MR AND ITS TIMED AUTOMATA MODELING

In this paper, we consider MR architectures organized according to a data flow-oriented scheme, from sensors to the actual result produced by the rendering loops. As represented in figure 2, data are produced by sensors (cameras, GPS, motion trackers,...), then they are processed by processing units (for example, in charge of noise filtering, image processing,...) and stored in a shared memory where they are picked up by the rendering loops. A rendering loop is a pipeline which processes such data and transforms them to the actual rendering result (for example, images on a screen or force feedback according to the rendering device).

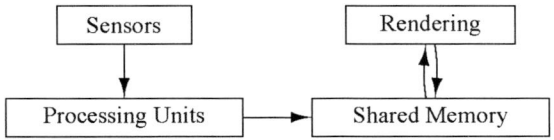

Fig. 2. A decomposition scheme of an MR software architecture.

Each element of this functional decomposition including different sorts of sensors, different sorts of processing units, a shared memory and possibly various rendering loops, will be represented as timed automata, as well as the logical glue which links them together. In the following, we present the concrete timed automata models of each element of our data flow architecture (notice that due to the page limitation, only automata used in our case study are presented).

Sensors. The sensors are devices that capture data from the environment. Data can be images coming from cameras, motions captured from trackers or events like mouse clicks or keyboard entries. We assume that each sensor has a unique output (possibly multiplexing sensor's data). The temporal characteristics of the sensors allow us to roughly classify them into two categories:

- *Periodic sensors* capture data at periodic time according to a well defined cycle. This periodicity can be expressed by time constraints, corresponding to the minimum and the maximum time of data-gathering. Other time constrains have to be taken into account namely their minimal delay of initialization and of shutting down. A generic timed automaton for a periodic sensor is represented in figure 3(a).
- *Aperiodic sensors* collect data only when an asynchronous event occurs. They may be a representation of a physical sensor (like a switch, any warning device) as well as an abstraction of any system using an event interface (typically graphical user interface). In general, such sensors have to respect a minimum delay between two events, which can be expressed as a time constraint and is also assumed to be the same as a minimal shutting down delay. A generic timed automaton for an aperiodic sensor is represented in figure 3(b).

Data processing units. A data processing unit (PU) processes data received from sensors. An example of data processing can be an extraction of position and orientation of a camera with respect to its workspace using image processing techniques. A data processing unit has time constraints corresponding to the minimum and the maximum time of processing, and may have in general several inputs and outputs. We only consider four basic kinds of PUs, that assume the number of entries is two at most and one at least. This is not a limitation: a PU with more than two entries may be modeled using a composition of several basic ones. Each basic processing unit may have several outputs $n\_outputs$, with $n\_outputs \geq 1$. As for sensors, the outputs are triggered sequentially according to their predefined rank comprised between 0 and $n\_outputs - 1$.

We consider four basic sorts of data processing units, one unary reception and three binary ones. Each of them abstracts actual processing (devoted to a specific automaton - see figure 1 for a simplified version) and focuses on durations (time constraints):

- A *Unary* PU starts processing the data as soon as it is received on its unique input,
- An *AtLeast* PU starts processing if data are received at least on one of its two inputs,
- A *Both* PU has two buffered inputs and starts processing if data are ready in both buffers,
- A *Priority* PU has one master input and one buffered slave input. It starts processing if data are ready in master input and possibly uses buffered data from the slave input.

Shared memory and rendering loops. We assume that the memory is composed of registers supporting reading and writing operations, which are mutually exclusive, see figure 3(c).

The rendering loops are periodic and their processing is decomposed in two phases corresponding to the processing and rendering time constraints. Since rendering loops are reading their data into memory, the allowed interval of reading has to be taken into account. If too much time is spent in these two phases, the user may observe phenomena like jerky images or jerky force feedback.

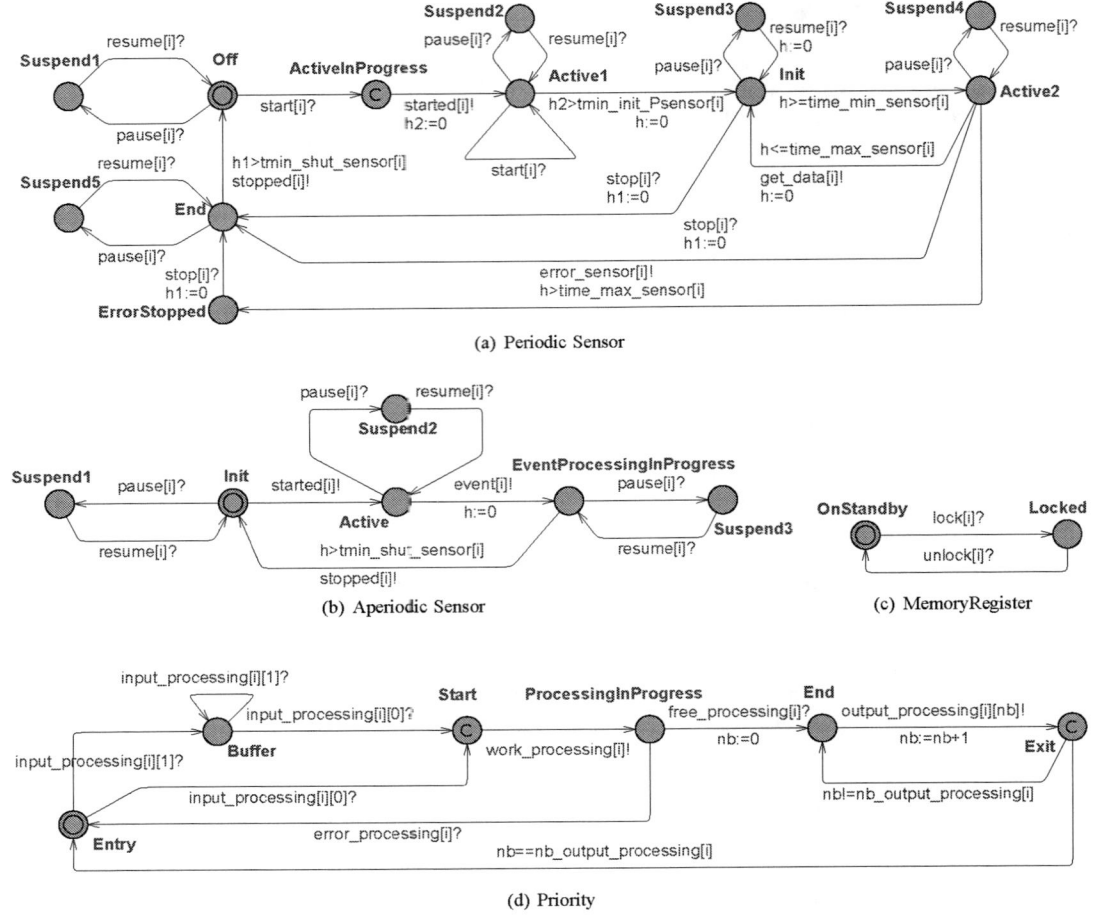

Fig. 3. Periodic sensor, Aperiodic sensor, Memory Register and Priority processing modeling.

Controllers. The timed automata corresponding to the previously introduced elements need to be composed together in order to model an MR application. These compositions may be realized directly through synchronizations on transitions (like between the memory and rendering loops), or through controllers (like between generic automata for sensors, PU and the memory). A scheme of such synchronizations for an example of application is represented in figure 4, where the collector controllers $cCo1$ and $cCo2$ link sensors with $Priority$ PU, the memory writer controller cMW links PU with memory, and cGr and cVg are respectively group and void group controllers. A controller links a given output to a given input, as indicated by integer labels.

- A *Collector controller* broadcasts sensor's data to associated data processing unit,
- A *Processing controller* connects two PUs,
- A *Memory Writer controller* connects a PU output to memory. This controller has a particularity to take into account the allowed interval of writing down data into memory,
- A *Group controller* manages conflicts between a sensor and a group of sensors. In other words, if the main sensor is activated, it deactivates all sensors from the group it supervises,

- A *Void group controller* is necessary to manage sensors that are not supervising any group,
- An *Initialization controller* is needed to start the periodic sensors of the application,
- An *Error management controller* intercepts error messages from periodic sensors and properly reinitializes them.

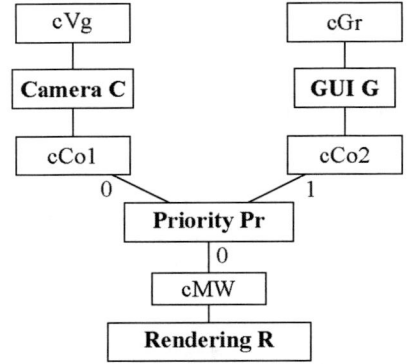

Fig. 4. An abstract view of an MR application.

Each of the above entities gives rise to a template (a parameterized timed automaton), which can be instantiated and used in the specification of an MR application. The complete specification, which is a parallel composition of the instanciated timed automata composing the application, can be then analyzed and simulated before any actual implementation. The following simple case study illustrates this approach.

5. A SIMPLE CASE STUDY: FORMAL ANALYSIS AND SIMULATION

We consider a simple classical setup for an MR application involving a camera (C) and a graphical user interface (G) with one visual rendering loop, cf. figure 4. C is modeled as a periodic sensor while G as an aperiodic one. We assume that the camera C may be suspended by the graphical user interface G supervising it.

The resulting model comprises 13 timed automata (including a initializations and error management for periodic sensors) synchronizing through channels and including a large number of time constraints.

Model-checking is the process of checking whether a given model, in our case the obtained set of timed automata, satisfies a given property, usually expressed in a query language. The state space of such a system comprises all the states reachable from the initial one by firing transitions at some dates. To check a property, the model-checker will explore all the reachable states of the system and answer if the property is satisfied or not. In UPPAAL, the query language allows to express, for instance:

- Reachability properties - denoted E<> p (does exist a state where p is satisfied reachable from the initial one?),
- Invariant properties - denoted A[] p (is p satisfied in all states reachable from the initial one?),
- Temporal implication properties - denoted p --> q meaning that if p is satisfied, then q is satisfied eventually.

It is well known that even for small systems, the number of reachable states may be huge, (in our case study it is about 200,000 states), which is making it impossible to analyze it by hand whereas model-checking tools can do it automatically and exhaustively, therefore furnishing a proof.

Table 1 summarizes the queries performed on our model as well as the results returned by the model checker. We should notice that these queries are also used during the modeling process to check if the system has the expected behavior, validating step by step the building of the model.

The first property we typically want to check is the absence of a deadlock, which is the case (cf. Table 1, line 1). However, the absence of a global deadlock does not guarantee that some parts are not locked. For example, if we change the timing constraints in the rendering loop and make an obvious error like having a rendering algorithm spending more time than the period of the rendering loop, then this locks the rendering loop whereas the other parts of the system are still running (cf. line 2).

Therefore, we should check the behavior of our model for example :

- Is the rendering loop able to perform its task even if the periodic sensor is disabled (cf. line 3) ?
- When we lock the memory either for reading or writing, will the memory eventually unlock (cf. line 4) ?
- Can we write and read in memory at the same time (cf. line 5) ?
- When the aperiodic sensor is activated, will the periodic sensor eventually be suspended (cf. line 6) ? Since we don't know in which state the periodic sensor will be suspended, we will have to check all suspended states,
- Can the system write data in memory (cf. line 7) ?

The answers to all the above requests confirm the expected behavior of the system.

When a reachability property is answered by true or an invariant property is answered by false, UPPAAL can also produce traces (paths) proving its answer. The model-checker can also look for traces that are the shortest in time of execution or the fastest in terms of steps composing the trace. Using traces it is also possible to start having some numerical values attached to the properties we want to check. For example, by attaching a clock initialized at the startup of the periodic sensor, we are able to know the minimum amount of time needed before the first value is written in memory.

However, query languages cannot express all sorts of questions we need. Typically, we cannot know how many times a given state will be reached. That is why model-checker tools are also coupled with simulation tools, the latter ones giving finer indications on the dynamic behavior of the system.

Simulation consists in making automata evolve step by step (manually or randomly) and observing several variables and states of the system. We carried out some simulations and the obtained result confirm the expected behavior of the system component at work. It was possible to analyze particular scenarios, like the activation of the aperiodic sensor suspending the periodic sensor. Up to some extent, the simulation provides also some quantitative results, for example the balance between reading and writing accesses to memory.

6. CONCLUSION AND PERSPECTIVES

We introduced in this paper a compositional modeling for MR software architectures dedicated to specify and validate formally the real-time properties of such systems. The basic elements taken into account are periodic and aperiodic sensors, various kinds of processing units and controllers, possibly several registers in memory and possibly several rendering loops. In particular, it is possible to define groups of sensors which may be suspended once another sensor is running. The approach has been illustrated on a small example of an MR architecture, whose timed automata representation has been simulated in UPPAAL and checked against basic behavioral properties. It permitted to show in particular, that the system was deadlock free and was meeting its expected behavior.

In our future work we will be interested in developing a method and tools allowing to use the obtained timed automata model for automatically generating source code skeletons for an implementation on an MR platform.

REFERENCES

R. Alur and David L. Dill. A theory of timed automata. *Theoretical Computer Science*, 126(2):183–235, 1994.

M. Bauer, B. Bruegge, G. Klinker, A. MacWilliams, T. Reicher, S. Riss, C. Sandor, and M. Wagner. Design of a component-based augmented reality framework. In *Proceedings of the*

#	Query	Result
1	`A[] not deadlock`	true
2	`E<> Rendering(0).End`	false
3	`E<> (PeriodicSensor(0).End and Rendering(0).Reading)`	true
4	`MemoryRegister(0).Locked --> MemoryRegister(0).OnStandBy`	true
5	`E<> (MemoryWriterCtr(0).Write and Rendering(0).Reading)`	false
6	`APeriodicSensor(1).Active --> ( PeriodicSensor(0).Suspend1 or PeriodicSensor(0).Suspend2 or PeriodicSensor(0).Suspend3 or PeriodicSensor(0).Suspend4 or PeriodicSensor(0).Suspend5)`	true
7	`E<> MemoryWriterCtr(0).Write`	true

Table 1. Queries and results

International Symposium on Augmented Reality (ISAR), oct 2001.

B. Bayart and A. Kheddar. Haptic augmented reality taxonomy: haptic enhancing and enhanced haptics. In *EuroHaptics 2006*, pages 641–644, jul 2006.

J.Y. Didier, S. Otmane, and M. Mallem. A component model for augmented/mixed reality applications with reconfigurable data-flow. In *8th International Conference on Virtual Reality (VRIC 2006)*, pages 243–252, Laval (France), April 26-28 2006.

B. P. Douglass. *Real-Time UML: Developing Efficient Objects for Embedded Systems*. Addison-Wesley Longman Publishing Co., Inc., Boston, MA, USA, 1997.

C. Endres, A. Butz, and A. MacWilliams. A survey of software infrastructures and frameworks for ubiquitous computing. *Mobile Information Systems Journal*, 1(1), January-March 2005.

S. Feiner, B. MacIntyre, T. Hollerer, and A. Webster. A touring machine: Prototyping 3d mobile augmented reality systems for exploring the urban environment. In *ISWC '97: Proceedings of the 1st IEEE International Symposium on Wearable Computers*, page 74, Washington, DC, USA, 1997. IEEE Computer Society.

A. Fuhrmann. *Studierstube: a Collaborative Virtual Environment for Scientific Visualization*. PhD thesis, Institute of Computer Graphics and Algorithms, Vienna University of Technology, Favoritenstrasse 9-11/186, A-1040 Vienna, Austria, 1999.

M. Haller, J. Zauner, W. Hartmann, and T. Luckeneder. A generic framework for a training application based on mixed reality. Technical report, Upper Austria University of Applied Sciences, Hagenberg, Austria, 2003.

D. Harel. Statecharts: A visual formalism for complex systems. *Science of Computer Programming*, 8(3):231–274, June 1987.

D. Harel, A. Pnueli, J. P. Schmidt, and R. Sherman. On the formal semantics of statecharts. In David Gries, editor, *Proceedings of the Second Annual IEEE Symp. on Logic in Computer Science, LICS 1987*, pages 54–64. IEEE Computer Society Press, June 1987.

K. G. Larsen, P. Pettersson, and W. Yi. UPPAAL in a Nutshell. *Int. Journal on Software Tools for Technology Transfer*, 1(1-2):134–152, October 1997.

Z. Manna and A. Pnueli. *The Temporal Logic of Reactive and Concurrent Systems*. Springer-Verlag, New York, 1992.

C. Owen, A. Tang, and F. Xiao. Imagetclar: A blended script and compiled code development system for augmented reality. In *Proceedings of the International Workshop on Software Technology for Augmented Reality Systems*, 2003.

W. Piekarski and B. H. Thomas. An object-oriented software architecture for 3d mixed reality applications. In *ISMAR '03: Proceedings of the The 2nd IEEE and ACM International Symposium on Mixed and Augmented Reality*, page 247, Washington, DC, USA, 2003. IEEE Computer Society.

W. Reisig. *Petri Nets: An Introduction*, volume 4 of *Monographs in Theoretical Computer Science. An EATCS Series*. Springer, 1985.

G. Reitmayr and D. Schmalstieg. Opentracker-an open software architecture for reconfigurable tracking based on xml. In *VR '01: Proceedings of the Virtual Reality 2001 Conference (VR'01)*, page 285, Washington, DC, USA, 2001. IEEE Computer Society.

G.W. Roberts, A. Evans, A.H. Dodson, S. Cooper, R. Hollands, B. Denby, W. Hatton, M. Sen, D. Muller, A. Marchant, D. Tragheim, M. Shaw, and J. Jones. The use of augmented reality, gps and ins to visualize mining and geological data. In *Proceedings of the 16th International Technical Meeting of the Satellite Division of the Institute of Navigation*, Portland, Oregon, USA, September 2003.

G. Welch and E. Foxlin. Motion tracking: No silver bullet, but a respectable arsenal. *IEEE Comput. Graph. Appl.*, 22(6):24–38, 2002.

SESSION DI: PREDICTIVE CONTROL

Predictive Direct Torque Control for Induction Motor Drive

A. Benzaioua*, M. Ouhrouche* and A. Merabet**

Electric Machines Identification and Control Laboratory (EMICLab)
Department of Applied Sciences, University of Quebec at Chicoutimi
555 Blvd de l'Université, G7H 2B1 Chicoutimi (Qc), Canada
(e-mail: Ammar_Benzaioua@uqac.ca and Mohand_Ouhrouche@uqac.ca)
**Dalhousie University, Department of Electrical & Computer Engineering*
1360 Barrington Street, B3J 2X4 Halifax (NS), Canada
(e-mail: Adel.Merabet@dal.ca)

Abstract: A predictive control combined with the direct torque control (DTC) to induction motor drive is presented. A new switching strategy is used in DTC, where the constant switching frequency is taken constant, and the speed tracking is done by a predictive controller. The scheme control is applied to induction motor drive in order to perform the dynamic responses of electromagnetic torque, stator flux and speed. A comparison between the PI controller and predictive controller for speed tracking is done. Results of simulation show that the performance of the proposed control scheme for induction motor drive is accurately achieved.

Keywords: Induction motor, Direct torque control, Predictive Control.

1. INTRODUCTION

This advancement of power electronics and DSP technology allows improving control techniques for induction motor drive. This advancement has led to the increased use of adjusted speed in induction motor (IM) drives. A direct control torque is one from the control strategies for controlling the speed and torque in induction motor drive. It was introduced by Takahashi et al [1]. The idea behind the DTC is to exploit the fast stator flux dynamics and to directly manipulate the stator flux vector such that the desired torque is produced [2].

The DTC shows very high quality torque control without the need of tuning current controllers or using co-ordinate transformation. However, there are some drawbacks such as operation with variable switching frequency and large torque ripple, due to hysteresis comparators. Several works has been done to overcome these problems [3-10]. Since the implementation of control algorithms with constant switching frequency are often an advantage, a predictive DTC, with constant switching frequency and constant sampling time, proposed in [4, 5] combined with a predictive control strategy for speed tracking is applied to induction motor drive in this work.

In this paper, the application of the predictive direct control for induction motor drive is done. It is shown that this control scheme leads to good dynamic responses of the electromagnetic torque and speed compared with the classical DTC. The paper is organized as follows: in section 2, the design of the DTC by the new switching strategy [4, 5] is given for induction motor drive. In section 3, PI and predictive control strategies are carried out for speed control. Finally, to check the effectiveness of the controller, simulations have been done for different conditions.

2. CONTROL STRATEGY

The model of the motor in a fixed frame reference (α-β) can be written by the following equations:

$$\frac{d}{dt}\begin{bmatrix} I_{s\alpha} \\ I_{s\beta} \\ \varphi_{r\alpha} \\ \varphi_{r\beta} \end{bmatrix} = \begin{bmatrix} -\gamma & 0 & \frac{L_m}{\sigma L_s L_r T_r} & \frac{L_m}{\sigma L_s L_r}p\Omega \\ 0 & -\gamma & -\frac{L_m}{\sigma L_s L_r}p\Omega & \frac{L_m}{\sigma L_s L_r T_r} \\ \frac{L_m}{T_r} & 0 & -\frac{1}{T_r} & -p\Omega \\ 0 & \frac{L_m}{T_r} & p\Omega & -\frac{1}{T_r} \end{bmatrix} \begin{bmatrix} I_{s\alpha} \\ I_{s\beta} \\ \varphi_{r\alpha} \\ \varphi_{r\beta} \end{bmatrix} + \begin{bmatrix} \frac{1}{\sigma L_s} & 0 \\ 0 & \frac{1}{\sigma L_s} \\ 0 & 0 \\ 0 & 0 \end{bmatrix}\begin{bmatrix} V_{s\alpha} \\ V_{s\beta} \end{bmatrix} \quad (1)$$

with $T_r = \frac{L_r}{R_r}$, $\gamma = \frac{R_s + \frac{L_m^2}{L_r T_r}}{\sigma L_s}$ and $\sigma = 1 - \frac{L_m^2}{L_s L_r}$

The electromagnetic torque of the machine related to the stator is carried out by several expressions. The equation, used for the calculation, combines the two components of rotor flux with the two components of the stator flux:

$$T_e = p\frac{L_m}{L_r}\text{Im}[\varphi_r \cdot \varphi_s^*] = -p\frac{L_m}{\sigma L_s L_r}[\varphi_{s\alpha}\varphi_{r\beta} - \varphi_{s\beta}\varphi_{r\alpha}] \quad (2)$$

The expression of stator flux is given by:

$$\varphi_s(t) = \int_0^t (V_s - R_s)\, dt \quad (3)$$

The idea is to change the switching strategy used in classical DTC (Fig.1) by a new switching strategy proposed in [4, 5]. In this strategy, the switching instants of the active voltage phasor V_s are predicted (Fig.2). In the switching cycle of inverter, the suitable active vector of voltage is chosen to be applied to the motor for the required time and the zero-vector of voltage is applied when the electromagnetic torque reaches the border of the torque ripple calculated for the rest time of the inverter switching cycle. The torque returns to the minimum limit.

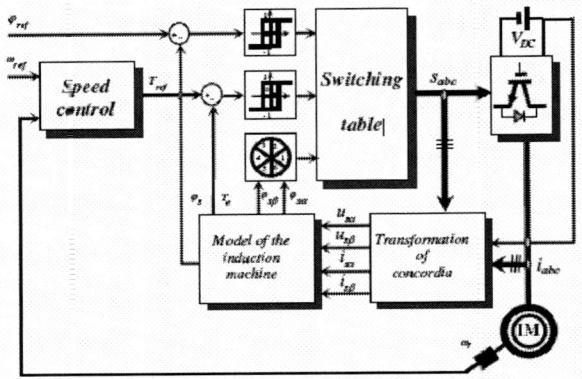

Fig. 1. Classical DTC

For the calculation of these switching instants, the derivative of the torque at each instant of sampling time must be calculated for the active voltage phasor, as well as for the zero-voltage phasor. The expression of the torque derivative is:

$$\frac{dT_e}{dt} = -P\frac{L_m}{L_c L_r}\left[\frac{d\varphi_{s\alpha}}{dt}\varphi_{r\beta} + \frac{d\varphi_{r\beta}}{dt}\varphi_{s\alpha} - \frac{d\varphi_{s\beta}}{dt}\varphi_{r\alpha} - \frac{d\varphi_{r\alpha}}{dt}\varphi_{s\beta}\right] \quad (4)$$

$L_c = \sigma L_s$ and P: poles number

The variation of rotor flux is small during the sampling interval because the time constant of rotor flux is large. Thus, the right hand side of equation (4) can be changed using the assumption [6]:

$$\frac{d\varphi_r}{dt} = \frac{d\varphi_{r\alpha}}{dt} = \frac{d\varphi_{r\beta}}{dt} = 0$$

Then, the variation of torque is given by:

$$\frac{dT_e}{dt} = -P\frac{L_m}{L_c L_r}\left[\frac{d\varphi_{s\alpha}}{dt}\varphi_{r\beta} - \frac{d\varphi_{s\beta}}{dt}\varphi_{r\alpha}\right] \quad (5.a)$$

$$\frac{dT_e}{dt} = -P\frac{L_m}{L_c L_r}\left[(V_{s\alpha} - R_s I_{s\alpha})\varphi_{r\beta} - (V_{s\beta} - R_s I_{s\beta})\varphi_{r\alpha}\right] \quad (5.b)$$

In steady state, a linear shape of torque is assumed during the switching interval T_{cycle} as shown in figure below:

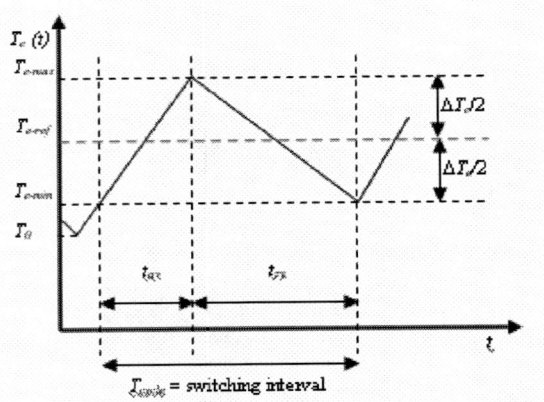

Fig. 2. Torque during the switching interval

From the equation (5), the derivative of torque for the zero-voltage ($V_S=0$) can be expressed as:

$$\frac{dT_e^-}{dt} = -P\frac{L_m}{L_c L_r}\left[(-R_s I_{s\alpha})\varphi_{r\beta} - (-R_s I_{s\beta})\varphi_{r\alpha}\right] \quad (6)$$

The other derivative of the torque for active voltage is equal to:

$$\frac{dT_e^+}{dt} = -P\frac{L_m}{L_c L_r}\left[V_{s\alpha}\varphi_{r\beta} - V_{s\beta}\varphi_{r\alpha}\right] + \frac{dT_e^-}{dt} \quad (7)$$

The ripple of the torque in steady state can be carried out by (see Fig.2):

$$\Delta T_e = -\frac{\dfrac{dT_e^+}{dt} \cdot \dfrac{dT_e^-}{dt}}{\dfrac{dT_e^+}{dt} - \dfrac{dT_e^-}{dt}} T_{cycle} \quad (8)$$

By the same geometric approach, to calculate the ripple in (8), the switching times for the active voltage are:

$$t_{av} = \frac{T_{eref} - T_0 - \dfrac{\Delta T_e}{2} - \dfrac{dT_e^-}{dt}T_{cycle}}{\dfrac{dT_e^+}{dt} - \dfrac{dT_e^-}{dt}} \quad (9)$$

and: $\quad t_{zv} = T_{cycle} - t_{av}$

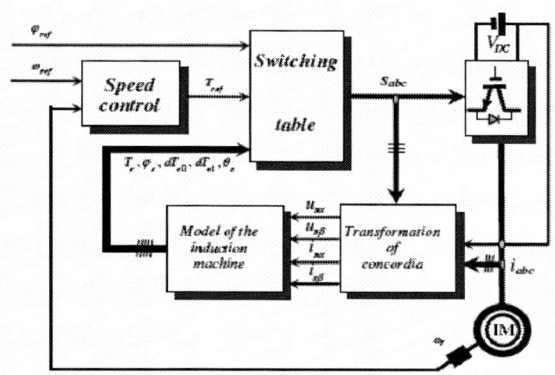

Fig.3. Application of predictive DTC to IM

The switching time corresponding to a given value of the torque reference and ripple can be calculated as a function of the voltages, the measured currents and the rotor position. Under the assumption of an ideal inverter, the instantaneous voltages can be only calculated by measuring the voltage of dc-link and the measured currents. The position of rotor is calculated by estimating the rotation speed with the mechanical equation:

$$J\frac{d\omega_r}{dt} + T_L = T_e - f\omega_r \quad (10)$$

A second criterion, which should not be neglected, is the sign of stator flux φ_s at the beginning and at the end of the switching interval. From the equation (3), the sign can be known. The value of flux can be predicted at each switching interval to decide which stator voltage V_S is ideal to have a minimum ripple of the electromagnetic torque and to ensure a good trajectory of stator flux.

3. SPEED CONTROL STRATEGIES

For the rotor speed control, two kinds of controller are used. One is the classical PI controller. Then, a predictive controller is proposed.

3.1 Classical PI Speed Controller

The following figure shows the diagram of speed control ω_r by the classical PI.

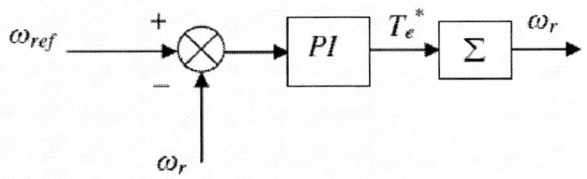

Fig. 4. PI speed control

3.2 Predictive Speed Controller

The diagram of the speed control (PC) with a predictive controller is presented in the following figure.

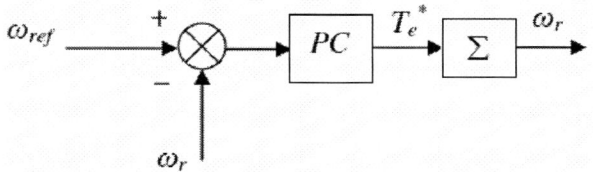

Fig. 5. Predictive speed control

The mechanical dynamic of the motor is described by:

$$\dot{\omega}_r(t) = -\frac{f_r}{J}\omega_r(t) + T_e(t) - \frac{1}{J}T_L(t) \quad (11)$$

$T_e = p\dfrac{L_m}{JL_r}(\varphi_{r\alpha}i_{s\beta} - \varphi_{r\beta}i_{s\alpha})$ is the electromagnetic torque, which is the control input for the system, and T_L is the load torque. The objective of the predictive controller is to find a control law in order to track the reference speed at the next time $(t+T)$, where $T > 0$ is the prediction horizon, through minimization of the cost function defined as:

$$J = \frac{1}{2}Q(\omega_r(t+T) - \omega_{ref}(t+T))^2 + \frac{1}{2}RT_e^2(t) \quad (12)$$

where
  Q and R are the weighting terms.

The predicted speed $\omega_r(t+T)$ is approximated by Taylor series expansion:

$$\begin{aligned}\omega_r(t+T) &= \omega_r(t) + T\dot{\omega}(t) \\ &= \omega_r(t) + T\left(-\frac{f_r}{J}\omega_r(t) + T_e(t) - T_L(t)\right)\end{aligned} \quad (13)$$

The same method is used to carry out the predicted reference speed $\omega_{ref}(t+T)$.

$$\omega_{ref}(t+T) = \omega_{ref}(t) + T\dot{\omega}_{ref}(t) \quad (14)$$

The optimal control is carried out by putting $\partial J / \partial T_e = 0$, and given by:

$$T_e(t) = -\frac{QTp/J}{Q\left(\dfrac{Tp}{J}\right)^2 + R}((\omega_r(t) - \omega_{ref}(t)) + \\ T(-\frac{f_r}{J}\omega_r(t) - \dot{\omega}_{ref}(t)) - \frac{Tp}{J}T_L)) \quad (15)$$

4. SIMULATION RESULTS

A SIMULINK model was developed in order to test the method presented above. The simulations are done to check the performance of the controller, and the comparison between the traditional PI control and predictive control for speed tracking is done.

The load torque profile during the whole simulation period is shown in Fig.6 and Fig.7. The details of the IM motor with nominal values used in simulation are given in appendix.

The model of the motor is run with a sample time of 1 µs, the inverter feeding the machine has a cycle period of $T_{cycle}=100$ µs. The sampling time of the controllers (PI and PC) $T_s=100$ µs and the prediction period is chosen as $T=85*T_s$ for the speed predictive controller. Fig.6 and Fig.7 give respectively the responses of the different states of the motor with PI and predictive controllers. It can be seen that the performances of the tracking responses, for speed, torque and stator flux, is successfully achieved.

As a comparison between the PI and predictive controllers, it can be seen that the flux, speed and the torque reach their references with the predictive controller faster than the PI controller.

Then, in order to test the effectiveness of the method for different steps of speed as shown in (Fig.8), while the machine asynchronous turns at the nominal torque. It can be noticed the system response is fast and the performance the proposed control system is successfully satisfied.

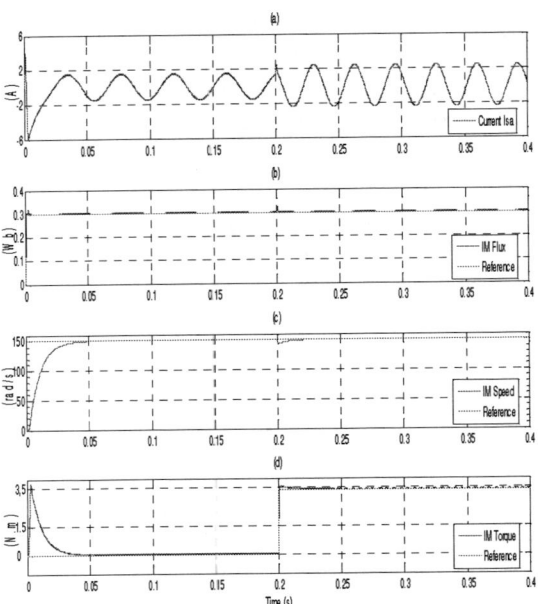

Fig.7 Responses to step torque reference with predictive speed controller; (a) current response, (b) flux response, (c) speed response, (d) torque response.

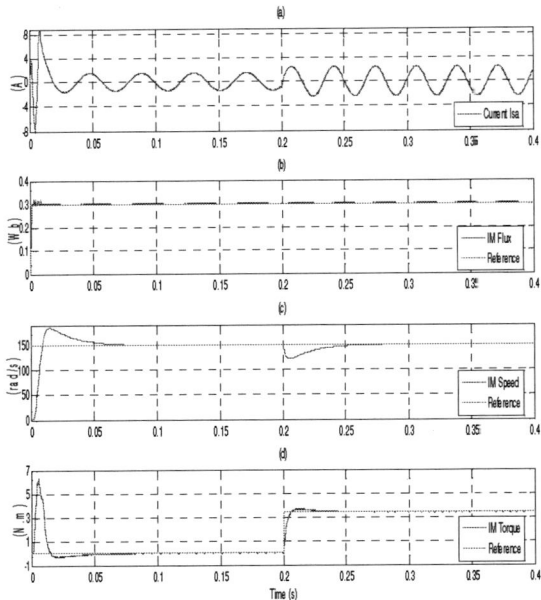

Fig. 6. Responses to step torque reference with PI speed controller; (a) current response, (b) flux response, (c) speed response, (d) torque response.

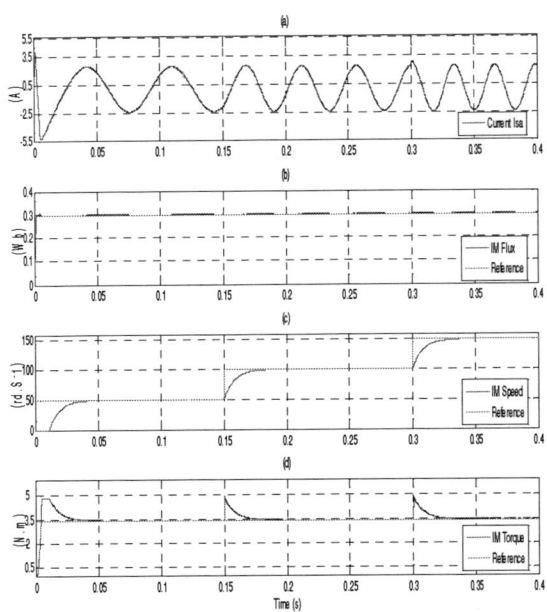

Fig.8 Responses for different steps of speed reference with predictive speed controller; (a) current response, (b) flux response, (c) speed response, (d) torque response.

5. CONCLUSION

In this paper, the predictive DTC control combined with the speed predictive controller is applied to induction motor drive. It leads to an acceptable dynamic for the motor and it can be a real alternative to the classical DTC with PI speed controller. A completely predictive structure for DTC is suggested to have a good performance. The torque ripple and the switching table can be predicted, when a constant inverter frequency is used. This method can overcome the drawbacks of the traditional DTC.

Simulation results show the effectiveness of the proposed control scheme for induction motor drive.

REFERENCES

[1] Takahashi I. and T. Noguchi (1986). A new quick-response and high-Efficiency control strategy of an induction motor, *IEEE Transactions on Industry Applications,* **Vol. IA-22, No. 5,** pp. 820-827.
[2] Papafotiou G., T. Geyer and M. Morari (2004). Optimal direct torque control of three-phase symmetric induction motors, *Proceedings of the 43rd IEEE Conference on Decision and Control,* Atlantis, Bahamas, pp. 1860-1865.
[3] Rodriguez, J. J. Pontt, C. Silva, R. Huerta and H. Miranda (2004). Simple direct torque control of induction machine using space vector modulation, *Electronics Letters,* **Vol. 40, No. 7**.
[4] Flasch E., R. Hoffmannand P. Mutscher (1997). Direct mean torque control of an induction motor, *Proceedings 7th European Conference Power Electronics and Applications,* Tronheim, Norway, pp. 672-677.
[5] Pacas M. and J. Weber (2005). Predictive direct torque control for the PM synchronous machine, *IEEE Transactions on Industrial Electronics,* **Vol.52, No.5,** pp. 1350-1356.
[6] Kaboli S., M.R. Zolghadi, D. Roye, J. Guiraud and J-L. Schanen (2004). Design and implementation of a predictive controller for reducing the torque ripple in direct torque control based high frequency induction motor drives, *35th Annual IEEE Power Electronics Specialists Conference,* Aachen, Germany, pp. 1169-1174.
[7] Noguchi T., M. Yamamoto, S. Kondo and I. Takahashi (1997). High frequency operation of PWM inverter for direct torque control of induction motor, *IEEE Industry Applications, annual meeting,* New Orleans, USA, pp. 775-780.
[8] Mutschler P. and E. Flach (1998). Digital implementation of predictive direct control algorithms for induction motors, *33rd IAS Annual Meeting, IEEE Industry Applications Conference,* **Vol. 1,** pp. 444 – 451.
[9] Rodriguez J., J. Pontt, C. Silva, S. Kouro and H. Miranda (2004). A novel direct torque control scheme for induction machines with space vector modulation, *35th Annual IEEE Power Electronics Specialists Conference,* Aachen, Germany, pp. 1392-1397.
[10] Tripathi A., A.M. Khambadkone and NS.K. Panda (2001). Space-vector based constant frequency, direct torque control and dedbeat stator flux control of AC machines, *The 27th Annual Conference of the IEEE Industrial Electronics Society,* pp. 1219-1224.

Appendix A.

Electrical parameters of the simulated circuit:

Rated power	P_r=500 W
DC voltage	V_c=312 V
Rated Torque	T_r=3.41 N.m
Number of poles	P=4
Stator Resistance	R_s=4.495 Ω
Stator Inductance	L_s=0.165 H
Magnetising Inductance	L_m=0.149H
Rotor Resistance	R_r=5.365 Ω
Rotor Inductance	L_r=0.162H.
Rotor Moment of Inertia	J=0.00095 Kg.m^2
Damping coefficient	f=0.0004 N.m.s
Switching Cycle	T_{cycle}=100 μs

Parameters of PI:

K_p=0.1.
K_i=4.8.

Weighting terms of predictive controller:

Q=1e^6.
R=1e^{-2}.

Parameter Tuning of Fractional $PI^\lambda D^\mu$ Controllers With Integral Performance Criterion

K. Bettou., A. Charef

Département d'Electronique
Université Mentouri Route Ain El-bey-25000 - Constantine Algérie
(e-mail: bettou_kh@yahoo.com)

Abstract: A new technique for tuning fractional $PI^\lambda D^\mu$ controllers is presented in this paper. This technique is based on the solution of an optimisation problem. The basic ideas of the proposed technique are based on the minimum Integral Squared Error (ISE) criterion, in the first place, for setting the parameters of the fractional $PI^\lambda D^\mu$ controller for $\lambda=1$ and $\mu=1$ which means setting the parameters of the conventional PID controller, and in the second place, for setting the fractional integration action order λ and the fractional differentiation action order μ. The performance of this technique is compared with other techniques already available in the literature, and the results show a significant improvement in the closed-loop responses and more robustness using the parameters obtained by the proposed approach.

Keywords: PID tuning, Fractional $PI^\lambda D^\mu$, ISE Optimization.

1. INTRODUCTION

Fractional calculus is a generalization of integration and derivation to non-integer order fundamental operator ${}_aD_t^\alpha$, where a and t are the limits of the operation. The two definitions used for the general fractional differintegral are Grunwald definition and Riemann-Liouville definition (Oldham and Spanier1974).

In recent years we observe an increasing number of studies related with the application of the fractional calculus (FC) theory in many areas of science and engineering(Miller and Ross 1993), (Podlubny 1999a) and (Hilfer 2000). This fact is due to a better understanding of the FC potentialities revealed by many phenomena. In what concerns the area of automatic control systems (Oustaloup 1995) and (Vinagre *et al.*, 2000) the application of the FC concepts is still scarce and only in the last two decades appeared the first applications.

Despite the development of more advanced control strategies, the majority of industrial control systems still use PID controllers because they are standard industrial components, and their principle is well understood by engineers (Aström and Hägglund 1995) and (Cheng-Ching 1999). Design and tuning of PID controllers have been a large research area ever since Ziegler and Nichols presented their method in 1942 (Ziegler and Nichols 1942).

One of the possibilities to improve PID controllers is to use fractional order controllers with fractional order differentiation and integration parts. The first who really introduced a fractional order controller was Oustaloup. He developed the so-called Commande Robuste d'Ordre Non Entier (CRONE) controller and applied it in various fields of control systems (Oustaloup 1999). More recently, Podlubny proposed a generalization of the PID controller, namely the fractional order $PI^\lambda D^\mu$ controller, involving an integration action of a fractional order λ and differentiation action of a fractional order μ (Podlubny 1999b). Since, many researchers have been interested in the use and tuning of this fractional $PI^\lambda D^\mu$ controller (Petras *et al.*, 1998) and (Barbosa *et al.*, 2003). The interest of this kind of controllers is justified by a better flexibility, since it has two more parameters which are the fractional integration action order λ and the fractional differentiation action order μ. These parameters can be used to fulfil additional specifications for the design or other interesting requirements for the controlled system, than in the case of a conventional PID controller ($\lambda=1$, $\mu=1$). Further research activities are running in order to develop new tuning rules for fractional order controllers, studying previously the effects of fractional order of the derivative and integral parts to design a more effective controller (Monje *et al.*, 2004) and (Bettou *et al.*, 2007).

In this paper we propose the design of the fractional $PI^\lambda D^\mu$ controller of a classical unity feedback control system that minimizes an ISE performance index, where the controller is the fractional order $PI^\lambda D^\mu$ controller whose transfer function is given as:

$$C(s) = K_p \left(1 + \frac{1}{T_I s^\lambda} + T_D s^\mu \right) \quad (1)$$

with K_P is the proportional constant, T_I is the integration constant, T_D is the differentiation constant, λ is the fractional integration action order such that $0<\lambda<1$ and μ is the fractional differentiation action order such that $0<\mu<1$.

The equation for the fractional order $PI^\lambda D^\mu$ controller's output in time domain is:

$$u(t) = K_p \left(e(t) + \frac{1}{T_I} . D^{-\lambda} e(t) + T_D . D^\mu e(t) \right) \quad (2)$$

The basic ideas of the proposed technique are based on the minimum Integral Squared Error (ISE) criterion, in the first place, for setting the parameters of the fractional $PI^\lambda D^\mu$ controller for $\lambda=1$ and $\mu=1$ which means setting the parameters of the conventional PID controller, and in the second place, for setting the fractional integration action order λ and the fractional differentiation action order μ. We apply the proposed strategy to first order plus delay time (FOPDT) model. Several techniques are known for approximating plant step responses by this type of transfer function. The performance of this technique is compared with other techniques already available in the literature, and the results show a significant improvement in the closed-loop responses and more robustness using the parameters obtained by the proposed technique.

Bearing these ideas in mind, the paper is organized as follows. Section 2 introduces the basic ideas and the derived formulations of the new design strategy of the fractional order $PI^\lambda D^\mu$ controller. In section 3, an illustrative example is presented to demonstrate the control enhancement of the tuning method. Finally, section 4 draws the main conclusions and addresses some perspectives of future works.

2. FRACTIONAL $PI^\lambda D^\mu$ CONTROLLER DESIGN

We propose the design of the fractional order $PI^\lambda D^\mu$ controller of a classical unity feedback control system shown in Fig. 1:

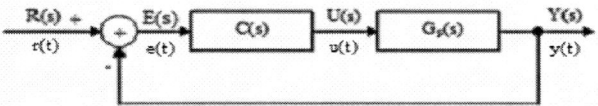

Fig. 1. Classical unity feedback control system

The plant's transfer functions $G_p(s)$ considered is a first order plant with a time delay:

$$G_p = \frac{K_0}{(1+\tau s)} e^{-Ls} \qquad (3)$$

and C(s) is the transfer function of the controller. If the mathematical model of controller includes fractional order derivation and integration, the controller is the fractional order controller. A. Oustaloup (Oustaloup 1999) developed the so-called CRONE controller for fractional order systems. Igor Podlubny (Podlubny 1999b) proposed a generalization of the PID controller, which is called the $PI^\lambda D^\mu$ because of involving an integrator of fractional order λ and differentiator of fractional order μ. The transfer function of the fractional order $PI^\lambda D^\mu$ controller is defined in (1).

Taking $\lambda=1$ and $\mu=1$, we obtain a conventional integer order PID controller. If $\lambda=1$, $\mu=0$, we obtain PI controller. If $\lambda=0$ and $\mu=1$, we have PD controller. All these conventional types of PID controllers are the particular cases of the fractional order $PI^\lambda D^\mu$ controller given by (1). Because the orders λ and μ can be arbitrary real number, the fractional order $PI^\lambda D^\mu$ controller is more flexible and given an opportunity to better adjust the dynamical properties of systems.

Let now consider the general form of a conventional PID controller

$$C(s) = K_p \left(1 + \frac{1}{T_I s} + T_D s\right) \qquad (4)$$

If we consider fractional order integration and derivation, the previous relation assumes the form of (1).

If $\lambda \neq \mu$ the relation of (1) becomes:

$$C(s) = \frac{K_p}{T_I} \frac{(T_D T_I s^{(\lambda+\mu)} + T_I s^\lambda + 1)}{s^\lambda} \qquad (5)$$

In the final expression of (5) it can be noted the presence of zeros that are related to (1). Taking into account the previous discussion, the asymptotic magnitude Bode diagram of the fractional order $PI^\lambda D^\mu$ controller in (1) can be obtained. Figure 2 shows an example of $PI^\lambda D^\mu$ magnitude asymptotic plot.

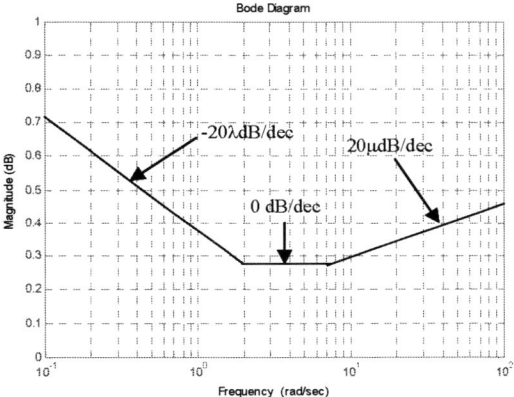

Fig. 2. An example of $PI^\lambda D^\mu$ magnitude asymptotic plot

The exact diagram of the $PI^\lambda D^\mu$ (magnitude and phase) are reported in Fig. 3.

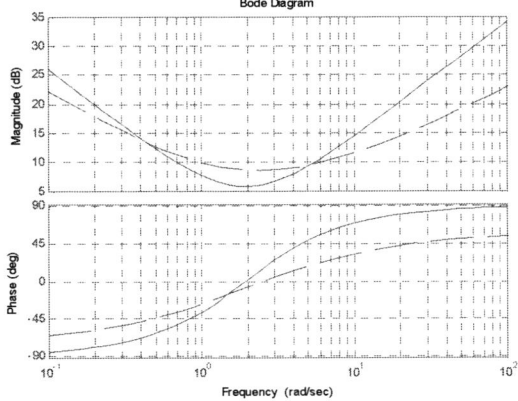

Fig. 3. An example of PID (solid line) and $PI^\lambda D^\mu$ (dashed line, with $\lambda=0.8$, $\mu=0.7$) magnitude and phase plot

2.1 Integral performance criteria

Over the years, many formulas for the optimum design of the controllers have been published. One common approach is to minimize a performance index. The popular performance indexes are the integral square error (ISE), the integral absolute error (IAE), and the integral time absolute error (ITAE) indexes. The ISE method has some advantages such as simple computation in the frequency domain.

When the plant transfer function is known, the parameters of the conventional PID and the fractional order $PI^\lambda D^\mu$ controllers may be optimised by minimising an integral performance criterion. Several optimization criterion are applied for tuning process controllers, the integral of squared error (ISE) is one of the most well known criteria since the relevant integral can easily be evaluated in the frequency domain and analytical solution obtained (Borne 1993).

The integral square error (ISE) is given as:

$$J = \int_0^\infty [e(t)]^2 dt = \int_0^\infty [r(t) - y(t)]^2 dt \qquad (6)$$

Where $e(t) = [r(t) - y(t)]$ is the error signal. Here $y(t)$ and $r(t)$ are the output and its set point, respectively. According to Laplace transform properties the integral J can be written as (Borne 1993):

$$J = \frac{1}{2\pi j} \int_{-j\infty}^{+j\infty} E(s) E(-s) ds \qquad (7)$$

Then, for $E(s) = \frac{N_E(s)}{D_E(s)}$ a rational function in s, the complex integral J will be:

$$J = \frac{1}{2\pi j} \int_{-j\infty}^{+j\infty} \frac{N_E(s) N_E(-s)}{D_E(s) D_E(-s)} ds \qquad (8)$$

From Fig. 1, the error signal $E(s)$ is given as:

$$E(s) = \left(\frac{1}{1 + C(s) G_p(s)}\right) R(s) = \left(\frac{1}{1 + C(s) G_p(s)}\right) \left(\frac{1}{s}\right) \qquad (9)$$

We calculate the corresponding ISE index J using the Hall-Sartorius method given in (Borne 1993).

2.2 Design of the parameters K_P, T_I and T_D

Our tuning technique is based, in the first place, on ISE optimization for setting the parameters K_P, T_I and T_D of the fractional order $PI^\lambda D^\mu$ controller for $\lambda=1$ and $\mu=1$ which means setting the parameters of a simple conventional PID controller.

2.3 Design of the parameters λ and μ

The proposed method consists of using the parameters K_P, T_I and T_D obtained in the first step for setting the parameters λ and μ minimizing the integral square error (ISE) of the classical unity feedback control system of Fig.1 for a unit step input.

The settings of the fractional integration action order λ and the fractional differentiation action order μ of the fractional order $PI^\lambda D^\mu$ controller consists in finding these two parameters that minimize the ISE index J of (6). We calculate the corresponding ISE index J using the Hall-Sartorius method given in (Borne 1993). In order to calculate the complex integral J using the Hall-Sartorius method (Borne 1993), $E(s)$ must be a rational function.

But the fractional order $PI^\lambda D^\mu$ controller's transfer function $C(s)$ given in (1) is an irrational function and the plant's transfer function $G_p(s)$ is also an irrational function because of the time delay. To circumvent this problem, the time delay of the plant's transfer function $G_p(s)$ is approximated by a rational function using the Padé approximation method as:

$$e^{-Ls} = \frac{\left(1 - \frac{L}{2}s\right)}{\left(1 + \frac{L}{2}s\right)} \qquad (10)$$

And the irrational function of the fractional order $PI^\lambda D^\mu$ controller $C(s)$ of (1) is also approximated by a rational function (Charef *et al.*, 1992) and (Charef *et al.*, 2006). With λ and μ are such that $0 < \lambda < 1$ and $0 < \mu < 1$.

The rational function approximation of $C(s)$, in a given frequency band of practical interest $[\omega_L, \omega_H]$, is given as:

$$C(s) = K_p \left(1 + \frac{1}{T_I s^\lambda} + T_D s^\mu\right)$$

$$= K_P \left(1 + \frac{K_I}{T_I} \frac{\prod_{i=0}^{N_I - 1}\left(1 + \frac{s}{z_{Ii}}\right)}{\prod_{i=0}^{N_I}\left(1 + \frac{s}{p_{Ii}}\right)} + T_D K_D \frac{\prod_{i=0}^{N_D}\left(1 + \frac{s}{z_{Di}}\right)}{\prod_{i=0}^{N_D}\left(1 + \frac{s}{p_{Di}}\right)}\right) \qquad (11)$$

The poles p_{Ii}'s, the zeros z_{Ii}'s, and the parameters K_I and N_I of the rational function approximation of the fractional order integrator, and the zeros z_{Di}'s, the poles p_{Di}'s, and the parameters K_D and N_D of the rational function approximation of the fractional order differentiator can be calculated as given in (Charef *et al.*, 1992) and (Charef *et al.*, 2006).

3. ILLUSTRATIVE EXAMPLE

This section shows the application of the results obtained for the design of the fractional order $PI^\lambda D^\mu$ controller for one selected plant.

Consider the plant model given by

$$G_p = \frac{K_0}{(1 + \tau s)} e^{-Ls} = \frac{1}{(1 + s)} e^{-0.5s} \qquad (12)$$

A conventional PID controller ($\lambda=1$, $\mu=1$) is used to control the above plant. Using the ISE optimization tuning method, the PID controller's parameters K_P, T_I and T_D are found to be $K_P = 1.9516$, $T_I = 0.9891$, and $T_D = 0.2642$. Hence, the conventional PID controller's transfer function $C_1(s)$ is given as:

$$C_1(s) = 1.9516\left(1 + \frac{1}{0.9891s} + 0.2642s\right) \quad (13)$$

Using the parameters K_P, T_I and T_D found above, the fractional order $PI^\lambda D^\mu$ controller's transfer function $C_2(s)$ is:

$$C_2(s) = 1.9516\left(1 + \frac{1}{0.9891s^\lambda} + 0.2642s^\mu\right) \quad (14)$$

The smallest ISE index J is obtained for the couple $(\lambda, \mu) = (0.9, 0.9)$. Then the fractional order $PI^\lambda D^\mu$ controller's transfer function $C_2(s)$ required is given as:

$$C_2(s) = 1.9516\left(1 + \frac{1}{0.9891s^{0.9}} + 0.2642s^{0.9}\right) \quad (15)$$

The fact of being $\lambda < 1$ makes the output converge to its final value more slowly than in the case of an integer controller. Furthermore, the fractional effects need to be band-limited when it is implemented. Therefore, the fractional order integrator must be implemented as $\frac{1}{s^\lambda} = \frac{1}{s}s^{1-\lambda}$, ensuring this way the effect of an integer integrator $1/s$ at very low frequency. So the fractional $PI^\lambda D^\mu$ controller's transfer function $C_2(s)$ required is given as:

$$C_2(s) = 1.9516\left(1 + \frac{1}{0.9891s}s^{0.1} + 0.2642s^{0.9}\right) \quad (16)$$

In this example of application, the fractional order integral and derivative parts have been implemented by Charef approximation of the fractional integrator and differentiator (Charef *et al.*, 1992) and (Charef *et al.*, 2006)., choosing a frequency band from $0.01\omega_c$ to $100\omega_c$, with ω_c is the unity gain crossover frequency of the open loop transfer function $C(s)G_P(s)$ when $C(s)$ is the conventional PID controller tuned by ISE criteria.

Figure 4 shows the step responses of the closed loop system with both controllers.

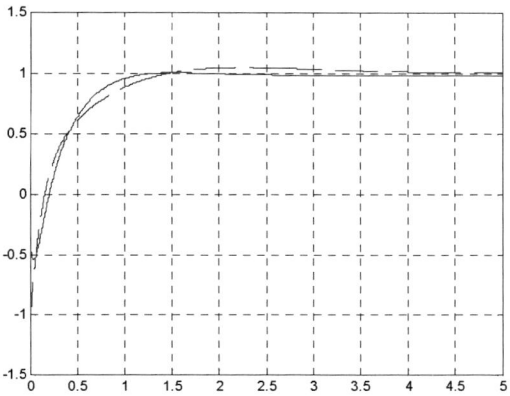

Fig. 4. Step responses of the closed loop system with C(s) a conventional PID controller (dashed line) and C(s) a fractional $PI^{0.9}D^{0.9}$ controller (solid line)

In order to evaluate the control quality enhancement using the fractional order $PI^\lambda D^\mu$ in comparison with the conventional PID controller, we have summarized some performances characteristics in Table 1 for the feedback control system with both controllers, in terms of the rise time t_r, the settling time t_s and the overshoot OS.

As it can be seen from Table 1, the rise time, the settling time and the overshoot with the fractional order $PI^{0.9}D^{0.9}$ controller are smaller than with the conventional PID controller.

Table 1. Performances characteristics for the conventional PID and the fractional $PI^{0.9}D^{0.9}$ controllers

Controller	t_r(s)	t_s(s)	OS(%)
PID	0.771	2.560	04.000
$PI^{0.9}D^{0.9}$	0.605	1.070	0.197

There could be variation or uncertainty in the process parameters. If the gain and the time delay of the process are 10% higher

$$G_p = \frac{1.1}{(1+s)}e^{-0.55s} \quad (17)$$

Then the step responses of the closed loop system with both controllers are given in Fig. 5.

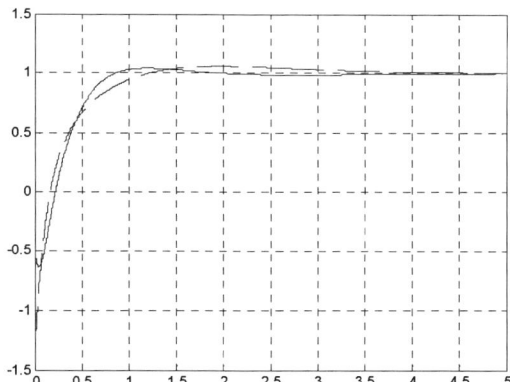

Fig. 5. Step responses of the closed loop system with C(s) a conventional PID controller (dashed line) and C(s) a fractional $PI^{0.9}D^{0.9}$ controller (solid line)

From the results obtained in the case of study, it can be concluded that the compensated system using the proposed fractional order $PI^{0.9}D^{0.9}$ controller is robust to changes in static gain and time delay, since variations of the performance characteristics are lower for the fractional order $PI^{0.9}D^{0.9}$ controller as it can be observed in Table 2.

In short, it can be said that the use of the fractional order $PI^\lambda D^\mu$ controller provide better responses and robust system.

Table 2. Performances characteristics for the conventional PID and the fractional $PI^{0.9}D^{0.9}$ controllers

Controller	$t_r(s)$	$t_s(s)$	OS(%)
PID	0.619	2.000	5.000
$PI^{0.9}D^{0.9}$	0.435	1.320	3.600

6. CONCLUSIONS

In this paper a new technique for tuning fractional order $PI^\lambda D^\mu$ controllers using performance index criteria have been investigated.

We are concerned with the design of fractional order $PI^\lambda D^\mu$ controller which involves fractional order integrator and differentiator. The three controller gains and two real orders of the fractional order $PI^\lambda D^\mu$ are determined to minimize an integral square error (ISE) performance index. Design example is given to compare the performance of the optimal fractionalorder $PI^\lambda D^\mu$ controller with the optimal integer order PID controller in controlling a first order with time delay plant.

In the software implementation the relevant integral has been evaluated using Padé approximation for time delay, simple manipulation and Hall-Sartorius integral algorithm.

Through one type of plant model, the fractional order $PI^\lambda D^\mu$ controller design by the proposed method has been demonstrated. The simulation results illustrate that fractional order $PI^\lambda D^\mu$ controller achieves better control performance with the proposed design method. The compensated system using the proposed fractional order $PI^\lambda D^\mu$ controller is robust to error model.

Our new tuning technique will be very suitable for already tuned PID controllers because in order to implement the fractional $PI^\lambda D^\mu$ controller we use the already existing conventional PID controller with a given fractional order differentiator and a given fractional order integrator.

Our further research efforts include: experiment on real plants, exploration of non-minimum phase, open-loop unstable systems, and oscillatory plants.

REFERENCES

[1] Aström, K. and T. Hägglund (1995). *PID Controllers: Theory, Design and Tuning.* Instrument Society of America, North Carolina.

[2] Barbosa, R.S., J.A.T. Machado, and I. M. Ferreira (2003). A Fractional Calculus Perspective of PID Tuning. *Proceedings of Design Engineering Technical Conferences and computers and Information in Engineering Conference.* Chicago. USA.

[3] Bettou, K., A. Charef and F. Mesquine (2007). Conception of Fractional $PI^\lambda D^\mu$ Controller. *Conference on Systems and Control.* Marrakech, Marroco.

[4] Borne, P. (1993). *Analyse et Régulation des Processus Indutriels.* Tome 1, Régulation Continue, Editions TECHNIP, Paris.

[5] Charef, A., H. Sun, Y. Tsao, Y. and B. Onaral (1992). Fractal System as Represented by Singularity Function. *IEEE Transaction on Automatic Control.* Vol. 37, N°9, pp. 1465-1470.

[6] Charef,A., (2006). Analogue Realization of Fractional Order Integrator, Differentiator and Fractional $PI^\lambda D^\mu$ Controllers. *IEE Proceedings on Control Theory Applications.* Vol. 135, N°6, pp. 714-720.

[7] Cheng-Ching, Y., (1999). *Autotuning of PID Controllers, Advances in Industrial Control.* Springer-Verlag.

[8] Hilfer, R., (2000). *Applications of fractional calculus in physics.* World Scientific. Singapore.

[9] Miller,K.S. and B. Ross (1993). *An introduction to the fractional calculus and fractional differential equations.* Wiley. New York.

[10] Monje, C.A., B.M.Vinagre, Y.Q. Chen, V. Feliu, P. Lanusse and J. Sabatier (2004). Proposals for Fractional $PI^\lambda D^\mu$ Tuning. *Proceedings of the 1st IFAC Workshop On Fractional Differentiation and its Applications FDA'04.* Bordeaux, France.

[11] Oldham, K.B. and J. Spanier (1974). *The fractional calculus.* Academic Press. New York.

[12] Oustaloup, A., (1995). *La dérivation non entière.* HERMES. Paris.

[13] Oustaloup, A., (1999). *La Commande CRONE: Commande Robuste d'Ordre Non Entier.* HERMES, Paris.

[14] Petras, I., L. Dorcak, and I. Kostial (1998). Control Quality Enhancement by Fractional Order Controllers. *ACTA Montanistica Slovaca.* Rocník 3, Vol. 3, N°2, pp. 143-148.

[15] Podlubny, I., (1999a). *Fractional equations.* Academic Press. San Diego, California.

[16] Podlubny, I., (1999b). Fractional order systems and $PI^\lambda D^\mu$ controllers. *IEEE Transaction on Automatic Control.* Vol. 44, N°1, pp. 208-214.

[17] Vinagre, B. M., I. Podlubny, A. Hernandez and V. Feliu (2000). 'Some Approximations of Fractional Order Operators Used in Control Theory and Applications', *FCAA fractional calculus and applied analysis.* 3(3), 231-248.

[18] Ziegler, J.G., and N.B. Nichols (1942). Optimum Settings for Automatic Controllers. *Transactions of the A.S.M.E.* Vol. 64, N° 8, pp. 759-768.

Preparation Model Based Control System For Hot Steel Strip Rolling Mill Stands

S.E. Bouazza*, H.A. Abbassi* and A.K. Moussaoui**

*Automatic & Signals Laboratory (LASA),
Université Badji Mokhtar BP 12 - 23000 - Annaba Algérie
(e-mail: h_a_abbassi@yahoo.fr and bouazza_s@yahoo.com)
**Electrical Engineering Laboratory of Guelma (LGEG),
BP.401, University of Guelma, 24000, Algeria (e-mail: a_k_moussaoui@yahoo.fr).

Abstract: As part of a research project on El-hadjar Hot Steel Rolling Mill Plant Annaba Algeria a new Model based control system is suggested to improve the performance of the hot strip rolling mill process. In this paper off-line model based controllers and a process simulator are described. The process models are based on the laws of physics. these models can predict the future behavior and the stability of the controlled process very reliably. The control scheme consists of a control algorithm. This Model based Control system is evaluated on a simulation model that represents accurately the dynamic of the process. Finally the usefulness to the Steel Industry of the suggested method is highlighted.

Keywords: Hot Rolling Mill Train, Rolling Mill Stands, Hot Steel Strip, Static model, Dynamic Model, Mill Stands Set-up, Thickness Control, Thickness prediction, Tension Control, Temperature prediction, Model based control, Feedforward control, Feedback control.

1. INTRODUCTION

In hot steel strip rolling process both feedforward and feedback control are necessary to reduce the effects of rapid strip thickness variations due to skid marks, roll eccentricity and other factors such as supports deformation, bearings ... as well as long term or slow variations of thickness in the six stands finishing train [2-13-5].

The consideration of model based control in hot steel strip rolling mill in which there is thickness, Tension and temperature control, involves the resolution of many problems.

The main problem is that both good static and dynamic mathematical models must be found so that this allows the investigation and observation of many phenomena and parameters than we could measure in a real situation in the Steel Rolling mill plant. This of course permits the synthesis of good feedforward or feedback controllers.

2. PROCESS DESCRIPTION

In a hot strip rolling mill, Finishing Mill Train is a kind of process that makes strips from thick slabs which have been produced by continuous casting. A typical process shown in Fig.1 consists of rolling stands where the attached rolls are used to press the steel slab. After a slab is reheated to re crystallization temperature in the furnace, it is reduced in several passes in a Roughing Mill before being rolled in the Finishing hot rolling mill Train.

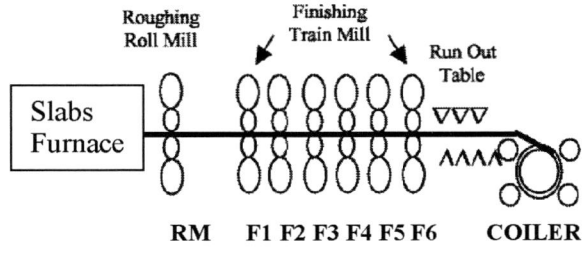

Fig.1. Hot Steel Strip Rolling Mill Train

2.1. Hot Strip Thickness Control System

As shown in Fig.2, the hot strip thickness control system called also Automatic Gage Control (AGC), is intended to maintain the thickness (the gage) of the strip constant at a predetermined value. To satisfy this condition, finishing train stands must feature roll positioning systems (hydraulic screw drives) capable of High speed operational response under full rolling load and should be equipped with load cells for the measurement of rolling force and a position measuring device [1]. Because of the pure time delay involved when making the measurement of the thickness down the rolling line this leads to the necessity of using a mathematical model of the system to be controlled and as it is well known the Smith predictor is often used in these sorts of situations.

The classical Smith Predictor [11] certainly improves the response of the overall system but does not combat disturbances. The method that avoids the effects of the pure time delay introduced by the thickness measuring device and gets the Stand exit thickness from the force on the Stand Work Rolls is known as BISRA [8-10] Control Method. In what follows we will improve the Smith Predictor control system and the BISRA Gages by using evolved mathematical Models derived from the industrial process nature using basic laws of physics.

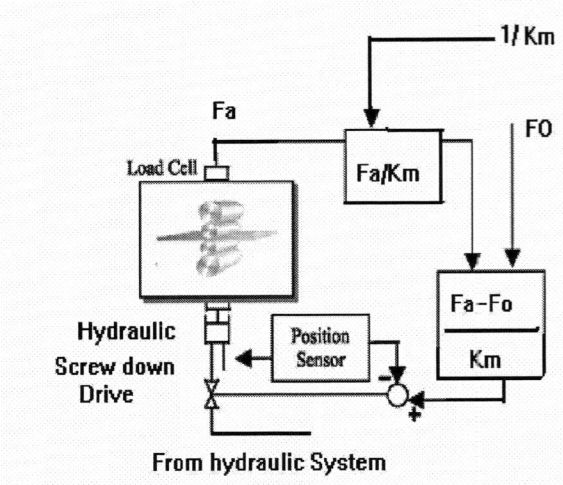

Fig.2. Hot Strip Thickness Basic (BISRA) Control System.

3. MATERIALS AND METHODS

3.1. Static Roll Gap Model Solution for Rolling Mill Set-up prediction

The fundamental Orowan's differential equation for a Rolled strip [10] is:

$$\frac{dF_H}{d\beta} = 2R * P(SIN\beta \pm uCOS\beta) \quad (1)$$

Where :

F_H = Horizontal Force applied to a strip unit section.
β = Angle of no slip point
R = roll radius
P = pressure applied to a strip unit section.
u = friction coefficient

The solution of the above differential equation (1) by many Authors : Sims, Bland, Ford, Siebel, Ellis, ... with different physical hypotheses and approximations led to the prediction of the results below :

3.2. Predictions of Force, Thickness, Temperature, Torque, Powers, Forward Slip, Strip entry and exit Speeds and Rolling pressure

$$V(t) = Vo + 2R'(1+COS\beta) \quad (2)$$

thickness at any point and time predicted by equation (2) in the roll Gap.

R'= the deformed work roll Radius calculated by equation 9 below.

From [3-12] and the schematic representation given in Fig.2, when ignoring the chatter e(t) due to eccentricity of the rolls and other less important factors (oil film, thermal camber, etc.) the desired finishing thickness V_d and the attained thickness V are expressed by:

$$V_d = Zo + (Fo / Km) \quad (3)$$

$$Zo = V - (Fa / Km) - \Delta Z \quad (4)$$

$$V = Zo + (Fa / Km) + \Delta Z \quad (5)$$

Where:
Zo: Unloaded roll gap (Set-up Roll Gap), Vo incoming thickness, V is the Outgoing thickness (output or actual thickness), Fo: Reference roll force, Fa: Actual rolling force, Km: Mill Spring Constant (Mill Modulus), ΔZ : change in roll gap by AGC (Correction based on X-Rays thickness meter).

The thickness deviation is then :

$$\Delta V = Vo - V = ((Fa - Fo)/Km) + \Delta Z \quad (6)$$

As in the ideal case ΔV must tend to zero, so that the thickness would reach the desired value.
The basic control equation is:

$$\Delta Z = -(Fa - Fo)/Km \quad (7)$$

For a measured or predicted material temperature of a given Steel grade and calculated rolling speed, the actual rolling force is calculated by Sims/Orowan efficient formula [4-7] :

$$Fa = 1.15.b.k1.Qp\sqrt{R'i(Vo - V)} \quad (8)$$

with

$$Qp = 0.7924 + 1.778 * \exp(-2.148 * A)$$

where

A = L(bite contact length) / V.
R'_i = Hitchcock's Roll Radius equation [8]

$$R'_i = R_i\left[1 + C_i \frac{F_a}{\Delta V}\right] \quad (9)$$

R' is the loaded Stand Work Roll Radius.
Ri is the unloaded work Roll Radius
Ci is a constant depending on Roll's Material elasticity, usually between 0.00021 mm²/N and 0.00025 mm²/N for hard steel rolls.

Where also:
T: Strip temperature, k1: deformation resistance, b: Strip width, Ri': Loaded work roll radius, Vo: Input thickness, V: output thickness.
k1 [kgf / mm^2] = K2*K3 is calculated by the use of MISAKA'S [9] deformation resistance Prediction Equation:

$$K_2 = \exp[0.126 - 1.75 c^2 + (\frac{2851 + 2968 c - 1120 c^2}{T})]$$
$$K_3 = \varepsilon^{0.21} [\frac{d\varepsilon}{dt}]^{0.13} \quad (10)$$

C: Slab Steel carbon content [weight %]
T: Rolled Steel Absolute Temperature [K]
ε: True Strain of deformation in mm per unit length

$\frac{d\varepsilon}{dt}$: Strain Rate of deformation (1/s)

t : Time [s]

The Mean Strain Rate is:

$$\text{Mean}\left(\frac{d\varepsilon}{dt}\right) = \frac{V_{roll} \ln\left(\frac{V}{Vo}\right)}{R'\sqrt{V-Vo}} \quad (11)$$

Usually taken as: 10^{-2} to $200 (1/s)$

For Mild steel, high carbon and low-alloy steel., the Steel Strip Temperature is predicted by:
knowing the temperatures of the strip measured by Pyrometer at the entry point of stand F1 and at the exit point of the last stand F6, the temperature of the strip at each stand could be predicted by the following expression :

$$T_i = T_{(i-1)} - \Delta T_i(irr) - \Delta T_i(desc) - \Delta T_i(csr) + \Delta T_i(defo). \quad (12)$$

Tirr=irradiation temperature Drop ; Tdesc = descaling temperature Drop ; Tcsr = contact strip /roll temperature Drop, Tdefo = deformation temperature Gain.

$\Delta T_i(defo) = k1 \varepsilon / \rho C_p$ = Temperature Gain. (13)

Cp =steel heat capacity
k1 = deformation resistance as above.
ε = True Strain of deformation
ρ = steel's density

All Temperatures could be calculated by a temperature model calibrated to the considered rolling mill train by (14) Formula:

$$T_i = T_w - (T_{i-1} - T_w)\exp\left(\frac{-2\alpha_F L_1}{C\rho V_F V_{ST}}\right) \quad (14)$$

T_w = Cooling Water Temperature
L_1 = Distance between Stands
V_{ST} = Strip threading speed (meters/h)
V_F = Finishing Mill delivery thickness in meters.
α_F = constant
C ' Specific Heat (kcal/kg °C degrees centigrade)
ρ = Density (kg/ square meter)

The total rolling Torque is predicted by :

$$\text{TRM} = (2 F_a L_a)/1000 \quad (15)$$

L_a = (C1*C2*exp(C3*D/2*V)*L' = The lever arm

Where L' is the contact bite length, D is roll diameter C1,C2,C3 are known constants.

The Torque at the motor shaft is expressed by:

TMSH = TRM/gear ratio (gr)*efficiency(η) (16)

The electrical input power is given by:

KW = TRM*Wroll(rd/s)/η
 = TRM* Vroll (rpm)*1000/R η (17)

The forward slip Ratio Model is given as follows:

fos =(Vstrip – Vroll) / Vroll (18)

The Slip model allows the prediction of the Strip input and output speeds of any Stand starting from the Roll Speed of the considered Stand.

The Angle ß of no Slip point can be calculated [10] by:

$$\beta = \frac{\sqrt{V}}{\sqrt{R'}} \tan\left[\frac{\pi}{8}\left(\frac{\sqrt{V}}{\sqrt{R'}}\right)\ln(1-r) + \frac{1}{2}\tan^{-1}\left(\frac{\sqrt{r}}{\sqrt{r-1}}\right)\right] \quad (19)$$

Thickness Reduction r = Vo – V

Vexi = exit speed from stand Fi is predicted by

$$V_{exi} = V_{roll}\left[1 + \left(\frac{2R'}{V}\cos\beta - 1\right)(1 - \cos\beta)\right] \quad (20)$$

The Elongation of the Strip EL must be kept constant in order to get a constant strip tension between stand Fi and F i+1 or Fi-1:

EL = (Vexi – Vientry) / Vientry   (21)

Rolling Speed at Stand Fi :

Vrsi = (1+ fos6)V6*Vrs6) / (1+ fosi)Vi   (22)

The last F6 stand rolling speed Vrs6 is directly introduced by the operator at the start of every rolling campaign.

The Rolling pressure (Ekelund Formula) :

Ps = (Ka* n*u)*(1+m)   (23)

Where :

Ka = (14-0.01t)(104+C+Mn+0.3Cr)
n = 0.01(14-0.01t)Cv

$$u = \frac{2000J\sqrt{2\Delta V/d}}{Z_o + V_o}$$

Zo = set-up gap
$\Delta V = V_0 - V$ = thickness deviation
J = 0.5(Vrollsup – Vrollinf)
u = friction coefficient = 1.05 – 0.0005t
d = work roll diameter
$Cv = 1.0942\, e^{-0.03J}$

4. RESULTS AND DISCUSSION

Table 1.[14]: EL- Hadjar Hot Steel Rolling Mill Finishing Train Annaba ,Algeria , plant specifications .

Finishing mill Type	6 stands equipped with hydraulic screw-downs
Nominal Power	2 x 3000 HP, D.C.
Rolling Thickness Range	1.5 to 15 mm
Strip Width Range	600 to 1350 mm
AGC (Autom. Gage control)	Hydraulic
Gauge meter	X Ray
Rolling speed	49 to 407 rpm
Mean Mill Stretch	1000 tons/mm
Rolling temperature	860 to 880 °C
Production capacity	1. 5 Million Tons per Year

The calculation of the Set-up (Fig:3) is in reality executed on-line in the real time available after the slab leaves the Roughing mill and before it enters the first Stand of the Finishing Train which has the main specifications shown in table.1.

In order to verify that the Simulator satisfies the time constraints and gives good prediction of Rolling Variables, Off-line Simulations using real process data is introduced into the Simulator program.

To keep the thickness deviation (Fig.5) to the specified minimum dictated by the client the simulator developed makes two sorts of adaptations one is between Stand (i) and Stand (i+1) in the Feed forward control Mode and the second adaptation is done in the Feedback Mode or something that could be called (Learning Mode) using data from the X.Ray Monitor & Pyrometer (Fig.3) of one coil and correcting the Rolling variables in the next run during the Rolling of the next Slab. If we look at Fig.3, the words (Operator Trims) are still in the control system, this is because there is always a restricted precision of the process Model but nevertheless the accuracy is getting better every day and the operators are needed less and less than before in modern Rolling Mill plants .

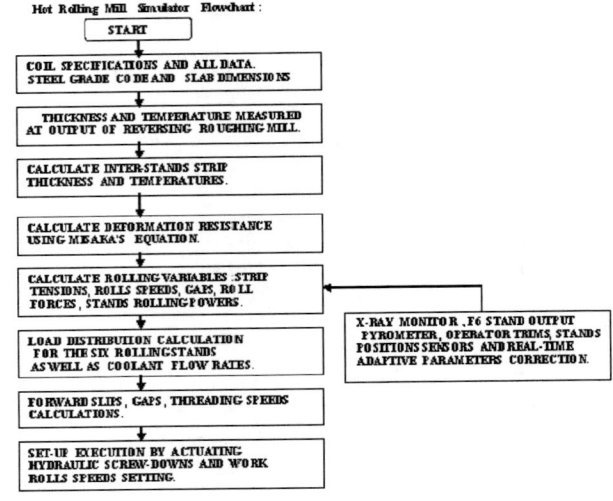

Fig.3. Simplified hot rolling mill finishing train simulator flowshart.

The simulation results of fig.4 representing the different rolling forces applied to stands F4, F5, F6 ,
Fig.6 and fig.7 represent the simulated results for the speed and tension at stand 3 . These simulated results are more or less the same as the practically measured by (force ,speed and tension) sensors in a practical rolling mill of the same size as the El-Hadjar finishing rolling mill simulated in this project.

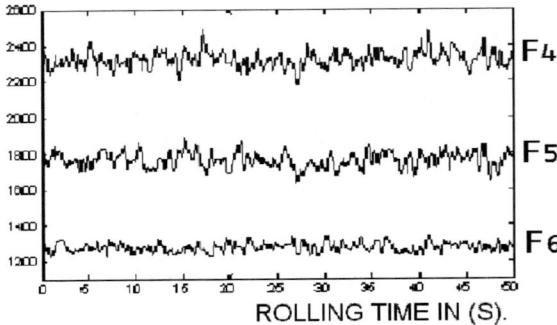

Fig.4. Evolution of calculated stands roll Forces

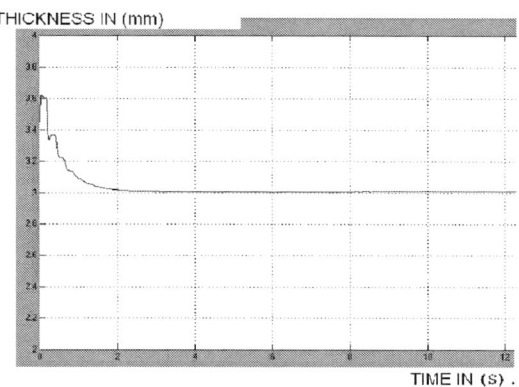

Fig.5. Evolution of calculated thickness

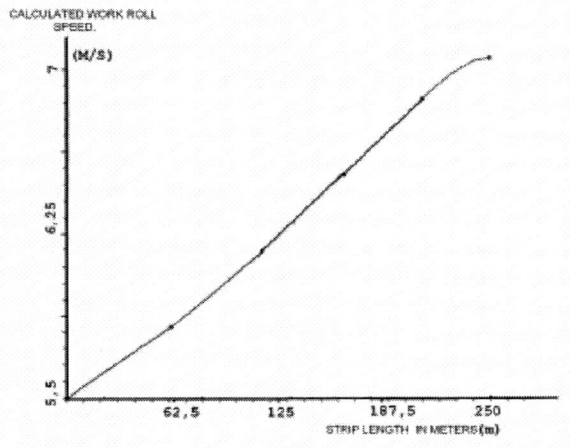

Fig.6. Evolution of calculated stand 3 rolling speed

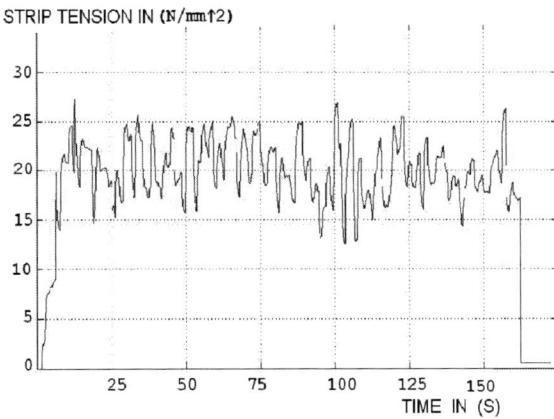

Fig.7. Calculated strip tension

5. CONCLUSION

An off-line Hot Steel Strip Rolling Mill Simulator was elaborated and tested under mainly Fortran language because of the types of calculations involved on the level two mainframe (Vax) process computer (formulae type) using off-line plant Data from the Arcelor-Mittal, El-Hadjar Annaba Algeria six stands finishing train.

From the simulation results fig.4: to fig.7, the usefulness and effectiveness of the suggested System and models were proved. Moreover, it is well shown that mathematical model based control of industrial plants is very reliable because of their real physical modeling and their predictable stability and behavior [6,7,12].

The new Simulator presented has good performance in terms of computing time and accuracy. The results obtained up to now are quite promising for future implementation practically on-line in the industrial finishing train.

ACKNOWLEDGEMENTS

The authors would like to express their appreciations and sincere thanks to the staff of URASM/CSC (ex. DRA, El Hadjar), Algeria, for their support and for the help, tolerance, encouragements and fruitful suggestions given and shown to us during the present research project.

We also acknowledge the superb service offered by both the members of staff of Badji Mokhtar University Annaba Algeria and Arcelor-Mittal, El-Hadjar Complex Technical Libraries.

REFERENCES

[1]. Wada.B,H.Sumi,I.Ueda , K.Suzuki ,Automatic Gage Control for heavy plate mill pp.465-470 La Revue de Metallurgie –CIT , Mars , 1994.

[2]. Konno.Y , Shioya.M , and T.Ueyama , Development and application of Dynamic system Simulator , Nippon Steel Report N°67 , pp.63-68.

[3]. Dairiki.O. , H.Mabuchi , I.Degawa , H. Nakamura , Progress of AGC utilization techniques at plate mill of Oita Works , Nippon Steel Technical Report , N°42 , July , 1989 , pp.38-47.

[4]. Ford, H. and J.M. Alexander, 1963-64, " Simplified Hot Rolling Calculations ", Journal of the institute of Metals, Vol.92, 1963-64, pp.397-403.

[5]. Min Huang , Xiaoke Fang , Jianhui Wang , Shusheng Gu , Roll Eccentricity compensation control for strip rolling mill based on wavelets packet de-noising theory . IEEE Proceeding of the fifth congress on intelligent control and automation ,june 15-19 , 2004, Hangzhou , P.R China.

[6]. L.M. Pederson . Modeling and Control of Plate Mill Processes .pp.25 & pp.40 Phd Thesis Lund Institute of Technology dept of Automatic Control Sweden 1999.

[7]. Gomez, C., " Modélisation et Simulation d'un Train de Laminage à Chaud, Thèse de Doctorat, présentée à l'Université de Paris-Sud, Centre d'Orsay, France, December, 1980.

[8]. Eustace.C. LARKE The Rolling of Strip,Sheet and Plate Chapman and Hall Ltd , London 1963.
[9]. Misaka. Y. et al. Journal of the Japan Society for the Technology of Plasticity 8,79,1967-1968,pp.414-422.
[10]. W.L . Robert .Hot Rolling Of Steel , Marcel Dekker Newyork U.S.A 1983.
[11]. Walter Ungerer , Ulrich Muller , Mohieddine Jelali , Andreas Wolff. Advanced control strategies for rolling mills MPT International N°3 ,2001.pp.54-57.
[12]. J.Pittner , Marwan A.Simaan. State-Dependent Riccati Equation Approach for Optimal Control of a Tandem cold Metal Rolling Process.IEEE Transactions on industry applications vol.42, N°3 pp.836-843 May/June 2006.
[13]. Siegfried Shreiber , Dr Rudiger Doll , Alfred Dummler Automation and process models at the new Wuhan N°2 Wide hot strip mill. MPT International N°6 , 2004.pp32-37.
[14]. Document des spécifications techniques: laminoir à chaud. Société Nationale de Sidérurgie (Annaba El-Hadjar) SNS (1979).

Adaptive Fuzzy Hysteresis Band Current Controller for Four-Wire Shunt Active Filter

F. Hamoudi *, A. Chaghi *, H. Amimeur *, E. Merabet *

* Group LSTIE-Research Laboratory, Department of Electrical Engineering, University of Batna. Street Chahid Mohamed El hadi Boukhlouf, 05000, Batna, Algeria (email : f_hamoudi@yahoo.fr; az_chaghi@yahoo.fr; amimeurhocine@yahoo.fr; merabet_elkheir@yahoo.fr).

Abstract: This paper presents an adaptive fuzzy hysteresis band current controller for four-wire shunt active power filters to eliminate harmonics and to compensate reactive power in distribution systems in order to keep currents at the point of common coupling sinusoidal and in phase with the corresponding voltage and the cancel neutral current. The conventional hysteresis band known for its robustness and its advantage in current controlled applications is adapted with a fuzzy logic controller to change the bandwidth according to the operating point in order to keep the frequency modulation at tolerable limits. The algorithm used to identify the reference currents is based on the synchronous reference frame theory ($dq\gamma$). Finally, simulation results using Matlab/Simulink are given to validate the proposed control.

Keywords: Four-wires shunt active filter, Synchronous reference frame, Fuzzy hysteresis band control.

1. INTRODUCTION

ACTIVE filters have came into view around 1970, after that, the concept has been successfully developed, tested and assisted by the power electronics technology, and put in practical uses . In term of topology, three-phase filters can be divided into three-wire and four-wire active filters [1, 2, 3, 4, 5, 6, 7]. The first one is well-developed and put in industrial applications, however, in last years, researches are more oriented to the second one for its advantage in four-wire distribution systems [5, 6, 7].

In this paper the four-wires three-phase shunt active filter for current compensation, using the conventional voltage three legs source inverter as illustrated in Fig.1 is studied.

The active filter control can be divided into two different parts; the first is the algorithm used to identify the suitable currents to be injected in the network to achieve compensation objectives, while the second part is the current control or regulation, which forces the shunt active filter to synthesize the desired current with minimized error. For this part, the hysteresis band current control is very suitable for several reasons such as: simplicity, easy implementation, very fast response and good accuracy. However, the hysteresis band current control presents an undesirable feature in its high commutation frequency, which varies with the operating point. From the theoretical point of view, increasing switching frequency permits a very good reproduction of the reference currents especially in harmonics current control, but in practical uses, this frequency is limited by the nature of the semi-conductors and the passive elements.

This paper focuses on the optimization of the hysteresis bandwidth using fuzzy logic concept, thus, the resulting fuzzy adaptive hysteresis band makes compromise between the switching frequency and the robustness of the conventional one.

In this paper, first, we have presented briefly the four wires active filter configuration and the reference current identification

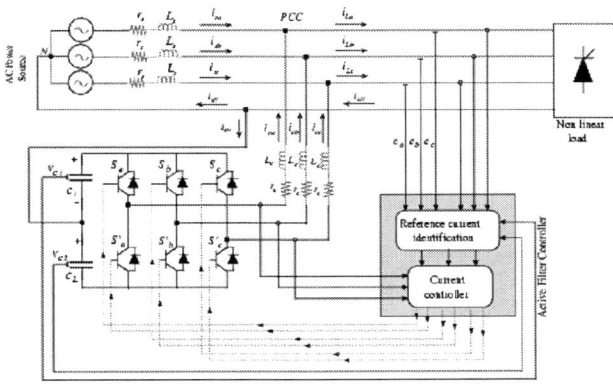

Fig. 1. Configuration of the four-wires shunt active filter.

based on the synchronous reference frame. Next, the adaptive fuzzy hysteresis band current controller is described. Finally, simulation results are given followed by the conclusion.

2. FOUR-WIRE SHUNT ACTIVE FILTER

The general configuration of the four wires shunt active filter presented in this paper is shown in Fig.1. The DC bus is constituted of two capacitors C_1 and C_1 with a midpoint connected to the neutral wire of the network. The AC side of the active filter is connected to the network through a first order passive inductive filter. This topology permits to compensate not only phase currents but also to cancel the neutral current.

The model of the three-leg four-wire voltage source inverter (1) permits to show that the output voltage of the k-leg depend only of the switching function d_k ($k = a, b, c$) of the same leg, this mean that this topology can be assimilated to three decoupled single phase active filter.

$$v_k = d_k V_{C_1} - \bar{d}_k V_{C_2} \qquad (1)$$

The interaction between the shunt active filter and the network can be described by the following state equation (2) with e_a, e_b and e_c are the network voltages.

$$[\dot{X}] = [A][X] + [B] \qquad (2)$$

Where;

$$A = \begin{bmatrix} -\dfrac{r_c}{L_c} & 0 & 0 & \dfrac{d_a}{L_c} & -\dfrac{d_a}{L_c} \\ 0 & -\dfrac{r_c}{L_c} & 0 & \dfrac{d_b}{L_c} & -\dfrac{d_b}{L_c} \\ 0 & 0 & -\dfrac{r_c}{L_c} & \dfrac{d_c}{L_c} & -\dfrac{d_c}{L_c} \\ -\dfrac{d_a}{C} & -\dfrac{d_b}{C} & -\dfrac{d_c}{C} & 0 & 0 \\ \dfrac{d_a}{C} & \dfrac{d_b}{C} & \dfrac{d_c}{C} & 0 & 0 \end{bmatrix},$$

$$X = \begin{bmatrix} i_{ca} \\ i_{cb} \\ i_{cc} \\ V_{C_1} \\ V_{C_2} \end{bmatrix}, \text{ and } B = \begin{bmatrix} -\dfrac{e_a}{L_c} \\ -\dfrac{e_b}{L_c} \\ -\dfrac{e_c}{L_c} \\ 0 \\ 0 \end{bmatrix}$$

3. REFERENCE CURRENTS IDENTIFICATION

The synchronous reference frame used to identify the reference compensating currents consists to transform the instantaneous load currents i_{Lk} in a turn coordinate system by using Park transformation, in order to make easy the separation of the undesired components from the measured currents [8].

$$\begin{bmatrix} i_d \\ i_q \\ i_\gamma \end{bmatrix} = \sqrt{\dfrac{2}{3}}.T(\hat{\theta}) \begin{bmatrix} i_{La} \\ i_{Lb} \\ i_{Lc} \end{bmatrix} \qquad (3)$$

Where;

$$T(\hat{\theta}) = \begin{bmatrix} \cos(\hat{\theta}) & \cos(\hat{\theta} - 2\pi/3) & \cos(\hat{\theta} - 4\pi/3) \\ -\sin(\hat{\theta}) & -\sin(\hat{\theta} - 2\pi/3) & -\sin(\hat{\theta} - 4\pi/3) \\ 1/\sqrt{2} & 1/\sqrt{2} & 1/\sqrt{2} \end{bmatrix}$$

In fact, if the rotating angle $\hat{\theta}$ correspond to the fundamental frequency, then currents in $dq\gamma$-axis are composed by a DC component (related with fundamentals) and AC components (with harmonics), which can be separated easily by using high-pass filter.

$$\begin{aligned} i_d &= \bar{i}_d + \tilde{i}_d \\ i_q &= \bar{i}_q + \tilde{i}_q \\ i_\gamma &= \bar{i}_\gamma + \tilde{i}_\gamma \end{aligned} \qquad (4)$$

Note that i_d correspond to the reactive, i_q correspond to the active power and i_γ to the homopolar power flowing through the neutral wire.

In general, the rotating angle $\hat{\theta}$ used in the Park transformation is given by a Phase Locked loop (PLL) since network voltages.

3.1 PLL System

The PLL system used to extract the instantaneous fundamental angle $\hat{\theta}$ is shown in Fig.2(a), it is based on the *pq-theory* [9]. The instantaneous real power in the PI-controller input is:

$$p_{PI} = e_\alpha i'_\alpha + e_\beta i'_\beta \qquad (5)$$

Where e_α and e_β are the measured voltages transformed in (α, β) coordinate, and the feedback signals i'_α, i'_β are the cosine and the sine of the angle at the PLL circuit.

This circuit can reach the stable point of operation only if the output p_{PI} of the PI-controller has zero average value ($\bar{p}_{PI}=0$) and minimized alternative value ($\tilde{p}_{PI} \approx 0$). Otherwise, the average power in three-phase system is given by:

$$P_{avg} = 3E_1^+ I_1^+ \cos(\phi_{E_1^+} - \phi_{I_1^+}) \qquad (6)$$

This means that the stable point of operation is found only $\hat{\omega}$ correspond to the fundamental frequency of the system and the auxiliary signals i'_α and i'_β become orthogonal to the measured network voltages e_α and e_β respectively.

3.2 DC Voltage Regulation

It is well known that the voltage of the DC bus should be maintained at a predefined design value to achieve correctly the compensation. As a particular characteristic for four-wire three-leg active filter is that the DC voltage regulation is constituted of two control loops; the first loop keeps the voltage $V_C = V_{C_1} + V_{C_2}$ constant around its reference V_C^*. This is achieved by a PI-controller, which generate an active current I_{VC} added to the active filter reference current in order to inject or to absorb an additional active current in or from the network, which keeps V_C constant. The second loop permits to balance the voltages V_{C_1} and V_{C_2}, for this, the deference $I_{VC_\gamma} = V_{C_1} - V_{C_2}$ is added through a low pass filter as an homopolar current in the active filter reference current with the same effect as the regulation [7].

3.3 Reference Current calculus

The reference compensating currents are calculated by the non desired parts of the currents i_d, i_q and i_γ to achieve compensation objectives, if we wan to compensate exclusively harmonics, then $\tilde{i}_d$, $\tilde{i}_q$ and $\tilde{i}_\gamma$ are used to calculate reference compensating currents, but in general the fundamental reactive power should be compensated. In this case, the DC part of i_d is included, therefore, the reference compensating current are given in the abc-coordinate using the inverse Park transformation.

$$\begin{bmatrix} i_{ca}^* \\ i_{cb}^* \\ i_{cc}^* \end{bmatrix} = \sqrt{\dfrac{2}{3}}.T^{-1}(\hat{\theta}) \begin{bmatrix} i_d \\ \tilde{i}_q + I_{VC} \\ i_\gamma + I_{VC_\gamma} \end{bmatrix} \qquad (7)$$

Finally, the complete block diagram of the compensating current identification is illustrated in fig2.

4. THE ADAPTIVE FUZZY HYSTERESIS BAND CURRENT CONTROLLER

As it was mentioned in the introduction, the hysteresis band current control technique presents very good characteristics for current control applications. In the case of a voltage source inverter with a midpoint in the DC bus the width of the hysteresis band HB_k is given for each phase in [10] by:

$$HB_k = \dfrac{V_{DC}}{8f_m L_c} \left[1 - \dfrac{4L_c^2}{V_{DC}^2} \left(\dfrac{e_k(t)}{L_c} + \dfrac{di_{ck}^*(t)}{dt} \right) \right] \qquad (8)$$

Where f_m is the modulation frequency, $\dfrac{di_{ck}^*}{dt(t)}$ is the slope of the reference current for the k-phase, e_k is the instantaneous

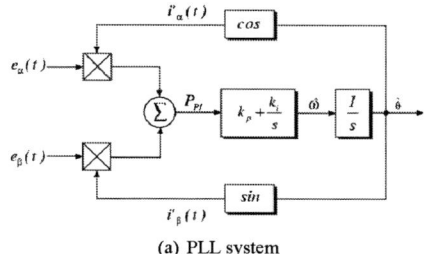

(a) PLL system

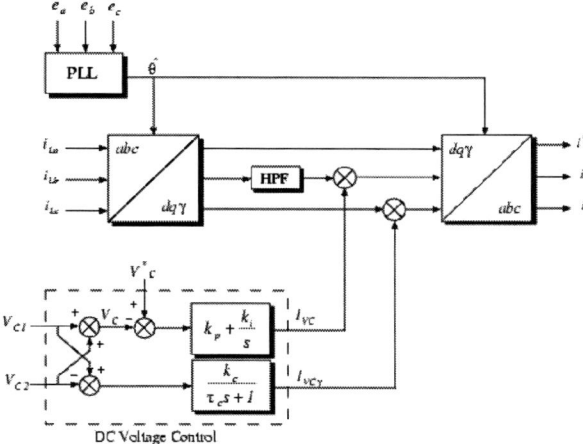

(b) Compensating current identification.

Fig. 2. Complete Block diagram of the reference current identification.

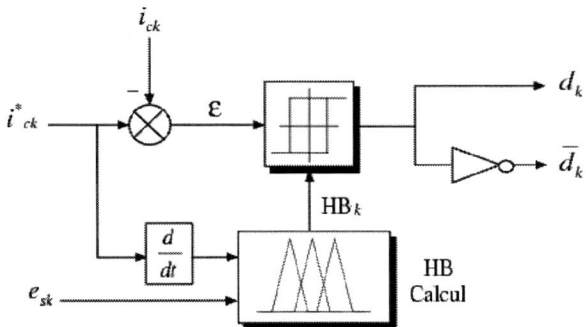

Fig. 3. The adaptive fuzzy hysteresis band current controller.

voltage at the common coupling point of the same phase, L_c and V_C are respectively the decoupling inductance and the DC bus voltage.

Regarding the equation (8), we can remark that in order to keep the modulation frequency f_m constant it is necessary to regulate the bandwidth HB_k at different operating points, following the variations of the different parameters in (8).

To improve good performances without precise knowledge of the active power filter parameters, the bandwidth HB_k is controlled using a fuzzy logic Controller [11] with e_k and $\frac{di^*_{ck}}{dt(t)}$ as input variables and HB_k as the output variable as shown in Fig.3.

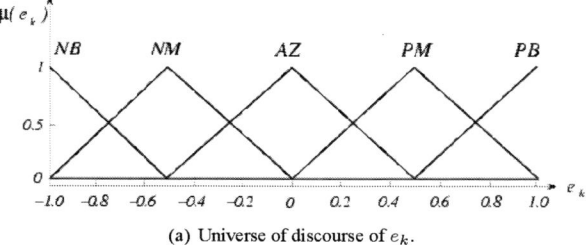

(a) Universe of discourse of e_k.

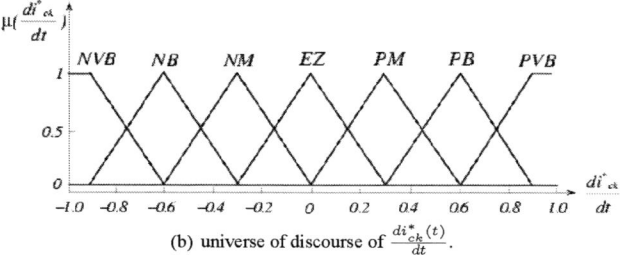

(b) universe of discourse of $\frac{di^*_{ck}(t)}{dt}$.

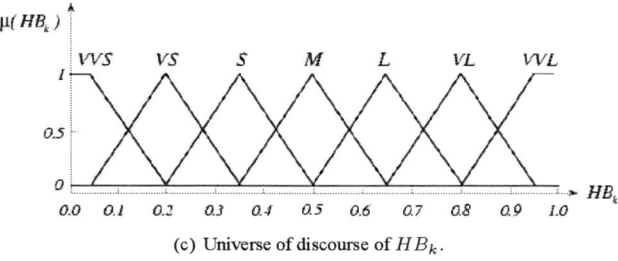

(c) Universe of discourse of HB_k.

Fig. 4. Universes of discourse for the inputs and output variables.

In reality, we can take in account the DC voltage as a third input variable but this can complicate the fuzzy logic controller without a substantial contribution if we suppose that the DC voltage regulation is correctly achieved.

The fuzzy logic controller is constituted of three principal elements which are: the fuzzification, rule base, and the defuzzification [12].

The fuzzification converts the real values into linguistic size with assigning to each variable a set of fuzzy subsets. For this step, five fuzzy subsets are chosen for e_k: NB is negative big, NM is negative medium AZ is almost zero, PM is positive medium and PB is positive big (Fig.4(a)). However, in order to increase the dynamic performances of the fuzzy logic controller, seven fuzzy subsets are chosen for the reference current slope: NVB is negative very big, NB is negative big, NM is negative medium, AZ is around zero, PM is positive medium, PB is positive big, and PVB is positive very big (Fig.4(b)). The universes of discourse of these variables are fixed between -1 and 1 by introducing a simple gain for each variable.

Otherwise, the output variable HB_k is fuzzed also with seven fuzzy subsets: VVS is very very small, VS is very small, S is small, M is medium L is large and VL is very large, and VVL is very very large (Fig.4(c)). The universe of discourse of the variable HB_k is fixed between 0 and 1, so, it is necessary to introduce a gain G_{HB_k} at the fuzzy logic controller output.

Table 1. Fuzzy rules for the hysteresis bandwidth

HB_k		e_k				
		NB	NM	AZ	PM	PB
di^*_{ck}/dt	NVB	VVS	VS	VVL	VVL	VL
	NB	VS	M	VVL	VL	M
	NM	S	L	VVL	L	S
	AZ	S	VL	VVL	VL	S
	PM	S	L	VVL	L	S
	PB	M	VL	VVL	L	VS
	PVB	L	VL	VVL	S	VVS

Table 2. Main parameters of the simulated model

Power Source	$230Vrms, 50Hz$ $R_s = 50m\Omega$ $L_s = 0.15mH$	
Non linear Load	• Six-pulse current-source converter bridge with a firing angle=15. • Four-pulse current-source converter bridge between the b-phase and the neutral wire. • Single phase diode bridge between the c-phase and the neutral wire.	
Active Filter	• Capacitors C_1 and C_2	$5mF$
	• DC voltages V_{C_1} and V_{C_2}	$500V$
	• Inverter side inductance	$3mH$
	• Switching frequency	$10kHz$

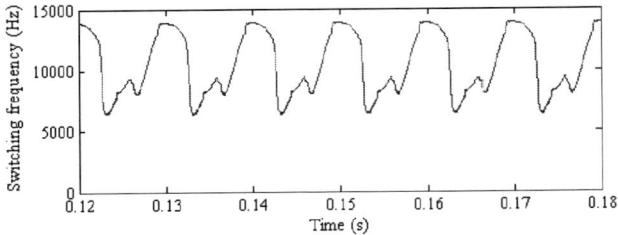

(a) Switching frequency (Conventional hysteresis band controller).

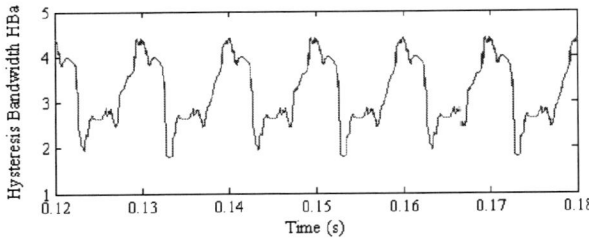

(b) Hysteresis bandwidth (Fuzzy hysteresis band controller).

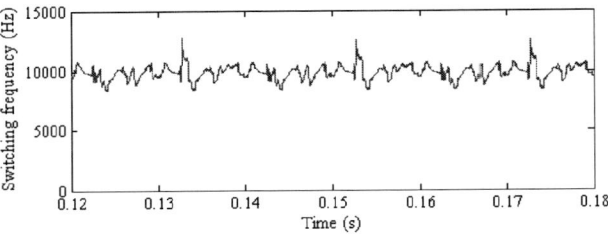

(c) Switching frequency (Fuzzy hysteresis band controller).

Fig. 5. Instantaneous switching frequency of the adaptive fuzzy hysteresis band current controller compared with a conventional fixed hysteresis band controller.

Thus, the rule base is expressed as a set of 35 $\mathcal{IF}...\mathcal{THEN}$...rules with the use of Mamdani's implication, these rule consigned in the Table.1.

Finally, the defuzzification which gives the real value of the instantaneous hysteresis band is realized by the centroid method.

5. SIMULATION RESULTS

The complete model of the system shown in Fig.1 is implemented and simulated using Matlab/Sumilink. The fuzzy logic controller is established using the Fuzzy Inference System Editor.

The main parameters of the simulated circuit are recapitulated in the Table2.

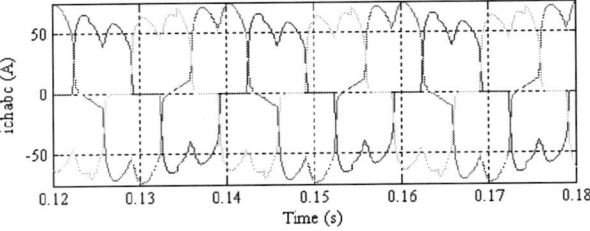

(a) The three-phase load current.

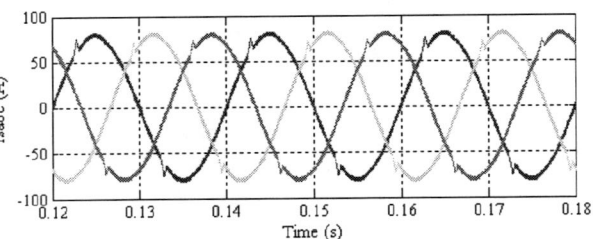

(b) The three-phase source current.

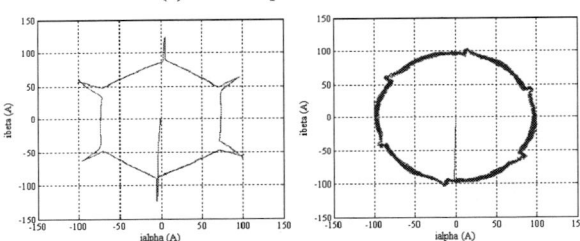

(c) Load current equivalent phasor.   (d) Source current equivalent phasor.

Fig. 6. Three-phase current compensation and balance.

Firstly, the performances of the fuzzy logic controller are shown in Fig.5. It makes in evidence the advantage of the fuzzy hysteresis band controller compared with a conventional fixed bandwidth one. In fact, the Fig.5(a) shows that the variations in instantaneous switching frequency in the case of a fixed bandwidth are important and it is to note that these variations are increasing with decreasing the hysteresis bandwidth. The instantaneous bandwidth established by the fuzzy logic controller is shown in Fig.5(b), and the instantaneous resulting switching frequency in Fig.5(c), we can remark that with varying the instantaneous bandwidth, the instantaneous switching frequency is remained constant around $10kHz$. This frequency can be increased or decreased by decreasing or increasing the gain at the fuzzy controller output.

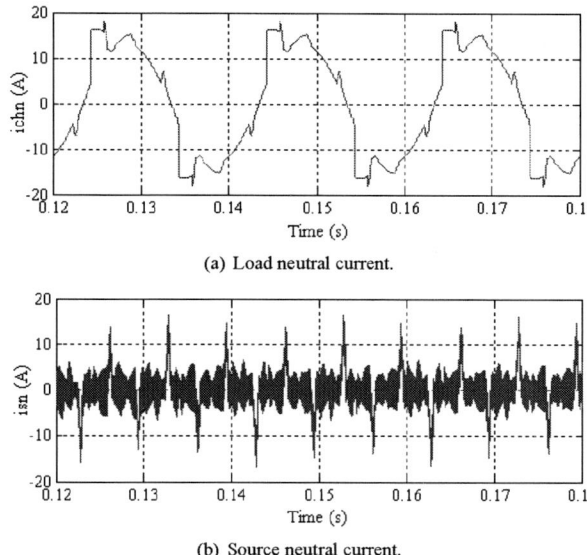

Fig. 7. Three-phase current compensation and balance.

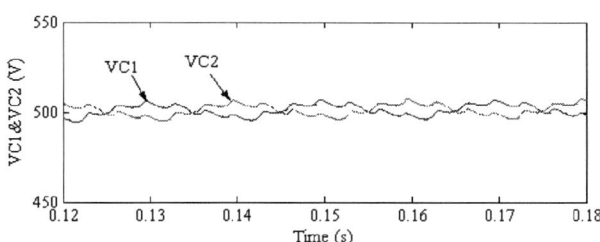

Fig. 8. *DC* Voltage regulation

Table 3. Main simulation results

	Load current		Source current	
	I_1	THD	I_1	THD
a-phase	74.55	32.17%	79.50	03.17%
b-phase	82.79	29.86%	79.41	02.49%
c-phase	90.49	26.02%	79.23	02.30%

The Fig.6 illustrate the waveforms of the load and the source current such as there corresponding equivalent phasors in (α, β) coordinate, this figure shows clearly that the harmonics and unbalance are correctly compensated, in fact the unbalance rates are 5% and 0.35% for the load and the source current respectively, therefore, the current in the neutral at source side is almost canceled as shows Fig.7.

Otherwise, for the DC voltage control, we can sea in Fig.8 that the voltage in the two capacitors are balanced and remained constant.

To show more clearly the performances of the proposed control, Fig.9 represents the harmonics spectrum of the load and source currents for each phase, and in Table.3, we resume the fundamental magnitudes and the total harmonic distortion ($THDs$) for each phase.

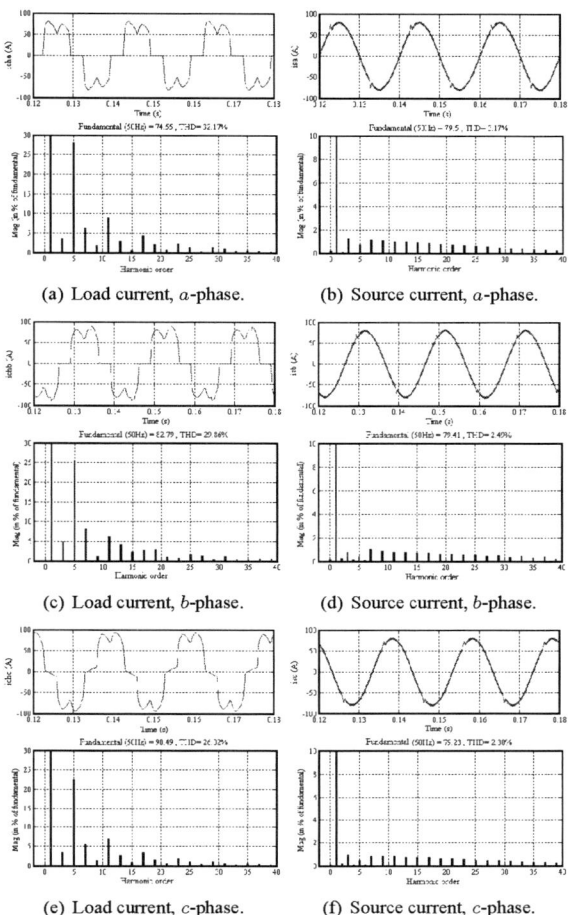

Fig. 9. Harmonics current spectrum of the load and source current.

6. CONCLUSION

In this paper, an adaptive fuzzy hysteresis band current controller for four-wire shunt active filter is proposed in order to achieve robust control with a moderate switching frequency. The fuzzy hysteresis controller was firstly described theoretically, and after verified and validated by simulation. In fact, the proposed fuzzy hysteresis controller presents very satisfactory results to compensate harmonics and to cancel neutral current, but its principal advantage is that in practical uses, this controller permits to optimize active filter parameters, therefore, to reduce active losses in the passive elements. However, it is important to note that a fuzzy hysteresis controller taking on account the DC voltage permits to improve the dynamic characteristics of the control and especially the transient state.

REFERENCES

[1] B.Singh, K.Al-Haddad and A.Chandra. A Review of Active Filters for Power Quality Improvement. *IEEE Transaction on Industrial Electronics*, vol. 46, no. 5, October 1999

[2] J.L.Afonso, H.J.Ribeiro da Silva and J.S.Martins. Active Filters for Power Quality Improvement. *2001 IEEE Porto PowerTech*, ISBN 0 7803 7139 9 Porto 2001.

[3] H. Akagi. Active Harmonic Filters. *Proceeding of the IEEE*, Vol.93, No 12, pp 2128-2141. December 2005.

[4] J. Tlust, P. Santarius, V. Valouch, J. Skramlyk. Optimal control of shunt active power filters in multibus industrial power systems for harmonic voltage mitigation. *Mathematics and Computers in Simulation*, 71 (2006) 369-376

[5] M. Aredes, J. Hfner, and K. Heulmann. Three-Phase Four-Wire Shunt Active Filter Control Strategies. *IEEE Transaction on Power Electronics*, Vol12, No.2, pp 311-318 March 1997.

[6] M. Ucar, E. Ozdemir. Control of a 3-phase 4-leg active power filter under non-ideal mains voltage condition. *Electric Power Systems Research*, xxx (2007) xxx-xxx

[7] B. R. Lin, H. K. Chiang and K. T. Yang. Shunt Active Filter with Three-phase Four-Wire NPC Inverter. *The 47th IEEE International Midwest Symposium on Circuits and Systems 2004*.

[8] D. Graovac, Vladimir K. Kati, and A. Rufter. Unified Power Quality Conditioner Based on Current Source Converter Topology., *EPE*, Graz 2001.

[9] L. G. Barbosa Rolim, D. R da Costa and M. Aredes. Analysing and Software Implimentation of a Robust Synchronizing PLL Circuit Based on the pq Theory. *IEEE Transaction on Industrial Electronics*, Vol 53 N6 December 2006.pp 1919-1926.

[10] M. Kale, E. Ozdemir. An adaptive hysteresis band current controller for shunt active power filters. *Electric Power System Research 73*, pp 113-119. 2005.

[11] B. Mazari, F. Mekri. Fuzzy hysteresis control and parameter optimization of a shunt active filter. *Journal of Information Science and Engineering 21*, pp 1139-1156.2005.

[12] M.N. Cirstea, A. Dinu, J.G. Khor. Neural and Fuzzy Logic Control of Drives and Power Systems. *Newnes edition First published 2002*.

Model Predictive Control of the Permanent Magnet Synchronous Motor in State Space with Input Constraints

Lh.Arab (*), A.Belemhedi (**), M. Aït Ahmed (***), N.Habani (*)

*Université Mouloud Mammeri de Tizi-Ouzou,
Département d'électrotechnique BP 17 RP, Algérie (e-mail: Arab_lh@yahoo.fr)
**Université de Bejaia, Algérie
***IREENA Université de Nantes, France*

Abstract: In this paper, speed control of permanent magnet synchronous motor (PMSM) using constrained model predictive in state space is presented. This model predictive control (MPC) minimizes a cost function which depends on tracking errors of speed, electrical current and control signal. The regulator thus synthesis is to deal with the system limits using the Lagrange multiplier approach.

Keywords: model predictive control, constraints, permanent magnet synchronous motor, Lagrange multiplier.

1. INTRODUCTION

During the control of the PMSM, to follow a certain required speed by industrial processes, and at the same time worked with a certain strategy (maximum of torque, to minimize the joules losses ...) represented by the current i_d in direct axis which is generally constant, we must respect the technology limits.

The electric actuator which is PMSM work in a field limited by many constraints (saturation of the magnetic circuit, commutation frequency of semiconductor, limited gradient speed and limited current...).

To deal with the constraint due to technological limits, anti-windup PI controller has been developed. Unfortunately this technique presents limitations in certain applications. Thus, we are interested by using an optimal control based on a criterion which takes account of the future trajectory. This will enable us to predict possible violation of the constraints, and to take them into account.

Therefore, we can formulate the control as being a problem of optimization of a quadratic criterion based on tracking errors of speed, electrical current and control signal [2].

2. MATHEMATICAL MODEL OF THE PMSM

In this paper, we consider the permanent magnet synchronous motor (PMSM). It is a three-phase salient stator and permanent magnet rotor. This kind of motors operates on the same principle as classical winding rotor synchronous motors. The mathematical model is based on dq–axis (Fig.1) and Park's transformation [1].

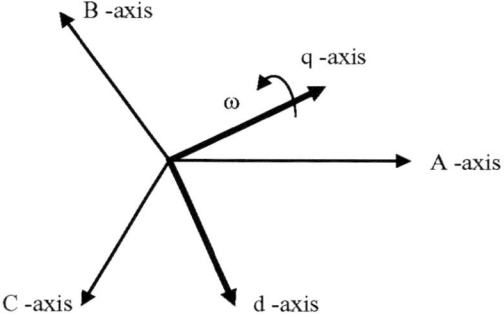

Fig. 1. Park's transformation

The (d) and (q) components of the current vector are calculated through a state transformation of the stator currents phases i_a, i_b and i_c, and of the position θ:

$$\begin{pmatrix} i_d \\ i_q \end{pmatrix} = \sqrt{\frac{2}{3}} \begin{pmatrix} \cos(\theta) & -\sin(\theta) \\ \sin(\theta) & \cos(\theta) \end{pmatrix} \begin{pmatrix} 1 & -\frac{1}{2} & -\frac{1}{2} \\ 0 & \frac{\sqrt{3}}{2} & -\frac{\sqrt{3}}{2} \end{pmatrix} \begin{pmatrix} i_a \\ i_b \\ i_c \end{pmatrix}$$

Where: $\omega = \dfrac{d\theta}{dt}$

The electrical equations of the permanent magnet synchronous motor in the rotor frame are as follows:

$$V_q = Ri_q + d\varphi_q/dt + P\Omega\varphi_d \qquad (1)$$

$$V_d = Ri_d + d\varphi_d/dt - P\Omega\varphi_q \qquad (2)$$

$$\varphi_d = L_d i_d + \Phi \qquad (3)$$

$$\varphi_q = L_q i_q \qquad (4)$$

Where V_d, V_q, i_d, i_q, R, L_d, L_q, Φ, P and Ω voltages and currents in the d-q reference frame, resistance, inductances, permanent magnetic flux linkage, number of pole pairs, and rotor speed respectively.

The mechanical equations are given by [4, 5].

$$C_{em} = P(\varphi_d i_q - \varphi_q i_d) \quad (5)$$

$$C_{em} - C_r = J\frac{d\Omega}{dt} + f\Omega \quad (6)$$

Where f is viscous friction coefficient, C_r is the load torque and J is the inertia.

By choosing (i_d, i_q, Ω) as state variables, the PMSM system can be written in state space as follows:

$$\begin{bmatrix} \frac{di_d}{dt} \\ \frac{di_q}{dt} \\ \frac{d\Omega}{dt} \end{bmatrix} = \begin{bmatrix} -\frac{R}{L_d} & P\frac{L_q}{L_d}\Omega & 0 \\ -\frac{L_d}{L_q}P\Omega & -\frac{R}{L_q} & -P\frac{\Phi}{L_q} \\ \frac{P}{J}(L_d - L_q)i_q & \frac{P\Phi}{J} & -\frac{f}{J} \end{bmatrix} \begin{bmatrix} i_d \\ i_q \\ \Omega \end{bmatrix}$$
$$+ \begin{bmatrix} \frac{1}{L_d} & 0 & 0 \\ 0 & \frac{1}{L_q} & 0 \\ 0 & 0 & -\frac{1}{J} \end{bmatrix} \begin{bmatrix} V_d \\ V_q \\ C_r \end{bmatrix} \quad (7)$$

In this paper, the available measurements are assumed to be the rotor speed (Ω), and the currents i_d and i_q.

To obtain a linear model, we can use a decoupling method by compensating the terms of coupling.
Then we can write:

$$\begin{cases} u_d = V_d - V_{dc} \\ u_q = V_q - V_{qc} \end{cases} \quad (8)$$

Where (u_d, u_q) are the news outputs and (V_{dc}, V_{qc}) the terms of coupling.

$$\begin{cases} V_{dc} = -P\Omega L_q i_q \\ V_{qc} = P\Omega L_d i_d \end{cases} \quad (9)$$

Finally we can write the linear continuous state equation of the PMSM (supposing that the salient is negligible) as follows:

$$\begin{bmatrix} \frac{di_d}{dt} \\ \frac{di_q}{dt} \\ \frac{d\Omega}{dt} \end{bmatrix} = \begin{bmatrix} -\frac{R}{L_d} & 0 & 0 \\ 0 & -\frac{R}{L_q} & -P\frac{\Phi}{L_q} \\ 0 & \frac{P\Phi}{J} & -\frac{f}{J} \end{bmatrix} \begin{bmatrix} i_d \\ i_q \\ \Omega \end{bmatrix} + \begin{bmatrix} \frac{1}{L_d} & 0 \\ 0 & \frac{1}{L_q} \\ 0 & 0 \end{bmatrix} \begin{bmatrix} u_d \\ u_q \end{bmatrix} \quad (10)$$

To obtain a discrete model, we use the Euler approximation by the introduction of sample time (T):

$$\frac{dx(t)}{dt} \rightarrow \frac{x(k+1) - x(k)}{T}$$

We can write the state discrete model of the PMSM as follows:

$$\begin{cases} x(k+1) = Ax(k) + Bu(k) \\ y(k) = Cx(k) \end{cases} \quad (11)$$

Where:

$$A = \begin{bmatrix} 1 & 0 & 0 \\ 0 & 1 & 0 \\ 0 & 0 & 1 \end{bmatrix} + T \begin{bmatrix} -\frac{R}{L_d} & 0 & 0 \\ 0 & -\frac{R}{L_q} & -\frac{P\Phi}{L_q} \\ 0 & \frac{P\Phi}{J} & -\frac{f}{J} \end{bmatrix};$$

$$B = T \begin{bmatrix} \frac{1}{L_d} & 0 \\ 0 & \frac{1}{L_q} \\ 0 & 0 \end{bmatrix} \text{ and } C = I = \begin{bmatrix} 1 & 0 & 0 \\ 0 & 1 & 0 \\ 0 & 0 & 1 \end{bmatrix}$$

3. BASIC PREDICTIVE CONTROL OF THE PMSM IN STATE SPACE

Linear model predictive control (MPC) is a generic term for computer control algorithms that utilize an explicit process model to predict future. An optimal input is computed by solving an open-loop optimal control problem over a finite time horizon, i.e. for a finite number of future samples. From the calculated input signal only the first element is applied to the system. This is done at every time step. The idea is thus to go one step at a time and check further and further ahead.

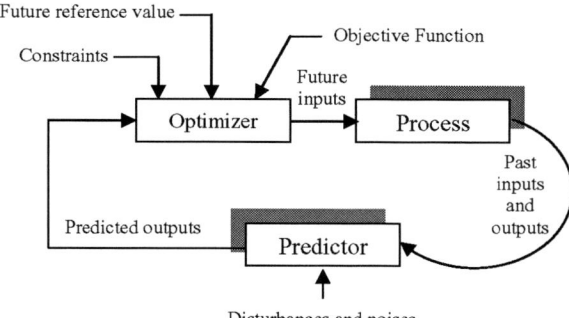

Fig. 2. Basic structure of the predictive control

These are the two most important advantages in the idea of predictive control [5]:
- The ability to take accounts of the explicit constraints under states and input variables.
- The ability to generalize to multi-variable systems.

The control $u(k)$ which must be applied to the system is calculated in order to have just after a sample time, the output $y(t:k+1) = y_r(t:k)$, expressed by the objective criteria (J) to be minimized where:

$$J = [y_r(k+1) - y(k+1)]^2 \quad (12)$$

The number of samples one looks ahead is called the prediction horizon N_2. In our MPC formulations a difference is made between prediction horizon and control horizon N_u. The control horizon is then the number of samples times that the optimal input is calculated for. With a shorter control horizon than prediction horizon the complexity of the problem can be reduced. Then the quadratic criterion to minimize becomes [3]; [4]

$$J = \sum_{j=1}^{j=N_2} [y_r(k+j) - y(k+j)]^2 \\ + R \sum_{j=1}^{j=N_u} [u(k+j-1) - u(k+j-2)]^2 \quad (13)$$

According to this criterion it is necessary to calculate the predicted output based on the model of the system and the reference in the future until the prediction horizon N_2. Rewriting of the output (y), until the moment ($t:k+N_2$), in the following matrix form with the performance criterion

$$Y = C_p.x(k) + G.U \quad (14)$$

$$Y = \begin{bmatrix} y(k+1) \\ y(k+2) \\ \ldots \\ \ldots \\ y(k+N_2) \end{bmatrix}; U = \begin{bmatrix} u(k) \\ u(k+1) \\ \ldots \\ \ldots \\ u(k+N_u) \end{bmatrix}$$

$$C_P = \begin{bmatrix} CA \\ CA^2 \\ \ldots \\ \ldots \\ CA^{N_2} \end{bmatrix}; G = \begin{bmatrix} CB & 0 & \ldots & \ldots & 0 \\ CAB & CB & 0 & \ldots & 0 \\ \ldots & \ldots & \ldots & \ldots & \ldots \\ \ldots & \ldots & \ldots & \ldots & \ldots \\ CA^{N_2-1}B & CA^{N_2-2}B & \ldots & \ldots & CB \end{bmatrix}$$

$$J(U) = [C_p x(k) + GU - Y_r]^t [C_p x(k) + GU - Y_r] \\ + U^t RU \quad (15)$$

$$R = \begin{bmatrix} r_1 & 0 & 0 \\ 0 & \ldots & 0 \\ 0 & 0 & r_{N_u} \end{bmatrix}$$

By minimizing the performances criterion (15) we obtain in matrix form the following input control:

$$U = (G^t.G + R)^{-1}.G^t (Y_r(k+1) - C_p.x(k)) \quad (16)$$

Knowing that $u(k+j) = 0; \forall j \geq N_u$

4. PREDICTIVE CONTROL DESIGN WITH CONSTRAINTS

In the following, we review the development of the constrained controller. Constraints are incorporated into the performance objective optimization. Generally we consider two types of the input constraints (17).

$$u_{\min} \leq u(k) \leq u_{\max} \quad (17)$$

$$-\Delta u_{\max} \leq \Delta u(k) \leq \Delta u_{\max} \quad (18)$$

If the constraints are respected then the real control at discrete time (k) is given by the first set of optimal control in equation (16). But if the constraints become active then they act at the optimum like equality constraints. Then we rewrite the constraints (17) and (18) in equality form as follows:

$$\begin{cases} g_1(U) = 0 \\ g_2(U) = 0 \end{cases} \quad (19)$$

In practice there are two types of constraints:
- Hard Constraints: they must be respected at any moment.
- Soft constraints: they must be respected if possible, if not they will be relaxed in the control algorithm.

In PMSM application the hard constraints, are the effective voltage limit V_l and the effective current limit I_l. We consider only input constrained in this work. So, without loss of generality we can define for our case the principal constraints equality:

$$g_1(U) = (V_q^2 + V_d^2) - U_l^2 = 0 \quad (20)$$

Where:

$$U_l = \sqrt{V_{q\max}^2 + V_d^2}$$

V_d remain within the acceptable limits because it fix a control torque strategy ($i_d = 0$). The optimal control problem then becomes minimization of the cost criterion (15) subjects to the constraint equality (20). From the objective function (15), we can construct an augmented cost objective L based on Lagrange multiplier [6]

$$L(x,U,\lambda) = J(x,U) + \lambda^t g_1(x,U) \quad (21)$$

Where $\lambda \in IR^\beta$ called the vector of the Lagrange multipliers, which is to be determined, β the number of constraints (in our case $\beta = 1$).

The Lagrange multipliers is a technique which makes it possible, to transform a problem of optimization under constraints, into another problem of a higher order but without constraints [7]; [8].

$$\begin{cases} \min_U J(x,U) \\ g_1(x,U) \end{cases} \Leftrightarrow \min_{U,\lambda} L(x,U,\lambda) \quad (22)$$

The constraint minimum is obtained when the gradient of L equals to zero, i.e.

$$\begin{cases} \dfrac{\partial J}{\partial U} = -\lambda^t \dfrac{\partial g_1}{\partial U} \\ U^2 = U_l^2 \end{cases} \quad (23)$$

From (23), the resultant optimal control with rate constraints can be derived as:

$$U = (G^t G + R + \lambda I_R)^{-1} G^t (Y_r(k+1) - C_p x(k)) \quad (24)$$

With I_R is matrix identity of dimension ($N_u \times N_u$) identical to R. The term λI_R acts exactly like a control cost weighting factor, penalizing the control not exceed the limiting value ($|U_l|$).

One thus finds the significance of the Lagrange multiplier, more it is high, more it far from the optimum without constraint.

Table 1. Parameters of the Machine

R	0.60	Ω
L_d	0.0014	H
L_q	0.0028	H
Φ	0.1194	Wb
J	0.0011	$kg.m^2$
f	0.0014	$Kg.m^2.s^{-1}$
P	4	

5. CONCLUSION

In this paper a various simulation tests have been executed to verify the quality of the proposed control and to prove validity and robustness of the controller. For the controller parameters, we set:

$$\begin{cases} N_2 = 500; N_u = 2, \lambda = 0.01 \\ r_1 = r_2 = 10^{-8}, T = 0.05ms \end{cases}$$

The speed response is shown in Fig.3 and Fig.4 which tracks the given trajectory. The current i_d remains practically equal to zero whereas i_q, which is the image of the torque, remains within the fixed limits.

Fig.5 illustrates some results of simulation by considering that all the system parameters vary from 100%.

The results of constrained controller (Fig.6) confirm that the limits are not violated. Whereas from the Fig.4 and Fig.5 we observe that the control signal of the unconstraint controller violates the amplitude limits this leads to larger overshoot by variation of the machines parameters and introducing the load torque with velocity inversion, severe mode with an important overshoot for manipulated variables and controls variables. Thus, it makes that the system remains in the zone of linearity. If not, the performances of the control decrease and even damage the machine. To guarantee the system to remain within acceptable limits for the input control we observe a little degradation of the performance in Fig.6.

Simulation results have shown that the proposed control strategy is effective in reducing overshoot and smother control input. Possible directions for future research include switching control based MPC where the decision must be taken to inject for the system two controls, one with constraints, another without constraints, which permit us to obtain a good dynamic performances, however we must take some precautions during transitions in order to ensure a system stability.

REFERENCES

[1] B.K. Bose (1990). *Power Electronics and Drives.* Prentice Hall, New Jersy.

[2] Lian-Bing Li, He-Xu Sun, Jian-Dong Chu, Guo-Liang Wang (2003). *The Predictive Control of PMSM Based on State Space.* Proceedings of the Second International Conference on Machine Learning and Cybernetics, IEEE.

[3] Kenneth Robert Muske, B.S. Che, M.S. (1995). *Linear Model Predictive Control of Chemical process.* Phd of the University of Texas at Austin.

[4] D.W. Clarke, C. Mohtadi and P.S Tuffs (1987). *Generalized Predictive control, Part I. Part II,* Automatica Vol.23 N°2 pp 137- 160.

[5] D.Q.Mayne, J.B. Rawling, C.V.Rao, P.O.M.Scokaert (2000). *Constrained model predictive control: Stability and optimality,* Automatica 36 pp 789-814.

[6] Andrey Alexandrovich Tyagunov (2004). *High-Performance Model Predictive control for process industry,* Phd, Eindhoven.

[7] Kay-Soon Low (2000). *Robust Model Predictive Control and Observer for Direct Drive applications,* IEEE Transactions on power electronics, Vol.15, N°3.

[8] Kay-Soon Low (2000). *Robust Model Predictive Control of Motor Drive with Control input Constraints,* IEEE Transactions on power electronics, Vol.15, N°3.

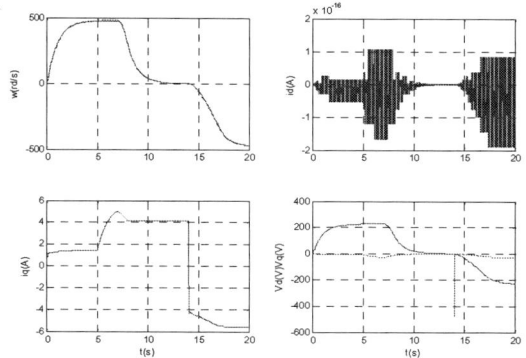

Fig. 4. Performances of the unconstrained predictive control (Cr=2Nm at t=5s).

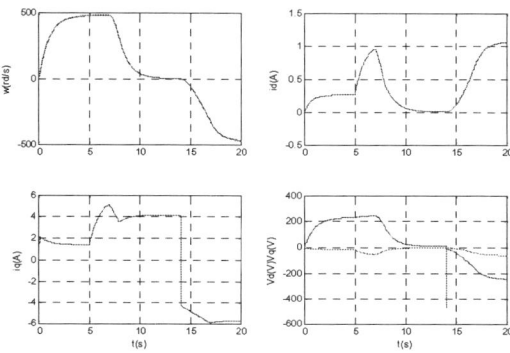

Fig. 5. Performances of the unconstrained predictive control (100% variation of the value parameters).

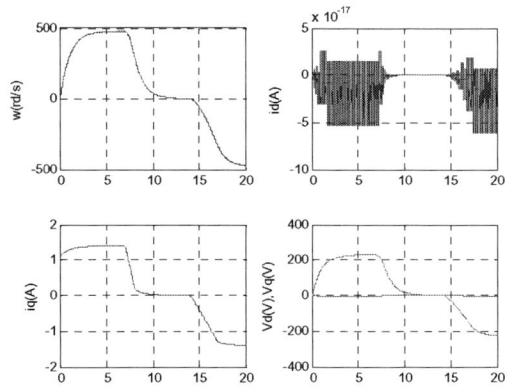

Fig. 3. Performances of the unconstrained predictive control without load.

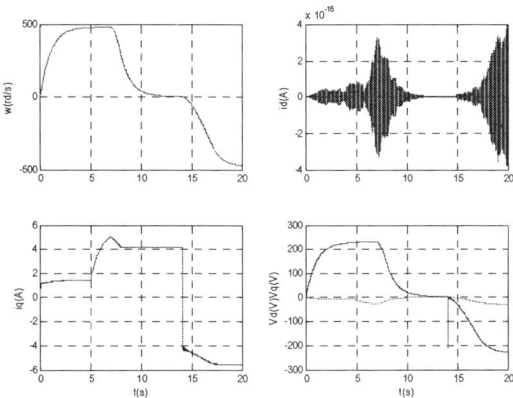

Fig. 6. Performances of the constrained predictive control.

SESSION D2: INTELLIGENT CONTROL

FPGA Implementation of Multilayer Perceptron for Modeling of Photovoltaic panel

H. MEKKI*, A. MELLIT**, H. SALHI***, and K. BELHOUT****

* Department of Electronics, RASIC laboratory BLIDA UniversityBLIDA, Algeria (Mekkihamza@yahoo.fr).
** Department of Electronics, Control laboratory BLIDA UniversityBLIDA, Algeria (a.mellit@yahoo.co.uk).
*** Department of Electronics, Control laboratory BLIDA UniversityBLIDA, Algeria (h_hassan@yahoo.com).
**** Department of Electronics, RASIC laboratory BLIDA UniversityBLIDA, Algeria (bk_elec@yahoo.fr).

Abstract: The Number of electronic applications using artificial neural network-based solutions has increased considerably in the last few years. However, their applications in photovoltaic systems are very limited. This paper introduces the preliminary result of the modeling and simulation of photovoltaic panel based on neural network and VHDL-language. In fact, an experimental database of meteorological data (irradiation, temperature) and output electrical generation signals of the PV-panel (current and voltage) has been used in this study. The inputs of the ANN-PV-panel are the daily total irradiation and mean average temperature while the outputs are the current and voltage generated from the panel. Firstly, a dataset of 4x364 have been used for training the network . Subsequently, the neural network (MLP) corresponding to PV-panel is simulated using VHDL language based on the saved weights and bias of the network. Simulation results of the trained MLP-PV panel based on Matlab and VHDL are presented. The proposed PV-panel model based ANN and VHDL permit to evaluate the performance PV-panel using only the environmental factors and involves less computational efforts, and it can be used for predicting the output electrical energy from the PV-panel.

Key Word: FPGA, MLP, VHDL, photovoltaic panel.

1. INTRODUCTION

Solar energy conversions has various advantages such as short time duration of installation and long life of exploitation, circuit simplicity, no need of moving part and realize a salient, safe, not pollutant an renewable source of electricity. The wide acceptance and utilization of the photovoltaic (PV) generation of electric power depends on reducing the cost of the power generated and improving the energy efficiency of PV systems. In recent years, it has been shown that artificial neural networks (ANN) have been successfully employed in solving complex problems in various fields of applications including pattern recognition, identification, classification, speech, vision, prediction and control systems [1]. Today ANNs can be trained to solve problems that are difficult for conventional computers or human beings. ANNs, overcome the limitations of the conventional approaches by extracting the desired information directly from the experimental (measured) data. However, the application of the ANN in PV-systems are very limited, in literature there are several models that are proposed for modeling and simulation of PV-panel for evaluating the performance and the power quality in the PV-panel [2,3]. The performances of a PV module are strictly dependent on the environmental factors such as irradiance and cell temperature. These two parameters make the PV module maximum power change and especially for PV-integrated power system dispatchers it is necessary to be able to predict accurately the PV output [3]. It is very difficult to develop an accurate PV-panel model based on analytical models or numerical simulations; due to the influencing of the environmental factors. In fact, some very recent works were developed for modeling a PV-system based on ANN [4-10].

The main objective of this work is to use of the ANN and VHDL-language for modeling and simulation PV-panel. The proposed model allows us to evaluate the performance of the PV-panel based only on the meteorological data such as mean average temperature and daily total solar radiation. This paper is organized as follows: the next section introduces the PV-panel and presents the database used in this study, section III provides the ANN architecture used for modeling of the PV-panel, also, the same section describe the simulation of the PV-panel based on ANN and VHDL. Results and discussion are presented in section 4.

2. PHOTOVOLTAIC PANEL AND DATABASE

The efficiency of solar energy conversion is related both to the optimal design and to the maximum power extraction from PV system. The PV-array used in this simulation consists of 16 ITALSOLAR modules. Each module comprises 30 square single crystal silicon cells. The total peak power of each PV array is 720 We, the PV-array voltage is 40 V (max) and the PV-array current is 20 A (max). The area of PV generator is 6 m². The traditional I-V characteristics of a solar array, when neglecting the internal shunt resistance, are given by the following equation [2]:

$$I_0 = I_g - I_{sat}\left\{\exp\left[\frac{q}{AKT}(V_0 + I_0 R_s)\right] - 1\right\} \quad (1)$$

where I_o, V_o are the output current and output voltage of the solar array, I_g is the generated current under a given insolation, I_{sat} is the reverse saturation current, q is the charge of an electron, k is the Boltzmann constant, A is the ideality factor for a P-N junction, T is the array temperature, R_s and R_{sh} are the intrinsic series and shunt resistances of the solar array.

Figure 1 shows the different signal recorded form the above PV-panel, there signals are the mean temperature, daily irradiation, the output PV-current and PV-voltage generated.

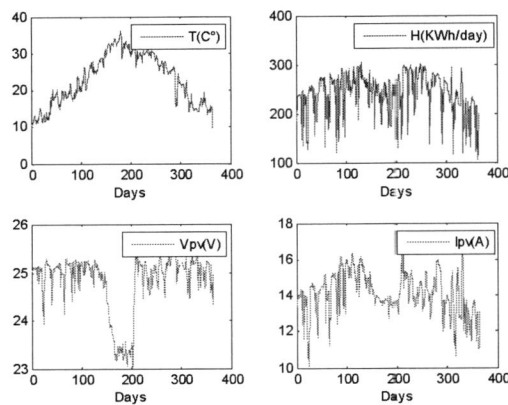

Fig. 1. the database of T, H, V_{pv} and I_{pv}

3. METHODOLOGY

3.1 Multi-layer Perceptron (MLP) network

Back-propagation (BP) has been widely adopted as a successful learning rule to find the appropriate values of the weights for NNs. The MLP consists of various layers: an input and output ones between which lie one or several hidden ones whose outputs are not observable. These layers are based upon some processing unit (neurons) interconnected by means of feed-forward pondered links (figure 2-a) [11]. All these processing units carry out the same operation (figure 2-b): i.e. the sum of their weighed inputs (See "(2)"). Then they apply the result to a non-linear function named activation function and generally based upon the sigmoid function (see "(3)").

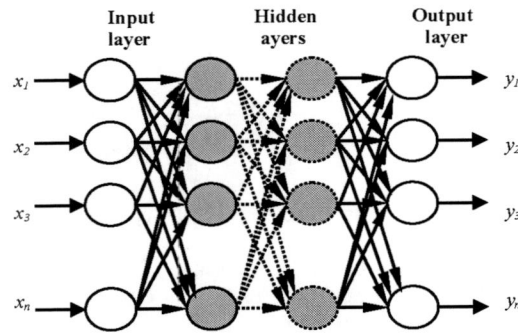

Fig. 2.a. Schematic diagram of a multi-layer feed-forward neural network.

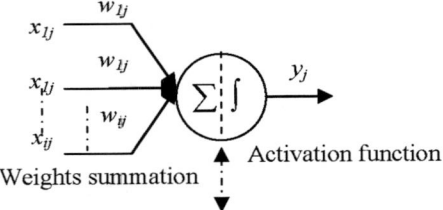

Fig. 2.b. Information processing in a neural network unit.

$$y_j = f\left[\left(\sum_i w_{ij}.x_{ij}\right) - bj\right]. \quad (2)$$

$$f(x) = \frac{1}{(1 + e^{-x})}. \quad (3)$$

where y_j, is the output of the processing unit w_{ij}, the synaptic weight coefficient of the i-th input of the processing unit b_j, is the bias.

3.2 MLP-Based modeling and simulation of PV-panel

Figure 3 shows the MLP configuration proposed for modeling and simulation of the PV-panel, the input of this model are the daily total solar radiation (H) and the mean average temperature (T) while the output are the generated current (I_{pv}) and voltage (V_{pv}). The input and the output are fixed initially however the number of hidden layers and the neurons within these layers are optimized during the learning process based on the good performance of MSE. A database of 365*5 patterns for each signals are divided in two parts, a dataset of 365*4 (4-years) patterns used for training the proposed MLP-PV panel model and a dataset of 365*4 (1-year) patterns used for testing and validation of the model. A soft computing program has been implemented for MLP-PV panel based on the LM-algorithm, the Matlab Ver 7.1b are used in this simulation.

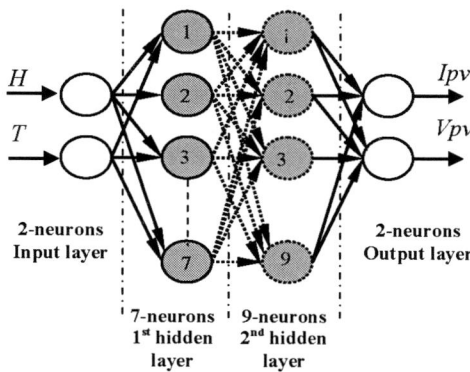

Fig. 3. MLP-PV panel model

3.3 VHDL-implementation of the MLP

Once the MLP-PV panel model is optimized, the weights and bias are saved in order to be used for implementation of the MLP based on VHDL language. The structural description of a neuron with VHDL allows during the compiling step to specify generically some characteristics like the input numbers and the data size, and even to modify the type of some arithmetic operators like the adder or the multiplier and eventually the activation function unit if one wants to modify the activation function. The neuron computes the product of its inputs, with the corresponding synaptic weights, which are memorized in an ROM, and then the results are added. The result is presented to a comparison unit designed to represent an appropriate sigmoid function. Then the output of this comparison unit is used as a controller for a multiplexing unit, which delivers they output of the neuron (See fig. 4) [11].

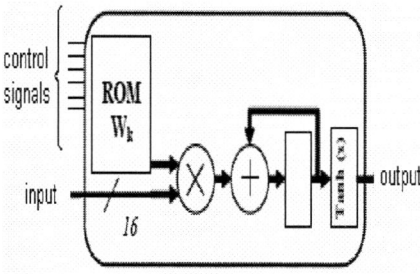

Fig. 4. Neural processor architecture

4. RESULTS AND DISCUSSION

Firstly, we present the results of MLP-PV panel model based on Matlab simulation. Figure 5 shows a comparison between measured and estimated output generating current and voltage of the PV-panel.

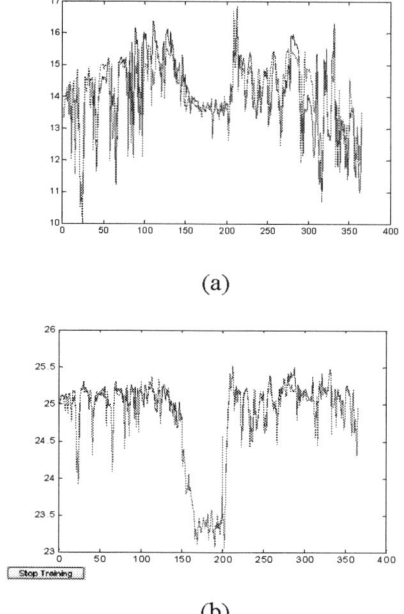

(a)

(b)

Fig. 5. Comparison between Measured and estimated output signal of
a. The current, b. the voltage

It should be noted there is a very good agreement between measured and estimated signal (I_{pv}, V_{pv}) in addition the coefficient of determination R^2 is 97% which is very satisfactory. In order to simulate the PV-panel model on VHDL language, the data was coded on 18 bits in fixed-point [12].

Function activation is very difficult in its known expression Tansig) "(3)". In order to simplify function expression, it was linearized on several intervals [C_i, C_{i+1}] and its value is evaluated using two constants (a_i and b_i) corresponding to this intervals [13] "(4)"and "(5)".

$$f(x_i) = a_i.x_i + b_i \quad \text{For } x_i \in [c_i, c_{i+1}]. \tag{4}$$

$$f(x) = 1 \quad \text{for x > 3.} \tag{5}$$

Figure 6.a presents the result of the activation function simulation. The obtained results by MODELSIM has been tested and compared with Matlab, which gives a good accurate result (See fig.6.b).

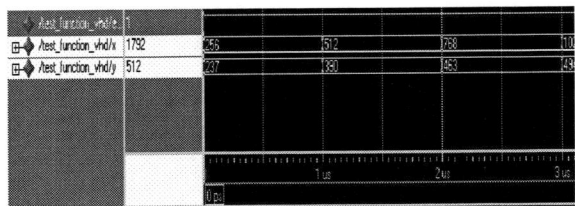

Fig. 6-a. VHDL Simulation of the activation function

The MODELSIM has been used for simulating the proposed architecture based on the VHDL, once the architecture is simulated. Figure 8 shows a comparison between Matlab and VHDL (ModelSim) estimation. the next step consists the implementation of this architecture on FPGA hardware.

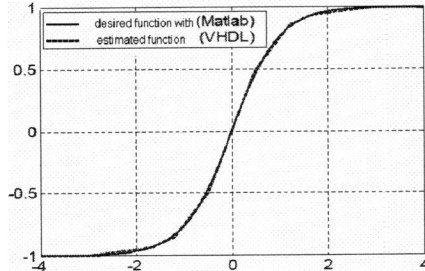

Fig. 6-b. Comparaison between activation function calculated by Matlab and simulated by MODELSIM (VHDL)

The network is composed of several elements of basis that are the neurons; this network is presented in fig. 7.a

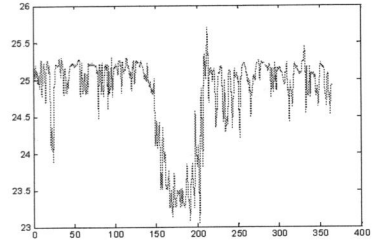

Fig. 8.a. comparison between MATLAB and VHDL estimation of a voltage

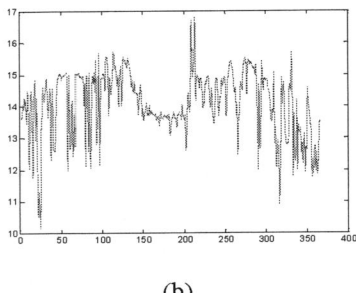

(b)

Fig. 8.b. Comparison between MATLAB and VHDL estimation of a. current

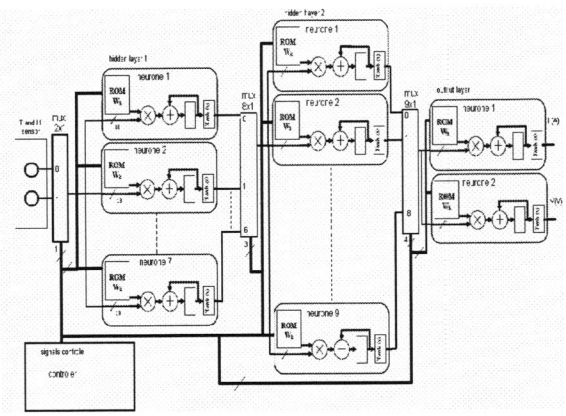

Fig. 7.a. Global architecture of the neural network

The simulation result based on VHDL is presented in figure 7.b, where X1 is the mean average temperature, X2 is the daily total solar radiation, Y1 is the output voltage and Y2 is the output current from the PV-panel.

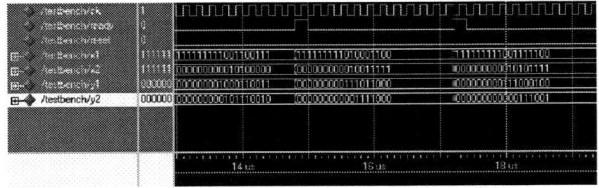

Fig. 7.b. Simulation results based on VHDL

5. CONCLUSION

In this paper, a PV-panel has been modeled and simulated based on ANN and VHDL-language. The accuracy and the generalization of the neural model in the system prediction are demonstrated by comparing test results with actual data. The advantage of the proposed model that it can be used for estimating the performance of PV-panel and it is able to predict the output electrical energy generation from the PV-panel based only on the environmental data. In addition, the using of the VHDL language for PV-panel simulation has been demonstrated. The modelsim has been used for simulate this network. Actually, it remains the implementation of this architecture on a FPGA hardware, that will be our objective in the future paper.

REFERENCES

[1] S.A. Kalogirou "Artificial neural networks in renewable energy systems applications: a review". Renewable and Sustainable Energy Reviews, vol.5, pp.373-401, 2000.

[2] A. Gow, C.D. Manning, "Development of a photovoltaic array model for use in power electronics simulation studies", IEE Proceedings on Electric Power Applications, Vol. 146, No 2, March pp. 193 -200. 1999.

[3] W. T. Jewell and T. D. Unruh: "Limits on clouds induced fluctuation in photovoltaic generation". IEEE Transactions on energy conversion, Vol. 5, N. 1 ,March, pp 8-14, 1990.

[4] TF. Elshatter M.T, Elhagree M.E Aboueldahab and. A.A, Elkousry "Fuzzy modeling and simulation of photovoltaic system". In 14th European photovoltaic Solar Energy Conference, 1997.

[5] M. Abdulhadi A. M, Al-Ibrahim and G.S Wirk. "Neuro-fuzzy based solar cell models". IEEE Transactions on Energy Conversion, vol.19 n3,pp. 619-629,.2004.

[6] L. Zhang and Y. Fei Bai "Genetic algorithm-trained radial basis functions neural networks for modelling photovoltaic panels". Engineering application of artificial intelligence;Vol.18: pp.833-844. 2005.

[7] A. Mellit, M Benghanem. A. Hadj Arab and A Guessoum. "Prediction and modeling signals from the monitoring of stand-alone PV systems using an adaptive neural network model". In Proceedings of The 5th ISES European Solar Conference. 2004, vol.3 :pp.224-230.

[8] A. Mellit and M., Benghanem "Modelling and simulation of stand-alone photovoltaic power system using artificial neural network". in Proceedings of the Solar Word Congress ISES 2005 August 6-12 2005 Florida, USA 2005.

[9] A. Mellit and S.A. Kalogirou "Neuro-fuzzy based modelling for photovoltaic power supply (PVPS) system". In First International Power and Energy Coference PECon IEEE, Nov. 28-29, 2006, Malaysia Vol.1, pp-88-93

[10] A Mellit, M, Benghanem and. S.A. Kalogirou "Modelling and simulation of a stand-alone photovoltaic system using an adaptive artificial neural network". Renewable Energy, vol.32, pp.285-313. 2007

[11] Y. Taright, M.l Hubin "FPGA implementation of a multilayer Perceptron Neural Network using VHDL" In Proceedings of ICSP pp-1311-1314, 1998

[12] E.M. Ortigosa a , A. Canas a, E. Ros a, P.M. Ortigosa b, S. Mota a, J. Dı́az a "Hardware description of multi-layer perceptrons with different abstraction levels " Elsevier .Microprocessors and Microsystems 30 (2006) 435–444

[13] F. Smach, M. Atri, J. Mitéran and M. Abid "Design of a Neural networks Classifier for Face Detection" transactions on engineering, computing and technology v5, 2005 issn 1305-5313

Hybrid Approach to Reinforcement Learning

Brahim BOULEBTATECHE, Mourad FEZARI, Mohamed BOUGHAZI

Electronics Department, Faculty of Engineering,
Badji Mokhtar University, Annaba BP.12, Annaba, 23000, Algeria
bbouleb@yahoo.fr, mohamed.fezari@gmail.com, boughazi@leri.univ-reims.fr

ABSTRACT: Reinforcement Learning (RL) is a general framework in which an autonomous agent tries to learn an optimal policy of actions from direct interaction with the surrounding environment (RL). However, one difficulty for the application of RL control is its slow convergence, especially in environments with continuous state space. In this paper, a modified structure of RL is proposed to speed up reinforcement learning control. In this approach, supervision technique is combined with the standard Q-learning, a model-free algorithm of reinforcement learning. The a priori information is provided to the RL by an optimal LQ-controller, used to indicate preferred actions at intermittent times. It is shown that the convergence speed of the supervised RL agent is greatly improved compared to the conventional Q-Learning algorithm. Simulation work and results on the cart-pole balancing problem are given to illustrate the efficiency of the proposed method.

Keywords: Supervised Reinforcement Learning, Autonomous Agents, LQ-controller, Machine Learning.

1. INTRODUCTION

In reinforcement learning paradigm, the agent learns its environment through trial-an-error interactions (Sutton et al.,1998). Learner is not told which actions to take, but gets reward/punishment from environment and learns the action to perform the next time. For each action it executes the environment returns a reward indicating how appropriate the action was in the given situation. This paradigm is well suited for learning on many domains where it is inappropriate to specify in an explicit way how to perform a task, e.g. navigating in unknown environment. After each action, it receives from the environment a scalar signal called reinforcement (reward / punishment) signal that inform on the appropriateness of taking a particular action in a given state. The goal of RL is to construct an optimal policy of actions for the agent to follow based on observed interactions with the environment. The agent is, thus, trained so that the long-term return of the expected sum of instantaneous reinforcement rewards is maximized.

However, standard RL algorithms are faced with a fundamental problem known as the curse of dimensionality. Although, many tasks defined over a finite state space can be dealt with successfully in this framework, in real applications, it would take an enormous amount of time for these algorithms to converge towards a suitable solution. There are two major approaches that address the problem of slow convergence in large finite state space or that try to find solutions to problems that seem intractable in complex environments with infinite set of states. The first approach is to apply generalization techniques, which involve approximations of the value function or some tiling of the state space (Sutton et al.,1998) (Bertsekas et al.,1995). The second approach is to provide the agent with a priori information about the environment. We can incorporate such knowledge either by modifying the reward function as in the reward shaping techniques (Dorigo et al.,1997) or we can create macro-actions from primitive ones as in (Mc Govern et al.,1997). A third approach, based on the use of supervision techniques, has been followed by a number of researchers to overcome some of the difficulties that may arise in previous methods. For instance, many techniques that combine the two concepts of supervised learning and reinforcement learning are well established in robotics and social sciences. Among them, we may cite imitation learning, LQ controller induction, and learning by demonstration, (Price and Boutilier, 2003) (Huber and Grupen, 1997) (Dixon et al., 2000). Despite the many successful implementations, none of these methods combines both kinds of learning instantaneously. Either supervised learning precedes RL during a separate training phase, or else the supervisory information is used to modify a value function rather than a policy.

In this paper, we propose a supervised approach to the classical Q-Learning algorithm of the RL paradigm (Watkins, 1989) (Watkins, 1992). The structure of the learning problem is modified by incorporating some a priori knowledge provided by a supervisor during the training phases. The supervisor adds structure to the classical framework for RL by integrating on-line a priori information or advices to the learning agent and that may help speeding-up learning tasks. An example is described to illustrate ideas behind our algorithm by using a feedback LQ controller that is easily designed yet sub-optimal to help an RL controller selecting appropriate actions for balancing a nonlinear system (an inverted pendulum on a moving cart). The remaining part of the paper is organized as follows: Section 2 provides a succinct review of the RL principles used to solve typical class of problems represented by MDPs models. Section 3 focuses on the use of supervisory techniques as applied to RL. This structure seeks to improve the learning rate by incorporating added knowledge to Q-learning algorithm during the training process, i.e., on-line. In sections 4 is described in more details the insertion of an LQ supervisor into the RL structure for learning how to balance an inverted pendulum on a cart. Finally, a conclusion is drawn in section 5 and avenues for future work are identified, therein, as well.

2. REVIEW OF REINFORCEMENT LEARNING

RL is the problem of learning an optimal behavior from direct interaction with an environment. Following is a succinct account of a reinforcement learning framework described from a Markov Decision Process (MDP) perspective.

The interaction between the learner and the environment having a MDP structure can be fully described by a finite set of states S , a finite set of actions A , and a real valued reward function, r (s ,a) : S × A → R . At some discrete time step t ∈ T the learner is in some state s_t ∈ S where it can choose to perform an action a_t ∈ A_s, where A_s ⊂ A is the set of available actions in state s. Upon execution of action a_t , it may result a state transition whereby the learner will find itself in state s_{t+1} with probability P(s_t , a_t , s_{t+1}), which is referred to as the transition probability. Arriving in state s_{t+1} the learner receives a reward r_t from the environment. Whereby the reward function gives an indication of the immediate utility of taking action a in state s and then following some policy. A policy π is defined as a mapping from states to actions. An optimal policy is a policy that optimizes some function of reward (either maximizing gain or minimizing cost) in the long run. Furthermore, it is helpful to define a real valued function Q (s, a): S × A → R, named the action-value function. The goal of RL is to derive the optimal action-value function, Q^* (s , a) from these interactions, for taking action a in state s and terminating in state s':

$$Q^*(s, a) = E\{r(s,a) + \lambda \max_{a'} Q^*(s', a')\} \quad (1)$$

where λ ∈ [0 , 1]is the discount factor, and E is the expectation operator. An iterative version of the optimal action-value function is given by the Q-Learning algorithm (Watkins, 1992). All state-action pairs are stored in a table, and their update takes place based on experiences (s, a , r , s') according to :

$$Q(s,a) \leftarrow \alpha(r(s,a) + \lambda \max_a Q(s',a')) + (1-\alpha)Q(s,a) \quad (2)$$

where α is the learning rate and gives a trade-off between a new observation and the present approximation . When the convergence of the Q-Learning algorithm is reached, the policy that it defines (referred to as the greedy policy) is simply obtained by taking actions with maximum value for the current state, given by :

$$\pi(s) = \operatorname{argmax}_a Q(s, a) \quad (3)$$

However, there are some problems associated with using Q-Learning on complex environments defined by infinite state/action space. Learning how to act in such domains is not guaranteed to converge to an optimal policy. Therefore, a modified structure of RL is well needed in order to effectively adapt to complex problems. As stated earlier, among many techniques investigated to alleviate the RL deficiencies, methods that, somehow, try to combine supervisory information and RL form a natural trend in machine learning. This will be considered with more details in the next section.

3. SUPERVISED REINFORCEMENT LEARNING

Most work on reinforcement learning has concentrated on the tabular case, in which the agent has little or no prior knowledge when it begins to act and learn in a finite state space. It is well known that in all branches of machine learning that, without some significant bias (or prior knowledge), learning cannot be efficient or effective. In reinforcement learning, the problem is even worse: if an agent starts with no knowledge at all and begins to act at random, it may take an extremely long time for the agent to even encounter the parts of its environment from which it can learn. For this reason, we must find ways to introduce additional knowledge into reinforcement learning agents.

One of the most appealing methods, which is widely used in nature, is for an agent to learn by watching and imitating other agents. This is a very practical way for an agent to gain additional knowledge and it has been the basis for some very successful robot learning programs. However, perception remains a big difficulty; most robots do not have sufficiently advanced perception to be able to sense what other robots are doing and whether they are succeeding.

The least attractive, but perhaps most immediately practical, method is for prior knowledge to be given directly to the robot. This knowledge might be in the form of partial or incorrect programs, program decompositions, or local reinforcement function. This initial knowledge, even if it is partial or sub-optimal, may guide the robot to behave well enough initially to learn effectively from its environment.

Supervised reinforcement learning as another method for improving the effectiveness of learning is a rule rather than the exception in natural life. With this approach, a supervisor adds structure to a learning problem and supervised learning makes that structure part of the reinforcement learning framework. The supervisor agent or trainer may intervene at different levels of abstraction during the learning process: at state level by identifying some special states as sub-goals (Dietterich,2000), at reward level by modifying the reward function as in reward shaping techniques, or at a more abstracted level by providing macro-actions or partial behaviors to the learning agent (Precup,2000). Figure 1 illustrates the general idea behind the supervised RL paradigm. Initially, the learning agent starts exploring the environment by executing some actions based on trial-and-error interaction. After each action taken, it receives from the environment a scalar signal that indicates the value of that action with respect to the given task goal to be accomplished. To speed-up the learning phase and reducing the exploration time needed to learn about the environment, a supervisor is used to provide the learning agent with pertinent information. Thus, the learning agent is instructed to take some privileged actions in particular situations, as specified by the supervising agent. In doing so, integrating a priori information into the learning agent policy leads to effectively directing search during learning and allowing relatively quick convergence to successful control policies.

A key feature of reinforcement learning is the exploration/exploitation trade-off. The choice of actions, between exploration of new actions or exploitation of previous optimal actions, can have significant effect on the behavior of the learner.

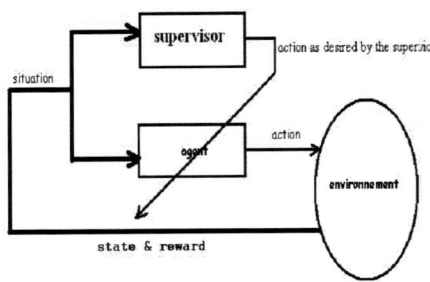

Figure 1. Supervised RL structure

In our case, the supervised reinforcement learning is carried out by indicating to the RL agent, through a human operator, some interesting parts of the environment state space , known as 'way-point ' states, for guidance purpose . Whenever the agent reaches one of these states it breaks down its usual way of selecting actions (for instance, ε-greedy policy is commonly used) and proceed its search by following related actions fixed by the supervisor agent. Another method that we also implemented, here, is based on a direct training of the learning agent by an LQ-controller (Linear Quadratic controller) used as a teacher. The learning agent try to imitate direct actions of the controller by passively observing the control signal of the latter. Thus, the agent's action policy follows, loosely speaking, the same trend as that of the controller. The reason for choosing this form of supervision is that incorporating the trainer's action, somehow, directly into the learning agent's building policy will not affect the underlying structure of the reinforcement learning process. Furthermore, due to the generalization capability of the RL, it is believed that the RL agent will well generalize to a wide range of conditions than the LQ-controller which can only perform robustly under limited conditions (i.e. , constrained to work only near around the equilibrium state of the linearized model of the pendulum-cart system). The conventional RL is slightly modified at the level of the search procedure in the action space: it uses ε-greedy policy, i.e. an action leading to the best estimate of $Q(s, a)$ function is chosen most of the time (with a probability equal to $1 - \varepsilon$). However, a small fraction of the time, ε , an action is given by the LQ-controller instead of being selected randomly from the action space. The following section describes in more detail the application domains used to implement and illustrate the effectiveness of the proposed approaches.

4. EXPERIMENTAL WORK AND RESULTS

The objective of these experiments is to compare the performance of supervised RL as implemented by our methods with that of conventional Q-Learning algorithm. The standard problem we used for this purpose is often utilized for testing learning algorithms performance on physical systems: the balancing task of an inverted pendulum on a moving cart.

4.1 Inverted Pendulum Balancing Task

In this experiment, we apply our algorithm to the standard cart-pole learning task, a.k.a. inverted pendulum problem, which involves learning to stabilize the upward equilibrium state of the inverted pendulum mounted on a mobile cart by applying appropriate forces to the cart, see Figure 5. The moving cart carries a pendulum that can swing freely around its origin. This system is described by the following nonlinear dynamic state equation:

$$\ddot{\theta} = \frac{g.\sin\theta + \cos\theta.\left[\dfrac{-F - m_p.l.\dot{\theta}^2.\sin\theta + \mu_c\, sign(\dot{x})}{m_c + m_p}\right] - \dfrac{\mu_p.\dot{\theta}}{m_p.l}}{l.\left[\dfrac{4}{3} - \dfrac{m_p.\cos^2\theta}{m_c + m_p}\right]} \quad (4)$$

$$\ddot{x} = \frac{F + m_p.l.\left[\dot{\theta}^2.\sin\theta - \ddot{\theta}.\cos\theta\right] - \mu_c\, sign(\dot{x})}{m_c + m_p} \quad (5)$$

The position and velocity of the cart (x , $\dot{x}$) and the pole angle and angular velocity (θ , $\dot{\theta}$) represent the state of the cart-pole system. The specification for the simulated pole and cart are as follows: length of the track, L = 4.0 m ; mass of the cart , m_c = 1.0 kg ; mass of the pole : m_p = 0.1 [kg] ; half length of the pole : l = 0.5 [m] ; gravity : g = 9.8 [m/s2]; coefficient of friction of cart on track : μ_c = 0.0005 ; coefficient of friction of pendulum on cart : μ_p = 0.000002

Only three actions F = {- 10 N, 0, + 10 N} were available to the RL agent for this task.

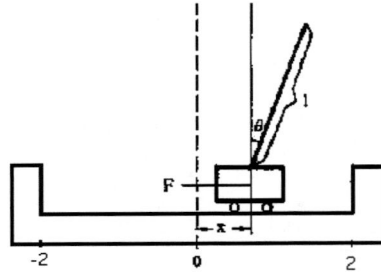

Figure 2. The cart-pole system

The proposed algorithm uses an LQ-Controller for supervisory purposes. This controller is applied to the linearized model of the cart-pole system by using Euler approximation technique with a time step of 0.02 s and intervene only during the balancing task of the pendulum (for $|\theta| \leq 15°$). The state space is partitioned into boxes of equivalent states (12705 boxes were used for the swinging up task and 270 boxes for the balancing task).

The whole system (RL agent , pole-cart model and LQ-controller) was implemented in Matlab. The Q-Learning algorithm utilizes the same parameters as indicted in the first experiment. At the beginning of each episode, the pendulum is initialized in downward position (angle $\theta = 180°$) with zero angular velocity. The agent has three available actions corresponding to forces of - 10, 0, +10. These actions are chosen small enough so that the only way to move the pendulum higher and higher is by swinging back and forth. The objective is to find a policy that will drive the pendulum past the upward position (the unstable equilibrium state) and maintaining the pendulum in that position for as long as possible.

As is illustrated by the learning curves depicted in Figures 3 through 6, the supervised RL agent gives better results in terms of convergence speed than those obtained without any supervision, and in terms of robustness to perturbation compared to the LQ-Controller. Figure 3 shows that the learning performance of the supervised agent is higher compared to the unsupervised case. The obtained curves are plotted by averaging produced success during 50000 learning episodes. Results for the pole balancing learning problem are depicted in Figure 4 where it is shown that the proposed agent has more success in maintaining the inverted pendulum in the vertical position within an angle less than 15°. If we looked at the magnitude of command signal that are issued by the RL agent and the LQ-controller we can easily notice on Figure 5 that the latter produce action signals of very high magnitude (much more greater than ± 25 N at the beginning of the balancing task) whereas actions of the former are limited to ± 10 N only. Thus, if the command signal is constrained to acceptable values, i.e. ± 10 N for instance, it will take a longer time for the LQ-controller to stabilize the inverted pole.

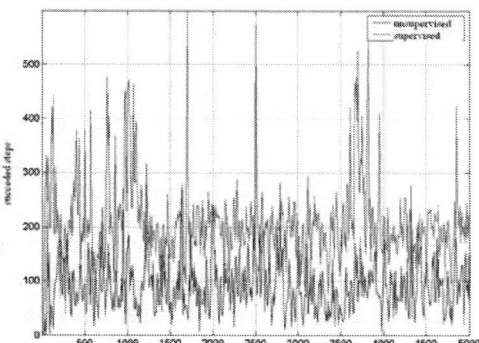

Figure 4. Balancing learning task. Curves show number of successful time steps in balancing the pendulum in function of learning episodes.

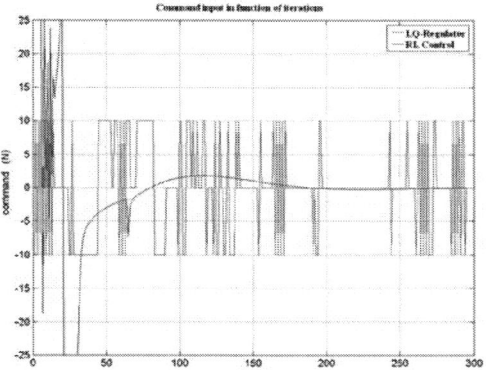

Figure 5. Command signal evolution along iterations, produced during the balancing task by our RL algorithm (green) and LQ-controller (blue)

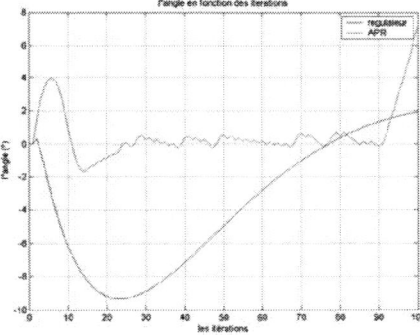

Fig. 6 : response to a perturbation of $\theta'=75°/s$.

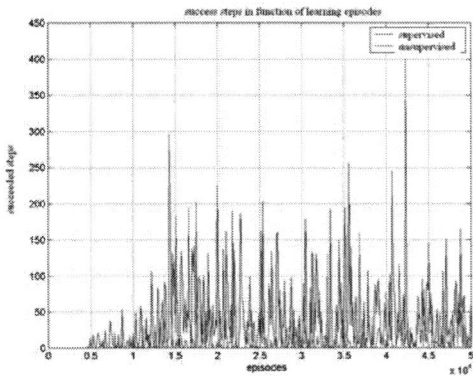

Figure 3. Swing up task. Curves show number of successful steps for swinging up the pendulum versus learning episodes. Supervised (blue curve) and unsupervised RL (green curve)

The supervised RL agent gives better results in terms of robustness to perturbation compared to the LQ-Controller. Comparing, in figure 6, the reaction of the two controllers to an impulsive perturbation (the pole has been given an impulse of 75° / s) we may observe that : the RL-based controller was very rapid in, effectively, responding to this perturbation by driving and stabilizing the pole to its vertical position (the RL agent takes just 30 iterations , i.e. 30 × 0.02s = 0.6 s) than it is done by the LQ-controller (which takes 250 iterations, i.e. 250 × 0.02s = 5 s). Due to its generalization property, the RL agent demonstrates a better ability to adapt to strong perturbations.

5. Conclusion

Scaling-up reinforcement learning techniques to complex environments is a challenging task for the AI community. It was our aim here to propose a simple method to accelerate convergence to an optimal policy by using a priori knowledge about the environment to be learned or by following advice proffered by an external controller. This is a modified version of the standard unsupervised RL algorithms built by incorporating supervision within this structure which suffers from low convergence speed in domains with continuous state space. The proposed approach has been implemented on two different environments: one is learning to freely navigate in a grid world containing obstacle and the second task is to learn solving the cart-pole balancing problem. Obtained results are encouraging and, more importantly, they show that supervisory information can really help improving RL performance. However, it should be noted that further work is needed to elucidate all aspects of balancing the need for a priori information integration and the need for sufficient autonomy and exploration by the RL agent in order to not degrade the learning agent's performance.

ACKNOWLEDGEMENT

This project was funded by the Laboratory of Automatic and Signal of Annaba, LASA, at Badji Mokhtar University, Algeria

REFERENCES

[1] Bertsekas, D. P. and Tsitsiklis, J. N., *Neuro-dynamic programming*, Athena Scientific, Belmont, MA, 1995.

[2] Dietterich, T.G. ,"Hierarchical reinforcement learning with the maxQ value function decomposition", *Journal of Artificial Intelligence Research*, 13: 227 –303, 2000

[3] Dixon, K. R., Malak, R. J. and Khosla, P. K., *Incorporating Prior Knowledge and Previously Learned Information into Reinforcement Learning Agents*, Technical Report, Institute for Complex Engineering Systems, CMU, 2000.

[4] Dorigo, M. and Colombetti, M., *Précis of " Robot Shaping : An Experiment in Behaviour Engineering"*, *Adaptive Behavior*, 5 (3 – 4)., Précis of the book from MIT Press, Oct. 1997.

[5] Huber M. and Grupen R. A., *A feedback control structure for on-line learning tasks*, Robotics and Autonomous Systems, 22 (3 – 4), pp. 303 - 315 ,1997

[6] Mc Govern A., Sutton, R. S. and Fagg A. H. "Roles of macro-actions in accelerating reinforcement learning", *in Proceedings of the 1997 Grace Hopper Celebration of Women in Computing*, pages 13 -18, 1997

[7] Mc Govern A., *Autonomous discovery of temporal abstractions from interaction with an environment*, PhD thesis, U. of Massachusetts, Amherst, MA, 2000

[8] Precup D., *Temporal abstraction in reinforcement learning*, PhD thesis, U. of Massachusetts, Amherst, MA, 2000.

[9] Price B., and Boutilier C., "Accelerating reinforcement learning through implicit imitation", *Journal of Artificial Intelligence Research*, 19, pp. 569 – 629, 2003

[10] Sutton, R. S. and Barto, A. G., *Reinforcement Learning : An Introduction.* MIT Press, Cambridge, MA, 1998.

[11] Watkins, C. J. C. H., *Learning from Delayed Rewards*, PhD thesis, Cambridge University, 1989

[12] Watkins, C. J. C. H. and Dayan, P., *Q-Learning*, Machine Learning, 8, pp. 279 – 292, 1992

Neural and Fuzzy Adaptive Control of Induction Motor Drives

Y. Bensalem, L.Sbita and M.N. Abdelkrim

*Research unit of Modelisation, Analyse, Command of systems MACS
(e-mail: bensalem.yemna@yahoo.fr).
**6029 Université High School of Engineering-Gabès-Tunisia
(e-mail:Lasad.Sbita@enig.rnu.tn)
(e-mail:Naceur.abdelkrim@enig.rnu.tn)

Abstract: This paper proposes an adaptive neural network speed control scheme for an induction motor (IM) drive. The proposed scheme consists of an adaptive neural network identifier (ANNI) and an adaptive neural network controller (ANNC). For learning the quoted neural networks, a back propagation algorithm was used to automatically adjust the weights of the ANNI and ANNC in order to minimize the performance functions. Here, the ANNI can quickly estimate the plant parameters and the ANNC is used to provide on-line identification of the command and to produce a control force, such that the motor speed can accurately track the reference command. By combining artificial neural network techniques with fuzzy logic concept, a neural and fuzzy adaptive control scheme is developed. Fuzzy logic was used for the adaptation of the neural controller to improve the robustness of the generated command. The developed method is robust to load torque disturbance and the speed target variations when it ensures precise trajectory tracking with the prescribed dynamics. The algorithm was verified by simulation and the results obtained demonstrate the effectiveness of the IM designed controller.

Keywords: Adaptive neural network, Adaptive Fuzzy logic Mechanism, ANNI and ANNC, Induction motor.

1. INTRODUCTION

The IM is a nonlinear system and in addition many of its parameters vary with time and operating condition. Conventional IM dynamic control method such as vector control and direct-self-control (DSC) attempt to reduce the complex nonlinear dynamic structure into a linear structure, in order to enable the application of linear design techniques. With its size and its nonlinear nature worsened by irreducible changes in its internal characteristics in operational mode, the IM constitutes a system of great complexity. So, the elaboration of an advanced command has been the subject of multiple efforts [11], [12], [13], [14].

Realizing an efficient command for nonlinear systems or whose dynamics are partially or totally unknown has always been an attempt to perform a less sophisticated linearization as the case of gain scheduling, adaptive control, self tuning, etc. To alleviate the disadvantages of adaptive control, for example the sensitivity of the model, researchers have developed the notion of neurocontrol which consists of the instillation of the system's dynamic behavior to an artificial neural network (ANN) trough the back propagation learning operation [1], [2]. The proposed neural control scheme is implemented in the field oriented control (FOC) which is widely used for IM drives. By providing decoupling of torque and flux control demands, the vector control can navigate an AC motor drive similar to a separately excited DC motor drive without sacrificing the quality of the dynamic performance [5]. In addition, both conventional methods are relatively complex and require the use of relatively highly trained commissioning personnel.

In the past few years, neural networks have been used in some power electronic applications, such as inverter current regulation, dc motor control, flux estimation, and observer based control of IM [3], [4]. Evidently, neural network technique is showing promise as a competitive method of signal processing for power electronics applications [12]. They have the advantages of extremely fast parallel computation, immunity from input harmonic ripple, and fault tolerance characteristics due to distributed network intelligence [2].

Neural network is well known for its learning ability and approximation to any arbitrary continuous function. Recently, it is proposed in the literature that neural networks can be applied to parameter identification and state estimation of IM control systems [6], [7]. The interest generated by artificial neural networks nowadays is justified by some of the fascinating properties that they have, such as adaptiveness, generalization, etc., which give a powerful tool to solve problems related to the control of nonlinear complex processes and to overcome the limitations of commonly used techniques [15], [16]. Many different types of ANN have been proposed, however this paper considers sigmoidal feedforward ANN exclusively. Because of the necessity for adaptive abilities in a network learning process, applying neural networks to system identification and control dynamics has become a promising alternative to process control [9], [10].

In this research, an adaptive neural network control method is developed for IM speed control which constitutes a dynamic mapping. Neural network can be applied to control and identify the nonlinear systems since they approximate any desired degree of accuracy with a wide range of nonlinear model [8]. In addition, the two neural network structures can be implemented in parallel. Here, a multilayer neural network using back-propagation learning algorithm was applied to identify the process model. The command is produced corresponding to the model reference desired response and the identified plant neural model. The fuzzy logic adaptive mechanism is used to improve the speed loop properties and to incorporate performance specifications.

The proposed control scheme was implemented and the computer simulation results demonstrate the effectiveness of the proposed control scheme.

2. INDUCTION MOTOR MODEL

Assuming linear magnetic circuits, equal mutual inductances and neglecting iron losses; the dynamic model of an IM can be represented according to the usual d-axis and q-axis components in a synchronous frame as [5]:

$$V_{ds} = R_s i_{ds} + \frac{d\phi_{ds}}{dt} - \omega_s \phi_{qs} \quad (1)$$

$$V_{qs} = R_s i_{qs} + \frac{d\phi_{qs}}{dt} + \omega_s \phi_{ds} \quad (2)$$

$$0 = R_r i_{dr} + \frac{d\phi_{dr}}{dt} - \omega_{sl} \phi_{qr} \quad (3)$$

$$0 = R_r i_{qr} + \frac{d\phi_{qr}}{dt} - \omega_{sl} \phi_{dr} \quad (4)$$

where ω, ω_s and ω_{sl} are respectively the electrical rotor, synchronous and slip speeds; V_{ds} and V_{qs} are the d, q axis voltages; i_{ds} and i_{qs} are the d, q stator axis currents; i_{dr} and i_{qr} are the d, q rotor currents; ϕ_{ds} and ϕ_{qs} are the d, q stator fluxes; ϕ_{dr} and ϕ_{qr} are the d, q rotor fluxes; R_s and R_r are the stator and the rotor resistances.

The stator and rotor flux are expressed as:

$$\phi_{ds} = L_s i_{ds} + M i_{dr} \quad (5)$$

$$\phi_{qs} = L_s i_{qs} + M i_{qr} \quad (6)$$

$$\phi_{dr} = L_r i_{dr} + M i_{ds} \quad (7)$$

$$\phi_{qr} = L_r i_{qr} + M i_{qs} \quad (8)$$

Where L_s and L_r are the stator and rotor self inductances; M is the mutual inductance.

As known, the torque produced by an IM can be written as follows:

$$T_e = n_p \frac{M}{L_r} (i_{qs} \phi_{dr} - i_{ds} \phi_{qr}) \quad (9)$$

Where T_e is the electromagnetic torque; n_p is the number of pole pairs.

The electromagnetic torque equation and the rotational rotor speed are related by the well known mechanical equation:

$$J \frac{d}{dt} \Omega + f \Omega = T_e - T_l \quad (10)$$

where $\Omega = \frac{\omega}{n_p}$; J is the rotor moment of inertia, f is the friction constant and T_l is the load torque.

3. ROTOR FIELD ORIENTED CONTROL

The IM can produce good performances using field-oriented vector control strategy. The main idea of the vector control is to monitor the torque and the flux separately. This technique is based on the orientation of the flux vector along the d axis [5], which can be expressed by considering:

$$\phi_{dr} = \phi_r \text{ and } \phi_{qr} = 0 \quad (11)$$

Using (11), we eliminate all the terms with quadratic rotor flux so we can express the synchronous angular speed as:

$$\omega_s = \omega + \frac{M i_{qs}}{\tau_r \phi_r} \quad (12)$$

The rotor flux vector orientation and the state space equation lead to the following V_{ds}, V_{qs} and the electromagnetic torque expressions:

$$V_{ds} = \sigma L_s \frac{di_{ds}}{dt} + (R_s + R_r \frac{M^2}{L_r^2}) i_{ds} - \omega_s \sigma L_s i_{qs} - \frac{M}{L_r^2} R_r \phi_r \quad (13)$$

$$= V_{ds1} - e_{ds}$$

$$V_{qs} = \sigma L_s \frac{di_{qs}}{dt} + \omega_s \sigma L_s i_{ds} + (R_s + R_r \frac{M^2}{L_r^2}) i_{qs} + \frac{M}{L_r^2} n_p \Omega \phi_r \quad (14)$$

$$= V_{qs1} - e_{qs}$$

$$\tau_r \frac{d\phi_r}{dt} + \phi_r = M i_{ds} \quad (15)$$

$$\omega_s = n_p \Omega + \frac{M i_{qs}}{\tau_r \phi_r} \quad (16)$$

The d and q compensations terms are respectively:

$$e_{ds} = \omega_s \sigma L_s i_{qs} + \frac{M}{L_r^2} R_r \phi_r \ , \ e_{qs} = -\omega_s \sigma L_s i_{ds} + \frac{M^2}{L_r \tau_r} i_{qs}$$

where τ_r is the rotor time constant, $\sigma = 1 - \frac{M^2}{L_s L_r}$ is the total leakage constant.

These expressions form the control algorithm to carry out the IM field orientation strategy. The voltages components V_{ds} and V_{qs} should act on the current i_{ds} and i_{qs} separately and consequently the flux and the torque. Actually, the field-oriented control technique is widely used in high performance motion control of IM. Because of torque/flux decoupling, it achieves a good dynamic response and accurate motion control as separately excited DC motors.

4. THE PROPOSED CONTROL SCHEME

A speed control system with an adaptive neural network controller and adaptive neural network identifier is shown in Fig. 1. The adaptive neural network identifier (ANNI) was used to determine an adaptive nonlinear model of the unknown motor dynamics. The adaptive neural network controller (ANNC) was used to produce an adaptive control to lead the motor speed to accurately track the reference model. The learning algorithm applied is the back-propagation which has been used successfully for a wide variety of applications because it offers a promising way of handling complex control problems [4]. It is used for the ANNI and the ANNC to automatically adjust their parameters.

The ANNC output which represents the control signal is added to the output signal of the fuzzy controller to produce the actual input command U. The control objective is that the speed output plant ω track the reference model output ω_m. In fact, the reference model represents the system dynamics with nominal parameters which in general represent a difficult task especially for nonlinear and complex systems. For this reason, the reference model is replaced by a transfer function which describes the desired performances related to the rise time, overshoot and steady-state error.

where g_1 is the nonlinear function which will be identified, ω and i_{qs} are the output and the input of the plant; n and m are the order of ω and i_{qs}. The neural network structure used in the ANNI is indicated in Fig. 2, it has three layers: input layer, hidden layer and output layer. The input and output layers have neurons equal to the respective number of signals, the number of neurons in each hidden layer depend on the system dynamics and the desired degree of accuracy [6]. The network is fully connected; the output of each neuron is connected to all the neurons in the forward layer through a weight. Besides, a bias is coupled to all the neurons of the hidden and output layers through a weight.

The feedforward neural network is usually trained by back-propagation training scheme. With the network initially untrained and its weights selected at random, so an output signal is obtained for a given input pattern. The actual output is compared with the desired output and the weights are adjusted by the supervised back-propagation training algorithm until the errors become acceptably small [7].

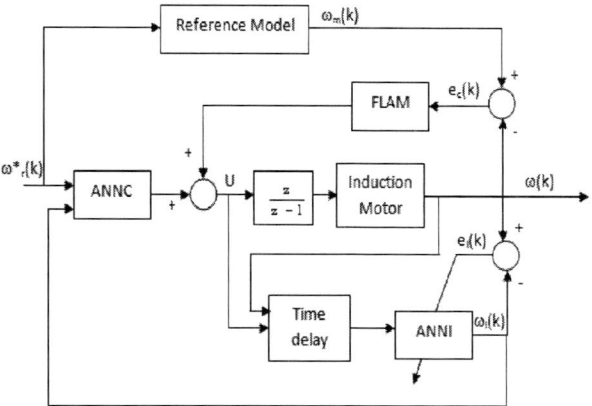

Fig. 1. Block diagram of the field oriented IM with the adaptive controller and identifier

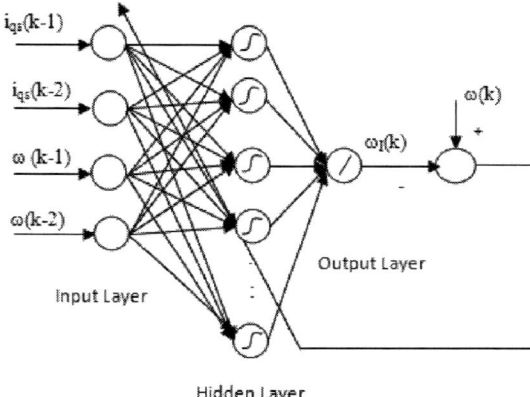

Fig. 2. The ANNI neural network architecture

4.1 Adaptive Neural Network Identifier

The ANNI is a neural network which calculates the output ω_I using the input-outputs values. The standard model which can be represented by the various nonlinear discrete systems is the NARMA model (Nonlinear Autoregression-Moving Average) which is used to approximate the input/output IM model:

$$y(k) = N[y(k), y(k-1), \ldots, y(k-n+1), u(k), u(k-1), \ldots, u(k-m+1)] \tag{17}$$

where $u(k) = i_{qs}(k)$ is the input, and $y(k) = \omega(k)$ is the output.

To estimate the unknown parameters, the plant dynamic is assumed to be represented by the expression follow:

$$\omega_I(k) = g_1(\omega(k-1), \ldots, \omega(k-n), i_{qs}(k-1), \ldots, i_{qs}(k-m)) \tag{18}$$

Training Procedure: A learning procedure of a neural network with M layers and n inputs using a back propagation algorithm is defined. In this description, the index i corresponds to a neuron in the output layer, the index j to a hidden layer. During the learning procedure, we search to minimize the couple (input, desired output) by modifying the weights ω_{ij}^m :

- Initialize all the network weights to small random numbers.
- Propagate the input forward through the network

$$x_i^m(t) = f\left[E_i^m(t)\right] \tag{19}$$

$$E_i^m(t) = \sum_j w_{ij}^m(t) x_j^{m-1}(t) \tag{20}$$

where

$x_i^m(t)$: output of the neuron i and the layer m,

$x_j^m(t)$: output of the neuron j of the layer m (m: 1, …, M),

223

w_{ij}^m : weights relating the neuron j of the layer (m-1) to the neuron i of the layer m.

-For each network output unit i, calculate the error term of the output layer:

$$\delta_i^M(t) = f'\left[E_i^M(t)\right]\left[\Delta\omega_{si}(t) - \Delta\omega_{mct}^M(t)\right] \quad (21)$$

where $d_i(t)$ the desired output of the neuron i of the output layer.

-For each hidden unit h, calculate the error term by propagating the error:

$$x_i^{m-1}(t) = f'\left[E_i^{m-1}(t)\right]\sum_j w_{ji}^m(t)\delta_j^m(t) \; ; \quad (22)$$

m = M, M-1, ..., 2

- Update the network weights W_{ij}:

$$(w_{ij}^m)_{new} = (w_{ij}^m)_{old} + \Delta w_{ij}^m \quad (23)$$

With:

$$\Delta w_{ij}^m = \eta \delta_i^m x_j^{m-1} \quad (24)$$

η is the learning rate of the back propagation algorithm
These steps are repeated so that to minimize the function J:

$$J = \frac{1}{2}\sum_t\sum_i\left[\Delta\omega_{si}(t) - \Delta\omega_{sinet}^M(t)\right]^2 \quad (25)$$

where t is the pattern number.

4.2 The adaptive neural network controller

The objective of the ANNC is to contribute with fuzzy controller to develop the adaptive command U such that the output of the system $\omega(k)$ can accurately track the reference command $\omega_r^*(k)$. So, it is necessary to develop the nonlinear controller as follow:

$$U_{net}(k) = g_2(\omega(k), \omega(k-1), ..., i_{qs}(k-1), i_{qs}(k-2), i_{qs}(k-m), \omega_r^*(k)) \quad (26)$$

Where g_2 is the unknown nonlinear function which will be identified, n and m are respectively the order of ω and i_{qs}.

4.3 The Fuzzy Logic Adaptation Mechanism

Fig. 3 shows the block diagram of the fuzzy adaptation mechanism FLAM, which constituted of two loops: a speed loop control which contains the neural network controller and an adaptation loop composed of a reference model and an adaptation mechanism.

For a speed reference ω_r^*, the reference model generates a desired output ω_m. In fact, the reference model is used in order to define the desired performances corresponding to the robust criteria as: rise time, overshoot and steady state error. Then, the reference model output will be compared with the plant output ω to produce a correction signal.

The FLAM has two inputs: the first input is a speed error which is the difference between the reference model output and actual speed; the second one is the speed error variation. The inputs are given the following equations:

$$e(k) = \omega_m(k) - \omega(k) \quad (27)$$
$$Ce(k) = e(k) - e(k-1) \quad (28)$$

These inputs will be treated by the adaptation mechanism using the fuzzy logic rules in order to produce an adaptive signal which will be added to the desired fuzzy controller output as shown in Fig. 3. So, the variation command obtained is given by the equation as follow:

$$Ci_{qs}^*(k) = Ci_{qs1}^*(k) + Ci_{qs2}^*(k) \quad (29)$$

The reference model command is obtained by integrating the following signal:

$$i_{qs}^*(k) = i_{qs}^*(k-1) + i_{qs}^*(k) = i_{qs}^*(k-1) + [Ci_{qs1}^*(k) + Ci_{qs2}^*(k)] \quad (30)$$

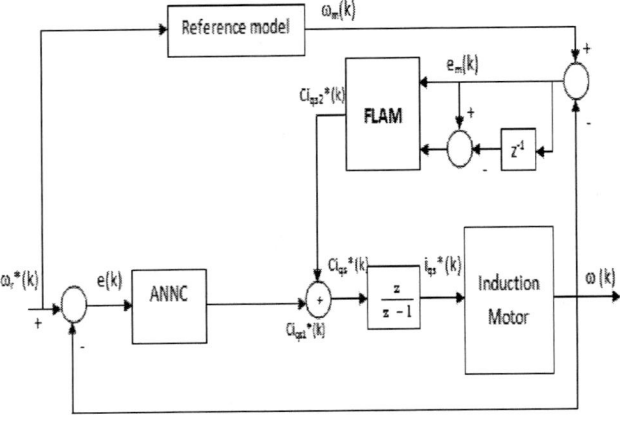

Fig. 3. Block diagram with the fuzzy logic adaptation mechanism FLAM

The structure of this mechanism is described by the diagram block of the Fig. 4. It is based on four well known stages: the fuzzification stage, the fuzzy rule base, the fuzzified inputs are combined using these rules in the third part to produce the fuzzy inference engine. The final stage constituted by the defuzzifier which produces a crisp output from the combined fuzzy output set. For the FLAM, the universe of discourse is partitioned into seven linguistic variables: NB, NM, NS, EZ, PS, PM and PB. Triangular membership functions are chosen to represent the linguistic variables. Fig. 5 shows the input and output membership functions. The used fuzzy rules are summarized in table 1.

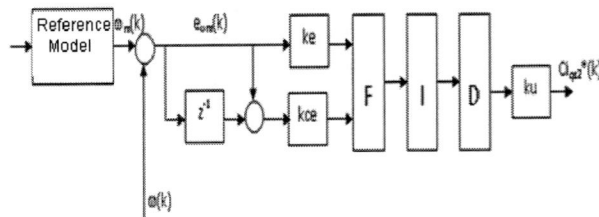

Fig. 4. The Fuzzy Logic Adaptation Mechanism Structure

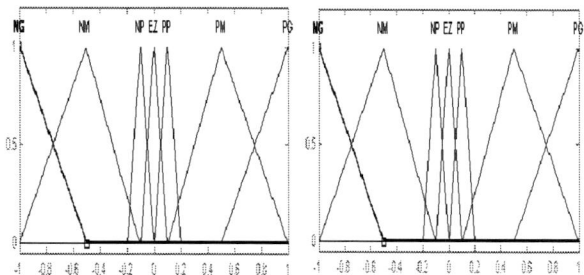

Fig. 5. Input and output membership functions

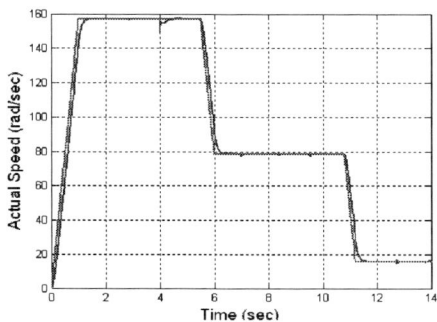

Fig. 6. The speed response with a variable target speeds

Table 1. FLAM rule base

e \ Δe	NG	NM	NP	EZ	PP	PM	PG
NG	NG	NG	NG	NG	NM	NP	EZ
NM	NG	NG	NG	NM	NP	EZ	PP
NP	NG	NG	NM	NP	EZ	PP	PM
EZ	NG	NM	NP	EZ	PP	PM	PG
PP	NM	NP	EZ	PP	PM	PG	PG
PM	NP	EZ	PP	PM	PG	PG	PG
PG	EZ	PP	PM	PG	PG	PG	PG

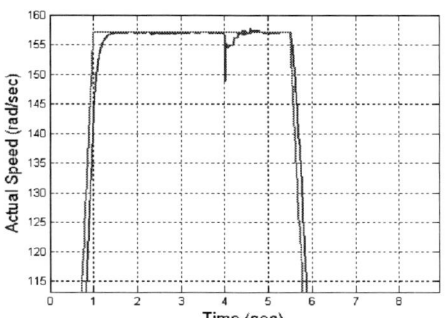

Fig. 7. A zoom of the speed response

4.4 Model Reference Choice

The reference model choice depends on many criteria: the system parameters, limitations of devices used. For example, if the load inertia moment is 5 to 10 times more than of the machine, it is not practical to impose to the IM to track a reference model with a very small rise time because the output controller can saturate to start up limited value. Here, the reference model is chosen in order to have the following desired performances: rise time: $\tau_m = 0.972$, time constant $\tau = 0.534$ sec, response time $t_r = 1.186$ sec, overshoot = 0%.

5. SIMULATION RESULTS

In this section, simulation results are presented to evaluate the effectiveness of the proposed control scheme for different operating conditions. The software environnement used for the simulation in Matlab 6.5 with simulink package.

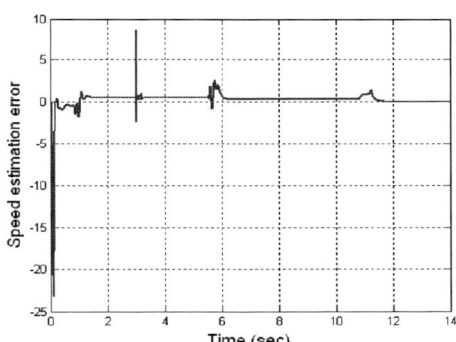

Fig. 8. The speed estimation error

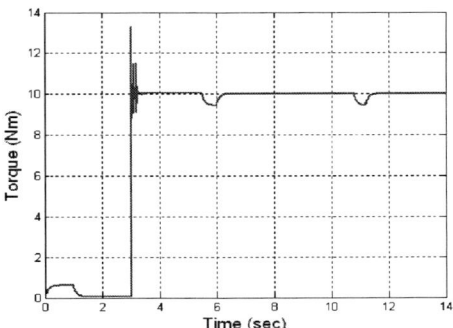

Fig. 9. The electromagnetic Torque

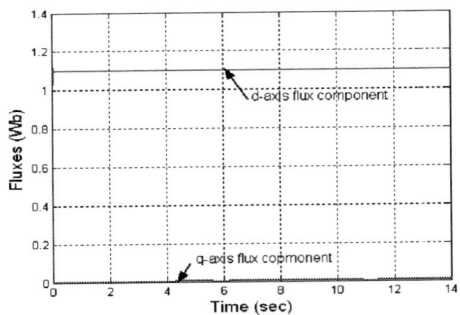

Fig. 10. The d and q axis flux components response

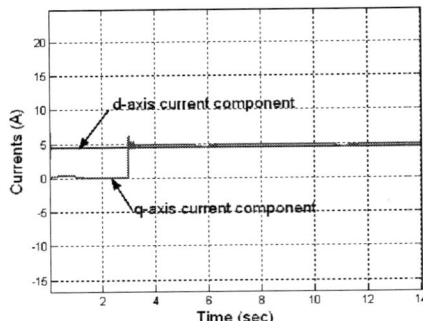

Fig. 11. The d and q axis current components response

The first test in Fig. 6 concerns a load application with a variation of the reference; a load torque disturbance ($T_l = 10$ Nm) is then applied at t = 4 sec and this figure show that the neural network controller compensate it very well. The motor shaft is subjected to a change of the reference speed at t = 5 sec (78.5 rad/sec) and at t = 11 sec (15.7 rad/sec). Fig. 8 shows the speed estimation error, which is difference between the actual output ω and the neural network identified output ω_I, of the proposed control IM drive. We can show that the flux tracks its reference value 1.1 Wb for all speed ranges. Note that the speed and the flux reach the desired reference values, and there is no effect of the load torque variation on the flux; which means that the speed and the flux are decoupled even when that the flux had to conserve the reference value 1.1 Wb, these results show that the ideal variables decoupling are established despite the application of a load torque. Fig. 11 shows the d and q current components i_{ds} and i_{qs} which respectively are the image of the flux and the torque and this is confirmed in this figure.

6. CONCLUSION

This paper presented an adaptive speed controller scheme based on fuzzy and neural network for an IM drive. An adaptive neural network identifier ANNI was designed using a back-propagation algorithm in order to provide on line adaptive estimation of the induction motor's model. In addition, neural network controller was used to contribute with the adaptive fuzzy logic mechanism in generating a robust control force so that the motor speed could accurately track the reference command. Theoretical analysis and simulation results demonstrated that the proposed speed control scheme could predict sufficiently the motor dynamics, confirm the feasibility of the proposed scheme and allows one to better understanding of the application of neural network to speed tracking control in IM.

REFERENCES

[1] Michael T. W. and Ronald G. H. (1995). Identification and Control of Induction Machines using Artificial Neural Networks. *IEEE trans. On Indust. Applic.*, **vol. 31**, no. 3.

[2] Ba-razzouk A., Cheriti A., Olivier G. (1997). A Neural Networks Based Field Oriented Control Scheme for Induction Motors. *IEEE Indust. Applic. Society*, New Orelans, Louisiana.

[3] Shi K. L., Chan T. F. (2001). direct self control of induction motor based on neural network. IEEE Trans. On Indust. Applicat., **vol. 37**, No. 5.

[4] Theocharis J. and Petridis V. (1994). Neural Networks observer for Induction Motor Control. *IEEE Control Systems*, pp 26-37.

[5] P. Vas. (1990). *Vector Control of AC Machines*. London, U. K.: Oxford Univ. Press.

[6] Zerikat M., Bendjebbar M. and Benouzza N. (2005) Dynamic Fuzzy-Neural Network Controller for Induction Motor *Drive. Trans. On Engineering, Computing and Technology*, **Vol. 10**.

[7] Bensalem Y and Sbita L. (2007) A robust speed sensorless induction motor Drives. *International Conference on Electrical Engineering Design and technology ICEEDT'07*.

[8] Tsai-Juin R., Tien-Chi C. (2006). Robust Speed-Controlled induction motor drive based on recurrent neural network. *Electric Power Systems Research*, pp. 1064-1074.

[9] Chen T. T., Sheu T. T. (2002). Model reference neural network controller for induction motor speed control. *IEEE trans. Energy Convers.* **Vol. 17**, pp. 157-16.

[10] Lin F. J, Wang and Huang P. K. (2004). RFNN controlled sensorless induction spindle motor drive. *Electr. Power Syst. Res.* **Vol. 70**, pp 211-222.

[11] Mouloud A. D., SID Ahmed A. Fuzzy and neural control of an induction motor, *Int. J. Appl. Math. Comput. Sci.*, 2002, **vol. 12**, pp. 221-233.

[12] Wishart M. A. and Harely R. G. (1995). Identification and control of induction machines using artificial neural networks. *IEEE Trans. Ind. Applicat.* **vol. 31**, pp. 612-619.

[13] Yang G., Chin T. H. (1993). Adaptive-speed identification scheme for a vector-controlled speed sensorless inverter-induction motor drive, *IEEE Trans. Ind. Appl.* 29, pp. 820-825.

[14] Chauder C. (1993). Adaptive speed identification for vector control of induction motors without rotational transducers. *IEEE Trans. Ind. Appl.* pp: 1054-1061.

[15] Kojabadi H.M., Chang L. (2002). Model Reference adaptive system pseudoreduced-order flux observer for

very low speed and zero speed estimation in sensorless induction motor drives. *IEEE Annual Power Electronics Specialists Conference,* Australia, **vol. 1**, pp. 301-308.

[16] Hiyama, T., M. Ikeda, T. nakayama, (2000). Artificial neural network based induction motor design. *IEEE Power Engineering Society Winter Meeting,* **vol. 1**, pp: 264-268.

Appendix

Induction motor parameters: Rated power: 3 kW, Rated voltage: 380/220V, Rated frequency: 50 Hz, rated speed = 1430 r/min, $R_s = 2.3\Omega$, $R_r = 1.55\Omega$, $L_s = L_r = 0.261$ H, $M = 0.249$ H, $\sigma = 0.0898$, $f = 0.0007$ Nm.s/rad, $J = 0.02$ Kg.m^2, $n_p = 2$.

Neural Network Identification For a C5 Parallel Robot

M. E. Daachi *, B. Achili **, B. Daachi *** and D. Chikouche *

*Université Ferhat Abbès Sétif Algérie
(e-mail : m_daachi@yahoo.fr).
** Laboratoire d'Informatique Avancées de Saint Denis (LIASD)
2, rue de la liberté 93526 Saint Denis Cedex - France
(e-mail : achili@ai.univ-paris8.fr).
*** Laboratoire Images, Signaux et Systmes Intelligents (LISSI)
122, rue Paul Armangot 94400 Vitry/Seine - France
(e-mail : daachi@univ-paris12.fr).

Abstract: This paper presents the design and analysis of a neural network-based identification of the inverse dynamic model of a C5 parallel robot. The identification structure is designed using the black box form (the dynamic model is completely unknown). This identification uses real data acquired on the C5 parallel robot by applying a nominal control scheme (PD). The desired trajectories of this scheme are based on Fourier series and the coefficients are chosen in a heuristic way. We have used this type of desired trajectories to obtain exciting trajectories for identification procedure. Three identification schemes are tested and compared. The comparison is performed based on the number of parameters used in each architecture and the quality of the generalization error. The used neural network is of MLP type and composed of one hidden layer.

Keywords: Identification, Neural Networks, Parallel robot.

1. INTRODUCTION

The identification of nonlinear systems has been widely considered in literature. The used techniques are diverse and varied: fuzzy logic, neural networks, least squares, to name but a few [7][5][6][3][2]. Every technique has its advantages and its drawbacks. A priori knowledge can be added in the identification procedure. In case of manipulator robots, one can add the structure of the dynamic model (a priori known) in the neural networks-based identification. A good identification facilitates the design of stable controllers. The identification is based on data obtained on the system. In practice, data collected from actual plants are corrupted by a noise.To obtain a database representing the actual dynamic system, two steps are to be carried. The first step consists to choose and apply a typical controller on the system and to collect result measures. The second step consists to pre-process these data to eliminate noise.

The properties of the neural networks were largely studied in the literature. Hornik and Funahashi [4][5] proved that if the network contains a sufficient number of hidden neurons, it will be able to approach any nonlinear function. The number of hidden neurons is given in a heuristic way and the approximation error is obviously not null. In the literature, the neural method do not consider the peculiarity of every system. Indeed, the nature and the number of entries of the neural network can be studied to reduce the number of parameters and to improve the quality of identification.

For black box identification of robot inverse dynamic, the possible inputs are positions, velocities and accelerations. In this case, it would be interesting to compare the results given by various neural architectures.

In this paper, the considered problem is the identification of the inverse dynamic of a C5 parallel robot. We use in this identification, neural networks of one hidden layer (MLP). Three neural architectures are studied and compared. The first architecture is a classical neural identification. In the second architecture, we will try to prove by practical data, that it is not possible to dissociate dynamics of the robot axes. In this approach, we use one neural network (MLP) for each axis. The inputs of every MLP are joint variables of the concerned axis. The third architecture is a modification of the second architecture with the same inputs for all neural networks. The Marquardt-Levenberg algorithm is used to optimize the neural parameters.

This paper consists of five sections. The next section describes the mechanical architecture of the C5 parallel robot. In the third section, we present the identification schemes to be applied. Fifth section is dedicated to the presentation and analysis of the experimental results. The last section is for conclusion and perspectives.

2. PRESENTATION OF THE C5 PARALLEL ROBOT

Parallel architectures were first used for building flight simulators and tire testers [9], [10]. More recently, parallel robots appeared in the medical field [11] . The latter requires the design of very precise parallel machines performing in a limited workspace. The C5 parallel robot was developed at the LIIA (Laboratory of Industrial Computing and Automatic). This robot has architecture allow it to make very precise movements. This robot consists of a static part and a mobile part connected together by six actuated links. Each segment is embedded to the static part at point A_i and linked to the mobile part through a spherical joint attached to two crossed sliding plates at point B_i (Fig. 1 and Fig. 2).

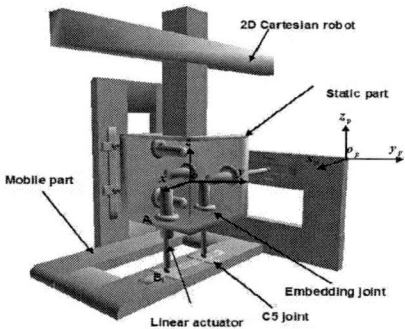

Fig. 1. Parallel robot.

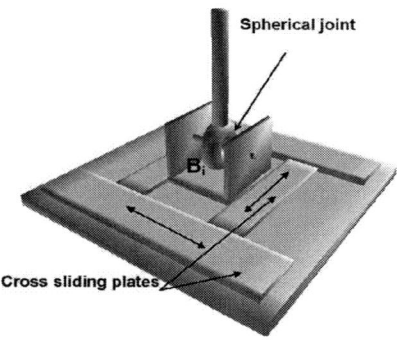

Fig. 2. Detail of the C5 joint

Theoretical study concerning this architecture has been presented in [8]. The C5 links parallel robot is equipped with six linear actuators; each of them is driven by a DC motor. Each motor drives a ball and screw arrangement. The position measurements are obtained from six incremental encoders, which are tied to the DC motors. In order to implement a force feedback control, the robot has been equipped with six strain gauge force sensors. Each sensor is serially displayed between the linear actuator end and the swivel (spherical joint). Hence, the contact force vector is computed from the information acquired from the force sensors.

The experimental setup shown in Fig. 3 is composed of a 2D Cartesian robot linked to the C5 parallel robot. The letter acts as an active force controlled wrist of the Cartesian robot.

3. NEURAL ARCHITECTURE FOR IDENTIFICATION

In the identification, the neural architectures which are used are different. We can talk about grey box and black box architectures. When we have a priori knowledge of the system (structure of the model etc.) to identify, it is interesting to associate them to the neural networks. We discuss in this case about grey box identification. Black box identification of the system is the identification when only available information is its inputs and its outputs. In the present work, we realize the black box identification to handle all possible scenarios.

The identification consists in obtaining a model representing the dynamic behavior of a system. In our case we use neural

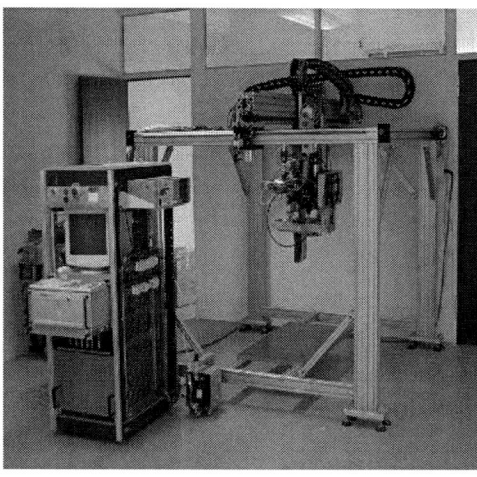

Fig. 3. Experimental setup.

network of the type MLP to identify the dynamic behavior of the C5 parallel robot.

We shall try through three different neural architectures, to realize the identification. The obtained results will be compared. The identification procedure consists of a stage of learning and of other one of the generalization.

3.1 Trajectory generation for identification

The robot movement can be defined by the interpolated joint displacements. This interpolation is done by a polynomial. The principle consists in calculating polynomial coefficients by non linear optimization, which minimizes an exciting criterion under constraints of position, velocity and acceleration.

The exciting trajectories give data rich in information about the system dynamics [1]. In our case, calculation of the exciting (modulated) trajectory is based on heuristics. The database representing this trajectory contains torques and positions relative to the 6 axes of the robot.

3.2 Identification method

The figure 4 summarizes identification phases. In this paper the pack propagation method with Levenberg-Marquardt algorithm is used to calculate the neural parameters. Three different neural architectures are applied to identify the system dynamics. The first identification diagram is composed of only one neural network. The network inputs are the positions, velocities and accelerations. In the second identification diagram, a neural network is reserved for each axis. The neural network inputs are the position, the velocity and the acceleration of the concerned axis (figure 6). For each neural network, parameters are updating using the partial error of the appropriated axe. In the third identification diagram, we also use a neural network per axis. Contrary to the second diagram of identification, the six networks have the same inputs (positions, velocities and accelerations of the 6 axes of the robot). For parameters updating, the neural networks are independent (figure 7). To implement these three identification diagrams, and as we use a supervised training, the following phases are to be achieved :

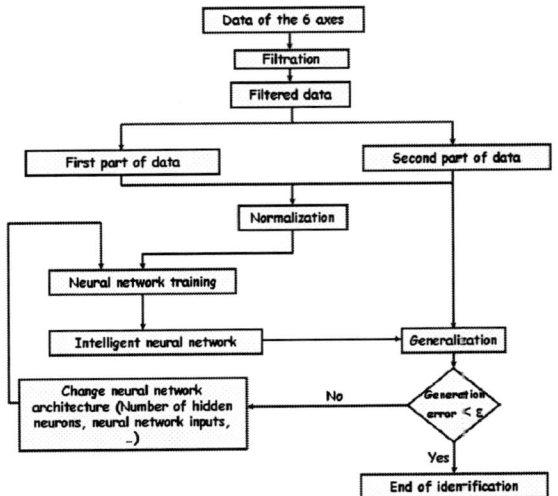

Fig. 4. Neural identification principle.

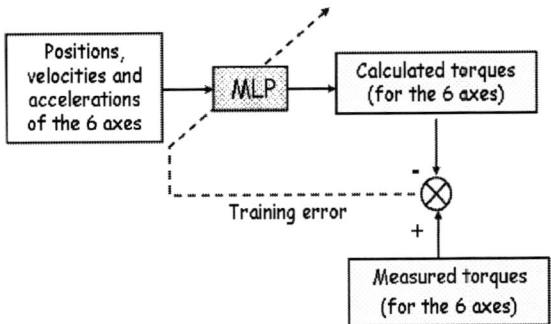

Fig. 5. First identification scheme (1 MLP)

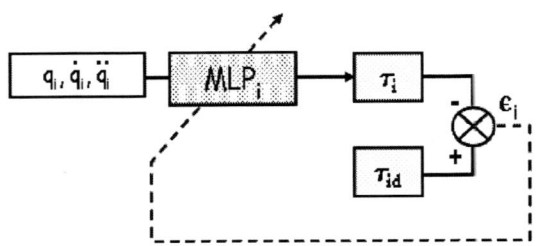

Fig. 6. Second identification scheme (1 MLP for each torque, i = 1, 6).

(1) Make measures on the real robot by using a nominal controller of type PD (figure 8). In the goal of avoiding any unexpected behavior of the robot, data generation is made in closed loop. The proportional and derivative gains of this controller (PD) are given as follows:
(2) As the obtained data are affected by noise, it is necessary to filter them. The used filter is of Butterworth.
(3) As the robot has not the sensors of velocity and acceleration, we apply derivation to joint position.

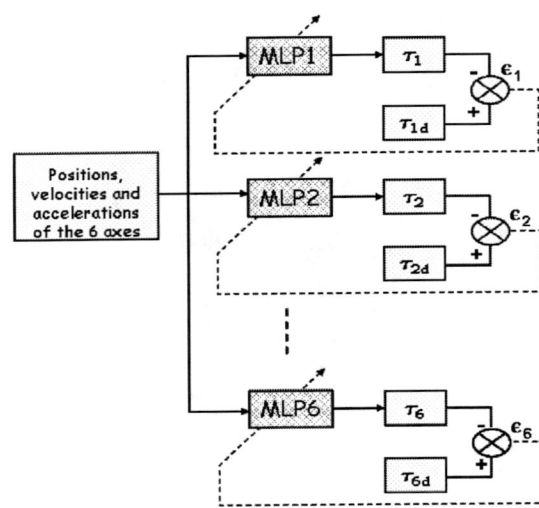

Fig. 7. Third identification scheme.

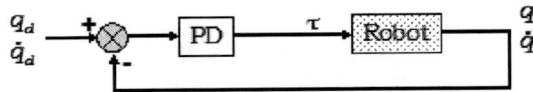

Fig. 8. Nominal controller for data generation.

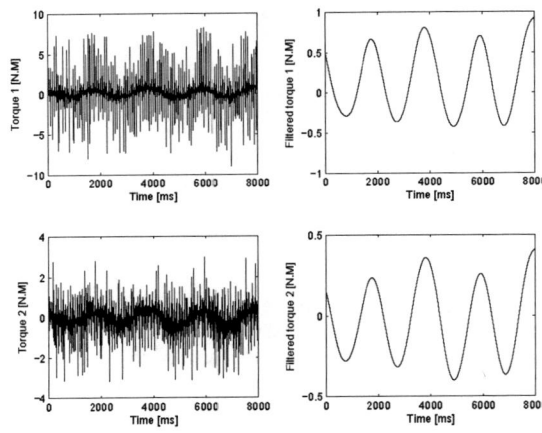

Fig. 9. Torque applied on the first joint.

4. EXPERIMENTAL RESULTS

For training and generalization, we used data generated by the same manipulation. For the training phase, we consider the first part of data and we use the second part for the generation phase. For an effective identification, it is very important to have for learning, exciting trajectories (trajectories that represent goodly the system dynamics). The dynamics to be identified is the inverse dynamics. The table given by figure (15) summarizes the main characteristics of each method. The parameters reserved and presented in this table (15) give the best identification results concerning each method.

In training phase, we obtained similar results (good results) for the three implemented architectures. The illustrations given in

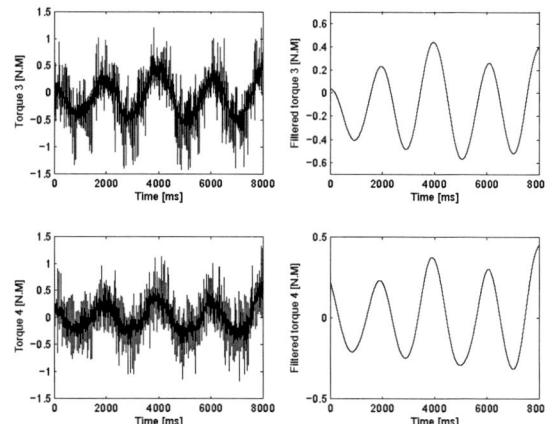

Fig. 10. Torques applied on the third and the fourth joint.

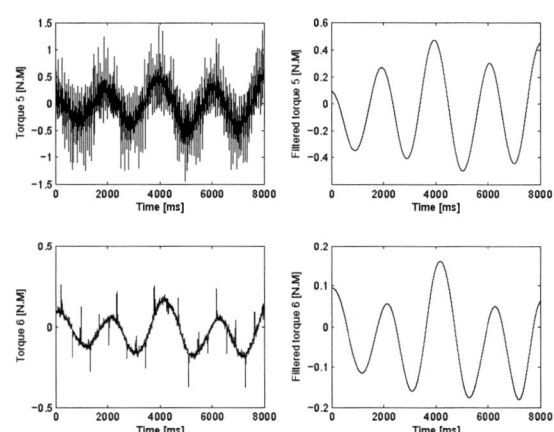

Fig. 11. Torques applied on the fifth and the sixth joint.

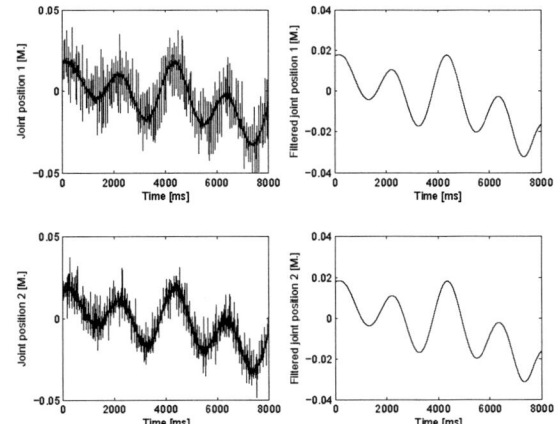

Fig. 12. Torque 1.

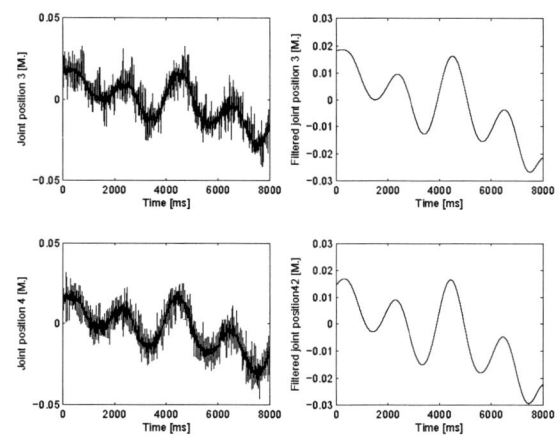

Fig. 13. Third and fourth joint positions.

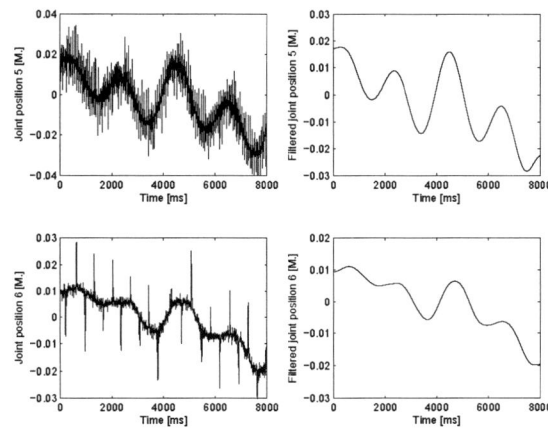

Fig. 14. Fifth and sixth joint positions.

Identification scheme / Number of	1	2	3
Parameters	144	144	114
Neural Networks	1	6	6
Hidden neurons in each MLP	6	6	1
Inputs for each MLP	18	3	18
Outputs for each MLP	6	1	1

Fig. 15. Comparison table.

this paper, concern only the generalization phase. The results of three architectures are presented on the same figure. With a simple examination of figures (16, 17, 18, 19, 20, 21) we exclude the second method. We can then notice that every torque can be correctly reconstituted only with the joint variables of the 6 axes of the robot. In this case, only the first and third architectures are to be retained for the identification of the C5 parallel robot inverse dynamics. Regarding the parameters number, the third architecture is the best.

Finally, we can have the certainty that even with neural networks; it is not possible to separate the dynamics of the 6 axes of the robot.

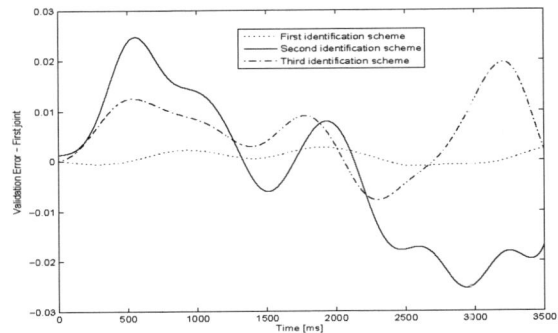

Fig. 16. Generation error concerning the first torque.

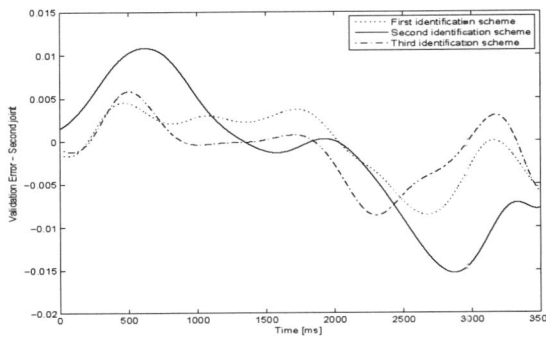

Fig. 17. Generation error concerning the second torque.

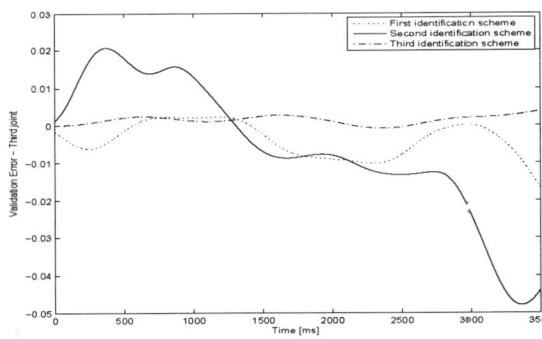

Fig. 18. Generation error concerning the third torque.

5. CONCLUSION

In this paper, we introduced the experimental identification of the inverse dynamic of a C5 parallel robot. One hidden layer neural networks were used for this identification. Trajectories used for identification are both modulated in frequency and amplitude. This modulation characteristic is used so as to cover the major part of the work space and to provide exciting trajectories. Results analysis suggests opting for the first or the third architecture. In our case the first architecture requests fewer parameters.

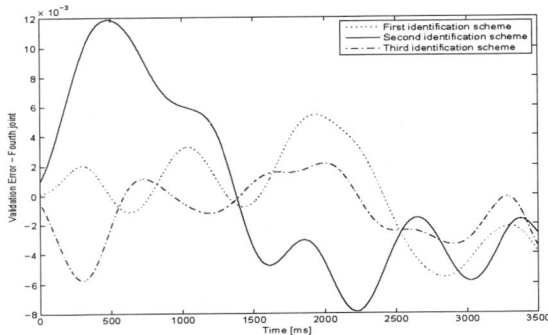

Fig. 19. Generation error concerning the fourth torque.

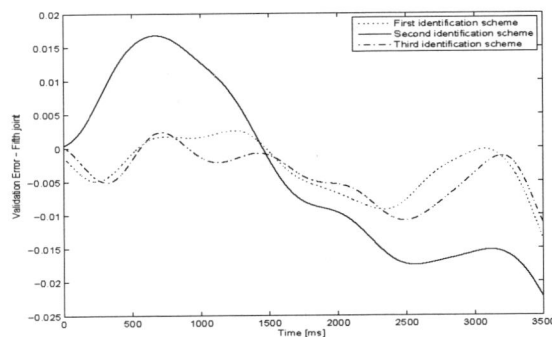

Fig. 20. Generation error concerning the fifth torque

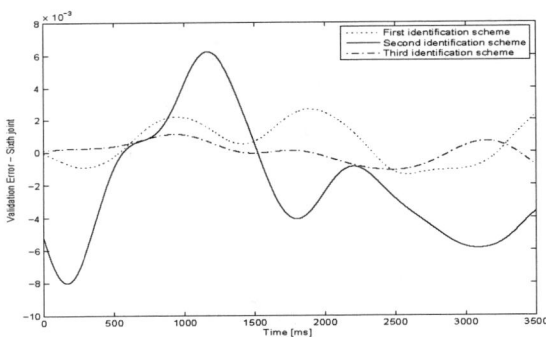

Fig. 21. Generation error concerning the sixth torque.

REFERENCES

[1] Swevers J., Ganseman C., Bilgin D., De Schutter J., Van Brussel H., 1997. Optimal robot excitation and identification. IEEE Transactions on Robotics and Automation, 13(5):730–740.

[2] M. Gautier and Ph. Poignet, Extended Kalman filtering and weighted least squares dynamic identification of robot. In Control Engineering Practice 9, pp. 1361–1372, (2001).

[3] M.Oussalaha, J. de Schutter and H. Bruyninckxb. Fuzzy-based approach for contact identification in force-controlled robot tasks : experimental results and modelling influence. In Engineering Applications of Artificial Intelligence 16, pp. 691–707, (2003).

[4] K. Hornik. Feedforward networks are universal approximators. In Neural Networks 2, pp.359:366. 1989.

[5] K. Funahashi. On the approximate realization of continuous mappings by neural networks. In Neural Networks 2, pp. 183:192. 1989.

[6] K. S. Narendra and K. Parthasarathy. Identification and control of dynamical systems using neural networks IEEE Transactions on Neural Networks 1, pp. 4:27. 1990.

[7] M. O. Efe and O. Kaynak. A comparative study of neural network structures in identification of nonlinear systems. In Mechatronics 9, pp. 287:300. 1999.

[8] M. Dafaoui, Y. Amirat, C. François, J. Pontnau, "Analysis and design of a six dof parallel robot. Modeling, singular configurations and workspace", IEEE Transactions on Robotics and Automation, Vol 14, N 1, February 1998, pp 78-92.

[9] V. E. Gough, 1957. *Contribution to discussion of papers on research in automobile stability, control and tyre performance*, Proceedings Auto Div. Inst. Mech. Eng.

[10] Baret, M., 1978. *Six degrees of freedom large motion for flight simulation*, AGARD conference proceeding n249, Piloted aircraft environment simulation techniques, Bruxelles, pp 22-1/22-7.

[11] Merlet, J. P., 1997. *Miniature in parallel positionning MIPS for minimally invesive surgery*. In Proc. Of World Congress on Medical Physics and Biomedical Engineering, Nice, France, 14-19 September

Extended Kalman Filter Based Neural Networks Controller For Hot Strip Rolling mill

A.K. Moussaoui*, H.A. Abbassi** and S. Bouazza**

*Electrical Engineering Laboratory of Guelma (LGEG),
BP.401, University of Guelma, 24000, Algeria (e-mail: a_k_moussaoui@yahoo.fr).
**Université Badji Mokhtar BP 12 - 23000 - Annaba Algérie
(e-mail: h_a_abbassi@yahoo.fr and bouazza_s@yahoo.com)

Abstract: The present paper deals with the application of an Extended Kalman filter based adaptive Neural-Network control scheme to improve the performance of a hot strip rolling mill. The suggested Neural Network model was implemented using Bayesian Evidence based training algorithm. The control input was estimated iteratively by an on-line extended Kalman filter updating scheme basing on the inversion of the learned neural networks model. The performance of the controller is evaluated using an accurate model estimated from real rolling mill input/output data, and the usefulness of the suggested method is proved..

Keywords: Neural Networks Control, Extended Kalman Filter, Hot Rolling Mill, Bayesian Evidence.

1. INTRODUCTION

Neural Networks control method is one of the most promising control schemes for non-linear processes because of their ability to handle non-linearities. Neural networks represent a simple but powerful technique for process modelling because their flexibility makes them able to discover complex relationships between process input/output data [1]. The Neural Networks models are developed quickly using only real data, and they show better precision than other empirical modelling techniques [2]. Several neural networks models and their learning strategies, particularly multilayered feed-forward neural networks with the back-propagation learning algorithm, have been suggested and applied successfully to the identification and the control of non-linear dynamic processes [3]. Unfortunately, the lack of instrumentation, data acquisition capabilities, and computer facilities in several industrial processes hinder the on-line application of neural networks models. For example, if process conditions change from those used when training the neural network, data must once again be collected, analysed, and used for retraining the system.

In hot rolling process, feed-forward control is necessary to reduce the effects on the strip thickness of skid-marks and roll eccentricity the two largest disturbances responsible for the deviation of strip thickness in the finishing train. It is difficult to determine this control law in a complex process such as hot rolling mill [4-6].

The efficiency of a hot strip mill can be increased if the amount of rejected material is reduced. A strip is considered as rejected material if it does not meet the requirements of the customer and thus has to be sold as lower quality or has to be re-melted. This last option implies a tremendous amount of extra materials handling and energy costs.

The present paper deals with the optimization of the efficiency of the hot strip rolling mill by the reduction of the amount of rejected materials, which will save considerably the production costs. To fulfil this requirement, an effective thickness control able of reducing variations in thickness of the hot strip is suggested. The control strategy consists of an adaptive feed-forward on-line control law based on off-line forward Neural Networks model learning [7] and a Neural Network based extended Kalman filter as an on-line nonlinear controller. The control input is estimated on-line to make the process output to track a given reference signal.

2. HOT STRIP ROLLING PROCESS DESCRIPTION

A simplified schematic diagram of a steel rolling mill for the production of coil plate is presented in Figure 1. It shows the transformation stages of slabs from entry at the reheat furnace to their exit at the coiler at the end of the mill.

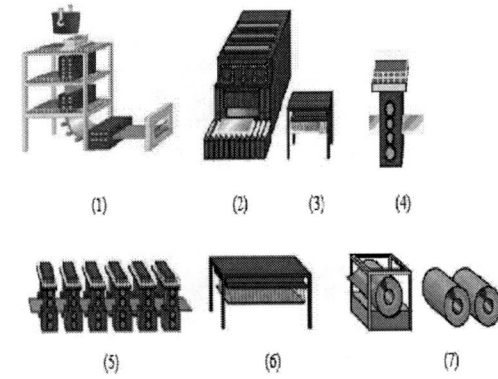

Fig.1. Hot rolling line layout

The feed stocks for the rolling mill are slabs produced by the continuous casting process in a steel plant (1). These are normally supplied at ambient temperature. The purpose of the reheat furnace (2) is to raise the temperature of the whole slab to the around 1250 °C (the re-crystallization temperature).

On exit from the reheat furnace, there is a build-up of scale on the surface of the slab, due to oxidation, which is detrimental to surface quality. This is removed within the de-scaling box (3), which consists of jets of high pressure water (140 bars).

After the de-scaling stage, the roughing mill (4) produces a breakdown bar (the product between the roughing mill and the finishing mill) by rolling the slab through a series of forward and reverse passes, typically reducing the slab thickness from 200 to 30 mm. The finishing mill (5) is designed to reduce the gauge (thickness) of the breakdown bar to that of the finished coil, while maintaining the desired width. The finishing mill control system is critical as constant mass flow must be maintained in all stands to ensure continuous production [8].

On exit from the finishing mill, the rolled strip is still at elevated temperatures, typically (> 800°C), which is above the phase transformation of the coil. Critical quality parameters, such as the mechanical properties and other metallurgical properties, of the finished coil are significantly affected by the cooling process applied in the run-out table (6). On exit from the mill / run-out table cooling system, the hot strip typically has a velocity of up to 15 m/s and can be hundreds of metres in length. The down coiler (7) allows the strip to be converted into a coil of dimensions that can be easily transported. The main characteristics of the considered hot strip rolling mill are given in Table 1 [9].

Table 1: Main characteristics of Arcelor Mittal hot strip mill, El-Hadjar, Algeria

Finisher mill Type	6 stand 4-High mill equipped with hydraulic screwdowns
Nominal Power	2 x 3000 HP, D.C.
Rolling Thickness Range	1.5 to 15 mm
Strip Width Range	600 to 1350 mm
AGC	Hydraulic
Gauge meter	X Ray
Rolling speed	49 to 407 rpm
Rolling temperature	860 to 880 °C
Maximum rolling Force	30000 KN
Maximum exit speed	15 m/sec

3. HOT STRIP ROLLING FORCE EMPIRICAL MODEL

The desired finishing thickness h_d and the attained thickness h_a are expressed by

$$h_d = S_o + \frac{F_o}{M} \quad (1)$$

$$h_a = S_o + \frac{F_a}{M} + \Delta S \quad (2)$$

where S_o is the unload roll gap, F_o is the reference roll separating force, F_a is the actual roll separating force, M is Mill stretch Constant (Mill Modulus), and ΔS is the change in roll gap by (AGC). The thickness deviation Δh is then

$$\Delta h = h_a - h_o = \frac{F_a - F_o}{M} + \Delta S \quad (3)$$

It is required that $\Delta h \to 0$ to bring the thickness to the desired value. The control equation is then

$$\Delta S = -\frac{F_a - F_o}{M} \quad (4)$$

For a given strip Temperature, a steel grade and a rolling speed, the actual separating rolling force can be calculated by the model of Alexander-Ford [10]:

$$F_a = 1,15\, b\, k_m \sqrt{R(h_e - h_a)}\, Q_p \quad (5)$$

where b is the plate width, R is the work roll radius, k_m is the mean constrained flow stress in the roll bite for plane strain conditions (strain resistance), expression under square root is the contact length, h_e is the incoming thickness, h_a is the exit thickness and Q_p is a geometric factor which is strongly affected by the geometry of roll bite and interface conditions between the rolls and rolled strip.

The term k_m can be calculated in function of a set of hot rolling parameters, like plate temperature T, viscous friction coefficient μ and rolling speed v.

In practice, there is no accurate physical model which describes efficiently the relationship between these hot rolling parameters. They are generally determined by empirical regressions. In fact, many authors proposed different equations for the calculation of the geometrical factor Q_p, based on their own experimental rolling data.

An analogue situation stands for the hot flow strength k_m, where there are many formulas that permit its evaluation from values of temperature, strain and strain rate [11].

A commonly used approximation of the term k_m is given by Misaka formula [12,13]:

$$k_m = \exp\left[0.126 - 1.75C + 0.594C^2 + \frac{(2851 + 2968C - 1120C^2)}{T}\right] k_r^{0.21} \left(\frac{dk_r}{dt}\right)^{0.13} \quad (6)$$

Where:
k_m: Steel Hot Strength [kgf/mm²],
C: Carbon content [%],
T: Absolute Temperature [°K],
k_r: True Strain,
t: Time [s].

As mentioned above, the mill operating parameters are generally determined from mathematical models based on expert metallurgical and mechanical knowledge [14]. Unfortunately, the rolling process involves many additional factors that still cannot adequately describe the deformation process in the roll gap. In this sense, the mathematical model is far from perfect.

To improve the thickness model accuracy, we suggest approximating the rolling process control model by the construction of a forward neural network using only measured input/output rolling mill data. The main reason for choosing neural networks is the fact that they are capable of reliably characterising non-linear functional relationships and require less process knowledge than the development of a phenomenological model.

4. NEURAL NETWORK TRAINING MODEL

To emulate the behaviour of the process, we consider the training of a Neural Network using Bayesian evidence approach [15].

The training of MLP may be classified into batch learning and pattern learning. In batch learning, the weights of the neural network are adjusted after a complete sweep of the entire training data, while in pattern learning the weights are updated at every time step in an on-line fashion. Batch learning has a better mathematical validity in the sense that the gradient descent method can be exactly implemented. Pattern learning, usually derived as batch learning approximations, can be used to modify the network weights on-line so that the model can track a time varying process dynamics.

In this work, batch learning is used for off-line training of the network to obtain the initial weights, and pattern learning is used for on-line updating of the weights.

According to Fig.2, the learning algorithm adjusts the weights in all connecting links and thresholds in the nodes so that the actual output $Y(t)$ and the target output $T(t)$ are minimized for all given training patterns.

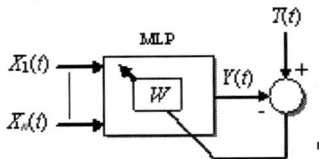

Fig.2. Learning Algorithm Scheme

The non-linear capability of the network was implemented using the *tanh* transfer function between the first and second layers. The Bayesian Evidence based training is investigated. When implementing Bayesian Evidence training [15], the error function was given by:

$$E_{BE} = \frac{\beta}{2}\sum_{n=1}^{N}(t_n - y_n)^2 + \frac{\alpha}{2}\sum_{i=1}^{W}w_i^2 \quad (7)$$

Where E_{BE} was the (Bayesian Evidence) error function and β the parameter describing the inverse variance of the noise model for the target data (predicted hot rolling force). If interest lies only in minimizing the error for a particular weight vector, then the effective value of the regularization parameter depends only on the ratio α/β.

Besides accommodating regularization in a consistent framework, the Bayesian approach has the additional advantage of providing a mechanism to generate confidence bounds on the output prediction values. Assuming that the posterior distribution of the weight matrix is Gaussian in nature, it is possible to find the variance corresponding to the mean output $y(x, w_{MP})$, i.e. the standard prediction output for the most probable weight distribution. This variance is given by [15]:

$$\sigma^2 = \frac{1}{\beta} + g^T A^{-1} g \quad (8)$$

Where A was the Hessian matrix defining the second derivatives of the error function and g was the gradient of the error function.

The standard deviation σ of the predictive distribution can be interpreted as an error bar on the mean value y_{MP} which has two contributions. The first arises from the intrinsic noise in the target data, the second from the posterior distribution of the network weights. The ease of implementation of these powerful network training paradigms was a major consideration in employing the NETLAB toolbox to realize the network training [15].

For the Bayesian Evidence training an initial value of $\alpha=0.01$ was employed along with an initial inverse noise variance parameter $\beta = 100$. During the Evidence update procedure of the network training, these hyper-parameters were re-evaluated iteratively.

5. ON-LINE ADAPTIVE CONTROL LAW

5.1. Extended Kalman Filter based control scheme

Extended Kalman filter can be used as a learning algorithm for forward multilayered neural networks weights estimation, or as a non-linear iterative on-line controller. For control purpose, the control input is estimated on-line to make the process output to track a given reference signal. Consider a class of non-linear discrete time dynamical system described by the regressive equation

$$y(k) = f(y(k-1), y(k-2), \ldots, y(k-p), u(k-1), u(k-2), \ldots, u(k-q)) \quad (9)$$

Where $y(k) = [y_1(k), y_2(k), \ldots, y_n(k)]^T \in R^n$ and $u(k) = [u_1(k), u_2(k), \ldots, u_m(k)]^T \in R^m$ are the process output and input vectors, respectively, p and q are the process orders and $f(\ ,\ )$ is an unknown vector function $f(x)=[f_1(x), f_2(x), \ldots f_n(x)]^T$. The process in equation (9) can be represented in a compact form as $y(k)=f(X(k-1))$, where

$X(k-1)=[u(k-1)^T, \ldots, u(k-q)^T, y(k-1)^T, \ldots, y(k-p)^T]^T \in R^{np+qm}$ is a vector consisting of the process inputs and outputs.

Fig.3 shows a n_i–input n_o–output feed-forward neural network with one hidden layer.

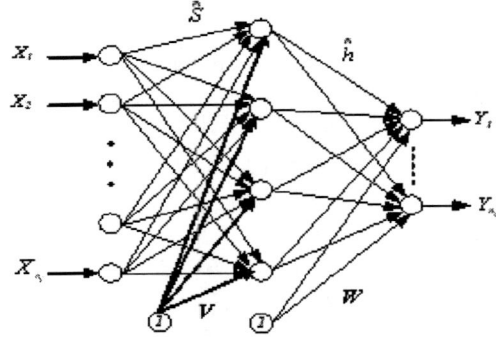

Fig.3. Feed-forward Neural Network model

The output of the hidden units can be represented as

$$\hat{h}_j = \rho(\hat{s}_j) = \rho(V_j^T I) \quad j=1,2,\ldots,n_h \quad (10)$$

Where $V_j^T = [V_{j0}, V_{j1}, ..., V_{jn_i}]$, $I^T = [1, X^T]$, $X^T = [x_1, x_2, ..., x_{n_i}]$ and $\rho(x) = \dfrac{1}{(1+e^{-x})}$ are sigmoid functions.

The output of the output layer neurons can be represented as

$$\hat{y}_i = W_i^T \hat{h}, \quad i=1,2,...,n_o \quad (11)$$

Where: $W_i^T = [W_{i1}, W_{i2}, ..., W_{in_h}]$ and $\hat{h} = [\hat{h}_1, \hat{h}_2, ..., \hat{h}_{n_h}]^T$

The non-linear process dynamics of equation (9) can be approximated by the network through properly determined parameters V and W. The process predictive output can be obtained using the neural network as

$$\hat{y}(k+1) = NN(X(k), V, W) \quad (12)$$

Where $X(k) \in R^{np+mq}$ is the network input and $\hat{y}(k+1) \in R^n$ is the process output representing the process output prediction.

5.1.1. EKF estimate of control input

For an output tracking control, the controller is designed to produce a proper control input $u(k)$, based on the information available at the time instant k, such that the process output $y(k+1)$ is made as close as possible to the desired $y_d(k+1)$, if the process is modelled with sufficient accuracy by the neural network prediction model.

5.1.2. Estimate Algorithm

At any time instant k, $y(k), y(k-1), ..., y(k-p+1), u(k-1), ...,$ and $u(k-q+1)$ are known. The neural network output can, therefore, be represented by:

$$\hat{y}(k+1) = NN(u(k)) \quad (13)$$

and the controlled process output is given by

$$y(k+1) = \hat{y}(k+1) + \xi(k) = NN(u(k)) + \xi(k) \quad (14)$$

where $\xi(k)$ represents the modelling error.

For a given reference signal, $y_d(k+1)$, the control input is determined to satisfy the following equation:

$$y_d(k+1) = NN(u(k)) + \xi(k) \quad (15)$$

The EKF is a state estimation method which can be applied in the case of this work if the control input vector $u(k)$ of equation (15) is viewed as a state vector. From equation (15), the following non-linear system equations can be obtained for the time instant k:

$$u_{i+1} = u_i \quad (16)$$

$$y_d = NN(u_i) + \xi_i \quad (17)$$

In the above equations, the time index k is omitted for simplicity and the subscript i denotes the number of the iterations during the kth control period.

Following the convention in the literature [16], ξ_i is assumed to be a white noise vector with a covariance matrix R_i. The application of the EKF to equations (16) and (17) gives the following discrete iterative estimation algorithm:

$$\hat{u}_i = \hat{u}_{i-1} + K_i(y_d - NN(\hat{u}_{i-1})) \quad (18)$$

$$K_i = P_{i-1} A_{i-1}^T (A_{i-1} P_{i-1} A_{i-1}^T + R_i)^{-1} \quad (19)$$

$$P_i = (I - K_i A_{i-1}) P_{i-1} \quad (20)$$

Where: $A_{i-1} = \left. \dfrac{\partial NN(u)}{\partial u} \right|_{u = \hat{u}_{i-1}}$.

In terms of the neural networks parameters, A_{i-1} is expressed as follows:

$$A_{i-1} = W^T \Lambda_{i-1} \bar{V} \quad (21)$$

where $W = [W_1, W_2, ..., W_n]$,

$\Lambda_{i-1} = diag\{\rho(s_{1,i-1})(1-\rho(s_{1,i-1})), ..., \rho(s_{n_h,i-1})(1-\rho(s_{n_h,i-1}))\}$

$\bar{V} = [\bar{V}_1, ..., \bar{V}_m]$ and $\bar{V}_l = [V_{1,l}, ..., V_{n_h,l}]$, $l=1,...,m$.

Starting from the initial estimate $u_0 = u(k-1)$, which is the control input at the last control instant, the above estimate procedure comprising equations (18) to (20) can be performed iteratively until the termination condition $\|y_d - NN(\hat{u}_i)\| \leq \delta$ is satisfied, where δ is a pre-specified error-tolerance level. To prevent excessive iterations in real-time implementation, a maximum iteration number i_{max} may be set. If $(i > i_{max})$ and the termination condition is still not satisfied, the estimate procedure breaks with the latest estimate $u_{i_{max}}$ as the control input, i.e.,

$$\begin{cases} u(k) = \hat{u}_i & \text{if } \|y_d - NN(\hat{u}_i)\| \leq \delta \\ u(k) = \hat{u}_{i_{max}} & \text{otherwise} \end{cases} \quad (22)$$

The control algorithm is executed in real-time according to the following steps:

1- Set $i=1$, $\hat{u}_0 = u(k-1)$, and $P_0 = \lambda I$ ($\lambda > 0$),
2- Calculate A_{i-1} using equation (21),
3- Compute $\hat{u}_i$ from equation (18) and (19) and update P_i using equation (20),
4- Apply $\hat{u}_i$ to the network and compute the tracking error; if $\|y_d - NN(\hat{u}_i)\| \leq \delta$ or $i \geq i_{max}$ go to step 6,
5- Set $i = i+1$; go to step 2,
6- Set $u(k) = \hat{u}_i$ and output $u(k)$ to the process.

The covariance matrix R_i in the equation (19) is usually chosen as a constant for the EKF.

In the case of this work, R_i represents the covariance of the model error; it is thus estimated by the following recursion:

$$R_i = R_{i-1} + \dfrac{1}{i}\left[(y_d - NN(\hat{u}_{i-1}))^T - R_{i-1}\right] \quad (23)$$

With an initial value $R_0 = \sigma I$ ($\sigma > 0$).

6. SIMULATION RESULTS

The EKF based control scheme developed in section 5, is shown in Fig.4. The overall control input u consists of the feed-forward on-line adaptive EKF based controller u_{ff} and a feed-back controller u_{fb}, where u_{ff} is determined by the iterative estimation algorithm described above and u_{fb} can be designed to be a conventional digital PID controller to compensate modelling errors.

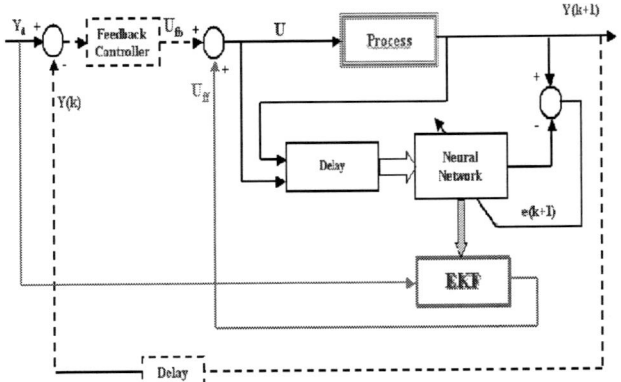

Fig.4. On-line EKF based Control Scheme

We consider Δh_p as the process output thickness deviation, Δh_n as the neural Network model output thickness deviation, ΔH_r as the reference thickness deviation and u as the control signal variation. The principles disturbances are: w_e (the roll eccentricity disturbance) and w_i (the input thickness disturbance).

6.1. Data acquisition and pre-processing

Measurements of several finishing rolling mill variables were recorded for a five days period of production using IbaAnalyser© (dedicated embedded real-time logging software). For our study, the manufacture of a single grade of steel coils (with constant width and exit thickness varying from 1.2 mm to 4.00 mm) was considered. The patterns were grouped into training, validation and test sets. The output values of Δh_n ranged from 0 to 1.

In order to fully utilise the maximum dynamic range of the *tanh* transfer function of the ANN networks, The input values were normalised to lie in the range -1 to +1 before presentation to the networks. Note that all results presented in this paper are plotted as the original (un-normalised) data. A sample of the most significant input/output variables used in our study for one rolling coil is shown in the curves of Fig. 5.

6.2. Discussion

A NNARX (non-linear based neural networks auto-regressive with exogenous input) topology is considered for modelling the process behaviour.

$$\Delta H_n(k+1) = NN(\Delta H_p(k), \Delta H_p(k-1), \Delta H_p(k-2), u(k), \quad (24)$$
$$u(k-1), u(k-2), T(k), w_i(k), w_e(k))$$

Where T(k) is the actual temperature of the hot strip.

The batch off-line learning of the model of equation (24) is conducted using a multilayered perceptron MPC and the real process data presented in section 6.1.

Bayesian training method was implemented to construct a series of MPC structures to model hot rolling process. According to training errors, it is shown from Fig.6., that the Bayesian training algorithm tended to produce smoother overall fitting functions to the training data [17].

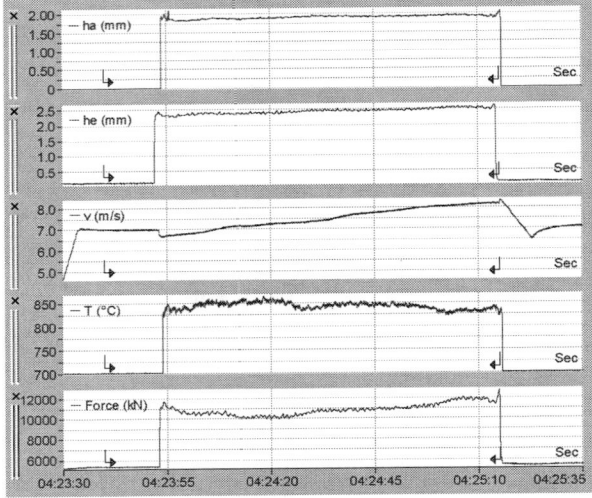

Fig.5. Hot rolling input/output samples

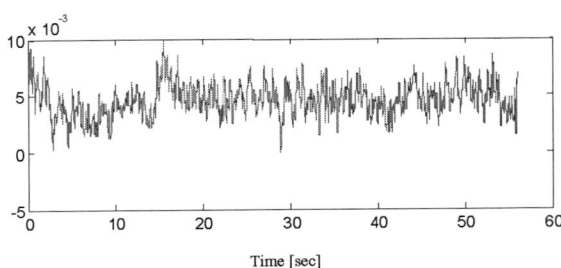

Fig.6. Evolution of Learning errors

With the learned neural network model, the suggested EKF based algorithm with an error tolerance of 0.02 and the maximum iteration number T_{max} =5 is applied control the hot rolling process. The process output is controlled to track a prescribed reference signal (from quality consideration).

With the application of the 2 principles hot rolling disturbance signals shown in Fig.7 (eccentricity and input disturbances), the simulation results, as shown in Fig.8, indicate that the suggested algorithm has a satisfactory reference tracking ability and disturbance rejection for such a non-linear considered process.

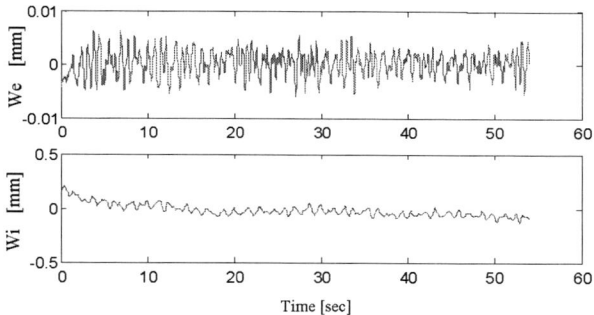

Fig.7. Eccentricity and Input Disturbances

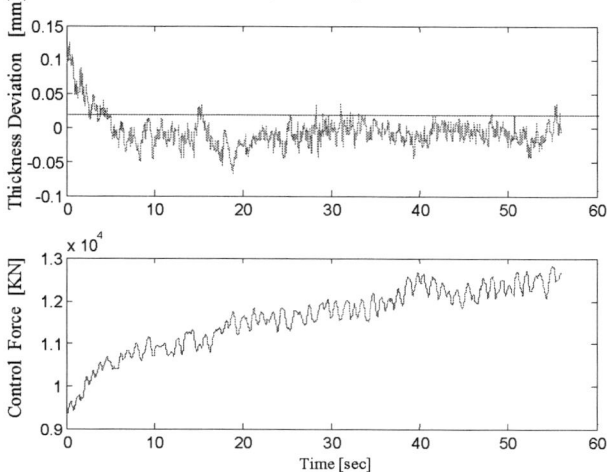

Fig.8. Evolution of Output and Control signals

CONCLUSION

Extended Kalman filter based Neural network, as a feed-forward non-linear iterative on-line controller, is considered. The control scheme is based on the on-line estimation of the control input to make the process output to track a given reference signal.

Using real input/output process data, Bayesian training method is implemented to construct a series of ANN structures to select a prediction model for the considered process.

From simulated results, it is proved that the suggested approach presents a good efficiency. Nevertheless, the control scheme can be enhanced by the consideration of a more sophisticated feed-back controller.

REFERENCES

[1] Neelakantan R., and J. Guiver (1998), Applying Neural Networks, *Hydrocarbon Processing*, Gulf Publishing Company, Houston, p.91-96.

[2] Bissessur Y., E.B. Martin, A.J. Morris and P. Kitson, (2000), Fault Detection in Hot Steel Rolling Using Neural Networks and Multivariate statistics, *IEE Proc. Control Theory Appl.*, **Vol 147, No. 6**, pp.633-640.

[3] Young-Sang, K., Y. Bong-Jin and K. Min, (2001), Robust design of artificial neural network for roll force prediction in hot strip mill, neural networks. Proc. IJCNN'01, Intl. Joint Conf., 4, pp.2800-2804.

[4] Zietsman J.H., S. Kumar, J.A. Meech, I.V. Samarasekera, & J.K. Brimacombe, (1998), Taper Design In Continuous Billet Casting Using Artificial Neural Networks, *Ironmaking and Steelmaking*, **25 (6)**, pp.476-483.

[5] Vermeulen W., A. Bodin, & S. Van Der Zwaag, (1997), Prediction of the Measured Temperature after the last Finishing Stand Using Artificial Neural Networks, *Steel Research*, **68(1)**, pp.20-26.

[6] Dairiki O., H. Mabushi, I. Degawa, & H. Nakamura, (1989), Progress of AGC Utilisation Techniques at Plate Mill of Oita Works, *Nippon Steel Technical Report*, **No.42**, pp.38-47.

[7] Hagan M.T., Menhaj M., (1994), Training Feed-forward Networks with the Marquardt Algorithm, *IEEE Transactions on Neural Networks*, **5(6)**, pp.989-993.

[8] Oda T., N. Satou, & T. Yabuta, (1995), Adaptive Technology For Thickness Control of Finisher Set-up on Hot Strip Mill, *ISIJ International*, **35(1)**, pp.42-49.

[9] Moussaoui, A.K., H.A. Abbassi, S.E. Bouazza and A. Kondratas, (2003), Neural-networks-based adaptive non-linear controller for hot strip rolling mill. *Mechanika*, **No. 6**, pp.55-61.

[10] Ford, H. and J.M. Alexander, (1963-64), Simplified hot rolling calculations. *J. Institute of Metals*, **92**, pp.397-403.

[11] Fumio, Y., S. Kunio, T. Masashi, A. Yoshiharu, A. Yasushi, F. Charles, G. Maurice and C. Trevor, (1991), Hot strip mathematical models and set-up calculation. *IEEE Trans. Industry Applications*, **27**, pp.131-139.

[12] Misaka, Y. *et al.*, (1967-1968), Formulatization of mean resistance to deformation of plain C steels at elevated temperature, *J. Japan Society for the Technology of Plasticity*, **8**, pp.414-422.

[13] Royzman, S.E., (1996), Thermal stresses in slab solidification, *Asia Steel*, pp.158-162.

[14] Watanabe, T., H. Narazaki, A. Kitamura, Y. Takahashi, H. Hasegawa, (1997), A new mill-setup system for hot strip rolling mill that integrates a process model and expertise, systems, man and cybernetics, computational cybernetics and simulation. *IEEE Intl. Conf.*, **3**, pp.2818-2822.

[15] Nabney I.T. (2002), *Netlab-Algorithms for Pattern Recognition*, Springer-Verlag.

[16] Bahera L., M. Gopal, and S. Chaudhury, (1995), Inversion of RBF Networks and Applications to Adaptive Control of nonlinear systems. *IEE. Proc.-Control Theory Appl.* **Vol. 142, No. 6**, pp.617-624.

[17] Moussaoui A.K., Y. Selaimia and H. A. Abbassi, (2006), Hybrid Hot Strip Rolling Force Prediction using a Bayesian Trained Artificial Neural Network and Analytical Models, *American Journal of Applied Sciences*, **Vol. 3, No. 6**, pp.1885-1889.

SESSION E1: COMPUTER VISION

A Direct Frequency-based Phase Algorithm for Motion Estimation

M. Boughazi*, B.Boulebtateche*, M. Fezari*, L. Zouaoui**, N. Bonnet***

*LASA, Electronic Department, Annaba University, Algeria.
**Physic Department, Setif University, Algeria.
***CReSTIC-LERI, Reims University, France.
boughazi@leri.univ-reims.fr

Abstract: In this paper, we propose an efficient direct method that estimates pure translation motion parameters in dynamic scenes. We develop an algorithm that computes, directly, phase information in Fourier domain. The computations are locally performed and limited to small patches of the image. The local measurement of motion parameters are then combined in a more global interpretation by using parametric images. We demonstrate the performance of our algorithm on synthetic and real image sequences and compare our results with those of two other algorithms based on the Hough transform and the inverse Fourier transform.

Keywords: Motion, Optical Flow, Phase, Hough Transform

1. INTRODUCTION

Motion estimation is a fundamental problem for the analysis of image sequence. The importance of motion in visual perception is mostly due to the role that motion stimulus plays in many aspect of human and machine vision, among them the perception of depth, segmentation of dynamic scenes and the estimation of the motion of objects in the world. In addition, there are a great variety of applications that depend on motion analysis in image sequences. Such applications cover a large spectrum of everyday life. These include motion detection for surveillance, image data compression, image understanding mobile robot navigation, aerial image registration and so on. However, determining the relative motion between an observer and his environment is a major problem in computer vision.

Motion estimation consists of measuring 2-D projection on the image plane of a 3-D scene due to the moving objects in the scene and to the displacement of the camera. The result is known as the optical flow field, a distribution of apparent velocities of moving brightness patterns in an image.

Optical flow, besides tracking techniques, allows estimation and analysis of object motion or regions of interest within image series. Since the influential paper on optical flow first published by Horn and Schunk [1] three decades ago, a great deal of effort has focused on finding ways to compute optical flow more efficiently and more accurately.

Most machine vision techniques that try to extract image flow utilize just two successive frames from an image sequence, either by matching features from one image to the next one or by calculating the intensity change between such successive frames along the image gradient direction.

The measurement of image velocities has been approached from many aspects. Cross-correlation techniques or global matching models try to extract image flow by performing a matching over some large regions or patches in the image [2]. This is accomplished by sliding the image from one frame to match the image in the next frame in an optimal way. The difficulty with these approaches is that their performance is greatly reduced when dealing with complex motions in a scene.

Feature extraction and matching is another way to compute the flow field [3]. Such methods are unable to deal with highly textured images because of the inherent problem of correspondence.

A third approach that avoids the problem of correspondence is based on calculating the gradient in spatial and temporal dimensions [4][5]. The derived methods use the brightness constancy assumption to constrain the space of solution and thus to express the relationship between the temporal derivative of the sequence and the spatial derivatives in a simples form. They use the computation of spatio-temporal derivatives of light intensity and stipules that the brightness intensity of an image pixel corresponding to a fixed point on the object does not change after motion. However, such constraint would only permit to compute one component of the local velocity vector because of the aperture problem. The accuracy of the corresponding discrete operators is limited and greatly affected by high frequency noise. Some regularization techniques have been proposed to avoid such limitation by achieving a balance between the brightness constraint and smoothness of the flow field by minimising a cost function. However, the solution thus obtained is not guaranteed to converge to a global minimum because of the nonlinear aspect of the optimization problem as formulated by these techniques [1] [6] [7].

Frequency-based methods, known as spatio-temporal filters, have also been used. They use a bank of 2-D or 3-D filters such as Gabor filters, where each filter is sensitive to a separate direction of motion. The main drawback of such operator is their sensitivity to illumination changes [8][9].

The use of the phase information has been proved to be well suited to deal with the problem of motion estimation. As pointed out by many researchers, phase information may lead to better result with respect to performance and robustness [10] [11] [12].

This is due to the fact that the phase component of a filtered band pass signal is less sensitive to the changes of illumination than the component of the amplitude [13]. More over, the phase of the Fourier transform is rich in essential information. For instance, an image can still be identified even if the information conveyed by the amplitude is not considered. Many techniques based on extracting motion parameters using phase information have been proposed in recent years. A technique that computes the phase information by the use of inverse Fourier Transform is given in [14]. The problem of phase-based motion estimate has also been formulated as a parametric model characterization in the Hough Transform space [15].

Our contribution, in this paper, use another approach based on a direct estimation of phase component in the frequency space.

In addition, an algorithm is also given to solve the problem of phase wrapping because in the frequency space is can only obtain a relative value of the phase i.e. modulo 2π (or $\pm\pi$). Therefore an algorithm for phase unwrapping is needed to compute the absolute value of the phase [16]. And finally, a novel technique which uses the concept of parametric images in order to infer global motion parameters is presented.

In the following sections, the problem of optical flow computation is formulated and some background is given in section 2. Then, the use of phase information in motion estimation is presented in more details and the approach that we use to directly estimate phase values in frequency space is fully described in section 3. This approach is actually applied to a pure translation motion in a dynamic scene. In section 4, experimental works and results obtained by our algorithm for both synthetic and real sequences are presented and compared with those obtained by the Hough transform and the inverse Fourier transform based algorithms. And finally, conclusions and perspectives for future research are discussed in section 5.

2. OPTICAL FLOW ESTIMATION PROBLEM

The apparent 2-D motion that results from projection of a 3-D scene onto the image plane is only accessible via the analysis of spatial and temporal variation of image sequences under some assumption on the brightness variation. The well-known brightness constraint stipules that the image brightness of a pixel is constant over time. As a result of this, the total derivative of its intensity with respect to the spatio-temporal coordinates is zero. According to this strong assumption, we thus obtain that:

$$I(x, y, t) = I(x + v_x \Delta t, y + v_y \Delta t, t + \Delta t) \quad (1)$$

$$\frac{dI}{dt}(p,t) = \nabla I(p,t)v(p) + \frac{\partial I(p,t)}{\partial t} = 0 \quad (2)$$

$$v_x I_x + v_y I_y + I_t = 0$$

where I_x, I_y, and I_t are the spatio-temporal image derivates and v_x, v_y are the components of flow along the x and y directions. This constraint describes the equation of a line.

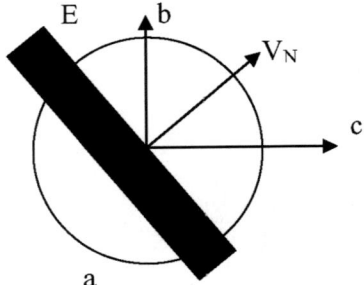

Fig. 1: The aperture problem.

Only the component of motion V_N in the direction perpendicular to the moving edge E can be computed though the aperture A.

This equation is called the equation of the optical flow and makes it possible to estimate the component V_N of the velocity vector oriented in the same direction as the gradient. The component V_N is given by the following expression:

$$V_N = -\frac{I_t}{\|\Delta I\|} \quad (3)$$

where $I_t = \frac{\partial I}{\partial t}(p,t)$ denotes the temporal variation of $I(p,t)$ and V_N is the projection of the velocity vector on the gradient vector of the image ∇I.

The relation (1) shows an inherent problem which is proper to local motion estimation. The assumption of brightness conservation when applied to local parts of the image allows finding only the velocity component that is perpendicular to moving contours seen through a small window in the image (See Fig. 1.). This indetermination is known under the name of aperture problem. Moreover, the estimate is impossible if $\nabla I = 0$. This is the case of uniform un-textured surface. Therefore, the whole set of optical flow estimation techniques must integrate some additional information on spatial and sometimes spatio-temporal domain in order to compute local measurement of the motion. The use of phase information may greatly help dealing with the issue of local motion estimation.

3. DIRECT FREQUENCY-BASED PHASE ESTIMATION ALGORITHM

The generalization of phase information for optical flow was emphasized by many researchers. For instance, it is claimed in [13] that the phase component of band pass filters is less sensitive to noise than the amplitude component when small deviation from image translation are considered in 3D scenes. However, they also show that phase singularities and hence the phase may become instable. In our implementation, the computation of phase information follows many stages. First, a pre-processing procedure is needed to smooth out unwanted noise that may occur in region of interest within each pair of successive images. For this purpose, moving Gaussian filters of various widths are used. This step will hopefully reduce the effect of noise and consequently it will control to some extent phase instabilities. Thus, each local part of the image is convolved with a spatial Gaussian window moving over the entire image the sampling rate of the image by such a window is taken to be an integer number of pixels (i.e. the centre of window moves by steps, of say 4, 8 or 10 pixels).

In addition, sampled parts of the image may overlap and this will avoid loss of information and assure correlation at boundaries of neighbouring regions. The Fourier transform is then computed for each pair of corresponding smoothed local region of the two successive images. The computation is performed locally and applied over the entire image at predefined steps. Afterwards, the phase information is derived for each such local region in the following manner:

Let $F(u,v)$ be the 2-D Fourier transform of an image $I(x,y)$, where x and y are the spatial coordinates, we may write:

$$\Im[I(x,y)] = F(u,v) = \sum_x \sum_y I(x,y) e^{j2\pi(ux+vy)} \quad (4)$$

where $\Im[I(x,y)]$ stands for the Fourier transform of a 2-D function, u and v denote the spatial frequency along the horizontal and vertical direction respectively. If we introduce the time variable as the third coordinate in the 3-D space, an image at time t can be represented as a 3-D function $I(x,y,t)$ and its Fourier transform is written as $\hat{I}(u,v,w)$ and w is the temporal frequency variable. Let v_x and v_y denote the horizontal and vertical components of the velocity vector for a pure translation motion.

Let $I_1(x,y,t)$ and $I_2(x,y,t)$ be two successive images in a sequence which undergoes a pure translation motion. We may thus, write that:

$$I_2(x,y,t) = I_1(x \pm v_x \Delta t, y \pm v_y \Delta t, t + \Delta t) \quad (5)$$

and according to the shift property of Fourier transform we obtain that:

$$\Im[I_2(x,y,t)] = \Im[I_1(x \pm v_x \Delta t, y \pm v_y \Delta t, t + \Delta t)]$$
$$= \Im[I_1(x,y,t)] e^{j2\pi(\pm v_x u \pm v_y v)} \quad (6)$$
$$= |\hat{I}(u,v,w)| e^{j(\phi_1 + \Delta\phi)}$$

where ϕ_1 is the phase component of the first image and $\Delta\phi$ is the amount of phase variation produced by the pure translation motion of the image. We observe that the phase difference between the two successive images is given as a linear expression in terms of Fourier component, and for a pure translation towards increasing value of x and y spatial variables this can be written as:

$$\Delta\phi = |\phi_1 - \phi_2| = v_x u + v_y v \quad (7)$$

There are many ways to derive the component v_x and v_y of the velocity vector at an (x, y) location when utilizing the difference in phase information. Among them we may use the Hough transform to compute v_x and v_y by constructing the Hough space for line parameters in v_x and v_y (or (u and v). This method is however very costly in computation time. The inverse Fourier transform can also be used to infer the velocity vector (v_x, v_y) by detecting peaks in the inverse transform of the normalized cross power spectrum of the two images which is equal to $e^{\pm(v_x u + v_y v)}$. This expression corresponds to the Fourier transform of a Dirac delta function centered at $\pm(v_x \Delta t, v_y \Delta t)$ in the image domain; Δt is the temporal sampling rate of the sequence. However, this technique is also computationally expensive.

In order to speed up necessary computation for extracting phase information we preferred to directly work in the Fourier domain and hence the extra time for calculating the inverse Fourier transform is avoided. In fact, we know that the phase difference between the two images can be expressed from the Fourier transform, which is written in complex form, as follows:

$$\Delta\phi = arctg\left(\frac{X_1 \cdot R_2 - X_2 \cdot R_1}{R_1 \cdot R_2 + X_2 \cdot X_1}\right) \quad (8)$$

where (R_1, X_1) and (R_2, X_2) are the complex Fourier transform of image I_1 and I_2 respectively.

However, the Fourier transform provides only a relative value of the phase, i.e. the phase spectrum is wrapped modulo 2π for all the spatial frequencies. Therefore, it is necessary to apply an unwrapping algorithm to derive the absolute value of the phase component. The results obtained from the unwrapping process are simply displayed in form of a set of graded bars arranged in both horizontal and vertical directions. The number of such bands depends on the velocity components v_x and v_y, and also on the dimension of the image domain, as shown in [17]. The components v_x and v_y are respectively computed by considering the cumulated sum of the differences between values of last and first band for both horizontal and vertical directions. The corresponding values will be denoted by V_{xx} and V_{yy} respectively and the velocities components are easily derived such that:

$$v_x = \frac{V_{xx}}{2\pi \times \omega} \quad (9)$$

$$v_y = \frac{V_{yy}}{2\pi \times \omega} \qquad (10)$$

where ω is the width of the local region of interest and (v_x, v_y) represents the velocity vector for that region centered at x and y.

4. RESULTS AND DISCUSSION

In this section, the performance of our algorithm is evaluated on synthetic and real image sequences and comparison of the obtained results with those of two other techniques, namely the ones based on the Hough Transform and the inverse Fourier transform is presented as well. In our implementation, the computation of phase information consists of, firstly, smoothing each image by applying a spatial Gaussian window of a given size. In our case, we have used three different windows of size (32*32, 64*64 and 128*128) pixels respectively. The standard deviation of the corresponding filters take on values of σ = (2, 4 and 8). The image is scanned over by the smoothing window at regular steps. The center of the window is moved by steps of (4, 8 and 10) pixels. The computation are, thus, carried out at lower resolution and limited only to overlapping small windows. The phase information is computed directly from the Fourier transform for each sub-region delimited by such windows. By using this scheme, it can be easily seen that the local implementation of our algorithm may result in great savings in computation time. Our algorithm was firstly run on synthetic sequences. The main advantage of using synthetic data is that the 2-D motion fields and the experimental environments can be controlled and tested on in a systematic manner. In particular, the true 2-D motion field is already known and therefore it will be easy to quantify the performance of implemented algorithms to some extent. And consequently, this can be employed as an optimistic bound on the measure of the expected errors with real image sequences test. The set of synthetic image inputs being used include the following test sequences:

- Translating square: this sequence contains a dark square (with a width of 40 pixels) translating over a bright background at uniform velocity along three directions (See Fig. 2.). This type of inputs helps to illustrate the aperture problem.

For the simulation with real image sequences, we have tested our algorithm on the following real sequences:

- Trees SRI Sequence: This sequence, described in the reference [2], was filmed by a camera moving in a plan parallel to the trees at the velocity vector $(v_x, v_y) = (-1, 0)$ pixels per frame.
- A Tornado sequence: the successive images of this sequence show up the effect of a tornado translating in an oblique direction defined by the velocity vector $(v_x, v_y) = (3, 2)$ pixels per frame.

For the comparative purpose, and following the work given in [17] by Barron et al., we have employed an angular measure of the error between the real velocity vector and the one estimated by the implemented algorithms. The angular error is defined as follows:

$$\phi_{err} = Arc\cos(V_c, V_e) \qquad (11)$$

where V_c is the correct velocity vector at any location and V_e is the estimated optical flow vector at that location. The angular error ϕ_{err} is estimated by calculating the following expression:

$$\phi_{err} = Arc\cos\left[\frac{V_c \cdot V_e}{\|V_c\| \cdot \|V_e\|}\right] \qquad (12)$$

The obtained performance is expressed as a measure of this error and is summarized in Table I for the three algorithms. As it is reported on the table, it can be seen that the proposed algorithm yields satisfactory results compared with those obtained by using the Hough transform or the Inverse Fourier transform.

5. CONCLUSION

We have examined the performance of the proposed algorithm that extracts phase information directly from the frequency domain and thereby gives an insight on the translation motion that may arise in dynamic scene. The results obtained by our algorithm are encouraging compared with those obtained by the use of the Hough transform which suffer from the problem of an excessive computational time or those produced by the inverse Fourier transform-based approach. The latter approach is less costly in computation time but suffers of imprecision in dealing with sub pixels displacements. Our algorithm, however, shows up same performance but performed in shorter time. On the other hand, the problem of phase wrapping has been solved in a simpler way. The algorithm works equally well for simply structured or complex textured images and gives good results in both cases. Moreover, a representation by parametric images is used to better characterize moving objects in a sequence of images. However, the implemented algorithm is only limited to pure translation motions and some further research is needed to deal with other type of motion, namely rotation and scaling. This will be our future work.

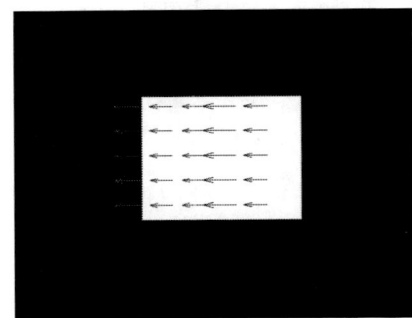

Fig. 2: Translating Square horizontal motion at a speed of (-1.5 , 0) pixels/frame.

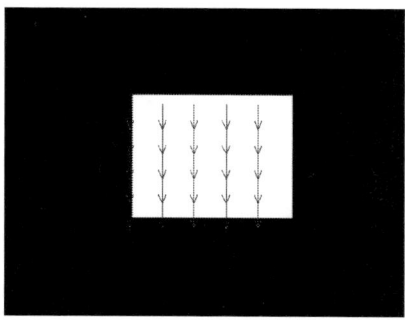

Fig. 3: Translating Square. Vertical motion at a speed of (0, 1.85) pixels/frame.

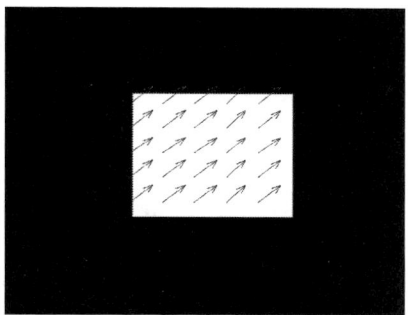

Fig. 4: Translating Square. Diagonal motion at a speed of (1.33, -1.33) pixels/frame.

Fig. 5: Outdoor Sequence Horizontal Camera motion at a speed of (1pixel/frame).

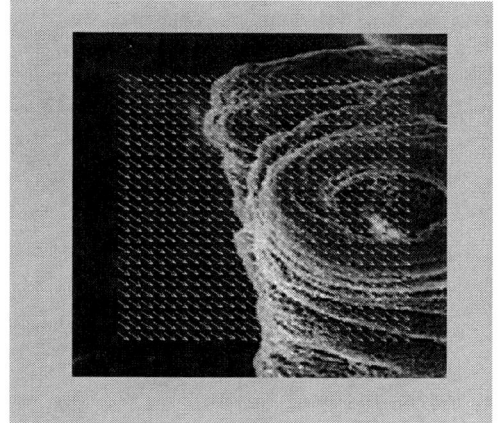

Fig. 6: Tornado Sequence. Oblique motion at speed of (3, 2) pixels/frame.

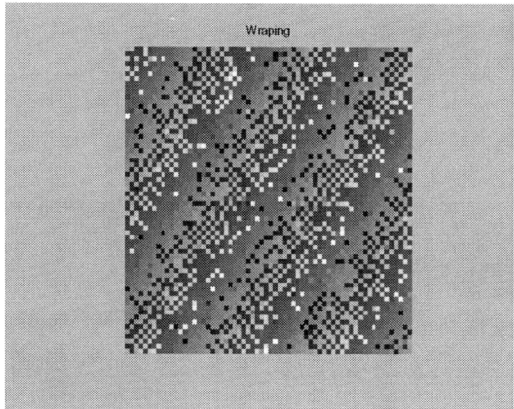

Fig. 7: Wrapped relative phase component for Tornado Sequence.

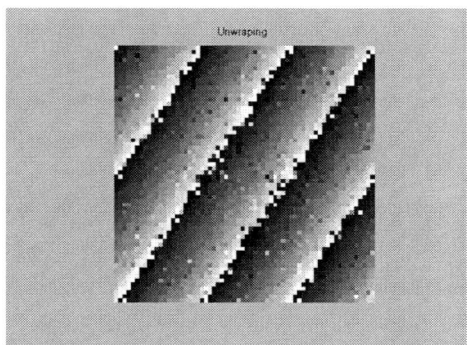

Fig. 8: Wrapped relative phase component for Tornado Sequence obtained by our algorithm.

Fig. 9: Unwrapped phase component along horizontal direction.

Fig. 10: Unwrapped phase component along vertical direction.

frames	Method	Mean angular error	Density
21	Fleet & Jepson	0.32°	74.5%
10	Inverse Fourier method	1.50°	100%
2	Hough transform (Vernon)	3.53°	100.0%
2	Proposed Algorithm	0.,78°	100%

Table 1 Results Measured angular error and performance obtained by different methods. Entries for first column indicate the number of successive images necessary for each method to compute the optical flow.

REFERENCES

[1] B.K.P. Horn and B.G. Schunck, "Determining Optical Flow," Artificial Intelligence, vol.17, pp.185-203, 1981

[2] M.A. Arredondo, K. Lebart and D. Lane, "Optical flow using textures", Pattern Recognition Letters, vol. 25, pp. 449-457, 2004.

[3] M. Pingault, D. Pellerin, "Motion estimation of transparent 84, pp. 709 – 719, April 2004.

[4] A.R.Bruss and B. K. Horn, "Passive navigation," Computer Graphics and Image Processing, Vol.21, pp. 3–20, 1983.

[5] Li, M., Biswas, M.: DCT-based phase correlation motion estimation, ICIP 2004, 24-27 Oct. 2004, p.445 – 448.

[6] K.Prazdny, "On the information in optical flows," Computer Graphics and Image Processing, Vol. 22, pp. 239–259, 1983.

[7] F. Girosi, A. Verri and V. Torre "Constraints for the computation of optical flow," Proc. IEEE Workshop on Visual Motion, Irvine, pp. 116-124, 1989.

[8] E.H. Adelson and J.R. Bergen, "Spatiotemporal Energy Models for the Perception of Motion," J. Opt. Soc. Am. A, Vol.2, pp.284-299, 1985.

[9] D.J. Heeger, "Model for the Extraction of Image Flow," J. Opt. Soc. Am. A, Vol.4, n°8, pp.1455-1471, 1987.

[10] A D Jepson and D J Heeger, "Linear subspace methods for recovering translation direction," Spatial Vision in Humans and Robots, pp.39–62. Cambridge University Press, New York, 1993.

[11] N. Bonnet and P. Vautrot, "Image analysis : is the Fourier transform becoming obsolete Microsc," Microanal. Microstruct Vol. 8, 59-75. 1997.

[12] S.S. Beauchemin and J. L. Barron, "The Computation of Optical Flow", ACM Computing Surveys, vol. 27, No.3, pp. 433-467, September 1995.

[13] D.J. Fleet and A. D. Jepson, "Stability of phase information," IEEE Trans. PAMI, Vol.15, n°12, pp.1253-1268, 1993.

[14] M. Boughazi, N. Bonnet, M. Bedda and L. Zouaoui, "Improvement and Comparative Study of Two Methods for the Estimation of the Optical Flow," Asian Journal of Information Technology, Grace Publications, Vol. 4, n°.7, pp. 694-7003, 2005.

[15] D .Vernon, "Computation of instantaneous optical flow using the phase of Fourier components," IEEE Trans. PAMI Vol. 14 , n° 3, pp. 346-352, 1998.

[16] M. Costantini, "A novel phase unwrapping method based on network programming," IEEEE Trans.Geosci. Remote Sens. Vol. 36, pp. 813-8, 1998.

[17] J.L.Barron, D.J. Fleet and S.S. Beauchemin, "Performance of Optical Flow Techniques," International Journal of Computer Vision, Vol..12, n°1, pp. 43-77, 1994.

MRI Images Compression Using Curvelets Transforms

M. Beladgham[*], I. Boucli Hacene[*], A. Taleb-Ahmed[**], and M. Khélif[*]

(*) Geni Biomedical laboratory, Departemet of computer science
Abou bekr Belkaid university, Tlemcen, 13000
(**) Biomecanic Laboratory, Valecienne, France
(e-mail: beladgham@yahoo.fr)

Abstract: In the field of medical diagnostics, interested parties have resorted increasingly to medical imaging, it is well established that the accuracy and completeness of diagnosis are initially connected with the image quality, but the quality of the image is itself dependent on a number of factors including primarily the processing that an image must undergo to enhance its quality. We are interested in MRI medical image compression by Curvelets, of which we have proposed in this paper the compression algorithm FDCT using the wrapping method. In order to enhance the compression algorithm by FDCT, we have compared the results obtained with wavelet and Ridgelet transforms. The results are very satisfactory regarding compression ratio, and the computation time and quality of the compressed image compared to those of traditional methods.

Keywords: Compression, MRI, Wavelets, Ridgelets, Curvelets, FDCT

1. INTRODUCTION

The massive use of numerical methods in medical imaging (MRI, X scanner, nuclear medicine, etc....) today generates increasingly important volumes of data. The problem becomes even more critical with the generalisation of 3D sequence. So it is necessary to use compressed images in order to limit the amount of data to be stored and transmitted.

Many compression schemes by transformation have been proposed, we can cite the standards JPEG images, MPEG 1 and 2 for compressing video. All of these standards are based on the discrete cosine transform (DCT). (Chappelier, 2005)

Over the past ten years, the wavelets (DWT), have had a huge success in the field of image processing, and have been used to solve many problems such as compression and restoration of images (Mallat, 1989). However, despite the success of wavelets in various fields of image processing such as encoding, weaknesses have been noted in its use in the detection and representation of the objects' contours. The wavelets transform and other classical multi resolutions decompositions seem to form a restricted and limited class of opportunities for multi-scale representations of multidimensional signals.

Recent work has shown that it is possible to define the conceptual context in larger multi scale representations giving rise to more interesting new transform that are more suited to the extraction of smooth and continuous geometric structures, such as objects' contour. One example is the new family of Curvelet transform. These are multi scale decompositions which operate under a variety of frequency guidelines that provide a good compromise between sparse representation (or compact) of characteristic features and perceptual quality of the reconstructed image whose effectiveness in image processing has already been proved. In this work we propose the FDCT algorithm for MRI image compression. For this reason, this paper is divided into three parts: the first of which is devoted to a representation of the curvelet transform, then we present the FDCT via wrapping method.

In order to enhance the image compression algorithm by FDCT, we compare the PSNR results obtained with the existing techniques namely the wavelets and Ridgelets

2. CURVELET TRANSFORM

The curvelets have been proposed by E. Candes & D. Donoho 1999 (Candes, and al,1999), (Boubchir, 2005). They are a new family of geometric wavelet frames that are more effective than traditional transforms. They are intended to represent the contours in a parsimonious way.

For example, on Fig.1 (a), the wavelet would take lot of coefficients to represent precisely such a contour. Compared with wavelets, curvelets may represent a smooth contour with fewer coefficients with the same precision (Fig.1). (Boubchir, 2005)

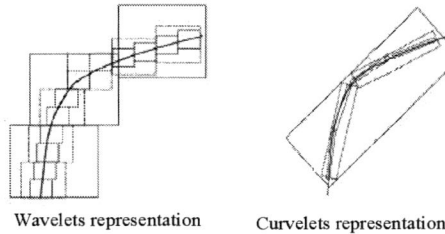

Wavelets representation   Curvelets representation

Fig.1. Comparison of the performance of the non-linear approximation using wavelets and curvelets

In the technical and scientific literature, we find two types of curvelet transform. The first continuous curvelet transform was built by E. Candes and Donoho in 1999 on the basis of ridgelet transform (Candes, and al,1999). This mode enables to obtain a multi scale analysis in sub-bands on the

Frequency crowns $f \in \left[0, \frac{1}{2^{2s}}\right]^2 \setminus \left[0, \frac{1}{2^{2s+2}}\right]^2$, where $s \in N$ represents the scale.

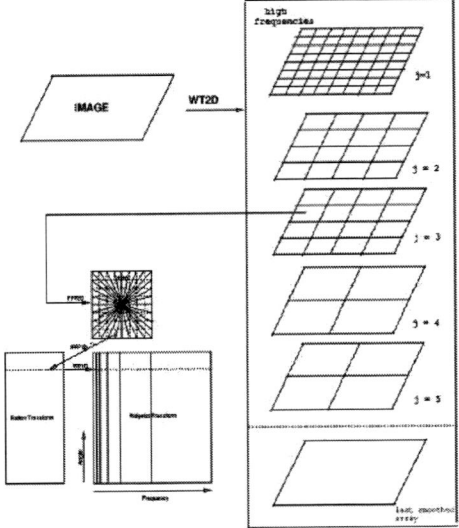

Fig. 2. Organizational structure of the Curvelets transform
- Decomposition of the original image into sub-bands
- Space Division in blocks of each sub-bands
- Ridgelets transform is applied to each block

Based on work carried out by scientists in the field of noise and image restoration, this implementation is also unnecessary.

Furthermore a delicate point is the computation of the Fourier transform in polar coordinates. Indeed this calculation requires 2D interpolations, which causes a lot of numerical errors.

These researchers have proposed replacing the last step (Ridgelet transform) by a clever combination of Fourier coefficients (wrapping method) (Candes and al, 2006), (Demanet, 2006). The construction of the spatial grid is based on the theory of frames. The analysis frame is built directly from a 2D high frequency mother function C according to one of the axes and low frequency depending on the other (typically the tensor product of a wavelet function and a scale function). The corresponding family of curvelets $(C_{l,n,\theta})_{l \in N, n \in Z^2, \theta \in \frac{2\pi}{2^l} Z / 2^l Z}$ is then given by

$$(C_{l,n,\theta}) = 2^{\frac{3l}{2}} C(D_l R_\theta t - n) \quad (1)$$

$$D_l = \begin{pmatrix} 2^{2l} & 0 \\ 0 & 2^l \end{pmatrix}$$

(Let us note in passing that $2^{\frac{3l}{2}}$ corresponds to the square root of its determinant), R_θ is the matrix angle rotation θ and n indicates the position of the curvelet. This transform has been used successfully in the denoising context (Minh, 2002), (Starck and al, 2002).

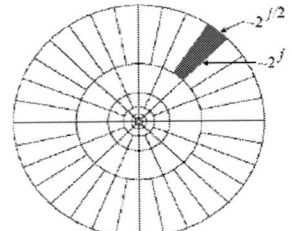

Fig.3. Curvelet tiling in the frequency

3. DISCRET CURVELET TRANSFORM VIA WARPPING METHOD

Our choice in this article was done by the algorithm of Laurent Demanet who has used only 2 types of rectangle $R_{j,l}$ for each scale. We will thus distinguish then curvelets with a vertical orientation rather than a horizontal one. This corresponds to the implementation of the "wrapping" that is introduced in reference (Candes and al, 2005).
One then obtains more elements in this new frame. However, it will require much less interpolation to translate an image from these curvelets coefficients.
Figure 4 shows the spectral partitioning caused by the FDCT (Candes and al. 2006)

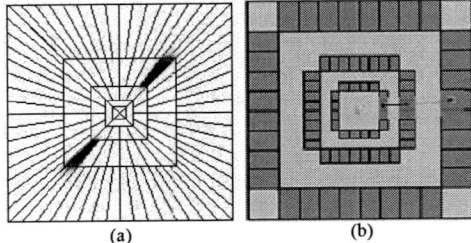

Fig.4. Oriented pyramidal decomposition by FDCT
(a) FDCT spectral partition (b) FDCT of Lena image

3.1 FDCT algorithm via wrapping

Discrete curvelet transform, as presented in reference (Candes and al, 2005) has been implemented in four steps:
1. Apply 2DFFT to the image.
2. Divide the interval of the frequency by cancelling a dyadic on concentric squares. **(Fig. 5 /a)** and obtain the Fourier sample $\hat{f}[n_1, n_2] - n/2 \leq n_1, n_2 < n/2$ (2)
3. Each crown is divided into trapezoidal region.
4. For each tile in the crown: every angle l and scale j
   A. Originally translate this tile. (Fig.5/a); by forming the product $\tilde{U}_{j,l}[n_1, n_2] \hat{f}[n_1, n_2]$
   B. wrapping parallelogram support which is formed by a tile around a centre rectangle at the origin **(Fig.5/b)**, and obtained

$$\hat{f}_{j,l}[n_1, n_2] = W\left(\tilde{U}_{j,l} f\right)[n_1, n_2] \quad (3)$$

Where a scale for n_1 et n_2 is : $0 \leq n1 < L_{1,j}$ et $0 \leq n2 < L_{2,j}$ (for θ in the scale $\theta \in (\pi/4, 3\pi/4)$);
C. Apply FFT 2D inverse in wrapped tiling
D. Gathering these coefficients of discrete curvelet $c^D(j,\ell,k)$

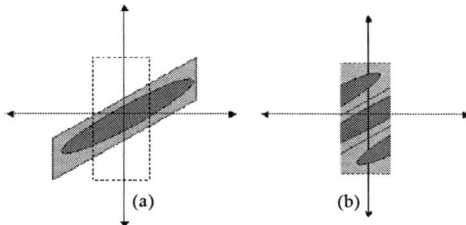

Fig.5. Corner tiling: (a) before the package. (b) after the package

3.2 Curvelets properties

We inscribe now some of the properties of curvelet transform: (Candes and al, 2005), (Candes and al, 2002).

1. Tight frame. Much like in an orthonormal basis, we can easily expand an arbitrary function $f(x_1, x_2) \in L^2(\mathfrak{R}^2)$ as a series of curvelets: we have a reconstruction formula

$$f = \sum_{j,l,k} <f, \varphi_{j,l,k}> \varphi_{j,l,k} \quad (4)$$

Using Parseval identity, we can also write it as follows

$$\sum_{j,l,k} |<f, \varphi_{j,l,k}>|^2 = \|f\|^2_{L^2(\mathfrak{R}^2)}, \forall f \in L^2(\mathfrak{R}^2) \quad (5)$$

2. Parabolic scaling. The curvelets are tied to the spatial localization scale. The atoms are anisotropic responding to a parabolic scale law where Heisenberg rectangles satisfy the notion of the tiling scale which is different in this case; obeying the (non dyadic) law:

$height = length^2$. $height \approx 2^{-j/2}, width \approx 2^{-j}$.

3. Oscillatory behaviour: the mother curvelet $\varphi_j(x_1, x_2)$ is oscillating with null average according to x_1 (in width), whereas it behaves as a low pass according to the direction x_2 (in length).

4. Vanishing moments. The curvelet template φ_j is said to have q vanishing moments when

$$\int_{-\infty}^{\infty} \varphi_j(x_1, x_2) x_1^n dx_1 \quad pour\ toutes\ 0 \leq n < q\ et\ \forall x_2$$

5. Isometric and Inversion:
In practice, the coefficients of curvelet are standardized as follows,

$$C^{D,N}(j,l,k) = \frac{n}{\sqrt{L_{1,j} L_{2,j}}} C^D(j,l,k) \quad (6)$$

Where $L_{1,j}, L_{2,j}$ are the side parallelogram with the length of $P_{j,\ell}$. Because of this property of isometry, transform curvelet is invertible. We can easily present the algorithm of inverse FDCT via wrapping in "flipping" all direct operations of the transform.

4. COMPRESSION QUALITY EVALUATION

The Peak Signal to Noise Ratio (PSNR) is the most commonly used as a measure of quality of reconstruction in image compression. The PSNR were identified using the following formulae (Sivakumar, 2007):

$$MSE^2 = \frac{1}{MxN} \sum_{i=1}^{i=N} \sum_{j=1}^{j=M} (I(i,j) - \hat{I}(i,j))^2 \quad (7)$$

Mean Square Error (MSE) which requires two $m \times n$ grey-scale images I and $\hat{I}$ where one of the images is considered as a compression of the other is defined as:

The PSNR is defined as:

$$PSNR = 10 \log_{10}\left(\frac{(Dynamics\ of\ image)^2}{MSE}\right) \quad (8)$$

Usually an image is encoded on 8 bits. It is represented by 256 gray levels, which vary between 0 and 255, the extent or dynamics of the image is 255.

5. RESULTS AND DISCUSSION

In this paper we applied the curvelets transform to compress medical images, this is why we have chosen two slices of brain (sagittal and axial) with a sizes of 256x256 (gray scale) encoded on 8bpp recorded by means of an MRI scanner (GE System).

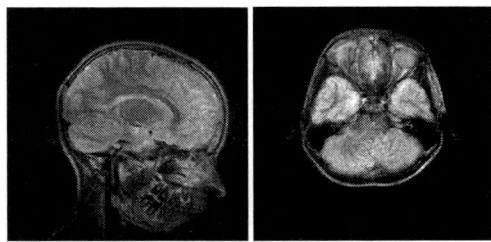

Fig.6. Medical image (sagittal, axial)

Application1:
As a first step we apply the transform curvelet to compress these images for different values of bitrate with a level of decomposition (S = 1)
Our objective appears particularly interesting to reduce the rate in which the image quality remains acceptable. For any image, estimation and judgment of the quality of the compressed image is given by the PSNR.
Figure (7) shown below shows the change in PSNR depending on the bit rate. We note that the FDCT gives an important PSNR for the two slices (axial and sagittal) for a decomposition level S=1. This result implies a quasi-perfect reconstruction of the image after compression.

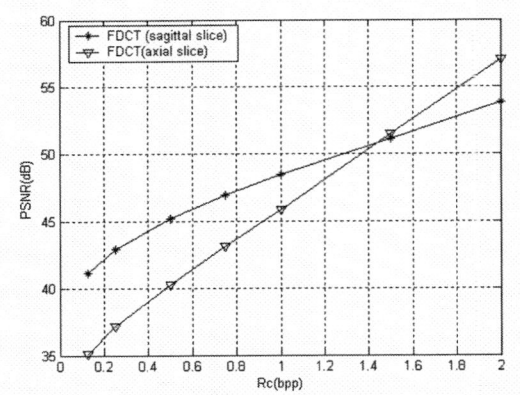

Fig.7. PSNR variation using FDCT method

These images present the results obtained after applying the FDCT.

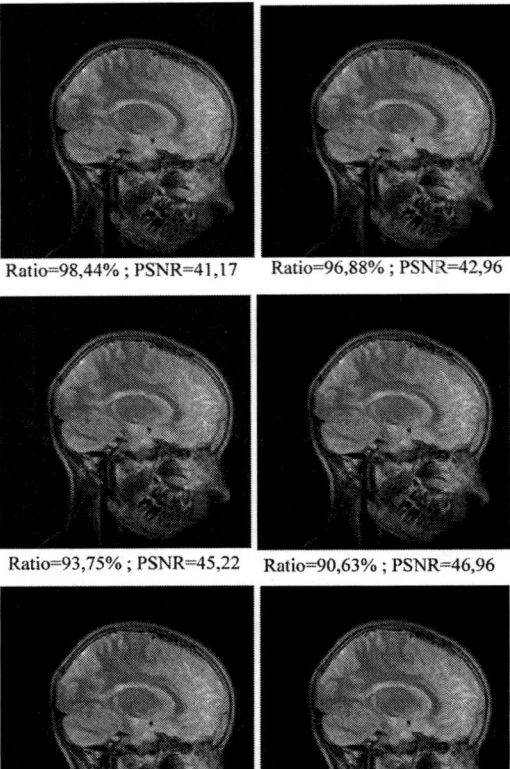

Ratio=98,44% ; PSNR=41,17   Ratio=96,88% ; PSNR=42,96

Ratio=93,75% ; PSNR=45,22   Ratio=90,63% ; PSNR=46,96

Ratio=87,50% ; PSNR=48,46   Ratio=81,25% ; PSNR=51,14

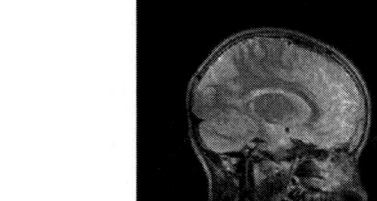

Ratio=75% ; PSNR=53,82 dB

Fig.8. Compressing of a sagittal slice with FDCT (S=1)

Application 2:

After demonstrating the performance of the curvelet transform, we shall now make a comparison between the different types of transforms (JPEG, Discret Wavelet Transform (DWT), Fine Ridgelet Transform (FRIT) and Fast Discret Curvelet Transform (FDCT)), to study the influence of the choice of the Wavelet on PSNR. For this we choose the Wavelet of coif 5 to compress these images, with a level of decomposition (N = 3) for DWT and FRIT, for the curvelet transform we choose S=1. For each application we vary the bit rate from 0.125 to 2bpp, and we calculate PSNR.

We choose for this application a sagittal slice with a size of 256x256 encoded on 8bpp.

The table presents the results of compression obtained by the (JPEG, DWT, FRIT, and FDCT)

Table1.

Rc(bpp)	PSNR (dB) JPEG	PSNR (dB) DWT	PSNR (dB) FRIT	PSNR (dB) FDCT	MSE FDCT
0.125	24.13	13.62	25.40	41.17	4.96
0.25	25.16	18.05	27.07	42.96	3.28
0.5	27.91	27.14	29.28	45.22	1.95
0.75	29.47	30.72	30.96	46.96	1.31
1	30.88	32.94	32.39	48.46	0.92
1.5	32.36	36.28	34.95	51.14	0.50
2	33.48	39.15	37.33	**53.82**	**0.26**

Note that typical a value of PSNR for images of good quality varies between 30 and 50 dB, where higher is better.

We note from the figure (9) that the values of PSNR found by FDCT for different bit rate values are very important in relation to other techniques. So the curvelet transform gives very satisfactory results.

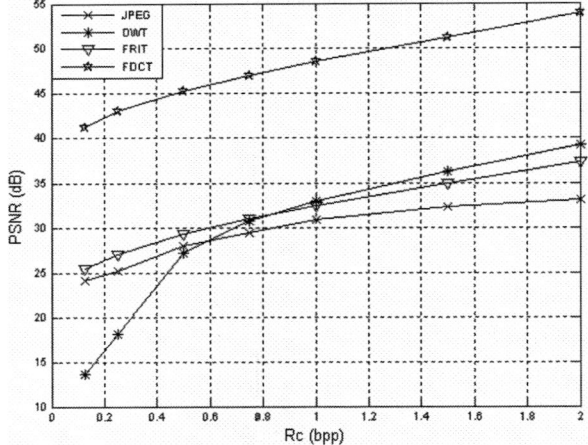

Fig.9. PSNR variation using different methods

The following images present the results obtained after the application of various types of transform (JPEG; DWT; FRIT and FDCT) on various slices. These results are obtained with a bit rate of 0.3 and 0.5 bpp in (Fig.10) and (Fig.11) successively.

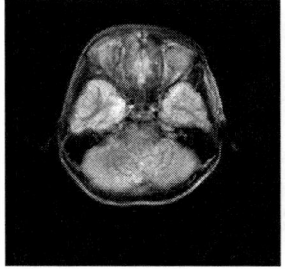
Ratio=96,25% ; PSNR=37,78 dB
MSE=10,85
FDCT compression

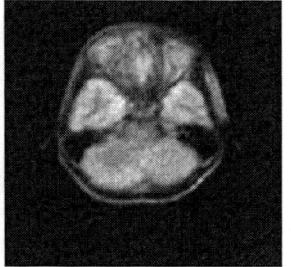
Ratio=96,25% ; PSNR=26,08 dB
MSE=160,53
FRIT compression

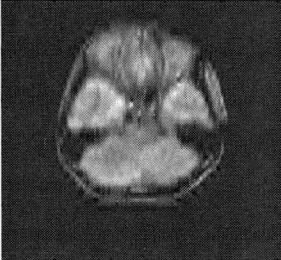
Ratio=96,25% ; PSNR=24,46 dB
MSE=233,79
DWT compression

Fig.10. Axial slice (256x256)

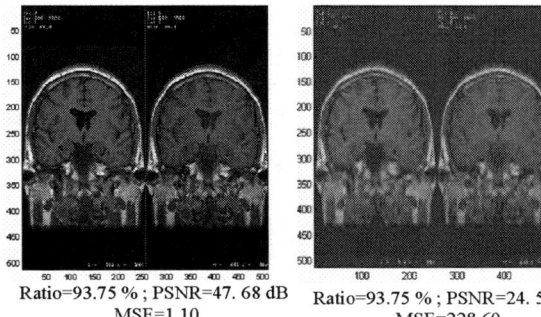
Ratio=93.75 % ; PSNR=47.68 dB   Ratio=93.75 % ; PSNR=24.54 dB
MSE=1.10   MSE=228.60
FDCT compression   FRIT compression

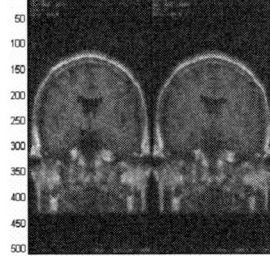
Ratio=93.75 % ; PSNR=21.53 dB
MSE=457.17
DWT compression

Fig.11. Coronal slice (512x512)

6. CONCLUSION

As a reminder, one of the objectives of this paper is undoubtedly improving the quality of Medical images after the compression step. The latter is regarded as an essential tool for storage or transmission of medical images. We have used the compression by Curvelet. After various applications, we have found that this transform gives better results than the other compression techniques.

In order to exploit this compression technique by Curvelets, we have applied this technique on different types of medical images (MRI). We have noticed that with a bit rate 0.5 bpp, the technique gives a very important value of PSNR of the MRI images. Therefore, we conclude that the results are very satisfactory in terms of compression ratio, and quality of the compressed image.

REFERENCES

[1] Chappelier.V (2005) . Codage progressif d'images par ondelettes orientées . *Ph.D. Thesis. Rennes University* . December.

[2] Mallat. S (1989) .A theory for multiresolution signal decomposition: The wavelet representation. *IEEE Transaction.* On Pami, Vol. 11, No.7.

[3] Candes, E.J., D.L. Donoho (1999) . Curvelets - A surprisingly effective nonadaptive representation for objects with edges. *Curve and Surface Fitting*, Vanderbilt Univ.

[4] Boubchir, L., J.M. fadili (2005) . Modélisation statistique multivarié des images dans le domaine de la transformée de curvelet . *IEEE.* 0-7803-9243-4/05.

[5] Candes, E.J., L. Demanet, D.L. Donoho and L. Ying (2006) . Fast discrete curvelet transforms. *SIAM J. on Mult. Model. Simul.* 5(3):861–899.

[6] Demanet.L (2006) .Curvelets, Wave Atoms, and Wave Equations. *Ph.D. Thesis.*

[7] Minh N. DO (2002) . Directional Multiresolution Image Representations . *Ph.D. Thesis*, Polytechnique coledge of Lausanne.

[8] Starck, J.L., E.J. Candès, and D L. Donoho (2002) .The Curvelet Transform for Image Denoising . *IEEE Transactions On Image Processing.* 670. VOL. 11, NO.6, June.

[9] Candes, E.J., L. Demanet, D.L. Donoho and L. Ying (2005) .Fast Discrete Curvelet Transforms. *Technical Report. Cal Tech.*

[10] Sivakumar, R. (2007) .Denoising Of Computer Tomography Images Using Curvelet Transform. *ARPN Journal of Engineering and Applied Sciences.* vol. 5. no. 1, february.

[11] Candes, E.J., D.L. Donoho (2002) . New Tight Frames of Curvelets and Optimal Representations of Objects with Smooth Singularities. *Technical Report. Stanford University.*

http://www.gemedicalsystems.com

Recognition of Handwritten Arabic words using a neuro-fuzzy network

Abdelhak Boukharouba*. Abdelhak Bennia**

*Département de Génie électrique, Université 08 Mai 45 de Guelma, Algérie
(e-mail : boukharouba_abdelhak@hotmail.com).
**Département d'Electronique, Université Mentouri de Constantine, Algérie

Abstract: We present a new method for the recognition of handwritten Arabic words based on neuro-fuzzy hybrid network. As a first step, connected components (CCs) of black pixels are detected. Then the system determines which CCs are sub-words and which are stress marks. The stress marks are then isolated and identified separately and the sub-words are segmented into graphemes. Each grapheme is described by topological and statistical features. Fuzzy rules are extracted from training examples by a hybrid learning scheme comprised of two phases: rule generation phase from data using a fuzzy c-means, and rule parameter tuning phase using gradient descent learning. After learning, the network encodes in its topology the essential design parameters of a fuzzy inference system.

The contribution of this technique is shown through the significant tests performed on a handwritten Arabic words database.

Keywords: Neural networks, Fuzzy inference, MLP, Neuro-fuzzy, Segmentation, Recognition

1. INTRODUCTION

One type of modeling with imprecise data is fuzzy modeling, whose objective is to extract a model in the form of fuzzy inference rules. Fuzzy modeling based on numerical data, which was first explored systematically by (Takagi and Sugeno., 1985), has found numerous successful applications to complex system modeling. Considerable work has been done to integrate the excellent learning capability of neural networks with fuzzy inference systems (Wang and Mendel., 1992), (Lin and Lu., 1996) and (Linkens and Chen., 1999), resulting in neuro-fuzzy modeling approaches that combine the benefits of these two powerful paradigms into a single network and provide a powerful framework to extract fuzzy rules from numerical data. Neural networks can learn from data, but cannot be interpreted; they are black boxes to the user. Fuzzy systems consist of interpretable linguistic rules. We use learning algorithms from the domain of neural networks to create fuzzy systems from data. The learning algorithms can learn both fuzzy sets, and fuzzy rules, and can also use prior knowledge.

The major challenge in the Arabic writing recognition systems comes from the cursive nature of the data. Generally, there are two approaches to tackle the problem of cursiveness in Arabic script: the global approach and the analytical approach. The global approach treats the words as whole. Features here are extracted from unsegmented word and compared to a model (Khorsheed., 2003). The analytical approach decomposes the word into smaller units (Miled et al., 1997) or primary and secondary strokes (Goraine et al.,). (Lorigo and Govindaraju., 2006) presented a comprehensive review of different methods. It was the first survey to focus on Arabic handwriting recognition and the first Arabic character recognition survey to provide recognition rates and descriptions of test data for the approaches discussed. It included background on the field, discussion of the methods, and future research directions.

(Amin and Al-Sadoun., 1996) proposed a structural approach for recognizing handwritten Arabic characters. The binary image of the character is first thinned and then the skeleton of the image is traced from right to left in order to build a graph to represent the character. Features like straight lines, curves and loops are then extracted from the graph. Finally, a five layer artificial neural network is used for the character classification. Each character is classified in term of the segments used in the system such as dot, hamza, line, curve and loop. (Altuwaijri and Bayoumi., 1994) introduced a system for recognizing printed Arabic words using artificial Neural Networks (NN). The system can be described into three different steps: first the Arabic input word is segmented into characters. Next, six moments are used for extracting features from the segmented characters feeding it to the neural network. Finally, a multi-layer perceptron network with back-propagation learning with one hidden layer is used to classify the characters. (Amin and Mansoor., 1997) used artificial neural networks for recognizing printed Arabic text. The technique can be summarized into three major steps: the first step is pre-processing in which the original image is transformed into a binary image and then forming the connected component. Second, global features of the input Arabic word are then extracted such as number of sub-words, number of peaks within the sub-word, number and position of the complementary character, etc. Finally, a three layer artificial neural network is used for the word classification.

(Sano *et al.*, 1996) introduced a structural approach using fuzzy relations for recognizing handwritten isolated Arabic characters. Each input pattern is divided into sub-patterns (strokes) by feature points; end points, branch points, intersections and maximum curvatures point, etc. The sub-patterns are then represented in terms of similarity to primitive elements (straight line, circle and diacritical point). The algorithm has been tested on a small number of handwritten samples. Finally, (Bouslama., 1997) adopted an algorithm based on structural technique and fuzzy logic for recognizing isolated printed Arabic characters. The structural technique is used to extract features from the input character such as number of strokes, the position of the center of gravity of each sub-segment, the black pixel ratio of the sub-segment with respect to the total number of black pixels in the skeleton, the chain code, the length ratio of the distance between end points and the total length of each sub-segment, etc. Fuzzy logic concepts are used to model any variations or uncertainties in the feature values to allow a better and more realistic representation of these features. Moreover, Fuzzy rules are also used for characters classification.

In the first phase of this work, a fuzzy c-means clustering algorithm is used as an unsupervised clustering method to establish the structure and parameters of a fuzzy rule base. These parameters are considered as initial parameters of the second learning phase. In the second learning phase, final parameters of the fuzzy rules are optimized via a gradient descent technique with an effort to improve the performance of the derived fuzzy model. Once the learning is completed, the network architecture encodes the knowledge learned in the form of fuzzy rules and processes data following fuzzy reasoning principles. This paper is organized as follows. Section 2 recalls the Fuzzy C-means Clustering method. Section 3 details the architecture of the neuro-fuzzy network on which is based our system. Section 4 gives the supervised learning process and the estimation of the neuro-fuzzy network parameters. Section 5 is devoted to the segmentation and feature extraction. Section 6 deals with the application of our model to handwriting Arabic characters and presents the experiments performed to validate the approach. Finally, in section 7 we present some concluding remarks and perspectives.

2. FUZZY C-MEANS CLUSTERING

Fuzzy C-means Clustering (*FCM*) employs fuzzy partitioning such that a data point can belong to all groups with different membership grades between 0 and 1. This method (Bezdek., 1981) is frequently used in pattern recognition. The aim of *FCM* is to find cluster centers (centroids) that minimize a following objective function:

$$J_m = \sum_{i=1}^{c} \sum_{j=1}^{n} u_{ij}^m \|x_i - c_j\|^2 \quad (1)$$

With

$$\sum_{i=1}^{c} u_{ij} = 1, \forall j = 1,...,n \quad (2)$$

where m is any real number greater than 1, u_{ij} is the degree of membership of x_i in the cluster j, x_i is the ith measured data, c_j is the center of the cluster, and $\|*\|$ is any norm expressing the similarity between any measured data and the center.

Fuzzy partitioning is carried out through an iterative optimization of the objective function shown above, with the update of membership u_{ij} and the cluster centers c_j by:

$$c_i = \frac{\sum_{j=1}^{n} u_{ij}^m x_j}{\sum_{j=1}^{n} u_{ij}^m} \quad (3)$$

$$u_{ij} = \frac{1}{\sum_{k=1}^{c} \left(\frac{\|x_i - c_j\|}{\|x_i - c_k\|}\right)^{2/(m-1)}} \quad (4)$$

This iteration will stop when $max_{ij}\left\{\left|u_{ij}^{(k+1)} - u_{ij}^{(k)}\right|\right\} < \varepsilon$, where ε is a termination criterion between 0 and 1. This procedure converges to a local minimum or a saddle point of J_m.

By iteratively updating the cluster centers and the membership grades for each data point, *FCM* iteratively moves the cluster centers to the "right" location within a data set.

3. ARCHITECTURE OF THE NEURAL NETWORK

The architecture of the network proposed here realizes the inference mechanism of a zero-order Takagi-Sugeno fuzzy model, based on a collection of H rules of the form:

R_k: *If* (x_1 *is* A_1^k, x_2 *is* A_2^k,......, x_n *is* A_n^k) *then* (y_1 *is* v_{k1}, y_2 *is* v_{k2},...... ,y_m *is* v_{km}).

Where R_k is the kth rule ($1 \leq k \leq H$), $\{x_i\}$ $i=1: n$ are the input variables, $\{y_j\}$ $j=1:m$ are the output variables, A_i^k are fuzzy sets defined on the input variables, and v_{kj} are fuzzy singletons defined on the output variables.

Fuzzy sets A_i^k are defined by bell-shaped (Gaussian) membership functions:

$$\mu_{ik}(x_i) = e^{-(x_i - w_{ik})^2 / \sigma_{ik}^2} \quad (5)$$

Where w_{ik} and σ_{ik} are the center and the width of the Gaussian function, respectively.

By adopting singleton fuzzification, product rule inference and center average defuzzification, the inferred crisp output value of this fuzzy system for any input $(x_{10}, x_{20},....,x_{n0})$ is calculated as:

$$y_j^0 = \frac{\sum_{k=1}^{H} \mu_k(x^0).v_{kj}}{\sum_{k=1}^{H} \mu_k(x^0)} \quad (6)$$

Where $\mu_k(x^0) = \prod_{i=1}^{n} \mu_{ik}(x_i^0)$ is the activation strength of the kth rule.

To realize the described fuzzy inference mechanism, we propose a neural network with three layers.

1. Input layer: Units in this layer receive the input values ($x_1, x_2,....,x_n$) and act as fuzzy sets representing the terms of the corresponding input variable. Nodes in this layer are arranged into H groups; each group representing the If-part of a fuzzy rule. Each node receives the input variable concerned and computes the membership value $\mu_{ik}(x_i)$ that specifies the degree to which the input value xi belongs to the fuzzy set A_i^k. Hence, the output of node ik is in the range [0, 1] and is computed by the following function:

$$f_{ik}^{(1)}(x_i) = e^{-(x_i-w_{ik})^2/\sigma_{ik}^2} \quad (7)$$

2. Hidden layer: The number of nodes in this layer is equal to the number of fuzzy rules. A node in this layer represents a fuzzy rule; for each node, there are n fixed links from the input term nodes representing the If-part of the fuzzy rule. The kth node performs the And operation for precondition matching of the kth rule by Larsen product operator; thus the output of this node is:

$$f_k^{(2)}(x) = \prod_{i=1}^{n} f_{ik}^{(1)}(x) \quad (8)$$

3. Output layer: Nodes in this layer represent the output variables of the system. Each node j acts as a defuzzifier and computes the output values according to:

$$f_j^{(3)}(x) = \frac{\sum_{k=1}^{H} f_k^{(2)}(x).v_{kj}}{\sum_{k=1}^{H} f_k^{(2)}(x)} \quad (9)$$

This neuro-fuzzy network encodes a set of fuzzy rules in its topology, and processes information in a way that matches the fuzzy reasoning scheme adopted.

The weights of the network correspond to the Gaussian membership function parameters $\{w_{ik}\}$, $\{\sigma_{ik}\}$ and to the consequent singletons $\{v_{kj}\}$.

The architecture of this neuro-fuzzy inference network is depicted in figure 1, where nodes representing the premise part of a fuzzy rule are enclosed in a gray circle.

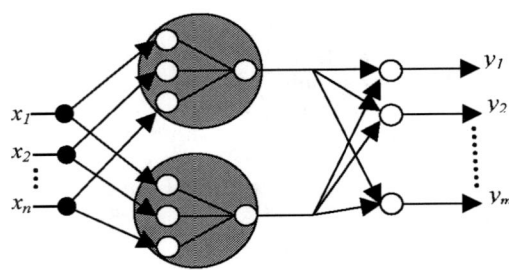

Fig.1. Neuo-fuzzy network.

4. SUPERVISED LEARNING PHASE

After a structure and initial weights of the network are established using a fuzzy c-means algorithm, the network enters the second learning phase to optimally adjust the parameters based on the same training data. A gradient method performing the steepest descent on a surface of the instantaneous error in the network weight space is used. Given the training set, the goal is to adjust weights so as to minimize an error function:

$$E_p = \frac{1}{2}\sum_{j=1}^{m}(y_j^p - f_j^{(3)}(x^p))^2 \quad (10)$$

Where $f_j^{(3)}(x^p)$ is the jth output of the neuro-fuzzy network for the current sample x^p, and y_j^p is the corresponding desired output. Weight update is performed repeatedly for total patterns P until the total error $e = \sum_{p=1}^{P} E_p$ is smaller than a predefined threshold value. For the sake of simplicity, the subscript p indicating the current sample will be dropped in the following. The general update formula for a generic weight α is $\Delta\alpha = -\eta\partial E/\partial\alpha$ where η is the learning rate. Starting at the first layer, a forward pass is used to compute the activity levels of all the nodes in the network to obtain the current output values. Then, starting at the output nodes, a backward pass is used to compute $\partial E/\partial\alpha$ for all the nodes. The complete learning algorithm is summarized below.

Supervised learning algorithm

0. Initialize weights using the fuzzy c-means algorithm

1. Input: Select the next sample (x, y) from S.

2. Forward step: propagate x through the network and determine the output values $f_j^{(3)}$, $j = 1,...,m$.

3. Backward step: compute error terms for units $j \in L3$, $k \in L2$ et $i_k \in L1$ in the order:

$$\delta_j^{(3)} = -\frac{\partial E}{\partial f_j^{(3)}} = y_j - f_j^{(3)}$$

$$\delta_k^{(2)} = -\frac{\partial E}{\partial f_k^{(2)}} = \sum_{j=1}^{m} \frac{\partial E}{\partial f_j^{(3)}} \frac{\partial f_j^{(3)}}{\partial f_k^{(2)}} = \frac{\sum_{j=1}^{m} \delta_j^{(3)} (v_{kj} - f_j^{(3)})}{\sum_{t=1}^{H} f_t^{(2)}}$$

$$\delta_{ik}^{(1)} = -\frac{\partial E}{\partial f_{ik}^{(1)}} = -\frac{\partial E}{\partial f_k^{(2)}} \frac{\partial f_k^{(2)}}{\partial f_{ik}^{(1)}} = \delta_k^{(2)} \cdot \frac{\partial f_k^{(2)}}{\partial f_{ik}^{(1)}}$$

4. Adjustment: update weights $\{v_{kj}\}, \{w_{ik}\}$ and $\{\sigma_{ik}\}$ respectively according to the update quantities:

$$\Delta v_{kj} = \eta \delta_j^{(3)} \cdot \frac{f_k^{(2)}}{\sum_{t=1}^{K} f_t^{(2)}}.$$

$$\Delta w_{ik} = \eta \delta_{ik}^{(1)} \left[\frac{2(x_i - w_{ik})}{\sigma_{ik}^2} \right] f_{ik}^{(1)}.$$

$$\Delta \sigma_{ik} = \eta \delta_{ik}^{(1)} \left[\frac{2(x_i - w_{ik})^2}{\sigma_{ik}^3} \right] f_{ik}^{(1)}.$$

$$v_{kj} = v_{kj} + \Delta v_{kj}.$$

$$w_{kj} = w_{kj} + \Delta w_{kj}.$$

$$\sigma_{kj} = \sigma_{kj} + \Delta \sigma_{kj}$$

5. If $E < \varepsilon$ then go to step 6. Else go to step 1.

6. End.

5. SEGMENTATION AND FEATURE EXTRACTION

This process segments an image of handwritten script into letters. Here, the segmentation is performed in three levels: line segmentation, word and sub-word segmentation, and character segmentation. The method based on horizontal histogram is used to segment the text image into line image. The segmentation procedure of lines into words and sub-words consist of identification and classification of connected components. After a detection of connected components in the input text line, these components are classified in three classes: sub-words, isolated characters, and stress marks including dots, see Fig. 5.

Most segmentation methods that are used classify the connected components using information on their sizes and positions. Any connected component whose size is less than a threshold is regarded as a stress mark. The algorithm used to evaluate the stress marks is presented in (Al-Yousefi and Udpa., 1992).

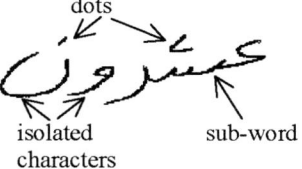

Fig. 2. Different types of connected components forming the word.

To improve segmentation efficiency, we adopted to remove stress marks like dots from characters. Their original position is remembered and reintroduced only in the recognition phase. The Arabic sub-word, which consists of one isolated character, does not need to be segmented. So simple some rules are used to check whether a sub-word includes only one character (Liying *et al.*, 2004). Only sub-words are considered by the next segmentation phase. The segmentation is carried out with the help of local minima on the upper contour see Fig. 4. Thus, every valley produces a descending type break (cut-off) hypothesis around this minimum (Miled *et al.*, 1997).

Fig. 3. Upper contour of sub-word عسر.

This procedure gives the Primary Segmentation Points (PSPs). Heuristic rules, such as the distance of the cut from the base line, the distance between the minima by using an average filtering criterion to eliminate cut-off hypotheses likely to be non potential, and the number of white-black or black white transitions encountered along the cut, are then applied for each PSP in order to validate the Decisive Segmentation Points (DSPs). Finally, the latter are used to segment the input image see Fig. 4

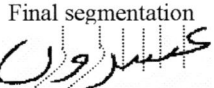

Fig. 4. Segmentation of the word in graphemes.

After the segmentation into graphemes, the image is composed of a sequence of patterns corresponding to graphemes. Each grapheme is either only one character or a portion of a character; the character س is segmented into three portions, Fig. 4, and each portion of character سـ has the same class as the character بـ of word سبعة

Arabic alphabet contains 28 characters; most characters (17 out 28) have a dot, two dots, three dots, hamza, or zigzags (secondaries) associated with character and can be above, below, or inside the character. Each character has two to four different forms depending on its position in the word that increases the classes to be recognized from 28 to 100 different classes. Isolation of the secondaries and recognition them separately reduces the number of patterns to be recognized from 100 to only 56. This helps in improving the recognition process since secondaries can be placed anywhere above or below the character, which introduces a great deal of variation between characters of the same class. By examining segmented characters, it is found that the main body of some characters in different positions is the same. For example, there are characters (ط, ظ) that have the same shape regardless of their position in a word. Therefore, such characters are included in the same main class. Consequently, the number of different main shapes (classes) reduces to 25. The characters can be recognized by identifying their main body shapes, their stress marks and their position in a word. Identifying the position of character in a word and the stress marks was done during the pre-processing and segmentation phase.

In this level, each character (grapheme) is described by a feature vector. In our application graphemes are described by two families of features. Each family is called a sub-vector.

The first sub-vector contains topological features that correspond to human perception: loops, dots, the relative size (width, height) of the grapheme, and the relative position compared to the base line.

The second sub-vector contains normalized moments and Fourier descriptors (El-Dabi et *al.*, 1990) and (El-sheikh and Guindi., 1988). Finally, we obtain a feature vector of 36 components per grapheme.

6. EXPERIMENTAL RESULTS

The validity of our approach to fuzzy inference and rule extraction has been tested on the Arabic handwritten characters. This network has as many outputs as possible class characters. In all, 25 letters are used. Consequently, it is a network of 25 classes that we need to build in this case; for each letter a set of 36 classic primitives are used. Neuro-fuzzy network architecture with 36 inputs and 25 outputs, corresponding to the 25 classes, was considered. There are 520 samples for this problem; each class is written by different writers. The whole data set was divided into two equally sized parts. One part was used as a test set for the network trained with the remaining 260 samples.

Table 1 summarizes the classification rate of the neuro-fuzzy network and Fuzzy C-means algorithm (FCM) classifier. The fuzzy rule bases generated by our approach provide good performance, in terms of classification rate, when compared with Fuzzy C-means algorithm (FCM) classifier.

Table 1. Recognition rates and network errors.

	Recognition rate	Error rate
Fuzzy C-means	94.62%	05.38%
Neuro-fuzzy network	96.15 %	03.85%

7. CONCLUSIONS

In this paper, we presented a new system for recognizing handwritten Arabic words. A neural network is proposed to resolve the main problem of fuzzy modeling, i.e. structural and parametric identification of a fuzzy rule base. The main features and advantages of the developed hybrid network are:

It is a general framework that combines two models, namely neural networks and fuzzy systems.

The inside of the neural network can be explained in concept of a fuzzy model and hence it can be easily understood.

The network encodes in its structure the essential design parameters to assemble a fuzzy model and processes data according to a fuzzy reasoning mechanism.

The network determines both structure and parameters of the fuzzy rule base via learning from data.

Simulations on a handwritten characters classification verify the effectiveness of the proposed network.

REFERENCES

[1] Altuwaijri, M., Bayoumi, M. (1994). Arabic text recognition using neural networks. *Proc. Int. Symp. On Circuits and Systems – ISCAS'94. 415-418.*
[2] Al-Yousefi, H., Udpa, S.S. (1992). Recognition of Arabic characters. *IEEE Trans. Pattern Anal. Mach. Intell. 14, 853-857.*
[3] Amin, A., Al-Sadoun, H. (1996). Handprinted Arabic character recognition system using an artificial neural network. *Pattern Recognition 29, 663-675.*
[4] Amin, A., Mansoon, W. ,(1997). Recognition of printed Arabic text using Neural networks. *Proc. 4th Int. Conf. on Document Analysis Recognition, Ulm, Germany.*
[5] Bezdek J. C. (1981). Pattern Recognition with Fuzzy Objective Function Algorithms. *Plenum Press, New York.*
[6] Bouslama, F. (1997). Arabic character recognition by Fuzzy techniques. *Proc. 5th European Congress on Intelligent Techniques and Soft Computing, Aachen, Germany.*
[7] El-Dabi, S., Ramsis, R., Kamel, A. (1990). Arabic character recognition system: A statistical approach for recognizing cursive typewritten text. *Pattern Recognition, Vol. 23, N°5, 485-495.*
[8] El-sheikh, T., Guindi R. (1988). Automatic recognition of isolated Arabic characters. *Signal processing 14, North-Holland.*
[9] Goraine, H., Usher, M., Alemami, S. Online Arabic character recognition. *IEEE Comput. J., 71-74.*
[10] Khorsheed, M. S. (2003). Recognising handwritten Arabic manuscripts using a single hidden Markov model. *Pattern Recognition Lett. 24 (14), 2235-2242.*

[11] Lin, C., Lu, Y. (1996). A neural fuzzy system with supervised learning. *IEEE Trans. Syst., Man, and Cyb. Part B, vol. 26. 744-763.*

[12] Linkens, D.A., Chen M.Y. (1999). Input selection and partition validation for fuzzy modeling using neural network. *Fuzzy Sets and Systems, vol. 107 299-308.*

[13] Liying, Z., Hassin, A. H., Tang, X. (2004). A new algorithm for machine printed Arabic character segmentation. *Pattern Recognition Letters, Vol. 25, No.15, 1723-1739.*

[14] Lorigo, L.M., Govindaraju, V. (2006). Offline Arabic Handwriting Recognition: A Survey. *IEEE Trans. Pattern Anal. Mach. Intell. vol. 28(5), 712-724.*

[15] Miled, H., Olivier, C., Cheriet, M., Lecourtier, Y. (1997). Coupling observations/letters for a markovian modeling applied to the recognition of the Arabic handwriting. *4th IAPR International Conference on Document Analysis and Recognition, ICDAR'97, Ulm, Germany 580-583.*

[16] Sano, M., Kosaki T., Bouslama F. (1996). Fuzzy structural approach for recognition of handwritten Arabic characters. *Proc. Int. Conf. on Robotics Vision and Parallel Processing for Industrial Automation, Ipon, Malaysia, 252-257.*

[17] Takagi, T., Sugeno, S.M. (1985). Fuzzy identification of systems and its application to modeling and control. *IEEE Trans. Syst., Man, and Cybernetics*, vol. 15. 116-132.

[18] Wang, L.X., Mendel, J. (1992). Generating fuzzy rules by learning from examples. *IEEE Trans. Syst., Man, and Cyb., vol. 22. 1414-1427.*

Indexing color images using color band moments and a relevance feedback

AYAD Mohammed[1], BESSAID Abdelhafid[1], BECHAR Hassane[1], TALEB AHMED Abdelmalik[2]

[1]*Biomedical engine laboratory, university of Tlemcen, Algeria*
[2]*LAMIH - UMR - CNRS University of Valencienne and Hainaut-Cambrésis, France*
E-mail: ayad144@yahoo.fr

Abstract: In most content based image retrieval systems there is the problem of localizing the spatial information. To resolve this problem we have proposed in this paper to split the picture into horizontal and vertical bands and characterize each band by their statistical moments of each color component. But the satisfaction of low level characteristics don't implicate that the semantic needs are satisfied, so the human interaction with the system is necessary. For this reason we have integrated in our system a relevance feedback loop, on which we have proposed a method which gives us the similarity degree in function of the positive and negative examples which are manually denoted.

Keywords: CBIR(Content Based Image Retrieval), spatial information, splitting into bands, relevance feedback

1. INTRODUCTION

As a result of computer science development, many problems have emerged concerning the data organization. The visual information has becoming widely used, what makes necessary to design the indexing systems. The new techniques of indexing image are based on the visual feature of the image. We can find for example the method which uses the color histogram as an image's descriptor 6, and the statistical moment's method [2]. We can also find other techniques based on the extraction of the texture feature like the co-occurrence matrix method. Most of these techniques suffer for the disadvantage of absence of the spatial information, which is necessary describe the distribution of the different colors on the picture regions. We can find for example images having similar feature vector, while there contents are completely different. Many approaches have been proposed to localize the spatial information [8]. We can find for example the approach of splitting the image into five areas: the middle and four side areas. We can also find other methods based on the segmentation like the CRT (Composite Region Templates) method [7] which consist of segmenting the image and counting the different transition between the different segmented areas.

Because of the structure of the natural pictures, which includes regions extended along the entire image, we have proposed to characterize the image by the signal of mean of the different lines; we have accomplished this description by the second and the third moments.

In spite of that localizing the spatial information can enhance the performances of our system; we can find many pictures having the same distribution of the colors on the different regions, so satisfying the low level features doesn't imply that the images have the same semantic. We can so find many systems which improve the outcome by the interaction with the user. This can be done dint the relevance feedback technique which we have used in our system. To estimate the similarity of an image, we have proposed to use the positives and negatives examples at the same time.

So In this paper, we have proposed two originalities:

- Splitting the image into horizontal and vertical bands.
- A feedback method based on the minimal distance between the positive and negative examples.

In section 2 we have presented the method of spiting the image, I section 3 we have presented our relevance feedback, and we have terminated by a conclusion the section 5.

2. Splitting into bands

In most of the CBIR systems which are based on the low level characteristics, the spatial information is almost absent. For example, the first CBIR system uses as characteristic vector the color histogram and similarity is given in function of the distance between these vectors. So we can find many distance's formulas. The commonly used one is the distance given by the norm L_P.

2.1 Disadvantages of the color histogram

One of the disadvantages of color histogram is that it is very sensible to the quantization. Many compression methods entrain a quantization of the colors either in the spatial domain or in the spectral domain with different degrees. A quantization in the spatial domain causes a high distortion on the color histogram, where many components are grouped into one,

Among the encoding systems which entrain a color quantization in the spatial domain, the GIF (Figure 1).

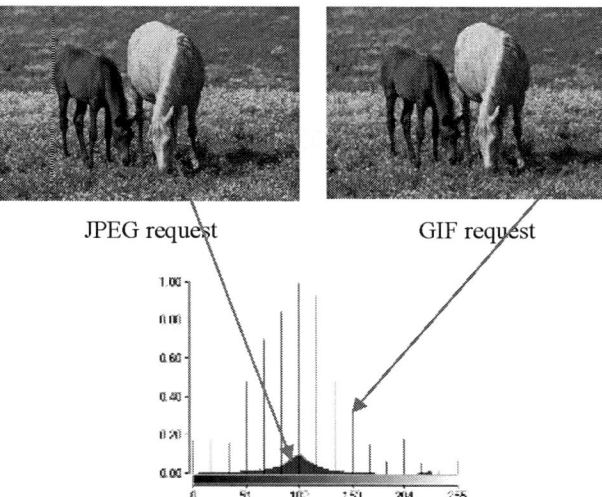

Figure 1: the quantization effect on the color histogram comparison

2.2 The statistical moments

Another feature which we can find in the CBIR systems is that using the statistical moments of order 1, 2, and 3 of each color channel. These moments are given by the formula below:

$$\mu_i = \frac{1}{N}\sum_j p_{i,j} \qquad \sigma_i = \sqrt{\frac{1}{N}\sum_{j=1}^{N}(p_{i,j} - \mu_i)^2}$$

$$s_i = \left(\frac{1}{N}\sum_{j=1}^{N}(p_{i,j} - \mu_i)^3\right)^{\frac{1}{3}}$$

The first moment μ_1 represent the mean color of the image. Despite that this characteristic is very important; it doesn't suffice to describe colors presented in the image. (For example, we can find many images having different color distribution).

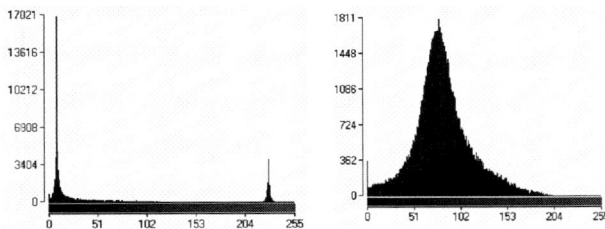

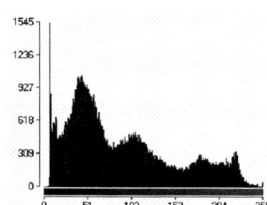

Figure 2: Example of three histograms having the same mean (61)

We see well that the distribution of the colors around the mean value is almost nil in the first histogram. In the second histogram we see that the colors are focused around the mean value, while it is almost uniformly distributed in the third one.

The first histogram is characterized with a high variance, so the second one has a low variance, so we can say that the second moment gives us information about the distribution of the colors, around the mean value, so it denotes if image is contrasted or not.

The third moment characterize the homology of the color histogram

The distance between two images can be defined as a weighted sum of differences between these quantities taken for each color channel.

$$d_{mom}(I,H) = \sum_{i=1}^{3} w_{i1}\left|\mu_i^I - \mu_i^H\right| + w_{i2}\left|\sigma_i^I - \sigma_i^H\right| + w_{i3}\left|s_i^I - s_i^H\right|$$

where:

$w_{i,j}$: weights corresponding to the moment m_j

and the index i means the color channel (R, G, B or H, S, V)

The weights w_{i1}, w_{i2}, w_{i3} represent indexes allowing to base our research on some features, and to slight others. They can be useful if we want for example to slight the brightness's differences where pictures are taken in different light conditions.

2.3 Our proposition

The color statistical moments technique is also one of the methods which they don't safeguard the spatial information. To compare two images caring of the spatial information; we can compare them for example pixel by pixel after normalizing their size. This operation can take long time, where we can not realize an On-Line application. To minimize this time we can take for example only 4 or 5 blocs, but this proposition depends highly on the structure of the image.

We have remarked that most of the natural images have a structure of uniform bands which can extend along the entire image, so we have proposed to characterize an image by the signal of the means of the different lines.

Figure 3: Example of horizontal and vertical bands

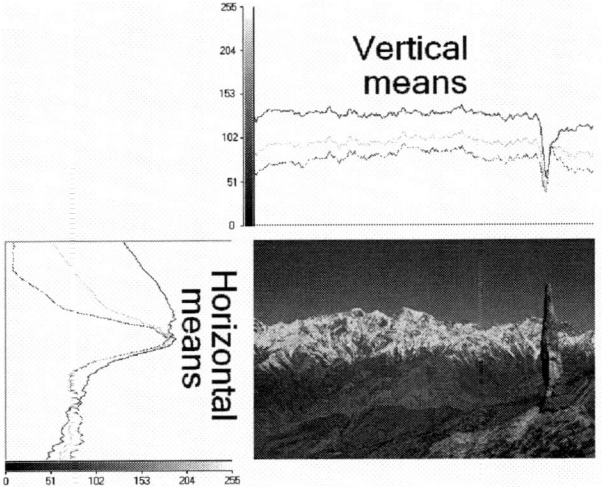

Figure 4: the means of rows and columns of an image

We have used also as feature the second and the third statistical moments. To normalize the dimensions we have collected these features as we keep just a limited component count, so the image will be split into horizontal and vertical bands characterized by their 3 statistical moments for each color component.

Why splitting the image into horizontal and vertical bands?

Saving the all the spatial information can cause some problems because when object change their positions, so we have propose to use bands because if the position of an object change, almost it stays in the same band.

2.3.1 Choice of the band's number

The problem which can face now is the choice of band's number. This problem relies substantially on the areas which we can find in the image. It is clear that as we increase this number as the precision increases. But this remark is not always true, so increasing the bands number may make our system very sensible for the small spatial changes which they are inevitable, like the translation of the objects. Therefore, we have study the effect of the band's number on the research result, so we have remarked that it depends highly on the image structure.

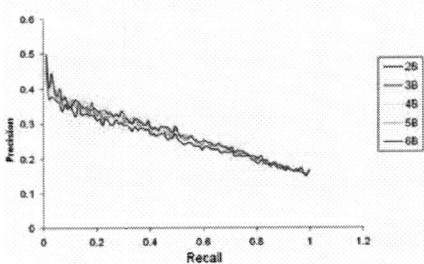

Figure 5: Recall and precision curve for different band's number

Figure 1 shows the mean recall and precision curve for different band's number taken from 1 to 6 (applied on Wang database) considering that the horizontal band's number is the same as vertical.

Therefore, we can see that as the band's number increases as we obtain best result. But this enhancement doesn't increases highly as we increase the band's number. Therefore, we can suppose that a number of 3 or 4 can be sufficient for the natural images.

3. Relevance feedback

One of the major problems which we can face in a CBIR is the problem of the semantic. We can present for example to our system a picture of pasturage, and don't b surprised if it give as the picture of green car. Therefore, it is difficult to make differences between images if we base our research only on the low-level characteristics. So it is recommended to give the user possibility to interact with the system by the technique of relevance feedback. In CBIR system based on the relevance feedback, we present a request image to our system, but when we obtain the result; we denote some image as relevant and others as not. When we reset our research, the system takes care of our opinion, so the new result will be more similar to the positive examples and far from the negative examples. We can repeat this operation until becoming satisfied.

These systems can be divided into two classes [5]: the systems, who introduce their correction on the request vector, and the systems who introduce their correction on the weights associated to the elements of the feature vector. Therefore, these two kinds accord well with vectorial model.

In the first kind, the new request vector is formed by the former request and the positive and negative examples. Generally, this new request is formed according the formula below:

$$Q' = \alpha \cdot Q + \beta \left(\frac{1}{N_P} \sum_{i=1}^{N_P} DP_i \right) - \gamma \left(\frac{1}{N_N} \sum_{i=1}^{N_N} DN_i \right)$$

Where

Q : The old request

Q' : The new request

α, β, γ : Suitable constancies.

DP_i: feature vector of the i relevant image

DN_i: feature vector of the i non relevant image

This technique was used in the MARS system [5] to replace the document vector with visual feature vectors. Experiments show that retrieval performance can be improved by using these relevance feedback approaches.

The re-weighting method enhances the importance of a feature or parts of it that help retrieve relevant images and reduce that of the features or the parts of the features that hinder this process. This is achieved by assigning weights to the feature vectors or to its entries and using a weighted distance metric.

For example the distance between two images can be defined as the norm L_1:

$$D = \sum_i w_i |x_i - y_i|$$

Where:

x_i, y_i: The components of the two vectors X and Y respectively

w_i Weights used to emphasize the distance.

So the weight for a feature component, w_i, is updated as follows:

$$w_i' = w_i(1 + \overline{\delta} - \delta_i)$$

Where

$\delta_i = f(Q_i - DP_i)$ is a function of the difference between the request vectors and the relevant image vector.

$\overline{\delta}$: mean of δ_i

2.3 Our approach

Another approach consist of considering the distance of an image (form the database) as the minimal distance between this image and all the images considered as positive example [6]. This method has the advantage of that we will never obtain as result an image which is very far from the entire positive image set.

This method considers only the positive examples, so we have proposed to introduce a formula which take care of the positive and the negative examples at the time. Below we explain well this method:

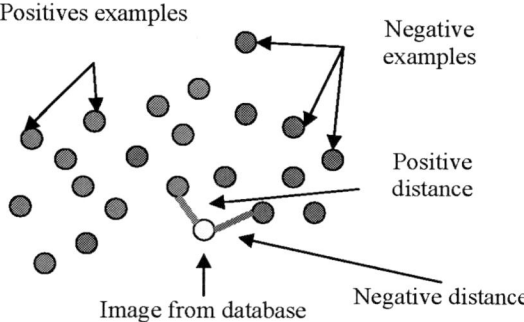

Figure 6: Computing the distance using the positive and the negative distance

Image form database After specifying the positive and the negative examples, and to compute the distance of an image from the database, we search its nearest positive example (manually denoted) and its nearest negative example, so the final distance can be a ratio between the positive and the negative distance. We can define this distance as:

$$D = \frac{D_P}{D_P + D_N}$$

Where:

D : The searched distance of database image.

D_p: The distance between the database image and its nearest positive example.

D_N: The distance between the database image and its nearest negative example.

Therefore, for a total similarity the distance D equals to 0 and for a total dissimilarity the distance D equals to 1.

Instead of specifying positive and negative examples, we have choose to specify a relevance degree or weight taken from -2 to 2 to specify if this example is very similar, similar, different, or completely different to our needs. This process is similar to a space where we can find many electrical charges, and every one tries to attract the free charges which represent in our case the database images where we try to find their distance.

So to find the positive distance the request's weight must widen the influence field of the example image, so positive distance can be defined as:

$$D_P = \min\left(\text{Dist}(Im_i^B, Im_j^P)/w_j\right)$$

With the same mean we can define the negative distance as:

$$D_N = \min\left(\text{Dist}(Im_i^B, Im_j^N)/|w_j|\right)$$

So the final distance is defined as:

If $D_P < D_N$ then $D = \left(\dfrac{D_P}{D_P + D_N} - 0.5\right) \times w_P$

Else $D = -\left(\dfrac{D_P}{D_P + D_N} - 0.5\right) \times w_N$

4. Experimental results:

4.1 splitting into bands

To validate our results we have done our experiences on the Wang image database [10]. It consists of 1000 image where we can find 10 classes.

We can see that splitting image becomes more efficient when the image contain separate areas.

Figure 7 show some results obtained when we have split the image into 3 horizontal and 3 vertical bands.

Figure 7: results obtained using 3 horizontal and 3 vertical bands.

We can see that most of the result images contain a blue sky at the top o we can consider that this method has efficiently safeguarded the spatial information, in opposite of methods which describe the entire image (histogram comparison method) where the images obtained in the result have not the same structure as the request image (Figure 8).

Figure 8: Result obtained using the histogram comparison method.

In other way, splitting the image into horizontal and vertical bands remains inefficient when the image contain many telescopic areas like in images which represent Africa in Wang database. So we have studied the influence of the bands number on the performance of the system, and we have seen that as we increase the bands number as we improve the outcome, but if the image is complex this number become least important.

4.2 Relevance feedback

The aim of using our method of relevance feedback is to evade obtaining results which are far from the positive examples selected by the user. Figure 9 shows some results obtained using the mean feature of positive and negative examples.

a: manual denotation showing the system the positive and the negative examples

b: result obtained using the relevance feedback

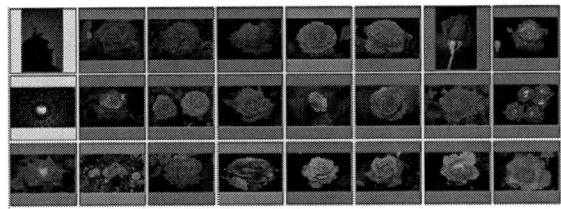

Result obtained after many interactions.
Figure 9: result obtained using the mean feature of the positive and the negative examples.

We can see that this method can improve the result, but it may give a very bad result if interact many time with the system while our method gives the highest priority to the positive examples and their neighbourhood. So don't care of obtaining a very bad result. We can see that method gives good result just with 2 or 3 interactions.

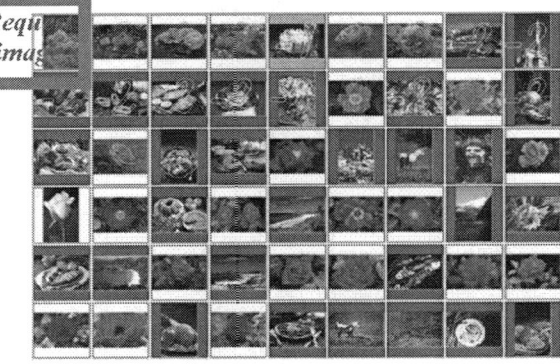

a: first result

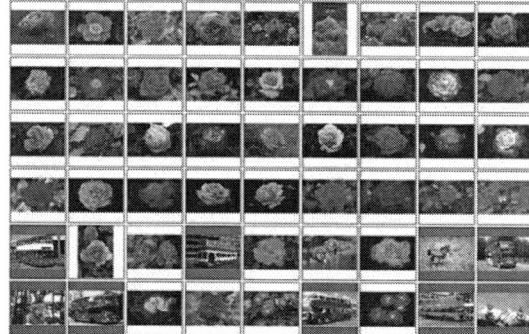

b: result obtained just with one interaction

Figure 10: relevance feedback based on the minimal distance

5. Conclusion and perspectives

Most of content-based image retrieval systems are based on the low-level characteristics. In most cases, these features are global descriptors so they give global statistics on the entire image like the color histogram. This problem forces us to think to use other methods to characterize the different areas and safeguard the spatial information. Localizing the spatial information can be an advantage like it can be a disadvantage. In most natural images, we have separate regions like sky, herbs, flowers…etc. so splitting the image can be very interesting. The unique problem in this case is the shifting between the different areas so we think that we adapt the different region using the concept of correlation. For the relevance feedback, we think that it is a permitting to define the limits between the relevant images and non relevant, so we propose to use classifying algorithms based on the learning process, and which can be adapted with each user.

6. References

[1] M.STRINKER M.SWAIN, The capacity of histogram an the sensitivity of color histogram indexing, Technical report 94-05 Zurich Switzerland, March1994

[2] N. KEEN, Color101 Moments, ID: 0341091, February 10, 2005.

[3] CARLOS JOEL, RIVERO MORENO, Contribution à la caractérisation des images par transformée polynomiale: Application à l'indexation des images et des vidéos, Thèse de doctorat, Institut national des sciences appliquées de Lyon, octobre 2005.

[4] OLIVIER ALATA, Caractérisation de textures par coefficients de réflexion 2-D Application en classification et segmentation, Thèse de doctorat, Université de Bordeaux, école, Janvier 1998.

[5] FAOUZI ALAYA CHEIKH, MUVIS: A System for Content-Based Image Retrieval, Tampere University of Technology, Finland, March 2004

[6] LE THI LAN, Interface de visualisation avec retour de pertinence pour la recherche d'images, Projet ORION, INRIA, 2004

[7] JOHN R. SMITH AND CHUNG-SHENG LI, Image Classification and Querying using Composite Region Templates, IBM T.J. Watson Research Center, 1999

[8] JÉRÔME FOURNIER, INDEXATION D'IMAGES PAR LE CONTENU ET RECHERCHE INTERACTIVE DANS LES BASES GÉNÉRALISTES, Thèse de doctorat Université de Cergy-Pontoise, octobre 2002

[9] M.CAMPEDEL, B.LUO, H.MAÎTRE, E.MOULINES, M.ROUX, I.KYRGYZOV, Indexation des images satellitaires Détection et évaluation des caractéristiques de classification, Département Traitement du Signal et des Images, Décembre 2004

[10] BASE DE WANG: http://wang.ist.psu.edu/IMAGE

SESSION E2: INTELLIGENT AND FLEXIBLE MANUFACTURING

On-line scheduling of Automatics and flexible Manufacturing System using SARSA technique

N. Aissani*. B. Beldjilali**

*,**Université d'Oran BP 1524 El M'nouer -Oran Algérie
*(e-mail: aissani.nassima@yahoo.com)
**(e-mail: bouzianebeldjilali@yahoo.fr)

Abstract: In this paper context, we will show what will be the best organization of decision entities in flexible manufacturing system, but also show our approach steps to achieve a manufacturing control system which is more reliable insofar as it has responding to queries in online. With this intention, we use a multi-agent system of which the decisions taken by the system are the result of those agents group work, these agents ensure in the same time manufacturing scheduling solution and a continuously improvement of their quality thanks to the reinforcement learning technique and particularly SARSA algorithm which was introduced to them. This technique of learning makes it possible the agents to be adaptive and to learn the best behavior in their various roles (answer the requests, self-organization, plan…) without attenuating the system on-line. A computer implementation and experimentation of this model are provided in this paper to demonstrate the contribution of our approach compared to a famous metaheuristic: tabu search, widely used for scheduling in complex manufacturing systems.

Keywords: On-line scheduling, Reinforcement learning, multi-agent system, Heterarchical organization

1. INTRODUCTION

The most popular automated systems are those subjected to a continuous closed loop control. As seen in (Fig. 1), a manufacturing control system is inspired from closed-loop system in automatics and robotics (Prouvost, 2004).

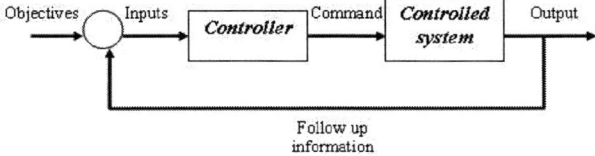

Fig. 1. Retroaction control system

This will form the basis of our approach principle. In this paper we are focusing the controller in a big scaled format of automated system, which is a flexible manufacturing system. It is a system in which all functions are automated, planning and scheduling (the allocation of resources) are done by an automated control system, a lot of researches are conducted to develop tools and models for designing this kind of systems.

2. CONTEXT AND PROBLEMATIC

In this part we propose an analysis of the pertinent literature in the field of the heterarchical or distributed control to generate on-line scheduling solutions.

2.1 Heterarchical system organisation

In (Bousbia and Trentesaux, 2002) an analysis of the whole of these points (adaptive control and improvement of performances) was carried out and in particular was highlighted the interest to adopt an approach of self-organized and heterarchical control system. The heterarchy term describes a relation between of the same hierarchical level entities. Initially, proposed in the medical biology field (McCulloch 1945) then experimented in several domains (Prabhu, 2003; Haruno and Kawato, 2006). The Heterarchical organization offers a great potential in communication inter decision entities of the same level that encourages greater reaction system on-line, in addition, it keeps a hierarchical organization to do long-term planning, in a nutshell, heterarchical is the ideal organization for a system control in on-line.

This term is closely linked to distribution in multi-agent systems (Although this is not always the case) and multi-agent systems are widely used in conception of manufacturing control systems (Chen, *et al.*, 2003; Albadawi, *et al.*, 2006). So that we could propose a control system based on entities (agents…) which is dynamic, reactive and which would to be self-organized and able to process in on-line the data concerning resources: the control structure is composed of agents unit which cooperate to make a decision and which, in this case, are also able to learn according to their experiment using reinforcement learning (RL) (Watkins, 1989). RL is used for learning in stochastic environment which is difficult or even impossible to model (Charton, *et al.*, 2003; Marthi, *et al.*, 2005, Dongbing and Yang, 2007). The objective of this system is to be ongoing collaboration with the controlled manufacturing system (feedback and reactivity) to produce a high-performance behavior improves as providing solutions for better scheduling.

2.2 Control system and scheduling

Scheduling constitute the most important function in manufacturing control systems. In this paper we focus reactive/on-line scheduling.

On-line scheduling is called *reactive*. The resource allocation process actually evolves and more information becomes available in order to make decision in real-time (Pujo and Brun-Picard, 2002; Csaji and Monostori, 2006; Aissani et al., 2008). Reactive scheduling can also be associated at the beginning to an off-line scheduling solution it is called *predictive* scheduling which assume a deterministic environment. This junction is called *rescheduling* But, when it is an unforeseeable environment and when there are some data (e.g., the tasks durations) that will only be available during the execution of the plan. The given scheduling solution is a *proactive* scheduling. A proactive solution allocates the operations to resources without precise starting times because the durations are uncertain (e.g. stochastic job-shop scheduling) (Bidot *et al.*, 2007).

Naturally, a reactive solution is not a simple objective function, but instead a resource allocation policy (a mapping from states to actions) which controls the process.

In the next section, we will represent some works using reinforcement learning (a reactive decisional and learning technique) in manufacturing systems researches.

2.3 Reinforcement learning in manufacturing systems

The machine learning has been widely introduced in the design of production control systems and even in the system decision making process itself. Let's talk about (Katalinic and Kordic 2004) who have dealt with the scheduling problem in a very expensive electric motor production system. The aim was to optimize resource use. (Katalinic and Kordic 2004) considered these units as insect colonies able to organize themselves to carry out a task, allowing production risk problems to be solved more easily. The evolutionist algorithms were also used in the resolution of resources allowance problems (Bousbia and Trentesaux, 2004), and the minimization of the products latency to carry out their tasks from where improvement of the system performances.

The objective here is to provide a manufacturing control system that generates on-line scheduling solutions (resource allocation at the job request time), but respects an overall goal at the same time (minimizing the overall production time, minimization of the total cost, etc.)

Reinforcement learning is learning to act by trial and error. Reinforcement learning is learning by trial and error. In this paradigm, agents can perceive their individual states and perform actions for which numerical rewards are given. The goal of the agents is thus to maximize the total reward they receive over time. This technique was often used in robotics, in order to teach a robot a proper behavior face of goals and obstacles. (Dongbing and Yang, 2007) used reinforcement learning algorithm to control cooperative multi-robots in a leader-follower robotic system and a flocking system, they demonstrate by simulation that the control performance can be improved by learning. Other searchers used this technique in several domains: to handle the changing dynamics of the complex traffic processes within the network, (Choy et al., 2003) used an online reinforcement learning module to update the knowledge base and inference rules of the agents. The most used reinforcement learning algorithm is Qlearning. Wei and Zhao (2005) extended this algorithm by using a reward function based on EMLT (Estimated Mean LaTeness) scheduling criteria, but in this paper, we are exploring a more developed algorithm "SARSA algorithm" in a heterarchical organisation of agents.

In conclusion, we are trying to experiment reinforcement learning by using **SARSA** algorithm to conceive an **adaptative** and **reactive** manufacturing control system based on **heterarchical** multi-agent architecture.

In the next section, we will present our system architecture and motivating our choices.

3. CONTRIBUTION AND PROPOSED CONTROL SYSTEM

A multi-agent system is a distributed system with local-ized decision-making and interaction among agents. An agent is an autonomous entity with its own value system and the means to communicate with other such entities. For a general survey of the application of multi-agent systems in manufacturing, see the review by Aissani et al. (2008). In order to develop multi-agent system with a reactive decision capability in an uncertain environment, they may be modelled as Markov Decision Process (MDP) (Russell and Norvig, 1995). And to improve the system performances and learn optimal policy in Markov environment, If the transition function T (modelling the system's evolution from state to state) is unknown while an objective can be identified a learn-by-trial process such as RL (Russell and Norvig, 1995), (Singh and Sutton, 1996) can be designed. This proposition is resumed in (Fig. 2).

3.1 The proposed manufacturing control system

Our control system must be closely related to the manufacturing system, which requires the presence of many sensors, which provide accurate data at the requested time. Then, our control system is composed of agents who are entities decision system, including: **resource agent** to represent the system resources and **part agent** to represent the manufactured products in this system. And in order to meet a long-term goal (produced in the shortest time) performed by the convergence of agent's decisions to meet specific needs (placing jobs on resources) to a common objective (reducing the production time or minimize machines downtime, etc.) **Observer agent** has been introduced (see Fig. 3); it had a view on the whole of system (agents, system, external user...). For more details about observer see (Aissani *et al.* 2008).

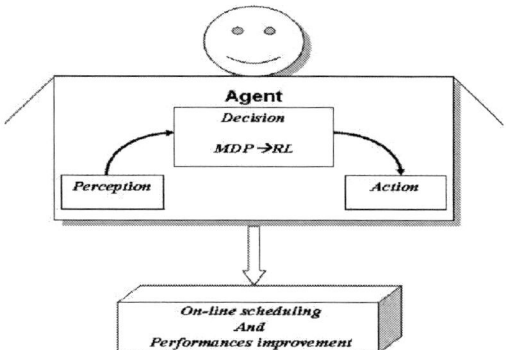

Fig. 2. MDP→ RL →improvement of on-line scheduling performances

The relevant idea underlying this approach is to consider both resources and physical parts as active entities modelled according to Alaadin modelization (Ferber and Gutknecht, 1998) these agent entities have certain properties, roles and groups. Initially, they must have knowledge about their properties (e.g., resource names, feasible tasks, and characteristics), a list of their own role (i.e., tools) and their group (e.g., list of resources able to realize the same production tasks).

In addition, we equipped these agents with three other modules: after the system state is coded (St) by the Perception module, an agent selects with the Learning module an action (At) to be executed by the Action module, and depending on what this action yields, the agent receives a numerical recompense, which may be positive or negative (Rt), to reward or punish the executed action (t is always a given instant).

3.2 SARSA (Stat, Action, Reward, new Stat, new Action) algorithm to resolve on-line scheduling problem

MDPs (Russell and Norvig, 1995) are useful for studying a wide range of optimization problems solved via dynamic programming and reinforcement learning. A Markov Decision Process is a tuple $(S, A, P(*,*), R(*))$, where:

- S is the State space,
- A is the action space,
- $P_a(s, s') = P(s_{t+1} = s' \mid s_t = s, a_t = a)$ is the probability that action a in state s at time t will lead to state s' at time $t+1$,
- $R(s)$ is the immediate reward (or expected immediate reward) received in state s.

The goal is to maximize a cumulative some of rewards (see Fig. 4). The Markov Decision Process solution is a policy π, which gives the action to take for a given state, regardless of prior history. Markov decision process is combined with a policy in this way; this fixes the action for each state.

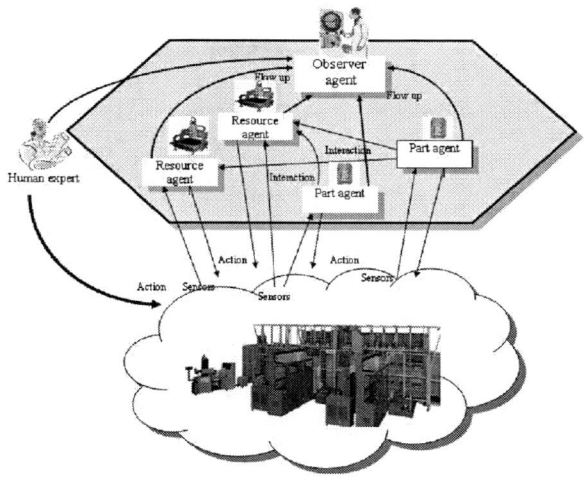

Fig. 3. Our Manufacturing Control System architecture

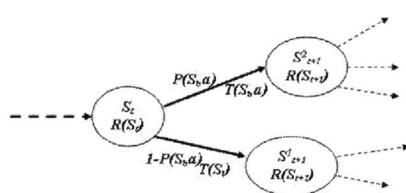

Fig. 4. MDP progression

The standard algorithms to calculate the policy requires storage for two arrays indexed by state: value V, which contains real values, and policy π which contains actions. At the end of the algorithm, π will contain the solution and $V(s_0)$ will contain the discounted sum of the rewards to be earned (on average) by following that solution. The algorithm then has the following two kinds of steps.

$$\pi(s) = \arg\max_a \sum_{s'} P_a(s,s')V(s')$$
$$V(s) = R(s) + \gamma \sum_{s'} P_{\pi(s)}(s,s')V(s') \quad (1)$$

Where γ is the discount rate and satisfies: The sum is to be maximized.

As long as no state is permanently excluded from either of the steps, the algorithm will eventually arrive at the correct solution.

But if the probabilities are unknown, the problem is one of reinforcement learning. For this purpose it is useful to define a further function, which corresponds to taking the action a and then continuing optimally (or according to whatever policy one currently has):

$$Q(s,a) = R(s) + \gamma \sum_{s'} P_a(s,s')V(s') \quad (2)$$

In Q-learning algorithm (2) become:

$$Q_{t+1}(s_t, a_t) = R(s_t, a_t, s_{t+1}) + \gamma \max_{a' \in A} Q_t(s_{t+1}, a') \quad (3)$$

Where t is instant and a' is the supposed next chosen action. For more details see (Watkins, 1989). Q-learning is the most famous reinforcement learning algorithm, but in this paper, we are investigating SARSA algorithm (Rummery and Niranjan, 1994). Essentially, this algorithm is performing a gradient descent through the reward function to maximize return on the path to the goal. As the goal is reached many times, the values along the path from start state to goal converge to their true values. Once every action has been experience in every state sufficiently, the optimal policy is to act greedily with regard to the value function. Formally, SARSA replaces the max over Q-values with the Q-value of the state action pair that is actually observed on the next step. Since updates are based on the actions that are actually taken, rather than on the best possible action, SARSA based modules discover Q-values that are closer to the true expected return under the composite policy. Then, (3) became (4) which will be used in our agent's decision module.

$$Q_{t+1}(s_t, a_t) = R(s_t, a_t, s_{t+1}) + \gamma Q_t(s_{t+1}, a') \qquad (4)$$

In the next figure (Fig. 5) we will see SARSA algorithm steps

```
1. In the current state s, select the action to be
executed which the Q value is raised and according
to a Boltzmann probability distribution
2. Execute the selected action a, which leads to
the new state s'
3. Update each module according to the update
formula (2)
4. Set s' to current state s and goto step 1
```

Fig. 5. SARSA algorithm

For an on-line scheduling problem, this algorithm data are in Fig. 6.

3.3 Agent's behavior

Our system agents are equipped of a decision-making process running the SARSA algorithm, their behavior is as follows: An agent perceive the resource or product stat (Working, Stopped, request for manufacturing (message from another agent), etc.) S_t, then, according to this state he choses the action with the maximum *Q value* to execute (sending to the system) *a*, then, it updates *Q (s_t, a)* according to (4) and perceive the new state S'_t (Waiting new message).

The Observer agent will have a total sight on the system, it will observe the indicators of the total performance. This agent will observe all decision making centre to learn from their behavior (agents, human expert and IDSS). It is a learning centred interaction. The Observer agent plays triple role:

Stats
S1 Working
S2 Broken down
S3 Stopped
S4 Working and Launch a task
S5 Broken down and Launch a task
S6 Stopped and Launch a task
S7 Working and end of task
S8 Broken down and end of task
S9 Stopped and end of task

Actions
A1 Put on standby
A2 Stop machine
A3 Change machine
A4 Starting

Rewards
R1 Working and task in progress 0
R2 Starting and task in progress +4
R3 Stop and end in progress +5
R4 Stop and task in progress -5

Fig. 6. SARSA algorithm data

- The first, is the interpretation and the digitalization of the perceived state or received information
- The second is the chart of the received messages.
- The third is the calculation of the system learning level evaluation criteria.

In the next section, we will show, how we did to realise the above described model and present also en experimentation to valid this approach.

4. IMPLEMENTATION AND EXPERIMENTS

Our model was simulated in the Borland Jbuilder environment because of its potential for facilitating communication and thread programming and because of its compatibility with the chosen MADKIT platform architecture for SMA development (visit http:// www.madkit.org/ downloads).

One of the advantages of the reinforcement learning algorithms is that they allow evaluation during learning. To permit this evaluation, we selected the following criteria.

4.1 Experiences and investigation

As an experimental test bench, we chose the data set used by Hurink et al. (1994), which was extended by Dauzère-Pérès and Paulli (1997) and applied to the gene-ralized multiprocessor jobshop scheduling problem.

This benchmark is composed of the data sets, edata, rdata, sdata and vdata, and contains File M x P files (i.e., files with M resources and P parts problems). L01 is 5×10, L10 is 5×15, L20 is 10×10, and L30 is 10×20; edata comprises the original problems used by Hurink et al. (1994) and Adams et al. (1988). Each set Mi represents the machine to which operation i is assigned in the original problem, plus any of the other machines with a given probability.

From this probability, Hurink et al. generated three sets of test problems: edata, rdata, and vdata. The first set edata contains the problems with the least amount of flexibility (close to the original job-shop scheduling problems). The average size of Mi is equal to 2 in rdata and m/2 in vdata. (For more details, see Hurink et al., 1994).

4.2 Improvement of Cmax in scheduling solution

We used first the data set m10 from *edata* to see how our approach can improve the quality of scheduling solutions to prove our approach ability of improving the **Cmax** of the generated scheduling solutions.

The first graph (Fig. 7) shows the *Cmax* evolution, before 6000 iterations, where one iteration is a new launch of the scheduling problem, the scheduling solutions are given with a large *Cmax*, up to 2500, in reinforcement learning this phase is called an **exploration** phase where actions are carried out by chance according to certain probability, here we used the Boltzmann distribution, the motivation of this choice is in (Aissani *et al.*, 2008). While the solutions suggested towards and after the 10000 iterations, *Cmax* is very interesting 670. Moreover, we notice a convergence of suggested solutions: the *Cmaxs* interval is [900,670]. This phase is called an **exploitation** phase, where the choice of actions is based on Q values.

We can point out as well, a significant improvement in the quality of the proposed scheduling solutions in terms of total time production *Cmax*. These results show that our system is able to learn how to establish a continuously improving optimal control policy.

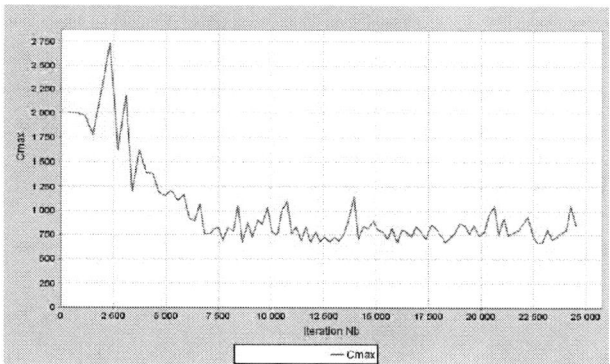

Fig. 7. Cmax Graph of 10 x 10 problem experimentation

Now, we will compare our best given solution (considering *Cmax*) to the best solutions given by Hurink *et al.*, (1994) on the same Benchmark (*edata*, *rdata* and *vdata*). The following (Table. 1) table will show these results.

Firstly, we can see that results obtained with *edata* are worse then the obtained results with *rdata* which are less good then the obtained with *sdata* and *vdata*, it can be explained by the flexibility. In other words, the increased number of resources able to execute a job in *rdata* and *vdata* allowed the system more flexibility.

We can also point out that ,our approach shortly provide even better solutions than traditional approaches (tabu search Hurink *et al.*, (1994)), in 33% of case our *Cmax* is the best and in 33% case they are equivalent.

This experiment has shown that our approach gives scheduling solutions with the wire of time which means a dynamic and reactive scheduling (responding to on-line resource allocation requests), even more so, our approach improves the quality of its solutions after several iteration of the learning algorithm (in exploitation phase). Finally, by comparing our approach with classical scheduling methods (tabu search (Hurink *et al.*, (1994)) on the same benchmark, our approach was quite successful and very promising.

Table 1. Comparison with prior results on the Hurink et al. data. Boldfaced values indicate the best solution values.

		Best Cmax according to...	
		Our approch	[Huring et al.], [Dauzeres-Peres et al.]
edata	L01	**605**	609
	L10	866	866
	L20	1343	**857**
	L30	**1264**	1147
	M06	**47**	55
	M10	**673**	871
rdata	L01	**421**	570
	L10	804	804
	L20	**774**	756
	L30	1068	1068
	M06	47	47
	M10	**675**	679
vdata	L01	**461**	570
	L10	804	804
	L20	**810**	756
	L30	1068	1068
	M06	49	**47**
	M10	660	**655**

5. CONCLUSIONS AND FUTURE WORKS

In this paper, we gave an overview of the work done in the field of manufacturing control and we focus on the **dynamic and reactive scheduling**, we gave the reason for choosing a **heterarchical** organisation of the control system by proposing a model based on **multi-agents system** with a reactive decision-process with a continuous improvement which is **reinforcement learning**, and more specifically **SARSA** algorithm.

Simulation results have been presented for a benchmark representing a variety of flexible workshops. These simulations demonstrate that this model can generate scheduling solution in on-line and improving their quality (Cmax) with mire of time.

By comparing these results to conventional scheduling approaches notably tabu search, our approach gives very promising results.

Nevertheless, we think to work on the perceived system stats representation in order to optimize the Q table entries. We think also test our approach on other Benchmarks using another heuristics and metaheuristics.

REFERENCES

[1] Adams, J., E. Balas, and D. Zawack, (1988), The shifting bottleneck procedure for job-shop scheduling, *Management Science*, **Vol 34**, p 391-401.

[2] Aissani. N, D. Trenteseaux and B. Beldjilali. (2008). Use of Machine Learning for Continuous improvement of the Real Time Manufacturing control

system performances. *IJISE: International Journal of Industrial System Engineering*, **Vol 3, No 4**, , p 474-497

[3] Albadawi. Z, B. Boulet, R. DiRaddo, P. Girard, A. Rail and V. Thomson, (2006), Agent-based control of manufacturing processes, *International Journal of Manufacturing Research (IJMR)*, **Vol. 1, No. 4**, 2006, p 466-481

[4] Bousbia, S and D. Trentesaux, (2002), Self-Organization in Distributed Manufacturing Control: state-of-the-art and future trends, *IEEE International conference on Systems, Man & Cybernetics*, **Vol 5**, 6 p.

[5] Bousbia S. and D. Trentesaux, (2004), Towards an intelligent heterarchical manufacturing control system for continuous improvement of manufacturing systems performances, *5th International Conference on Integrated Design and Manufacturing in Mechanical Engineering – IDMME'2004*, Bramley, Culley, Dekoninck, McMahon, Medland, Mileham, Newnes & Owen (eds), CD-Rom, ISBN: 1-85790-129-0, Hadleys Ltd – Essex (University of Bath, UK, Avril 2004), p. 125 (abstract), paper #123, 10 p.

[6] Charton. R, A. Boyer and F. Charpillet, (2003). Learning of Mediation Strategies for Heterogeneous Agents Cooperation, *15th IEEE International Conference on Tools with Artificial Intelligence - ICTAI'2003, Sacramento, Californie, USA, IEEE Computer Society*, B. Werner editor, p. 330-337, Nov, 2003.

[7] Chen. R., K. Lu and C. Chang, (2003), Application of the multi-agent approach in just-in-time production control system', *International Journal of Computer Applications in Technology (IJCAT)*, **Vol. 17, No. 2**, p 90-100

[8] Choy. M. C, R. L. Cheu, D. Srinivasan and F. Logi, (2003), Real-time coordinated signal control through use of agents with online reinforcement learning, *Transportation Research Board*, **No 1836**, p 64-75

[9] Csaji B. C and L. Monostori, (2006), Adaptive algorithms in distributed resource allocation. *Proc of the 6th International Workshop on Emergent Synthesis (IWES-06)*, August 18–19, The University of Tokyo, Japan, p. 69-75

[10] Dongbing. G. and E. Yang, (2007), Fuzzy Policy Reinforcement Learning in Cooperative Multi-robot Systems, *Journal of Intelligent and Robotic Systems*, **Vol 48, No 1**, p 7-22

[11] Ferber. J and O. Gutknecht, (1998) Un méta-modèle organisationnel pour l'analyse, la conception et l'execution de systèmes multi-agents, *Proc of 3rd International Conference on Multi-Agent Systems ICMAS'98*, p 128-135

[12] Haruno. M, and M. Kawato, (2006), Heterarchical reinforcement-learning model for integration of multiple cortico-striatal loops: fMRI examination in stimulus-action-reward association learning', *Neural Networks*, **Vol 19, Special Issue (2006)**, p 1242–1254

[13] Hurink, J., B. Jurisch, and M. Thole, (1994), Tabu search for the job-shop scheduling problem with multi-purpose machines, *OR Spektrum*, **No 15**, p 205-215.

[14] Katalinic. B and V. Kordic, (2004), Bionic assembly system: concept, structure and function, *the 5th International Conference on Integrated design and Manufacturing in Mechanical Engineering, IDMME* 2004, Bath, UK, April 5-7, 2004

[15] Marthi. B, S. Russell, D. Latham and C. Guestrin, (2005), Concurrent Hierarchical Reinforcement Learning, *International Joint Conference on Artificial Intelligence, IJCAI-05*, p 779-785

[16] McCulloch, W. S, (1945), A Heterarchy of values Determined by the Topology of Nervous Nets, *Bull. math. biophys.*, **Vol. 7**, pp. 89-93.

[17] Prabhu, V.V, (2003), Stability and Fault Adaptation in Distributed Control of Heterarchical Manufacturing Job Shops, *IEEE Transactions on Robotics and Automation*, **Vol. 19, No. 1**, p. 142-147.

[18] Prouvost, P. (2004), *Automatique Contrôle et régulation*, Dunod, France

[19] Pujo. P and D. Brun-Picard, (2002), Pilotage sans plan prévisionnel ni ordonnancement préalable , *Méthodes du pilotage des systèmes de production*, Hèrmes, 2002. p 129- 162.

[20] Rummery. G and M. Niranjan, (1994), On-line q-learning using connectionist systems, *Technical Report CUED/F-INFENG/TR 166*, Cambridge, University Engineering Department.

[21] Russell S. and P. Norvig, (1995), Artificial Intelligence: A Modern Approach, *The Intelligent Agent Book. Prentice Hall Series in Artificial Intelligence*, 1995.

[22] Wang, Y.C. and Usher, J.M., (2005), Application of reinforcement learning for agent-based production scheduling, *Engineering Applications of Artificial Intelligence*, **Vol. 18, No 1**, P 73-82

[23] Watkins, C. J. C. H, (1989), Learning from delayed rewards, *PhD thesis*, Cambridge University, Cambridge, England

[24] Wei Y-Z and Zhao M-Y, (2005), A reinforcement learning-based approach to dynamic Job-shop scheduling, *Acta automarica sinica*, **Vol 31, No 5**, p 765-771.

Intelligent Production Monitoring and Control based on Three Main Modules for Automated Manufacturing Cells in the Automotive Industry

Ulrich Berger*, Ralf Kretzschmann* and Algebra Veronica Vargas A.*

*Brandenburg University of Technology Cottbus,
Chair of Automation Technology
Siemens-Halske-Ring 14, 03046 Cottbus Germany
{ulrich.berger, ralf.kretzschmann, veronica.vargas @tu-cottbus.de}

Abstract: The automotive industry is distinguished by regionalization and customization of products. As consequence, the diversity of products will increase while the lot sizes will decrease. Thus, more product types will be handled along the process chain and common production paradigms will fail. Although Rapid Manufacturing (RM) methodology will be used for producing small individual lot sizes, new solution for joining and assembling these components are needed. On the other hand, the non-availability of existing operational knowledge and the absence of dynamic and explicit knowledge retrieval minimize the achievement of on-demand capabilities. Thus, in this paper, an approach for an Intelligent Production System will be introduced. The concept is based on three interlinked main modules: a Technology Data Catalogue (TDC) based on an ontology system, an Automated Scheduling Processor (ASP) based on graph theory and a central Programmable Automation Controller (PAC) for real-time sensor/actor communication. The concept is being implemented in a laboratory set-up with several assembly and joining processes and will be experimentally validated in some research and development projects.

Keywords: process control, process monitoring, ontology, semantic net.

1. INTRODUCTION

The current situation in the automotive industry is characterized by increasing requirements from the customer side on quality and individualization of products and, at the same time, upcoming pressure on product prices. Car manufacturers create new product segments and enrich existing segments with more possibilities for individualization like regionalization and customization of products. The product diversification is combined with ongoing reduction of product life cycle time and an acceleration of innovation (Kuhn, et al., 2002). Furthermore, the automotive industry is characterized by enabling innovations in light-weight vehicle structure, energy efficient power-train solutions and assistance systems. New manufacturing paradigms for automotive structures and components force the automotive industry to continuously promote the development of cost-efficient and innovative vehicles, with high-added customer value, increased personalization capabilities and environmental sustainability. According to the recent Global Technology Revolution 2020 report, issued by the RAND corporation, there is foreseen a strong trend for on-demand manufacturing of components and small products according to personal or corporate specifications (Silberglitt, 2006). Therefore the whole product lifecycle hast to be customized. The development and process planning of diversified products is supported by powerful CAD/CAM and PLM systems. Nevertheless, the production will be the bottleneck along the lifecycle. On the other hand, the knowledge about technologies, materials and data engineering in general, that is normally generated at the shop-floor, is not documented and consequently it is lost. The non-availability of ubiquitous operational knowledge and the absence of dynamic and explicit knowledge recuperation procedures minimize the achievement of on-demand capabilities. In this context, the Japanese NISTEP report no. 99 from 2005 identifies with high priority the establishment of a technology for converting implicit knowledge on manufacturing and manufacturing technique into explicit knowledge (NISTEP, 2005).

It is proposed in this article that the handling of different products in the production line and the retrieving, sharing, processing and structuring of relevant information related to these products are still unsatisfactory. Therefore, a new intelligent production monitoring and control for automated manufacturing cells in the automotive industry will be introduced.

2. FRAME OF REFERENCE

2.1 Rapid Manufacturing (RM) for diversified components

Besides already existing manifold efforts in the sector of Rapid Manufacturing (RM), the introduction of such principles in automated manufacturing cells is still in the beginning (Park, 2006). RM technologies are based on linear thermo-mechanical and chemical processes, varying ingot material properties, recipes and process parameters, whereas automated manufacturing tasks are based on complex operational condition, real-time monitoring and heuristic decision features.

2.2 Adaptation of Manufacturing Cells for Regionalization and Customization of Products

As shown in figure 1, a current used manufacturing cell consists of several components. The main components are the industrial robot with the Robot Control (RC), the Human-Machine-Interface (HMI) and additional programming devices.

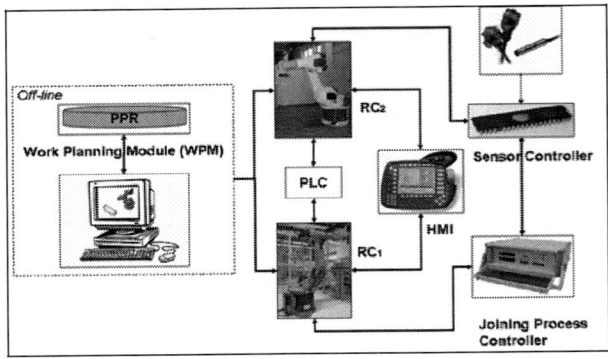

Fig. 1. Standard approach of a production cell

These devices can be classified in on-line and off-line devices. It is a common practice to generate robot programs with an off-line work planning module based on Product, Process and Resources (PPR) information. Furthermore, the robot is connected via the robot control and Programmable Logic Controller (PLC) with supplementary devices such as further Robot Controllers and sensor controllers. These sensors have to determine diversified assembly and joining tasks in order to modify the robot program and the paths so as to fulfil the corresponding operations like adhesive bonding, mechanical joining as riveting and welding. Thus, each sensor controller has to be connected to the RC. As results of this standard approach a lot of interfaces have to be kept. Furthermore, the actual implementation has low capability, high downtimes of the automated machinery and consequently high costs of trouble shooting.

2.3 Managing engineering data

The product development process can be defined as a cross-functional, inter-company and market oriented process which requires constant interaction based on the exchange of data to respond to customer needs (Calabrese 1999). Therefore, the product development process involves distribution of explicit knowledge within the company through formal communication channels whereas implicit knowledge requires co-operation (Calabrese, 1997).

According to Van der Bij, et al., (2003), higher levels of knowledge dissemination and information exchange leads to a better understanding of technology capabilities and trends, market- customer and competitors- and competitive actions, which are essential information to design the manufacturing process and determine the product features and specifications. Higher levels of knowledge dissemination and information exchange leads, as well, to a considerable decrease of marketing and technical uncertainties in the innovation process. Moreover, greater levels of information exchange within an organization increase the likelihood that the new product will be positioned in the right market segments and will be introduced at an optimal time improving the chances of product success.

It has been suggested that information can be exchanged formally within the boundaries of defined mechanisms, such as structured methods and formal processes; or informally; and both horizontally, e.g. cross-functional, and vertically within the organization (Perks 2000, p. 182; Van der Bij, et al., 2003, p. 164; Calabrese, 1999, p. 440).

Management can determine knowledge sharing by implementing formal procedures for guiding information flows; moreover, there are mechanisms which can originate such process (Berends, et al., 2006, pp. 88-91):

1) Diffusion- members of an organization select and communicate existing information without being oriented towards a particular problem.

2) Information retrieval- someone who needs a particular piece of information obtains it by asking someone who has it.

3) Information pooling- members of an organization working together pool information; it is transferred not only factual information but also questions, suggestions and instructions.

4) Collaborative problem solving- new information is developed with regard to a shared problem.

5) Pushing- someone chooses to provide someone else with the existing information. It involves thinking that the other person needs to know something, or that certain information might be useful for his research activities.

6) Thinking along- someone developed new ideas with regard to someone else's problem. It may yield new ideas, hypotheses or questions.

7) Self-suggestion- in the same way as one can think about someone else's problem, one can also think about one's own problem during interaction. The need to explain one's own problem stimulates one to come up with new explanations, solutions, arguments and conclusions.

Nevertheless, sharing knowledge among members of a big organization may be a complex activity. And as long as the knowledge is not shared, can not be exploited by the organization (Choo, 1996).

Research based on organizational memory indicates that there are sophisticated computerized tools for externalizing personal knowledge such as idea managing tools, expert systems, among others; nevertheless, there is still a danger that tacit knowledge may remain as personal knowledge due to the complexity to share information throughout distributed or big companies (Klammer & Mathias, 1998).

According to Basson, Bonnema and Liu (2004), there are also many tools, such as CAD systems, that are available to manage the latter type of design information -the final product definition-. In contrast, much of earlier type of design information is currently not captured in structured or systematic ways due to its level of abstraction; it is supposed that a high level of abstraction is related to less detailed information.

Further, the access to information has been solved, by large-scale computer networks, but the processing and interpretation of retrieved information remain a problem due to heterogeneity of the data. There can be distinguished three main heterogeneity problem categories (Stuckenschmidt and van Harmelen, 2005):

1) Syntactical problems, e.g. data format heterogeneity

2) Structural problems, which are originated because the same objects and facts can be described in different ways using homonyms (the use of the same word with different meaning), synonyms (the use of different words with the same meaning), etc.

3) Problems of semantics, which refer to intended meaning of terms in a particular context or application. They occur due to the inherent context dependency of information that can be only understood in the context of their original source and purpose.

The first kind of problem can be solved through standards that are used as interfaces to integrate different information sources. Structural problems and semantic conflicts can be partially solved by one to one structural mappings; if structural mappings do not apply, like in the case of large-scale information sharing, the semantics of the information has to be taken in account in order to decide how different information items relate to each other. It is proposed the use of ontologies as technology for approaching the problem of explicating semantic knowledge about information (Stuckenschmidt and van Harmelen, 2005).

Many authors have addressed the definition of ontologies; some of them will be here mentioned to broaden the understanding of this subject. An ontology "is a set of formal terms, usually with a hierarchical organization, with associated formal definitions that specify their relationships with the other formal terms, and a set of constraints about their use in the knowledge representation of the domain studied" (Goossenaerts and Pelletier, 2001).

Similarly, ontology can also be defined as an "explicit specification of a conceptualization" (Gruber, 1993a and Gruber, 1993b).

3. APPROACH FOR AN INTELLIGENT PRODUCTION SYSTEM

To overcome the limitations above explained, an intelligent production monitoring and control of automated manufacturing cells will be introduced and experimentally validated. The concept will be used for monitoring and controlling diversified components and/or products, which are previously manufactured by RM technologies. The concept is based on three interlinked main modules, (i) a Technology Data Catalogue (TDC) based on an ontology system, (ii) an Automated Scheduling Processor (ASP) based on graph theory, and (iii) a Programmable Automation Controller (PAC) for real-time sensor/actor communication.

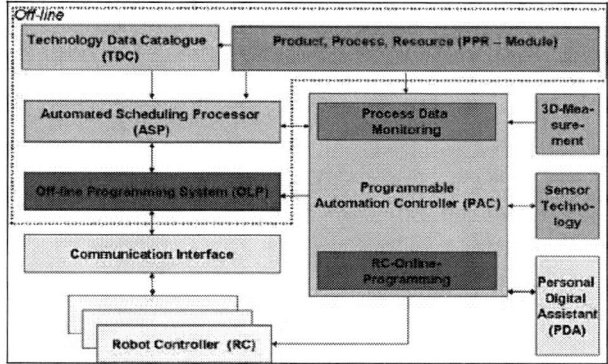

Fig. 2. Architecture of the intelligent production system

As shown in figure 2 the concept is enriched with further modules. The modules will be divided in on-line and off-line modules. The off-line modules, the TDC (i) and the Product Process Resource (PPR) – module, will supply the ASP (ii) with all relevant engineering data. The ASP will generate a process work plan which will match to the given production task. Afterwards, an off-line programming system (OLS) will transfer the work plan as robot program via a communication interface to the RC, while the on-line modules will monitor and control the current joining and/or assembling processes. Moreover, the innovative PAC (iii) will be introduced as central intelligent cell controller. All relevant on-line information, which will be measured by a 3D-measurement and a sensor technology, will be gathered and pre-processed by the PAC. Thus, the PAC may influence the execution of the robot program directly, through one interface to the RC; consequently, it is achieved reduction of interfaces to the RC and reduction of off-line programming complexity. A new kind of direct interaction with the shop-floor technician is warranted through the new human-machine-interaction paradigm, a Personal Digital Assistant (PDA) and the access to several process planning information systems (Berger and Kretzschmann, 2007).

3.1 Module I: Technology Data Catalogue

With the aim of retrieving, sharing, processing and structuring relevant engineering data, the development of a Technology Data Catalogue (TDC) is proposed. Following the ideas of Stuckenschmidt and van Harmelen (2005), to overcome data format heterogeneity, it is suggested the use of STEP (the international standard neutral file format); this will allow gaining information from CAD/CAE/CAM, Product Data Management (PDM) and Enterprise Resource Planning (ERP) systems. Moreover, in order to workout problems of semantics, an ontological system is suggested. Specifically, this model will contain (1) a database with shared vocabulary -which will allow the use of terminology most appropriate to the particular context-, (2) translators –which will determine the relation among concepts coming from different sources- and (3) filters –which will enable sorting out information that better match the requested technology (Lepratti, 2005; Basson, Bonnema and Liu 2004; Goossenaerts and Pelletier, 2001).

The TDC will also fulfil other requirements, highlighted in the literature as general requirements for information systems (Basson, Bonnema and Liu; 2004); some of them are for example:

1. Facilitate the extraction of information selectively, avoiding redundancies
2. Improvement and increment information by recording informal statements, generated for example at the shop floor: avoiding that they remain as personal hidden knowledge.
3. Ensuring consistency of the information

The last component of the TDC, a semantic net, defined in the literature as "a graphic notation for representing knowledge in patterns of interconnected nodes and arcs" (Sowa, 1992), will facilitate the dynamic navigation through the information, the information visualization and the sharing of information through formal procedures. Mechanisms which may originate the knowledge sharing (Berends, et al., 2006) will be also accomplished to assure that operational knowledge will not remain personal knowledge.

General TDC outputs will be relevant process parameters, such as: cutting speed or pressure force for a specific technology. The next figure (Fig. 3) shows the Technology Data Catalogue structure above discussed.

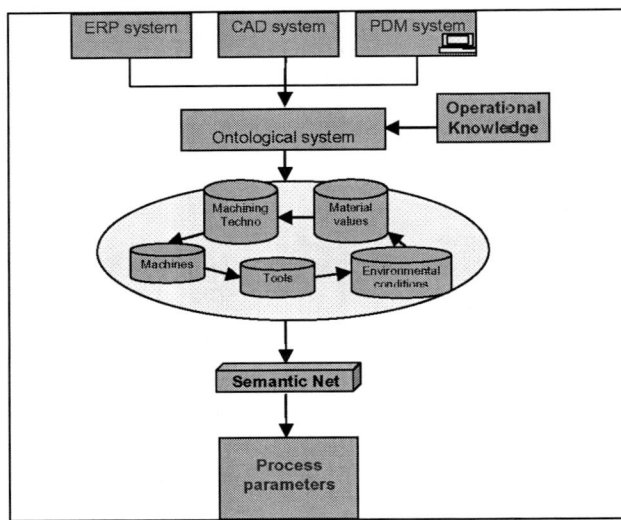

Fig. 3. Technology Data Catalogue Model

3.2. Module II: Automated Scheduling Processor (ASP)

The ASP has to create an optimized and adaptable work plan based on feature technology, which will be used in the process planning (Berger, Cai, Weyrich, 2005). See Fig. 4.

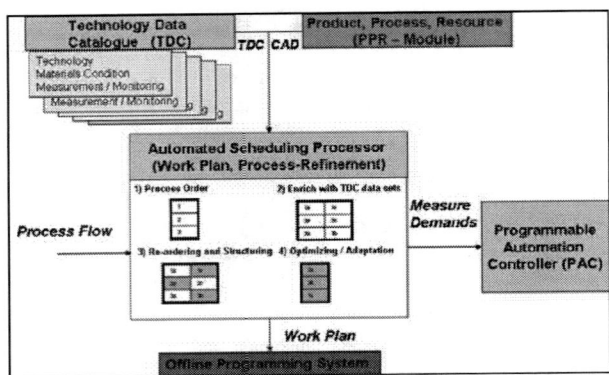

Fig. 4. Architecture and interfaces of the ASP

The ASP has access to the TDC to get all information about applicable technologies for joining and assembling with the corresponding materials conditions, measuring and monitoring conditions for these technologies. As well, the ASP has manifold access to the PPR-module to get information about the produced product and other relevant information such as materials and tolerances. The ASP creates an optimized work plan and defines measurement demand to control this work plan at the shop-floor. Then, the given information stock and the given process flow will be used. The module's main aim is to create a work plan, which can be adapted to current production task corresponding to the current assembly and/or joining processes detected by the PAC. It is proposed that, the workflow to create the optimized work plan has four steps (as shown in Figure 5).

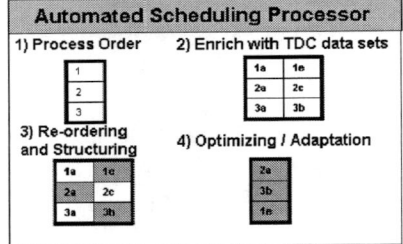

Fig. 5. Workflow for creating the optimized work plan

First, the normal process order will be given by the process flow (1). This process order defines joining and/or assembling operations with the work piece. The next step will be the enrichment of the process order with alternative technologies provided by the TDC. With the help of re-ordering and structuring algorithms, known from the graph theory, the alternatives technologies of each process step will be eliminated. Additionally, the process will be ordered in an optimized way. Finally, each process step will be adapted to the corresponding process flow and the whole work plan will be transferred to the shop-floor.

3.3. Module III: - Programmable Automation Controller (PAC)

Part of the intelligent control unit is the automated monitoring and processing of assembly and joining parameters in real-time. The PAC is further connected to standard automation devices like the robots control or PLC based manufacturing cells (as shown in figure 6). Intelligent production control will be achieved by integration of production simulations; intelligent sensors and behaviours, able to recognize their environment; and novel feedback loops, which link the simulation and the practical validation.

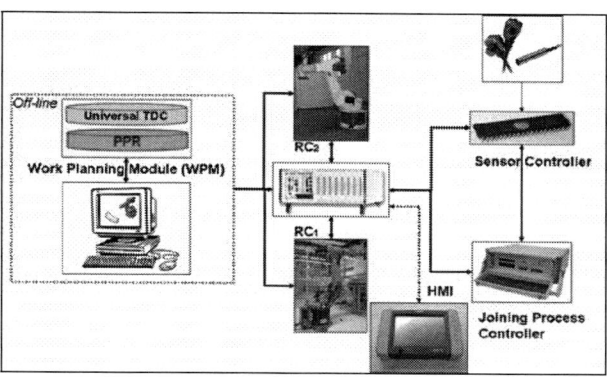

Fig. 6. Architecture and interfaces of the new approach

The PAC has several interfaces and can access all information provided by the TDC and PPR-module. Additionally, the PAC has access to the ASP to get the process planning information. The PAC is the central device to gather all sensor data for the cell. The PAC is connected to the RC in order to influence the execution of the robot program. The central scheme is to pre-process all sensor data in a powerful CPU provided by the PAC. Thus, the PAC can calculate new robot paths by segmentation of detected curves and paths (Berger, Noack and Kretzschmann, 2006) in order to control the robot program corresponding to the process planning information via one interface. Furthermore, the technician can influence the intelligent process control via the new Human-Machine Interface (HMI) communication device called Personal Digital Assistant (PDA). The powerful PAC architecture offers interfaces to further data bases at the PAC to store best practice decision of the technician in order to re-use these experiences later on. The PAC represents a central process control in order to pre-process sensor data with high-level algorithm. Therefore, the complexity of the robot programs can be reduced.

4. TECHNICAL REALIZATION

The Technology Data Catalogue is still an ongoing process. While the ontology system is already validated, the semantic net is not yet developed. In order to determine if a program to develop the desired semantic net is already available in the market or if this net will be programmed, research and benchmarking activities has been achieved.

The Automated Scheduling Processor (ASP) and the Programmable Automation Controller (PAC) are running in a laboratory set-up (shown in Figure 7) with distinct assembly and joining processes and will be experimentally validated in research and development projects.

One of these projects is the EU-FP6 IP FUTURA, which will pursue innovative concepts that will facilitate the integration of multifunctional materials in the automotive industry according to customized demands. These concepts will include modular, scalable, hybrid body and chassis structures. The scalability will enable an accommodation of size and design requirements for different models, while the modularity concept will facilitate an easy derivation of variants and a flexible assembly and/or joining process.

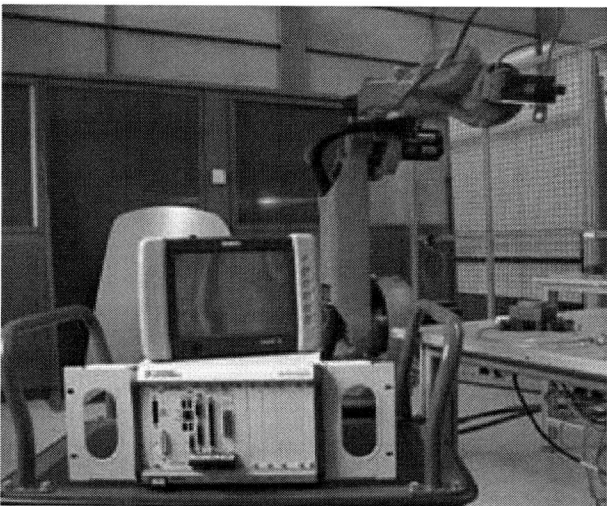

Fig. 7. Laboratory set-up (HMI, PAC, RC)

The implementation is still an ongoing process. The actual development status enables the communication between the PAC and the connected devices. The TDC will provide the PAC with technology process data in order to be processed by the PAC.

5. CONCLUSIONS

The adaptation of manufacturing cells in order to handle diversified products is a key issue in the successfully automated assembly and joining operations of customized parts. The production of high diversified products can be divided into technologies for manufacturing diversified components and for joining and/or assembling them into complex products.

On the other hand, the non-availability of ubiquitous operational knowledge and the absence of dynamic and explicit knowledge recuperation procedures minimize the achievement of on-demand capabilities. In this context, the Japanese NISTEP report no. 99 from 2005 identifies with high priority the establishment of a technology for converting implicit knowledge on manufacturing and manufacturing technique into explicit knowledge

Therefore a new intelligent production monitoring and control for automated manufacturing cells will be introduced in the automotive industry. Moreover, an approach for controlling the manufacturing cell in real-time was proposed. The main idea is the use of a central device, the Programmable Automation Controller (PAC), in order to pre-process information provided by a central Technology Data Catalogue in the manufacturing cell and to control the industrial robot in a very effective way. Thus, the robot control will be supported. The next steps are the achievement of further validations and the formal implementation in a research and development project. It is planned that the concept implementation results will be presented in further articles.

Although the approach discussed in this article was initially developed for automated manufacturing cells in the automotive industry, it is foreseen its application in other industries.

REFERENCES

[1] Basson, Anton H.; Bonnema, G. Maarten; Liu, Yang (2007). *A flexible Electro-Mechanical Design Information System*. In: Tools and Methods of Competitive Engineering, Imre Horváth and Paul Xirouc hakis (ed.), v.2, pp. 879- 889, Millpress Rotterdam Netherlands, 2004.

[2] Berends, Hans; Van der Bij, Hans; Debackere, Koenraad; Weggeman, Mathieu (2006). *Knowledge sharing mechanisms in industrial research*. In: R&D Management, **Vol. 36**, 1, 85-95.

[3] Berger, Kretzschmann (2007). *Aufbau einer werkstattgerechten Informationsversorgung*. In: Industrie-Management: Zeitschrift für industrielle Geschäftsprozesse; 23. Jahrgang, Ausgabe 4.

[4] Berger, Lepratti, Cai, Weyrich (2006). *Toward the Knowlegde-based Enterprise*. In: IFIP International Federation for Information Processing, **Volume 183**, Jan 2005, pp. 351 – 361.

[5] Berger, Noack, Kretzschmann (2006). *Automatic Generation of Robot Paths from CAD-Data Based on Linear and Circular Approximation*. In: Proceedings of the 4th IFAC-Symposium on Mechatronic Systems.

[6] Berger, Cai, Weyrich (2005). *Ontological Machining Process Data Modelling for Powertrain Production in Extended Enterprise*. In: Journal of Advanced Manufacturing System (JAMS), **Vol. 4**, No. 1: 69-82.

[7] Calabrese, Giuseppe (1997). *Communication and co-operation in product development: a case study of a European car producer*. In: R&D Management, **Vol. 27**, 3, 239-252.

[8] Calabrese, Giuseppe (1999). *Managing information in product development*. In: Logistics Information Management, **Vol. 12**, 6, 439-450.

[9] Choo, C. W. (1996). *The knowing organization: How organizations use information to construct meaning, create knowledge and make decisions*. In: International Journal of Information Management, **Vol. 16**, 5, 329-340.

[10] Goossenaerts, J.B.M and Pelletier, C. (2001). *Enterprise Ontologies and Knowledge Management*. In: K.-D. Thoben, F. Weber and K.S. Pawar (ed.) Proceedings of the 7th International Conference on Concurrent Enterprising: "Engineering the Knowledge Economy through Co-operation" Bremen, Germany, pp. 281-289.

[11] Gruber, Thomas R. (1993a): *A Translation Approach to Portable Ontology Specifications*. Knowledge Acquisition, 5, (2): 199-220.

[12] Gruber, Thomas R. (1993b): *Toward Principles for the Design of Ontologies Used for Knowledge Sharing*. In: International Journal Human-Computer Studies 43, pp. 907-928.

[13] Klamma Ralf, Mathias Jarke (1998): *Driving the Organizational Learning Cycle: The Case of Computer-Aided Failure Management*. In: Baets, Walter R. J. (ed.): Proceedings of the 6th European Conference on Information Systems (ECIS'98), Aix-En-Provence, France, **Vol. 1**. Granada, Euro-Arab Management School.

[14] Kuhn, Wiendahl, Eversheim, Wiesinger (2002). *Schneller Produktionsanlauf von Serienprodukten, Ergebnisbericht der Untersuchung `fast ramp-up`*. Verlag Praxiswissen. Dortmund.

[15] Lepratti, Raffaello (2005). *Ein Beitrag zur fortschrittlichen Mensch-Maschine-Interaktion auf Basis ontologischer Filterung*. Logos Verlag Berlin.

[16] NISTEP (2005). *Nistep Report No. 99, Science and Technology Foresight Center National Institute of Science and Technology Policy (NISTEP)*. Ministry of Education, Culture, Sports, Science and Technology.

[17] Park (2006). *Rapid Manufacturing Today*. http://www.rm-platform.com, Rev. 2007, 08-31.

[18] Silberglitt (2006). *Global Technology Revolution 2020, In-depth Analysis*. Bio/Nano/materials/information, Rand Corp.

[19] Perks, Helen (2000). *Marketing Information Exchange Mechanisms in Collaborative New Product Development, the Influence of Resource Balance and Competitiveness*. In: Industrial Marketing Management, **Vol. 29**, 179-189.

[20] Sowa, John F. (1992). *Semantic Networks*. In: Shapiro Stuart C., Wiley (ed.) Encyclopedia of Artificial Intelligence.

[21] Stuckenschmidt, Heiner; van Harmelen, Frank (2005). *Information Sharing on the Semantic Web*. Springer-Verlag Berlin Heidelberg Germany.

[22] Van der Bij, Hans; Song, Michael X.; Weggeman, Mathieu (2003). *An Empirical Investigation into the Antecedents of Knowledge Dissemination at the Strategic Business Unit Level*. In: Journal of Product Innovation Management, **Vol. 20**, 163-179.

Synthesis of a Discrete Controller Based on Grafcet

Hamdi Hocine *, Alla Hassane **, El Ani Ines *

Automatic and Robotics Laboratory of Constantine
University Mentouri Constantine - Road of Ain El Bey BO 133 -25017 Constantine Algeria
(e-mail: hhamdi@hotmail.com)
GIPSA-Lab Department of Automatic
BO 46 - 38402 St Martin d' Heres Cedex - France (email: Hassane.Alla@inpg.fr)

Abstract: In the supervised control of discrete events systems (DES), the uncontrollable transitions are the source of prohibited states. This article presents an approach for checking the Grafcet properties (presenting synchronization transitions) and its validation compared to the set of conditions. It relies on the extraction of the equivalent automat to the grafcet (graph of accessible states) and uses the control by supervision theory to determine the controller.

Key words: controller, discrete event system, grafcet, graph of accessible states, prohibited state, linear constraint.

1. INTRODUCTION

The supervision theory of the discrete events systems (DES) provides techniques that permit to guarantee, a priori, that the operation of the supervised system will respect specifications imposed. This theory, initiated by work of Ramadge and Wonham (RW) in the beginning of the eighties, is based on the use of the automats models and formal languages.

A supervisor is a DES that makes it possible to modify the operation of a process by prohibiting the generation of certain events by the process. The techniques that permit to synthesize it in a systematic way are based on the concept of controllability, which is the key of the RW theory [7].

The existence of uncontrollable transitions in the DES causes the existence of prohibited states. Many results were obtained, by using the healthy Petri nets [2], to prohibit the access to these forbidden states. In this study one re-uses with a grafcet approach, the already defined concepts for the Petri nets. The main part of the properties used in this article is demonstrated in [3]. Instead of remaking the demonstration of the theorems and properties already established, we will adapt it to the grafcet, by relying us on the definitions necessary to comprehension. An algorithm based on Grafcet is proposed. It presents an implementation technique that makes effective the synthesis of the controller.

The method used to obtain the supervisor provides control stages that prevent the validation of the transition leading to a forbidden state.

2. GRAFCET OF THE UNSUPERVISED PROCESS

Our support example concerns a process made up on the one hand of a balance for scaling and mixing the products, and on the other hand of a transport wagon and product draining. It is represented schematically by Fig. 1.
In *the initial state the* tank C1 is empty (zero of the rocker) and the wagon W is in left position detected by x.

The constraints of operation are expressed as:
- the valve opening (order Vc) of the tank C1 can be done only if the wagon W is in left position ;

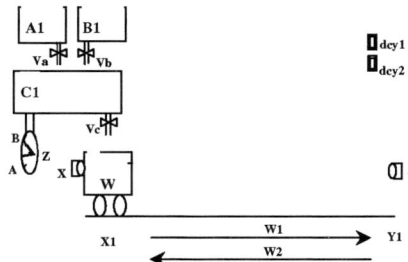

Fig. 1. Representation of the process

- the displacement of the wagon to the right (W1) can be done only if the empty tank C1, is detected by the zero z of the rocker.

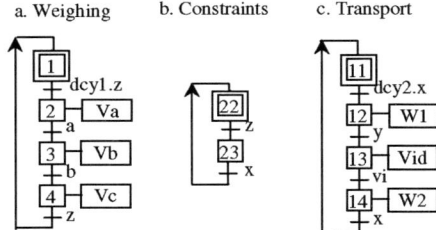

Fig. 2. Grafcets of control

The systems of weighing and transport are modelled separately (cf. fig. 2). Then one carries out the synchronization by merging the transitions conditioned by the same events, to reach the global model of the process Fig. 3.

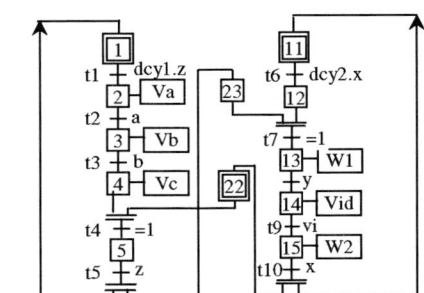

Fig. 3. Grafcet of the unsupervised process

3. GRAPH OF ACCESSIBLE STATES AND BASIC DEFINITIONS

One directly builds the graph of accessible states starting from the initial state of the grafcet (fixed by the specifications).

A state change of a grafcet stage (activation or deactivation), following the crossing of a transition, causes a change in the state of the graph.

To build this graph one adopts the *fundamental assumption of the sequential systems: not correlated events never occur at the same time.*

The tops of the graph correspond to the active stages of the grafcet at this moment, and the transitions are represented by arrows. Being a problem of supervision, there are three different representations of arcs corresponding to three types of transitions.

Controllable transition or with controllable event →
Uncontrollable transition or with uncontrollable event →
Forbidden transition or with forbidden event --▸

Definition 1[4]
Transitions are *controllable* (in other words the associated events are controllable), when the crossing of these transitions, if they are validated by the activity of their stage(s) upstream, can be authorized or prohibited by a control action that authorize or not to take into account the occurrence of the event that is associated with them.

Definition 2
The transitions are known as *not controllable* (in other words the associated events are uncontrollable), when the crossing of these transitions, if they are validated by the activity of their stage(s) upstream, on occurrence of the event that is associated to them, can not (in no case) be prohibited.

In the case of our example Fig. 3, only the events {b, vi, (= 1).$_{4.5}$, (= 1).$_{12.13}$} are uncontrollable.

An uncontrollable event (corresponding to the crossing of an uncontrollable transition) can pose problem on the synchronization level, corresponding to an AND convergence represented by a simultaneous sequence exit. If the under system forces [1] the passage of the transition without awaiting the synchronization information brought by the specification, we end up in a prohibited situation.

Definition 3
It will be said that a *transition* (with uncontrollable event) *is forbidden* if it is a synchronization transition leading to a prohibited situation.

Definition 4
A *forbidden event* is an uncontrollable event corresponding to the crossing of a partially validated synchronization transition (absence of external information to the process coming from specifications).

---
[1] This concept of a transition "forced crossing" is to be brought closer that of Quasi Petri Net introduced by Dideban into [3].

As one can see it on the Fig. 4, between the states $E_{4.12.22}$ and $E_{5.12}$, the transition t_4 is not forbidden because information's necessary to its crossing are all present. On the other hand between the states $E_{4.13}$ and $E_{5.13}$, although it is about the same transition, t_4 is prohibited because it misses the activation information of the stage E_{23} to cross the transition t_4.

The Fig. 4 represents the graph of the accessible states associated with the model of the Fig. 3. It's an equivalent automat that describes the successive phases by which the system will pass, and represents the possible physical behaviors of the process.

To prohibit the access to the transitions preceding the prohibited states, one will prohibit the states located upstream of these transitions, which we call thereafter *forbidden states*.

Definition 5
A *forbidden state* is a state that has downstream at least one forbidden transition (e.g. states $E_{4.14}$ and $E_{2.12.22}$ of the Fig. 4).

Definition 6
A *state border* is
 Either a prohibited state that has a controllable transition upstream (e.g. state $E_{2.12.22}$ of the Fig. 4);
 Or a state located upstream of a forbidden state and that has at least a controllable transition upstream (e.g. state $E_{3.14}$ of the Fig. 4).

Definition 7
An *acceptable state* is a state of the accessible states graph that is neither a state border nor a forbidden state.

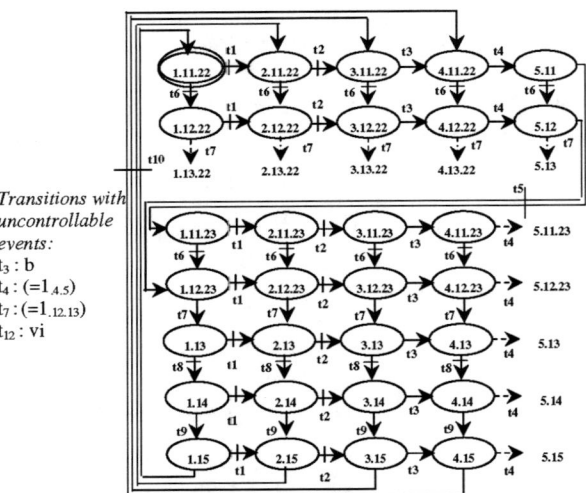

Fig. 4. Accessible states graph of the unsupervised

4. ALGORITHME

To prevent the crossing of the uncontrollable transitions that lead to situations prohibited, one prohibits the access to the forbidden states (preceding the "undesirable" transitions) from the authorized states immediately upstream. That results in the crossings prohibition of a controllable transition.

The method derived from the formalism of the Petri nets consists in adding stages of control [5] preventing the transition from being "sensitized" or "validated".

The algorithm is based on the prohibition of the states borders:
- Constraints determination by the states borders.
- Number of constraints reduction by the method of *over states*.
- Calculation of the supervisor.
- Model validation of the supervised process.

Before applying the algorithm, one search at first the sets of the prohibited states $\mathcal{E}_I$, borders $\mathcal{E}_B$ and acceptable $\mathcal{E}_A$, starting from the graph of the Fig. 3.

$$\mathcal{E}_I = \{ E_1E_{12}E_{22}, E_2E_{12}E_{22}, E_3E_{12}E_{22}, E_4E_{12}E_{22}, E_5E_{12}, \\ E_4E_{11}E_{23}, E_4E_{12}E_{23}, E_4E_{13}, E_4E_{14}, E_4E_{15}\}. \tag{1}$$

$$\mathcal{E}_B = \{ E_1E_{12}E_{22}, E_2E_{12}E_{22}, E_3E_{12}E_{22}, E_4E_{12}E_{22}, E_5E_{12}, \\ E_3E_{11}E_{23}, E_3E_{12}E_{23}, E_3E_{13}, E_3E_{14}, E_3E_{15}\} \tag{2}$$

$$\mathcal{E}_A = \{E_1E_{11}E_{22}, E_2E_{11}E_{22}, E_3E_{11}E_{22}, E_4E_{11}E_{22}, E_5E_{11}, \\ E_1E_{11}E_{23}, E_2E_{11}E_{23}, E_1E_{12}E_{23}, E_2E_{12}E_{23}, E_1E_{13}, \\ E_2E_{13}, E_1E_{14}, E_2E_{14}, E_1E_{15}, E_2E_{15}\}. \tag{3}$$

5. PROHIBITION OF THE BORDER STATES BY CONTROL STAGES

5.1 Transformation of the border states into linear constraints [2]

If $E_1 E_2 \ldots E_n$ is a state border made up of n stages, to prohibit it means that one must have at least one inactive stage. One can express that in the form of linear constraint by

$$\sum_{i=1}^{n} E_i \leq n - 1 \quad (E_i = 1 \text{ corresponds to the activation of the grafcet stage } E_i).$$

If one expresses the 10 states borders of the example in the form of linear constraints, one obtains:

$$\begin{aligned}
E_1 + E_{12} + E_{22} &\leq 2 \quad \text{(two active stages out of three)} \\
E_2 + E_{12} + E_{22} &\leq 2 \\
E_3 + E_{12} + E_{22} &\leq 2 \\
E_4 + E_{12} + E_{22} &\leq 2 \\
E_3 + E_{11} + E_{23} &\leq 2 \\
E_3 + E_{12} + E_{23} &\leq 2 \\
E_3 + E_{13} &\leq 1 \quad \text{(one active stage out of two)} \\
E_3 + E_{14} &\leq 1 \\
E_3 + E_{15} &\leq 1 \\
E_5 + E_{12} &\leq 1
\end{aligned} \tag{4}$$

The number of control stages will depend on the number of linear constraints. A stage of control (noted E_c) is a stage added to the initial grafcet that makes it possible to satisfy the constraints.

5.2 Reduction of the linear constraints number by the method of over states

Definition 8
Suppose $E_i = \{E_{i1}E_{i2}\ldots E_{in}\}$ an accessible state, $E_j = \{E_{j1}E_{j2} \ldots E_{jm}\}$ ($m \leq n$) is an *over state* of E_i if :

$$\forall E_{jk} \in E_j, k \in [1,..m], \exists E_{ik} \in E_i \text{ such as } E_{ik} = E_{jk}$$

Remark
- The set of the stages forming the state E_j is included in the set of the stages forming the state E_i.
- The set of the states covered by E_j contains the state E_i.

It will be said for example that the state $E_{12}E_{23}$ is an over state of $E_1E_{12}E_{23}$. Stages E_{12} and E_{23} forming the state $E_{12}E_{23}$ are included in the set of the stages forming the state $E_1E_{12}E_{23}$.
Thus the prohibition of the state $E_{12}E_{23}$ interdicts automatically the set of the states covered by him.

The number of constraints reduction steps consists of the following:
- to build the sets of borders and acceptable over states,
- to simplify the set of over states borders: initially by removing from it the elements that belong to the set of acceptable over state S, then by removing the non minimal over states,
- to reduce this set by a simplification technique of the combinatory logic,
- and finally to merge the constraints by seeking the common elements.

5.2.A Set of border and acceptable over states

Definition 9
Suppose $E_i = \{E_{i1}E_{i2} \ldots E_{im}\}$ a state of the system. The set of the E_i over states noted $\mathcal{E}^{sur}_i$ is equal to the set of the E_i parts without the empty set.
Thus for the state $b_i = E_1E_2E_3$, the set of its over states is given by $b_i^{sur} = \{ E_1, E_2, E_3, E_1E_3, E_1E_2, E_2E_3, E_1E_2E_3 \}$.

---
[2] Yamalidou and Kantor [8] formalized, in the case of the Petri nets, transformation of boolean constraints into linear inequalities.

The union of b_i^{sur} form the global set $\mathcal{B}_1$ of border over states:

$$\mathcal{B}_1 = \bigcup_{i=1}^{card(\mathcal{E}_B)} b_i^{sur} \qquad (5)$$

The set of the 10 border over states of our process is:

$\mathcal{B}_1 = \{$ $E_1, E_2, E_3, E_4, E_5, E_{11}, E_{12}, E_{13}, E_{14}, E_{15}, E_{22}, E_{23},$
$E_1E_{12}, E_1E_{22}, E_2E_{12}, E_2E_{22}, E_3E_{11}, E_3E_{12}, , E_3E_{13}, E_3E_{14},$
$E_3E_{15}, E_3E_{22}, E_3E_{23}, E_4E_{12}, E_4E_{22}, E_5E_{12}, E_{11}E_{23},$
$E_{12}E_{22}, E_{12}E_{23}, E_4E_{22}, E_5E_{12}, E_1E_{12}E_{22}, E_2E_{12}E_{22},$
$E_3E_{12}E_{22}, E_4E_{12}E_{22}, E_3E_{11}E_{23}, E_3E_{12}E_{23}\}$ (6)

The global set of the 15 acceptable over states is given by:

$$\mathcal{A}_1 = \bigcup_{i=1}^{15} a_i^{sur} \qquad (7)$$

$\mathcal{A}_1 = \{E_1, E_2, E_3, E_4, E_5, E_{11}, E_{12}, E_{13}, E_{14}, E_{15}, E_{22}, E_{23}, E_4E_{11}, E_1E_{12},$
$E_1E_{13}, E_1E_{14}, E_1E_{15}, E_1E_{22}, E_1E_{23}, E_2E_{11}, E_2E_{12}, E_2E_{13},$
$E_2E_{14}, E_2E_{15}, E_2E_{22}, E_2E_{23}, E_3E_{11}, E_3E_{22}, E_4E_{11}, E_4E_{22},$
$E_5E_{11}, E_{11}E_{22}, E_{11}E_{23}, E_{12}E_{23}, E_1E_{11}E_{22}, E_2E_{11}E_{22},$
$E_3E_{11}E_{22}, E_4E_{11}E_{22}, E_1E_{11}E_{23}, E_2E_{11}E_{23}, E_1E_{12}E_{23},$
$E_2E_{12}E_{23}\}$ (8)

5.2.B Suppression of the acceptable over states from the set of borders over states

By construction of the set $\mathcal{B}_1$, one can have introduced acceptable over state belonging to $\mathcal{A}_1$. They should then be removed.

$\mathcal{B}_2' = \mathcal{B}_1 \setminus (\mathcal{B}_1 \cap \mathcal{A}_1)$: one remove from $\mathcal{B}_1$ list all the elements that belong to $\mathcal{A}_1$.

$\mathcal{B}_2' = \{$ $E_3E_{12}, E_3E_{13}, E_3E_{14}, E_3E_{15}, E_3E_{23}, E_4E_{12}, E_5E_{12}, E_{12}E_{22},$
$E_1E_{12}E_{22}, E_2E_{12}E_{22}, E_3E_{12}E_{22}, E_4E_{12}E_{22}, E_3E_{11}E_{23},$
$E_3E_{12}E_{23}\}$ (9)

5.2.C Suppression of non minimal states

When a state is already covered by another over state of smaller cardinal, it is said that it is not minimal and it must be removed from the list $\mathcal{B}_2'$.

$$\mathcal{B}_2 = \mathcal{B}_2' \setminus \{b_{2j} \in \mathcal{B}_2' / \exists\ b_{2i} \in \mathcal{B}_2' \text{ telque } b_{2i} \subset b_{2j}\} \quad (10)$$

It is necessary to remove b_{2j} from the list $\mathcal{B}_2'$ if b_{2i} is a subset of b_{2j} that is included in the set $\mathcal{B}_2'$.

The final set of border over states obtained is given by :

$\mathcal{B}_2 = \{$ $E_3E_{12}, E_3E_{13}, E_3E_{14}, E_3E_{15}, E_3E_{23}, E_4E_{12}, E_5E_{12}, E_{12}E_{22}\}$ (11)

5.2.D Reduction of the number of over states

One builds initially a table of over states and constraints: over states given by (11) on line and constraints given by (2) in column. Then by adopting a simplification technique close to that of Mc Cluskey in combinatory logic, one builds the final reduced set of over states C_2, that contains the over states allowing to cover all the constraints while eliminating the redundancy.

The construction algorithm of the set C_2 is as follows:

- Reading the tables $\mathcal{B}2$ (over states) and $\mathcal{E}_B$ (constraints).

- Constructing the table D
  - To take an element of $\mathcal{B}2$ and to write it in D. To associate to him (i.e. to write in D) all the elements of $\mathcal{E}_B$ covered by this over state.
  - To remake the same work for whole elements of $\mathcal{B}2$.

- Constructing the table C_2
  - To see the first element of D and to put it in C_2.
  - To take the following element of D. To compare its associated components with the components of the elements present in C_2:
    * if a constraint is already covered (redundant) by one preceding over state, one removes it from the over state that has the minimum of constraints (one keeps it for over state that covers the greatest number of constraints);
    * then to check
      ** if the over state in progress has zero constraints, it is eliminated,
      ** if not it is recorded in C_2.
  - To remake the same work for all the elements of D.

Remark

If there are states borders that are covered by no constraint, there is no solution with the problem of synthesis, because one is brought to remove acceptable states.

Table 1. Table of reduction Over state-Constraints

	E1E12E22	E3E12E22	E5E12	E3E12E23	E3E14	E2E12E22	E4E12E22	E3E11E23	E3E13	E3E15
E3E12										
E3E13										√
E3E14										√
E3E15										√
E3E23										√
E4E12										
E5E12										√
E12E22										√

The reduced set of over states is:

$C_2 = \{$ $E_3E_{13}, E_3E_{14}, E_3E_{15}, E_3E_{23}, E_5E_{12}, E_{12}E_{22}\}$ (12)

5.2.E Merging of constraints by search for common elements

Let $\{E_iE_{i1}, E_iE_{i2}, ..., E_iE_{in}\}$ be elements of C_2.

If none of the elements of the unit $\{E_{i1}E_{i2}, E_{i1}E_{i3},..., E_{i1}E_{in}, E_{i2}E_{i3}, E_{i2}E_{i4},..., E_{i2}E_{in},..., E_{in-1}E_{in}\}$ belongs to the set of the acceptable states $\mathcal{A}1$ (it means that the stages $E_{i1}, E_{i2},..., E_{in}$ are never active at the same time), then one can write [3] :

$$\left.\begin{array}{l} E_i + E_{i1} \leq 1 \\ E_i + E_{i2} \leq 1 \\ \text{------------} \\ E_i + E_{in} \leq 1 \end{array}\right\} \leftrightarrow E_i + E_{i1} + E_{i2} + \cdots + E_{in} \leq 1 \qquad (13)$$

According to the expression of C_2 given by (12), one has two sets of linear constraints:

$E_3 + E_{13} \leq 1,\ E_3 + E_{14} \leq 1,\ E_3 + E_{15} \leq 1,\ E_3 + E_{23} \leq 1$ (14)
$E_{12} + E_5 \leq 1,\ E_{12} + E_{22} \leq 1$ (15)

---
[3] That was demonstrated in [3]

In (14) the stage E_3 is common to all the constraints. Like when combining two to two the stages constituting the constraint, any state obtained isn't an acceptable state, one can then replace the constraints in (14) by only one expressed in the form :

$$E_3 + E_{13} + E_{14} + E_{15} + E_{23} \leq 1 \qquad (16)$$

By applying the same property to (15), one obtains:
$$E_5 + E_{12} + E_{22} \leq 1 \qquad (17)$$

The set of border over states after merging is:
$$C_1 = \{(E_3, E_{13}E_{23}E_{14}E_{15}), (E_{12}, E_5E_{22})\}^4 \qquad (18)$$

Remark
In the original method of Dideban [3], the merging of constraints is applied to the set $\mathcal{B}2$ of over states, and then one reduces the number of over states. Whereas here one carries out initially the choice of the states that cover the constraints (thus elimination of non useful states), then one applies the merging to the reduced set C_2. That permits to avoid useless calculations and to reduce the number of equations (in our case 2 instead of 4).

5.3 Calculation of the control stages

One can translate the constraints on the model by stages of control.
The accessible states graph represents all the states by which the system will pass, like their order of occurrence. Thus let us start by eliminating the border and prohibited states in the states graph, then observe the occurrence of the common stages in the constraints of the expression (18).
Equation (18) indicates that the activation of E_3 can occur only if all the stages E_{13}, E_{23}, E_{14} and E_{15} are deactivated.

On the Fig. 5, let us examine the predecessors of the common stage E_3, appearing among the four stages constituting the constraint. One finds upstream E_3 only the stage E_{15}. One thus will have a control stage $E_{c15.3}$: the deactivation of the stage E_{15} allows the activation of the stage E_3 via the stage $E_{c15.3}$.

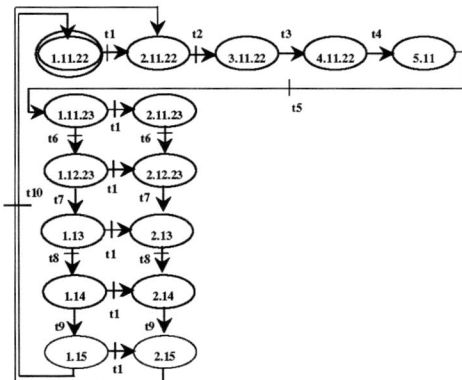

Fig. 5 Simplified accessible states graph of the unsupervised process

[4]This writing has the merit to store the information on the common stages

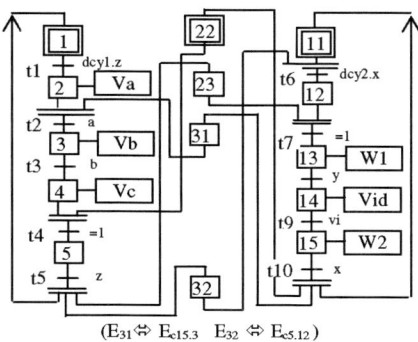

($E_{31} \Leftrightarrow E_{c15.3} \quad E_{32} \Leftrightarrow E_{c5.12}$)

Fig. 6. Grafcet of the supervised process with control stages

By applying the same principle to the second constraint of (18) where E_{12} represents the common stage, there is a stage of control $E_{c5.12}$. One obtains finally the diagram Fig. 6.

6. GRAFCETS VALIDATION OF THE SUPERVISED PROCESS

While modeling by states graph to describe the system is a step where one doesn't consider all the possible combinations of the entry variables, it is necessary to validate the model obtained, by checking in particular that the model translates the set of conditions correctly.

For that let us carry out the graph of accessible states from the new grafcet with its two stages of control (cf. Fig. 7).
It is noticed that this new graph of states comprises neither state border nor prohibited state. The prohibited states and their arcs upstream and downstream were automatically removed, and the specifications of operation are satisfied.

Indeed in the supervised grafcet, the control stage $E_{C15.3}$ guarantees that the passage from the stage E_4 to E_5 can be effective only if E_{22} is active, because the activity of E_4 is subordinated to the activity of E_3 controlled by the stage $E_{C15.3}$ (activated at the same time as E_{22}). Consequently the valve will be open only if the wagon is present under the tank.

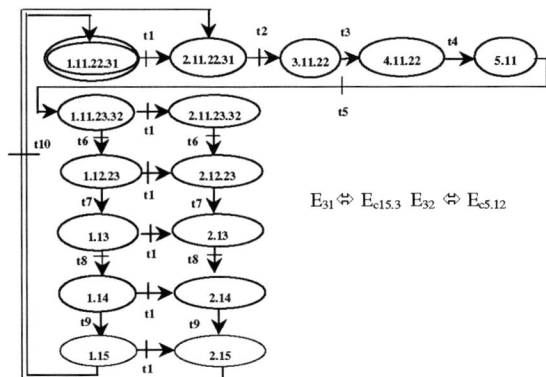

$E_{31} \Leftrightarrow E_{c15.3} \quad E_{32} \Leftrightarrow E_{c5.12}$

Fig. 7 Accessible states graph of the supervised process

It is the same for the wagon start (action W1 at stage 13) that can take place before the end of the tank draining in the unsupervised grafcet, and that is inhibited in the supervised grafcet thanks to the control stage $E_{c5.12}$.

However a meticulous observation of the supervised process grafcet, emphasizes a real problem during the ordering, and that did not exist in the original grafcet of the Fig. 3.

Indeed the addition of the control stage $E_{c15.3}$, like condition of validation (or sensitizing) of the controllable transition t_2, will penalize the actions that are carried out at the stage upstream. The actions of the stage E_2 will be maintained without taking account of the real condition of stop, as long as the stage $E_{c15.3}$ is not active[5].

To mitigate this difficulty, one can:
   o   transform the actions of the stage number 2 into conditional actions (cf. Fig. 8.b), which permits to respect the specifications of the process (by observing the stop condition of the action associated at the stage E_2);
   o   or create a waiting stage downstream from the stage number 2 (cf. Fig. 8.a), that will wait until the control stage $E_{c15.3}$ will be active. So that the exit transition from the simultaneous sequence stays always controllable, the associated information will correspond to the event actions end of the stage E_2.

The two structures are identical from the operation point of view, and their programs on the PLC will be exactly the same one. However the structure Fig. 8.a has the merit to be very explicit.

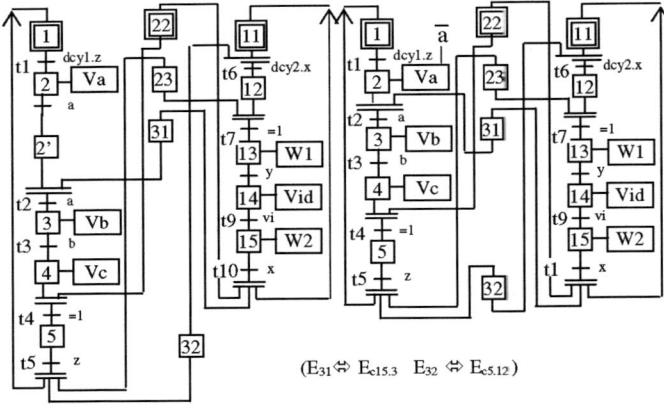

a. By waiting stage   b. By conditional action

Fig. 8. Adaptation of the grafcet structure

7. CONCLUSION

There are many calculations in the developed method, but it is systematic and thus easy to implement on computer. With the simplification technique based on the chronology of the states occurrence, the number of control stages is reduced: each constraint equation is modeled by only one stage.

The controller that one reaches is maximum permissive.

However, it is required to make the calculation of the accessible states graph off line, or to use a systematic computation software like that developed by Roussel in [6].

It can happen that the system is not controllable; it means there exist a sequence of uncontrollable crossings S, that leads to a prohibited state, starting from the initial state. In this case the specification and no more restrictive under specification are then uncontrollable.

REFERENCES

[1] F. Charbonnier (1996). "Commande supervisée des systèmes à événements discrets", *Doctorate Thesis of the National Polytechnique Institute of Grenoble*, 1996.

[2] R. David, H. Alla (2005). *Discret, Continous, and hybrid Petri Nets*, Springer, 2005.

[3] A. Dideban (2005). "Synthèse optimale d'un contrôleur par construction de l'ensemble minimal de contraintes", *RS-JESA-39/2005.MSR'05*, pp. 127-141.

[4] J.L. Ferrier, J.L. Boimond (2004). "Modèles et Systèmes: un tour d'horizon", Monograph, *Laboratory of Engeneering of the Automated Systems, University of Angers*, June 2004, http://www.istia.univ-angers.fr/LISA.

[5] L.E. Holloway, B.H. Krogh (1990). "Synthesis of feedback logic for a class of controlled Petri nets", *IEEE Trans. Autom. Control*, vol 35, n°5, pp. 514 -523,1990.

[6] J.M. Roussel, "analyse de grafcets par génération logique de l'automate équivalent", *Doctorate Thesis of the Higher Teacher Training School of Cachan*, LURPA/ENS of Cachan, 16 December 1994.

[7] W.M. Wonham, E.S. Rogers (2006). "Supervisory control of discrete event systems", Monograph ECE 1636f/1637S 2006-07, *Dpt. of Electrical & Computer Engineering, University of Toronto*, 1 july 2006.

[8] E. Yamalidou, J.C. Kantor (1991). "Modelling and optimal control of discrete-event chemical processing using Petri nets", Comput. Chemic. Engineering, vol15, pp.503-519, 1991.

---

[5]That recalls the cause of the creation of waiting stages at the end of the simultaneous sequences.

Decentralized method for complex task allocation in massive MAS

Zaki Brahmi[*], M.M. Gammoudi[**]
Research Unit (URPAH)

[*]Faculty of Sciences of Tunis
[**]High Institute of Computer Sciences and Management of Kairouan
e-mail : Zaki.Brahmi@isigk.rnu.tn, mohamed.gammoudi@fst.rnu.tn

Abstract: Task allocation is still a fundamental problem in Multi-Agents System (MAS). It allows the formation a coalition of agents to cooperate together in order to carry out a complex task. Generally, the process of task allocation requires calculating the value of all the possible allocations, then determining which optimal. In the context of Massive Multi-Agent Systems (MMAS), one agent generates all the possible coalitions and then calculates the value of each, is inefficient. Moreover, the traditional approaches based on the negotiation between agents, are impractical because of the complexity of communications between agents. In this paper we propose a decentralized method, based on grouping agents using the Formal Concepts Analysis (FCA) approach for the task allocation in MMAS. In our model, agents are fully cooperative and each task is composed of several subtasks. The proposed solution is based on two steps: i) computing groups of agents having similar characteristics in the objective to distribute the task allocation process among agents and minimize the communication between agents. ii) Finding the optimal allocation by sharing the task allocation process among groups of agents.

Keywords: Task Allocation, MMAS, FCA, Group, Concept, Concepts lattice.

1. INTRODUCTION

A Multi-Agent System (MAS) is defined as a set of agents in interaction in order to complete a global goal. A form of interaction is *coalition formation*. In [7], the coalition formation is defined as the process for which a group of agents meets and agrees to coordinate and cooperate in order to carry out a given task. The problem of task allocation to agents in the Multi-Agents Systems is a fundamental problem, which attract much research in this field. It is used in several fields like: the E-commerce [11], distributed routing of vehicles [15], etc.

In [6], the author presents a classification of the task allocation problem according to two levels. The first is the distribution level of the mechanism of task allocation, which can be decentralized [10] or be centralized. The second is the level of co-operation between agents, in which the agents can be completely co-operative [4, 5] or self-interest agents.

This work is placed in the context of Massive Multi-Agents Systems (MMAS). According to [9], a MMAS is a very large MAS, which contains tens or hundreds of thousands agents in interactions. In particular, we are interested in the mechanism of tasks allocation in a cooperative environment. Generally, the process of tasks allocation is the following:

1. Generate the set of possible coalitions among the set of agents implied in task allocation process (noted L).
2. For each coalition $l \in L$, compute *it coalition value V (l)*.
3. Determine the *optimal coalition* expressed by the minimal value of the set values $V(l_i)$, obtained in 2).

In the context of MMAS, using such task allocation process is inefficient. This is because of the following problems:

1. Great flow of communication between agents during the task allocation problem, for example through the negotiation.
2. Complexity and inefficiency of task allocation if one allocator is used during the allocation process. This is caused by: i) the subdivision of a task in several sub-tasks, ii) significant number of agents and tasks and the iii) exponential number, according to the number of agents of the possible allocations.
3. Times and memory capacity used to store all the possible allocations and to determine which one is optimal are significant. Indeed, they must be minimal.

In order to deal with these problems, we propose a decentralized task allocation method in a MMAS where the agents are completely, cooperative. Within the framework of this work, we consider that the tasks allocation could be carried out as follows:

1. Each task is subdivided into several elementary sub-tasks
2. The achievement of a task requires the assignment of its various sub-tasks to the adequate agents.

This article is organized as follows: the following section presents work related to our problems. In section 3, we present the formalisms on which we will base our proposal. In section 4, we will detail our contribution for complex task allocation in MAMS. In section 5, we will study the communication between agents. In section 5, a comparative study is carried out. Lastly, we conclude our work and we give some prospects.

2. RELATED WORKS

The task allocation to the agents in a system Multi-Agents represents a problem which remains until our days of topicality. Indeed, several pieces of work are interesting to solve the problem, which can be expressed as follows: *how to find in a reasonable time, the optimal allocation by reducing the communication between agents* [4, 14, 10, 8, 1]. Indeed, some of them are interesting in the regrouping of agents according to their common properties, such as that the work of E MALVILLE [14]. The goal of the author is to minimize the flow of communication between agents. The solution suggested is based on clusters of agents sharing a set of characteristics. The size of a cluster is fixed by the user. The task allocation amounts to a customer to find the relevant provider. In the context of a MMAS, the research space becomes very complex. Moreover, this solution is centralized and does not answer our problem since it treats only simple tasks. Other similar work which is interesting in coalition formation in a cooperative environment, we give as an example the work of Talal Rahwan and Al [7, 1]. The authors propose an algorithm for coalition forming named DCVC (*Distributing Coalitional Value Calculations among cooperating agents*) in a cooperative environment. The algorithm is decentralized. It distributes the calculation of the values of possible coalitions among agents. The number of the coalitions is exponential according to the number of agents. Moreover the solution, suggested is not applicable to our complex task allocation problem.

In our knowledge, the only work which exists in the literature and which is interesting to the allocation of complex tasks in a cooperative environment is proposed by E. Manisterski, and Al [4]. These authors propose a solution based on the bipartite graph. Indeed, the idea is to find the perfect bonds balanced with a minimal weighting in a bipartite graph. The nodes of the graph represent agents and sub-tasks. For a problem of N tasks, the algorithm generates 2^N-1 sets of tasks. For each set of tasks, the algorithm provides a bipartite graph in order to find the optimal task allocation belonging to this unit. From the set of generated allocation, the algorithm finds the optimal one. Thus, the complexity of the algorithm is exponential according to the number of tasks. Moreover, computing the optimal allocation is done by one allocator agent. The unicity of allocator agent (or the centralization of the processes allocation) is a limit [1], for reasons of congestion of the number of agents or for a technical reason, like the breakdowns and the memory capacity necessary to store all the possible allocations.

To minimize communication between agents, Danny Weyns and Al [2] propose a field-based approach is for task assignment in an AGV transportation system. This approach based on the *emissions of fields* in the environment by the agents composing the system. But, in the context of massive MAS, the fields sent by the agents and the tasks, overload the network. To find the optimal allocation, the majority of the solutions suggested generate all the possible allocation, which presents a problem of complexity of treatment and thus a heaviness of execution time.

In order to palliate these limits, we propose a decentralized group-based solution, using the Formal Concepts Analyse (FCA) approach. This grouping allows:

1. To reduce the communications between agents. Indeed, the communication is between the persons in charge of the groups. An agent plays the role of a responsible for the group where it belongs. It is named *local allocator*.
2. Distribute the task allocation process by sharing the computing among agents.

Before giving our solution in detail, we recall in the following section the description of task allocation problem defined in [4] and the basic concepts FCA approach.

3. TASKS ALLOCATION PROBLEM DESCRIPTION AND FCA APPROACH

3.1 Problem description

The task allocation problem can be formulated as bellow:

- Consider a set T of n tasks $T = \{T_1, T_N\}$. Where, each task T_I is composed of a set of sub-tasks.
- Each task T_i has an execution cost $C(T_i)$. The costs of the T_i sub-tasks are not given.
- A set MAS of m agents $SMA = \{Ag_1, Ag_m\}$. Each agent Ag_i, have a set of capacities to carry out several sub-tasks. The capacities of each agent are known by the *allocator agent*.

The goal is to find the *feasible allocation* that grouped a set of agents in order to execute the set or sub-set of tasks T. The following definitions will be used.

Definition 1: *an allocation Al is feasible, if it satisfies the following conditions:*

C1: Each agent is assigned to at least one sub-task.

C2: Any task is either fully allocated (one agent per sub-task) or it is unallocated.

C3: For each task $T_i \in Al$, the total cost proposed by the agents to perform its constituent subtasks does not exceed the task's total payment $C(T_i)$.

Definition 2: *a task is feasible, if it satisfies C1, C2 and C3 conditions.*

Definition 3: *the cost of a feasible allocation Al is the sum of the costs suggested by the agents in order to carry out the sub-tasks belonging to Al:*

$$V(Al) = \sum_{ST_i \in Al} C_{Agi}^{STi} \cdot (1)$$

Where, C_{Agi}^{STi} is the cost proposed by Ag_i agent in order to execute the sub-task ST_i.

2. Basic concepts of FCA approach

We point out the bases mathematics of the Formal Concepts Analysis (FCA) approach as they are necessary for this article. For more detail, the reader can see [12].

Formal context: A formal context is a triplet (O, A, I) for which O is a set of objects (or entities), A set of attributes (or properties) and I ($\subseteq O \times A$) a binary relation between O and A. I associate an object to a property: $(o, a) \in I$; when "o has the property a" or "the property a applied to the object o".

Correspondence of Galois: Let be $k = (O, A, I)$ a formal context, we can define two functions f and g make it possible to express the correspondences between the subsets of objects $P(O)$ and the subsets of attributes $P(A)$ induced by relation I, in the following way:

- f is the application which with any element o of O associates $f(o) = \{a \in A / (o,a) \in I\}$.
- g is the application which with any element a of A associates $g(a) = \{o \in O / (o,a) \in I\}$.

These two applications constitute the *Galois correspondence* of the context.

Concept: Given a context $k = (O, A, I)$ and $O1 \subseteq O, A1 \subseteq A$, the couple $C = (O1, A1)$ is called a *concept* of K if and only if $f(O1) = A1$ and $g(A1) = O1$. $O1$ is called *extension* of the concept; $A1$ is called the *intention* of the concept.

Galois lattice (Lattice of the concepts): A Lattice is a mathematical structure $F = <F, \leq, \vee, \wedge, 0_F, 1_F>$, F is a set of a partial order by the relation $\leq$, with 1_F is the broadest element, 0_F is the smallest element, and $\wedge, \vee$ are laws of composition interns "supremum" and "infimum". Are $C1 = (A1, B1)$ and $C2 = (A2, B2)$ a couple of concepts then $C1 \vee C2 = (\{A1 \cup A2\}, \{B1 \cap B2\})$ and $C1 \wedge C2 = (\{A1 \cap A2\}, \{B1 \cup B2\})$ are *concepts*. The set L of all the concepts of a context (O, A, I), provided with the relation of order $\leq$, has the mathematical structure of lattice and is called *Galois lattice* (or the concepts) of the context (O, A, I).

Several algorithms are proposed for the generation of a Galois lattice, as it is known, Godin [6], Nourine [13], etc. The last algorithm appeared in the literature is in [3] which was the faster one in runtime and the complexity by comparison with the quoted algorithms. For this reason, we decide to use it in our work.

In the follows, we present our decentralized group-based method for the task allocation problem. It is based on an algorithm called GROUPAL which will be presented in the following section.

4. OUR PROPOSITION: METHOD OF TASK ALLOCATION BASES ON THE GROUPAL ALGORITHM

Before presenting our task allocation method, we recall some significant definitions.

Definition4: A group of agent or concept G is composed by the set AG of p agents which share the same capacities defined by the couple: $(\{Ag1, Ag2, ...Agp\}, \{c1, c2, ...cl\})$.

Where, C_i indicates the common competence to all the agents AG to carry out an elementary sub-task ST_{ij}. To facilitate the comprehension of our step, we associate the execution of a given sub-task by an agent to its capacity.

Definition5: An agent is responsible for its group if it is the first in the list of agents which compose this group.

Definition6: An allocator agent is an agent responsible for its group. A global allocator is the agent which belongs to the group or supremum concept of the Galois lattices in FCA approach. A local allocator is a responsible for a given group in the Galois lattices.

Definition7: An allocation $Al = \{<Ag_i, ST_{jk}> / Ag_i \in SMA$ and $ST_{jk} \in STj\}$ is a distinct set of assignment. The coupl $<Ag_i, ST_{jk}>$ is the assignment of the Ag_i agent to the ST_{jk} sub-task, which belongs to the set of sub-tasks which composed the STj task.

Definition8: A conflict exists between two assignments Af_i an Af_j, if they do not check the C1condition.

The proposed task allocation method in MAS where agents are cooperative is decentralized. It distributes the task allocation process by the distribution of calculation between the various responsible agents (local allocator). We find the global allocation by the union of the set of the local allocation, after resolution of conflict situations if they exist.

The idea of the method is based on two task allocation phases: *local allocation*, which is related to sub-tasks (component task) and *global allocation* which is related to composite tasks. These allocations are based on a process of regrouping of agents according to their capacities to carry out a given task. This regrouping is carried out by an algorithm based on FCA approach and defined by [3]. Each local allocator, proposes a local allocation by solving the *local conflict* situations between agents. These proposals will be communicated to an agent responsible for the allocation. This, the latter proposes an *optimal allocation*, after resolution of the global conflict situations. The Galois lattices studied in the FCA approach, allows to organize the groups of agents called also (concepts) according to a partial order relation.

In following, we will give in detail the two types of allocation (local allocation and global allocation).

4.1 Local allocation

It is the execution of a sub-process belongs to task allocation process. It is named local if the assignment of the elementary sub-tasks is done with the agents of the same group. A local allocation is the set of assignments of each sub-task ST_{kj} to an Ag_i agent. The Ag_i agent is which has the minimum cost (C_{Agi}^{STkj}) to execute ST_{kj}. The idea to choose the agent having the minimum cost guarantees that the global allocation, found, is optimal. During the local allocation process, we identified two following conflict situations, S1 and S2:

S1: The cardinality of sub-tasks belongs to a task assigned to an agent exceeds value 1. In this case, the agent takes the sub-task, which has the minimal cost.

S2: The cardinality of sub-tasks belongs to two different tasks $T1$ and $T2$ assigned to the Ag_i agent exceeds value 1. In this case, Ag_i takes the cost which can be like this:

$$C_{Agi}^{STi}/C(T1) < C_{Agi}^{STj}/C(T2) \quad (2)$$

Where, $C(T_1)$ and $C(T_2)$ are, respectively, the execution cost of task T_1 and the execution cost of the task T_2.

4.2 Total allowance

The global allocation is the union of all the local allocation. Formally, we define a global allocation AlG by:

$$AlG = \bigcup_{1 \leq i \leq nbrG} AlLi \quad (3)$$

$nbrG$ indicates the number of groups. Indeed, each local allocator communicates his local allocation to the global allocator. The latter, determines the possible conflict situations between these allocations. In order to solve these conflict situations, it uses the solution suggested in the section 4.1).

In the following, we present the GROUPAL algorithm of the solution suggested for tasks allocation in MMAS.

4.3 GROUPEAL Algorithm

The proposed algorithm has an objective to search for an optimal allocation. The principle of the algorithm is as follows:

1. Computing the set agents groups (*ListeG*), after simplification of the Galois lattices T calculated among the formal context defined by the triplet *(SMA, P, I)*, where P is the whole properties belongs to all agents. I is the binary relation between *SMA* and *P*. The Galois lattices T is calculated by the algorithm presented in [3]. It is the role of the agent *globa allocator* agent.
2. Each responsible for the group G_i proposes a local allocation.
3. The global allocator agent determines the global allocation after resolution of conflict situations if they exist.

The following pseudo-code presents the GROUPAL algorithm:

```
Initialisation
 Allocation = ø
 MAS = {A₁, A₂,…,Agₙ}// n agents
 Tasks = {T₁, T2,…, Tₘ}// m tasks
 SetG = ∅ // empty set of group.
Start GROUPEAL
 1. Computing groupes of agents: SetG =
    CreationG(MAS, Tasks)
 2. For each group Gᵢ ∈ SetG
       Al = ProposeAllocation(Gi)
       If (Al.nomAgent = Alⱼ.nomAgent)Then
          ResolutionConf(Allocation,
       Tasks,MAS,Al,Alⱼ)
         else
            Allocation.AddElement(Al)
         En If
    End For
 3. Feasible allocations:
    For each Alᵢ ∈ Allocation
       If CheckFaisible(Alᵢ,Tasks,MAS)=
    false) Then
          Allocation.romoveElement(Alᵢ)
       End If
    End For
End GroupeAl
```

Fig.1. GROUPEAL Algorithme

The *ProposeAllocation(Gi)* function presents the behaviour of the local allocator agent, responsible for the G_i group.

ResolutionConf(Allocation, Tasks, MAS, Al$_i$, Al$_j$) is a function which solves conflict between a new Al_i allocation and an already existing Al_j allocation in the allocation list (*Allocation*). Indeed, before adding a new allocation proposed by a local allocator, the agent global allocator checks if there is a conflict. If it is the case, it chooses the allocation which will be modified. Then, the second will be inserted in the list *Allocation*. By recursive call of the function, the process is repeated until a conflict exists. Indeed, each responsible for a group uses the pseudo following code to solve a conflict between two allocations Al_i and Al_j:

```
If Alᵢ ∩ Alⱼ = Agk  Then
  a) Choose the modified allocation Al
     based on S1 and S2.
  b) If ∃ Ag₁∈ G1\{Agₖ} Then Al.Agent =Ag₁
     Else modify the second allocation
        If ∃ Agᵣ∈ G2\{Agₖ}Then Al.Agent=Agᵣ
        Else mark ST_Al not assigned.
        End If
     End If
```

Fig.2. Pseudo code for conflict resolution

Where, Ag_l and Ag_r are agents which replace the agent Ag_k in the allocations which will be modified. And G_i is either the group which proposed the allocation Al (Al_i or Al_j), or one of the sub-group of the group which proposed Al, if $G \setminus \{Ag_k\} = \phi$.

Computing the set of groups (concepts) of agents according to their capacities: CreationG(MAS, Tasks)

In order to illustrate our proposal, we use the following example (Example1):
$MAS = \{Agent1, Agent2, Agent3, Agent4, Agent5\}$
Tasks = {T1, T2}. Where;

- *T1* is composed by *ST11* and *ST12* sub-tasks, and
- *T2* is composed by *ST21* and *ST22* sub-tasks.
- The cost to perform *T1* is *C(T1) = 500* and the cost to perform *T2* is *C(T2) = 12*.

The following table (Table1); shows the agent's capacities and theirs proposed cost:

Table 1. Capacities and cost

	ST11	ST12	ST21	ST22
Agent1	50			1
Agent2	5		5	
Agent3		5		5
Agent4	30		10	
Agent5		30		10

The creation of clusters of agents follows three steps:

1. *Generation of the binary relation*: A system Multi-Agent is defined by a set *MAS* of agents. Each Ag_i agent has a set of properties (competences) p_i. From the set of agents MAS, we define the formal context by the triplet *(MAS, P, I)* where *P* is the set of properties of the agents and *I* is the binary relation between *MAS* and *P*.

The binary relation of the *examlpe1* is as following:

Table 2. Binary matrix (agent, compétences)

	ST11	ST12	ST21	ST22
Agent1	1			1
Agent2	1		1	
Agent3		1		1
Agent4	1		1	
Agent5		1		1

2. *Generation of Galois lattice*: After the definition of the binary relation, the algorithm [3] generates a Galois lattice from this binary matrix (agent, competences). The generated concepts correspond to clusters of agents having the same characteristics. the Galois lattice corresponds to *Examlpe1* is as follows :

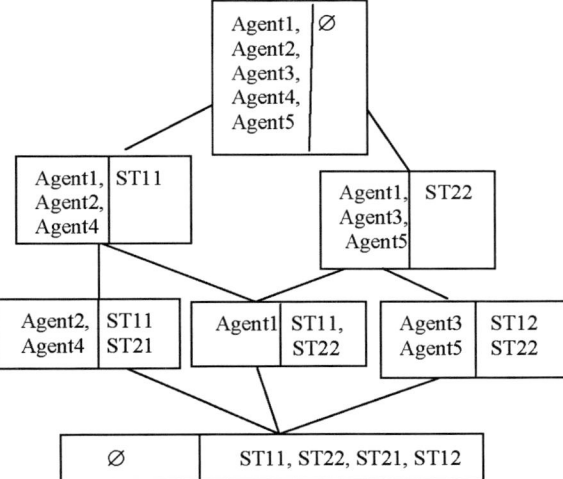

Fig.3. Galois lattice T which corresponds to exemple1

1. *Simplification of the Galois lattice*: The nodes (concepts) of the Hasse's diagram obtained by the algorithm [3] contain redundant elements. There are redundant elements between two nodes *Ni* and *Nj*, where *Nj* is the successor of *Ni*, if *P* is a property to be removed from the node *Nj*. In the preceding lattice we have *({Agent1, Agent2, Agent4}, {ST11})* ∩ *({Agent2, Agent4}, {ST11, ST22})* = *{ST11}*, then *ST11* will be removed from the second node. The removal of *P* guarantees that *P* will not be treated by the same agents representing the extensions of nodes *Ni* and *Nj*. Indeed, the set of agents representing the extensions of *Nj* are included in the set of agents representing the *Ni* extension. This makes it possible to check the *C1* condition which consists of: *a sub-task is assigned only to one agent*. If a concept is empty it will be removed. In our example the node corresponds to the concept *({Agent1}, {ST11,ST22})* will be removed because *ST11* and *ST22* belong to these predecessors. The following function simplifies the Galois lattices:

```
SimplifiyLattice(Lattice T)
  For each node N_i ∈ T
    For each successeur N_j of Ni
      If Ni ∩ Nj = P then
        Delete(N_j,P)//delete P from N_j
        If N_j is empty then
            delele(N_j)
        End If
      End If
    End For
  End For
```

Fig.4. Simplification function of the Galois lattice T

The obtained graph contains all groups of agents. A node represents a group and the arcs represent the bonds of communications between the clusters of agents. An agent can communicate only with the agents belonging to the higher or lower cluster. Indeed, this makes it possible to reduce the time of communication between agents. The following graph (Fig2) illustrates the lattice *T* simplified:

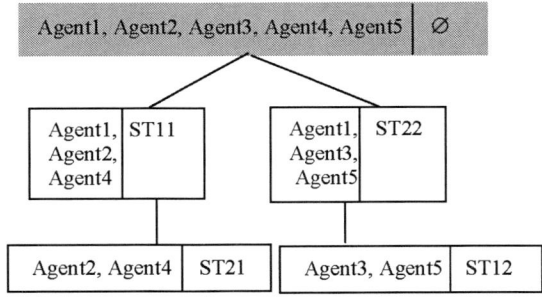

Fig.5. Graph of groups

Computing the optimal allocation

The proposed solution is differed from the other task allocation methods in a cooperative environment by: the optimal allocation is obtained by the union of the optimal local allocation proposed by the responsible agents. Indeed, we don't generate all possible allocation. Thus, the cost of the global allocation (*AlG*) is obtained by

$$V(AlG) = \sum_{i=1}^{K} V(Ali) \quad (4)$$

Where, *K* indicates the number of agents groups, and *V(Ali)* the cost of the local allocation *Ali*. Given that each G_i group proposes the optimal local allocation then, necessarily, the cost of the global allocation *V(AlG)* is minimal.

The result of the last stage on the Example1 is a set of agent clusters. This set is obtained from the simplification of the Galois lattice. This set consists of four clusters:
Groupe1 = ({Agent1, Agent2, Agent4}, {ST11}), Groupe2 = ({Agent1, Agent3, Agent5}, {ST22}), Groupe3 = ({Agent2, Agent4}, {ST21}) and Groupe4 = ({Agent3, Agent5, ST12}).

The local allocations of the Examlpe1 are the following couples: *Al1 = (Agent2, ST11), Al2 = (Agent1, ST22), Al3 = (Agent2, ST21)* and *Al4 = (Agent3, ST12)*. The detection of conflicts between groups consists in checking the intersection between two different allocations. In our case, we have $Al1 \cap Al3 = Agent2$, which means that *Agent2* is assigned to more than one sub-task.

Indeed, *Agent2* increases the optimality of the global allocation if it carries out *ST11* sub-task. The solution, then, is to keep the local allocation *Al1* and modify the local allocation *Al3*, by assigning *ST21* sub-task to an agent belongs to the group *Groupe3\{Agent2}*; which has the minimal cost. Thus, the *Al3* allocation will be: *Al3 = (Agent4, ST21)*.

Thus, the global allocation (*GA*) obtained is composed of the locals allocations *Al1*, *Al2*, *Al3* and *Al4*. This allocation checks the condition *C1*, *C2* and *C3* defined in *section 2*), it is feasible and it is optimal; because the value of this allocation is *V(GA)* = 5+1+10+5 = 21. This value is the minimum of all values related to the remaining allocations.

5. INTERACTION BETWEEN AGENTS

The interaction between the agents is based on interaction protocol, which describes the rules of conversation between agents. These conversations are based on the following communication primitives:

1. *proposeC({<sous-tâche,cost>})*: this message indicates that an agent proposes the set of the couple *<sub-task, cost>*. Where, *cost* is the cost suggested by the agent to carry out the sub-task *sub-task*.
2. *proposeAll(Alli, ListeAg$_i$)*: this message is sent by a local allocator agent. Where, *Alli* indicates the local allocation, and *ListeAgi* indicates the list of the agents which composed its group.
3. *joinGroupe(ListeAg)*: this message is sent by the agent global alloctor agent. It indicates for each $Ag_i \subset SMA$ to join the group composed by the set *ListeAg* of agents.
4. *result(Coalition)* : this message is sent by global allocator agent. It indicates the result (*Coalition*) of task allocation process.

Each message sent has the following structure: *typeMessage(IDE, IDR, <Primitives >)*, Where:

- *typeMessage* can be *send* or *receive*.
- *DE* and *IDR* indicate, respectively, identifier of the transmitting agent and the receiving agent.
- *Primitive* indicates the contents of the message which can be *proposeC, proposeAll* or *joinGroupe*.

The idea to distribute the task allocation process allows minimizing the flow of communication between agents. Indeed, we can summarize the interaction protocol between agents as following:

1. The first agent which has the set of the tasks which will be carried out is named global allocator. We assume that the other agents know already, the set of tasks to be carried out.
2. The other agents send the message *send(IDi IDAlG, proposeC({<STij, Cij >}))*. Where, IDi indicates the identifier of an agent, *IDAlG* identifies the global allocator agent, and *{<STij, Cij>}* indicates the set of the couples < *sub-task, cost* >. Thus, |*SMA*| messages are sent.
3. The global allocator agent generates the Galois lattice, which represents the groups of agents, as well as the bonds of communication between them. Then, it informs each agent on its future group. Thus, the message sent to each agent identified by *IDi*, is *send(IDAlG,IDi, joinGroupe(ListeAg)*.
4. Each responsible for a group *Gi* proposes a local allocation and sends it to globa allocator. The message is: *send(IDri,IDAlG,proposeAll(Alli, ListeAgi)*. Where, *IDri* indicates the identifier of the agent responsible for the group *Gi*.
5. After resolution of the conflict situations, if they exist, the global alloctor sends the following message to the agents: *send(IDAlG,IDi, result(Coalition))*. Thus, the number of messages sent will be to the maximum the number *NST* of all the sub-tasks.

Thus, the number *NbM* of messages communicated between agents is calculated as follows:

$$NbM = 2*|MAS| + NbRep + NST. \quad (5)$$

Where, *NbRep* indicates the number of local allocator or the number of concepts in the filtered Galois lattice. |*MAS*| represents the cardinality of the society of agents. This number is multiplied by two because the global allocator agent sends to all the agents the message *rejoinGroupe*, and each agent sends to this last the message *proposeC*.

6. IMPLEMENTATION AND EVALUATION OF OUR METHOD

In order to compare our method with the method developed by E. Manisterski and Al [4], we implemented them in language JAVA. The evaluation of the results according to the run-time and the number of agents is presented as follows:

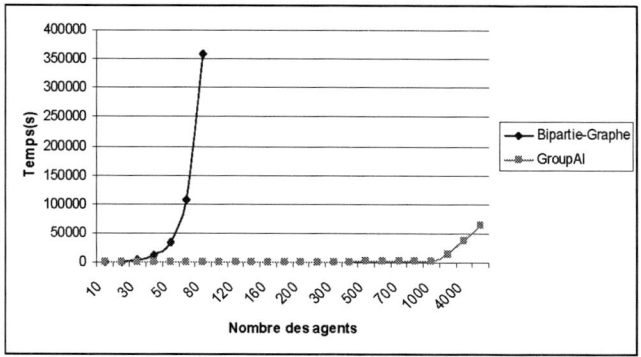

Fig. 6. Comparison between Bipartite-Graph and ClusterAll

The result shows (see fig6) that our solution takes less time than that developed by E. Manisterski et al. [4] when the number of agents exceeds value 15. This is due to three factors:

1. The distribution of computing the optimal allocation among local alloctor, and without generating all possible allocations. Indeed, in [4], only one allocator agent calculates all the possible allocations, using a graph bipartite. By increasing the number of agents, the number of nodes increases.
2. The research of the optimal allocation is computed simultaneously with the computation of the local allocation. For [4] it generates all the possible allocations, and then it proceeds to search the best allocation.

7. CONCLUSION

In this work, we proposed a method for tasks allocation problem in a massive MAS. The solution suggested is based on two basic ideas: 1) clustering agents by using the Formal Concepts Analysis approach in order to exploit the order relation between clusters (concepts) of the Galois lattice. The idea allows minimizing, on the one hand, the flow of communication between agents and on the other hand the categorization of agents. Thus, the structure of Galois lattice allows us to make a better cooperation between agents. In order to find the optimal allocation in a reasonable time, all created clusters propose a local allocation. The optimal solution is computed by resolution of the conflict situations between clusters of agents by exploiting their bonds of communication. As a future work, we plan to use this method in the field of information retrieval.

BIBLIOGRAPHIE

[1] Rahwan, T., and Jennings, N.R. (2007). An Algorithm for Distributing Coalitional Value Calculations among cooperating Agents. *Artificial Intelligence Journal* 2007, 171(8-9) pp. 535-567.

[2] Jong, S.; Tuyls, K. and Sprinkhuizen-Kuyper , I. (2006). Gradient Field-Based Task Assignment in an AGV Transportation System. *AAMAS'06* May 8–12 2006, Hakodate, Hokkaido, Japan.

[3] Vicky Choi (2006). Faster Algorithms for Constructing a Concept (Galois) Lattice. *arXiv:cs.DM/0602069 v2*, 1 Jun 2006.

[4] Manisterski, E., David, E., Kraus, S. and Jennings, N. R. (2006). Forming efficient agent groups for completing complex tasks. In: *5th Int. Conf. on Autonomous Agents and Multi-Agent Systems*, Hakodate, Japan.

[5] C. Sombattheera, et A. Ghose (2006). A Prune-Based Algorithm for Computing Optimal Coalition Structures in Linear Production Domains.

[6] David Sarne and Sarit Kraus (2005).Solving the Auction-Based Task Allocation Problem in an Open Environment. In*: Proceedings of Twentieth National Conference on Artificial Intelligence (AAAI)*, pp. 164-169.

[7] Rahwan, T. Jennings, N. R. (2005). Distributing Coalitional Value Calculations among Cooperative Agents. In: *American Association for Artificial intelligence*. AAAI Press, pp. 152-157.

[8] Oliveira, D. and Ferreira, P. and Bazzan, Ana L. C., (2004). A Swarm Based Approach for Task Allocation in Dynamic Agents Organizations. In: *Proceedings of the 3rd International Joint Conference on Autonomous Agents and Multi Agent Systems*, AAMAS. vol.3. July.

[9] P. T. Tosic et G. A. Agha (2004). Maximal Clique Based Distributed Group Formation For Task Allocation in Large-Scale Multi-Agent Systems. In: *The First International Workshop on Massively Multi-Agent Systems, December 2004*.

[10] Sander, P., Peleshcuk, D. and Grosz, B. (2002). A scalable, distributed algorithm for efficient task allocation. In: *AAMAS2002*, 1191–1198, Italy.

[11] Maksim Tsvetovat and Katia P. Sycara and Yian Chen and James Ying (2000). Customer coalitions in the electronic marketplace. In: *Proceedings of the Seventeenth National Conference on Artificial Intelligence and Twelfth Conference on Innovative Applications of Artificial Intelligence*, pp.1133-1134.

[12] Ganter, B. and Wille, R. (1999). Formal Concept Analysis, Mathematical Foundation. Berlin: Springer Verlag. 1999.

[13] Lhouari Nourine and Olivier Raynaud (1999). A fast algorithm for building Lattices. In: *Information Processing Letters*, pp.199-204.

[14] Eric MALVILLE (1999). L'auto-organisation de groupes pour l'allocation de tâches dans les Systèmes Multi-Agents : Application à CORBA. *PhD thesis*, Université de Savoir, 25 mars 1999.

[15] Tuomas W. Sandholm and Victor R. Lesser (1997). Coalitions among computationally bounded agents. In*: artificial Intelligence* 94 pp. 99–137.

SESSION E3: PRODUCT DESIGN & ROBOTICS

Antenna Automation For NOAA Satellite Images Reception

W.L.Rahal, N.Benabadji, A.H.Belbachir

University of Sciences and Technology of Oran
Laboratory of Analysis and Application of Radiations
B.P. 1505, El M'nouar, Oran, Algerie
Phone: 213-71-992-536   Fax: 213-41-530-461
(wassilaleila@hotmail.com, benanour2000@yahoo.com, ahbelbachir@yahoo.com)

Abstract: In this paper, we present a novel, precise and efficient software tool (LAAR-TRACK) for Low Earth Orbit (LEO) Satellites orbit determination. It's based on using orbital elements, which are given by the NORAD (North American Aerospace Defence) by taking into considerations orbital perturbations due to the atmospheric drag, the influence of the moon and the sun and the geopotential field.
The LAAR-TRACK gives the azimuth and the elevation that must have the antenna for pointing in real time the LEO satellites. This software is loaded on a computer directly connected, via the parallel port, to the tracking interface that we have developed, and which will be detailed in this paper. By this way the antenna can be automatically directed for receiving NOAA (National Oceanic and Atmospheric Administration) HRPT (High Resolution Picture Transmission) pictures.

Key words: LEO satellites, tracking interface, NOAA HRPT images, prevision and tracking satellites, TLE, SGP4 model.

1. INTRODUCTION

The problem of satellite orbit determination is a very important one, especially for defiling satellites such as NOAA series.
General perturbations element sets are generated by NORAD for all resident space objects. These element sets are periodically refined so as to maintain a reasonable prediction capability on all space objects, and can be used to predict position and velocity of Earth-orbiting objects. To do this one must be careful to use a prediction method which is compatible with the way in which the elements were generated.
The orbit determination problem consists of two basic parts: Propagation of the state estimated forward in time and updating the state estimate based upon new measurements of parameters which are function of the states.
As we are interested by the HRPT images of NOAA satellites, we've choose to predict the position of the satellites by using the SGP4 model, applied for near-Earth satellites.

2. THE PREDICTION MODELS

Satellite orbits around the Earth are not the perfect ellipses from Newtonian mechanics. The oblateness of the Earth, the irregular gravitational field, atmospheric drag, the pull of the Moon and Sun, and solar light pressure all effect satellites. Accurate models take into account these perturbations to predict where a satellite will be and what path it will follow.

When orbital elements are generated from observations a particular model is used. The elements released are the mean elements and are meant to be used with a particular model. Blindly converting to a different format or using a different model results in decreased accuracy. The most accuracy is achieved when using the same model as used to generate the elements.
The accuracy of element sets also depends on time. The perturbations change the orbit in non-predictable ways over long periods of time. This means that accuracy of the predictions from an element set decrease over time. The most accuracy requires getting fresh element sets. Atmospheric drag is the least predictable factor. This affects lower satellites more and makes their orbits less predictable which mean that the elements need to be updated more frequently to be accurate.

2.1 The SGP4 propagator

SGP4 (Simplified General Perturbations Satellite Orbit Model 4) is a NASA/NORAD algorithm of calculating near earth satellites. Any satellite with an orbital time of less than 225 minutes should use this algorithm. Satellites with orbital times greater than 225 minutes should use the SDP4 [1] or SDP8 algorithms. The choice of 225 minutes for selecting the propagation model (near-Earth or Deep-Space) appears somewhat arbitrary, but is thought to relate to the original range of the NORAD tracking radar system.

TLE data should be used as the input for the SGP4 algorithm. The accuracy of SGP4 is typically 0.1° longitude and 0.1° latitude from the ground. TLE data older than 30 days is considerably inaccurate due to perturbations in the orbit.

The SGP4 model was developed by Ken Cranford [2]. This model was obtained by simplification of the more extensive analytical theory of Lane and Cranford [3] which uses the solution of Brouwer [4] for its gravitational model and a power density function for its atmospheric model [5].

For SGP4 the mean motion is first recovered from its altered form and the drag effect is obtained from the SGP4 drag term (B*) with the pseudo-drag term being ignored.

To summarize, the SGP4 uses a geopotential model of 4th order [6] [7] [8]:

$$V = -\frac{GM}{r}\left(1 - \sum_{n=2}^{4} J_n \left(\frac{r_e}{r}\right)^n P_n(\sin\phi)\right)$$

To model:
- Equatorial pad (J2).
- Form pear (J3).
- Additional deviation (J4).

It models also the atmospheric trail which slows down the satellites on their trajectory.

3. THE TWO LINES ELEMENTS

This is the format used by NASA to distribute satellite elements in their NASA Prediction Bulletin [9] [10].

The format consists of groups of 3 lines: One line containing the satellite's name, followed by the standard Two-Line Orbital Element Set Format identical to that used by NASA and NORAD [11]. Data for each satellite consists of three lines in the following format:

```
NOAA 18
1 28654U 05018A   07011.61594517 +.00000288 +00000-0
+18419-3 0 07954
2 28654 098.8143 315.6637 0014444 354.5392 005.5612
14.11027555084775
```

Line 0 is a twenty-four character name.
Lines 1 and 2 are the standard Two-Line Orbital Element Set Format.

3.1 Decoding the Two Lines Elements

For decoding the TLE files, we've developed software, shown in the figure1, which extracts the following orbital elements:

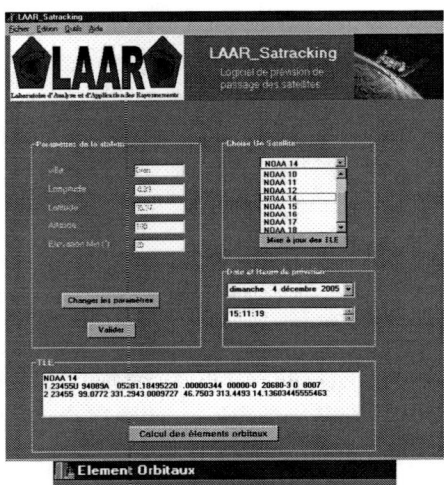

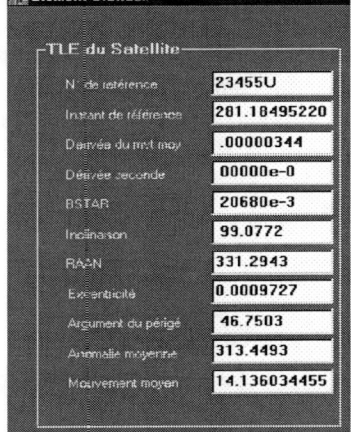

Fig. 1. Decoding TLE files interface

- Epoch : Specifies the time at which the set of orbital elements was taken.
- Orbital Inclination : the angle between the orbital and the equatorial plane
- Right Ascension of Ascending Node (R.A.A.N.): An angle, in the range 0 to 360 degrees, measured at the centre of the earth, from the vernal equinox to the ascending node.
- Argument of Perigee : The angle between the line-of-apsides and the line of nodes.
- Eccentricity : In the Keplerian orbit model, the satellite orbit is an ellipse. Eccentricity determines the shape of the ellipse: when e=0, the ellipse is a circle. When e is very near 1, the ellipse is very long and skinny.
- Mean Motion : A number which indicates the complete number of orbits a satellite makes in one day
- Mean Anomaly : An angle that increases uniformly with time, starting at perigee, use to indicate where a satellite is located along its orbit.

4. POSITIONNING ANTENNA CALCULATION

Disposing of the orbital parameters, it is possible now to call upon the SGP4 propagator which will provide us the position and the speed of the satellite in the ECI coordinate system.

The ECI coordinate system, shown in the figure2, is defined as a Cartesian coordinate system, where the coordinates (position) are defined as the distance from the origin along the three orthogonal axes [12].

- The z axis runs along the Earth's rotational axis pointing North,
- The x axis points in the direction of the vernal equinox,
- The y axis completes the right-handed orthogonal system.

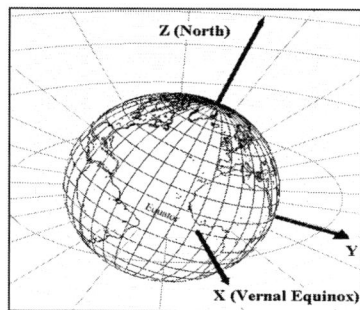

Fig. 2. The ECI coordinate system

The vernal equinox is an imaginary point in space which lies along the line representing the intersection of the Earth's equatorial plane and the plane of the Earth's orbit around the Sun or the ecliptic. Another way of thinking of the *x* axis is that it is the line segment pointing from the centre of the Earth towards the centre of the Sun at the beginning of spring, when the Sun crosses the Earth's equator moving North. The *x* axis, therefore, lies in both the equatorial plane and the ecliptic. These three axes defining the Earth-Centred Inertial coordinate system are fixed in space and do not rotate with the Earth.

We have also to define the ground station position in same reference mark (ECI) to be able to calculate the vector outdistances between the satellite and the ground station $[r_X, r_E, r_Z]$.

The coordinates of this vector must be converted in the topocentric coordinate system shown in the figure 3:

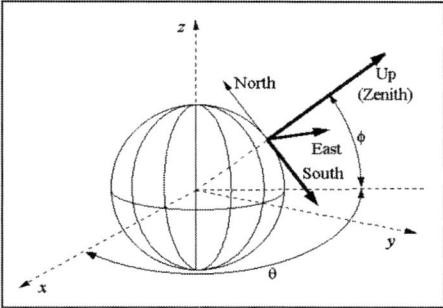

Fig. 3. Topocentric coordinates system

The conversion is done by [13]:
- A rotation of θ angle around the z axis.
- A rotation of φ angle around the y axis.

The coordinates of $[r_x, r_y, r_z]$ became:

$$r_S = \sin \varphi \cos \theta \, r_x + \sin \varphi \sin \theta \, r_y - \cos \varphi \, r_z \quad (1)$$

$$r_E = -\sin \theta \, r_x + \cos \theta \, r_y \quad (2)$$

$$r_Z = \cos \varphi \cos \theta \, r_x + \cos \varphi \sin \theta \, r_y + \sin \varphi \, r_z \quad (3)$$

The distance between the satellite and the ground station is calculated by:

$$r = \sqrt{r_S^2 + r_E^2 + r_Z^2} \quad (4)$$

The elevation is given by:

$$El = \sin^{-1}\left(\frac{r_Z}{r}\right) \quad (5)$$

And the azimuth by:

$$Az = \tan^{-1}\left(\frac{r_E}{r_S}\right) \quad (6)$$

5. DESCRIPTION OF THE TRACKING HARDWARE PART

5.1 Synoptic diagram of satellite tracking

The tracking can be summarized in three steps:
- Upturn of the antenna position.
- Comparison with the values calculated by the prevision software.
- Control of the antenna.

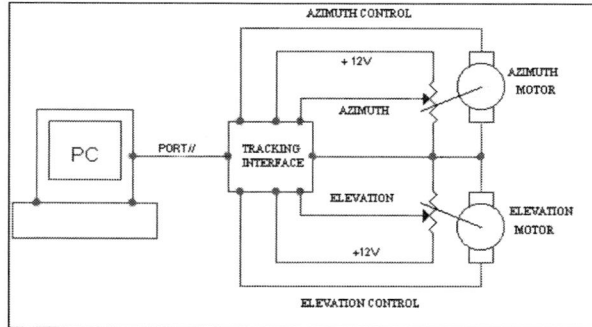

Fig. 4. Synoptic diagram of satellite tracking

We recover the antenna position information via the control panel of the rotor.

While comparing the antenna position with the values calculated by the prevision software, the same interface orders the rotor so that the values read on the potentiometers correspond to those calculated by the software.

5.2 The tracking interface:

The tracking interface guarantees the communication between the hardware part and the prevision software, by ensuring the data flow between the antenna rotor and the computer which contains the prevision algorithm.

5.3 Tracking interface implementation

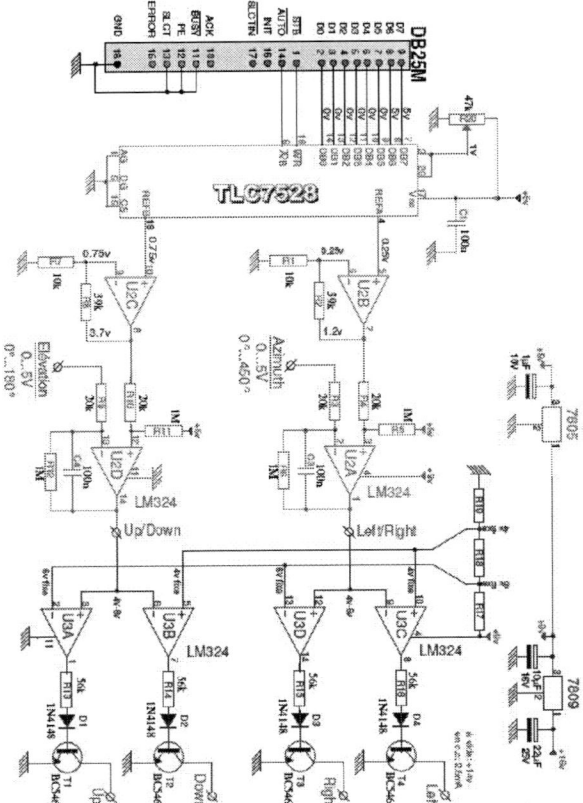

Fig. 5. Electronic scheme of the tracking interface

The interface is connected to the PC where the prevision software is installed, through the parallel port.

The antenna elevation is recovered through the pin1 of the rotor:
- When the potential = 0 volt → elevation = 0°
- When the potential = 5 volt → elevation = 180°

The antenna azimuth is recovered through the pin6 of the rotor:
- When the potential = 0 volt → Azimuth = 0°
- When the potential = 5 volt → Azimuth = 450°

These data are compared with the elevation and azimuth values generated by the prevision software, by using differential amplifiers (LM324).

Transistors (BC546) are used to amplify the control currents of motors, Azimuth (Left, Right) and elevation (Up, Down), through pins 2, 3, 4 and 5 of the DIN card located on the back face of the control panel, in such way that the difference at the exit of the differential amplifiers is null.

6. CONCLUSIONS

The satellite then goes into an inclined orbit, requiring earth stations to have intelligent tracking mechanism. The aim of the proposed work is to provide a tracking system for LEO satellites, composed of a software part and a material one.
Automatic pointing of satellite antennas helps to improve signal quality and also frees the operator to focus on the satellite contact. There are three components in an automatic pointing system; azimuth and elevation rotors, satellite track software loaded on a computer, and an interface that connects the rotors to the computer
The satellite track software, that we have developed, calculates at any time the position of the Polar Orbiting satellites and where the antennas should be pointed in space. It is based on the SGP4 model that takes into consideration the disturbing forces that constantly cause the satellite to move from its assigned position, such as gravitational attraction of the sun and the moon and the earth gravitational field.
The hardware part ensures the control of the rotor antenna (Yeasu G5500). The tracking interface reads the antenna position voltages from the rotor and converts them to a digital word value that is compared with satellite position data information (calculated with the satellite track software) and send appropriate commands to move the rotors.

REFERENCES

[1] Lainy V. (2002) « *Théorie dynamique des satellites galiléens* » Thèse de doctorat présentée à L'observatoire De Paris. Spécialité: Dynamique des systèmes gravitationnels

[2] Lane, M.H. and Hoots, F.R. (1979) *"General Perturbations Theories Derived from the Lane Drag Theory"*, Project Space Track Report No. 2, Aerospace Defense Command, Peterson AFB, CO.

[3] Lane, M.H. and Cranford, K.H., (1969) *"An Improved Analytical Drag Theory for the Artificial Satellite Problem"*, AIAA Paper N° **69** p 925.

[4] Brouwer, D. (1959) *"Solution of the Problem of Artificial Satellite Theory without Drag"*, Astronomical. Journal **64**, p.378-397,.

[5] Lane, M.H., Fitzpatrick, P.M., and Murphy, J.J. (1962) *"On the Representation of Air Density in Satellite Deceleration Equations by Power Functions with Integral Exponents"*, Project Space

Track Technical Report No. APGC-TDR-62-15., Air Force Systems Command, Eglin AFB, FL.

[6] Hilton, C.G. and Kuhlman, J.R. (1966) *"Mathematical Models for the Space Defense Center"*, Philco-Ford Publication No. **U-3871**, p.17-28,.

[7] Hoots, F.R. (1980) *"A Short, Efficient Analytical Satellite Theory"*. AIAA Paper **N°.80** p.1659.

[8] Hoots, F.R., *"Theory of the Motion of an Artificial Earth Satellite"*, accepted for publication in Celestial Mechanics.

[9] Kozai, Y. (1959) *"The Motion of a Close Earth Satellite"*, Astronomical Journal **64**, p.367-377.

[10] Defense Command Space Computational Center Program Documentation (1977), DCD 8, Section 3, p.82-104.

[11] : Felix R. Hoots and Ronald L. Roehrich (1988) *"Models for Propagation of NORAD Element Sets"* SPACETRACK REPORT N° 3.

[12] Hujsak, R.S. (1979) *"A Restricted Four Body Solution for Resonating Satellites with an Oblate Earth"*, AIAA Paper **N°79** p.136.

[13]: T.S. Kelso (1995) *"Orbital Coordinate Systems"*, Satellite Times Journal.

On the Partial Attitude Control of Axi-Symmetric Rigid Spacecraft

Chaker Jammazi

Ecole Polytechnique de Tunisie
Laboratoire d'Ingénierie Mathématique
Chaker.Jammazi@ept.rnu.tn

Abstract: This paper studies the partial asymptotic stabilization of underactuated axi-symmetric rigid spacecraft with two controllers. We have shown that axi-symmetric rigid spacecraft is not controllable, and cannot be asymptotically stabilizable by continuous pure state feedback laws. In order to overcome these limitations we treat the stabilization problem in the sense of partial asymptotic stabilization.

Keywords: Partial asymptotic stabilization, Axi-symmetric rigid spacecraft, Brockett's condition, Attitude control.

1. INTRODUCTION

The attitude control of rigid spacecraft with only two controllers has been the subject of numerous research, Tsiotras [1994], Tsiotras and Longuski [1993], Tsiotras et al. [1995], Crouth [1984], Coron [1999], Morin [2004], Keraï [1995], Coron and Keraï [1996], Morin et al. [1995], Jammazi and Abichou [2006], Zuyev [2001]. This type of system is used to illustrate several aspects of nonlinear controllability. Let us mention the results of Bonnard [1982] and Crouth [1984] proved that the system is globally controllable in large time, Keraï [1995] proved that the system satisfies Sussmann's condition, and is small time locally controllable. The attitude stabilization is studied by the work of Morin et al. [1995], Morin [2004] where the concept of time-varying feedback laws stabilizing locally the system is explicitly derived by using center manifold theory combined with averaging techniques. Coron and Keraï [1996] establish the time-varying homogenous and periodic feedback laws stabilizing the rigid spacecraft with two controllers. The method describes by Morin and Samson [2000] where they studied the attitude of underactuated rigid spacecraft. The authors have shown that the rigid spacecraft is locally, exponential stabilized with respect to a given dilation. In addition the obtained controllers was periodic, time-varying and non-differentiable at the origin and the construction relied on the proprieties of homogenous systems.

In this paper, we will study the stabilization problem of axi-symmetric rigid spacecraft with two inputs. For the axi-symmetric rigid body there exists a wide range of results. However, stabilization is only possible for the restricted case of zero spin rate the unactuated axis. Spin-stabilization with two control torques is addressed in Tsiotras [1994] based on new formulation for the attitude dynamics. The new attitude formulation, described in Tsiotras et al. [1995] was subsequently derived a time-invarying discontinuous control law that achieves arbitrary reorientation of the rigid spacecraft. In a more recent work of Tsiotras and Luo [2000], bounded feedback laws are designed for stabilization and tracking underactuated spacecraft.

In the present paper we propose other control strategies; these strategies are based on the partial asymptotic stabilizability Jammazi and Abichou [2006]. The partial asymptotic stabilization considered in the present work, is the stabilization with respect to the major components of the system and the rest converges to the same position which depends on the initial conditions.

This theory is a natural extension of the classical concept of stabilizability in Lyapunov sense. Stability (respectively, stabilizability) with respect to part of the state called also "partial stability (respectively, partial stabilizability)" has been intensively studied within last 50 years. Basic results in this field belong to Rumyantsev, Rouche et al. [1977], Chellaboina and Haddad [2001], Fradkov et al. [2001], Haddad et al. [2003], Vorotnikov [1998], the founder of the theory of partial stability for systems of ordinary differential equations with continuous right side and had demonstrated the applicability of this results to problems of stability of more general models of distribute-parameter systems. Subsequently, a large number of researchers have contributed to the development of theory and methods for studying partial stability and stabilization and resolved several important problems; see, for instance, Vorotnikov [1998].

Moreover, our partial asymptotic stability (respectively, stabilizability) definition is different from the definitions given and used in Chellaboina and Haddad [2001], Fradkov et al. [2001], Haddad et al. [2003], Vorotnikov [1998], Zuyev [2001], Zuiev [1999], in that the authors have been occupied with the part of the system and supposes that the rest is bounded. The definition used in this paper takes in to consideration the complete stability of the system, the asymptotic stability with respect to part of the sate and the convergence of the rest. The latter property is necessary since it removes the possible oscillations of the system. This partial asymptotic stability is applied in many engineering fields as the stabilization problems of rigid spacecraft with two controls, the ship Jammazi and Abichou [2006], the underwater vehicle Jammazi and Abichou [2007] and the nonholonomic systems.

In our approach, we have constructed two smooth feedback laws that makes the four first components of the system describing the rigid spacecraft asymptotically stable and the remaining component converges. This work improves Tsiotras et al. (Tsiotras [1994]) result, and improves the regularity of the feedback

---

* This work is sponsored by the Polytechnical School of Tunisia

controllers, indeed, this special stabilization is obtained by C^∞ feedback laws.

The paper is organized as follows. In section 2 we review the notion of partial asymptotic stabilizability and we give some related results. Section 3 presents the partial stabilization of the underactuated axi-symmetric rigid spacecraft. Section 5 contains the conclusion.

2. PRELIMINARIES

We begin with a review of the concept of partial asymptotic stability (respectively, partial asymptotic stabilizability), and let R denotes the set of real numbers, let R^n denotes the set $n \times 1$ real column vectors and $|.|$ denotes the Euclidean vector norm. Finally $f \in L^1([0, +\infty))$ implies that $\int_0^{+\infty} |f(t)|\, dt < \infty$. Consider the following system on the form:

$$\begin{cases} \dot{x}_1 = f_1(x_1, x_2) \\ \dot{x}_2 = f_2(x_1, x_2). \end{cases} \quad (1)$$

where $f = (f_1, f_2)$ is supposed to be class $C^\infty(R^p \times R^{n-p})$, $x_1 \in R^p$, $x_2 \in R^{n-p}$ and p integer such that $0 < p \leq n$. In the equation given by x_1, we can see x_2 as a parameter and we can adopt as Lin et al. [1995], Zuyev [2001] the following hypothesis

$$f_1(0, x_2) = 0,\ \forall x_2 \in R^{n-p},\ f_2(0, 0) = 0. \quad (2)$$

Definition 1. (**Partial asymptotic stability**) The system (1) is said to be partially asymptotic stable if the following properties are satisfied:
a) The system (1) is stable;

$$\forall\, \epsilon > 0, \exists\, \eta > 0 : (|x_1(0)| + |x_2(0)| < \eta) \\ \Rightarrow (|x_1(t)| + |x_2(t)| < \epsilon\ \forall\, t \geq 0). \quad (3)$$

b) The system (1) is asymptotically stable with respect to x_1 and x_2 converges;

$$\exists\, r > 0 : (|x_1(0)| + |x_2(0)| \leq r) \Rightarrow (\lim_{t \to +\infty} x_1(t) = 0),\ (4)$$

and there exists a constant vector $\alpha(x(0))$ depending on initial conditions such that

$$\lim_{t \to +\infty} x_2(t) = \alpha(x(0)). \quad (5)$$

Definition 2. (**Partial asymptotic stabilizability**) The control system

$$\begin{cases} \dot{x}_1 = f_1(x_1, x_2, u) \\ \dot{x}_2 = f_2(x_1, x_2, u), \end{cases} \quad (6)$$

where $u \in R^m$ is the control and $x = (x_1, x_2) \in R^p \times R^{n-p}$ is said to be partially asymptotic stabilizable if there exists a continuous function $\phi : R^p \times R^{n-p} \longrightarrow R^m$, $\phi(0, x_2) = 0$ such that the system (8) in closed-loop:

$$\begin{cases} \dot{x}_1 = f_1(x_1, x_2, \phi(x)) \\ \dot{x}_2 = f_2(x_1, x_2, \phi(x)), \end{cases} \quad (7)$$

is partially asymptotically stable in the sense of definition 1.

Theorem 3. (Jammazi and Abichou [2006]) If the system

$$\begin{cases} \dot{x}_1 = f_1(x_1, x_2, u) \\ \dot{x}_2 = f_2(x_1, x_2, u), \end{cases} \quad (8)$$

is partially asymptotically stabilizable by a C^1 feedback laws, then the augmented system

$$\begin{cases} \dot{x}_1 = f_1(x_1, x_2, y) \\ \dot{y} = u \\ \dot{x}_2 = f_2(x_1, x_2, y), \end{cases} \quad (9)$$

is partially asymptotically stabilizable by a C^0 feedback laws in the following sense: (x_1, y) is asymptotically stabilizable and x_2 converges.

The following lemma is well known, and justifies the limit of differentiable functions near the infinity.

Lemma 4. Let $f : [0, +\infty) \to R$ derivable function such that

$$\int_0^{+\infty} |f'(t)|\, dt < \infty,$$

then $\lim_{t \to +\infty} f(t)$ exists.

3. AXI-SYMMETRIC RIGID SPACECRAFT

Various control algorithms for controlling axi-symmetric rigid spacecraft have appeared in the literature. Tsiotras [1994], Tsiotras et al. [1995] and Tsiotras and Longuski [1993] was the first to control such system. It was shown that discontinuous feedback controllers at the equilibrium can achieve the stabilization of all axi-symmetric rigid spacecraft. Contrarily to works cited in Tsiotras and Longuski [1993], Tsiotras [1994], Tsiotras et al. [1995] that are focalized on discontinuous feedback laws and in order to overcome the limitations imposed by Brockett's condition, we propose a smooth feedback laws that can assure the asymptotic stability of four components and the convergence of the remaining one.

It is important to mentioning that the terms axi-symmetric means the symmetric of the inertia matrix not the geometric symmetry.

All rigid spacecraft with two control can be described by the following systems (Coron [2007])

$$\begin{cases} \dot{\omega} = J^{-1} S(\omega) J \omega + \sum_{i=1}^{2} u_i e_i \\ \dot{\eta} = A(\eta) \omega, \end{cases} \quad (10)$$

where

$$J = \begin{pmatrix} j_1 & 0 & 0 \\ 0 & j_2 & 0 \\ 0 & 0 & j_3 \end{pmatrix},$$

is the inertia matrix of the rigid spacecraft.
S is skew-symmetric matrix

$$S = \begin{pmatrix} 0 & \omega_3 & -\omega_2 \\ -\omega_3 & 0 & \omega_1 \\ \omega_2 & -\omega_1 & 0 \end{pmatrix}.$$

The matrix $A(\eta)$ is given by:

$$A = \begin{pmatrix} 1 & sin\phi\, tg\theta & cos\phi\, tg\theta \\ 0 & cos\phi & -sin\phi \\ 0 & \dfrac{sin\phi}{cos\theta} & \dfrac{cos\phi}{cos\theta} \end{pmatrix},$$

$e_1 = (1, 0, 0)'$, $e_2 = (0, 1, 0)'$, and $cos\theta \neq 0$.

Definition 5. The rigid spacecraft described by the system (10) is said to be axi-symmetric if
$$c_3 = \frac{j_1 - j_2}{j_3} = 0 \Rightarrow \dot{\omega}_3 = 0;$$
i.e. in the inertia matrix J, we have $j_1 = j_2 \Rightarrow \dot{\omega}_3 = 0$.

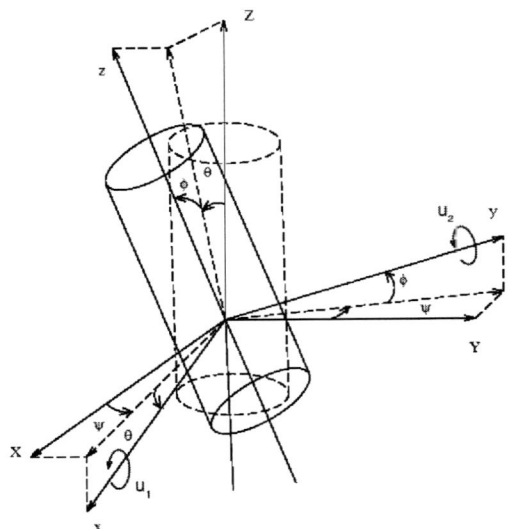

Fig. 1. Axi-symmetric satellite

In the sequel, we will be interested to stabilization problem of axi-symmetric rigid spacecraft. To simplify our task we assume as in Tsiotras [1994] that $\omega_3(0) = 0$, thus we obtain

$$\begin{cases} \dot{\omega}_1 = u_1 \\ \dot{\omega}_2 = u_2 \\ \dot{\phi} = \omega_1 + \omega_2 \sin\phi\, tg\theta \\ \dot{\theta} = \omega_2 \cos\phi \\ \dot{\psi} = \frac{\omega_2 \sin\phi}{\cos\theta}. \end{cases} \quad (11)$$

Proposition 6. The system (11) is not small time locally controllable (STLC).

Proof The system (11) is an example of system with drift, it can be written in the form: $\dot{x} = f_0(x) + u_1 f_1(x) + u_2 f_2(x)$, where

$$f_0(x) = \begin{pmatrix} 0 \\ 0 \\ \omega_1 + \omega_2 \sin\phi\, tg\theta \\ \omega_2 \cos\phi \\ \frac{\omega_2 \sin\phi}{\cos\theta} \end{pmatrix}, f_1(x) = \begin{pmatrix} 1 \\ 0 \\ 0 \\ 0 \\ 0 \end{pmatrix}, f_2(x) = \begin{pmatrix} 0 \\ 1 \\ 0 \\ 0 \\ 0 \end{pmatrix}$$

It is easy to see that, $[ad^k f_0, f_i] = 0_{R^4}$, for all $k \geq 2$ and $i = 1, 2$, and $span\{f_1(0), f_2(0), [f_0, f_1](0), [f_0, f_2](0)\} \neq R^4$, then the strong Lie algebra condition at equilibrium zero is not satisfied (Coron [2007]). Thus the system (11) is not small time locally controllable (STLC).

Proposition 7. The system (11) does not satisfy the Brockett necessary condition for the stabilizability.

Proof For $\phi \in]-\eta, \eta[$ ($\eta > 0$ small enough), all point of the form $(0, 0, *, 0, \epsilon)'$ ($\epsilon \neq 0$) is not in the image of the map f given by

$$f(x, u) = \begin{pmatrix} u_1 \\ u_2 \\ \omega_1 + \omega_2 \sin\phi\, tg\theta \\ \omega_2 \cos\phi \\ \frac{\omega_2 \sin\phi}{\cos\theta} \end{pmatrix}.$$

Indeed, the equation $\omega_2 \cos(\phi) = 0$ with ϕ small enough, implies $\omega_2 = 0$ and then $\epsilon = 0$. Contradiction.

Remark 8. The proposition 6 show that no continuous or discontinuous static-state feedback law exists which makes the origin of the axi-symmetric system asymptotically stable, because the latter system, is an example of system with drift, cannot satisfy Brockett's condition for stabilizability (Brockett [1983]) and Coron and Rosier conditions (Coron and Rosier [1994]) for stabilizability by discontinuous feedback laws.

4. PARTIAL ASYMPTOTIC STABILIZATION OF AXI-SYMMETRIC SPACECRAFT

We have shown that the system (11) is not asymptotically stabilizable by means of continuous pure state feedback laws. The stabilization should be treated in the sense of partial asymptotic stabilization.

Our objective is to conceive two smooth feedback controllers such that $(\omega_1, \omega_2, \phi, \theta)' = (0, 0, 0, 0)'$ is asymptotically stable and the state ψ converges.

In this context we will apply the theorem of backstepping in partial stabilization (Theorem 3). It is then sufficient to stabilize partially the following reduced system:

$$\begin{cases} \dot{\phi} = u_1 + u_2 \sin\phi\, tg\theta \\ \dot{\theta} = u_2 \cos\phi \\ \dot{\psi} = \frac{u_2 \sin\phi}{\cos\theta}. \end{cases} \quad (12)$$

Let the feedback transformation
$$v_1 := u_1 + u_2 \sin\phi\, tg\theta$$
$$v_2 := u_2. \quad (13)$$

Then the system (12) take the form
$$\begin{cases} \dot{\phi} = v_1 \\ \dot{\theta} = v_2 \cos\phi \\ \dot{\psi} = \frac{v_2 \sin\phi}{\cos\theta}. \end{cases} \quad (14)$$

In this case, the state ϕ is described by
$$\dot{\phi} = v_1.$$
If we choose
$$v_1 = -\phi, \quad (15)$$
we get
$$\dot{\phi}(t) = -\phi(t),$$
this implies the exponential stability of the equilibrium $\phi = 0$.
Let the Lyapunov candidate function V given by:
$$V(\theta, \phi) = \frac{1}{2}\theta^2 + \frac{1}{2}\phi^2, \quad (16)$$
then
$$\dot{V} = \theta\, v_2 \cos\phi - \phi^2. \quad (17)$$

If we choose
$$v_2 = -\theta \cos\phi, \tag{18}$$
we get
$$\dot{V} = -\theta^2 \cos^2\phi - \phi^2, \tag{19}$$
and we have
$$\dot{V} = 0 \Leftrightarrow -\theta^2 \cos^2\phi - \phi^2 = 0 \Leftrightarrow (\phi, \theta) = (0, 0).$$
We conclude by direct Lyapunov methods that $(\phi, \theta) = (0, 0)$ is asymptotically stable.
The dynamic of ψ is given by
$$\dot{\psi} = \frac{-\theta \cos\phi \sin\phi}{\cos\theta},$$
the state θ and ϕ are asymptotically stable, then the expression, $\frac{-\theta \cos\phi}{\cos\theta}$ is bounded by a reel $k > 0$.
Moreover, we have
$$|\sin\phi| \leq |\phi|.$$
then in the neighborhood of $+\infty$, we get
$$|\dot{\psi}| \leq k|\phi| \leq k|\phi(0)|e^{-t}. \tag{20}$$

the inequality (20) show that $\dot{\psi}$ is integrable on $[0, +\infty)$, therefore, by lemma 4, the state ψ converges.

Proposition 9. With our feedback laws (15) and (18) the angle θ is exponentially stable.

Proof: With (18), the state θ satisfy the equation
$$\dot{\theta} = -\theta \cos^2(\phi), \tag{21}$$
we integer (21) and taking account of (15) we obtain
$$\theta(t) = \theta(0) e^{-\int_0^t \cos(\phi_0 e^{-s}) ds}. \tag{22}$$
In the neighborhood of $+\infty$, we have $\phi_0 e^{-s} \to 0$, as $s \to +\infty$, then we have
$$\cos(\phi_0 e^{-s}) = 1 - \frac{\phi_0^2 e^{-2s}}{2} + o(e^{-2s}), \tag{23}$$
we get
$$\int_0^t \cos(\phi_0 e^{-s}) ds = t + \frac{\phi_0^2 (e^{-2t} - 1)}{4} + o(e^{-3t}), \tag{24}$$
and
$$e^{-\int_0^t \cos(\phi_0 e^{-s}) ds} = e^{-t} e^{-\frac{\phi_0^2(e^{-2t}-1)}{4} + o(e^{-3t})} \leq c\, e^{-t}. \tag{25}$$
With $c > 0$ is the upper bound on $[0, +\infty)$ of the expression
$$e^{-\frac{\phi_0^2(e^{-2t}-1)}{4} + o(e^{-3t})}.$$
The inequality (25) implies
$$|\theta(t)| \leq c|\theta(0)|e^{-t}, \tag{26}$$
then $\theta = 0$ is exponentially stable.

Comments

(1) With the feedback (15), ϕ is exponentially stable, then for t near $+\infty$ and $|\phi(0)| < \epsilon$ we have $\cos\phi \sim 1$, thus in the dynamic of θ we obtain $\dot{\theta} = v_2$, we can take for example $v_2 = -\theta$ to have the exponential stability of θ.

(2) The feedback controllers that stabilizing the extended system are inspired by the work of Morin and Samson [1997], the difference here is that the reduced system is not homogenous of degree zero with respect to any dilation, because the reduced system includes trigonometric terms. Then, to show the stabilization propriety of the augmented system we proceed by linearization techniques. The following proposition gives these feedback

Proposition 10. Let $k \geq 4$. The feedback controllers
$$\tau_1 = -k(\omega_1 - v_1(x)) \tag{27}$$
$$\tau_2 = -k(\omega_2 - v_2(x)) \tag{28}$$
assured the partial asymptotic stabilization of the system (11) in the sense:
$(\omega_1, \omega_2, \phi, \theta)' = (0, 0, 0, 0)'$ is asymptotically stable and ψ converges.
$v_1(x)$ and $v_2(x)$ are given respectively in (15) and (18).

Proof
• $(\omega_1, \omega_2, \phi, \theta)' = (0, 0, 0, 0)'$ is asymptotically stabilizable?
With the feedback (27) and (28), the system (11) in closed loop is given by
$$\begin{cases} \dot{\omega}_1 = -k(\omega_1 + \phi) \\ \dot{\omega}_2 = -k_2(\omega_2 + \theta \cos\phi) \\ \dot{\phi} = \omega_1 + \omega_2 \sin\phi \, tg\theta \\ \dot{\theta} = \omega_2 \cos\phi \\ \dot{\psi} = \frac{\omega_2 \sin\phi}{\cos\theta}. \end{cases} \tag{29}$$

Let the subsystem
$$\begin{cases} \dot{\omega}_1 = -k(\omega_1 + \phi) \\ \dot{\omega}_2 = -k(\omega_2 + \theta \cos\phi) \\ \dot{\phi} = \omega_1 + \omega_2 \sin\phi \, tg\theta \\ \dot{\theta} = \omega_2 \cos\phi. \end{cases} \tag{30}$$

The linearized of (30) around the equilibrium is given as follow
$$\begin{cases} \dot{\omega}_1 = -k(\omega_1 + \phi) \\ \dot{\omega}_2 = -k(\omega_2 + \theta) \\ \dot{\phi} = \omega_1 \\ \dot{\theta} = \omega_2. \end{cases} \tag{31}$$

In (31), with the dynamic of ω_1 and ω_2, the angles ϕ and θ respectively satisfies the second order equations
$$\ddot{\phi} = -k\omega_1 - k\phi$$
$$= -k\dot{\phi} - k\phi. \tag{32}$$
$$\ddot{\theta} = -k\omega_2 - k\theta$$
$$= -k\dot{\theta} - k\theta. \tag{33}$$

Thus, we choose the gain k such that the polynom $P = x^2 + kx + k$ is Hurwitz. All roots of P are given by:
$$x_1(k) = \frac{-k - \sqrt{k^2 - 4k}}{2}, \quad x_2(k) = \frac{-k + \sqrt{k^2 - 4k}}{2}. \tag{34}$$
It is clear that $k \geq 4$, roots x_1 and x_2 are strictly negatively. Therefore, with the choice of the gain k, the states ϕ and θ solutions respectively of (32) and (33) are given by
$$\phi(t) = a_1 e^{x_1(k)t} + a_2 e^{x_2(k)t}$$
$$\theta(t) = b_1 e^{x_1(k)t} + b_2 e^{x_2(k)t}. \tag{35}$$

For $i = 1, 2$, a_i and b_i are constants depends on initials conditions.

It is clear that ϕ and θ are exponentially stable.

Since $x_1(k) < x_2(k)$ then

$$|\phi(t)| \leq (|a_1| + |a_2|)e^{x_2(k)t}, \tag{36}$$

and

$$|\theta(t)| \leq (|b_1| + |b_2|)e^{x_2(k)t}. \tag{37}$$

For the angular velocity ω_1, we have $\dot{\phi} = \omega_1$ then,

$$\omega_1(t) = a_1 x_1(k) e^{x_1(k)t} + a_2 x_2(k) e^{x_2(k)t}, \tag{38}$$

i. e.

$$\begin{aligned}|\omega_1(t)| &\leq |a_1 x_1(k)| e^{x_1(k)t} + |a_2 x_2(k)| e^{x_2(k)t} \\ &\leq (|a_1| + |a_2|)|x_1(k)| e^{x_2(k)t}.\end{aligned} \tag{39}$$

From (39), we deduce the exponential stability of ω_1. by the similar idea, we show that

$$|\omega_2(t)| \leq (|b_1| + |b_2|)|x_2(k)| e^{x_2(k)t}. \tag{40}$$

Finally, the linearized system (31) is globally exponentially stable, then by classical result (see Khalil [2002], Zabczyk [1992]), the subsystem (30) is locally exponentially stable.

• ψ converges ?.

Since ψ satisfy the equation $\dot{\psi} = \dfrac{\omega_2 \sin\phi}{\cos\theta}$, we have then $|\dot{\psi}| \leq r|\omega_2||\phi|$, with r is upper bounded of $\dfrac{1}{\cos(\theta(t))}$, consequently, $\dot{\psi} \in L^1[0, +\infty)$, then, ψ converges (Lemma 4).

Comments In Tsiotras et al. [1995], the authors proposed the following feedback laws:

$$u_1 = -k\frac{\sin\phi\cos\theta}{1 + \cos\phi\cos\theta}, \ k > 0, \tag{41}$$

$$u_2 = -k\frac{\sin\phi}{1 + \cos\phi\cos\theta}. \tag{42}$$

These feedback (41) and (42) assured the local asymptotic stabilization of (11) such that the set

$$S := \{(\phi, \theta, \psi) : z(\phi, \theta, \psi) = 0\},$$

an invariant manifold for system (11), with z is the function defined by

$$z(\phi, \theta, \psi) = \psi + \arcsin(p\cos\phi) - \arcsin(p)$$

$p = \dfrac{a}{\sqrt{1+a^2}}$ and $a = \dfrac{tg\theta}{\sin\phi}$, the inconvenient of these feedback laws that present a singularity for example in the point $(\pi, 0, 0)$. The advantage of our feedback laws that are of class C^∞ and the stabilization is exponential.

4.1 Numerical simulation

In this subsection we present numerical simulations to valid our results. The feedback given in (27) and (28) with $x = (\omega_1, \omega_2, \phi, \theta, \psi)'$ are:

$$\begin{aligned}\tau_1(x) &= -5(\omega_1 - \alpha_1(x)), \\ \tau_2(x) &= -10(\omega_2 - \alpha_2(x)), \\ \alpha_1(x) &= -\phi - \alpha_2 \sin\phi \, tg\theta, \\ \alpha_2(x) &= -\theta \cos\phi.\end{aligned}$$

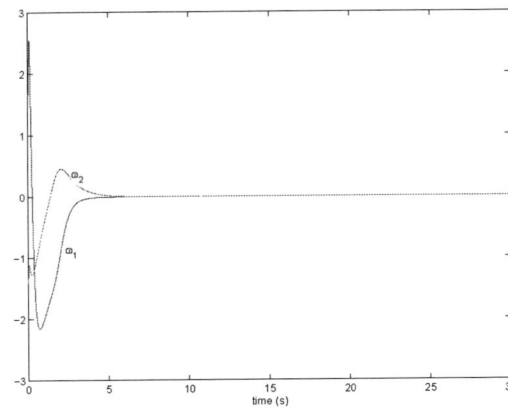

Fig. 2. Trajectories of the velocities ω_1 and ω_2

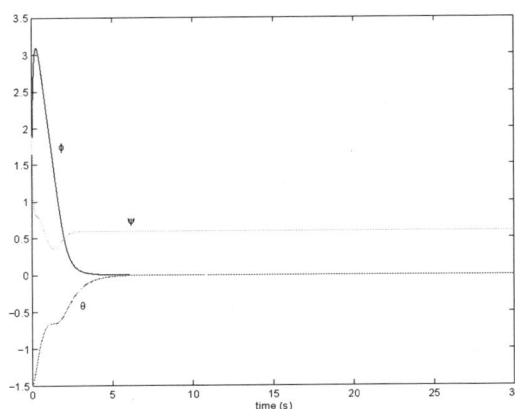

Fig. 3. Comportement of the angles θ, ϕ and ψ

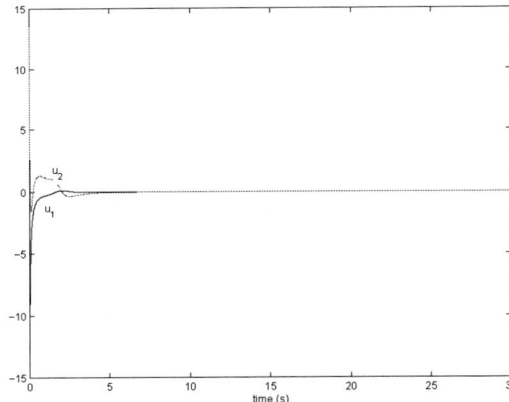

Fig. 4. Trajectories of applied torques u_1 and u_2

these simulations shown the action of the feedback laws (27) and (28) and assured the partial asymptotic stabilization of axi-symmetric rigid spacecraft with respect to four states and one converge.

5. CONCLUSION

In this paper the problem of angular velocity and the attitude stabilization of a axi-symmetric rigid spacecraft are exposed. The main result showed that is impossible to stabilize the system by means of continuous pure state feedback laws the system. In order to overcome this limitations we have proposed the partial asymptotic stabilization approach. This special stabilization is assured by indefinitely differentiable feedback laws, and without restriction on the initial conditions (contrarily to Tsiotras et al. [1995], Tsiotras and Longuski [1993]). Our results improved the Tsiotras (Tsiotras et al. [1995]) feedback laws.

Acknowledgements. The author would like to thank Professor Azgal abichou and Professor Jean-Michel Coron for several interesting discussions and helpful comments. We also thanks the reviewers for useful comments and suggestions.

REFERENCES

B. Bonnard. Contrôle de l'attitude d'un satellite rigide. *R.A.I.R.O Automatique/Systems Analysis and Control*, 16: 85–93, 1982.

R. W. Brockett. Asymptotic stability and feedback stabilization. *Differential geometric control theory, Progress in Math.*, 27: 181–191, 1983.

V. Chellaboina and W. M. Haddad. Teaching time-varying stability theory using autonomous partial stability theory. *Proc. IEEE Conf. Dec. Contr.*, Orlando, FL, pages 3230–3235, December 2001.

J.-M. Coron. On the stabilization of some nonlinear control systems: results, tools, and applications. *Kluwer Academic Publishers*, pages 307–367, 1999.

J.-M. Coron. *Control and Nonlinearity*, volume 136. Mathematical Surveys and Monographs, 1 edition, May 18 2007.

J.-M. Coron and E. Y. Keraï. Explicit feedbacks stabilizing the attitude of a rigid spacecraft with two torques. *Automatica*, 32:669–677, 1996.

J.-M. Coron and L. Rosier. A relation between continuous time-varying and discontinuous feedback stabilization. *J. Math. Systems Estimation and Control*, 4:67–84, 1994.

P. E. Crouth. Spacecraft attitude control and stabilization: Applications of geometric control theory to rigid body models. *IEEE Trans. Autom. Control*, AC, 29(4):321–331, 1984.

A. L. Fradkov, I. V. Miroshnik, and V. O. Nikiforov. *Nonlinear and adaptive Control of Complex systems*. Kluwer Academic, 2001.

W. Haddad, V. Chellaboina, and S. Nersesov. A unification between partial stability of state-dependent impulsive systems and stability theory for time-dependent impulsive systems. In *Proc. Amer. Contr. Conf*, pages 4004–4009, June 2003.

C. Jammazi and A. Abichou. Partial stabilizability of cascaded systems. Applications to partial attitude control. In *Proc in 3rd Inter. Conf on information in control, Automation and Robotic*, volume 2, pages 650–656, Portugal, 1-5 August 2006.

C. Jammazi and A. Abichou. Partial stabilizability of an underactuated autonomous underwater vehicle. In *Proc in International Conference "System Identification and Control Problems" SICPRO '07*, volume 7, pages 976–986. Moscow Institute of Control, January 29- February 1 2007.

E. Keraï. Analysis of small time local controllability of rigid body model. *In IFAC Conf. on System Structure and Control*, pages 645–650, 1995.

H. K. Khalil. *Nonlinear Systms*. Prentice Hall, third edition, 2002.

Y. Lin, E. D. Sontag, and Y. Wang. Input to state stabilizability for parameterized families of systems. *Intern. J. Robust and Nonlinear Control*, 5:187–205, 1995.

P. Morin. *Stabilisation de systèmes non linéaires critiques et application la commande de véhicules*. Habilitation à diriger des recherches, L'université de Nice-Sofia Antipolis, 2004.

P. Morin and C. Samson. Time-varying exponential stabilization of a rigid spacecraft with two control torques. *IEEE Trans. on Automatic Control*, 42:528–534, 1997.

P. Morin and C. Samson. Control of nonlinear chained systems. from the Routh-Hurwitz stability criterion to time - varying exponential stabilizers. *IEEE, Trans. on Automatic Control*, 45:141–146, 2000.

P. Morin, C. Samson, J-B. Pomet, and Z-Ping Jiang. Time-varying feedback stabilization of the attitude of a rigid spacecraft with two controls. *Systems and Control Letters*, 25: 375–385, 1995.

N. Rouche, P. Habets, and P. Laloy. *Stability theory by Lyapunov's direct method*. Applied Mathematical Sciences, springer edition, 1977.

P. Tsiotras. New control laws for the attitude stabilization of rigid bodies. In *13th IFAC Symposium on Automatic Control in Aerospace*, Palo Alto, California, pages 316–321, Sept. 12-16 1994.

P. Tsiotras, M. Corless, and J. M. Longuski. A novel approach to the attitude control of axi-symmetric spacecraft. *Automatica*, 31(8):1099–1112, 1995.

P. Tsiotras and J. M Longuski. On attitude stabilization of symmetric spacecraft with two control torques. In *Proceedings, American Control Conference*, San Francisco, California, pages 46–50, June 2-4 1993.

P. Tsiotras and J. Luo. Control of underactuated spacecraft with bounded inputs. *Automatica*, 36:1153–1169, 2000.

V. I. Vorotnikov. *Partial Stability and Control*. Birkhäuser, 1998.

J. Zabczyk. *Mathematical control theory*. Birkhäuser, Boston, Basel, Berlin edition, 1992.

A. L. Zuiev. On Brockett's condition for smooth stabilization with respect to part of variables. In ECC, editor, *Proc. European control conference ECC'99, (Karlsruhe, Germany, 1999)*, pages 1–6, 1999.

A. L. Zuyev. On partial stabilization of nonlinear autonomous systems: Sufficient conditions and examples. In *Proc. of the European Control Conference, Porto (Portugal)*, pages 1918–1922, 2001.

VLSI Cells Placement Using the Neural Networks

Hacène AZIZI (*), Lamri Zouaoui (**), Salah Mokhnache (***)
Université ferhat Abbas, Faculté des Sciences
Laboratoire Optoélectronique et Composants
(*) Email : Azizi_hacene@yahoo.fr
(**) Email : Lam_zou@yahoo.fr
(***) Email : M_Salsh@yahoo.fr

Abstract: The artificial neural networks have been studied for several years. Their effectiveness makes it possible to expect high performances. The privileged fields of these techniques remain the recognition and classification. Various applications of optimization are also studied under the angle of the artificial neural networks. They make it possible to apply distributed heuristic algorithms. In this article, a solution to placement problem of the various cells at the time of the realization of an integrated circuit is proposed by using the KOHONEN network.

Key words : neural Networks, Cells Placement, KOHONEN networks

1. INTRODUCTION

The development of the CAD tools made possible the whole automation of the design process. This is possible with the use of the style of design (logical gates, cells standards, ...) coupled with an effective whole of programs (Software) for the placement operation and automatics routing.

The problem of the placement [3] [6] can be defined as follows. Either an electric circuit made up of modules with preset inputs/outputs, it is necessary to build a layout indicating the positions of the modules so that the estimate length of connections and surface of the circuit are minimized. The entry of the problem corresponds to the description of the modules (size) and the file of connection (netlist) describing the interconnections enter the terminals of these modules. The output is a list of co-ordinates (x, y) of all the modules. Figure 1, gives an example of placement where the diagram of the circuit of figure 1 (a) is placed, according to the style of design standard cells, is shown in figure 1 (b).

The principal goal of the algorithms of placement is to minimize the total surface of the circuit as well as the estimate length of connections. For an acceptable placement, another criterion is to be taken into account is the fact that it should be realizable physically [1] [8]; i.e., (1) the modules should not cover, (2) they should not exceed the borders of the circuit, (3) standard cells should be confined with the lines in predetermined positions, and (4) the gates should be assigned to the lines in points of grid [5]. In practice, it is a question of defining a performance index or an objective function, which includes/understands the sum of all these criteria, the estimate length of connection, surfaces total circuit, and so on. The goal of the algorithm of placement is to determine a placement with a possible minimum of cost [9] [4].

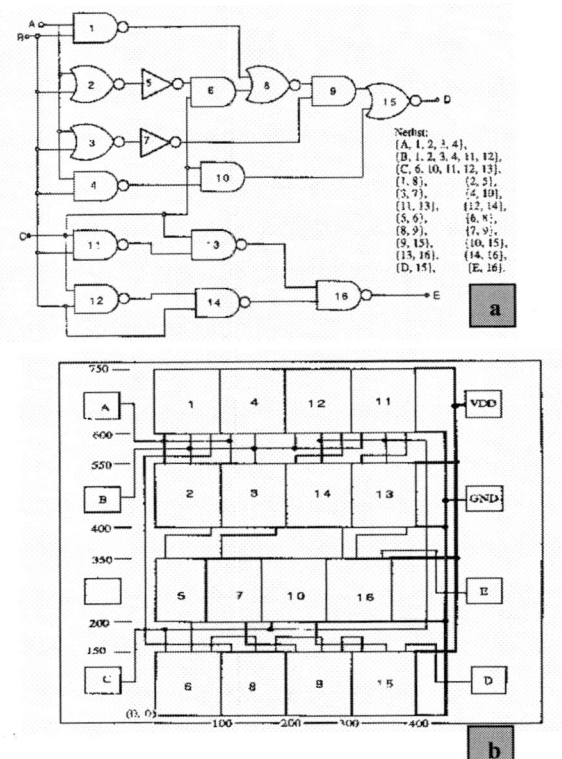

Fig. 1. Cells Placement Problem

(a) Input: file of connection
(b) Output: Positions (x,y) of the modules.

Studies showed that a particular type of networks of artificial neural networks, the KOHONEN network [5], can be used to position cells of identical size and form on a plan of mass without rails of connections. In this article, we presented an algorithm of placement of cells VLSI with simplifying assumptions.

2. ANALOGY PLACEMENT SELF-ORGANIZATION

In this part, we will present the analogy which exists between the problem of the cells placement and the the self-organization problem in neural networks.

Definition 1: sm_{ij} the similarity between the output neuron i and the output neuron j, and d_{ij} the distance (Difference) between i and j, the Problem of the self-Organization consists in minimizing the following cost function:

$$C_{SO=} \sum_{i=1}^{m} \sum_{j=i+1}^{m} d_{ij} \cdot sm_{ij}$$

Definition 2: cm_{ij} the connectivity between cell i and the cells j, and d_{ij} the distance between cell i and j then, the Problem of the placement consists in minimizing the following cost function:

$$C_{WL=} \sum_{i=1}^{m} \sum_{j=i+1}^{m} d_{ij} \cdot cm_{ij}$$

It is noticed that the two functions present certain similarities and consequently the cell placement problem can be dealt with like a Self-organization problem.

We will use the KOHONEN networks to solve this problem and we will proceed as follows:
- a network of N neurons arranged on a grid of dimension 2, where each neuron represents a possible site for a cell.
- a input neuron e connected to each neuron of the network.
- a stimulus vector at time T $x(t) = (x_1(t), x_2(t), ..., x_c(t))$
- a synaptic weights vector which characterize a neuron i at time t $w_i(t) = (w_{i1}(t), w_{i2}(t), ... w_{ic}(t))$.

3. ALGORITHM

The problem is to minimize the function of energy, by minimizing the contribution of each Ck cell. The Function of energy is

$$\sum_{k \in K} \sum_{l \in K} d_{kl} cm_{lk}$$

Contribution of the cell is

$$c_k = \sum_{l \in K} d_{kl} cm_{lk}$$

The general idea of the algorithm is to place each one of these cells at the best possible position, to optimize this sum. First of all, there is a empty neural network, after a pre-treatment and initializations, a cell is presented to the network to be placed.

An iterative loop is started, with each iteration, a cell which is presented until one reaches a stable state, i.e. until each cell is associated with a neuron which produces the smallest contribution. Therefore the algorithm proceeds in the 4 following stages [2]:

3.1 PRE-TREATMENT

The goal of the pre-treatment is to transform the matrix of connections in order to reveal indirect connections which exist between cells which are not connected directly.

Let us suppose that one has a cell c1 which is strongly connected to two other cells c2 and c3 which is not connected between them (figure 2), then by transitivity if the cells are so close to c1 then they are inevitably close one to the other. There is thus an implicit connection between the cell c2 and c3.

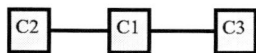

Fig.2 Implicit connection between the cells

<u>Algorithm</u>
Begin
For k=0; k < nmb_cel; k++
For l=k+1; l < nmb_cel; l++
max_d = 0
If cm_{kl} = 0 Then
For m =0; m < nmb_cel; m++
If $cm_{ml} \neq 0$ and $cm_{ml} \neq 0$ Then dist(k,l) = max(cm_k)/cm_{km} + max(cm_m)/cm_{ml}
If max_d < dist (k,l) Then max_d = dist (k,l)
End_If
End_If
End_For
cm_{kl} = max(cm_k)/max_d
End_If
End_For
End_For
End.

3.2 CREATION OF THE INPUT SEQUENCE

The creation of the input sequence consists in placing all the cells, the ones after the others while presenting after each cell, the cells which are connected there in the order descending of the intensity of connection. The algorithm of the input sequence is as follows:

<u>Algorithm</u>
Begin
k = {1,...,c}
k = Random (nmb_cel)
For each cell K to make
S_k = < k,l_1,l_2,l_3,...,l_L> tel que $cm_{kl1} \geq cm_{kl2} \geq ... cm_{klL} > 0$
Sequence = Sequence $\subset S_k$
End.

Where the symbol ⊂ represents the concatenation of two sequences.

3.3 INITIALIZATION OF THE SYNAPTIC WEIGHTS MATRIX

The synaptic weights matrix is initialized by positive low values which one will call Rnd_val.

3.4 ITERATION

It is the most important stage in this algorithm, we have three principal phases [4].

a. PRESENTATION OF A STIMULUS
We will define time as being a complete treatment of the input sequence. The presentation of a stimulus is done as follows:
k ← seq[1]; beginning time
k ← seq[t]; at time t

b. CALCULATION OF THE ANSWERS AND SEARCH FOR A WINING NEURON
The minimization of the energy function is obtained by combined effect of the computation of the answers and the law of adaptation of the synaptic weights. A good placement for a cell ck minimizes the contribution of this cell i.e. contribution of ck

$$\sum_{l \in K} d_{kl} \, cm_{lk}$$

The response of a neuron must have the following form:
Response of the neuron nor to the stimulus ck is:

$$\sum_{l \in K} w_{il} \, cm_{lk}$$

With w_{il} inversely proportional to d_{il}

The response of each neuron is obtained then by scalar product between the synaptic weights vector neuron and the stimulus vector presented in input. A neuron ni can give the same response for two different cells ck and ck', therefore it is necessary to determine which of these two cells will occupy this neuron. To be done, we will put these two cells in competition. The wining cell is given by comparing the response of the neuron ni with the stimuli ck and ck', abstraction made of the interconnections between these two cells.

The cell which gives the strongest answer is placed in ni, the losing cell is removed from the network, and synaptic weights is re-initialized with rnd_val.

Algorithm
Response calculation of each neuron:
For i = 1 to N Do
$$rik = \sum_{l \in K} w_{lk} \, cm_{lk}$$
End_For

The file of the neurons is ordered by descending order of each neuron response of i to the cell k In the case of several neurons having the same response for the cell k, we take initially the neuron occupied by the cell k, then the free neurons, and finally the neurons occupied by the other cells, to prevent that the cell unnecessarily does not change a position which it occupies or to pursue of the cells, whereas there is a free position having an equal answer.

c. ADAPTATION OF THE WEIGHTS SYNAPTIC
The neurons close to the wining neuron, will have a value of weight higher than that of the neurons distant from this one, and this after the adaptation. After each determination of the wining neuron, the weights vector is updated so as to increase the wining neuron response to the stimulus, like that of its neighbours. The adaptation of the weights is done as follows:

Algorithm
Begin
i: neurons set
For i=1 to N Do
If i≠γ Then
$w_{ik} = w_{ik} + \text{cost} \; \Gamma_{i\gamma} \, (\max \, (cm_k)/d_{i\gamma} - w_{ik})$; /* I is not a wining neuron */
If i = γ then /* I is a wining neuron */
$w_{ik} = w_{ik} + \text{cost} \; \Gamma_{i\gamma} \, nb\_p/nmb\_cel \, (\max(CM_k) - w_{ik})$;
End.

Nb_p : A number of placed cells.
Cost : static gain (parameter of low value of about 0.001).
Γ : Topological function of vicinity.
Γ 1, si dis(n_i,n_j) ≤ vois_min
Γ_{ij} : {
        0, si dis(n_i,n_j) > vois_min

vois_min: minimum vicinity of a neuron ni with which is associated the cell ck as being the smallest vicinity which can contain all the cells connected to the cell k, it is equal in general to 2.

4. STOPING CONDITIONS
The iterative loop stops when the placement stabilizes, after having to try to place each cell at least once. This situation is reached before the convergence of the matrix of the weights is not total. When a fixed position is allotted to each cell, the solution of the placemen problem is found. The effect of the later iterations would be to modify the level of attraction of a cell on its neighbours.

5 RESULTS AND TESTS

Let us consider an example of 9 cells illustrated by figure 3, and see the behaviour of the algorithm with this example.

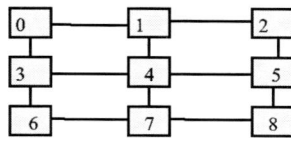

Fig. 3 Circuit with 9 cells

The interconnections between cells are given by the following matrix of connections (table 1):

	0	1	2	3	4	5	6	7	8
0	0	1	0	1	0	0	0	0	0
1	1	0	1	0	1	0	0	0	0
2	0	1	0	0	0	1	0	0	0
3	1	0	0	0	1	0	1	0	0
4	0	1	0	1	0	1	0	1	0
5	0	0	1	0	1	0	0	0	1
6	0	0	0	1	0	0	0	1	0
7	0	0	0	0	1	0	1	0	1
8	0	0	0	0	0	1	0	1	0

Table 1: Matrix connection for circuit fig.1

The optimal solution is known and corresponds to the diagram of figure 3. The matrix of connections after the pre-treatment becomes (table2).

0.0	1.0	0.5	1.0	0.5	0.0	0.5	0.0	0.0
1.0	0.0	1.0	0.5	1.0	0.5	0.0	0.5	0.0
0.5	1.0	0.0	0.0	0.5	1.0	0.0	0.0	0.5
1.0	0.5	0.0	0.0	1.0	0.5	1.0	0.5	0.0
0.5	1.0	0.5	1.0	0.0	1.0	0.5	1.0	0.5
0.0	0.5	1.0	0.5	1.0	0.0	0.0	0.5	1.0
0.5	0.0	0.0	1.0	0.5	0.0	0.0	1.0	0.5
0.0	0.5	0.0	0.5	1.0	0.5	1.0	0.0	1.0
0.0	0.0	0.5	0.0	0.5	1.0	0.5	1.0	0.0

Table 2 Matrix connection after the pre-treatment.

As for the input sequence of the cells it will take the following form:

Input sequence =   { 4, 1, 3, 5, 7, 0, 2, 6, 8,
2, 1, 5, 0, 4, 8,
1, 0, 2, 4, 3, 5, 7,
7, 4, 6, 8, 1, 3, 5,
0, 1, 3, 2, 4, 6,
5, 2, 4, 8, 1, 3, 7,
6, 3, 7, 0, 4, 8,
3, 0, 4, 6, 1, 5, 7,
8, 5, 7, 2, 4, 6.   }

At the end of the study of the fundamental case, a first conclusion, the neuronal approach is applicable for the placement of V.LS.I cells. The property of classification of the KOHONEN networks is adapted to this type of problems. The results obtained are now acceptable, though the computing time is too important for problems of big sizes (table 3)

Cells Number	network	Iteration Number
100	20 x 20	211
144	24 x 24	302
256	32 x 32	532
289	34 x 34	600

Table 3 Results of the tests

We will test our algorithm improved and adapted on several problems, to evaluate its performances (table 4). We will carry out these tests on problems of 25 and 35 cells (figure 4), and finally known a problem of 100 cells analogue to that of 36 cells. Then, we will then carry out a series of tests on a problem which has a form of hypercube (figure 5), on an asymmetrical problem (figure 6).
The network size will be taken as agreed four times larger than the size of the problem

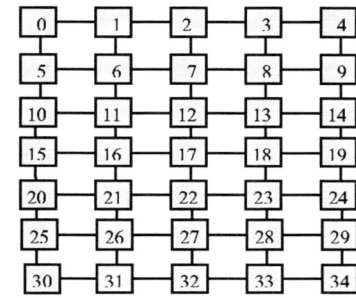

Fig.4 optimal Solution to a problem of 35 cells

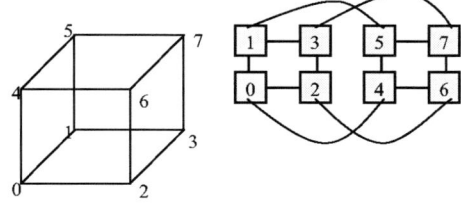

Fig.5 Hyper cube of 3 dimension

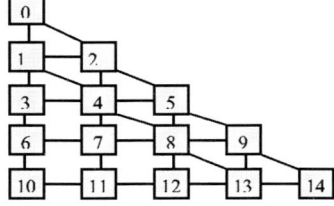

Fig.6 triangular Problem

Cellules	1ère cellule en entrée	Placement Neurone centre	Itérations
25	12	oui	54
36	14	oui	115
100	44	oui	211
Hyper cube (8)	2	oui	34
Asymétrique (15)	5	oui	113

Table 4 Results of the tests

6. CONCLUSIONS

At the end of this work several conclusions can be drawn. With regard to the problem of placement, it should be said that the fact of using the neural networks to optimize the cells placement proves to be an interesting and "original" idea. Therefore, the neuronal approach is applicable for the optimization of the VLSI cells placement.

In spite of the very restrictive assumptions posed, the starting algorithm gave only unsatisfactory results. We improved this work and of the tests on true circuits can be considered. As prospect, the pretreatment must take into account all the possibilities of placement and establishes bonds between cells only if those contribute to structure the unit of the circuit. Also, envisage tests on benchmarks circuits

REFERENCES

[1] Pinaki Mazumder; *VLSI Cell Placement Techniques.* ACM Comput. Surv. 23(2) pp 143-220, 1991

[2] Mehmet Can Yildiz, Patrick H. Madden; *Global objectives for standard cell placement.* ACM Great Lakes Symposium on VLSI 2001 68-72

[3] Guofang Nan, Minqiang Li, Dan Lin, Jisong Kou, *Adaptative Simulated Annealing for Standard Cell Placement.* ICNC (3) pp 943-947. 2005

[4] S. Goto. *An Efficient Algorithm for the Two-Dimensional Placement Problem* in Electrical Circuit Layout. IEEE Transactions on Circuits and Systems, CAS-28, Jan. 1981.

[5] M. A. Breuer, "*Min-Cut Placement*", Journal of Design Automation and Fault Tolerant Computing, Vol. 1, No. 4, 1977, pp. 343-362.

[6] W. Swartz and C. Sechen. *New Algorithms for the Placement and Routing of Macro Cells.* In ICCAD, 1990.

[7] K. Shahookar and P. Mazumder, "*A Genetic Approach to Standard Cell Placement Using Meta-Genetic Parameter Optimization,*" IEEE Trans. on Computer-Aided Design, Vol. 9, No. 5, May 1990, pp. 500-513.

[8] Kohonen T; *Self-Organization and Associative Memory.* 1989. 3rd Edition. Springer-Verlag. NY USA

[9] Hemani A; Postula A; *Cell placement by self-organization.* Neural Networks.Vol.3,. pp. 377-383, 1990.

3D Curves With a Prescribed Curvature and Torsion for a Flying Robot

Yasmina BESTAOUI*

*Informatics, Integrative Biology and Complex Systems, University of Evry
40 Rue de Pelvoux, 91020 Evry Cedex, France (e-mail: bestaoui@iup.univ-evry.fr).

Abstract: The objective of this paper is to generate a desired flight path to be followed by an flying robot. A curve with discontinuous curvature and torsion is not appropriate for smooth motions for any vehicle architecture. Three different classes of curves are presented. First, constant curvature and torsion followed by a linear variation versus the curvilinear abscissa then a quadratic variation. Finally, the problem of maneuvers between two trim helices of different curvature and torsion is tackled with.

Keywords: Flying Robot, Path Generation

1. INTRODUCTION

Recent advances in guidance technologies have enabled some flying robots to execute simple mission tasks without human interaction. Many of these tasks are pre-planned using reconnaissance or environment information. The actual trend is towards more autonomy. An autonomous flying robot will be suitable for applications like search and rescue, surveillance and remote inspection. Task planning and safe trajectory solutions are essential to the survivability and success of an autonomous system [1-5, 7, 9-10, 14].

This article is concerned with methods of computing a trajectory in 6 degrees of freedom space that describes the desired motion. Classical differential geometry curve theory is a study of 3D space curves with orthogonal coordinate systems attached to moving points on the space curve. It is well known from the local theory of curves in differential geometry, that if two different curves parameterized by the curvilinear abscissa have the same torsion and curvature functions, they are the same curves up to a rigid motion (translation and/or rotation). The intrinsic representation of curvature and torsion is thus used in the sequel.

Classically, in motion planning and generation, methods such as continuous optimization and discrete search are sought [2-5]. [7] presented a randomized motion planning algorithm by employing obstacle free guidance system as local planners in a probabilistic roadmap framework. In [15], path planning of autonomous fixed wing aircraft is based on a learning real-time A* search algorithm, considering only the motion on a horizontal plane. A family of trim autonomous helicopter trajectories in level flight were used in all these references to construct paths. As stated in [1], "Despite these results, significant research efforts are still needed to advance the state of the art of trajectory planning for autonomous aerospace vehicles." In the literature, the proposed trajectories are based on classical trim helices and maneuvers are not considered. In this paper, a trajectory planning method is presented for a generic flying robot where maneuvers of first and second order are introduced. The reference [13] discussed the shortest 3D paths with a prescribed curvature bound, in a theoretical way.

This paper consists of 4 sections. Section 2 introduces the basic concepts. Path generation results are discussed in Section 3 and finally some concluding remarks are given in section 4.

2. BASIC CONCEPTS

As in this study, we are interested by the curvilinear abscissa s instead of the time, let's consider the curve $C(s)$ representing the motion of this vehicle in $\Re^3$, the tangent to this curve is

$$T(s) = C'/\|C'\| \qquad (1)$$

This tangent is constrained to have unity norm ('represents derivation versus the curvilinear abscissa s). Now taking the derivative of the above equation with respect to s, gives

$$K(s) = \frac{\|C' \times C''\|}{\|C'\|^3} \qquad (2)$$

where $K(s)$ is called the curvature of the motion. Curvature measures how quickly the curve is pulling away from the tangent:

$$T'(s) = K(s)n(s) \qquad (3)$$

where $n(s)$ is the unit normal vector to the curve $C(s)$. The binormal to the curve is

$$b(s) = T(s) \times n(s) \qquad (4)$$

Let the torsion of the motion be

$$\tau(s) = \frac{(C' \times C'').C'''}{\|C' \times C''\|^2} \qquad (5)$$

it measures how quickly the curve is pulling out of the plane defined by $n(s), T(s)$. Since $T(s), n(s), b(s)$ are all orthogonal to each other, the matrix with these vectors as its columns is an element of $SO(3)$, thus

$$\phi(s) = (T(s) \quad n(s) \quad b(s)) \in SO(3) \quad (6)$$

$$g(s) = \begin{pmatrix} T(s) & n(s) & b(s) & C(s) \\ 0 & 0 & 0 & 1 \end{pmatrix} \quad (7)$$

In the above equations, K, τ completely determine the frame behavior. At this point, it is worth mentioning the fundamental existence and uniqueness theorems of 3D curves. These theorems are proved as a consequence of the fundamental theorems of ordinary differential equations [8, 10-12]:

<u>Theorem 1</u> (existence) Let $K(s)$ and $\tau(s)$ be continuous functions of a real variable s then there exists a curve $C(s)$ for which $K(s)$ is the curvature function, $\tau(s)$ is the torsion function and s a natural parameter.

<u>Theorem 2</u> (uniqueness) If two curves $C_1(s)$ and $C_2(s)$ with the same curvature functions $K_1(s)$ and $K_2(s)$ and torsion functions $\tau_1(s)$ and $\tau_2(s)$ then $C_1(s)$ and $C_2(s)$ can be made to coincide by an appropriate rotation and translation.

The shape of a 3D curve can be completely captured by its curvature and torsion functions. Hence let's consider $K(s)$ and $\tau(s)$ to be a set of intrinsic and complete shape features of the curve $C(s)$. The length of the Darboux vector, also called total curvature, includes both of the above features

$$\Omega(s) = \tau(s)T(s) + K(s)b(s) \quad (8)$$

Ω indicates how the entire frame rotates, making it the measure of the structural variation in C. The entire frame rotates about Ω at the angular rate of $\|\Omega\|$.

3. PATH GENERATION

The classical approach in dealing with Frenet-Serret frame is to calculate the curvature and the torsion from a given path [2]. In this paper, the opposite point of view is taken. Let's propose the curvature K and torsion τ as continuous functions of the curvilinear abscissa s then find the corresponding path. Polynomial functions are a good example of continuous easy to deal with functions. So, let's take the torsion and curvature as :

$$\tau(s) = \sum_{i=0}^{n} a_i s^i \quad K(s) = \sum_{j=0}^{m} b_j s^j \quad (9)$$

n and m represent respectively the order of the torsion and curvature polynomials, we restrain our study to $0 \leq n, m \leq 2$.

The differential equation [8] describing the evolution of ϕ (Frenet Formulas for parameterized curves in $\mathbb{R}^3$ for a unit speed curve) can be written as :

$$\frac{d\phi(s)}{ds} = \phi(s)\hat{\sigma} \quad (10)$$

where

$$\sigma = (\tau \quad 0 \quad K)^T \quad (11)$$

Integrating this matrix differential equation gives

$$\phi(s) = \exp(\hat{\varpi}(s)) \in SO(3) \quad (12)$$

where

$$\varpi(s) = \left(a_{-1} + \sum_{i=0}^{n} \frac{a_i s^{i+1}}{i+1} \quad 0 \quad b_{-1} + \sum_{j=0}^{m} \frac{b_j s^{j+1}}{j+1} \right)^T \quad (13)$$

Using Rodrigues Formula to write a closed form solution

$$\phi(s) = I + \frac{\sin\|\varpi(s)\|}{\|\varpi(s)\|}\hat{\varpi} + \frac{1-\cos\|\varpi(s)\|}{\|\varpi(s)\|^2}\hat{\varpi}^2 \quad (14)$$

The Tangent vector of the Frenet-Serret frame is given by

$$T(s) = \begin{pmatrix} 1 - \varpi_3^2(s) \frac{1-\cos\sqrt{\varpi_1^2(s)+\varpi_3^2(s)}}{(\varpi_1^2(s)+\varpi_3^2(s))} \\ \varpi_3(s) \frac{\sin\sqrt{\varpi_1^2(s)+\varpi_3^2(s)}}{\sqrt{(\varpi_1^2(s)+\varpi_3^2(s))}} \\ \varpi_1(s)\varpi_3(s) \frac{1-\cos\sqrt{\varpi_1^2(s)+\varpi_3^2(s)}}{(\varpi_1^2(s)+\varpi_3^2(s))} \end{pmatrix} \quad (15)$$

while the normal vector is eq 1

$$n(s) = \begin{pmatrix} -\varpi_3(s) \frac{\sin\sqrt{\varpi_1^2(s)+\varpi_3^2(s)}}{\sqrt{(\varpi_1^2(s)+\varpi_3^2(s))}} \\ 1-(\varpi_1^2(s)+\varpi_3^2(s))\frac{1-\cos\sqrt{\varpi_1^2(s)+\varpi_3^2(s)}}{(\varpi_1^2(s)+\varpi_3^2(s))} \\ \varpi_1(s) \frac{\sin\sqrt{\varpi_1^2(s)+\varpi_3^2(s)}}{\sqrt{(\varpi_1^2(s)+\varpi_3^2(s))}} \end{pmatrix} \quad (16)$$

and the binormal vector is

$$b(s) = \begin{pmatrix} \varpi_1(s)\varpi_3(s) \frac{1-\cos\sqrt{\varpi_1^2(s)+\varpi_3^2(s)}}{(\varpi_1^2(s)+\varpi_3^2(s))} \\ -\varpi_1(s) \frac{\sin\sqrt{\varpi_1^2(s)+\varpi_3^2(s)}}{\sqrt{(\varpi_1^2(s)+\varpi_3^2(s))}} \\ 1-\varpi_1^2(s) \frac{1-\cos\sqrt{\varpi_1^2(s)+\varpi_3^2(s)}}{(\varpi_1^2(s)+\varpi_3^2(s))} \end{pmatrix} \quad (17)$$

Doing a parallel with planar curves, three cases are studied
1. constant curvature and torsion (planar curve : line or circular path)
2. curvature and torsion linear functions of s (planar curve : cornu spiral or clothoid)
3. curvature and torsion quadratic function of s (planar curve : cubic spiral [9])

The three curve classes are useful for motion planning for 3D autonomous vehicles (aerial or underwater). However, once these trajectories characterized, we have to choose those respecting the kinematical and dynamical constraints. As dynamical constraints depend on the dynamic model, an admissible trajectory for an airplane robot may not be an admissible trajectory for a helicopter or an airship robot, and vice-versa.

Constant Curvature and Torsion

The trajectories characterized by constant curvature and torsion are in general cylindrical helices (see figure 1). Particular cases are straight lines and circles arcs. Cylindrical helices are classically known in aeronautical science to be trim conditions. A trimmed flight condition is defined as one in which the rate of change of magnitude of the flying robot state vector is zero (in the body fixed frame) and the resultant of the applied forces and moments are zero. In a trimmed maneuver, the flying robot will be accelerated under the action of non zero resultant aerodynamic and gravitational forces and moments: effects such as centrifugal and gyroscopic inertial forces and moments will balance these effects.

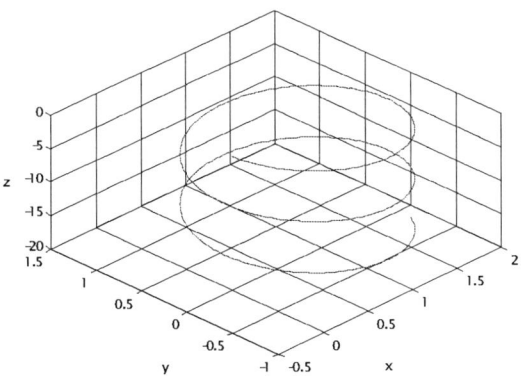

Figure 1 : A cylindrical Helix

The following results are found for constant curvature $b_0 = \alpha$ and torsion $a_0 = \beta$. The position is given by

$$x(s) = \frac{\beta^2 s\sqrt{\alpha^2+\beta^2} + \alpha^2 \sin\left(s\sqrt{\alpha^2+\beta^2}\right)}{(\alpha^2+\beta^2)\sqrt{\alpha^2+\beta^2}}$$

$$y(s) = \alpha \frac{-1 + \cos\left(s\sqrt{\alpha^2+\beta^2}\right)}{(\alpha^2+\beta^2)}$$

$$z(s) = \alpha\beta \frac{s\sqrt{\alpha^2+\beta^2} + \sin\left(s\sqrt{\alpha^2+\beta^2}\right)}{(\alpha^2+\beta^2)\sqrt{\alpha^2+\beta^2}}$$

(18)

while the tangential vector is obtained by :

$$T(s) = \begin{pmatrix} \dfrac{\beta^2 + \alpha^2 \cos\left(s\sqrt{\alpha^2+\beta^2}\right)}{(\alpha^2+\beta^2)} \\ -\alpha \dfrac{\sin\left(s\sqrt{\alpha^2+\beta^2}\right)}{(\alpha^2+\beta^2)} \\ \alpha\beta \dfrac{1 - \cos\left(s\sqrt{\alpha^2+\beta^2}\right)}{(\alpha^2+\beta^2)} \end{pmatrix}$$

(19)

Curvature and torsion linear functions of the curvilinear abscissa s

In this subsection and the following, the proposed paths can be used as maneuvers between two trim helices of different curvature and torsion. Let's consider a curvature and a torsion given respectively by :

$$\kappa(s) = \alpha s \qquad \tau(s) = \beta s \qquad (20)$$

All integration constants are considered to be zero.

The position is given for $\alpha = 1; \beta = 1$ by

$$x(s) = \frac{1}{2} + \frac{\sqrt{\pi}}{4} 2^{3/4} \mathrm{FresnelC}\left(\frac{2^{1/4}}{\sqrt{\pi}} s\right)$$

$$y(s) = \frac{\sqrt{\pi}}{4} 2^{1/4} \mathrm{FresnelS}\left(\frac{2^{1/4}}{\sqrt{\pi}} s\right)$$

$$z(s) = \frac{-\sqrt{\pi}}{4} 2^{3/4} \mathrm{FresnelC}\left(\frac{2^{1/4}}{\sqrt{\pi}} s\right) + \frac{s}{2}$$

(21)

while the tangential vector is obtained by :

$$T(s) = \begin{pmatrix} \dfrac{\alpha^2 + \beta^2 \cos\left(\dfrac{s^2}{2}\sqrt{\alpha^2+\beta^2}\right)}{(\alpha^2+\beta^2)} \\ -\beta \dfrac{\sin\left(\dfrac{s^2}{2}\sqrt{\alpha^2+\beta^2}\right)}{(\alpha^2+\beta^2)} \\ \alpha\beta \dfrac{1 - \cos\left(\dfrac{s^2}{2}\sqrt{\alpha^2+\beta^2}\right)}{(\alpha^2+\beta^2)} \end{pmatrix}$$

(22)

The functions FresnelS and FresnelC represent respectively the sin and cosine integrals

$$FresnelS(x) = \int_0^x \sin\left(\frac{\pi}{2}t^2\right)dt$$
$$FresnelC(x) = \int_0^x \cos\left(\frac{\pi}{2}t^2\right)dt \qquad (23)$$

Figures 2 and 3 present the particular case
$a_{-1} = 0 \quad a_0 = 0 \quad a_1 = 0.5 \qquad a_2 = 0$
$b_{-1} = 0 \quad b_0 = 0 \quad b_1 = 0.1 \qquad b_2 = 0$.

Figure 2 shows the norm of the darboux vector versus s and Figure 3 presents the 3D curve.

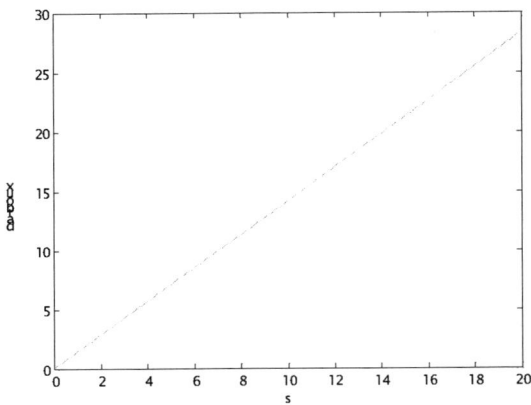

Figure 2 : The norm of darboux vector versus s

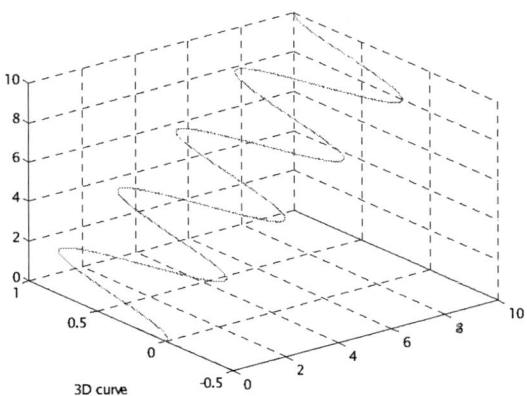

Figure 3 : Path in 3D

Curvature and torsion quadratic functions of the curvilinear abscissa s

Let's consider a curvature and a torsion given respectively by: $\kappa(s) = \alpha s^2 \qquad \tau(s) = \beta s^2 \qquad (24)$

All integration constants are considered to be zero.
The tangential vector is obtained by:

$$T(s) = \begin{pmatrix} \dfrac{\alpha^2 + \beta^2 \cos\left(\dfrac{s^3}{3}\sqrt{\alpha^2 + \beta^2}\right)}{(\alpha^2 + \beta^2)} \\ \beta \dfrac{\sin\left(\dfrac{s^3}{3}\sqrt{\alpha^2 + \beta^2}\right)}{(\alpha^2 + \beta^2)} \\ \alpha\beta \dfrac{1 - \cos\left(\dfrac{s^3}{3}\sqrt{\alpha^2 + \beta^2}\right)}{(\alpha^2 + \beta^2)} \end{pmatrix} \qquad (25)$$

The value of the position can be obtained by a numerical integration method. Figures 4, 5 present the particular case
$a_{-1} = 0 \quad a_0 = 0 \quad a_1 = 0 \quad a_2 = 0.5$
$b_{-1} = 0 \quad b_0 = 0 \quad b_1 = 0 \quad b_2 = 0.1$

Figure 4 shows the norm of Darboux vector versus s, while figure 5 shows the 3D curve.

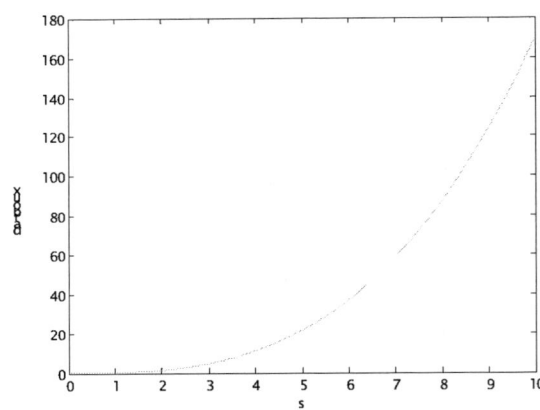

Figure 4: Norm of Darboux vector versus s

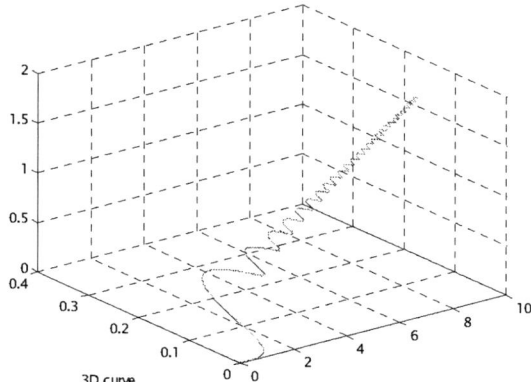

Figure 5 : Trajectory in 3D

Maneuvers Between Two Different Trim Trajectories

The problem of maneuvers between two different trim trajectories is an important problem in motion planning for an flying robot. Let be two cylindrical helices of different curvatures and torsion respectively (κ_i, τ_i) for the initial one and (κ_f, τ_f) for the final one. The length ℓ of the path is supposed known. The easiest way for a maneuver is to propose a linear variation of the curvature and the torsion

$$\begin{aligned} &\mathrm{K}(s) = a_\kappa \left(\frac{s}{\ell}\right) + b_\kappa \\ &b_\kappa = \mathrm{K}_i \quad a_\kappa = \mathrm{K}_f - \mathrm{K}_i \\ &\tau(s) = a_\tau \left(\frac{s}{\ell}\right) + b_\tau \\ &b_\tau = \tau_i \qquad a_\tau = \tau_f - \tau_i \end{aligned} \quad (26)$$

If we adopt a quadratic variation, we can add the constraint of maximum torsion and curvature

$$\begin{aligned} &\mathrm{K}(s) = a_\kappa \left(\frac{s}{\ell}\right)^2 + b_\kappa \left(\frac{s}{\ell}\right) + c_\kappa \\ &c_\kappa = \mathrm{K}_i \quad a_\kappa + b_\kappa = \mathrm{K}_f - \mathrm{K}_i \\ &\tau(s) = a_\tau \left(\frac{s}{\ell}\right)^2 + b_\tau \left(\frac{s}{\ell}\right) + c_\tau \\ &c_\tau = \tau_i \qquad a_\tau + b_\tau = \tau_f - \tau_i \end{aligned} \quad (27)$$

while

$$\mathrm{K}_{max} = -\frac{b_\kappa^2}{4 a_\kappa} + \mathrm{K}_i \qquad \tau_{max} = -\frac{b_\tau^2}{4 a_\tau} + \tau_i \quad (28)$$

The Frenet Serret frame is similar to the generalized stability axis system in atmospheric flight mechanics in which the T vector is along the velocity vector and the N vector is along the normal lift direction of the vehicle. This coordinate system will be used to study the shape of a space curve.

4. CONCLUSIONS

This paper addresses the problem of characterizing continuous paths on the group of rigid body motions in 3D. Paths with given curvature and torsion are investigated. Three particular cases are studied : constant, linear and quadratic variation versus the curvilinear abscissa. Then, maneuvers between two different trim trajectories are proposed.

The role of the trajectory generator is to generate a feasible time trajectory for the flying robot. Once the path has been calculated in the Earth fixed frame, motion must be investigated and reference trajectories determined taking into account actuators constraints (inequality constraints) and the under-actuation (equality constraints) of an flying robot. Two differential algebraic equations must be solved if there is six degrees of freedom and four inputs.

Moreover, it is desirable that the plan makes optimal use of the available resources to achieve the goal optimizing some 'cost' measure: the time required for the execution of the trajectory, its length, the deviation from a reference trajectory, control effort or energy. Flying robots are required to achieve maximum performance in severe environments. This is the subject of our actual research

REFERENCES

[1] G. Avanzini 'Frenet based algorithm for trajectory prediction' AIAA Journal of Guidance, Control and Dynamics, vol. 27, #1, 2004, pp. 127-135.

[2] J. Angeles, A. Rojas, C. S. Lopez-Cajun 'Trajectory planning in robotics continuous path applications' IEEE Journal of Robotics and automation, vol. 4, #4, 1988, pp. 380-385

[3] C. Belta, V. Kumar 'An SVD based projection method for interpolation on SE(3)' IEEE Transactions on Robotics and Automation, vol. 18, #3, 2002, pp 334 –345

[4] Y. Bestaoui, S. Hima 'Some insights in path planning of small autonomous blimps' Archives of Control Sciences, Volume 11 (XLVII), 2001, N° 3-4, pp. 21-49.

[5] Y. Bestaoui, S. Hima, C. Sentouh 'Motion planning of a fully actuated unmanned air vehicle' AIAA Conference on Navigation, Guidance and Control, Austin, Texas, 2003.

[6] A. M. Bloch 'Non holonomic mechanics and control' Springer-Verlag, 2003

[7] N. Faiz, S. Agrawal, R. M. Murray 'Trajectory planning of differentially flat systems with dynamics and inequality' AIAA J. Of Guidance, Control and Dynamics, vol. 24, #2, 2001, pp. 219-227

[8] R. Grimshaw 'Non linear ordinary differential equations' CRC Press, 1993.

[9] Y. Kanayama, N. Miyake 'Trajectory generation for mobile robots' Robotics research, vol. 3, MIT Press, pp. 333-340, 1986

[10] Z. Li, J.F. Canny 'Nonholonomic motion planning' Kluwer Academic Publishers, 1992

[11] R. J. Murray, S. Sastry, Z. Li 'A mathematical introduction to robotic manipulation' CRC Press, 1994

[12] B. O'Neil 'Elementary differential geometry' Academic Press, 1997

[13] H. J. Sussmann 'Shortest 3 dimensional paths with a prescribed curvature bound' 34th IEEE Conference on Decision and Control, New Orleans, 1995, pp. 3306-3312

[14] J. A. Thorpe 'Elementary Topics in Differential Geometry' Springer, 1979

[15] O. A. Yakimenko 'Direct method for rapid prototyping of near optimal aircraft trajectory' AIAA Journal of Guidance, Control and Dynamics, vol. 23, #5, 2000, pp. 865-875

Fuzzy Visual Path Following by a Mobile Robot

A. HAMISSI *, A. BAZOULA**

* Military Polytechnic School P.O. Box 17 – 16111 Bordj El Bahri
Algiers Algeria (e-mail: mariposaa77@yahoo.com).
** Military Polytechnic School P.O. Box 17 – 16111 Bordj El Bahri
Algeirs Algeria (e-mail: abdelouahab.bazoula@gmail.com)

Abstract: We present in this work a variant of a visual navigation method developed for path following by a nonholonomic mobile robot moving in an environment free of obstacles. Only an embedded CCD camera is used for perception.

The integration of perception and action leads us to develop firstly a method of extraction of the useful information from each acquired image, secondly a control approach using fuzzy logic.

Key word: path following, mobile robot, fuzzy control, HSI transformation.

1. INTRODUCTION

Robotic system is a mechanism equipped with means of perception, reasoning and action which enable it to interact with its environment because robotics is a field of research which is at the crossroads of the artificial intelligence, automatic and perception by computer.

We are interested in the autonomous mobile robotics which more specifically aims the autonomy of displacement. We find direct applications in the field of the automobiles or the robotics of service for example.

This work concerns the integration of perception and action in a robotic system. For this we have dealt with the problem of path following using images acquired by a color CCD camera mounted on the robot «Pioneer II», in other words, to deal with the problem of robot control on a reference trajectory using only the visual information. This problem is regarded as an initial stage for other more complex tasks like tracking or navigation in an urban environment.

Several methods of systems control were developed in mobile robotics; we can classify them in two categories: those based on total planning and those using a local control. The global methods, based on a precise knowledge of the robot and environment, have the advantage of proving the existence of an optimal solution making it possible to the robot to achieve its goal and then build a map of free space. But, they have a disadvantage that requires an exact model of the environment, so, we cannot take into account any variation of the world model. The local methods can be considered as reactive methods or actions reflexes which are based only on the interaction robot/environment. In our case the robot has only a reduced vision of the environment; the global methods cannot guarantee the success of the mission in all circumstances, for this, in this work we take into consideration local techniques (action-reflex).

The use of the vision sensor is particularly interesting because of the amount of information which a camera can provide and because of the large variety in tasks that makes it possible to be realized (identification, inspection, localization etc). The realization of our work requires the analysis of the image to extracted useful information, so it will be used for the conception of a controller using fuzzy logic in order to ensure the path following by a nonholonomic robot.

2. PROBLEM STATEMENT

The robot «Pioneer II» is a nonholonomic mobile robot, the trajectory is a marking on the floor, the color CCD camera used is positioned ahead with a tilt angle. As we apply a zoom to include only the close zones with the robot and to make the trajectory more dominant in the acquired image.

Fig.1. Robot « Pioneer II » follows the path

We are in front of the autonomy displacement of a mobile robot problem. It consists in determining at every moment which orders must be sent to the effectors, knowing firstly, the goal to achieve, and the measurements provided by the different sensors. It is a question of determining the relationship existing between perception and action knowing the goals to reach. For that, the robot must follow the diagram corresponding to the paradigm Perceive-Decide-Act [1].

Our approach to resolve this problem is summarized as follows: when the robot is on the trajectory, it must be in the middle of the acquired image or, there would be an error of orientation (noted E_θ) between the middle axis of the image and the reference trajectory. This error is the input variable of the designed fuzzy controller which generates the lateral control (noted ω_r). The translation velocity of the robot is constant (noted v_r).

Our strategy of control have two crucial steps: the first consists in extracting from each acquired image the useful information to the controller, the second is the lateral velocity estimation.

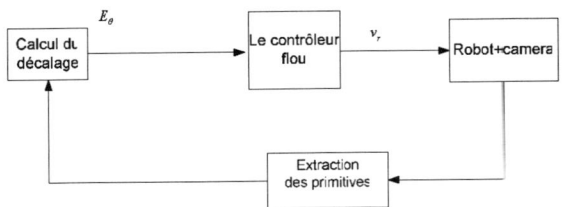

Fig.2. Control diagrams

3. EXTRACTION OF THE USEFUL INFORMATION TO THE CONTROL

The angular error analysis requires the extraction of the starting and the ending points of the trajectory; this is realized by the following algorithm which is based primarily on the detection of edge:

1. Change of space image representation;
2. Image resizing;
3. Edge detection;
4. Edge chaining and extraction of the representative points.

3.1 Change of the image representation space

The acquired images are in RGB format we apply an HSI transformation « Hue-Saturation-Intensity » or color-saturation-brightness given by the following equations [2]:

$$I = \frac{(R+G+B)}{3} \quad (1)$$

$$S = 1 - [\min(R,G,B)]/I \quad (2)$$

$$H = \begin{cases} w, & B \leq G \\ 360 - w, & B > G \end{cases} \quad (3)$$

Where $w = \cos^{-1}\left[\dfrac{[(R-G)+(R-B)]/2}{[(R-G)^2 + (R-B)(G-B)]^{1/2}}\right]$

We use only the component S, where we can easily extract marking on the ground to reduce both the memory capacity occupied during the analysis and time.

 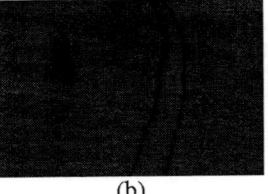

(a) (b)

Fig.3. (a) acquired floor image, (b) image (S) component

In this picture we present the acquired image Fig.3 (a), and only its (S) component after the application of HIS transformation Fig.3. (b).

3.2 Image resizing

To reduce the image size we replace a number of pixels by only one representative which its intensity is equal to the sum balanced to the pixels intensities of its vicinity.

The weighting coefficient is selected so that only the dominant intensity which will be allotted to the representative pixel, the following diagram explains this procedure.

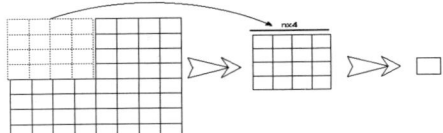

Fig. 4. Image resizing principle

The aim of this operation is to decrease the research field and to be more precise in the trajectory primitive's extraction also to solve the problem due to the ground nature. The result of this phase is given on the Fig. 5.

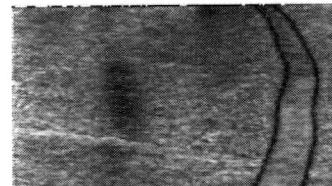

Fig.5. Image resulting from the change of format

3.3 Edge detection

At this stage the image is ready for an edge detection operation more effective and fast made by the optimization approach of Canny as shown in Fig.6.

Fig.6 Image after edges detection

3.4 Edge chaining and extraction of the representative point.
The chaining operation consists in gathering the edges pixels present in matrix form in whole chained lists. Each list is then comparable with a geometrical entity (line segment, conical, etc) we used a function which gathers the pixels in lists lengths is higher or equal to a preset value. The longest list represents marking on the ground where the two ends are the required primitives.

Fig.7. Trajectory primitive (two points)

4. DESING OF THE FUZZY CONTROLLER

Our choice to design a controller based on fuzzy logic is justified by its capacity to treat the unclear, the dubious and vagueness. It tries to imitate the man capacity to decide and act in a relevant way in spite of the knowledge blur. It was introduced with an aim of approaching the human reasoning using an adequate representation of knowledge. Also, the success of the fuzzy control finds is origin from its capacity to translate a control strategy of an operator qualified into a whole linguistic rules "if…then" easily interpretable. The use of the fuzzy control is particularly interesting when we haven't precise mathematical models of the process to control or when this process presents too strong nonlinearities or inaccuracies [3].

A fuzzy controller does not differ so much from a traditional regulator. We finds a block of treatment, a block of entry (quantification, preliminary calculations…) and a terminal block (for the determination of the control). Two additional blocks appear in the case of a fuzzy controller: a block of fuzzyfication and a block of defuzzyfication.

The block of fuzzyfication constitutes the interface between the physical world and inference subsets (Inference Engine) and a base of rules (Rules Base). The block of fuzzyfication will convert the values of the entries into fuzzy subsets. The inference engine will activate the rules whose premises will be checked. Each activated rule will give place to a subset in output. It will remain with the block of defuzzyfication to incorporate those and to extract from it a precise and realizable action control. The following diagram fig.8 explains what we are saying [3], [4].

The most important part in a fuzzy controller is the control set rules connected by the concepts of fuzzy implication and composition, and the fuzzy inference rules. So a fuzzy controller represents an algorithm which can convert a formal strategy (linguistic) of control based on knowledge of an expert into an automatic control strategy. However, there is not a systematic procedure for the fuzzy controller's design due to the great choices parameters diversity and operators in the algorithm of fuzzy control. The general procedure of fuzzy controller's design [3], [4], [5] comprises the stages below:

1. Input and output choice;
2. Membership functions definition;
3. Definition of the fuzzy controller behavior ;
4. Selection of defuzzyfication method.

In our work we will make the design of two fuzzy controllers one with only one entry (angular error) the other has an entry in more (variation of the angular error).

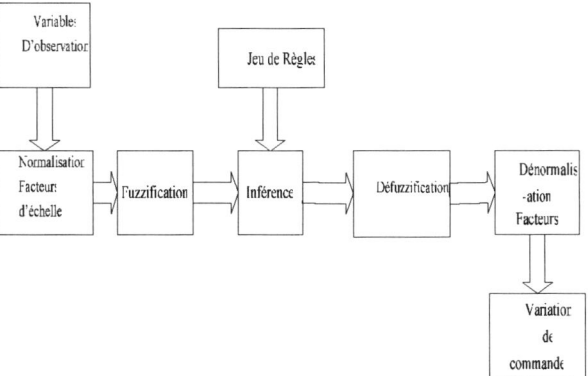

Fig.8. Internal configuration of a fuzzy logic controller

4.1. Design of one input fuzzy controller

Before designing the controller we must define membership functions definition, which consists in specifying the variation field of the variables which is divided into intervals. This distribution, which consists in fixing the number of these values and to distribute them on the field, is made on the base of the knowledge of the system and according to the desired precision.

Our choice of the strong fuzzy distributions at the beginning is justified by the absence of information about the robot behavior with respect to a control which is based, much more, on the experimentation. The orientation error is limited between (- 90°, and +90°), given on Fig.9.

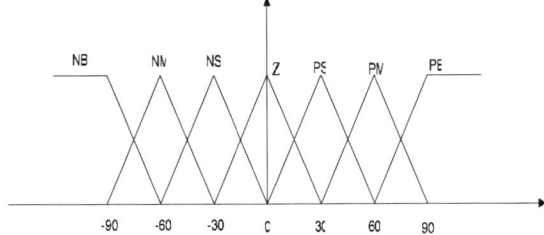

Fig.9. Membership functions of the orientation error

Such as:
NG: Negative Great   **NM**: Negative Means   **NP**: Negative Small   **EZ**: Approximately Zero   **PP**: Positive Small
PM: Positive Means   **PG**: Positive Great.

The output of this controller is the angular velocity it is limited between -1rad/s and +1 rad/s. The repartition of the membership function is done on fig.9.

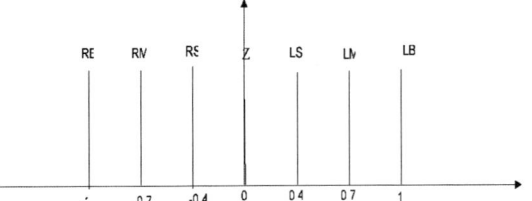

Fig.10. Membership function of angular velocity v_r

Such as:
RB: Right rotation Big   **RM**: Right rotation means
RS: Right rotation Small **Z**: Zero
LS: Left rotation Small   **LM**: Left rotation Means
LB: Left rotation Big

4.2 Fuzzy controller behavior definition

In this stage we develop the base of the rules, to define the behavior expected by the robot. For each value of the input, there is a situation for which we must associate an action for the variable of output. Thus, we obtain the table of the rules situation/corresponding Action:

Situation	NB	NM	NS	Z	PS	PM	PB
Action	LB	LM	LS	Z	RS	RM	RB

Table.1. Roles of the first fuzzy controller

The use of this table is illustrated for this example:
If E_θ is **NB**, **Then** v_r is **LB**

After the definition of membership functions and the rules which defines the behavior of the controller we go to the defuzzyfication method selection, this stage makes it possible to transform the values of fuzzy control towards the real field (physical variables). It means to define precisely which must be the action on the process [4].

This choice is generally conditioned by a compromise between facility of implementation and performance of calculation.

In this work, we used the method of the gravity center of the resulting membership function $\mu_{res}(x_R)$, the coordinate μ of this gravity center gives the value of the order to apply and can be given by the following general relation:

$$u = \frac{\int_{-1}^{1} x_R \mu_{res}(x_R) dx_R}{\int_{-1}^{1} x \mu_{res}(x_R) dx_R}$$

4.3 Two inputs fuzzy controller design

The first alternative of control used is the variation angle; we add now its variation for more robustness, reliability and general information [5]. It raises a fuzzy controller with two inputs, the angular error and its variation to have the lateral control in output.

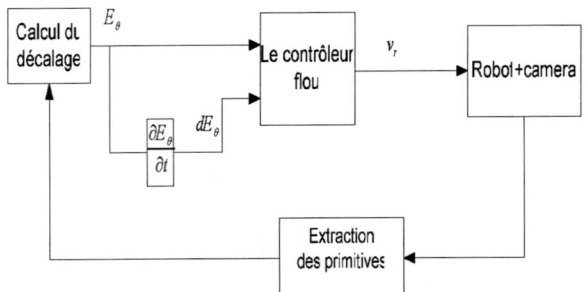

Fig.11. Fuzzy controller with tow input diagram

a- Membership Functions

In this second controller the distribution of the input in error membership functions is the same as the first one. And that of the error variation is a strong fuzzy partition given on fig. 12

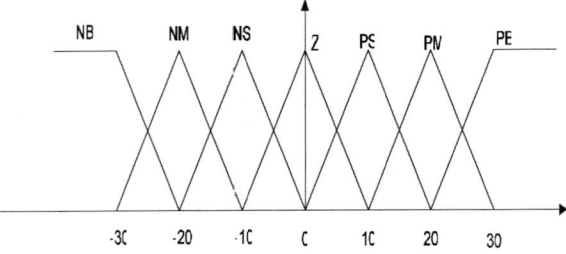

Fig.12. Membership Function of the error variation dE_θ

For the inference engine, we use the same operators and the same method of defuzzyfication as previously, the rules of control are gathered in (table2), it is the resultant of the experimental tests, and the significance of the linguistic terms is also the same one.

E_θ \ dE_θ	NB	NM	NS	ZE	PS	PM	PB
PB	ZE	RS	RM	RB	RB	RB	RB
PM	LS	ZE	RS	RM	RB	RB	RB
PS	LM	LS	ZE	RS	RM	RB	RB
ZE	LB	LM	LS	ZE	RS	RM	RB
NS	LB	LB	LM	LS	ZE	RS	RM
NM	LB	LB	LB	LM	LS	ZE	RS
NB	LB	LB	LB	LB	LM	LS	ZE

Table 2. Roles of The second fuzzy controller

5. EXPERIMENTATION

The experimentation was carried out on the robot «Pioneer II», shown Fig.13. The results are given for each controller we use the same method of the useful information to the control extraction.

In our work parameters controller are tuned by the tray-error method; in each experience we took the results than we do a comparison so we change the different parameters of our controller.

Finally this controller can compensate the failures of the first because the angle variation calculation will give us more information on dynamics of the system, for controlling best.

Fig.13. Robot Pioneer II

5.1. Test1

The results of implementation of the first controller with only one input are presented by tracing its real trajectory and the trajectory executed by the robot.

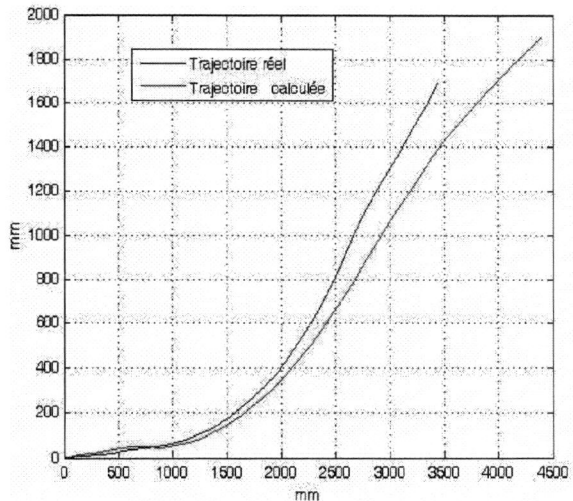

Fig.14. Real trajectory and the trajectory executed by the robot

5.2. Test2

We will test the second controller designed at the base of the calculated angle and its variation, the following figure shows the behavior of the robot.

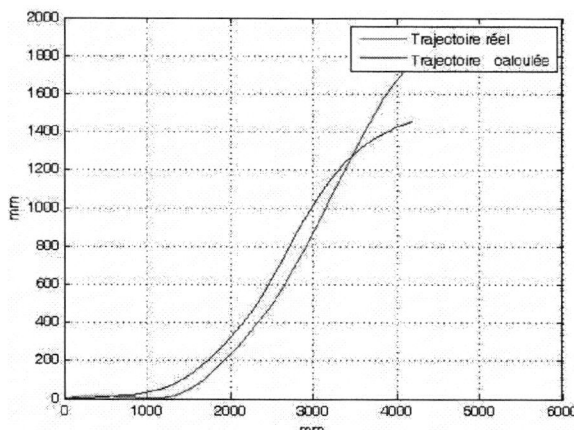

Fig.15. Real trajectory and the trajectory executed by the robot.

6. CONCLUSION

Work presented is in the field of the nonholonomic mobile robots fuzzy control. Our objective was to control robot «Pioneer II» on a reference trajectory, we used only one embedded color camera like tool for perception.

The experimental results obtained are satisfactory; nevertheless, we can always fear that the rules deduced from a simple human expertise are more or less largely under optimal. We propose to use on line optimization methods in order to adjust the distributions of the membership functions.

REFERENCES

[1]. R.A. Brooks, « A Robust Layered Control System for a Mobile Robot», IEEE J. of Robotics and Automation, RA-2, PP.14-23, April 1986.

[2]. Chengjun Ding et al «Mobile Robot's Road Following Based on Color Vision and FGA Control Strategy », School of Mechanical Engineering, Hebei University of Technology, China, 2006

[3]. A. Benmakhlouf « Contrôleur Flou pour la navigation d'un robot mobile d'intérieur » Thèse de Magister, laboratoire d'Electronique Avancée (LEA), Algeria Batna, 2006

[4]. Gregor Klancar, Drago Matko, Saso Blazic «Mobile Robot Control on a Reference Path» Proceedings of the 13th Mediterranean Conference on Control and Automation Limassol, Cyprus, June 27-29, 2005 pp 1343-1348

[5]. N.ouadah « Implémentation d'un contrôleur flou pour la navigation d'un robot mobile de type voiture » Algeria CDTA, 2005

Integration of additional constraints to Inverse the differential kinematic model for a nonholonomic and redundant mobile manipulator

Isma Akli [1,2], Noura Achour [1]

1-Université Houari Boumediène, BP 32 USTHB 16111 Bab-Ezzouar Alger, Algérie
2-Centre de Développement des Technologies Avancées, Cité 20 août 1956 Baba Hassen, Alger, Algérie
(e-mail : iakli@cdta.dz, nachour@usthb.dz)

Abstract: This article presents a differential kinematic study for a car-like mobile system carrying a four degrees of freedom manipulator. The generalized coordinates and velocities of the mobile manipulator are required, when the position and the velocity of the end-effector are imposed in the cartesian space. Our approach consists of planning the motion of the mobile platform with make the onboard manipulator able to follow the cartesian trajectory. The resulted generalized coordinates are exploited to calculate the Forward Differential Kinematic Model. Since the mobile manipulator is redundant regarding to the task, we profit from the system characteristics to augment the jacobian matrix, while integrating additional constraints, to inverse the differential kinematic model.

Keywords: Redundancy, Mobile Manipulator, Inverse Differential Kinematic Model

1. INTRODUCTION

A mobile manipulator system consists of coupling manipulation represented by an articulated system, and locomotion with using a mobile robot [2] ; this system is presented as a vehicle carrying a manipulator. This alliance has some advantages because it extends the arm workspace, whereas, the arm offers much operational functionality [14]. The integration of the manipulator to the mobile system, However, gives rise to some difficulties, since, the vehicle adds degrees of freedom (d.o.f) responsible to create a redundancy.

This paper focuses on the study of the inverse differential kinematic modeling for redundant and nonholonomic mobile manipulator.

Hu and Guo[1] proposed to model a system including a three d.o.f arm with revolute joints, and a nonholonomic platform (a car-like mobile system). They started by the presentation of the forward kinematic model and they calculate the inverse kinematic model analytically thanks to the forward model with imposing the orientation of the platform. The authors derivate the forward kinamatic equations to calculate the differential kinematic model. The same approach that is decoupling the combined system has been adopted by Papadopoulos and Poulakakis [2], two kinds of mobile manipulators have been considered, so, they studied the most commonly available mobile platforms which use either a differential drive or a car-like drive. Each of them has its own differential kinematic model, even if they are equipped with the same two-link manipulator. The authors presented a planning methodology for such systems allowing them to follow simultaneously desired end-effector trajectory and a platform path, without violating nonholonomic constraints. The executed task is a robotic crack-sealing, while the end-effector must follow some crack on the pavement, and a mobile platform is required to follow a given path, while respecting the condition that the distance between the end-effector and a reference point on the platform is within the reach of the manipulator.

Foulon [7] derived the forward differential kinematic model with leaving the wheels of the platform out of account and he deduced the reduced differential kinematic model without violating the nonholonomic constraints. The author used the augmented jacobian matrix in order to calculate the inverse differential kinematic model. The considered mobile manipulator is a differential-drive vehicle (Hilare H2bis) carrying a six d.o.f manipulator with revolute joints (GT6A).

Bayle [6] presents a Kinematic study of mobile manipulators built based on various platform designs, that differ by the drive mechanism employed, it consists on the most commonly available systems (holonomic mobile platforms, differential-drive platforms and car-like mobile systems). He presents the forward reduced differential kinematic model dedicated to the control with tacking into account the vehicle's drive system.

In this paper, we present an approach considering the inversing of the kinematic model for a redundant manipulator mounted on a car-like mobile system, in order to calculate the configuration of the system, with imposing a three dimensional cartesian path to the end-effector. Next, we focus on the inversion of the differential kinematic model for the performed task, using a kinematic approach requiring the insertion of additional constraints.

1.1. The studied system

Consider the mobile manipulator depicted in Fig .1. the vehicle is a car-like mobile platform with two steering front wheels, and two rear wheels, parallel to the main axis of the platform. The onboard articulated system has four revolute joints. In our case, we have considered the sub-chain

$Ang=[q_{b1}\ q_{b2}\ q_{b3}\ q_{b4}]^T$ responsible of positioning the end-effector of the seven d.o.f Mitsubishi PA10 7CE arm [5]. This manipulator is placed in the centre of the front wheels axis.

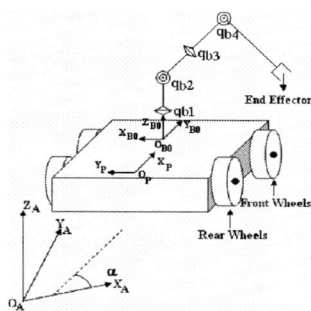

Fig. 1. The Mobile manipulator

1.2. Presentation of the reference frames

The kinematic and differential kinematic models of the mobile manipulator need to focus on three main reference frames.

Absolute Reference Frame: $R_A = (O_A, \vec{x}_A, \vec{y}_A, \vec{z}_A)$ indicates the notation in which must be expressed the position coordinates $[A_1\ A_2\ A_3]^T$ and the cartesian velocities $[\dot{A}_1\ \dot{A}_2\ \dot{A}_3]^T$ of the end-effector, according respectively to $\vec{x}_A$, $\vec{y}_A$ and $\vec{z}_A$ axis, as well as, its orientations $[A_4\ A_5\ A_6]^T$ and orientation velocities $[\dot{A}_4\ \dot{A}_5\ \dot{A}_6]^T$ with respect to the same three reference axis. R_A is chosen orthonormal, direct, and $\vec{z}_A$ is perpendicular to the surface on which moves the wheeled platform [8].

Platform Reference Frame: The mobile robot must have a reference frame $R_P = (O_P, \vec{x}_P, \vec{y}_P, \vec{z}_P)$, which is represented in R_A by the vector $A_P = [X_P\ Y_P\ \alpha]^T$ ($[X_P\ Y_P]$ is the position of O_P, and α is the orientation of the platform). The velocity vector $\dot{A}_P = [\dot{X}_P\ \dot{Y}_P\ \dot{\alpha}]$ is also presented in the frame R_P.

Manipulator Reference Frame: The reference frame $R_{B0} = (O_{B0}, \vec{x}_{B0}, \vec{y}_{B0}, \vec{z}_{B0})$ is assigned to the manipulation basis where must be calculated the position coordinates $[X_E\ Y_E\ Z_E]^T$ and cartesian velocities $[\dot{X}_E\ \dot{Y}_E\ \dot{Z}_E]^T$ of the end-effector, and its orientations $[\psi\ \theta\ \phi]^T$ and orientation velocities $[\dot{\psi}\ \dot{\theta}\ \dot{\phi}]^T$ given with Euler angles[6].

2. INVERSE KINEMATIC MODEL

Before presenting the differential kinematic model of the mobile manipulator, we briefly discuss its Inverse Kinematic Model. Define the configuration vector:

$$q=[q_1\ q_2\ q_3\ q_4\ q_5\ q_6\ q_7]^T \quad (1)$$

Where $q_1, q_2, q_3,$ and q_4 denote the manipulator joint angles q_{b1}, q_{b2}, q_{b3} and q_{b4} respectively, and q_5, q_6 and q_7 denote the cartesian position of O_P $[X_P\ Y_P]^T$ and the heading angle α respectively.

The prescribed end-effector trajectory (or operational trajectory) consists of a succession of some points close to one another, the aim of this representation is to allow the end-effector to follow a smoothed imposed trajectory. Those points will be called "*step points*", every step point has its own cartesian coordinates $[Xc\ Yc\ Zc]^T$ in R_A. The inverse kinematic model of the four d.o.f manipulator mounted on the nonholonomic platform was developed in detail in [16]. Therefore, we present in this article a resume of our inverse kinematic modelling approach.

The end-effector must reach the series of step points in the robot configuration space that require that the vehicle moves. The path planning methodology generates a succession of connected segments representing a smooth path that the mobile system follows, without violating the nonholonomic constraints. The main aim of this motion generation is to find the time-based coordinates of O_P, and the rotation angle α of the mobile robot considering the car rate v and the steering front wheels angle φ [13].

The task is decoupled into the vehicle moving into region containing the next step point in the workspace of the arm, assuming that the vehicle is fixed and the articulated system moving. An iterative method is used to calculate the configuration Ang_c of the manipulator[9][11]with imposing the cartesian coordinates for the current step point. Fig.2. shows the most important parameters in the motion planning and modelling of the mobile system. However, the steering wheels are grouped into one wheel.

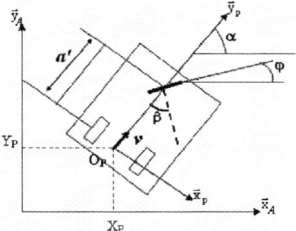

Fig.2. Representation of a car-like mobile platform

In the following sections, the vector of the operational velocities is $\dot{A} = [\dot{A}_1\ \dot{A}_2\ \dot{A}_3]^T$.

3. FORWARD DIFFERENTIAL KINEMATIC MODEL

The jacobian equation representing the forward differential kinematic model is defined as:

$$\dot{A} = J.\dot{q} \quad (2)$$

Where $\dot{q} = [\dot{q}_1\ \dot{q}_2\ \dot{q}_3\ \dot{q}_4\ \dot{q}_5\ \dot{q}_6\ \dot{q}_7]^T$ is the vector of the joints velocities, assuming that $\dot{Ang} = [\dot{q}_1\ \dot{q}_2\ \dot{q}_3\ \dot{q}_4]^T$ and $\dot{A}_P = [\dot{q}_5\ \dot{q}_6\ \dot{q}_7]^T$

Equation (3) shows the jacobian matrix J for the nonholonomic vehicle carrying the redundant manipulator.

Denote a' the distance between O_P and O_{B0} presented in Fig.2.

$$\begin{bmatrix} J_{11}C_\alpha - J_{21}S_\alpha & J_{12}C_\alpha - J_{22}S_\alpha & J_{13}C_\alpha - J_{23}S_\alpha & J_{14}C_\alpha - J_{24}S_\alpha \\ J_{11}S_\alpha + J_{21}C_\alpha & J_{12}S_\alpha + J_{22}C_\alpha & J_{13}S_\alpha + J_{23}C_\alpha & J_{14}S_\alpha + J_{24}C_\alpha \\ J_{31} & J_{32} & J_{33} & J_{34} \\ 1 & 0 & -\{(a'+X_E)S_\alpha - (Y_E)C_\alpha\} & \\ 0 & 1 & \{(a'+X_E)C_\alpha - (Y_E)S_\alpha\} & \\ 0 & 0 & 0 & \end{bmatrix} \quad (3)$$

The J_{ij} (i=1,2,3 and j=1,2,3,4) terms are the elements of the fixed base jacobian of the manipulator employed. The symbols C and S have been used instead of cosines and sinus.

3.1 The Reduced forward differential kinematic model

Rolling contact between the wheels and ground generally causes a non integrable kinematic constraints[15], reducing the number of parameters describing the vector $\dot{A}_F$.

$$\begin{bmatrix} \dot{X}_P & \dot{Y}_P & \dot{\alpha} \end{bmatrix}^T = \begin{bmatrix} -LC_\alpha S_\beta & -LS_\alpha S_\beta & C_\beta \end{bmatrix}^T [\eta_p] \quad (4)$$

Equation (4) shows the reduction of the vector $\dot{A}_p$ into one parameter for the car-like drive mobile system, where β is the rotation angle of the front wheels (Fig.2), and η_p is the velocity due to the driving wheels at O_{B0} [2].

After inserting the kinematic constraints, the kinematic equation representing the RDKM is given as:

$$\dot{A} = \bar{J} \cdot \dot{q}_{cfg} \quad (5)$$

Where $\dot{q}_{cfg} = \begin{bmatrix} \dot{q}_{b1} & \dot{q}_{b2} & \dot{q}_{b3} & \dot{q}_{b4} & \eta_p \end{bmatrix}^T$ is the vector of generalized velocities. Therefore, the reduced jacobian matrix $\bar{J}$ of the case study is written as follows:

$$\begin{bmatrix} J_{11}C_\alpha - J_{21}S_\alpha & J_{12}C_\alpha - J_{22}S_\alpha & J_{13}C_\alpha - J_{23}S_\alpha \\ J_{11}S_\alpha + J_{21}C_\alpha & J_{12}S_\alpha + J_{22}C_\alpha & J_{13}S_\alpha + J_{23}C_\alpha \\ J_{31} & J_{32} & J_{33} \\ J_{14}C_\alpha - J_{24}S_\alpha & -LC_\alpha S_\beta - \{(a'+XE)S_\alpha + (YE)C_\alpha\}C_\beta \\ J_{14}S_\alpha + J_{24}C_\alpha & -LS_\alpha S_\beta + \{(a'+XE)C_\alpha - (YE)S_\alpha\}C_\beta \\ J_{34} & 0 \end{bmatrix} \quad (6)$$

The $\bar{J}$ matrix depends on the configuration q of the mobile manipulator. Once the configuration q^{sp} deduced from the inverse kinematic model, with imposing the cartesian coordinates $[Xc\ Yc\ Zc]^T$, for the current step point, $\bar{J}$ is calculated.

Assuming that every step point is identified by its cartesian $[\dot{X}c\ \dot{Y}c\ \dot{Z}c]^T$ velocities, we must inverse the kinematic model to result in the generalized velocities vector $\dot{q}_{cfg}$.

4. INVERSE DIFFERENTIAL KINEMATIC MODEL

The mounted manipulator is redundant regarding to the operational task, hence, the integration of the mobile system to the fixed-base arm confirm this redundancy. We denote the mobility plus the degrees of freedom of the onboard manipulator by DN (with $DN=5$), and working cartesian space dimension by DM (with $DM=3$). We have $DN>DM$, then the degree of redundancy DR is calculated in what follow:

$$DR = DN - DM = 5 - 3 = 2 \quad (7)$$

Therefore, to resolve the redundancy, we can apply two additional constraints equations [15] in cartesian or joint space to reflect the additional task [3].

Equation (8) presents the kinematic equation of the mobile manipulator with integrating the DR constraints, where J_{a1} and J_{a2} are the introduced extra-lines to the $\bar{J}$ matrix. The matrix J_t is named, the augmented jacobian matrix [15].

$$\underbrace{[\dot{A}\ \dot{W}_{a1}\ \dot{W}_{a2}]^T}_{DN \times 1} = \underbrace{[\bar{J}\ J_{a1}\ J_{a2}]^T}_{J_t\ DN \times DN} \underbrace{[\dot{q}_{cfg}]}_{DN} \quad (8)$$

4.1. Insertion of the additional task

In what follow, we are going to present the additional constraints introduced for obtaining the augmented jacobian matrix for the car-like mobile robot carrying the four d.o.f manipulator.

4.1.1 First kinematic constraint $\dot{W}_{a1}$

The first additional constraint is deduced from the inverse kinematic model, where the velocity of the platform is imposed. Hence, the first extra line is:

$$[\eta_p] = [0\ 0\ 0\ 0\ 1] \quad (9)$$

Otherwise, if this extra-line was not deduced from the inverse kinematic model, the following of the operational trajectory could not be insured [7].

4.1.2. Second kinematic constraint $\dot{W}_{a2}$

In this section, three propositions $\dot{W}_{a2}^1$, $\dot{W}_{a2}^2$ and $\dot{W}_{a2}^3$ for the second additional constraint are presented because only the velocity of the mobile system is imposed in the inverse kinematic model.

- *The additional constraint $\dot{W}_{a2}^1$*

While the mobile manipulator system can control four velocities $\dot{A}_1$, $\dot{A}_2$, $\dot{A}_3$ and $\dot{A}_4$ in operational space, thanks to joints velocities of the manipulator and the rate η_p, the task is performed in cartesian space. Therefore, the extra-line representing the second constraint is:

$$[\dot{A}_4] = [J_{41} \ J_{42} \ J_{43} \ J_{44} \ C_\beta] \quad (10)$$

- The additional constraint $\dot{W}_{a2}^2$

The velocity $\dot{X}_E$ (the velocity of the end-effector according to the $\vec{x}_{B0}$ axis) is a proposition representing the second additional constraint. The integrated extra-line is as follows:

$$[\dot{X}_E] = [J_{11} \ J_{12} \ J_{13} \ J_{14} \ 0] \quad (11)$$

- The additional constraint $\dot{W}_{a2}^3$

Since the articulated system is redundant regarding to the task, the third proposition for an additional constraint is a joint velocity. Thus, the extra-line can be defined as:

$$[\dot{q}_{b1}] = [1 \ 0 \ 0 \ 0 \ 0] \quad (12)$$

5. SIMULATIONS

To experiment the theretical concepts, the simulation study was carried out. Different additional constraints were tested, assuming that the end-effector must follows a helicoidally form, in the cartesian space Fig.3. The operational trajectory is discretized into 360 step points, and the task is performed in approximately 14 seconds. The car-like mobile system follows the prescribed planned path with initialising the velocity of the platform v into 1m/s, and the maximum steering angle value of the front wheels into φ_{max}=0.35 radians.

The intervals of velocities for every joint of the onboard manipulator are: $[\dot{q}_{b1}^{min}, \dot{q}_{b1}^{max}] = [-1,1]$, $[\dot{q}_{b2}^{min}, \dot{q}_{b2}^{max}] = [-1,1]$, $[\dot{q}_{b3}^{min}, \dot{q}_{b3}^{max}] = [-2,2]$, $[\dot{q}_{b4}^{min}, \dot{q}_{b4}^{max}] = [-2,2]$.

The resulting trajectory of the mobile system is different from the planned path (Fig.4). This is due to the nonholonomic constraints.

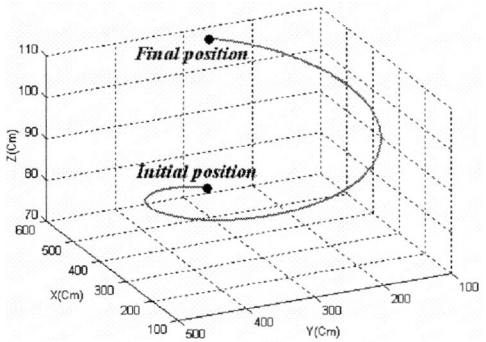

Fig.3. Imposed operational trajectory

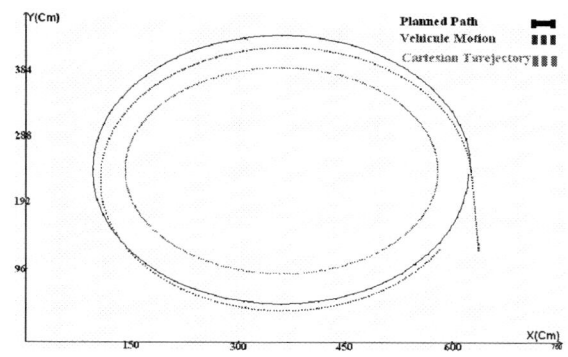

Fig.4. Motion planning of the mobile system

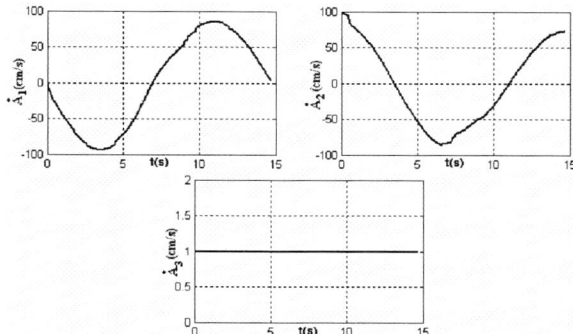

Fig.5. Imposed cartesian velocities

The given operational velocities are shown in Fig.5. where the vector $[\dot{X}c \ \dot{Y}c \ \dot{Z}c]^T = [\dot{A}_1 \ \dot{A}_2 \ \dot{A}_3]^T$.

In the following part, the additional constraints $\dot{W}_{a2}^1, \dot{W}_{a2}^2$ and $\dot{W}_{a2}^3$ are experimented, and their influence on the joint velocities of the carried arm is tested, considering the first additional constraint $\dot{W}_{a1}$.

- The imposed $\dot{W}_{a2}^1$ constraint

Fig.6 shows the joint velocities with imposing the velocity $\dot{A}_4$ =-0.4 rad/s. It can be seen that the joint velocity $\dot{q}_{b3}$ is outside the interval of the admitted velocities between t=7s and t=9s.

However, with imposing the velocity $\dot{W}_{a2}^1 = \dot{\Psi}$ as a second additional constraint (where the extra-line $[J_{41} J_{42} J_{43} J_{44} 0]$ is integrated), the resulting joints velocities presented in Fig.7 are in the interval of the admitted velocities.

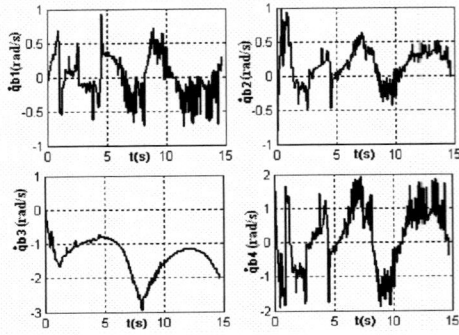

Fig.6. Generalized velocities with imposing $\dot{A}_4 = -0.4$

The insertion of $\dot{\Psi}$ as an additional constraint presents an advantage, because this rotation velocity does not depend on the rate of the mobile system, and can be controlled only with controlling the motion of the mounted manipulator.

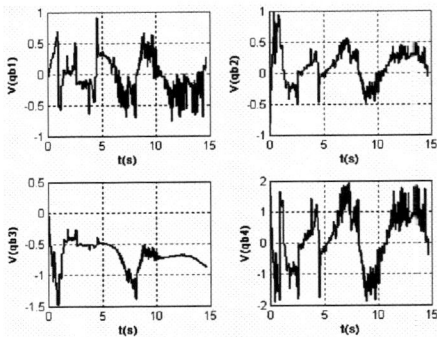

Fig.7. Generalized velocities with imposing $\dot{\Psi} = -0.4$ rad/s

- *The imposed $\dot{W}_{a2}^{2}$ constraint*

The augmented Jacobian matrix is singular for all the step points with integrating the kinematic constraint $\dot{W}_{a2}^{2}$ (Fig.8).

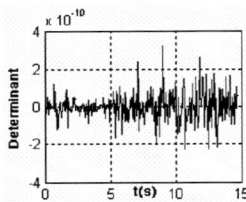

Fig.8. Determinant of J_t for the additional constraint $\dot{W}_{a2}$

Hence, the resulted generalized velocities of the manipulator shown in Fig.9 present very significant values.

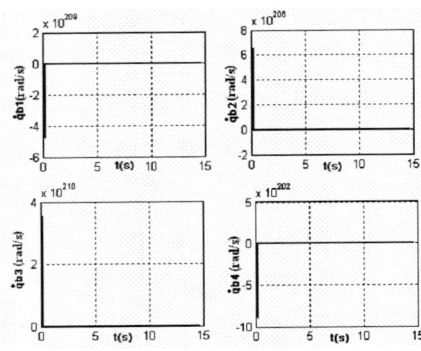

Fig.9. Generalized velocities with imposing $\dot{X}E = -0.4$ rad/s

- *The imposed $\dot{W}_{a2}^{3}$ constraint*

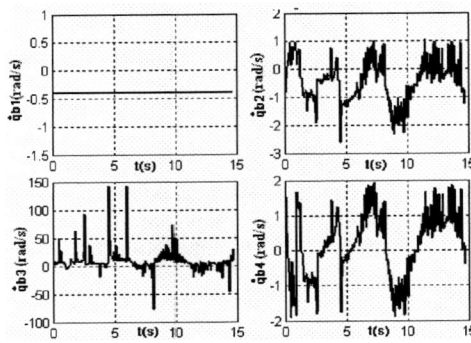

Fig.10. Generalized velocities with imposing $\dot{q}_{b1} = -0.4$ rad/s

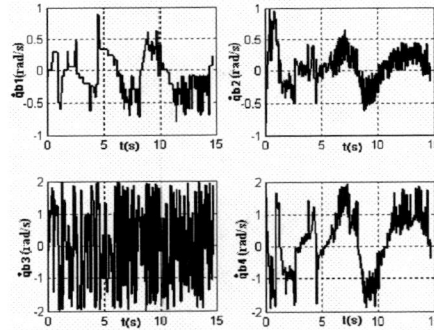

Fig.11. Imposed operational velocities with imposing $\dot{q}_{b1}$

For the imposed additional task $\dot{W}_{a2}^{3}$ (with considering $\dot{q}_{b1} = 0.4$ rad/s), the resulted joints velocities presented in Fig.10 are corrects, except $\dot{q}_{b3}$ which presents some extreme variations exceeding the admitted velocity limits $\left[\dot{q}_{b3}^{min}, \dot{q}_{b3}^{max}\right]$. However, when the velocity $\dot{q}_{b1}$ is correctly adjusted Fig.11, or with changing the additional constraint into $\dot{W}_{a2}^{3} = \dot{q}_{b3}$ (Fig.12) and the second extra-line to [0 0 1 0 0], the resulting velocities for all the joints become correct.

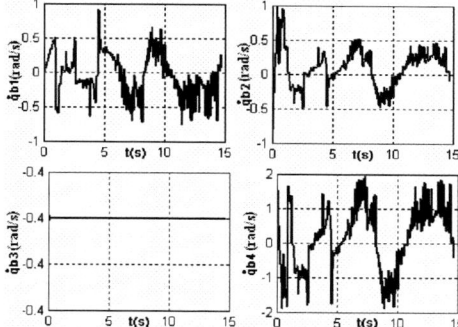

Fig12. Imposed operational velocities with imposing $\dot{q}_{b3}$

6. CONCLUSIONS

The mobile manipulation systems are able to accomplish tasks requiring mobility and manipulation. For the case study, the four degrees of freedom manipulator is mounted on a car-like mobile system. Since the system is redundant, two additional constraints are necessary to augment the reduced jacobian matrix, and inverse the differential kinematic model.

The experimented method to inverse the differential kinematic model depends on the accomplished task, and the studied system.

While integrating the velocity of the mobile system, as a first constraint, we exploited the redundancy of the manipulator regarding to the task to insert the second additional constraint. When the velocity constraint is defined in the joint space of the manipulator, the resulted generalized velocities are correct and are in the interval of the admitted joints velocities. However, while considering the integrated constraints in the task space, it is preferable to prescribe velocities regarding to the arm reference frame for a better control of the system. Therefore, the augmented jacobian matrix must be non-singular and the velocities representing the additional tasks can be adjusted to have a correct resulting joints velocities.

For future work, it would be interesting to consider the rotation rate of the steering front wheels angle for the redundant and nonholonomic mobile manipulator. The other alternative could be also to calculate the pseudo inverse of the system by minimizing a norm of the joint velocities.

REFERENCES

[1] Y.M.HU and B.H.Guo, "Modelling and Motion Planning of a three link wheeled mobile Manipulator" *8th Int Conf. Control, Automation, Robotics and Vision*, China, 2004.

[2] E.Papadopoulos and J.Poulakakis, "Planning And Model-Based Control for Mobile Manipulators" in *Proc IROS 2000 Conf. Intelligent Robots and Systems*, Takamatsu, Japan, 2000.

[3] H. Seraji: "Configuration Control of redundant Manipulators: Theory and Implementation", in *Proc IEEE Trans. Robotics and Automation*, **VOl 5. NO 4**, pp.472-490, August 1989

[4] D.Xu, H.HuCarlos, A.A.Calderonand and M. Tan, "Motion Planning for a Mobile Manipulator with Redundant DOFs"in *Proc ICIC'05. Int. Conf. Intelligent Computing*, China, 23-26 August, 2005.

[5] D.N.Nenchev,Y. Tsumaki and M. Takahashi "Singularity-Consistent Kinematic Redundancy Resolution for the S-R-S Manipulator" in *Proc 2004 IEEE/RSJ Int. Conf. Intelligent Robots and Systems*, Sendai, Japan, September 28 October 2, 2004.

[6] B. Bayle, "Modélisation et Commande Cinématiques des Manipulateurs Mobiles à Roues", PhD. Thesis, LAAS-CNRS, Paul Sabatier Univ, France, 2001.

[7] G.Foulon, "Génération de Mouvements Coordonnés pour un Ensemble constitué d'une Plate-forme Mobile à Roues et d'un bras manipulateur", PhD. Thesis, LAAS–CNRS, Institut National des Sciences Appliquées, France, 1998.

[8] V.Padois "Enchainenemts Dynamiques de Tâches pour des Manipulateurs Mobiles à Roues", PhD. Thesis, Laboratoire de Génie de Production, Ecole Nationale d'Ingénieurs de Tarbes, Institut national polytechnique de Toulouse, 2005

[9] B.Gorla and M.Renault, "Modèles des Robots Manipulateurs : Application à leurs Commandes", Cepadues, Ed. 1984.

[10] L.Flückiger, "Interface pour le pilotage et l'analyse des robots basée sur un générateur de cinématique", PhD. Thesis, école polytechnique fédérale de Lausanne, Suisse, November1998.

[11] C.Pholsiri, "Task based decision making and control of robotic manipulator", PhD. Thesis, Texas University, Austin, USA, December 2004.

[12] Y.Yamamoto and X.Yun, "Coordinating Locomotion and Manipulation of a Mobile Manipulator", in *Proc. IEEE Int. Conf. Robotics and Automation*, USA, pp157-181, May 1993.

[13] A.Pruski, "Robotique mobile: La planification de trajectoire", Hermès Science Publications, Collection traité des nouvelles technologies, 1996.

[14] J.B.Mbede, P.Ele, C.M. Mveh-Abia, Y.Toure,V.Graefe, S.Ma, "Intelligent mobile manipulator navigation using adaptive neuro-fuzzy systems", *elsevier trans. Infirmation Sciences*, **VOL 171**, 447-474, 2005.

[15] M.H.Korayem and H.Gharibllu, "Maximum allowable load on wheeled mobile manipulator imposing redundancy constraints", *Elsevier trans. Robotics and Autonomous Systems*, **VOL 44**, 151-159, 2003.

[16] I.Akli, N.Achour and R.Toumi, "Path Planning for redundant Mobile Manipulator", *4th Int Conf. Computer Integrated Manufacturing CIP'2007*, 03-04 November, 2007.

[17] I.Akli, "Elaboration d'une stratégie de coordination de mouvements pour un manipulateur mobile redondant", M.S Thesis, USTHB, Algeria, Jully, 2007.

SESSION F1: MOTION & ROBUST CONTROLS

Numerical Scheme for Viability Computation Using Randomized Technique with Linear Programming

Badis Djeridane, *IEEE Member* *

*ETH-Zurich, Automatic Control Laboratory, 8092 Zurich, Switzerland,
djeridane@control.ee.ethz.ch*

Abstract: We deal with the problem of computing viability sets for nonlinear continuous or hybrid systems. Our main objective is to beat the *curse of dimensionality*, that is, we want to avoid the exponential growth of required computational resource with respect to the dimension of the system. We propose a randomized approach for viability computation: we avoid griding the state-space, use random extraction of points instead, and the computation of viable set test is formulated as a classical feasibility problem. This algorithm was implemented successfully to linear and nonlinear examples. We provide comparison of our results with results of other method.

Keywords: Viability Theory, Randomized Algorithm, Feasibility Problem.

1. INTRODUCTION

Because of their importance in application ranging from engineering to biology and economics, questions of reachability, viability and invariance have been studied extensively in the dynamics and control literature. Most recently, the study of these concepts has received renewed attention through the study of safety problems in hybrid systems. Reachability computations have been used in this context to address problems in the safety of ground transportation systems Lygeros et al. [1998], air traffic management systems Livadas et al. [2000], flight control Lygeros et al. [1999], etc.

Indirect approach to viability questions is using optimal control methods. In this case, the reachable, viable or invariant sets are characterized as level sets of the value function of an appropriate optimal control problem. Using dynamic programming, the value function can in turn be characterized as solution of partial differential equation Lygeros [2004].

Alternative, direct characterization of viability concepts is one of the topics addressed by viability theory Aubin [1991]. The development of computational tools to support the numerous viability theory concepts is an going effort Cardaliaguet et al. [1999]. Methods for directly addressing viability questions have been proposed in the hybrid systems literature. For example, for certain classes of continuous systems.

It is well known that, frequently, the complexity for computing the viability kernel is very high even for low state space dimensions. Hence, it appears natural to seek approximate method involving suitable discretization to facilitate computer work. The computationally efficient full discretization of state space has been a challenge for researchers for many decades. To overcome this difficulty, for example Chow and Tsitsiklis [1988] propose a multi-grid method adapted to a class of discrete time, continuous state, discounted, infinite horizon dynamic programming problems to improve the computational complexity for this class of system. Another method called cell-mapping uses the basic idea of considering the state space not as continuous but rather as a collection of large number of state cells with each cell being taken as a state entity Hu and Chiu [1986a,b]. This method has been successfully applied to optimal control problem Bursal and Hsu [1989] by representing all the admissible controls and their duration application as finite set. Then, the process of extracting optimal control results from the family of controlled mappings becomes a matter of systematic search.

All these methods suffer from an exponential computational complexity. Recently, new idea have emerged that could find solution at "most of time" for particular problem with "high confidence" that the candidate solution is the true solution. Randomized algorithms are gaining popularity among control theory community Ariola et al. [2003], and have been applied successfully to compute a reachable set using neural networks Djeridane et al. [2007] and to system identification of ARMA Model Vidyasagar and Karandikar [2002]. Another application is the identification of a piecewise affine system presented in Prandini [2004].

In this paper, we present an improved algorithm of our previous work Djeridane et al. [2008]. This approach is motivated by learning theory Vidyasagar [1998] that aims to beat the curse of dimensionality for viability kernel computation by generating points randomly instead of gridding over the whole state-space. Once we have all sample points, we start by generating our sample points according to Halton sequence scheme and afterward all our operations are based on this new representation of our points. And formulated the problem of finding the appropriate control to determine the viability kernel set as a classical feasibility problem.

The paper is organized into sections as follows: Section II deals with a problem statement and provides some background material on the viability theory. Section III presents our randomized algorithm used to compute the viability kernel. In section VI we show many results where our method has been applied successfully from linear to nonlinear system and finally we

---

* Work supported by the European Commission under the HYCON Network of Excellence, IST-511368

conclude our paper with conclusion V and some directions for our future research.

2. VIABILITY FRAMEWORK

This section is devoted to the presentation of results of viability theory. In particular, we recall the definition and the geometrical characterization of the *Viability Kernel* and *Invariance Kernel*.

2.1 Basics results on viability theory

Differential inclusions Consider the control system
$$\dot{x} = f(x, u) \quad (1)$$
with $u \in U$.

It is almost classical result that control system (1) can be represented by the following differential inclusion
$$\dot{x} \in F(x) \quad (2)$$
where $F : X \to X$ is the set-valued map defined by
$$\forall x \in X, F(x) = \{f(x, u), u \in U\}$$
The system (1) and (2) have the same absolutely continuous solutions.

The Viability kernel Let us consider a closed nonempty set $K \subset X$. We shall say that a solution $x(.)$ to (1) (or equivalently to (2)) is *viable* in K if and only if $x(t) \in K$ for any $t \geq 0$.

Definition 1 Let K be closed subset of X. The viability kernel of K for F is the set
$$\{x_0 \in K | \exists x(.) \in S_F(x_0), x(t) \in K, \forall t \geq 0\}$$
We denote it by $Viab_F(K)$.

Let us notice that this set is empty if and only if any solution, starting from K, leaves K in finite time. For computing $Viab_F(K)$ without computing any trajectory, we need to characterize $Viab_F(K)$ in geometrical way Aubin [1991].

The viability kernel of K for a set-valued map F consists in of the set of initial positions x_0 of K from which at least one solutions start $x(.) \in S_F(x_0)$ which remains in K. It is quite natural to consider the set of initial conditions x_0 of K such that any solution $x(.) \in S_F(x_0)$ remains in K.

Definition 2 Let K be closed subset of X. The invariance kernel of K for F is the set
$$\{x_0 \in K | \forall x(.) \in S_F(x) : x(t) \in K, \forall t \geq 0\}$$
We denote it by $Inv_F(K)$.

As for the viability kernel, it is possible to characterize the invariance kernel by the mean of geometric conditions Aubin [1991].

2.2 Approximation of Viability kernel

To approach the viability kernel, we first replace the initial differential inclusion system by a finite difference inclusion. Secondly, we replace the state space X by an integer lattice X_h of X.

The semi-discrete Viability Kernel Algorithm Let us consider F_ϵ some approximation of F and define
$$G_\epsilon = x + \epsilon F_\epsilon(x)$$
The choice of F_ϵ depends in general on the regularity of the dynamic of F. The discretization dynamic corresponding to the Euler scheme is
$$x_{k+1} \in G_\epsilon(x_k) \quad (3)$$
where F_ϵ is an approximation of F which satisfies the following properties:

(1) F_ϵ is upper semi-continuous with convex compact nonempty values.
(2) $Graph(F_\epsilon) \subset Graph(F) + \phi(\epsilon)B$ where $\lim_{\epsilon \to 0^+} \phi(\epsilon) = 0^+$
(3) $\forall x \in X, \cup_{\|x-y\| \leq M\epsilon} F(y) \subset F_\epsilon(x)$

where B is unit ball in $\Re^n$.

Next algorithm provides a constructive computation of the viability kernel in discrete-time scheme

Algorithm 1 Semi-discrete Viability Kernel Algorithm
$K^0 = K$
$K^{k+1} = \{x \in K^k | G(x) \cap K^k \neq \emptyset\}$

And Viability Kernel is defined by $Viab_G(K) = K^\infty$.

The fully discrete Viability kernel Algorithm To implement this algorithm we have to associate with G_ϵ suitable finite set-valued maps on finite sets. For that purpose, we introduce a grid X_h of X, associated with any $h \in \Re$, satisfying

- the set X_h has a finite intersection with any compact of X
- $\forall h > 0, \forall x \in X, \exists x_h \in X, \|x - x_h\| \leq h$

Now we are dealing with systems which are not only discrete but finite. With any closed set K we associate its "projection onto the grid" defined by
$$K_h = (K + hB) \cap X_h$$
The next algorithm is similar to Algorithm (1) for finite sets. Let $\Gamma_{\epsilon,h} : X_h \to X_h$ be a set-valued map with finite nonempty values and K_h a finite subset of X_h. Let $\{K_{\epsilon,h}^{k+1}\}$ be the sequence of the subsets of K_h defined as follows

Algorithm 2 Fully-discrete Viability Kernel Algorithm
$K_{\epsilon,h}^0 = K_h$
$K_{\epsilon,h}^{k+1} = \{z_k \in K_{\epsilon,h}^k | \Gamma_{\epsilon,h}(z_h) \cap K_{\epsilon,h}^k \neq \emptyset\}$

Then, $\exists k \in N : Viab_{\Gamma_{\epsilon,h}}(K_h) = K_{\epsilon,h}^k$

3. COMPUTATION PROCEDURE

3.1 Settings problem

To obtain an approximation of the viability kernel, the collocation method used. First, we have randomly samples point over the set K. Now, we comment on how we compute x_{k+1} (successor of x_k). In the algorithm is stated that

$$x_{k+1} = x_k + \epsilon F_\epsilon(x_k)$$

where $F(x)_\epsilon = F(x) + Ml\epsilon B$ with $l = \sup_x \frac{\partial f(x,u)}{\partial x}$ for $\forall u \in U$, $M = \sup_u f(x,u)$ $\forall u \in U$ and B is a ball.

Consider the set K is defined by polytopes

$$K = \{x \in \Re^n : Ax \leq 0\}$$

where $A \in \Re^{p \times n}$.

and the constraints set U is defined too by polytope set

$$U = \{u \in \Re^m : Du \leq 0\}$$

where $D \in \Re^{q \times m}$.

3.2 Generating samples

Assume that we are trying to minimize a function over the unit cube ($[0,1]^d$), and the minimum point is exactly at the center of the hypercube. Also consider a smaller hypercube inside of the original one with sides equal to $1 - \frac{\epsilon}{2}$. Now, extract a random point inside the hypercube according to an uniform distribution probability, then the results in the probability of sampling inside the smaller cube is

$$(1-\epsilon)^d$$

where d is the dimension of the hypercube Tempo et al. [2005]. Then, the probability of the sampling inside the smaller cube tends to zero as d goes to infinity, hence the clustering effect on the surface is observed as we go into higher dimensions. We would like to replace the random samples required for computing a viability set with deterministic samples that posses certain regularity condition, i.e. they are regularly spread within the sampling space. This method is also independent of the sampling space. It has shown its superiority over classic Monte Carlo methods in the calculation of certain integrals Gentle [1998] and robust control problem Hokayem et al. [2003].

There are many sequences in the literature used to generate a sampling points, which include the property of "evenly distributed", such as Halton, Hammersley, Sobol sequences. In this paper, we are interested in Hatlon sequence which has low discrepany sequence (measure of how the samples set is equidistributed within integration domain) and efficient computation technique. Based on prime numbers, these pseudo-random distribution is uniform and irregular, but lack point in close proximity, i.e., they have a minimum resolution that increases as the number of points in the sample increases. The gives the algorithm for computing Halton points in up to ten dimension. The function, $p_2(n,d)$, takes the number of points and parameter as input and returns a (rational) number belonging to the interval $[0,1)$. Wong et al. [1997]

Figure (1) compares the uniform random Halton distribution in two dimension. The Halton distribution is more even, but the random distribution displays a wider range of difference vector magnitudes.

Algorithm 3 Algorithm for Generating Halton

$prime = \{2, 3, 5, 7, 11, 13, 17, 19, 23, 29\}$
choose p_1 from $prime$
$p_2 = p_1$
$\phi = 0$
repeat
$\quad a = n \mod p_1$
$\quad \phi = \phi + \frac{a}{p2}$
$\quad n = int(\frac{n}{p1})$, where $int(x)$ returns the integer part of x.
$\quad p_2 = p_2 p_1$
until $n \leq 0$

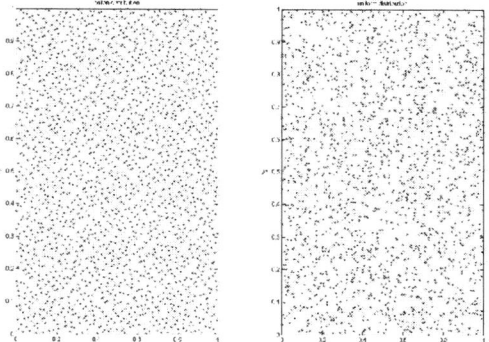

Fig. 1. 500 points distributed with random uniform distribution (left) and according to a two-dimensional Halton point set (right)

3.3 Feasibility problem

In a feasibility problem, we typically have a system of inequalities defined by functions $l(u) : \Re^m \rightarrow \Re^p$. We are then interested in the set

$$P = \{u \in \Re^m | l(u) \leq 0\}$$

Typically we would like to know if the set P is non-empty; that is, it there exists $u \in \Re^m$ which simultaneously satisfies all of the inequalities. Such a point is called a *feasible point* and the set P is then called *feasible*. If P is not feasible, it is called *infeasible*.

Since we are dealing with linear constraints, the feasibility problem could be turn it as follows:

Find $u \in \Re^m$ such that

$$\exists u_k \in U : Ax_{k+1} \leq 0$$

where

$$x_{k+1} = f(x_k) + g(x_k)u_k$$

can be stated as

$$\exists u_k \in U : Af(x_k) + Ag(x_k)u_k \leq 0$$

Towards that end, we reformulate the quantifier free formula as an optimization problem:

$$\min_{u_k} 0$$

with

$$[Af(x_k)] + [Ag(x_k)]u_k \leq 0$$
$$Du_k \leq 0$$

The problem of finding u such the successor stay inside the polytope K is turn into a classical feasibility problem, and we could use an efficient numerical tools to solve such as CPLEX.

3.4 Outline of the algorithm

Instead of using a regular griding of the state space we approximate the set K using a finite number of points generated randomly. Let $\{x^i\}_{i \leq N}$ denote these points. And we define the *state discretization index* h. As usual with a numerical explicit scheme, the space and time discretization steps are linked up. Therefore the time-step ϵ is determined after the random sampling of K in order to ensure consistency of the approximation.

The procedure for computing an approximation of the viability kernel is relatively simple once you can check if a point is *locally viable* in a set K, that is, if it has a successor in K. In order to check *locally viable* of x_k for all value of u, a simple idea is enumerate all possible values of u belongs to the control constraints U and test if x_{k+1} with different values of u_k still belongs to the set K. However this naive approach needs a huge amount of computing resources if the number of inputs is large. In order to handle this problem, we propose a technique based on formulating the problem as feasibility problem. The principle is explained in the section.

This procedure is formulated in the following algorithm

Algorithm 4 Local viability test for x^i to K^p

Pick $x_k \in K^p$
Solve feasibility problem state in section
if $P \neq \emptyset$ **then**
    x_k is not viable point
else {otherwise}
    x_k is viable point
end if

The overall viability kernel approximation procedure is summarized in algorithm 5.

Algorithm 5 Computation of the viability kernel

Generate randomly N points x^i over K
Compute the state discretization index h
Initially set $K^0 = \{x^i : i \in \leq N\}$
Initially set $p = 0$
repeat
    for all $i = 1$ to N **do**
        Do the local viable test of x^i in K^p
    end for
    M number of points x^i removed from K^p
    Compute the new set $K^P = \{A_p x \leq 0\}$
    $N = card(K^p)$
    $p = p + 1$
until $N = 0$ or $M = 0$

The computation of the new polytope is done by using geometrical tools developed in Multi-Parametric Toolbox IfA [2005] We should stress that when using the local viability test described below, this algorithm provides an upper approximation of the viability kernel *with a certain confidence*. Indeed, our approximation method guarantees that a point the successor of which in inside the real kernel is not eliminated only if

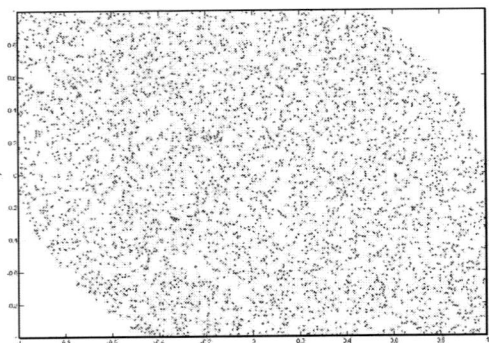

Fig. 2. Viability set for 2D

our initial random sampling is *good enough*. Therefore, the randomization of the methods loses the guarantee of over-approximation of Cardaliaguet et al. [1999].

4. EXAMPLES

To validate the algorithm presented in this paper, we have worked on linear system and nonlinear systems and we have compared our results with MPT Toolbox which is used to compute the viability set for linear system Baric et al. [2005]. We have perform all the computations on the a Opteron 64 processor running on Linux with 4 GB memory.

4.1 Linear system

We have applied our approach to linear system with canonical controllability form. And we have computed the exact solution using Level Set Toolbox with high accuracy Mitchell [2005].

2D example   Consider the double integrator

$$\dot{x}_1 = x_2$$
$$\dot{x2}_2 = u$$

with $u \in U = [-1, 1]$ and $x \in K = [-1, 1] \times [-1, 1]$.

3D example
$$\dot{x}_1 = x_2$$
$$\dot{x2}_2 = x_3$$
$$\dot{x2}_3 = u$$

with $u \in U = [-1, 1]$ and $x \in K = [-1, 1]^3$.

To compare, first we have computed the "exact" solution for our example using level set toolbox with high accuracy, and afterward we have used the technique developed (Method 1) in this paper with tolerance of 5% of error from the exact solution. And also we have computed the viability set using MPT Toolbox with same accuracy of 5 % to show that our proposed technique provide better results even in the linear case.

	2D	3D	4D	5D
Number of vertices Method 2	14	440	–	–
Number of facts Method 2	14	330	–	–
n Method 1	20000	50000	100000	280000

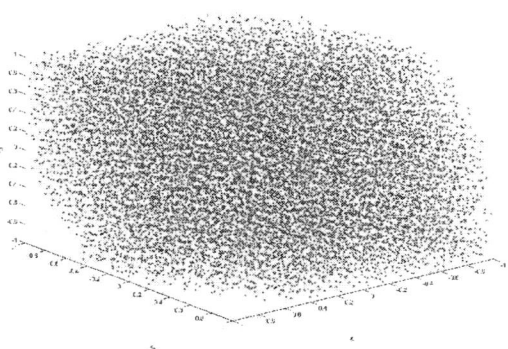

Fig. 3. Viability set for 3D

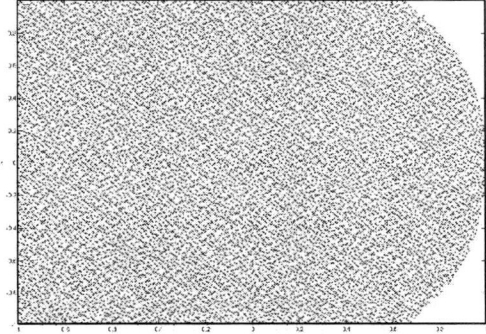

Fig. 4. Viability set for Nonlinear system

4.2 Nonlinear system

To show that our technique works too for the case of nonlinear system, we have worked in simple nonlinear system defined as follow

$$\dot{x}_1 = x_2^2$$
$$\dot{x2}_2 = u$$

with $u \in U = [-1, 1]$ and $x \in K = [-1, 1] \times [-1, 1]$.

We have computed an approximation of the viable set using 20000 samples within 5% error from the "exact" solution. Which make clear that our method perform well in linear case as well in nonlinear case.

5. CONCLUSION

The goal of this paper was to propose an approximation method that circumvents the curse of dimensionality encountered in viability computations. We have used our algorithm on a set of different examples with good results. In current work we are trying to extended this trial solution to more complicated system. A direct advantage of our method is that the size of the state space to be explored can be fixed in advance as the number of points taken randomly in the initial set. We are currently working on a multi-step refinement process to obtain fine approximation with a limited number of points. The hope is that the computation required to achieve a certain accuracy will grow polynomially with the dimension of the system.

The real advantages introduced by our randomized method is could handle on efficient way the problem with multi-inputs system. The probability that the output of our algorithm is a *bad* approximation of the real viability kernel is strictly positive. We are working on a convergence proof in order to provide a statistical guarantee of convergence in a confidence interval. The proof is rather challenging since it requires mixing elements from randomized techniques together with non-smooth analysis. Another features of our algorithm

REFERENCES

M. Ariola, C.T. Abdallah, and V. Koltchinskii. Applications of statistical-learning control in system and control. In *IFAC Workshop*, Villa Erba, Cernobbio-Como, Agust 29-31 2003. on Adaptation and Learning in Control and Signal Processing.

J. P. Aubin. *Viability theory*. Bikhauser, Boston, 1991.

M. Baric, P. Grieder, M. Baotic, and M. Morari. Optimal control of pwa systems by exploiting problem structure. In *IFAC World Congress*, Prague, july 2005.

F.H. Bursal and C.S. Hsu. Application of a cell-mapping method to optimal control problems. *International Journal of Control*, (5):1505–1522, 1989.

P. Cardaliaguet, M. Quincampoix, and P. Saint-Pierre. Stochastic and differential games: Theory and numerical methods. In *Annals of the International Society of Dynamic Games*. Birkaser, 1999.

C. Chow and J.N. Tsitsiklis. An optimal multigrid algorithm for continuous state discrete time stochastic control. In *Conference on Decision and Control*, pages 1908–1912, Austin, Texas, December 1988.

B. Djeridane, E. Cruk, and J. Lygeros. A learning theory approach: to the computation of reachable sets. In *European Control Conference*, Kos, July 2007.

B. Djeridane, E. Cruck, and J. Lygeros. Randomized algorithm: A viability computation. In *submitted to IFAC World Congress*, Seoul, July 2008.

J.E. Gentle. *Random Number Generation and Monte Carlo Methods*. Springer-Verlag, New York, 1998.

P. F. Hokayem, C.T. Abdallah, and P. Dorato. Quasi-monte carlo methods in robust control desgin. In *IEEE Mediterranean Conference on Control and Automation*, Rhodes, Greece, June 2003.

C.S. Hu and H.M. Chiu. A cell mapping method for nonlinear deterministic and stochastic systems -part i: The method of analysis. *Trans. ASME J. Appl. Mech.*, 53(3):702–710, 1986a.

C.S. Hu and H.M. Chiu. A cell mapping method for nonlinear deterministic and stochastic systems -part ii: Examples of application. *Trans. ASME J. Appl. Mech.*, 53(3):695–701, 1986b.

IfA. *Multi-Parametric Toolbox*. http://control.ee.ethz.ch/ mpt/, 2005.

C. Livadas, J. Lygeros, and N. Lynch. High-level modeling and analysis of the traffic alert and collision avoidance system (tcas). *Proceeding of the IEEE*, 88(7):926–948, July 2000.

J. Lygeros. On reachability and minimum cost optimal control. *Automatica*, 40:917–927, 2004.

J. Lygeros, D.N. Godbole, and S. Sastry. Verified hybrid controllers for automated vehicles. *IEEE Transactions on Automatics Control*, 43(3):522, 1998.

J. Lygeros, C. Tomlin, and S. Sastry. Controllers for reachability specifications for hybrid systems. *Automatica*, 35:349–370, 1999.

I. Mitchell. *A Toolbox of Level Set Methods version 1.1*. http://www.cs.ubc.ca/ mitchell/ToolboxLS/index html, 2005.

M. Prandini. Priecewise affine systems identification: a learning theoretical approach. In *Conference on Decision and Control*, pages 3844–3849, Bahamas, December 2004.

R. Tempo, G. Calafiore, and F. Dabbene. *Randomized Algorithms for Analysis and Control of Uncertain Systems*. Springer, New York, 2005.

M. Vidyasagar. Statistical learning theory and randomized algorithms for control. *IEEE Control Systems*, pages 69–85, December 1998.

M. Vidyasagar and R.L. Karandikar. A learning theory approach to system identification. In *Technical committe on robust control*, Portugal, July 2002. Workshop of IFAC.

T. Wong, W. Luk, and P. Heng. Sampling with hammersley and halton points. *Journal of Graphics Tools*, (2):9–24, 1997.

Nonlinear Control of the Doubly Fed Induction Motor with Copper Losses Minimization for Electrical Vehicle

S. DRID*, M.-S. NAIT-SAID*, M. TADJINE** and A. MAKOUF*

*Laboratory of Electromagnetic Induction and Propulsion Systems "LSPIE", Electrical engineering department; University of Batna, Rue M.E.H Boukhlof, Algeria, Tel/Fax: +213.33.81.51.23 e-mail: s_drid@yahoo.fr.
** Control Laboratory, Electrical engineering department, ENP Algiers, 10 Av. Hassen Badi, B182. Algiers, Algeria

Abstract: There is an increasing interest in electric vehicles due to environmental concerns. Recent efforts are directed toward developing an improved propulsion system for electric vehicles applications with minimal power losses. This paper deals with the high efficient vector control for the reduction of copper losses of the doubly fed motor. Firstly, the feedback linearization control based on Lyapunov approach is employed to design the underlying controller achieving the double fluxes orientation. The fluxes controllers are designed independently of the speed. The speed controller is designed using the Lyapunov method especially employed to the unknown load torques. The global asymptotic stability of the overall system is theoretically proven. Secondly, a new Torque Copper Losses Factor is proposed to deal with the problem of the machine copper losses. Its main function is to optimize the torque in keeping the machine saturation at an acceptable level. This leads to a reduction in machine currents and therefore their accompanied copper losses guaranteeing improved machine efficiency. The simulation results in comparative presentation confirm largely the effectiveness of the proposed DFIM control with a very interesting energy saving contribution.

Keywords: Doubly Fed Induction Machine, Vector Control, Flux Orientation, Lyapunov Function, Nonlinear Feedback Control, Optimization, Copper Losses

1. INTRODUCTION

The electric vehicle (EV) was conceived in the middle of the previous century. EV's offer the most promising solutions to reduce vehicular emissions. EV's constitute the only commonly known group of automobiles that qualify as zero-emission vehicles. These vehicles use an electric motor for propulsion, batteries as electrical-energy storage devices and associated with power electronics, microelectronics, and microprocessor control of motor drives.

The doubly fed induction motor (DFIM) is a wound rotor asynchronous machine supplied by the stator and the rotor from two external source voltages. This machine is very attractive for the variable speed applications such as the electric vehicle and the electrical energy production [1-5]. Consequently, it covers all power ranges. Obviously, the requested variable speed domain and the desired performances depend of the application kinds [1-7].

Sponsor and financial support acknowledgment goes here

The use of DFIM offers the opportunity to modulate power flow into and out of the rotor winding in order to have, at the same time, a variable speed in the characterized super–synchronous or sub–synchronous modes in motor or in generator regimes. Two modes can be associated to slip power recovery: sub–synchronous motoring and super-synchronous generating operations. In general, while the rotor is fed through a cycloconverter, the power range can attain the MW order which presents the size power often reserved to the synchronous machine. [1-10].The DFIM has some distinct advantages compared to the conventional squirrel-cage machine. The DFIM can be controlled from the stator or rotor by various possible combinations. The disadvantage of two used converters for stator and rotor supplying can be compensated by the best control performances of the powered systems [3]. Indeed, the input–commands are done by means of four precise degrees of control freedom relatively to the squirrel cage induction machine where its control appears quite simple. The flux orientation strategy can transform the non linear and coupled DFIM-mathematical model into a linear model leading to one attractive solution for generating or motoring operations [14].

It is known that the motor driven systems account for approximately 65% of the electricity consumed in the world. Implementing high efficiency motor driven systems, or improving existing ones, could save over 200 billion kWh of electricity per year. This issue has become very important especially following the economic crisis due to the oil prices raising, the new energy saving technologies are appearing and developing rapidly in this century [4], [12] and [15-18].

In this framework, the DFIM continues to find great interest since the birth of the idea of the double flux orientation [19-20]. The philosophy of this idea is to get a simpler machine model expression (ideal machine) [19]. Consequently, in the same time, we can solve a non linear problem presented by the DFIM control and step up from many digital simulations toward the experimental test by the use of the system dSPACE-1103. This method gives entire satisfaction and consolidates our theory, especially using the *Torque Optimization Factor* TOF strategy [20]. Always the search for a solution has more optimal, us nap leans towards the minimization of the copper losses in the DFIM.

In this paper we developed a new optimization factor *Torque Copper Losses Optimization* TCLO. The article will be organized as follows. The DFIM mathematical model is presented in section III. In section IV, the feedback linearization is exposed. Section V concerns the two energy torque optimization strategies TOF and TCLO. In the section VI, simulation results are exposed and comparative illustration shows the performances in energy saving between TOF and TCLO

2. THE DFIM MODEL

Its dynamic model expressed in the synchronous reference frame is given by

Voltage equations:

$$\begin{cases} \bar{u}_s = R_s \, \bar{i}_s + \dfrac{d\bar{\phi}_s}{dt} + j\omega_s \bar{\phi}_s \\ \bar{u}_r = R_r \, \bar{i}_r + \dfrac{d\bar{\phi}_r}{dt} + j\omega_r \bar{\phi}_r \end{cases} \quad (1)$$

Flux equations:

$$\begin{cases} \bar{\phi}_s = L_s \, \bar{i}_s + M \, \bar{i}_r \\ \bar{\phi}_r = L_r \, \bar{i}_r + M \, \bar{i}_s \end{cases} \quad (2)$$

From (1) and (2), the state-all-flux model is written like:

$$\begin{cases} \bar{u}_s = \dfrac{1}{\sigma T_s}\bar{\phi}_s - \dfrac{M}{\sigma T_s L_r}\bar{\phi}_r + \dfrac{d\bar{\phi}_s}{dt} + j\omega_s \bar{\phi}_s \\ \bar{u}_r = -\dfrac{M}{\sigma T_r L_s}\bar{\phi}_s + \dfrac{1}{\sigma T_r}\bar{\phi}_r + \dfrac{d\bar{\phi}_r}{dt} + j\omega_r \bar{\phi}_r \end{cases} \quad (3)$$

The electromagnetic torque is done as

$$C_e = \dfrac{PM}{\sigma L_s L_r} \Im m\!\left[\bar{\phi}_s \bar{\phi}_r^*\right] \quad (4)$$

The copper losses are giving as:

$$P_{cl} = R_s i_s^2 + R_r i_r^2 \quad (5)$$

The motion equation is:

$$C_e - d = J \dfrac{d\omega}{dt} \quad (6)$$

In DFIM operations, the stator and rotor mmf's (magneto motive forces) rotations are directly imposed by the two external voltage source frequencies. Hence, the rotor speed becomes depending toward the linear combination of theses frequencies, and it will be constant if they are too constants for any load torque, given of course in the machine stability domain. In DFIM modes, the synchronization between both mmf's is mainly required in order to guarantee machine stability [7]. This is the similar situation of the synchronous machine stability problem where without the recourse to the strict control of the DFIM mmf's relative position, the machine instability risk or brake down mode become imminent.

3. NONLINEAR VECTOR CONTROL STRATEGY

3.1. Double flux orientation:

It consists in orienting, at the same time, stator flux and rotor flux. Thus, it results the constraints given below by (7). Rotor flux is oriented on the *d*-axis, and the stator flux is oriented on the *q*-axis. Conventionally, the *d*-axis remains reserved to magnetizing axis and *q*-axis to torque axis, so we can write [19-20]

$$\begin{cases} \phi_{sq} = \phi_s \\ \phi_{rd} = \phi_r \\ \phi_{sd} = \phi_{rq} = 0 \end{cases} \quad (7)$$

Using (7), the developed torque given by (4) can be rewritten as follows:

$$C_e = k_c \phi_s \phi_r. \quad (8)$$

where, $k_c = \dfrac{PM}{\sigma L_s L_r}$

ϕ_s Appears as the input command of the active power or simply of the developed torque, while ϕ_r appears as the input command of the reactive power or simply the main magnetizing machine system acting.

3.2. Vector control by Lyapunov feedback linearization

Separating the real and the imaginary part of (3), we can write:

$$\begin{cases} \dfrac{d\phi_{sd}}{dt} = f_1 + u_{sd} \\ \dfrac{d\phi_{sq}}{dt} = f_2 + u_{sq} \\ \dfrac{d\phi_{rd}}{dt} = f_3 + u_{rd} \\ \dfrac{d\phi_{rq}}{dt} = f_4 + u_{rq} \end{cases} \quad (9)$$

Where f_1, f_2, f_3 and f_4 are done as follows :

$$\begin{cases} -f_1 = \gamma_1 \phi_{sd} - \gamma_2 \phi_{rd} - \omega_s \phi_{sq} \\ -f_2 = \gamma_1 \phi_{sq} - \gamma_2 \phi_{rq} + \omega_s \phi_{sd} \\ -f_3 = -\gamma_3 \phi_{sd} + \gamma_4 \phi_{rd} - \omega_r \phi_{rq} \\ -f_4 = -\gamma_3 \phi_{sq} + \gamma_4 \phi_{rq} + \omega_r \phi_{rd} \end{cases} \quad (10)$$

With: $\gamma_1 = \dfrac{1}{\sigma T_s}$; $\gamma_2 = \dfrac{M}{\sigma T_s L_r}$; $\gamma_3 = \dfrac{M}{\sigma T_r L_s}$; $\gamma_4 = \dfrac{1}{\sigma T_r}$

Tacking into account of the constraints given by (7), one can formulate the Lyapunov function as follows

$$V = \frac{1}{2}\phi_{sd}^2 + \frac{1}{2}\phi_{rq}^2 + \frac{1}{2}(\phi_{sq} - \phi_s)^2 + \frac{1}{2}(\phi_{rd} - \phi_r)^2 > 0 \quad (11)$$

From (11), the first and second quadrate terms concern the fluxes orientation process defined in (7) with the third and fourth terms characterizing the fluxes feedback control. Where its derivative function becomes

$$\dot{V} = \phi_{sd}\dot{\phi}_{sd} + \phi_{rq}\dot{\phi}_{rq} + (\phi_{sq} - \phi_s)(\dot{\phi}_{sq} - \dot{\phi}_s) + (\phi_{rd} - \phi_r)(\dot{\phi}_{rd} - \dot{\phi}_r) \quad (12)$$

Substituting (9) in (12), it results

$$\dot{V} = \phi_{sd}(f_1 + u_{sd}) + \phi_{rq}(f_4 + u_{rq}) + (\phi_{sq} - \phi_s)(f_2 + u_{sq} - \dot{\phi}_s) + (\phi_{rd} - \phi_r)(f_3 + u_{rd} - \dot{\phi}_r) \quad (13)$$

Let us define the following law control as [21]:

$$\begin{cases} u_{sd} = -f_1 - K_1 \phi_{sd} \\ u_{rq} = -f_4 - K_2 \phi_{rq} \\ u_{sq} = -f_2 + \dot{\phi}_s - K_3 (\phi_{sq} - \phi_s) \\ u_{rd} = -f_3 + \dot{\phi}_r - K_4 (\phi_{rd} - \phi_r) \end{cases} \quad (14)$$

Hence (14) replaced in (13) gives:

$$\dot{V} = -K_1 \phi_{sd}^2 - K_2 \phi_{rq}^2 - K_3 (\phi_{sq} - \phi_s)^2 - K_4 (\phi_{rd} - \phi_r)^2 < 0 \quad (15)$$

The function (15) is negative one. Furthermore, (14) introduced into (9) leads to a stable convergence process if the gains K_i (i=1, 2,3, 4) are evidently all positive, otherwise:

$$\begin{cases} \lim_{t \to +\infty} \phi_{sd} = 0 \\ \lim_{t \to +\infty} \phi_{rq} = 0 \\ \lim_{t \to +\infty}(\phi_{rd} - \phi_r^*) = 0 \\ \lim_{t \to +\infty}(\phi_{sq} - \phi_s^*) = 0 \end{cases} \quad (16)$$

In (16), the first and second equations concern the double flux orientation constraints applied for DFIM-model which are define above by (7), while the third and fourth equations define the errors after the feedback fluxes control. This latter offers the possibility to control the main machine magnetizing on the d-axis by ϕ_{rd} and the developed torque on the q-axis by ϕ_{sq}.

4. ENERGY OPTIMIZATION STRATEGY

In this section we will explain why and what is the optimization strategy used in this work. Fig. 1 illustrates the problem which occurs in the proposed DFIM vector control system when the machine magnetizing excitation is maintained at a constant level.

4.1. Why the energy optimization strategy?

Considering an iso-torque-curve (hyperbole form), drawn from (8) for a constant torque in the (ϕ_s, ϕ_r) plan and lower load machine (Fig.1), on which we define two points **A** and **B**, respectively, corresponding to the two machine magnetizing extreme levels. Theses points concern respectively an excited machine ($\phi_r = 1 Wb = Const$) and an under excited machine ($\phi_r = 0.1 Wb = Const$). Both points define the steady state operation machine or equilibrium points. The machine rotates to satisfy the required reference speed acted by a given slope speed acceleration $\alpha = \dfrac{d\Omega}{dt} = Const$. So, the machine in both magnetizing cases must develop a transient torque such as:

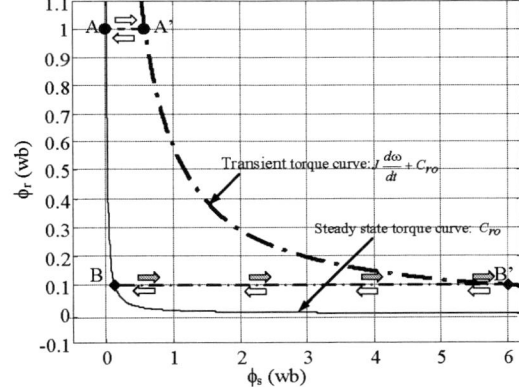

Fig.1 Illustration of the posed problem in the DFIM control system with constant excitation.

$$C_{eT} = C_{r0} + J\frac{d\Omega}{dt} = C_{ro} + J\alpha \quad (17)$$

On the same graph, we define a second iso-torque-curve C_{eT}=Const in the (ϕ_s, ϕ_r) plan. This curve is a transient one on which we place two transient points **A'** and **B'**. Here we distinguish the first transitions **A–A'** and **B–B'** due to the acceleration set, respectively for each magnetizing case. Both transitions are rapidly occurring in respect to the adopted control.

Once the machine speed reaches its reference, the inertial torque is cancelled ($\alpha = 0$), then the developed torque must return immediately to the initial load torque C_{ro}, characterized by the second transitions **A'–A** and **B'–B** towards the preceding equilibrium points **A** and **B**. One can notice that during the transition **B–B'**, corresponding to the under excited machine, the stator flux can attain very high values greater than the tolerable limit ($\phi_{s\max}$), and can tend to infinite values if the load torque C_{ro} tends to zero. So the armature currents expressed by the following formula deduced from (2) and (7) are strongly increased and can certainly destruct the machine and their supplied converters.

$$\begin{aligned}\bar{i}_s &= \lambda.\phi_r + j\gamma.\phi_s \\ \bar{i}_r &= \chi.\phi_r + j\lambda.\phi_s\end{aligned} \quad (18)$$

Where, $\lambda = -\dfrac{M}{\sigma.L_s.L_r}$; $\gamma = \dfrac{1}{\sigma.L_s}$; $\chi = \dfrac{1}{\sigma.L_r}$

In the other hand, for the case **A** (excited machine), if the **A–A'** transition remains tolerable, the armature currents can present prohibitory magnitude in the steady state operation due to the orthogonal contribution of stator and rotor fluxes at the moment that the machine is sufficiently excited. The steady state armature currents can be calculated by (18), where we can note the amplification effect of the coefficients λ, γ and χ.

4.2. Torque Optimization Factor (TOF) design

In the previous sub-section, the problem is in the transient torque, especially when the machine is low loaded. So it becomes very important to minimize the torque transition such as [20]:

$$\frac{dC_e}{dt} \rightarrow 0 \quad (19)$$

where, $dC_e = \dfrac{\partial C_e}{\partial \phi_s}d\phi_s + \dfrac{\partial C_e}{\partial \phi_r}d\phi_r$ (20)

This condition should be realized respecting the stator flux constraint given by

$$\phi_s \leq \phi_{s\max} \quad (21)$$

In this way the rotor and stator fluxes, though orthogonal, their modulus will be related by the so-called TOF strategy which will be designed from the resolution of the differential equations (19-20) with constraint (21) as follows:

$$\begin{cases}\dot{\phi}_s\phi_r + \dot{\phi}_r\phi_s = 0 \\ \phi_s \leq \phi_{s\max}\end{cases} \quad (22)$$

from (21) we can write

$$-\dot{\phi}_s\phi_r = \dot{\phi}_r\phi_s \leq \dot{\phi}_r\phi_{s\max} \quad (23)$$

thus,

$$-\frac{\dot{\phi}_s}{\phi_{s\max}} \leq \frac{\dot{\phi}_r}{\phi_r} \quad (24)$$

the resolution of (24) leads to

$$-\frac{\phi_s}{\phi_{s\max}} + C \leq \ln \phi_r \quad (25)$$

where C is an arbitrary integration constant, therefore

$$\phi_r \geq e^{(\frac{\phi_s}{\phi_{s\max}} - C)} \quad (26)$$

Since, the main torque input-command in motoring DFIM operation is related to the stator flux, it becomes dependent on the speed rotor sign and thus we can write

$$\phi_{sq} = \phi_s \, \text{sgn}(\Omega) = \begin{cases} +\phi_s & if \quad \Omega > 0 \\ -\phi_s & if \quad \Omega < 0 \end{cases} \quad (27)$$

with (27), (26), the rotor flux may be rewritten as follows

$$\phi_r = e^{(\frac{|\phi_{sq}|}{\phi_{s\max}} - C)} \quad (28)$$

The resolution of (24) gives place to the arbitrary integration constant C from which the TOF-relationship (28) can be easily tuned. This one can be adjusted by a judicious choice of the integration constant, while figure 2 presents TOF effect on armature DFIM currents with C-tuning. Note that this method offers the possibility to reduce substantially the magnitude of the armature currents into the machine and we can notice an increase in energy saving. Hence using TOF strategy, we can avoid the saturation effect and reduce the magnitude of machine currents from which the DFIM efficiency could be clearly enhanced.

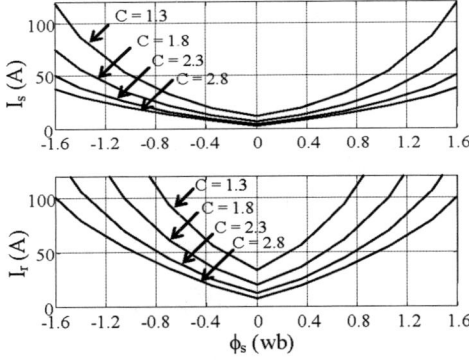

Fig. 2 TOF effect on armature DFIM currents

4.3. Torque–Copper Losses Optimization (TCLO) design

The objective is to find the relation between fluxes which can optimize the compromise between torque and copper losses in steady state as well as in transient state, (i.e. for all $\{C_e\}$ find (ϕ_s,ϕ_r) *let* min$\{P_{cl}\}$). The *Rolle's Theorem* is the key result behind applications of the derivative to optimization problems. The second derivative test is used to finding

minimum point. From (5), (8) and (18), the torque and copper losses can be to written as:

$$\begin{cases} C_e = k_c.\phi_r.\phi_s \\ P_{cl} = a_1\phi_r^2 + a_2\phi_s^2 \end{cases} \quad (29)$$

with :
$$a_1 = (\frac{R_r}{(\sigma.L_r)^2} + \frac{R_s M^2}{(\sigma.L_r.L_s)^2})$$
$$a_2 = (\frac{R_r M^2}{(\sigma.L_r.L_s)^2} + \frac{R_s}{(\sigma.L_s)^2})$$

The figure 3 represents the layout of (29) for a constant level of torque and copper losses in the (ϕ_s, ϕ_r) plan. These curves present respectively a hyperbole for the iso-torque and ellipse for iso-copper-losses.

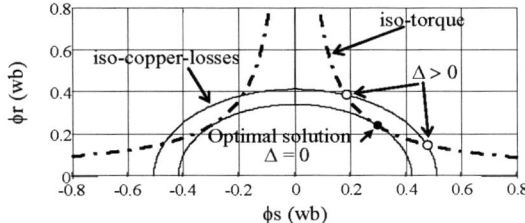

Fig. 3 The iso–torque curves and the iso–losses curves in the plan (ϕ_s, ϕ_r)

Figure 4 illustrates a general block diagram of the suggested DFIM control scheme. Here, we can note the placement of optimization block, the first estimator-block which evaluates torque and the second estimator-block which evaluates firstly the modulus and position fluxes, respectively ϕ_s, ϕ_r, ρ_s and ρ_r, from the measured currents using (2) and secondly the feedback functions f_1, f_2, f_3, f_4 given by (10). Optimization process allows adapting the main flux magnetizing defined by rotor flux to the applied load torque characterized by the stator flux. With the analogical switch we can select the type of the reference rotor flux. The switch position 1, 2 gives respectively TCLO and TOF for optimized operation and the position 3 for a magnetizing constant level.

5. SIMULATION RESULTS

The Figure 5 shows the speed response versus time according to its desired profile drawn on the same figure. Figure 6 illustrate the fluxes trajectory of the closed–loop system. It moves along manifold toward the equilibrium point. We can notice the stability of the system. Figures 7 and 8 show respectively the stator and the rotor input control voltages versus time during the test. Figure 9 present the copper losses according to the stator flux variations in steady state operation and we can see the contribution of the TCLO compared to the TOF.

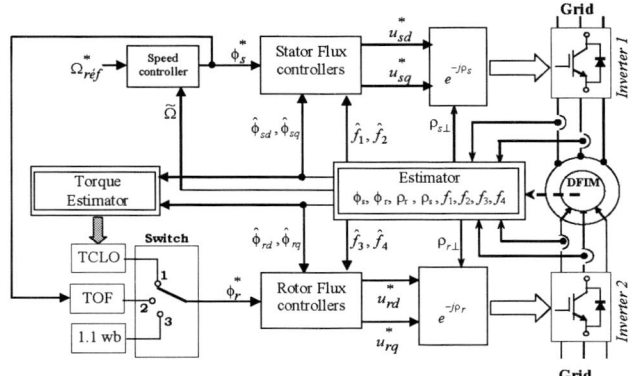

Fig. 4 General block diagram of control scheme

Finally figure 10 present the dissipated energy versus time from which we can observe clearly the influence of the three switch positions on the copper losses in transient state. We can conclude that the TCLO is the best optimization.

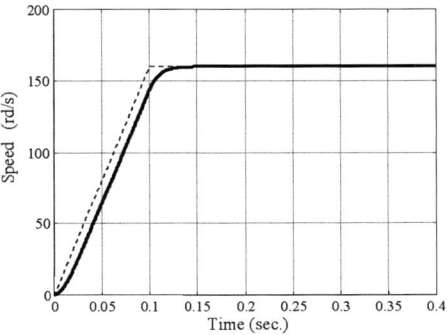

Fig. 5 Speed response

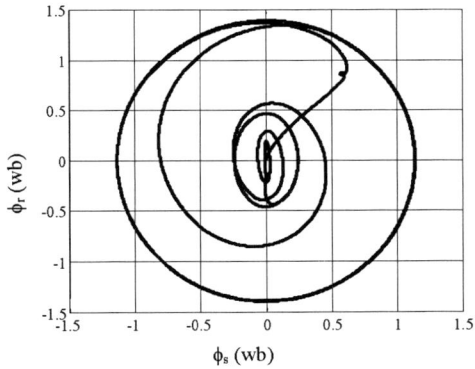

Fig. 6 Fluxes trajectories of the closed–loop system

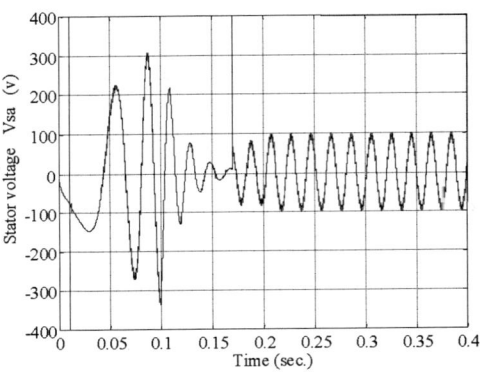

Fig. 7 The input control stator voltage response in the stator reference frames with TCLO

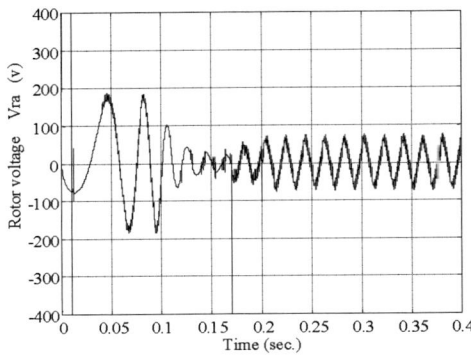

Fig. 8 The input control rotor voltage response in the stator reference frames with TCLO

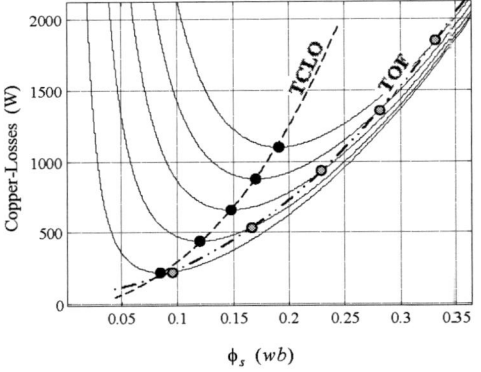

Fig. 9 Minimized copper losses in steady state operation with TOF and TCLO

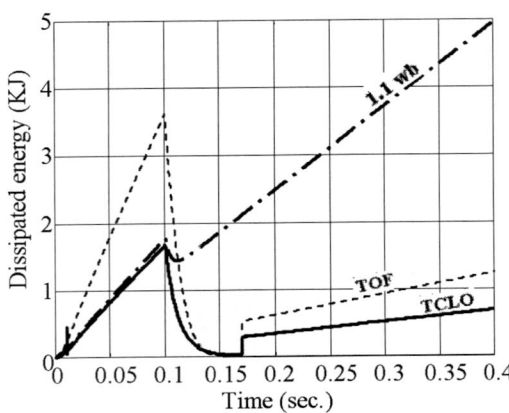

Fig. 10 Total copper losses versus time during test for the three switch positions (Energy saving illustration)

6. CONCLUSION

In this paper was presented a vector control intended for doubly fed induction motor (DFIM) mode. The use of the state-all-flux induction machine model with a flux orientation constraint gives place to a simpler control model. The stability of the nonlinear feedback control is proven using the Lyapunov function.

The simulation results of the suggested DFIM system control based on double flux orientation which is achieved by the proposed DFIM control demonstrates clearly the suitable obtained performances required by the references profiles defined above. The speed tracks its desired reference without any effect of the load torque. Therefore the high control performances can be well affirmed. To optimize the machine operation we chose to minimize the copper losses. The proposed TCLO factor performs better than the already designed TOF. Indeed, the energy saving process can be well achieved if the magnetizing flux decreases in the same way as the load torque. It results in an interesting balance between the core losses and the copper losses into the machine, so the machine efficiency may be largely improved. The simulation results confirm largely the effectiveness of the proposed DFIM control system.

REFERENCES

[1] Michael S. Vicatos, and John A. Tegopoulos, "A Doubly-Fed Induction Machine Differential Drive Model for Automobiles" *IEEE Transactions on Energy Conversion*, vol.18, N°: 2, June 2003 pp: 225-230.

[2] Akagi, H.; Sato, H., "Control and Performance of a Flywheel Energy Storage System Based on a Doubly-Fed Induction Generator-Motor", *IEEE Power Electronics Specialists Conference*, 1999. PESC 99. 30th Annual vol1, 27 June-1 July 1999, pp32-39.

[3] Debiprasad Panda, Eric L. Benedict, Giri Venkataramanan and Thomas A. Lipo "A Novel Control Strategy for the Rotor Side Control of a Doubly-Fed Induction Machine" 2001 *IEEE*, pp: 1695-1702.

[4] W. Leonhard, "Control Electrical Drives ", Springier verlag Berlin Heidelberg 1997, Printed in Germany.

[5] Wang S. and Ding Y. " Stability Analysis of Field Oriented doubly Fed induction Machine drive Based on Computed Simulation", *Electrical Machines and Power Systems* (Taylor & Francis), 1993.

[6] L. Morel et al. "Double-fed induction machine: converter optimisation and field oriented control without position sensor", *IEE Proc. Electr. Power Appl.* Vol. 145. N°4 July 1998.

[7] B. Hopfensperger et al. "Stator flux oriented control of a cascaded doubly fed induction machine", *IEE Proc. Electr. Power Appl.* Vol. 146. N°6 November 1999.

[8] B. Hopfensperger et al. "Stator flux oriented control of a cascaded doubly fed induction machine with and without position encoder ", *IEE Proc. Electr. Power Appl.* Vol. 147. N°4 July1999, pp241-250.

[9] H.M.B. Metwally et al. "Optimum performance characteristics of doubly fed induction motors using field oriented control", *Energy conversion and Management* 43 (2002) 3-13.Elsevier Science.

[10] Hirofumi Akagi and Hikaru Sato. "Control and Performance of a Doubly fed induction Machine Intended for a Flywheel Energy Storage System", *IEEE Transactions on Power Electronics*, Vol. 17 No 1 January 2002.

[11] Djurovic M. et al. "Double Fed Induction Generator with Two Pair of Poles". *Conferences of Electrical Machines and Drives (IEMDC)*, 11-13 September 1995, Conference Publication, N°412 IEE 1995.

[12] Leonhard W., "Adjustable-Speed AC Drives" Invited Paper, Proceedings of the IEEE, vol. 76, Issue: 4, April 1988, pp.455-471.

[13] Longya Xu, Wei Cheng "Torque and Reactive Power control of a Doubly Fed Induction Machine by Position Position Sensorless Scheme". *IEEE Transactions on Industry Applications*, vol. 31,n° 3, May/June 1995, pp 636-642.

[14] Sergei Peresada, Andrea Tilli, and Alberto Tonielli, "Indirect Stator Flux-Oriented Output Feedback Control of a Doubly Fed Induction Machine," *IEEE Trans. On control Systems Technology*, vol. 11, pp. 875–888, Nov. 2003.

[15] Wang, D.H.; Cheng, K.W.E., "General discussion on energy saving Power Electronics Systems and Applications," 2004. Proceedings. 2004 First International Conference on 9-11 Nov. 2004 pp298-303.

[16] Li Zang, Hasan K H, "Neural Network Aided Energy Efficiency control for a Field Orientation Induction Machine Drive," Proceeding of Ninth International conference on Electrical Machine and Drives, Conference publication N.468, IEE 1999.

[17] David E. Rice, "A Suggested Energy-Savings Evaluation Method for AC Adjustable-Speed Drive Applications," *IEEE Trans. on Industry Applications*, vol. 24, N.6 , Nov/Dec. 1988 pp1107-1117

[18] Jaime Rodriguez et al, "Optimal Vector Control of Pumping and Ventilation Induction Motor Drives," *IEEE Trans. on Electronics Industry*, vol. 49, N. 4, AUGUST 2002 889-895.

[19] S Drid, M.S. Nait_Said and M. Tadjine, "Double flux oriented control for the doubly fed induction motor," *Electric Power Components & Systems Journal*, vol. 33, N.10, October 2005, USA.

[20] S Drid, M. Tadjine and M.S. Nait_Said, "Nonlinear Feedback Control and Torque Optimization of a Doubly Fed Induction Motor," *JEEEC Journal of Electrical Engineering Elektrotechnický časopis*, vol. 56, N.3-4, 2005, pp:57-63, Slovakia.

[21] Khalil H., "Nonlinear systems" Prentice–Hall, 2ed edition 1996, Printed in USA

Appendix

The machine parameters are:

Rs =1.2 Ω; Ls =0.158 H; Lr =0.156 H; Rr =1.8 Ω; M =0.15 H; P =2 ; J = 0.07 Kg.m² ; Pn = 4 Kw ; 220/380V ; 50Hz ; 1440tr/min ; 15/8.6 A ; cosφ = 0.85.

NOMENCLATURE

s, r	Rotor and stator indices.
d, q	Direct and quadrate indices for orthogonal components
$\bar{x}$	Variable complex such as: $\bar{x} = \Re e[\bar{x}] + j.\Im m[\bar{x}]$.
$\bar{x}$	It can be a voltage as $\bar{u}$, a current as $\bar{i}$ or a flux as $\bar{\phi}$
$\bar{x}^*$	Complex conjugate
R_s, R_r	Stator and rotor resistances
L_s, L_r	Stator and rotor inductances
T_s, T_r	Stator and rotor time-constants ($T_{s,r}=L_{s,r}/R_{s,r}$)
σ	Leakage flux total coefficient ($\sigma =1- M^2/L_rL_s$)
M	Mutual inductance
θ	Absolute rotor position
P	Number of pairs poles
δ	Torque angle
ρ_s, ρ_r	Stator and rotor flux absolute positions
ω	Mechanical rotor frequency (rd/s)
Ω	Rotor speed (rd/s)
ω_s	Stator current frequency (rd/s)
ω_r	Induced rotor current frequency (rd/s)
J	Inertia
d	Unknown load torque
C_e	Electromagnetic torque
$\sim$	Symbol indicating measured value
$\wedge$	Symbol indicating the estimated value
$*$	Symbol indicating the command value
DFIM	Doubly Fed Induction Machine
TOF	Torque Optimization Factor
TCLO	Torque Copper Losses Optimization

Sub-Optimal Motion Planner of Wheeled Mobile Manipulators With Under-Actuated Platform

M. Haddad*, T. Chettibi*, H. E. Lehtihet*, W. Khalil**, F. Boyer**

*Laboratory of Structure Mechanics, E.M.P., B.P. 17, Bordj El-Bahri,16111, Algiers, Algeria.
(e-mail: Haddadmoussa2003@yahoo.fr he.lehtihet@gmail.com).
** IRCCyN- UMR CNRS 6597, Ecole Centrale de Nantes B.P. 92 101- 44321 Nantes, France.
(e-mail: Wisama.Khalil@irccyn.ec-nantes.fr)

Abstract: We propose a stochastic optimization scheme for point-to-point trajectory planning of nonholonomic wheeled mobile manipulators with under-actuated platform. The problem is known to be complex, particularly if dynamic constraints are considered. The proposed method consists of extending to under-actuated systems the random-profile approach recently applied to wheeled mobile robots. This versatile method handles constraints on: *geometry* (obstacle avoidance, bounds on joint positions and path curvature), *kinematics* (bounded velocities and accelerations), *dynamics* (bounded torques, stability issues). It may be applied using a cost function involving travel time, efforts and power. A solution is presented for a planar wheeled manipulator with under-actuated platform undertaking a point-to-point generalized task.

Keywords: wheeled mobile manipulator, under-actuated systems, trajectory planning, stochastic optimization, random profile approach.

1. INTRODUCTION

Under-actuated systems are mechanical systems with fewer actuators (i.e. controls) than degree of freedom of the system and subjected to second-order nonholonomic constraints. Trajectory planning and control for such systems is currently an active field of research due to their broad applications in robotics, biomechanical system, aerospace vehicles, marine vehicles, etc.

A mechanical system may be under-actuated for different reasons. For example:
(i) Dynamics of the system: aerospace vehicles (Shirazi and Ghaari-Saadat, 2004), undulatory locomotion (Ostrowski et al., 1995);
(ii) Study of the dynamic behavior of animals (Hirose 1993), (Henning et al., 1998), (Prausch et al., 2000), (Chernousko, 2004);
(iii) Actuator failure (Arai and Tachi, 1991), etc.

Here, we consider the generalized point-to-point trajectory-planning problem for a Wheeled Mobile Manipulator with Under-Actuated Platform (WMM-UAP). The method we propose is an adaptation of the Random Profile Approach (RPA), which has been applied to WMMs (Haddad et al., 2006) and to wheeled platforms (Haddad et al., 2007). The RPA is a broad-spectrum technique that was initially proposed in (Chettibi and Lehtihet, 2002) for trajectory planning of fixed-base manipulators. It uses the concept of time-scaling (Hollerbach, 1983) and the concept of trajectory profiles mapping real trajectories over the unit interval. It converts non-tight kinodynamic constraints to bounds on the travel time and reduces the dynamic problem to a constrained-profile optimization subsequently solved via a stochastic scheme. Although it is expected to yield sub-optimal solutions, the RPA is remarkably flexible. It is applicable in various types of problems (including those with discontinuous dynamic models involving friction efforts) and for various types of cost functions and constraints.

Let us first review some relevant papers. Oriolo and Nakamura (1991) derived conditions for systems with nonholonomic acceleration constraint and proved that an equilibrium point cannot be stabilized using any smooth state feedback in the absence of potential forces. Later, the stabilization of a planar 2R robot with a passive elbow joint has been obtained in (Nakamura et al., 1997) where a time-varying feedback is designed via Poincaré map analysis. In (De-Luca et al., 2000) an iterative steering technique is used to design a non-smooth feedback that guarantees robust convergence to a desired configuration for an under-actuated manipulator. Aria et al. (1998) presented a simple method to plan and control point-to-point transfer of a 2 d.o.f. under-actuated manipulator. It is based on time scaling of active joint and a bi-directional motion planning from the desired limit configurations. Olfati-saber (2001) proposed to use an explicit change of coordinates and control, which transform the under-actuated system into a cascade of nonlinear systems with structural properties and make the control design easier. This decomposition is based on the concept of normalized generalized momentums. The author exposes a large number of applications for various under-actuated systems (flexible arms, helicopters, acrobat, pendubot…).

2. DYNAMIC MODEL

Let $\Re = (O, x, y, z)$ be the fixed frame of the world coordinates (Fig. 1). The WMM-UAP considered here is a series-chain multi-link manipulator mounted on an under-actuated wheeled platform. We assume that the motion of the platform is confined to the $(O, \mathbf{x}, \mathbf{y})$ plan and that the main part of this platform is a rigid chassis with non-deformable wheels. All the wheels are assumed free (not power-driven and not equipped with brakes). The type of the platform is not specified. It may belong to any of the nonholonomic wheeled robots discussed in Campion et al. (1996).

Let $^{MP}\mathfrak{R} = (^{MP}O, {}^{MP}\mathbf{x}, {}^{MP}\mathbf{y}, {}^{MP}\mathbf{z})$ and $^{EE}\mathfrak{R} = (^{EE}O, {}^{EE}\mathbf{x}, {}^{EE}\mathbf{y}, {}^{EE}\mathbf{z})$ be the moving frames linked, respectively, to the mobile platform and to the end-effector's. The operational coordinates are, in the most general case, described by a 6-vector $\mathbf{U}_e = [x_e, y_e, z_e, \theta_e, \beta_e, \varphi_e]^T$ that defines the situation of the end-effector's (Cartesian coordinates of ^{EE}O and Euler's orientation angles of $^{EE}\mathfrak{R}$ in $\mathfrak{R}$). The generalized coordinates are described by the n-vector $\mathbf{q} = [\mathbf{q}_p^T, \mathbf{q}_a^T]^T$, where the n_p-vector $\mathbf{q}_p$ and the n_a-vector $\mathbf{q}_a$ define, respectively, the configuration of the platform in $\mathfrak{R}$ and that of the arms in $^{MP}\mathfrak{R}$ (so that $n = n_p + n_a$).

The situation of the platform in $\mathfrak{R}$ is described by $\mathbf{U}_p = [x_p, y_p, h_p, \theta_p, 0, 0]^T$; but since h_p is the constant distance between ^{MP}O and O measured along the $\mathbf{z}$-axis, the situation of the platform is completely specified by the 3-vector $\mathbf{X}_p = [x_p, y_p, \theta_p]^T$. Thus, we may write $\mathbf{q}_p = [\mathbf{X}_p^T, \mathbf{q}_w^T]^T$, where $\mathbf{q}_w$ is the n_w-vector of coordinates associated to the rotation/orientation of the wheels in $^{MP}\mathfrak{R}$ (so that $n_p = 3 + n_w$). However, as an arbitrary number of choices of $\mathbf{q}_w$ may be associated to any given $\mathbf{U}_e$, it will be most useful to consider the partial generalized coordinates of the system (Bayle et al., 2003). These are given by the $(3 + n_a)$-vector $\mathbf{\Omega} = [\mathbf{X}_p^T, \mathbf{q}_a^T]^T$ that actually affects the end-effector's situation.

The platform is subjected to m nonholonomic constraints ($m < n_p$) that can be written in the following compact form (Campion et al., 1991):

$$A(\mathbf{q}_p)\dot{\mathbf{q}}_p = \vec{0} \quad (1)$$

where $A(\mathbf{q}_p)$ is a full-rank $(m \times n_p)$-matrix. Using the Lagrangian approach while considering (1), we can model the dynamics of the system as follows (Bloch et al., 1992):

$$M_{pp}(\mathbf{q})\ddot{\mathbf{q}}_p + M_{ap}(\mathbf{q})\ddot{\mathbf{q}}_a + \mathbf{c}_p(\mathbf{q},\dot{\mathbf{q}}) = A^T(\mathbf{q}_p)\Lambda \quad (2a)$$

$$M_{pa}(\mathbf{q})\ddot{\mathbf{q}}_p + M_{aa}(\mathbf{q})\ddot{\mathbf{q}}_a + \mathbf{c}_a(\mathbf{q},\dot{\mathbf{q}}) + \mathbf{g}(\mathbf{q}_a) = \tau_a \quad (2b)$$

where $M(\mathbf{q}) = \begin{bmatrix} M_{pp} & M_{ap} \\ M_{pa} & M_{aa} \end{bmatrix}$ and $\mathbf{c}(\mathbf{q},\dot{\mathbf{q}}) = \begin{bmatrix} \mathbf{c}_p \\ \mathbf{c}_a \end{bmatrix}$ are, respectively, the symmetric positive-defined $(n \times n)$ inertia matrix and the n-vector of centrifugal and Coriolis forces in which generalized velocities appear under a quadratic form. $\mathbf{g}(\mathbf{q}_a)$ is the n_a-vector of gravity forces. Λ is the m-vector of Lagrange's multipliers associated to nonholonomic constraints. τ_a is the n_a-vector of external forces and torques applied to the manipulator by the actuators.

Following Bloch et al. (1992), we write $A(\mathbf{q}_p) = [A_v \ A_u]$, where A_v is a $m \times (n_p - m)$-matrix and A_u is a non singular $(m \times m)$-matrix. Next, if we define the full-rank matrix:

$$R(\mathbf{q}_p) = \begin{bmatrix} I \\ -A_u^{-1} A_v \end{bmatrix},$$

where I is the $(n_p - m) \times (n_p - m)$ identity matrix, then it follows that $A(\mathbf{q}_p) R(\mathbf{q}_p) = \vec{0}$. Thus, we may pre-multiply (2a) by $R^T(\mathbf{q}_p)$ to eliminate Λ and to extract the $(n_p - m)$ second-order nonholonomic constraints of the under-actuated platform as follows:

$$R^T(\mathbf{q}_p)\left[M_{pp}(\mathbf{q})\ddot{\mathbf{q}}_p + M_{ap}(\mathbf{q})\ddot{\mathbf{q}}_a + \mathbf{c}_p(\mathbf{q},\dot{\mathbf{q}})\right] = \vec{0} \quad (3)$$

Expressions (1), (2b) and (3) give a full description of the dynamics of the system.

3. PROBLEM STATEMENT

The WMM is required to move freely from an initial configuration $\mathbf{\Omega}^{START} = [(\mathbf{X}_p^S)^T, (\mathbf{q}_a^S)^T]^T$ to a final configuration $\mathbf{\Omega}^{GOAL} = [(\mathbf{X}_p^G)^T, (free)]^T$. In other words, as shown in Figure 1, the final configuration $\mathbf{q}_a^G$ of the arms is not specified. We must find the trajectory $\mathbf{q}(t)$, the time history of the vector $\tau_a(t)$ of actuator efforts and the traveling time T so that boundary conditions are matched, constraints are respected and a given cost function is minimized.

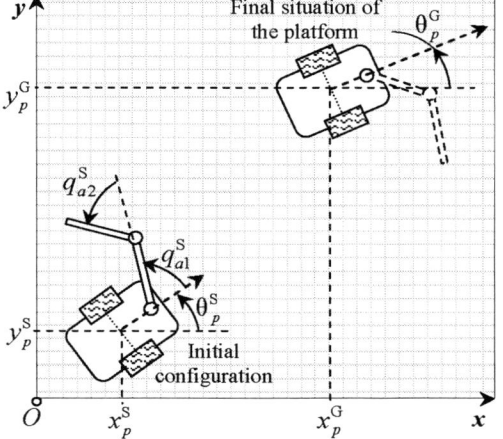

Fig. 1: Point-to-point trajectory planning problem for a WMM with under-actuated mobile platform

3.1 Constraints

In addition to nonholonomic constraints (1) and (3), the set of feasible trajectories is restricted by other constraints that must be satisfied during the travel from $\mathbf{\Omega}^{START}$ to $\mathbf{\Omega}^{GOAL}$. These constraints concern the boundary conditions and the physical limitations on the kinodynamic performances of the system.

a) Boundary conditions

The following conditions must be considered:
- *Position/Orientation*

$$\mathbf{\Omega}(t=0) = \mathbf{\Omega}^{START} \quad \text{and} \quad \mathbf{\Omega}(t=T) = \mathbf{\Omega}^{GOAL} \quad (4a)$$

- *Velocity*

$$\dot{\mathbf{q}}(t=0) = \vec{0} \quad \text{and} \quad \dot{\mathbf{q}}(t=T) = \vec{0} \quad (4b)$$

b) Physical limitations

Additional constraints that may have to be verified at each instant $t \in [0, T]$ represent other limitations such as those on:

- *Joint positions*

$$|q_i(t)| \leq q_i^{\max} \qquad i = 1 \ldots n \qquad (4c)$$

- *Active-joint velocities:*

$$|\dot{q}_{ai}(t)| \leq \dot{q}_{ai}^{\max} \qquad i = 1 \ldots n_a \qquad (4d)$$

- *Active-joint accelerations:*

$$|\ddot{q}_{ai}(t)| \leq \ddot{q}_{ai}^{\max} \qquad i = 1 \ldots n_a \qquad (4e)$$

- *Active-joint torques:*

$$|\tau_{ai}(t)| \leq \tau_{ai}^{\max} \qquad i = 1 \ldots n_a \qquad (4f)$$

3.2 Performance index

The goal function J represents the travel cost. For simplicity, we will adopt the following balance between the travel time T and the quadratic average of actuator efforts:

$$J = (1-\alpha)T + \alpha \int_0^T \sum_{i=1}^{n_a} \left(\frac{\tau_{ai}(t)}{\tau_{ai}^{\max}}\right)^2 dt \qquad (5)$$

α is a weight coefficient selected in [0, 1] according to the significance one would like to give to the minimization of T. The case $\alpha = 0$ corresponds to the minimum-time problem (i.e.: $J \equiv T$).

4. PROPOSED METHOD

We begin with an outline of the RPA as applied in (Haddad et al., 2006, 2007) to a WMM and to a wheeled platform. Then we will extend the RPA to the case of a WMM-UAP.

4.1 Overview of the RPA

Using the following time-scaling transformation, a trajectory $q(t)$ can be uniquely characterized by a travel time T and a time-evolution shape $\mathbf{Q}$:

$$q(t) = \mathbf{Q}(\xi(t)) = \mathbf{Q}(\xi) \circ \xi(t) \qquad (6)$$

The time-scale function is defined as $\xi(t) \equiv t / T$ so that meaningful values of ξ will belong to the unit interval. The *trajectory profile* $\mathbf{Q}(\xi)$ gives the shape of the time history of q from the start of the motion ($\xi = 0$) to its end ($\xi = 1$). Hereafter, the *prime* symbol will be reserved to indicate derivatives with respect to ξ.

A profile $\mathbf{Q}$ may be viewed also as a trajectory template that uniquely defines a class of trajectories having the same shape but distinctive travel times. The key point is that, for any given class $\mathbf{Q}$, we can apply a straightforward *clipping process* that will account for kinodynamic constraints and that will extract both the travel cost $J_\mathbf{Q}$ and the travel time $T_\mathbf{Q}$ of the most competitive member of this class. Thus, the difficult task of finding the optimal trajectory $q(t)^{\text{best}}$, with unknown T^{best}, can be reduced to that of finding only the class $\mathbf{Q}^{\text{best}}$ of this desired trajectory.

The RPA strives to achieve this goal using a nested master/slave optimization. The slave routine is the above-cited clipping. Given an input class $\mathbf{Q}$, the travel cost J is rewritten in terms of the single variable T. In addition, non-tight kinodynamic constraints, including stability considerations, are rewritten in terms of bounds on T. Thus, the clipping simply solves a one-dimensional minimization problem within some admissible time window(s).

This inner-level deterministic routine actually plays the role of a workhorse fitness function that is repeatedly called at the outer level by a master stochastic routine. The master routine explores the available class space using a simulated-annealing scheme (Hajek, 1985). Its built-in Metropolis algorithm randomly targets promising classes among those that would most likely lead to the global minimum of the cost function. In other words, the aimed output of the RPA is the most competitive member of the best class reachable.

Naturally, this difficult functional optimization is converted, as described next, to a more tractable parametric problem via a discrete representation of trajectory profiles. Namely, each random profile is generated from a finite set of free control nodes; continuity is achieved by fitting a smooth model that accounts for geometric constraints and for boundary conditions. Since a trajectory may be written in terms of a geometric path and a motion on this path, it is then convenient to express a trajectory profile as follows:

$$\mathbf{Q}(\xi) = \mathbf{P}(\lambda(\xi)) = \mathbf{P}(\lambda) \circ \lambda(\xi) \qquad (7)$$

$\mathbf{P}(\lambda)$ is a time-independent continuous sequence of configurations that defines completely the geometry of the robot path in the generalized-coordinate space as λ varies in [0, 1]. The shape of the time history of such a sequence is described by a monotonically increasing motion function $\lambda(\xi)$. Clearly, a feasible path $\mathbf{P}(\lambda)$ must match boundary configurations and must satisfy geometric constraints for any λ in [0, 1]. Whereas, a valid motion function must verify the following conditions:

$$\lambda(\xi = 0) = 0 \quad \text{and} \quad \lambda(\xi = 1) = 1 \qquad (8a)$$

$$\lambda'(\xi) > 0 \qquad \forall \xi \in {]0, 1[} \qquad (8b)$$

$$\lambda'(\xi = 0) = 0 \quad \text{and} \quad \lambda'(\xi = 1) = 0 \qquad (8c)$$

Conditions (8c) simply insure null limit velocities. This class partition into path and motion components offers some practical advantages. In particular, each part can make use of a specific fitting model as well as a separate set of control nodes. Indeed, a random class $\mathbf{Q}$ can be built using altogether three distinct sets S_m, S_a and S_p of control nodes. S_m is made up of control nodes in the (ξ, λ) plan (fig. 2) while S_a and S_p contain control nodes placed respectively, in the arm and in the platform configuration spaces.

Using S_m, a trial motion function can be generated by fitting a smooth curve through the $N_m + 2$ nodes of this set. As shown in Figure 2, the first and last nodes are held fixed according to (8a) but the N_m interior nodes are freely positioned while matching (8b) and (8c). Here, a clamped cubic-spline model (i.e.: user-defined end-point slopes) is well adapted to handle the first-order boundary conditions (8c). It will also insure continuity up to second-order while remaining computationally efficient.

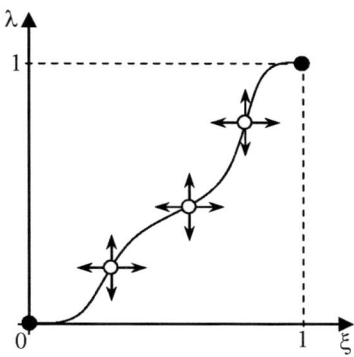

Fig. 2. A random motion function $\lambda(\xi)$ through $N_m + 2$ nodes. Free (fixed) nodes are shown in white (black)

Similarly, using the other two sets S_a and S_p, we can build a trial path $\mathbf{P}(\lambda)$ in the generalized-coordinate space. Here however a parametric B-spline model of order four or higher will be more suitable to match boundary conditions imposed on the robot position/orientation and to insure the continuity of joint positions, and later on, that of velocities, accelerations and torques.

In summary, the RPA is a stochastic technique that explores the available solution space directly at the class level. In fact, the whole trajectory problem boils down to finding the optimal positions of some randomly disrupted control nodes of the fitting models. Evidently, although the simulated-annealing scheme is not easily attracted by poorer local extrema, it may fail to find the global optimal solution of the problem. The major drawback, however, is that the search space itself is restricted to the sub-space reachable via the fitting models. Moreover, if obstacles are present in the workspace, the calculation scope will be further restricted to a user-defined corridor (Haddad et al., 2007). Therefore, solutions are expected to be sub-optimal. On the other hand, the RPA is a versatile tool that may provide reasonably good approximate solutions to a variety of practical dynamic problems.

4.2 Adaptation of the RPA to a WMM-UAP

In the case of a WMM-UAP, the profile $\mathbf{Q}_p$ of the platform is completely tied up to the profile $\mathbf{Q}_a$ of the arms through second-order nonholonomic constraints. Using the time-scaling (6), constraints (1) and (3) become:

$$A(\mathbf{Q}_p)\mathbf{Q}'_p = \vec{0} \quad (9a)$$

$$R^T(\mathbf{Q}_p)\left[M_{pp}(\mathbf{Q})\mathbf{Q}''_p + M_{ap}(\mathbf{Q})\mathbf{Q}''_a + c_p(\mathbf{Q},\mathbf{Q}')\right] = \vec{0} \quad (9b)$$

with the following boundary conditions:

$$\mathbf{X}_p(\xi=0) = \mathbf{X}_p^S \quad \text{and} \quad \mathbf{X}_p(\xi=1) = \mathbf{X}_p^G \quad (9c)$$

$$\mathbf{Q}'_p(\xi=0) = \vec{0} \quad \text{and} \quad \mathbf{Q}'_p(\xi=1) = \vec{0} \quad (9d)$$

It is clear that the RPA cannot be applied directly as given in the previous section. Now a valid trajectory profile $\mathbf{Q} = [\mathbf{Q}_p^T, \mathbf{Q}_a^T]^T$ must also satisfy (9).

The proposed modification consists of building the profile $\mathbf{Q}$ from $\mathbf{Q}_a$. Namely, for each randomly-generated profile $\mathbf{Q}_a$ of the arms, the corresponding profile $\mathbf{Q}_p$ of the platform will be deduced by a numerical integration of (9).

Similarly to (7), here again we can carry out a partition of $\mathbf{Q}_a(\xi)$ into a path and a motion component:

$$\mathbf{Q}_a(\xi) = \mathbf{Q}_a(\lambda(\xi)) = \mathbf{P}_a(\lambda) \circ \lambda(\xi) \quad (10)$$

A feasible path $\mathbf{P}_a$ must satisfy geometric constraint (4c) for any λ in [0, 1]. Moreover, it must match conditions (4a) imposed on the start configuration, namely:

$$\mathbf{P}_a(\lambda=0) = \mathbf{q}_a^S \quad (11)$$

A random profile $\mathbf{Q}_a(\xi)$ is then generated using two distinct sets S_m and S_a of free nodes. The set S_m, having $N_m + 2$ control points, is used as previously illustrated in figure 2 to build a motion profile $\lambda(\xi)$. The set S_a, having $N_a + 1$ control points, is used to generate a random fourth-order B-spline path $\mathbf{P}_a(\lambda)$ in the manipulator C-space. The first node is fixed according to (11) while the N_a other nodes are freely disrupted within the limits imposed by (4c). A typical example is given in the following Figure 3, which illustrates one single component of a random path $\mathbf{P}_a(\lambda)$.

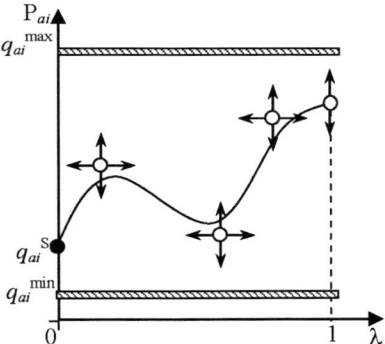

Fig. 3. A component of a trial path $P_a(\lambda)$ of the arms through $N_a + 1$ control points. Free (fixed) points are shown in white (black).

However, given a random profile $\mathbf{Q}_a(\xi)$, the numerical integration of (9b) starting at $\xi = 0$ will not, in general, yield a solution $\mathbf{Q}_p(\xi)$ that satisfies conditions (9c) and (9d) imposed at $\xi = 1$. In order to force the stochastic process to converge to a valid solution $\mathbf{Q}_p(\xi)$, we penalize in the cost function J any discrepancy observed at $\xi = 1$ between the calculated solution and the desired goal conditions. In other words, we treat the boundary-value problem using a shooting technique imbedded within the stochastic process that minimizes the travel cost.

Before we give an example, we need first to explain how to rewrite the travel cost as a function of the travel time T and how to convert non-tight kinodynamic constraints into bounds on T.

4.3 Evaluation of the cost function

Using the time-scaling (6), we rewrite (2b) in terms of T and derivatives of q with respect to ξ as follows:

$$\tau_a = \frac{D_\tau(q, q', q'')}{T^2} + G_\tau(q) \quad (12)$$

where $G_\tau(q) = g(q_a)$,
$D_\tau(q, q', q'') = M_{pa}(q)q_p'' + M_{aa}(q)q_a'' + c_a(q, q')$.

Then, given a trajectory profile $Q(\xi)$, the cost function (5) becomes:

$$J = T \cdot \left(C_0 + \frac{C_2}{T^2} + \frac{C_4}{T^4} \right) \quad (13)$$

where $C_0 = (1-\alpha) + \alpha \int_0^1 \sum_{i=1}^{n_a} \overline{G_{\tau i}}^2 \, d\xi$,

$C_2 = 2\alpha \int_0^1 \sum_{i=1}^{n_a} \overline{D_{\tau i}} \, \overline{G_{\tau i}} \, d\xi$, $C_4 = \alpha \int_0^1 \sum_{i=1}^{n_a} \overline{D_{\tau i}}^2 \, d\xi$.

$\overline{D_{\tau i}} = D_{\tau i}(q, q', q'')/\tau_{ai}^{max}$, $\overline{G_{\tau i}} = G_{\tau i}(q)/\tau_{ai}^{max}$

C_0, C_2 and C_4 are real coefficients that do depend implicitly on the trajectory profile Q but that do not depend on T. Note also that C_0 and C_4 are always positive. Expression (13) represents a family of curves whose general shape, for any feasible profile is shown in Figure 4. The minimum of each of these curves is reached when T takes on the value T_m given by:

$$T_m = \left(\frac{C_2 + \sqrt{C_2^2 + 12 C_0 C_4}}{2 C_0} \right)^{1/2} \quad (14)$$

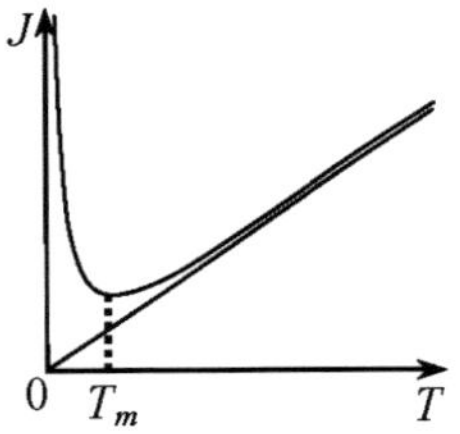

Fig. 4. General shape of the cost function J vs. T for a given trajectory profile Q

4.4 Clipping process

Given a candidate trajectory profile Q, most constraints simply translate to bounds on admissible values of the travel time of that candidate. Following (Chettibi and Lehtihet, 2002) we group constraints in three categories.

The first concerns geometric constraints. They will not yield any restriction on T. For example, any candidate that infringes (4c) will be rejected early in the process of selection as it will not lead to a feasible solution.

The second concerns kinematic constraints (4d) and (4e). Using a chain-rule differentiation, they translate to an explicit lower bound on T:

$$T \geq \max(T_V, T_A) \quad (15)$$

where $T_V = \max_i \left[\frac{\max |q'_{ai}(\xi)|}{\dot{q}_{ai}^{max}} \right] \quad \xi \in [0\ 1]$,

$T_A = \max_i \left[\frac{\max |q''_{ai}(\xi)|}{\ddot{q}_{ai}^{max}} \right]^{1/2} \quad \xi \in [0\ 1]$

As to dynamic constraints (4f), since they may be rewritten via (12) under the form:

$$-\tau_{ai}^{max} \leq \frac{D_{\tau i}}{T^2} + G_{\tau i} \leq \tau_{ai}^{max} \quad i = 1\ldots, n_a,$$

they usually translate to bracketing bounds on T:

$$T \in [T_L, T_R] \quad (16)$$

Intersecting (15) and (16) along the trajectory for ξ in $[0, 1]$ will yield a final window $\Pi = [T_1, T_2]$. Consequently, for any random trajectory profile Q, the corresponding optimal travel T_Q is found by minimizing J within Π. As to the value of T_m given by (14) in the unconstrained case, it will coincide with T_Q only if it belongs to Π. Otherwise, T_Q will coincide either with T_1 or T_2.

5. EXAMPLE

We consider the minimum-time trajectory-planning problem for a 2-link planar manipulator (Fig. 5) with two vertical parallel revolute axes, mounted on a two independently fixed wheel mobile platform whose characteristics are listed in Table 1.

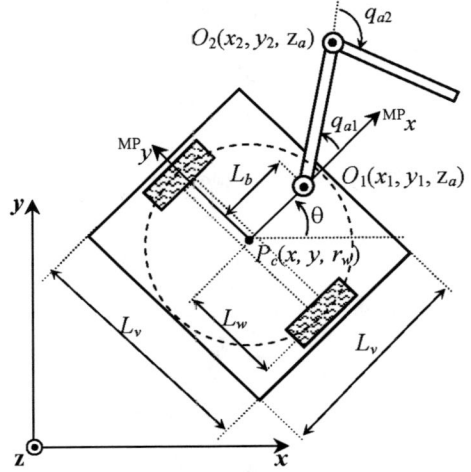

Figure 5: Planar mobile manipulator

The constraints on joint positions, velocities acceleration and driving torques are given as follows:

For $i = 1, 2$.
$$q_{ai}^{max} = -q_{ai}^{min} = 160°, \quad \dot{q}_{ai}^{max} = 3.0 \ rd/s,$$
$$\ddot{q}_{ai}^{max} = 8.0 \ rd/s^2, \quad \tau_{ai}^{max} = 10.0 \ Nm.$$

The platform is completely under-actuated. It has the form of a $(L_v \times L_v)$ square. r_w is radius of the wheel. L_1, L_2 are the length of the links. L_h is the distance between the middle of the wheel axis and the first joint axes. m_p, m_w, m_1 and m_2 are the masses of, respectively, the platform, the wheels, link 1 and link 2. I_{zp}, I_{zw}, I_{z1} and I_{z2} are the inertia moment about the ^{MP}z-axis of, respectively, the platform, the wheels, link 1 and link 2. I_{yw} is the inertia moment of the wheel about the ^{MP}y-axis. All inertia moments are given with respect to the center of mass of the corresponding bodies.

Table 1. Parameters of the two 2-link planar WMM

I	m_p = 30.0 kg	L_v = 0.60 m
I_{zw} = 0.05 kgm^2	m_w = 3.0 kg	r_w = 0.15 m
I_{yw} = 0.10 kgm^2		L_w = 0.25 m
I_{z1} = 0.30 kgm^2	m_1 = 15.0 kg	L_1 = 0.50 m
I_{z2} = 0.30 kgm^2	m_2 = 15.0 kg	L_2 = 0.50 m
z_{Gp} = 0.20 m	z_a = 0.65 m	L_h = 0.10 m

The workspace is a $8m \times 6m$ flat floor (Fig. 6a). The motion starts at (2.0, 2.0, 0°, 0°, 0°) and ends at (6.0, 6.0, 90°, free, free). For this problem we have adopted a fourth-order B-spline model with $N_a = 7$ control points and a clamped cubic-spline model with $N_m = 3$ interpolation points. Computations have been carried out on a 2.4 GHz P4. Results obtained are shown in Figure 6. The score of the calculated solution is $T = 46.40$ seconds for a runtime of 286 seconds. The remaining discrepancy between the calculated and the desired goal configuration of the platform is negligible (relative error of less than 10^{-5} in the final position/orientation and linear/angular velocity of the platform). Similar results can be obtained also in the case in which the start and goal orientations of the platform are completely inverted.

6. CONCLUSIONS

We have proposed a modification of the random-profile approach to handle a trajectory-planning problem for a wheeled mobile manipulator with an under-actuated platform. The modification consists of generating trial trajectory profiles for the arms and then deducing the corresponding profiles for the platform by integrating numerically the second-order nonholonomic constraints. Moreover, a penalty has been introduced in the cost function to steer the stochastic optimization process towards the desired goal configuration of the platform.

The result is a flexible sub-optimal stochastic shooting technique that is again based, as the ordinary RPA, on a random disruption of a few control nodes. This technique remains applicable to problems with stability and obstacle-avoidance considerations. Additional tests are necessary to assess its performance for high-dimension problems.

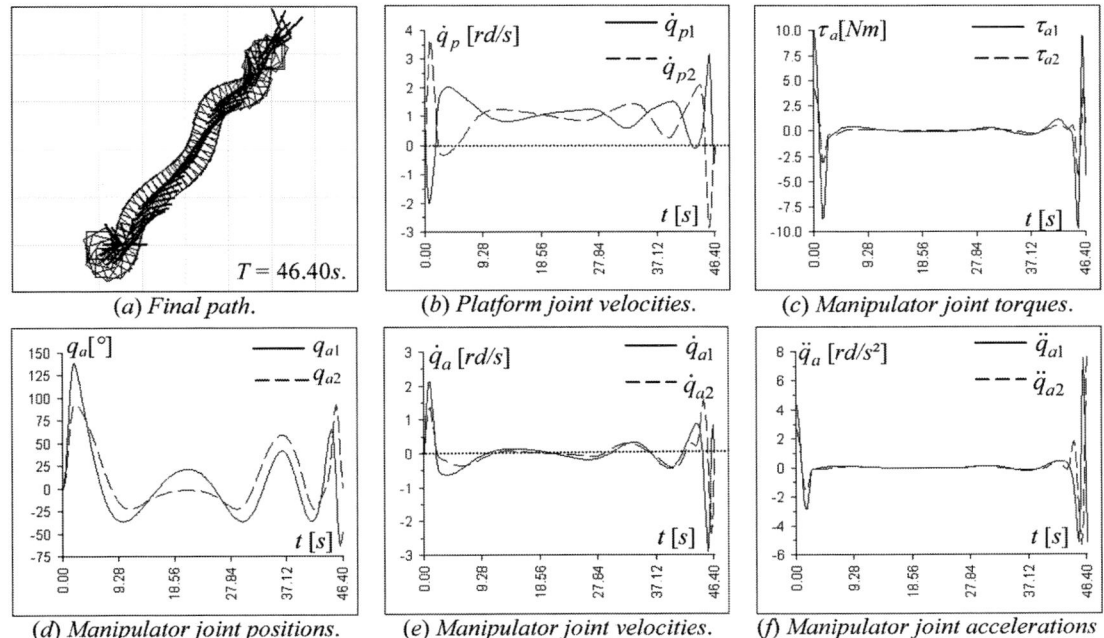

(a) Final path. (b) Platform joint velocities. (c) Manipulator joint torques.
(d) Manipulator joint positions. (e) Manipulator joint velocities. (f) Manipulator joint accelerations

Fig. 6: Results obtained using $N_m = 3$ and $N_a = 7$

REFERENCES

Arai, H., and Tachi, S., 1991. Position Control of a Manipulator with Passive Joints Using Dynamic Coupling. *In IEEE Transactions on Robotics and Automation*, **Vol. 7, No. 4**, pp. 528-534.

Aria, H., Kazuo, T., and Naoji, S., 1998. Time-scaling control of an underactuated manipulator. *Journal of Robotics Systems*, **Vol. 15, No. 9,** *pp.* 525-536.

Bayle, B., Fourquet, J.-Y. and Renaud, M., 2003. From Manipulation to Wheeled Mobile Manipulation : Analogies and Differences. 7th *Symposium on Robot Control (SYROCO'03)*, **Vol. 1**, pp. 97-104.

Bloch, A. M., Reyhaniglu, M., and McClamroch, N. H., 1992. Control and Stabilization of Nonholonomic Dynamical Systems. *In IEEE Trans. on Automatic Control*, **Vol. 37, No.11**, *pp.*1746-1757.

Campion, G., d'Andrea-Novel, B., Bastin, G., 1991. Modelling and State Feedback Control of Nonholonomic Mechanical Systemss. *In IEEE Proc.* 30th *Conf. on Decision and Control, Brighton, England,* pp. 1184-1189.

Campion, G., Bastin, G., and d'Andrea-Novel, B., 1996. Structural Properties and Classification of kinematics and Dynamic Models of Wheeled Mobile Robots. *In IEEE Trans. on Robotics and Automation,* **Vol. 12, No.1**, pp. 47-62.

Chettibi, T., and Lehtihet, H. E., 2002. A new approach for point to point optimal motion planning problems of robotic manipulators. In *Proc. Of 6th Biennial Conf. on Engineering System Design and Analysis*, APM10.

Chernousko, F.L., 2004. Modelling of snake-like locomotion. *In Applied Mathematics and Computation*, **Vol. 164 No.2**, pp.415-434

De-Luca, A., Mattone, R., and Oriolo, G., 2000. Stabilization of an underactuated planar 2R manipulator. *In International Journal of Robust and Nonlinear Control*, **Vol. 10**, pp. 181-198.

Haddad, M., Chettibi, T., Saidouni, T., Hanchi, S., and Lehtihet, H. E., 2006. Sub-Optimal Motion Planner of Mobile Manipulators in Generalized Point-to-Point Task With Stability Constraint. *In Proc. Of* 16th *CISM-IFToMM Symposium*, pp. 171-178.

Haddad M., Chettibi T., Hanchi S., and Lehtihet H. E., 2007. A Random-profile Approach for trajectory planning of Wheeled Mobile Robots. *European J. of Mechanics, A/Solid* **No. 26 (2007)**, pp. 519-540.

Hajek, B., 1985. A Tutorial survey of Theory and Application of Simulated Annealing. *In Proceeding of* 24th *Conference on Decision and Control*, pp. 755-760.

Henning, W., Hickman, F., and Choset, H., 1998. Motion Planning for Serpentine Robots. *ASCE.98, Space and Robotics Albuquerque, New Mexico.*

Hirose, S., 1993. Biologically Inspired Robots (Snake-like Locomotor and Manipulator), *Oxford University Press*; *ISBN*: 0 19 856261 6.

Hollerbach, J. M., 1983. Dynamic Scaling of Manipulator Trajectories. *M.I.T.A.I. Lab Memo* 700.

Nakamura, Y. Suzuki, T., and Koinuma, M., 1997, Nonlinear behavior and control of a nonholonomic free-jointmanipulator. *In IEEE Transactions on Robotics and Automation*, **Vol. 13, No. 6**, pp. 853-862.

Olfati-Saber, R., 2001. Nonlinear Control of Underactuated Mechanical Systems with Application to Robotics and Aerospace Vehicles. *PhD Thesis, Department of Electrical Engineering and Computer Science, Massachusetts Institute of Technology, Cambridge, MA.*

Oriolo, G., and Nakamura, Y., 1991. Free joint manipulators: Motion control under second order Nonholonomic constraints. *In IEEE/RSJ Int. Work. on Intelligent Robots and Systems*, pp. 1248-1253.

Ostrowski, J., Burdick, J., Lewis, A., and Murray, R., 1995. The Mechanics of Undulatory Locomotion: the Mixed Kinematic and Dynamic Case. *In IEEE International Conference on Robotics and Automation.*

Prautsch, P., Mita, T., and Iwasaki, T., 2000 . Analysis and Control of a Gait of Snake Robot. *In Transaction IEE of Japan*, **Vol. 120-D, No. 3**, *March* 2000.

Shirazi, K. H., and Ghaari-Saadat M. H., 2004. Chaotic motion in a class of asymmetrical Kelvin type gyrostat satellite. *International Journal of Non-Linear Mechanics*, **Vol. 39, pp.** 785 – 793.

Minimum time trajectory planning for a micro quadrotor aerial robot

Y. Bouktir, T. Chettibi

Laboratory of Structure Mechanics
E.M.P., Bordj El Bahri, BP 17, 16111, Algiers, Algeria (Bouktir.yasser, tahachettib@yahoo.fr).

Abstract: A simple direct method able to generate minimum time trajectories for micro quadrotor aerial robots is presented. It is based on modeling the quadrotor trajectory as a composition of a parametric form **P**(λ), defining the quadrotor path, and a monotonically increasing function λ(t), specifying the motion on this path. The optimal evolutions of **P**(λ) and λ(t), which are approximated using algebraic spline functions, are found using a nonlinear optimization technique. The proposed method accounts for the most important constraints inherent to the system behavior, such as underactuation and limits on actuator torques and speeds. Furthermore, this method may be used for generating optimal reference dynamic motions in the presence of obstacles.

Keywords: quadrotor, dynamic model, trajectory optimization

1. INTRODUCTION

Autonomous or tele-operated vehicles are considered as an excellent cost-effective and harmless way to substitute human operators in both civilian and military applications, particularly in hazardous environments or specific tasks. For this reason, many vehicles have been developed including, ground, underwater and aerial robots. Nowadays, thanks to developments recorded in various technological fields these systems are getting cheaper, more effective, miniaturized, flexible, accurate and more autonomous.

Unmanned air vehicles (UAVs) are self-propelled aerial robots, equipped with high technology sensors, capable of conducting autonomous operations or they are remotely controlled. The family of UAVs embraces various models ranging from micro aerial vehicles to big inhabited combat aircraft vehicles, and some models are being sold as RC toys. UAVs are mainly used to accomplish military defensive or offensive missions such as bombing, spying, surveillance, etc. But, they are also used for numerous civilian applications such as search and rescue operations, terrain and utilities inspection, disaster monitoring, environmental surveillance, sport broadcasting, and more.

Among existing small UAVs, we find quadrotors which are Vertical Take-Off and Landing (VTOL) four rotor helicopters (fig.1). They are becoming a standard platform for UAV research, due to the simplicity of their construction and maintenance. They are controlled simply by changing the rotation speed of the four rotors. The front and rear rotors (2, 4) rotate in a clockwise direction while the left and right rotors (1, 3) rotate in a counter-clockwise direction to balance the torque created by the spinning rotors. The up/down motion is achieved by increasing/decreasing the rotors speed while maintaining an equal individual speed. The forward/backward, left/right motions are achieved through a differential control strategy of rotors speed. Thanks to this configuration, quadrotors are able to hover, takeoff, and land in small areas and enable them to perform tasks that fixed-wing craft are unable to do. Although the mechanical design of the quadrotor is simple, its particular dynamics makes the vehicle control relatively difficult. This is due principally to the fact that the system is underactuated: there are only four rotors which generate four inputs thrusts (T_i) to control the six degrees of freedom of the crossing body during fly (fig.1).

In recent years, there have been a number of papers dealing with various problems inherent to the exploitation of quadrotor like systems. They generally deal with three main topics: modeling, trajectory planning and control problems. For example, Mistler *et al.* developed in [1] the dynamic model of a four rotors helicopter and gave a dynamic feedback controller. Altug *et al.* proposed a visual feedback control method of a quadrotor using a camera equipped on the UAV as the main sensor for attitude estimation [2]. Hamel et al. proposed in [3] a vision based visual servo controller for performing trajectory tracking tasks of a four rotor VTOL aerial vehicle. McKerrow obtained the dynamic model for theoretical analysis of a Draganflyer [4]. In reference [5], Mokhtari and Benallegue developed a dynamic model of a quadrotor for state variable control based on Euler angles and an open loop position state observer, emphasized attitude control rather than the translational motion of the UAV. Castilo *et al.* achieved autonomous takeoff and hovering and landing control of a quadrotor by synthesizing a controller using the Lagrangien model based on the Lyapunov analysis [6]. Bouabdellah *et al.* [7] discussed the design, dynamic modeling, sensing and control of the OS4 quadrotor and proposed a tractable and effective dynamic model. They also discussed preliminary control results. Hamel et al. proposed a vision based controller which performs visual servo control by positioning a camera onto a fixed target for the hovering of a quadrotor [8]. In reference [9], the problem of constrained nonlinear tracking control for small fixed-wing UAVs is addressed. This paper deals with the issue of tracking control for UAV kinematic models with physically motivated heading rate and velocity constraints. The problem is treated using a constrained control Lyapunov function based approach. Yang et al. treated in [10] the problem of time-optimal control of a quadrotor by using a nonlinear programming method coupled with a genetic algorithm. They succeeded to generate in simulation minimum time point-to-point trajectories under various technological constraints. Bouabdallah and Siegwart suggested in [11] two nonlinear control techniques: a backstepping and a sliding-mode control applied on the OS4 quadrotor. Waslander et al. [12] presented STARMAC which is an operational multivehicle quadrotor platform capable of autonomous outdoor flight, without tethers or motion guides. Furthermore, they did a comparison of several control design techniques, specifically for outdoor altitude control, in and above ground effect. They showed that integral sliding mode and reinforcement learning control are able to accommodate the nonlinear disturbances of the aircraft. Altug el al. [13] introduced a vision-based stabilization and output tracking control. They also presented specific pose estimation algorithms based on the use of two cameras: one onboard the quadrotor and another on the ground. Successful experiments showing the efficiency of the proposed approach were done on a commercially quadrotor called HMX-4. Tayebi et al. proposed in ref. [14] a quaternion-based feedback control scheme for exponential attitude stabilization of a quadrotor aircraft. The proposed controller is based upon the compensation of the Coriolis and gyroscopic torques and the use of a PD feedback structure. Experiments were conducted on a modified version of the Draganflyer quadrotor. Cowling et al. presented in [15] an optimal trajectory planner with a linear control scheme to follow a reference trajectory. Authors exploited the differential flatness of the quadrotor to address the optimization problem within the output space. This scheme has been validated in simulation using a full dynamic model of the quadrotor. In reference [16], a Model Predictive Control Based Trajectory Tracking system for small unmanned helicopters is presented. It is based on a linear model predictive controller and showed in simulations a good robustness to parameter uncertainty. In reference [17], authors discussed the nonholonomic constraints which characterize the underactuation of the quadrotor and highlighted potential applications of these constraints in generating feasible reference trajectories. Bouadi et al. [18] developed an efficient sliding mode control based on backstepping approach and presented numerical simulations showing the perfect tracking of prescribed trajectories. Chettibi and Haddad [28] presented a detailed dynamic model of the quadrotor using a multi-body approach and accounting for the main forces acting on the system. Simplification hypotheses were also introduced in order to get a tractable model for trajectory generation and system control problems.

In this paper, we propose a simple numerical method to treat the problem of generating minimum time trajectories for a quadrotor aerial robot under various constraints. It is based on previous works developed at our laboratory [19, 20] dealing with the problem of trajectory generation for serial manipulators and mobile robots. For this purpose, the dynamic model of the quadrotor robot is first developed using results of reference [28]. Then, the trajectory generation problem is stated as a nonlinear optimization problem by parameterizing both of the robot path and the associated motion profile using algebraic spline functions.

The resulting problem may be solved using classical optimization techniques such as the sequential quadratic programming method (SQP). Finally, simulation results corresponding to point-to-point tasks in free and encumbered environments are presented to illustrate the efficiency of the proposed approach.

2. QUADROTOR MODELING

2.1 Kinematics

The quadrotor aerial robot is a multibody dynamic system composed of mainly five rigid bodies (figure 1): four rotors ($i = 1,...,4$) connected to a crossing body frame ($i = 0$) by means of four actuated revolute joints. The kinematic modeling of the quadrotor involves the use of the following frames (fig.1): an inertial earth frame $\{R_E\}\,(O, \mathbf{x}, \mathbf{y}, \mathbf{z})$ where $\mathbf{z}$ is oriented upwards, a body-fixed frame $\{R_0\}\,(O, \mathbf{x}_0, \mathbf{y}_0, \mathbf{z}_0)$, and four rotor-fixed frames $\{R_i\}\,(O, \mathbf{x}_i, \mathbf{y}_i, \mathbf{z}_i)$ $i = 1,...4$.

Furthermore, a set of ten parameters should be selected in order to define completely the configuration of the system: (*i*) **six** parameters to describe the six degrees of freedom (*dof*) of the crossing body frame in $\{R_E\}$ (3 *dof* of translation and 3 *dof* of rotation); (*ii*) **four** parameters to describe the motion of rotors in $\{R_0\}$.

The translation motion of the crossing body ($i = 0$) can be studied in the inertial frame using O_0 Cartesian coordinates, i.e. (x, y, z). Whereas, the rotation may be described using three independent angles e.g.: α, β, γ. The parameters

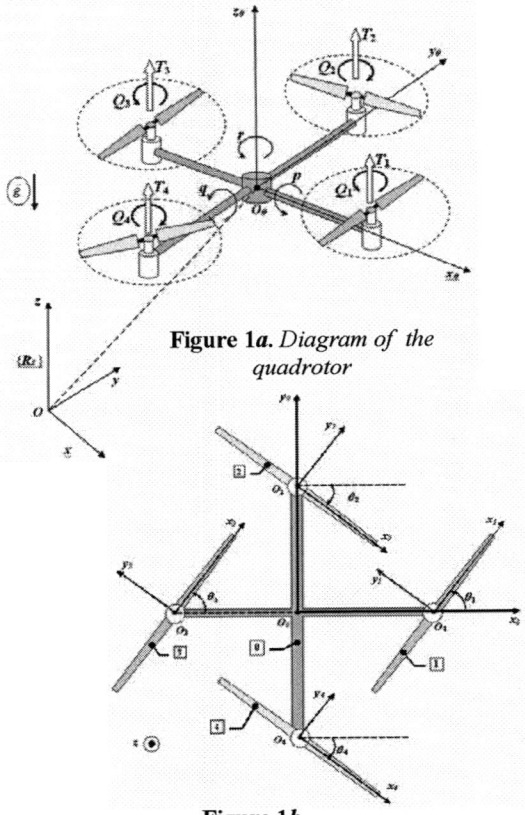

Figure 1a. *Diagram of the quadrotor*

Figure 1b.
Top view of the quadrotor

Figure 1. Description of the quadrotor motion

$(x, y, z, \alpha, \beta, \gamma)$ describe completely the six *dof* of the crossing body frame. These parameters and their time variations may be estimated from measurements delivered by a combination of various sensors like inertial measurement units (IMUs), GPS sensors, gyros, etc [12, 21]. For instance, it is possible to measure the angular velocities of the crossing body in the inertial frame $\{R_0\}$, i.e. ${}^0\mathbf{\Omega}_0 = [p\ q\ r]^T$, using three gyros. Then, from ${}^0\mathbf{\Omega}_0$ the body attitude is deduced simply by integration. However, it is necessary to establish the relationship between the time variation of Euler angles and the components of ${}^0\mathbf{\Omega}_0$ in the following way:

$${}^0\mathbf{\Omega}_0 = \mathbf{H}\dot{\mathbf{\Theta}} \quad (1)$$

where $\mathbf{\Theta} = [\gamma\ \beta\ \alpha]^T$ and $\mathbf{H}$ is a 3×3 matrix that depends on the adopted Euler angles.

The rotation matrix ${}^E\mathbf{R}_0$ that defines the orientation of $\{R_0\}$ relative to the earth frame $\{R_E\}$ can be written in a general form as follows:

$${}^E\mathbf{R}_0 = \begin{bmatrix} r_{11} & r_{12} & r_{13} \\ r_{21} & r_{22} & r_{23} \\ r_{31} & r_{32} & r_{33} \end{bmatrix} \quad (2)$$

The orientation of the frame $\{R_0\}$ is described hereafter using the **roll-pitch-yaw** angles (*X-Y-Z fixed* angles). The three rotations take place about the axes of the inertial frame $\{R_E\}$ in the following order: rolling $rot(\gamma, \mathbf{x})$, then pitching $rot(\beta, \mathbf{y})$ and finally yawing $rot(\alpha, \mathbf{z})$. Also, they verify the inequalities: $-\pi \leq \alpha \leq \pi$, $-\pi/2 \leq \beta \leq \pi/2$, $-\pi \leq \gamma \leq \pi$. Using the following shorthand notation ($Cos = c$, $Sin = s$), the components r_{jk} of the matrix ${}^E\mathbf{R}_0$ are:

$$\begin{array}{lll} r_{11} = c\alpha c\beta & r_{12} = c\alpha s\beta s\gamma - s\alpha c\gamma & r_{13} = c\alpha s\beta c\gamma + s\alpha s\gamma \\ r_{21} = s\alpha c\beta & r_{22} = s\alpha s\beta s\gamma + c\alpha c\gamma & r_{23} = s\alpha s\beta c\gamma - c\alpha s\gamma \\ r_{31} = -s\beta & r_{32} = c\beta s\gamma & r_{33} = c\beta c\gamma \end{array} \quad (3)$$

The angular velocity of the body frame "0" may be obtained using the following formula [22]:

$$S({}^0\mathbf{\Omega}_0) = {}^E\mathbf{R}_0^T\ {}^E\dot{\mathbf{R}}_0 \quad (4)$$

where $S({}^0\mathbf{\Omega}_0)$ is the skew-symmetric matrix associated to the vector ${}^0\mathbf{\Omega}_0$. From this matrix equation, one can extract the components of the rotation vector as follows:

$$\begin{bmatrix} p \\ q \\ r \end{bmatrix} = \begin{bmatrix} -\dot{\alpha}s\beta + \dot{\gamma} \\ \dot{\alpha}c\beta s\gamma + \dot{\beta}c\gamma \\ \dot{\alpha}c\beta c\gamma - \dot{\beta}s\gamma \end{bmatrix}$$

Therefore, the matrix $\mathbf{H}$ is:

$$\mathbf{H} = \begin{bmatrix} 1 & 0 & -s\beta \\ 0 & c\gamma & c\beta s\gamma \\ 0 & -s\gamma & c\beta c\gamma \end{bmatrix} \quad (5)$$

The linear velocity of the mass center of the crossing body O_0, expressed in $\{R_E\}$, is:

$${}^E\mathbf{V}_0 = [\dot{x}\ \dot{y}\ \dot{z}]^T \quad (6)$$

The orientation of the frame $\{R_i\}$ attached to the ith rotor relative to $\{R_0\}$ is given by the following matrix:

$${}^0\mathbf{R}_i = \begin{bmatrix} c\theta_i & -s\theta_i & 0 \\ s\theta_i & c\theta_i & 0 \\ 0 & 0 & 1 \end{bmatrix} \quad i=1, ..., 4 \quad (7)$$

The angular velocity of the ith rotor calculated in $\{R_i\}$ is obtained by composition:

$${}^i\mathbf{\Omega}_i = {}^i\mathbf{R}_0\ {}^0\mathbf{\Omega}_0 + \dot{\theta}_i \mathbf{z}_i = \begin{pmatrix} pc\theta_i + qs\theta_i \\ -ps\theta_i + qc\theta_i \\ r + \dot{\theta}_i \end{pmatrix} \quad i=1, ..., 4 \quad (8)$$

The linear velocity of the mass center O_i of the ith rotor calculated in $\{R_E\}$ is:

$${}^E\mathbf{V}_i = {}^E\mathbf{V}_0 + \overrightarrow{O_0O_i} \times {}^E\mathbf{\Omega}_0 \quad i = 1, ..., 4 \quad (9)$$

2.2 Dynamics

We denote by $\mathbf{I}_i$ and m_i ($i = 0, ..., 4$) the inertia matrix and the mass of the ith body. The inertia matrix $\mathbf{I}_i$ is defined with respect to each body proper frame $\{R_i\}$. Point O_i is supposed coincident with the mass center of the ith body.

The dynamics of the ith body under the action of an external wrench of force $\mathbf{F}_i^{ext}$ and moment $\mathbf{M}_i^{ext}$, applied to its mass center, can be described using the Newton-Euler formalism [23] as follows:

$$\mathbf{F}_i^{ext} = \frac{D}{Dt}(m_i \mathbf{V}_i) = m_i\ {}^E\dot{\mathbf{V}}_i \quad (10a)$$

$$\mathbf{M}_i^{ext} = \frac{D}{Dt}(\mathbf{I}_i\ {}^i\mathbf{\Omega}_i) = \mathbf{I}_i\ {}^i\dot{\mathbf{\Omega}}_i + {}^i\mathbf{\Omega}_i \times (\mathbf{I}_i\ {}^i\mathbf{\Omega}_i) \quad (10b)$$

The vectors $m_i \mathbf{V}_i$ and $\mathbf{I}_i \mathbf{\Omega}_i$ define respectively the linear and angular momentum of the considered body. The operator D/Dt indicates the time derivative in the inertial frame.

Note that for convenience, the translation dynamics equations are given in the earth frame while the rotation dynamics equations are projected in the body proper frame.

External Forces and moments acting on the ith rotor, considered at its mass center O_i, comprise: (*i*) the weight $\mathbf{W}_i$, (*ii*) the aerodynamic forces induced by the rotor rotation which are designated hereafter by the resultant $\mathbf{F}_i^a$ and the moment $\mathbf{M}_i^a$, (*iii*) the action of the crossing body, modelled by the force $\mathbf{F}_{0/i}^v$ and the moment $\mathbf{M}_{0/i}^v$, which is transmitted through the corresponding actuated joint. $\mathbf{F}_{0/i}^v$ and $\mathbf{M}_{0/i}^v$ account for actuator efforts, joint friction and cohesion forces. Hence, we have for $i = 1, ..., 4$:

$$\mathbf{F}_i^{ext} = \mathbf{W}_i + \mathbf{F}_i^a + \mathbf{F}_{0/i}^v, \quad \mathbf{M}_i^{ext} = \mathbf{M}_i^a + \mathbf{M}_{0/i}^v \quad (11)$$

Using relations (10) and (11), it is possible to calculate the elements of the wrench characterizing the action of the body frame on each rotor as follows; *for* $i = 1, ..., 4$:

$$\mathbf{F}_{0/i}^v = \frac{D}{Dt}(m_i \mathbf{V}_i) - \mathbf{W}_i - \mathbf{F}_i^a, \quad \mathbf{M}_{0/i}^v = \frac{D}{Dt}(\mathbf{I}_i\ {}^i\mathbf{\Omega}_i) - \mathbf{M}_i^a \quad (12)$$

According to the momentum theory of rotors [24, 25], each rotor produces mainly a trust T_i and a drag moment Q_i about its axis of rotation. Both of them are proportional to the square of the rotor speed. Hence, the aerodynamic resultant

$\mathbf{F}_i^a$ and the aerodynamic moment $\mathbf{M}_i^a$ can be written, in the body frame $\{R_0\}$ (note that $\mathbf{z}_0 = \mathbf{z}_i$), as follows; $i = 1, ..., 4$

$$^i\mathbf{F}_i^a = {}^0\mathbf{F}_i^a = T_i\,\mathbf{z}_0 = b\,\dot{\theta}_i^2\,\mathbf{z}_0 \qquad (13a)$$

$$^i\mathbf{M}_i^a = {}^0\mathbf{M}_i^a = (-1)^i Q_i \mathbf{z}_0 = (-1)^i d\,\dot{\theta}_i^2\,\mathbf{z}_0 \qquad (13b)$$

where $b > 0$ and $d > 0$ are coefficients depending upon the blade's Reynolds Number, Mach number and angle of attack as well as other factors (see [26] for more details). They are generally identified experimentally as done in ref. [27, 29].

The crossing body ($i = 0$) is moving under the action of the following efforts: (*i*) the weight $\mathbf{W}_0$ applied at its mass center O_0, (*ii*) the aerodynamic forces exerted by the air at the body mass center O_0, they are characterized by the resultant $\mathbf{F}_0^a$ and the moment $\mathbf{M}_0^a$, (*iii*) the actions of the four rotors across the four joints which are defined, at points O_i ($i = 1, ..., 4$), by the forces $\mathbf{F}_{i/0}^v = -\mathbf{F}_{0/i}^v$ and the moments $\mathbf{M}_{i/0}^v = -\mathbf{M}_{0/i}^v$. Thus, we have:

$$\mathbf{F}_0^{ext} = \mathbf{W}_0 + \mathbf{F}_0^a + \sum_{i=1}^{4}\mathbf{F}_{i/0}^v \qquad (14a)$$

$$\mathbf{M}_0^{ext} = \mathbf{M}_0^a + \sum_{i=1}^{4}\left(\mathbf{M}_{i/0}^v + \overrightarrow{O_0 O_i} \times \mathbf{F}_{i/0}^v\right) \qquad (14b)$$

The body "0" is performing mainly translation motions with small angular rotations. So, it undergoes a drag force $\mathbf{F}_0^a$ applied at the pressure center and in a direction $\mathbf{e}_w$ opposite to its relative velocity with the oncoming wind ($\mathbf{e}_w = -\mathbf{V}_0/\|\mathbf{V}_0\|$). The aerodynamic moment $\mathbf{M}_0^a$ appears if the pressure center P is distant of the mass center O_0 (i.e. $\mathbf{M}_0^a = \overrightarrow{O_0 P} \times \mathbf{F}_0^a$).

The rotational dynamics needs to be developed for Euler angles. This can be done by introducing relation (1) in (10b). First, we have:

$$^0\dot{\mathbf{\Omega}}_0 = \dot{\mathbf{H}}\dot{\mathbf{\Theta}} + \mathbf{H}\ddot{\mathbf{\Theta}} \qquad (15a)$$

Then, the second member of relation (10b) becomes:

$$\mathbf{I}_0\,{}^0\dot{\mathbf{\Omega}}_0 + {}^0\mathbf{\Omega}_0 \times (\mathbf{I}_0\,{}^0\mathbf{\Omega}_0) = \mathbf{I}_0\dot{\mathbf{H}}\dot{\mathbf{\Theta}} + \mathbf{I}_0\mathbf{H}\ddot{\mathbf{\Theta}} + \mathbf{H}\dot{\mathbf{\Theta}}\times(\mathbf{I}_0\mathbf{H}\dot{\mathbf{\Theta}}) \qquad (15b)$$

By introducing relations (14b) and (15b) in (10b) we get:

$$\ddot{\mathbf{\Theta}} = (\mathbf{I}_0\mathbf{H})^{-1}\left(\mathbf{M}_0^a + \sum_{i=1}^{4}\left(\mathbf{M}_{i/0}^v + \overrightarrow{O_0 O_i} \times \mathbf{F}_{i/0}^v\right) - \mathbf{I}_0\dot{\mathbf{H}}\dot{\mathbf{\Theta}} - \mathbf{H}\dot{\mathbf{\Theta}}\times(\mathbf{I}_0\mathbf{H}\dot{\mathbf{\Theta}})\right) \qquad (15c)$$

Finally, The dynamics equations of the whole system are obtained by regrouping relations (10), (12), (14) and (15) and accounting for the fact that $\mathbf{F}_{i/0}^v = -\mathbf{F}_{0/i}^v$ and $\mathbf{M}_{i/0}^v = -\mathbf{M}_{0/i}^v$. Given the inertial and geometrical parameters of the quadrotor, the complete model of the quadrotor aerial robot is:

$$\begin{cases}
{}^0\mathbf{\Omega}_0 = [p\,q\,r]^T,\ S({}^0\mathbf{\Omega}_0) = {}^E\mathbf{R}_0^T\,{}^E\dot{\mathbf{R}}_0 \\
\mathbf{\Theta} = [\gamma\,\beta\,\alpha]^T,\ {}^0\mathbf{\Omega}_0 = \mathbf{H}\dot{\mathbf{\Theta}} \\
{}^i\mathbf{\Omega}_i = {}^i\mathbf{R}_0\,{}^0\mathbf{\Omega}_0 + \dot{\theta}_i\,\mathbf{z}_i & i=1,...,4 \\
{}^E\mathbf{V}_0 = [\dot{x}\,\dot{y}\,\dot{z}]^T \\
{}^i\mathbf{V}_i = {}^i\mathbf{V}_0 + \overrightarrow{O_0 O_i}\times{}^i\mathbf{\Omega}_0\quad {}^0\mathbf{V}_i = {}^iR_0^T\,{}^i\mathbf{V}_i & i=1,...,4 \\
\mathbf{F}_{0/i}^v = m_i\dot{\mathbf{V}}_i - \mathbf{W}_i - \mathbf{F}_i^a & i=1,...,4 \\
\mathbf{M}_{0/i}^v = \mathbf{I}_i\,{}^i\dot{\mathbf{\Omega}}_i + {}^i\mathbf{\Omega}_i\times(\mathbf{I}_i\,{}^i\mathbf{\Omega}_i) - \mathbf{M}_i^a & i=1,...,4 \\
\dot{\mathbf{V}}_0 = \frac{1}{m_0}\left(\mathbf{W}_0 + \mathbf{F}_0^a - \sum_{i=1}^{4}\mathbf{F}_{0/i}^v\right) \\
\ddot{\mathbf{\Theta}} = (\mathbf{I}_0\mathbf{H})^{-1}\left(\mathbf{M}_0^a - \sum_{i=1}^{4}(\mathbf{M}_{0/i}^v + \overrightarrow{O_0 O_i}\times\mathbf{F}_{0/i}^v) - \mathbf{H}\dot{\mathbf{\Theta}}\times(\mathbf{I}_0\mathbf{H}\dot{\mathbf{\Theta}}) - \mathbf{I}_0\dot{\mathbf{H}}\dot{\mathbf{\Theta}}\right)
\end{cases} \qquad (16)$$

The dynamic model (16) is relatively difficult to handle and to use for the synthesis of control laws and the generation of reference trajectories. Therefore, we propose to adopt the following simplification assumptions, in order to get a more tractable model:

- The linear and angular momentums of rotors are neglected since their mass and inertia matrix are very small comparatively to those of the frame body. Hence, by putting $m_i \approx 0$ and $I_i \approx 0$, relation (12) become:

$$\begin{cases} \mathbf{F}_{0/i}^v \approx -\mathbf{F}_i^a \\ \mathbf{M}_{0/i}^v \approx -\mathbf{M}_i^a \end{cases} \qquad (17a)$$

- Joints relating rotors to the crossing body are supposed perfect (no friction). Hence, $\mathbf{M}_{0/i}^v \cdot \mathbf{z}_i = \tau_i$ represents practically the amplitude of the input actuator torque. In consequence, using (13b) and (17a) we get:

$$\tau_i = -(-1)^i d\,\dot{\theta}_i^2 = (-1)^{i+1} d\,\dot{\theta}_i^2 \qquad (17b)$$

- The crossing body is supposed symmetrical and the corresponding inertia matrix $\mathbf{I}_0$ is given in $\{R_0\}$ by:

$$\mathbf{I}_0 = diag(I_{xx}, I_{yy}, I_{zz}).$$

- The translation velocity of the quadrotor is small, therefore the aerodynamic forces $\mathbf{F}_0^a$ and the corresponding moment $\mathbf{M}_0^a$ are neglected, i.e.

$$\mathbf{F}_0^a \approx 0, \quad \mathbf{M}_0^a \approx 0 \qquad (17c)$$

In consequence of the above mentioned assumptions (17a-c), the main dynamic equations are those related to the crossing body motion, i.e.:

$$\dot{\mathbf{V}}_0 = \frac{1}{m_0}\left(\mathbf{W}_0 + \sum_{i=1}^{4}\mathbf{F}_i^a\right) \qquad (18a)$$

$$\ddot{\mathbf{\Theta}} = (\mathbf{I}_0\mathbf{H})^{-1}\left(\sum_{i=1}^{4}\left(\mathbf{M}_i^a + \overrightarrow{O_0 O_i}\times\mathbf{F}_i^a\right) - \mathbf{H}\dot{\mathbf{\Theta}}\times(\mathbf{I}_0\mathbf{H}\dot{\mathbf{\Theta}}) - \mathbf{I}_0\dot{\mathbf{H}}\dot{\mathbf{\Theta}}\right) \qquad (18b)$$

By projecting relation (18a) in $\{R_E\}$ and accounting for (13a) we get:

$${}^E\dot{\mathbf{V}}_0 = [\ddot{x}\,\ddot{y}\,\ddot{z}]^T = -g\,\mathbf{z} + \frac{1}{m_0}\left({}^E\mathbf{R}_0 \cdot \sum_{i=1}^{4}T_i\mathbf{z}_0\right) \qquad (19a)$$

or more simply:

$$\ddot{x} = r_{13}u_1,\quad \ddot{y} = r_{23}u_1,\quad \ddot{z} = -g + r_{33}u_1 \qquad (19b)$$

where:

$$u_1 = \frac{1}{m_0}\sum_{i=1}^{4}T_i \qquad (19c)$$

In many papers dealing with control problems of the quadrotor, e.g. ref. [7, 11, 18], the hovering and near hovering conditions are generally adopted in order to simplify the relationship between measured angular velocity ${}^0\mathbf{\Omega}_0$ and time variation of Euler angles $\dot{\mathbf{\Theta}}$. In fact, under this assumption, Euler angles (α,β,γ) are considered small and this leads to:

$$\mathbf{H} \approx \begin{bmatrix} 1 & 0 & 0 \\ 0 & 1 & 0 \\ 0 & 0 & 1 \end{bmatrix} \Rightarrow {}^0\mathbf{\Omega}_0 \approx \dot{\mathbf{\Theta}} \qquad (20)$$

Thus, relation (18b) becomes:

$$\ddot{\mathbf{\Theta}} = \mathbf{I}_0^{-1}\left(\sum_{i=1}^{4}\left(\mathbf{M}_i^a + \overrightarrow{O_0 O_i}\times\mathbf{F}_i^a\right) - \dot{\mathbf{\Theta}}\times(\mathbf{I}_0\dot{\mathbf{\Theta}})\right) \qquad (21a)$$

The projection of relation (21a) in the local frame $\{R_0\}$ gives:

$$\begin{cases} \ddot{\gamma} = u_2 + \dot{\alpha}\,\dot{\beta}\,I_1 \\ \ddot{\beta} = u_3 + \dot{\alpha}\,\dot{\gamma}\,I_2 \\ \ddot{\alpha} = u_4 + \dot{\gamma}\,\dot{\beta}\,I_3 \end{cases} \quad (21b)$$

where $\begin{cases} u_2 = l(T_2 - T_4)/I_{xx} \\ u_3 = l(T_3 - T_1)/I_{yy} \\ u_4 = (Q_4 + Q_2 - Q_1 - Q_3)/I_{zz} \end{cases}$ (22)

and $\begin{cases} l = \|\overrightarrow{O_0 O_i}\| \quad i = 1,\dots,4 \\ I_1 = (I_{yy} - I_{zz})/I_{xx},\; I_2 = (I_{zz} - I_{xx})/I_{yy},\; I_3 = (I_{xx} - I_{yy})/I_{zz} \end{cases}$

Note that points O_i are supposed to belong to the same plan. Also, the quantities u_i can be seen as equivalent control inputs of our system since they are a linear combination of effective input torques τ_i, $i = 1,\dots,4$ (see rel. (28)).

Finally, the simplified direct dynamic model of the quadrotor is composed of relations (19b) and (22). The inverse form is also possible. By noting that $r_{13}^2 + r_{23}^2 + r_{33}^2 = 1$, we get:

$$\begin{cases} u_1 = \sqrt{\ddot{x}^2 + \ddot{y}^2 + (g+\ddot{z})^2} & u_3 = \ddot{\beta} - \dot{\alpha}\,\dot{\gamma}\,I_2 \\ u_2 = \ddot{\gamma} - \dot{\alpha}\,\dot{\beta}\,I_1 & u_4 = \ddot{\alpha} - \dot{\gamma}\,\dot{\beta}\,I_3 \end{cases} \quad (23)$$

3. STATEMENT OF THE MINIMUM TIME TRAJECTORY PLANNING PROBLEM

The problem of trajectory planning can be stated in a simple form as follows. The quadrotor is required to move freely from an initial configuration P_i to a final one P_f, both of which are characterized by null velocities. In addition to solving for the transfer time T, we must find the robot trajectory $\Im(t) = (x(t), y(t), z(t), \alpha(t), \beta(t), \gamma(t))$ and the corresponding inputs controls $\Gamma(t) = (\tau_1(t),\dots,\tau_4(t))$ (or $U(t) = (u_1(t),\dots,u_4(t))$) such as the initial and final states are matched, constraints are respected and the following objective function is minimized:

$$F^{obj} = \int_0^T dt \quad (24)$$

The following boundary conditions inherent to the achievement of the desired task must be taken into account:
- Position $\quad \Im(0) = \Im^{ini}$ and $\Im(T) = \Im^{fin}$ (25a)
- Velocity $\quad \dot{\Im}(0) = \vec{0}$ and $\dot{\Im}(T) = \vec{0}$ (25b)

The other constraints that may have to be satisfied during the quadrotor fly are:
- *bounds on the quadrotor configurations*
$$\Im^{min} \le \Im(t) \le \Im^{max} \quad (26a)$$
- *bounds on the actuator velocities*:
$$\dot{\theta}_i^{min} \le \dot{\theta}_i(t) \le \dot{\theta}_i^{max} \quad i = 1,\dots,4 \quad (26b)$$
- *bounds on the actuator torques*:
$$\tau_i^{min} \le \tau_i(t) \le \tau_i^{max} \quad i = 1,\dots,4 \quad (26c)$$
- *obstacles avoidance*:
$$Col(\Im(t)) = false \quad (26d)$$

The constraint (26a) traduces the fact that the quadrotor will move in a limited space and its orientation should be compatible with the simplification hypotheses (small angles) adopted in relation (20). The constraint (26b) arises from the fact that each rotor should turn in a specific direction and the corresponding motor has a limited speed. The limited power of actuators implies also bounds on input torques as indicated in (26c). These limits are systematically projected on the amplitude of the equivalent control inputs u_i, $i = 1,\dots,4$. When obstacles are present in the workspace, the constraint (26d) will hold during the quadrotor fly. The function *Col* indicates whether the robot at a given configuration is in collision with an obstacle or not.

Moreover, the quadrotor has only four motors actuating the four rotors whereas the system has six *dof*. This means that the system is underactuated. The exploitation of such a system involves the identification of the dependency existing between the six *dof*, i.e. between the elements of $\Im(t)$.

From relation (19b), it is possible to write ($u_1 \ne 0$):

$$r_{13} = \ddot{x}/u_1 \qquad r_{23} = \ddot{y}/u_1 \qquad r_{33} = (g+\ddot{z})/u_1;$$

but, from (3) we have: $\begin{cases} c\alpha\, r_{13} + s\alpha\, r_{23} = s\beta c\gamma \\ s\alpha\, r_{13} - c\alpha\, r_{23} = s\gamma \end{cases}$.

Thus, $\begin{cases} \ddot{x} c\alpha + \ddot{y} s\alpha - (\ddot{z}+g)tg\beta = 0 \\ \ddot{x} s\alpha - \ddot{y} c\alpha - \sqrt{\ddot{x}^2 + \ddot{y}^2 + (g+\ddot{z})^2}\, s\gamma = 0 \end{cases}$ (27a)

Relation (27a) represents two nonholonomic constraints of second order traducing the dependency existing between the quadrotor kinematic parameters. Using this relation, it is possible to deduce two parameters from a non linear combination of the four other parameters, e.g.:

$$\beta = \arctan\left(\frac{\ddot{x} c\alpha + \ddot{y} s\alpha}{\ddot{z}+g}\right),\; \gamma = \arcsin\left(\frac{\ddot{x} s\alpha - \ddot{y} c\alpha}{\sqrt{\ddot{x}^2 + \ddot{y}^2 + (g+\ddot{z})^2}}\right) \quad (27b)$$

Note that during the quadrotor fly, the condition $u_1 \ne 0$ is generally verified. The unique case for which u_1 is null is that of the free fall motion which is discarded in our study.

In addition to previous constraints, there is another constraint arising from the relationship existing between the quantities u_i and the effective input torques τ_i, $i = 1,\dots,4$, which are proportional to $\dot{\theta}_i^2$. In fact, from relations (13a-b), (17b), (19c) and (22), it is easy to establish that:

$$\begin{aligned}(u_1\; u_2\; u_3\; u_4)^T &= \Lambda(\dot{\theta}_1^2\; \dot{\theta}_2^2\; \dot{\theta}_3^2\; \dot{\theta}_4^2)^T \\ &= -\frac{1}{d}\Lambda(\tau_1\; \tau_2\; \tau_3\; \tau_4)^T\end{aligned} \quad (28)$$

where: $\Lambda = \begin{pmatrix} \dfrac{b}{m_0} & \dfrac{b}{m_0} & \dfrac{b}{m_0} & \dfrac{b}{m_0} \\ 0 & -\dfrac{lb}{I_{xx}} & 0 & \dfrac{lb}{I_{xx}} \\ -\dfrac{lb}{I_{yy}} & 0 & \dfrac{lb}{I_{yy}} & 0 \\ -\dfrac{d}{I_{zz}} & \dfrac{d}{I_{zz}} & -\dfrac{d}{I_{zz}} & \dfrac{d}{I_{zz}} \end{pmatrix}$

Hence, the feasibility of any set of inputs u_i, $i = 1,\dots,4$, is conditioned by the following constraint:

$$\Lambda^{-1}(u_1\; u_2\; u_3\; u_4)^T > \vec{0} \quad (29)$$

The satisfaction of inequality (29) ensures the existence of the vector $(\dot{\theta}_1^2\; \dot{\theta}_2^2\; \dot{\theta}_3^2\; \dot{\theta}_4^2)^T$, thus the corresponding input torques may be achieved by the quadrotor actuators. In contrast, violating constraint (29) means that the requested trajectory involves unfeasible control inputs, and thus should be modified.

The problem defined by relations (24-29) is a generic optimal control problem and may be solved using either direct or indirect methods. In the following section, we present a simple direct numerical method able to treat this problem.

4. THE PROPOSED APPROACH

It is clear that using non-holonomic constraints (27), it is possible to express all elements of the optimization problem as a function of the time evolution of only four configuration parameters, e.g. $\{x(t), y(t), z(t), \alpha(t)\}$ and their time derivatives. Figure (2) highlights the way how this process may be achieved. The definition of a trajectory candidate is done by defining the evolution of $\{x(t), y(t), z(t), \alpha(t)\}$, for $t \in [0\ T]$, while accounting for boundary conditions (25) and bounds (26a) on the quadrotor configurations. Using relation (27), the time evolution of the two other configuration parameters are deduced. By application of the inverse dynamic model (23), the input controls are calculated. After that, the remaining constraints (26b, 26c, 26d, 29) are easily checked.

The remaining question is how to generate the trajectory candidates? For this purpose, we propose to consider any trajectory candidate as a composition of a parametric form $\mathbf{P}(\lambda) = (x(\lambda), y(\lambda), z(\lambda), \alpha(\lambda)), \lambda \in [0\ 1]$, defining the quadrotor path, and a monotonically increasing function $\lambda(t)$, $t \in [0\ T]$, specifying the motion on this path. The optimal evolutions of $\mathbf{P}(\lambda)$ and $\lambda(t)$, which are approximated using algebraic polynomial splines [30], are found using a nonlinear optimization technique. In what follows we propose to adopt the following choices:

- generate $\mathbf{P}(\lambda), \lambda \in [0\ 1]$, by means a quintic B-spline model and using a set of control points generated within the admissible workspace defined by (26a) and accounting for constraints (25a, 26d)
- build the motion profile $\lambda(t)$ on the interval $[0\ T]$ using a 7th order polynomial function that account for the following boundary conditions:

$$\begin{cases} \lambda(0) = 0 \\ \dot{\lambda}(0) = 0 \\ \ddot{\lambda}(0) = 0 \end{cases} \text{ and } \begin{cases} \lambda(T) = 1 \\ \dot{\lambda}(T) = 0 \\ \ddot{\lambda}(T) = 0 \end{cases}$$

These conditions ensure the respect of constraints (25b).

The design parameters of $\mathbf{P}(\lambda)$ and $\lambda(t)$, which are the positions of control points of the B-spline functions and the value of T, become the unique unknowns of the trajectory generation problem. Their optimal values may be found easily using for example the sequential quadratic programming technique [31].

5. SIMULATION RESULTS

Hereafter, two cases will be discussed. Firstly, the quadrodor is required to move from the initial configuration $\mathfrak{I}^{in} = (5,5,5,0)$ to the final one $\mathfrak{I}^{fn} = (40,50,30,0)$, in a free space and under constraints given in Table 1. The Bspline curve representing $\mathbf{P}(\lambda)$ is generated using five free control points. The minimum transfer time found in simulation is T = 10.6seconds. The evolution of a number of selected kinematic and dynamic parameters is given in figure 3. Secondly, the quadrotor is asked to move from $\mathfrak{I}^{in} = (1,1,0,0)$ to $\mathfrak{I}^{fn} = (12,6,9,0)$ in an encumbered environment. The initial solution is given manually by specifying the initial position of the control points. The optimization process enhances the quality of this solution until that depicted on figure 4. The transfer is executed in 17.995seconds. Note that both transfers are executed in a smooth way and no abrupt variations are recorded in any kinematic or dynamic motion parameter. This is due principally to the high order continuity functions employed for generating the quadrotor path and motion profile.

Table 1. Simulations parameters

m = 0.5 kg	I_{xx} = 0.0622 $kg.m^2$	z^{min} = 0; z^{max} =100m
L = 0.2 m	I_{yy} = 0.0733 $kg.m^2$	α^{max} =-α^{min} = 15°
b =4.74E-5	I_{zz} = 0.0964 $kg.m^2$	β^{max} =-β^{min} = 15°
d =2.35E-7	I_r= 1E-4 $kg.m^2$	γ^{max} =-γ^{min} = 15°
g=9.81$m.s^{-2}$	x^{min} = 0; x^{max} =100m	$\dot{q}_i^{max}$ =180$rd.s^{-1}$
τ^{max} =- τ^{min} =0.05Nm	y^{min} = 0; y^{max} =100m	

6. CONCLUSION

In this paper, we have first developed the dynamic model of an aerial quadrotor robot using a multibody approach. After that, the problem of minimum time trajectory planning has been addressed accounting for the main constraints inherent to the system and its environment. Then, a simple direct numerical method, based on an adequate parametrization of the quadrotor trajectory and using a nonlinear optimization technique, has been introduced. Finally, two numerical examples have been exposed to illustrate the efficiency of the proposed approach. As a future work, other optimization techniques and further constraints inherent to the actuators design will be considered. Also, implementation on a realistic quadrotor is envisaged.

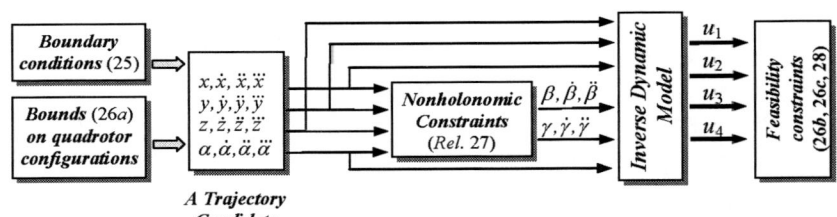

Fig. 2. Synoptic diagram of the construction and the evaluation of a quadrotor trajectory candidate

REFERENCES

[1] V. Mistler, A. Benallegue and N.K. M'Sirdi, "Exact linearization and noninteracting control of a 4 rotors helicopter via dynamic feedback," *Proceedings of IEEE Intrnaltional Workshop on Robot and Human Interactive Communication*, 2001, pp. 586~593.

[2] E. Altug, J.P. Ostrowski, and R. Mahony, "Control of a Quadrotor Helicopter Using Visual Feedback," *Proceedings of the 2002 IEEE International Conference on Robotics & Automation*, Washington, DC, 2002, pp. 72~77.

[3] T. Hamel, R. Mahony, and A. Chriette, "Visual servo trajectory tracking for a four rotor VTOL aerial vehicle," *Proceedings of the 2002 IEEE Int. Conf. on Robotics & Aut.*, Vol.3, Washington, DC, 2002, pp. 2781~2786.

[4] P. McKerrow, "Modeling the Draganflyer for-rotor helicopter," *Proceedings of the 2004 IEEE Inter. Conf. on Robotics & Automation*, New Orleans, LA, 2004, pp. 3596~3601.

[5] A. Mokhtari and A. Benallegue, "Dynamic Feedback Controller of Euler Angles and Wind parameters estimation for a Quadrotor Unmanned Aerial Vehicle," *Proceedings of the 2004 IEEE Inter. Conf. on Robotics & Automation*, New Orleans, LA, 2004, pp. 2359~2366.

[6] P. Catillo, R. Loranzo and A. Dzul, "Stabilization of a mini-rotorcraft having four rotors," *Proceedings of IEEE/RSJ Inter. Conf. on Intelligent Robots and Systems*, Sendai, Japan, 2004, pp. 2693~2698.

[7] S. Bouabdallah, P. Murrieri, and R. Siegwart, "Design and Control of an Indoor Micro Quadrotor," *Proceedings of the 2004 IEEE International Conference on Robotics & Automation*, New Orleans, LA, 2004, pp. 4393~4398.

[8] T. Hamel and R. Mahony, "Pure 2D Visual Servo control for a class of under-actuated dynamic system," *Proceedings of the 2002 IEEE Inter. Conf. on Robotics & Aut.*, New Orleans, LA, 2004, pp. 2229~2235.

[9] W. Ren, and R. W. Beard, "Trajectory Tracking for Unmanned Air Vehicles with Velocity and Heading Rate Constraints," *IEEE TRANSACTIONS ON CONTROL SYSTEMS TECHNOLOGY*, 2004.

[10] C. C. Yang, L.C. Lai, and C. J. Wu, "Time optimal control of a hovering quad-rotor helicopter," Int. Conf. on Syst. & Signals, IEEE-ICSS 2005, pp295-300.

[11] S. Bouabdallah and R. Siegwart, "Backstepping and Sliding-mode Techmiques Applied to an Indoor Micro Quadrotor," *Proceedings of the 2005 IEEE International Conference on Robotics & Automation*, Barcelona, Spain, April 2005, pp. 2259~2264.

[12] S. L. Waslander, G. M. Hoffmann, J. S. Jang, C. J. Tomlin, "Multi-Agent Quadrotor Testbed Control Design: Integral Sliding Mode vs. Reinforcement Learning," *IEEE/RSJ International Conference on Intelligent Robots and Systems*, 2005.

[13] E. Altug, J. P. Ostrowski, and C. J. Taylor, "Quadrotor Control Using Dual Camera Visual Feedback," *The International Jour. of Robotics Research* Vol. 24, No. 5, pp. 329-341, May 2005.

[14] T. Abdelhamid and M. Stephen, "Attitude Stabilization of a VTOL Quadrotor Aircraft," *IEEE Transactions On Control Systems Technology*, VOL. 14, NO. 3, May 2006.

[15] I. D. Cowling, J. F. Whidborne, A. K. Cooke, "Optimal Trajectory Planning And LQR Control For A Quadrotor UAV," in *Proc. Of the int. conf. control*, Glasgow, Scotland, august 30, September 2007.

[16] C. L. Castillo, W. Moreno, K. P. Valavanis, "Unmanned Helicopter Waypoint Trajectory Tracking Using Model Predictive Control," *Proc. of the 15th Mediterranean conf. on Cont. & Aut.*, July 27-29, 2007, Athens –Greece.

[17] S. Yazir, M. Mekki and T. Chettibi, "Modélisation dynamique directe et inverse d'un quadrotor," *4th Conference on electrical engineering*, Algiers, April 2007.

[18] H. Bouadi, M. Bouchoucha, and M. Tadjine, "Sliding Mode Control based on Backstepping Approach for an UAV Type-Quadrotor," *Inter. Jour. Of Applied Math. & Computer Sciences*, Vol.4, N°2, 2007.

[19] T. Chettibi, H. E. Lehtihet, M. Haddad and S. Hanchi, "Minimum cost trajectory planning for industrial robots", *European J. of Mechanics/A*; pp703-715, 2004.

[20] M. Haddad, T. Chettibi, S. Hanchi and H. E. Lehtihet, "A Random Profile Approach for Trajectory Planning of Wheeled Mobile Robots", *European Journal of Mechanics A/Solids, 26* (2006), pp 519-540.

[21] N. Metni, J.-M. Pflimlin, T. Hamel, P. Souères, "Attitude and gyro bias estimation for a VTOL UAV," *Control Engineering Practice*, Vol 14, pp 1511–1520, 2006.

[22] J. J. Craig, "Introduction to robotics, mechanics and controls," 3rd edition, Pearson Education 2005.

[23] M. D. Ardema, "Newton-Euler Dynamics," springer 2005.

[24] J. Seddon, "*Basic Helicopter Aerodynamics*," Blackwell Science, Osney Mead, Oxford, 1996.

[25] R. W. Prouty, "Helicopter Performance, Stability, and Control," Krieger Publishing Company, 2002.

[26] D. Honnery, "*Introduction to the Theory of Flight*," Gracie Press, Northcote, Victoria, 2000.

[27] M. Mekki, S. Yazir, "Contribution à la conception et la réalisation d'un mini-drone quadrotor," *Projet de fin d'études*, EMP, Algiers, 2007.

[28] T. Chettibi, M. Haddad, "*Dynamic modelling of a quadrotor aerial robot*," Journées D'études Nationales de Mécanique, Batna, Algérie 2007.

[29] L. Derafa, T. Madani and A. Benallegue, "*Dynamic Modelling and Experimental Identification of Four Rotors Helicopter Parameters*" international conference on industrial technology, Munbai, India 2006.

[30] T. Chettibi, M. Haddad, S. Hanchi, "Application of algebraic polynomial splines for trajectory generation problems," 5ème journées de la mécanique, EMP, Alger, 2006, Algérie.

[31] T. Coleman, M. A. Branch, A. Grace, "*Optimization Toolbox, for use with Matlab*," The MathWorks Inc., 1999.

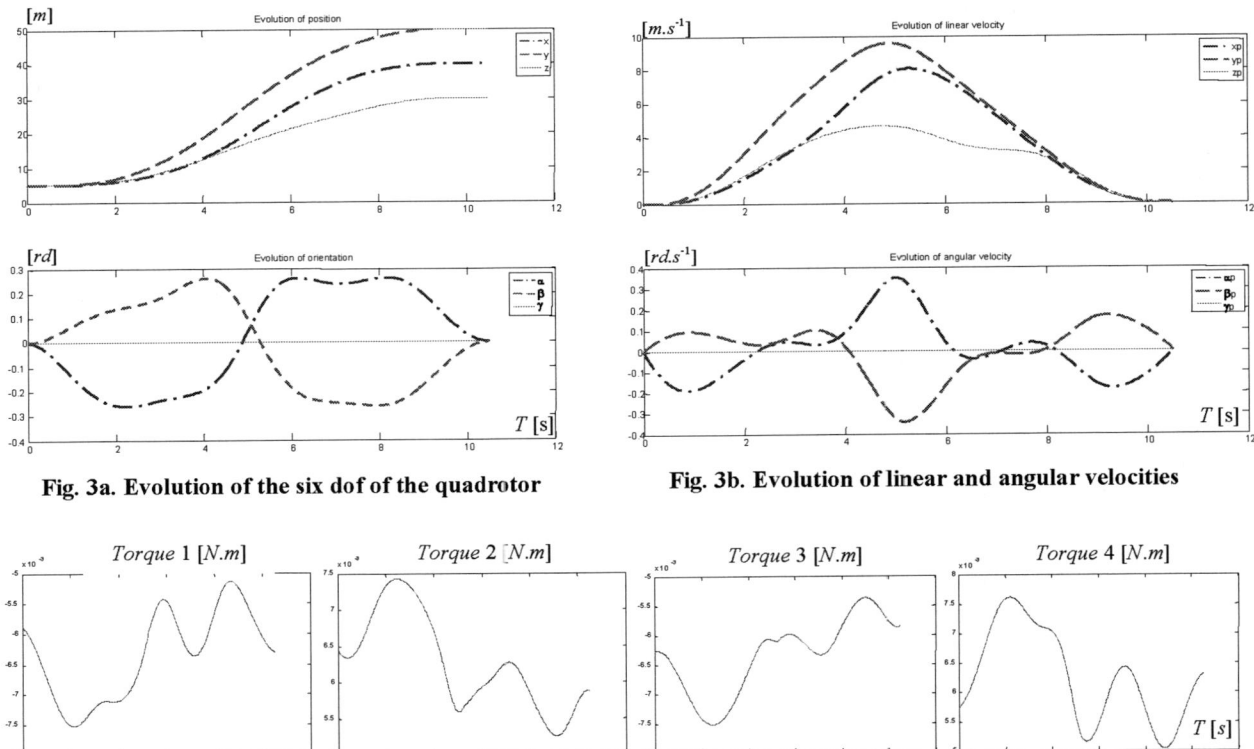

Fig. 3a. Evolution of the six dof of the quadrotor

Fig. 3b. Evolution of linear and angular velocities

Fig. 3c. Evolution of actuator torques

Fig. 3. Optimal Trajectory Generation Between two Configurations.

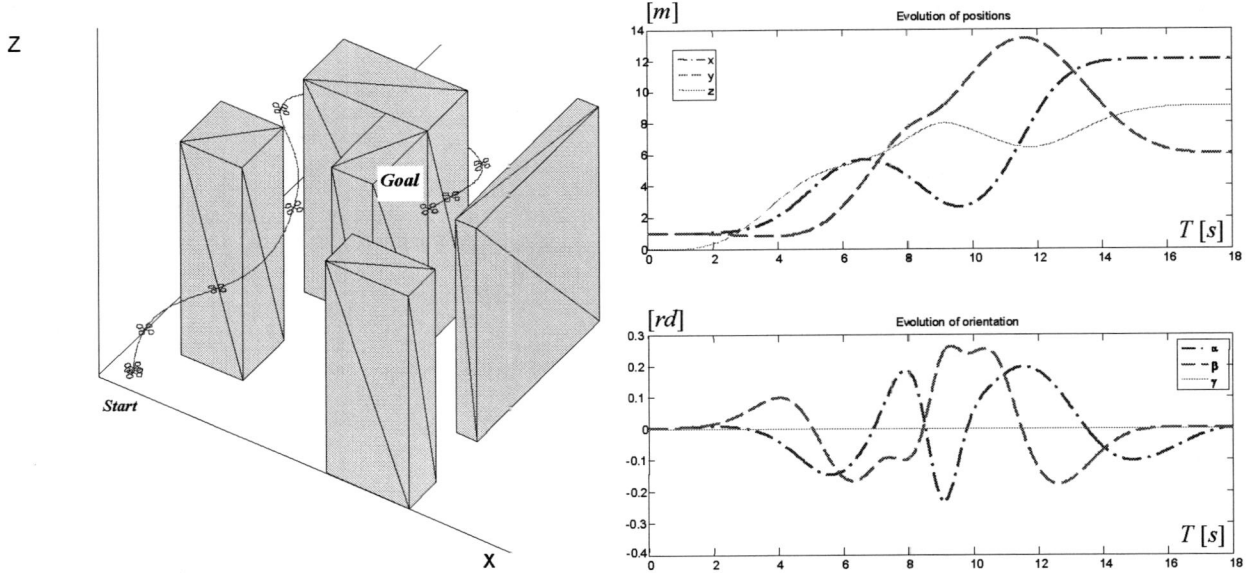

Fig. 4a. Optimal trajectory through obstacles

Fig. 4b. Optimal time evolution of the six quadrotor *dof*.

Fig. 4. Trajectory generation in the presence of obstacles

SESSION F2: ADAPTIVE CONTROL & FAULT DETECTION

Comparative Study of Adaptive Type-1 and Type-2 Fuzzy Controls for Nonlinear Systems under Uncertainty

S. Mokaddem, F. Khaber.

Laboratoire QUERE, Département d' Electrotechnique, Université Ferhat Abbas 19000 - Sétif- Algérie
mokdsana@yahoo.fr, Farid.Khaber@ieee.org

Abstract: This work presents a development of adaptive type-1 and type-2 fuzzy controls for uncertain nonlinear systems. Using the adaptive type-1 fuzzy control, the dynamic of the nonlinear systems is approximated with type-1 fuzzy systems whose parameters are adjusted by appropriate law adaptation. For adaptive type-2 fuzzy control, the dynamic of the nonlinear systems is approximated with interval type-2 fuzzy systems. The use of this type-2 control requires an additional operation witch is the type reduction, in comparing with typ-1 control. The closed-loop system stability is guaranteed by the Lyaponov synthesis. To show the performance of the developed controls, a comparative study is realized through the application of these controls so that an inverted pendulum tracks a given trajectory in presence of disturbances.

Keywords: Type-1 fuzzy system, Adaptive control, Type-2 fuzzy system, Type reduction, Nonlinear uncertain system.

1. INTRODUCTION

Uncertainty is an inherent part in control systems used in real world application. This uncertainty appears in a number of different forms and results of some information deficiency, which may be incomplete, imprecise not fully reliable, or deficient in some other way. The framework of fuzzy reasoning allows handling much of this uncertainty, fuzzy systems employ type-I fuzzy sets (T1 FSs), which represents uncertainty by numbers in the range [0, 1]. However, it is not reasonable to use an accurate membership function for something uncertain like a measurement. In this case we use fuzzy sets which are able to handle these uncertainties, the so called type-2 fuzzy sets (T2 FSs). So, the amount of uncertainty in a system can be reduced by using type-2 fuzzy logic because it offers better capabilities to handle linguistic uncertainties by modeling vagueness and unreliability of information. The structure of the type-2 fuzzy system (T2 FLS) that remains similar at this of the type-1 fuzzy system (T1 FLS) including a supplementary module charged of the operation of type reduction. Several methods exist to achieve this type reduction among which: Center of sets, centroid, center of sums, and height type reduction.

Currently, the T2 FLS are used for control systems in different domains. In fact, in [7] and [6] a combination of T2 FLS and neural networks is used to control the dynamic of nonlinear system. In other interesting studies, that used the T2 FLS in order to develop a structure of control conferring a good behavior of robots especially in the non structured environments [1].

This paper is organized as follows:

In the second section a detailed description of interval type-2 fuzzy system (IT2 FLS) is presented. The basic definitions of these systems are given. The operations of fuzzification, inference, type reduction and defuzzification are explained for the case of T2 FLS. In the third section an algorithm of adaptive fuzzy control uses the T1 FLS and IT2 FLS is exposed. To enable the two types of the controls developed an example of simulation of the nonlinear system with uncertainty is treated in section four.

2. DESCRIPTION OF TYPE-2 FUZZY SYSTEM

A type-2 fuzzy set in universal set X is denoted as $\tilde{A}$ which is characterized by a type-2 membership function in (1).

$$\tilde{A} = \int_{x \in X} u_{\tilde{A}}(x)/x = \int_{x \in X} \left[\int_{u \in J_x} f_x(u)/u \right]/x \qquad (1)$$
$$J_x \in [0,1]$$

The $\mu_{\tilde{A}}(x)$ can be referred as a secondary membership function (MF), which is a type-1 fuzzy set in [0, 1]. In (1), $f_x(u)$ is a secondary grade, which is the amplitude of a secondary MF; i.e., $0 < f_x(u) < 1$ (fig.1).

The domain of a secondary MF is called the primary membership of x noted J_x, where $u \in J_x \subseteq [0,1]$ for $\forall x \in X$; u is a fuzzy set in [0,1], rather than a crisp point in [0, 1].

When, $f_x(u) = 1 \ \forall u \in J_x \subseteq [0,1]$, then the secondary MFs are interval sets such that $\mu_{\tilde{A}}(x)$ in (1) can be called an interval type-2 MF. Therefore the type-2 fuzzy set can be re-expressed as

$$\tilde{A} = \int_{x \in X} u_{\tilde{A}}(x)/x = \int_{x \in X} \left[\int_{u \in J_x} 1/u \right]/x \qquad (2)$$
$$J_x \in [0,1]$$

IT2 FSs reflect a uniform uncertainty at the primary memberships of x. This type of MF is the most often used in T2 FLS for their simplicity of calculation especially in the type reduction.

Also, a Gaussian primary MF with uncertain mean and fixed standard deviation having an interval type-2 secondary MF can be called an interval type-2 Gaussian MF (3), will be adopted in this paper. It can be stated as

$$\mu_{\tilde{A}}(x) = \exp\left[-\frac{1}{2}\left(\frac{x-m}{\sigma}\right)^2\right], \quad m \in [m_1, m_2] \quad (3)$$

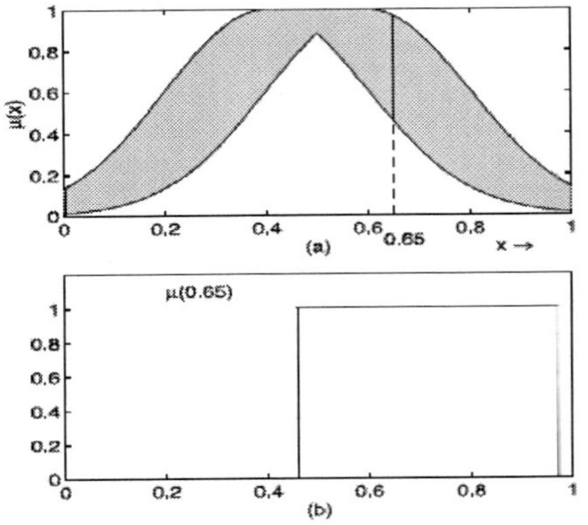

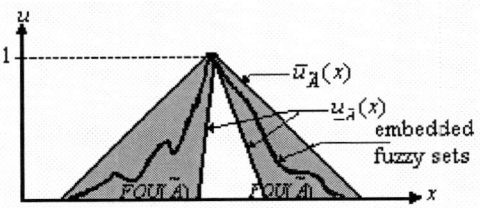

Fig.1. (a) Pictorial representation of interval type-2 fuzzy set with a Gaussian primary MF, (b) the secondary MF is interval type-1 set in $x=0.65$

Uncertainty about $\tilde{A}$ is conveyed by the union of all the primary memberships, which is called the footprint of uncertainty (FOU) of $\tilde{A}$ (fig.2), i.e.

$$FOU(\tilde{A}) = \bigcup_{\forall x \in X} J_x = \{(x,u) : u \in J_x \subseteq [0,1]\} \quad (4)$$

The upper MF and a lower MF of $\tilde{A}$ which are denoted as $\overline{\mu}_{\tilde{A}}(x)$ and $\underline{\mu}_{\tilde{A}}(x)$, respectively. Both of them are two type-1 MFs that bound the FOU (fig.2). Hence, (4) can be re-stated as

$$FOU(\tilde{A}) = \bigcup_{\forall x \in X} [\overline{\mu}_{\tilde{A}}(x), \underline{\mu}_{\tilde{A}}(x)] \quad (5)$$

Fig.2 Type-2 membership function

An IT2 FLS, which is a FLS that uses at least one IT2 FS, contains five components- fuzzifier, rule base, inference engine, type-reducer and defuzzier- that are inter-connected, as shown in fig.3. In the following sections, we will introduce each block in the type-2 FLC.

A. Fuzzifier

The fuzzifier maps a crisp input into output fuzzy sets. These fuzzy sets can, in general, be type-2 fuzzy input sets $\tilde{A}$. However, in this paper, we consider only singleton fuzzification, for which the input fuzzy set has only a single point of nonzero membership.

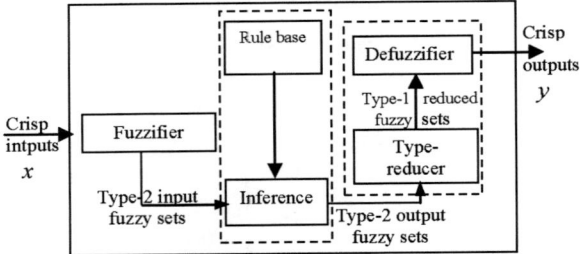

Fig.3. Structure of a type-2 fuzzy system

B. Rule Base

The rules will remain the same as in type-1 FLS but the antecedents and/or the consequents will be represented by IT2 FSs. The ith rule in the T2 FLS can be written as follows:

$$"R^i : \text{IF } x_1 \text{ is } \tilde{F}_1^i \text{ and } x_2 \text{ is } \tilde{F}_2^i \text{ and}\ldots\text{and } x_p \text{ is } \tilde{F}_p^i \\ \text{THEN } y \text{ is } \tilde{G}^i" \quad (6)$$

where x_n's are the inputs, $\tilde{F}_n^i$'s are the antecedent sets ($n = 1 \ldots p$), y is the output, and $\tilde{G}^i$'s are the consequent sets.

C. Fuzzy Inference Engine

The inference engine combines rules and gives a mapping from input type-2 sets to output type-2 sets [3]. In the inference engine, multiple antecedents in the rules are connected using the Meet operation. The membership grades in the input sets are combined with those in the output sets using the extended sup–star composition. Multiple rules are combined using the Join operation.

Each rule in a fuzzy rule base with M rules having p inputs $x_1 \in X_1, x_2 \in X_2, \ldots, x_p \in X_p$ and one output $y \in Y$, is described by the MF:

$$\mu_{\tilde{F}_1^i \times \ldots \times \tilde{F}_p^i \to \tilde{G}^i}(x,y) = \mu_{A^i \to G^i}(x,y) \quad (7)$$

where $x = \{x_1, x_2, \ldots, x_p\}$, this MF can be written as

$$\mu_{A^i \to G^i}(x,y) = \mu_{\tilde{F}_1^i}(x_1) \cap \ldots \cap \mu_{\tilde{F}_p^i}(x_p) \cap \mu_{\tilde{G}^i}(y) \\ = \left[\bigcap_{n=1}^{p} \mu_{\tilde{F}_n^i}(x_i)\right] \cap \mu_{\tilde{G}^i}(y) \quad (8)$$

As we are using the Singleton fuzzification, then the type-2 fuzzy input set $\tilde{X}'$ contains a single element x' and each $\mu_{\tilde{X}'}(x)$ is non zero only at one point $x = x'$, (input is modeled as perfect measurement). In our interval type-2 FLS, we will use the product t-norm. The result of the input and antecedent operations, is an interval type-1 set, called the firing set $\bigcap_{n=1}^{p} \mu_{\tilde{F}_n^i}(x_n) \equiv F^i(x')$, i.e.,

$$F^i(x') = \left[\underline{f^i}(x'), \overline{f^i}(x')\right] = \left[\underline{f^i}, \overline{f^i}\right]$$
$$= \left[\underline{\mu}_{\tilde{F}_1^i}(x_1')*...*\underline{\mu}_{\tilde{F}_p^i}(x_p'), \overline{\mu}_{\tilde{F}_1^i}(x_1')*...*\overline{\mu}_{\tilde{F}_p^i}(x_p')\right] \quad (9)$$

D. Type Reduction

Type reduction was proposed by Karnik and Mendel [3],[5]. It is called type-reduction because this operation takes us from the type-2 output sets of the inference engine to a type-1 set that is called the "the type-reduced set". This set provides an interval of uncertainty for the output of an IT2 FLS.

Five different type reduction methods are described in [4], which are based on computing the centroid of an IT2 FS. Computing the centroid of general T2 FS can be very intensive; however, for an IT2 FS, an exact iterative method for computing its centroid has been developed [2]. This was possible because the centroid of an IT2 FS is an IT1 FS, and such sets are completely characterized by their left and right end points; hence, computing the centroid of an IT2 FS set only requires computing those two end-points.

Center of sets, centroid, center of sums, and height type reduction can all be expressed as

$$Y_{TR} = [y_l, y_r]$$
$$= \int_{y^1 \in [y_l^1, y_r^1]} \cdots \int_{y^M \in [y_l^M, y_r^M]} \int_{f^1 \in [\underline{f^1}, \overline{f^1}]} \cdots \int_{f^M \in [\underline{f^M}, \overline{f^M}]} 1 / \frac{\sum_{i=1}^M f^i y^i}{\sum_{i=1}^M f^i} \quad (10)$$

where the multiple integral signs denote the union operation. The most widely used type reduction is centre of sets (COS) type reduction, which is adopted in this paper, for which: y_l^i and y_r^i are the left and right end points of the centroid of the consequent of the ith rule, $\underline{f^i}$ and $\overline{f^i}$ are the lower and upper firing degrees of the ith rule, computed using (9); and M is the number of fired rules.

In general, there are no closed form formulae for y_l^i and y_r^i; however, Karnik and Mendel have developed two iterative algorithms very simple and easy to implement (known as the Karnik-Mendel or KM Algorithms [2]) for computing these end-point exactly.

E. Defuzzification

From the type-reduction stage, we obtained a type reduced set determined by its left most point y_l^i and right most point y_r^i. We defuzzify the interval set by using the average of y_l^i and y_r^i, hence, the defuzzified crisp output is [5].

$$y(x') = (y_l(x') + y_r(x'))/2 \quad (11)$$

F. Approximation of function by type-2 fuzzy system

The class of the systems which use the singleton fuzzification, product inference and centroid déffuzifucation, are universal approximators [10]. They can approximate any continuous real function $y(x)$ on a compact set U [8].

$$y(x) = \sum_{i=1}^M \overline{y}^i \left(\prod_{n=1}^p \mu_{\tilde{F}_n^i}(x_n)\right) / \sum_{i=1}^M \left(\prod_{n=1}^p \mu_{\tilde{F}_n^i}(x_n)\right) \quad (12)$$

where $x = (x_1,...,x_p) \in U$ and $\overline{y}^i$ is a point in R where $\mu_{\tilde{G}^i}$ achieves its maximal value. The equation (12) can be rewritten under the following compact shape:

$$y(x) = \theta^T \xi(x) \quad (13)$$

where $\theta = (\overline{y}^1,...,\overline{y}^M)^T$ is parameters vector, $\xi(x) = (\xi^1(x),...,\xi^M(x))$ is a regressive vector, with $\xi^i(x)$ called fuzzy basis function (FBF). In the case of the IT2 FLS, the relation (13) remains valid with a supplementary phase of type reduction. Since COS type reduction represented by (10) is used in this work, observe that each set (10) is an interval type-1 set, hence, $y_{TR} \equiv y_{COS}$ is also an interval type-1 set. So, to find y_{COS}, we just need to compute the two end-points of this interval. Unfortunately, no closed-form formula is available for y_{COS}.

For any value, $y \in y_{COS}$ can be represented as

$$y = \sum_{i=1}^M f^i y^i / \sum_{i=1}^M f^i \quad (14)$$

The maximum value of y is y_r and the minimum value of y is y_l. From (14), we see that y is a monotonic increasing function with respect to y^i; so y_r is associated only with y_r^i and, similarly, y_l is associated only with y_l^i. In the COS type reduction method, Karnik and Mendel [2]; have shown that the two end points of y_{COS}, y_r and y_l depend only on a mixture $\overline{f^i}$ or $\underline{f^i}$ values, since $f^i \in F^i = \left[\underline{f^i}, \overline{f^i}\right]$. In this case, each on of y_l and y_r can be represented as a FBF expansion [5], i.e,

$$y_l = \sum_{i=1}^M f_l^i y_l^i / \sum_{i=1}^M f_l^i = \sum_{i=1}^M y_l^i p_l^i \quad (15)$$

$$p_l^i = f_l^i / \sum_{i=1}^M f_l^i \quad (16)$$

where $\overline{f}^i$ denotes the firing strength membership grade [either $\overline{f}^i$ or $\underline{f}^i$] contributing to the left-most point y_l, and p_l^i is the FBF. Similarly

$$y_r = \sum_{i=1}^M f_r^i y_r^i / \sum_{i=1}^M f_r^i = \sum_{i=1}^M y_r^i p_r^i \quad (17)$$

$$p_r^i = f_r^i / \sum_{i=1}^M f_r^i \quad (18)$$

where f_r^i denotes the firing strength membership grade [either $\overline{f}^i$ or $\underline{f}^i$] contributing to the right-most point y_r

and p_r^i is another FBF.

Note that whereas a T1 FLS is characterized by a single FBF expansion [5], an IT2 FLS is characterized by two FBF expansions. A general T2 FLS is characterized by a huge number of FBF expansions; hence, we have shown that by choosing secondary MFs to be interval sets, the complexity of a general T2 FLS is vastly reduced.

3. DESIGN OF ADAPTIVE FUZZY CONTROL

Consider the nth order nonlinear system [9] of the form

$$\begin{cases} x^{(n)} = f(x,\dot{x},\ldots,x^{(n-1)}) + g(x,\dot{x},\ldots,x^{(n-1)})u \\ y = x \end{cases} \quad (19)$$

where f and g are unknown continuous functions, $u \in R$ and $y \in R$ are the input and output respectively, and $x = (x,\dot{x},\ldots,x^{(n-1)})^T \in \Re^n$ is the state vector of the system which is assumed to be available for measurement.

According to the analyses developed in [9] of adaptive fuzzy control, we can summarize the step of conception of adaptive fuzzy controller by the following algorithm:

Step 1: off-line processing

*Specify the gain of control k, such as the matrix A_c is stable.

*Specify a positive definite $n \times n$ matrix Q, and solve the Lyapunov equation, to obtain a symmetric matrix $P>0$.

*Specify the design parameters M_f, M_g, ε and $\bar{V}$ based on practical constraints.

Step 2: Initial controller constrain

*Define m_i type-1 (type-2) fuzzy sets $\tilde{F}_p^{i_p}$ whose MFs $\mu_{\tilde{F}_p^{i_p}}$ where $i_p=1,2,\ldots,m_p$, $p=1,2,\ldots,n$. we require that the $\tilde{F}_p^{i_p}$'s include the antecedent sets in the fuzzy rules of $\hat{f}$ and $\hat{g}$.

*Construct the fuzzy rule bases for the fuzzy systems $\hat{f}(x/\theta_f)$ and $\hat{g}(x/\theta_g)$, according to (6).

*construct the fuzzy basis functions in order to constraint the fuzzy systems as

$$\hat{f}(x/\theta_f) = \theta_f^T \xi(x) \quad (20)$$
$$\hat{g}(x/\theta_g) = \theta_g^T \xi(x) \quad (21)$$

Step 3: On-line adaptation

*Apply the following feedback control $u = u_c + u_s$ to the plant (19), where $u_c = \frac{1}{\hat{g}(x/\theta_g)}\left[-\hat{f}(x/\theta_f) + y_m^{(n)} + k^T e\right]$ is the certainty equivalent controller and $u_s = I_1^* \text{sgn}(e^T P \ b_c) \times \frac{1}{g_L(x)}\left[|\hat{f}| + f^U + |\hat{g} \ u_c| + |g^U u_c|\right]$ is the supervisory control

with $I_1^* = \begin{cases} 1 & si \ V_e > \bar{V} \\ 0 & si \ V_e > \tilde{V} \end{cases}$, $\hat{f}$ and $\hat{g}$ is given by (20) and (21) respectively [9].

*use the adaptation law developed in [9] to adjust the parameter vector θ_f and θ_g.

4. SIMULATION

In this section, we make a comparison of the performance obtained by the application of adaptive type-1 and type-2 fuzzy approaches to control a nonlinear system (inverted pendulum) in presence of the disturbance. Four cases are studies. In the first two cases (T1 FS and IT2 FS) our adaptive fuzzy control is developed in ideal conditions (without perturbation), for the last two cases, the same control is developed in presence of Gaussian noise. A comparative study is based on calculate the integral of the absolute value of the tracking error (IAE) as:

$$IAE = \int_0^\infty |e(t)| \ dt \quad (22)$$

The objective of the developed control is to guarantee a good tracking of a sinewave trajectory $y_m(t) = \pi/30 \sin(t)$, from initial condition $x(0) = [-\pi/6 \ 0]$.

The dynamic equations of the inverted pendulum system are

$$\begin{cases} \dot{x}_1 = x_2 \\ \dot{x}_2 = f(x) + g(x)u \end{cases} \quad (23)$$

with

$$f(x) = \left(g \sin x_1 - \frac{m l x_2^2 \cos x_1 \sin x_1}{m_c + m}\right) \bigg/ l\left(\frac{4}{3} - \frac{m \cos^2 x_1}{m_c + m}\right) \quad (24)$$

$$g(x) = \left(\frac{\cos x_1}{m_c + m}\right) \bigg/ l\left(\frac{4}{3} - \frac{m \cos^2 x_1}{m_c + m}\right) \quad (25)$$

where $x_1 = \theta$ and $x_2 = \dot{\theta}$, $g=9.8 m/s^2$ is the acceleration due to gravity, m_c=1Kg is the mass of cart, m=0.1Kg is the mass of pole, l=0.5m is the length of pole, u is the applied force (control).

*1st case: T1 FS without disturbances.

The antecedent sets are defined by five type-1 Gaussian MFs centred in $\{-\pi/6, -\pi/12, 0, \pi/12, \pi/6\}$ with standard deviation $\sigma = \pi/24$ on the interval $|x_i| \leq \pi/6$ i=1,2. The fuzzy rule bases of $\hat{f}$ and $\hat{g}$ consist of the following rules:

"R_f^i: IF x_1 is $F_1^{i_1}$ and x_2 is $F_2^{i_2}$ THEN $\hat{f}$ is $G^{(i_1,i_2)}$"
"R_g^i: IF x_1 is $F_1^{i_1}$ and x_2 is $F_2^{i_2}$ THEN g is $H^{(i_1,i_2)}$"

where $G^{(i_1,i_2)}$ et $H^{(i_1,i_2)}$ are the fuzzy sets in R corresponding to $\bar{y}^i$ parameters (parameters vectors θ_f, θ_g) which have a random initial values in a certain bounded region [9].

The simulation results show the good tracking of desired trajectory (fig.5), with a tracking error quantified through the calculation of the (IAE) criteria, which value is illustrate in table.1.

*2^{end} case: IT2 FS without disturbance.

In this case, the same fuzzy rules are used except the antecedent sets are represented only by three type-2 Gaussian MFs, which are constructed from the type-1 antecedent sets (1^{rst} case), by adding a region of uncertainty to the mean of type-1 fuzzy sets $m \in [-0.1, +0.1]$, while keeping a fixed standard deviation. To facilitate the manipulation of these sets, each of them is represented by its upper and lower MFs, $\bar{\mu}_{\tilde{A}}(x)$ and $\underline{\mu}_{\tilde{A}}(x)$, respectively as

$$\bar{\mu}_{\tilde{A}}(x) = \begin{cases} e^{-\frac{1}{2}\left((x_k - m_{k_1}^l)/\sigma_k^l\right)^2} & x_k < m_{k_1}^l \\ 1 & m_{k_1}^l < x_k < m_{k_2}^l \\ e^{-\frac{1}{2}\left((x_k - m_{k_2}^l)/\sigma_k^l\right)^2} & x_k > m_{k_2}^l \end{cases} \quad (26)$$

$$\underline{\mu}_{\tilde{A}}(x) = \begin{cases} e^{-\frac{1}{2}\left((x_k - m_{k_2}^l)/\sigma_k^l\right)^2} & x_k \leq (m_{k_1}^l + m_{k_2}^l)/2 \\ e^{-\frac{1}{2}\left((x_k - m_{k_1}^l)/\sigma_k^l\right)^2} & x_k > (m_{k_1}^l + m_{k_2}^l)/2 \end{cases} \quad (27)$$

Since $\hat{f}$ and $\hat{g}$ are IT2 FLS, theirs type reduction sets $\hat{F}_{cos}$ and $\hat{G}_{cos}$ are IT1 FS, then $\hat{f} = (\hat{f}_l + \hat{f}_r)/2$, $\hat{g} = (\hat{g}_l + \hat{g}_r)/2$

$$\hat{f} = \left(\theta_f^T \xi_{fl} + \theta_f^T \xi_{fr}\right)/2 = \theta_f^T \left[(\xi_{fl} + \xi_{fr})/2\right] = \theta_f^T \xi_f \quad (28)$$

$$\hat{g} = \left(\theta_g^T \xi_{gl} + \theta_g^T \xi_{gr}\right)/2 = \theta_g^T \left[(\xi_{gl} + \xi_{gr})/2\right] = \theta_g^T \xi_g \quad (29)$$

We constructed the average of FBF $\xi_f(x)$ and $\xi_g(x)$ of IT2 FLS, by computing $\xi_{fr}(\xi_{gr})$ according (16), and $\xi_{fl}(\xi_{gl})$ according (18).

The consequent sets are T1 FLS; their centroid is the adjustable parameters, which have random initial value in certain bounded region.

Table1: Comparison of performance criteria for T2 FLS and T1 FLS after 2000 samples

Performance criteria		IAE
T1 FLS with 5 MFs	System without disturbance	9.6861
	System with disturbance	19.9559
IT2 FLS with 3 MFs	System without disturbance	6.9122
	System with disturbance	16.9152

The results of simulation obtained by application of adaptive type-2 fuzzy control are presented in fig.6. We observed that the tracking is similar for the one obtained in the 1^{rst} case. But a value of performance criteria in table 1 (fig.7), indicate that the use of the IT-2 FLS gives a lower tracking error than T1 FLS, although the IT2 FLS use only three T2 FSs, which reduce the number of used rules to 9 rules comparing with T1 FLS that used 25 rules

*3rd case: T1 FLS with disturbances.

We consider the same conditions of the 1^{rst} case; with addition of disturbance from the system (19) by the introduction of a random noise with normal distribution according to the schema of fig.4. The noise will be add to the system output $y(t)$ as follows

$$y(t) = y(t) + 0.001 \cdot \text{randn} \quad (30)$$

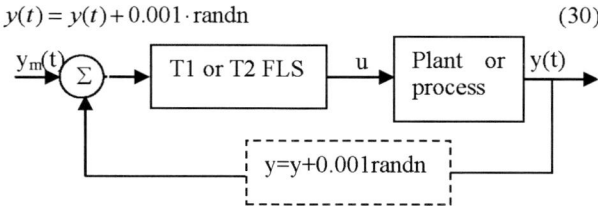

Fig.4. System with additive disturbance

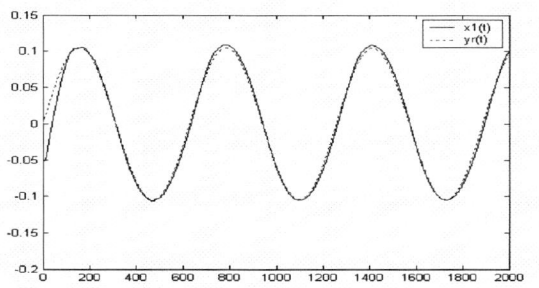

Fig.5. Output of ideal system $x_1(t)$ and reference trajectory $y_r(t)$: using T1 FLS with 5 MFs

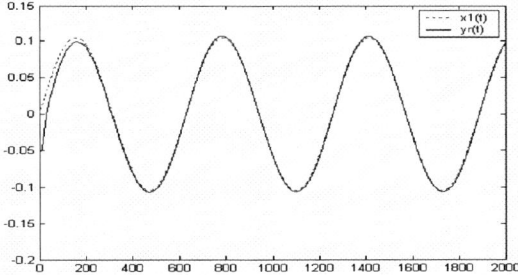

Fig.6. Output of ideal system $x_1(t)$ and reference trajectory $y_r(t)$: using IT2 FLS with 3 MFs

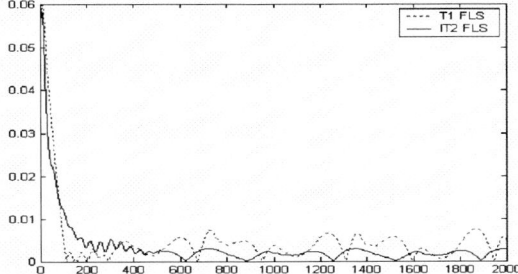

Fig.7. Criteria IAE without disturbances.

The simulation results show that the output system $x_1(t)$ represented in the fig.8; achieve acceptable tracking of desired trajectory, although the system is perturbed. The value of IAE criteria indicate that the tracking error is bigger than the one obtained in the two precedent cases (without disturbances) fig.10.

*4th case: T1 FLS with disturbances.

We consider the same situation as in the 3rd case, except that we use the IT2 FLS of the 2nd case.

The simulation results which are represented in fig.9, show that the output track the desired trajectory. The value of IAE in the table 1 shows that the tracking error in this case is more important than the error obtained in non disturb case (cases 1st, 2^{end} case), but weaker than the one obtained by using T1 FLS in presence of disturbances (fig.10). This affirms the capacity of the T2 FLS to handle uncertainties.

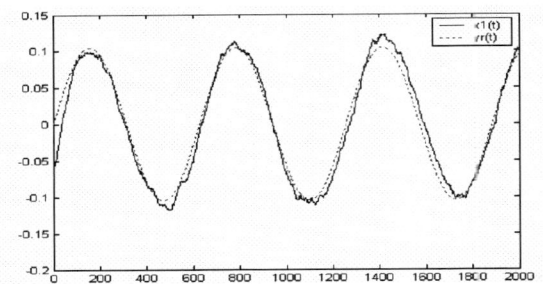

Fig.8. Output of perturbed system $x_1(t)$ and reference trajectory $y_r(t)$: using T1 FLS with 5 MFs

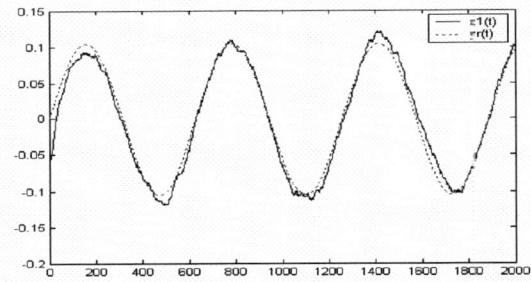

Fig.9. Output of perturbed system $x_1(t)$ and reference trajectory $y_r(t)$: using IT2 FLS with 3 MFs

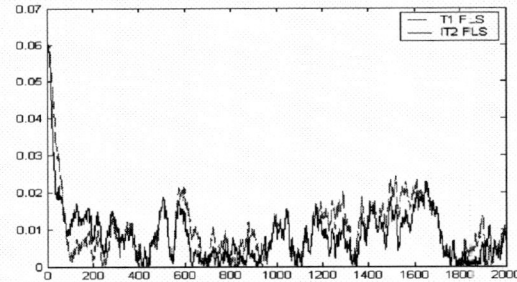

Fig.10. Criteria IAE with disturbances.

5. CONCLUSION

In this work we have developed two adaptive type-1 and type-2 fuzzy controls. For the adaptive type-1 fuzzy control we have used an adaptive T1 FLS to approximate the dynamic of the systems. For adaptive type-2 fuzzy control the T1 FLS are changed by T2 FLS, which need a supplementary operation of type reduction. The T2 FLS used, are of type interval, and the type reduction operation is accomplished by using a centre of sets method. To enable the performance of the two developed controls, we have used an example of simulation in order to elaborate a comparative study. The simulation results obtained by an implementation in MATLAB environment have allowed us to validate the two laws of adaptive type-1 and type-2 fuzzy control. The evaluation of IAE performance criteria, based on the tracking error, shows that T2 FLS can handle uncertainties in better way because they provide more parameters, more design degrees of freedom, and furthermore the time computation needed for IT2 FLS implementation is of the same order of size of time needed of T1 FLS implementation.

REFERENCES

[1] Hegras. H. A, (2004) "hierachical type-2 fuzzy logic control Architecture for autonomous mobile robots", *IEEE Trans. Fuzzy Syst*, **vol. 12, no. 4**, pp 524-539.

[2] Karnik. N. N, J. M. Mendel, (2001) " Centroid of a type-2 fuzzy logic systems", *Information sciences*, **Vol.132**, pp.195-220.

[3] Karnik. N. N, J. M. Mendel, and Q.liang, (1999),"Type-2 fuzzy logic systems", *IEEE Trans. Fuzzy Syst*, **Vol. 7**, pp. 643-658.

[4] Karnik. N. N, J. M. Mendel (October 1998),"Type-2 fuzzy logic systems: Type-reduction", *IEEE Syst*, Man, Cybern. Conf, Sandiago, CA.

[5] Liang. Q, J. M. Mendel, (2002),"Interval type-2 fuzzy logic systems: Theory and design", *IEEE Trans. Fuzzy Syst*, **Vol. 8, no. 5**, pp. 535-550.

[6] Melin. P and O. Castillo, (2003) "A new method for adaptive modelbased control of non-linear plants using type-2 fuzzy logic and neural networks", *Proceedings of the 12th IEEE International conference on Fuzzy Sys*.

[7] Melin. P, and O. Castillo, (2002)"Intelligent control of non-linear dynamic plants using type-2 fuzzy logic and neural networks", *Proceedings of the Annual Meeting of the North American Fuzzy Information Processing Society*.

[8] Mendel. J. M, (1995) " Fuzzy logic systems for engineering: A titorial" *Proc. IEEE*, **Vol. 83, no. 3**, pp. 345-377.

[9] Wang. L. X, (1996) "Stable adaptive fuzzy controllers with application to inverted pendulum tracking" *IEEE Trans. On sys, Man, Cybern-Part B: Cybernetics*, **Vol. 26, no.5**, pp.677-691.

[10] Wang. L. X, J. M. Mendel, (1992) "Fuzzy basis function, universal approximation, and orthogonal least squares learning", *IEEE Trans. Neural Networks*, **Vol. 3, no. 5**, pp. 807-814.

Control of perturbed systems in the frequency Domain

GHEDJATI. Keltoum, ABDELAZIZ. Mourad

*Automatic Laboratory,
Department of Electrotechnics,
University of Sétif.
Cite Maabouda, Setif, 19000.
ALGERIA
ghedjati_r@yahoo.fr   abde_m@yahoo.fr*

Abstract: One present in this paper the design of a robust direct model reference adaptive control (DMRAC) algorithm which will be applied to systems with unmodeled dynamics. These unmodeled dynamics are represented as either additive or multiplicative norm bounded perturbations in the transfer function. Frequency domain design conditions for a feedforward compensator have been developed. The validity of the algorithm is illustrated by means of examples.

Keywords: Adaptive control, Strictly positive real, Asymptotic stability, Lyapunov function MRAC.

1. INTRODUCTION

Model reference adaptive methods may be classified according to three different approaches. First is the full state access method described by Landau [10], which assumes that the state variables are measurable. Second is the input-output method originating from Monopoli's augmented error signal concept [13]. In this latter approach, adaptive observers are incorporated with the controller to estimate missing states. Third is the simple adaptive control approach originated by Kaufman et. al [13]. This approach is an output feedback method which requires neither full state feedback nor adaptive observers, Asymptotic stability is guaranteed if the plant is ASPR; that means, if there exists a feedback gain K_e such that the resulting closed-loop transfer function is strictly positive real (SPR). This gain need not be physically realized during implementation.

We know that most real systems are not ASPR, the algorithm was extended by Bar-Kana and Kaufman [3,8] to the class of non-ASPR plants.

Kaufman and Neat [9] suggested a further modification incorporating supplementary feedforward into the reference model in such a manner that asymptotic tracking of the augmented plant and model outputs implies asymptotic tracking of the original plant and model outputs.

We develop in this paper, the conditions of output stability for plants in the presence of structural perturbations.

Barkana [1,2] gives more investigations in DMRAC and more studies about the convergence of the adaptive gains. This paper is organised as follows :

The formulation of the DMRAC algorithm is discussed in section 2, robust feedforward compensator design is formulated in section 3, and simulation results are given in section 4. Finally, conclusions are drawn in section 5.

2. FORMULATION OF THE DMRAC

The linear time invariant model reference adaptive control pro problem is considered for a plant given by .

$$\begin{aligned} \dot{x}_p(t) &= A_p x_p(t) + B_p u_p(t) \\ y_p(t) &= C_p x_p(t) \end{aligned} \qquad (1)$$

Where $x_p(t)$ is the $(n \times 1)$ state vector, $u_p(t)$ is the $(m \times 1)$ control vector, $y_p(t)$ is the $(q \times 1)$ plant output vector, and A_p, B_p and C_p are matrices with appropriate dimensions. The range of the plant parameters is assumed to be bounded as defined by

$$\underline{a}_{ij} \le a_p(i,j) \le \overline{a}_{ij}, i = 1..n, j = 1..n \qquad (2)$$

$$\underline{b}_{ij} \le b_p(i,j) \le \overline{b}_{ij}, i = 1..n, j = 1..m \qquad (3)$$

Where $a_p(i,j)$ is the $(i,j)^{th}$ element of A_p, and $b_p(i,j)$ is the $(i,j)^{th}$ element of B_p.

The design objective is to find, without explicit knowledge of A_p and B_p, some control $u_p(t)$ such that the plant output vector $y_p(t)$ follows the output of the reference model

$$\begin{aligned} \dot{x}_m(t) &= A_m x_m(t) + B_m u_m(t) \\ y_m(t) &= C_m x_m(t) \end{aligned} \qquad (4)$$

The model incorporates desired plant behaviour and in many cases $\dim[x_p(t)] \gg \dim[x_m(t)]$.

The adaptive control algorithm being presented is based upon the command generator tracker concept (CGT) developed by O'Brien and Broussard [5]. In the CGT method, it is assumed that there exists an ideal plant with ideal state and control trajectories, $x_p^*(t)$ and $u_p^*(t)$, respectively, which corresponds to perfect output tracking (i.e., when $y_p(t) = y_m(t)$ for $t \geq 0$). By definition, this ideal plant satisfies the same dynamics as the real plant, and the ideal plant output is identically equal to the model output. Hence, when perfect tracking occurs, the real plant trajectories become the ideal plant trajectories. The ideal control law $u_p^*(t)$, generating perfect output tracking and the ideal state trajectories is assumed to be a linear combination of the model states and model input:

$$\begin{bmatrix} x_p^*(t) \\ u_p^*(t) \end{bmatrix} = \begin{bmatrix} S_{11} & S_{12} \\ S_{21} & S_{22} \end{bmatrix} \begin{bmatrix} x_m(t) \\ u_m(t) \end{bmatrix} \quad (5)$$

Where the S_{ij} sub matrices satisfy the following conditions

$$\begin{aligned} S_{11}A_m &= A_pS_{11} + B_pS_{21} \\ S_{11}B_m &= A_pS_{12} + B_pS_{22} \\ C_m &= C_pS_{11} \\ 0 &= C_pS_{12} \end{aligned} \quad (6)$$

The adaptive control law based on this CGT approach is given as [8]

$$u(t) = K_e(t)(y_m(t) - y(t)) + K_x(t)x_m(t) + K_u(t)u_m(t) \quad (7)$$

Where $K_e(t)$, $K_x(t)$ and $K_u(t)$ are gains concatenated into a single matrix

$$K(t) = [K_e(t), K_x(t), K_u(t)] \quad (8)$$

Defining a vector $r(t)$ as

$$r(t) = \begin{bmatrix} y_m(t) - y_p(t) \\ x_m(t) \\ u_m(t) \end{bmatrix} \quad (9)$$

The control $u_p(t)$ can be written in compact form as

$$u_p(t) = K(t)r(t) \quad (10)$$

The adaptive gains are obtained as a combination of the following integral and proportional gains [8]

$$K(t) = K_p(t) + K_i(t)$$
$$K_p(t) = (y_m(t) - y(t))r^T T_p, T_p \geq 0 \quad (11)$$
$$\dot{K}_i(t) = (y_m(t) - y(t))r^T T_i, T_i > 0$$

Sufficiency for asymptotic tracking is achieved if:

1. There exists a solution to the CGT problem, eq. (6).

2. The plant is ASPR; that is, there exists a gain matrix K_e, not needed for implementation, such that the closed-loop transfer function $G_c(s) = [1 + G_p(s)K_e]^{-1}G_p(s)$ is SPR.

In general, the ASPR conditions are not satisfied by most real systems. Bar-Kana and Kaufman [3] have remedied this situation by showing that a non-ASPR plant of the form $G_p(s) = C_p(sI - A_p)^{-1}B_p$ can be augmented with a feedforward compensator $H(s)$ such that the augmented plant transfer matrix $G_a(s) = G_p(s) + H(s)$ is ASPR.

It was shown in [8] that the resulting adaptive controller will, in general, result in a model following error that is bounded, but not zero, in steady state. To improve upon this, the modification, incorporating the supplementary feedforward into the reference model output, has been developed [9]. In [9], asymptotic model following was achieved using a Strictly proper stable feedforward compensator. However, it is also possible using a proper but not strictly proper stable feedforward compensator [14].

3. ROBUST COMPENSATOR DESIGN

In this section, we derive frequency domain design conditions for the feedforward compensator using the so-called Q parameterization. We first give the ASPR lemma [1], which will be needed in the development of a design procedure for a feedforward compensator.

Lemma 1 [12]

Let $G_p(s)$ be any transfer function of arbitrary relative degree. $G_p(s)$ not necessarily stable or minimum phase.

Let $C(s) = H^{-1}(s)$ be any dynamic stabilizing controller as shown in Figure 1, then

$$G_a(s) = G_p(s) + H(s) \quad (12)$$

is ASPR if the relative degree of $G_a(s)$ is zero or 1

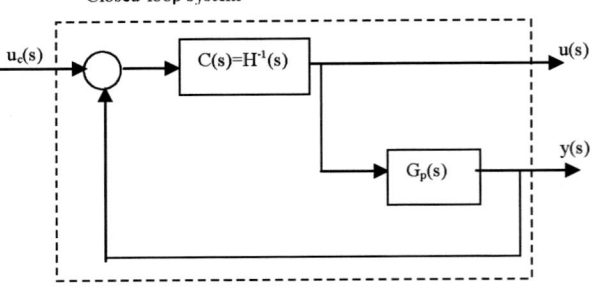

Fig. 1. Augmented closed-loop system

3.1 Stable Nominal Plant

For now, consider only a non-ASPR nominal plant without any perturbation. Suppose that the controller C(s) is given in terms of a parameter Q(s) as follows [6].

$$C(s) = \frac{Q(s)}{1 - Q(s)G_{p0}(s)} \quad (13)$$

Where $G_{p0}(s)$ is the non-ASPR nominal plant transfer function. Then, the closed-loop system in Figure 1 is

$$G_{cl}(s) = [1 + C(s)G_{p0}(s)]^{-1}C(s) \quad (14)$$

Substituting (13) into (14) gives

$$G_{cl}(s) = Q(s) \quad (15)$$

From 15, we see that the closed-loop system is stable if and only if $Q(s)$ is stable. Now, the following lemma gives the ASPR condition for the augmented plant $G_a(s)$.

Lemma 2 [12]

The augmented plant $G_a(s)$ is ASPR if and only if $Q^{-1}(s)$ is ASPR.

Now consider the nominal plant with additive plant perturbation, i.e.

$$G_p(s) = G_{p0}(s) + \Delta_a(s) \quad (16)$$

Or with multiplicative plant perturbation, i.e.,

$$G_p(s) = G_{p0}(s)[1 + \Delta_m(s)] \quad (17)$$

Assuming that the actual perturbation is not exactly Known but there exists a known rational function $W(s) \in RH_\infty$, such that

$$|W(j\omega)| \geq \max|\Delta(j\omega)|, \forall \omega \quad (18)$$

Where $\Delta(j\omega)$ is either an additive or a multiplicative plant perturbation in the transfer function, depending upon the perturbation model considered. Then, the following assumptions are made for the plant

Assumption 1

1. The nominal plant is stable

2. $\Delta(s) \in RH_\infty$ and satisfies (17)

Then, the following theorem gives the ASPR condition of the augmented plant $G_a(s)$.

Theorem 1 [12]

If the following design conditions are satisfied, then the augmented plant $G_a(s) = G_p(s) + H(s)$ will be ASPR in the presence of plant perturbations.

1. $Q(s)$ is designed, so that $Q^{-1}(s)$ is ASPR

2. $\|R\|_\infty < 1$

Where $R(s) = Q(s)W(s)$ for additive perturbations and $R(s) = Q(s)G_{p0}(s)W(s)$ for multiplicative perturbations.

With regard to the design conditions for a feedforward compensator given by Theorem 1, the following optimization procedure is proposed to determine the coefficients of $Q(s)$.

$$\min_q \|R(j\omega)\|_\infty$$

$$\text{Subject to: } [roots(\eta_Q(s))] < 0 \quad (19)$$

Where q is a vector whose elements are the coefficients of $Q(s)$ and $n_Q(s)$ is the numerator polynomial of $Q(s)$. The constraint given in the above optimization ensures the stability of the feedforward compensator [12].

3.2 Unstable Nominal Plant

In this section, we extend the above results for plant with an unstable nominal part.

We first use the fact [11] that every proper transfer function can be written in the form

$$G_{p0}(s) = \frac{N(s)}{M(s)} \quad (20)$$

Where $N(s)$ and $M(s)$ are stable coprime transfer function.

A useful characterisation of coprimeness is given in the following by Bezuot's theorem [4,7]

Theorem 2 [5]

$N(s)$ and $M(s)$ are coprime if and only if there exists two other stable functions $X(s)$ and $Y(s)$ such that

$$N(s)X(s) + M(s)Y(s) = 1 \quad (21)$$

Now suppose that the controller $C(s)$ is given in terms of the parameter Q as follows

$$C(s) = \frac{X(s) + M(s)Q(s)}{Y(s) - N(s)Q(s)} \quad (22)$$

Then, the closed-loop system given in Fig 1 is

$$G_{cl}(s) = [1 + C(s)G_{p0}(s)]^{-1}C(s) \quad (23)$$

Substituting (22) into (23) gives

$$G_{cl}(s) = M(s)(X(s) + M(s)Q(s)) \quad (24)$$

From (29), since M(s) and X(s) are all stable, we see that the closed-loop system is stable if and only if $Q(s)$ is stable. Then, the following lemma gives the ASPR condition of the augmented plant $G_a(s)$.

Lemma 3 [12]

The augmented plant $G_a(s)$ is ASPR if

1. Q(s) is stable,

2. its relative degree

$$\gamma_Q = \begin{cases} 0, 1 \text{ or } -1, & \text{if } \gamma_x = 0 \\ 0 \text{ or } -1, & \text{if } \gamma_x \geq 1 \end{cases}$$

Where γ_x is the relative degree of $X(s)$

Now, consider the nominal plant with additive plant perturbation, i.e,

$$G(s) = G_{p0}(s) + \Delta_a(s) \quad (25)$$

or with multiplicative plant perturbation, i.e.,

$$G(s) = G_{p0}(s)[1 + \Delta_m(s)] \quad (26)$$

Assuming that the actual perturbation is not exactly known but there exists a known rational function $W(s) \in RH_\infty$ such that

$$|W(j\omega)| \geq \max|\Delta(j\omega)|, \forall \omega \quad (27)$$

Where $\Delta(j\omega)$ is either an additive or multiplicative plant perturbation in the transfer function depending on the perturbation model considered. Then, the following assumptions are made for the plant

Assumption 2

1. The nominal plant is unstable

2. $\Delta(s) \in RH_\infty$ and satisfies (26)

Then, the following theorem gives the ASPR condition for the augmented plant $G_a(s)$.

Theorem 3 [12]

if the following design conditions are satisfied then the augmented plant $G_a(s) = G_p(s) + H(s)$ will be ASPR in the presence of plant perturbations.

1. $Q(s)$ is designed according to Lemma 3

2. $\|R\|_\infty < 1$

Where

$$R(s) = M(s)(X(s) + M(s)Q(s))W(s)$$

for additive perturbations and

$$R(s) = N(s)(X(s) + M(s)Q(s))W(s)$$

for multiplicative perturbations

With regard to the design conditions for a feedforward compensator given above, the following optimization procedure is proposed to determine the coefficients of $Q(s)$.

$$\underset{q}{\text{mimimise}} \|R(j\omega)\|_\infty$$

Subject to: $[roots(p(s))] < 0 \quad (28)$

Where q is a vector whose elements are the coefficients of $Q(s)$ and $p(s)$ is a polynomial that appears in the denominator of $H(s)$. The constraint given in the above optimization guarantees the stability of the feedforward compensator $H(s)$ [12].

4. SIMULATION RESULTS

Example1 In this example, we apply the proposed method to a stable nominal plant with modeled additive uncertainty. The nominal plant is assumed to be of the form,

$$G_{p0}(s) = \frac{-0.1s + 0.8}{s^2 + 5s + 8}$$

The reference model considered is

$$G_m(s) = \frac{y_m(s)}{u_m(s)} = \frac{1}{s+1}$$

With u_m being a square wave with a magnitude of 1 unit and period of 60 seconds. The plant perturbation is of the form

$$\Delta_a(s) = \frac{s}{\mu s + 1}, \quad \mu = 0.01, \ldots, 1$$

$W(s)$ satisfying (17) is set to

$$W(s) = \frac{2.55}{1 + 0.5s}$$

$Q(s)$ was determined using Theorem 1 and the optimization procedure given by (18) to be

$$Q(s) = \frac{0.05s + 0.1}{0.14 + 0.3}$$

Since $H(s) = C^{-1}(s)$, we get $H(s)$ as

$$H(s) = \frac{3.75}{1 + 0.5s}$$

As seen from Figure 2, reasonable model following is observed. The input signal is shown in Figure 3, which is a smoothing signal with bounded values. Figure 4 shows the amplitude of both $|\Delta_{max}(jw)|$ and $|W(jw)|$ for 100 frequency's values

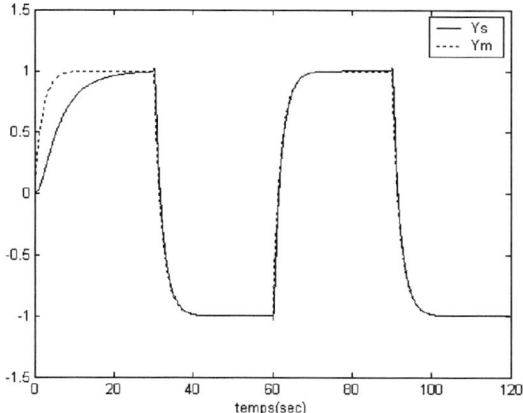

Fig. 2. Model and Actual Plant Response, $\mu = 0.5$

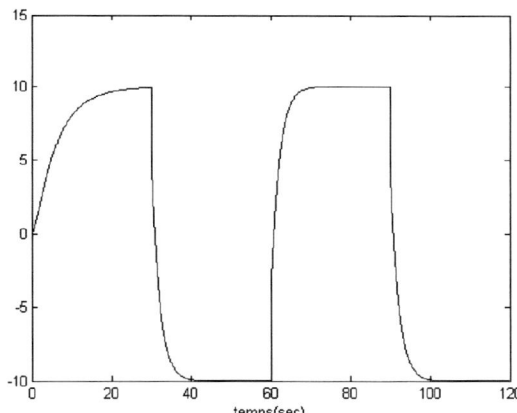

Fig. 3. Input to system, $\mu = 0.5$

Example 2

In this example, we will consider an unstable nominal plant with multiplicative perturbation in the transfer function $G(s) = G_{p0}(s)[1 + \Delta_m(s)]$. Consider the unstable nominal plant as

$$G_{p0}(s) = \frac{s+1}{(s+2)(s+3)}$$

The reference model was as defined in the previous example.

Since the nominal plant is unstable, the coprime factorization of the nominal plant using the procedure given in [4] results in

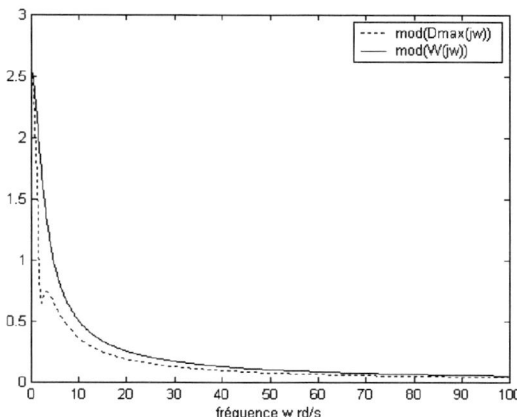

Fig. 4. $|\Delta_{max}(jw)|$ and $|W(jw)|$

for 100 frequency's values

$$N(s) = \frac{s+1}{(s+2)^2} \quad M(s) = \frac{(s-2)(s-3)}{(s+2)^2}$$

$$X(s) = \frac{7.9s+1.4}{s+2} \quad Y(s) = \frac{s+1.1}{s+2}$$

The plant perturbation is assumed to be

$$\Delta_m(s) = \frac{s}{\mu s + 1}, \quad \mu = 0.01,....,1$$

$W(s)$ satisfying (26) is set to

$$W(s) = \frac{3.41}{1+0.8s}$$

$Q(s)$ was determined using Theorem 3 and the optimization procedure given by (27) to be

$$Q(s) = \frac{12.25s^2 + 43.36s + 9.2}{11.5s + 12.36}$$

Since $H(s) = C^{-1}(s)$, we get H(s) as

$$H(s) = \frac{2.32s^3 + 11.9s^2 + 36.2s + 22.5}{3.3s^4 + 11.2s^3 + 4.25s^2 + 25.62s + 1.52}$$

Also, we see from Figure 5, that the actual plant follows reasonable well the model. The input is shown in Figure 6.

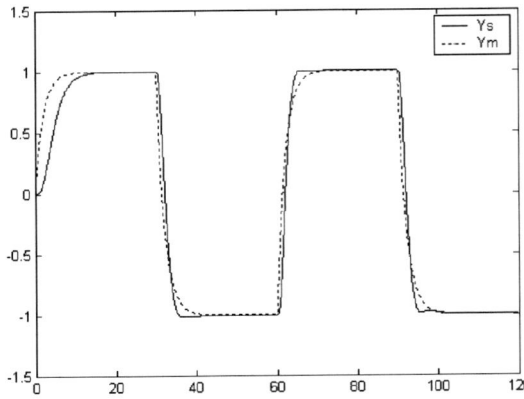

Fig. 5. Model and Actual Plant Response, $\mu = 0.5$

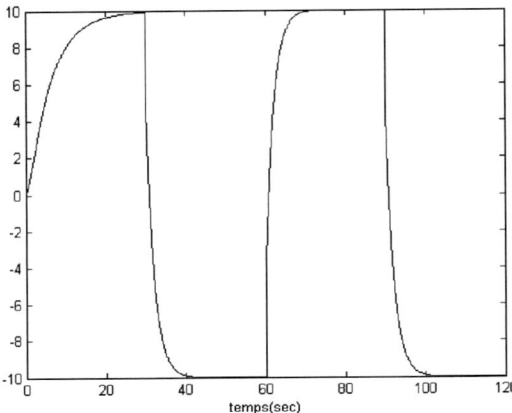

Fig. 6. Input to system, $\mu = 0.5$

5. CONCLUSION

We have presented in this paper a procedure for determining the DMRAC feedforward compensator needed for satisfying the positive real conditions which is necessary for the convergence of the tacking errors. This development enables the augmented plant to satisfy the ASPR condition in the presence of plant unmodeled dynamics. Simulation results show the efficiency of the algorithm for most stable and instable unmoded systems.

REFERENCES

[1] Barkana, I.. Gain Condition and Convergence of Simple Adaptive Control, International Journal of Adaptive
Control and Signal Processing,19, pp.13-40, 2005

[2] Barkana, I. On Output Feedback Stability and Passivity in Discrete Linear Systems, Proceedings of The 16th, Triennial IFAC World Congress, Prague, Czech Republic, July 2005.

[3] I.Bar-Kana and H. Kaufman. Global stability and performance of a simplified adaptive control algorithm. Int. J. Control, 42(6):1491-1505, 1985.

[4] S. P. Bhattacharyya, H. Chapellat, and L. H. Keel. Robust Control, The Parametric Approach. Prentice Hall, 1995.

[5] J. Broussard and 0. O'Brien. Feedforward control to track the output of a forced model. 17th IEEE, CDC, volume 42, pages 1149-1155, San Diego, CA, 1979.

[6] J. C. Doyle, B. A. Francis, and A. R. Tannenbaum. Feedback Control Theory. Macmillan, 1992.

[7] T. Kailath. Linear Systems. Prentice Hall, 1980.

[8] H. Kaufman, I. Bar-Kana, and K. Sobel. Direct Adaptive Control Algorithms. Springer-Verlag, 1994.

[9] H. Kaufman and G. Neat. Asymptotically stable mimo direct mrac for processes not necessarily satisfying a positive real constraint. Int. J. Control,58:1011-1031,1993.

[10] I. D. Landau. A survey of model reference adaptive techniques : Theory and applications. Automatica, 10:353- 379,1974.

[11] J. M. Maciejowski. Multivariable Feedback Design. Addison Wesley, 1994.

[12] I. Mizumoto and Z. Iwai. Simplified adaptive model output following for plants with unmodeled dynamics. Int J. Control, 64(1):61-80, 1996.

[13] R. V. Monopoli. Model reference adaptive control with an augmented error signal. IEEE Trans. on Automatica, Control, AC-19:474-484, 1974.

[14] S. Ozcelik. Design of Robust Feedforward ,Compensators for Direct Model Reference Adaptive, Controllers. PhD thesis, Rensselaer Polytechnic Institute, Troy, NY, 1996.

FOCOVE: Formal Concurrency Verification Environment for Complex Systems

Djamel Eddine Saïdouni*, Adel Benamira*, Nabil Belala* and Farid Arfi*

*LIRE Laboratory, University of Mentouri, 25000 Constantine, Algeria
(e-mail: saidounid@hotmail.com, a.benamira@yahoo.fr, nbelala@gmail.com, arfi_f@hotmail.com)

Abstract: This paper presents a tool to exploit a true concurrency model, namely the Maximality-based Labeled Transitions Systems. We show the use of this model in model checking technique. Three techniques are implemented in order to solve the state space combinatorial explosion problem: the reduction modulo α-equivalence relation, the joint use of covering steps and maximality semantics, and the maximality-based symbolic representation.

Keywords: Maximality semantics, Partial order semantics, Maximality-based symbolic representation, Model checking, LOTOS specification language.

1. INTRODUCTION

FOCOVE (for Formal Concurrency Verification Environment) is an integrated environment designed to edit Basic LOTOS [3] behavior expressions which describe reactive systems and to generate and analyze Maximality-based Labeled Transitions Systems structures (MLTS) [11]. FOCOVE offers various implementations of some true-concurrency based techniques: maximality-based model checking [14], maximality-based symbolic representation, reduction modulo equivalence relation [11] and partial order semantics [2,7,8,16].

The paper is organized as follows. Section 2 introduces maximality-based model checking technique for complex and reactive systems formal verification. α-equivalence relation, partial order technique and maximality-based symbolic representation are presented respectively in sections 3, 4 and 5. Some FOCOVE screenshots are given in Section 6. The paper is enclosed by appendix recalls some definitions related to the maximality-based semantics of Basic LOTOS as presented in [6,11].

2. MAXIMALITY-BASED MODEL CHECKING

Formal verification approach supported by FOCOVE is based on models. In this approach, the application to be verified firstly specified by means of the formal description technique Basic LOTOS [3]. This specification will be translated in an operational way towards an underlying model represented by a graph called Maximality-based Labeled Transition System (MLTS) [11]. The expected properties of the system are written in Computation Tree Logic CTL [5] logic and they are verified by means of the model checking approach.

In spite of temporal logics facilitate the specification of systems to be verified, model checking approach is limited by the state graph combinatorial explosion problem, particularly when the specification model underlying semantics is the interleaving one. Such semantic is characterized, on one hand by the action temporal and structural atomicity hypothesis and on the other hand by the interpretation of parallel execution of two actions as their interleaving executions in time. To escape the action atomicity hypothesis imposed by interleaving semantics, new semantics, said true concurrency semantics, were defined. Among these semantics, we can quote a variant of the maximality semantics [6,11]; its principle consists in using the dependence relations between actions occurrences and by associating to every state of the system the actions, which are potentially in execution.

In [14], is has been shown that model checking algorithms proposed in the literature, which are based on interleaving semantics, may be adapted easily to true concurrency semantics for the verification of new properties classes related to simultaneous progress of actions at different states. This was mainly led by the presence of maximality information in states and transitions. In particular, this information makes easy writing atomic propositions expressing properties to be verified. A particular attention was concerned the natural and intuitive reading of these properties.

3. REDUCTION MODULO α-EQUIVALENCE RELATION

The α-equivalence relation [11] is purposed to put in correspondence MLTSs describing the same behavior of which the only difference resides in the choice of event names.

For example, both MLTSs of Fig. 1 describe the same behavior, i.e. the parallel (or true-concurrent) execution of actions a and b. We can obtain MLTS of Fig. 1.(a) from that of Fig. 1.(b) by substituting event names e by x and event name z by y.

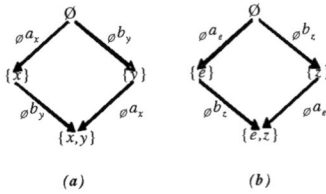

Fig. 1. Two α-equivalent MLTSs

A reduction consists in eliminating redundancy via certain relations by preserving properties to be checked. In this section, we will use the α-relation as a criterion of redundant behaviors.

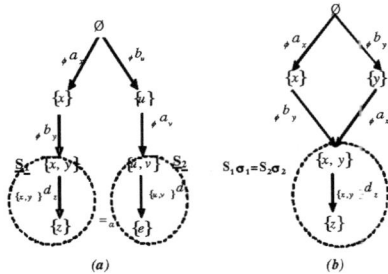

Fig. 2. α-reduction

As illustration, MLTS of Fig. 2.(a) represents the behavior of the LOTOS expression $a;d;stop|[d]|b;d;stop$. It was generated by a mapping of LOTOS maximality-based operational semantics [11]. Both sub-MLTSs S_1 and S_2 of Fig. 2.(a) are α-equivalent. Indeed, it exists two functions of substitution $\sigma_1=\{x/x,y/y,z/z\}$ and $\sigma_2=\{x/v,y/u,z/e\}$ such as $S_1\sigma_1 \equiv S_2\sigma_2$. To remove such a redundancy, we must, initially, apply the substitution function $\sigma_1 \cup \sigma_2$ to the MLTS of Fig. 2.(a), group the start stats of S_1 and S_2, and then, we remove $S_1\sigma_1$ or $S_2\sigma_2$. As a result, we obtain the MLTS of Fig. 2.(b).

Table 1 shows results obtained with the MLTS α-reduction construction on the same examples, where $A=a;A$.

Table 1. Results of MLTS construction

	MLTS		MLTS$_{/=\alpha}$	
	States	Transitions	States	Transitions
A|||A	5	10	4	8
A|||A|||A	16	48	8	24
A|||A|||A|||A	65	260	16	64
A|||A|||A|||A|||A	326	1630	32	150

4. PARTIAL ORDER TECHNIQUE

Inspiring from the covering steps [17] technique, we do not consider all possible interleaving sequences. On the other hand, we build, under certain conditions, a step allowing directly reaching the final state which would have been reached by each interleaved sequence. Fig. 3 shows the obtained benefit in the case of the derivation of three parallel actions a, b and c in the presence of differed conflict.

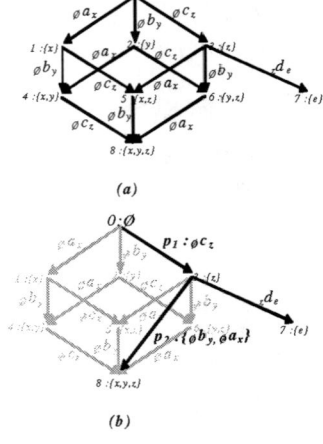

Fig. 3. An MLTS and its maximality-based step graph

The graph of Fig.3.(b) is the Maximality-based Step Graph (MSG) of the MLTS of Fig.3.(a) in which all interleaving runs were converted into two steps (p_1 and p_2); the first step expresses the start of c and the second one expresses true-concurrent execution of a and b. The built step graph covers the initial MLTS via the Mazurckiewicz's traces equivalence [9]. It is proved that our approach preserves deadlock states and liveness property, so its on-the-fly generation is thus possible.

Table 2 compares results by consideration of MLTS and MSG construction on the system of Fig. 4.

Table 2. Results of MSG construction

	MLTS$_{/=\alpha}$		MSG	
n	States	Transitions	States	Transitions
4	33	81	4	3
6	129	449	4	3
8	513	2305	4	3
10	2049	11265	4	3

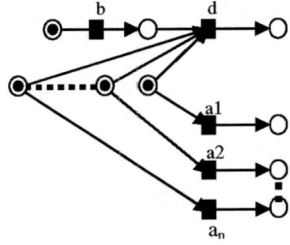

Fig. 4. Example system

5. MAXIMALITY-BASED SYMBOLIC REPRESENTATION

Anther method to tackle the state space explosion problem is to use a symbolic representation based on binary decision diagrams (BDDs) for implicitly representing state spaces of the model [4]. A trivial solution is to build the MLTS of source algebraic specification and then convert the MLTS to BDDs. However this kind of solution, which has already been adopted to get BDD representations from ATP specification [10], is not satisfactory, in that the usefulness of BDD representations stands mainly in the ability to manage specifications for which the MLTS is too big to be represented or even traversed.

To remedy this problem, a more satisfactory solution is implemented in our tool, which exploits compositionality of process algebras so as to build BDD representations without enumerating all states and transitions [15]. Proposed technique consists in building small MLTS to represent basic building blocks of the specification. These are converted to BDDs which in turn are combined together, according to the various process algebras operators to obtain the overall BDD. This kind of generation is called *on-the-fly generation*.

Tab. 3. shows comparative results between MLTS and symbolic-based MLTS where $B=b;b;b;stop$. Size is given by Megabytes and total memory used is 1 Gigabyte.

Table 3. Results of symbolic-based MLTS construction

	Symbolic-based MLTS			MLTS		
	Size	States	Trans	Size	States	Trans
B|||B	1	25	36	1	25	36
B|||B|||B	2	226	477	3	226	477
B|||B|||B|||B	10	2713	7536	63	2713	7536
B|||B|||B|||B|||B	63	40696	140175	Out of memory		

6. USER INTERFACE

The FOCOVE toolbox, available for download on website `http://www.focove.new.fr`, is built in a modular way. Available modules can be used independently or in combination. Modules include:
- An editor for Basic LOTOS behavior expressions.
- A compiler of Basic LOTOS behavior expressions.
- A maximality-based model checker.
- A tool for building α-reduced MLTS.
- A tool for building LTS and MLTS structures.
- A tool for building MSG.
- A tool for building maximality-based symbolic representation MSR.
- A viewer for MLTS, LTS, MSG and MSR.

After compiling a Basic LOTOS description, the generation of MLTS, LTS, MSG and MSR is possible. The result of generation can be obtained in textual output format or graphical output format. The screenshot of Fig. 5.(*a*) represents a Basic LOTOS expression input, and the second screenshot is a CTL formula input. Fig. 5.(c) illustrates an MLTS.

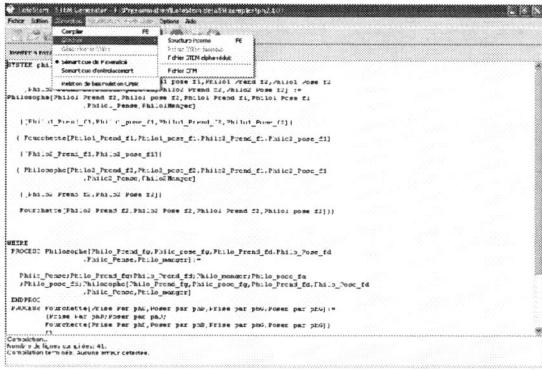

(*a*)

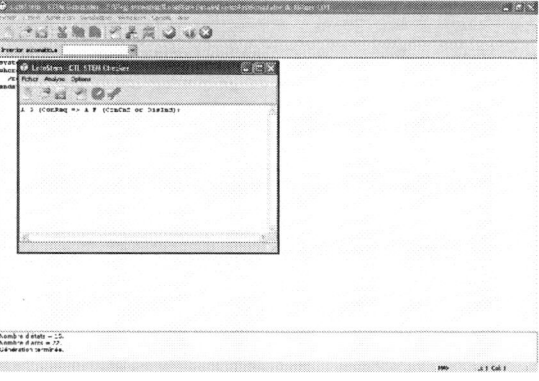

(*b*)

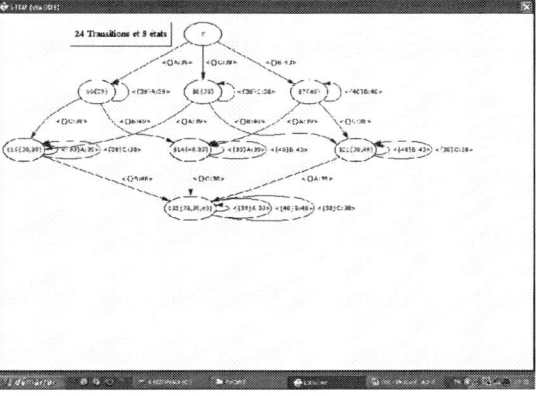

(*c*)

Fig. 5. Some screenshots of FOCOVE environment

7. CONCLUSION

This paper presents FOCOVE, an integrated environment to edit Basic LOTOS behavior expressions and generate and analyze Maximality-based Labeled Transitions Systems structures.

Different techniques are taken into account to solve state space combinatorial explosion problem. In FOCOVE, three techniques are supported, a maximality-based symbolic representation, a reduction modulo equivalence relation, a partial order semantics and a maximality-based model checking technique to verify MLTSs.

As a perspective, It is interesting to extend FOCOVE to take into account time [1,12] and conformance testing in the context of maximality semantics [13].

REFERENCES

[1] N. Belala and D. E. Saïdouni, Non-Atomicity in Timed Models, in Proceedings of *ACIT'2005*, 2005.

[2] A. Benamira and D. E. Saïdouni, Consideration of the Covering Steps in the Maximality-Based Labeled Transitions Systems, In Proceedings of *International Arab Conference on Information Technology (ACIT'2006)*, Yarmouk University, Irbid, Jordan, December 19-21st, 2006.

[3] T. Bolognesi and E. Brinksma, Introduction to the ISO Specification Language LOTOS, *Computer Networks and ISDN Systems*, volume 14, 1987.

[4] J. R. Burch, E. M. Clarke, K. L. McMillan, D. L. Dill and L. J. Hwang. Symoblic model checking: 10^{20} states and beyond. In Proceeding of *5th IEEE Symposium on Logic in Computer Science*, 1990.

[5] E. M. Clarke, E. A. Emerson, and A. P. Sistla, Automatic Verification of Finite State Concurrent Systems Using Temporal Logic Specifications, *ACM Transactions on Programming Languages and Systems*, volume 8, no 2, pages 244-263, 1986.

[6] J. P. Courtiat and D. E. Saidouni, Relating Maximality-Based Semantics to Action Refinement in Process Algebras, *in D. Hogrefe and S. Leue, Editors, {IFIP} {TC/WG6}.1, 7th Int. Cof of Formal Description Techniques (FORTE'94)"*, pages 293-308, Chapman &Hall, 1995.

[7] P. Godefroid, Using Partial Orders to Improve Automatic Verification Methods, in Proceedings of *CAV'90*, volume 3, pages 321-340, ACM, DIMACS, 1990.

[8] P. Godefroid and P. Wolper, A Partial Approach to Model Cheking, in Proceedings of the *6th Symp. On Logic in Computer Science*, volume 531, pages 406-415, Amsterdam 1991.

[9] A. Mazurckiewicz, Trace Theory. In Petri Nets: Applications and Relationships to Other Model of Concurrency, in *Advances in Petri Nets 1986, Part {II;} Proceedings of an Advanced Course*, pages 279-324, Springer Verlag, LNCS 255, 1986.

[10] X.Nicollin, J. Sifakis and S.Yovine, Compiling Real-Time Specifications into Extended Automata, *IEEE Trans. on Software Engineering*, volume 18,794-804, 1992.

[11] D. E. Saidouni, Sémantique de Maximalité: Application Au Raffinement D'actions Dans LOTOS, *Phd thesis*, LAAS, Univerity of Paul Sabastier, Toulouse,1996.

[12] D. E. Saïdouni and N. Belala, Actions Duration in Timed Models, in Proceedings of *International Arab Conference on Information Technology (ACIT'2006)*, Yarmouk University, Irbid, Jordan, December 19-21st, 2006.

[13] D. E. Saidouni and A. Ghenai, Intégration des refus temporaries dans les graphes de refus, in proceeding of *NOTERE '2006*, Toulouse, France, 2006.

[14] D. E. Saïdouni and N. Belala. Using Maximality-Based Labeled Transition System Model for Concurrency Logic Verification, *The International Arab Journal of Information Technology (IAJIT)*, Vol. 2, No. 3, pages 199-205. July 2005. Zarka Private University, P. O. Box 2000, Zarka 13110, Jordan, ISSN: 1683-3198.

[15] R. Sisto, A method to build symbolic representations of LOTOS specification. *Protocol specification, Testing and verification*, pages 323-338, 1995.

[16] A. Valmari, Sets for Reduced State Space Generation, in Proceedings of the *Tenth International Conference on Application and Theory of Petri Nets*, volume II, Bohn 1989.

[17] F. Vernadat, P. Azéma and F. Michel, Covering Step Graph, in Proceedings of *Application and Theory of Petri Nets 96*, volume LNCS 1091, Springer Verlag, 1996.

Appendix A. MAXIMALITY SEMANTICS

A.1 Intuition [6,11]

The semantics of a concurrent system can be characterized by the set of the system states and the transitions by which the system passes from a state to another. In the approach based on the maximality, the transitions are events which represent only the beginning of actions execution. To distinguish each action execution, an identifier is associated to its start. In a state, an event is said to be *maximal* if it corresponds to the beginning of the execution (start) of an action which can possibly be always under execution in this state.

To illustrate the maximality principle, let us consider the LOTOS behavior expressions $E \equiv a;stop |||b;stop$ and $F \equiv a;b;stop[]b;a;stop$. In the initial state, no action is under execution, therefore, the set of maximal events is empty, from where following initial configurations associated to E and F are $\emptyset[E]$ and $\emptyset[F]$. From the configuration $\emptyset[E]$, starting the execution of actions a and b leads to the following transitions:

$$_\emptyset[E] \xrightarrow{a_x} {}_{m\{x\}}[stop]|||_\emptyset[b;stop] \xrightarrow{b_y} {}_{m\{x\}}[stop]|||_{\{y\}}[stop]$$

x (respectively y) is the event name identifying the beginning of the action a (respectively b). Since nothing can be concluded about the termination of both actions a and b in the configuration $_{\{x\}}[stop]\ |||\ _{\{y\}}[stop]$, x and y are then maximal in this configuration. Let us note that x is also maximal in the intermediate state represented by the configuration $_{\{x\}}[stop]\ |||\ _{\emptyset}[b;stop]$.

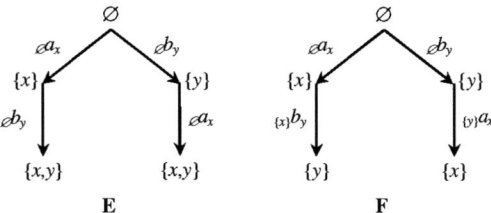

Fig. 6. MLTS structures of E and F expressions

For the initial configuration, associated to the behavior expression F, the following transition is possible: $_{\emptyset}[F] \xrightarrow{\emptyset a_x} _{\{x\}}[b;stop]$. As previously, x identifies the beginning of the action a and it is the unique maximal event name in the configuration $_{\{x\}}[b;stop]$. It is clear that, within sight of the action prefixing operator semantics, the beginning of the action b is possible only if the action a terminates its execution. Consequently, x does not remain maximal any more when the action b begins its execution; the unique maximal event in the resulting configuration is y which identifies the beginning of execution of action b. Thus the following derivation $_{\{x\}}[b;stop] \xrightarrow{\{x\}b_y} _{\{y\}}[stop]$.

The configuration $_{\{y\}}[stop]$ is different from the configuration $_{\{x\}}[stop]\ |||\ _{\{y\}}[stop]$, because the first has only one maximal event (identified by y), whereas the second has two (identified by x and y). The derivation structures of the behavior expressions E and F obtained by the application of the maximality semantics principle are represented in Figure 5. These structures are called Maximality-based Labeled Transition System (MLTS).

A.2. Related definitions

In this section, we recall some related definitions of Basic LOTOS. The complete presentation of Basic LOTOS maximality-based operational semantics may be found in [6,11].

Syntax of Basic LOTOS: Let PN be the set of processes ranged over by P and let G be the set of gates ranged over by g. A particular observable action $\delta \notin G$ is used to notify the successful termination of the processes. L indicates any subset of G, the internal action is noted by i. The set of all actions is indicated by $Act = G \cup \{i,\delta\}$. $\mathcal{B}$, ranged over by E, F, ... denotes the set of behavior expressions whose syntax is:

$E ::=\quad Stop\ |\ exit\ |\ E[L]\ |\ g;E\ |\ i;E\ |\ E[\ ]\ E$
$\qquad E|[L]|E\ |\ hide\ L\ in\ E\ |\ E>>E\ |\ E[>E$

Given a process P which have the behavior E, the definition of P is expressed by $P:=E$.

The set of event names is a countable set indicated by M. This set is ranged over by $x,y,...$ $M,N,...$ indicate finite subsets of M. The set of atoms of support Act is $Atm = 2^M_{fn} \times Act \times M \cdot 2^M_{fn}$ being the set of finite parts of M. For $M \in 2^M_{fn}$, $x \in M$ and $a \in Act$, the atom (M,a,x) will be noted $_M a_x$. The choice of an event name can be done in a deterministic way by the use of any function $get: 2^M-\{\emptyset\} \to M$ satisfying $get(M) \in M$ for any $M \in 2^M-\{\emptyset\}$.

Configuration: The set C of configurations of Basic LOTOS behavior expressions is the smallest set defined inductively as follows:

- $\forall E \in \mathcal{B}, \forall M \in 2^M_{fn} :_M[E] \in C$
- $\forall P \in PN, \forall M \in 2^M_{fn} :_M[P] \in C$
- If $\mathcal{E} \in C$ then $hide\ L\ in\ \mathcal{E} \in C$
- If $\mathcal{E} \in C$ and $F \in \mathcal{B}$ then $\mathcal{E}>>F \in C$
- If $\mathcal{E}, \mathcal{F} \in C$ then $\mathcal{E}\ op\ \mathcal{F} \in C$
  $op \in \{\ [],\ |||,\ ||\ ,|[\ ]|\ ,[>\}$
- If $\mathcal{E} \in C$ and $\{a_1,a_2, ..., a_n\}, \{b_1,b_2, ..., b_n\} \in 2^M_{fn}$ then $\mathcal{E}[a_1/b_1, a_2/b_2, ..., a_n/b_n]$

Given a set $M \in 2^M_{fn}$, $_M[...]$ is called embedding operation. This operation is distributive over the operations $[], |[L]|$, hide, $[>$ and the renaming gates operation. We also admit that $_M[E>>F] \equiv _M[E]>>F$. A configuration is known as canonical if it cannot be reduced any more by the distribution of the embedding operation on the other operators. Thereafter, we suppose that all configurations are in canonical form.

Any canonical configuration is under one of the following forms ($\mathcal{E}$ and $\mathcal{F}$ being canonical configurations):

$_M[stop]\ |\ _M[exit]\ |\ _M[a;E]\ |\ _M[P]\ |\ \mathcal{E}[\ ]\mathcal{F}\ |$
$hide\ L\ in\ \mathcal{E}\ |\ \mathcal{E}>>F\ |\ \mathcal{E}[>\mathcal{F}\ |\ \mathcal{E}[a_1/b_1, a_2/b_2,...,a_n/b_n]$

The function $\varphi: C \to 2^M_{fn}$, which determines the set of event names in a configuration, is defined inductively by:

$\varphi(_M[E]) = M$
$\varphi(\mathcal{E}[\ ]\mathcal{F}) = \varphi(\mathcal{E}) \cup \varphi(\mathcal{F})$
$\varphi(\mathcal{E}\ |[L]|\ \mathcal{F}) = \varphi(\mathcal{E}) \cup \varphi(\mathcal{F})$
$\varphi(\mathcal{E} >> F) = \varphi(\mathcal{E})$
$\varphi(hide\ L\ in\ \mathcal{E}) = \varphi(\mathcal{E})$
$\varphi(\mathcal{E}[>\mathcal{F}) = \varphi(\mathcal{E}) \cup \varphi(\mathcal{F})$
$\varphi(\mathcal{E}[b_1/a_1,...,b_n/a_n]) = \varphi(\mathcal{E})$

Let E be a configuration; $E \setminus N$ indicates the configuration obtained by removing the set of event names N from the configuration E. $E \setminus N$ is defined inductively as follows:

$(\iota_M[E])\backslash N = \iota_{M\text{-}N}[E]$
$(\mathcal{E}\ []\ \mathcal{F})\backslash N = \mathcal{E}\backslash N\ []\ \mathcal{F}\backslash N$
$(\mathcal{E}\ |[L]|\ \mathcal{F})\backslash N = \mathcal{E}\backslash N\ |[L]|\ \mathcal{F}\backslash N$
$(hide\ L\ in\ \mathcal{E})\backslash N = hide\ L\ in\ \mathcal{E}\backslash N$
$(\mathcal{E} >> \mathcal{F})\backslash N = \mathcal{E}\backslash N >> \mathcal{F}$
$(\mathcal{E}[>\mathcal{F})\backslash N = \mathcal{E}\backslash N\ [>\mathcal{F}\backslash N$
$(\mathcal{E}[b_1/a_1,...,b_n/a_n])\backslash N = \mathcal{E}\backslash N\ [b_1/a_1,...,b_n/a_n]$

The set of substitution functions of event names is noted *Subs* (i.e. $Subs = M \to 2^M_{fn}$); $\sigma, \sigma_1, \sigma_2,...$ are elements of Subs. Given $x, y, z \in M$ and $M \in 2^M_{fn}$, then

- The application of σ to x will be written by σx
- The substitution identity function ι is defined by $\iota x = \{x\}$
- $M\sigma = \cup_{x \in M}\ \sigma x$;
- $\sigma[y/z]$ is defined by

$$\sigma[y/z]x = \begin{cases} \{y\} & \text{if } z = x \\ \sigma x & \text{otherwise} \end{cases}$$

Let σ be a substitution function, the simultaneous substitution of all occurrences of x in $\mathcal{E}$ by σx, is defined recursively on the configuration $\mathcal{E}$ as follows:

$(\iota_M[\mathcal{E}])\sigma = \iota_{M\sigma}[\mathcal{E}]$
$(\mathcal{E}[]\mathcal{F})\sigma = \mathcal{E}\sigma\ []\ \mathcal{F}\sigma$
$(\mathcal{E}|[L]|\ \mathcal{F})\sigma = \mathcal{E}\sigma\ |[L]|\ \mathcal{F}\sigma$
$(hide\ L\ in\ \mathcal{E})\sigma = hide\ L\ in\ \mathcal{E}\sigma$
$(\mathcal{E} >> \mathcal{F})\sigma = \mathcal{E}\sigma >> F$
$(\mathcal{E}[>\mathcal{F})\sigma = \mathcal{E}\sigma\ [>\mathcal{F}\sigma$
$(\mathcal{E}[b_1/a_1,...,b_n/a_n])\sigma = \mathcal{E}\sigma[b_1/a_1,...,b_n/a_n]$

Data Reconciliation and Gross Error Detection: A Filtered Measurement Test

Y. HIMOUR

Université 20 Août 1955, BP 26 - 21000 - Skikda Algérie
(e-mail: himoury@yahoo.fr)

Abstract: Measured process data commonly contain inaccuracies because the measurements are obtained using imperfect instruments. As well as random errors one can expect systematic bias caused by miscalibrated instruments or outliers caused by process peaks such as sudden power fluctuations. Data reconciliation is the adjustment of a set of process data based on a model of the process so that the derived estimates conform to natural laws. In this paper, we will explore a predictor-corrector filter based on data reconciliation, and then a modified version of the measurement test is combined with the studied filter to detect probable outliers that can affect process measurements. The strategy presented is tested using dynamic simulation of an inverted pendulum.

Keywords: Dynamic data reconciliation, measurement test, EWAM filtering.

1. INTRODUCTION

Measurements obtained from sensors are generally corrupted with two types of errors: gross and random errors, thus provoking a violation of the constraints defined by process model equations. To improve the quality of measurements these errors must be eliminated. One of the most popular techniques used in this purpose is data reconciliation, which is a model based filtering technique introduced first by Kuehn and Davidson [4]. The difference between data reconciliation and other filtering techniques remains in the fact that data reconciliation explicitly uses process model equations as constraints in such way that the estimates obtained satisfies these constraints. The usual use of data reconciliation is the detection of gross errors, so that most famous works which one can find in the technique literature, are directed towards the task detection with regard to [2,3]; without neglecting of course the importance of research which aimed at the improvement of the stage optimization (i.e.) [5]. Taking the work of S. Bai et al [1] as the base references, we will try in the second paragraph a reformulation of data reconciliation problem, this reformulation enables us to use the principle of data reconciliation in the form of a predictor-corrector filter. Then, we will present the principle of the measurement test which is largely combined with data reconciliation to detect gross errors; an attempt to improve the detectability of the test by the proposal of an EWMA filtering is detailed. Finally, a simulation of an inverted pendulum is carried out to concretize the steps of the studied theory.

2. PREDICTOR-CORRECTOR FORM OF DATA RECONCILIATION

If the measurement errors follow a normal distribution, the problem of data reconciliation can be posed like an estimate of the variable reconciled by the likelihood method [1].

The relation between the measured values, the real values and the random errors affecting measurements is expressed by:

$$y = y^* + \varepsilon \quad (1)$$

Let us note V the covariance matrix of the error ε.

Still let us consider the following expression:

$$yp = y^* + e \quad (2)$$

Where, yp is the model prediction, and e is the prediction error, which is the result of several factors such as the inaccuracy of the model structure, the uncertainty of model parameters, as well as the inaccuracy of measurements of entries of the model. To simplify, it is supposed that e follows a normal distribution of zero mean and a covariance matrix R. According to our assumptions concerning measurements and the predictions; and with some mathematical manipulations we can obtain the predictor corrector form of the data reconciliation:

$$\hat{y} = yp + K(y - yp) \quad (3)$$

Where $K = (V^{-1} + R^{-1})^{-1} V^{-1} = (I + VR^{-1})^{-1} \quad (4)$

I: is the matrix identity.

3. IMPROVED MEASUREMENT TEST

The measurement test uses a statistical criterion Z_{MJ} based on the values r_M and Σ_M, such that:

$$r_M = QA^T(AQA^T)Ay \quad (5)$$

$$\Sigma_M = QA^T(AQA^T)^{-1}AQ \quad (6)$$

$$Z_{MJ} = \frac{r_{MJ}}{\sqrt{\Sigma_{MJJ}}} \quad (7)$$

r_M is the vector of residuals and Σ_M is covariance matrix of r_M. Supposing that the errors have a normal distribution, measurements with gross errors can be detected by comparison of Z_M with a threshold value Z_{MC}. If α is the level of confidence, then the threshold value Z_{MC} can be taken directly form the normal distribution table $x_{\alpha/2}$. To overcome the problem of the isolated peaks, we propose to use an Exponentially Weighted Moving Average filtering (EWMA) on the residuals of the measurement test. This filtering is regarded as a good solution against false alarm, but it suffers from a delay of detection which can be of a great influence on the process, particularly if the sampling time is sufficiently large. The general expression of the filter applied is given by:

$$\bar{r}_M(k) = (I - \beta)\bar{r}_M(k-1) + \beta r_M(k) \quad (8)$$

Where the term β represents the lapse of memory factors matrix, and I is the identity matrix. The initialization of the algorithm can be carried out by taking $\bar{r}(0) = 0$.

The matrix β can be adjusted according to the type of fault to detect. In general, a value close to the matrix identity supports the detection of the slow changes, while a value close to zero supports the detection of the abrupt changes.

The filtered test threshold is $\bar{Z}_{MC} = \dfrac{\beta}{2-\beta} Z_{MC}$.

4. SIMULATION EXAMPLE

To highlight the effectiveness of the theory studied, we considered the example of the inverted pendulum, where the measured variables and those estimated by data reconciliation with the presence of a defect on the variable x3 are illustrated in figure 1. Figure 2 shows respectively the evolution of the residuals of the measurement test in the classic case and the case with filtering suggested.

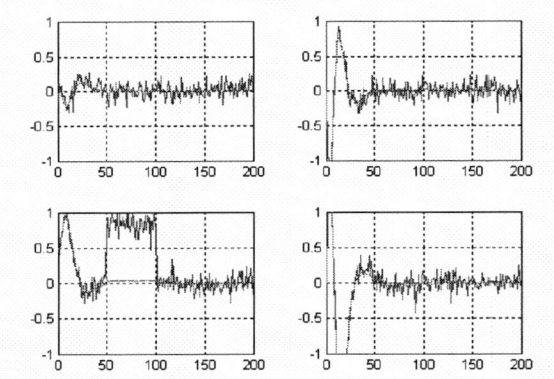

Fig.1. Process measured and reconciled data

Table.1. Measurement and reconciled errors variances

	Reconciled data errors variances	Measurement errors variances
X1	0.0009	0.0392
X2	0.0013	0.0440
X3	0.0059	0.0477
X4	0.0011	0.0394

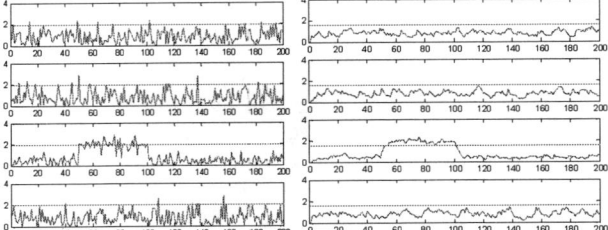

Fig.2. Measurement test residuals and measurement test filtered residuals evolution

5. CONCLUSION

A new approach of gross error detection by data reconciliation was proposed. Initially the reconciliation of the data is accommodated in a predictor-corrector filter form, with which the test of measurement is combined to achieve the task detection, and to improve this latter one, a filtering EWMA on the residuals of the test used is proposed; the results show a great efficacy of the filter to eliminate the random error effects on measured data (see table 1). Also by comparison of both figure 2 and figure 3, one can clearly see the improvement of detection by the introduction of the EWMA filtering.

REFERENCES

[1] Bai S., Thibault J. and McLean D.D (2006). Dynamic data reconciliation: Alternative to Kalman filter. *Journal of Process Control.* **Vol.6**, pages.485–498.

[2] C. Karlsson, E. Dahlquist and E. Dotzauer2004). Data Reconciliation and Gross Error Detection for the Flue Gas Train in a Heat & Power Plant. *International Conference on Probabilistic Methods Applied to Power Systems,* Iowa State University, Ames, Iowa, September 12-16, 2004.

[3] In-Won Kim, Mun Sik Kang, Sunwon Park and Thomas F. Edgar (1997). Robust data reconciliation and gross error detection: the modified MIMT unsing NLP. *Computers and Chemical Engineering*, **Vol. 21, No. 7,** pages 775-782.

[4] Kuehn, D. R., & Davidson, H(1961). Computer control II: Mathematics of control. *Chemical Engineering Progress,* **Vol. 57(6),** 44–47.

[5] Nikhil Arora and Lorenz T. Biegler (2001). Redescending estimators for data reconciliation and parameter estimation. *Computers & Chemical Engineering,* **Vol 25**, pages 1585-1599.

Intelligent Diagnosis of Degradation State under Corrosion *

Dorin Isoc* Aurelian Ignat-Coman** Adrian Joldiş***

*Technical University of Cluj Napoca, Romania, 400020 – Cluj Napoca, str.C.Dacoviciu nr.15 (e-mail: Dorin.Isoc@aut.utcluj.ro).
**Technical University of Cluj Napoca, Romania, 400020 – Cluj Napoca, str.C.Dacoviciu nr.15 (e-mail: Aurelian.Ignat@aut.utcluj.ro).
***Technical University of Cluj Napoca, Romania, 400020 – Cluj Napoca, str.C.Dacoviciu nr.15 (e-mail: Adrian.Joldis@com.utcluj.ro).

Abstract: The work presents an inter- and multi- disciplinary research where the diagnosis is treated by using the artificial intelligence means and the application the degradation state of buildings and urban power networks. A possible model of degradation process caused by the corrosion and the technical achievement manner is given. The notions of micro- and macro - modeling and model granularity are introduced and applied. For resulting model the specification of intelligent processing of information and further the knowledge for suggested model are prepared. As concluding remarks the results are analysed and interpreted and a generalized approach is suggested and argued.

Keywords: corrosion, degradation state, intelligent system, case-based reasoning, inference engine, intelligent diagnosis, case base.

1. INTRODUCTION

Diagnosis is an usual application in many technical fields, especially in numerical systems, communications, power plants, control systems (Venkatasubramanian et al. (2003), Chen (1995), Frank and Seliger (1991)).

Despite these results, the faults have specific shapes in any technical application. The fault effects are also different both as consequences and as the value of damages.

Interesting fields and less studied before are of buildings or built systems and infrastructures, especially of urban utility networks. Usually, the fault state is associated with the destruction and the fact that the destruction acts slowly in time is almost always neglected. The researchers treating the degradation processes of buildings and infrastructure have not the use to think their dynamics. So, they donŠt realize that here we can talk about prediction, and anticipating of effective preventive action.

It isn't to forget that eliminating the effects of building and urban grids degradation, as example, means large amount of investments and these investments need a careful planning before to be too late.

Many of used diagnosis approaches are based exclusively on physical and chemical phenomena provoking the fault or ageing of the used materials (Densley (2001), Hirano et al. (1988), Zhang et al. (1994)).

The work aims to integrate the knowledge collected by physical and chemical measurements with other non-homogeneous information on the supervised technical system and use them by means of other typical approaches of artificial intelligence. The integration has the goal to support the decision process which can allow the diagnosis by anticipating and predicting the degradation degree and fault state. The main, but not the unique cause of degradation is considered the corrosion which acts like a set of phenomena of natural and artificial, built environment.

The second part of the paper is dedicated to state the monitoring problem of degradation process of technical systems. The fault state is further in the third section introduced axiomatically as a study and description mean of degradation process. The fourth section approach the aspects of degradation process modeling in the built systems and in the urban utility grids, and further aims to analysis and selection of case-based reasoning with the results interpreted in the sixth section. The conclusions emphasize the results and possible extents of artificial intelligence for other similar applications.

2. STATEMENT OF MONITORING PROBLEM OF DEGRADATION PROCESS

2.1 Degradation as effect

Each technical system has a finality established by its designer or architect.

Following the finality nature, two large classes of systems can be identified: systems useful by their simple existence, and systems useful by their running manner. In the first class are involved the buildings, the utilities grids, the bridges, and in the second one are involved machines, equipments, electrical circuitries and so on.

---

* The research was supported by the Romanian National Authority for Science and Technology under the projects CEEX No.136/2006 DIRECTOR and CEEX No.264/2006 MATELIZ.

The time, the weather, the environment act on the technical systems with actions in order to diminish their usefulness. Both classes of technical systems previous identified endure differently the usefulness diminishing.

The first class reduces its usefulness till the complete cessation of this.

The second class of systems is affected in two stages. The first stage affects the running, as it is, of the technical system. The second stage of affecting of usefulness consists in its diminishing till the time moment when the costs to put the parametersŠ system inside the necessary borders are inadmissible.

The stage where the usefulness is diminished will be named *degradation*.

One finds that the degradation of complex process occurs inherently of the first time moment of any technical system.

2.2 Degradation cannot be directly measured, it can be only evaluated

The degradation of a technical system cannot be directly measured by using dedicated tools.

So, the degradation degree of a technical system can be estimated by means of the information set which can be associated to technical system existence and running, to environment actions against the system.

One can identify many categories of such information evaluating the degradation degree:

- historic information on the used materials, on the used technology, on the natural conditions existing at initial time moment of technical systemŠs achievement. Historical information assumes also the technical interventions destined to bring back the system to initial or new desired parameters.
- running information referring the sequences time effort associated to technical system. We understand here any kind of effort, non-named and typical to each system type.
- environment information referring to concretized actions of natural and/or artificial environment on the technical system.

A concise analysis of general information categories on the technical system points out that we have a non-homogeneous, difficult to be homogenized and collected information.

Despite these drawbacks it is to emphasize that:

- the historical information respects the strictness of some professional standards and rules and it is collected and archived quasi-systematically. The information is stored in public or private archives but is entirely accessible for control bodies.
- the working information is at least managed. The respected principle is of over-dimensioning in order to cover all the practical situations. We will assume always that the material choosing manner, existing already in historical information doesnŠt require a resuming.
- the environment information will be considered to be the most important in order to describe the main agents which causes the degradation.

2.3 Degradation process occurs by fault successions

The main hypothesis adopted by the authors assumes that the technical system degradation is a consequence of a fault set which occurs and evolves provoking effects which can be found.

By associating the fault to a state, that are a value set of a variable set associated to a time moment, the idea that the degradation can be studied by fault state knowledge will be reached.

2.4 The statement of degradation monitoring problem

In the above stated framework the problem of degradation monitoring and study can be put in a typical pattern as in Table 1.

Table 1. Typical pattern for degradation study.

Initial hypothesis:
• An existing system is considered.
• One assumes that the historical information regarding the manner to obtain or maintain the technical system is known or is accessible.
• One assumes that the working information is inside the foreseen ranges for a normal and sufficient lifecycle.
Working hypothesis:
• Information on the manner in which the natural or/and built environment aggresses the technical system can be collected.
• The amount of possible direct experiments on the technical system is relatively small.
• One has an important amount of information on similar technical systems working in somehow similar circumstances.
One requires:
• The degradation degree of studied technical system.
• The prediction of existing degradation degree for a set of known historical, working and environment information.

2.5 Premises of problem solving

The problem tackling will be considered in connection with two representative applications. The first application is regarding the buildings and the second aims the utilities grids, mainly the underground power grid.

In both situations the main degradation cause will be considered the corrosion.

For an uniform treatment we will assume that the degradation is known and can be anticipated if the fault state is known and interpretable.

3. ABOUT THE FAULT STATE

Any degradation process has as cause the occurrence and development of a generic state further named fault state.

The fault state consists of the overall value of variables and information which contribute to system output in the normal working regime. By working or running regime also the usefulness associated to the technical system like a building or infrastructure part will be understood.

So, as in (Isoc (2003)) the working fault state can be described by a feature set introduced as rules:

- it is a possible state of system.
- it is an abnormal work state according to the system objectives and goals.
- it is a stable state of the system.
- it has a time horizon during which it has a some transient evolution.
- it has working or structural specific reasons (but difficult to be specified).
- it could be always foreseen. When we are talking about the forecasting we will understand the fault nature and never the moment of its occurrence.
- it couldn't be modeled using only measurable information. To be described they need information known by the technical system beneficiary or user.
- it is always outside of the associated control systems possibilities.
- it doesn't affect always the controlled performances of the system. So it could affect only the steady state values, without to affect the dynamic part of variables.
- it doesn't direct necessarily to damages, but such prospective exists for a more important maintaining time of such state.

It is to emphasize that the disturbances cannot be assimilated to faults.

The fault state doesn't damage yet the component part of the system but it is a potential premise for future accidents.

The connection between the rule set defining the fault state occurs at least at the level of next statements:

- the degradation state of a technical system is treated by means of the concept of fault state.
- the fault state must be introduced by means of all measurable variables and known or accessible information.
- the fault state must allow to user an efficient intervention in classifying the abnormal nature of it face the current states of the system.

Once the corrosion, as environment agent was accepted, one raises the problem of studying and especially integrating the knowledge on the corrosion so that this information contributes to decision-making.

Systematic researches developed to deeply study of corrosion on the structures in reinforced concrete (Lingvay and Lingvay (2007)) pointed out that the main actions are:

- The intimate chemistry of phenomenon in different stages.
- The influences of environment agents, especially the temperature, humidity, concentration of corrosive agents.
- The influence of encouraging agents expressed as actions of strong electrical currents of any nature.
- The influence of micro-organisms existence encouraged by the physical and chemical environment of reinforced concrete.

Both the cited and other similar works emphasize in their concluding remarks the relevance of micro-phenomena typical to corrosion.

Such point of view cannot be accepted when one try to integrate the information of different sources in decision-making systems.

To avoid this drawback we made the option to build an overall quantitative criterion able to describe the corrosion state without to emphasize to close the details of phenomenon physical and chemical modeling.

This quantitative criterion was named *corrosivity degree* and it is built an a weighting mean of all agent values relevant for the corrosion phenomena. These values are obtained by technical measurements.

4. TWO CASE STUDIES

4.1 Building models for monitored systems

Once the main cause of degradation process identified and the quantitative degree able to characterize it accepted, the model building was approached.

The given definition of fault state and the complexity of information describing such system already suggested the way to build a knowledge based model.

Building the model of monitored systems of the point of view of degradation process, a special attention was paid to the fact that during the analysis we learned a coexistence of a large amount of information of different classes.

Especially the association of the information regarding the corrosion phenomenon surprised, its very close privacy at molecule level and the information regarding the technologies used to obtain the built structures or the spatial relations established between the buildings and power grids.

We will say that here are parts of micro-modeling regarding phenomena and processes situated under the dimension of visualization required usually by the human being and, in the same time, parts of macro-modeling situated much over the dimension of visualization required by the same.

The simultaneous treatment of both information inside the same model has as consequences an exaggerate complication of the model and the occurrence of situation where some information become insignificant for the description.

The technique selected to be applied was to define information granules that is small quantities of information which can be viewed as parts in the same manner.

So the corrosion phenomenon was replaced by the *corrosivity degree*.

Following the modeling type, that is micro- or macro-modeling, the corrosivity degree will be treated as a variable belonging to macro-modeling while the corrosion as phenomenon together with its modeling is a member of micro- modeling parts.

By introducing the corrosivity degree, the homogenizing of granularity of available information was done in order to build models able to be used inside the monitoring of degradation process of studied technical systems.

Synthesis of information which can describe a built structure of reinforced concrete has the possible attributes and values as in Table 2. A typical situation is described in Table 3 using the defined attributes.

Table 4 describes a pattern for an underground power grid and Table 5 is a situation using the introduced pattern.

4.2 Decision-making using the case-base reasoning

The complexity of built models with the aim to be used in degradation process monitoring and decision-making regarding to predict events or necessary interventions needs specialized methods of artificial intelligence.

The following premises have based the selection of decision-making technique:

- Each monitored system possesses a set of attributes having mainly verbal values.
- Not all the selected attributes to model monitored systems haves sure values at a given time moment.
- Each monitored system is a case which can be representative for the attributes associated to degradation degree.

In these circumstances the optimum method to elaborate the decision is the case-based reasoning.

For these systems (Kolodner (2006)) the working manner as in Table 6 is used.

As software tool using the case-based reasoning *jColibri* (Recio-García et al. (2007)) was selected and an user interface associated to degradation process monitoring for buildings and insulation of underground power grid was attached.

4.3 Preparing the information for application testing

In this stage, two are the problems to be solved for an intelligent software dedicated to decision-making using the case-based reasoning.

The first problem is to prepare the minimal information in order to test the attribute set and a second problem is to choose the representative cases set to test the decision-making quality.

The minimum set of cases which allows to test the correctness and sufficiency of selected attributes was obtained by successive trials.

On the ground of examples of values associated to attributes as in Table 3 and 5, we learned that the minimum number of cases inside of case base should be in the range of 15 ... 25 times the attribute accepted number but not less than un case for each value set for each accepted attributes.

By the developed experiment set we learned that the case set suits with the estimated amount calculated to test the correctness and sufficiency of selected attributes, then the

Table 2. Variables (attributes) and associated values.

Item	Attribute name	Values and symbols
Building details		
1.	Foundation	FCB - continuous foundation under concrete walls
		FCBA – continuous foundation under reinforced concrete walls
		FISBA – insulated foundation under reinforced concrete pillars
		FISM – insulated foundation under metalic pillars
2.	Foundation material	MfBaNN – isn't the case, MfBaDU – light reinforced concrete, MfBaDG – hard reinforced concrete
3.	Reinforcement	MfBaRNN – isn't the case, MfBaRO – regular reinforced concrete, MfBaRIR – high resistance reinforced concrete, MfBaRS – special reinforced concrete.
4.	Reinforcement type	MfArSlOB00 - with smooth surface OB 00, MfArSlOL38 - with smooth surface OL 38, MfArSlSTNS - with smooth surface STNB MfArPpPC52 - with periodic shaped PC 52,
5.	Reinforcement diameter	Diameter of reinforcement wire
6.	Reinforcement method	MfMArI - nets and core-frames of independent rods, MfMArS - nets and core-frames of welded rods
7.	Structure	SPP – structure with partition wall
		SDPoBa - structure with partition diaphragm of reinforced concrete
		SDPoPre - structure with prefab partition diaphragm
		SDPuBa - structure with carried diaphragm of reinforced concrete
		SDPuPre - structure with prefab carried diaphragm
		SCaBa – frame structure of reinforced concrete
		SCaPre – prefab frame structures
	Structure material density	MfBaNN – isn't the case, MfBaDU – light reinforced concrete, MfBaDG – hard reinforced concrete
	Structure material resistance	MfBaRNN – isn't the case, MfBaRO – regular reinforced concrete, MfBaRIR – high resistance reinforced concrete, MfBaRS – special reinforced concrete.
	Structure reinforcement type	MfArSlOB00 - smooth OB 00, MfArSlSTNS - smooth surface STNB MfArPpPC52 - periodical shaped PC 52,
	Structure reinforcement diameter	Diameter of reinforcement wire
	Structure reinforcement method	MfMArI - nets and core-frames of independent rods, MfMArS - nets and core-frames of welded rods
Achievement details		
8.	Achievement period	PRStart – start date, PRStop – end date
Environment details		
9.	Corrosivity degree	Grade between 1 and 10.

Table 4. Pattern to describe an underground power cable grid segment. Attributes, values, nature of values: SM =symbolic, multivalued; RP - real positive, L-logic; PI-positive, integer.

Part	Attribute	Subattribute	Nature	Values
Cable	Insulator material	Type	SM	EPR, XLPE, TRPE, H
		Connection	SM	C1, C2, C3
		Dielectric strength	RP	
Grid sector	Commercial name		SM	DC1, DC2, DC3, DC4
	Shielding		L	Yes, No
	Sector		SM	S1, S2, S3, S4, S5
	Length		RP	
	Installed load		RP	
	Consumption regime		SM	R1, R2, R3, R4
	Neighbouring consumers	Type	SM	CEV1, CEV2, CEV3
		Minimal distance, m	RP	
	Neighbouring utiliy grid	Type	SM	REU1, REU2, REU3
		Minimal distance, m	RP	
Laying environment	Corrosivity degree		RP	
Technological details	Laying date	Year	PI	
		Month	SM	
		Day	PI	
	Working team		SM	E1, E2, E3
	Last event date	Year	PI	
		Month	SM	
		Day	PI	
	Cathodic protection		SM	Yes, No

Table 5. A case set of urban power grid sectors descriptions. In some fields for the 'isn't the case' the '0' value was used.

Part	Attribute	Subattribute	S1	S2	S3	S4	S5
Power cable	Insulator material	Type	XLPE	XLPE	TRPE	H	TRPE
		Connection	C1	C1	C2	C1	C2
		Dielectric strength					
Grid sector	Commercial name		DC1	DC1	DC2	DC3	DC2
	Shield		Yes	Yes	Yes	No	Yes
	Sector		S1	S1	S3	S5	S1
	Length, m		200	350	280	400	230
	Installed load [kVA]		3000	1200	1000	1200	1700
	Consumption regime		R1	R2	R2	R1	R1
	Neighbouring consumers	Type	CEV1	CEV1	CEV2	CEV2	CEV1
		Minimal distance, m	10	5	12	14	10
	Neighbouring utiliy grid	Type	REU1	REU1	REU0	REU2	REU1
		Minimal distance, m	10	15	10	15	30
Laying environment	Corrosivity degree		5.0	6.4	7.4	6.1	5.6
Technological details	Laying date	Year	1998	2001	1974	1974	1999
		Month	March	March	January	November	October
		Day	3	13	6	23	21
	Working team		E1	E1	E3	E1	E1
	Last event date	Year	0	0	1988	0	0
		Month	No	No	January	No	No
		Day	0	0	23	0	0
	Cathodic protection		Yes	No	Yes	No	Yes
Power cable state			Faulty	Correct	Faulty	Correct	Correct

Table 3. Case pattern for a situation to be included in decision-making case base

Case No.1	
Problem attributes and values	
Foundation	FCBa
Foundation material	MfBaDU
Reinforcement	MfBaRO
Reinforcement type	MfArSlOB00
Reinforcement diameter	12
Reinforcement method	MfMArI
Structure	SDPuBa
Structure material density	MfBaDU
Structure material resistence	MfBaRO
Structure reinforcement type	MfArSlOB00
Structure reinforcement – diameter	12
Structure reinforcement method	MfMArI
Achievement period	1976, Jun. – 1977, Jan.
Corrosivity degree	7.5
Problem solution	
Degradation	Normal
Corrosion	Normal
Strengthen necessity	15 years
Major overhaul necessity	45 years

Table 6. General working algorithm for a system using case-based reasoning.

Step	Action
Step 1:	Each case is introduced during its treating in a collection named *case base* together with the solution identified as optimum in the given context.
Step 2:	When a new case is processed, the values of its attributes are considered to be as essential.
Step 3:	The new case is compared to all cases already solved.
Step 4:	One identifies in the case base the closest case to the new one to be treated.
Step 5:	One gives the solution of similar case to the studied one as a solution to the new case to be solved.

same set is enough and ensure the elaborated decision quality.

4.4 Results interpreting

After modeling the cases, selecting the decision-making technique and of software tool (environment) to carry out this work we learned that:

- The introduced attributes are enough and representative to describe the degradation process based on fault state knowledge.
- Prediction of degradation degree is possible in the range of the known degradation degree among the existing cases in case base.
- Reaching an enough credibility of prediction process is possible by increasing the amount and continuously updating of the case base.

5. CONCLUDING REMARKS

The reported research has a deep inter- and multi- disciplinary nature. That is proved by the diversity of reported aspects which is however an unity treatment of this kind of theme.

The main outcome of the research is the systematic manner to build a case base where the aggressive action of the environment is integrated in a corrosivity degree. Using such degree, one homogenizes the granularity degree between aspects modeled using the micro-modeling and the aspects modeled using the macro-modeling.

Using such model described by a collection of attributes with non-homogeneous values, the use of case-based reasoning has proved to be a very efficient way to get a decision oriented to state prediction.

The presented methodology has an degree enough of generality in order to be extent easily to other application where the artificial intelligence can find optimum condition for use.

REFERENCES

Y. Chen. A fuzzy decision system for fault classification under high levels of uncertainty. *Transactions of the ASME - Journal of Dynamic Systems, Measurement, and Control*, 117:108–115, 1995.

J. Densley. Ageing mechanisms and diagnostics for power cables – an overview. *IEEE Electrical Insulation Magazine*, 17:14–22, January 2001.

P.M. Frank and R. Seliger. Fault detection and isolation in automatic processes. *Control and Dynamic Systems*, 49:241–287, 1991.

N. Hirano, T. Tsujimura, N. Shimizu, and K. Horii. Diagnosis of the aged XLPE cable using frequency and temperature characteristics of tan /spl delta/ II. In *Proceedings of the Twenty-First Symposium on Volume*, pages 179–182, 1988.

D. Isoc. Faults, diagnosis, and fault detecting structures in complex systems. In *Study and Control of Corrosion in the Perspective of Sustainable Development of Urban Distribution Grids - The 2nd International Conference*, pages 5–12, Miercurea Ciuc, Romania, June 19–21 2003.

J.L. Kolodner. An introduction to case-based reasoning. *Artificial intelligence*, (6):3–34, 2006.

I. Lingvay and C. Lingvay. Complex investigation approach of degradation state by corrosion in structures of reinforced concrete (In Romanian). Technical Reports RTH 07007, INCDO-INOE2000 Research Institute in Analytical Instrumentation, Cluj Napoca Subsidiary, Cluj Napoca, November 2007.

J.A. Recio-García, B. Díaz-Agudo, and P. González Calero. jCOLIBRI2 Tutorial. Technical Report IT/2007/02, University Complutense of Madrid. Department of Software Engineering and Arňtiňfiňcial Intelligence, Cluj Napoca, October 2007.

V. Venkatasubramanian, R. Rengaswamy, and K. Yin. A review of process fault detection and diagnosis Part III: Process history based methods. *Computers and Chemical Engineering*, 27:327–346, 2003.

W. Zhang, Y. Zhu, B. Yang, and Y. Liu. Study on DC component method for hot-line XLPE cable diagnosis. In *Electrical Insulation. Conference Record of the 1994 IEEE International Symposium on Volume*, pages 95–98, June 5–8 1994.

SESSION F3: VISUAL SERVOING AND MODELING

Kinematic Visual Servo Controls of an X4-Flyer: Practical Study

Odile Bourquardez * Nicolas Guenard ** Tarek Hamel ***
François Chaumette * Robert Mahony **** Laurent Eck **

* IRISA - CNRS and INRIA, Campus de Beaulieu, 35042 Rennes cedex,
France (firstname.lastname@irisa.fr).
** CEA/List, Fontenay-Aux-Roses, France (firstname.lastname@cea.fr).
*** I3S, UNSA - CNRS, Sophia Antipolis, France (thamel@i3s.unice.fr).
**** Dep. of Eng., Australian Nat. Univ., ACT, 0200 Australia
(Robert.Mahony@anu.edu.au).

Abstract: Image moments provide an important class of image features used for image-based visual servo control. Perspective zeroth and rst order image moments provide a quasi linear and decoupled link between the image features and the translational degrees of freedom. Spherical rst-order image moments have the additional desirable passivity property. They allow to decouple the position control scheme from the rotation dynamics. This property is suitable to control an under-actuated aerial vehicle such as a quadrotor. In this paper a range of kinematic control laws using spherical image moments and perspective image moments are experimented on a quadrotor aerial vehicle prototype. The task considered is to reach a desired position with respect to a speci ed target. Three control schemes show excellent performances in practice whereas each one has different theoretical properties.

Keywords: Visual servo control, VTOL aircraft.

1. INTRODUCTION

VISUAL servo algorithms have been extensively developed in the robotics eld over the last ten years [1, 2, 3]. Visual servo control techniques have also been applied recently to a large variety of reduced scales aerial vehicles, such as quadrotors [4, 5], helicopters [6, 7, 8, 9], airships [10, 11] and airplanes [12, 13]. In this paper we consider visual servo control of a quadrotor aerial vehicle.

Much of the existing work in visual servo control of aerial robots (and particularly autonomous helicopters) has used position-based visual servo techniques [6, 14, 7, 8, 4, 5, 9]. The estimated pose can be used directly in the control law [4], or as part of a scheme fusing visual data and inertial measurements [9]. In this paper, we do not deal with pose estimation, but consider image-based visual servo (IBVS), similar to the approach considered in [10, 15, 11].

Image based visual servo control has been used for robotic manipulators [16, 17, 18] and for aerial vehicles [11, 19], by taking into account the system dynamics in the control law. Another approach is based on separating the control problem into an inner (attitude regulation) loop and an outer position control loop [8, 14]. The inner attitude loop is run at high gain using inputs from inertial sensors, rate gyrometers and accelerometers acquired at high data rate; while the outer loop is run at low gain using video input from the camera. The outer (visual servo) loop provides set points for the inner attitude loop and classical time-scale separation and high gain arguments can be used to ensure stability of the closed-loop system [4, 20, 19, 14]. In this paper, we take the inner/outer loop stability for granted (see [21] for details) and concentrate on the speci c properties of the outer loop image based visual servo control design. One of the interests of this approach is to

Fig. 1. The X4-flyer.

decouple the navigation part (considered in the inner loop) from high-level tasks, interacting with the environment. For example using an embedded camera which sends the images to a ground station implies time delays and then a slow image based control loop. It is thus necessary to have a lower-level loop to ensure stabilisation. An other advantage to consider the high-level loop is to enable easier re-use of the IBVS scheme, since it is not closed to the material equipment of the aerial vehicle.

Following earlier work [22, 15], we have chosen to use zero and rst order image moments as primary visual features for the control design. Perspective projection moments with suitable scaling along with a classical IBVS control design lead to satisfactory transients and asymptotic stability of the closed-loop system when the image plane remains parallel to the target. However, the system response may lack robustness for aggressive manoeuvres. In order to overcome this problem, new control schemes, based on spherical rst order image moments, have been proposed [23]. In [23], the experimental results had been obtained on a 6 degrees of freedom robot arm, whereas we present in this paper experimental results on a quadrotor. Note that [19] deals with dynamic control whereas we consider kinematics translation control.

The goal of this paper is to experiment and compare a range of kinematic image based control schemes with a quadrotor aerial vehicle named X4- yer (Fig.1), an omnidirectional VTOL (vertical take off and landing) vehicle ideally suited for stationary and quasi-stationary flight conditions. The task considered is to reach a desired position with respect to a specified target.

This paper is organized as follows: in Section 2 the X4-flyer is described, and the experimental conditions used in all experiments are given. In Section 3 perspective zeroth and first order image moments are used to control the translation kinematics of the X4-flyer prototype. In Section 4 spherical first order image moments and a range of related control schemes are presented. Experimental results are analysed and compared in each section.

2. EXPERIMENTAL CONDITIONS

2.1 Prototype description

The unmanned aerial vehicle (UAV) used for the experimentations is an X4-flyer (Fig.1), that is an omnidirectional VTOL vehicle ideally suited for stationary and quasi-stationary flight conditions. It consists of four fixed pitch propellers linked to an electrical motor at each extremity of a cross frame (Fig. 1). The vehicle is equipped with an avionics stack including an Inertial Measurement Unit (IMU) supplying the vehicle attitude and a controller board [24, 21]. A wireless link allows the transmission of the attitude command between the X4-flyer and a ground station (Pentium 4). A camera situated below the X4-flyer (Fig. 2.a) is embedded and observes a target on the ground, consisting of four black marks on the vertices of a planar rectangle (30 × 40 cm) (Fig. 2.b). A wireless analogue link transmits camera images to the ground station. A 3D estimation of the vehicle position with respect to the target is obtained by fusing the data of the embedded IMU and the visual data in a particle filter [25]. This estimate is used to provide an estimate of ground truth for the 3D behaviour of the vehicle and to provide an estimate of the linear velocity of the vehicle that is used by the high-gain controller of the airframe dynamics. In this paper, only 2D visual information is used in the outer IBVS control loop for position regulation. All the visual servo controls tested are implemented on the ground station. The outer IBVS control loop provides desired translational velocity. This velocity is considered as a set point for an inner control loop, which regulates the rotational dynamics of the vehicle. Time-scale separation and high gain arguments can be used to ensure stability of the closed-loop system.

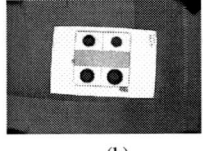

(a) (b)

Fig. 2. (a) The camera. (b) The target view from the camera.

2.2 Experimental protocol

In order to compare the proposed different kinematic visual servo controls, the initial conditions of the experiments were chosen identically. For each experiment, the X4-flyer was servo controlled to a specific initial position using a standard state-space controller deriving information from the task space position estimate. When the vehicle is stabilised at this position, the visual control is initiated and the 3D position, obtained from a particle filter, is recorded. This protocol ensures that the flight conditions are the same and allows the comparison between the different proposed controllers. The velocity demand is also saturated at 20 cm/s to ensure the vehicle remains in quasi-stationary flight regime [26]. The gains of different control laws have been tuned so that the X and Y positions converge in about 10 seconds.

3. VISUAL SERVO CONTROL USING PERSPECTIVE IMAGE MOMENTS

In this section, we use the perspective zeroth and first order image moments [22] to control the translational displacement of the X4-flyer. These image features provide a linear and decoupled link between the task space and the image space, which allows to ensure a good 3D behaviour.

Let $\mathcal{A}$ denote the inertial or task space reference frame and let $\mathcal{C}$ denote the camera or body-fixed reference frame.

Let us define the visual feature vector $\mathbf{s} = (x_n, y_n, a_n)$ such that

$$a_n = Z^*\sqrt{\frac{a^*}{a}}, \quad x_n = a_n x_g, \quad y_n = a_n y_g$$

where x_g and y_g are the centroid coordinates of the object in the image, a is the area of the object in the image, a^* is its desired value and Z^* is the desired depth between the camera and the target. The time derivative of $\mathbf{s}$ and the relative motion between the camera and the object can be related by the classical equation:

$$\dot{\mathbf{s}} = \mathbf{L}_v v + \mathbf{L}_\omega \omega$$

where v and ω are respectively the linear and angular velocity of the camera both expressed in the camera frame, and where $\mathbf{L}_v$ and $\mathbf{L}_\omega$ are respectively the parts of the interaction matrix related to the translational and the rotational motions. The desired image feature is denoted by $\mathbf{s}^* \in \mathcal{C}$, and the visual error is defined by $\mathbf{e} = \mathbf{s} - \mathbf{s}^*$.

Classical image based visual servo control design aims to impose linear exponential stability on the image error kinematics [1, 22] to ensure an exponential decoupled decrease for $\mathbf{e}$ ($\dot{\mathbf{e}} = -\lambda \mathbf{e}$, with λ a positive constant). Using $\mathbf{e}$ to control the translational degrees of freedom, the classical IBVS control input is:

$$v = -(\mathbf{L}_v)^{-1}(\lambda \mathbf{e} + \mathbf{L}_\omega \omega), \quad \lambda > 0. \quad (1)$$

Generally, the interaction terms $\mathbf{L}_v$ and $\mathbf{L}_\omega$ depend non-linearly on the state of the system and cannot be reconstructed exactly from the observed visual data. The visual feature $\mathbf{s} = (x_n, y_n, a_n)$ is of particular interest since $\mathbf{L}_v = -\mathbf{I}_3$ in the case where the camera image plane is parallel to the target plane [22]. In the application considered in this paper, the camera is mounted to point directly downward in the X4-flyer and the image and target plane are never more than a couple of degrees offset. As a consequence, the approximation $\mathbf{L}_v \approx -\mathbf{I}_3$ is valid. The control law is thus simplified to

$$v = \lambda \mathbf{e} + \mathbf{L}_\omega \omega, \quad \lambda > 0. \quad (2)$$

Since the link between image space and task space is almost linear and decoupled ($\mathbf{L}_v \approx -\mathbf{I}_3$), this control scheme is known to lead to satisfactory closed-loop behaviour for holonomic robot [22]. It is in fact equivalent to a position-based visual servo, but without any pose estimation required.

The motion of the X4-flyer is smooth and slow and the value of $\mathbf{L}_\omega \omega$ is small compared with the error $\lambda \mathbf{e}$ in (2). Thus, a reasonable approximation of (2) for the purposes of this paper is

$$v = \lambda \mathbf{e}, \quad \lambda > 0. \quad (3)$$

Equation (3) does not require the estimation of any 3D parameters and can be implemented based only on the observed image features $\mathbf{s}$. This control was implemented on the experimental platform.

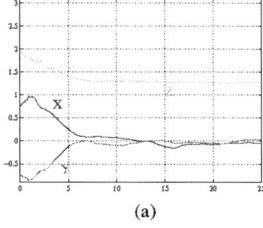
(a)

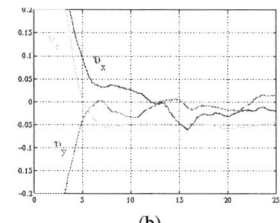
(b)

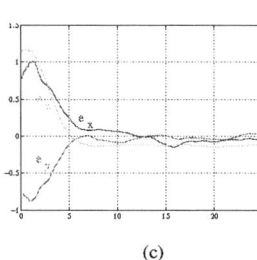
(c)

Fig. 3. Results obtained for $v = \lambda e$: time evolution (in seconds) of the real position in the task space (in meters) (a) with the velocity output of the visual servo control v (in meters per seconds) (b). The evolution of the visual error is plotted on (c).

The practical results are very satisfactory (see Fig. 3) : the vehicle has actually a very good behaviour in each direction. Moreover, the control law is a simple proportional control, and, as the visual error design needs only visual data, it is very easily implemented.

The limitation of this approach, however, lies in its dependence on the particular geometry of the application considered and the requirement to consider only smooth slow trajectories of the vehicle. If the vehicle undertakes aggressive manoeuvres, or the parallel target plane assumption is invalidated for a particular application, the approximation $\mathbf{L}_v \approx -\mathbf{I}_3$ will fail and more importantly the approximation $\mathbf{L}_\omega \omega \approx 0$ may also fail. This second issue introduces a significant dynamic disturbance in the system response that cannot be cancelled directly without the risk of introducing zero dynamic effects into the closed-loop response similar to those studied in recent works [20, 27]. The potential limitations of the classical IBVS control design based on perspective projection features motivate us to consider a class of spherical projection features and non-linear control design techniques.

4. VISUAL SERVO CONTROL USING SPHERICAL PROJECTION

4.1 Image based feature

In this section we use an un-normalised first order spherical image moment along with an inertial goal vector [15] that allows us to obtain the desirable passivity property. This passivity property is very interesting in order to control an under-actuated vehicle such as the X4-flyer [15].

Consider a point target consisting of n points $\{\mathbf{P}_i\} \in \mathcal{C}$ corresponding to image points on the spherical plane $\{\mathbf{p}_i\}$. The centroid of the target is defined to be

$$\mathbf{q} := \sum_{i=1}^{n} \mathbf{p}_i \in \Re^3. \quad (4)$$

For a point target comprising a finite number of image points the kinematics of the image centroid are easily verified to be [15]

$$\dot{\mathbf{q}} = -\omega \times \mathbf{q} - \mathbf{Q} v, \quad (5)$$

with $\mathbf{Q} = \sum_{i=1}^{i=n} \frac{\pi_{\mathbf{P}_i}}{|\mathbf{P}_i|}$ where $\pi_{\mathbf{p}} = (\mathbf{I}_3 - \mathbf{p}\mathbf{p}^T)$. Note that $\mathbf{Q}$ is a positive definite matrix if there are at least two different points $\mathbf{p}_i$ in the image space (see [15] for more details).

Let $\mathbf{b} \in \mathcal{A}$ denote the fixed set point for visual feature $\mathbf{q}$. The feature $\mathbf{q}$ is measured relative to the camera frame and not in the inertial frame, and it is necessary to map the desired set point into the camera frame before an image based error can be defined.

Let $\mathbf{q}^* := \mathbf{R}^\top \mathbf{b} \in \mathcal{C}$, where rotation matrix $\mathbf{R}$ between the camera frame and the inertial frame is obtained from the data supplied by the embedded IMU. The image based error considered is

$$\delta := \mathbf{q} - \mathbf{q}^*. \quad (6)$$

Since $\mathbf{q}^* \in \mathcal{C}$, it inherits dynamics from the motion of the camera: $\dot{\mathbf{q}}^* = -\omega \times \mathbf{q}^*$. Thus, the image error kinematics are [15]

$$\dot{\delta} = \delta \times \omega - \mathbf{Q} v \quad (7)$$

from which we deduce $|\dot{\delta}| = -\frac{\delta^\top \mathbf{Q} v}{|\delta|}$. Since $|\delta|$ is a function of position only, its behaviour will thus not be perturbed by the camera rotational motions. That is why the visual feature δ seems to be very interesting to control an X4-flyer.

In the following subsections, a range of control design for the translational motion of the X4-flyer based on the visual feature $\mathbf{q}$ is considered. Some of them have already been theoretically developed in [23], but we did not have experimented them on an aerial vehicle. In this paper experimental results using an X4-flyer prototype are provided. Moreover the results are compared with the classical IBVS control design based on perspective image moments presented in Section 3.

For each experimentation, the asymptotic value to reach for $\mathbf{Q}$ is

$$\mathbf{Q}^* = diag(2.35, 2.36, 0.057) \quad (8)$$

and we have $\mathbf{b} \cong (0, 0, 3.96)$. These values have been computed when the vehicle is situated at the desired position: approximatively above the center of the target at 1.4 m height of the ground.

4.2 Asymptotic compensation

Linearization at the set point. Using pure proportionnal feedback of the un-normalized centroid ensures global asymptotic stability, but does not give suitable behaviour [23]. The problem is that the eigenvalues of the matrix $\mathbf{Q}$ are not the same and in the general case $\lambda_{\min}(\mathbf{Q}) \ll \lambda_{\max}(\mathbf{Q})$ (where $\lambda_{\min}$ and $\lambda_{\min}$ are respectively the smallest and largest eigenvalues). This means that convergence rates of the components of δ are not the same and the component which is affected by the eigenvalue $\lambda_{\min}(\mathbf{Q})$ is more sensitive to perturbations. By computing matrix $\mathbf{Q}$ at the desired position ($\mathbf{Q}^*$), it follows that $\lambda_{\min}$ is the third eigenvalue of matrix $\mathbf{Q}$ (see (8)). The third component of $\mathbf{q}$ (or δ) is thus sensitive to perturbations. So, it is important that the control schemes designed compensate this sensitivity problem. The first idea to compensate the poor sensitivity is to use the inverse interaction matrix [23] as in classical IBVS.

Indeed the control law $v = k_\mathbf{Q} \mathbf{Q}^{-1} \delta$, $k_\mathbf{Q} > 0$ yields $\dot{\mathcal{L}} = -k_\mathbf{Q} \delta^\top \mathbf{Q} \mathbf{Q}^{-1} \delta = -k_\mathbf{Q} \delta^\top \delta$, where $\mathcal{L}$ is a storage function defined by $\mathcal{L} = \frac{1}{2}|\boldsymbol{\delta}|^2$. This choice guarantees global asymptotic stability and equal convergence rates. The problem is that the matrix $\mathbf{Q}^{-1}$ is not exactly known, since it depends on the 3D depths $|\mathbf{P}_i|$. Thus we can not use easily this control law.

The idea is then to use the *desired* interaction matrix $\mathbf{Q}^*$ [23] instead of the *current* interaction matrix $\mathbf{Q}$, as it is often done in classical IBVS:
$$\boldsymbol{v} = k_* \mathbf{Q}^{*-1} \boldsymbol{\delta}, \quad k_* > 0. \tag{9}$$

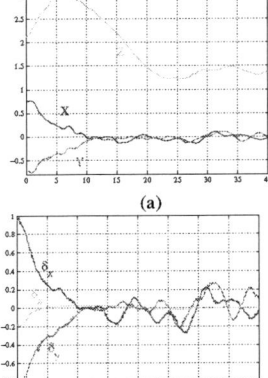

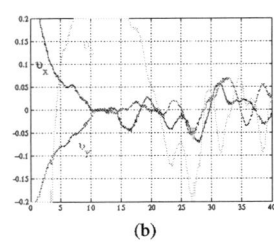

Fig. 4. Results obtained for $\boldsymbol{v} = k\mathbf{Q}^{*-1}\boldsymbol{\delta}$, configured as Figure 3.

As can be seen on Fig. 4, this control law enables the convergence of all the visual error components. However, as the matrix $\mathbf{Q}$ is never updated during the vehicle evolution and fixed to $\mathbf{Q}^*$, this control scheme is not adequate far from the desired position. Consequently, we can see that the convergence rate is not the same on the three components of the position (Fig. 4.a) and the Z component is not suitable. That is why we experiment another approach in the next subsection.

Partitioned control. A second idea for compensating the relative poor sensitivity in the control design is to modify the visual error term in keeping the passivity-like properties. Since difficulties observed in control designs presented in the previous section result from sensitivity in the z-axis, a possible solution is to use a partitioned approach by singling out the problematic component for a special treatment [28].

We separate the visual error term into two criteria with different sensitivity. Two new error terms are introduced in order to compensate the poor sensitivity of $\mathbf{q}$:
$$\boldsymbol{\delta}_{11} = \mathrm{sk}(\mathbf{q}_0^*)\mathbf{q}, \quad \delta_{12} = \mathbf{q}_0^{*\top}\boldsymbol{\delta}, \quad \text{with } \mathbf{q}_0^* = \frac{\mathbf{q}^*}{|\mathbf{q}^*|}.$$

Note that due to the properties of the skew symmetric matrix $\mathrm{sk}(\mathbf{q}_0^*)$, $\boldsymbol{\delta}_{11}$ and $\mathbf{q}_0^*\delta_{12}$ are orthogonal. δ_{12} is the projection of the error $\boldsymbol{\delta}$ along the $\mathbf{q}^*$ direction.

Deriving $\boldsymbol{\delta}_{11}$ and δ_{12}, it follows that
$$\dot{\boldsymbol{\delta}}_{11} = -\mathrm{sk}(\boldsymbol{\omega})\boldsymbol{\delta}_{11} - \mathrm{sk}(\mathbf{q}_0^*)\mathbf{Q}\boldsymbol{v}, \tag{10}$$
$$\dot{\delta}_{12} = -\mathbf{q}_0^{*\top}\mathbf{Q}\boldsymbol{v}. \tag{11}$$

Let us define as Lyapunov function $\mathcal{L}$ such that
$$\mathcal{L} = \frac{1}{2}(|\boldsymbol{\delta}_{11}|^2 + \lambda^2 \delta_{12}^2) \tag{12}$$

where λ is a constant chosen as shown below. It is straightforward to verify that $\mathcal{L} = \frac{1}{2}|\boldsymbol{\delta}_\mathbf{A}|^2$, with
$$\boldsymbol{\delta}_\mathbf{A} = \boldsymbol{\delta}_{11} + \lambda \mathbf{q}_0^* \delta_{12}. \tag{13}$$

Deriving (12), recalling (10), (11), and substituting for (13), one obtains
$$\dot{\mathcal{L}} = -\boldsymbol{\delta}_\mathbf{A}^\top \mathbf{A}(\mathbf{q}_0^*)\mathbf{Q}\boldsymbol{v} \tag{14}$$
where $\mathbf{A}(\mathbf{q}_0^*) = \mathrm{sk}(\mathbf{q}_0^*) + \lambda \mathbf{q}_0^* \mathbf{q}_0^{*\top}$. We define the following control input
$$\boldsymbol{v} = k_\mathbf{A} \mathbf{A}(\mathbf{q}_0^*)^\top \boldsymbol{\delta}_\mathbf{A}, \quad k_\mathbf{A} > 0. \tag{15}$$
Recalling (14) and substituting the control input $\boldsymbol{v}$ by its expression yields
$$\dot{\mathcal{L}} = -k_\mathbf{A} \boldsymbol{\delta}_\mathbf{A}^\top \mathbf{A}(\mathbf{q}_0^*)\mathbf{Q}\mathbf{A}(\mathbf{q}_0^*)^\top \boldsymbol{\delta}_\mathbf{A}.$$

Since $\mathbf{Q}$ is a positive definite matrix and $\mathbf{A}(\mathbf{q}_0^*)$ a non singular matrix, $\mathbf{A}(\mathbf{q}_0^*)\mathbf{Q}\mathbf{A}(\mathbf{q}_0^*)^\top > 0$ and therefore $\boldsymbol{\delta}_\mathbf{A}$ converges exponentially to zero. Consequently, $\boldsymbol{\delta}_{11}$ and δ_{12} converge exponentially to zero (see (13)). Exponential convergence of the initial error $\boldsymbol{\delta}$ to zero is guaranteed.

Note that the best choice of the gain λ is characterized by the following constraint: $\mathbf{A}(\mathbf{q}_0^*)\mathbf{Q}^*\mathbf{A}(\mathbf{q}_0^*)^\top \cong \mathbf{I}_3$. where the symbol $\cong$ means equality up to a multiplicative constant. This choice ensures asymptotically equivalent convergence rate for all the components of the error $\boldsymbol{\delta}_\mathbf{A}$. $\lambda = 6.44$ was used for the presented experimentation; it gave $\mathbf{A}(\mathbf{q}_0^*)\mathbf{Q}^*\mathbf{A}(\mathbf{q}_0^*)^\top \cong 2.35\mathbf{I}_3$.

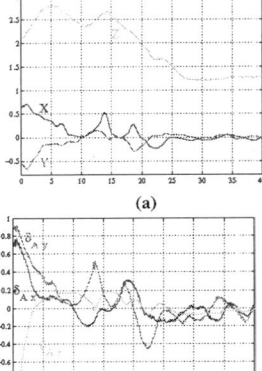

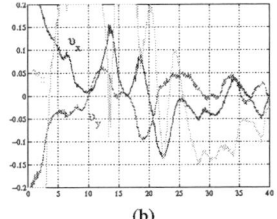

Fig. 5. Results obtained for $\boldsymbol{v} = k\mathbf{A}(\mathbf{q}_0^*)^\top \boldsymbol{\delta}_1$, configured as Figure 3.

At the view of the Fig.5, we can see that although the initial position is far from the set point, this control law enables the convergence of all the visual error components. However the dynamic behaviour of the Z component is strange and consequently, the control law is not perfectly suitable.

4.3 Global compensation using rescaled image feature

The previous control schemes use the desired position in order to equalize the dynamics of the control law: in Section 4.2.1 the desired interaction matrix $\mathbf{Q}^*$ was used, and in Section 4.2.2 the visual error was projected on the direction of the desired visual feature $\mathbf{q}^*$. However the asymptotic compensation is not suitable during the transient, and the behaviour of the X4-flyer is not satisfactory.

In [23], we have shown that a rescaled image feature allow to improve the results, with suitable transient and asymptotic behaviour. In the following subsections, we recall the basics and analyse the new experimental results.

Proportionnal control law with rescaled image feature. The visual error δ_f is defined as follows:

$$\delta_f = F(|\mathbf{q}|)\mathbf{q}_0 - F(|\mathbf{q}^*|)\mathbf{q}_0^*$$

It incorporates the normalised first order moments $\mathbf{q}_0 = \frac{\mathbf{q}}{|\mathbf{q}|}$ along with the scaled depth parameter $F(|\mathbf{q}|)$ defined by:

$$F(|\mathbf{q}|) := \frac{a|\mathbf{q}|}{\sqrt{n^2 - |\mathbf{q}|^2}} \quad (16)$$

where n is the number of points observed and a is the approximate radius of the target. This parameter ensures that the link between task space and image space is almost linear. Consequently the image based visual servoing will give similar behaviour in image space as in task space (see [23] for more details).

Thus we design the control law such that the convergence rates of the components of the visual error δ_f are very close.

Taking the time derivative of the storage function $\mathcal{L} = \frac{1}{2}|\delta_f|^2$ yields after developments: $\dot{\mathcal{L}} = -\delta_f^\top \mathbf{MQ}v$.

Note that the matrix $\mathbf{M}$ is such that $\mathbf{MQ} \simeq \mathbf{I}_3$ [23].

Thus an intuitive idea is to design the control law such that the convergence rates are given by the eigenvalues of $\mathbf{MQ}$. We choose

$$v = k_f \delta_f, \quad k_f > 0 \quad (17)$$

in order to obtain for the derivative of the storage function: $\dot{\mathcal{L}} = -k_f \delta_f^\top \mathbf{MQ}\delta_f$. This form of the storage function ensures the desired property, since the convergence rate of the components of the visual error δ_f are given by the eigenvalues of $\mathbf{MQ}$.

Theoretically this control scheme gives approximately the same convergence rate for the components of the visual error. Moreover, the image feature is chosen close to the 3D position, in order to have a good 3D behaviour with same convergence rate for the components of the 3D position.

As expected, the transient behaviour of the X4-flyer is very good and the three components converge at equal rates in image space (see Fig. 6.c) and in task space (see Fig. 6.a). Moreover, the asymptotic behaviour of the velocity control is less disturbed than previously (see Fig. 6.b, Fig. 4.b, and Fig. 5.b).

Its advantage is also that it is easily implemented, since the control law is a direct function of the visual error δ_f.

Since this control law preserves the passivity property, it is expected to be well-adapted for wide range of aerial vehicles and experimental conditions. However, similar to the perspective moments control design, the global asymptotic stability has not been demonstrated because we are not sure to have $\mathbf{MQ} > 0$ in all the task space.

Globally asymptotically stable control law with modified rescaled image feature. In [23] we proposed a control law based on the depth parameter $F(|\mathbf{q}|)$ (defined by (16)), which ensures suitable image space and task space convergence, in addition to global asymptotic stability.

The new visual error δ_g is defined as follows:

$$\delta_g = G(|\mathbf{q}|)\mathbf{q}_0 - G(|\mathbf{q}^*|)\mathbf{q}_0^* \quad (18)$$

where $G(|\mathbf{q}|) = \alpha(|\mathbf{q}|)\sqrt{|\mathbf{q}|F(|\mathbf{q}|)}$ and $\alpha(|\mathbf{q}|)$ is chosen such that $\alpha(|\mathbf{q}^*|) = 1$.

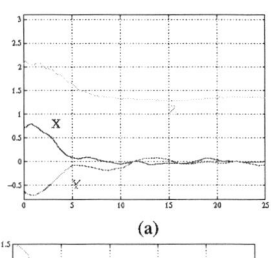
(a)

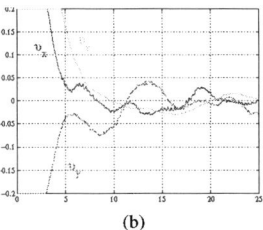
(b)

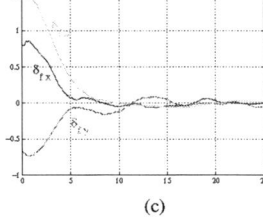
(c)

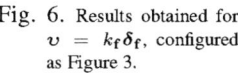

Fig. 6. Results obtained for $v = k_f \delta_f$, configured as Figure 3.

Taking the time derivative of the storage function $\mathcal{L} = \frac{1}{2}|\delta_g|^2$ yields after developments $\dot{\mathcal{L}} = -\delta_g^\top \mathbf{HQ}v$ with $\mathbf{H} = \alpha(|\mathbf{q}|)\sqrt{\mathbf{M}}$.

Thus, if we choose as control law

$$v = \frac{k_g}{\alpha(|\mathbf{q}|)^2}\mathbf{H}(\mathbf{q})\delta_g, \quad k_g > 0 \quad (19)$$

the derivative of the storage function becomes

$$\dot{\mathcal{L}} = -k_g \, \delta_g^\top \frac{\mathbf{H}(\mathbf{q})\mathbf{Q}\mathbf{H}(\mathbf{q})}{\alpha(|\mathbf{q}|)^2}\delta_g.$$

Since $\mathbf{Q}$ is a positive definite matrix, classical Lyapunov theory guarantees that δ_g converges exponentially to zero. Since $\alpha(|\mathbf{q}^*|) = 1$, we have $\frac{\mathbf{HQH}}{\alpha(|\mathbf{q}|)^2} \simeq \mathbf{I}_3$ and consequently good convergence rates in image space (see [23] for more details). Suitable task space behaviour is ensured by the visual feature choice.

As can be seen on Fig. 7, this control scheme leads to equal convergence rates of the visual error components, and equal convergence rates in the task space. The transient behaviour is acceptable.

This control law ensures good behaviour as well as the theoretical important properties of global asymptotic stability and passivity. However the linear link between task space and image space is destroyed, and this could lead to undesirable transient behaviour in some situations.

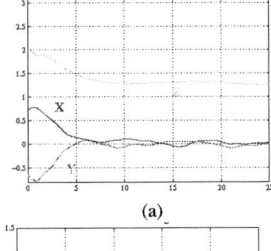
(a)

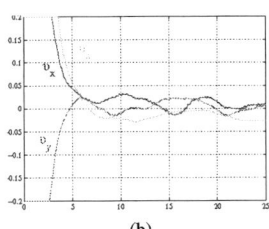
(b)

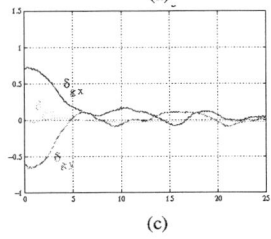
(c)

Fig. 7. Results obtained for $v = \frac{k_g}{\alpha^2}\mathbf{H}\delta_g$, configured as Figure 3.

5. CONCLUSION

This paper presented experimental results for kinematic IBVS control of an X4-flyer. Since they have desirable properties, image moments have been used as visual features. Experimental results have been shown, analyzed and compared for each proposed control scheme.

Using the well-known perspective image moments to design a classical IBVS translational control law leads to a good system behaviour in the undertaken experimental studies. However this control scheme does not ensure global asymptotic stability or passivity of the closed-loop system.

Using spherical first order image moments along with an inertial goal vector allows us to design translational control laws independent from the rotation motion. Global asymptotic stability can be obtained by using these visual features and a simple proportional feedback, but the behaviour on the z-axis is not acceptable. Asymptotic compensation by using classical linearization at the set point or partitionned control give better results but bad transient behaviour: while the X4-flyer has to go down, it starts to go up at the beginning of the control. Suitable feature rescaling allows to compensate globally the sensitivity problem and to improve this behaviour. However, one of the suitable control law does not ensure global asymtotic stability, and the other one does not preserve the linear link between task space and image space.

Finally, the perspective image moments control design, as well as the globally compensated control laws using spherical image moments lead to an acceptable behaviour of the system. None of these three control schemes can be said better than the others, but each one has different theoretical properties.

ACKNOWLEDGEMENTS

This work was supported by CNRS under the project ROBEA-Robvolint.

REFERENCES

[1] B. Espiau, F. Chaumette and P. Rives. A new approach to visual servoing in robotics. *IEEE Trans. on Rob. and Autom.*, 8(3):313 326, 1992.

[2] F. Chaumette and S. Hutchinson. Visual servo control, part I: basic approaches. *IEEE Rob. and Autom. Mag.*, 13(4):82 90, Dec. 2006.

[3] K. Hatano and K. Hashimoto. Image-based visual servo using zoom mechanism. *SICE 2003 Annual Conf.*, 3:2443 2446, Aug. 2003.

[4] E. Altug, J.P. Ostrowski and C.J. Taylor. Control of a quadrotor helicopter using dual camera visual feedback. *IJRR*, 24(5),329 341, Sage Publications, Inc., Thousand Oaks, USA, 2005.

[5] H. Romero, R. Benosman and R. Lozano. Stabilization and location of a four rotor helicopter applying vision. *ACC*, 3930 3936, Minneapolis, USA, June 2006.

[6] O. Amidi, T. Kanade and K. Fujita. A visual odometer for autonomous helicopter flight. *J. of Rob. and Aut. Sys.*, 28:186 193, Aug. 1999.

[7] K. Nordberg, P. Doherty, G. Farnebck, P.E. Forssn, G. Granlund, A. Moe and J. Wiklund. Vision for a UAV helicopter. *IROS, workshop on aerial robotics*, Lausanne, Switzerland, Oct. 2002.

[8] S. Saripalli, J.F. Montgomery and G.S. Sukhatme. Visually-guided landing of an UAV. *IEEE Trans. on Rob. and Autom.*, 19(3):371 381, June 2003.

[9] A.D. Wu, E.N. Johnson and A.A. Proctor. Vision-aided inertial navigation for flight control, *AIAA Guidance, Navigation, and Control Conf. and Exhibit*, San Francisco, USA, Aug. 2005.

[10] J.R. Azinheira, P. Rives, J.R.H Carvalho, G.F. Silveira, E.C. de Paiva and S. S Bueno. Visual servo control for the hovering of an outdoor robotic airship. *ICRA*, 3:2787 2792, Washington, USA, May 2002.

[11] H. Zhang and J.P. Ostrowski. Visual servoing with dynamics: control of an unmanned blimp. *ICRA*, 618 623, Detroit, USA, May 1999.

[12] P. Rives and J.R. Azinheira. Visual auto-landing of an autonomous aircraft. *Research Rep., INRIA-Sophia Antipolis*, n. 4606, Nov. 2002.

[13] O. Bourquardez and F. Chaumette. Visual servoing of an airplane for auto-landing. *IROS*, San Diego, USA, Oct. 2007.

[14] O. Shakernia, Y. Ma, T. Koo and S Sastry. Landing an UAV: Vision based motion estimation and nonlinear control. *Asian J. of Control*, 1(3):128 145, Sept. 1999.

[15] T. Hamel and R. Mahony. Visual servoing of an underactuated dynamic rigid-body system: An image based approach. *ITRO*, 18(2):187-198, Apr. 2002.

[16] P.I. Corke and M.C. Good. Dynamic effects in visual closed-loop systems. *IEEE Trans. on Rob. and Autom.*, 12(5):671 683, 1996.

[17] J. Gangloff and M. de Mathelin. Visual servoing of a 6 DOF manipulator for unknown 3D profile following. *IEEE Trans. on Rob. and Autom.*, Aug. 2002.

[18] R. Kelly, R. Carelli, O. Nasisi, B. Kuchen and F. Reyes. Stable visual servoing of camera-in-hand robotic systems. *IEEE/ASME Trans. on Mech.*, 5(1):39 48, Mar. 2000.

[19] N. Guenard, T. Hamel and R. Mahony. A practical Visual Servo Control for a UAV. *ITRO*, April 2007.

[20] E. Frazzoli, M.A. Dahleh and E. Feron. Real-time motion planning for agile autonomous vehicles. *J. Guidance Cont. and Dyn.*, 25(1):116-129, 2002.

[21] N. Guenard, T. Hamel and L. Eck. Control law for the tele operation of an UAV known as an X4-flyer. *IROS*, Beijing, China, Oct. 2006.

[22] O. Tahri and F. Chaumette. Point-based and region-based image moments for visual servoing of planar objects. *ITRO*, 21(6):1116-1127, Dec. 2005.

[23] O. Bourquardez, R. Mahony, T. Hamel and F. Chaumette. Stability and performance of image based visual servo control using first order spherical image moments. *IROS*, Beijing, China, Oct. 2006.

[24] N. Guenard, T. Hamel and V. Moreau. Dynamic modeling and intuitive control strategy for an X4-flyer, *Int. Conf.on Cont. and Autom.*, Budapest, Hongrie, June 2005.

[25] S. Arulampalam, S. Maskell, N.J. Gordon and T. Clapp. A tutorial on particle filters for on-line non-linear/non-gaussian bayesian tracking. *IEEE Trans. of Signal Processing*, 50(2):174-188, Feb. 2002.

[26] N. Guenard, T. Hamel, V. Moreau and R. Mahony. Design of a controller allowed the intuitive control of an X4-flyer. *Int. IFAC Symp. on Rob. Cont.*, Bologna, Italy, Sept. 2006.

[27] T. Hamel and R. Mahony. Image based visual servo-control for a class of aerial robotic systems. *To appear in Automatica*, 2007.

[28] P. Corke and S.A. Hutchinson. A new partitioned approach to image-based visual servo control. *IEEE Trans. on Rob. and Autom.*, 17(4):507-515, Aug. 2001.

Modelling and Control of Flexible Airship

S.BENNACEUR*. A.ABICHOU*, N.AZOUZ **

**Informatics, Integrative Biology and Complex Systems (IBISC),
40 Rue de Pelvoux, 91020 Evry Cedex, France (e-mail:Azouz@iup.univ-evry.fr).
*Laboratoire d'Ingénieurie Mathématique (LIM) BP2078 - Tunisie
(e-mail: bennaceur_selima@yahoo.fr, azgal.abichou@ept.rnu.tn,)

Abstract: Unmanned Aerial Vehicles (U.A.V.) have a need of a greater autonomy in their new missions. Autonomous U.A.V. flight control systems require a precise modeling of the dynamic behavior taking into account the effect of flexibility and the interaction with surrounding fluid. In this paper, we present an efficient modeling of the autonomous flexible blimps. These flying objects are assumed to undergo large rigid-body motion and small elastic deformations. The formalism used is based on the Newton-Euler approach. This one is frequently used for flying rigid objects. In this study we develop a method to generalize the existing Newton-Euler "rigid body" formalisms by including the effect of the flexibility without destroying the global methodology. The method is hybrid. It uses the Lagrange equations and the Eulerian variables. The flexibility appears in the global dynamical system by the way of few supplementary degrees of freedom. This method has the advantage of making easier the elaboration of algorithms of control, stabilization or generation of trajectories. The added mass phenomenon is also taken into account in the dynamical system. This phenomenon is important for big and light objects moving in a fluid such as airships. As validation we use the parameters of an AS-200 blimp belonging to the University of Evry.

Keywords: FlexibleBlimps, Small deformations, Stabilization.

1. INTRODUCTION

Autonomous aerial vehicles such as airships and drones have recently gained importance. The airships began these last decades to retrieve a new youth after a half century of hibernation. Capability of airships are expanding rapidly now, and the range of missions they designed to support is growing. We can give the examples of climate research, surveillance or even infiltration in war scenarios, and more closer to us we always saw them as means of advertising. A common denominator in all these situations is the impossibility or unwillingness to have human presence at the scene. In that case a precise dynamic model should be elaborated to permit an easy control, stabilization or navigation task of this autonomous object. In order to fulfill this requirement, it is necessary to introduce the effect of the structural flexibility in the dynamic model. It is important to note that several kinds of airships, usually called blimps, are mainly constituted of a balloon filled with gas. The only solid parts are the careen and the tail fins (for more details see [1]). The integration of the structural flexibility in the dynamic analysis is then useful; however it is now in an embryonic state and is only just emerging. Several researches were done by using the assumption of rigid body behaviour for airships [2],[3]. The flexibility effects are sometimes modelled as a perturbation. However in other flying objects, such as light aircrafts, the introduction of the flexibility in the dynamic model becomes essential [4,5]. Thus we try to contribute to the study of the deformation of the airships by introducing the effect of flexibility as non controlled supplementary degrees of freedom. The deformation of the blimp is not considered as a perturbation but rather acting on the motion of the airship.

The influence of structural flexibility on the dynamics of mechanical systems has become increasingly important in classical robotics [6,7,8], and recently in flying robots (i.e. Airships, drones…).

Several approaches, to study the problems of flexible bodies, have been proposed in the literature. These approaches can be classified into two groups. The first one uses the total Lagrangian method [9]. This consists in defining the motion relatively to a fixed reference frame, but this often leads to complex relations when describing stresses and strains in the flexible body subject to large displacements and small deformations. An Updated Lagrangian Method (U.L.M.) was proposed by Bathe & al.[10] and developed for deformable bodies that undergo large translational and rotational displacements. The resolution of the dynamic problem is incremental. The configuration and the motion of the body are identified using a moving reference configuration representing the position of the deformable body in the preceding step. Bennaceur & al[11] propose as reference configuration a rigid body configuration which follows the motion of the body without coinciding with it. This approach is convenient for a flying body with small deformations. The motion is given by coupled sets of rigid and elastic variables. The nonlinear equations are formulated in terms of a set of time invariant matrices expressed in a reference configuration (i.e. mass and stiffness matrices). Time-variant quantities appear in the nonlinear terms that represent the dynamic coupling between the rigid body modes and the elastic deformation. A suitable technique to actualize these terms using matrix partitioning and canonical decomposition is proposed, and the dynamic system is reduced and solved via a modal synthesis. The Newton-Euler description [12], which is an interesting method in regard of the time computation, was extensively studied by another important group of searchers in the case of rigid airships. This choice is mainly motivated by the facility to build control or stabilization algorithms based on this model. However the use of the Newton-Euler approach in the dynamic analysis of flexible complex structures is rare[12]. We propose through this paper to extend the classical rigid bodies' model to the deformable bodies, without destroying the general formalism obtained. However this requires a total reformulation of the spatial flexible structure modelization by the use of a hybrid method based on both Lagrangian and Eulerian description. The model obtained should be acceptable for control community and closer to the one used for the rigid bodies. On the other hand airships are also governed by the aerodynamic forces that have to be modelled. The basis to analyze the motion of a rigid body in a perfect fluid has been established in the 19th century and has been described by Lamb [13]. In his work, Lamb considered the case of simple displacement in a big infinite mass of fluid where the movement of this last is entirely due to the motion of a solid, and it is irrotational and acyclic. He proved that the kinetic energy of the fluid can be expressed as a quadratic shape of the six velocities of translation and rotation of the vehicle. The derivations given by Lamb will be used in the description of the airship, in a stationary uniform atmosphere. The terms depending on the acceleration (i.e. the added masses) come from the fact that the fluid considered perfect is accelerated. When an ellipsoid body moves in an incompressible and infinite inviscid fluid in order that the external flow is everywhere irrotational and continuous, the kinetic energy of the fluid produces an effect equivalent to an important increase of the mass and of the moments of inertia of the body [14], [15], [2]. The field of research on the control of under-actuated autonomous vehicles is potentially interesting. Thereafter, the main control techniques used in the literature are displayed. A class of under-actuated autonomous systems is examined by Abichou, Bestaoui and Beji [21], who presented an unsteady periodic and uniform stabilizing an under-actuated airship. Approximations homogeneous and return control of state is applied successfully to systems without drift by Morin and Samson. They presented a return to continuous and unsteady asymptotically stabilizing of an under actuated satellite. Hamel, Mahony, Lozano, and Ostrowski [22] have proposed a new method of control "X4flyer", in which they separated the dynamics of rigid device to the dynamic engine for the purpose of obtaining a high practice stability of the complete system.

Pettersen [23] have shown that broad class of vehicles can not be stabilized either by a continuous return or by a discontinuous return state. They considered a system with drift (boat) as a model to study and they showed a law returning state continuously, periodically and unsteady may stabilize.

2. Dynamics

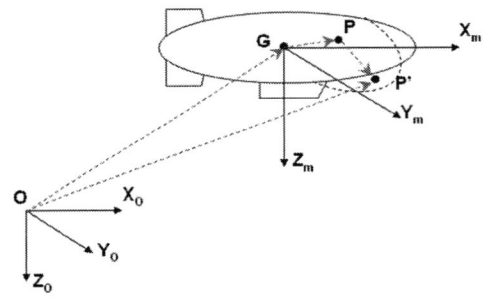

Fig. 1. Definitions of the position vectors

The position of a point in the blimp is given by:

$$OP' = OG + GP + PP' \quad (1)$$

Knowing that P' is a point of the deformed configuration. As in the rigid case, we will start with a Lagrangian description. Consequently all the dynamic parameters are expressed in R_0. Consequently:

$$r = \eta_1 + J_{1q}\underbrace{(u_0 + u_d)}_{U} = \eta_1 + J_{1q}U \quad (2)$$

By deriving the relation (1) with respect to time, and when using similar developments to those presented in (2), we obtain:

$$\dot{r} = \dot{\eta}_1 - J_{1q}\tilde{U}J_{2q}\dot{q} + J_{1q}\dot{u}_d \quad (3)$$

The displacement of deformation $u_d = u_d(s,t)$ is a function of time and space. It can be broken up into a sum of two separate functions:

$$\mathbf{u_d}(s,t) = \sum_{i=1}^{2} \mathbf{Y_d^i}(t) X^i(s)$$

Where:

$X^i(s)$: represent the ith eigenmode of the solid

$\mathbf{Y^i_d}(t)$: the associated modal amplitude.

This writing can be condensed in the following way:

$$\mathbf{u_d} = S\overline{\mathbf{Y}}_\mathbf{d}$$

S represents the two selected modes, and $\overline{\mathbf{Y}}_\mathbf{d}$ is the column matrix composed by the various Y_d^i.

We denote by $\overline{\eta} = \begin{pmatrix} \eta_1 \\ q \\ \overline{Y}_d \end{pmatrix}$ the total position vector of an arbitrary point of the flexible airship.

The modes of deformation S being constant compared to time, we can thus deduce that the speed of a given point from the blimp will be:

$$\dot{r} = \dot{\eta}_1 - J_{1q} \tilde{U} J_{2q} \dot{q} + J_{1q} S \dot{\overline{Y}}_d \qquad (4)$$

Or in a more compact form:

$$\dot{r} = \overline{C}\dot{\overline{\eta}} \qquad (5)$$

With:

$$\overline{C} = \begin{pmatrix} I & -J_{1q}\tilde{U}J_{2q} & J_{1q}S \end{pmatrix} \qquad (6)$$

The preceding form of the speed will be used for the development of the kinetic energy of a deformable body.

2. Expression of the Kinetic Energy of an arbitary Point **P'**

Similarly to the developments presented in § 2, we will have:

$$E_c = \frac{1}{2}\int_V \rho \dot{r}^T \dot{r} \, dV \qquad (7)$$

By using the expression of the velocity vector in the preceding equation, the kinetic energy is expressed as follows:

$$E_c = \frac{1}{2}\int_V \rho \dot{\overline{\eta}}^T \overline{C}^T \overline{C} \dot{\overline{\eta}} \, dV = \frac{1}{2}\dot{\overline{\eta}}^T M \dot{\overline{\eta}} \qquad (8)$$

M is recognized as the symmetrical matrix of mass and is defined by:

$$M = \int_V \rho \begin{pmatrix} I \\ (-J_{1q}\tilde{U}J_{2q})^t \\ (J_{1q}S)^t \end{pmatrix} \begin{pmatrix} I & -J_{1q}\tilde{U}J_{2q} & J_{1q}S \end{pmatrix} dV \qquad (9)$$

The mass matrix in (8) can be written in a symbolic form as follows:

$$M = \begin{pmatrix} m_{TT} & m_{TR} & m_{TD} \\ m_{TR}^T & m_{RR} & m_{RD} \\ m_{TD}^T & m_{RD}^T & m_{DD} \end{pmatrix}$$

It is clear that the two sub-matrices respectively associated with the translation and the elastic co-ordinates are constant. The other matrices depend on the generalized co-ordinates of the system, and consequently they are implicit functions of time.

By coinciding the centre of gravity of the machine and the centre of R_m, and by assuming that the position of this last does not vary significantly in R_m in presence of small deformations, we can thus consider that the sub-matrix m_{TR} is void.

We will also assume here that the terms of coupling translation-deformation m_{TD} are void.

The kinetic energy is then:

$$E_C = \frac{1}{2}(\dot{\eta}_1^T m_{TT} \dot{\eta}_1 + \dot{q}^T m_{RR} \dot{q} \qquad (10)$$
$$+ 2\dot{q}^T m_{RD} \dot{\overline{Y}}_d + \dot{\overline{Y}}_d^T m_{DD} \dot{\overline{Y}}_d)$$

By using the Lagrange's equations, we obtain the following relation $\quad M\ddot{\overline{\eta}} = \tau - K\overline{\eta} + Q_G \qquad (11)$

Where τ is the vector of the external forces and moments, and K is the matrix of stiffness given by:

$$K = \begin{pmatrix} 0 & 0 & 0 \\ 0 & 0 & 0 \\ 0 & 0 & K_{DD} \end{pmatrix}$$

The relation can be given in a matrix form as follows:

$$\begin{pmatrix} m_{TT} & 0 & 0 \\ 0 & m_{RR} & m_{RD} \\ 0 & m_{RD}^T & m_{DD} \end{pmatrix} \begin{pmatrix} \ddot{\eta}_1 \\ \ddot{q} \\ \ddot{\overline{Y}}_d \end{pmatrix}$$
$$+ \begin{pmatrix} 0 & 0 & 0 \\ 0 & 0 & 0 \\ 0 & 0 & K_{DD} \end{pmatrix} \begin{pmatrix} \eta_1 \\ q \\ \overline{Y}_d \end{pmatrix} = \begin{pmatrix} \tau_1 \\ \tau_2 \\ 0 \end{pmatrix} + \begin{pmatrix} Q_{G\eta_1} \\ Q_{Gq} \\ Q_{Gd} \end{pmatrix} \qquad (12)$$

Then $\quad M\ddot{\overline{\eta}} = \tau - \begin{pmatrix} 0 \\ 0 \\ K_{DD}\overline{Y}_d \end{pmatrix} + Q_G \qquad (13)$

Let us now analyze the vector of gyroscopic and Coriolis forces Q_G.

From the Lagrange's equation, we have:

$$Q_G = -\dot{M}\dot{\overline{\eta}} + \left[\frac{\partial}{\partial \overline{\eta}}(\frac{1}{2}\dot{\overline{\eta}}^T M \dot{\overline{\eta}})\right]^T \qquad (14)$$

Or in more explicit form:

$$Q_G = \begin{pmatrix} Q_{G\eta_1} \\ Q_{Gq} \\ Q_{Gd} \end{pmatrix} = \begin{pmatrix} \left[\frac{\partial}{\partial \eta_1}(\frac{1}{2}\dot{\overline{\eta}}^T M \dot{\overline{\eta}})\right]^T \\ -\dot{m}_{RR}\dot{q} - \dot{m}_{RD}\dot{\overline{Y}}_d + \left[\frac{\partial}{\partial q}(\frac{1}{2}\dot{\overline{\eta}}^T M \dot{\overline{\eta}})\right]^T \\ -\dot{m}_{RD}\dot{q} - \dot{m}_{DD}\dot{\overline{Y}}_d + \left[\frac{\partial}{\partial \overline{Y}_d}(\frac{1}{2}\dot{\overline{\eta}}^T M \dot{\overline{\eta}})\right]^T \end{pmatrix} \qquad (15)$$

The development of this relation leads to these following results:

The first component could be written as:

$$Q_{G\eta_1} = \left[\frac{\partial}{\partial \eta_1}(\frac{1}{2}\dot{\bar{\eta}}^T M \dot{\bar{\eta}})\right]^T = 0$$

For the second component Q_{Gq} we have:

$$Q_{Gq} = -\dot{m}_{RR}\dot{q} - \dot{m}_{RD}\dot{\bar{Y}}_d + \left[\frac{\partial}{\partial q}(\frac{1}{2}\dot{\bar{\eta}}^T M \dot{\bar{\eta}})\right]^T \quad (16)$$

The first term is:

$$-\dot{m}_{RR}\dot{q} = -\overbrace{(J_{2q})^T I_{RR} J_{2q}}\dot{q} \quad (17)$$

With I_{RR} the matrix of rotation inertia expressed in R_{ra}.

Basing on the fact that this analysis is intended for the airships subjected to small deformations, therefore we will keep the same assumptions used for the rigid case. Consequently the inertial terms expressed in the pointer are invariants.

Thus

$$-\overbrace{(J_{2q})^T I_{RR} J_{2q}}\dot{q} = -(\dot{J}_{2q})^T I_{RR} \mathbf{v_2} \quad (18)$$

and we can say for the second term that:

$$-\dot{m}_{RD}\bar{Y}_d = -(\dot{J}_{2q})^T I_{RD} \bar{Y}_d \quad (19)$$

For the third term:

$$\frac{\partial}{\partial q}(\frac{1}{2}\dot{\bar{\eta}}^T M \dot{\bar{\eta}}) = \frac{\partial}{\partial q}(\frac{1}{2}\dot{q}^T m_{RR} \dot{q}) \\ = \frac{1}{2}\frac{\partial}{\partial q}(\dot{q}^T (J_{2q})^T I_{RR} J_{2q}\dot{q}) \quad (20)$$

By using the relation (20), and after differentiating, we will have:

$$\left[\frac{\partial}{\partial q}(\frac{1}{2}\dot{\bar{\eta}}^T M \dot{\bar{\eta}})\right]^T = -(\dot{J}_{2q})^T I_{RR} \mathbf{v_2} \quad (21)$$

Thus:

$$Q_{Gq} = -2(\dot{J}_{2q})^T I_{RR} \mathbf{v_2} - (\dot{J}_{2q})^T I_{RD} \bar{Y}_d \quad (22)$$

For the third component Q_{Gd} we have:

$$Q_{Gd} = -\dot{m}_{RD}^T \dot{q} + \underbrace{\left[\frac{\partial}{\partial \bar{Y}_d}(\frac{1}{2}\dot{\bar{\eta}}^T M \dot{\bar{\eta}})\right]^T}_{0}$$

According to [9] the preceding notations we can obtain the following relation:

$$Q_{Gd} = -I_{RD}^T \underbrace{\dot{J}_{2q}\dot{q}}_{0} = 0$$

When collecting all these developments, could be expressed in this form:

$$\begin{pmatrix} I_{TT} & 0 & 0 \\ 0 & m_{RR} & m_{RD} \\ 0 & m_{RD}^T & I_{DD} \end{pmatrix} \begin{pmatrix} \ddot{\bar{\eta}}_1 \\ \ddot{q} \\ \ddot{\bar{Y}}_d \end{pmatrix} = \begin{pmatrix} \tau_1 \\ \tau_2 \\ -K_{dd}\bar{Y}_d \end{pmatrix} + \\ \begin{pmatrix} 0 \\ -2(\dot{J}_{2q}^{-1})^T I_{RR} \mathbf{v_2} - (\dot{J}_{2q})^T I_{RD} \bar{Y}_d \\ 0 \end{pmatrix} \quad (23)$$

In the Eulerian variables we will have:

$$\begin{pmatrix} I_{TT} & 0 & 0 \\ 0 & I_{RR} & I_{RD} \\ 0 & I_{RD}^T & I_{DD} \end{pmatrix} \begin{pmatrix} \dot{\mathbf{v}}_1 \\ \dot{\mathbf{v}}_2 \\ \ddot{\bar{Y}}_d \end{pmatrix} = \begin{pmatrix} \tau_1 \\ \tau_2 \\ 0 \end{pmatrix} + \begin{pmatrix} -I_{TT}(\mathbf{v}_2 \wedge \mathbf{v}_1) \\ -\mathbf{v}_2 \wedge (I_{RR}\mathbf{v}_2) - \mathbf{v}_2 \wedge (I_{RD}\bar{Y}_d) \\ -K_{dd}\bar{Y}_d \end{pmatrix} \quad (24)$$

Or in compact form

$$M_E \dot{\bar{\nu}} = \underline{\tau} + \bar{Q}_G \quad (25)$$

We thus obtain a Newton-Euler system equation similar to that defined in the rigid case but extended to the flexible blimps.

3. Aerodynamic Contribution:

We show here the influence of the surrounding air on the general behavior of the flexible blimps. Let us consider a simple dynamical model for the action of the air on a body. To present this model, we assume that the flow is quasi-steady, i.e. the distribution of the velocities of particles of the medium coincides with the distribution corresponding to the steady motion of the body. Thus the medium responds only to the current motion of the body and forgets its initial conditions. Therefore, within the framework of this hypothesis, the resultant force and torque acting on the body can be represented in the form of a function of the instantaneous distribution of velocities in this body. Thus we reach at the statement of the problem of the motion of a body in a dragging medium as a problem of classical dynamics.

3.1. Flow Representation:

To take into account the interaction of the airship with the surrounding fluid medium, a model of the flow is needed. Here, we rely on the potential flow theory corresponding to the following hypothesis:

the air can be considered as a perfect fluid with uniform density ρ_{air}, i.e. an incompressible gas with vanishing viscosity,

the flow is irrotational

Only the flow outside the airship contributes significantly to the aerodynamics. Denoting by **v** the velocity field in the fluid domain "air", the incompressibility and irrotational assumptions leads to:
$$\nabla \cdot \mathbf{v} = 0 \quad ; \quad \nabla \wedge \mathbf{v} = 0 \quad (26)$$

∇ is the gradient symbol, and the flow field may be described in terms of a potential F such as:
$$\mathbf{v} = \nabla \Phi$$

From the incompressibility constraint, it is easy to show that the potential obeys to the homogeneous Laplace equation:

$$\nabla^2 \Phi = 0 \quad \text{in } \Omega_{air} \quad (27)$$

with Newman boundary conditions:

$$\nabla \Phi \mathbf{n} = -\dot{\mathbf{q}} \, \mathbf{n} \text{ on the boundary } \left(\partial C^t\right) \text{ (careen)} \quad (28)$$

n is a unit vector, normal to $\left(\partial C^t\right)$.

Thus, one of the most important characteristic of this representation is that v only depends on the current boundary conditions, and not on the history of the flow: the model is quasi-steady. To solve the potential equation, we use the boundary integral representation of the Laplace equation, together with standard boundary element method. It consists in the determination of a piecewise constant distribution of singularities over $\left(\partial C^t\right)$ (see [19] for details on the numerical treatment).

3.2. Fluid Forces:

For this assumption, the pressure at any point in the fluid domain (including $\left(\partial C^t\right)$) is given by Bernoulli theorem:

$$P + \rho_{air}\left[\frac{1}{2}\mathbf{v}\cdot\mathbf{v} + \frac{\partial \Phi}{\partial t}\right] = P_\infty + \rho_{air} \cdot \frac{1}{2}\mathbf{v}_\infty \cdot \mathbf{v}_\infty \quad (29)$$

The subscript ∞ denotes the undisturbed conditions far from the airship. This pressure distribution over the airship surface can be integrated to compute the resulting forces and torques. At the end, and with the linear property of the Laplace equation, the generalized fluid forces vector can be rewritten as:

$$\mathbf{F_f} = -M_{ad}\,\overline{\dot{\mathbf{v}}} - B_f\,\overline{\mathbf{v}}$$

Where Mad is the matrix of the added masses (virtual masses), and Bf is a damping due to the flexibility of the hull. Taking into account the previous developments, the dynamic equation of structure becomes:

$$M'_E \, \overline{\dot{\mathbf{v}}} = \underline{\tau} + \overline{Q}'_G \quad (30)$$

we note $M'_E = M_E + M_{ad}$, $\overline{Q}'_G = \overline{Q}_G - B_f \mathbf{v}$. The effect of the fluid on the structure is then represented mainly by the adjunction of the added masses matrix Mad to the mass matrix of the structure.

For a quasi-ellipsoid airship the extra-diagonal terms of M_{ad} can be neglected (for more details about the constitutive terms of M_{ad}, the reader can see [2], [13]).

4. Stabilization and control:

With the previous assumptions, the dynamic and kinematics of an airship with small deformations can be written in the following compact form as:

$$M'_E \, \overline{\dot{\mathbf{v}}} = \underline{\tau} + \overline{Q}'_G \quad \text{and} \quad \begin{cases} \dot{x} = u \\ \dot{y} = v \\ \dot{z} = w \\ \dot{\phi} = p \\ \dot{\theta} = q \\ \dot{\psi} = r \\ \dot{y}_{d1} = \dot{y}_{d1} \\ \dot{y}_{d2} = \dot{y}_{d2} \end{cases} \quad (33)$$

$$M_E = \begin{pmatrix} m+m_{ad_1} & 0 & 0 & 0 & 0 & 0 & 0 & 0 \\ 0 & m+m_{ad_2} & 0 & 0 & 0 & 0 & 0 & 0 \\ 0 & 0 & m+m_{ad_3} & 0 & 0 & 0 & 0 & 0 \\ 0 & 0 & 0 & I_x & 0 & 0 & 0 & 0 \\ 0 & 0 & 0 & 0 & I_y+m_{ad_4} & 0 & 0 & 0 \\ 0 & 0 & 0 & 0 & 0 & I_z+m_{ad_5} & 0 & 0 \\ 0 & 0 & 0 & 0 & 0 & 0 & I_{d1}+m_{adf_1} & 0 \\ 0 & 0 & 0 & 0 & 0 & 0 & 0 & I_{d1}+m_{adf_2} \end{pmatrix}$$

In order to stabilize our system we write the system (33) as follows:

$$\overline{\dot{\mathbf{v}}} = M'^{-1}_E (\underline{\tau} + \overline{Q}'_G) \quad (34)$$

In more developed form we have:

$$\begin{pmatrix} \dot{u} \\ \dot{v} \\ \dot{w} \\ \dot{p} \\ \dot{q} \\ \dot{r} \\ \ddot{y}_{d_1} \\ \ddot{y}_{d_2} \end{pmatrix} = \begin{pmatrix} \frac{1}{m+m_{ad_1}} & 0 & 0 & 0 & 0 & 0 & 0 & 0 \\ 0 & \frac{1}{m+m_{ad_2}} & 0 & 0 & 0 & 0 & 0 & 0 \\ 0 & 0 & \frac{1}{m+m_{ad_3}} & 0 & 0 & 0 & 0 & 0 \\ 0 & 0 & 0 & L_1 & 0 & L_2 & L_3 & 0 \\ 0 & 0 & 0 & 0 & L_4 & 0 & 0 & L_5 \\ 0 & 0 & 0 & L_6 & 0 & L_7 & L_8 & 0 \\ 0 & 0 & 0 & L_9 & 0 & L_{10} & L_{11} & 0 \\ 0 & 0 & 0 & 0 & L_{12} & 0 & 0 & L_{13} \end{pmatrix} \times$$

$$\begin{pmatrix} \tau_{1x} \\ \tau_{1y} \\ \tau_{1z} \\ \tau_{2x} \\ \tau_{2y} \\ \tau_{2z} \\ 0 \\ 0 \end{pmatrix} + \begin{pmatrix} m(rv-qw)+X_u u \\ m(pw-ru)+Y_v v \\ m(qu-pv)+Z_w w \\ rqI_y - qrI_z - q\dot{y}_{d1} I_{RD_2} - r\dot{y}_{d2} I_{RD_1} + L_p p \\ prI_z - prI_x + p\dot{y}_{d1} I_{RD_2} + M_q q \\ qpI_x - qpI_y - p\dot{y}_{d2} I_{RD_1} + N_r r \\ -K_{dd_1} y_{d1} \\ -K_{dd_2} y_{d2} \end{pmatrix} \quad (35)$$

Theorem:

Let x_e an equilibrium point of the dynamic system $\dot{x} = f(x)$:

If x_e is asymptotically stable by the linerized system then there is for the initial system.

If x_e is not stable hen there is for the initial system.

Thus the linear model will be given by this expression

$$\begin{cases} \dot{u} = \frac{1}{m+m_{ad_1}} \varphi_1 \\ \dot{v} = \frac{1}{m+m_{ad_2}} \varphi_2 \\ \dot{w} = \frac{1}{m+m_{ad_3}} \varphi_3 \\ \dot{p} = L_1 \varphi_4 + L_2 \varphi_6 + L_3 \varphi_7 \\ \dot{q} = L_4 \varphi_5 + L_5 \varphi_8 \\ \dot{r} = L_6 \varphi_4 + L_7 \varphi_6 + L_8 \varphi_7 \\ \ddot{y}_{d_1} = L_9 \varphi_4 + L_{10} \varphi_6 + L_{11} \varphi_7 \\ \ddot{y}_{d_2} = L_{12} \varphi_5 + L_{13} \varphi_8 \\ \dot{x} = u \\ \dot{y} = v \\ \dot{z} = w \\ \dot{\phi} = p \\ \dot{\theta} = q \\ \dot{\psi} = r \\ \dot{y}_{d_1} = \dot{y}_{d_1} \\ \dot{y}_{d_2} = \dot{y}_{d_2} \end{cases} \quad (36)$$

With $$\begin{cases} \varphi_1 = \tau_{1x} + X_u u \\ \varphi_2 = \tau_{1y} + Y_v v \\ \varphi_3 = \tau_{1z} + Z_w w \\ \varphi_4 = \tau_{2x} + L_p p \\ \varphi_5 = \tau_{2y} + M_q q \\ \varphi_6 = \tau_{2z} + N_r r \\ \varphi_7 = -K_{dd_1} y_{d_1} \\ \varphi_8 = -K_{dd_2} y_{d_2} \end{cases} \quad (37)$$

In order to reach a desired position by a flexible airship, in this section, we present an explicit algorithm such as [20],[21]. Furthermore our system can be subdivided in six subsystems as:

$$\begin{cases} \dot{u} = \frac{1}{m+m_{ad_1}} \varphi_1 \\ \dot{x} = u \end{cases} \quad (38)$$

$$\begin{cases} \dot{v} = \frac{1}{m+m_{ad_2}} \varphi_2 \\ \dot{y} = v \end{cases} \quad (39)$$

$$\begin{cases} \dot{w} = \frac{1}{m+m_{ad_3}} \varphi_3 \\ \dot{z} = w \end{cases} \quad (40)$$

$$\begin{cases} \dot{\theta} = q \\ \dot{q} = L_4 \varphi_5 + L_5 \varphi_8 \\ \ddot{y}_{d_2} = L_{12} \varphi_5 + L_{13} \varphi_8 \\ \dot{y}_{d_2} = \dot{y}_{d_2} \end{cases} \quad (41)$$

$$\begin{cases} \dot{p} = L_1 \varphi_4 + L_2 \varphi_6 + L_3 \varphi_7 \\ \dot{\phi} = p \end{cases} \quad (42)$$

$$\begin{cases} \dot{r} = L_6 \varphi_4 + L_7 \varphi_6 + L_8 \varphi_7 \\ \ddot{y}_{d_1} = L_9 \varphi_4 + L_{10} \varphi_6 + L_{11} \varphi_7 \\ \dot{\psi} = r \\ \dot{y}_{d_1} = \dot{y}_{d_1} \end{cases} \quad (43)$$

Being given that the basic problem is to stabilize asymptotically neighbourhood the desired point $(x_d, 0, z_d, 0_{\Re^5})$.

Hence we define the error systems given by:

$$\begin{cases} \dot{u} = \frac{1}{m+m_{ad_1}} \varphi_1 \\ \dot{\xi}_1 = u \end{cases} \quad (44) \text{ and } \begin{cases} \dot{w} = \frac{1}{m+m_{ad_3}} \varphi_3 \\ \dot{\xi}_2 = w \end{cases} \quad (45)$$

The subsystems (44), (45), (39) and (42) are simple to be stabilized by using the Backstepping method.

4.1. Stabilization:

First Step: Commandability

The study of stabilization will be given in two parts.

In the first part we study the command ability of the system by using the theorem of KALMAN given by:

Theorem (Kalman):

It is said that the linear system $\dot{x}=f(x,u)=Ax+Bu$ with A a linear application from $\Re^n$ to $\Re^n$ and B a linear application from $\Re^n$ to $\Re^m$ can be commendable if and only if the matrix of controllability $C=(B \quad AB \quad .... A^{n-1}B)$ is of rank n.

Further more the linear system (44) can be written as:

$$\begin{cases} \ddot{\xi}_1 = \dfrac{X_u}{m+m_{ad_1}}\dot{\xi}_1 + \underbrace{\dfrac{1}{m+m_{ad_1}}\tau_1}_{\tau_{12}} \\ \dot{\xi}_1 = \dot{\xi}_1 \end{cases} \quad (46)$$

Thus the system in the equation (44) takes the form of

$$\dot{x}=Ax+B\tau_{12} \quad \text{With} \quad A = \begin{pmatrix} \dfrac{X_u}{m+m_{ad_1}} & 0 \\ 1 & 0 \end{pmatrix} \quad \text{and} \quad B = \begin{pmatrix} 1 \\ 0 \end{pmatrix}$$

We find the matrix of commandability as
$C = (B \quad AB) = \begin{pmatrix} 1 & \dfrac{X_u}{m+m_{ad_1}} \\ 0 & 1 \end{pmatrix}$ with a rank equal to 2.

Second Step: Construction of Command

By using the Backstepping.

The structure of our system using some techniques based on the Backstepping to calculate the command recursively.

In seeking to reach a desired altitude we pose an intermediate control τ_1' in our system so we obtain:

$$\begin{cases} \dot{u} = \tau_1' \\ \dot{\xi}_1 = u \end{cases} \quad (47)$$

By using the Backstepping method we obtain this system which is obtained by taking the virtual control τ_1''

$$\dot{\xi}_1 = \tau_1'' \quad (48)$$

thus we pose $\tau_1'' = -k_1\xi_1$ with k_1 the gain strictly positive.

In order to prove that the system is asymptotically stable we choose a function of Lyapunov $\dot{V}(x) < 0$.

Theorem:

If we pose $x = 0$ a point of balance of the system $\dot{x}=f(x,u)$ and if there exist a Lyapunov function $V(x)$, thus the origin x=0 is stable. If more $\dot{V}(x) < 0$ in $\Re^n - \{0\}$ thus x=0 is asymptotically stable.

Proof:

We pose a Lyapunov function $\dot{V}(x) < 0$ definite positive
$V(u,\xi_1) = \dfrac{1}{2}\xi^2{}_1 > 0$ so $\dot{V}(\xi_1) = -k_1\xi_1 < 0$ definite negative.

This choice ensures us the stabilization of the system (48) near zero, hence the stabilization of the system (47) near (0,0) by

$$\tau_1' = -k_2(u+k_1\xi_1)$$

thus we obtain $\tau_1 = -k_2(m+m_{ad_1})(u+k_1\xi_1) - X_u u$

By replacing ξ_1 by its expression we finally obtain

$$\tau_1 = -k_2(m+m_{ad_1})(u+k_1(x-x_d)) - X_u u$$

the gain k_1 and k_2 are big enough.

We use the same method for the system (39), (40), and (42).

So for the system (41) and (43) will be stabilized by the Backstepping method.

In order to reduce our system we write

$$\begin{cases} \ddot{\theta} = L_4\varphi_5 - L_5 K_{dd_2} y_{d_2} \\ \ddot{y}_{d_2} = L_{12}\varphi_5 - L_{13} K_{dd_2} y_{d_2} \end{cases}$$

We verify that the system is commandable by Kalman.

So in order to stabilize our system we make some changes:

$$\begin{cases} \ddot{\theta} = L_4\varphi_5 - L_5 K_{dd_2} Y_1 - \dfrac{L_{12}}{L_4} L_5 K_{dd_2}\theta \\ \ddot{Y}_1 = (-L_{13} - \dfrac{L_{12}}{L_4} L_5 K_{dd_2})Y_1 - (\dfrac{L_{12}L_{13}}{L_4} - \dfrac{L_{12}^2}{L_4^2} L_5 K_{dd_2})\theta \end{cases} \quad (49)$$

We pose $Y_1 = y_{d_2} - \dfrac{L_{12}}{L_4}\theta$

By using a Bacckstepping as shown in the first model:

$$\begin{cases} \ddot{\theta} = \varphi_5' \\ \ddot{Y}_1 = \alpha Y_1 - \beta\theta \end{cases} \quad (50)$$

So we pose $Y_2 = \dot{Y}_1$ thus we obtain

$$\begin{cases} \dot{q} = \varphi_5' \\ \dot{\theta} = q \\ \dot{Y}_1 = Y_2 \\ \dot{Y}_1 = \alpha Y_1 + \beta \theta \end{cases} \Rightarrow \begin{cases} \dot{\theta} = v \\ \dot{Y}_1 = Y_2 \\ \dot{Y}_1 = \alpha Y_1 + \beta \theta \end{cases} \quad (51)$$

We pose $\varphi_5' = -k_1(q-v)$ with $k_1 > 0$

Then by applying another Backstepping

$$\begin{cases} \dot{Y}_1 = Y_2 \\ \dot{Y}_2 = \alpha Y_1 + \beta v_1 \end{cases} \text{ with } v = -k_2(\theta - v_1) \quad k_2 > 0$$

thus we obtain

$$\begin{cases} \dot{Y}_1 = Y_2 \\ \dot{Y}_2 = v_2 \end{cases} \text{ with } v_2 = \alpha Y_1 + \beta v_1$$

By applying another Backstepping with $k_3 > 0$

We have $\dot{Y}_1 = v_3$ with $v_2 = -k_3(Y_2 - v_3)$ thus we choose

$v_3 = -k_4 Y_2$ with a gain $k_4 > 0$.

Finally we obtain a control witch stabilize the angular state and the deformations given by

$$\tau_5' = \frac{-k_1(q + k_2(\theta - k_3 \frac{\dot{Y}_1}{\beta} - \frac{k_3 k_4}{\beta} Y_1) + \frac{\alpha}{\beta} Y_1)}{L_4} \quad (52)$$
$$-M_q q + \frac{L_5 K_{dd_2}}{L_4} Y_1 + \frac{L_{12}}{L_4^2} L_{13} \theta$$

We stabilize the system (43) by the same method.

5. Simulations results:

To illustrate this Eulerian approach we study the blimp belonging to the LSC-IBISC and having the following characteristics:

-The envelope: Length: 6.25 m., Diameter: 1.52 m.,

Volume: 7.48 m3., -Mass of the airship: 5.8 Kg.

-Payload: 1.58 Kg.

The blimp is thrust by two contrarotating propellers. In the first step we apply to the blimp two opposite forces in the propellers to generate a yaw motion. The flexible airship should rotate about 90° around the z-axis. A Proportional-Integrated-Derivative (P.I.D) controller will impose this task, and we study the behavior of the airship during this maneuver. The number of deformable modes kept is nd = 2. This number seems to give an acceptable approximation of the flexible behavior. We were guided for that by the modal masses of these two modes which represent roughly 70% of the total mass of the airship. We applied the force on the point P (2, 0, 0) of the blimp. In this simulation we applied around the yaw angle a torque consigns with: K_d = 800, K_v = 50, ψ_d : The desired angular position and ψ : The angular position of the airship. One notices in figure 5 and 6 that the amplitude of the deviation of the yaw angle decreases significantly in a few seconds, it stabilizes oneself while merging with the instruction.

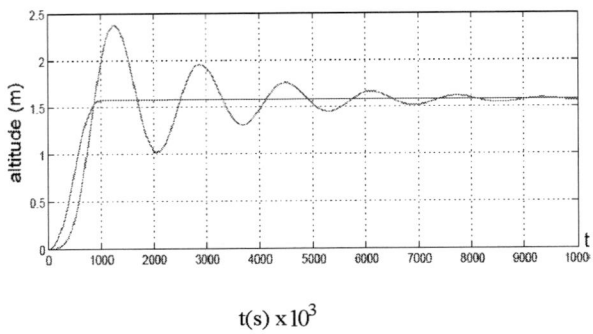

t(s) x 10^3

Fig. 2. Superposition of the motions of the yaw and desired trajectory.

This represents a complete aerial motion of the airship. The propellers were oriented adequately to assure a combined motion along the moving x-axis and z-axis and the tails were oriented to give a homogeneous yaw motion.

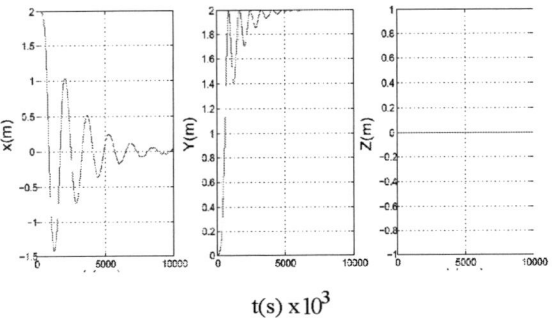

t(s) x 10^3

Fig.3. Position of the tip of the airship.

We superimpose total displacement along the X-axis of the rigid airship and the flexible device. It is noticed that the flexible device continues to oscillate which proves the impact of flexibility on displacement. For the rigid behavior we eliminate the two deformable modes and keep only the rigid motion.

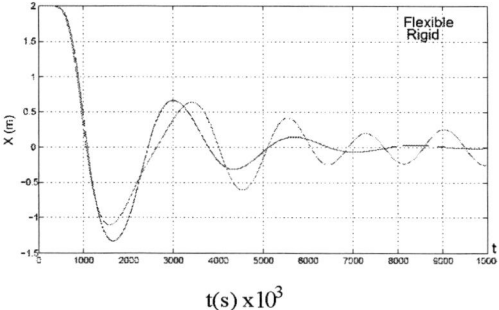

Fig.4. Superposition of Flexible and Rigid airship.

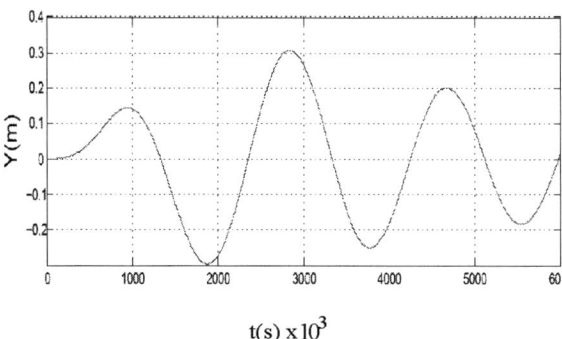

Fig. 5. The deformation's displacement.

The deformations are about 0.15m what is more or less significant considering one has small deformations. In this simulation we visualized the prow of the airship, and we see well the oscillation of this later with non-negligible amplitude.

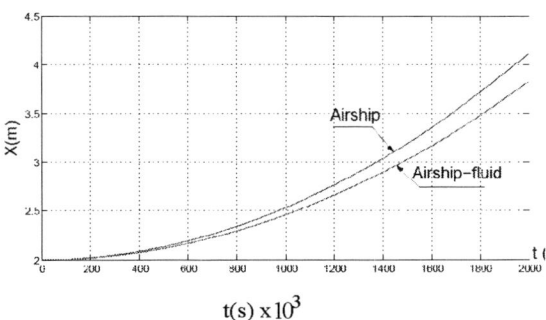

Fig. 6. Superposition of the motions of the airship with and without fluid.

Finally we show the influence of the added masses on the total displacement of the airship. It is noticed that the immersed airship has a delay compared to the body alone and this is because of the addition of the virtual masses representing the influence of the displacement of the mass of air around the airship when this last accelerates. After that we simulate the result of the control of the rigid and flexible system we obtain:

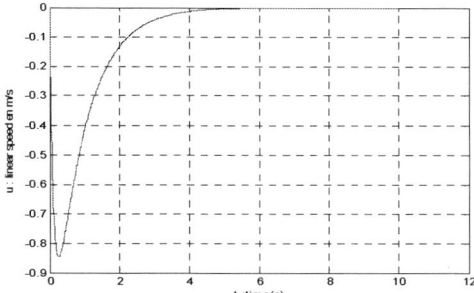

Fig. 7. The linear speed of the system without deformations.

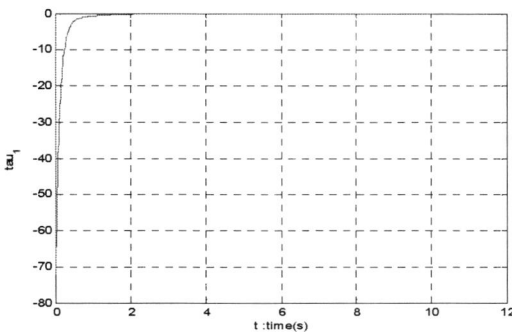

Fig. 8. The control of the rigid system.

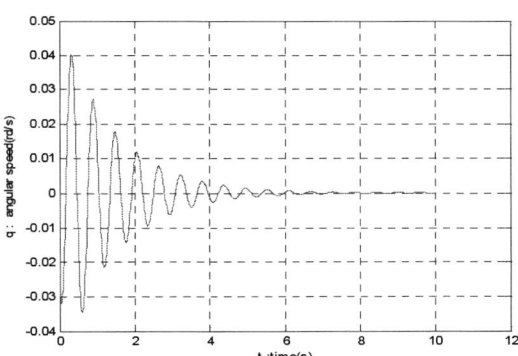

Fig. 9. The angular speed of the system with flexibility.

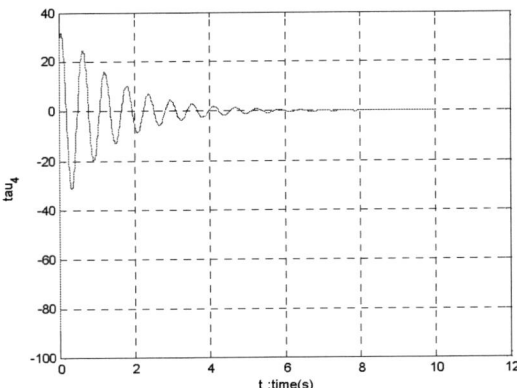

Fig. 10. The control witch stabilizes the deformation and the angular variables of state.

So we can see in the figure of the control the impact of the flexibility in the system, the flexible model oscillates compared to the rigid one.

So in the rigid case, the model stabilizes rapidly but in the flexible case we show some vibrations, which create a delay compared to the rigid case.

6. CONCLUSIONS

In this paper we presented a description of the flexible airships by an Eulerian method. We introduced the flexibility of the airship as an extension of a rigid body algorithm based on Eulerian description and taking into account the coupling between the rigid body motion and the deformation. A hybrid methodology is used, based on both Lagrangian and Eulerian theory. The final goal of developing this model is to enable an easy implementation of control and stabilization algorithms for the flexible flying objects .We prove in this work the impact of the flexibility in the model to be controlled however, we prove also that the extend model is easy to be stabilized.

The model built presents also an interesting ratio precision computational time. Simulation results prove that the integration of the flexibility in the dynamic system of the airship is very important and could not be neglected. The added mass phenomenon is also taken into account. However we neglected in this study the added masses issued from the vibration of the airship in the air and their coupling with those issued from the overall rigid body motion. This point is under investigation and will be presented in future works.

This work proves that in our model, the deformation is not a sample perturbation around the main rigid motion, but it acts on this motion. Experimental data will be available in few months to validate the results of our incremental scheme.

REFERENCES

[1] Khoury G.A.,Gillet J.D.,1999 *"Airship Technology"*, Cambridge Univ. Press.

[2] Fossen T., 1996. "Guidance and control of ocean vehicles", J. Wiley press, Chichester,

[3] Bestaoui Y.,Hamel T. 2000 "Dynamic modeling of small autonomous blimps" Proc. Of Conference on Methods and Modeles in Automation and Robotics (MMAR00), Miedzyzdroje, Poland,pp.579-584.

[4] Calise A.,Nakwan K., 2002 " Adaptive compensation for flexible dynamics". AIAA Guidance, navigation and control conference. Montery, Canada.

[5] Bianchin M.,Quaranta G.,Mantegazza P., 2003 "State space reduced order models for static aeroelasticity and flight mechanics of flexible aircrafts". 17th national conference AIDAA,Italy,.

[6] Simo J.C., 1987"The role of non-linear theories in transient dynamic analysis of flexible aircrafts», Journal of Sound &Vibration, 119, pp. 487-508.

[7] Pascal. M. 1991, "Vibrations analysis of flexible multibody systems", Dyn. & Stability of systems, Vol.6, N° 3.

[8] SchiehlenW, Cuse N. and Seifried R.,2006 "Multibody dynamics in computational mechanics and engineering applications", Computer Methods in Applied Mechanics and Engineering , Volume 195, Issues 41-43, 15 August 2006, pp. 5509-5522.

[9] Shabana A., 1988 "Dynamics of multibody systems". Edition Springer-Verlag.

[10] Bathe K.J & al. 1975 "Finite Elements for large deformation dynamic analysis" INT.Jour. For Num. Meth In Eng., Vol 9, pp.353-386.

[11] Bennaceur S., Azouz N., Boukraa D. 2006 *"An efficient modelling of flexible Airships: Lagrangian approach"*. Proceeding of the ESDA'06 ASME International Conference. ESDA2006-95741. Torino, Italy, July.

[12] Boyer F., Coiffet Ph. 1996 *"Generalization of Newton-Euler model for flexible manipulators"*, Int. Jour. For Robotics Research 17(3), pp. 282- 293.

[13] Lamb H., 1945 *"On the motion of solids through a liquid. Hydrodynamics"*, Dover, New York, 6th edition.

[14] Thomasson P.G 2000 *"Equation of motion of a vehicle in a moving fluid"*, Jour. of Aircraft, Vol.37, N°4, pp. 630-639.

[15] Ortega J.,Rosier L.,Takahashi T 2006 " On the motion of a rigid body immersed in a bidimensional incompressible perfect fluid", Annales de l'institut Poincare (c) Non linear Analysis .

[16] Goldestein 2001 "Classical Mechanics", Ed. Addison Wesley Publishing Company; Boston, 3rd edition.

[17] Gibert R-J *"Vibrations des structures. Interactions avec les fluides"*, Ed. Eyrolles, Paris, 1988, pp. 137-140.

[18] Clough R-W., PENZIEN J., 1993 *"Dynamics of Structures"*, McGraw; International edition.

[19] Katz J., Plotkins A. 1991 *"Low-speed Aerodynamics, from Wings Theory to Panel Methods"*, McGraw-Hill Book Co, New York.

[20] L. Beji, A. Abichou, Y. Bestaoui 'Stabilization of a nonlinear underactuated autonomous airship – A combined averaging and backstepping approach' 3rd IEEE Workshop on Robot Motion and Control, Bukowy Dworek, Nov. 2002.

[21] L. Beji, A. Abichou, Y. Bestaoui 'Stabilization of a nonlinear underactuated autonomous airship – A combined averaging and backstepping approach' 3rd IEEE Workshop on Robot Motion and Control, Bukowy Dworek, Nov. 2002.

[22] T.Hamel, R.Mahony, R.Lozano and J.Ostrowski, 'Dynamic Modelling and Configuration Stabilization for an X4-flyer, Proceedings of the International Federation of Automatic Control Symposium , Barcelona, Spain, 2002.

[23] K.Y.Pettersen and H. Nijmeijer,'Underactuated ship tracking control: theory and expriments,' Int.J.Control,2001,vol.74,N14,435-14.

A Technique Combining Optimal Filtering and Periodicity

Soraya. Zenati *, Abdelhani. Boukrouche**

*University of Guelma, Box 401 Guelma,
24000, Algeria (e-mail: blzenague@yahoo.fr).
** Faculty of Sciences and Engineering Box 401 Guelma, 24000, Algeria
Laboratory of Automatic and Computer of Guelma L.A.I.G.
(e-mail. hani.Boukrouche@gmail.com)

Abstract: The proposed method is a derivative of the iterative deconvolution with constraint and optimal filtering introduced by Neveux & al. [2000] [10], method which offers the advantage of using the impulse response of the distortion process includes a filtering step in a deconvolution algorithm with constraint. In the following it is shown that this method can be adapted to restore 1D or 2D signals without ever using the causality property. Our approach combines the method of periodicity suggested by Blanc-Feraud & al. [1988] [7].

Keywords: Blur, Convolution, Periodicity

INTRODUCTION

The proposed method is a derivative of the iterative deconvolution with constraint and optimal filtering introduced by Neveux & al. [2000] [10], method which offers the advantage of using the impulse response of the distortion process, includes a filtering step in a deconvolution algorithm with constraint. This approach was presented by Sekko & al. [1999] [11, 12] and involves the transfer function of the distortion process from the state space model and can be considered as a modified version of classical regularization techniques. The procedure described by [11, 12] can be applied to 1D as well as 2D signals. Authors have taken care to use only concepts that do not use the property of transfer causality. The other hand this is not the case for the method given by [10] where the unknown signal u to be restored is considered causal. In the following it is shown that this method can be adapted to restore 1D or 2D signals without ever using the causality property. Our approach combines the method of periodicity suggested by Blanc-Feraud & al. [1988] [7].

In general, the impulse response can be represented by a circulant matrix which is easily diagonalizable by a Fast Fourier Transform (FFT). It's diagonalization facilitates the restoration, the filtering and decreases the number of undertaken operations as well as the size of memory space used.

The performances of the proposed method are illustrated through two examples. The first example deals with the estimation of the heat rate in a reaction calorimeter, while the second one focuses on the image restoration.

1.1 Approach

Various methods of image (signal) restoration are described in literature. This diversity reveals the importance of the problem and its great difficulty. As a matter of fact, signal restoration is an ill posed problem [13], that is to say: A bounded perturbation on the observed data gives an unbounded perturbation on the restored signal.

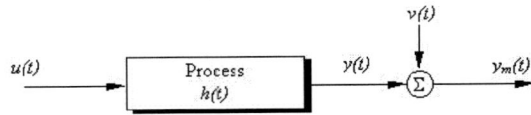

Fig. 1. Block diagram of the signal distortion

In control theory, the response of a process to an input signal is described by a convolution product of the form:

$$y(t) = (h * u)(t) \qquad (1)$$

Where * denote 1D or 2D convolution, h is the impulse response of the degradation process or the point spread function (PSF) in image processing, and u is the original signal. The recorded data $y_m(t)$ is given by:

$$y_m(t) = y(t) + v(t) \qquad (2)$$

where v is the noise measurement.

The inverse operation consisting in the reconstruction of the signal u knowing y_m and h, is called deconvolution. This technique has a wide array of application (such as astronomy, seismology, spectroscopy, etc ...) and also a wide range of solutions [3, 4, 9] already existing. As one may find in the literature, this problem is known to be an ill-posed one. This is due to the fact that the recorded signal $y_m(t)$ is corrupted by an additive noise so that the inverse procedure is hardened.

Various methods of deconvolution are described in the literature. This diversity reveals the importance of the problem and its great difficulty. As a matter of fact, signal restoration is an ill-posed problem [13], that is to say: a bounded perturbation on the observed data yields an unbounded perturbation on the restored signal. The main technique used in order to transform this ill-posed problem into a well-posed one is the so called 'regularizarion procedure'.

Recorded images, or more generally, recorded signals obtained by sensors, are degraded versions of original signals. In most cases, this degradation can be modeled by a convolution procedure followed by noise addition.

The fundamental issue in signal restoration is to find a signal $\hat{u}$ as close as possible to the original unknown signal u from available information on the degradation process. That is to say, estimate u, using the impulse response h (or any model of the distorting process), stochastic knowledge of the measurement noise, and of course the observed data y_m.

In the following, h is supposed to be known as in [2].

Usually, the measured signal is previously filtered in order to lessen the level of noise, then, this filtered version is introduced in the deconvolution algorithm, Sekko & *al.* [11, 12] proposed, in the frequency domain, to introduce the filtering step in the deconvolution algorithm.

2. NOTATIONS

Let u be the non causal unknown of the 1D signal to be restored and y_m the noisy measured output of a distortion process supposed to be linear and described by the impulse response h (see Figure 2).

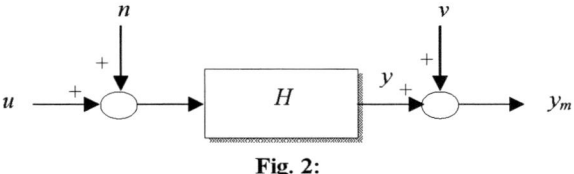

Fig. 2:

Therefore, after discretization, the input-output relation can be written as follow:

$$y_m(i) = \sum_{k=-\infty}^{+\infty} h(i)[u(k-i) + n(k-i)] + v(k) \quad (3)$$

where $h(i)$, $u(i)$, $n(i)$, $y_m(i)$ and $v(i)$ are the discretized counter part of $h(t)$, $u(t)$, $n(t)$, $y_m(t)$ and $v(t)$ respectively. We suppose that $n(t)$ and $v(t)$ be independent Gaussian white noise with variance σ_n et σ_v, respectively. In addition, the signals u, n, and v are supposed to be mutually uncorrelated.

The input-output relation is given by:

$$y = H(u+n) \quad (4)$$

Where H is the Fourier transform of h.

2.1 1D signal case

The impulse response given by
$$RI = [h_m \cdots h_0 \cdots h_{-m}]$$
RI is symmetric and impairs.

We construct now the convolution matrix of RI with $u(t)$

$$y(t) = \sum_{j=-m+1}^{j=m+n} h_{i-j} u_j \quad \text{with} \quad i = 1 \text{ to } n \quad \text{and} \quad m \in N, n \in N$$

if h is finished dimension then $(2m+1)$

$$y_1 = h_m u_{-m+1} + h_{m-1} u_{-m+2} + h_{m-2} u_{-m+3} \cdots h_{-m} u_{m+1}$$
$$y_2 = h_m u_{-m+2} + h_{m-1} u_{-m+3} + h_{m-2} u_{-m+4} \cdots h_{-m} u_{m+2}$$
$$\vdots$$
$$y_n = h_m u_{-m+n} + h_{m-1} u_{-m+n+1} + h_{m-2} u_{-m+n+2} \cdots h_{-m} u_{m+n}$$

what translated into

$$y(1\ldots n) = \begin{bmatrix} h_m & \cdots & h_0 & \cdots & h_{-m} & 0 & \cdots & 0 \\ 0 & \ddots & & & & \ddots & \ddots & \vdots \\ \vdots & \ddots & \ddots & & & & \ddots & 0 \\ 0 & \cdots & 0 & h_m & \cdots & h_0 & \cdots & h_{-m} \end{bmatrix} \begin{bmatrix} u_{-m+1} \\ u_{-m+2} \\ \vdots \\ u_0 \\ \vdots \\ u_n \\ u_{n+1} \\ \vdots \\ u_{n+m} \end{bmatrix}$$

under this form it is not invertible.

Following [6], we approach the matrix H by

$$\hat{H} = [H_{left} + H_0 + H_{right}]$$

where

$$H_{left} = \begin{bmatrix} 0 & \cdots & \cdots & \cdots & \cdots & 0 \\ \vdots & \ddots & & & & \vdots \\ 0 & & \ddots & & & \vdots \\ h_{-m} & & & \ddots & & \vdots \\ \vdots & \ddots & & & \ddots & \vdots \\ h_{-1} & \cdots & h_{-m} & 0 & \cdots & 0 \end{bmatrix}$$

is the matrix nxn with h_{-m} on its diagonal

$$H_{right} = \begin{bmatrix} 0 & \cdots & 0 & h_m & \cdots & h_1 \\ \vdots & \ddots & & \ddots & \ddots & \vdots \\ \vdots & & & & \ddots & h_m \\ \vdots & & & & & 0 \\ \vdots & & & & & \vdots \\ 0 & & & & & 0 \end{bmatrix}$$

is the matrix nxn with h_m on its diagonal

and

$$H_0 = \begin{bmatrix} h_0 & h_{-1} & \cdots & h_{-m} & 0 & \cdots & \cdots & 0 \\ h_1 & \ddots & \ddots & & \ddots & 0 & & \vdots \\ \vdots & \ddots & \ddots & & & \ddots & \ddots & \vdots \\ h_m & & & & & & \ddots & \vdots \\ 0 & \ddots & & & & & & 0 \\ \vdots & \ddots & h_m & \cdots & h_1 & h_0 & h_{-1} & \cdots & h_{-m} \\ \vdots & & & 0 & \ddots & & & \ddots & h_{-1} \\ 0 & \cdots & & & 0 & h_m & \cdots & h_1 & h_0 \end{bmatrix}$$

is the matrix nxn with h_0 on its diagonal

Which gives the following circulant matrix:

$$\hat{H}_{1D} = \begin{bmatrix} h_0 & h_{-1} & \cdots & h_{-m} & 0 & \cdots & 0 & h_m & \cdots & h_1 \\ h_1 & \ddots & & & & & & & & \vdots \\ \vdots & \ddots & & & & & & & & h_m \\ h_m & & & & & & & & & 0 \\ 0 & & & & & & & & & \vdots \\ \vdots & & & & & & & & & 0 \\ 0 & \cdots & 0 & h_m & \cdots & h_1 & h_0 & h_{-1} & \cdots & h_{-m} \\ h_{-m} & & & & & & & & & \vdots \\ \vdots & & & & & & & & & h_{-1} \\ h_{-1} & \cdots & h_{-m} & 0 & \cdots & 0 & h_m & \cdots & h_1 & h_0 \end{bmatrix} \quad (5)$$

the matrix nxn with h_0 on its principle diagonal $\hat{H}_{1D}$ is easily diagonalizable by a Fast Fourier Transform (FFT).

2.2 2D signal case

In this case, the impulse response *RI* is a symmetric matrix in all sens and impair. The determinant of *RI* is zero for the general case non causal, to overcome the singularity one goes to construct a convolution matrix whose dimension is equal to the dimension of the image. Then the impulse response centered symmetrically in all direction feels them impair will be given by:

$$\begin{bmatrix} . & & h_m & & . \\ & & \vdots & & \\ h_m & \cdots & h_0 & \cdots & h_m \\ & & \vdots & & \\ . & & h_m & & . \end{bmatrix}$$

We suppose that *RI* operate line by line therefore the central line allows to construct H_0, the same manner as in the 1D case.

$$[H_m \quad \cdots \quad H_0 \quad \cdots \quad H_m]$$

Then the convolution matrix:

$$H_{2D} = \begin{bmatrix} H_m & \cdots & H_0 & \cdots & H_{-m} & 0 & \cdots & 0 \\ 0 & \ddots & & & & \ddots & & \vdots \\ \vdots & & \ddots & & & & \ddots & 0 \\ 0 & \cdots & 0 & H_m & \cdots & H_0 & \cdots & H_{-m} \end{bmatrix}$$

is the matrix (2m+1)xn and each block is(2m+1) xn.

Finally following Biemond [5] it is approached by the block circulant matrix which is it also circulant matrix.

$$\hat{H}_{2D} = \begin{bmatrix} \hat{H}_0 & \hat{H}_{-1} & \cdots & \hat{H}_{-m} & 0 & \cdots & 0 & \hat{H}_m & \cdots & \hat{H}_1 \\ \hat{H}_1 & \ddots & & & & & & & & \vdots \\ \vdots & \ddots & & & & & & & & \hat{H}_m \\ \hat{H}_m & & & & & & & & & 0 \\ 0 & & & & & & & & & \vdots \\ \vdots & & & & & & & & & 0 \\ 0 & \cdots & 0 & \hat{H}_m & \cdots & \hat{H}_1 & \hat{H}_0 & \hat{H}_{-1} & \cdots & \hat{H}_{-m} \\ \hat{H}_{-m} & & & & & & & & & \vdots \\ \vdots & & & & & & & & & \hat{H}_{-1} \\ \hat{H}_{-1} & \cdots & \hat{H}_{-m} & 0 & \cdots & 0 & \hat{H}_m & \cdots & \hat{H}_1 & \hat{H}_0 \end{bmatrix}$$

This matrix is the dimension nxn invertible diagonalizable with transformation by FFT. (Each block is construct as shown in the matrix (5)).

3. PROPPOSED TECHNIQUE

3.1 Optimal filter design

In this section, the signal u is supposed to be known. Then, by the knowledge of u, y_m and the variance of the noises, it is possible to define an optimal filter if its gains minimize a specified criterion representing the fidelity of the filtered signal to the ideal one. As a matter of fact, filtering is an operation used to reduce the noise component of a measured signal. Hence, the following criterion is defined by:

$$J = E\{(\hat{\underline{y}} - \underline{y})^t (\hat{\underline{y}} - \underline{y})\} \quad (6)$$

$E\{.\}$ is the average operator.
where the structure of the searched optimal linear filter is

$$\hat{\underline{y}} = F \underline{y}_m + G \underline{u} \quad (7)$$

Theorem: The minimization of criterion (6) leads to the following expression for the gain matrices F and G:

$$\begin{cases} F = (\hat{H}\hat{H}^t + \beta.I)^{-1} \hat{H}\hat{H}^t & \beta = \dfrac{\sigma_v}{\sigma_n} \\ G = (I - F)\hat{H} \end{cases} \quad (8)$$

Proof: The criterion (6) can be written as follows:
$$J = E\{(\hat{\underline{y}} - \underline{y})^t (\hat{\underline{y}} - \underline{y})\} = tr\{E(\underline{y} - \hat{\underline{y}})(\underline{y} - \hat{\underline{y}})^t\} \quad (9)$$
The error estimation is defined by
$$\underline{\varepsilon} = \hat{\underline{y}} - \underline{y} \quad (10)$$
Using (4), (7) and (10), we obtain the expression
$$\underline{\varepsilon} = [(F - I)\hat{H} + G]\underline{u} + [(F - I)\hat{H}]\underline{n} + F\underline{v} \quad (11)$$
By virtue the expression (11) the relation ship (9) becomes

$$J = tr\left\{ \begin{array}{l} [Z + G]E\{\underline{uu}^t\}[Z + G]^t + [Z]E\{\underline{n.n}^t\}[Z]^t \\ + FE\{\underline{v.v}^t\}F^t \end{array} \right\}$$

with $Z = (F - I)\hat{H}$ \quad (12)
The estimate should be unbiased, then is turns out that
$$E(\underline{\varepsilon}) = [(F - I)\hat{H} + G]E\{\underline{u}\} + [(F - I)\hat{H}]E\{\underline{n}\} + FE\{\underline{v}\} = 0$$
Optimal values of matrices F and G are such that

$$\dfrac{\partial J}{\partial G} = 0 \qquad \dfrac{\partial J}{\partial F} = 0 \quad (13)$$

From (12) and (13), we obtain:

$$\dfrac{\partial J}{\partial G} = 2[(F - I)\hat{H} + G](E\{\underline{uu}^t\})^t \quad (14)$$

So we obtain that $\hat{G}$
$$\hat{G} = (I - F)\hat{H} \quad (15)$$
By introducing (15) in the equation (12), we obtain a new expression for the criterion (10).
$$J = tr\{(F - I)\hat{H}]E\{\underline{nn}^t\}(F - I)\hat{H}]^t + FE\{\underline{vv}^t\}F^t\} \quad (16)$$
and from equations (11) and (12), we have :

$$\dfrac{\partial J}{\partial F} = 2[(F - I)\hat{H}](E\{\underline{nn}^t\})^t H^t + 2FE\{\underline{vv}^t\} = 0 \quad (17)$$

So
$$\hat{F} = [\hat{H}(E\{\underline{nn}^t\})^t \hat{H}^t + (E\{\underline{vv}^t\})^t]^{-1} H(E\{\underline{nn}^t\})^t \hat{H}^t \quad (18)$$
As
$$E\{\underline{nn}^t\} = \sigma_n^2 I \quad \text{and} \quad E\{\underline{vv}^t\} = \sigma_v^2 I$$
one obtains:
$$\hat{F} = [\hat{H}\hat{H}^t + \beta.I]^{-1} \hat{H}\hat{H}^t \quad (19)$$

3.2 Optimal estimate u

After designing an optimal filter in order to lessen the level of measurement noise, an algorithm giving the best estimate of the signal u has to be designed. In order to use the optimal linear filtering presented in the previous section, in an input restoration algorithm, the following technique is proposed. In the filtering step, the unknown input signal u is replaced by its estimate.

As it has been shown in [11, 12], the optimal estimate of u is such that:

$$\hat{\underline{u}} = \arg\min\left\{\left\|F\underline{y_m} - F\hat{H}\right\|^2 + \alpha\left\|L\underline{u}\right\|^2\right\} \quad (20)$$

where L is used to penalize the light-frequency spectrum of $\hat{\underline{u}}$ which results in a smoothing constraint on the restored signal; and α is the regularization parameter.

The optimal estimate of $\hat{\underline{u}}$ is given by [1]

$$((F.\hat{H})^t(F.\hat{H}) + \alpha.L^t.L)\hat{\underline{u}} = (F.\hat{H})^t F.\underline{y_m} \quad (21)$$

The equation (21) can be solved by direct inversion or with an iterative method.

4. EXAMPLES

In order to illustrate the performance of the proposed restoration technique, we applied our results for reconstruction of 1D and 2D signal.

4.1 1D signal case

We apply the described method to estimate the heat reaction rate $q_r(t)$ of acetic anhydride hydrolysis in a calorimeter.

Let us write the energy balance over a batch reaction calorimeter [8]:

$$\begin{cases} C_r \dfrac{dT_r}{dt} = UA(T_j - T_r) + q_r \\ \dfrac{dT_j}{dt} = \dfrac{D_j}{V_j}(T_{ji} - T_j) + \dfrac{UA}{\rho_j V_j C_j}(T_j - T_r) \end{cases} \quad (22)$$

Where T_r and T_j respectively denote the temperature inside the reactor and the jacket, A the heat exchange surface, T_{ji} the inlet jacket temperature, U the global heat exchange coefficient, C_r the reactor heat capacity, q_r the heat reaction rate, D_j, ρ_j, V_j and C_j respectively denote the jacket flow rate, fluid density, volume and fluid heat capacity.

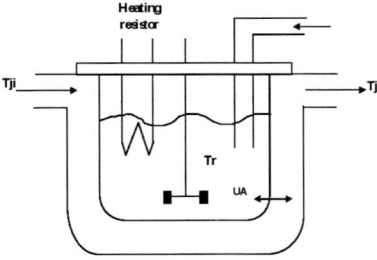

Fig.3: The reactor

We shall deal with the estimation of q_r from the measurements of T_r and T_j for a first order reaction. The thermal measure exploitation allows then to evaluate the rate of conversion of chemical reaction via the estimation of q_r in function of the time.

Hypothesis:
- Stirred-tank reactor
- The reactional volume is constant
- The jacket is perfectly agitated
- The heat transfers are defined by global coefficient heat transfer
- Neither the reactor nor the jacket does not exchange heat with the exterior

The system (22) becomes

$$\begin{aligned}\dfrac{dx}{dt} &= Ax + B_1.q_r + B_2.T_{ji} \\ y &= Cx\end{aligned} \quad (23)$$

where

$$x = \begin{bmatrix} T_r \\ T_j \end{bmatrix}, \quad B_1 = \begin{bmatrix} \dfrac{1}{C_r} \\ 0 \end{bmatrix}, \quad B_2 = \begin{bmatrix} 0 \\ \dfrac{D_j}{V_j} \end{bmatrix}$$

$$A = \begin{bmatrix} -\dfrac{UA}{C_r} & \dfrac{UA}{C_r} \\ \dfrac{UA}{\rho_j V_j C_j} & -(\dfrac{D_j}{V_j} + \dfrac{UA}{\rho_j V_j C_j}) \end{bmatrix}$$

and $C = \begin{bmatrix} 1 & 0 \end{bmatrix}$

In our approach we suppose UA and C_r are known by calibration method.

We can write:

$$T_r = k_1 * q_r + k_2 * T_{ji} + v \quad (24)$$

and

$$T_r - k_2 * T_{ji} = k_1 * q_r + v \quad (25)$$

where $*$ represents the convolution operator.

Let k_1 the transfer function from q_r to T_r and k_2 the transfer function from T_{ji} to T_r. Where k_1 and k_2 are respectively the impulse response of the both transfer function from q_r to T_r and from T_{ji} to T_r.

Let

$$T_s = T_r - k_2 * T_{ji} \quad (26)$$

We obtain:

$$T_s = k_1 * q_r + v \quad (27)$$

The previous equation (27) is well known as an convolution equation. Our goal is to estimate q_r from knowledge of T_s, k_1 and some informations about the error measurement v.

If we consider that there is a random signal n added to the input signal q_r.

We have

$$T_s = T_r + v \quad (28)$$

with

$$T_r = h * [q_r + n] \quad (29)$$

We apply the described method to estimate the heat reaction rate q_r from the pseudo temperature T_s (see equation (27)).

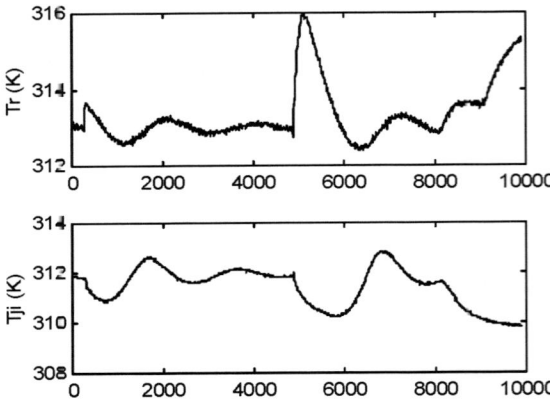

Fig 4: recorded signals T_r and T_{ji}

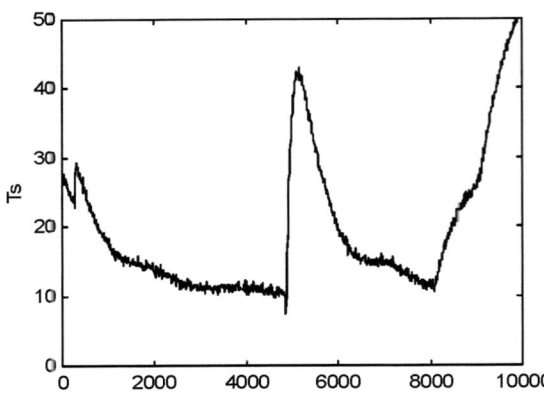

Fig 5: recorded signals T_s

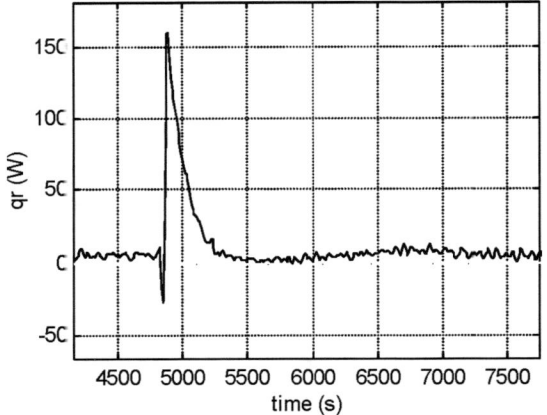

Fig 6 : Estimation of the heat reaction rate $q_r(t)$. with periodicity

Again, in order to illustrate the performances of the proposed method (fig. 6), The result of the restoration carried out without periodicity was bad, but interesting with periodicity. We have compared the corresponding results (fig. 6) with the heat reaction rate $q_r(t)$ resulting from the numerical derivation (fig. 7).

$$q_r = -k_0.n.\exp\left(\frac{-E}{R(T_r + 273)}\right)\Delta H_r$$

The expression of the heat reaction rate $q_r(t)$ has been chosen as follows :

$$\frac{dn}{dt} = -k_0.n.\exp\left(\frac{-E}{R(T_r + 273))}\right)$$

The heat transfer capacity UA and the reactor heat capacity C_r were assumed to be constant (UA= 6 W/K and C_r = 5000 J/K).

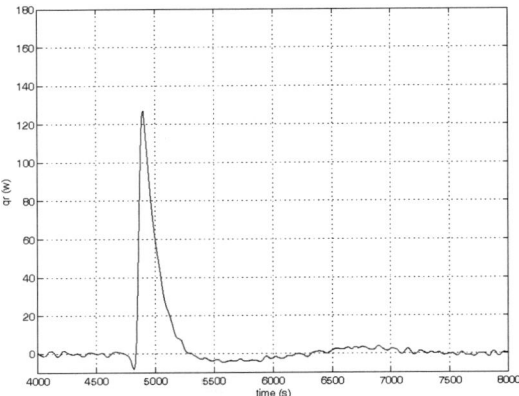

Fig 7: $q_r(t)$ obtained with numerical derivation
(Filtering with Butterwort filter)

Inconvenients of the direct method:

- The direct calculation presents an handicap at the equation 8, which necessitates a high memory space.
- It is not possible to find I (identity matrix) conform to the size of the image; because it exceeds the capacity of the machine (matlab).
- The product of the impulse response matrix by its transposed matrix modifies the size and the sum will be never executed (see equation 8).

Advantages of the periodicity:

- Eliminates the transposition of frequencies that is due to the utilization of FFT.
- Facilitates the calculation by diagonalization of the impulse response matrix.
- Possibility to construct a convolution matrix from the impulse response for any size of the image.

4.1 2D signal case

We applied our results for reconstruction of Lena image degraded by blur and noise SNR=10 dB.

We have used noniterative method to determine $\hat{U}$ from (21), because when the number of iterations increases, the noise effect can be amplified. At second we suppose that the regularization can be replaced by a direct inversion and the problem can be resolved.

Fig. 8:

Fig. 9:

Fig. 10:

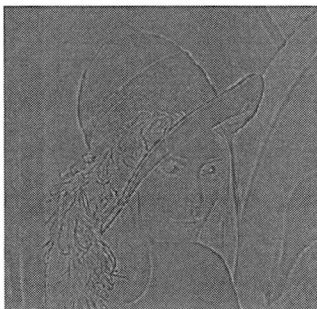

Fig. 11:

The fig.8 represented the original image; the fig.9 the image degraded by a blur and a noise, fig.10 the restored image, and the fig.11 the error of restoration. That represents the quality of the restoration. Noting that a test of restoration 2D without periodicity has given the weak result.

5. CONCLUSIONS

The new proposed technique can be applied to 1D or 2D signals. As a matter of fact, the causality property is never used.

REFERENCES

[1] Badeva, V., and V. Morozov (1991): Problème Incorrectement poses, Théorie et Application (Masson. (Ed)).

[2] Katsaggelos, A. (1991). *Digital Image Restoration.* New York, Springer-Verlay.

[3] P. Bolzern, P. Colaneri, and G. De Nicolao (1994) On computation of upper covariance bounds for perturbed linear systems, *IEEE Trans. On Auto Control*, **Volume 39, N°3,**pp.623-626.

[4] P. Bolzern, P. Colaneri, and G. De Nicolao (1996) optimal robust filtering with time-varying parameters uncertainty. *INT. J. Control*, **Volume 63, N°3,** pp.557-576.

[5] Biemond, J., Rieske Jelle, and Jean Gerbands. (1983). A Fast Kalman Filter for Images Degrades by BothBlur and noise in A.F. Round, editor. *IEEE Transactions on Acoustics Speech and Signal Processing*, **Volume ASSP, 31, N° 5.**

[6] Biemond, J. (1982). A Fast Kalman Filter for Images Degrades by BothBlur and noise. *IEEE Volume ASSP,* **31,** pp. 1146-1149.

[7] Blanc-Feraud, L., and al. (1988). Amélioration de la restauration d'images floues. *Traitement du signal*, **Volume 5, N° 4.**

[8] Fiaty, K., A. Accary, C. Jallut and O. Croulé. (1991). Calibration of a discontinuous calorimeter a set of specific transient experiments. *Termochimica Acta ,* **Volume 188,** pp. 191-200.

[9] T. H. Hopp and W. E. Schmitendorf (1990) Design of linear controller for robust tracking and model following. *Trans. Of the ASME, Volume* **12, N°3,** pp.552-528.

[10] Neveux, Ph, E. Sekko, and G. Thomas. (2000) A constrained iterative deconvolution technique with an optimal filtering: application to a hydrocarbon concentration sensor. *IEE trans. On Instrument and measurement ,* **Volume 49,N° 4,** pp. 852-856.

[11] E. Sekko, A. Boukrouche, P. Sarri, and G. Thomas. (1997) Comparaison de deux methodes de Restauration d'Images: Application à l'astronomie. *16eme Colloque GRETSI, 15-19 Septembre Grenoble*, pp. 837-839.

[2] E. Sekko, G. Thomas and A. Boukrouche. (1999) Deconvolution Technique Using Optimal Wiener Filtering and Regularization. *Signal processing*, **Volume 72, N°1,**pp.23-32.

[13] Tikhonov A.N. and V.Y. Arsenin. (1977): *Solution of ill-posed problem.* (New York, Winston, Wiley).

Telerobotics Using a Gestural Servoing Interface

H. Abdelmoumene and N.E. Berrached

*LAboratory of REsearch in Intelligent Systems,
USTO- BP. 1505 Oran El M'naouer –31 000 ORAN
E-mail : infvrf6@gmail.com, nasr1berrached@yahoo.fr*

Abstract: Man-Machine interaction through hand gestures is a rich, natural, and intuitive tool to control virtual and real environment. This paper proposes a vision-based hand gesture interface (VBHGI) to remotely control a robot arm through the web. A VBHGI requires real time and robust hand detection and gesture recognition. This recognition is carried out in three phases: acquisition, segmentation and identification of the hand posture. Since we are not using gloves or markers, we propose appropriate motion detection and segmentation. For the identification phase, we opted for principal component analysis in order to better represent the classes of gesture in reduced spaces. Once the gesture is recognized, it is analyzed to be used as an articulation command to remotely control a robot arm end effector.

Keywords: VBHGI, Telerobotics, Hand gesture recognition, Vision based command.

1. INTRODUCTION

In the area of robotics, current developments are used to make the Man-Machine interaction more intuitive and more natural. Gestural interaction using hand postures, is of interest due to its richness related to the number of degrees of freedom that involves the hand.

The implementation of a VBHGI for robot control requires the design of a system whose first phase is the detection and the acquisition of 2D or 3D hand images. In our application, we used a single Webcam coupled to computer vision and image (video) processing.

The second phase is hand segmentation to isolate the hand. The complexity of this phase is due to the scene complexity, and variability in backgrounds and lighting condition. This is accentuated in our application, since to make the interaction more natural, we used a bare hand without the use of gloves or markers to distinguish the hand from background. To do that, we have introduced a technique based on motion detection, considering that the hand is the only moving object in the scene with a fixed camera.

This technique is widespreading in several areas, and it is adapted to the objective of each application.

Most algorithms motion detection in the literature are presented as methods of removal of a reference image (background). The change in the implementation of these algorithms is presented in the proposed solutions for improving their results.

Among theses solutions; a step relaxation markovian is applied in [1] on noisy images resulting, to improve their quality.

Another solution offered by [2], is to apply a 5 × 5 Gaussian filter on the image difference.

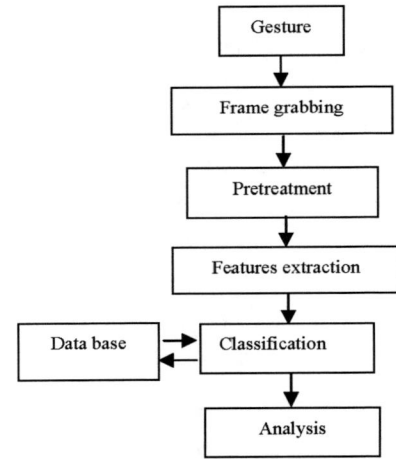

Fig.1. The proposed hand gesture recognition system.

While [3] proposes an improvement based on the Median and Average filter to eliminate remote zones. On the other hand [4] applied dilation followed by erosion on the result of subtraction.

The major drawback of the difference technique with background image is the need for the updating of the reference image, for example in case of light variation.

After image extraction from the background, the next step is the identification of the gesture by applying an appropriate pattern recognition method.

Many methods have been developed for such recognition. For example in [2], they distinguished between two types of gestures, pointing and pinch using the number of peaks obtained from the contour of the detected hand. [5] and [6] have used Freeman coding characteristics (gravity center and the surface) to recognize six gestures.

The approach used in [7] is based on three dimensions defined by the fist center hand gesture to identify four gestures.

In [8], the method is based on artificial neural networks, to recognize six gestures and [9] used Hidden Markov Models to recognize eight gestures.

[10] using fuzzy clustering (Fuzzy C-Means clustering) for the identification of 12 gestures.

In our application, we propose to use Principal Component Analysis to identify 12 postures dedicated to control the Mentor robot arm.

2. GOAL

So our goal, as we have mentioned before, is to achieve a gestural interface for the control of the robot Mentor, in teleoperation and in real-time. To do this, we must define a vocabulary of 12 hand gestures given in Fig. 2.

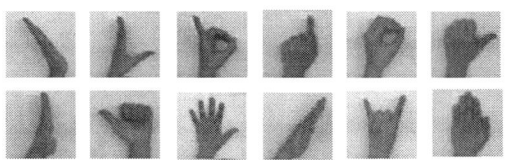

Fig.2. Gestural Vocabulary

The principle of our system is represented in the synoptic shown in Fig. 3.

The design of such recognition and analysis system requires the study of several fundamentals problems, whose difficulties are mainly:

- The detection and segmentation of the hand, without the use of gloves or markers of the 2D scene.
- The gesture identification, while considering time of execution.
- The analysis of the last commands to the robot.

For the segmentation step, we used an algorithm for motion detection to separate the hand from the background, principal component analysis is the method proposed for the stage of recognition, and we will explain a method of command for the robot based on the position of the effector by direct and inverse geometric models. Much later, we will show some results, as well as the assumptions taken into account.

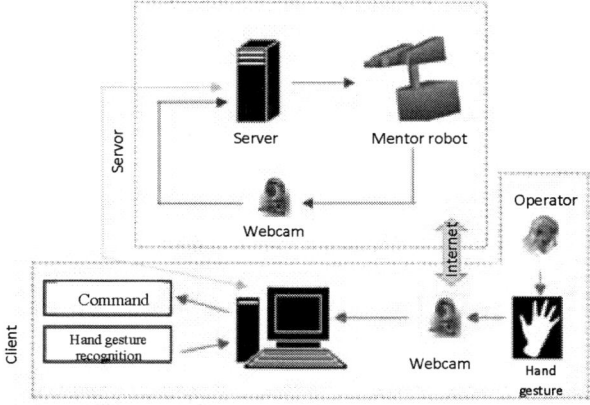

Fig.3. Control system of the Mentor robot in teleoperation

3. IMPLEMENTATION

3.1. Motion detection method :

As we mentioned previously, the movement is defined by temporary changes between two different times. At each image in the sequence is subtracted images known as reference image (image from the background). These difference images are very noisy, which is why it is useful to carry out a filtering step. We have applied an average filter.

The principle of this method is shown in the following diagramm:

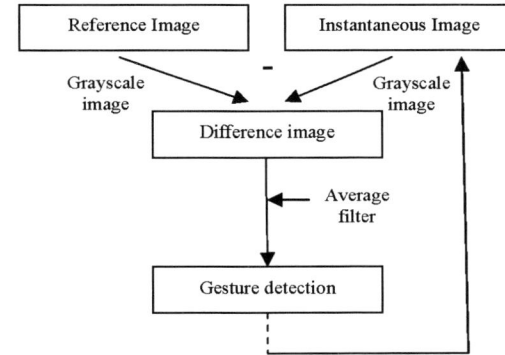

Fig.4. Motion detection method

The disadvantage of this method is the updating of the reference image. In our application, updating the background is done manually, because the automation of this process requires a journey, which will penalize our processing system in time and this is not in our interest because the conduct of our system is in real time.

3.2. Hand gesture recognition :

This step is to identify the hand tracked configuration, by the movement, and to assign it to a class among the 12 gestures classes.

The approach taken in this achievement is the principal component analysis for the gestures classification because it allows:

- To give interesting results in a reasonable time,

- To reduce in a very important way initialization

3.2.1. Learning phase:

3.2.1.1. Construction of the image database:

The first phase is the construction of the database of images to make learning. To do this, we took for each hand configuration, specified in our vocabulary gesture, 10 binarized images, standardized, normalized and reduced on size 32×32. Images in the same class have the same configuration with different appearances.

3.2.1.2. Covariance matrix:

This conducted to a matrix of N vectors l_i of $N \times M$ dimensions, of M images of light intensity, represented in the columns form. Each vector l_i is standardized by the subtraction of the average image $\bar{l}$ such as:

$$\bar{l} = \frac{1}{m}\sum_{i=1}^{i=m} l_i \qquad (1)$$

The standardized images provide a new matrix B such as:

$$B = [\hat{i}_1, \hat{i}_2, \hat{i}_3, \ldots, \hat{i}_m] \qquad (2)$$

The $N \times N$ covariance matrix is defined by:

$$Cov = \frac{1}{m} B.B^t \qquad (3)$$

After computing the covariance matrix, it is necessary to calculate the eigenvalues and the eigenvectors by solving the following linear system:

$$Cov.U = U.\lambda \qquad (4)$$

Where U represents the N eigenvectors matrix and λ is the diagonal matrix of eigenvalues.

3.2.2. Classification:

Classification consists in projecting the new configuration in all spaces and calculating the distance to these spaces, while seeking the index of the class which minimizes the error of reconstruction for all the class Ci :

$$i = \arg_i \min e_i \qquad (5)$$

Knowing that:

$$e_i = \left\| T^*(T(l)) - \bar{l} \right\| \qquad (6)$$

$$\phi = T(l) = E^t(l - \bar{l}) \qquad (7)$$

$$T^*(\phi) = E.\phi + \bar{l} \qquad (8)$$

The transformations T and T* denote projection and partial reconstruction of an image.

4. MENTOR ROBOT ARM CONTROL

After identifying the gesture tracked meaning, we used it to control the robot. The first proposal was to assign to each articulation two gestures, the first to increase the rotation angle and the second to decrease it by a fixed step. So the we have an articular command in this case. See Fig. 5

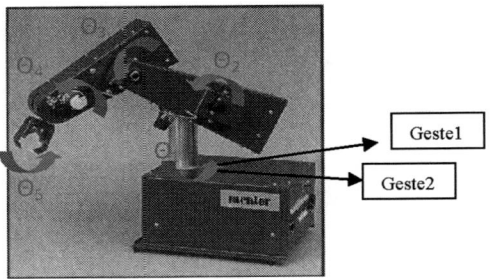

Fig. 5. Articular Command

For instance:

Gesture1 and gesture2 are awarded to the first articulation and θ_1 is its rotation angle:

If gesture1 is recognized then $\theta_1 = \theta_1 + step$.

Else If gesture2 is recognized then $\theta_1 = \theta_1 - step$.

This solution is too slow, and does not reflect the natural look of the command.

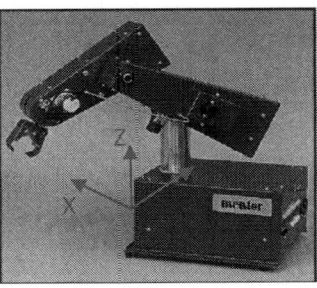

Fig. 6. Operational Command

The second proposal is based on an operational command. See Fig. 6. In this case, each axis of the operational space, was controlled by two gestures, the first gesture is to move one direction and the second gesture for the opposite direction. As for the movement of the gripper (elevation, rotation, opening and closing), we kept the same side that previous solution, as shown in the following representation:

For instance:

Gesture1 and gesture2 are awarded to the first axis of the effector over base:

If gesture1 is recognized then $X = X + step$.

Else If gesture2 is recognized then $X = X - step$.

Indeed, it is obvious to calculate the position of the terminal organ. The appropriate model which allows us to perform this calculation is the inverse geometrical model; this model involves calculating the direct geometrical model.

The geometrical models are based on the Denavit – Hartenberg convention [11].

4.1. The direct geometrical model (DGM):

The DGM allows to obtain the coordinates (x, y, z) position of the effector over base, according to the parameters of translations and/or rotations given by the system.

$$T(i-1)(i) = \begin{bmatrix} Cos\theta_i & -Sin\theta_i & 0 & a_{i-1} \\ Sin\theta_i Cos\alpha_{i-1} & Cos\theta_i Cos\alpha_{i-1} & -Sin\alpha_{i-1} & -r_i Sin\alpha_{i-1} \\ Sin\theta_i Sin\alpha_{i-1} & Cos\theta_i Sin\alpha_{i-1} & Cos\alpha_{i-1} & r_i Cos\alpha_{i-1} \\ 0 & 0 & 0 & 1 \end{bmatrix} \quad (9)$$

The position of the final organ of robot manipulator is described by the following equation:

$$X_i = f(q_i) \quad (10)$$

With x_i: operational variable (position and orientation).
      q_i: articular variable

4.2. The inverse geometrical model (IGM):

The inverse geometrical model consists on calculating the articular coordinates which bring the final organ in a wished position, specified by its operational coordinates. The inverse geometrical model of therobot arm is written as:

$$X_i = f^{-1}(q_i) \quad (11)$$

With: f^{-1}: the inverse of function f.

5. SYSTEM DESCRIPTION

We installed our application on a Client station, and a Server application on a station which is connected to the Mentor robot. The hand is segmented by the motion detection method that is based on the subtraction of the background image. See Fig. 7.

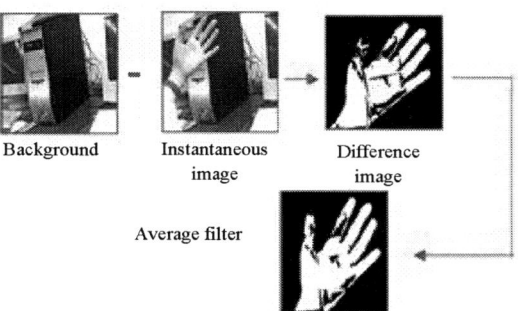

Fig.7. Hand Acquisition.

The principle of the real time Mentor's robot control is based on the gestural identification, which is done by projection on 12 spaces. This makes the execution expensive in term of time (including the various image processing time).

We fixed a threshold error T in order to avoid any confusion between gestures. The following step is to translate the gesture into an order according to the associated gesture for each axis from operational space. See Fig.8.

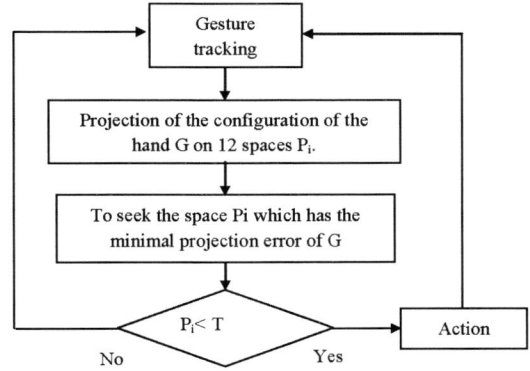

Fig.8. Flow diagram of the robot control.

Control of the robot is done by the operator who perceives the robot end effectuor position (through the visual feedback), and send a command using a hand gesture which is interpreted using the learned gestural vocabulary. The identification of the posture creates a specification of the axis of articulation or concerned by the gesture, as well as its direction of motion. So, the gesture information and returns received from the server generate a new command to the robot arm. But, according to our second proposed solution, we must calculate the direct geometrical model, in order to take the position of the terminal organ.

Then, in order to be transmitted to the robot, the information must be in the form of rotation angles, as this is the way exploitable by the arm through a RJ45 connection. So, we need to calculate the inverse geometrical model of the robot arm.

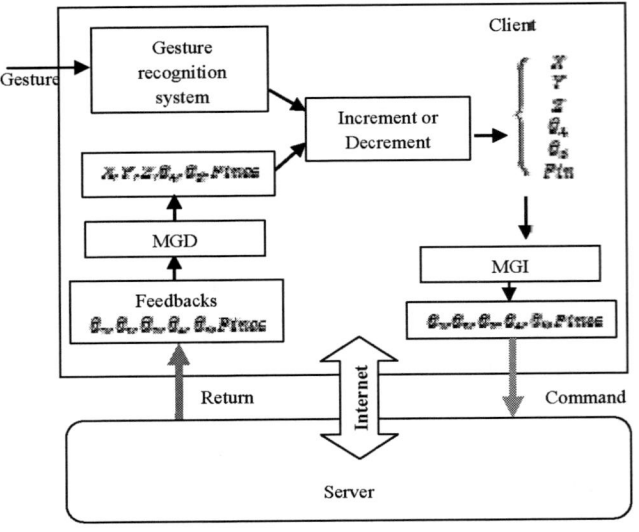

Fig. 9. Control system.

6. RESULTS

We designed interface, developed in Builder C++, for the Mentor robot in teleoperation and in real time. Initially, as shown below, we tested our system in a virtual model of the robot designed in OpenGl, and then we made control in reality.

Entering object is an elementary task of robotics and several works that are involved in this field, mainly as regards the role of assisting the disabled.

To perform this task in loop controlled by the gesture of the hand, we realized the following loop: Fig 10

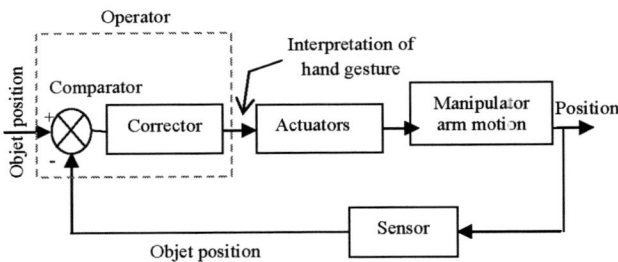

Fig.10. Gestural servoing system.

The results of our application concerning both virtual and real models of the robot, were almost similar, the only difference is in the type of transmission used for the data transfer to the robot. This data transfer is direct as it is the case in the virtual control on the same station, and it is made through a Server in the real case.

During segmentation, we had good results with a reasonable speed except in a few cases. Segmented or object does not reflect a pattern consistent and it was due to lighting and shadows on the back of other objects or even for the hand itself.

The following diagram (Fig. 11) includes all the units used in the realization of such an interface, as well as the applications used such as the software "Webcam32", for the visual return and a FTP Servre, "TYPSoft ftp server" for its transmission.

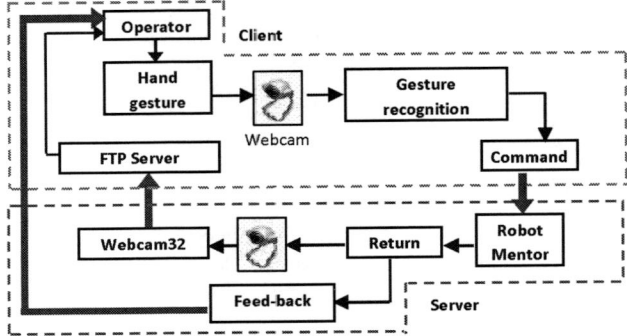

Fig. 11. Command System of robot Mentor.

We tested the classification on 480 examples, containing examples of the database and examples of 6 different hands. We could obtain the following recognition rate of each gesture:

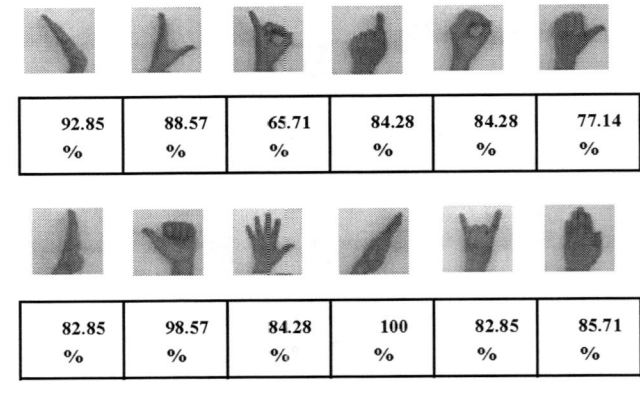

7. CONCLUSION

We have designed a a VBHGR interface to remotely control through the web a robot arm. Using hand gestures provides a natural communication interface between humans and robots. The results of the system were encouraging.

The inclusion of themotion has improved the performance of our system.

The principal components analysis was used for the gestural identification. We had satisfactory results reaching 85.59% for the average recognition rate.

REFERENCES

[1] J. Martin. (2000). « *Reconnaissance de Gestes en Vision par Ordinateur* », Ph.D, INP Grenoble, France,

[2] M. Shahzad, « *Real-time Hand Tracking and Finger Tracking for Interaction* », CSC2503F Project Report, 2003,

[3] R.Ramdani, (2003), «*Détection et Classification des objets en mouvement dans une séquence d'images* », University of Sciences and Technology of Oran, Algeria,

[4] M. Ekinci, E. Gedikli, (2005), «*Silhouette Based Human Motion Detection and Analysis for Real-Time Automated Video Surveillance*», Dept. of Computer Engineering, Karadeniz Technical University, Trabzon, TURKEY, Turk J Elec Engin, VOL.13, NO.2 2005.

[5] A. Ahmed Blaha et L.Bouhalassa, (2006), «*Commande gestuelle d'un bras manipulateur*», University of Sciences and Technology of Oran -LARESI, Algeria,

[6] M. Zalegh et A. Kerroumi, (2006), «*Asservissement gestuel d'un robot virtuel*», University of Sciences and Technology of Oran -LARESI, Algeria,

[7] N. Kawarazaki, T.Yoshidime, and K. Nishihara, (2002), «*An Assistive Robot System Using Gesture and Voice Instructions*», 2nd Cambridge Workshop on Universal Access and Assistive Technology 1Dept. of Welfare Systems Engineering, Kanagawa Institute of Technology, Japan.

[8] G. Heidemann, H. Ritter, (2003), «*Learning to Recognise Objects and Situations to Control a Robot End-Effector*», Kunstliche Intelligenz, 2:24-29, 2003,

[9] G. Heidemann, H. Ritter, «*Real time hand tracking and 3d gesture recognition for interactive interfaces using hmm*», Computer Engineering Dept. Boğaziçi University.

[10] W. Juan and K. Uri, (2001), « *Hand Gesture Telerobotic Systems using Fuzzy clustering algorithms* », Ben Gurion University, Israel.

[11] E. Dombre et W. Khalil, (1988) «*Modélisation et commande des robots*», Ed. Hermes.

Driving Forces Study in the Ultrasonic Motor

M.Djaghloul*[a], Z.Boumous*[b], S.Belkhiat*[c].

* Ferhat Abbas University, Electrotechnic Department, DAC laboratory, Sétif, 19000, Algeria.
*[a](e-mail:Djaghloul_mehdi@yahoo.fr).
*[b](e-mail: Zsid3@yahoo.fr).
*[c](e-mail: Belsa_set@yahoo.fr).

Abstract: This work presents an overview on ultrasonic motors of type traveling wave (TWUSMs). The mathematical model for the TWUSMs is developed where two additional cases, neglected in the literature, concerning the dynamic behavior of the rotor and the stator, are detailed. The major utility of this study is to detail the various additional cases, describing more the functional work of the motor, by exposing particular cases which represent our contributions and also by putting in value our results of simulation, implemented and realized in the environment Matlab/Simulink. The refined model resulting of this study is implemented on software Matlab/Simulink. This mathematical model describes the dynamic behavior of ultrasonic motor AMW90. The implementation is based on the aptitude of the different blocks of simulation in to be adapted according to our needs. The obtained simulation results in comparison with manufacturer measures are satisfactory, and confirm the robustness of the refined model.

Keywords: Implementation on Matlab/Simulink, Mathematical model, Forces in ultrasonic motor, Control of USM motor, Traveling wave ultrasonic motors (TWUSMs).

1. INTRODUCTION

As novel motors with a new principle, new mechanism and new material, ultrasonic motors (USM) is a special type motor which is driven by the ultrasonic vibration force of piezoelectric elements. It has an excellent performance and many other useful features that traditional electromagnetic ones do not possess, in particularly the traveling wave ones [1].

This type of ultrasonic motor has excellent functional characteristics. The strong torque of maintains without alimentation , the high torque at low speed (supporting the direct drives), the absence of parasitic magnetic fields, and its small size make of this motor an ideal actuator [2].

These motors have large application domains, as simple actuator in automobiles windows, as precise actuator in Canon camera, and recently it has been used as direct drive actuators for articulated robots, control valves and a positioning table of machine tools. All these applications require a quick response and precise position control of actuators which can be found in this motor [1], [3], [4].

Several studies, considers these motors as application domain, and try to develop estimating of their different characteristics, in order to find the optimal ones, and to make of these motors an ideal actuators for various applications [5],[6],[7].

In this view way, the constructors test their products by experimental tests, and hope to interpret the experimental results on mathematical model, which describe exactly the functional behaviors of their products [8],[9].

The search for a mathematical description proves very difficult and complex, but at the same time necessary. For describing better the functional behavior of these motors, this model must represents a partial and total reflection of the dynamic associated to them [5],[6].

The tribological phenomena and the high frequency vibrations acting in the contact area between stator and rotor are the base of the principle of work, which make the presence of the strongly nonlinear work characteristics, and moreover, vary in time. The concepts of transfer function, spaces states, very useful in the dimensioning of the regulators, can't be applicable generally to our system [5],[10].

These phenomena are only the physical interpretation of the various effects of the structure components in the motor, which can be mutually between them or on their works and their influence on functional description [10].

The tribological influences of the phenomena are concretized by the manner of contacts and its interpretation by the existence of the points of calculates (x_l, x_r)[11].

A multitude of works study the influence of these points on the forces existence, their appearance and calculate, by the distingue of four studied essential cases [12],[5],[6]. The majorities of the studies neglects or eliminate other cases which characterize, better, the dynamic of the motor.

About the control of this kind of motor, and as it represent a non linear system, a particular type of control is used which is a nonlinear control based generally on an artificial intelligence approaches. Some references for these approaches are given in the section three.

This work presents an overview on ultrasonic motors of type traveling wave (TWUSMs). The mathematical model for the TWUSMs is developed where two additional cases, neglected in the literature, concerning the dynamic behavior of the rotor and the stator, are detailed. The major utility of this study is to detail the various additional cases, describing more the functional work of the motor, by exposing particular cases which represent our contributions and also by putting in value our results of simulation, implemented and realized in the environment Matlab/Simulink.

We exposed, so in this paper, in first the principle of function and their general structure. The detailed studies on the various cases of the emerged forces and their evolution are presented in the second section. Finally in thirst we exposed, some references concerning the proposed control approaches in the literature, and the simulation results which are compared with manufacturer measures.

2. FUNCTIONAL DESCRIPTION OF COMPONENT BEHAVIOUR IN THE MOTOR

The model of the traveling wave ultrasonic motor used in this study is composed of four principal modules or blocks:

- Excitation or the voltage source,
- Piezoelectric ceramic stator,
- Stator-rotor interface contact,
- Rotor.

In the fallowing sections we propose more detail on each blocks having influence on the dynamic of the motor.

2.1 Stator

The stator is the part of the motor where the electromechanical conversion of energy will take place. Applied piezoelectric ceramics, constituting the stator, according to its deformation produces the traveling wave.

The stator model may be mathematically represented as an input/output system, which has as input variables:

- Electric tensions of excitation,
- Modal forces of the stator-rotor interface,

And as outputs:

- Modal displacements,
- Temporal derivatives of modal displacements.

The equation (1) represents the state representation describing the stator behavior [13],[14].

$$\begin{bmatrix} \dot{w}_1 \\ \ddot{w}_1 \\ \dot{w}_2 \\ \ddot{w}_2 \end{bmatrix} = \begin{bmatrix} 0 & 1 & 0 & 0 \\ \frac{-c_{s1}}{m_{eff}} & \frac{-d_{s1}}{m_{eff}} & 0 & 0 \\ 0 & 0 & 0 & 1 \\ 0 & 0 & \frac{-c_{s2}}{m_{eff}} & \frac{-d_{s2}}{m_{eff}} \end{bmatrix} * \begin{bmatrix} w_1 \\ \dot{w}_1 \\ w_2 \\ \dot{w}_2 \end{bmatrix} + \begin{bmatrix} 0 & 0 & 0 & 0 \\ \frac{A_1^*(1-\varepsilon_1)}{m_{eff}} & \frac{A_1^*\varepsilon_2}{m_{eff}} & \frac{1}{m_{eff}} & 0 \\ 0 & 0 & 0 & 0 \\ \frac{A_2^*\varepsilon_1}{m_{eff}} & \frac{A_2^*(1-\varepsilon_2)}{m_{eff}} & 0 & \frac{1}{m_{eff}} \end{bmatrix} * \begin{bmatrix} U_{p1} \\ U_{p2} \\ F_{s1} \\ F_{s2} \end{bmatrix}$$

$$\begin{bmatrix} y1 \\ y2 \\ y3 \\ y4 \end{bmatrix} = \begin{bmatrix} 1 & 0 & 0 & 0 \\ 0 & 1 & 0 & 0 \\ 0 & 0 & 1 & 0 \\ 0 & 0 & 0 & 1 \end{bmatrix} * \begin{bmatrix} w_1 \\ \dot{w}_1 \\ w_2 \\ \dot{w}_2 \end{bmatrix} + \begin{bmatrix} 0 & 0 & 0 & 0 \\ 0 & 0 & 0 & 0 \\ 0 & 0 & 0 & 0 \\ 0 & 0 & 0 & 0 \end{bmatrix} * \begin{bmatrix} U_{p1} \\ U_{p2} \\ F_{s1} \\ F_{s2} \end{bmatrix} \quad (1)$$

2.2 Stator-rotor interface

The model of the interface stator-rotor is the most complex part in the ultrasonic motor model. It is supposed that the stator is rigid and its vibration profile does not change after the contact with the rotor, knowing that this one has a conform layer of contact [5],[14].

This part is the one where the interne functional behavior is assured by the existence of some forces [13]. These forces depend on comparison between the displacements speeds of the stator and the rotor respectively [14].

One of manners to describe the mechanics of contact is to employ the model of contact zone showed in figure1 [5]. This model supposes that the stator is rigid and the rotor has a contact layer specified as a linear spring with an equivalent rigidity in the axial and tangential direction c_N [15].

We present a new system (two freedom degrees) having the same coordinates, which moves with the speed and in the same direction of the traveling wave in the equations (2) and (3).

$$\tilde{x} = x - \frac{\omega}{k}t \quad (2)$$

$$\tilde{z} = z \quad (3)$$

The velocity distribution of points of the surface stator is expressed by formula (4):

$$v_{hor}(\tilde{x}) = v_0 \cdot \cos(k\tilde{x}) \quad (4)$$

In the contact layer, the stator exerts a normal pressure along of the surface contact. Figure 1 and 2, showed the overlapping between the stator and the rotor, and the forces details in the stator-rotor interface [13]. The layer contact is expressed by equation (5):

$$\Delta W = W_0 \cdot (\cos(k\tilde{x}) - \cos(kx_k)) \quad (5)$$

x_k is the half of length value of the contact surface.

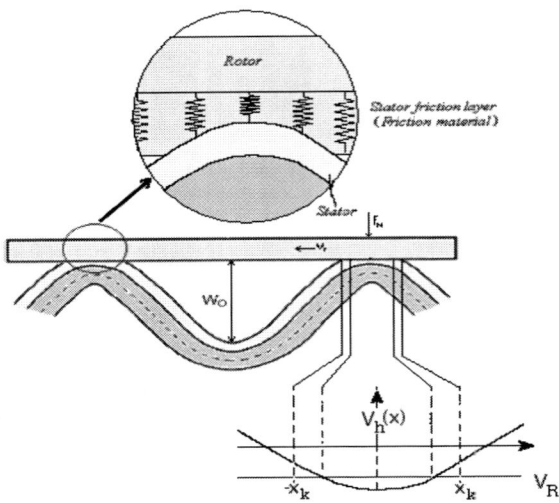

Fig.1. Overlapping between the stator surface and the contact layer of the rotor.

The force field according to [12] is expressed as follows:

$$f_z = c_N \cdot \Delta W = c_N \cdot W_0 \cdot (\cos(k\tilde{x}) - \cos(kx_k)) \quad (6)$$

c_N is the equivalent stiffness of the contact layer, [15].

The normal force of the working rotor can be calculated by integrating of f_z corresponding to the contact zone:

$$F_n = n \int_{-x_k}^{x_k} f_z(\tilde{x}) d\tilde{x} \quad (7)$$

n represents the peaks number of the wave.

Comparing the speed of points of the stator surface in along the contact zone with the speed of the rotor, we can see that there are zones where the speed of the rotor is higher than the speed of points of stator surface and zones where the speed of the rotor is lower, and also points where these two speeds are equals (no-slip point).

Since Coulomb's friction is considered as friction force (F_{antr}); this one can be expressed by equation (8):

$$F_{antr} = n\mu \int_{-x_k}^{x_k} \text{sign}(v_{hor}(\tilde{x}) - v_R) \cdot f_z(\tilde{x}) d\tilde{x} \quad (8)$$

The direction of this force depends on sign of the relative speed between the stator speed the rotor one. The tangential forces of stator acting on rotor are exposed in Figure 2.

The friction forces active on the rotor in the zones where the speed of rotor is higher than speeds of the external points of the stator which are directed in the same direction as the force of load (torque). Thus, they contribute to the slowing down effect of the rotor. The force of traction F_{antr} depends on breaking zones (surface of negative sign).

From the point of change of the sign (figure 2) the force of friction can be of the traction or braking type. The points where the speed of the rotor is equal to the speed of the surface of the stator are called the no-slip points x_L, x_R.

The position of these points determines the width of the traction zone (of positive sign), and depends on the load applied to the rotor. The traction torque must have the same value as the applied load torque, in order to maintain equilibrium.

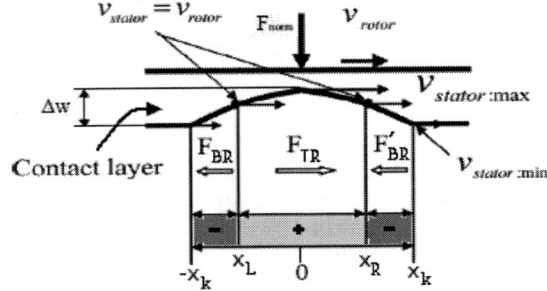

Fig.2. Breaking and traction forces in the contact zone.

To study the general case of the traveling wave, a whole speculation is realized [14], which we benefit of the results and formulas [5],[14].

Since force calculates is based primordially on the terminals in associated integral, like showed in equation (8), we must imperatively define them to assure the evolution of system in time, these terminals are the x_L, x_R and x_k.[6],[13].

Forces existence in the stator-rotor interface and their study

Different case of forces existence emerges in the behavior study of stator-rotor contacts and which are interpreted by the position of the no-slip points in the length of the contacts zone (- x_k to x_k).

Four cases are distinguished in the literature [5],[6],[8]:

- All the two no-slip points are placed at interior of the contact zone ($x_L <= x_k$, and $x_R <= x_k$),
- The left no-slip points is outside and the right no-slip points is inside the zone of contact ($x_L > x_k$ and $x_R <= x_k$),
- The traction zone is on the right and the braking one on the left side ($x_L <= x_k$, and $x_R > x_k$),
- Only the traction zone exists in the whole of contact zone ($x_L > x_k$ and $x_R > x_k$).

Every one of the exposed cases is associated to a distribution of portions, indicating the existence and the direction (sign) of forces, and imposing its effect in traction or braking influence [5],[14], practically by identical manner of the representation in figure 2.

Particular cases

All the studies made on the forces existence cases in the contact zone, in the literature [5],[6],[8] neglected completely two cases, which don't even depend on the position of the no-slip points nor of their existence.

In our previous work we have deduce two additional cases to complete the combination of the cases, which depend on the speed of rotation and of the stator points one, by associating a whole mathematical description [13].

These two cases characterize the forces state in permanent working mode of the ultrasonic motor. Existence of these two cases doesn't depend on the non slip points. They are defined as follows:

- When the speed of rotor can be higher than the maximum horizontal speed of the stator surface points above the whole contact zone, we have only braking forces.
- When the speed of rotor can be lower than the minimal horizontal speed of the stator surface points above the whole contact zone, we have only traction forces.

The intervening forces, associated to these cases are:

Driving force for the first case noted (BR):

$$(BR)F_{antr} = -nc_N W_0 \mu \cdot \xi(x_k) \quad (9)$$

Feedback tangential forces for (BR) case:

$$(BR)F_{fbtg} = -2n\mu a \frac{c_N}{k}[\Xi_2(x_k)] \quad (10)$$

And for the second case noted (TR):

$$(TR)F_{antr} = -(BR)F_{antr} \quad (11)$$

$$(TR)F_{fbtg} = (BR)F_{fbtg} \quad (12)$$

With simplification equations (13), (14) and (15):

$$[\Xi_2] = \begin{bmatrix} W_{2m}\xi_1 \\ -W_{1m}\xi_1 \end{bmatrix} \quad (13)$$

w_{1m} and w_{2m} are modal displacements.

$$\xi_1(\tilde{x}) = -\frac{1}{4}\cos(2\tilde{x}) + \cos(kx_k)\cos(2\tilde{x}) \quad (14)$$

$$\xi(\tilde{x}) = \sin(k\tilde{x}) - k\tilde{x}\cos(kx_k) \quad (15)$$

The flowing figure (Figure 3) represents the second case (just driving forces):

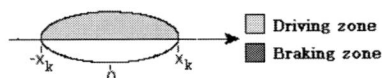

Fig.3. Only traction forces.

2.3 Rotor

The rotor of the studied ultrasonic motor is modeled according to two freedom degrees:

- Vertical movement
- Rotation

The space-state representation (16) describes the vertical move of the rotor. F_{norm} is the applied axial force, m_{rot} is the equivalent mass of the rotor, d_{rot} and c_{rot} are the equivalent dumping and stiffness in the rotor (figure.4).

$$\begin{bmatrix} \ddot{w}_{rot} \\ \dot{w}_{rot} \end{bmatrix} = \begin{bmatrix} \frac{d_{rot}}{m_{rot}} & \frac{c_{rot}}{m_{rot}} \\ 1 & 0 \end{bmatrix} \cdot \begin{bmatrix} \dot{w}_{rot} \\ w_{rot} \end{bmatrix} + \begin{bmatrix} \frac{1}{m_{rot}} \\ 0 \end{bmatrix} \cdot (F_n - F_{norm})$$

$$y = \begin{bmatrix} 0 & 1 \end{bmatrix} \cdot \begin{bmatrix} \dot{w}_{rot} \\ w_{rot} \end{bmatrix} + [0] \cdot (F_n - F_{norm}) \quad (16)$$

The rotational move of the rotor is described by a simple transfer function (17):

$$G_{ra}(s) = \frac{w_{rot}}{M_{antr} - M_{Load}} = \frac{1}{J_{rot}S} \quad (17)$$

M_{antr} : Traction torque.
M_{load} : Load torque.

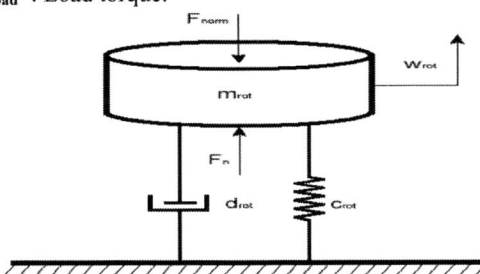

Fig.4. The equivalent mechanical model for the vertical move of the rotor.

3. CONTROL APPROACHES

In this section we will interest to the functional aspect of the USM motor and his influence on its application domain, and the new needs to use this type motor, by presenting its control approaches.

3.1 Domains application

An USM is a newly developed motor. It is a device that transforms vibration and wave motions of solids into progressive or rotational motions by means of contact frictional forces. The USM has some excellent performances and useful features, such as high torque at low speeds, compactness in size, no electromagnetic interference, short start–stop times, and many others [1],[5]. Owing to the advantages mentioned above, the USM has been used in many practical applications [3][4], such as MEMS, robots [16][17], medical instruments, cameras and aeronautics.

All this domains application reflect the precise use aspect of this type of motor, then a good modeling and robust control is a necessity.

3.2 Model of TWUSM motor

The traveling wave motor is generally described as a multi variable system, with four inputs (frequency of control, amplitudes of the two excitation electric voltages, difference in phase between these voltages), and two outputs (speed and torque) [14].

But due to piezoelectric characteristics source of functioning principle of the ultrasonic motor which are strongly nonlinear and moreover vary as a function time because of drifts of some parameters, which change, with the increase of the motor working temperature. The concepts of transfer function, state spaces, very useful for the regulators dimensioning, is thus not applicable generally to our system, because it is nonlinear and change in time [5].

This view pushes us to think about a particular kind of control, which considers our system just as a black box and tack care of it nonlinear behavior.

3.3 Short overview on control approaches for USM motor

The TWUSM is a peculiar motor whose driving principle is different from that of other electromagnetic-type motors, It has strong nonlinear characteristics.

In recent years, some models of the USM have been proposed [5],[9] with their simulation [15], but most models are too complex to apply to practical applications. Therefore, it is also difficult to control the USM using the conventional algorithms.

Several approach control for Usm motor exist in the literature [18] , [19] , [20] , [21] , [23] , [24] , [25] , [26] ,[27], based on identification or on direct control and with different manner of implementation. Ones use electrical approach like (Güngör Bal) works [19],[20] , other use artificial intelligent approaches (X.Xu)[21], (KenjiroTakemura) [22], (Hong-Wei Ge) [23], (Y.C.Liang) [24], (Zhangfan) [25] and other combine the two kind approaches [26].

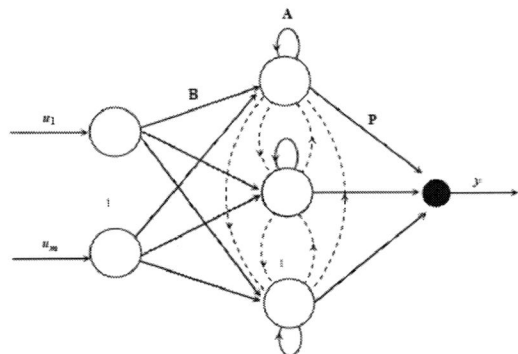

Fig.5. One of Architecture recurrent neural network.

Our interest type of control is the neural network based intelligent artificial control which represent an ideal control approach for our system [21] ,[27] and satisfies our condition of implementation on Matlab/Simulink software.

For a simple description, this approach is based on a recurrent neural network [21], may be showed as a network in figure 5,'ui' are the input of the network and 'y' represents its output.

4. SIMULATION AND DISCUSSION RESULTS

After the exhibition of the used analytical model, with mathematical approaches, our task is to implement this model on Matlab/Simulink, by benefiting from the technique of modularity granted in this software, to treat the refined and reduced version of the equations of each part in the motor.

All noted parameter used to ensure the good simulation of the refined model [14], based in general on the proposed model in [5],[6], are gathered in the table 1.

This part of study is based on our previous work on the TWUSM motor [13],[14].

Table 1. Parameters of AMW 90-X Motor [6],[14]

NAME	Symbol and value
Resistances of entries	Rp_1= 5 Ω Rp_2= 5 Ω
Ceramics capacity	Cp_1= 7.8e-9 F Cp_2= 7.87e-9 F
Capacity of stator	Cp_{s1}= 0.42e-9 Cp_{s2}= 0.428e-9
Inertia of Rotor	J_{Rot}=3.4367e-004 Kgm^2
Ray	R_w = 40.5e-3 m
Effective mass	m_{eff} = 40.5 Kg
Mass of rotor	m_{Rot} = (m_{eff}+22.8+3)*1e-3 Kg
The rigidity of Rotor	c_{Rot}=300e3 N/m
Attenuation of Rotor	d_{Rot}= 50*1e3 Ns/m
Coefficient of Coulomb of friction	μ = 0.21
The distance enters the points of surface of stator	a=4.5e-3 m
Peak of wave numbers	n=11
Frequency of Resonance	Wres2=wres1 wres1=2*pi*43.365*1e3 Hz
Wavelength	λ=2*pi*R_w/n m
Wave numbers	k=2*pi/λ
The rigidity of the zone of contacts	c_N=8500*1e6 N/m^2
Frequency of Antiresonance	W_{ant2}=W_{ant1} W_{ant1}=2*pi*43.425e3 Hz
Modal mass of Stator	m=0.082 Kg
Report/ratio of transfer	A1=($m*C_{p1}*(w_{ant1})^2-w_{res1})^2))^{1/2}$(kgFs^{-2})$^{1/2}$ A2=($m*C_{p2}*(w_{ant2})^2-w_{res2})^2))^{1/2}$(kgFs^{-2})$^{1/2}$
Rigidity of the stator	c_{S1}=$(w_{res1})^2$*m N/m c_{S2}=$(w_{res2})^2$*m N/m
Factor of disturbance	ε1= ε2,ε1 =0.02
Damping ratios	d_{S2}=d_{S1} d_{S1}=10 Ns/m

4.1 Simulation results

Matlab/Simulink has been used to simulate the dynamic behavior of the motor. The tools such as Ode45 and Ode23 and the associated methods allow calculating the integrals and, giving the evolution values of outputs (rotation speed and torque).

Considering the parameter values (see table 1) used for the simulation [12],[15], the figure 6 shows the evolution of the rotation speed as a function of different loads applied. The first curve represents the motor speed without load. From a time of $3,5*10^{-3}$ seconds, the motor is connected to loads of 1Nm, 2Nm, 3Nm, 4Nm and 5Nm. We can see that the speed decreases as a function of the load. For the first four loads (1, 2, 3 and 4Nm), a time of 1 to $1,5*10^{-3}$ seconds has been necessary to stabilize the speed. For a load of 5Nm, the speed tends towards zero.

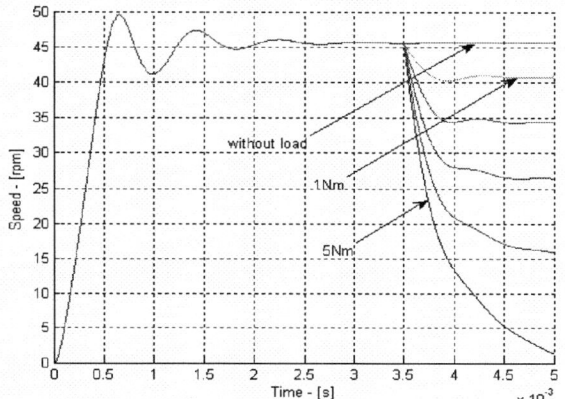

Fig.6. Motor speed as a function of time for different loads.

Figure 7 and 8 represent respectively the evolution of traction force $\mathbf{F_{antr}}$ and feedback tangential forces $\mathbf{F_{fbtg}}$ as functions of time, simulated without load and with 3Nm as applied load.

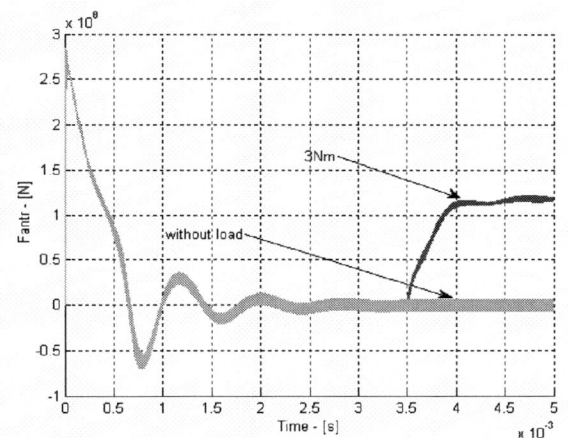

Fig.7. $\mathbf{F_{antr}}$ as a function of time for different loads.

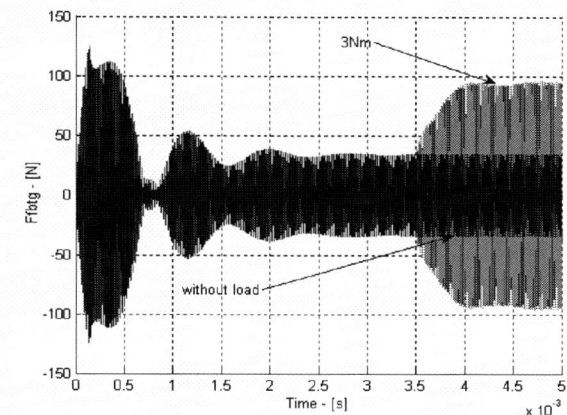

Fig.8. $\mathbf{F_{fbtg}}$ as functions of time for different loads.

4.2 Results validation

By comparison of obtained results with the data of the manufacturer [28], we can say that our implementation, on the software Matlab/Simulink, of refined model reflects the true behavior of the motor, and it is shown in figure 9.

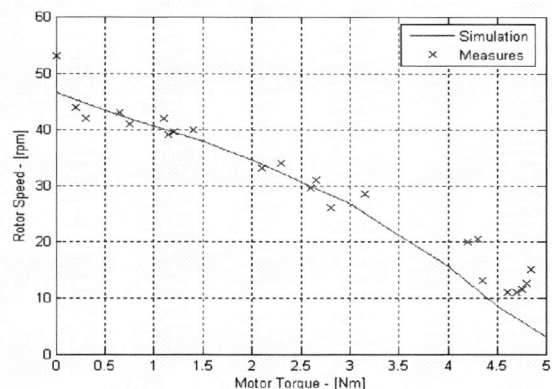

Fig.9. Measured and simulated curve correspondence speed-torque of motor AMW 90-X with various loads [28].

On figure 10, the (x) points represent the points of measurements of the manufacturer, and the solid lines the interpolation ensured by the points extracted from the simulation results.

5. CONCLUSION

In this paper, we have studied the various forces cases existing which ensure the dynamic of the ultrasonic motor with detailing more the missed cases in the majority of the studies in the literature.

We have presented a short overview on control approaches applied for the USM motors by detailing their kind of implementation. A control of motor using artificial intelligent approaches (X.Xu) will be presented in the next paper.

We have realized the implementation of the refined analytical model of the ultrasonic motor (TWUSM), with the software Matlab/Simulink.

In order to ensure this implementation, the different parts of motor have been interpreted in Matlab/Simulink by parameterized blocks. The simulation results of the rotation speed and of torque showed a good functional behavior of this ultrasonic motor (TWUSM).

The results of simulation have been compared with experimental values [28]. This comparison validates the success of, this implementation and our study.

REFERENCES

[1] T. Sashida, T. Kenjo: "An Introduction to Ultrasonic Motors", Clarendon, Oxford, 1993.

[2] MS Tsai, CH Lee, and SH Hwang : "Dynamic modeling and analysis of a bimodal ultrasonic motor", IEEE Trans. Ultrason., Freq. Contr., vol. 50, no. 3, pp. 245-256, 2003.

[3] Ueha, S. Tomikawa : "Ultrasonic Motors Theory and Applications", Clarendon, Oxford, 1993.

[4] Flynn, A.M: "Piezoelectric Ultrasonic Micromotors" MIT Artificial Intelligence Laboratory, December, 1997.

[5] Mattec.B. Mémoire de doctorat : " Modélisation et commande de moteurs piézoélectriques à onde progressive ", Université Lausanne 2005.

[6] G. Kandare and J. Wallaschek : "Derivation and validation of a mathematical model for traveling wave ultrasonic motors" Smart Mater. Struct.,vol.11, pp. 565-574, 2002.

[7] T. Sashida : "Trial construction and operation of an ultrasonic vibration driven motor," OYO BUTURI, (in Japanese) vol. 51, no. 6, pp. 713-720,1982.

[8] N. Hagood and A. J. McFarland : "Modelling of a piezoelectric rotary ultrasonic motor" IEEE Trnns. Ultmson., Fermeleet . Freq., Gontrr., vol. 42, no. 2, pp. 210-224, 1995.

[9] J M. Fernandez, M. Krummen, Y.Perriard : "Analytical and Numerical Modeling of an Ultrasonic Stepping Motor Using Standing Waves", Ultrasonics Symposium,0-7803-8412-1/04 IEEE, 2004.

[10] H. Storck , W. Littmann, J. Wallaschek, M. Mracek : "The effect of friction reduction in presence of ultrasonic vibrations and its relevance to travelling wave ultrasonic motors", Ultrasonics vol. 40 pp. 379–383. 2002.

[11] L. Huafeng Z. Chunsheng G. Chenglin ; "Study On The Contact Model Of Ultrasonic Motor Considering Shearing Defomation" ,Journal of ELECTRICAL ENGINEERING, VOL. 55, NO. 7-8, pp. 216-220. 2004.

[12] X. Cao und J. Wallaschek : "Dynamische Kontaktprobleme bei Schwingungsantrieben", DFGAbschlußbericht, DFG-Projekt: Wa 564/6, September 1996.

[13] M.Djaghloul ,Z.Boumous, : " Structure, Function, Existing Forces in the Ultrasonic Motor ", will be published, International conference on Sciences and Techniques of Automatic control STA ,2007.

[14] Z.Boumous, M.Djaghloul, Z.E.Kheribeche, S.Boumous, S.Belkhiat, "Simulation of Ultrasonic Piezoelectric Motor ", will be published, International Conference on Electrical Engineering Design and Technologies ICEEDT ,2007.

[15] A.M. Flynn : "Torque Production in Ultrasonic Motors." ,MIT Artificial Intelligence Laboratory 1993.

[16] Zhijun Sun , Rentao Xing , Chunsheng Zhao , Weiqing Huang: "Fuzzy auto-tuning PID control of multiple joint robot driven by ultrasonic motors", Ultrasonics 46 ,pp303–312, 2007

[17] Ikuo Yamano Takashi Maeno; "Five-fingered Robot Hand using Ultrasonic Motors and Elastic Elements", International Conference on Robotics and Automation ,Barcelona, Spain, April 2005

[18] Kenjiro Takemura, and Takashi Maeno;"Design and Control of an Ultrasonic Motor Capable of Generating Multi-DOF Motion", IEEE/ASME Transactions on Mechatronics, Vol. 6, No. 4, pp. 499-506, 2001.

[19] Güngör Bal, Erdal Bekiroglu ,:"Servo speed control of travelling-wave ultrasonic motor using digital signal processor", Sensors and Actuators A 109 ,pp 212–219, 2004.

[20] Güngör Bal: "A Digitally Controlled Drive System for Travelling-wave Ultrasonic Motor", Turk J Elec Engin, VOL.11, NO.3 , c TUBITAK, 2003.

[21] X. Xu, Y.C. Liang, H.P. Lee, W.Z. Lin, S.P. Lime, X.H. Shi,:"A stable adaptive neural-network-based scheme for dynamical system control", Journal of Sound and Vibration 285 ,pp 653–667 ,2005.

[22] Kenjiro Takemura, Takashi Maeno :"Control of Multi-DOF Ultrasonic Motor using Neural Network based Inverse Model", Conference on Intelligent Robot and Systems October 2002.

[23] Hong-Wei Ge , Yan-Chun Liang , Maurizio Marchese ,:"A modified particle swarm optimization-based dynamic recurrent neural network for identifying and controlling nonlinear systems" , Computers and Structures 85 , pp1611–1622,2007.

[24] Y.C.Liang ,X.H.Shi,H.P.Lee, X.Xu, W.Z.Lin S.P.Lim K.H.Lee "A neurol-network based-methode on speed control of ultrasonic motor" , , Vol. 2, ISBN 0-9728422-1-7 ,Nanotech 2003.

[25] Zhangfan, Chen Weishan, Liu Junkao, Zhao Xuetao:"Control of an ultrasonic transducer to realize low speed driven", Ultrasonics 44 e569–e574,2006.

[26] Güngör Bal, Erdal Bekiroglu, Sevki Demirbas , Ilhami Colak "Fuzzy logic based DSP controlled servo position control for ultrasonic motor" , Energy Conversion and Management 45 3139–3153, 2004

[27] Hongwei Ge, Wenli Du, Feng Qian, Zhencheng Ye ,"Speed Identification of Ultrasonic Motors Based on Evolutionary Elman Network",2006.

[28] Daimler-Benz AG, Druckschrift TE/ P67052104000791. 1991.

A Behavior-Based Visual Servoing Control Law

Mohammed Marey [1] & François Chaumette

IRISA/INRIA-Rennes, Campus de Beaulieu, 35042 Rennes, France.
(e-mail: Firstname.Name@irisa.fr)

Abstract: In this paper, we analyze and compare four image-based visual servoing control laws. Three of them are classical while a new one is proposed. This new control law is based on a behavior controller to adjust the movement of the camera. It can also be used to switch between the classical methods. An analytical study of all control schemes when translational motion along and rotational motion around the optical axis is also presented. Finally, simulation and experimental results show that the new control law with a behavior controller has a wider range of success than the other control schemes and can be used to avoid local minima and singularities.

Keywords: visual servoing, robotics control, singularities, local minima.

1. INTRODUCTION

Visual servoing is a well known approach to increase the accuracy, the versatility and the robustness of a vision-based robotic system [11, 5]. Two main aspects have a great impact on the behavior of any visual servoing scheme: the selection of the visual features used as input of the control law and the form of the control scheme. As for the visual features, they can be selected in the image space (point coordinates, parameters representing straight lines or ellipses, moments,... [8, 12, 6, 9, 4]), in the Cartesian space (pose, coordinates of 3D points,... [16, 18]), or composed of a mixture of both kinds of features attempting to incorporate the advantages of both image-based and position-based methods [13, 7, 2]. As for the choice of the control law [8, 14, 5], it affects the behavior of the selected visual features (local or global exponential decrease, second order minimization, ...) and may lead, or not, to local minima and singularities [3].

This paper is not concerned with the choice of the visual features, but with the analysis of different control schemes. That is why we will consider the most usual and simple features, that are the Cartesian coordinates of image points. As for the control schemes, we consider three classical control laws and we also propose in this paper a new control law that follows an hybrid strategy. It is based on a behavior parameter that can be used to tune the weight of the current and the desired interaction matrix in the control law. We will see that in some configurations where all other control schemes fail, this new control law allows the system to converge. The paper also includes an analysis of the control laws with respect to translational motion along and rotational motion around the optical axis. As we will see, a singularity of the control law proposed in [14] will be exhibited thanks to this analysis.

The paper is organized as follows: In Section 2, classical control schemes are recalled from which the control law with a behavior controller is proposed. In Section 3, an analysis of the control laws in the presence of rotation and translation w.r.t. the camera optical axis is presented. Finally, experimental and simulation results are presented in Section 4.

---
[1] Mohammed Marey is granted by the Egyptian Government.

2. NEW CONTROLLER WITH A BEHAVIOR PARAMETER

Let $\mathbf{s} \in \mathbb{R}^k$ be the vector of the selected k visual features, $\mathbf{s}^*$ their desired value and $\mathbf{v} \in \mathbb{R}^6$ the instantaneous velocity of the camera. Most classical control laws have the following form:

$$\mathbf{v} = -\lambda \widehat{\mathbf{L}_\mathbf{s}}^+ (\mathbf{s} - \mathbf{s}^*) \qquad (1)$$

where λ is a gain and $\widehat{\mathbf{L}_\mathbf{s}}^+$ is the pseudoinverse of an estimation or an approximation of the interaction matrix related to $\mathbf{s}$ (defined such that $\dot{\mathbf{s}} = \mathbf{L}_\mathbf{s} \mathbf{v}$ where $\mathbf{v} = (v, \omega)$ with v the translational velocity and ω the rotational one). Different forms for $\widehat{\mathbf{L}_\mathbf{s}}$ have been proposed in the past [8, 14, 5]. For simplicity, we consider that all values can be computed accurately, leading to the following choices

$$1) : \widehat{\mathbf{L}_\mathbf{s}} = \mathbf{L}_{\mathbf{s}^*} \qquad (2)$$

$$2) : \widehat{\mathbf{L}_\mathbf{s}} = \mathbf{L}_{\mathbf{s}(t)} \qquad (3)$$

$$3) : \widehat{\mathbf{L}_\mathbf{s}} = (\mathbf{L}_{\mathbf{s}^*} + \mathbf{L}_{\mathbf{s}(t)})/2. \qquad (4)$$

In the first case, $\widehat{\mathbf{L}_\mathbf{s}}$ is constant during all the servo since it is the value of the interaction matrix computed at the desired configuration. In the second case, $\widehat{\mathbf{L}_\mathbf{s}}$ changes at each iteration of the servo since the current value of the interaction matrix is used. Finally, in the third case, the average of these two values is used [14]. These three usual choices for $\widehat{\mathbf{L}_\mathbf{s}}$ when used with (1) define three distinct control laws, that we will denote D, C and A (for desired, current and average respectively) in the remainder of the paper.

As explained in [17], it is possible to improve the behavior of control law A by using:

$$\widehat{\mathbf{L}_\mathbf{s}} = (\mathbf{L}_{\mathbf{s}^*}{}^{c^*}\mathbf{T}_c + \mathbf{L}_{\mathbf{s}(t)})/2$$

where ${}^{c^*}\mathbf{T}_c$ is the spatial motion transform matrix to transform velocities expressed in the desired camera frame to the current camera frame. However, we will not consider this supplementary control scheme in the following.

On one hand, near the desired pose where the error $\mathbf{s} - \mathbf{s}^*$ is low, the same behavior is obtained whatever the choice of $\widehat{\mathbf{L}_\mathbf{s}}$

since we have in that case $\mathbf{L}_{s(t)} \approx \mathbf{L}_{s^*}$. On the other hand, as soon as $\mathbf{s} - \mathbf{s}^*$ is large, it is well known that the choice of $\widehat{\mathbf{L}_s}$ induces a particular behavior of the system since we thus have $\mathbf{L}_{s(t)} \neq \mathbf{L}_{s^*}$. This motivates the current research on the determination of visual features such that the interaction matrix is constant in all the configuration space of the camera, but it is clearly still an open problem, and, as already said, not the subject of this paper.

From (2), (3) and (4), a general form for $\widehat{\mathbf{L}_s}$ can easily be written by introducing a behavior controller $\beta \in \mathbb{R}$

$$\widehat{\mathbf{L}_s} = \mathbf{L}_\beta = (\beta \mathbf{L}_{s^*} + (1-\beta) \mathbf{L}_{s(t)}). \quad (5)$$

Using (5) in (1), we obtain a new control law, denoted G in the following (for "general"). Control laws D, C, and A are known to be locally asymptotically stable only [5]. The same is also true for control law G. Of course, if $\beta = 1$, we find again control law D, if $\beta = 0$, we obtain control law C, and if $\beta = 1/2$ we obtain control law A. Control law G could thus be used to switch between the different control schemes during the execution of the task. Switching strategies have already been proposed in [10, 1] but, in these works, switching is performed between image-based and position-based approaches, that is between different features, while here the features are the same but their control would be different.

In this paper, we are not interested in designing a possible strategy to switch between the different control laws. We are looking if particular values of β provide a better behavior of the system. Indeed, the main interesting property of control law G obtained using (5) is that the behavior of the system changes gradually from the behavior using control law C to the behavior using control law A when β varies from 0 to 1/2, and similarly, the behavior changes gradually from the behavior using control law A to the behavior using control law D when β varies from 1/2 to 1. Hence, this new control scheme allows us to adapt the behavior of the system based on the selected value of β. We will see in Section 4.1 that particular values of β indeed allows the system to converge while the other control schemes fail for some configurations.

Let us finally note that in case of modeling or calibration errors, the matrices $\mathbf{L}_{s^*}$ and $\mathbf{L}_{s(t)}$ have to be respectively replaced by approximations $\widehat{\mathbf{L}_{s^*}}$ and $\widehat{\mathbf{L}_{s(t)}}$, but that does not change the general properties of the control schemes as long as the approximations are not too coarse.

3. MOTION ALONG AND AROUND THE OPTICAL AXIS

This section presents an analytical analysis of the control laws described previously when the camera displacement is a combination of a translation t_z and a rotation r_z w.r.t. the camera optical axis. As usually done in IBVS, we have considered an object composed of four points forming a square.

The study includes two cases in which the movement along z-axis is from Z to Z^* and where $r_z = 90°$ in the first case and $r_z = 180°$ in the second case. In both cases, the object plane is parallel to the image plane.

The coordinates of a 3D point in the camera frame are denoted (X, Y, Z) and the coordinates of that point on the image plane are given by $\mathbf{x} = (x, y)$ with $x = X/Z$ and $y = Y/Z$. It is well known that the interaction matrix related to $\mathbf{x}$ is given by

$$\mathbf{L_x} = \begin{bmatrix} -\frac{1}{Z} & 0 & \frac{x}{Z} & xy & -(1+x^2) & y \\ 0 & -\frac{1}{Z} & \frac{y}{Z} & 1+y^2 & -xy & -x \end{bmatrix}$$

Using four points, the visual feature vector $\mathbf{s}$ is $\mathbf{s} = (x_0, x_1, x_2, x_3, y_0, y_1, y_2, y_3)$ whose desired value is $\mathbf{s}^* = (x_0^*, x_1^*, x_2^*, x_3^*, y_0^*, y_1^*, y_2^*, y_3^*)$.

Case 1: $r_z = 90°$ & $t_z = (Z \to Z^*)$ The coordinates of the four points w.r.t. the camera frame at the initial and the desired poses are denoted $p_{i0} = (-L, -L, Z)$, $p_{i1} = (-L, L, Z)$, $p_{i2} = (L, L, Z)$, $p_{i3} = (L, -L, Z)$, $p_{d0} = (-L, L, Z^*)$, $p_{d1} = (L, L, Z^*)$, $p_{d2} = (L, -L, Z^*)$ and $p_{d3} = (-L, -L, Z^*)$. Let $l = L/Z$ and $l^* = L/Z^*$. The initial value of $\mathbf{s}$ is then $\mathbf{s_i} = (-l, -l, l, l, -l, l, l, -l)$, the desired value is $\mathbf{s}^* = (-l^*, l^*, l^*, -l^*, l^*, l^*, -l^*, -l^*)$ and $\mathbf{s_i} - \mathbf{s}^* = (-l+l^*, -l-l^*, l-l^*, l+l^*, -l-l^*, l-l^*, l+l^*, l^*-l)$ is the error vector. Using the analytical form of $\mathbf{L_x}$, it is possible to compute the analytical form of $\mathbf{L_\beta}$ defined in (5) and then its pseudoinverse $\mathbf{L_\beta^+}$. Using $Z = l^* Z^*/l$, we obtain after computations and simplifications

$$\mathbf{L_\beta^+} = \begin{bmatrix} -c_0 & -c_0 & -c_0 & -c_0 & -c_1 & c_1 & -c_1 & c_1 \\ -c_1 & c_1 & -c_1 & c_1 & -c_0 & -c_0 & -c_0 & -c_0 \\ -c_3 & c_4 & c_3 & -c_4 & c_4 & c_3 & -c_4 & -c_3 \\ -c_5 & c_5 & -c_5 & c_5 & 0 & 0 & 0 & 0 \\ 0 & 0 & 0 & 0 & c_5 & -c_5 & c_5 & -c_5 \\ c_7 & c_6 & -c_6 & -c_7 & c_6 & -c_7 & -c_6 & c_7 \end{bmatrix}$$

where, when $\beta \in [0; 1]$,

$$c_0 = \frac{l^* Z^*}{4(\beta l^* + (1-\beta)l)}$$

$$c_1 = \begin{cases} 0 & \text{if } \beta l^{*2} = (1-\beta)l^2 \\ c_0 \frac{\beta(1+l^{*2}) + (1-\beta)(1+l^2)}{(\beta l^{*2} - (1-\beta)l^2)} & \text{else.} \end{cases}$$

$$c_3 = \frac{l^* Z^* (\beta l^* + (1-\beta)l)}{8((1-\beta)^2 l^3 + \beta^2 l^{*3})}, \quad c_4 = \frac{l^* Z^* (\beta l^* - (1-\beta)l)}{8((1-\beta)^2 l^3 + \beta^2 l^{*3})}$$

$$c_5 = \begin{cases} 0 & \text{if } \beta l^{*2} = (1-\beta)l^2 \\ \frac{-1}{4(\beta l^{*2} - (1-\beta)l^2)} & \text{else.} \end{cases}$$

$$c_6 = \frac{\beta l^{*2} + (1-\beta)l^2}{8((1-\beta)^2 l^3 + \beta^2 l^{*3})}, \quad c_7 = \frac{\beta l^{*2} - (1-\beta)l^2}{8((1-\beta)^2 l^3 + \beta^2 l^{*3})}$$

Using the value of $\mathbf{s_i} - \mathbf{s}^*$, the initial velocity $\mathbf{v_i}$ is easily deduced from (1) as

$$\mathbf{v_i} = (0, 0, v_z, 0, 0, \omega_z) \quad (6)$$

where

$$v_z = \frac{\lambda Z^* l^* (\beta l^{*2} - (1-\beta)l^2)}{\beta^2 l^{*3} + (1-\beta)^2 l^3}, \quad \omega_z = \frac{\lambda l l^* (\beta l^* + (1-\beta)l)}{\beta^2 l^{*3} + (1-\beta)^2 l^3}$$

As expected, the initial camera motion consists in performing a translation combined with a rotation whose value only depends on image data and on the chosen value for β and λ. We can note that $\mathbf{L_\beta}$ is singular if $\beta l^{*2} = (1-\beta)l^2$. For instance, such a singularity occurs when $l = l^*$ (i.e. $Z = Z^*$) and $\beta = 1/2$, which is very surprising. The control law A proposed in [14] is thus singular for a pure rotation of $90°$, which had not been exhibited before as far as we know. In fact, the only way to avoid this singularity whatever the value of l and l^* is to select $\beta = 0$ or $\beta = 1$. As can be seen on (6), this singularity has no effect on the computed velocity in perfect conditions, but, as we will see in Section 4.1, a quite unstable behavior is obtained in the presence of image noise or for configurations near that singularity (such that for instance the object plane is almost parallel to the image plane).

When $Z = Z^*$ then $l = l^*$ and the initial velocity $\mathbf{v_i}$ becomes
$$\mathbf{v_i} = \left(0, 0, \frac{\lambda Z^*(2\beta - 1)}{2\beta^2 - 2\beta + 1}, 0, 0, \frac{\lambda}{2\beta^2 - 2\beta + 1}\right).$$

In that classical case, the velocity $\mathbf{v_i}$ contains an unexpected translation whose direction depends on the value of β ($v_z < 0$ if $\beta < 1/2$ and $v_z > 0$ if $\beta > 1/2$). The only way to avoid this nonzero translation is to select $\beta = 1/2$ as already shown in [14], but $\mathbf{L}_\beta$ is singular in that case...

Coming back to the more general case and setting $\beta = 1$ in $\mathbf{L}_\beta^+$, the initial velocity $\mathbf{v_i}$ using control law D is given by
$$\mathbf{v_i} = \left(0, 0, \lambda Z^*, 0, 0, \frac{\lambda l}{l^*}\right). \quad (7)$$

Whatever the value of Z, that is even when $Z < Z^*$ in which case the camera has to move backward, the initial camera motion contains a forward translational term. This surprising result extends the same property obtained when $Z = Z^*$ [5].

Setting $\beta = 0$, the initial velocity $\mathbf{v_i}$ using the control law C is now
$$\mathbf{v_i} = \left(0, 0, \frac{-\lambda l^* Z^*}{l}, 0, 0, \frac{\lambda l^*}{l}\right). \quad (8)$$

In that case, the initial camera motion contains a backward translational term whatever the value of Z, that is even when $Z \geq Z^*$. We can even note that, more l is small, i.e. more Z is large, more the initial backward motion is large, which is even more surprising than the result obtained for $\beta = 1$. These results extend thus largely the property exhibited in [6] when $Z = Z^*$. By comparing (7) and (8), we can also note that the amplitude of the rotational motion using control laws D and C is surprisingly not the same as long as $l \neq l^*$, that is as soon as $Z \neq Z^*$.

Setting $\beta = 1/2$, the velocity $\mathbf{v_i}$ using control law A is
$$\mathbf{v_i} = \left(0, 0, \frac{2\lambda Z^* l^*(l^{*2} - l^2)}{l^{*3} + l^3}, 0, 0, \frac{2\lambda l l^*(l + l^*)}{l^3 + l^{*3}}\right).$$

In that case, a good behavior is obtained since the translational motion is always in the expected direction ($v_z < 0$ when $l^* < l$, that is when $Z < Z^*$, $v_z > 0$ when $l^* > l$ ($Z > Z^*$), and, as already said, $v_z = 0$ when $l = l^*$ (where $Z = Z^*$ but where $\mathbf{L}_\beta$ is singular).

Case 2: $r_z = 180^o$ & $t_z = (Z \to Z^*)$ We now consider the more problematic case where the camera displacement is composed of a translation and of a rotation of 180° around the camera optical axis. In that case, $\mathbf{s_i} - \mathbf{s}^* = (l+l^*, -l-l^*, -l-l^*, l+l^*, -l-l^*, -l-l^*, l+l^*, l+l^*)$ and $\mathbf{L}_\beta^+$ is given by

$$\mathbf{L}_\beta^+ = \begin{bmatrix} -c_0 & -c_0 & -c_0 & -c_0 & -c_1 & c_1 & -c_1 & c_1 \\ -c_1 & c_1 & -c_1 & c_1 & -c_0 & -c_0 & -c_0 & -c_0 \\ -c_3 & c_3 & c_3 & -c_3 & c_3 & c_3 & -c_3 & -c_3 \\ -c_4 & c_4 & -c_4 & c_4 & 0 & 0 & 0 & 0 \\ 0 & 0 & 0 & 0 & c_4 & -c_4 & c_4 & -c_4 \\ c_5 & c_5 & -c_5 & -c_5 & c_5 & -c_5 & c_5 & c_5 \end{bmatrix}$$

where, when $\beta \in [0; 1]$,

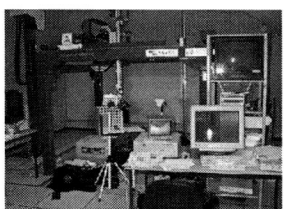

Fig. 1. Afma6 robot

$$c_0 = \frac{l^* Z^*}{4(\beta l^* + (1-\beta)l)}$$
$$c_1 = c_0 \frac{\beta(1 + l^{*2}) + (1-\beta)(1 + l^2)}{\beta l^{*2} + (1-\beta)l^2}$$
$$c_3 = \begin{cases} 0 & \text{if } \beta l^{*2} = (1-\beta)l^2 \\ \dfrac{l^* Z^*}{8(\beta l^{*2} - (1-\beta)l^2)} & \text{else} \end{cases}$$
$$c_4 = \frac{1}{4(\beta l^{*2} + (1-\beta)l^2)}$$
$$c_5 = \begin{cases} 0 & \text{if } \beta l^* = (1-\beta)l \\ \dfrac{1}{8(\beta l^* - (1-\beta)l)} & \text{else} \end{cases}$$

Proceeding as before, we obtain using the value of $\mathbf{s_i} - \mathbf{s}^*$
$$\mathbf{v_i} = (0, 0, v_z, 0, 0, 0)$$
$$\text{where } v_z = \begin{cases} 0 & \text{if } \beta l^{*2} = (1-\beta)l^2 \\ \dfrac{\lambda Z^* l^*(l + l^*)}{\beta l^{*2} - (1-\beta)l^2} & \text{else.} \end{cases}$$

In all cases, no rotational motion is produced while a translational motion is generally obtained, but when $\beta l^{*2} = (1-\beta)l^2$ in which case $\mathbf{L}_\beta$ is singular, leading to a repulsive local minimum where $v_z = 0$. Such a case occurs for instance when $Z = Z^*$ (i.e. $l = l^*$) and $\beta = 1/2$, which corresponds to the control law proposed in [14]. Another singularity occurs when $\beta l^* = (1-\beta)l$, which is also the case when $l = l^*$ and $\beta = 1/2$.

Of course, when $Z = Z^*$, we find again the results given in [3]: a pure forward motion is involved when $\beta = 1$ and a pure backward motion is involved when $\beta = 0$. More generally, for $\beta = 1$ and $\beta = 0$, the direction of motion is the same (i.e. forward or backward) whatever the value of l and l^*, that is whatever the value of Z with respect to Z^*. For any other value of β, the direction of motion depends on the relative value of Z with respect to Z^*, but unfortunately, there does not exist any value of β that will give a good behavior in that case since no rotational motion is computed by the control law.

4. RESULTS

In this section, experimental and simulation results are given. They have been obtained using the ViSP library [15] in which the new control schemes have been implemented.

4.1 Experimental results : Singularities

The experimental results have been obtained on a six degrees of freedom robot as shown in Fig. 1. They allow to validate the analysis presented in the previous section about the motion along and around the optical axis. Note that the velocities are saturated to forbid the application of too high values which may be computed near a singularity. More precisely, all velocity

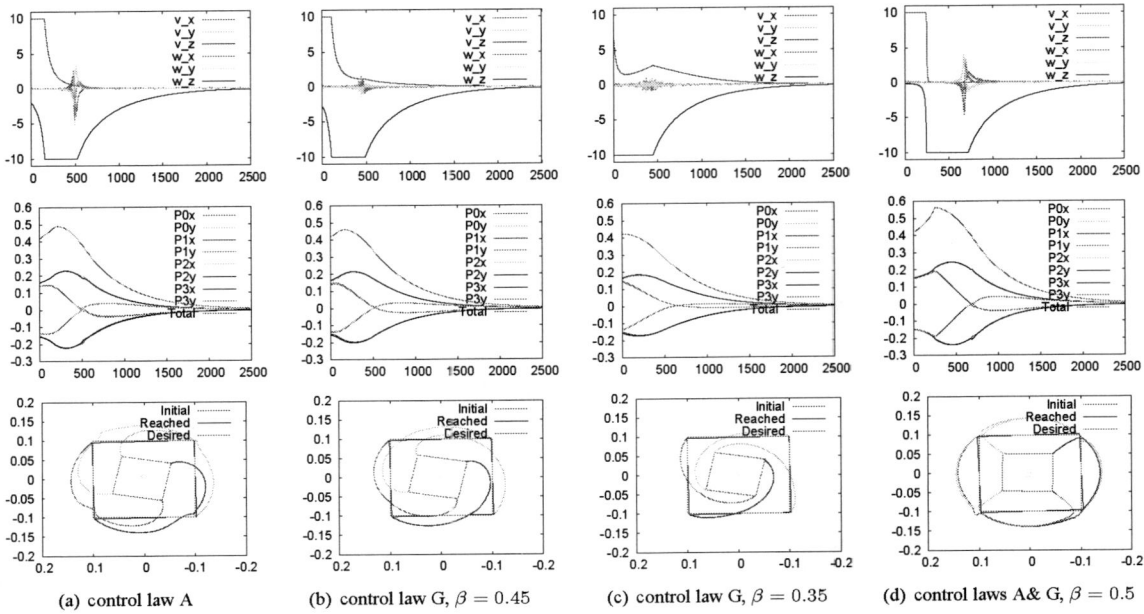

Fig. 2. Experimental results. Case A ($r_z = 170°$ and $t_z = 0.5$ m) in (a), (b) and (c); case B ($r_z = 180°$ and $t_z = 0.5$ m) in (d). Top line: camera velocity components (in m/s and rad/s), middle line: visual features error components and global error, bottom line: image points trajectories.

components are normalized when needed so that the maximal one is not more than 10 cm/s or 10 deg/s.

Case A In this first case, the required camera motion is composed of a rotation of $170°$ around the optical axis combined with a translation of 0.5 m along the optical axis toward the object (a square once again). As usual, gain λ has been set to 0.1. As expected unfortunately, control law D makes the points leave the camera field of view due to a forward motion, while control law C makes the robot reach its joints limits due to a backward motion. As can be seen in Fig. 2.a, control law A starts with high value of v_z toward the object, while ω_z increases until the translational motion is almost finished. As demonstrated in the analytical study, since the pure rotation $r_z = 90°$ corresponds to a singularity of control law A, the behavior of the system is quite unstable near this configuration, that is from iterations 800 to 1200, as can be observed in Fig. 2.a. As can be seen in Fig. 2.b, using control law G with $\beta = 0.45$ enables to decrease significantly the effect of the singularity near $r_z = 90°$, while its effect completely disappears for $\beta = 0.35$ (see Fig. 2.c).

Case B In this second case, the task is still to perform a translation of 0.5 m toward the object but combined now with a rotation of $180°$. Figure 2.d shows the results obtained for control law A (that is G with $\beta = 0.5$). The velocity components show that the motion of the camera starts with a pure translation toward Z^*. From the analytical study, no rotational motion should occur. However, due to small image noise and to the use of a real robot, that is a non perfectly calibrated system, the robot moves away from the repulsive local minimum and starts to rotate. The effect of the singularity at $90°$ is clearly visible, but after its crossing, the system converges to the desired pose.

4.2 Simulation Results : Optical Axis Studies

A general description of the camera behavior when the required movement is along the optical axis with all possible values of r_z is now given. It has been obtained through extensive simulations. As for the experimental results, we have set $L = 0.1$, the initial camera pose is $(0, 0, 1, 0, 0, r_z)$, and the desired camera pose is $(0, 0, 0.5, 0, 0, 0)$ so that the square appears as a centered square in the image with $l^* = 0.2$ and $l = 0.1$. We have also set $\lambda = 0.1$ and saturation terms on the velocity components have been introduced.

Applying control law D, the camera rotates and translates toward the desired pose without any additional movement as soon as $r_z \leq 78°$. When $r_z > 78°$, the camera continues its translational motion after reaching $Z = Z^*$ and then moves back toward the desired pose. The translation increases as r_z increases (see Fig. 3.a). When $r_z \geq 155°$, the control law fails since the camera reaches the object plane where $Z = 0$. Finally, v_z reaches its saturated maximal value at the first iteration of the control scheme while ω_z reaches its saturated value after several iterations (see Fig. 3.a).

Applying control law C, the camera rotates and translates correctly as long as $r_z \leq 61°$. When $r_z > 61°$, the camera starts moving backward and then translates forward. The backward translation increases as r_z increases (see Fig. 3.b). We can note on Fig. 3.b that the maximal rotational velocity is reached and saturated when the translational motion changes from backward to forward. The number of iterations required to reach the desired pose increases rapidly when $r_z > 150°$. Finally, when $r_z \geq 178.6°$, the backward translation is so large that the camera is not able to reach the desired pose.

Control law A converges with a perfect behavior (that is without any supplementary translation) as long as $r_z < 180°$ (see

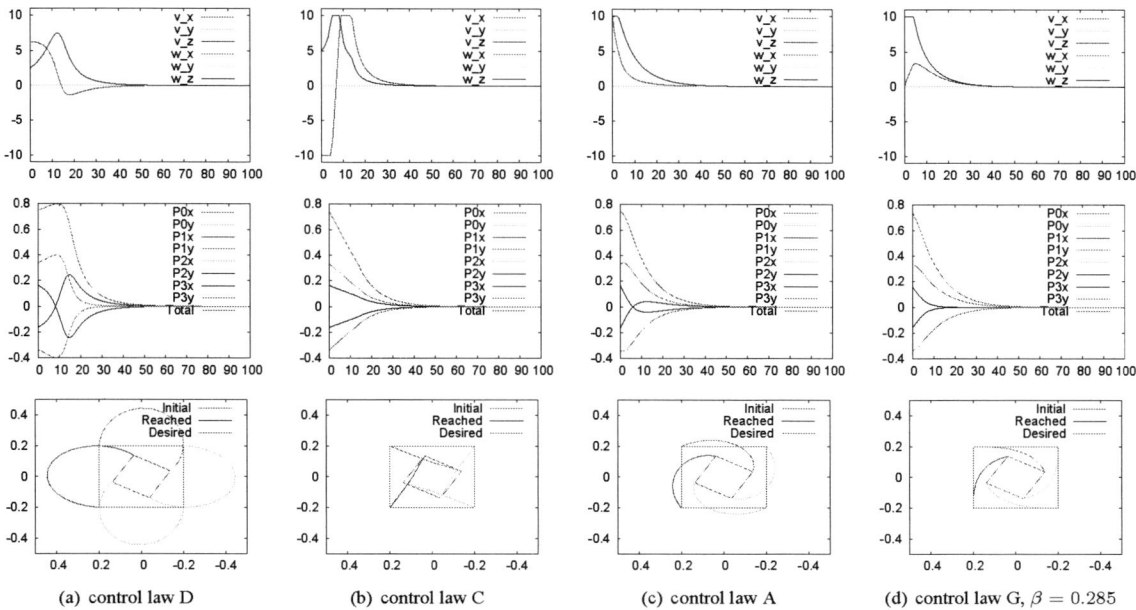

(a) control law D (b) control law C (c) control law A (d) control law G, $\beta = 0.285$

Fig. 3. Simulation results for optical axis studies obtained when $t_z = 0.5$ m and $r_z = 120°$.

Fig. 3.c). As discussed before, control law A has a singularity when $r_z = 180°$, that is why the velocity components are saturated at the beginning of the servo for large values of r_z.

Applying the new control law G, different behaviors are obtained based on the value selected for β. When the value of β is near to 0, 1 and 1/2, the behavior of the control law approaches the behavior of control laws C, D and A respectively. Best selection of β leads to enhance the behavior of the control law for a given displacement. For example, when $r_z = 120°$, control law G allows the camera to reach its desired pose when $\beta \in [-0.08, 1.19]$ with the best behavior obtained when $\beta = 0.285$ (see Fig. 3.d). In that case, the rotational velocity ω_z reaches its maximum value at the first iteration. The error on each point coordinates starts also to decrease at the first iteration. When $r_z = 170°$, the camera reaches its desired pose as long as $\beta \in [0.33, 0.85]$ with best behavior obtained when β is between 0.35 and 0.4.

4.3 Simulation results : Local Minima

Now, we consider a difficult configuration and compare the results obtained with the different control schemes described previously. A pose is denoted as $\mathbf{p} = (\mathbf{t}, \mathbf{r})$ where $\mathbf{t}$ is the translation expressed in meter and $\mathbf{r}$ the roll, pitch and yaw angles expressed in degrees. The desired camera pose is given by $(0, 0, 1, 45, -30, 30)$ which means that the desired position of the image plane is not parallel to the object. The initial camera pose is given by $(0, 0, 1, -46, 30, 30)$. As can be seen on Fig. 4.a, using control law D, the camera is first motionless, as in a local minimum, and then starts to diverge so that the points leave the camera field of view. Even if we do not consider this constraint (we are here in simulation where an image plane of infinite size can be assumed), the camera then reaches the object plane where $Z = 0$, leading of course to a failure. From the results depicted in Fig. 4.b and 4.c, we can see that control laws C and A both fail in a local minimum. As for control law A, it is the first time, as far as we know, that such a local minimum problem is exhibited. Finally, control law G is the only one to converge to the desired pose as soon as $0.515 < \beta < 0.569$ (see Fig. 4.d). The oscillations observed in the camera velocity and in the points coordinates allow the camera to go out from the workspace corresponding to the attractive area of the local minimum for the other control schemes.

5. CONCLUSIONS

The control laws used in image-based visual servoing have their respective drawbacks and strengths. In some cases, a control law is not able to converge while the others succeed. In other cases, all classical control laws may fail. Different behaviors may explain these failures. For example, the camera moves to infinity, the camera moves to be too near to the object, the camera reaches a local minimum or a singular configuration. In this paper, new configurations have been exhibited, for the first time as far as we know: a local minimum for all classical control schemes, especially for the control law proposed in [14]. A singularity of the control scheme proposed in [14] has also been exhibited and its effects have been emphasized through experiments obtained on a 6 dof robot. New surprising results have also been obtained for the other classical control schemes for motion combining translation along and rotation around the optical axis. Finally, a new control law based on a behavior controller has also been proposed. Setting $\beta = 0, 1$, or $1/2$ would allow to switch between the three most classical schemes but we have prefered to analyse the behavior of the control scheme for all possible values of this parameter. In all considered cases (difficult configurations subject to local minima for all classical schemes, motion along and around the optical axis), it has always been possible to determine values of this parameter that provide a satisfactory behavior of the control scheme. In fact, the suitable values of the behavior controller rely on the displacement that the camera has to realize. Future work will

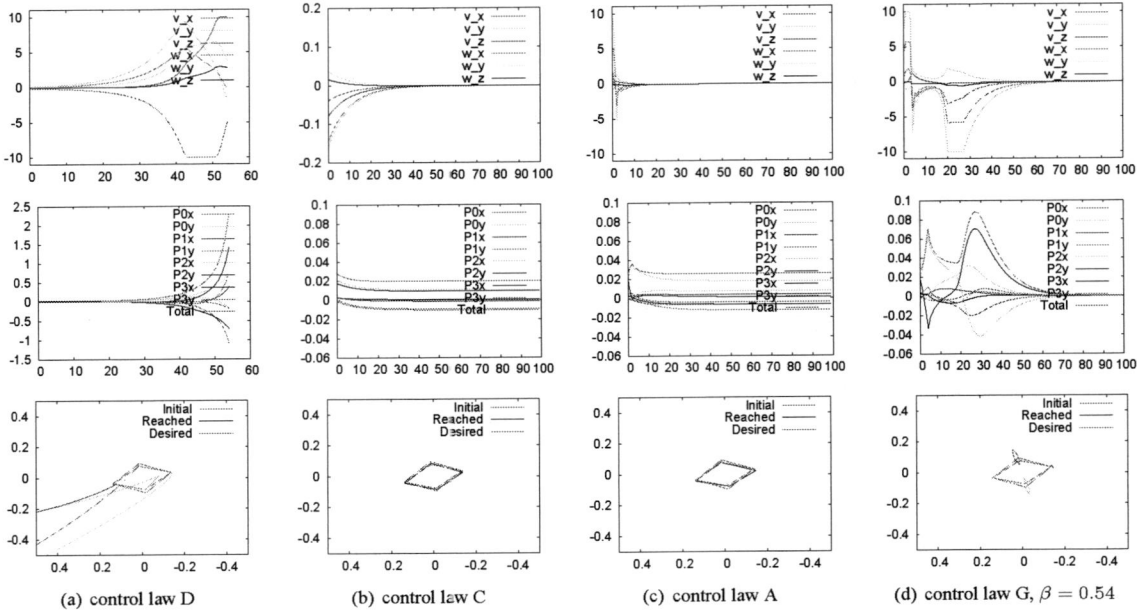

Fig. 4. Simulation results for local minima situation

thus be devoted to determining how to select automatically the value of the behavior controller to obtain a good behavior in all cases. Modifying on line the value of the behavior controller during the task execution will be also studied.

ACKNOWLEGMENTS

The authors would like to thank Seth Hutchinson for his comments on this research and earlier draft of this paper.

REFERENCES

[1] A. H. Abdul Hafez and C.V. Jawahar. Visual servoing by optimization of a 2D/3D hybrid objective function. ICRA'07, pp. 1691-1696, Apr. 2007.

[2] E. Cervera, A. Pobil, F. Berry and P. Martinet. Improving image-based visual servoing with three-dimensional features. Int. Journal of Robotics Research, 22:821-839, Oct. 2003.

[3] F. Chaumette. Potential problems of stability and convergence in image-based and position-based visual servoing. The Conference of Vision and Control. LNCIS 237, pp 66-78, Springer Verlag, 1998.

[4] F. Chaumette. Image moments: a general and useful set of features for visual servoing. IEEE Trans. on Robotics and Automation, 20(4):713-723, Aug. 2004.

[5] F. Chaumette and S. Hutchinson. Visual servo control Part I: basic approaches. IEEE Robotics and Automation Magazine, 13(4):82-90, Dec. 2006.

[6] P. Corke and S. Hutchinson. A new partitioned approach to image-based visual servo control. IEEE Trans. on Robotics and Automation, 17(4):507-515, Aug 2001.

[7] L. Deng, F. Janabi-Sharifi and W. Wilson. Hybrid strategies for image constraints avoidance in visual servoing. IROS02, Lausanne, pp 348-353, Oct. 2002.

[8] B. Espiau, F. Chaumette and P. Rives. A new approach to visual servoing in robotics. IEEE Trans. on Robotics and Automation, 8(3):313-326, June 1992.

[9] N. Gans, S. Hutchinson and P. Corke. Performance tests for visual servo control systems, with application to partitioned approaches to visual servo control. Int. Journal of Robotics Research, 22:955-981, Oct. 2003.

[10] N. Gans and S. Hutchinson. Stable visual servoing through hybrid switched-systems control. IEEE Trans. on Robotics, 23(3):530-540, June 2007.

[11] S. Hutchinson, G. Hager and P. Corke. A tutorial on visual servo control. IEEE Trans. on Robotics and Automation, 12(5):651-670, Oct. 1996.

[12] F. Janabi-Sharifi and W. Wilson. Automatic selection of image features for visual servoing. IEEE Trans. on Robotics and Automation 13(6):890-903, Dec. 1997.

[13] E. Malis, F. Chaumette and S. Boudet. 2 1/2 D visual servoing. IEEE Trans. on Robotics and Automation, 15(2):238-250, Apr. 1999.

[14] E. Malis. Improving vision-based control using efficient second-order minimization techniques. ICRA'04, pp 1843-1848, New Orleans, Apr. 2004.

[15] E. Marchand, F. Spindler, F. Chaumette. ViSP for visual servoing: a generic software platform with a wide class of robot control skills. IEEE Robotics and Automation Magazine, 12(4):40-52, Dec. 2005.

[16] P. Martinet, J. Gallice and D. Khadraoui. Vision based control law using 3d visual features. Second World Automation Congress, Vol. 3, pp. 497-502, Montpellier, France, May 1996.

[17] O. Tahri, Y. Mezouar. On the efficient second order minimization and image-based visual servoing ICRA'08, Pasadena, May 2008.

[18] W. Wilson, C. Hulls and G. Bell. Relative end-effector control using Cartesian position based visual servoing. IEEE Trans. on Robotics and Automation, 12(5):684-696, Oct. 1996.

Stochastic Wireless Channel Modeling, Estimation and Identification from Measurements

Mohammed M. Olama*. Yanyan Li**. Seddik M. Djouadi**. Charalambos D. Charalambous. ***

*Oak Ridge National Laboratory, 1 Bethel Valley Road, Oak Ridge, TN 37831, USA
(E-mail: olamahussemm@ornl.gov).
**University of Tennessee, 1508 Middle Dr., Knoxville, TN 37996, USA
(E-mail: yli24@utk.edu, djouadi@ece.utk.edu).
***University of Cyprus, 75, Kallipoleos Street, P.O.Box 20537, 1678, Nicosia, Cyprus
(E-mail: chadcha@ucy.ac.cy).

Abstract: This paper is concerned with stochastic modeling of wireless fading channels, parameter estimation, and system identification from measurement data. Wireless channels are represented by stochastic state-space form, whose parameters and state variables are estimated using the expectation maximization algorithm and Kalman filtering, respectively. The latter are carried out solely from received signal measurements. These algorithms estimate the channel inphase and quadrature components and identify the channel parameters recursively. The proposed algorithm is tested using measurement data, and the results are presented.

Keywords: Identification, Expectation Maximization, Kalman Filter, Stochastic Modeling, Wireless Channels

1. INTRODUCTION

Stochastic wireless channel models capture both the space and time variations of wireless systems, which are due to the relative mobility of the receiver, transmitter and/or scatterers [1]-[3]. The majority of research papers in this field such as in [4]-[6] use time-invariant (static) models for wireless channels. In time-invariant models, channel parameters are random but do not depend on time, and remain constant throughout the observation and estimation phase. This contrasts with stochastic models, where the channel dynamics become stochastic processes [1]-[3]. This paper is concerned with the development of stochastic wireless channel models based on system identification algorithms to extract various channel parameters using received signal measurement data.

In [1] and [3], the stochastic channel parameters are estimated from approximating the Doppler power spectral density (DPSD) of the wireless fading channel. However, in reality one can not have access to the TV DPSD at all times during the estimation process. We propose to estimate channel parameters as well as the inphase and quadrature components directly from the received signal measurements, which are usually available at the receiver. The expectation maximization (EM) algorithm [7] and Kalman filtering [8] are employed in the estimation process.

The paper is organized as follows. In Section 2, the TV wireless fading channel mathematical model is introduced. In Section 3, the EM algorithm together with the Kalman filter, to estimate the channel parameters as well as the inphase and quadrature components from received signal measurements, are developed. In Section 4, numerical results are presented. Finally, Section 5 provides the conclusion.

2. MATHEMATICAL MODELS FOR WIRELESS CHANNELS

The general time-varying (TV) model of a wireless channel is typically represented by the band-pass impulse response [5]

$$H(t;\tau) = \sum_{j=1}^{J(t)} \begin{pmatrix} I_j(t,\tau)\cos(\omega_c t) \\ -Q_j(t,\tau)\sin(\omega_c t) \end{pmatrix} \delta(\tau - \tau_j(t)) \quad (1)$$

where $H(t;\tau)$ is the band-pass response of the channel at time t, due to an impulse applied at time $t-\tau$, $J(t)$ is the random number of multipath components, ω_c is the carrier frequency, and the set $\{I_j(t,\tau), Q_j(t,\tau), \tau_j(t)\}_{j=1}^{J(t)}$ describes the random TV inphase component, quadrature component, and arrival time of the different paths, respectively. Let $s_l(t)$ be the low pass equivalent representation of the transmitted signal, then the band-pass representation of the received signal is given by

$$y(t) = \sum_{j=1}^{J(t)} \bigl(I_j(t,\tau)\cos(\omega_c t) - Q_j(t,\tau)\sin(\omega_c t)\bigr) s_l(t-\tau_j(t)) \\ + v_I(t)\cos(\omega_c t) - v_Q(t)\sin(\omega_c t) \quad (2)$$

where $\{v_I(t)\}_{t\geq 0}$ and $\{v_Q(t)\}_{t\geq 0}$ are two independent and identically distributed (iid) white Gaussian noise processes.

It is shown in [1] and [3] that the DPSD of the wireless fading channel, denoted by $S(s)$, can be approximated by an even, stable, rational, and factorizable transfer function, $\tilde{S}(s) = H(s)H(-s)$, where $H(s)$ is

$$H(s) = \frac{b_{n-1}s^{n-1} + \ldots + b_1 s + b_0}{s^n + a_{n-1}s^{n-1} + \ldots + a_1 s + a_0} \quad (3)$$

Consequently, the inphase and quadrature components can be realized using the following stochastic observable canonical form (OCF) state space representation [11]

$$\begin{aligned} dX_{I,j}(t) &= A_I X_{I,j}(t)dt + B_I dW_j^I(t) \\ I_j(t) &= C_I X_{I,j}(t) + f_j^I(t) \\ dX_{Q,j}(t) &= A_Q X_{Q,j}(t)dt + B_Q dW_j^Q(t) \\ Q_j(t) &= C_Q X_{Q,j}(t) + f_j^Q(t) \end{aligned} \quad (4)$$

where

$$X_{I,j}(t) = \left[X_{I,j}^1(t), X_{I,j}^2(t), \ldots, X_{I,j}^n(t) \right]^T,$$
$$X_{Q,j}(t) = \left[X_{Q,j}^1(t), X_{Q,j}^2(t), \ldots, X_{Q,j}^n(t) \right]^T,$$

$$A_I = A_Q = \begin{bmatrix} 0 & 1 & 0 & \cdots & 0 \\ 0 & 0 & 1 & \cdots & 0 \\ \vdots & \vdots & \vdots & \ddots & \vdots \\ 0 & 0 & 0 & \cdots & 1 \\ -a_0 & -a_1 & -a_2 & & -a_{n-1} \end{bmatrix}, \quad (5)$$

$$B_I = B_Q = \begin{bmatrix} b_{n-1} \\ \vdots \\ b_1 \\ b_0 \end{bmatrix}, \quad C_I = C_Q = \begin{bmatrix} 1 & 0 \cdots & 0 \end{bmatrix}$$

$\{W_j^I(t)\}_{t \geq 0}$, $\{W_j^Q(t)\}_{t \geq 0}$ are independent standard Brownian motions, which correspond to the inphase and quadrature components of the jth path respectively, $f_j^I(t)$ and $f_j^Q(t)$ are arbitrary functions representing the line-of-sight (LOS) of the inphase and quadrature components for the jth path respectively. Without loss of generality, we consider the case of flat fading, in which the fading channel has purely a multiplicative effect on the signal and the multipath components are not resolvable. Thus, it can be considered as a single path [5]. We also consider the non-line-of-sight (NLOS) case, i.e., $f_j^I(t) = f_j^Q(t) = 0$, which represents an environment with large obstructions. The state space model described in (4) has a solution given by [11]

$$X_L(t) = e^{A_L(t-t_0)} X_L(t_0) + \int_{t_0}^{t} e^{A_L(t-u)} B_L dW_L(u) \quad (6)$$

where $L = I$ or Q. Therefore, the mean of $X_L(t)$ is

$$E[X_L(t)] = e^{A_L(t-t_0)} E[X_L(t_0)] \quad (7)$$

and the covariance is

$$\Sigma_L(t) = e^{A_L(t-t_0)} Var[X_L(t_0)] e^{A_L^T(t-t_0)} + \int_{t_0}^{t} e^{A_L(t-u)} B_L B_L^T e^{A_L^T(t-u)} du \quad (8)$$

It can be seen in (7) and (8) that the mean and variance of the inphase and quadrature components are functions of time. Thus the statistics of the inphase and quadrature components, and therefore the statistics of the ad hoc channel, are times varying.

Similarly, following the state space representation in (4) and the received signal in (2), the fading channel can be represented using general stochastic state space representation of the form

$$\begin{aligned} dX(t) &= A(t)X(t)dt + B(t)dW(t) \\ y(t) &= C(t)X(t) + D(t)v(t) \end{aligned} \quad (9)$$

where

$$X(t) = \begin{bmatrix} X_I(t) & X_Q(t) \end{bmatrix}^T,$$
$$A(t) = \begin{bmatrix} A_I(t) & 0 \\ 0 & A_Q(t) \end{bmatrix},$$
$$B(t) = \begin{bmatrix} B_I(t) & 0 \\ 0 & B_Q(t) \end{bmatrix}, \quad (10)$$
$$C(t) = \begin{bmatrix} \cos(\omega_c t)C_I & -\sin(\omega_c t)C_Q \end{bmatrix},$$
$$D(t) = \begin{bmatrix} \cos(\omega_c t) & -\sin(\omega_c t) \end{bmatrix}$$
$$v(t) = \begin{bmatrix} v_I(t) & v_Q(t) \end{bmatrix}^T,$$
$$dW(t) = \begin{bmatrix} dW^I(t) & dW^Q(t) \end{bmatrix}^T$$

In this case, $y(t)$ represents the received signal measurements, $X(t)$ is the state variable of the inphase and quadrature components.

In [1] and [3], the channel parameters $\{a_{n-1}, \ldots, a_0, b_{n-1}, \ldots, b_0\}$ are obtained from approximating the DPSD. However, in reality one can not have access to the DPSD at all times during the estimation process. In this paper, channel parameters as well as inphase and quadrature components are estimated directly from received signal measurements, which are usually available or easy to obtain in any wireless network. The EM algorithm and Kalman filtering are employed in the channel parameter and state estimation, respectively. These algorithms are introduced in the next section.

3. WIRELESS CHANNEL ESTIMATION VIA THE EM ALGORITHM AND KALMAN FILTERING

This section describes the procedure employed to estimate the channel model parameters and states associated with the state space model in (9), using the EM algorithm [7] combined with Kalman filtering [8]. However, since the estimation process is carried out in discrete time we consider the discrete-time version of (9) given by

$$\begin{aligned} x_{t+1} &= A_t x_t + B_t w_t \\ y_t &= C_t x_t + D_t v_t \end{aligned} \quad (11)$$

where $x_t \in \Re^n$ is a state vector, $y_t \in \Re^d$ is a measurement vector, $w_t \in \Re^m$ is a state noise, and $v_t \in \Re^d$ is a measurement noise. The noise processes w_t and v_t are assumed to be independent zero mean and unit variance Gaussian processes.

The unknown system parameters $\theta_t = \{A_t, B_t, C_t, D_t\}$ as well as the system states x_t are estimated through the received signal measurement data, $Y_N = \{y_1, y_2, ..., y_N\}$. The methodology employed is recursive and based on the EM algorithm together with the Kalman filter. The Kalman filter is introduced next.

3.1 Channel State Estimation: The Kalman Filter

The Kalman filter estimates the channel states x_t for given system parameter θ_t and measurements Y_t. The Kalman filter is described by the following equations [8]

$$\begin{aligned} \hat{x}_{t|t} &= A_t \hat{x}_{t-1|t-1} + P_{t|t} C_t^T D_t^{-2} \left(y_t - C_t A_t \hat{x}_{t-1|t-1} \right) \\ \hat{x}_{t|t-1} &= A_t \hat{x}_{t-1|t-1}, \quad \hat{x}_{0|0} = m_0 \end{aligned} \quad (12)$$

where $t = 0, 1, 2, ..., N$, $D_t^2 = D_t D_t^T$, and $P_{t|t}$ is given by

$$\begin{aligned} \overline{P}_{t|t}^{-1} &= P_{t-1|t-1}^{-1} + A_t^T B_t^{-2} A_t \\ P_{t|t}^{-1} &= C_t^T D_t^{-2} C_t + B_t^{-2} - B_t^{-2} \overline{P}_{t|t} A_t^T B_t^{-2} \\ P_{t|t-1} &= A_t P_{t-1|t-1} A_t^T + B_t^2 \end{aligned} \quad (13)$$

and $B_t^2 = B_t B_t^T$. The channel parameters $\theta_t = \{A_t, B_t, C_t, D_t\}$ are estimated using the EM algorithm which is introduced next.

3.2 Channel Parameter Estimation: The EM Algorithm

The EM algorithm uses a bank of Kalman filters to yield a maximum likelihood (ML) parameter estimate of the state space model. It is an iterative scheme for computing the ML estimate of the system parameters θ_t, given the data Y_t. Specifically, a single iteration of the EM algorithm consists of two steps: The expectation step and the maximization step [9]. The filtered expectation step only uses filters for the first and second order statistics. The algorithm yields parameter estimates with nondecreasing values of the likelihood function, and converges under mild assumptions [10]. The expectation step evaluates the conditional expectation of the log-likelihood function given the complete data, which is described by [9]

$$\Lambda(\theta_t, \hat{\theta}_t) = E_{\hat{\theta}_t} \left\{ \log \frac{dP_{\theta_t}}{dP_{\hat{\theta}_t}} | Y_t \right\} \quad (14)$$

where $\{P_{\theta_t}; \theta_t \in \Theta\}$ denotes a family of probability measures induced by the system parameters θ_t, and $\hat{\theta}_t$ denotes the estimated system parameters at time step t. The maximization step finds

$$\hat{\theta}_{t+1} \in \arg\max_{\theta_t \in \Theta} \Lambda(\theta_t, \hat{\theta}_t) \quad (15)$$

The expectation and maximization steps are repeated until the sequence of model parameters converge to the real parameters. The EM algorithm is given by [9]

$$\begin{aligned} \hat{A}_t &= E\left(\sum_{k=1}^{t} x_k x_{k-1}^T | Y_t \right) \times \left[E\left(\sum_{k=1}^{t} x_k x_k^T | Y_t \right) \right]^{-1} \\ \hat{B}_t^2 &= \frac{1}{t} E\left(\sum_{k=1}^{t} \left((x_k - A_k x_{k-1})(x_k - A_k x_{k-1})^T \right) | Y_t \right) \\ &= \frac{1}{t} E\left(\sum_{k=1}^{t} \left(\begin{array}{c} (x_k x_k^T) - A_k (x_k x_{k-1}^T)^T \\ -(x_k x_{k-1}^T) A_k^T + A_k (x_{k-1} x_{k-1}^T) A_k^T \end{array} \right) | Y_t \right) \\ \hat{C} &= E\left(\sum_{k=1}^{t} y_k x_k^T | Y_t \right) \times \left[E\left(\sum_{k=1}^{t} x_k x_k^T | Y_t \right) \right]^{-1} \\ \hat{D}_t^2 &= \frac{1}{t} E\left(\sum_{k=1}^{t} \left((y_k - C_k x_k)(y_k - C_k x_k)^T \right) | Y_t \right) \\ &= \frac{1}{t} E\left(\sum_{k=1}^{t} \left(\begin{array}{c} (y_k y_k^T) - (y_k x_k^T) C_k^T \\ -C_k (y_k x_k^T)^T + C_k (x_k x_k^T) C_k^T \end{array} \right) | Y_t \right) \end{aligned} \quad (16)$$

where $E(\cdot)$ denotes the expectation operator. The system parameters $\{\hat{A}_t, \hat{B}_t^2, \hat{C}_t, \hat{D}_t^2\}$ can be computed from the following conditional expectations as [9]

$$\begin{aligned} L_t^{(1)} &= E\left\{ \sum_{k=1}^{t} x_k^T Q x_k | Y_t \right\}, \\ L_t^{(2)} &= E\left\{ \sum_{k=1}^{t} x_{k-1}^T Q x_{k-1} | Y_t \right\}, \\ L_t^{(3)} &= E\left\{ \sum_{k=1}^{t} \left[x_k^T R x_{k-1} + x_{k-1}^T R^T x_k \right] | Y_t \right\}, \\ L_t^{(4)} &= E\left\{ \sum_{k=1}^{t} \left[x_k^T S y_k + y_k^T S^T x_k \right] | Y_t \right\} \end{aligned} \quad (17)$$

where Q, R and S are given by

$$Q = \left\{\frac{e_i e_j^T + e_j e_i^T}{2}\right\},$$
$$R = \left\{\frac{e_i e_j^T}{2}\right\}, \quad (18)$$
$$S = \left\{\frac{e_i e_i^T}{2}\right\}$$

where $i, j = 1, 2, \ldots n$; $l = 1, 2, \ldots d$, and e_i is the unit vector in the Euclidean space; that is $e_i = 1$ in the ith position, and 0 elsewhere. For instance, consider the case $n = d = 2$, then

$$E\left(\sum_{k=1}^{t} x_k x_{k-1}^T | Y_t\right) = \begin{bmatrix} L_t^{(3)}(R_{11}) & L_t^{(3)}(R_{21}) \\ L_t^{(3)}(R_{12}) & L_t^{(3)}(R_{22}) \end{bmatrix} \quad (19)$$

where $R_{ij} = \{e_i e_j^T / 2; i, j = 1, 2\}$. The other terms in (16) can be computed similarly.

The conditional expectations $\{L_t^{(1)}, L_t^{(2)}, L_t^{(3)}, L_t^{(4)}\}$ can be estimated from measurements Y_t as follows [9]:

1) Filter estimate of $L_t^{(1)}$:

$$L_t^{(1)} = -\frac{1}{2} Tr\left(N_t^{(1)} P_{t|t}\right) - \frac{1}{2} \sum_{k=1}^{t} Tr\left(N_{k-1}^{(1)} \bar{P}_{k|k}\right)$$
$$-\frac{1}{2} \sum_{k=1}^{t} \begin{pmatrix} -2 x_{k|k}^T P_{k|k}^{-1} r_k^{(1)} + 2 x_{k-1|k}^T P_{k-1|k}^{-1} r_{k-1}^{(1)} - x_{k|k}^T N_k^{(1)} x_{k|k} \\ + x_{k-1|k}^T B_k^{-2} A_k \bar{P}_{k|k} N_{k-1}^{(1)} \bar{P}_{k|k} A_k^T B_k^{-2} x_{k-1|k} \end{pmatrix} \quad (20)$$

where $Tr(\cdot)$ denotes the matrix trace. In (20), $r_k^{(1)}$ and $N_k^{(1)}$ satisfy the following recursions

$$\begin{cases} r_k^{(1)} = \left(A_k - P_{k|k} C_k^T D_k^{-2} C_k A_k\right) r_{k-1}^{(1)} + 2 P_{k|k} Q x_{k-1} \\ \quad - P_{k|k} N_k^{(1)} P_{k|k} C_k^T D_k^{-2} \left(y_k - C_k x_{k|k-1}\right) \\ r_{k|k-1}^{(1)} = A_k r_k^{(1)} \\ r_0^{(1)} = 0_{m \times 1} \end{cases} \quad (21)$$

$$\begin{cases} N_k^{(1)} = B_k^{-2} A_k \bar{P}_{k|k} N_{k-1}^{(1)} \bar{P}_{k|k} A_k^T B_k^{-2} - 2Q \\ N_0^{(1)} = 0_{m \times m} \end{cases}$$

2) Filter estimate of $L_t^{(2)}$:

$$L_t^{(2)} = E\left\{\sum_{k=1}^{t} x_{k-1}^T Q x_{k-1} | Y_t\right\} = E_\theta\{x_0^T Q x_0 | Y_t\}$$
$$+ E_\theta\left\{\sum_{k=1}^{t} x_k^T Q x_k | Y_t\right\} - E_\theta\{x_t^T Q x_t | Y_t\} \quad (22)$$

Therefore, $L_t^{(2)}$ can be obtained from $L_t^{(1)}$.

3) Filter estimate of $L_t^{(3)}$:

$$L_t^{(3)} = E\left\{\sum_{k=1}^{t} \left(x_k^T R x_{k-1} + x_{k-1}^T R^T x_k\right) | Y_t\right\}$$
$$= -\frac{1}{2} Tr\left(N_t^{(3)} P_{t|t}\right) - \frac{1}{2} \sum_{k=1}^{t} Tr\left(N_{k-1}^{(3)} \bar{P}_{k|k}\right) \quad (23)$$
$$-\frac{1}{2} \sum_{k=1}^{t} \begin{pmatrix} -2 x_{k|k}^T P_{k|k}^{-1} r_k^{(3)} + 2 x_{k-1|k}^T P_{k-1|k}^{-1} r_{k-1}^{(3)} - x_{k|k}^T N_k^{(3)} x_{k|k} \\ + x_{k-1|k}^T B_k^{-2} A_k \bar{P}_{k|k} N_{k-1}^{(3)} \bar{P}_{k|k} A_k^T B_k^{-2} x_{k-1|k} \end{pmatrix}$$

In this case, $r_k^{(3)}$ and $N_k^{(3)}$ satisfy the following recursions

$$\begin{cases} r_k^{(3)} = \left(A_k - P_{k|k} C_k^T D_k^{-2} C_k A_k\right) r_{k-1}^{(3)} - P_{k|k} N_k^{(3)} P_{k|k} C_k^T D_k^{-2} \\ \left(y_k - C_k x_{k|k-1}\right) + \left(2 P_{k|k} R + 2 P_{k|k} B_k^{-2} A_k \bar{P}_{k|k} R^T A_k\right) x_{k-1|k-1} \\ r_{k|k-1}^{(3)} = A_k r_k^{(3)} \\ r_0^{(3)} = 0_{m \times 1} \end{cases} \quad (24)$$

$$\begin{cases} N_k^{(3)} = B_k^{-2} A_k \bar{P}_{k|k} N_{k-1}^{(3)} \bar{P}_{k|k} A_k^T B_k^{-2} - 2R \bar{P}_{k|k} A_k^T B_k^{-2} \\ \quad - 2 B_k^{-2} A_k \bar{P}_{k|k} R^T \\ N_0^{(3)} = 0_{m \times m} \end{cases}$$

4) Filter estimate of $L_t^{(4)}$:

$$L_t^{(4)} = E\left\{\sum_{k=1}^{t} \left(x_k^T S y_k + y_k^T S^T x_k\right) | Y_t\right\}$$
$$= \sum_{k=1}^{t} \left(x_{k|k}^T P_{k|k}^{-1} r_k^{(4)} - x_{k-1|k}^T P_{k-1|k}^{-1} r_{k-1}^{(4)}\right) \quad (25)$$

where $r_k^{(4)}$ satisfy the following recursions

$$\begin{cases} r_k^{(4)} = (A_k - P_{k|k} C_k^T D_k^{-2} C_k A_k) r_{k-1}^{(4)} + 2 P_{k|k} S y_k \\ r_{k|k-1}^{(4)} = A_k r_k^{(4)} \\ r_0^{(4)} = 0_{m \times 1} \end{cases} \quad (26)$$

Using the filters for $L_t^{(i)}$ ($i = 1, 2, 3, 4$) and the Kalman filter described earlier, the system parameters $\theta_t = \{A_t, B_t, C_t, D_t\}$ can be estimated through the EM algorithm described in (16). Numerical results that show the applicability of the above algorithm in estimating the channel parameters as well as the inphase and quadrature components from measurements are discussed in the next section.

4. NMERICAL RESULTS

In this section, the accuracy of the EM algorithm together with the Kalman filter to estimate the channel parameters as well as the inphase and quadrature components from received signal measurements, is determined. We consider a 4th order channel model as described in (9) and (10). Therefore, the system parameters $\theta_t = \{A_t, B_t, C_t, D_t\}$ can be represented as

$$A_t = \begin{bmatrix} 0 & 1 & 0 & 0 \\ a_1 & a_2 & 0 & 0 \\ 0 & 0 & 0 & 1 \\ 0 & 0 & a_3 & a_4 \end{bmatrix}, B_t = \begin{bmatrix} b_1 & \delta_{12} & \delta_{13} & \delta_{14} \\ b_2 & \delta_{22} & \delta_{23} & \delta_{24} \\ \delta_{31} & \delta_{32} & b_3 & \delta_{34} \\ \delta_{41} & \delta_{42} & b_4 & \delta_{44} \end{bmatrix}, \quad (27)$$
$$C_t = \begin{bmatrix} \cos(\omega_c t) & 0 & -\sin(\omega_c t) & 0 \end{bmatrix}, D_t = \begin{bmatrix} d_1 & d_2 \end{bmatrix}$$

The measurement data are provided by the Canadian Communication Research Center (CRC) and include samples for the inphase and quadrature components and received signal level. Using the measurement data, the inphase and quadrature components as well as channel parameters are estimated and then compared to the ones obtained by viewing the measurements as corrupted by white noise sequences. Fig. 1 shows the measured and estimated inphase and quadrature components as well as the received signal using the EM algorithm together with the Kalman filter for 400 sampled data taken from the measurements of one channel chosen at random. The system parameters are estimated as

$$\hat{A} = \begin{bmatrix} 0 & 1 & 0 & 0 \\ -0.0756 & -0.0474 & 0 & 0 \\ 0 & 0 & 0 & 1 \\ 0 & 0 & -0.6638 & 0.0717 \end{bmatrix},$$
$$\hat{B}^2 = \begin{bmatrix} 0.0484 & -0.0029 & -0.0453 & 4.0686*10^{-4} \\ -0.0029 & 0.0462 & 0.0013 & -0.0438 \\ -0.0453 & 0.0013 & 0.0573 & 0.0047 \\ 4.0686*10^{-4} & -0.0438 & 0.0047 & 0.0564 \end{bmatrix}, \quad (28)$$
$$\hat{C} = \begin{bmatrix} \cos(\omega_c t) & 0 & -\sin(\omega_c t) & 0 \end{bmatrix}, \hat{D}^2 = [0.0119].$$

From Fig. 1, it can be noticed that the inphase and quadrature components of the wireless fading channel as well as the received signal have been estimated with very good accuracy. It can also be noticed that the estimation error decreases as the number of samples increases; this is because the algorithm is recursive and the channel parameters converge to the actual values as more samples are being estimated. Fig. 2 shows the received signal estimates root mean square error (RMSE) for 100 runs. It can be noticed that it takes just few iterations (less than 15) for the filter to converge, and the steady state performance of the proposed channel estimation algorithm is excellent. Since we consider 4th order channel model, the computational cost of the proposed estimation algorithm is moderate and can be implemented on-line. Moreover, the filters of the expectation step are recursive and decoupled and hence are easy to implement in parallel on a multi-processor system [9].

5. CONCLUSION

This paper develops a general scheme for extracting mathematical STF channel models from noisy received signal measurements. The proposed estimation algorithm is recursive and consists of filtering based on the Kalman filter to remove noise from data, and identification based on the EM algorithm to determine the parameters of the model which best describe the measurements. Performance of the latter is investigated through a numerical example that shows excellent results. Therefore the proposed algorithms have good potential for real-time applications.

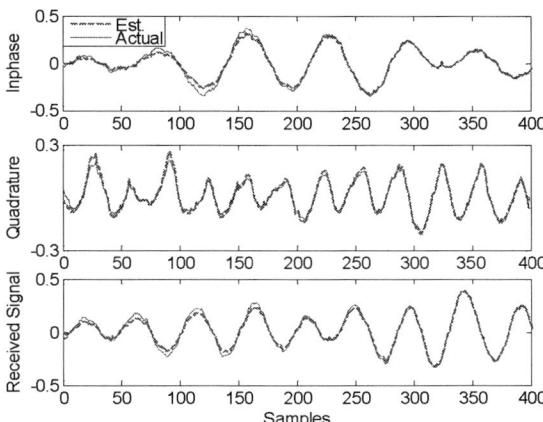

Fig. 1. Measured and estimated inphase and quadrature components, and received signals for 4th order channel using the EM algorithm combined with the Kalman filter.

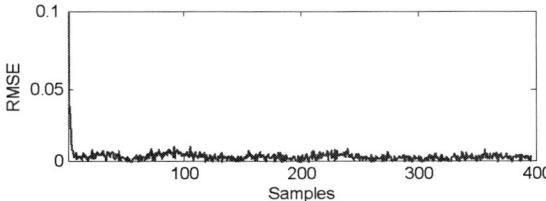

Fig. 2. Received signal estimates RMSE for 100 runs using the EM algorithm together with the Kalman filter.

REFERENCES

[1] C.D. Charalambous, S.M. Djouadi, and S.Z. Denic, "Stochastic power control for wireless networks via SDE's: Probabilistic QoS measures", *IEEE Trans. on Information Theory*, vol. 51, No. 2, pp. 4396-4401, December 2005.

[2] M.M. Olama, S.M. Djouadi, and C.D. Charalambous, "Stochastic power control for time-varying long-term fading wireless networks," *EURASIP Journal on Applied Signal Processing*, vol. 2006, Article ID 89864, 13 pages, 2006.

[3] M.M. Olama, S.M. Shajaat, S.M. Djouadi, and C.D. Charalambous, "Stochastic power control for time-varying short-term fading wireless channels", *Proceedings of the 16th IFAC World Congress*, Prague, Czech Republic, July 2005.

[4] W. Jakes, *Microwave mobile communications*, IEEE, Inc. NY, 1974.

[5] J.G. Proakis, *Digital communications*, Fourth Edition, McGraw Hill, New York, 2000.

[6] T.S. Rappaport, *Wireless communications: Principles and practice*, Prentice Hall, 2nd Edition, 2002.

[7] C.D. Charalambous and A. Logothetis, "Maximum-likelihood parameter estimation from incomplete data via the sensitivity equations: The continuous-time case", *IEEE Transaction on Automatic Control*, vol. 45, no. 5, pp. 928-934, May 2000.

[8] G. Bishop and G. Welch, *An introduction to the Kalman filters*, University of North Carolina, 2001.

[9] R.J. Elliott and V. Krishnamurthy, "New finite-dimensional filters for parameter estimation of discrete-time linear Guassian models," *IEEE Trans. On Automatic Control*, vol. 44, no. 5, pp. 938-951, 1999.

[10] C.F.J. Wu, "On the convergence properties of the EM algorithm," *Annals of Statistics*, vol. 11, pp. 95-103, 1983.

[11] B. Oksendal, *Stochastic differential equations: An introduction with applications*, Springer, Berlin, Germany, 1998.

A hybrid approach for MILP partitioning problem

Boudour, R*. Farfar, D.*Kimour, Mohamed T. *

*Department of Computer Science,
University of Annaba, Bp. 12, Annaba, Algeria
Tel / Fax: 213-38-87-24-36
e-mail: {racboudour, kimour}@yahoo.fr*

Abstract: Splitting a large system into smaller and more manageable units has become an important problem and a challenging task for some fields of research.The question of generating a good partitioning into smaller modules becomes a minimization problem for the number of parts being called by other parts.Two different and complementary approaches turned out to be promising to tackle the problem. First, we used a new approach, known as extreme partitioning, where hw/sw partitioning is based on profiling. Nevertheless, this one didn't guarantee hard timing constraints. To overcome this weakness, we coupled it with a MILP partitioning approach and used to reach this solution strategy, by efficient reformulations and a clever implementation. The advantage of the first one is to reduce particularly the time to market and the second one is to verify timing constraints by scheduling and to keep the cost down by clustering the nodes. We illustrate this method on two examples. The results are encouraging.

Keywords: hardware/software partitioning, embedded systems, Mixed Integer Linear Programming, branch and bound method, timing constraints, extreme partitioning

1. INTRODUCTION

The Hardware/software partitioning is a key issue in the design of embedded systems when performance constraints have to be met and some criteria are critical. Automatic partitioning is one key issue. Many problems in codesign are optimization problems under constraints. Generally problems in codesign are integer programming (IP) problems with many booleans. There are a number of approaches to solving mixed integer linear programming (MILP) problems, including branch and bound algorithms, implicit enumeration algorithms, cutting plane algorithms, etc. We try to explore this problem towards a new method called extreme partitioning (Some may call it overkill, others may call it wasteful)., which can defined as the use of extra hardware, often multiple processors dedicated to specific tasks, to greatly simplify the software-development effort. This latter provide a fine grain architecture but didn't guarantee the hard timing constraints. To surmount that, we coupled it with a well-known MILP approach [1]. We will concentrate our attention on both mixed methods in order to solve MILP partitioning problem with offering a low cost in reasonable time and mitigating risk at a minimum.

The rest of the paper is organized as follows. In section 2, we give an overview of related work. In section 3, we present briefly extreme partitioning approach and MILP method. Therfore, section 4 describe our approach. In section 5, we provide experimental results to demonstrate the effectiveness of our technique. Finally, we draw conclusions and gave future work.

2. RELATED WORKS

A key phase in the design of an embedded system is hardware/software partitioning that refers to the partitioning of the application into separate hardware and software modules. Traditional approaches to this problem as highlighted in [2][3][4][5][6] have been to initiate the process after the system specifications have been translated into code. The input to such partitioning approaches is thus the source code of the application, a binary implementation, or an internal format generated from the source code during analysis [7][8][9] as seen in Figure 1.

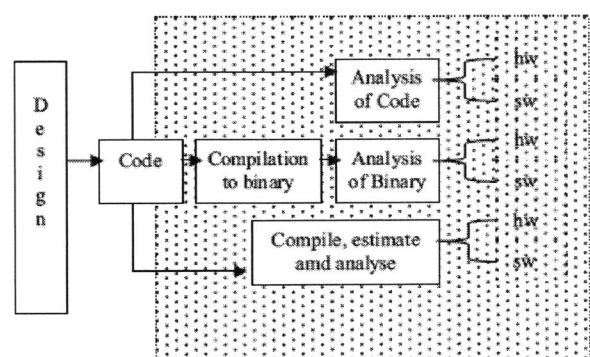

Fig. 1. Traditional Partitioning approach

An exception to the above is a work based on UML design specification [10] that uses function point analysis and COCOMO to compare different design alternatives at an early stage of analysis. A major assumption in most of these approaches, the target architecture is fixed and the user spent his time to optimize system specification on this latter.

3. APPROACHES PRESENTATION

3.1 Extreme partitioning

In recent years, the capabilities of information technology have increased tremendously. At the same time, large communication systems in today's organizations such as banks, health care providers, or government agencies, have become costly to obtain. To reduce the interfaces costs, the systems need to be split into smaller, more manageable modules. Each module can then be assigned to a separate team of designers. A partitioned system needs interfaces for the communication between modules; the number of interfaces is the main cost factor. In a good partitioning, the size of each module is restricted and the total size of the interface is minimized. To find such a partitioning is a job for a trained expert, but when the system is large an automated suggestion for a partitioning into modules becomes useful. To reach this goal, we use an appropriate approach, called extreme partitioning. The goal is to speed-up a system by incorporating hardware. A key feature is a memory allocation method which minimizes the interface traffic between hardware and software. The disadvantage of this approach is that hard timing constraints can not be guaranteed because the cost model is based on profiling [11].

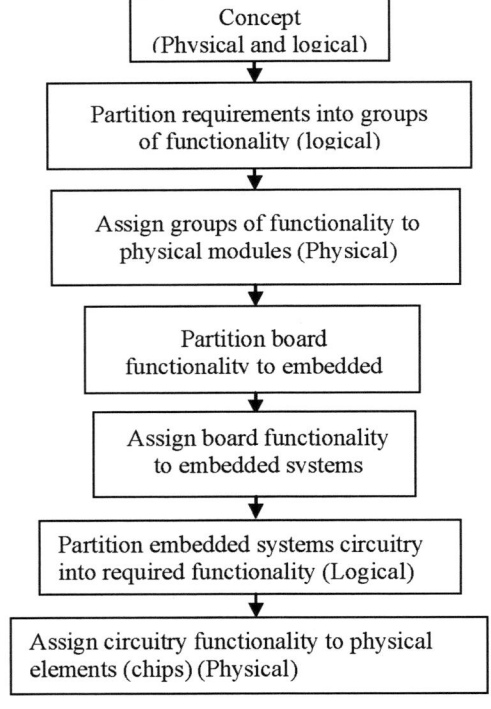

Fig. 2. Extreme partitioning generalized process model

Figure 2 shows the seven stages for the generalized process for extreme partitioning. The process begins with a combined logical/physical concept and then alternates between logical and physical partitioning until a final physical outcome is generated. Each stage is usually iterative, requiring several cycles of consideration before a final outcome is acceptable and the next stage can be entered. The model does not show any feedback mechanisms, although certainly they exist. For example, if a physical stage is entered, and the designers determine that a logical element is missing, the process must revert to at least the previous stage. In some cases the process must revert to the beginning concept stage.

3.2 MILP partitioning approach

Usually, to formulate a system that has to be partitioned on hw/sw parts with the help of ILP model with variables taking values only in {0,1}, the designer has to define the following [1][12][13]:
1. The target technology has to be specified by defining the set of processors for the software parts and the components library for synthesizing the hardware parts of the embedded system.
2. The system has to be defined in systemC in our case, as a set of interconnected instances of components
3. The design constraints have to be defined, including performance constraints (timing) and resource constraints (area-memory).

After the system has been specified with systemC, the specification is compiled into an internal syntax graph model. For each component of this model, software source code and hardware source code is generated. The software parts are compiled and the hardware parts are synthesized. The results are values for software metrics and values for hardware metrics for the components [14][15].

Then, a partitioning graph is created in which each node of the graph represents an instance of a component in the system. Edges of the partitioning graph represent the wires between these instances. Nodes are weighted with hardware and software costs, edges are weighted with interface costs which reflect the cost of hardware/software interfaces. The partitioning graph is then transformed into an IP-model. Afterwards, the model can be solved for example by standard commercial software such as CPLEX[16], or using the LINDO package[17], LINDO starts with a feasible linear programming solution and searches for optimal integer solutions using the branch-and-bound method, where optimal solution for used metrics is found, sw parts mapped onto the same processor are clustered into one new node and evaluated together, which results in a new sw cost metric [18][19][20]. We repeated the process until no more components can be moved. Parts of the system may be implemented in hardware or in software.

Software parts may be sequentialized and executed on the processor and hardware parts may be executed in parallel on the hardware components.

4. HYBRID SOLUTION BY ILP

After having applied the seven steps of the process shown in figure 2, we obtain a first solution of partitioning problem with a fine grain specification system and target architecture (behaviours and components) [21][22]. This solution constitute the input of the MILP process. We will first formulate the problem as a graph partitioning problem - a bicoloured control data flow graph (CDFG) - and then translate it into an ILP problem with 0-1 variables, without forgetting information particularly the timing constraints, needed for partitioning as is given in figure 3.

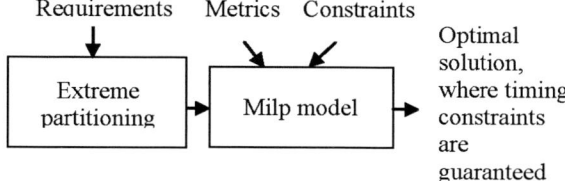

System specification and target architecture

Fig. 3. Main steps of our mixte approach

Our basic idea to reduce costs and risks is presented through the rules in the following subsections:

4.1 Timing constraints

The timing costs cannot be calculated by accumulating the execution time of the nodes, because two nodes v_1, v_2 can be executed in parallel if they do not share the same resources and if there is no path from v_1 to v_2 and vice versa. To determine the starting time and ending time for each node, scheduling has to be performed. The execution time of v_i is either a hardware or a software execution time. The ending time of v_i is the sum of starting time and execution time

$$\forall j \in J \qquad T^E_J = T^S_J + T^D_J \qquad (1)$$

where T^E_J is the ending time, T^S_J is the starting time and T^D_J the execution time of the node j.

The system execution time C^t is the maximum MAX^t of the ending times of all nodes v_i and may not violate the global design timing constraint.

$$\forall j \in J \qquad T^E_J \leq C^t \leq MAX^t \qquad (2)$$

4.2 Interfacing

An interface has to be realized for an edge e = (v_{i1}, v_{i2}), if v_{i1} and v_{i2} are realized on different target technology components.

4.3 Scheduling

Two nodes which can be executed in parallel have to be sequentialized, if
- v_{i1} and v_{i2} are executed on the same processor or
- v_{i1} and v_{i2} share the same hardware instance on the same hardware component.
To sequentialize two nodes v_{i1}, v_{i2} the binary decision variable is used, idem for the edges.

The implemented software nodes are scheduled by an algorithme ALAP in a preprocessing step.

4.4 Heuristic Scheduling

Resource constrained scheduling is a NP-complete problem [9][16]. Therefore, it is that clear solving scheduling problem optimally can not be done efficiently. For this reason, we have developed an algorithm using integer programming that solves the partitioning problem while iterating the following steps:
1. Solve an IP-model for the hardware-software mapping with help of approximated time values.
2. Solve an IP-model for calculating a valid schedule with nodes mapped to hardware or software.
3. If the resulting total time violates the timing constraint, repeat the first two steps with a timing constraint that is tighter than the approximated total time of step 1.

```
//Exploration grah
Begin
//the maximum number of iteration graph is equal to 2^instances
//if any component has one instance
// Iteration number = all solutions = Π (Instances number for
     each component + 1)
For i=1 until all_solutions do
    Convert in binary number
    If (solution not repeated) and (solution is importante)
    then
    Affect all metrics and parameters at solution
        Shedule it
        If solution is feasible then
            If solution is optimal the
                -  store the number I,
                -  store solution parameters
            endif
    endif
endfor
end
```

5. EXPERIMENTS AND TESTS

In our approach, the target architecture, the system, and the design constraints are specified by using a graphical user interface provided by our tool, implemented in Visual C++ environment. We have experimented our technique on two examples : A 4-band equalizer and a subsytem of currency distributor called returned currency.

Exemple 1 : A 4-band equalizer [19]

The 4-band-equalizer consists of 3 system components: an FIR-Filter, a multiplier and an adder. The equalizer is specified by using 4 instances of the FIR-filter, 4 instances of the multiplier and 3 instances of the adder. This 4-band equalizer is a well suited example to demonstrate the scheduling, hardware sharing and interfacing problem.

(a)

Comp.	Num_Inst	Successor	Surf_Com	Time_Com
FIR	1	5	7	4
	2	6	9	5
	3	7	13	10
	4	8	11	2
MUL	5	9	5	5
	6	9	4	3
	7	10	10	4
	8	10	6	9
ADD	9	11	8	12
	10	11	6	2
	11			

(b)

Comp.	Surface Mat.	Time d'exé_mat.	Espace mém st.	Espace mém d.	Temps d'exé_log.
FIR	22	2	20	25	5
MUL	12	2	30	30	3
ADD	17	1	15	40	2

(c)

Contrainte de conception		Facteurs d'amélioration (%)
Surface matérielle totale	120	30
Temps d'exécution total	30	40
Espace mém. statique	60	15
Espace mém. dynamique	80	15

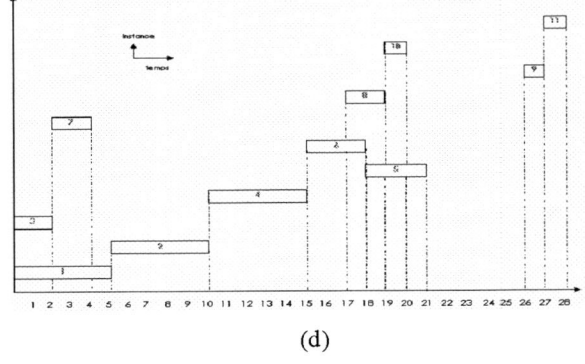

(d)

Fig. 4. (a) Specification, (b) metrics, (c) constraints and (d) scheduling of 4-band equalizer

Objective function minimization
Min $30(22(4-x1) +12(4-x2) +17(3-x3)) + 40(7x1+5x2+3x3) + 15(20x1+30x2+15x3) +15(25x1+30x2+40x3) +$

+ Sur_temps_com   (3)

Partitioning results :

Hardware area   =   117
Execution time   =   28
Static memory size =   50
Dynamic memory size =   55

Exemple 2 : Subsystem of returned currency

Objective function :
Min $25(5(1-x1) +12(5-x2) +12(2-x3) +12(1-x4) +5(1-x5) + 10(-1x6) +7(2-x7) +6(1-x8)) + 25(13x1+26x2+14x3+21x4+10x5+20x6+11x7+13x8) + 25(2x1+4x2+5x3+4x4+2x5+2x6+2x7+4x8) + 25(10x1+15x2+10x3+8x4+4x5+8x6+4x7+6x8) + \Delta$ (4)

Partitioning results :

Hardware area   =   89
Execution time   =   97
Static memory size =   17
Dynamic memory size =   43

The computing execution time is performed after the scheduling operation (ALAP) of parts system.

6. CONCLUSION

This paper presents a new approach of hardware/software partitioning supporting multi-processor systems, interfacing and hardware sharing. A hybrid approach has been proposed, which is able of solving the hardware/software partitioning problem using extreme partitioning and integer programming with no loss in optimality. In contrast to other approaches, our approach allows the design goals of guaranting timing constraints, quickly delivering prototype units, to make use of parallel development efforts and easy to debug. This helps to keep the development time, risk, and costs down. The presented results are very promising, are calculated in short time. Future work will deal with further refinement of the MILP-model for target architecture extraction from output extreme partitioning and design studies of other system level examples. More importantly, our approach is very flexible and can readily extend to the partitioning problems with various objectives and constraints, which makes the ILP formulations superior alternatives to the partitioning problems.

REFERENCES

[1] R. Niemann and P. Marwedel. HW/SW Partitioning based on Mixed Integer Linear Programming. 2000.

[2] R. Ernst, J. Henkel, and T. Benner. Hardware / software cosynthesis for Microcontrollers. *IEEE Design and Test*, **Vol.12**, p. 64-75, 1993.

[3] P. Eles, Z. Peng, and A. Doboli. VHDL System level Specification and Partitioning in a Hardware/Software CoSynthesis Environment. Third International Workshop on Hardware/Software Codesign, Grenoble, p. 49-55, 1994.

[4] R. K. Gupta, C. Coelho and G. De Micheli. Synthesis and Simulation of Digital Systems Containing Interacting Hw and Sw Components. 29th

IEEE Design Automation Conference, 225-230, 1992.

[5] Z. Peng and K. Kuchcinski. An Algorithm for Partitioning of Application Specific Systems. *Proceedings of the European Conference on Design Automation EDAC*, p. 316-321, 1993.

[6] A. Kalavade and E.A. Lee. The Extended Partitioning Problem, Hardware/Software Mapping and Implementation-Bin Selection. *Proceedings of the 6th International Workshop on Rapid Systems Prototyping*, 1995.

[7] A. Jantsch, P. Ellervee, J. Oberg, A. Hemani, and H. Tenhunen. Hardware/Software Partitioning and Minimizing Memory Interface Traffic. *European Design Automation Conference EURO-DAC*, p. 226-231, 1994.

[8] A. Kalavade and E.A. Lee. A Global Critically-Local Phase Driven Algorithm for the Constrained Hardware-Software Partitioning Problem. Third International Workshop on Hardware/Software Codesign, Grenoble, pages 42-48, 1994.

[9] F. Vahid, J. Gong, and D. Gajski. A Binary Constraint Search Algorithm for Minimizing Hardware during Hardware/Software Partitioning. *European Design Automation Conference EURO-DAC*, p. 214-219, 1994.

[10] W.Fornaciari, P.Micheli, F. Salice, L.Zampella, A First Step Towards Hw/Sw Partitioning of UML Specifications, *Proceedings of the Design, Automation and Test in Europe (DATE'03)*, 2003

[11] M. S. Alexander. Extreme partitioning. Embedded.com, 2005.

[12] R. Niemann and P. Marwedel. Hardware/Software Partitioning using Integer Programming. European Design and Test Conference ED & TC, p. 473-479, 1996.

[13] R. Niemann Hardware/Software Co-Design for Data Flow Dominated Embedded Systems. Hardcover, 244 pages, 2004.

[14] F. Vahid and D. Gajski. Closeness metrics for systemlevel functional partitioning. In Proceedings of the European Design Automation Conference, p. 328-333, 1995.

[15] D. Henkel, J. Herrmann, and R. Ernst. An Approach to the Adaption of Estimated Cost Parameters in the COSYMA System. Third International Workshop on Hardware/Software Codesign, Grenoble, p. 100-107, 1994

[16] CPLEX homepage, http://www.cplex.com.

[17] LINDO:Linear Interactive and Discrete Optimizer for linear, integer,and quadratic programming Problems. LINDO Systems, Inc.,1999.

[18] B. Knerr, M. Holzer, and M. Rupp. HW/SW Partitioning Using High Level Metrics. Published in the proceedings of the International Conference on Computing, Communications and Control Technologies (CCCT), pp. 33-38, Austin, 2004

[19] J. Peng, S. Abdi and D. Gajski. A Clustering Technique to Optimize Hardware/Software Synchronization. ASP-DAC, p. 965-968, 2005.

[20] W. Ahmed, D. Myers.Design Refinement for Efficient Clustering of Objects in Embedded Systems. Proceedings of the Design, Automation and Test in Europe Conference and Exhibition (DATE'05), 2005

[21] R. H.Bisseling, J. Byrkay, S.C. Erbas, N.Gvozdenovi, M. Lorenz, R. Pendavingh,

[22] C.Reeves, M.Röger,A. Verhoeven. Partitioning a Call Graph. Technical report, 2005.

POSTER SESSION

(Organizer: J. Y. Didier)

Classical Control System Design: A non-Graphical Method for Finding the Exact System Parameters

Mohammed Tawfik Hussein

Electrical & Computer Engineering Department
Islamic University of Gaza
Po Box 108, Gaza City, Palestine
(e-mail: mhussein@iugaza.edu or mhussein_98@hotmail.com)

Abstract: The Root Locus method of control system design was developed in the 1940's. It is a set of rules that helps in sketching the path traced by the roots of the closed loop characteristic equation of the system, as a parameter such as a controller gain, k, is varied. The procedure provides approximate sketching guidelines. Designs on control systems using the method are therefore not exact. This paper aims at a non-graphical method for finding the exact system parameters to place a pair of complex conjugate poles on a specified damping ratio line. The overall procedure is based on the exact solution of complex equations on the PC using numerical methods.

Key words: Root Locus, poles, controller gain, damping ratio.

1. INTRODUCTION

Faced with poor root-solving capabilities in his era, Walter Evans developed a set of rules in the mid-1940's by which the path traced by the closed-loop characteristic equation roots, could be sketched to reasonable accuracy as the gain varies. This plot, known as the Root Locus, is used by design engineers to determine the value of a variable parameter which places the poles of the closed loop system in a particular location. Here, a dominant pair of poles is located and this enables second-order system tendencies to be exploited or used as a starting point for pole placement and controller design.

The root-Locus technique is not confined only to the study of control systems. It is commonly formulated for the simple negative feedback loop of figure 1, which has an equivalent transfer function of [1]

$$F(s) = KG(s) / 1 + KG(s) H(s)$$

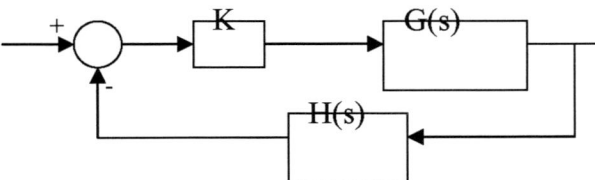

Fig. 1. System configuration usually used in root locus plots

In general, the method can be applied to study the behavior of roots of any algebraic equation with one or more variable parameters [2]. The basic equation employed in plotting the Root Locus is [3].

$$1 + K G(s) H(s) = 0. \quad (1)$$

Equation 1 is usually obtained from the characteristic equation, CE, of the system. For instance, given the CE

$$S^4 + 8S^3 + 24 S^2 + (32 + K) S + 6 K = 0$$

Where K, for example the gain, is a variable parameter of interest. We can transform this equation into

$$1 + K(S+6)/ (S^4 + 8S^3 + 24S^2 + 32S) = 0 \quad (2)$$

Which takes the form of Equation 1.

Procedures for calculating values of K and subsequent closed loop poles locations are based on Equation 1. Routh-Hurwitz criterion [3,4] applied to the CE can yield jw-axis crossings of the Root Locus, if the crossing points do exist. References 2 and 3 give the rules and examples for the Root Locus plot. Once the Root Locus is sketched, a graphical approach is used in determining closed-loop poles, as intersection point of the locus with a specified constant damping ratio or zeta line [5,6]. Values of the poles thus obtained are approximate.

Efforts to improve the accuracy of this procedure have been hampered by the difficulties in approximation methods for solving complex equations by numerical techniques. However, recently, the advent of Personal Computers with the proliferation of software that aid in computations of design problems, calls for re-examination of graphical techniques currently used in the system analysis and design Rather than employing these methods as solution procedures they need to play a different role of assisting in providing physical insight in the design procedure.

This article presents an analytical approach for finding the specific gain value K that will place a pair of system closed-loop poles on a desired constant damping ratio or zeta line.

2. FINDING K FOR A SPECIFIED ZETA (THE GRAPHICAL APPROACH)

The Root Locus plot for the control system described by equation 2 is shown in Figure 1. Suppose it is desired to find the gain K which places a pair of dominant complex conjugate closed loop poles such that the system's response to a step input will yield a percent peak overshoot of 20 corresponding to a damping ratio or zeta of about 0.46.

The usual approach is to superimpose a zeta line of 0.46 on the Root Locus plot. This translates into a line that subtends a positive angle of theta = (180^{deg} - $\cos^{-1} 0.46$) as shown in figure 2. The intersection of this line with the Root Locus gives the value(s) of the complex frequency, s, which can be used to evaluate K by means of Equation [2,3]. It is clear from the outlined procedure that the accuracy of the method depends on how the Root Locus is plotted. Lacking this however, graphical evaluation of K becomes a trail and error technique.

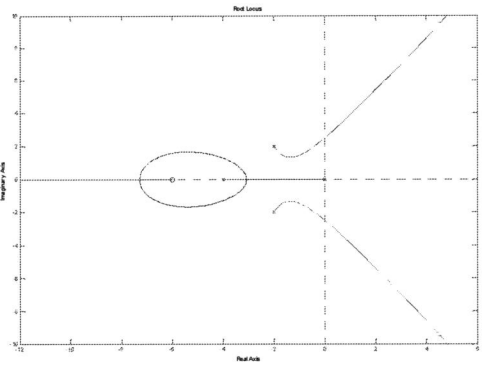

Fig.2. Root Locus and The Graphical Method

3. THE PROPOSED METHOD

As equation 1 defines the Root Locus, the approach adopted here is to evaluate this equation at a value of the complex frequency, s, which lies on the specified constant damping ratio or zeta line. This results in a set of two simultaneous equations in two unknown parameters, the gain K and the natural frequency w_n of the system. In all cases that occur, K is a linear parameter and so can be easily eliminated. This results in a single equation, a polynomial in the unknown w_n. This can be solved using any root finding technique.

4. AN EXAMPLE OF THE PROPOSED METHOD

Consider the system described by equation 2. With reference to figure 1, along the constant zeta line in the s-plane, the complex frequency s can be expressed as

$$S = w_n e^{j\theta} \quad (3)$$

Where the angle θ is as defined in the previous section. Substituting for S in Eq. 2 and simplifying yields the following:

$w_n^4 e^{j4\theta} + 8 w_n^3 e^{j3\theta} + 24 w_n^2 e^{j2\theta} + w_n^4 e^{j4\theta} + (32 + K) w_n e^{j\theta} + 6K = 0$ \quad (4)

For a specified value of zeta = 0.456, for example, the value of θ is 117.129 deg. Using the identity

$e^{j\theta} = \cos(\theta) + j \sin(\theta)$

Eq. 4 becomes

$- 0.32 w_n^4 + 7.91 w_n^3 - 14.02 w_n^2 + (-14.59 - 0.46) w_n + 6K = 0$ \quad (5a)

for real part

$0.95 w_n^4 - 1.2 w_n^3 - 19.48 w_n^2 + (28.48 + 0.89) w_n = 0$ \quad (5b)

for the imaginary part.

Eliminating K in Equation 5a and 5b results in the single equation

$-.15 w_n^5 - .80 w_n^4 + 14.17 w_n^3 - 116.88 w_n^2 + 170.87 w_n = 0$ \quad (6)

The roots of Eq. 6 using Newton's method on PC are $w_n = 0, -15.25, 1.802, 4.04 + j 5.02$

The presence of only one positive real root of 1.0802 implies the existence of only one value of K for which the roots are on the specified constant zeta line. Substituting for $w_n = 1.802$ in either Eq. 5a or Eq. 5b gives a value of the gain K as 5.58.

5. COMPUTER IMPLEMENTATION

Consider the numerator and denominator of G(s) H(s) as N(s) and D(s) respectively. Then Eq. 2 can be generally expressed as:

$$1 + KN(s)/D(s) = 0 \qquad (7)$$

From Eq. 7, the corresponding expressions for equations 5(a) and 5 (b) are

$$Re[D(s)] + K\,Re[N(s)] = 0 \qquad (8a)$$

And

$$Im[D(s)] + K\,Im[N(s)] = 0 \qquad (8b)$$

Where Re is the real part and Im denotes the imaginary part. Eliminating K yields
$$Re[N(s)]\,Im[D(s)] - Re[D(s)]\,Im[N(s)] = 0 \qquad (9)$$

Assume $N(s) = a + jb$ and $D(s) = c + jb$ which are complex quantities. Then, indicating the conjugate expression by a^*, as follows:
$[N(s)]^* = N(s^*)$
Therefore
$\{[N(s)]^*\}\,D(s) = N(s^*)\,D(s)$
$\qquad = [(a+jb)^*][c+jd]$
$\qquad = (a-jb)(c+jd)$
$\qquad = (ac+bd) + j(ad-bc) \qquad (10)$

It is worth noting that the left hand side of Eq. 9 is identical with the imaginary part of Eq. 10. Thus
$$Im[N(s^*)\,D(s)] = 0 \qquad (11)$$

Equation. 10 is more amenable to computer implementation. It can be solved using the software MATLAB[7] for w_n. The vector K of gain values corresponding to each value of w_n can also be computed in the same matlab environment by using Eq. 8a or 8b.

The following lines of matlab codes was used in the previous example control system:
For a kth order N(s) with (k+1) coefficients and an mth order D(s), denoting theta by th and w_n by wn.
N = [1   6]
D = [1   8   24   32   0]
Eq. 10 becomes

```
k  = length(N) – 1;
Nw = N .* exp(-j* th .* [k: -1: 0];
m  = length(D) –1;
Dw = D .* exp(j*th*[m:-1:0]);
wn = roots(imag(conv(Nw,Dw)));
wn = wn ((real(wn)>0)&(abs(imag(wn))<1e-5));
Nval = polyval(N, wn .*exp(j*th));
Dval = polyval(D, wn exp(j*th));
K = - real(Dval)/real(Nval);
```

6. CONCLISION

A non-graphical method for finding the exact system parameters to place a pair of complex conjugate poles on a specified damping ratio line has been presented. The procedure is based on exact solution of complex equations on the PC. The technique is non-iterative and can provide an alternative approach to design using the Root Locus method.

REFERENCES

[1] Tomas O. Silva, *Automatic Generation of Root Locus Plots*, Revista Do Detua, Vol. 2, No 3, September 1998.

[2] Benjamin C. Kuo, Farid Golnaraghi, *Automatic Control Systems*, 8th ed., John Wiley& Sons, Inc.,2003.

[3] Wang Z., etl., *"Robust Hurwitz Stability Test for Linear Systems"*, IEEE Transaction on Automatic Control, Vol. 49, No. 8, August 2004.

[4] Dorf R. C., *Modern Control Systems*, 6th ed., Addison- Wesley Publishing Co. Inc., NY, 1992.

[5] Shinners S. M, *Modern Control System: Theory and Application*, 2nd ed, Addison-Wesley Publishing Co. Inc. NY, 1978.

[6] Van De Vegte, J., *Feedback Control Systems*, 3rd ed., Prentice Hall, Englewood Cliffs, New Jersey, 1994.

[7] Shahian B., Hassul M., *Control System Design using MATLAB*, 1st ed, Prentice Hall, New Jersey, 1993.

Application of GMMs to Speech Recognition using very short time series

S. FRIHA & (*) N. MANSOURI (**)

(*)Département d'Electronique
Centre Universitaire Chikh Laarbi Tébessi
Route de Constantine, Tébess,ALGERIA. E-mail: souad_fri@yahoo.com
(**)Laboratoire d'Automatique et de Robotique
Département d'Electronique, Université Mentouri Constantine
Route Ain EL Bey, Constantine,ALGERIA. Email: nor_mansouri@yahoo.fr

Abstract: This paper reports on some recent results in speech recognition using the state of the art GMMs modeling. This is done over a reconstructed multidimensional attractor that is obtained via an embedding procedure into a phase space.
The novelty is being the use of very short time series of 20ms of speech for both the training data base and the test samples.
Classification accuracies reached 75.99% when four phoneme classes are concerned and 100% when there are only two phoneme classes. Experiments over two monosyllabic words gave an accuracy of 89.55%. Application of GMMs to speaker recognition, without using the traditional MFCC parameters was performed too and has resulted in an accuracy of 68.51%.

Keywords: GMM, short time series, speech recognition.

1. INTRODUCTION

Spectral Auto-Correlation, Peak-to-Value-Ratio, Linear Prediction Coding and many other linear techniques have been extensively used for the last years in speech processing and have resulted in many successful speech applications [1][6]. However, their shortcoming is that they are built upon linear assumptions of the vocal tract while there are strong evidences of the nonlinear nature of the speech production [1][5][6].

As an alternative, interest has emerged in applying nonlinear methods to speech. These nonlinear methods use in their majority dynamical systems models in the time domain rather than the spectral domain [5].

Our approach to capture the nonlinearities of the speech production is based on a dynamical system method called Phase Space Reconstruction.

The reconstructed phase space is a plot of the time-lagged vectors of signal, which are used to represent the nonlinear structure. Reconstructed phase spaces are topologically equivalent to the original system, if the embedding dimension is large enough [1][5][6].

Structural patterns, commonly referred to as trajectories or attractors, occur in this phase space. They can be quantified through direct models of the phase space distribution.

State of the art speech recognition systems use Gaussian Mixture Models (GMM) [5][8]. This is justified by the ability of linear combinations of gaussians to represent a large amount of distributions especially those emanating from the real world.

In this paper, we investigate the usefulness of GMMs to characterize and classify very short speech signals. The modeling is done not on the scalar speech signal but on its reconstructed multidimensional attractor that is obtained via an embedding procedure into a phase space.

The paper is organized as follows. In Section 2, the embedding of the system into phase space is presented. In Section 3, we report the theoretical background of GMMs. The experimental procedure and the main results are presented in section 4.

2. EMBEDDING AND ATTRACTOR RECONSTRUCTION

Phase space reconstruction techniques are founded on underlying principles of dynamical system theory and have been applied to a variety of time series analysis and nonlinear signal processing applications [1][5][6].

On both theoretical and practical levels, there are three benefits of chaos theory for system analysis. First, it applies to highly nonlinear systems and naturally accounts for all important system dynamics and secondly, it uncovers system information and relationships without having to uncover the laws and equations of the underlying dynamics [2].

One of the most popular representations of the chaotic nature of signals can be attained via Takens' embedding theorem, which states that the reconstruction of a state space representation is topologically equivalent to the original state space. Nonlinear dynamic progression of speech can be observed as a vector, which travels along a space trajectory, where the coordinates of the point are the degrees of independence of the system.

We assume that the speech production system can be viewed as a nonlinear but finite dimensional dynamical system. Given a speech signal segment s(n), n =1 ,...,N and according to the Taken's theorem, the vector:

$$X(n)=[s(n), s(n+Td), s(n+2Td), ..., s(n+(De-1)Td)] \quad (1)$$

formed by samples of the original signal delayed by multiples of a constant time delay Td, defines a motion in a reconstructed De dimensional space that has many common aspects with the original phase space.

X(n) reproduces the major part of the dynamics of the original system. Thus, by studying the constructible dynamical system we can uncover useful information about the original unknown dynamical system.

The geometric representation of equation (1) is called an attractor. Fig. 1 provides an illustrative reconstructed phase space with trajectory information for the French word "oui" (lag = 7).

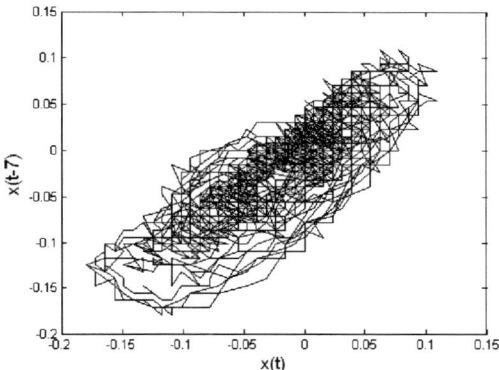

Fig. 1. *Attractor in a 2-D phase space of an uttered french word 'oui' (in order to be fully unfolded, dimension should be greater or equal to 8)*

The theorem does not specify a method to determine the required parameters (Td, De) but it only sets constraints on their values. Hence, procedures to estimate good values of these parameters are essential.

Td is related to the correlation or mutual information among speech samples. It must be neither too small so that the original time series is mimicked nor too large so that s (n) and s (n-Td) are independent.

In order to achieve a compromise one can use the first minimum of the average mutual function:

$$I(T) = \sum_{n=1}^{N-T} P(s(n), s(n+T)) \cdot \log_2 \left[\frac{P(s(n), s(n+T))}{P(s(n))P(s(n+T))} \right] \quad (2)$$

where P(.) denotes probability. I(T) measures the average mutual information between samples that are Td positions apart.

After fixing Td, the next step is to select the embedding dimension of the reconstructed vectors. The dimension is found by increasing its value until the percentage of false neighbours reaches zero or is minimized [3][7].

After choosing Td and De the embedding of the speech signal into a multidimensional phase space and the construction of its attractor can be accomplished as shown in Fig. 1.

3. GAUSSIAN MIXTURE MODELS

GMMs are a generalization of Gaussian distribution functions. Gaussian probability density function (pdf) are ideal because mathematical analysis of Gaussian's is simple, and because of the central limit theorem, which states that in the limit, the sum of random variables with identical distributions of any type has a Gaussian distribution.

A Gaussian pdf of an independent random variable x is given by :

$$p(x) = \frac{1}{\sqrt{2\pi\sigma^2}} e^{\frac{(x-\mu)^2}{2\sigma^2}} \quad (2a)$$

where μ is the mean and σ^2 is the variance of the Gaussian.

The multivariate Gaussian function of the random vector x is given by:

$$p(x) = \sum_{m=1}^{M} \omega_m p_m(x) = \sum_{m=1}^{M} \omega_m N(x, \mu_m, S_m) \quad (3)$$

Where ω_m is the mixture weights ($\sum \omega_m = 1$) M is the number of mixtures and $N(x, \mu_m, S_m)$ is a normal distribution with mean μ_m and covariance matrix Sm.

The weights, means and covariance matrix of the GMM are estimated using the Expectation Maximization (EM) algorithm [8] The method begins with initial values for each parameter, then iterates through the available data to find the Maximum Likelihood (ML) estimate.

The (ML) classifier uses the estimates of the distribution from the direct statistical modelling of each RPS (Reconstructed Phase Space). This classifier computes the conditional probabilities of the different classes given the phase space and then selects the class with the highest likelihood. Each RPS (Reconstructed Phase Space) signal is assigned a log likelihood $\hat{\omega}$ for the ith class according to :

$$\hat{\omega} = \arg\max_{i=1...C} \{p_i(x)\} \quad (4)$$

With pi(x) is the GMM probability distribution of the ith class and C is the number of classes. The probabilities implied in the above formula are obtained from the estimates of the probability mass function using training data.

The formulas used for the estimation are:

$$\mu_m = \frac{\sum_{n=1+(d-1)\tau}^{N}(p_m(x_n)x_n)}{\sum_{n=1+(d-1)\tau}^{N}p_m(x_n)}$$

$$\Sigma_m = \frac{\sum_{n=1+(d-1)\tau}^{N}(p_m(x_n)(x_n - \mu_m))}{\sum_{n=1+(d-1)\tau}^{N}p_m(x_n)}$$

$$\omega_m = \frac{\sum_{n=1+(d-1)\tau}^{N}p_m(x_n)}{\sum_{n=1+(d-1)\tau}^{N}\sum_{m=1}^{M}p_m(x_n)}$$

(5)

Fig.2 illustrates a GMM based modeling of the word 'oui' from Fig.1. The set of GMMs approximately models the distribution of the data.

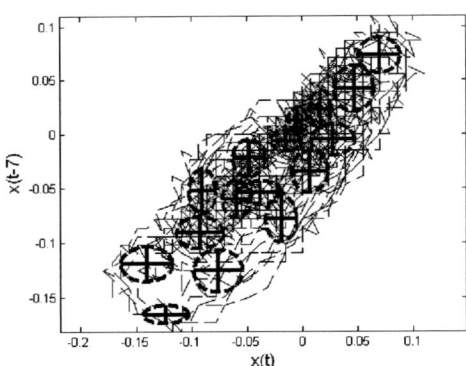

Fig. 2. *Data and covariance. The principal axes of each ellipse are labeled for each GMM.*

4. EXPERIMENTAL RESULTS

In a first time, we investigate the usefulness of GMMs to characterize two classes of monosyllabic french words 'oui' and 'non'. Each word is uttered 25 times by seven speakers, three males and four females. Then in a second time, GMMs are applied to the following french phonemes, each of them uttered ten times by three different speakers:

Vowels: 'a', 'o', 'e', 'i', 'u'

Nasals: 'm', 'n,'

Stops: 't', 'p'

Fricatives: 's', 'z'.

In all cases, signals are sampled at 22Khz, with 16 bits resolution; they are pre-filtered and silence parts are eliminated. For the construction of the phase space, we choose the values of 7 for the time lag and 10 for the embedding dimension. GMMs with full covariance matrices are used because they achieve a better fit to the distribution of training data than do GMMs with other covariance matrices. The resulting additional calculation is overcome when the size of the training data is limited.

A total of seven fricatives, seven vowels, five stops and five nasals are selected for the test.

Tables [1-4] illustrate the classification average accuracy obtained for different numbers of GMMs.

Table 1. Classification accuracy versus the number of GMMs within the first data base ("oui"/ "non")

Classification parameters	Accuracy
# GMMs = 2	89.14%
# GMMs = 5	**89.55%**

Table 2. Classification accuracy versus the number of GMMs within vowels and nasals

Classification parameters	Accuracy
# classes=2 (vowel 'e'/nasal 'n')	
# GMMs = 2	**100%**
# GMMs = 5	**100%**

Table 3. Classification accuracy within the second data base (4 classes are considered).

Classification parameters	Accuracy
# clclasses=4 (vowels, nasals, stops and fricatives)	
# GMMs = 2	71.44%
# GMMs = 3	69.98%
# GMMs = 4	71.04%
# GMMs = 5	73.59%
# GMMs = 6	74.31%
# GMMs = 7	74.18%
# GMMs = 8	74.94%
# GMMs = 9	**75.99%**
# GMMs = 10	73.94%
# GMMs = 12	72.57%
# GMMs = 14	71.66%
# GMMs = 16	70.52%

Table 4. Classification accuracy within the second data base (3 speakers are considered).

Classification parameters	Accuracy
# classes=3 (three speakers)	
# GMMs = 2	64.98%
# GMMs = 3	66.22%
# GMMs =4	67.37%
# GMMs =5	60.78%
# GMMs =6	60.97%
# GMMs =7	64.91%
# GMMs =8	62.21%
# GMMs =10	67.46%
# GMMs =12	**68.51%**

5. CONCLUSIONS

This study has shown the effectiveness of using very short time series of speech for recognition tasks in a GMM-based framework.

Application of GMMs to speaker recognition, without using the traditional MFCC parameters has resulted in an accuracy of 68.51%.

Although GMMs are in most cases employed in combination with a MFCC analysis for speaker recognition, experimentation showed that they can be used successfully in speech recognition too. Good recognition rates are obtained, 89.55% for the YES/NO data base and 75.99% for the four classes phonemes.

When classes are too dissimilar rates of 100% could be obtained (Table 2).

A big number of GMMs does not imply necessarily good recognition rates. We obtained an accuracy of 71.44% with only 2 GMMs (Table 3) while in [1] only 38.06% are obtained with 128 GMMs.

In future work we will consider more closely the problem of choosing the convenient number of GMMs and also look at increasing the recognition rates by some combinations between GMMs and linear or nonlinear methods.

REFERENCES

[1] Andrew C. Lindgren, Michael T. Johnson, Richard J. Povinelli, (2003). Speech Recognition using Reconstructed Phase Space Features. *IEEE, ICASSP*.

[2] Christopher Frazier, Kara M. Kockelman (2004). Chaos theory and Transportation Systems: An Instructive Example. 83rd Annual Meeting of the Transplantation Research Board, Washingtone DC.

[3] Iasonas Kokkinos and Petros Maragos (2005). Nonlinear Speech Analysis Using Models for Chaotic Systems. *IEEE Transactions on Speech and Audio Processing*, **Vol. 13, No. 6**., pp.1098-1109.

[4] Jérôme Louradour, (2007). Noyaux de Séquences pour la Vérification du Locuteur par Machines à Vecteurs de Support. Université de ToulouseIII.

[5] Kevin M. Indrebo, Richard Povinelli, Michael T. Johnson (2004a). A comparison of Reconstructed Phase Spaces and Cepstral Coefficients for Multi-Band Phoneme Classification. Speechlab.eeece.mu-edu/papers/Indrebo_icsp04.pdf.

[6] Kevin M. Indrebo, Richard Povinelli, Michael T. Johnson (2004b). A combined Sub-band and Reconstructed Phase Space Approach to Phoneme Classification. Speechlab.eeece.mu-edu/papers/Indrebo_ nolisp03.

[7] Rainer Hegger, Holger Kantz and Thomas Shreiber. (1998). Practical Implementaion of Nonliner Time Series Methods: the TISEAN Package. arXiv: chao-dyn/9810005 **vol. 1**, pp.1-2

[8] Yassine Mami, (2003). Reconnaissance de Locuteurs par Localisation dans un espace de locuteurs de Références. Ecole nationale supérieure des télécommunications, France.

Particle Swarm Optimization for Image Deblurring

A. Toumi*, A. Taleb-Ahmed**, K. Benmahammed***, N. Rechid ****

*University Mohamed Khidher Biskra Algérie
** LAMIH UMR CNRS UVHC 8530, le Mont Houy 59313 Valenciennes Cedex 9, France
***Université Farhat Abbas Setif Algérie
****University Mohamed Khidher Biskra Algérie
* abida_ba@yahoo.fr,
** taleb@univ-valenciennes.fr
*** khierben@yahoo.fr
**** Rechidnaima@yahoo.fr

Abstract: Within the framework of this first study we suggest the use of the Particle Swarm Optimization technique (PSO) in the image restoration field. In our knowledge, we did not still find works concerning the image deblurring (restoration) using the PSO. So, in this paper, we present the use of the PSO in two manners: (i) by taking as hypothesis the degraded image as the entire swarm and pixels as particles; (ii) the degraded image is taken as particle and we generate a population of images to buildup a swarm of variable size. The first results which we give were validated on real images degraded by a Gaussian blur only, and degraded by a Gaussian blur and an additive Gaussian noise. We finish with the comparison of our results with some classical restoration methods.

Keywords: Particle swarm optimizations, Image restoration, image deblurring, SNR, Gaussian blur, Gaussian noise

1. INTRODUCTION

Image restoration constitutes an important step in image processing process. The images which we analyze are in most part of cases degraded not only by an additive noise but also by a degradation of blur type. Several techniques were developed to resolve this problem [1, 5, 9-12, 15]. The objective of the restoration techniques is to reconstruct original image from that degraded based on a priori knowledge about the conditions of degradation. In that, the image restoration techniques are numerous and can be classified in two classes: deterministic approaches and stochastic approaches.

1.1 Deterministic approaches [9-12]

They are techniques based on the results of the one-dimensional signal processing, since image is taken as a signal in two dimensions. We distinguish:

a) Direct methods: the inverse filtering, the pseudo- inverse filtering, the Wiener filter.
b) Iterative methods: the Lucy- Richardson method, the successive estimates method of Tikhonov- Miller.
c) Methods based on constrained quadratic error: The maximum likelihood methods, the methods with Bayesian approach, the Wiener estimation, the Kalman Filter ...etc.

The direct techniques are not always feasible because of the inversion problem of the degradation matrix. The same for the techniques based on constrained quadratic error, we find the problem of matrix inversion.

For the iterative techniques which were conceived to resolve the inversion problem, in these, obtained images are solutions of a certain cost function where constraints are imposed. This introduction required the use of a regularization parameter, the choice of this parameter has a big influence on the restoration process executed by these methods.

1.2 Stochastic approaches [11]

These techniques are interested in the non stationary spatial nature of the image, in the variation case of the degradation function, in case when the degradation function is unknown (the blind Restoration).

For stochastic methods, we notice their sensibility to noise, their need to use supplementary algorithms or methods to finish a treatment, the complexity of their implementation, and the choice of their initialization which can not ensure the convergence toward the desired solution.

1.3 Model of the degradation process [9-12]

Principle

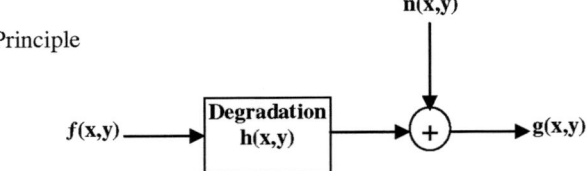

Fig.1. Principle diagram of an image degraded by a blur

$$g(x, y) = h(x, y) * f(x, y) + n(x, y) \quad (1)$$

Where h: represent the impulse response of the degradation system.

G, f and n represent: degraded image, original image, and the additive noise introduced by the system, respectively.

In discrete notation, the convolution is formulated as fellows:

$$g(x,y) = \sum_{k=1}^{M}\sum_{l=1}^{N}[h(x-k, y-l) \times f(k,l)] + n(x,y) \quad (2)$$

In matrix notation, the process became:

$$g = H \times f + n \quad (3)$$

Note that, g, f and n are NxM matrix, and H is the degradation matrix.

Within the framework of this article, we suggest for the first time, to our knowledge, the application of the PSO in the case of the image restoration. For which we are giving two possible ways of the use of the PSO:

1) PSO in restoration by pixel, where the swarm particles are all degraded image pixels;

2) PSO in global restoration, where the degraded image in total is taken as particle within a swarm of similar images.

The results which we propose are given when the image undergone a Gaussian blur degradation. In a case the image is degraded by the blur alone, and in other case degraded by a blur and a noise which are the two Gaussians.

We compare our results with those of classical restoration (deterministic methods and stochastic methods).

The comparison criterions we used are the SNR and calculus time.

Our article is planned as follows: in section 2, we remember and detail the PSO approach, in section 3, we present our contribution, in section 4, the results obtained with discussion, and at last we end by a conclusion and some perspectives related to this work.

2. PARICLE SWARM OPTIMIZATION

The evolutionary computation field is often considered to comprise four major paradigms: genetic algorithms, evolutionary programming, evolution strategies, and genetic programming. In a similar way to these evolutionary paradigms, particle swarm optimization uses "a population" of candidate solutions to develop an optimal solution of the problem. The degree of optimality is measured by a fitness function defined by the user [2, 4, 6, 7, 16].

Particle swarm Optimization (PSO), which has roots in artificial life and social psychology as well as engineering and computer science, differs from evolutionary computation methods in that the population members called "particles", are scattered in the space of the problem [6, 7].

Particles swarm Optimization method has been born in 1995 in the United States under the name of: Particle Swarm Optimization (PSO). His two designers, Russel Eberhart and James Kennedy [6], tried to model social interactions among "agents" to reach an objective given in a common space of search. Every agent has a certain capacity of memorization and data processing. The basic rule was that it did not there have to have any leader, or even any knowledge by the agents of the information set, only local knowledge. A simple model was then elaborated [3, 6-8]. The PSO can be arranged under the class of iterative methods as well as within the stochastic techniques.

2.1 Informal description [7, 8, 13, 14]

The historic version can be easily described from a Particle point of view. The algorithm starts with a randomly distributed swarm in the space of Search. Also, every particle has a random speed. Then, at any time: (**fig.2**)

Each particle is able to estimate the quality of its position and to keep in memory its best performance, that is the best position which it reached up to here (which can in fact be sometimes the current position) and the quality (value in this Position of the function to be optimized).

Each particle is capable of interrogating certain number of her congeners (her advisers, of whom it even) and to obtain from each of them its own better performance (and concerned quality).

At any step of time, each particle chooses the best of the best performances of which it has knowledge, modifies its speed according to this information and to its own data, and then moves in consequence.

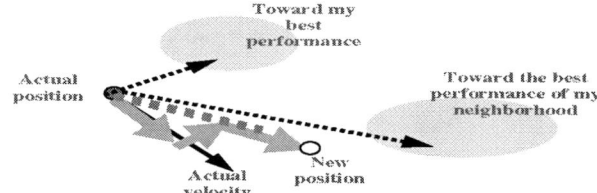

Fig.2. Principle diagram of a particle movement.

To realize the next Movement, every particle combines three tendencies: follows the appropriate speed, returns towards the best performance, and goes towards the best performance of her advisers.

Best time discovered adviser, the modification of the speed is a simple linear combination of three tendencies, by means of reliable coefficients:

"Adventurous" tendency, consisting in continuing according to the current speed,

"Conservative" tendency, returning more or less towards the best already found position,

"Panurgian" tendency, directing approximately to the best adviser,

Terms "more or less" or "approximately" make reference to the fact that hazard plays a role, due to a random and limited modification of the confidence coefficients, what favours the exploration of the search space.

The (**Fig.3**) presents a diagram summarizing explanations above.

Fig.3. Simple local application rules used in a particle swarm

2.2 Particles interactions [7]

The effectiveness of the PSO comes from the interactions of particles with their neighbours. When one particle discovers a local optimum, *it* becomes the best among her neighbours. And *t*hose are attracted towards this optimal region. During their movement towards the new optimum, particles can met a new better optimum. In that case, they are going to attract the first better particle towards the best position which they discovered.

In a typical particle swarm, neighbourhood size can vary from three (the particle and her two neighbours) to the total population size (in that case, there is a single neighbourhood).

2.3 Neighbourhood [7, 8]

The neighbourhood constitutes the structure of the social network. Particles inside a neighbourhood communicate with each other. Various neighbourhoods were studied [7]

Star topology (**fig.4**): the social network is complete. Each particle is attracted towards the best particle noted *gbest* and communicates with the others.

 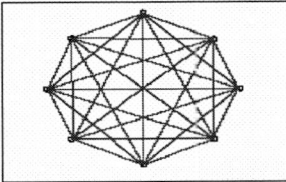

Fig.4. star Neighbourhood

Ring topology (**fig.5**): each particle communicates with n (n=3) immediate neighbours. Each particle tends to move towards the best particle in its local neighbourhood (*lbest*).

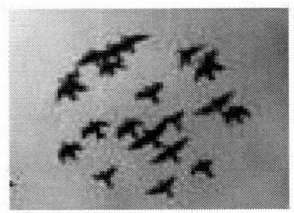 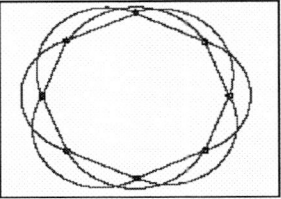

Fig.5. Ring neighbourhood

Beam (Von Neumann) topology (**fig.6**): a central particle is connected to all others. Only this central particle adjusts its position towards the best, if it provokes improvement, information is propagated to the others.

 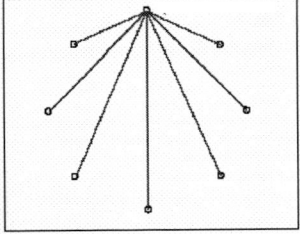

Fig.6. Beam neighbourhood

The choice of the neighbourhood topology of has a big effect in the propagation of the best solution found with the swarm. Using the *gbest* model propagation is very fast (that is all the particles in the swarm will be affected by the best solution found in the iteration t, at once in the iteration t+1). However, using the ring and beam topologies will slow down the convergence rate because the best found solution has to propagate through several Neighbourhoods before affecting all particles in the swarm. This slow propagation will allow the particles to explore more areas in the search space of the problem.

2.4 PSO algorithm [6-8]

Each particle in the swarm is represented by the followed characteristics:

x_i : The current position of the particle *i*.

v_i : The current velocity of the particle *i*.

y_i : The best personal position of the particle *i*.

$\hat{y}_i$: The best neighbourhood position of the particle i.

The update of the personal best position of a particle is as follows:

$$y_i(t+1) = \begin{cases} y_i(t) & si \quad f(x_i(t+1)) \geq f(y_i(t)) \\ x_i(t+1) & si \quad f(x_i(t+1)) < f(y_i(t)) \end{cases} \quad (4)$$

The position of the global best particle is then given by:

$$\hat{y}(t) \in \{y_0, y_1, ..., y_s\} = \min\{f(y_0(t)), f(y_1(t)), ..., f(y_s(t))\} \quad (5)$$

S: denotes the size of the swarm.

So the velocity of the particle *i* is updated using the following equation:

$$v_{ij}(t+1) = wv_{ij}(t) + r_1 c_1 (y_{i,j}(t) - x_{ij}(t)) + r_2 c_2 (\hat{y}_j(t) - x_{ij}(t)) \quad (6)$$

Where: *w* is the inertia weight

c_1 and c_2 are acceleration constants

r_1 and r_2 are uniformly distributed variables.

j=1: *D*, where *D*: the dimension of the search space of the considered problem.

The position of the particle i is updated by the equation:

$$x_{ij}(t+1) = x_{ij}(t) + v_{ij}(t+1) \quad (7)$$

It is the velocity vector which drives the search process and reflects the "sociability" of particles.

We envisaged the application of this tool of optimization in restoration of image to profit of advantages which it brings.

3. APPLICATION

To be able to use the PSO as an optimization tool, we convert the image restoration problem on an optimization problem. In this case a cost function, closed to this problem, which has to be optimized is presented. The solution which gives the optimal value constitutes the desired image.

Our problem cost function is the following:

$$J(\hat{f}) = \frac{1}{2}\|g - H \times \hat{f}\|^2 + \frac{1}{2}\lambda\|C \times \hat{f}\|^2 \qquad (8)$$

It can be presented in other way by:

$$J(\hat{f}) = \frac{1}{2}\hat{f}^T \times (H^T \times H + \lambda C^T \times C) \times \hat{f} - g^T \times H \times \hat{f} + \frac{1}{2}\|g\|^2 \qquad (9)$$

Called Constrained Least Square Error (CLS).
Where:
C: an MxN matrix associated to the high-pass operator c(x,y), which represent the imposed noise constraint.
λ: regularization operator, $\lambda \in [0, 1]$.
$\hat{f}$: is the estimated or desired image.

In this section we present the application of this tool, the PSO, to the supervised image restoration where the noise and the blur are known, and both are taken Gaussian.
The Gaussian blur is often due to the atmospheric turbulences. The impulse response of the Gaussian degradation filter is as follows:

$$h(x, y) = \frac{1}{\sigma_x \sigma_y}\exp\left\{\frac{x^2 + y^2}{2\sigma_x \sigma_y}\right\} \qquad (10)$$

For this purpose we proceeded in two manners:
1. In the first one, the image is taken as the entire swarm, and the pixels are the particles. So a local restoration is made.

The followed procedure is as follows:
- The image represented by a matrix is converted on line vector;
- Search space dimension is taken $D=1$, because our variable in this case is the pixel which is an one-dimensional variable;
- The swarm size is the total number of the image pixels, so it's the length of the image vector;
- The fitness function is applied to each element of the vector (pixel) as showed in the followed function:

$$J(\hat{f}(i)) = \frac{1}{2}\hat{f}^T(i) \times (H^T \times H + \lambda C^T \times C)(i,j) \times \hat{f}(i)$$
$$- g^T(i) \times H(i,j) \times \hat{f}(i) + \frac{1}{2}\|g\|^2 \qquad (11)$$

- The velocity and position update equations of the PSO become:

The velocity:
$$v_i(t+1) = wv_i(t) + r_1c_1(y_i(t) - x_i(t)) + r_2c_2(\hat{y}(t) - x_i(t)) \qquad (12)$$
The position:
$$x_i(t+1) = x_i(t) + v_i(t+1) \qquad (13)$$

In this case, $x_i(t)$ is the current pixel, $y_i(t)$ is the best personal performance of the pixel, $\hat{y}(t)$ is the best global pixel, and $v_i(t)$ is the pixel's velocity.
The PSO has been applied for one-dimensional problems. Our contribution in this case is the use of the pixel as particle which can move possessing a changeable velocity and position.

2. In the second one, a swarm of images is generated from the degraded one; in this case it's a global restoration that is made.

The followed procedure is as follows:
- Similarly, the image represented by a matrix is converted on line vector;
- Search space dimension is taken D=the length of the image vector;
- The swarm size is variable. In our application we take it 20; The fitness function is applied to each particle which is in this case the image vector as follows:

$$J(\hat{f}) = \frac{1}{2}\hat{f}^T \times (H^T \times H + \lambda C^T \times C) \times \hat{f} - g^T \times H \times \hat{f} + \frac{1}{2}\|g\|^2 \qquad (14)$$

- The velocity and position update equations of the PSO become:

The velocity:
$$v_i(t+1) = wv_i(t) + r_1c_1(Y_i(t) - X_i(t)) + r_2c_2(\hat{Y}(t) - X_i(t)) \qquad (15)$$
The position:
$$x_i(t+1) = x_i(t) + v_i(t+1) \qquad (16)$$

In this case, $X_i(t)$ is the current image, $Y_i(t)$ the best personal performance of the image, $\hat{Y}(t)$ the best global image obtained, and $v_i(t)$ is the image's velocity.
The PSO has also been applied for N-dimensional problems. Our contribution in this case is the use of the entire image as particle which can move possessing a changeable velocity and position which can reach.

To test the two methods we used a test image often used in this problem, which is the cameraman image.
We undergone this image to a Gaussian blur degradation with variance $\sigma_f = 5$, and a mean $\mu_f = 0$.
First of all we restore only blur degraded image;
In second place we restore blurred and noised image. The noise is white noise with zero mean, $\mu_n = 0$, and variance $\sigma_n = 0.002$.

(a) (b)

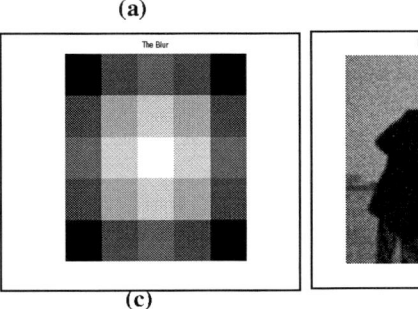

(c) (d)

Fig.7. (a) original image, (b) degraded image with a Gaussian blur, $\sigma_f = 5$ and SNR = 4.0994, (c) the Gaussian blur, (d) blurred image and noised by a white noise with mean $\mu_n=0$ and variance $\sigma_n=0.002$ and a SNR = 0.7026.,

The obtained results after application of the PSO on the blurred image, and blurred and noisy image gave us the following images:

3.1 Pixel restoration

(a)   (b)

Fig.8. Restored image with the PSO (the image is the swarm), (a) blurred SNR= 3.9761, (b) blurred and noised with SNR = 2.9731.

3.2 Global restoration

(a)   (b)

Fig.9. restored image with the PSO (swarm of 20 images), (a) blurred SNR= 2.0203, b) blurred and noisy with SNR = -0.8030.

4. RESULTS DISCUSSION

To show the effectiveness of the PSO in image restoration, we have degraded the original image **fig.7.a** with a Gaussian blur, the resulted image **fig.7.b** gave an SNR= 4.0994. For more validity we suggested the application of this tool not only to the blurred image, but also noised by a Gaussian noise **fig.7.d**.

Algorithms were implemented under Matlab7, in a computer Intel, Pentium 4, of processor 2.80 GHz, and 512 MØctet of RAM.

The simulation results which we obtained show the possibility of using the PSO in the image restoration field. Besides, obtained images **fig.8** and **fig.9** show the efficiency of the application of this tool in this domain.

By comparing results obtained from signal to noise report (SNR) point of view, we notice that the case in which image is taken as entire swarm **fig.8.a**, the SNR of the restored image has value SNR = 3.9761. Whereas in the case when the image is taken as a particle in a swarm of 20 images **fig.9.a**, the SNR of the restored image has value SNR = 2.0203.

So we notice that restoration by pixel, or we can call local restoration, gives a better result than restoration by image, or as well as we can call global restoration.

Similarly for the obtained results in the case of blurred and noised image, the restored images **fig.8.b** and **fig.9.b** gave both an SNR: 2.9731 and -0.8030 respectively.

Now, we compare the results from execution time point of view. For the results of **fig.8.a** and **fig.9.a** the execution time is approximately 10min and 3min respectively. For the results of fig.8.b and fig.9.b the time is estimated 13min and 5min respectively.

We notice that the local restoration took more time than the global restoration in both cases, either for the blurred image only, or for the blurred and noisy image.

So, the advantage which gives the one is itself the drawback of the other one. Either at time or in quality, represented with the SNR.

A comparison of the simulation results of some classical techniques of image restoration (determinist and stochastic), with obtained results by using the PSO is summarized in table **1**.

Table.1. Summarizing Table of some techniques of image restoration resulted SNR

Restoration methods			SNR	
			blurred	blurred and noisy
Inverse filtering			1.1860	1.1860
pseudo Inverse filtering			1.1860	1.1860
direct CLS filter			4.4986	-5.9177
iterative CLS filter			4.0994	0.7337
Wiener filter			-5.3706	-21.8618
Tikhonov- Miler method			-17.3017	-17.3021
Lucy- Richardson method			3.7169	-2.9828
Hopfield neural network			-3.1260	-3.8150
Markov Random Field(MRF)			-3.0262	10.6521
Wavelet (Daubechies)	Lucy- Richardson		-16.4836	-26.9108
	Wiener filter		-5.3706	-22.0004
	denoising	Lucy		-9.5221
		Wiener		-7.0339
PSO in local restoration			3.9761	2.9731
PSO in global restoration			2.0203	-0.8030

5. CONCLUSION

Our contribution consisted of the application of the PSO algorithm in the image restoration field. This tool which was not still applied, to our knowledge, in this domain, gave results which can be esteemed as good according to those given by the classic techniques image restoration (**table 1**).

So we can conclude that this tool (the PSO), proved its efficiency in the field of image restoration as it proved it in numerous other Ingeneering domains.

Besides the results that the PSO gave, we noticed its simplicity of implementation on computer unlike the other evolutionary techniques (as genetic algorithm for example).

Finally, we can say that the OEP is a very powerful optimization tool. And the results which we obtained after its use are very satisfactory. This tool gave sense and direction to our contribution.

Perspectives

Although the PSO finds good solutions in a time much shorter than the other evolutionary algorithms, the improvement of the quality of solutions can not be guaranteed by increasing the number of iteration.

THE PSO and other stochastic search algorithms have two main inconveniences:

The first inconvenience is that the swarm can prematurely converge;

The second inconvenience is that stochastic approaches have a problem of dependence; any change of one of their parameters can have an effect in the functioning of the algorithm as in the obtained solution.

Several variants of the PSO were developed to remedy these inconveniences [8].

As future perspectives we can intend to conceive a supplementary tool to remedy the problem of local optimum. This tool can be crossed with the PSO to have an even more powerful and more effective tool of optimization.

REFERENCES

[1] Al Bovik (2000). *Hand Book of Image and Video Processing*, Academic Press.

[2] M. Clerc and J. Kennedy (2001). The Particle Swarm: Explosion, Stability and Convergence in a Multi-Dimensional complex Space. IEEE Transactions on Evolutionary Computation, vol. 6.

[3] J. Dréo & P. Siarry (2003). Métaheuristiques pour l'optimisation difficile, Edition Eyrolles.

[4] A. Dutot et D. Olivier (2002). Optimisation par essaim de particules Application au problème des n- Reines. Laboratoire Informatique du Havre, Université du Havre.

[5] Gilles Burel (2001). *Introduction au Traitement d'images : Simulation sous Matlab.* Hermes Science.

[6] J. Kennedy and R. Eberhart (1995). Particle Swarm Optimization. *In Proceedings of IEEE International Conference on Neural Networks*, Perth, Australia, vol. 4.

[7] J. Kennedy & R.C. Eberhart (2001). *Swarm Intelligence*, Morgan Kaufman Publishers, Academic Press.

[8] M.G.H. Omran (2004). Particle Swarm Optimization Methods for Pattern Recognition and Image Processing, PhD Thesis, University of Pretoria, November.

[9] Rafael C. Gonzalez & Richard E. Woods (2002). *Digital Image Processing*, Second Edition, Prentice Hall.

[10] Russ John C. (1999). *The image processing handbook*, John C. Russ, Third Edition, CRC Press LLC.

[11] Stéphane Mallat (1999). *A Wavelet Tour Of Signal Processing*, Second Edition, Academic Press, Elsevier (USA).

[12] Steven K Smith (2003). *Digital Signal Processing, A Practical Guide for Engineers and Scientists*, Newnes.

[13] F. Van den Bergh (2002). An Analysis of Particle Swarm Optimizers, PhD Thesis, Department of Computer Science, University of Pretoria, South Africa.

[14] G. Venter and J. Sobieszczanski- Sobieski (2002). "Particle Swarm Optimization", In the 43rd AIAA/ ASME/ ASCE/ AHA/ ASC Structures, Structural Dynamics and Materials Conference, Denver, Colorado, USA.

[15] William K. Pratt (2001). *Digital Image Processing: PIKS Inside*, Third Edition, John Wiley & Sons, Inc.

[16] Wu, Q.H., Ji, T.Y. and Lu, Z. (2007). A Particle Swarm Optimizer Applied to Soft Morphological Filters for Periodic Noise Reduction, Evo Workshops 2007, LNCS4448,.

Improved Topology Control Algorithm for MANETs

Hatem Hamad*

*Department of Electrical and Computer Engineering
The Islamic University of Gaza, Palestine
hhamad@mail.iugaza.edu

Abstract: A mobile wireless ad hoc network is formed dynamically without a need for a pre-existing infrastructure. Frequent Topology changes leads to more processing and hence more power consumption. Reducing power consumption during node's life time is a challenging task. Adaptive Self-Configuring sEnsor Networks Topologies (ASCENT) is a topology control technique for reducing the power consumption during the node lifetime. In ASCENT, each node assesses its connectivity and adapts its participation in the multi-hop network topology based on the operating region. In this paper, I study some of the problems in ASCENT algorithm and propose a modified state diagram, which adaptively adjusts the states of individual nodes, to reduce redundancy. This helps to achieve the optimum number of Active nodes in the network. I implement the modified state diagram, and simulation results highlight that the improved ASCENT state diagram is able to achieve better performance than the original ASCENT algorithm.

Keywords: Topology Control, ad hoc networks, Power Consumption

1. INTRODUCTION

Topology control is an important way to reduce the consumption of power, that's why many algorithms (Chen et al. 2002, Deb et al. 2005, Hu. 1993, Ahmed et al. 2005, Wattenhofer et al. 2005, Ramanathan et al. 2000) were developed to get the best performance of the mobile ad hoc networks (MANETs). This paper discusses the node scheduling scheme by improving the performance of ASCENT algorithm (Cerpa et al. 2004). In general, coordination between nodes reduces the redundancy in the high density networks. This helps in extending the overall system lifetime. Also having too many nodes deployed in a wide range area is very difficult to manage and configure manually, design-time pre-configuration is precluded because of the environmental dynamics. That's why self-configuration is used to achieve the desired topology control.

ASCENT algorithm depends on the concept of node scheduling to achieve topology control (Cerpa et al. 2004, Deb et al. 2005). It has four states to represent node's active/inactive state. Active states are:
- *Active:* Forwards data and routes packets.
- *Test:* Sends neighbour announcement message, monitors network for neighbours and data loss rates and forwards data and routes packets.

Inactive states are:
- *Passive:* Monitors network for neighbours and data loss rates, also periodically checks if necessary to become active.
- *Sleep:* Turns radio off and goes to sleep.

ASCENT consists of several phases. When a node first initializes, it enters into a listening-only phase called neighbour discovery phase, where each node obtains an estimate of the number of neighbours actively transmitting messages based on local measurements. Upon completion of this phase, nodes enter into the join decision phase, where they decide whether to join the multi-hop diffusion sensor network. During this phase, a node may temporarily join the network for a certain period of time to test whether it contributes to improved connectivity.

If a node decides to join the network for a longer time, it enters into an active phase and starts sending routing control and data messages. If a node decides not to join the network, it enters into the adaptive phase, where it turns itself off for a period of time, or reduces its transmission range (Cerpa et al. 2004).

2. STATE OF THE ART

My work has been informed and influenced by a variety of other research efforts. There has been a great deal of work in the area of topology control, mostly using theoretical analysis or simulation, and involving MAC (IEEE Standards 1994) and power control mechanisms (Agarwal et al. 2001, Chen et al. 2002).

There have been several important theoretical evaluations of topology control. Most of this work focuses on node scheduling algorithms. They are used to control topology by reducing number of routers especially in dense network. There are a lot of works related to this field.

Alberto Cerpa and Deborah Estrin proposed self-configuring algorithm (Cerpa et al. 2004) that uses node scheduling concept. This algorithm is the basic of my work. it states that not all nodes are supposed to work as routers in the dense network. Chen proposes a localized algorithm called SPAN for node scheduling (Chen et al. 2002). Using simulations they proved that the node-scheduled topology does not suffer too much in terms of latency and capacity of the network while reducing redundant power consumption. However none of the papers consider the impact of node scheduling on the overhead of reliable transmissions for ad hoc networks. Another important work I referred to during my work was the analysis of algorithms for distributed construction of a connected dominating set (CDS) of the corresponding unit-disk graph and the routing strategies using the CDS backbone (Wan. et al. 2002, Bharghavan et al. 1997).

3. ASCENT ALGORITHM

ASCENT adaptively elects "Active" nodes from all nodes in the network. Active nodes stay awake all the time and perform multi-hop packet routing, while the rest of the nodes remain "passive" and periodically check if they should become active. Initially, only some nodes are active. The other nodes remain passively listening to packets but not transmitting. This situation is depicted in Fig. 1(a). The source starts transmitting data packets toward the sink. Because the sink is at the limit of radio range, it gets very high packet loss from the source. This situation is called a communication hole; the receiver gets high packet loss due to poor connectivity with the sender. The sink then starts sending help messages to signal neighbours that are in listen-only mode, also called passive neighbours, to join the network. When a neighbour receives a help message, it may decide to join the network. This situation is illustrated in Fig.1(b). When a node joins the network it starts transmitting and receiving packets, i.e. it becomes an active neighbour. As soon as a node decides to join the network, it signals the existence of a new active neighbour to other passive neighbours by sending a neighbour announcement message. This situation continues until the number of active nodes stabilizes on a certain value and the cycle stops (see Fig. 1(c)). When the process completes, the group of newly active neighbours that have joined the network makes the delivery of data from source to sink more reliable. The process will re-start when some future network event (e.g. node failure) or environmental effect (e.g. new obstacle) causes packet loss again (Cerpa et al. 2004).

3.1 ASCENT state transitions

In ASCENT, nodes are in one of four states: sleep, passive, test, and active. Fig. 2 shows a state transition diagram. Initially, a random timer turns on the nodes to avoid synchronization. When a node starts, it initializes in the test state.

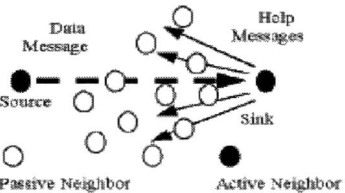

(a) Communication Hole

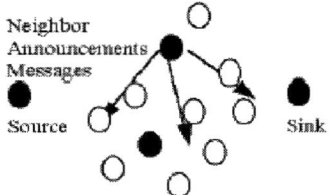

(b) Transition

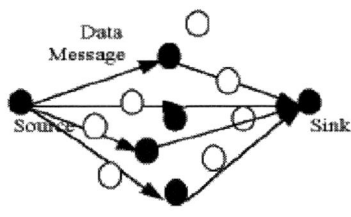

(c) Final State

Figure 1: Network self-configuration

Nodes in the test state exchange data and routing control messages. In addition, when a node enters the test state, it sets up a timer Tt, and sends neighbour announcement messages. When Tt expires, the node enters the active state.If before Tt expires the number of active neighbours is above the neighbour threshold (NT), or if the average data loss rate (DL) is higher than the average loss before entering in the test state, then the node moves into the passive state. If multiple nodes make a transition to the test state, then I use the node ID in the announcement message as a tie breaking mechanism (higher IDs win). The intuition behind the test state is to probe the network to see if the addition of a new node may actually improve connectivity.

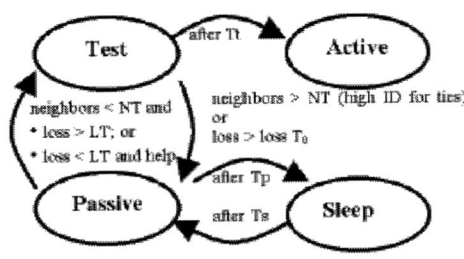

Figure 2: ASCENT state transitions

When a node enters the passive state, it sets up a timer Tp and sends new passive node announcement messages. This information is used by active nodes to make an estimate of the total density of nodes in the neighbourhood. Active nodes transmit this density estimate to any new passive node in the neighbourhood. When Tp expires, the node enters the sleep state. If before Tp expires the number of neighbours is below NT, and either the DL is higher than the loss threshold (LT) or DL is below the loss threshold but the node received a help message from an active neighbour, it makes a transition to the test state. While in passive state nodes have their radio on, and are able to overhear all packets transmitted by their active neighbours (even if the packets are not addressed to the passive node, since the radio is in promiscuous mode). No routing or data packets are forwarded in this state, since this is a listen-only state. The intuition behind the passive state is to gather information regarding the state of the network without causing interference with the other nodes. Nodes in the passive and test states continuously update the number of active neighbours and data loss rate values. Energy is still consumed in the passive state, since the radio is still on when not receiving packets. A node that enters the sleep state turns the radio off, sets a timer Ts and goes to sleep. When Ts expires, the node moves into passive state. Finally, a node in active state continues forwarding data and routing packets until it runs out of energy. If the data loss rate is greater than LT, the active node sends help messages (Cerpa et al. 2004).

4. PROBLEMS

Active nodes are responsible for routing, they construct the backbone of the network and form its topology (Santi 2005). Having too many Active nodes consumes a lot of unnecessary energy.

The original state diagram has no returning path from Active state, as shown in Fig. 2. There are many cases and scenarios that may cause this problem. For example, if the Active node changes its position due to the mobility of the network. The new position may have no routing tasks which means the node will be Active with no tasks (Idle node). There is no need for this node to be Active, so it is better to return to Passive again. Another problem can arise if the new position has enough Active nodes, which means they will exceed Neighbour Threshold (NT) (Cerpa et al. 2004). Too many Active nodes cause a lot of overhead in the network. When a node sends help message due to packet loss, there might be several responded nodes. This may cause to extra numbers of Active nodes eventually. A similar scenario happens when an Active node is lost because of the mobility or node's failure. The gap will be detected and a help message will be sent to find a replacement. Then, several nodes will replace the failed node.

My solution suggests that it is better to drain an area of nodes more slowly than to drain a node completely. I solved this problem by adding Active withdrawal state to the original state diagram as discussed in the next section.

5. SOLUTION

As stated in the previous section, the original state diagram doesn't fix the problem of the unnecessary number of nodes entering Active states, it has no mechanism to check the need for these Active nodes. Also the node that enters the Active state remains Active until its power is drained. To solve the previous problems a new modified state transition diagram is presented which will handle the additional issues provided previously. A new state was added to the original diagram called "Active Test". It is responsible for testing the necessity of the node to stay in the Active state or not, therefore reducing the power consumption in the network. The purpose for this modification is to reduce the unnecessary number of Active nodes. To achieve this, the node will move to *ActiveTest* state if it has no routing tasks, Fig 3. This should be done with regard to the network-wide information, particularly the number of Active neighbours, in order to guarantee the connectivity between nodes in the network and to ensure that only the unneeded nodes returns to passive. During *ActiveTest* state, the node has no routing tasks and ready to go to Passive state if possible. Being in Active state consumes energy in sending Neighbour Announcement messages. To maintain network's connectivity the node will check number of Active neighbours to insure that it will not affect the connectivity.

The transition from Active to ActiveTest happens when the node has no routing tasks for Δt, I assumed $\Delta t = Tt$ in the original algorithm. This is an indication that the node might be useless for the network; it should go through several tests as described next to check if the node is needed to stay in Active state or not. This is done in the ActiveTest state. During ActiveTest state, the node will check the number of its Active Neighbors. Having too many Active nodes makes the network very crowded, the interference is increased due to the heavy message exchange between nodes. This will affect the performance of the network (Gang et al. 2006). If number of Active Neighbors > Neighbors Threshold (NT), the node will return to Passive state, Fig. 3.

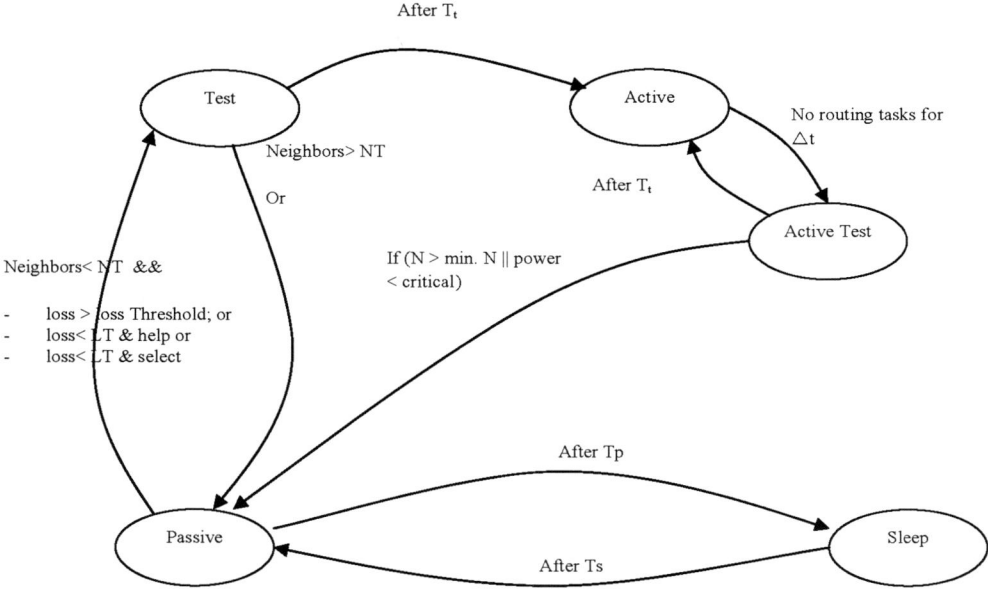

Figure 3: Modified state diagram

Such partition of the network is very crowded and has too many Active nodes. Redundancy between Active nodes can be reduced by removing the unnecessary ones with no routing tasks in order to maintain the routing paths. This situation happens due to the mobility in the network.

Another important measure is the power level of the node. If a node is in critical power which I assume after experiments to be 5%, it should not work as a router for other nodes consuming its remaining power. This could lead to a sudden death of that node. One way to solve this issue is to return to Passive state after selecting one of its Passive neighbours to go to Active. A unicast message is sent by that node to the selected Passive node. The selection criteria is the power level of that node, the neighbour Passive node with the highest power level will receive the message "Select Message" and move to Test state as described in Fig. 3.

After Tt, the node will return to Active if non of the previous conditions were achieved. This indicates that the node is still needed in the network and has a role in the topology formation and future routing tasks, Fig. 3. In general, this modified state diagram solves the problems of achieving the optimum number of Active nodes by reducing their numbers and withdrawing the unneeded to the Passive state.

The resulted redundancy will be reduced as well as the power consumption of the useless Active nodes. Another advantage is reducing the sudden gap in the network which happens because of the sudden death of the node after its power is drained.

6. SIMULATION AND RESULTS

I used the JiST-SWANS simulator (Barr 2006a,b) to simulate the proposed solution. My simulation compares the performance of the original ASCENT and the modified algorithm. The results show the changes of performance in the Network life time and Delivery ratio (network connectivity). My evaluation is based on the simulation of different node densities in the field. For measuring connectivity I used 20 node in different field range 100x100 to 3000x3000 Km^2 with transmission range of each node equals to 625 meters, two-ray ground propagation channel is assumed with a data rate of 1 Mbps. For measuring network life I took multiple values during simulation time at 0 to 1000 seconds with 50 second increment in 1500x1500 meters field, repeating the results for 20, 40 and 80 nodes. The data traffic simulated is constant bit rate (CBR) traffic (Perkins 2001). 50% of nodes, CBR sources, generate ten 128-byte data packets every (20-25) second. Random waypoint mobility model was used in my experiments with a maximum node speed of 2 m/s and a pause time 500 ms. With this approach, a node travels towards a randomly selected destination in the network. After the node arrives at the destination, it pauses for the predetermined period of time and travels towards another randomly selected destination. Simulation time is 2000 seconds and each simulation scenario is repeated several times to obtain steady-state performance metrics. In order to provide a fault situation, I used Uniform packet loss model that drops packets at certain probability. I used AODV as a routing protocol between the nodes (Perkins et al. 2000)

6.1 Average Network life time

This is a measure of the network life time. The value is measured by summing the power percent of each node in the network divided by the number of nodes.

Avg. Network Life time = $((P1+P2+.....+Pi)/i)$ (1)

where i is the number of nodes in the network.

The simulation was implemented in several densities, 40 and 80 nodes in 1500x1500 meters (18 and 36 nodes/Km2). The results are as shown in the figures below, Fig. 4, Fig. 5. In these figures, I can see clearly that improved algorithm extends the networks lifetime comparing to the actual algorithm. This is very useful and cost-effective for different network applications. I can observe more improvement in dense network (80 node). The improvement happened because of the Active withdrawal and optimum number of Active nodes, draining power in the improved algorithm comes from an area of nodes which is more slow and efficient than draining one node completely.

6.2 Success Delivery ratio

This parameter indicates the effect of my algorithm on network's connectivity. The previous work causes less number of nodes to be active, which might decrease the network connectivity. The success delivery ratio (SDR) is computed by dividing the number of received packets over the number of sent packets.

$$SDR = \frac{(number\_of\_received\_packets)}{(number\_of\_sent\_packets)} x100\%$$ (2)

The simulation was implemented in several densities by increasing the field range. I repeated the simulation first on static network then on mobile network with Random Way point model explained previously. In the static mobility, I observe no significant change in the success Delivery ratio compared to my modified algorithm, Fig. 6. This is because there is no often error case and path broken in the static network, which reduces the chances of my algorithm to take effect on the overall performance. After all I can see a little improvement in the high density network. In the dynamic mobility I can observe the increasing of the performance on the success delivery ratio especially at low density, Fig. 7. This is because my algorithm maintains a minimum number of Active nodes and in the case of Active node's failure, the node selects a replacement which will help in improving the connectivity.

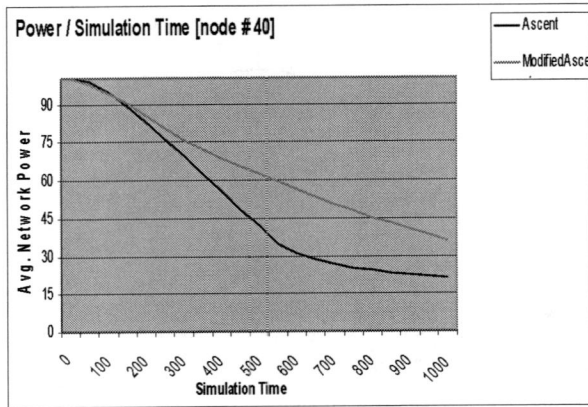

Figure 4: Network life time at 40 nodes

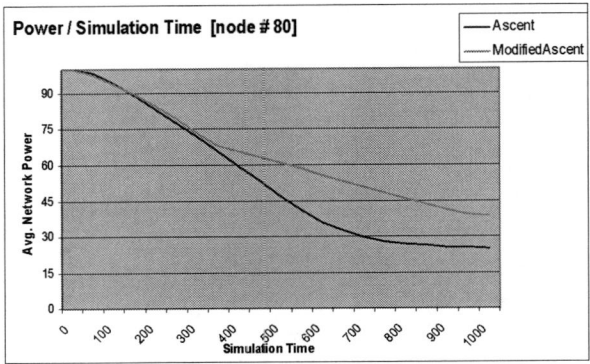

Figure 5: Network life time at 80 nodes

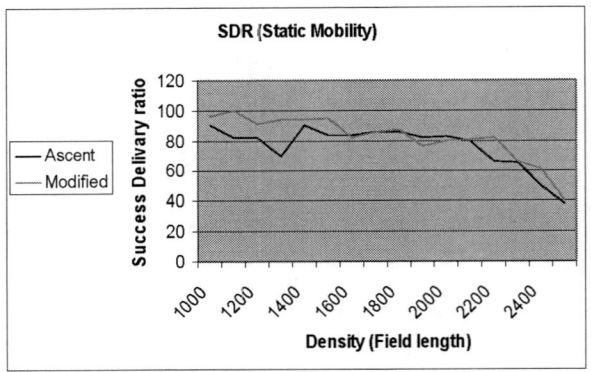

Figure 6: SDR at static Mobility

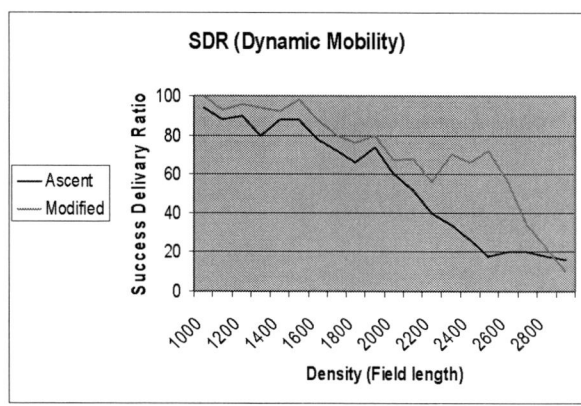

Figure 7: SDR at Dynamic Mobility

7. CONCLUSION

In this paper I presented a solution to some problems in ASCENT topology control algorithm. The solution is a modified state diagram that provides the following advantages to the original algorithm: (a) achieves optimum number of Active nodes by adding "Active withdrawal" stage, (b) reduces the redundancy resulted by a crowded network and (c) reduces the gap in the network which is resulted by the sudden death of the node after its power is drained. I hope in the future work to solve other open issues related to ASCENT Algorithm, such as the problem of Neighbor Threshold value. In the original ASCENT the value is fixed and depends on the application. In general, it is better to have a dynamic value of threshold for different environments and different types of nodes (in case of heterogeneous network). Another open issue is that broadcasting help message will cause nodes to go from passive to test states needlessly in areas where help is not needed, increasing collisions in that area temporarily. My solution handles this issue after detecting it, but this could be improved by not allowing these nodes to enter Active state from the start.

ACKNOWLEDGEMENT

The author would like to thank the research assistant Eng. Mohammed H. Alafifi for his assist in doing this work.

REFERENCES

[1] Agarwal S., Krishnamurthy S., Katz R., Dao S. (2001). *Distributed Power Control in Ad-hoc Wireless Networks*. PIMRC01.

[2] Ahmed A., Homayoun B. (2005). *Topology discovery for network fault management using mobile agents in Ad-Hoc networks*. IEEE 0-7803-9305-8/05.

[3] Barr R. (2004). *JiST– Java in Simulation Time User Guide*, Department of Computer Science, Cornell University, Ithaca, NY 14853.

[4] Barr R. (2004). *JiST SWANS– Scalable Wireless Ad hoc Network Simulator User Guide*, Department of Computer Science, Cornell University, Ithaca, NY 14853.

[5] Bharghavan V. and Das B. (1997). *Routing in Ad Hoc Networks Using Minimum Connected Dominating Sets*. International Conference on Communications' 97, Montreal, Canada.

[6] Cerpa A., Estrin D. (2004). *ASCENT: Adaptive Self-Configuring sEnsor Networks Topologies*. IEEE 1536-1233/04.

[7] Chen B., Jamieson K., Balakrishnan H., Morris R. (2002). *Span: An Energy Efficient Coordination Algorithm for Topology Maintenance in Ad Hoc Wireless Networks*. ACM

[8] Chen B., Jamieson K., Balakrishnan H., Morris R. (2002) *Distributed Construction of Connected Dominating Set in Wireless Ad Hoc Networks*. IEEE 0-7803-7476-2.

[9] Deb. B., Nath B. (2005). *On the Node-Scheduling Approach to Topology Control In Ad Hoc Networks*. Proceedings of the 6th ACM international symposium on Mobile ad hoc networking and computing. International Symposium on Mobile Ad Hoc Networking & Computing.

[10] Gang Z., Stankovic J., Son S. (2006), *Crowded Spectrum in Wireless Sensor Networks*, unpublished.

[11] Hu L. (1993). *Topology control for multihop packet radio networks*. IEEE Trans. Communications, **vol. 41, no. 10**.

[12] IEEE Standards Department (1994). Wireless LAN Medium Access Control (MAC) and Physical Layer (PHY) Specifications. IEEE Standard 802.11_1997.

[13] Perkins C. (2001). *Ad Hoc Networking*, Addison-Wesley.

[14] Perkins C., Royer E., and Das S. (2000). *Ad Hoc On Demand Distance Vector (AODV) Routing*. IETF Internet Draft, draft-ietf-manet-aodv-05.txt.

[15] Ramanathan R., Rosales-Hain R. (2000). *Topology control of multihop wireless networks using transmit power adjustment*. IEEE INFOCOM.

[16] Santi P. (2005). *Topology Control In Wireless Ad Hoc And Sensor Networks*, John Wiley & Sons.

[17] Wattenhofer. R., Li L., Bahl P., Wang Y. (2005). *Distributed topology control for wireless multihop ad-hoc network*. IEEE INFOCOM.

Training RBF Networks using DDA Algorithm Combined with Genetic Algorithm

G. KHENSOUS[1,2], B. MESSABIH[1] and N. BENAMRANE[1]

[1] *Computer Science Department, USTO-MB University, Oran, Algeria.*
[2] *Computer Science Department, University of Mostaganem, Mostaganem, Algeria.*
gh.khensous@yahoo.fr

Abstract: This article presents a new system for training Radial Basis Function Networks (RBF networks) using heuristic training techniques and more precisely the DDA Algorithm (short for *"Dynamic Decay Adjustments"*) combined with a Genetic Algorithm. This system is called GA-DDA. The GA is used in the pre-training stage to find the initial centroids of the network. The performance of the proposed system is evaluated on PSSP problems (*Protein Secondary Structure Prediction*) and the achieved results are compared with those gotten by the DDA Algorithm.

Keywords: RBF Network, DDA Algorithm, Genetic Algorithm, PSSP.

1. INTRODUCTION

The Radial Basis Function Networks are Feed-forward networks with one hidden layer. Their performances are greatly affected by their topology; i.e. the choice of centroids making up the hidden layer.

To find the best RBF network architecture, we used the DDA algorithm (DDA is the acronym of *"Dynamic Decay Adjustments"*); afterwards we improved the performance of the network by introducing the Genetic Algorithms in the pre-training stage. We tested both the DDA algorithm and the new system named GA-DDA on PSSP problems.

In section 2, the different training techniques of RBF networks are enumerated. The section 3 is a description of the proposed approach. The experimental results are presented in section 4. Finally, the conclusion and future work are given in section 5.

2. RBF NETWORKS

An RBF network consists of 3 layers. Each layer is fully connected to the following one:

- the input layer: retransmits the inputs without distortion.

- the RBF layer: hidden layer that contains the RBF neurons.

- the output layer: simple layer that contains a linear function.

Each RBF neuron contains a Gaussian which is centered on one point of the input space.

The Gaussian function of the RBF neurons is [1]:

$$\hat{y} = \sum_{i=1}^{k} \lambda_i e^{\left(\frac{\|x-c_i\|}{2\sigma_i^2}\right)} \quad (1)$$

Where x is the input vector, $\hat{y}$ is the scalar output of the RBF network, the c_i represent the Gaussian functions' centers, σ_i their respective widths, and λ_i are the network's weights. These weights correspond to the relative importance of each core in the network's output $\hat{y}$.

There are 4 principal parameters to adjust in a RBF network [2,3,4]:

- the number of the RBF neurons (number of neurons in the unique hidden layer).

- the position of the Gaussian centers of each neuron.

- the width of these Gaussians.

- the weight of the connections between the RBF neurons and the output(s) one(s).

Any alteration of one of these parameters carries away directly a change of the network's behaviour.

We can classify the training techniques of RBF networks in three groups [2,3,4]:

(1) *Supervised techniques* that rely on the principle of minimizing the quadratic error,

(2) *Two stages training techniques*, these techniques estimate the RBF network parameters in two stages: the first one determines the centers and the standard deviation of the basis functions. In this stage, only the input vectors are used and the training is not supervised. Whereas the second stage is aiming to calculate the connections' weights of the hidden layer towards the output layer (supervised training).

(3) *Heuristic techniques*, the purpose of these techniques is to determine the network's parameters in an iterative manner. Generally, we start by initializing the network on a center with an initial influence ray (μ_0, σ_0). The centeroids μ_i are created during the presentation of the training vectors. The following step has as purpose to modify the influence rays and the connections' weights (σ_i, w_i) (only the weights between the intermediate layer - Gaussian neurons - and the output one).

Two algorithms were established for an heuristic training: the RCE Algorithm "*Restricted Coulomb Energy*" and the DDA Algorithm.

The DDA algorithm is partially extracted from the RCE Algorithm; it is used for applications in classification (discrimination). Its principle is to introduce two thresholds $\theta-$ and $\theta+$ in instead of one in the RCE algorithm, with the aim of reducing the conflict zones between prototypes (essential problem in the RCE algorithm).

To carry out our experimentation, our choice was made on the heuristic technique. We have chosen the DDA algorithm for its many advantages: the authors tested it on several databases and compared the DDA performances with other training techniques as well as with the performances of the MLP network [5]:

First of all, the results of the DDA seem clearly better than the others especially in term of iterations' number before that the training converges. As an example, for an application on the problem of two spirals (classification type problem), RBF network "boosted" by the DDA technique converged at the end of 4 periods (a period represents a cycle of presentation of all the vectors of the training base) whereas the MLP network trained with the retro-propagation algorithm converged at the end of 40000 periods. On the other hand, all the vectors belonging to the training base were correctly classified with the RBF-DDA algorithm, result which is not necessarily gotten with the MLP network.

The following pseudo code presents a training iteration of a vector x from the class c [5]:

// Initialize weights at zero

for each prototype i of class k μ_i^k **do**
$$w_i^k = 0.0$$
end

// Training loop
for each training vector x of class c **do**

if $\exists \mu_i^c : \phi_i^c(x) \geq \theta^+$ **then** $w_i^c += 1.0$

else
//Create a new prototype
add a new prototype $\mu_{m_c+1}^c$ with:

$$\mu_{m_c+1}^c = x$$

$$\sigma_{m_c+1}^c = \max_{k \neq c \wedge 1 \leq j \leq m_k} \{\sigma : \phi_{m_c+1}^c(\mu_j^k) < \theta^-\}$$

$$w_{m_c+1}^c = 1.0$$
$$m_c += 1$$
end
//Adjust conflict areas
for each $k \neq c, 1 \leq j \leq m_k$ **do**

$$\sigma_j^k = \max \{\sigma : \phi_j^k(x) < \theta^-\}$$
end
end

3. GA-DDA

Like any other ANN "*Artificial Neural Networks*", the major problems encountered during the implementation of the RBF-DDA algorithm are essentially the initialization of the prototypes.

The implementation of the DDA algorithm requires an initialization of a representative of each class. Therefore, the choice of a better prototype is a crucial problem in this algorithm.

To solve this problem, we resorted to using the GA "*Genetic Algorithms*" in order to find the initial centroids. Unlike the contribution proposed by the authors of the paper [6] who choose the two stages training, we opted for the heuristic training technique. The GAs are then used in the pre-training stage to find the initial centroids, whereas the other RBF network's parameters are calculated via the DDA algorithm. We call the resulting system 'GA-DDA' short for "*Genetic Algorithms –Dynamic Decay Adjustment*".

We choose to optimize the position of the centers for two reasons [7]:

- it is often found in literature that the effectiveness of the classification of a network is generally influenced by the number of centroids and their positions in each class.

- the performance of the RBF networks is extremely affected by their topology, i.e.; the choice of the centers forming the hidden layer.

Although these centers can be found using the KMeans algorithm, this latter has many drawbacks. We try therefore to find these centers by using GAs which are well adapted to our problem since the centers are amino acids coded as binary vectors.

In general, to implement systems combining RBF networks and GAs, the following points should be derived [6]:

1. a suitable representation of the genotype (choice of coding).

2. a measurement of fitness (performance evaluation).

3. a method to produce new genotypes from old ones (selection and reproduction).

4. a training method to find the number of Gaussian, their centers, their widths and the weights of the outputs (RBF networks training).

3.1. Coding

The binary coding is one of the common methods of coding in GAs. This coding can be direct or indirect [6]. For the first, little effort is required to decode the chromosome, while the second requires more complicated techniques of decoding.

Direct coding is used in our work where the size of each chromosome is given by:

$$size(chromosome) = i \times j \qquad (2)$$

with i: the number of centers, it is equal to 4 (4 output neurons, therefore 4 hidden neurons initially).

j: the dimension of the input space, it is equal to 21 (each amino acid is a binary vector of 21 bits).

The following figure illustrates the structure of a chromosome.

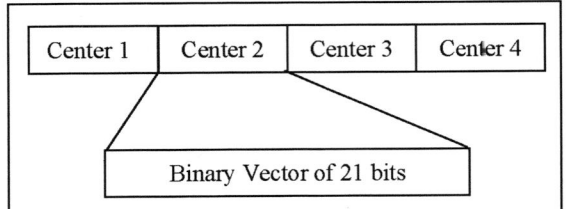

Fig. 1. Structure of chromosome

3.2. Fitness

The training of neural networks implies usually the minimization of an objective function which is typically MSE "*Mean Squared Error*". However, GAs are usually applied to maximize a fitness function [6].

Consequently to apply GAs, a method is required to convert the objective function into fitness one. A common approach is to use the following transformation:

$$f(X)=1/1+MSE(X) \qquad (3)$$

where *X* indicates the training set, *f* the fitness function.

3.3. Selection and Reproduction

In our work, a simple genetic algorithm is used in the selection and reproduction process. An initial population of 40 individuals is created by randomizing the genetic chains of size 84 (21×4), this population represents obviously random centers of the RBF network.

A new population is created by choosing pairs of individuals among the 20 best individuals of the current population which produce the new offspring. The latter is added to the new population. This process is reiterated until the size of the new population is equal to that of the initial one.

With the aim of selecting the best individuals, the Roulette Wheel Selection method is employed.

The evolution is assured by the application of the selection process as well as the use of the simple crossover operator (one cut point), and the mutation operator. The evolution process is then repeated until the population does not evolve any more.

3.4. Optimization of the Other Parameters

The widths and the number of neurons in the hidden layer as well as the weights of connections between hidden neurons and the output ones are optimized by the DDA algorithm.

The following pseudo code presents the various mentioned stages for the pre-training phase of RBF network:

Simple Genetic Algorithm
{
  //*Initialization*
  t = 0;
  //*Initialization of binary vectors of 84 bits*
  Initialization (P(t));

  //*Computing the fitness for each individual*
  Evaluation (P(t));

  //*Iteration*
  While not end **Do**
  //*Selection of the 20 best individuals*
  P' = Selection (P(t));
  //*Generation of the 20 new individuals*
  CrossOver (P ');
  Mutation (P ');
  //*Determination of the new population*
  P = Survival (P, P ');
  //*Evaluation of the new population*
  Evaluation (P);
  t = t + 1;
}

4. EXPERIMENTAL RESULTS

To test the GA-DDA system, we used the data set of Rost and Sander who had experimented the MLP network to predict protein secondary structures from its sequence of amino acids (20 amino acids). This problem is a discrimination task with four classes (Helix, Sheet, Turn or Random Coil) consists on the assignment of a specific class (secondary structure) to each amino acid of the input sequence.

To qualify the prediction effectiveness, the Q_3 value is calculated. This value, often mentioned in literature, represents the success rate of the predictions compared to the assignments [8,9,10]. It is given by:

$$Q_3 = \frac{\sum_{i \in SS} a_{ii}}{N} \qquad (4)$$

$SS = \{H, E, T, C\}$: set of secondary structure states.

a_{ii} = a number of residues observed in the state *i* and predicted in the state *i*.

N = a total number of residues.

To qualify the results gotten using the RBF-DDA network as well as the GA-DDA system, we calculated the Q_3 value. By doing the training of 120 proteins of the Rost and Sander Base and testing the remaining sequences, we reached the Q_3 scores gathered in the table below.

Table 1. Q_3 gotten

	RBF-DDA	GA-DDA
Q_3	63.26%	67.34%

Indeed, we could improve the Q_3 score which is equal to 67.34% by introducing the Genetic Algorithms. Note that the satisfactory scores for the results obtained for other approaches vary between 56% and 75%.

5. CONCLUSION AND FUTURE WORK

In our contribution to improve effectiveness of the DDA algorithm, we introduced the GA in the pre-training stage to predict a protein secondary structure from its amino acids' sequence.

So, the RBF-DDA approach was applied to a more or less large number of amino acids; this allowed to assign a specific class to each amino acid and to know their structures within the protein.

To qualify the results gotten using the RBF-DDA network, the Q_3 value is calculated. Indeed, we could reach a Q_3 score which is equal to 63.26 %, this score is better than the one gotten by Rost and Sander using MLP networks without introducing the MSA "*Multiple Sequences Alignments*" (Q_3=61.7%).

As mentioned above (see table 1), the introduction of the GA in the pre-training stage improved our score. The Q_3 value gotten using the GA-DDA system is equal to 67.34%. Note that the satisfactory scores for the results obtained by other approaches vary between 56% and 75%.

Our future interests are tended to implement a parallel training RBF network in order to improve its execution time.

REFERENCES

[1] DABLEMONT Simon, GEOFFROY Simon, LENDASSE Amaury, RUTTIENS Alain, VERLEYSEN Michel: *prédiction de séries temporelles financières par double carte de KOHONEN et modèles RBFNS locaux, application à la prédiction de l'indice boursierDAX30*. (2003).

[2] ZEMOURI Mohamed Ryad: *Contribution à la surveillance des systèmes de production à l'aide des réseaux de neurones dynamiques : Application à la e - maintenance.* (2003).

[3] Mourad ABERBOUR, "*Architecture d'un Système Hétérogène pour la Reconnaissance de Formes*", thèse Doctorat, Université Paris VI, Septembre 1999.

[4] Julien ROS, "*Représentations structurelles parcimonieuses et monodimensionnelles des singularités d'une image: Application à la classification d'images naturelles*", thèse Doctorat, Institut National des Sciences Appliquées de Lyon, Décembre 2006.

[5] Michael R. BERTHOLD and Jay DIAMOND, "*Boosting the Performance of RBF Networks with Dynamic Decay Adjustment*", Advances in Neural Information Processing, July 1995.

[6] Man Wai MAK and Kin Wai CHO, "*Genetic Evolution of Radial Basis Function Centers for Pattern Classification*", Hong Kong Polytechnic University, Hong Kong.

[7] Ben BURDSALL and Christophe GIRAUD-CARRIER, "*GA-RBF: A Self-Optimising RBF Network*", ENST de Bretagne, FRANCE.

[8] CHAKROUN Guillaume, "*prédiction de la structure d'une protéine*", Soluscience, 2004.

[9] BLIN Laurent, "*prédiction de la structure secondaire de protéines par apprentissage automatique symbolique numérique*", IFSIC, Mars 1996.

[10] Yann GUERMEUR, "*Combinaison de Classifieurs Statistiques, Application à la Prédiction de la Structure Secondaire des Protéines*", thèse Doctorat, Université Paris 6, Décembre 1997.

A Sliding Mode Controller Using Nonlinear Sliding Surface Improved With Fuzzy Logic: Application to the Coupled Tanks System

A. Boubakir [*], F. Boudjema [**], C. Boubakir [***]

[*] *Laboratoire Contrôle et commande, U.E.R Automatique, EMP, Bordj-EL-Bahri, Alger, 16111, Algérie (e-mail: ah_boubakir@yahoo.fr)*

[**] *Laboratoire de Commande des Processus, Département de génie électrique, Ecole Nationale Polytechnique, 10, Avenue Pasteur, Hassen Badi, El-Harrach, Alger, Algérie.*

[***] *Laboratoire d'Etudes et de Modélisation en Electrotechnique, Faculté des Sciences de l'ingénieur Université de Jijel, B.P. 98, Ouled Aissa, 18000, Jijel, Algérie.*

Abstract: This paper proposes an approach of hybrid control that is based on the concept of combining fuzzy logic and the methodology of sliding mode control (SMC). In the present works, a first-order nonlinear sliding surface is presented, on which the developed control law is based. Mathematical proof for the stability and convergence of the system is presented. In order to reduce the chattering in sliding mode control, a fixed boundary layer around the switch surface is used. Within the boundary layer, since the fuzzy logic control is applied, the chattering phenomenon, which is inherent in a sliding mode control, is avoided by smoothing the switch signal. Outside the boundary, the sliding mode control is applied to driving the system states into the boundary layer. Experimental studies carried out on a coupled Tanks system indicate that the proposed fuzzy sliding mode control (FSMC) is a good candidate for control applications.

Keywords: Sliding mode, Fuzzy logic, Coupled Tanks system

1. INTRODUCTION

Variable structure systems with a sliding mode were discussed first in the Soviet literature (Emel'yanov,1967; Utkin, 1978), and have been widely developed in recent years. Sliding mode control has found several applications in various fields such as power electronics (Boudjema et al., 1990), power electrical systems (Aggoune et al., 1994), and robot manipulators (Boukhetala et al.,2003; Yeung and Chen, 1988). A SMC law is designed such that the representative point's trajectories of the closed-loop system are attracted the sliding surface and once on the sliding surface they slide towards the origin. As the sliding surface is hit, the system response is governed by the surface; consequently, the robustness to the uncertainty or disturbance is achieved. However, it is known that there is a major drawback in the sliding mode control approach: undesired phenomenon of chattering due to high frequency switching, which will often excite undesired dynamics.

Several methods of chattering reduction have been reported. One approach (Slotine and Sastry, 1983) places a boundary layer around the switching surface such that the relay control is replaced by a saturation function. Another method (Spurgeon, 1991) replaces a max-min-type control by a unit vector function. These approaches, however, provide no guarantee of convergence to the sliding mode and involve a trade-off between chattering and robustness. Reduced chattering may be achieved without sacrificing robust performance by combining the attractive features of fuzzy control with SMC (Kim and Lee, 1995).

Fuzzy logic, first proposed by Zadeh (Zadeh, 1965), has proven to be a potent tool for controlling ill-defined or parameter-variant plants.

In this paper, a novel design method of sliding mode controller for coupled Tanks system based on fuzzy logic is proposed. In fist time, we develop a nonlinear sliding surface and we search the properties that must be fulfilled in order to achieve our control objective. Then, we design the control law to make the developed surface globally attractive and invariant. In order to reduce the chattering phenomenon, a fuzzy logic controller is used to approximate the corrective control. In the procedure, a fixed boundary layer is adopted and fuzzy logic control is applied within the boundary. Outside the boundary, the sliding mode control is applied to driving the system states into the boundary layer. By means of tuning the input center point in the fuzzy logic, the system tracking error can converge to a smaller neighborhood of zero than the conventional sliding mode control with boundary layer. Finally, experimental results are given to show the effectiveness and feasibility of the proposed control strategy.

2. SYSTEM MODELING

We had used the CE105 Coupled Tanks Apparatus, which is produced by TecQuipment limited for teaching system dynamics and control engineering principles (Wellstead, 1993). It consists of two separated vertical tanks that have sensors converting the liquid level to voltage for feedback control (Fig.1). Both tanks are interconnected by a flow channel (channel 1) that a rotary valve may be used to vary the sectional area of the channel and, hence, change the flow characteristics between the tanks. The model of the coupled tanks system can be written as (Wellstead, 1993),

$$\begin{cases} \dfrac{dh_1}{dt} = \dfrac{1}{A}\left(-s_1 \cdot a_{12}\sqrt{2g(h_1-h_2)} + K_p \cdot u\right) \\ \dfrac{dh_2}{dt} = \dfrac{1}{A}\left(s_1 \cdot a_{12}\sqrt{2g(h_1-h_2)} - s_2 \cdot a_0\sqrt{2gh_2}\right) \\ y = K_s \cdot h_2 \end{cases} \quad (1)$$

where,

$$u = \begin{cases} u_{max} & \text{if } u \geq u_{max} \\ 0 & \text{if } u \leq 0 \end{cases} \quad (2)$$

Where s_1 and s_2 are the sectional area of channel 1 and channel 2, g the acceleration caused by gravity, h_1 the liquid level in tank 1, h_2 the liquid level in tank 2, a_{12} and a_0 are the discharge coefficient of valve 1 and valve 2 respectively, K_p is the pump gain and K_s is the sensor gain, A the tank sectional area.

Some parameters (both nominal and experimental) are shown in table 1.

The model in (1) can be written in the state-space form representations as,

$$\begin{cases} \dot{x}_1 = f_1(x) \\ \dot{x}_2 = f_2(x) + Ka \cdot u \\ y = K_s \cdot x_1 \end{cases} \quad (3)$$

Where, $x = [x_1, x_2]^T = [h_2, h_1]^T$, and

$$\begin{aligned} f_1(x) &= \beta_1 \cdot \sqrt{x_2 - x_1} - \beta_2 \cdot \sqrt{x_1} \\ f_2(x) &= -\beta_1 \cdot \sqrt{x_2 - x_1} \end{aligned} \quad (4)$$

The coefficients (β_1, β_2, Ka) are given by, $Ka = \dfrac{Kp}{A}$;

$\beta_1 = \dfrac{s_1 \cdot a_{12}\sqrt{2g}}{A}$; $\beta_2 = \dfrac{s_2 \cdot a_0 \sqrt{2g}}{A}$.

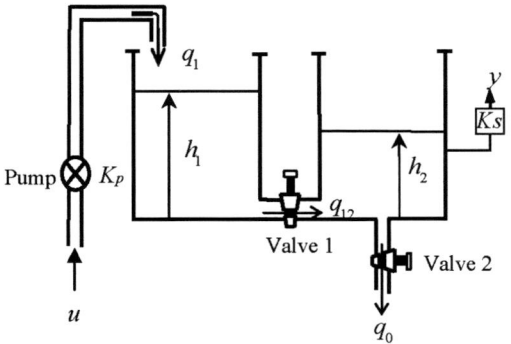

Fig. 1. Layout of the coupled tanks system

Table 1. Parameters of the coupled tanks system

Tank sectional area	A	$9350 \cdot 10^{-6}$ m^2
Channel sectional area	$s_{1\,max}$	$78.5 \cdot 10^{-6}$ m^2
	$s_{2\,max}$	$78.5 \cdot 10^{-6}$ m^2
Discharge coefficient	a_{12}	~1
	a_0	~1
Max. liquid level	h_{max}	0.25 m
Max. entry voltage	u_{max}	10 v
Pump gain	K_p	7.5 m^3/s.v
Sensor gain	K_s	40 v/m
Gravity constant	g	9.8 m/s^2

3. DESIGN OF SLIDING MODE CONTROLLER

In this section, the first goal is to characterize a class of manifold on which control objective is achieved. We recall that, the sliding mode control objective consists of designing a suitable manifold $\Psi(x,t) \in R^m$ defined by: $\Psi = \{x \in R^n / S(x) = 0\}$; so that restricting the state trajectories of the plant to this manifold result in the desired behavior such as tracking, regulation and stability. Then, determine a switching control law, $u(x,t)$, that is able to drive the state trajectory to this manifold and maintain it on it, once intercept, for all subsequent time. That is, $u(x,t)$ is determined such that the selected manifold $\Psi(x,t)$ is made attractive and invariant. In (Slotine and Lee, 1991) the authors give a form of this sliding manifold which is a Hurwitz polynomial of the error and its derivatives up to $\hat{r}-1$, where $\hat{r}$ is the relative degree of the output.

From the fact that, the output $y = K_s \cdot x_1$ are of relative degree two and in order to obtain static feedback we define the manifold $\Psi(e)$ as follows:

$$\Psi(e) = \{x \in R^2 / S(e) = \dot{e} + \Lambda(e) = 0\} \quad (5)$$

With $e = y - y_d$ and $y_d = K_s \cdot h_{d_2}$ is the tracking error, $\Lambda(\cdot)$ is any given class C_1 function whose property will be derived below. One has the following result:

3.1 Proposition 1

Consider the manifold Ψ defined in (5), and assume that $\Lambda(\cdot)$ is a continuous function such that $e \cdot \Lambda(e) > 0 \,\forall e \neq 0$. Then, on the manifold Ψ the output error e converges at least asymptotically to zero.

3.2 Proof 1

Due to the form of manifold Ψ, one has:

$$\dot{e} = -\Lambda(e) \quad (6)$$

Let us use the Lyapunov function given by $V = \frac{1}{2}e^2$. Its derivative is then:

$$\dot{V} = -e \cdot \Lambda(e) \quad (7)$$

In order to make $\dot{V}$ negative definite, it is enough that $e \cdot \Lambda(e) > 0 \; \forall e \neq 0$. Hence, the output error e is bounded and more over it tends at least asymptotically to zero.

As an example, for the function $\Lambda(\cdot)$, it can be taken as the sigmoid function (Yeganefar, Dambrine and Kokosy, 2004).

Hence, Ψ is a suitable manifold for our control system, since the control objective is achieved on it. Let us now, design the control law u that makes Ψ attractive and invariant.

3.3 Proposition 2

Consider the manifold Ψ defined in (5) and let the control signal u be given by,

$$u = u_{eq} + u_c \quad (8)$$

Where

$$u_c = -A^{-1}(x) \cdot m \cdot sign(S) \quad (9)$$

$$u_{eq} = -A^{-1}(x) \cdot [B(x) + C(x)] \quad (10)$$

With $m > 0$ and,

$$\begin{cases} \Lambda(x) = \dfrac{2}{1+e^{-\mu x}} - 1 \\ and, \qquad\qquad\qquad , \mu > 0 \\ \dfrac{d\Lambda(x)}{dx} = \dfrac{\mu}{2} \cdot \left[1 - \Lambda(x)^2\right] \end{cases} \quad (11)$$

$$\begin{cases} B(x) = K_s \cdot \left[\beta_1 \cdot \dfrac{f_2(x) - f_1(x)}{2 \cdot \sqrt{x_2 - x_1}} - \beta_2 \cdot \dfrac{f_1(x)}{2 \cdot \sqrt{x_1}}\right] \\ C(x) = (K_s \cdot f_1(x) - \dot{y}_d) \cdot \dfrac{\mu}{2} \cdot \left[1 - \Lambda(e)^2\right] - \ddot{y}_d \\ A(x) = \dfrac{K_s \cdot \beta_1 \cdot K_a}{2 \cdot \sqrt{x_2 - x_1}} \end{cases} \quad (12)$$

Where $f_i(x)$ for $i = 1, 2$ are given in (4) and the function $\Lambda(\cdot)$ is characterized in proposition 1. Then, Ψ is globally attractive and invariant.

3.4 Proof 2

Let us consider the following Lyapunov function $V = \frac{1}{2}S^2$, its time derivative is then:

$$\dot{V} = S \cdot \dot{S} \quad (13)$$

Where $\dot{S} = B(x) + C(x) + A(x) \cdot u$ with the control law given by,

$$u = -A^{-1}(x) \cdot [B(x) + C(x) + m \cdot sign(S)] \quad (14)$$

the surface dynamic $\dot{S}$ can be rewritten under the form:

$$\dot{S} = -m \cdot sign(S) \quad (15)$$

with relation (15), the expression (13) takes the form:

$$\dot{V} = -m \cdot S \cdot sign(S) \quad (16)$$

In order to make $\dot{V}$ negative $\forall S \neq 0$, it is sufficient to take $m > 0$, this condition makes $(S=0)$ and hence Ψ is globally attractive. Furthermore, if $\dot{S} = 0$, Ψ is invariant.

The control input signal can be given as, $u = u_{eq} + u_c$ where u_{eq} is the equivalent control given in (10) and u_c is the corrective control defined as,

$$u_c = -K \cdot sign(S) \quad (17)$$

with $K > 0$

In order to make $\dot{V}$ negative $\forall S \neq 0$, it is sufficient to take $K > \max |m/A(x)|$.

4. FUZZY SMOOTHING SWITCH CONTROL

In the proposed structure, the corrective term in the sliding controller is approximated by a continuous Fuzzy logic. The controller in (8) results with high frequency oscillations in its outputs, causes a problem known as chattering. Chattering is undesirable because it can excite the high frequency dynamics of the system. To eliminate chattering, a continuous fuzzy logic control u_{fuzzy} is used to approximate u_c.

The overall system with the proposed controller is given in Fig. 2.

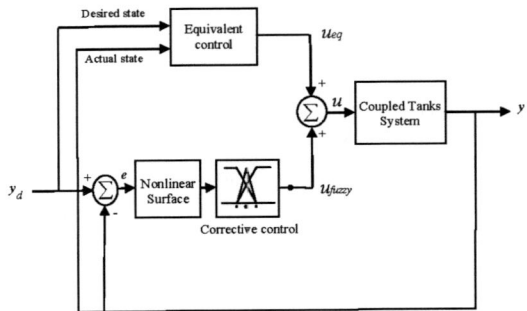

Fig. 2. The structure of FSMC

The design of the fuzzy controller begins from extending the crisp sliding surface $S = 0$ to the fuzzy sliding surface defined by linguistic expression (Liu, Zhao and Zhang, 2004):

$$\tilde{S} \text{ is ZERO} \quad (18)$$

Where $\tilde{S}$ is the linguistic variable for S and ZERO is one of its fuzzy sets. In order to partition the universe of discourse of S, the following fuzzy sets are introduced:

$$T(\tilde{S}) = \{NB, NM, ZR, PM, PB\} = \{F_S^1, \cdots, F_S^5\} \quad (19)$$

Where $T(\tilde{S})$ is the term set of $\tilde{S}$, and NB, NM, ZR, PM and PB are labels of fuzzy sets, which are negative big, negative medium, zero, positive medium, and positive big, respectively. For the control output u_{fuzzy}, its term set and labels of the fuzzy sets are defined similarly by,

$$T(\tilde{u}_c) = \{NB, NM, ZR, PM, PB\} = \{F_u^1, \cdots, F_u^5\} \quad (20)$$

The membership functions of these fuzzy sets are depicted in Fig.3. In Fig.3 (a), $r \in (0,1]$ is a coefficient to be used to adjust the input center point and Φ is the defined boundary layer around the switch surface.

From these two term sets, we can build the following fuzzy rules (Kim and Lee, 1995):

R^1 : If S is NB then u_{fuzzy} is PB.

R^2 : If S is NM then u_{fuzzy} is PM.

R^3 : If S is ZR then u_{fuzzy} is ZR.

R^4 : If S is PM then u_{fuzzy} is NM.

R^5 : If S is PB then u_{fuzzy} is NB.

Or equivalently, R^l : If S is F_s^l then u_{fuzzy} is F_u^{6-l} where $l = 1, \cdots, 5$.

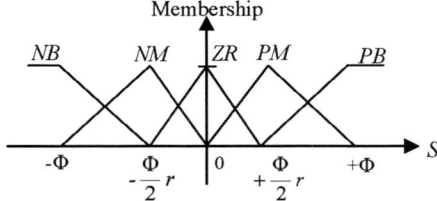

(a) Fuzzy partition of the universe of discourse of S

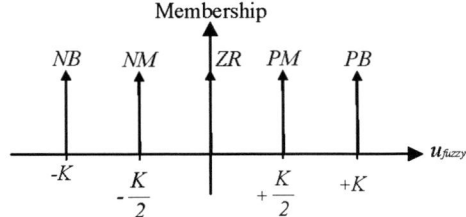

(b) Fuzzy partition of the universe of discourse of u_{fuzzy}

Fig. 3. Diagrammatic representation of term sets, $T(\tilde{S})$ and $T(\tilde{u}_c)$

Once the membership functions and fuzzy rules have been determined, the final step is the defuzzification, which is the procedure to determine a crisp control for u_{fuzzy}. There are many defuzzification strategies such as the maximum criterion, the mean of maximum, the centre of area and the weighted average method. We use the weighted average method to get the crisp control for u_{fuzzy}. Then,

$$u_{fuzzy} = \frac{\sum_{i=1}^{5} C_i \cdot \mu_i(S)}{\sum_{i=1}^{5} \mu_i(S)} \quad (21)$$

Where C_i is the associated singleton membership function of u_{fuzzy}.

In Fig. 4, we show the results of the influence of the fuzzy rules with different r value.

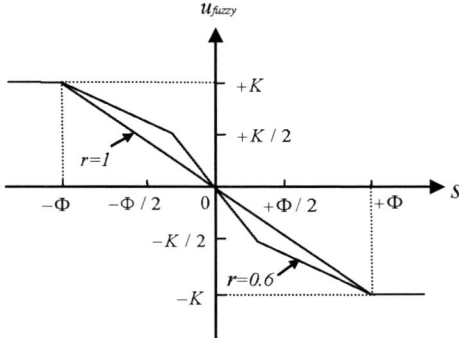

Fig. 4. Results of the influence of the fuzzy rules

From Fig.4, we see that the value of r plays an important role in the shape of this function, and if $r = 1$ then this function is a saturation function.

5. EXPERIMENTAL STUDIES

In order to verify the effectiveness and the efficiency of the proposed *FSMC*, an application to the Coupled Tanks system have been conducted. The reference level h_{d_2} used to evaluate the response of the *FSMC* applied to Coupled Tanks system is a pulse train whose amplitude is of 2.5 to 7.5 *cm* and changes every $T/2 = 200 \sec.$, with duty cycle $\tau = 0.5$. The plant and controller parameters used for experimental are given in table 1 and table 2, respectively. The parameter s_1 is kept at 100%, the channel 1 is completely opening, and the parameter s_2 is varying as (Fig. 5). In practice, the sectional area of the outflow channel is manually changed with the rotary valve.

Table 2. Parameters of the Controller and Sliding Manifold used in Experimentation

Sliding Surface	μ (gradient of sigmoid function)	0.5
Fuzzy Controller	Φ (boundary layer)	0.1
	r (coefficient used to adjust the centre point)	0.8
	K (feedback gain)	8

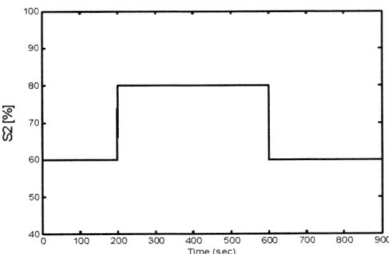

Fig. 5. Section area of the outflow channel

473

The control law was implemented on a PC Pentium II at 200 MHz, equipped with a dSPACE DS1102 controller board., using Matlab 5.3.0. And Simulink 3.0.1 with sampling time 0.2 sec. The sliding manifold that represents a desired system dynamics is given by (5) and (11). In the case of classical sliding mode controller, the corrective control is given by (17) and equivalent control by (10) and (12). In the proposed *FSMC*, the corrective control is computed by (21). For practical reasons, two Chebyshev low-pass filters are used in order to reduce the amount of noise caused by the oscillation of the liquid in the tanks provoked by some internal liquid flows and turbulence.

For each filter, the design parameters are a cut-off frequency of 0.5 [Hz] with a cut-off gain of –20 [dB].

Fig. 6. Experimental test bench

6. COMPARATIVE STUDY

The results of using conventional sliding mode controller are shown in Fig.7. and the results obtained when the fuzzy smoothing switch control has been used are shown in Fig.8. From the experimental results, we can find that the control result of conventional sliding mode controller produces a serious chattering phenomenon, Figs.7. (a), (b) and (c). On the contrary, the chattering phenomenon of the controlled system is suppressed in the case of FSMC, Figs.8. (a), (b) and (c). Moreover, the proposed controller is a robust controller since the variation of the sectional area of the outflow channel hasn't influence on the control performances.

To compare the performance of FSMC with SMC, we define tow cost functions (table 3): $J_1 = \frac{1}{2} \cdot \sum_{i=1}^{p} e^2$ and $J_2 = \frac{1}{2} \cdot \sum_{i=1}^{p} u^2$ with $p = \text{int}(t_{max}/\Delta t)$, t_{max} is the running time and Δt is the sampling period. The results for each performance index are given as:

Table 3. Comparative study with p= 4500

Controller		FSMC	SMC
Cost function	J_1	633.53	699.20
	J_2	1.35×10^5	2.44×10^5

Comparing the experimental results, it can be said that the proposed control strategy, FSMC, gave better performance than using the conventional sliding mode controller.

7. CONCLUSION

In this paper, a Fuzzy Sliding Mode Controller is proposed for a Coupled Tanks system and experiment results are presented. Firstly, a general class of manifolds for sliding mode control of Coupled Tanks system is developed. The proprieties of sliding surface, ensuring the control objective, are derived. Secondly, the SMC, using the proposed nonlinear sliding surface, is designed by selecting a Lyapunov function. The design yields an equivalent control term plus an addition control term. Thirdly, it is surveyed how the fuzzy controller is used to compute the corrective control; such that to alleviate the chattering phenomenon, a continuous fuzzy logic control is used to approximate the corrective control. The experiment results presented in this paper indicate that the suggested approach has considerable advantages compared to the classical sliding mode control. These characteristics make it a promising approach for motion control applications.

REFERENCES

[1] Aggoune, M.E., F. Boudjema, A. Bensenousi, A. Hellal, M.R. Elmesai and S.V. Vadari. (1994). Design of variable structure voltage regulator using pole assignement technique, *IEEE Trans. Autom. Control*, vol. 39, no. 10, pp. 2106-2110.
[2] Boudjema, F. and J.L. Abatut. (1990). Sliding-Mode : A new way to control series resonant converters. *(1990). IEEE Conf. Ind. Electron. Society*, Pacific Grove, CA, pp. 938-943.
[3] Boukhetala, D., F. Boudjema, T. Madani, M.S. Boucherit and N.K. M'Sirdi. (2003). A new decentralized variable structure control for robot manipulators. *Int. J. of Robotics and Automation*, vol. 18, no. 1, pp. 28-40.
[4] Emel'yanov, S.V. (1967). *Variable Structure Control Systems*, Moscow. Nouka.
[5] Kim, S.W. and J.J. Lee. (1995). Design of a fuzzy controller with fuzzy sliding surface. *Fuzzy Sets and Systems*, vol. 71, no. 3, pp. 359-367.
[6] Liu, J. Z., W. J. Zhao, and L. J. Zhang. (2004). Design of Sliding Mode Controller Based on Fuzzy Logic. *Proc. 3rd IEEE Conf. on Machine Learning and Cybernetics*, Shanghai, pp. 616-618.
[7] Slotine, J.J. and S.S. Sastry. (1983). tracking control of nonlinear systems using sliding surfaces with application to robot manipulators. *Int. J. Control*, vol. 38, no. 2, pp. 465-492.
[8] Slotine, J.J. and W. Lee. (1991). *Applied Nonlinear Control*, Prentice Hall.
[9] Spurgeon, SK. (1991). Choice of discontinuous control component for robust sliding mode performance. *Int. J. Control*, vol. 53, no. 1, pp. 161-179.
[10] Utkin, V. I. (1978). *Sliding Modes and Their Application in Variable Structure Systems* (in Russian), Moscow, U.S.S.R.: Nauka.
[11] Wellstead, P. (1993). *TecQuipment CE105 Coupled Tanks Apparatus*, Control Systems Centre, Manchester, U.K.
[12] Yeganefar. N., M. Dambrine and A. Kokosy. (2004). Stabilisation pratique par modes glissants pour un système linéaire à retard. *Conférence Internationale Francophone d'Automatique*, CIFA2004, Tunisie.
[13] Yeung, K.S. and Y.P. Chen. (1988). A new controller design for manipulators using the theory of variable structure systems. *IEEE Trans. Autom. Control*, vol. 33, no. 2, pp. 200-206 .
[14] Zadeh, L. (1965). Fuzzy sets. *Information Control*, vol.8, no.1, pp.338-353.

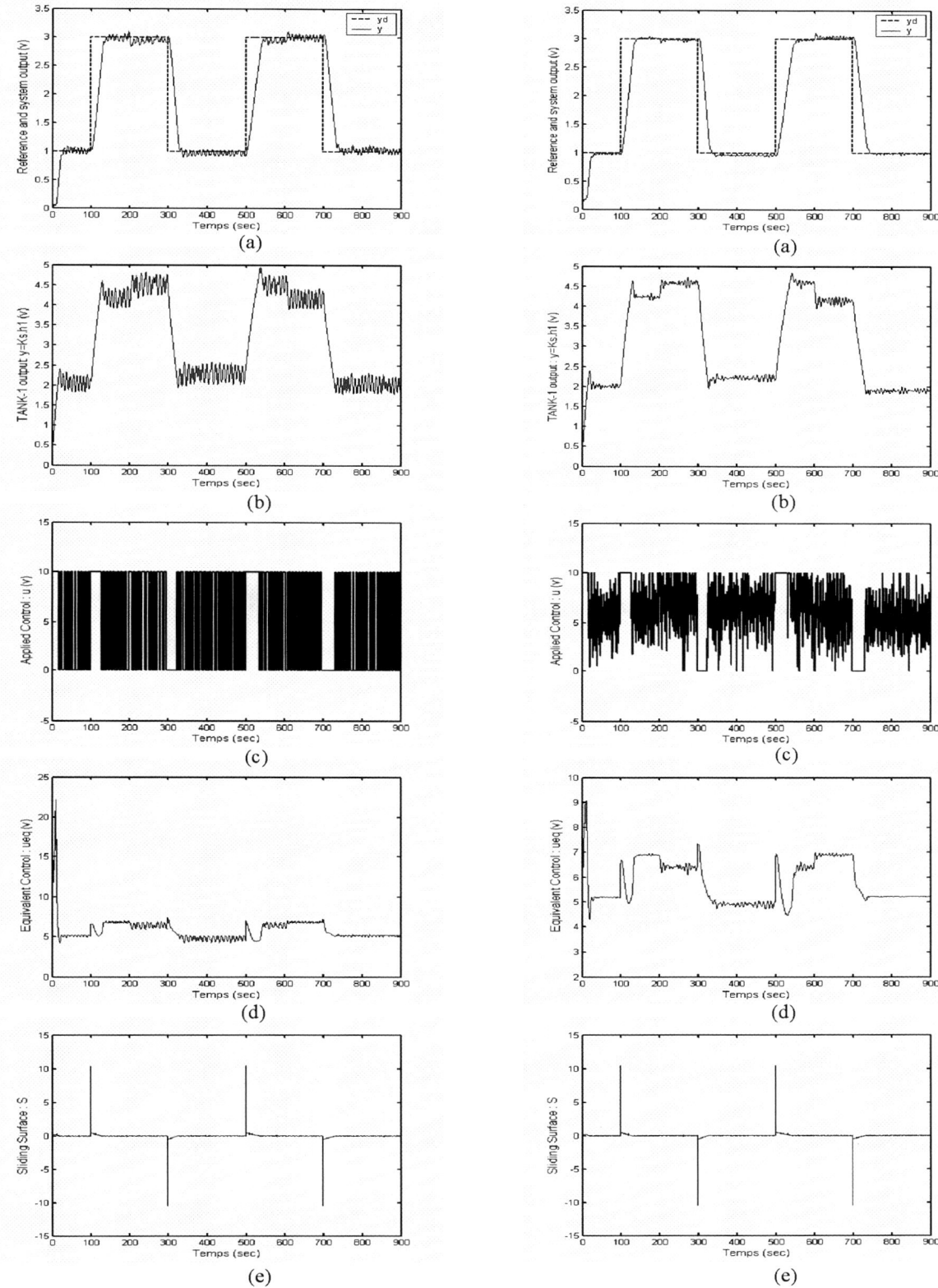

Fig. 7. Experimental results of SMC

Fig. 8. Experimental results of FSMC

Presentation of a GIS for road network

B. BENARBIA   **N. BERRACHED**

Intelligent Systems Research Laboratory
Sciences and Technology University of Oran
BenarbiaBenaissa@Gmail.com, Laresi_Usto@Hotmail.com

Abstract: We propose GISRE, a Geographical Information System for Road network Extraction, management and updating. This GIS includes the following mean modules:

- Database management and handling module,
- Database analysis and interrogation module,
- Data updating module,
- Road network extraction from satellite image module,
- Road network extraction from maps module,
- Registration module.

The proposed GIS is dedicated at this time to Western Algeria Road Network and will extend to the whole Algerian Road network in the near future.

Key Words: GIS, Maps, Road network, Satellite Image.

I. INTRODUCTION:

Roads play a strategic role in economic development of a country by its essential role in the transportation of travelers and merchandises as well. The road infrastructure is one of the main strategic elements for the socioeconomic development of a region.

Let's note that the national road inheritance account of the thousands of kilometers of different types of roads, especially with the new projects in horizon, and in particular the Algerian East-West highway. These infrastructures change in the course of time. Besides, informations concerning the management of these roads, that are numeric and alphanumeric, are managed in a heterogeneous manner. Therefore, the decision making concerning a simple positioning operation is time consuming and rises several difficulties of management, especially after the natural disasters. For example after the alert to flu aviaire, in a French department, three days were necessary to achieve the plan of crisis. To theses problems are added problems of data collection for the scheduling and the construction of news roads, and supplied tools for their management do no longer respond to user's needs. Therefore the development of a GIS (Geographic Information System) then becomes a necessity.

It is facing all these realities and starting from the old works in the laboratory of research in intelligent systems (LARESI, USTO), that we propose " GISRE: Geographical Information System Road for network Extraction " a GIS to extract, to analyze, to visualize, to plan and to manage the road network.

II. STATE OF ART OF GIS:

Due to their necessity and their importance presented previously several GIS products are developed and are commercialised. We can mention:

- ArcView of ESRI [www.esri.com], it proposes a complete set of functionalities for the intyroduction, the updating, the representation, the interrogation, the analysis and the geographical data impression.

- MapInfo [www.mapinfo.com], leader of GIS; used in several countries permits to manage the graphic and alphanumeric data and notably to publish, from SQL requests, varied thematic cartographies.

- Grass [www.grass.itc.it], is today the only complete free GIS software. The software arranges an extended functional whole notably in terms of raster processing and 3D modelling.

- Idrissi [www.clarklabs.org], of the Clark Lab®, is a GIS in raster mode, as well as a system of image processing. Since its apparition in 1987, Idrissi is used in more than 120 countries

About roads GIS, we find the ImaRoute© software package [www.esri.com], an application of management of the road network. Completely integrated in the ArcGIS© environment, and WebRoute©, an application server (local, intranet, internet) for the consultation of the road network, it completes the road architecture in complement of the ImaRoute© product.

III. ROAD DATABASE CONCEPTION:

A GIS project is an application that addresses to a class of spatial data management problems oriented to no expert user [1].

The modelling constitutes a previous essential step in the development of computer applications. It remained truly with the geographical information systems whose data must be organized and structured. Several methods of conception have been proposed, since the beginning of years 90.

Currently, several studies show that the "entity-relation" formalisms or "oriented object" can be spread in order to facilitate the conceptual modelling of information manipulated by the GIS [2][3][4][5][6].

Among these methods we can mention MODUL_R, MECOSIG, the Perceptory method and its Geo-UML, MADS and the Project MurMur, and POLLEN, etc.

a. III.A. Conception method of GISRE:

About our GIS, we used the MODUL_R method that is a method based on an extension of Merise. Some specific notations of spatial data are integrated to the conceptual models of data (CMD).

The way adopted by this method remains that recommended by Merise for all the phases of the development cycle [7].

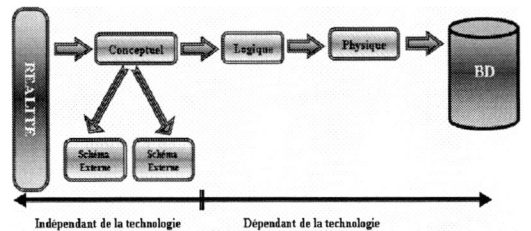

Fig.1 Different types of data models

The formalism of this method named « wide individual formalism" is a formalism of the family "entity-relation", developed within the framework of the method Merise [8].

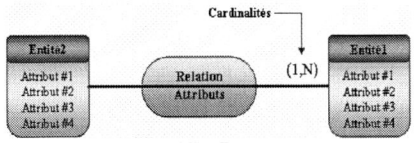

Fig.2. MODUL_R components.

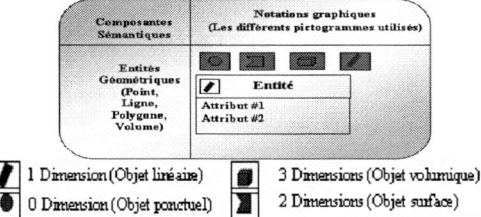

Fig.3 Geometric entities and spatial reference pictograms of MODUL_R

Our choice was made on method MODUL_R for the following reasons: the step of Merise adopted by MODUL_R is well adapted to the design of GIS, and the majority of currently available tools GIS are based on the relational model and integrate this method easily.

b. III.B. Conceptual model of data (CMD):

Our prototype contains the six following principal entities: **Road**, **Section** (which represents a segment of the road), **Repair** (which indicates reparation on the road), **Accident**, **Obstacle**, and **City**. The conceptual model of data of our GIS is as follows:

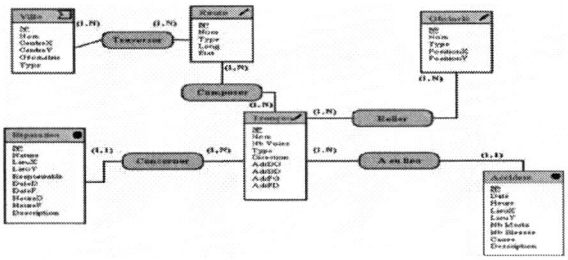

Fig.4 Le modèle conceptuel de données de notre SIG.

c. III.C. Logical model of data (LMD):

According to the CMD and by observing the rules of passage associated with the Merise method, we deduce the logical model of data:

City (N°, Name, CentreX, CentreY, Geometry, Type)

Road (N°, Name, Type, Width, State)

Obstacle (N°, Name, Type, PositionX, PositionY)

Section (N°, Name, Nb_Ways, Type, Direction, AdrDG, AdrDD, AdrFG, AdrFD)

Reparation (<u>N°</u>, Nature, LieuX, LieuY, Responsable, SDate, EDate, SHour, EHour, Description, N°_Section*)

Accident (<u>N°</u>, Date, Hour, PositionX, PositionY, Nb_deaths, Nb_Casualties, Cause, Description, N°_Section*)

Cross (<u>N°Road, N°City</u>, Width).

Compose (<u>N° Road, N° Section</u>)

Relier (<u>N° Section, N° Obstacle</u>)

IV: FUNCTIONALITIES OF GISRE:

For the implementation of our database, we used the standard database of Borland. For the development of the modules of the application we used Borland C++ Builder, Version 6. Our application integrates the following principal modules:

IV.1. Road extraction module:

This module has as goal the extraction of road network from the satellite and cartographic images, by using several methods. (in construction)

```
Begin
o  Dilate the initial image I; Let D this dilated image.
o  Erode D; Let E this eroded image.
o  Erode E; Let F this eroded image.
o  Dilate F; Let G this dilated image.
o  Cut off the result from the opened of the closed G of the
      initial image I, Let H this difference.
o  For all points H (i, j) do
   Fix a threshold S;
   If H (i, j) >S then  I (i, j) ←255
                  Else I (i, j) ← 0
   Endif
EndFor
End.
```

Fig.6 .Input Image.

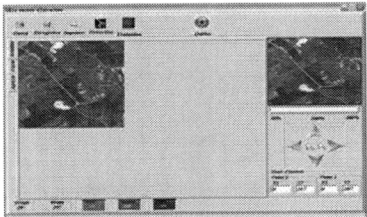

Fig.5 .Road extraction module.

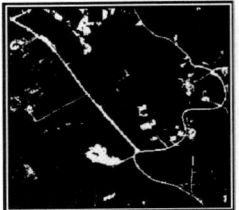 

Fig.7 Result of extraction with Thresh=20   **Fig.8 Result of extraction with Thresh=30**

We used as method of extraction of road from satellite images:
- The top hat algorithm
- Hough transformation.
- Extraction algorithm based on basic image processing operators of Kalyan Takru **[11]**.

IV.1.1. Top hat algorithm:

We obtain roads by a treatment in two parts: the top hat then thresholding. The top hat consists in cutting off the result from open from closed (a closing of the original image followed by an opening) to the initial image; this makes it possible to select pixels candidates with the membership of portions of roads. A thresholding on this image leads to the road network **[9]**. The algorithm of extraction by using the top hat is as follows:

IV.1.2. Hough Transformation:

Hough Transformation was developed by Paul Hough in 1962. It allows the detection of right-hand sides, circles or ellipses in a traditional way. It can also be extended to cases of description of more complex objects. The Hough transformation is a family of transformations which makes pass from an image (or any other signal with 1 or several dimensions) to a space of parameters. This space of parameters is the space of the possible parameters for a curve which we seek to detect **[10]**. For the extraction of the road network, the followed algorithm is:

- Each point of contour "Votes" for each parameter which defines a curve which crosses it.
- Find the stakes of parameters which are majority.

- Calculation of Hough parameters (ρ,θ), for each bipoint, we takes into account only the pixel and its eight neighbors.
- Elimination of all the pixels, and the insulated bipoints.
- Memorizing of all bipoints having the same parameters of Hough.
- Elimination of the primitives segments having the number of pixels lower than Kmin, i.e. elimination of the nonmaximum local, with Kmin: the number of Co-linears points.
- View the detected line portions.

 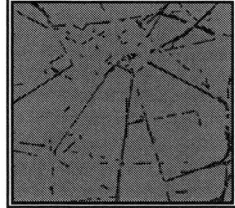

Fig.9. Input Image   **Fig.10. Application of HT**

IV.1.3. Road Extraction by the method of K. Takru:

This method uses basic functions of image processing to recover the road network by exploiting the basic characteristics of the road network. The developed method is composed of three principal steps:

- The pre-processing.
- The elimination of the not road segments.
- The reconstruction of the road network.

A. The pre-treatment:

Among the principal characteristics of the satellite images, the presence of the details, which can distort detection. To solve this problem we apply a high pass filter on the image.

$$\begin{pmatrix} -1 & -1 & -1 \\ -1 & +8 & -1 \\ -1 & -1 & -1 \end{pmatrix}$$

Fig.11 The high pass filter.

After application of the filter, we apply a thresholding to the image. It is supposed that the class of road pixels is separable to pixels in the immediate neighbourhoods of the roads. A range of values of pixels is indicated like foreground and a range of pixels with intense values is classified like background by:

$$Output(i,j) = \begin{cases} 0, LowerThreshold \leq input(i,j) \leq UpperThreshold \\ 1, Otherwise \end{cases} \quad (1)$$

Note: If a pixel value is *Output (I, J) =0*, it belongs to the road network.

Where, input(i , j) is the pixel value of the input image at coordinate i and j, and *Output(x, y)* is the result image.

It should be noted that the binary image contains an extremely high level of noise which is defined as any pixel which does not belong to the road.

The upper and lower thresholds are automatically calculated using the Mean and standard deviation:

$$Mean = \frac{\sum_{j=1}^{N}\sum_{i=1}^{M} input(i,j)}{N \times M} \quad (2)$$

Where:

$N = MaxColumn$
$M = MaxRow$

$$StdDeviation = \sqrt{\frac{\sum_{j=1}^{N}\sum_{i=1}^{M}(input(i,j) - Mean)^2}{(M \times N) - 1}} \quad (3)$$

LowThreshold=Mean-(StdDeviation)   (4)
UpperThreshold=Mean+(StdDeviation)   (5)

To allow flexibility in the thresholding process, another parameter called GreyThreshFactor is incorporated into the equation to calculate UpperThreshold and LowerThreshold. Essentially, the GreyThreshFactor parameter allows the user to increase or decrease the region of interest.

LowThreshold=Mean-(StdDeviation×GreyThresg Factor)   (6)
UpperThreshold=Mean+(StdDeviatio×GreyThresg   Factor)   (7)

Where: $0 \leq GreyThreshFactor \leq 1$

B. Road Detection Step:

Noise Removal is a very important step because it removes non-road particles from the image. The algorithm to remove noise by connected component analysis is contributed by Mr Kalyan Takru.[11]

This Removal method firstly groups all connected pixels into groups and weeding out groups with low pixel count. The method assumes roads are usually connected together and the pixel count of these groups are very large compared to unconnected noise object groups. Therefore, groups with low pixel count can be regarded as nonroad structures.

$$m'''(x,y) = 1 \quad \text{Si} \quad \Delta C_i < L \wedge \Delta R_i < L \quad (8)$$

Where ΔC_i and ΔR_i represent respectively the width and the height of the component.

C. Road reconstruction process:

Due to the complexity of automatic road network extraction, there are no algorithms available that can guarantee 100% accuracy in the road network extraction. This implies that parts of the road network will be lost in the extraction process. [12].

A method to rebuild the network of road lost caused by the process of extraction is used, as testing every point of segment end on the network of road extracted for test the possibility to rebuild lost sections.

Fig.12 Input Image   **Fig.13 The result of extraction**

IV.3. Database management module:

This module has for function the management and the analysis of tables of the data base.

It is composed of two parts:
1) **Data management:**

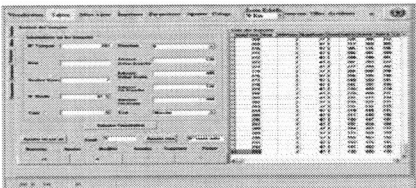

Fig.14 Tables management module.

2) **The graphic visualization.**

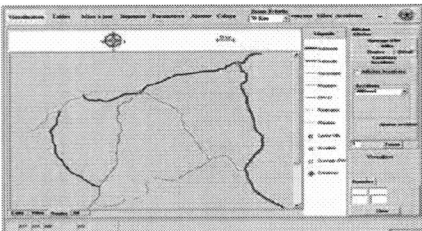

Fig.15. Graphic visualization.

IV.4. Module of database analysis:

This module has the interrogation of the data base according to the most effective criteria:
For roads:
   Active roads
   Roads in realization
   Roads according to the type (Highway, National road...)

For accidents:
   Accidents on a given road.
   Accidents according to their dates.
   Accidents according to their gravities.

For Reparations :
   Current reparation.
   Reparation on a given road.
   … etc.

IV.5. Updating data module:

This module updates the road database with descriptive data introduced by the user, or with data which come from the module of road extraction from the satellite images or maps.
To update the database we fellow the algorithm:
- Loading result image of extraction.
- Thinning of the image.
- Cross point detection.
- Apply the CCA algorithm (Connected component Analysis).
- Update database with the result and descriptive data of the user.

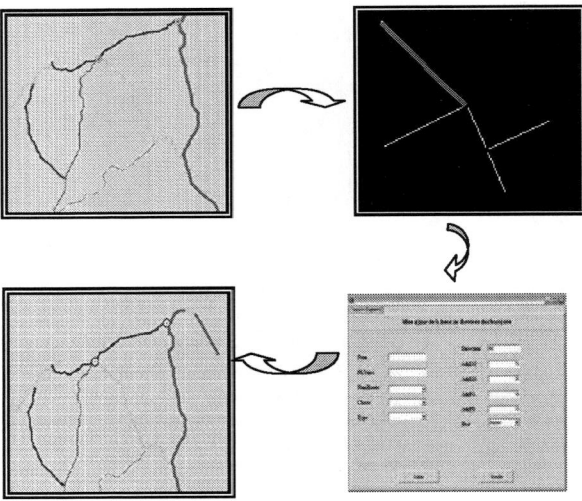

Fig.16. Datbase update module

V. CONCLUSION

During this study, we used several data-processing tools to develop our prototype GIS for the road network which meets the needs for the users of management of the roads. In particular, we exploited the Modul_R method to design our road data base, as well image processing methods, to extract and update the roads, and in particular mathematical morphology, the transform of Hough, CCA, and other basic operators. This prototype of GIS is composed of several modules, which are the module of management of the data base, the module of analysis and of interrogation, the module of extraction of the road from the maps, and the satellite images, the graphic visualization module of data, and updating data module.

VI. REFERENCES:

[1] A.Lbath, « *Aigle : un environnement visuel pour la conception et la génération automatique d'applications géomatiques* », Thesis of doctorate in Data processing, INSA Lyon, 1997.

[2] F.Golay, « *Méthodes de conception de SIRS, Modélisation des données* » Lesson n° 8,9,10, EPFL, Lausanne, 2001.

[3] C.Caron, Y.Bedard, and P.Gagnon, « *Modul_R, un formalisme individuel adapté pour les SIRS* ». International magazine of Geomatique, vol n°3, p283-306, 1993.

[4] A.Lbath, « *Aigle : un environnement visuel pour la conception et la génération automatique d'applications géographiques* », International magazine of Geomatique, Vol.5. n°2 , p.179-195, 1995.

[5] G.Kosters, B.U..Pagel, H.W.Six, « *GeoOOA: Object-Oriented Analysis for GIS-Applications* » International Journal of Geographical Information Science, vol 11 n°4, p307-335, 1997.

[6] D.Pantazis, T.P Donnay, «*La conception des SIG* », Hermès, Paris, 1996

[7] Y.Bédard, J.Pagea, Nsantenne, « *Introduction aux bases de données spatiales* », Masson, 1993.

[8] J.P.Matheron, « *Comprendre Merise, outils conceptuels et organisationnels* », Ed. Eyrolles, 1994.

[9] Z.Nougrara, « Un système d'extraction de réseaux routier à partir d'images satellitaires », Master thesis, USTO, 2000.

[10] L.Meddeber, « Un système d'inférence floue et possibiliste pour l'extraction du réseau routier et le recalage Markovien en imagerie satellitaire », Master thesis, USTO, 2005.

[11] K.Takru, T.Bretschneider, G.Leedham, «*Unsupervised detection of roads in high resolution panchromatic satellite images*», Nanyang Technological University, Singapore 2004.

[12] L Sung Yiau, «*Extracting Road Networks from High-Resolution Satellite Images*» Submitted in Partial Fulfillment of the Requirements for the Degree of Bachelor of Computer Engineering, Nanyang Technological University, Singapore, 2003/2004.

Solving Capelin Time Series Ecosystem Problem Using Hybrid ANN- GAs Model and Multiple Linear Regression Model

Karam M. Eghnam, Alaa F. Sheta

Information Technology Department,
Prince Abdullah Bin Ghazi Faculty of Science and Information Technology,
Al-Balqa Applied University Al-Salt, Jordan

(e-mail: it_karam@bau.edu.jo, asheta2@bau.edu.jo)

Abstract: Development of accurate models is necessary in critical applications such as prediction. In this paper, a solution to the stock prediction problem of the Barents Sea capelin is introduced using Artificial Neural Network (ANN) and Multiple Linear model Regression (MLR) models. The Capelin stock in the Barents Sea is one of the largest in the world. It normally maintained a fishery with annual catches of up to 3 million tons. The Capelin stock problem has an impact in the fish stock development. The proposed prediction model was developed using an ANNs with their weights adapted using Genetic Algorithm (GA). The proposed model was compared to traditional linear model the MLR. The results showed that the ANN-GA model produced an overall accuracy of 21% better than the MLR model.

Keywords: Forecasting, Capelin stock, Neural Networks, Genetic Algorithm, Ecosystem.

1. INTRODUCTION

Capelin (pelagic fish species) is an Arctic salmon fish that spends most of the year swimming around in the Arctic Ocean (Fig.1). In the Atlantic, the capelin is located in the Barents Sea. This problem is commonly attacked using many methods such as the virtual population analysis (VPA) or other techniques that apply fish harvest and estimates of natural mortality and growth to project stock development [1].

In this paper, we propose a solution to the stock prediction problem of the Barents Sea capelin using two distinct models the first one is Neural Network (NN) [2] and the second is Multiple Linear Regression Model (MLR). ANNs apply principles from neurology to find patterns in complex data and have successfully been used to predict yields of the Japanese sardine population [3] and African lake fisheries [4].

A NN is a powerful data modelling tool which is able to capture and represent complex input-output relationships. The motivation for the development of Neural Network technology stemmed from the desire to develop an artificial system that could perform "intelligent" tasks similar to those performed by the human brain. NN resemble the human brain in the following two ways: (1) A NN acquires knowledge through learning. (2) A NN's knowledge is stored within inter-neuron connection strengths known as synaptic weights. GA has been applied to a number of problems in NN [5]. This is significant for two reasons. Firstly, many of the problems in Neural Networks are important in their own right and do not presently have any wholly satisfactory means of resolution. A good example of this is the network weight optimization as we presented in this research. Secondly, the failure modes of the GA seen in NN applications are common to a broader class of problems, and their study can yield more general insights.

The models used here are based on factors that represent an ecological importance to the capelin stock. Ecosystems are spatially heterogeneous and spatial patterns and processes that represent an importance to ecosystem structure and function [6] for this reason the problem in capelins stock is influenced by the stock development, therefore we have motivated to do this research to contribute in solving the fishery estimation problem.

2. THE CAPELIN ECOSYSTEM

The capelin has a northerly circumpolar distribution, and it plays a key role in the arctic food [5] (see Fig. 1). Since 1979, the Barents Sea capelin fishery has been regulated by a bilateral fishery management agreement between Russia and Norway. During its autumn 2003 meeting the Mixed Russian Norwegian Fishery Commission decided that no fishing should take place on Barents Sea capelin for the winter season 2004.

Capelin overlaps spatially with cod, herring, and capelin itself at different stages of it is life history. Capelin is the most important food item for the cod (a large fish often lives close to the seafloor). The only food item of similar abundance and energy content is herring. Herring may replace capelin in the diet of cod during capelin collapses [7], for example herring present in the Barents Sea in part of the period when the capelin was practically absent [8]. When abundant in the Barents Sea, Herring often causes recruitment failure and eventually population crashes in the capelin stock [3]. The presence of 0-group herring has little effect on capelin recruitment compared to significant amounts of One-Year-old capelin and young herring, also indicating a negative interaction between these two species.

Since the biomass of capelin in the coming year will depend on the current one, abundance of capelin may be an important input factor in the model. The average weight of the two-year old capelin further indicates the current feeding conditions, which may impact on stock development [9]. The capelin Ecosystem input factors are shown in Table 1. These are the inputs used for our experiments.

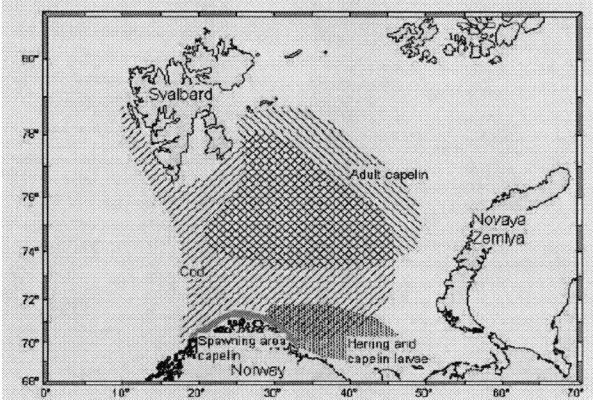

Fig. 1. Map of the Barents Sea with the main features of the distribution of various age groups of capelin as well as cod and herring. Predation by juvenile herring (1-3 years old) on larval capelin seems to have great impact on capelin recruitment at times when herring is abundant in the Barents Sea. Cod predation infers high mortality on the adult capelin, especially during capelin spawning in March [9].

Table 1. Input data in the models. T refers to the year of making the prognosis.

1.	0-group *T-1*
2.	Capelin 2 *T*
3.	Weight 2 *T*
4.	Herring *T-1*
5.	Herring *T*
6.	Cod *T*

3. ANN-GA MODEL

Selecting weights for a NN itself is an optimization problem. GA represents a solution to the weight optimization problem. We plan to use the inverse error as the measure of utility (fitness). Whitley and his co-workers [10, 11, and 12] have done much work in this area, and the study by Montana and Davis [13] is especially ingenious and noteworthy. Hybrid approaches have also been discussed in [14], and there have been studies in which GA have been used to tune the parameters of other training schemes, including initial weight configurations.

In this research, we applied the concept of hybrid model to combine both NN and GA to solve the problem of Capelin Ecosystem. Our proposed model considered the prediction of the next year biomass of capelin as a function of the input parameters given in Table 1. The proposed hybrid model is shown in Fig. 2. We used 8 neurons in the hidden layer. The following steps describe our approach:

1) **Training stage**
   - **Present data to the network.** The neural network proposed model is presented in Fig. 3.

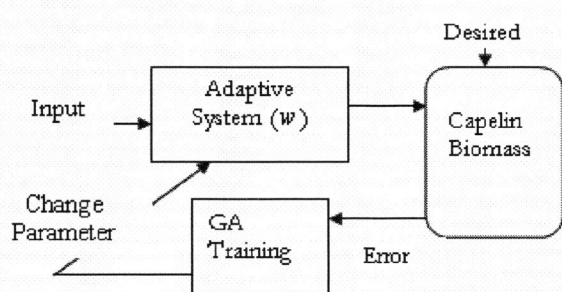

Fig. 2. Proposed Neural Network

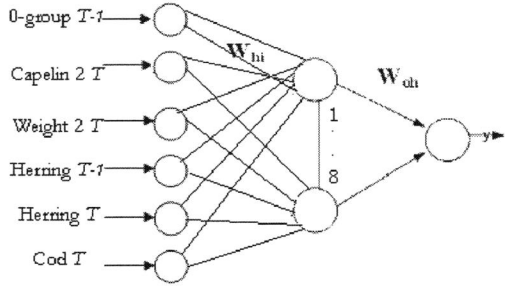

Fig. 3 NN and GA hybrid Model

- **NN weight optimization.** In this step the representation of weights is decided as shown in Table 2. This is the chromosome representation for GAs. W_{hi} is 6x8 matrix of synaptic weights connecting the inputs and hidden layers and W_{oh}, is 8x1 matrix of synaptic weights connecting the hidden and output layers. A floating point representation of connection weights will be used in our case for the ANN weights.

Table 2. Chromosome Representation

W_1	W_2	...	W_{n*p}	$W_{(n*p)+1}$	$W_{(n*p)+2}$	...	$W_{(n*p)+p}$

- **Start the evolutionary process.** Selection, crossover, and mutation operation are applied by GA. The best individuals survive to the next generation.
- **Compute the fitness function.** The fitness of these connection weights (chromosome) is computed by constructing the corresponding feed-forward neural network through decoding each chromosome and computing its fitness functions and Mean Square Error (MSE) function. In our case we adopted the Mean Square Error (MSE) as the fitness criterion as given in Equation 1.

$$MSE = \frac{1}{n}\sum_{I=1}^{n}\left(y - \hat{y}\right)^2 \quad (1)$$

The fitness of an individual is determined by the total MSE. The higher the error is the lower the fitness. y is the actual neurons output for the inputs training samples at the output layer, while $\hat{y}$ is the desired response.

- **Network computes an output.** A one-dimensional actual output vector $y = \{y_1, y_2, \ldots, y_8\}$, and for simplicity. We assume that $\hat{y}$ is the desired response of the ANN.
- **Networks output compared to desired Output.** This is implemented using Equation 1.
- **Weights modification.** Network weights are modified to reduce errors.

2. **Testing (Validation) Stage**
- **Present new data to the network.** The capelin Ecosystem input data (Table 1) was divided in two main subsets. Half of them were used for training the NN and the other were used for testing the NN.
- **Network computes an output based on its training.**

We run GA for 500 generations to obtain the optimal set of weights.

4. MLR MODEL

MLR was used to solve variety of prediction problems since many years [15]. It provides the scientist with powerful tool, allowing prediction future events to be made with information about past or present events. Regression analysis is one of the most widely used techniques for analyzing multifactor data. This is because of its ability to assess which factors to include and which to exclude, in order to develop alternate models with different factors.

In order to construct a regression model, the following steps can be undertaken:

(1) Both the information which is going to be used to make the prediction and the information which is to be predicted must be obtained from a sample of objects or individuals.

(2) The relationship between two pieces of information is then modeled with a linear transformation.

(3) In future, only the first information is necessary, and the regression model is used to transform this information in to the predicted (it is necessary to have information on both variables before the model can be constructed).

The MLR models can be represented using Equation 2.

$$y = \beta_0 + \beta_1 x_1 + \beta_2 x_2 + \ldots + \beta_6 x_6 + \varepsilon \quad (2)$$

Where,
- y is the price of the general index for,
- x_1 0-group T-1,
- x_2 Capelin 2 T,
- x_3 Weight 2 T,
- x_4 Herring T-1,
- x_5 Herring T,
- x_6 Cod T
- β_0 Intercept of the regression equation,
- β_i Regression coefficients, $i = 1, 2,3,4,5,6$ and
- ε Independent $N(0, \sigma^2)$.

We may write the sample regression model corresponding to Equation 3.

$$y = \beta_0 + \sum_{j=1}^{5}\beta_j x_{ij} + \varepsilon_i \quad (3)$$

5. EXPERIMENTAL RESULTS

The next year capelin biomass in 1979-1999 was collected from VPA for technical analysis of Capelins biomass. The first 13 years entries were used as training data. The rest of the 13 were used as a testing (validation) data.

The raw data was pre-processed using Min-Max Standardization. Six technical factors were selected as inputs of the model: 0-group T-1, Capelin 2 T, Weight 2 T, Herring T-1, Herring T, Cod T, and also the next year capelin biomass was predicted with different performance capability.

In order to evaluate the performance of our models, the Variance-Account-For (VAF) was chosen as an evaluation criterion in both cases, as given in Equation 3.

$$VAF = 1 - \frac{\text{var}(y - \hat{y})}{\text{var}(y)} \quad (4)$$

Model (1) provides a better fit with observations than model (2), the simple one (Fig.4a, B and Fig.5a, b). Model (1) has a better overall accuracy than model (2) (Fig. 4.a., b and Fig. 5a, b). On the other hand, Model (2) provided a good productivity in training case, but it predicted poorly in the testing (validation) case (See Table 3).

Table 3. Computed VAF

Model (1): NN-GA (VAF)	
Training Data	Testing Data
81%	77%
Model (2): MLR (VAF)	
Training Data	Testing Data
86%	56%

In model (1), Note that the curve sub-area (Fig. 4a) numbered by 2 to 7 represents data for the years 1979 to 1987. In this period of time the proposed model show a high degree of predict, while in the other half of the curve area has shown some miss prediction years such as in the years 1989 to 1991. In general, the total biomass of capelin is well predicted by model (1) with a high degree of VAF (77%). It performed 21% better than model (2) in the testing case.

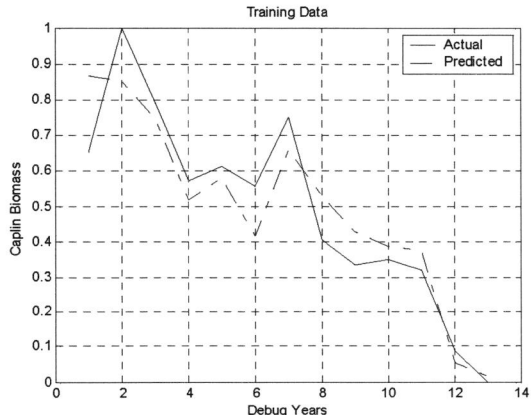

Fig. 4a. Actual and predicted Capelin biomass training case: Model (1)

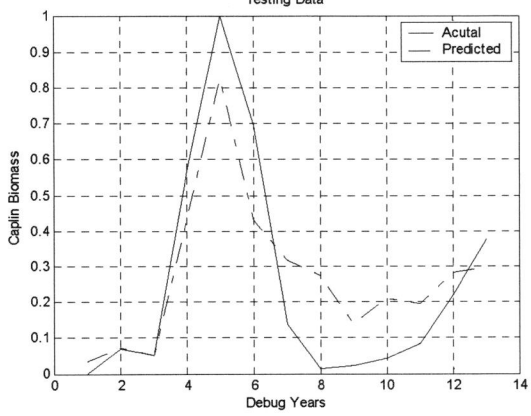

Fig. 4b. Actual and predicted Capelin biomass training case: Model (1)

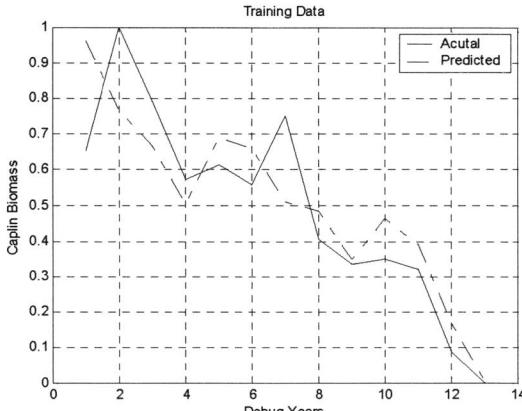

Fig. 5a. Actual and predicted Capelin biomass training case: Model (2)

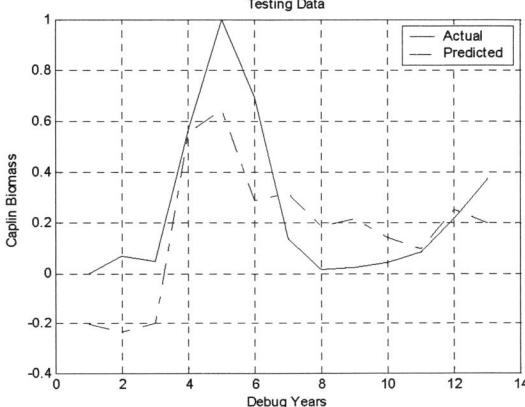

Fig. 5b. Actual and predicted Capelin biomass training case: Model (2)

6. DISCUSSION

The results from the developed models have shown relatively stable constant growth for the capelin biomass which has a good impact on the capelin stock. The proposed ANN-GA (Model 2) good predictive capabilities compared to Model 1. The following reasons are enhance the hybrid model results:

(1) The use of ANNs which are the most effective technique in financial problem is promoted mainly because its ability to discover nonlinear relationship and irregularities in input data makes them ideal for predicting the stock problems market.
(2) The use of GA was a suitable choice to overcome the problem of shortage data availability and to avoid the possibility that NN can stuck by local minimum.
(3) Using the GA procedures over generations an increasingly better solution to the problem.

7. CONCLUSIONS

In this paper, we presented a solution to the stock prediction problem of the Barents Sea capelin using a hybrid Artificial Neural Network and Genetic Algorithm Model. Experimental results show that it is possible to model Capelin stock based on capelin ecosystem factors by the proposed hybrid model. The developed model results were compared to the Multiple Linear Model Regression models. ANN-GA model provides a better fit with observations than the MLR Model (2).

REFERENCES

1. Seafood: http://www.seafood.no/Eff/eng/effandfacts. Sf/ May 12, 2005.
2. H.Yndestad, and A. Stene (2002). System dynamics of the Barents Sea capelin. *ICES Journal of Marine Science*, pp. 1155–1166
3. K. Eghnam and Alaa F. Sheta (2007), Training artificial neural networks using genetic algorithms to predict the price of the general index for Amman Stock Exchange, *In the Proceedings of the Midwest Artificial Intelligence and Cognitive Science Conference, DePaul University, Chicago, IL, USA, pp. 1-7.*
4. Vilhjálmsson, H. (1994). The Icelandic Capelin Stock. Journal of Marine Research Institute, Reykjavik, Vol. XIII, No. 1.
5. K. Richard, Belew, J. McInerny, and N. Schraudolph (1990). Evolving Networks: Using the Genetic Algorithm with Connectionist Learning. Technical Report CS90–174, UCSD (La Jolla).
6. Anon, (1999). Preliminary report of the international 0-group survey in the Barents Sea and adjacent waters in August- September. ICES Council Meeting.
7. B. Anon, Ressursoversikten. (1999). Bergen, Norway: Institute of Marine Research.
8. H. Holland (1975). Adaptation in natural and artificial systems. University of Michigan, pp.183.
9. H. Gjosaeter, H. Loeng (1987). Growth of the Barents Sea capelin, Mallotus villosus, in relation to climate. Environmental Biology of Fishes, no. 29pp. 293-300.
10. Darrell Whitley. Applying Genetic Algorithms to Neural Network Problems: A Preliminary Report.
11. D. Whitley and T. Hanson (1989). The Genitor Algorithm: Using Genetic Algorithms to Optimize Neural Networks. Technical Report, pp. S-89-107, Colorado State University.
12. D. Whitley, T. Starkweather, and C. Bogart (1989). Genetic Algorithms and Neural Networks: Optimizing Connections and Connectivity. Technical Report, pp. 89-117, Colorado State University.
13. D.Whitley and T. Hanson (1989). The genetic algorithm: Using genetic algorithms to optimize neural networks. Technical Report, pp. 89-107, University of Colorado state.
14. N.J. Radcliffe (1991). Genetic Set Recombination and its Application to Neural Network Topology Optimization. *Proceedings of the 4rth International Conference on Genetic Algorithms.* pp. 222–229.
15. N. J. Radcliffe (1990). *Genetic Neural Networks on MIMD Computers* Ph.D. Diss., PhD thesis, University of Edinburgh, UK.
16. A.S Chen, M.T. Leung, and H. Daouk (2003). Application of Neural Networks to an Emerging Financial Market: Forecasting and Trading the Taiwan Stock Index. Computers and Operations Research, pp. 901-923.

Automatic Generation of Observers for the Dala Robot with TTG [*]

Saddek Bensalem [*] Marius Bozga [*] Matthieu Gallien [**]
François Félix Ingrand [**] Moez Krichen [*] Stavros Tripakis [***]

[*] *Verimag Laboratory, Centre Equation 2, avenue de Vignate, 38610, Gières, France (e-mail: bensalem@imag.fr, bozga@imag.fr, krichen@imag.fr).*
[**] *LAAS Laboratory, 7, Avenue du Colonel Roche, 31077, Toulouse, France (e-mail: felix@laas.fr,mgallien@laas.fr).*
[***] *Verimag Laboratory and Cadence Berkeley Labs, 1995 University avenue, Suite 460, Berkeley, CA 94704, USA (e-mail: tripakis@cadence.com)*

Abstract: We report on the use of the timed test generator tool TTG as an automatic generator of observers for monitoring purposes for the Dala Robot (LAAS) case study. The starting point is a plan, which is taken to be a high-level specification. This plan is automatically translated into a network of timed automata written in IF language. From the latter an observer is automatically synthesized. The observer checks whether a sequence of observations conforms to the specification. We applied our method on a non-conforming trace of length 60. The non-conformance was detected using discrete-time semantics.

Keywords: TTG, monitoring, Dala Robot, plan, timed automaton, observer.

1. INTRODUCTION

The TTG tool [Krichen and Tripakis, 2004] is a timed test generator which may be used for monitoring purposes. It is built on top of the IF environment [Bozga et al., 2000]. The IF modeling language allows to specify systems consisting of many processes communicating through message passing or shared variables and includes features such as hierarchy, priorities, dynamic creation and complex data types.

In this work, TTG is used as an automatic generator of observers for the Dala Robot [Lemai et al., 2005] case study. We follow the same methodology as in [Bensalem et al., 2005]. Our methodology is illustrated in Figure 1. It consists of the following phases:

(1) Automatic generation of a timed-automaton specification from the plan.
(2) Automatic generation of an observer from the timed-automaton specification.
(3) Instrumentation of the system under test, that is, the execution platform.
(4) Execution and testing of the instrumented execution platform.

In the figure, solid arrows represent model and program transformations and dashed arrows represent data flow (output/input). We elaborate on each of the above phases in what follows.

The first step is to translate the plan in the form of a timed automaton, or a network of timed automata (TA). The translation must preserve the semantics of the plan, that is, the semantics of the TA and of the plan must be equivalent. It may also be the case that the TA *defines* the semantics of the plan in a formal way, as in [Akhavan et al., 2004].

[*] Work partially supported by the ANR ARA-SSIA AMAES Project.

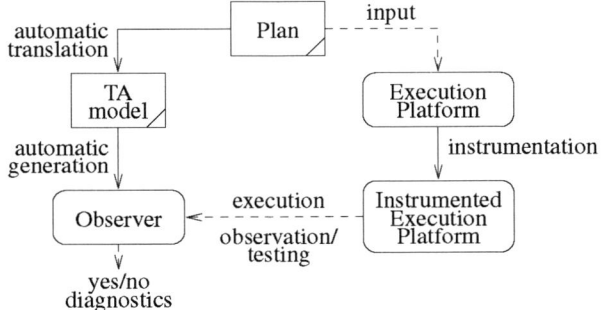

Fig. 1. Methodology [Bensalem et al., 2005].

Having obtained the TA specification A, the next step consists in generating automatically an *observer* for A. The observer is a testing device. It observes the system under test (SUT) and checks whether the trace generated by the SUT conforms to A. The observed traces are sequences of observable events and associated time-stamps. The accuracy of the time-stamps depends on the accuracy of the clocks of the observer.

There are mainly two types of observers (we follow the terminology of [Henzinger et al., 1992]). *Analog* observers, which can observe real-time precisely, and *digital* observers, which measure time with a clock ticking at a given period.

Digital-clock observers are clearly more realistic to implement, since in practice the observer will only have access to a finite-precision clock. However, analog-clock observers are still useful, for instance, when the implementation is discrete-time but its time step is not known a-priori. In this work, we restrict ourselves to the case of digital observers.

An observed trace conforms to A if it can possibly be generated by A. Notice that A is typically modeled as a network of timed automata, which induces non-determinism and internal communication between the automata. These are artifacts of the model, irrelevant to the external behavior and to the specification itself. Thus, we "hide" them, by considering them as unobservable events. This means that the observer checks if the observed trace is a possible observation resulting from some trace of A.

The third step is the instrumentation of the execution platform. It aims at interfacing the latter with the testing device (the observer). Two possibilities exist here. Either testing is performed *on-the-fly* (or *on-line*), that is, during execution of the platform, which is connected to the observer at real-time. Or it is performed *off-line*, that is, by first executing the platform multiple times to obtain a set of *log-traces*, then feeding these traces to the observer.

In both cases, the instrumented platform must be able to expose a set of observable events to the observer. In the case of testing off-line, the platform must also record the time-stamps of these events. For testing on-line, time-stamping can be done by the platform or by the observer. In the latter case, possible interfacing delays must be taken into account.

Instrumentation can be done manually or automatically. Depending on the complexity of the SUT, it can be a non-trivial task. Care should be taken so that the instrumentation does not itself alter the behavior of the system. For instance, the overhead of added code should be minimal, so as not to affect execution times of the tasks in the system. These are problems inherent in any instrumentation process, and are beyond the scope of this work.

The final step is the testing procedure per-se. The traces generated by the instrumented platform are fed to the observer, either in real-time (for on-the-fly testing) or off-line. The observer checks conformance of each trace. If a trace is found non-conforming to the specification, the SUT is non-conforming. Otherwise, no conclusion can be made. However, confidence to the correctness of the SUT is increased with the number of tests.

The main advantage of our method is that it is potentially fully-automatic. The considered plans can be automatically translated into networks of timed automata [Akhavan et al., 2004]. Observers for timed automata can be generated automatically, as we show here.

The remaining part of this paper is structured as follows. Section 2 gives a brief description about the Dala Robot architecture. Section 3 defines the simple temporal network (STN) model. Section 4 defines the timed automaton (TA) model. Section 5 explains how a simple temporal network is translated into a network of timed automata. Section 6 tells how to install and to use the TTG tool. Section 7 summarizes the obtained results. Finally, Section 8 concludes the paper and gives directions for future work.

2. DALA ROBOT

We applied this approach to the case of the robot Dala by LAAS [Lemai et al., 2005]. The LAAS robot architecture is decomposed into three levels.

- The *functional level*: it includes all the basic built-in robot action and perception capacities. These processing functions and control loops (e.g., image processing, obstacle avoidance, motion control, etc.) are encapsulated into controllable communicating modules. Each modules provide services which can be activated by the decisional level according to the current tasks, and posters containing data produced by the module and for other (modules or the decisional level) to use.
- The *decisional level*: this level includes the capacities of producing the task plan and supervising its execution, while being at the same time reactive to events from the functional level. The coexistence of these two features, a time-consuming planning process, and a time-bounded reactive execution process poses the key problem of their interaction and their integration to balance deliberation and reaction at the decisional level.
- The *execution control level*: at the interface between the decisional and the functional levels, lies an execution control level that controls the proper execution of the services according to safety constraints and rules, and prevents functional modules from unforeseen interactions leading to catastrophic outcomes. In recent years, we have used the R2C [Py and Ingrand, 2004] to play this role, yet it was programmed on the top of existing functional modules, and controlling their services execution and interactions, but not the internal execution of the modules themselves.

More details about the LAAS system architecture are to be found in [Alami et al., 1998].

We are given a plan of the robot as a simple temporal network (STN). The latter is translated into an equivalent network of timed automata. We want to check whether a given trace generated by the robot is accepted by the plan or not. For that purpose, we use TTG to generate an observer corresponding to the obtained network of TA. The observer may accept or reject the trace depending on whether it is a valid trace or not.

3. SIMPLE TEMPORAL NETWORKS

A *timepoint variable* is a temporal variable ranging over non-negative real values. Given a pair of timepoints (t_i, t_j), a *binary constraint* $c_{i,j}$ over (t_i, t_j) is a constraint of the form $l_{i,j} \leq t_j - t_i \leq u_{i,j}$, where $l_{i,j}$ and $u_{i,j}$ are real-values such that $l_{i,j} \leq u_{i,j}$.

Definition 1. A simple temporal network S is a pair (T, C) where T is a finite set of timepoint variables and C a finite set of binary constraints on these variables.

A solution of the STN S is a complete set of assignments for all the timepoint variables which satisfy all the constraints in C. An STN can be represented as a graph where nodes are the timepoint variables and the arcs are labeled with the binary constraints.

An example of an STN is given in Figure 3. The latter has three timepoint variables: t_1, t_2 and t_3. It should be interpreted as follows. Let e_1, e_2 and e_3 be the events corresponding to the timepoints t_1, t_2 and t_3, respectively. The event e_1 must happen at first. Events e_2 and e_3 must take place during the interval $[t_1 + 1, t_1 + 2]$ (but not necessarily at the same time). The binary constraint between t_2 and t_3 is the combination of two other constraints of the STN. A possible solution of this STN is $\{t_1 = 0, t_1 = 1, t_3 = 1.5\}$.

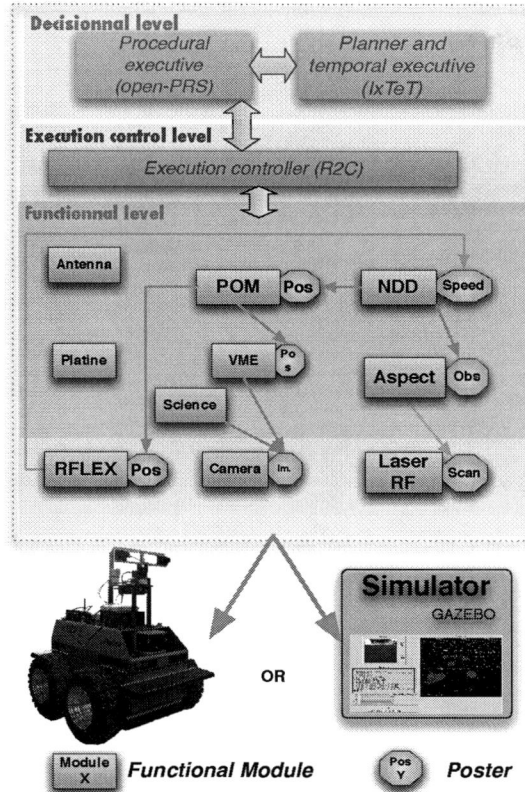

Fig. 2. The LAAS Dala Robot architecture.

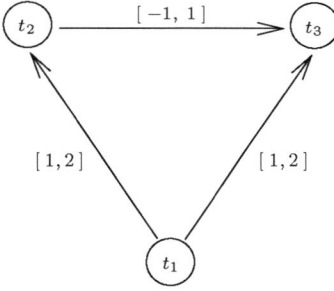

Fig. 3. A example of a simple temporal network (STN).

4. TIMED AUTOMATA

We use timed automata [Alur and Dill, 1994] with deadlines.

Definition 2. A *timed automaton over* Act is a tuple $A = (Q, q_0, X, \text{Act}, E)$, where:

- Q is a finite set of *locations*.
- $q_0 \in Q$ is the initial location.
- X is a finite set of *clocks*.
- E is a finite set of *edges*.

Each edge is a tuple (q, q', ψ, r, d, a), where:

- $q, q' \in Q$ are the source and destination locations.

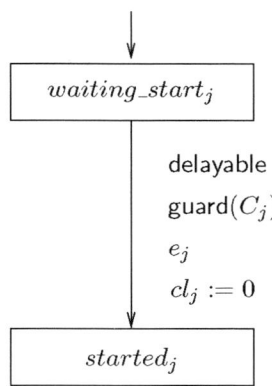

Fig. 4. The IF process P_j corresponding to the timepoint variable t_j.

- ψ is the *guard*, a conjunction of constraints of the form $x \# c$, where $x \in X$, c is an integer constant and $\# \in \{<, \leq, =, \geq, >\}$.
- $r \subseteq X$ is a set of clocks to *reset* to zero.
- $d \in \{\text{lazy}, \text{delayable}, \text{eager}\}$ is the *deadline*.
- $a \in \text{Act}$ is the action.

A timed automaton A defines an infinite *timed labeled transition system* (TLTS) which is denoted L_A. Its states are pairs $s = (q, v)$, where $q \in Q$ and $v : X \to \mathsf{R}$ is a clock *valuation*. 0 is the valuation assigning 0 to every clock of A. S_A is the set of all states and $s_0^A = (q_0, 0)$ is the initial state.

Discrete transitions are of the form $(q, v) \xrightarrow{a} (q', v')$, where $a \in \text{Act}$ and there is an edge (q, q', ψ, r, d, a), such that v satisfies ψ and v' is obtained by resetting to zero all clocks in r and leaving the others unchanged. Timed transitions are of the form $(q, v) \xrightarrow{t} (q, v + t)$, where $t \in \mathsf{R}, t > 0$ and there is no edge (q, q'', ψ, r, d, a), such that: either $d = \text{delayable}$ and there exist $0 \leq t_1 < t_2 \leq t$ such that $v + t_1 \models \psi$ and $v + t_2 \not\models \psi$; or $d = \text{eager}$ and $v \models \psi$.

5. TRANSFORMATION OF AN STN INTO A NETWORK OF TA

For pedagogical reasons, we first consider a restricted subclass of STN. For the considered STN, we assume that all the binary constraints are of the form $l_{i,j} \leq t_j - t_i \leq u_{i,j}$ such that $0 \leq l_{i,j} \leq u_{i,j}$ (e.g., $1 \leq t_j - t_i \leq 3$).

Consider such an STN $S = (T, C)$. We assume $T = \{t_1, t_2, \cdots, t_n\}$. For each $j = 1, \cdots, n$, let e_j be the event corresponding to the timepoint t_j. Also, let $C_j \subseteq C$ be the set of constraints $c_{i,j}$ such that $t_i \to t_j$ is an arc of S.

We associate an IF process P_j to each timepoint variable t_j. The process P_j has one clock cl_j and tow nodes $wait\_start_j$ (the initial node) and $started_j$. The process has only one edge. The latter is labelled with e_j, the event corresponding to t_j. The edge has the delayable guard $\text{guard}(C_j)$ such that

$$\text{guard}(C_j) = \bigwedge_{c_{i,j} \in C_j} (l_{i,j} \leq cl_i \leq u_{i,j}).$$

The clock cl_j is reset to 0 as soon as the transition is fired. The process P_j is depicted in Figure 4.

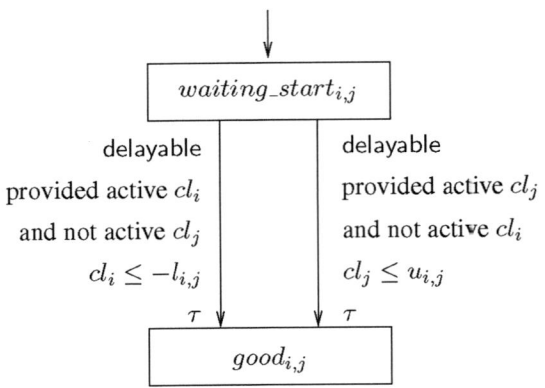

Fig. 5. The IF process $P_{i,j}$ corresponding to the binary constraint $c_{i,j} = l_{i,j} \leq t_j - t_i \leq u_{i,j}$ such that $(l_{i,j} < 0 \wedge 0 < u_{i,j})$.

Notice that at node $wait\_start_j$ the clock cl_j is not activated yet. It is activated as soon as the transition is fired. In the IF language, it is possible to label an edge with "provided active cl_j" to test whether the clock cl_j is active or not. Moreover, the edge is fired only if the clock is activated.

Now, we move to the general case where we may have constraints of the form $l_{i,j} \leq t_j - t_i \leq u_{i,j}$ such that $(l_{i,j} < 0 \wedge 0 < u_{i,j})$ (e.g., $-5 \leq t_j - t_i \leq 3$). For encoding this kind of constraints at the IF level we proceed as follows. For each constraint $c_{i,j}$ of this form, we define a new IF process $P_{i,j}$ which has two nodes $wait\_start_{i,j}$ (the initial node) and $good_{i,j}$. Moreover, $P_{i,j}$ has two edges labelled with the unobservable action τ. The conditions and guards of the two edges are depicted in Figure 5. The guards of the two edges are delayable in order to block time which, in turn, allows to prevent the binary constraints of the corresponding timepoints from becoming false.

6. HOW TO: TTG?

The TTG tool may be used for both test generation and monitoring purposes. Four generation modes are possible:

(1) *Interactive*: the user guides the test generation algorithm, resolving the non deterministic points (whether to issue an output or wait for an input, which output if many are possible, when to stop generating the test, etc.).
(2) *Random*: the non-deterministic points are resolved randomly.
(3) *Exhaustive*: all possible tests are generated up to a user-defined depth.
(4) *Coverage*: a set of tests that achieves a user-defined coverage criterion is generated.

Test generation is beyond the scope of this work. More details about the test generation mode are to be found in [Krichen and Tripakis, 2005].

TTG is written in C++. It works on linux iX86 platforms. It is built on top of the IF environment [Bozga et al., 2000]. The IF modeling language allows to specify systems consisting of many processes communicating through message passing or shared variables and includes features such as hierarchy, priorities, dynamic creation and complex data types.

Usage: TTG [option] argument1 argument2 ...	
TTG [-monit] out trace	read list of outputs from file <out>, read trace from file <trace> and check whether the trace is accepted or not
TTG [-help]	display help instructins on the standard output
TTG	allow to choose monitoring (or a test-generation) mode interactively

Table 1. Usage of TTG.

The IF tool-suite includes a simulator, a model checker and a connection to the untimed test generator TGV [Fernandez et al., 1996]. TTG is implemented independently from TGV. TTG uses the basic libraries of IF for parsing and symbolic reachability of timed automata with deadlines.

We give a brief description about how to install and to use TTG. A documentation about the tool and its binary code are available at:

http://www-verimag.imag.fr/ krichen/ttg/.

The different steps for installing TTG are the following:

(1) Download the IFx version of IF and install it.
(2) Download the TTG.tar package and unzip it into a directory of your choice.
(3) Set the "$TTG" environment variable to refer the directory where the TTG.tar package is unziped.
(4) Add into your "$PATH" variable the "$TTG/com" and "$TTG/bin/" directories.

For using TTG, one shall proceed as follows:

- Write your specification [1] in the IFx syntax and save into a text file with ".if" extension (e.g., "spec.if").
- Run command "runttg.sh spec.if" (or "runttg.sh spec" for short) to generate the "TTG" executable file.
- Run the executable file "TTG" as shown in Table 1.

7. RESULTS

We applied our method to a task plan of the robot Dala. The total number of timepoints of the considered plan is 62 (tp0, tp1, $\cdots$, tp61). Thus, the IF system we obtained has 62 clocks as well.[2] The obtained plan written in IF language is of length 1496 lines.

---

[1] specification = model of the SUT + model of the digital clock of the tester + priority rules.
[2] Notice that one other extra clock is used by the ticker process.

```
process tp52(1);

    var cl52 clock public;

    state wait_start52 # start ;
        deadline delayable;
        when
            ({tp0}0).cl0   ≥ 6136   and
            ({tp0}0).cl0   ≤ 8229   and
            ({tp33}0).cl33 ≥ 3735   and
            ({tp33}0).cl33 ≤ 6627   and
            ({tp34}0).cl34 ≥ 3335   and
            ({tp34}0).cl34 ≤ 6427   and
            ({tp41}0).cl41 ≥ 2933   and
            ({tp41}0).cl41 ≤ 5026   and
            ({tp42}0).cl42 ≥ 401    and
            ({tp42}0).cl42 ≤ 3693   and
            ({tp45}0).cl45 ≥ 6135   and
            ({tp45}0).cl45 ≤ 8228   and
            ({tp46}0).cl46 ≥ 4535   and
            ({tp46}0).cl46 ≤ 6628   and
            ({tp48}0).cl48 ≥ 4534   and
            ({tp48}0).cl48 ≤ 6627   and
            ({tp49}0).cl49 ≥ 2934   and
            ({tp49}0).cl49 ≤ 5027   and
            ({tp51}0).cl51 ≥ 1600   and
            ({tp51}0).cl51 ≤ 2000;
        output end11();
        set cl52 := 0;
        nextstate started52;
    endstate;

    state started52 ;
    endstate;
endprocess;
```

Fig. 6. An example of an automatically generated IF process encoding a timepoint of a plan of the Dala robot.

An example of an automatically generated IF process encoding a timepoint of the considered plan of the Dala robot is shown in Figure 6. The corresponding timepoint is "tp52". It has one clock "cl52" and two states "wait_start52" (the initial node) and "started52". The event corresponding to this timepoint is "end11". There are 10 time-constraints appearing within this process.

We tried to monitor a non-conforming trace. The trace we consider is given in Table 2. The trace is of length 60. The total duration of the trace is 94758 milliseconds (ms). The second colon of Table 2 gives the time separating two consecutive events of the third colon. The fourth colon gives the cumulated time corresponding to each depth.

We were unable to detect the non-conformance using analog-time semantics due to state explosion problems. The non-conformance was detected by using discrete-time semantics.

Depth	Time elapse in (ms)	Event	Cumulated time in (ms)
0	0	-	0
1	-	start0	0
2	47	-	47
3	-	start9	47
4	1621	-	1168
5	-	end9	1168
6	32	-	1700
7	-	start10	1700
8	23	-	1723
9	-	start3	1723
10	235	-	1958
11	-	end3	1958
12	1410	-	3368
13	-	end10	3368
14	52	-	3420
15	-	start7	3420
16	1948	-	5368
17	-	end7	5368
18	43	-	5411
19	-	start11	5411
20	1657	-	7068
21	-	end11	7068
22	39	-	7107
23	-	start12	7107
24	21	-	7128
25	-	start2	7128
26	240	-	7368
27	-	end2	7368
28	1400	-	8768
29	-	end12	8768
30	41	-	8809
31	-	start6	8809
32	2905	-	11714
33	-	start13	11714
34	505	-	12209
35	-	end6	12209
36	267	-	12476
37	-	start5	12476
38	842	-	13318
39	-	end13	13318
40	31	-	13349
41	-	start14	13349
42	24	-	13373
43	-	start1	13373
44	236	-	13609
45	-	end1	13609
46	701	-	14310
47	-	end5	14310
48	559	-	14869
49	-	start8	14869
50	139	-	15008
51	-	end14	15008
52	9801	-	24809
53	-	end8	24809
54	12652	-	37461
55	-	start4	37461
56	1858	-	39319
57	-	end4	39319
58	55439	-	94758
59	-	end0	94758

Table 2. Trace to monitor.

8. CONCLUSION

We used the timed test generator tool TTG as an automatic generator of observers for monitoring purposes for the Dala Robot (LAAS) case study. We start from a plan given as a simple temporal network (STN). This plan is taken to be a high-level specification. It is automatically translated into a network of timed automata (TA). The timed automata are written in the IF language.

An observer is automatically synthesized. The observer checks whether a sequence of observations conforms to the specification or not. We applied our method on a non-conforming trace of length 60. The non-conformance was detected using discrete-time semantics.

As a future work direction, it will be interesting to consider the analog-observer case. We may find another way to estimate our uncertainty about time measuring other than considering digital-clocks. It will be also interesting to consider on-line monitoring and to integrate it in the whole LAAS architecture in order to take into account in real-time the detected errors and failures.

Moreover, we are currently working on replacing the preexisting GENOM framework [Fleury et al., 1997] on which the current functional level of the Dala robot is based with the BIP component framework [Basu et al., 2006]. BIP is a software framework for modeling heterogeneous real-time components. A BIP component model is the superposition of three layers: behavior, connectors and priority rules. Some preliminary results about this work are to be found in [Basu et al., 2007]. It will be interesting to use monitoring-like techniques to validate this approach.

REFERENCES

A. Akhavan, S. Bensalem, M. Bozga, and E. Orfanidou. Experiment on verification of a planetary rover controller. In *4th International Workshop on Planning and Scheduling for Space (IWPSS'04), Darmstadt, Germany*, 2004.

R. Alami, R. Chatila, S. Fleury, M. Ghallab, and F. Ingrand. An architecture for autonomy. *International Journal of Robotics Research*, 17(4):315–337, 1998.

R. Alur and D. Dill. A theory of timed automata. *Theoretical Computer Science*, 126:183–235, 1994.

Ananda Basu, Marius Bozga, and Joseph Sifakis. Modeling heterogeneous real-time components in bip. In *SEFM '06: Proceedings of the Fourth IEEE International Conference on Software Engineering and Formal Methods*, pages 3–12. IEEE Computer Society, 2006.

Ananda Basu, Matthieu Gallien, Charles Lesire, Thanh-Hung Nguyen, Saddek Bensalem, Flix Ingrand, and Joseph Sifakis. In *2nd National Workshop on Control Architectures of Robots (CAR'07), Paris, France*, 2007.

S. Bensalem, M. Bozga, M. Krichen, and S. Tripakis. Testing conformance of real-time applications by automatic generation of observers. In *4th International Workshop on Runtime Verification (RV'04), Barcelona, Spain,*, volume 113 of *ENTCS*, pages 23–43. Elsevier, 2005.

M. Bozga, J.C. Fernandez, L. Ghirvu, S. Graf, J.P. Krimm, and L. Mounier. IF: a validation environment for timed asynchronous systems. In *Computer Aided Verification, 12th International Conference (CAV'00), Chicago, IL, USA*, volume 1855 of *LNCS*, pages 543–547. Springer, 2000.

J.C. Fernandez, C. Jard, T. Jéron, and G. Viho. Using on-the-fly verification techniques for the generation of test suites. In *Computer Aided Verification, 8th International Conference (CAV'96), New Brunswick, NJ, USA*, volume 1102 of *LNCS*, pages 348–359. Springer, 1996.

S. Fleury, M. Herrb, and R. Chatila. Genom: a tool for the specification and the implementation of operating modules in a distributed robot architecture. In *IEEE/RSJ International Conference on Intelligent Robots & Systems (IROS), Grenoble, France*, pages 842–848, 1997.

T. Henzinger, Z. Manna, and A. Pnueli. What good are digital clocks? In *Automata, Languages and Programming, 19th International Colloquium (ICALP'92), Vienna, Austria*, volume 623 of *LNCS*, pages 545–558. Springer, 1992.

M. Krichen and S. Tripakis. Black-box conformance testing for real-time systems. In *11th International SPIN Workshop on Model Checking of Software (SPIN'04), Barcelona, Spain*, volume 2989 of *LNCS*, pages 109–126. Springer, 2004.

M. Krichen and S. Tripakis. An expressive and implementable formal framework for testing real-time systems. In *The 17th IFIP Intl. Conf. on Testing of Communicating Systems (TestCom'05), Montreal, Canada*, volume 3502 of *LNCS*, pages 209–225. Springer, 2005.

S. Lemai, F. Ingrand, and M. Gallien. Embedded decision in the laas architecture. In *IEEE International Conference on Robotics and Automation (Workshop Principle and Practice of Software Development in Robotics: crafting modular and interoperable systems), Barcelona, Spain*. IEEE, 2005.

F. Py and F. Ingrand. Dependable execution control for autonomous robots. In *International Conference on Intelligent Robots and Systems (IROS), Takamatsu, Japan*. IEEE, 2004.

3D molecular modeling systems: State of art

M. Essabbah*. S. Otmane.*. M. Mallem.*

*Informatics, Integrative Biology and Complex Systems,
40 Rue de Pelvoux, 91020 Evry Cedex, France (e-mail: mouna.essabbah@ibisc.fr).

Abstract: Genomic sequences are initially known by their linear form. However, they have also a three-dimensional structure which can be useful for genomes analysis. This 3D structure representation brings a new point of view for the sequences analysis. It was established that the importance of molecular space structure has created increasingly a growing interest for 3D modeling molecules. Therefore, several studies have described the design of software for 3D molecular visualization. Some of these tools offer even 3D molecular manipulation through virtual models. However these 3D models are often based on predictive methods leading to a family of "predictive models". This constraint made that biologists doubt the effectiveness of these models, thinking they lack of structural realism and therefore functional one. The solution that we propose is the confrontation between virtual models and real data. This approach compares 3D virtual models and real microscopic images of the same molecule in order to validate and/or improve the 3D model.

Keywords: Molecules space structure, 3D modeling, 3D visualization, 3D manipulation.

1. INTRODUCTION

The 3D molecular modeling is a new technology that is increasingly attracting scientists interest. Its ability to simulate natural phenomena that are not exploitable experimentally offers great potential and opens doors to new research in the field. As a result, a large number of 3D modeling systems have emerged trying to be more precise as well as required. However, most software for 3D molecular modeling are based on predictive methods. These methods often use local conformation tables - generated by statistics relative to biological experiments - which restricts the 3D rending.

The advances made in the biology field and in parallel in 3D modeling, allows the implementation of 3D molecular modeling applications increasingly complex, dedicated to the study of molecular structures and molecular dynamics analysis. However 3D molecular modeling presents several problems that are still the subject of intense researches, in particular on the fundamental interests of this modeling and numerical simulation, the accuracy of 3D models, and so on.

This article presents a recent state of the art on 3D molecular modeling. It is divided into five sections: we begin with a focus on the importance of molecular space structure. Then, we will present different 3D molecular modeling systems. Finally, we present the solution that we propose in the conclusion.

2. IMPORTANCE OF MOLECULAR SPACE STRUCTURE

The importance of molecular space structure may seem obvious even to non-specialist. However, this section provides concrete evidence to real importance of modeling in molecular field.

2.1. Interest of the molecule spatial representation

Initially scientists were interested in sequence analysis for the study of its rich syntax. Gradually (between 1984 and 1987) molecular biology lived a fulgurating progress of its technical means which leads to automation and an increasingly advanced and refined miniaturization. The researchers are now studying the molecules behavior by spatial visualization and structures analysis. Actually, the three-dimensional structures inform us about the molecule functionality. This approach allows a global view of the molecule space structure as well as possibilities of interaction with other molecules. It also allows modeling phenomenon or simulates a biological mechanism.

Currently, molecular modeling is an essential research area. It helps scientists to develop new drugs against diseases. This sector assists also the genomes analysis. As the interest that was accorded to it is extremely important, many researchers became specialized in molecular modeling.

2.2. When biology becomes molecular

There are many researches in the literature that link molecular modeling in its biological context, based on three-dimensional models. These studies present various concepts used in the development of models for biological structures (DNA, RNA, proteins, etc.).

In the chemistry department at the University of Pennsylvania, Zou & al. studied the structure and dynamics of a three-dimensional protein (Amphiphilic, Metallo-Porphyrin-Binding Protein) via molecular dynamics simulations [Zou 06]. The simulation results match with the available experimental data, describing the structures in a lower resolution and a limited size.

In addition, a Polish group is working on forecasting (high resolution) of three-dimensional RNA structures (low resolution) to answer questions from the RNA molecular biology [Popenda 06].

The molecular structures are more interesting because of their complexity. Indeed, the molecular representation helped to make the connection between molecular complexity and its impact on the probability of discovering new drugs [Hann 01].

Structural analysis of Mu DNA transposition was successful using 3D reconstruction of images obtained by scanning transmission electron microscopy (TIGE) [Yuan 05].

Moreover, a three-dimensional structure was used as a model for the replication study based on curl degree analysis along the DNA axis. The structure was built by cryo-electron microscopy and simple-particle reconstruction techniques [Gomez-Lorenzo 03]. The same technique has been adopted for the reconstruction of a three-dimensional complex DNA-protein [Abu-Arish 04].

Those researches represent a rich structural basis for different areas such as the biochemical function, the study of various phenomena (replication, transcription), etc.

3. VISUALIZATION OF 3D STRUCTURES

The 3D molecules visualization has always been an important chemistry chapter. As a result, chemists have always tried to represent molecules and crystals using elementary ways (the wire DNA double helix of Watson and Crick) for teaching and research. Since then, molecular modeling has become a growing discipline, benefiting from the rapid development of hardware and software.

Moreover, visualization of a three-dimensional structure is one of the first tools developed for structures analysis, but also one of the first tests that do wish biologist.

We can classify 3D molecular representation systems into two categories: 3D visualization systems and 3D interactive modeling. Each of these two categories can be devided into two athers classes: predictive systems and those based on 3D data (crystallographic data, PDB: Protein Data Bank).

3.1. Molecular visualization software

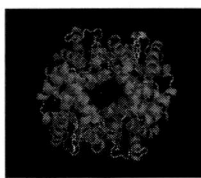
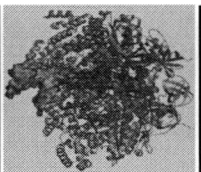
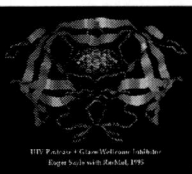

| Figure 1 Hemoglobin structure by PyMol | Figure 2 Mitochondrie F1-ATPase bovin by PyMol | Figure 3 HIV Protease by RasMol |

Due to this evolution, several three-dimensional software for molecules visualization, simple and public, have emerged (gOpenMol, MoleKel, ViewMol, MolMol, RasMol (Figure 3), PyMol (Figure 1 and 2), etc.). Other molecules viewers are online for immediate structure reconstruction (CBS-Metaserver, GENO3D, Swiss-Model, Biomer, etc.).

3.2. Interactive systems for molecular modeling

Gradually the visualization interest does not stop any more to the molecules observation, but it extends to the structure analysis and functionality interpretation of molecules by their forms. So a new wave of interactive viewers appears in order to study the molecule dynamics through its three-dimensional structure. These specific applications offer more complex modeling as molecules functionality.

VMD (Visual Molecular Dynamics) [Humphrey 96] is a molecular visualization program for displaying, animation and analysis of large molecular systems using three-dimensional graphics. It can also read standards PDB files and display structures. VMD provides a wide variety of methods to display and color molecules. It can also be used to animate and analyze the trajectory of molecular dynamics (MD) simulation. Its special feature is that it can be used as graphical interface for MD external program, displaying and animating a molecule that is simulated on a remote computer.

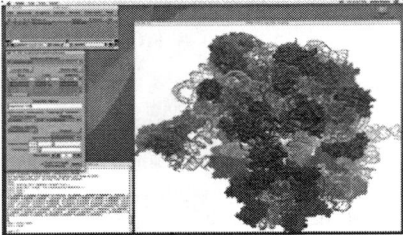

Figure 4 Screenshot of VMD 1.8.5

ADN-Viewer (DNA 3D modeling and stereoscopic visualization) [Hérisson 01] is part of the research conducted by the bioinformatics team of LIMSI-CNRS laboratory. The researchers were particularly interested in DNA spatial distribution.

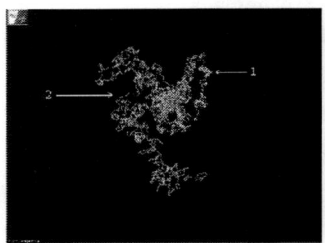
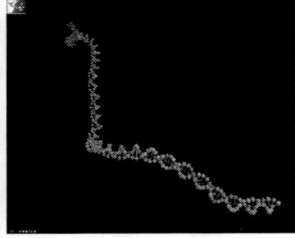

| Figure 5 Compact (zone 1) and relaxed (zone 2) DNA areas | Figure 6 Each colored sphere corresponds to a nucleotide |

The software offers the 3D reconstruction of DNA structure, which is based on a prediction table. This table has been established experimentally by physicists. The model is based on space conformation between two nucleotide of DNA textual sequence. The conformation table includes rotation angles that allow positioning two successive nucleotides.

This is called a predicted model. ADN-Viewer offers several 3D DNA sequences representations. There is a genomics representation (see Figure 5) and a gene representation (see Figure 6).

3DNA (Analysis and reconstruction of 3D nucleic acid structures) [Lu 03] was identified as American software for 3D nucleic acids structure analysis, reconstruction and visualization. 3DNA can handle double helices non-parallel and parallel, simple structures, triplex, and other quadrupled motifs found in complex DNA and RNA structures from a PDB file coordinates. The program uses a reference frame - recently recommended for the geometry description of nucleic acid base pairs - and a rigorous matrices arrangement to calculate parameters and local conformation to rebuild the structure of these parameters. Tools are provided to locate base pairs and regions in a helical structure and to reorient structures for an effective visualization. Helical regular models based on X-ray diffraction measurements of various repetition levels can also be handled by this program.

AMMP-Vis (a virtual environment for collaborative molecular modeling) [Chastine 05] is an immersive system that offers to biologists and chemists the possibility of manipulating molecular models through a natural gesture. It allows receiving and displaying real-time molecular dynamics simulation results. It allows also sharing adapted views and provides support for local and remote collaborative research. It is based on the molecular visualization system AMMP described in the next section.

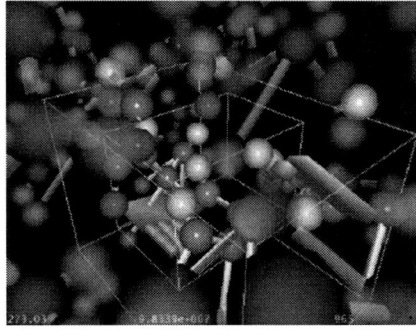

Figure 7 Example of a colaborative research

NAVRNA (Interactive system for structural RNA) [Bailly 06] is an interactive system to visualize, explore and edit the RNA molecules.

Figure 8 2D molecule (on the table) and 3D visualization (on the wall)

NAVRNA can be used to visualize at the same time the three-dimensional structure (3D) projected onto a white wall and the secondary structure (2D) also projected on a table. Both shows are strongly linked. It is a multi-surface collaborative system for RNA analysis.

Augmented reality with auto-made tangible models for molecular biology applications [Gillet 05]

The evolution of auto-made computer technology ("3D Printing") can now allows the production of physical models such as molecules and biological complex sets. It presents an application that demonstrates the use of tangible auto-made models and augmented reality for research and education in molecular biology, and to improve the environment for scientific collaboration and exploration. They have adapted an augmented reality system to allow 3D virtual representations (produced by the Python molecular viewer) to be overlaid on a real molecular model. User can easily change this superposition of information, switching between different molecule representations, the display of molecular properties such as electrostatics, or dynamic information. The physical model provides a powerful and intuitive interface for manipulating computer models, improving the interface between the human intention, the physical model and the computing activity.

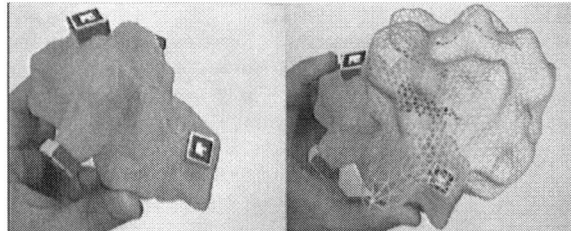

Figure 9 Ribosome with (right) and without (left) RA.

3.3. Molecular dynamics modeling

The molecular dynamics simulation [Leach 01] is the second essential direction after molecular modeling. More generally almost all researchers in chemistry or structural biology have been attracted by this experiment aspect. It allows calculating the evolution of a particles system over time. These simulations are used as structural and dynamic models for understanding experimental results.

This growing interest in the molecular dynamics simulation promoted the emergence of many modeling software. The description of some of the most popular is given in the following.

AMBER (Assisted Model Building with Energy Refinement) [Pearlmen 95] is a package i.e. a programs set developed from a program that was created in the end of the 70's. AMBER applies molecular mechanics, normal way analysis, molecular dynamics and calculating free energy to simulate the structural molecules and energetic properties. It now includes a program group representing a number of powerful tools of modern chemistry, focused on the molecular dynamics and free calculations of proteins energy, nucleic acids and carbohydrates [Case 05].

AMMP [Harisson 99] is a program that models complete and modern dynamics and molecular mechanics. It can handle small molecules and macromolecules including proteins, nucleic acids and other polymers. In addition to the standard common features molecular modeling software, AMMP has a potential flexible choice and a simple and powerful capability to manipulate molecules and analyze various energy limits.

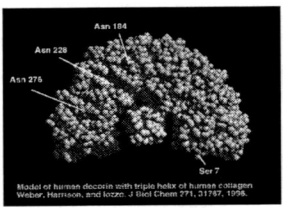 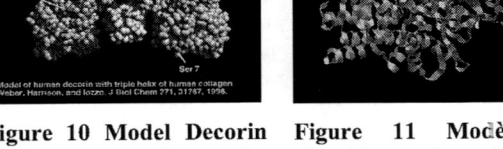

Figure 10 Model Decorin with fibrille collagen by AMMP

Figure 11 Modèle de glucokinase humain avec du glucose par AMMP

A main advantage over many other programs is that it is easy to present the non-standard links between polymer ligands unusual or non-standards residue. Furthermore it is possible to add the missed hydrogen atoms and implement partial structures, which is difficult for many other modeling tools.

GROMACS is a set of codes running molecular dynamics, initially developed by Herman Berendsen of Groningen University [Bekker 93]. It also has a lot of analysis tools including a trajectories viewer. It was initially designed for biochemical molecules such as proteins and lipids that have rich and complex interactions. However, since GROMACS is extremely fast to calculate the interactions [van der Spoel 05] many groups use it for research on non-biologic systems as polymers.

4. CONCLUSION

3D molecular models are approximated (i.e. discrete and simplified). We have often to be based on local or partial data. So the 3D molecular modeling is often predicted. We noticed that biologists manipulate those predictive programs as ADN-Viewer, with a certain reserve, finding that local predictions cannot be applied to a global structure. The prediction error would be cumulated, especially since it is obviously larger on a larger scale. In the other hand, computer scientists are struggling to create the three-dimensional shape of complex structures, such as DNA, from real images because those images are relatively crude. In addition, the molecule is a deformable object consisting of curvatures forms. So we decided to combine these two approaches, molecular modeling and real data (i.e. microscopic images). The goal is to estimate the 3D proposed models accuracy and so to improve them for more credibility. One proposed solution is to confront predict models to corresponding real data by matching techniques and image processing. Moreover, previous studies have shown that immersive virtual environments have unique advantages compared to desktops systems in the molecular visualization field. However, exploration and interaction in existing virtual environments of molecular modeling are often simply basic and limited to a single user, lacking of strong 3D natural interaction and collaboration support. In addition, scientists are often reluctant to adopt those systems because of their lack of availability. It would therefore be interesting to propose a confrontation tool between 3D models and real data while providing an immersive interaction paradigm. It will be a powerful and tailorable tool.

REFERENCES

[Hann 01] Hann, MM., AR. Leach, G Harper: Molecular Complexity and Its Impact on the Probability of Finding Leads for Drug Discovery. *Journal of Chemical Information and Computer Sciences* 41(3): 856-864 (2001)

[Zou 06] Zou, H., J. Strzalka, T. Xu, A. Tronin, and J.K. Blasie Three-Dimensional Structure and Dynamics of a de Novo Designed, Amphiphilic, Metallo-Porphyrin-Binding Protein Maquette at Soft Interfaces by Molecular Dynamics Simulations; *J. Phys. Chem. B*; 2007; *111*(7) pp 1823 – 1833.

[Popenda 06] M. Popenda, Ł. Bielecki and R. W. Adamiak, High-throughput method for the prediction of low-resolution, three-dimensional RNA structures, Nucleic Acids Symposium Series, 50, 67-68 (2006).

[Pearlmen 95] Pearlmen, D.A., Case, D.A., Caldwell, J.W., Ross, W.S., Cheatham, III, T.E., DeBolt, S., Ferguson, D., Seibel G. & Kollman, P. (1995) AMBER, a package of computer programs for applying molecular mechanics, normal mode analysis, molecular dynamics and free energy calculations to simulate the structural and energetic properties of molecules. *Comp. Phys. Commun.* 91, 1-41.

[Case 05] Case, D.A., T.E. Cheatham, III, T. Darden, H. Gohlke, R. Luo, K.M. Merz, Jr., A. Onufriev, C. Simmerling, B. Wang and R. Woods (2005). The Amber biomolecular simulation programs. *J. Computat. Chem.* 26: 1668-1688.

[Harisson 93] Harrison R.W. (1993), Stiffness and Energy Conservation in Molecular Dynamics: an Improved Integrator J. Comp. Chem. 14: 1112-1122.

[Harisson 99] Harrison, R.W (1999). Integrating Quantum and Molecular Mechanics, *J. Computat. Chem*, 20: 1618-1633.

[Bekker 93] Bekker, H., H.J.C. Berendsen, E.J. Dijkstra, S. Achterop, R. van Drunen, D. van der Spoel, A. Sijbers, H. Keegstra, B. Reitsma and M.K.R. Renardus (1993) Gromacs: A parallel computer for molecular dynamics simulations. *In Physics Computing* 92 (Singapore)

[Van der Spoel 05] Van der Spoel, D., E. Lindahl, B. Hess, G. Groenhof, A. E. Mark and H. J. C. Berendsen (2005): GROMACS: Fast, Flexible and Free, *J. Comp. Chem.* 26: 1701-1718.

[Gomez-Lorenzo 03] Gomez-Lorenzo MG, Valle M, Frank J, Gruss C, Sorzano CO, Chen XS, Donate LE, Carazo JM. (2003) Large T antigen on the simian virus 40 origin of replication: a 3D snapshot prior to DNA replication. *EMBO J. Dec* 1;22(23):6205-13.

[Abu-Arish 04] Abu-Arish, A., D. Frenkiel-Krispin, T. Fricke, T. Tzfira, V. Citovsky, S.G. Wolf and M. Elbaum, (2004) Three-dimensional reconstruction of *Agrobacterium* VirE2 Protein with Single-stranded DNA, Vol. 279, No. 24, Issue of June 11, pp. 25359–25363.

[Yuan 05] Yuan JF, Beniac DR, Chaconas G, Ottensmeyer FP. (2005), 3D reconstruction of the Mu transposase and the Type 1 transpososome: a structural framework for Mu DNA transposition. *Genes Dev.* Apr 1;19(7):840-52.

[Humphrey 96] Humphrey, W. Andrew Dalke, and Klaus Schulten. (1996) VMD - Visual Molecular Dynamics. *Journal of Molecular Graphics*, 14:33-38.

[Chastine 05] Chastine, J. W., Brooks, J. C., Zhu, Y., Owen, G. S., Harrison, R. W., and Weber, I. T. (2005). AMMP-Vis: a collaborative virtual environment for molecular modeling. In *Proceedings of the ACM Symposium on Virtual Reality Software and Technology* (Monterey, CA, USA, November 07 - 09, 2005). VRST '05. ACM Press, New York, NY, 8-15.

[Lu 03] Lu, XJ. and Wilma K. Olson, (2003) 3DNA: a software package for the analysis, rebuilding and visualization of three-dimensional nucleic acid structures, *Nucleic Acids Research*, , Vol. 31, No. 17 5108-512.

[Bailly 06] Bailly, G., Auber, D., and Nigay, L. (2006). From Visualization to Manipulation of RNA Secondary and Tertiary Structures. In *Proceedings of the Conference on information Visualization* (July 05 - 07, 2006). IV. IEEE Computer Society, Washington, DC, 107-116.

[Gillet 05] Gillet A, Sanner M, Stoffler D, and Olson A (2005), Tangible Interfaces for Structural Molecular Biology *Structure*: 13 p483–491.

[Leach 01] Leach, AR., *Molecular Modelling:Principles and Applications*, Prentice Hall (Pearson Education), Harlow; 2001; ISBN 0 582 38210 6; 744 pp.

[Hérisson 01] Hérisson J., Gherbi R., "Model-Based Prediction of the 3D Trajectory of Huge DNA Sequences Interactive Visualization and Exploration". BIBE 2001: 263-270

A STRATEGY FOR UNICYCLE'S FORMATION CONTROL BASED ON INVARIANCE PRINCIPLE

M.A. El Kamel * L. Beji ** A. Abichou *

*Mathematical Engineering Laboratory, Polytechnic School of Tunisia Rue El khawarizmi B.P. 743 - 2078 La Marsa. Tunisie
(e-mail: Anouar.elkamel@ibisc.univ-evry.fr)
(e-mail: azgal.abichou@ept.rnu.tn)
** Informatics, Integrative Biology and Complex Systems, 40 Rue de Pelvoux, 91020 Evry Cedex, France
(e-mail: Lotfi.Beji@iup.univ-evry.fr).

Abstract: In this work one presents a control strategy for n unicycles in formation called "formation control". This in order to achieve a target-capturing task in 2D space including collision-avoidance. The strategy developed here is based on the Invariance Principle (IP) where we construct a set which contains the target and rather big to contain all unicycles. This set will be attractive for the formation by developing a cooperative control law. The stability of the formation is realized when all unicycles are in the attractive set.

Keywords: multi-vehicle, cooperative control, invariance principle, collision avoidance, formation stability.

1. INTRODUCTION

Noticing that the study of formation of moving agents and its control is very significant in several applications inter alia in the medical field (behavior of a drug in the body), the study of movement of the cells, the road traffics, the migration of a group of animals(birds...), to military goals etc..., that's why we are interested in our work of the cooperation between a group of robots with avoidance of obstacles and collisions.

The formation control of vehicles in hostile environment is treated by several researchers by various methods according to the goal to achieve. As the formation is composed of several subsystems, the construction of a mathematical model is essential to facilitate the connection between its various agents. Among the models most used and most logical to think of, is using the kinematic distances model between robots[5, 3, 6, 7]. Another approach consists to group all subsystems in an only one and define a distributed control so that the distances between vehicles are bounded as in [2], also to avoid singularity of the kinematic equations in unicycle's formation, dynamic extension of the system was proposed in [8]. Concerning the pursuit of several similar robots we can quote the work realized in [9] which treats the cyclic pursuit of N agents described by a integrator $\dot{z}_i = u_i$ with a control $u_i = k(z_{i+1} - z_i)$ each agent i effectively pursues the next $i+1$ modulo N to exponentially converges to the centroid of N agents. This pursuit was also the object of several works like [3] for the unicycle and [6] for a target tracking in a 3D space. With regard to the formation stabilization, we can note, during the two last years, the special interest carried for the methods based on the graph theory which gives to the formation a quite precise configuration as in [3, 6, 10].

1.1 Problem description

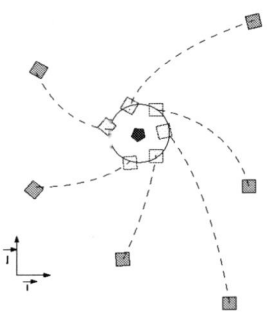

Fig. 1. Unicycles moving to the attractive set

We consider a fixed target in the plan and a group of n similar unicycles moving in a 2D environment without obstacles. Each unicycle has the following kinematics:

$$\begin{cases} \dot{x}_i = u_i \cos\theta_i \\ \dot{y}_i = u_i \sin\theta_i \\ \dot{\theta}_i = w_i \end{cases}, i \in I = \{1..n\} \qquad (1)$$

Where x_i, y_i are the cartesian coordinates, θ_i is the steering angle, u_i is the linear velocity and w_i is angular velocity of each unicycle i.

We assume that each vehicle knows the position of the others beside of the target's.

In our work, in opposite to the leader-follower trajectory tracking approach [2] and the cyclic pursuit [3], we conduct the whole formation without need neither to generate trajectories nor track a leader or even define a model based on relative distances as in [4, 3].

The main idea of our work is to conduct these n unicycles to a well defined "attractive" set containing a predefined target so that they will surround it.

We propose a new strategy for vehicle formation control. This will be developed in two steps :

In fact, in the section 2 we will define a function that we call "collision avoidance function". This function, injected into the control law's expression, will guarantee that the solutions of different systems under (1) will never collide over time. Then we obtain the kinematic system with a new control law. The last system will be controlled in section 3, so we can obtain convergence to a set defined by using the invariance principle Theorem.

2. COLLISION AVOIDANCE

Generally when we deal with the formation control of unicycles the first problem that must be solved is how to avoid collision between the different robots.

In order to solve this problem, we define a new control model such that each unicycle takes into consideration the other ones and guarantees that no collision occurs over time.

In the following proposition, we will use the notation L_{ik} defined as :

$$L_{ik} = \int_{t_0}^{t} [\frac{\prod_{j \neq i, \neq k}^{n} d_{ij}^2 \tilde{u}_i}{1 + \sum_{m \neq i}^{n} \prod_{j \neq i, \neq m}^{n} d_{ij}^2}((x_i-x_k)\cos\theta_i+(y_i-y_k)\sin\theta_i)$$
$$+$$
$$\frac{\prod_{j \neq k, \neq i}^{n} d_{kj}^2 \tilde{u}_k}{1 + \sum_{m=0, \neq i}^{n} \prod_{j \neq k, \neq m}^{n} d_{kj}^2}((x_i-x_k)\cos\theta_k+(y_i-y_k)\sin\theta_k)]ds$$

Where
$$d_{ik} = [(x_i-x_k)^2 + (y_i-y_k)^2]^{1/2}, \quad \forall i,k \in I$$
is the distance separating robot i from robot k.

Propostion 1. For $i \in I$, we define the avoidance function

$$ev_i = \frac{\prod_{j \neq i}^{n} d_{ij}^2}{1 + \sum_{m=0, \neq i}^{n} \prod_{j \neq i, \neq m}^{n} d_{ij}^2}$$

Under the following control laws
$$u_i = ev_i \tilde{u}_i$$

where $\tilde{u}_i$ is a new convenable control input of (1) such that L_{ik} is convergent, the distance between ith and kth vehicle
$$d_{ik} \neq 0, \quad \forall k \neq i$$
. □

Proof. Using the control law u_i given above the system become

$$\begin{cases} \dot{x}_i = ev_i \tilde{u}_i \cos\theta_i \\ \dot{y}_i = ev_i \tilde{u}_i \sin\theta_i \\ \dot{\theta}_i = w_i \end{cases} \quad (2)$$

For $i \neq k$ the differential of distance between two vehicles d_{ik}^2 with respect to t is given as follows

$$\dot{\overline{[(x_i - x_k)^2 + (y_i - y_k)^2]}} = [(x_i - x_k)^2 + (y_i - y_k)^2] *$$

$$[\frac{\prod_{j \neq i, \neq k}^{n} ((x_i - x_j)^2 + (y_i - y_j)^2)\tilde{u}_i}{[1 + \sum_{0=m \neq i}^{n} \prod_{j \neq i, \neq m}^{n} [(x_i - x_j)^2 + (y_i - y_j)^2]} *$$

$$((x_i - x_k)\cos\theta_i + (y_i - y_k)\sin\theta_i)$$
$$+$$
$$\frac{\prod_{j \neq k, \neq i}^{n} ((x_k - x_j)^2 + (y_k - y_j)^2)\tilde{u}_k}{[1 + \sum_{0=0=m \neq i}^{n} \prod_{j \neq k, \neq m}^{n} [(x_k - x_j)^2 + (y_k - y_j)^2]} *$$

$$((x_i - x_k)\cos\theta_k + (y_i - y_k)\sin\theta_k)]$$

The form of this equation is
$$\dot{z} = zf(t)$$
the solution of this model is
$$z = z_0 exp(\int_{0}^{t} f(s)ds)$$

if $\int_{0}^{\infty} f(s)ds$ is convergent
Then
$$[(x_i-x_k)^2+(y_i-y_k)^2] = [(x_{i0}-x_{k0})^2+(y_{i0}-y_{k0})^2]\exp[L_{ik}]$$

when $d_{ik0}^2 = [(x_{i0} - x_{k0})^2 + (y_{i0} - y_{k0})^2] \neq 0$ and L_{ik} is convergent.

So
$$d_{ik} = [(x_i-x_k)^2 + (y_i-y_k)^2]^{1/2} \neq 0$$
where d_{ik} designs the distance between robot i and robot k. □

3. CONTROL LAW CONSTRUCTION FOR TARGET CAPTURING

In this section we will demonstrate how we construct an attractive set and a control law that makes all the unicycles converge to this set and so surround the target as shown in $figure 1$.

Given the target position (a, b) in a 2D space, we define a circle centered in (a, b) of radius l rather big to contain all the unicycles in such a way that they surround the target.
In the following theorem we give a control law that ensures the convergence of all unicycle's position state to the circle defined above.

Theorem 1. Let
$$D = \{(x, y) \in \mathbb{R}^2 / k \geq (x-a)^2 + (y-b)^2 \geq l\}$$
and
$$M = \{(x, y) \in D / (x-a)^2 + (y-b)^2 = l\}$$

Consider n unicycles with kinematics (2) defined in D. By using the control inputs
$$\begin{aligned}\tilde{u}_i &= -[(x_i - a)\cos\theta_i + (y_i - b)\sin\theta_i] \\ &\quad [(x_i - a)^2 + (y_i - b)^2 - l](\frac{1}{(1+t)^2}) \\ w_i &= [(x_i - a)^2 + (y_i - b)^2 - l](\frac{1}{(1+t)^2})\end{aligned} \quad (3)$$

the solutions of (2) approaches M for every initial conditions in D. □

The proof of this theorem is based on the two lemmas that we will state before we prove it. Let us remind the first lemma known as the *Barbala's lemma*

Lemma 1. Let $\phi : \mathbb{R} \to \mathbb{R}$ be a uniformly continuous function on $[0, +\infty)$
Suppose that $\lim_{t \to \infty} \int_0^t \phi(s) ds$ exists and is finite. Then,
$$\phi(t) \to \infty \text{ as } t \to \infty$$
□

For the Proof of this lemma, the reader can refer to [1]

Now, here is the second lemma we need to achieve the proof of theorem 1

Lemma 2. let
$D = \{(x, y) \in \mathbb{R}^3 / k \geq (x-a)^2 + (y-b)^2 \geq l\}$
Every solution of (1) which starts in D remains for all future time in D i.e D is an invariant set for the kinematic equation (1). □

Proof. Now we shown that each solution starts in D remains for all future time in D
Assume that $(x_0, y_0) \in D$ and given
$$S(x_i, y_i) = (x_i - a)^2 + (y_i - b)^2 - l$$
. The differential of S with respect to t is given as follows:
$$\dot{S}(x_i, y_i) = -2ev_i[(x_i - a)\cos\theta + (y_i - b)\sin\theta]^2$$
$$((x_i - a)^2 + (y_i - b)^2 - l)\frac{1}{1+t^2}S(x_i, y_i)$$

The solution of this equation has the form
$$S(x_i, y_i) =$$
$$S(x_0, y_0)\exp(\int_{t_0}^t -2ev_i[(x_i - a)\cos\theta + (y_i - b)\sin\theta]^2$$
$$((x_i - a)^2 + (y_i - b)^2 - l)\frac{1}{1+t^2}ds)$$

and is positive when $S(x_0, y_0) \geq 0$ and
$$\int_{t_0}^t -2ev_i[(x_i - a)\cos\theta + (y_i - b)\sin\theta]^2$$
$$((x_i - a)^2 + (y_i - b)^2 - l)\frac{1}{1+t^2}ds$$
is convergent for $t \to \infty$ and $t_0 \in [0, +\infty[$.

This can be shown as follows:
We note $X = (x_i, y_i)$, $A = (a, b)$ and $I = (\cos\theta, \sin\theta)$.
we have
$$[(x_i - a)\cos\theta + (y_i - b)\sin\theta]^2 = [(X-A)^t I]^2 \leq \|X-A\|^2 \leq k$$
for $(x_0, y_0) \in D$ and $0 \leq ev_i \leq 1$.
Which implies that the integral
$$\int_{t_0}^t -2ev_i[(x_i - a)\cos\theta + (y_i - b)\sin\theta]^2$$
$$((x_i - a)^2 + (y_i - b)^2 - l)\frac{1}{1+t^2}ds$$
is convergent for $t \to \infty$ and $t_0 \in [0, +\infty[$

Moreover $S(x_0, y_0) \geq 0$ since $(x_0, y_0) \in D$.

Then
$$(x_i - a)^2 + (y_i - b)^2 \geq l \quad (4)$$

In the other side, let
$$L(x_i, y_i) = (x_i - a)^2 + (y_i - b)^2$$
Its differential with respect to t is given by :
$$\dot{L}(x_i, y_i) = -2ev_i[(x_i - a)\cos\theta + (y_i - b)\sin\theta]^2$$
$$((x_i - a)^2 + (y_i - b)^2 - l)\frac{1}{1+t^2}S(x_i, y_i)$$
for $(x_0, y_0) \in D$.

We have $\dot{L}$ negative, therefore L is decreasing.
Hence
$$(x_i - a)^2 + (y_i - b)^2 \leq (x_0 - a)^2 + (y_0 - b)^2 \leq k \quad (5)$$

Finally according to inequalities (4) and (5), one can conclude that $(x_i, y_i) \in D$ for all time when $(x_0, y_0) \in D$. □

Proof. *(Theorem 1)* We note $X = (x_i, y_i)$ and $A = (a, b)$.
We have
$$l \leq \|X - A\|^2 \leq k$$
then
$$l^{1/2} - \|A\| \leq \|X\| \leq k^{1/2} + \|A\|$$

since

$$\|X\| - \|A\| \leq \|X - A\| \leq \|X\| + \|A\|$$

(Cauchy inequality)
then D is a compact set.

Beside lemma2 ensure that $(x_i, y_i) \in D$ for all $t \geq t_0$.

Let

$$V(t, x_i(t), y_i(t)) = (x_i - a)^2 + (y_i - b)^2 + \theta_i$$

we have

$$\theta_i = \theta_{i0} + \int_{t_0}^{t} (x_i - a)^2 + (y_i - b)^2 - l \, d\tau$$

then θ_i is positive and convergent to finite value. This implies that V is positive.

The differential of V with respect to t has the following expression :

$$\dot{V}(t, x_i(t), y_i(t)) = [(x_i - a)^2 + (y_i - b)^2 - l]*$$

$$[-2ev_i((x_i - a)\cos\theta + (y_i - a)\sin\theta)^2 + \frac{2}{3}]$$

Since $V(t, x_i(t), y_i(t))$ is monotonically non increasing and bounded from below by zero, it converges as $t \to \infty$. Now,

$$\int_{t_0}^{t} S(x_i(\tau), y_i(\tau)) d\tau \leq \int_{t_0}^{t} \dot{V}(\tau, x_i(\tau), y_i(\tau)) d\tau \qquad (6)$$
$$= V(t_0, x_{i0}(t), y_{i0}(t))) - V(t, x_i(t), y_i(t))$$

Therefore, the integral $\int_{t_0}^{t} S(x_i(\tau), y_i(\tau)) d\tau$ exists and is finite. Since $(x_i, y_i) \in D$ for all $t \geq t_0$ and the write hand of equation (4) is locally lipschitz in (x_i, y_i), uniformly in t, we conclude that (x_i, y_i) is uniformly continuous in t on $[t_0, \infty[$. Consequently $S(x_i, y_i)$ is uniformly continuous in t on $[t_0, \infty[$, since $S(x_i, y_i)$ is uniformly continuous on the compact set D. Hence by lemma 1, we conclude that $S(x_i(t), y_i(t)) \to 0$ as $t \to \infty$. The limit $S(x_i(t), y_i(t)) \to 0$ implies that (x_i, y_i) approaches

$$M = \{(x, y) \in D / S(x_i(t), y_i(t)) = 0\}$$

4. SIMULATION

In this section we consider a group of three robots labeled $1, 2$ and 3 moving in the plane.
$Fig2$ represent, respectively, the relative distance between robots 1&2, 1&3 and 2&3.
We can verify that, using the control laws defined in Proposition 1, these distances are nonzero for all time t.

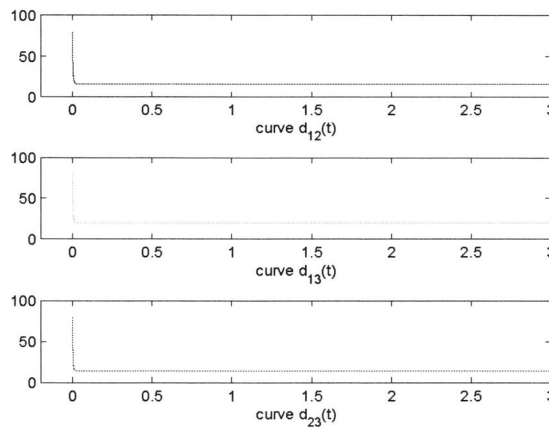

Fig. 2. Relative distances between unicycles 1,2 and 3

$Fig3$ shows the convergence of the formation to the attractive set and so fulfill the target capturing task without collision, using the control laws defined both in Thorem 1 and Proposition 1.
The initial conditions and parameters are chosen as follows :
$x_{10} = -40, y_{10} = 40, \theta_{10} = 2\pi - 0.1, x_{20} = -40, y_{20} = -40, \theta_{20} = 1.5, x_{30} = 30, y_{30} = 0, \theta_{30} = 2\pi - 0.5$, the target position is $(a = 0, b = 0)$ and the radius of attractive set is chosen as $l = \sqrt{5}$

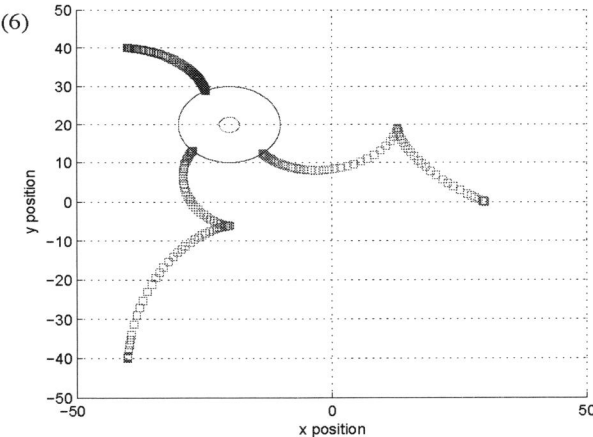

Fig. 3. Formation task achievement to target surrounding

5. CONCLUSION

As have been shown in the different sections of this paper, we have developed a method that make a formation of n unicycles converge to an "attractive" set in order to achieve a predefined target capturing task in a 2D plan. We have defined, as well, a function that we called "collision avoidance function" and which guarantee that no collision occurs over time between the different members of the unicycle's group. This result is based on the famous LaSalle's theorem that we remember in the appendix of the paper.

REFERENCES

[1] Hassan.K. Khalil. Nonlinear Systems. Macmillan Publishing Company, New York, 1992.

[2] N. Léchevin, C.A. Rabbath, and P.Sicard. Trajectory tracking of leader-follower formations characterized by constant line-of-sight angles . In *Automatica*, volume 42, pages 2131–2141, 2006.

[3] Joshua A. Marshall, Mireille E. Broucke, Bruce A.Francis Pursuit Formations of unicycles. *Automatica* , volume 42, pages 3–12, 2006.

[4] E.W. Justh., P.S. Krishnaprasad; Equilibria and steering laws for planar formations. *Systems & Control Letters*, volume 52, pages 25–38. 2004.

[5] Brian D.O. Anderson, Changbin Yu, Soura Dasgubta, A. Stephene Morse; Control of three-coleader formation in the plane. *system & control letters*, volume 56, pages 573-578, 2007.

[6] Tae-hyoung Kim, Toshiharu Sugie; Cooperative control for target-capturing task based on acyclic pursuit strategy. *Automatica*, volume 43, pages 1426-1431, 2007.

[7] A. Sinha, D; Ghose; Generalization of nonlinear cyclic pursuit. *Automatica* article in press 2007.

[8] Dussan M. Stipanovic, Gokhan Inalhana, Rodney Teo, Claire J Tomlin; Decentralized overlapping control of a formation of unmanned aerial vehicles. *Automatica*, volume 40, pages 1285-1296, 2004.

[9] Bruckstein A. M., Cohen N., Efrat A.; Ants, crickets and frogs in cyclic pursuit". Center for intelligent systems. Technical Report 9105, Technion-Israel Institute of Technology, Haifa, Israel. 1991

[10] Dimos V. Dimarogonas and Kostas J. Kyriakopoulos; A connection between formation control and flocking behavior in ninholonomic multiagent systems. *Robotics and Automation, 2006. ICRA 2006. Proceedings 2006 IEEE International Conference*

Development of Algorithms for the Classification of the benign and malignant tumors

L. Zouaoui[*], H. Azizi[*], M. Boughazi[**], H. Akdag[***]

[*]Faculty of Science, Department of Physics. Ferhat Abbas University.
[**]Faculty of Engineering, Department of Electronics. Annaba. Univ.
[***] CReSTIC Reims University, France

lam_zou@yahoo.fr

Abstract: The main objective of this paper is to develop and implement new algorithms of classification and show that the method of the nearest neighbors rule can be also applied successfully to deal with the medical classification problems. In this context, we developed two original algorithms of classification by the method of the nearest neighbors rule and we validated them by a real application in the field of classification for the assistance to the treatment of the breast cancer to detect possible benign or malignant tumors.

Keywords: Algorithm, classification, nearest neighbor, cancer.

1. INTRODUCTION

Many practical problems can be reduced to the assignment of various objects to predifined classes. For example in the case of the medical diagnosis, it is a question of recognizing the pathology of a given patient, the objects (characteristic) corresponding to the patients and the classes with various pathologies.

In general the methods of classification are made up in several steps. The most important step consists in working out rules of classification starting from knowledge available a priori; i e the phase of training. The latter uses the deductive or inductive training. The algorithms of inductive training lead to some rules of classification obtained from already classified examples [6].

The goal of these algorithms is to produce rules of classification in order to predict the class of assignment of a new case. Among the methods of classification using this type of training, let us quote the methods of the k-nearest neighbor rule, the method of Bayes, the method of discriminatory analysis, the approach of the neurons networks of and the method of the decision tree [3,4,7,10]. In the algorithms of deductive training, the rules of assignment are given a priori by the interaction with the decision maker, or the expert.

From these rules one determines the classes of assignment of these objects. Among the methods of classification using this type of training, let us announce by way of examples the expert systems and the approximate sets [2].

The cancer, which is already a major element of the load and morbidity in the world, will become a huge problem during the next decades. The need for the improvement of fight against this disease became paramount for several research teams in Biology and Industrial Informatics. Their objective lay the improvement of the strategies of prevention and the early detection of the disease [8].

The etiology of the breast cancer, which can be multicausale, remains unknown. The only mean to the medical staff to give the patients a maximum possibility of cure, is the early detection of the disease.

This early detection not only makes it possible to increase considerably the chances of cures but also to achieve this goal with average therapeutic.

Several works were interested in the assistance in the treatment of the breast cancer. One can quote for example [1,9,11].These works represent a computerized decision-making system in cancerology designed to solve problems of treatment of the breast cancer. It is a question of finding, according to certain characteristics of a given patient, the treatment which is best adapted. These characteristics are mainly its age, its sex, the size of the tumor, its localization in the breast, the presence of several hearths, and the presence of infected ganglia.

Our study has thus for principal objective to contribute to develop and implement new methods of classification and to show that the nearest neighbor method can be applied successfully to deal with the problems of medical classification.

2. PROPOSED METHOD

The method that we propose approaches the use of the classification rule by the method of the nearest neighbor. We describe in a first part an algorithm of partitioning of the space of training in adjacent hyper cubic cells non-empty followed by an algorithm which gives us an approximate nearest neighbor, then in the second part an algorithm of classification which gives us a real nearest neighbor using the transformation of the hyper cubic cells into hyper spherical cells.

2.1 Partition of the Space of Training

The method that we propose is based mainly on the fact that for an unknown sample, its nearest neighbor (NN) is not required in the space data, but only in the closest area. This method thus suggests the exploration of the partition concept of the training space in adjacent hyper cubic cells non-empty as shown in Figure 1.

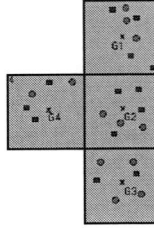

Fig. 1. Partition of the training space into adjacent hyper cubic cells

● Samples of class 1
■ Samples of class 2
G1, G2.,.,.,Centres of the different cells.

Algorithm of Partition of the training space
The organigram shown below describes the different steps of the algorithm used to identify the non-empty cells. These steps are:
1. Scale each axis of the data space by the factor 1/h such that an hyper-cubic cell of width h might be transformed into an hyper-cubic unit.
2. The first scaled design sample $x_1 = (x_{11}, x_{12},...,x_{1k})$ is used to create the first cell with center:
$G_1 = [Pe(x_{11}) + 0.5, Pe(x_{12}) + 0.5,..., Pe(x_{1k}) + 0.5]$
Where $Pe(x_{ik})$ is the integer part of the real number x_{ik}.
3. The second scaled sample $x_2 = (x_{21}, x_{22},...,x_{2k})$ may be put in the first cell, otherwise a second cell is created, with center:
$G_2 = [Pe(x_{21}) + 0.5, Pe(x_{22}) + 0.5,..., Pe(x_{2k}) + 0.5]$
4. The ith sample may be placed in one of the precedent cells, otherwise, it will be used to create another cell.

The Proposed Algorithm for Classification
The classification steps of an unknown sample are defined below:
1. Scale the components of $Z_i = (z_{i1}, z_{i2},...,z_{in})$ by 1/h.
2. Find the cell whose gravity center is closest to this sample.
3. Assign the unknown sample to the nearest neighbor class among the samples located in that cell.

The above procedure may lead to an approximate nearest neighbor as shown in Figure 2 where one can see that the nearest cell with Z is the cell 1.

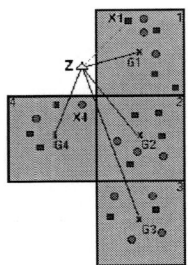

Fig. 2. Approximate classification of Z

The Improved Algorithm for Classification
Now if a real NN is wished, a certain additional pre-processing is necessary to transform the set of non-empty hyper cubic cells defining the data space into a set of hyper spheres.

Let $H(G_i)$ be the hyper-sphere with radius $R(G_i)$ and center G_i.

Where $R(G_i)$ is the maximal distance between G_i and a classified sample of C_j.

Under these new transformations, the partition of the data space is given by Figure 3.

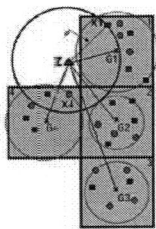

Fig. 3. Transformation of the data space into hyper-spherical cells

Thus, to find a real nearest neighbor, the steps of the algorithm involved in the classification phase are defined below:
1. Calculate
$D_1 = |Z - G_1|, D_2 = |Z - G_2|,...,..., D_{N_c} = |Z - G_{N_c}|$
and store the result.

2. Find the closest Cj hyper cubic cell to Z.

3. By using the design samples which are in the Cj cell, calculate the minimal distance D from Z and assign Z with the class of nearest neighbor corresponding.

4. Explore all the hyper spheres defined above and whose distance from the unknown sample Z satisfies the relation:
$D_j \langle D + R(G_j)_{j=1,2,3,...,N_c}$

5- By using the samples being in the corresponding hyper cubic cells, find the minimal distance D_1 to Z.

If $D_1 < D$, then Z is assigned to the class of the new sample corresponding to D_1. In Figure 2, it is clear that the nearest cell to Z is the cell with center G_1. Z is assigned to the class of its nearest neighbor among the samples located in that cell, which is the x_1 class.

Now, if a real nearest neighbor is wished, one must explore all the hyper spheres whose D_j distance satisfies the relation:

$$D_j \langle D + R(G_j)_{j=1,2,3,...,N_c}$$

Where D is the distance between x_1 and Z.
In the Figure 3, one can see that the cell of center G_4 might be explored, by considering the design samples inside it. One notices that the minimal distance (D_1) between these samples and Z is lower than D, therefore, Z will be assigned to the class of the sample corresponding to D_1 which is the x_4 class as shown in Figure 4.

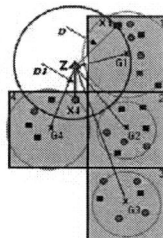

Fig. 4. Real classification of Z

3 APPLICATION TO THE ASSISTANCE WITH THE MEDICAL DIAGNOSIS

3.1 Implementation of the Results

The implementation was made on a basis of data "Breast-cancer-Wisconsin.data"[5]. This data base of the breast cancer was obtained at the teaching hospitals of Wisconsin. It is made up of 699 patients divided into two classes:
- The class benign cancer is made up of 458 samples.
- The class malignant cancer is made up of 241 samples.

Each example (vector) is composed of 11 attributes:
1- Sample code number.
2- Clump thickness.
3- Uniformity of cell size.
4- Uniformity of cell shape.
5- Marginal adhesion.
6- Single epithelial cell size.
7- Bare nuclei.
8- Bland Chromatin.
9- Normal Nucleoli.
10- Mitoses.
11- Class (2 for benign cancer and 4 for malignant cancer).

The first attribute is not other than a code allotted to each patient.

The other attributes (from 2 to 9) have values varying from 1 to 10.
The various parameters of our tests are:
- The width of the cell " h " which varies from 1 to 10.
- The rates of classification are given starting from a cross validation of order 10.
- The given rate of classification is an average rate starting from several bases of trainings.

3.2 Experimental Results

The results obtained by simulation are represented in the tables and the curves below:

h	A number of operations by the algorithm proposed	A number of operations by the algorithm improved
1	4535	5207
2	3881	4571
3	3340	3910
4	3030	4161
5	3483	4467
6	3100	4155
7	3192	4936
8	3327	5795
9	3253	6032
10	3295	6146

Table 1. Variation of the number of operations with respect to "h"

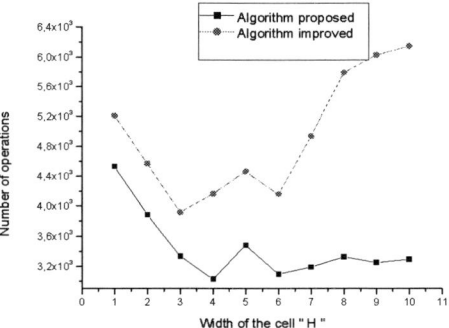

Figure 5. Variation of the number of operations according to the width of the cellule "h"

An analysis of the table shows that the number of operations is minimal for:
1- The proposed algorithm h=4.
2- The improved algorithm h=3.

It is clear that the improved version request a number of operations higher than the algorithm proposed.

h	Rate of classification by the algorithm proposed (%)	Rate of classification by the improved algorithm (%)
1	96	99
2	97	99
3	96	99
4	95	98

5	91	95
6	95	93
7	96	95
8	98	97
9	96	94
10	97	95

Table 2. Rate of classification by the improved algorithm

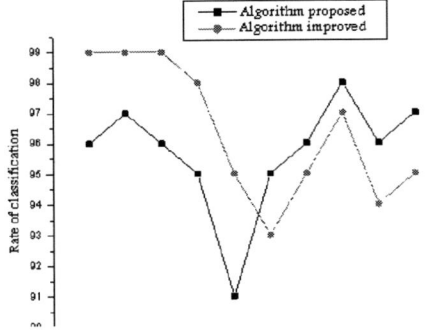

Figure 3.2 Variation of the rate of classification with respect to the cell width "h"

Figure 6. Variation of the rate of classification with respect to the cell width "h"

For low values of "h" one obtains a very important rate of classification of about 99% for the improved algorithm (a real nearest neighbor).

4. CONCLUSION

The work that we presented locates in the field of the classification and the recognition of the types of cancer (benign, malignant). We were interested in an algorithm of reference of the field, the algorithm of the nearest neighbors, which has very interesting properties of generalization and whose applications are numerous. Our contribution in the field of classification was initially to propose new methods of classification, then to apply them in the field of the assistance to the medical diagnosis. The results obtained by this application show that our method has a satisfying rate of classification.

5. REFERENCES

[1] B. Bressont and J. Lieber : "Classification pour l'aide au traitement du cancer du sein", Septième journée de la Société Francophone de Classification – SFC'99, PP 53-59. Septembre Nancy 1999.

[2] B. Chandrasekaran, A. Goel, et al. " From numbers to symbols to knowledge structures: Artificiel inlellegence perspectives on the classification task". IEEE Transaction on systems. Man and cybermetics 18.3: pp 415-425. 1989.

[3] R.D. Duda, P.E.Hart and D.G. Stork. « Pattern classification » John Wieley & sons,Ing, second edition 2001.

[4] R.o. Duda, P.E.Hart « Pattern classification and scene analysis».New York: Wiley 1973.

[5] ftp://ftp.ics.uci.edu/pub/machine-learning-databases

[6] N. Kaddeche et al. "Supervised Machine Learning by Generation of Rules: Optimization of the Classification Rate by Mixed Correlation", Asian Journal of Information Technology Vol. 4 N°2, pp152-161, 2005.

[7] GJ.Mc Lchlan « Discriminant analysis and statical pattern recognition ». Wiley et sons, Inc. 1992.

[8] J. Lieber, M. D'Aquin., P. Bey, B. Bresson, O. Croissant, P. Falzon., A. Lesur, J. Lévèque, V. Mollo, A. Napoli., M. Rios, C. Sauvagnac, " The Kasimir Project : Knowledge Management in Cancerology ", *Proc. of the 4th International Workshop on Enterprise Networking and Computing in Health Care Industry, HealthCom 2002*, June 2002.

[9] J. Lieber, M. d'Aquin, P. Bey and all. "Acquisition of Adaptation Knowledge for Breast Cancer Treatment Decision Support » , In 9th Conférence on Artificial Intelligence in Medicine in Europe 2003 – AIME 2003.

[10] D. Michie, DJ. Spiegelhlter, C. Laylor «Machine learning, Neural and statistical classification ». Ellis Horword series in artificiel intelligence; Ellis Horword. 1994.

[11] A. Napoli et al. " Acquisition et Modélisation de Connaissances d'Adaptation, une étude pour le traitement du cancer du sein », journée ingénierie des connaissances – IC'2001 pp. 409-426 (Grenoble, France).

Improvement of the performances of the genetic algorithms by using an adaptive search space reduction and the transformation.

L. Yousfi (1), N. Mansouri (2)

(1) : Department of Electronics, University of Tebessa
Road Constantine, 12002,Tebessa, Algeria
E-mail: yousfi_laatra@yahoo.fr
(2) : Laboratory of Automatic and Robotics,
Department of Electronics, Engineer science faculty, University of Constantine
Road Ain EL Bey, 25000 Constantine, Algeria
E-mail: nor_mansouri@yahoo.fr
Tel/Fax : 031-81-90-10

Abstract : The aim of this paper is the identification of the parameters in systems modeled by nonlinear differential equations. The proposed method is based on Genetic algorithms with domain's reduction and transformation strategies. The studied problems are successively solved using transformation technique, domain's reduction and a combination of the two strategies. The results obtained, using all these methods are comparables. The good results obtained by transformation seem to be related to the great degree of diversity that the mechanism introduces in population.

Keywords: Identification, Genetic Algorithms, Search domain reduction, Transformation.

1. Introduction

In recent years, many researchers have been interested by the development of new methods of parameters identification in non linear models. Different methods are proposed in the literature to solve this problem. The efficiency of all these algorithms is heavily dependent on the choice of initial parameters or population and there is no technique thoroughly reliable in terms of solution quality.

The classical techniques also called gradient methods base their search on the calculation of the first or second derivative of the cost function. Their convergence, to the global optimum is guaranteed only in the case where the initial algorithm values are chosen in the neighborhood of the solution.

Techniques such as evolutionary algorithms or simulated annealing are developed under different principles and are inspired from different sources. Each technique has its own properties, advantages and disadvantages. For example, evolutionary algorithms draw their inspiration from the hypothesis that the species evolve through a process of survival of the fittest individuals, whereas the simulated annealing method is slow and must be adapted with the problem being treated. The use of evolutionary algorithm requires very intensive computation and the performances depend on the choice of genetic initial population.

Genetic Algorithms are one of these algorithms. They have been widely used because of their performance and easy implementation. But as noted before, the performances depend mainly on the choice of the search space. Many researchers have been focused on finding most promising regions from an original large search space using successive reduction technique[1,2].

The traditional genetic operators used in Genetic Algorithms are crossover and mutation. In this paper, we introduce a biologically inspired recombination operator that occurs in the colonies of bacteria. The mechanism is called transformation and is responsible for the genetic variation and consequently the advantageous characteristics that some bacteria possess [3].

The proposed algorithm of parameters identification, is based on the transformation strategy mixed with search space strategy. The obtained results show the efficiency of the method.

2. Mechanism of genetic algorithms

Genetic algorithms are biological inspired search procedures that have been used to solve different hard problems. They are based on the neo-Darwinian idea of natural selection and reproduction.

These algorithms operate on coding of parameters rather than the parameters themselves and they are naturally formulated in term of maximization. The following differences make genetic algorithms distinct from the traditional search and optimization methods, help genetic algorithms to find global optimal points and make then more amenable for application to a wide variety of problems[2].

The mechanics of simple genetic algorithm is very simple, involving nothing more complex than copying strings and swapping partial strings. It starts with an initial population of individuals created at random. Then, this population is evolved through times by a string manipulation process based on three genetic operators: selection (reproduction), crossover and mutation[4,5,6,7].

These last years, and after having tested the efficiency of the genetic algorithms in many fields, many researchers have carried out extensive studies to understand several aspects of those algorithms, such as selection, space representation and the way of applying genetic operators.

The crossover operator promotes diversity in the individuals of the population. In this work, we choose to replace this standard operator by a biologically inspired genetic operator proposed by Simoies and Costa [3] and called transformation.

2.1 Transformation

In Evolutionary Algorithms, several authors have already used some biologically inspired mechanisms besides crossover and mutation. For instance, inversion, conjugation, translocation, transduction, transposition and transformation, were already used as the main genetic operators.

Transformation is an operator which is applied in every generation instead of the standard crossover operator. First, the individuals to be transformed are selected using the roulette-wheel selection method with a fixed probability of 70% and a gene segments pool is created randomly.

Second, a gene segment which consists in a binary string, is taken in the pool and replaced in the selected individual, after the transformation point. Part of the gene segment pool is updated every generation, using genetic information of the

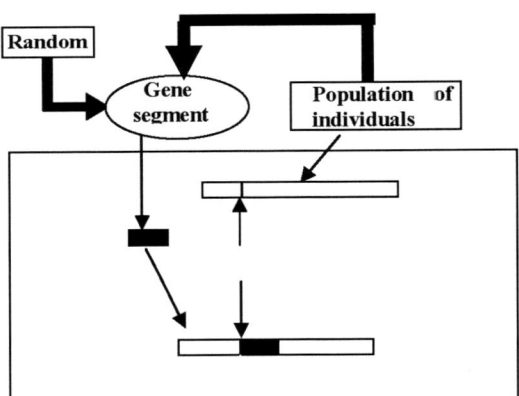

Fig.1. Transformation computation

individuals of the population.

Fig.1 illustrates the process of transforming an individual [3].

2.2 Strategy of domain's reduction

In a statistic sample from initial populations with size N, we will obtain, after a minimal number of generations of a Genetic Algorithm, different optimum values. If the sample is sufficiently dense, the best values of the objective function will be in the part of the search space where probability of finding the global optimum is great [8].

In this sense, the proposed algorithm works as follows: having launched the Genetic Algorithm 10 times with different initial populations and a reduced number of generations, we select the best individual of each launching. We deduce the dispersion of individuals σmax, using equation 1 (nber =10).

$$\sigma_{max} = \frac{\sqrt{\sum_{i=1}^{nber}(F_{best}-F_i)^2}}{nber} \quad (1)$$

Where *Fbest* represents the best value for the objective function

The new radius of the search domain is defined as follows:

$$r_i = \frac{d_i.\sigma_i}{|F_{best}-F_{worst}|} \quad (2)$$

With: $d_i = X_{max} - X_{min}$ (for each parameter)

And *Xmax, Xmin*: the limiting values for the parameter search domain

3. Algorithm of parameters identification

The identification problem considered here is the identification of the parameters λ*, which minimize a quadratic criterion φ(λ), i.e. which maximize a fitness function F(λ) defined as follows:

$$F(\lambda) = \frac{1}{1+\varphi(\lambda)} \quad (3)$$

With:

$$\varphi(\lambda) = 0.5\sum_{i=1}^{m}(y(t_i,\lambda)-Zd(t_i))^2 \quad (4)$$

Zd(t) are the observations and *y(t,λ)*, the outputs of the model.

The algorithm is tested on two systems. We present in this section some of the obtained results.

Example 1

The differential equation (5) is known as « problem of Bellman ». The parameters to be identified are p_1 and p_2:

$$y' = p_1(126.2-y)(91.2-y)^2 - p_2 y^2 \quad (5)$$

The observations are obtained by simulation using the following parameters values:

$$p_1 = 0,463.10^{-5}, \quad p_2 = 0,2434.10^{-3}, \quad y(1) = 0$$

Example 2

This example represents the Monod model, which describe growth and substrate consumption of a microorganism in batch cultivation. The model equations are as follows [9]:

$$\frac{dC}{dt} = \frac{k_1 S C}{k_2 + S} - k_D C \qquad C(0) = C_0$$
$$\frac{dS}{dt} = -\frac{1}{Y} \frac{k_1 S C}{k_2 + S} \qquad S(0) = S_0 \tag{6}$$

Where C is the biomass concentration; S the substrate concentration, Y the yield coefficient, k_D the death rate coefficient; k_2 the saturation constant and k_1 the maximum specific growth rate.

The parameters to be identified are: (k_1, k_2, k_D, Y).

It can be mentioned that the substrate measurement gives values that deviate from true values, i.e., the measurement is biased. The bias S_B is assumed to be constant and it must be identified.

The following parameters values were used for simulations:

$C_0 = 0.5 g/liter \quad S_0 = 20 g/liter \quad k_1 = 2.5 Day^{-1}$
$k_2 = 10 g/liter \quad k_D = 0.2 Day^{-1} \quad Y = 0.5 \quad S_B = 2 g/liter$

The Genetic Algorithm with transformation and search domain's reduction strategy is given in Fig. 2.

The genetic parameters used for the simulation are given in table 1.

Table 1

Genetic parameters	Value
Transformation rate	70%
Mutation probability P_m	0.03
Number of individuals	50
Number of generations	300
Number of launching	10

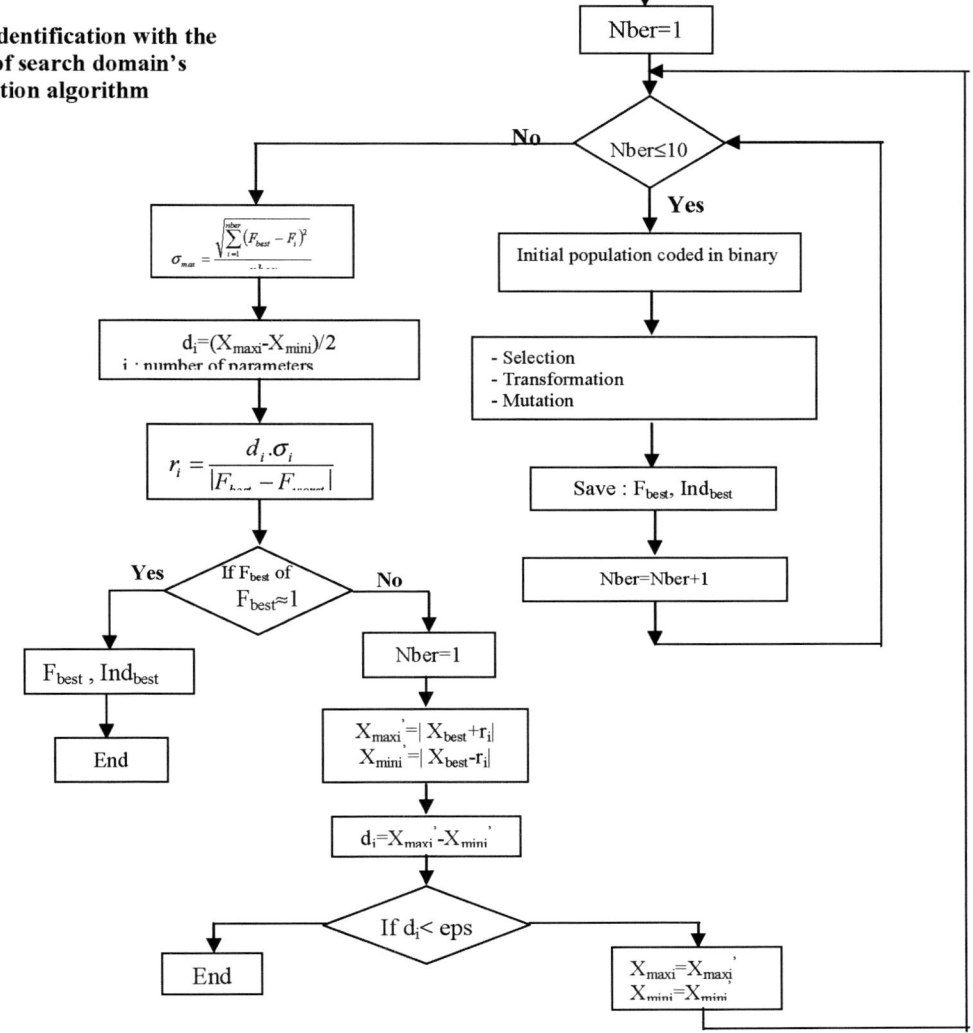

Fig. 2. The identification with the strategy of search domain's reduction algorithm

In the first part of the study, we have tested the influence of the transformation and the standard crossover on the effectiveness of the identification algorithm.

Fig. 3 and 4 shows the variation of the best solutions for different executions using transformation and standard crossover.

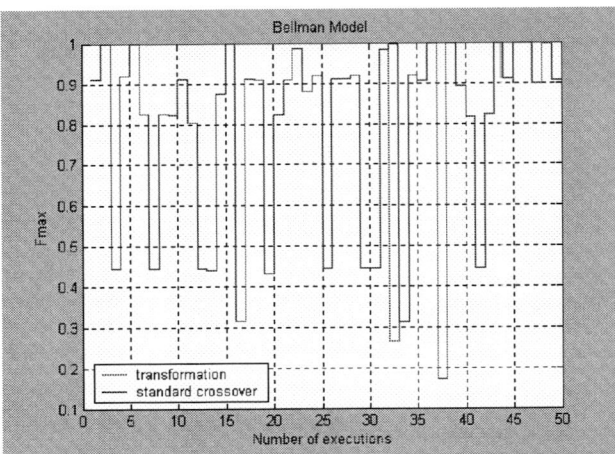

Fig.3. Variation of F_{best} using transformation and standard crossover for Bellman model

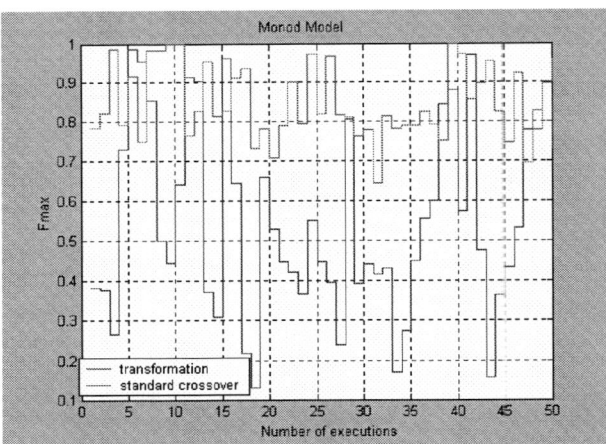

Fig.4. Variation of F_{best} using transformation and standard crossover for Monod model

If we observe the figures, we can see that with the transformation, the values of fitness function are greater and the optimal solution is founded.

We can conclude that the transformation promote diversity in the individuals of the population and a correct choice of transformation rate increase the quality of the results.

In the second part of the study, the performance of domain's reduction strategy is analyzed by comparing two cases :

- strategy of domain's reduction associated with a standard crossover.

- strategy of domain's reduction associated with the transformation.

The limits of the domain search parameters are shown in the table 2.

Table 2

Monod model	
Parameter	Search domain's
k_1	[0.5 5]
k_2	[8 20]
k_D	[0.1 0.3]
Y	[0.1 0.5]
S_B	[0.5 2]
Bellman model	
Parameter	Search domain's
P_1	[10^{-6} $0.8.10^{-5}$]
P_2	[10^{-4} $0.5.10^{-3}$]

The results of the two cases are shown in the table 3 and 4. The optimal parameter values are obtained after two successsive reduction of the search domain's.

As we can see, the identification results for the Bellman and Monod models using strategy of domain's reduction with the transformation were quite better than the results found by using the same strategy associated with standard crossover. The value of the fitness function is very near to the unity and the estimated parameters are the same as those used for the simulation.

We can also see, that the value of the fitness function for the Bellman model is greater than that found for the Monod model. This difference can be explained by the number of estimated parameters which is smaller in the Bellman model.

Table 3

	Bellman Model	Monod Model
λ^*	10^{-3}. [0.00463004841 0.24341044754]	[2.50046083753424 10.00184000516794 0.20004947931039 0.50010684797350 2.00003794067150]
F	0.99999964259482	0.99998610865343
φ	0.00000035740531	0.00001389153954

Table 4

	Bellman Model	Monod Model
λ^*	10^{-3}. [0.00463000381558 0.24340073592978]	[2.49869905874161 9.98952916932792 0.19989799394120 0.49989143537158 2.00020728759503]
F	0.99999999779499	0.99999368276500
φ	$2.205010254340383 \cdot 10^{-9}$	0.00000631727491

4. Conclusion

Previous work using the transformation operator showed that it is capable of preserving the population diversity during the entire evolutionary process and can promote convergence of the algorithm without the use of the strategy of domain's reduction.

The method of reduction suggested in this paper makes it possible to increase the value of the fitness function, especially when it is combined with the transformation. Thus, the genetic algorithms with adaptive domains are very effective for the identification of the parameters in nonlinear problems even if the number of parameters to be identified is important.

References

[1] M. Lyaon and K.C. Messa (1998), " Genetic algorithm model fitting", *Chapter 8*, pp.269-285.

[2] Z. Yong and N. Sannomiga (2001), "An improvement of genetic algorithms by search space reduction solving large-scale flow shop problems, *T.IEE Japan*, Vol.121-C, N°6.

[3] A. Simôes, and E. Costa (2002), "Parametric study to enhance genetic algorithm's performance when using transformation", *Proceeding of the Genetic and Evolutionary Computation Conference (GECCO'02). Morgan Kaufmann publishers*, New York,

[4] L. Oulladji and A. Janka (2004), "Aerodynamic optimization using hybrid genetic algorithms :application to sonic bang reduction", Rapport INRIA.

[5] V. Magnin (2003), " Genetic algorithms and optimization".

[6] G Renner (2003), " Genetic algorithms in computer aided design", *computer aided design*, Vol.35, No.8, pp.709-726.

[7] A. Simôes, and E. Costa, "Transposition: A biologically inspired mechanism to use with genetic algorithms "(1999), *Proceeding of the fourth international conference on neural networks and genetic algorithms (ICANNGA99)*, pp.178-186, Portoroz, Slovenia.

[8] J.A. Jiménéz, P.D. Cuesta and J.C Abderramon (2000), "Mixed strategy in genetic algorithms: Domain's reduction and multirecombination", *European congress on computational methods in applied sciences and engineering, ECCOMAS 2000*, Barcelona, pp. 11-14.

[9] M. Nihtila and J. Virkkunen (1977), " Practical identifiability of growth and substrate consumption models", *Proceeding of the Biotechnology and Bioengineering*, Vol.19, pp.1813-1850.

Indirect Adaptive Fuzzy Power System Stabilizer

Kamel Saoudi, Ziad Bouchama, Mohamed Naguib Harmas, and Khaled Zehar.

*Laboratory of Quality in Electrical Energy QUERE,
Université Ferhat Abbas -19000 - Sétif, Algérie*
(saoudi_k@yahoo.fr)

Abstract: A power system stabilizer based on adaptive fuzzy technique is presented. The design of a fuzzy logic power system stabilizer (FLPSS) requires the collection of fuzzy IF-THEN rules which are used to initialize an adaptive fuzzy power system AFPSS. The rule-base can be then tuned on-line so that the stabilizer can adapt to the different operating conditions occurring in the power system. The adaptation laws are developed based on a Lyapunov synthesis approach. Assessing the validity of this technique simulation of a power system is conducted and results are discussed.

Keywords: Adaptive fuzzy control, Indirect adaptive control, Power system stabilizer.

1. INTRODUCTION

Fuzzy logic control has been successfully applied to many commercial products and industrial problems, where: 1) no accurate mathematical plant under control are available ;and 2) human expert are available to provide linguistic fuzzy control rules or linguistic fuzzy descriptions about the systems.

Fuzzy logic power system is a technique of incorporating expert knowledge in designing a controller [1]. Past research of universal approximation theorem [6] shown that any nonlinear function over a compact set with arbitrary accuracy can be approximated by a fuzzy system. There have been significant research efforts on adaptive fuzzy control for nonlinear system [8], [10].

The proposed indirect adaptive fuzzy power system stabilizer (IAFPSS) is initialized using the fuzzy logic system for the approximations unknown nonlinear function of the model of power system [2], [3], [4]. By the Lyapunov synthesis approach, adaptation laws are developed to make the fuzzy logic systems adaptive to change in the different operating conditions occurring in the power system. The simulation of the proposed stabilizers concept for a one machine-infinite bus system shown that performs of adaptive fuzzy is better than the conventional (CPSS).

2. FUZZY-LOGIC DESIGN

The basic configuration of the fuzzy logic systems is shown in Fig.1. The fuzzy logic system performs a mapping from $u \subset R^n$ to R. It is assumed that $U = U_1 \times ... \times U_n$ where $U_i \subset R$, $i=1, 2,..., n$.

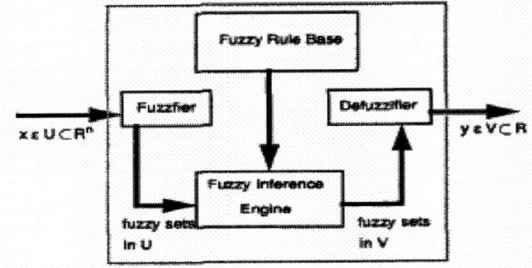

Fig.1. Basic structure of a fuzzy logic system.

The fuzzy rule base consists of a collection of fuzzy IF-THEN rules:

$$R(l): IF\ x_1\ is.F_1^l\ and...and.x_n\ is.F_n^l\ THEN\ y.is.G^l \quad (1)$$

Where $x = (x_1,........,x_n)^T \in U$ and $y \in R$ are the input and output of the fuzzy logic system, respectively, F_i^l and G^l are labels of fuzzy sets in U_i and R, respectively, where $l = 1, 2,..., M$. Each fuzzy IF-THEN rule of (l) defines a fuzzy implication [8], $F_1^l \times ... \times F_n^l \to G^l$, which is a fuzzy set defined in the product space $U \times R$. Based on generalization of implications in multi-value logic, many fuzzy implication rules have been proposed in the fuzzy logic literature [7].

For singleton fuzzification, max-product composition, product inference, and centroid defuzzification maps fuzzy sets in V to crisp values in R using the correlation product inference [9]. The crisp output is defined by:

$$y(\underline{x}) = \frac{\sum_{l=1}^{M} \theta_l \left(\prod_{i=1}^{n} \mu_{F_i^l}(x_i) \right)}{\sum_{l=1}^{M} \left(\prod_{i=1}^{n} \mu_{F_i^l}(x_i) \right)} \quad (2)$$

θ_l is the centre of gravity of the membership function of the output corresponding to the lth rule. Equation (2) can be rewritten as:

$$y(\underline{x}) = \underline{\theta}^T \underline{\xi}(\underline{x}) \quad (3)$$

Where $\underline{\theta}_l = [\underline{\theta}_1 ... \underline{\theta}_M]^T$ and $\underline{\xi}(\underline{x}) = [\xi^1(\underline{x}) ... \xi^M(\underline{x})]^T$ the fuzzy basis functions defined as [6]:

$$\xi_l(\underline{x}) = \frac{\prod_{i=1}^{n} \mu_{F_i^l}(x_i)}{\sum_{l=1}^{M} \left(\prod_{i=1}^{n} \mu_{F_i^l}(x_i) \right)} \quad (4)$$

3. INDIRECT ADAPTIVE FUZZY POWER SYSTEM STABILIZER

In order to represent the synchronous machine mathematically, let:
$x_1 = \Delta\omega$ = speed deviation and
$x_2 = \Delta P = P_m - P_e$ = accelerating power. It *is* possible to represent the system with the following nonlinear equations [4], [5]:

$$\dot{x}_1 = ax_2$$
$$a\dot{x}_2 = f(x_1, x_2) + g(x_1, x_2)u \quad (5)$$
$$y = x_1$$

where $a = 1/2H$ where H is the per unit inertia constant of the machine. $\underline{x} = [x_1, x_2]^T \in R^2$ is the state vector of the system and can be measured. The controlling signal u is the output of the PSS to be designed and f and g are nonlinear functions. Equation (5) represents the machine during a transient period after a major disturbance has occurred in the system. It has been assumed that two nonlinear functions f and g can be found such that:

$$\dot{P}_e = -2H[f(x_1, x_2) + g(x_1, x_2)u] \quad (6)$$

This equation is based on the fact that the governor time constant is large compared to the time constants of the synchronous machine and its exciter, so that during the first few seconds after the occurrence of severe disturbance the governor function can be ignored. Therefore the mechanical input power is constant during the transient interval.
It is known that a positive u will cause a positive change in $\dot{P}_e$, i.e $\dot{P}_e > 0$ whenever $u > 0$. This means that g is a negative function, i.e :

$$g(x_1, x_2) < 0, \text{ for all } x_1, x_2 \quad (7)$$

The control objective is to force restoration of the system equilibrium after a faulty operating condition. In the rest of this section the procedure to construct an indirect adaptive fuzzy controller [10], [11] to achieve the above control objective is discussed.

Let $\underline{e} = [e, \dot{e}]^T$ and $\underline{k} = [k_2, k_1]^T \in R^2$ be such that the two roots of the polynomial $h(s) = s^2 + k_1 s + k_2$ are in the open left half-plane.
If the functions f and g are known, then the control law

$$u^* = -\frac{1}{g(\underline{x})} \left[f(\underline{x}) - \ddot{y}_m + \underline{k}^T \underline{e} \right] \quad (8)$$

Applied to (3.1) using u^* instead of u, results in

$$\ddot{e} + k_1 \dot{e} + k_2 e = 0 \quad (9)$$

Which implies that $\lim_{t \to \infty} e(t) = 0$.

Since f and g are unknown, the ideal controller (8) cannot be implemented. However, the fuzzy IF-THEN rules (2) give estimates $\hat{f}(\underline{x}) = \hat{f}(\underline{x}|\underline{\theta}_f)$ and $\hat{g}(\underline{x}) = \hat{g}(\underline{x}|\underline{\theta}_g)$, where $\underline{\theta}_f \in R^{M_f}$ and $\underline{\theta}_g \in R^{M_f}$ are unknown parameter vectors in $\hat{f}(\underline{x})$ and $\hat{g}(\underline{x})$ respectively.
Thus, the fuzzy controller becomes

$$u_c = -\frac{1}{\hat{g}(\underline{x}|\underline{\theta}_g)} \left[\hat{f}(\underline{x}|\underline{\theta}_f) - \ddot{y}_m + \underline{k}^T \underline{e} \right] \quad (10)$$

By adding $\hat{g}(\underline{x}|\underline{\theta}_g)u_c$ to both sides of (9) and using (10), the following equation is derived:

$$\ddot{e} = -\underline{k}^T \underline{e} + [f(\underline{x}) - \hat{f}(\underline{x}|\underline{\theta}_f)] + [g(\underline{x}) - \hat{g}(\underline{x}|\underline{\theta}_g)]u_c \quad (11)$$

Since $\underline{e} = [e, \dot{e}]^T$, (11) can be written as

$$\dot{\underline{e}} = -A_c \underline{e} + \underline{b}_c \left([f(\underline{x}) - \hat{f}(\underline{x}|\underline{\theta}_f)] + [g(\underline{x}) - \hat{g}(\underline{x}|\underline{\theta}_g)]u_c \right) \quad (12)$$

Where

$$A_c = \begin{bmatrix} 0 & 1 \\ -k_2 & -k_1 \end{bmatrix}, b_c = \begin{bmatrix} 0 \\ 1 \end{bmatrix} \quad (13)$$

A_c is a stable matrix.
Therefore, there exists a unique symmetric positive definite matrix P which satisfies the Lyapunov equation

$$A_c^T P + P A_c = -Q \quad (14)$$

Where Q is an arbitrary positive definite matrix.
From (3), the estimates are of the form:

$$\hat{f}(\underline{x}|\underline{\theta}_f) = \underline{\theta}_f^T \underline{\xi}(\underline{x}) \quad (15)$$

$$\hat{g}(\underline{x}|\underline{\theta}_g) = \underline{\theta}_g^T \underline{\xi}(\underline{x}) \quad (16)$$

Next step is to adjust the parameter vectors $\underline{\theta}_f^T$ and $\underline{\theta}_g^T$ such that the tracking error $\underline{e} = [e, \dot{e}]^T$ and the parameter errors $\underline{\theta}_f - \underline{\theta}_f^*$ and $\underline{\theta}_g - \underline{\theta}_g^*$ are minimized. Using the following Lyapunov candidate function :

$$V = \frac{1}{2} \underline{e}^T P \underline{e} + \frac{1}{2\gamma_1} (\underline{\theta}_f - \underline{\theta}_f^*)^T (\underline{\theta}_f - \underline{\theta}_f^*) + \frac{1}{2\gamma_2} (\underline{\theta}_g - \underline{\theta}_g^*)^T (\underline{\theta}_g - \underline{\theta}_g^*) \quad (17)$$

Where $\gamma_1 = 2$ and $\gamma_2 = 20$ are positive constants which will be used as learning rate in the adaptation procedure. An adaptation law which minimizes the Lyapunov function is given by:

$$\dot{\underline{\theta}}_f = \gamma_1 \underline{e}^T P \underline{b}_c \xi(\underline{x}) \tag{18}$$

$$\dot{\underline{\theta}}_g = \gamma_2 \underline{e}^T P \underline{b}_c \xi(\underline{x}) u_c \tag{19}$$

Otherwise, if the parameter vectors are on the boundary of the constraint sets and moving toward the outside, we modify the adaptation law as[1] :

$$\dot{\underline{\theta}}_f = \gamma_1 \underline{e}^T P \underline{b}_c \xi(\underline{x}) - \gamma_1 \underline{e}^T P \underline{b}_c \frac{\underline{\theta}_f \underline{\theta}_f^T \xi(\underline{x})}{\left|\underline{\theta}_f\right|^2} \tag{20}$$

$$\dot{\underline{\theta}}_g = \gamma_2 \underline{e}^T P \underline{b}_c \xi(\underline{x}) u_c - \gamma_2 \underline{e}^T P \underline{b}_c u_c \frac{\underline{\theta}_g \underline{\theta}_g^T \xi(\underline{x})}{\left|\underline{\theta}_g\right|^2} \tag{21}$$

4. SIMULATION RESULTS

A nonlinear power system model that consists of a synchronous machine connected to a constant voltage bus through a double circuit of three phase transmission lines is chosen for simulation studies. A schematic diagram representation of the power system is shown in Fig.2. A three phase transformer is used between the synchronous machine and the transmission lines to boost the machine voltage.

A conventional (CPSS) controller structure Lead-lag network with gain K, washout time constant T_W, lead and lag time constants T_1 ,T_3 and T_2 , T_4 respectively is used for comparison. The transfer function of the stabilizer is given by:

$$U_{PSS} = \frac{sT_W(1+sT_1)(1+sT_3)}{(1+sT_W)(1+sT_2)(1+sT_4)} \tag{22}$$

A set of nonlinear differential equations has been used to simulate the synchronous machine. A simplified model of IEEE type ST 1A excitation system has been used.

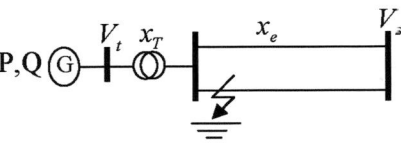

Fig.2. Single machine infinite bus power system.

Three different perturbed operating conditions a used to evaluate the soundness of the approach.

1) First Case: operating point
$P_0 = 0.9 \text{ pu}, Q_0 = 0.3 \text{ pu and } X_e = 0.2 \text{ pu}$
Fig.3. shown system response under a simulation three-phase fault to ground on the transmission line occurring at *t=0.2 sec* with a duration of *0.06 sec* for conventional, fuzzy PSS and adaptive fuzzy PSS is presented in Fig .3. The speed variation is rapidly suppressed by the adaptive stabilizer.

2) Second Case: operating point
$P_0 = 0.8 \text{ pu}, Q_0 = 0.8 \text{ pu and } X_e = 0.45 \text{ pu}$
Here again while the two approaches are lagging behind , the power system stabilizer based on the fuzzy adaptive technique rapidly eliminates oscillations present in the speed variation response of the synchronous machine as illustrated in Fig.4.

3) Third Case: operating point
$P_0 = 0.9 \text{ pu}, Q_0 = -0.3 \text{ pu and } X_e = 0.2 \text{ pu}$
In this last case, the fuzzy stabilizer is unable to damp oscillations while the AFFSS exhibit superior performance of its classical counterpart as shown in Fig.5.

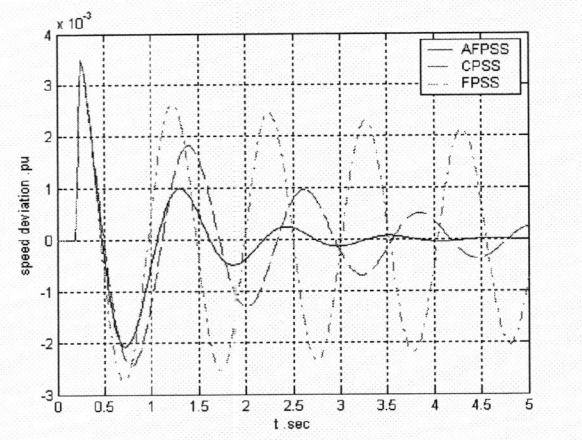

Fig.3. Speed deviation response for first case

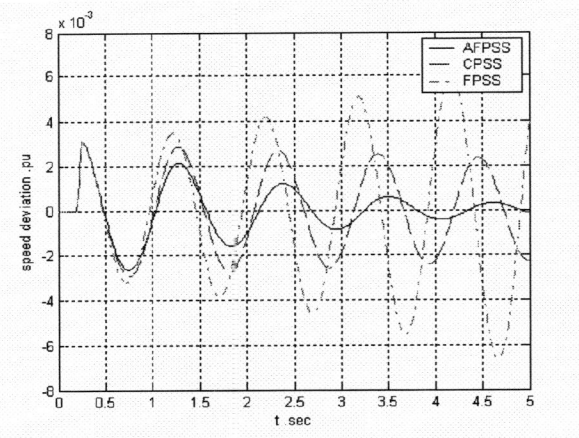

Fig.4. Speed deviation response for second case

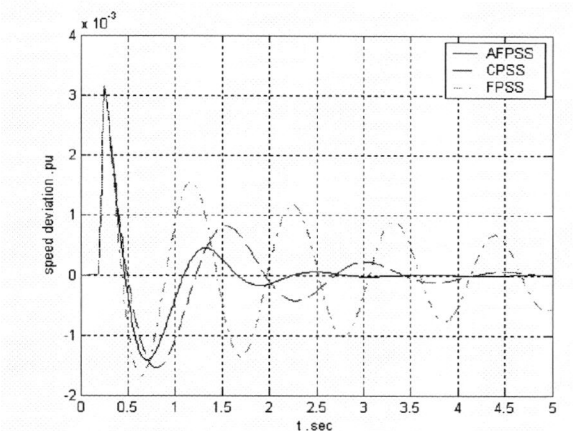

Fig. 5. Speed deviation response for third case

5. CONCLUSION

In this paper, an indirect adaptive fuzzy power system stabilizer based on the Lyapunov synthesis approach was designed by incorporating fuzzy rules describing the system, directly into the controllers. Using the fuzzy logic system for the approximation of unknown nonlinear functions of the power system model. The adaptation laws are developed to make the fuzzy logic systems adaptive.

It's evident from simulation studies that an indirect adaptive fuzzy power system stabilizer shows better response in a wide range of operating conditions over the initial fuzzy or conventional power system stabilizers.

REFERENCES

[1] El-Metwally, K., and Malik O., "Fuzzy logic power system stabilizer", IEE Proc. Generation, Transmission, and Distribution, Vol. 142, No. 3, pp. 277-281, 1995.

[2] Elshafei A., and El-Metwally K., ''Power System Stabilization Via Adaptive Fuzzy-Logic Control'', Proceedings of the 12th IEEE. International Symposium on Intelligent Control,16-18 July 1997, Istanbul, Turkey.

[3] Elshafei A., El-Metwally K., and Shaltout A., ''Design Analysis of a Variable Structure Adaptive Fuzzy-Logic Power System Stabilizer''. Proceedings of the American Control Conference Chicago, Illinois, June 2000.

[4] Hosseinzadeh N., Kalam A., ''A Direct Adaptive Fuzzy Power System stabilizer''. IEEE Transactions on Energy Conversion, Vol. 14, No. 4, December 1999.

[5] Zadeh N..H, Kalam A. , An indirect Adaptive Fuzzy Logic Power System Stabiliser, Elsevier Electrical Power and energy Systems 24, 837-842, 2002.

[6] Wang, L. and Mendel J., "Fuzzy basis functions, universal approximation, and orthogonal least squares learning ", IEEE Trans. Neural Networks, Vol. 3, pp. 807-814, 1992

[7] Lee C., " Fuzzy logic in control systems: Fuzzy logic controller, Parts I and 11", BEE Trans. Sys., Man, Cybernetic., Vol. 20, no. 2, pp. 404-435, 1990.

[8] Wang L.., Stable adaptive fuzzy control of nonlinear systems IEEE Trans. on Fuzzy Systems, Vol. 1, No. 2, 1993.

[9] Mendel J. M., Fuzzy logic systems for engineering, A tutorial, IEEE Proc., vol. 83, no. 3, , pp. 345-377, March 1995

[10] L. X. Wang, "Stable Adaptive Fuzzy Controllers with Application to Inverted Pendulum Tracking" IEEE transactions on systems, man, and cybernetics-part b: cybernetics, vol. 26, no. 5, October 1996.

[11] Tomsovic K and Chow M.Y, "Tutorial on Fuzzy Logic Applications in Power Systems", .IEEE-PES Winter Meeting in Singapore January, 2000.

Appendix.

System model.

$$\dot{\delta} = \omega_0 \Delta\omega$$

$$\dot{\omega} = (P_m - P_e)/M$$

$$\dot{E}'_q = (E_{fd} - (x_d - x'_d)i_d - E'_q)/T'_{do}$$

$$\dot{E}_{fd} = \frac{1}{T_A}(K_A(V_{ref} - V_t + U_{PSS}) - E_{fd})$$

$$V_d = V_s \sin\delta + R_e i_d - x_e i_q$$

$$V_q = V_s \cos\delta + R_e i_q + x_e i_d$$

$$V_t = \sqrt{V_d^2 + V_q^2}$$

$$T_e = E'_q i_q - (x'_d - x_q)i_d i_q$$

Parameters.

X_d=2.19 pu, X_q=1.01 pu, X_d'=0.18 pu, T_{do}'=4.14 sec, H=6 sec, *TA*=0.05, *KA*=50, *f*=50 Hz, -0.2 pu $\leq U_{PSS} \leq$ *0.2* pu, -2 pu $\leq E_{fd} \leq$ 6 pu.

Improvement of Arab Digits Recognition Rate Based in the Parameters Choice

C. Hadri, M. Boughazi, M. Fezari

Laboratoire d'Automatique et Signaux de Annaba LASA, Faculté des Sciences de l'Ingénieur,
Université de Badji Mokhtar, Annaba, Algérie.

(E-mail : hadricherif@yahoo.fr boughazi@leri.univ-reims.fr mouradfezari@yahoo.fr)

Abstract : Automatic speech recognition (ASR) is the process of automatically recognizing the speech on the basis of information obtained by acoustic features extracted from the speech signal. Because features extraction is the first component in ASR systems, the quality of the later component depends from the quality of feature extractor. The goal of this work is to study and implement features (representations) extraction, which are robust to the differences between the acoustic conditions of training and evolution. These features will be evaluated in an Automatic Arab digits recognition system. A particular attention will be taken to the robust features extraction methods (CMS, CGN, RASTAPLP, MBLPCC, and LPC MFCC).

Keywords: Speech Recognition, Cepstre, ASR System.

1. INTRODUCTION

The first stage in all algorithms ASR is the acoustic features extraction. The goal behinds features extraction process is to translate the information contained in the acoustic signals towards a representation of data which is appropriate for the statistical representation. ASR requires acoustic features which represent reliable phonetic information consistently, i.e. features must effectively describe the distinctive properties of speech sounds and which are reproducible. In the best of the cases, the acoustic features should clearly indicate the linguistically suitable differences between various speech sounds, while masking the acoustic variation of the signal which does not represent phonetic differences or which is not related to the speech events. Acoustic features extraction is an essence component in ASR system, because during the process a transition is made starting from the continuous signals to the most fundamental discrete speech recognition elements. A multitude of technique were proposed to produce acoustic features extraction effectively, for example LPC, LPCC, MFCC, PLP., the common goal of almost all these features representations is to describe the acoustic signals in terms of their short term spectral energy distribution. Information on the spectral energy distribution of a signal is usually extracted at regular time intervals. In addition to these so called static coefficients the first and second time derivatives of the static features also referred to as dynamic features are almost always included in the acoustic feature vectors. However the performance of ASR systems degrades significantly in noisy environments, for example it has been observed that additive white noise severely degrades the performance of mel frequency cepstral coefficients (MFCCs) based recognition systems. This performance degradation is attrib to unavoidable mismatch between the acoustic conditions of training and evolution [1]. The recognition algorithms for clean speech usually perform very unsatisfactorily for noisy speech. Even for a moderately noisy environment and a small vocabulary task, additional techniques are required to reach an acceptable recognition rat. The issue of robustness in ASR has mainly been translated into technical approaches to better accommodate into a speech recognizer various sours of variability introduced by corrupted speech signal. In this paper we study and implement features (representations), which are robust to the differences between the acoustic conditions of training and evolution. These features will be evaluated in an Automatic Arab digits recognition system. A particular attention will be lent to the methods of robust features extractions (CMS, CGN, RASTAPLP, MBLPCC, and LPC MFCC.

2. FEATURES EXTRACTION METHOD

The most commonly used features for speech recognition are the Mel Frequency Cepstrum Coefficients (MFCC, Davis and Mermelstein, 1980). The MFCCs are obtained by applying the cosine transform to the log energies of the outputs of a filter bank with filters regularly positioned along the mel frequency scale. The resulting coefficients can be liftered in order to equalize their range that can vary from low to high order coefficients. The principal advantage of the cepstral coefficients for speech recognition is that they are in general decorrelated, allowing the use of simpler statistical models [2]. Alternatives to the MFCCs exist that include more knowledge about speech perception. An example is Perceptual Linear Prediction (PLP, Hermansky, 1990;Junqua et al., 1993) [03], which weights the output of the filter banks by an equal loudness curve, estimated from perceptual experiments. These features are extremely similar to the MFCC. The principal difference between the two feature sets originates from the nature of the spectral smoothing that was

used to compute them. PLP combines several engineering approximations to selected characteristics of human hearing:
- Critical band (Bark) nonlinear frequency resolution, emulated by integrating the short-term Fourier spectrum of speech under increasingly wider trapezoidal curves (sometimes substituted by Mel-spaced triangular filters).,
- Asymmetries of auditory filters, emulated by a relatively steep (25 dB/Bark) slope of the trapezoidal curve towards higher frequencies and a more gradual (10 dB/Bark) slope towards lower ones,
- unequal sensitivity of human hearing at different frequencies, emulated by a fixed approximated Fletcher-Munson equal loudness curve.
- Intensity-loudness nonlinear relation, emulated by a cubic root compression, and
- broader than critical-band integration, hypothesized in perception of speech, emulated by an autoregressive all-pole model.

All these steps contribute to effectiveness of PLP analysis, the most important being the nonlinear warping of the frequency axis. Another popular feature set is the set of linear prediction cepstral coefficients (LPCCs). LPCC computes a LPC spectral envelope first, before converting into cepstral coefficients [1]. Linear predictive coefficients are very commonly used in speech recognition, as well as in speech coding and synthesis. LPC is implemented as an all pole autoregressive model of the spectrum that captures the vocal–tract properties of vowel-like sounds. The first steps of LPCC are the same as MFCC. However, instead of modeling the properties of human (with critical-band integration and compressing the spectral amplitudes in MFCC), LPCC takes the simpler approach of only smoothing the spectrum by an autoregressive filter (LPC). Then Durbin recursion solves for the autoregressive filter's coefficients or LPC coefficients from the autocorrelation coefficients. Finally, cepstral coefficients are solved from the LPC coefficients in a recursive manner.

3. ROBUST FEATURES EXTRACTION

A. Cepstral Mean Subtractions (CMS)

Cepstral Mean Subtraction is one of the earliest and the simplest methods used to remove channel distortion from signal. The principle behind this method is that a convolutional distortion in the time domain, such as a channel distortion, corresponds to an additive distortion in the cepstral domain [4]. If we denote by s(t) a speech signal, by w(t) the channel impulse response and by y(t) the speech signal transmitted through the channel we have the following equivalence :

$$y(t) = s(t) \otimes w(t) \Leftrightarrow C_y(i) = C_s(i) \cdot C_w(i)$$

Where $\otimes$ is the convolution operator $C_y(i)$, $C(i)$ and $C(i)$ are the cepstrum of respectively the transmitted w signal, the speech signal and the channel, now if we apply the expectation operator to the right side of the equivalence, we have: $\overline{C_y(i)} = \overline{C_s(i)} + \overline{C_w(i)}$

With the hypothesis that the channel distortion characteristics are constant and that the expectation of the speech is null (except for the 0th coefficient), we obtain:

$$\overline{C_y(i)} = C_w$$

Now by computing the long time average of the cepstrum of the transmitted speech, we have:

$$C_w = \frac{1}{N}\sum_{i=1}^{N} C_y(i)$$

It is now possible to subtract C_w from the observed cepstral $C_y(i)$ vectors in order to remove the channel effect. The Cepstral Mean Subtraction is a noise reduction technique very simple. The long time average of the cepstral coefficients is computed off-line and subtracted from each coefficient. This subtraction is done for all coefficients.

B. Cepstral Gain Normalizaion

When recognition training is executed in clean environment and recognition testing is evaluated in noisy environment, a difference of log-spectra between training and testing environments can be removed by adjusting the gain and the DC offsets [5]. The adjusted log-spectrum is obtained by the following equation:

$$\log S'(n,\omega) = \log E(n,\omega) - D_C(\omega)$$

In the noisy environment, logS'(n,w) can be obtained by canceling the gain G(w) and the DC offset DN(w) according to [05]. The DC offset DN(w) can be calculated from an average of log-spectra. The gain G(w) can be eliminated by normalizing both clean and noisy log-spectra gain GN(w)=1, GC(w)=1. These operations can be applied into cepstral parameters approximately. A series of procedures are summarized to the following two steps, which are applied to both training and testing data.

- Step1: Subtract an average of cepstral coefficients. The operation is known as cepstral mean subtraction (CMS)[5] (see section 3.1).

$$C'(n,k) = C(n,k) - \left(\sum_{n=1}^{L} C(n,k)\right)/L \quad for\ 1 \leq k \leq M$$

- Step2: Normalize gains by calculating the maximum and the minimum values of cepstral coefficients. It is called as cepstral gain normalization (CGN) in [5].

$$C''(n,k) = C'(n,k)/\left(\max_{1\leq n\leq k} C'(n,k) - \min_{1\leq n\leq k} C'(n,k)\right) \quad for\ 1 \leq n \leq k$$

where k is quefrency in M-order cepstrum. These steps are applied to delta cepstrum and delta-delta cepstrum.

C. RASTA PLP Coefficients

A generalization of CMS is RASTA (RealAtive SpecTrAl) filtering. In this approach, each cepstral coefficient is considered as the sample of a time signal, and this time signal is supplied to a filter that removes the low and high frequency modulations from this signal. The technique can be implemented with different filters and performed in different feature spaces (e.g. the log-power spectrum, the MFCC and the PLP space). The assumption is again that the

convolutional noise is quasi-stationary compared to the speech. The steps of RASTA-PLP are as follows [06]. For each analysis frame, do the following operations:
- Compute the critical-band power spectrum (as in PLP).
- Transform spectral amplitude through a compressing static nonlinear transformation.
- Filter the time trajectory of each transformed spectral component.
- Transform the filtered speech representation through expanding static nonlinear transformation.
- As in conventional PLP, multiply by the equal loudness curve and rise to the power 0.33 to simulate the power law of hearing.
- Compute an all-pole model of the resulting spectrum, following the conventional PLP technique.

The filter used is an IIR filter with the transfer function indicated in [06].

$$H(z) = 0.1z^4 \cdot \frac{2 + z^{-1} - z^{-3} + 2z^{-4}}{1 - 0.98.z^{-1}}$$

The low cut-off frequency of the filter determines the fastest spectral change of the log spectrum, which is ignored in the output, whereas the high cut-off frequency determines the fastest spectral change that is preserved in the output parameters. The high-pass portion of the equivalent band-pass filter is expected to alleviate the effect of convolutional noise introduced in the channel. The low-pass filtering helps to smooth some of the fast frame-to-frame spectral changes present in the short-term spectral estimate due to analysis artifacts.

D. Multiband Features based on Wavelet Transform

Based on time-frequency multi-resolution analysis, the effective and robust MBLPCC feature are used as the front end of the speech recognition system [7][8]. First, the LPCCs are extracted from the full-band input signal. Then the wavelet transform is applied to decompose the input signal into two frequency subbands: a lower frequency subband and a higher frequency subband. The recursive decomposition process enables us to easily acquire the multiband features of the speech signal. Based on this method, the number of MBLPCCs depends on the level of the decomposition process. If speech signals band limited from 0 to 4000 Hz are decomposed into two subbands, then three bands signals, (0-4000), (0-2000), and (0-1000) Hz, will be generated. Since the spectra of the three bands will overlap in the lower frequency region, it is clearly that the multiband feature extraction method focuses on the spectrum of the speech signal in the low frequency region similar to extracting MFCC features. In this paper, we use the orthonormal basis of DWT based on 16 coefficients of the quadrature mirror filters (QMF) introduced by Daubechies [1988].

E. Coupling between LPC and MFCC Analysis

Spectral smoothing certainly seems to be a desirable property of spectral estimators, because spectral smoothing translates into a reduction in spectral variance which, in turn, should yield a reduction in the variance of the acoustic models and therefore better class separability [09][10]. Spectral estimates can exhibit smoothness in two dimensions: frequency and time. FFT spectra are not inherently smooth. However, the application of (mel-scaled) filter banks reduces the variance in FFT spectra to a large extent, because the individual FFT coefficients within each mel band are averaged. LPC spectral estimates, on the other hand, are guaranteed to be smooth in the frequency dimension by their very nature. Intuitively one might expect that smoothness in the frequency dimension should translate into smoothness in time, due to the relatively slow changes in the spectral envelope of speech signals over time. However, in (extremely) noisy conditions, the relative contribution of the noise may be so large that the spectral properties of the input signal are primarily determined by the noise. Under these circumstances, the spectra of two adjacent frames may differ so much that a fixed-order LPC estimator may yield very different models for the two frames. A part of the frame-to-frame variance introduced by this phenomenon may be alleviated by mel-frequency averaging, but it remains to be seen whether the compensation is sufficient.

Based on LPC spectral smoothing LPCMFCC algorithm uses LPC analysis to replace the short term spectrum calculated by FFT by a new smooth spectrum calculated by LPC estimator. As known that LPC offers an estimate of the vocal tract based in an all pole system, while the analysis with MFCC parameters approaches the human audditive system. Therefore a coupling between LPC and MFCC group more knowledge concerning speech signal.

4. EXPERIMENTAL RESULTS

A. Experimental Conditions

Extracted feature vectors performance has been evaluated in an isolated Arabic digits speech recognition task. The speech database chosen for the experiments is taken from LASA (Laboratoire d'automatique et signaux Annaba) [11]. A speech data is sampled at 11.025 KHz and 16bit quantization. In speech analysis, MFCC, LPCC and PLP features are extracted after pre-emphasis and Hamming windowing, and converted to 39 dimensional feature vectors. Frame length and shift are 25ms and 10ms respectively. The feature vectors consist of 12 static features, 12 deltas, 12 delta-delta, log energy, delta log energy and delta-delta log energy. In training, we have created 10 HMMs models from 21 males and 21 females' (42 speaker's in the first half of the database), these models have 16 states and 1 mixture per states based on a continuous density function and a diagonal covariance matrix. In testing, clean speech is artificially added with noise at various SNRs. Recognition accuracy has been measured from 21 males and 21 females' (42 speaker's in the second half of the database) in recognition experiments.

B. Experimental Results

The evaluation of the feature performances under several noise conditions is summarized in the following table 1. First the CMS significantly increases recognition on clean and slowly noised speech (92.3913 to 95.2174 in SNR=20dB with MFCC). On average, for the tree common feature, using CMS yields a relative increase of recognition (81.5218 to 83.6087 with LPCC, 77.3478 to 78.9826 with PLP, 75.0000 to 74.5303 with MFCC). This result was expected since the data base is microphone isolated speech and since the CMS removes the convolutional channel noise, the combination of both CMS and CGN also gives significant improvement in noise conditions. On average, for the tree common feature, using CMS and CGN combination yields a relative increase of recognition (81.5218 to 89.00 with LPCC, 77.3478 to 86.7391 with PLP, 75.0000 to 85.8077 with MFCC).

Second, the combination with new robust feature increase recognition significantly. On average, of different feature combination methods, yields a good increase of recognition (90.0435 with MBLPCC-CMS, 88.1304 with RASTAPLP, 93.9565 with LPCMFCC-CMS-CGN). The results also show that the combination LPCMFCC-CMS-CMN are better than all other combination.

	Clean	SNR=30dB	SNR=20dB	SNR=10dB	SNR=05dB	Averag
39 LPCCs	96.7391	90.4348	86.0870	78.0435	56.3048	81.5218
39LPCCs+CMS	95.6522	95.8696	94.3478	82.3913	49.7826	83.6087
39LPCCs+CMS +CGN	96.9565	96.5217	93.4783	83.2609	74.7826	89.0000
39 MBLPCCs +CMS	98.0435	97.1739	95.6522	88.6957	70.6522	90.0435
39 .PLP	99.3478	93.9130	88.6957	67.3913	37.3913	77.3478
39 PLP+CMS	99.3478	98.0435	95.2174	68.9130	33.3913	78.9826
39 PLP +CMS +CGN	98.0435	97.8261	94.1304	78.6957	65.0000	86.7391
39 RASTA PLP	98.6957	99.1304	97.1739	86.0870	59.5652	88.1304
39 MFCCs	99.3478	99.1304	92.3913	58.4783	25.6522	75.0000
39 MFCCs +CMS	99.3478	98.0435	95.2174	58.2609	21.7820	74.5303
39 MFCCs+CMS+CGN	98.6057	97.6087	93.0435	76.5217	63.2609	85.8077
39 LPCMFCC +CMS	98.6957	98.2609	97.8261	82.6087	61.0870	87.6957
39LPCMFCC +CMS+CGN	98.4783	98.2609	97.3913	92.1739	83.4783	93.9565

Table 1: The evaluation of the feature performances under several noise

5. CONCLUSION

In this work, several features extraction techniques are implemented and tested on a speech recognition system of isolated Arabic digits. A particular attention have been given to noise reduction techniques like : CMS, CGN, RASTA-PLP, MBLPCC and LPCMFCC. Both additive noise and channel distortion can be removed by these techniques. Cepstral Mean subtraction and RASTA filtering can remove convolutional noise (channel distortion). Cepstral Gain normalisation is able to remove the additive noise. The combination of these techniques with three common feature vectors (LPCC, MFCC and PLP) improves robustness against noise, and the combination with a new robust feature vectors like RASTA-PLP, MBLPCC-CMS and LPCMFCC-CMS-CGN provides a good robustness.

The obtained results shows that the combination of LPCMFCC, CMS and CMN gives better recognition rate than all other techniques and combinations used in this work.

REFERENCES

[1] S. Stephane, An algorithm for automatic formant extraction using linear prediction spectra, IEEE transaction on acoustics, speech, and signal processing, vol.assp -22. April 1974.

[2] S. FURUI, Cepstral Analysis Technique for Automatic Speaker Verification, transaction on acoustics, speech, and signal processing, vol.assp -29, NO. 2, APRIL 1981

[3] H. Hermansky, B, Hanson and H. Wakita. Perceptually based linear predictive analysis of speech. Ch11-81851000-0509 1985 IEEE.

[4] C.Krmorvant, A comparison of noise reduction techniques for robust speech recognition. IDIAP-RR 99-10 july 1999.

[5] S. Yoshizawa, N. Hayasaka, N. Wada and Y. Miyanaga, Cepstral Gain Normalization for Noise Robust Speech Recognition, ICASP 2004.

[6] H. Hermansky N. Morgan, . RASTA processing of speech, IEEE Transection on Speech and Audio Processing , VOL .2.NO 4, October 1994.

[7] C .T. Hsieh, E . lai and Y. C . wang. Robust Speaker identification system based on wavelet transform and Gaussian Mixture Model , journal of information science and engineering A-267 -282(2003).

[8] W –Chen Chen, C .T . Hsieh and E. lai. Multiband approach to robust text –independent speaker identification.Natural Language Processing-IJCNLP 2004

[9] J.de veth,, B. cranenn , F.de wet , et L.boves (2002), A comparison of LPC and FFT – based acoustic feature for noise Robust automatic speech recognition, European project on Speech driven Multi –modal Automatic Directory Assistance (SMADA).

[10] F. de Wet, Automatic speech recognition in adverse acoustic conditions. PhD Departement of Language et Speech of the University of Nijmegen in the Netherlands.

[11] S. Chérifa, Conception d'un système de reconnaissance de mots isolés a base de l'approche stochastique en temps réel. Application : commande vocale d'une calculatrice Mémoire de magister, institut d'électronique, université d'Annaba, 2004.

Fuzzy neural order robust of the non-linear systems

F. Madour* and K. Benmahammed**

University Farhat abbas-Sétif Algeria
laboratory LSI (laboratory of the systems inteligents)
(e-mail: fouzia. madour@yahoo.fr).*
*(e-mail: Khierben@ieee.com)**.*

Abstract: This article introduces a controller at structure of a network multi-layer neurons specified by the fuzzy reasoning of Takagi-Sugeno (TS) order one [1], the weights of the network represent the standard deviations of the membership function. This controller is applied to the ordering of a reversed pendulum. Changes in the entries and the exit, as of the environment changes of operation are introduced in order to test the robustness of the designed controller.

Keywords: Fuzzy control, fuzzy neural network, robustness.

1. INTRODUCTION

The fuzzy order is a very powerful theory which is applicable to the badly mathematically modelled processes. The advantage of such a system is that only knowledge on the behaviour of the process to be ordered are sufficient for the laws orders synthesis [2].But the fuzzy controller is unable to learn knowledge contrary to the neurons network, another point of view; the neurons network need a minimum information [3]; of or utility of the hybridization of these two approaches.

Constraint major of this network is the optimization of the TS coefficients, that to avoid the local minimum that the network undergone during adaptation phase of the weights [4].A new technique was applied to overcome this problem and which proved is effectiveness. The objective of order is to maintain the system (reversed pendulum) inside a certain zone by a judicious choice of horizontal force F to apply to the pendulum reversed carriages, with respect to the robustness controller.

2. FUZZY CONTROLLER

The fuzzy controller can be seen like an expert system functioning starting from a knowledge representation based on the fuzzy set theory [2], its design requires the choice of the following parameters:

2.1 Definition of the system in term of its entries and exit variables

In our work, the entries variables are the angles (θ), and the angular velocity ($\dot{\theta}$), the exit variable is the force (F).

2.2 Choices of the fuzzy partitions

The choice of the fuzzy partitions consists in determining the number of terms in this unit, it depends on the problem to treat, if it requires the precision or the robustness of the system [5], the partition selected are:

(3) fuzzy partitions (N, Z, P) for each entry (θ),($\dot{\theta}$).

(9) fuzzy partitions (NG, NM, NP, ZN, Z, ZP, PP, PM, PG) for the exit (F).

2.3 Choices of the membership functions

The membership functions of entries variables and the exit are gaussian function, because it's only common function and rapid between the two approaches, neurons network and fuzzy logic [6].

The overlapping of these functions is 50%.

2.4 Conclusion of the base of rules

The fuzzy controller is defined by a whole of rules of the type:

IF x_1 is A_1 AND x_2 is B_1 THEN y is C_1.

Example:

IF θ is N AND $\dot{\theta}$ is P THEN F is ZN.

The base of rules contain the whole of the rules integrated in the controller.

Table. base of rules

θ $\dot{\theta}$	N	Z	P
N	NG	NM	ZN
Z	NP	Z	PP
P	ZP	PM	PG

2.5 Choices of the inference method and the defuzzyfication strategy

The Takagi-Sugeno method (TS) is the inference method used, because it convoy more information, and of this fact less rules are necessary to carry out a given spot; that is to say:

IF (x_1 is A1) **AND** (x_2 is A2) **AND**..... **THEN** (y = B(x)).

Where: B(x) is an analytical entries function x.

In our case there are two entries, the angle (θ) and angular velocity ($\dot{\theta}$) and an exit (Force), therefore:

$$F = a_1 \theta + a_2 \dot{\theta} + r \quad (2.1)$$

Where a_1 and a_2 represent (TS) coefficients.

Generally Fi (-) has a polynomial from :

$x_1, x_2, x_3, \ldots, x_m$.

$$F_i = (x_1, x_2, ..., x_n) = a_{0i} + a_{1i}x_1 + ... + a_{mi}x_m \quad (2.2)$$

This form is called Sugeno model of order one, weighted averages is the most defuzzyfication method used with this form of rules, it's given by :

$$y = \sum_{i=1}^{M} \mu_i F_i(x_1, x_2, ..., x_n) / \sum_{i=1}^{M} \mu_i \quad (2.3)$$

Where:

M: The number of rules.

μi: Activation degree of the rule number i.

$$\mu_i = Min(\mu_i(x_1), \mu_i(x_2), ..., \mu_i(x_n))$$

To simplify calculate them we can take:

$$\mu_i = \mu_i(x).\mu_i(x_2)...\mu_i(x_m) \quad (2.4)$$

with :

$$F_i(x_1, x_2, ..., x_n) = k_i F(x_i, x_2, ..., x_n) \quad (2.5)$$

Where: ki are the gravity centres from the exit fuzzy sets associated with each rule.

$$F_i(x_1, x_2, ..., x_n) = a_0 + a_1 x_1 + ... + a_n x_n \quad (2.6)$$

Thus:

$$\left(y = (\sum_{i=1}^{M} \mu_i k_i).f(x_1, x_1, ..., x_n) / \sum_{i=1}^{M} \mu_i \right)$$
$$\Rightarrow \left(y = (\sum_{i=1}^{M} \mu_i k_i).(a_0 + a_1 x_1 + ... + a_m x_m) / \sum_{i=1}^{M} \mu_i \right) \quad (2.7)$$

3. NEURAL FUZZY NETWORK

The network structure adapted to the Takagi-Sugeno method is that of Hamaifar, it acts of the type (333) controller.

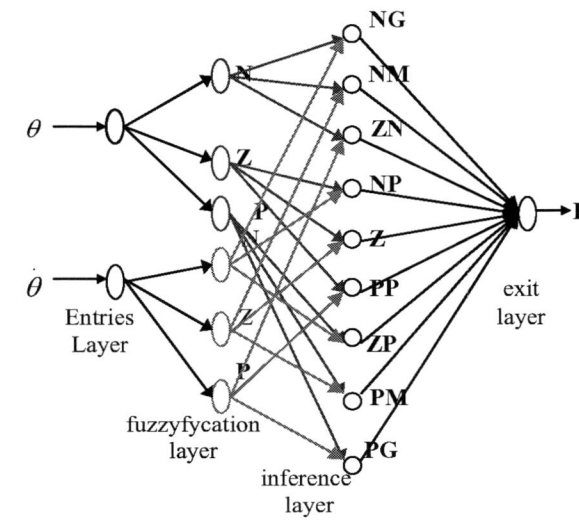

Fig. 1. Network controller structures

Composition of this structure is as follows:

3.1 Entries layer

This layer is passive which carries out the transfer direct of the input signals.

3.2 Fuzzyfication layer

In this layer we calculates the membership degree of each entry. The exit of each neuron is :

$$Si = e^{(x_i - c/\delta)^2} \quad (3.1)$$

xi: Entry of the neuron.

C: Center of the Gaussian function and (δ) its standard deviation.

3.3 Inference or basic of rules layer

each neuron calculate the activation degree of each rule, using the conjunction MIN operator [2]. The exit of a neuron is:

$$Si = MIN(\mu_i(x_1).\mu_i(x_2)) \quad (3.2)$$

The weights between this layer and that of fuzzyfication are fixed at 1.

3.4 Defuzzyfication layer

This layer calculates the digital display (y) by the weighted average method of we as quoted previously.

The weights between this layer and that basic of rules represent the consequences of the rules which are the centers of distribution fuzzy of exit of each rule, and the weights between the entry layer and that of fuzzyfication which represent the standard deviation associated with each linguistic variable are adapted by the retro-propagation method.

4. PRINCIPLE OF THE RETRO-PROPAGATION METHOD

The calculation of the error signal enables us to adjust the weights starting from this error, so that its current exit coincides with the desired answer. The law of training proposed is:

$$W_{ij}^k(t+1) = W_{ij}^k(t) - \mu(d_j(W)/dW_{ij}^k(t)) + \alpha(W_{ij}^k(t) - W_{ij}^k(t-1)) \quad (4.1)$$

Where:

$0 <= \alpha < 1$ is the moment, and (μ) is the step of training, (t) is index of iterartions.

The weights of connections are adjusted so that the performance index:

$$j(W) = \frac{1}{2}\sum_{i=1}^{n}(d_i - y_i)^2 \quad (4.2)$$

That is to say Minimized.

N: The number of the drive examples.

The derivative of the error compared to a W_{ij}^K weight is

$$d_j(W)/dW_{ij}^k(t) = (dj(W)/dy_j^k(t)).(dy_j^k(t)).(dW_{ij}^k(t)) \quad (4.3)$$

and:

$$(dy_j^k(t)/dW_{ij}^k(t)) = (df^k(S_j^k(t))/dW_{ij}^k(t))$$
$$(dy_j^k(t)/dW_{ij}^k(t)) = f^k(S_j^k(t)).(dS_j^k(t)).(dS_j^k(t)). \quad (4.4)$$

and:

$$(dS_j^k(t)/dW_{ij}^k(t)) = d\left[\sum_{i=1}^{i=nk-1} W_{ij}^k . y_j^{k-1}(t)\right]/dW_{ij}^k(t) = y_j^{k-1}(t) \quad (4.5)$$

Where:

$$(dy_j^k(t)/dW_{ij}^k(t)) = f^k(S_j^k(t)).y_j^{k-1}(t) \quad (4.6)$$

Thus:

$$dj(W)/dW_{ij}^k(t) = (dj(w)/dy_j^k(t)).f^k(S_j^k(t)).y_j^{k-1}(t) \quad (4.7)$$

While taking:

$$S_j^k(t) = -f^k(S_j^k(t)).(dj(W)/dy_j^k(t)) \quad (4.8)$$

Thus:

$$dj(W)/dW_{ij}^k(t) = -S_j^k(t).y_j^{k-1}(t) \quad (4.9)$$

if we replace

$(dj(W)/dW_{ij}^k(t))$ in (4.1) we find:

$$W_{ij}^k(t+1) = W_{ij}^k(t) + \mu S_j^k(t).y_j^{k-1}(t) + \alpha(W_{ij}^k(t) - W_{ij}^k(t-1)) \quad (4.10)$$

The large obstacle of this work is how to find the TS coefficients ?.

5. NEW TECHNIQUE TO FIND THE TAKAGI-SUGENO COEFFICIENTS

By observing the function of TS, $F = a_1 x_1 + a_2 x_2 + r$

we notice that there is a relation between (x1) and (x2) to find the force, it is the same thing with regard to the base of rules!

Thus the idea is that we can extract the coefficients from (TS) starting from the base of rules.

Example:

IF (θ is P) AND ($\dot{\theta}$ is Z) THEN (F is PM)

We notice that the force follows the angle and not the angular velocity thus the rule used is as follows:

- **If** the force follows the angle, then (a_1) receives a numerical value near to (1) and (a_2) receives a numerical value near to (0).

- **If** the force follows the angular velocity, then (a_2) receives a numerical value near to (1) and (a_1) receives a numerical value near to (0).

- **If** the force follows the angular velocity and the angle together, then (a_1) and (a_2) receive values numerical near to (1).

- **If** the force not follows the angular velocity nor the angle, then (a_1) and (a_2) receive numerical values near to (0).

This technique is simple rapid and proved its effectiveness.

6. DESCRIPTION OF THE PROCESS

the reversed pendulum is an unstable, non-linear and multivariable system, the ordering of this system must realize the regulation on the angle (θ) on the basis of a condition initial ranging between (-180°, +180°).

The pendulum length 2L and mass m, is related to the carriage of mass M which can move on a horizontal axis, the frictions located at the level of the rotation axis, and that due to the displacement of the carriage are neglected.

The state of the system is determined by two variables (θ, $\dot{\theta}$) respectively indicating the angular position reversed pendulum and its angular velocity.

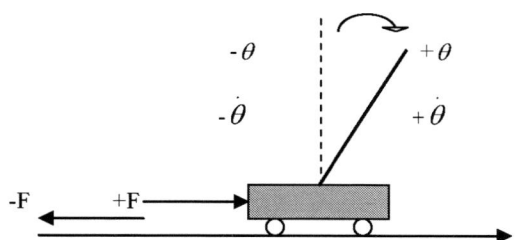

Fig. 2. Structure of the reversed pendulum

The dynamic of the system are characterized by:

$$\dot{x}_1 = x_2$$

$$\dot{x}_2 = \frac{g\sin(x_1) + \cos(x_1).(-F - m_p.L.(x_2)^2.\sin(x_1))/(m_c + m_p)}{L\left[\frac{4}{3}.(m_p.\cos^2(x_1)/(m_c + m_p))\right]}$$

Where:

G: Gravitational acceleration.

m_c: Mass carriage.

m_p: Mass pendulum.

L: Half length of the segment (pendulum).

x_1: The pendulum angle compared to the vertical axis.

x_2: Angular velocity of the pendulum.

7. RESULTS OF SIMULATION

For simulation we took the following initial conditions:

θ = 15 (deg), $\dot{\theta}$ = 0 (deg/s), F = 12 (N).

With the constraint: F <= 25 N.

And the numerical values used are:

m_c = 1 kg, m_p = 0.1kg, L = 0.5m, G = 9.8m/s²

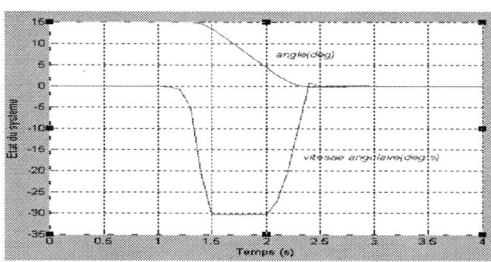

Fig. 3. Variation of the Angle and angular velocity

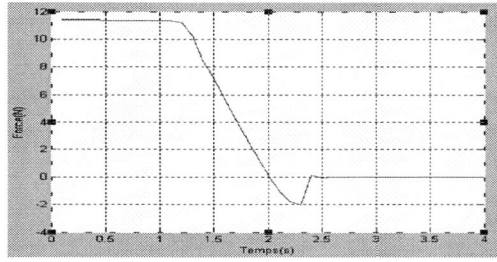

Fig. 4. Form of the force

we notice according to the preceding figures that the controller manages to stabilize to it clock around the position of balance during a time lower than 2.5s.

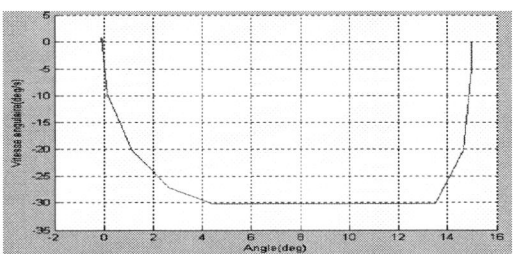

Fig. 5. Plan of phase

The figure shows how the trajectory approaches balance (0°,0°/ s) starting from the initial conditions

(15°,0°/ s).

8. TEST OF THE ROBUSTNESS

To test the robustness of our controller FLC (333) we must deviate from the normal conditions of use and see whether its aptitude reacts well. For that we tested our controller for several initial conditions:

{(12°, 0°/s, 10 N), (20°, 0°/ s, 16 N),

(25°, 0°/ s, 20 N), (30°, 0°/ s, 25 N)}.

Note:

If the angle increases we must increase the force to be able to raise the stem (the pendulum).

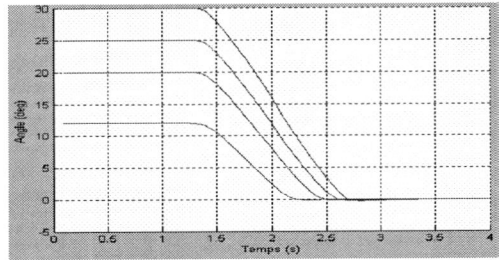

Fig. 6. Variations of the angles

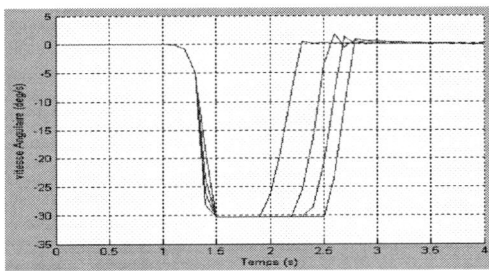

Fig. 7. Variations of the Angular velocities

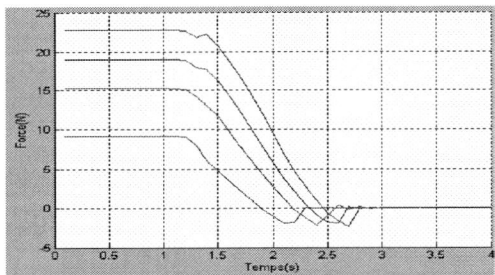

Fig. 8. Forms of the forces

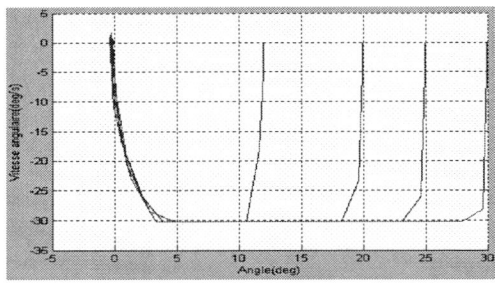

Fig. 9. Plan of phase

It is noticed that more one moves away from the balance points (0°,0°/s) more time to reach the balance point again is long, but does not exceed 3s.

The preceding figures show the robustness of our controller.

For better testing our controller, we make tests on the change stem lengths (0.5, 0.7, 0.9 and 1.1m) with the following initial conditions (20°,0°/s,15 N).

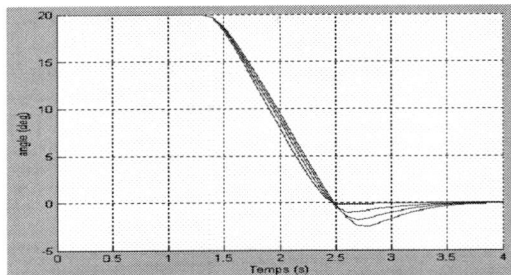

Fig. 10. Variations of the angles

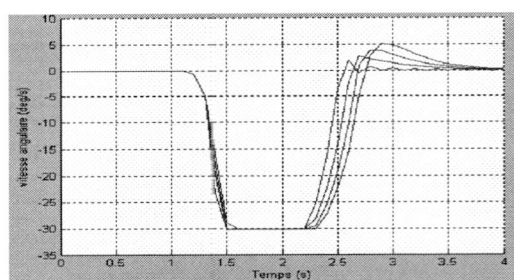

Fig. 11. Variations of Angular velocities

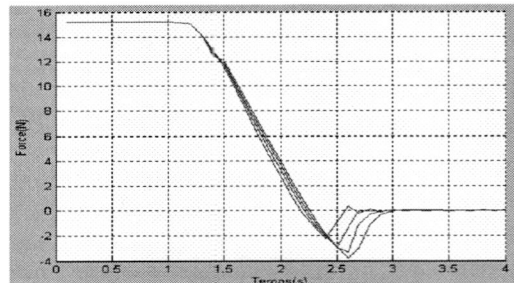

Fig. 12. Forms of the forces

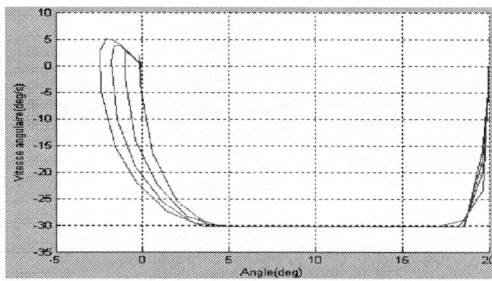

Fig. 13. Plan of phase

The figure shows how the trajectory approaches balance ($0°,0°/s$) starting from the initial conditions ($20°,0°/s,15$ N).

The figures show always the robustness of our controller.

9. CONCLUSIONS

The controller proposed FLC (333) applied to order the system of the reversed pendulum proved his effectiveness to bring back the pendulum to his position driving corresponding at an angle and a null velocity angular and without oscillations. This controller uses the physical field of the entries variations variables and the exit without standardization.

The robustness of the controller is justified by his capacity to react well opposite to the change of the internal parameters (angle and angular velocity of the reversed pendulum) characterizing the process to be ordered, like with the change of the environment of operation (the stem lengths and the force applied to the carriage).

The hybridization of the two approaches, consists an effective means to exploit the power and the flexibility of the elements handled by these various procedures and adds to the controller resulting an intelligent functionality which is the aim set by the intelligent controllers.

REFERENCES

[1] Hao Ying (Constructing non linear variable profit controllers via the Takagi Sugeno fuzzy control) *IEEE transaction Fuzzy systems*: flight.6, No 2, pp.226-233.May 1988.

[2] Pierre Gabriel (Introduction to fuzzy logic and the fuzzy order). Paris, 2000-2001.

[3] G. Dreyfus and J-m Martinez and Mr. Samue lides and M.B Cord and F.Badran and ST. hiniaL Hérault (Networks of neuron methodology and application). Paris 2002.

[4] Jean-François Joduin (networks of neurons: Principle and definition) Hermès, Paris 1994.

[5] C.Lee (Fuzzy logic in control systems: Fuzzy logic controller, part I and II) *Trans IEEE. sys*, Man, Cyben, flight 20, No.2, pp, 400-485, 1990.

[6] Jhy - Shing Roger Jang and Chuen - Tsai Sun, (Neuro fuzzy modeling and control) *proceedings of the IEEE*, vol. 83, No.3, March 1995, pp 378-406.

A Verbal Guidance System for Severe Disabled People

Abdelghani Redjati. Mounir Bousbia-Salah

*Badji Mokhtar University, Faculty of Engineering, Department of Electronics,
Laboratory of automatic and signals-Annaba, BP 12 - 23000 - Annaba, Algeria
(e-mail: redjati @yahoo.fr, bousbia.salah@univ-annaba.org)*

Abstract: The recent development in rehabilitation technology allows to significantly broaden the range of possible applications that support handicapped people in their daily lives. This paper presents a moral and physical support for the disabled. It consists in the development of a verbal guidance system based on a speech recognition development kit 'VD364'. This aid is intended to control a wheelchair and a manipulator arm for people with severe disabilities and who can speak. The study and design, conducted in the framework of this contribution have enabled an adaptation for a possible application and maximum exploitation of words that can be generated by a vocal module. The problem addressed is to allow a manipulator arm to compensate mechanically arm movements to give the handicapped satisfaction of his needs (for instance, drinking a glass of water). The objective is then to put forward a vocal command system that allows the arm to move in a well determined area to accomplish tasks that must be given by the user in addition to the displacement of the wheelchair.

Keywords: Speech, Vocal command, Robotic assistance

1. INTRODUCTION

In the last decades some research have been devoted to the development of technology for the increase of autonomy and independence of disabled people. In this context, robotics plays a major role thanks to the large number of aids it provides [1]. Many activities in this sector have been focused on telemanipulators controlled by users with disabilities and also on manipulators on board of electric wheelchairs to help the users in their work and in domestic activities [2]. Among the aids available is the "Senario" [3] which is an intelligent navigation system based on sensors that are easily ported between various electric vehicles. Another system is the "Movaid" [4] which is mainly focused on the development of a modular mobile robot interacting with the user by means of a friendly interface for the accomplishment of domestic tasks. In order to facilitate the integration of input–output devices for robotic assistive systems, the "M3S" project (Multi Master Multi Slave) [5] has been developed. This allows a flexible and reliable communication among devices made by different manufacturers.

Other systems are those related to smart powered wheelchairs. The "Vahm" project [6] was aimed at improving the control of powered wheelchairs by adding possibilities of autonomous mobility. Different operating modes have been defined, with different levels of autonomy, for adapting the system to various situations. The "NavChair" assistive wheelchair navigation system [7] is based on a commercial wheelchair system and employs different operating modes and has applications to the development of shared control systems where the machine can automatically adapt to human behaviours. The Hephaestus smart wheelchair [8] is an example of a modular system based on commercially available wheelchairs with various levels of autonomy. In all the above smart wheelchairs, the autonomy and security aspects are guaranteed by a set of sonar sensors.

In this area, locomotion and manipulation of objects by verbal command or guidance requires the use of a vocal module and a manipulator arm. The proposed configurations are varied depending on the size and location of the equipment that is attaching an arm on the wheelchair as for instance, the project "ARPH" [9].

In this paper, we propose a verbal system guidance for the disabled based on a vocal module to command a wheelchair which is conducted with a manipulator arm [10] used for the execution of various commands given by the user. A brief description of the VD364 [11] vocal module is then presented. This kit is an important element which allowed the realization of the prototype card for our tests. This voice direct 364 kit of sensory, summarizes the state of the art technology for automatic speech recognition [12]. It is perfect in a variety of consumer product and easily integrated into existing or new products. It can work alone as vocal recognition package, or as slave controlled by a microcontroller or a microprocessor. The developed card is primarily intended to control a wheelchair and a manipulator arm equipped with a control board containing essentially the module with its vocal elements [13]. In addition, the disabled gives orders through a microphone to his movements.

2. DESCRIPTION OF THE SYSTEM

The block diagram of the system is shown in figure 1 and includes the various elements necessary for the development of the aid. This system is based on a vocal module which is the VD364. The microcontroller PIC 16F876 [14] is used for the management of data and a power card common for the wheelchair and the manipulator arm.

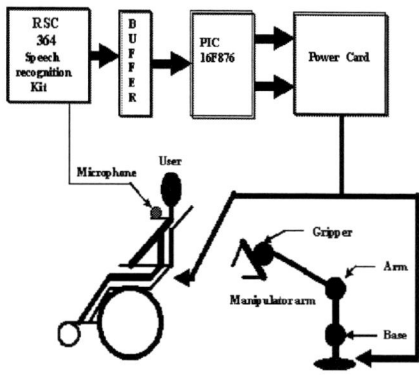

Fig. 1. Block diagram of the system

Recognition Word	Out 1	Out 2	Out 3	Out 4	Out 5	Out 6	Out 7	Out 8
Forward	x							
Backward		x						
Left			x					
Right				x				
Stop					x			
Basfor						x		
Basback							x	
Bastop	x							x
Armfor		x						x
Armback			x					x
Armstop				x				x
Grifor					x			x
Griback						x		x
Gristop							x	x

Table 1. Data table

If the object is located within a sufficient range of the body to reach, the reach and acquisition of this object can be done without using the locomotion, something which is necessary since the number of generated words by the vocal module does not allows the command of the wheelchair and the arm simultaneously. If the object is located beyond this reach, the locomotion is required to perform the action.
The selection mode of the VD364 Kit is shown in figure 2.

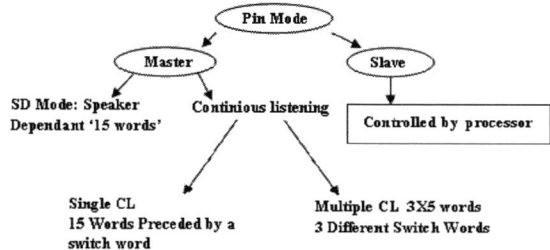

Fig. 2. Selection mode of the VD364

Among the various modes of training and recognition module, we have chosen the single continuous listening mode "SCL" as it is adapted to our application [15], and the exploitation of 15 words of the module.
The user is equipped with a microphone connected permanently to the card for the different commands. These commands are registered after training depending on the chosen mode. It is enough to state the activation mode followed by the different commands wanted by the handicapped. All these commands are shown in table 1.

3. PRINCIPLE OF OPERATION

When power is on, there is an automatic reset on the vocal module VD364. One LED is then on, indicating the system is waiting for word command. This word is common to all the commands depending on the chosen mode. Once this word is said, the module waits the pronounced action word. If it is accepted, it is reflected in the allocation of one or two bits on the data bus of vocal module, according to the algorithm of figure 3.

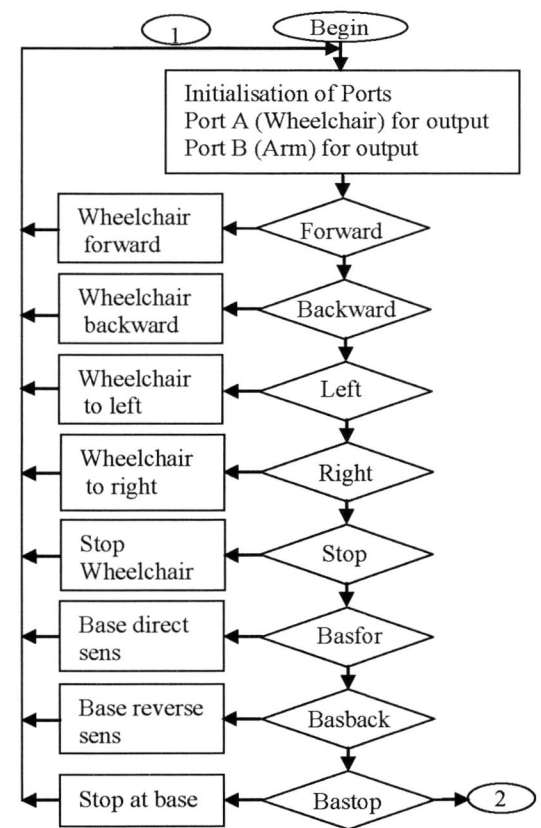

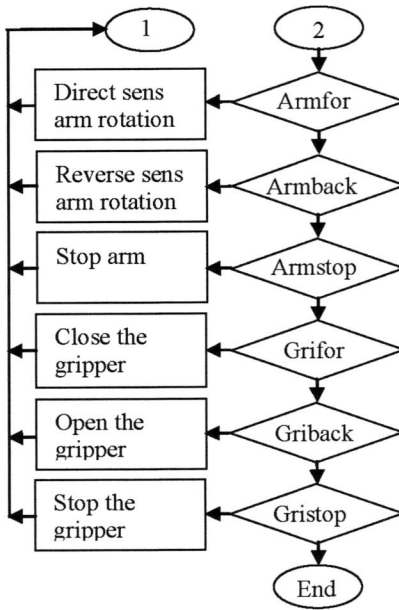

Fig. 3. Algorithm of the system

These bits will be translated by the microcontroller, providing the corresponding command to the wheelchair or to the arm. If you take for example, the word GRIFOR which corresponds to the closure of the gripper arm manipulator, the assignment will be out 5 and out 8. These two bits managed by the module during one second, will be transmitted to the microcontroller via a buffer line. This microcontroller reads in turn this information and then translates it by sending the corresponding action to the concerned stage giving then the wanted command by the handicapped. This information will be transmitted to the power amplifiers, which are directly connected to the engines of the arm manipulator and the wheelchair.

4. TESTS AND RESULTS

We first determine the recognition characteristics of the vocal module by varying the distance from the microphone and the energy of the vocal signal. This could be done by taking as an example, the word engine as a switch word and the words forward, backward, left, right and stop as command words. The second step is to inject a white noise to the speech signal and vary it while keeping the same energy of this signal. We then determine the signal-to-noise ratio and its gain in dB.

Activation Word: Engine
Energy : E1=166,5 ; E2=283,08 ; E3=347,8 ; E4=547,9

Command Words	Energy
Move forward	336,12
Move backward	393,68
Left	289,31
Right	274,68
Stop	183,17

Changes in the rate of recognition as a function of energy of the vocal signal and the distance from the microphone are given in table 2 and figure 4.

Distance from microphone [cm]	Energy of Signal	Recognition Rate
5	E1	10/10
10	E1	10/10
15	E1	8/10
20	E1	9/10
5	E2	9/10
10	E2	9/10
15	E2	8/10
20	E2	9/10
5	E3	9/10
10	E3	9/10
15	E3	10/10
20	E3	7/10
5	E4	10/10
10	E4	10/10
15	E4	9/10
20	E4	10/10

Table 2. Results of test 1

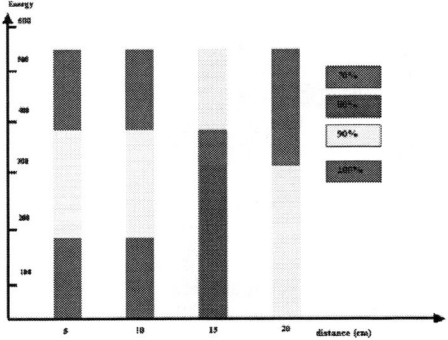

Fig. 4. Changes in the rate of recognition as a function of energy of the vocal signal and the distance from the microphone

On the other hand, when we inject a white noise to the speech signal with amplitude equal to 1 and we will also determine the signal-to-noise ratio (S/N).

Changes in the rate of recognition as a function of energy of the vocal signal and the distance from the microphone are shown in table 3 and figure 5.

Note: According to the tests, beyond 10% of injected noise to the speech signal and whatever the distance from the microphone to the speaker, the speech recognition module does not recognize the word, which means, the recognition rate is zero.

Distance from Micro [cm]	Energy	Noise	S/N Ratio [dB]	Recognition Rate
5	E4	100%	-38,99	0/10
10	E4	100%	-38,99	0/10
15	E4	100%	-38,99	0/10
20	E4	100%	-38,99	0/10
5	E4	10%	7,11	0/10
10	E4	10%	7,11	0/10
15	E4	10%	7,11	0/10
20	E4	10%	7,11	0/10
5	E4	5%	20,8	5/10
10	E4	5%	20,8	7/10
15	E4	5%	20,8	5/10
20	E4	5%	20,8	7/10
5	E4	2%	39,13	5/10
10	E4	2%	39,13	7/10
15	E4	2%	39,13	8/10
20	E4	2%	39,13	8/10
5	E4	1%	53	8/10
10	E4	1%	53	6/10
15	E4	1%	53	6/10
20	E4	1%	53	8/10
5	E4	0,2%	85,21	8/10
10	E4	0,2%	85,21	8/10
15	E4	0,2%	85,21	9/10
20	E4	0,2%	85,21	9/10

Table 3. Results of test 2

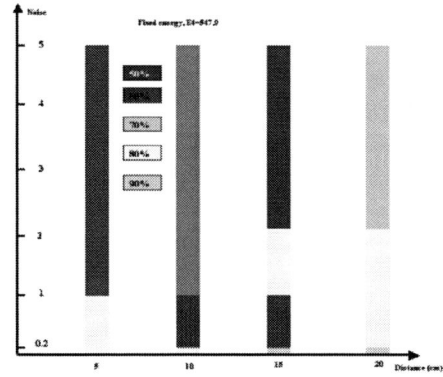

Fig. 5. Changes in the rate of recognition as a function of energy of the vocal signal and the distance from the microphone

5. CONCLUSIONS

In this paper, we have developed a very rich system in information which is the vocal command. This project can be considered in a number of areas such as vocal command for the severe disabled or the command of an AVG (wheelchair, manipulator arm, lift).

This development has allowed us to do a processing on real signals and the results demonstrate its reliability. Tests on the vocal module gave a meaning to this work and have contributed to the knowledge of the essential features of this module. The choice of this application allowed us to use the module with an interface based on a microcontroller to manage the training stages of recognition.

The current research is based on the development of a set of command modes (automatic or manual or shared) to give a human aspect to the manipulator arm. We also wanted to give in our application this aspect by using the vocal command because of the different techniques and recent research in the field of robotic assistance for the disabled and then adapt it to dynamic needs of the user.

The results obtained are encouraging and further testing shall be considered in the near future.

The aid is still under development as it is planned to add an obstacle detection system [16] in order to help the disabled to navigate safely and quickly among obstacles and other hazards. However, the problem of estimation of the disabled position, based on information from different sources, will be sorted out by using the approach known as particle filtering [17]. The particle clustering and convex region mapping techniques will be used to guarantee that at all times the position estimates are feasible.

Finally, this work serves as a powerful tool for vocal command. Its use can easily be extended by increasing the vocabulary, enhancing the robustness of the design and reducing noise, making then the system independent of the environment.

REFERENCES

[1] Cooper, R.A., M.L. Boninger, D.M. Spaeth, D. Ding, S. Guo, A.M. Koontz, S.G. Fitzgerald, R. Cooper, A. Kelleher and D.M. Collins (2006). Engineering Better Wheelchairs to Enhance Community Participation. *IEEE Transactions on neural neutwork systems and rehabilitation engineering*, **Vol. 14**, no. 4, pp. 438-455.

[2] Fioretti, S., T. Leo and S. Longhi (2000). A Navigation System for Increasing the Autonomy and the Security of Powered Wheelchairs. *IEEE Transactions on rehabilitation engineering*, **Vol. 8**, no. 4, pp. 490-498.

[3] Katevas, N. I., N. M. Sgouros, S. G. Tzafestas, G. Papakonstantinou, P. Beattie, J. M. Bishop, P. Tsanakas, and D. Koutsouris (1997). The autonomous mobile robot SENARIO: A sensor-aided intelligent navigation system for powered wheelchairs. *IEEE Robot. Automat. Mag.*, **Vol. 4**.

[4] Dario. P (1994). MOVAID (Mobility and activity assistance systems for the disabled). *Technical Annex, TIDE Technology and Development*, Project no. 1279.

[5] M3S Dissemination office-*TNO Delft*, vers. 2.0, Netherlands. [Online]. Available: http://www.tno.nl/m3s

[6] Bourhis G. and P. Pino (1996). Mobile robotics and mobility assistance for people with motor impairments: Rational justification for the VAHM Project. *IEEE Trans. Rehab. Eng.*, **Vol. 4**, no. 1, pp. 7–12.

[7] Levine, S. P., D. A. Bell, L. A. Jaros, R. C. Simpson, Y. Koren, and J. Borenstein (1999). The NavChair assistive wheelchair navigation system. *IEEE Trans. Rehab. Eng.*, **Vol. 7**, pp. 443–451.

[8] Simpson, R., D. Poirot and M. F. Baxter (1999). Evaluation of the Hephaestus smart wheelchair system. in *Proc. International Conf. Rehabilitation Robotics, CORR'99*, Stanford, CA, pp. 99–105.

[9] Hoppenot, P. Colle, E. Aider, O.A. and Rybarczyk, Y. (2001). ARPH-assistant robot for handicapped people-a pluridisciplinary project. *Proceedings of the10th IEEE InternationalWorkshop on Robot and Human Interactive Communication*, pp. 624-629.

[10] Alqasemi, R.M., E.J. McCaffrey, K.D. Edwards and R. Dubey (2005). Analysis, Evaluation and Development of Wheelchair-Mounted Robotic Arms. *Proceedings of the IEEE 9th International Conference on Rehabilitation Robotics*, Chicago, USA, pp. 469-472.

[11] Sensory Company (2003). *RSC-364 Datasheet*.USA.

[12] Jurafsky, D. and Martin, J.H. (2007). *Speech and Language Processing: An introduction to natural language processing,computational linguistics, and speech recognition*. Chapter 9, 2nd edition, Prentice-Hall, New Jersey, USA.

[13] A. Redjati (2005). *Commande vocale d'un AVG*. Master thesis, Badji Mokhtar University, Annaba, Algeria.

[14] Microchip Technology (2001). *Microcontroller PIC 16F876 Datasheet*. USA.

[15] S. Kibria (2005). *Speech Recognition for Robotic Control*. Master's Thesis, Umea University,Umea, Sweden.

[16] Shoval, S. and J. Borenstein (2001). Using Coded Signals to Benefit from Ultrasonic Sensor Crosstalk in Mobile Robot Obstacle Avoidance. *Proceedings of the 2001 IEEE International Conference on Robotics and Automation*, Seoul, Korea, pp. 2879-2884.

[17] **Gustafsson, F., F. Gunnarsson, N. Bergman, U. Forssell, J. Jansson, R. Karlsson and P.J. Nordlund** (2002). Particle Filters for Positioning, Navigation and Tracking. *IEEE Transactions on signal processing*, **Vol.** 50, no. 2, pp. 425-437.

Fabrication of Wall Shear Stress Sensor for Micro Flow Measurement

M. Laghrouche[*], A. Adane[*]   J.Boussey[**], S.Ameur[*] and D. Meunier[**] M.Tahanout[*]

[*] FGEI, university Mouloud Mammeri, BRP 017 H   hasnaoua Tizi ouzou; Algerie
[**] Institute of Microelectronics, Electromagnetism and Photonics
IMEP, ENSERG, 23 rue des Martyrs, 38016 Grenoble Cédex 1, France
(larouche_67@yahoo.fr)

Abstract: a new type of hot wire anemometer was developed by using surface micro machining techniques. The reduction of the microprobes section and the development of a cavity below the hot polycrystalline silicon film reduced considerably indirect thermal exchanges between the hot wire and its substrate enhancing the dynamic response time of these probes. This will make possible to measure fast fluctuations and turbulent phenomena in fluid mechanics.

Keywords: wall shear stress sensors, MEMS

1. INTRODUCTION

The control of turbulence in mechanics of fluids is a subject which interests several industrial fields; the economic stakes are enormous, mainly in the civil and military aviation. The use of the conventional anemometer (with Platinum wire) cannot quantify any swirling structure responsible for parietal friction. These structures are a few hundred microns of width and a few millimetres length [1] and have a very short lifespan, about the millisecond.
Microsystems, thanks to their integration capability and the optimization of geometrical structure, are of a great interest in this field and seem adapted to these needs. Several micro sensors of various geometries were produced by surface micro machining, using thermally sensitive materials like polycrystalline silicon.
In this article, we present the first results of electric characterization obtained from the first prototypes.

2. REALIZATION

Figure 1 shows the technological process adopted for the realization of the sensors. PECVD (Plasma Enhanced Chemical Vapor Deposition) silicon nitride was first deposited on p type 4" wafers to a thickness of 0.3 µm. Then the silicon nitride film was patterned, and wet thermal oxidation of silicon made through the nitride window. Oxidation time was adjusted with technological step simulation for achieving a flat surface. A 0.3 µm depth silica cavity was obtained. Polycrystalline silicon was deposited (LPCVD – thickness 0,5 µm) and doped by boron ion implantation (various doses were undertaken). After the thermal annealing, polycrystalline silicon was patterned into variable section wire. Electrical contacts were finally taken with chromium pads. Releasing of the wire was performed by wet-etching of the silica cavity after dicing the sensors.

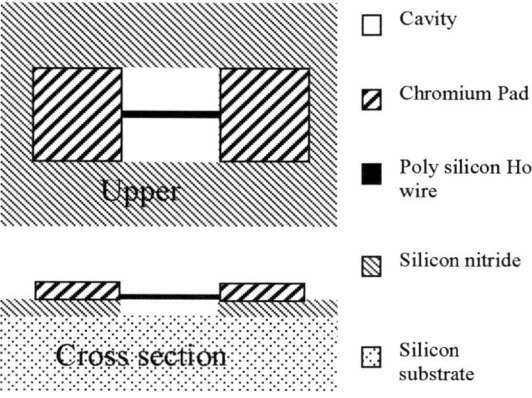

Fig. 1 Technological realization of the hot-wire probe suspended on a micro-cavity.

3. STATIC AND ELECTRIC SENSIBILITY

The measurement of temperature coefficient consists in plotting the hot wire resistance against temperature at different positions of the two-point probe and evaluating the slope of this curve. Under the same experimental conditions, the same a-values are obtained at any point of the upper surface of the sensors. This means that the temperature coefficient does not depend on the geometry of the hot wire resistance. As an example, Fig.2 illustrates the RKT curve obtained for a 50x2x0.5µm3 hot wire. This plot shows that the resistance variations are linear and a equals to 5.9.10-4/°C for this sensor. This result has also been obtained for the other polycrystalline silicon transducers. The TCR values do not depend of the hot wire geometry, so the TCR was relatively uniform on all the surface of the wafer.

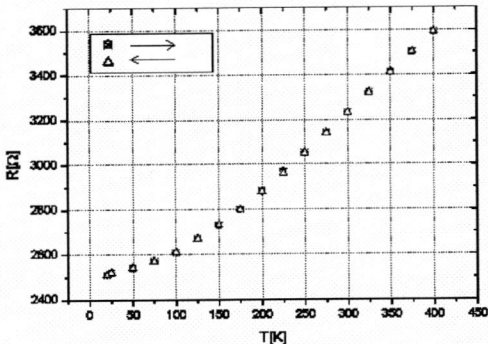

Fig. 2. Plot of resistance variations of against temperature for a 50x2x0.5µm3

The resistance measurement in function of the temperature for determining the TCR (Temperature Coefficient Resistance. The TCR values do not depend of the hot wire geometry, so the TCR was relatively uniform on all the surface of the wafer [4].

The voltage measurement in function of the polarization current. The variation of the resistance in function of the temperature is linear and in order to have an 70°C overheating, the polarization current was found on the order of 235µA.

4. DYNAMIC TIME RESPONSE

4.1 Measurements methods

Although static measurement is founded on a simple principle, it is now accepted that the frequency response of such sensors is in general complex because of convective and conductive heat exchanges. An electric circuit based on the internal heating method was designed and carried out for the dynamic response time determination of the sensors (Fig.3)

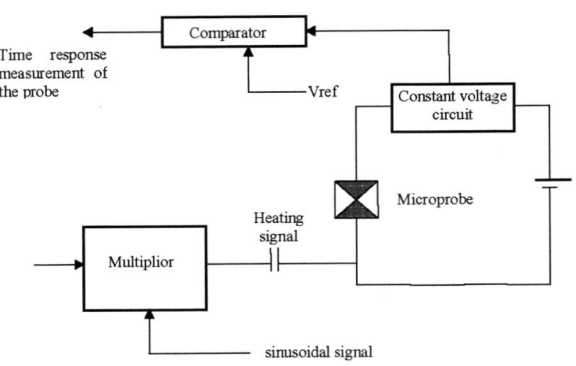

Fig. 3. Time response measurement circuit.

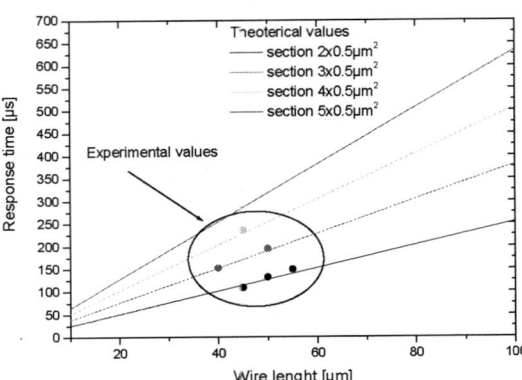

Fig. 4. Influence of the hot wire geometry on its dynamic performance: Model (straight lines) and experimental values.

The circuit makes possible to simulate, on the sensor level, the fluctuations of temperature by carrying out an internal modulation of heating of the probe created by a D.C. current (rated current of the probe). To this current, a AC current of 1 MHz frequency modulated by a square signal of 1 kHz frequency is superimposed.

Figure 5 shows the time response of a $50x2x0.5\mu m^3$ hot wire, which is about 150µs. So the bandwidth sensor lies between 0 and 1 KHz. This bandwidth is not good enough for detecting all turbulence phenomena associated with airflow over aircrafts in wind tunnels.

4.2. Time response study in function of the geometry

A study comparing the time response of various hot wire geometry was undertaken. The obtained results are in agreement with the analytical model [5,6] where the time response is calculated for constant current mode (CCM), with the next formula:

$$\tau_v = \frac{MC}{U(v_0)A - i_0^2 K_\theta} \quad (1)$$

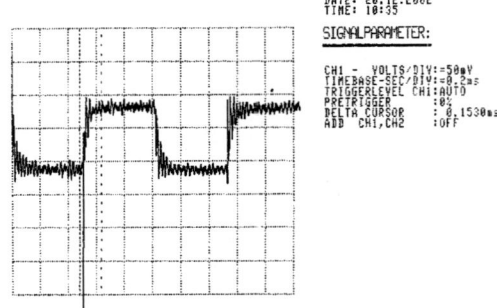

Fig. 5. Time response measurement of a $50x2x0.5\mu m^3$ wall shear stress sensor.

M is the sensor mass, C, the sensor specific heat, U the convection heat transfer coefficient between fluid and sensor, equal to 0,65 Wcm^{-2}°C^{-1}, A is the cross section, i$_0$ the current passing through the wire and $K_\theta = \dfrac{\Delta R_\theta}{\Delta T}$.

On Fig. 4, one can see that the time response is dependent on the cross section wire: the smaller is the width, the shorter is the response time.

We can also note that the experimental values are fitting very well the theoretical model.

The time response could be improved in mounting the hot wire in a Constant Temperature Mode (CTM) (see fig. 6).

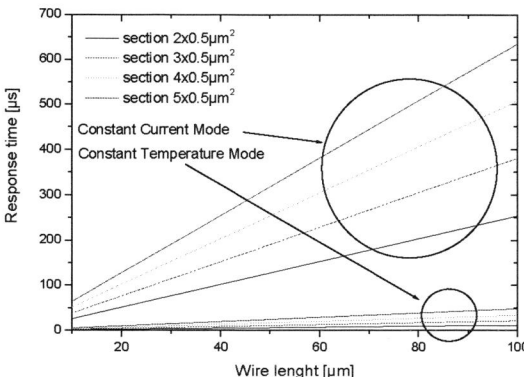

Fig. 7. Theoretical time response comparison between Constant Current Mode and Constant Temperature Mode.

4.3. The Conditioning circuit

The associated analogue circuit (fig.8) is divided into 3 stages: a Wheatstone bridge circuit, a CTA feed back circuit with a power stage and a differential amplifier circuit with an offset compensation network.

The Wheatstone circuit is formed with two precision resistors (0,01%) and a tunable resistor to control the initial voltage references. In the fourth branch we place the sensor near the resistors and the conditioned circuit in order to minimize the noise and external perturbation.

The AD620 circuit component, is a low cost, high accuracy instrumentation amplifier which requires only one external resistor to set main gains from 1 to 1000. This latter is working as preamplifier due to its low voltage noise.

In order to reduce the circuit size, we have pasted the sensor and the Wheatstone bridge on the ceramic support. The conditioning circuit is realized with a very small circuit under the ceramic circuit in which the CMS technology is used.

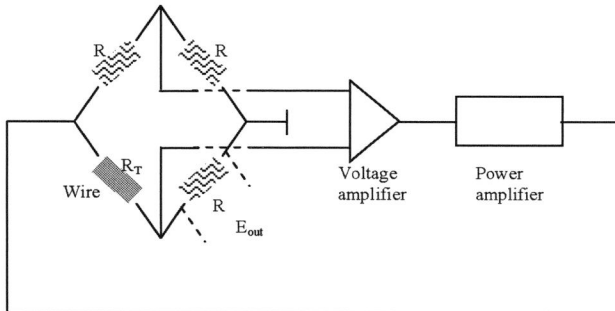

Figure. 6. Schematic diagram of hot wire mounted in Constant Temperature Mode.

The hot wire anemometer is mounted in a Wheatstone bridge: This is a self-balancing bridge which maintains the resistance R$_T$ of the sensor at a constant value R. An increase in fluid velocity v causes a decrease of T and R$_T$ values inducing an unbalancing of the bridge. This causes the amplifier output current and current through the sensor to increase thereby restoring T and R$_T$ to their required values.

The advantage of such CTM circuit is to minimize thermal inertia of the sensors, which was leading to rather bad time response.

It was demonstrated [5,6,7] that the time response in CTM becomes :

$$\tau_{CTA} = \dfrac{\tau_v}{1 + K_I K_A K_B} \qquad (2)$$

with $K_I = \dfrac{2K_\theta i_0 R_{\theta 0}}{[U_{v0}A - i_0^2 K\theta]}$, $K_A \approx -4$ and $K_B = \dfrac{1}{4}\dfrac{V_s}{R}$

where V$_s$ is the bridge supply voltage and R the variable resistance.

In Fig. 6, one can see the theoretical time response comparison between CCM and CTM. For our sensors, it means that the sensor time response will decrease of about 22 times. The bandwidth covered will lie between 0 and 25 kHz which is largely sufficient for covering most of turbulence measurement applications.

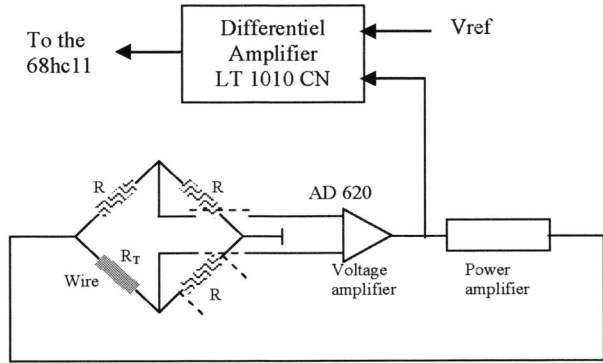

Fig. 8. Schematic diagram of hot wire anemometer mounted in Constant Temperature Mode.

The hot wire anemometer is mounted in a Wheatstone bridge: This is a self-balancing bridge which maintains the resistance R_T of the sensor at a constant value R. An increase in fluid velocity v causes a decrease of T and R_T values inducing an unbalancing of the bridge. This causes the amplifier output current and current through the sensor to increase thereby restoring T and R_T to their required values.

The advantage of such CTM circuit [6] is to minimize thermal inertia of the sensors, which was leading to rather bad time response.

4.4. The Data Acquisition System

The Data Acquisition System is composed of two main subsystems: the programmable digital data acquisition system and transducers with their associated circuits [5]. The requirement of data acquisition system dictated the use of the microcontroller. In our case the Motorola 68HC11 CMOS microcontroller integrated circuit was chosen for this purpose thanks to its low power consummation, small size and built-in analog to digital (A/D) converter with 8 input multiplexer.

The A/D converter of the microcontroller convert simple signals into digital values and transfers them to a personal computer with RF circuit via an RS 232 serial port (fig 9,10).

The RF module transmits the data through the AM modulation and 464 MHz frequency, so the consumed power of transmission is 13m W.

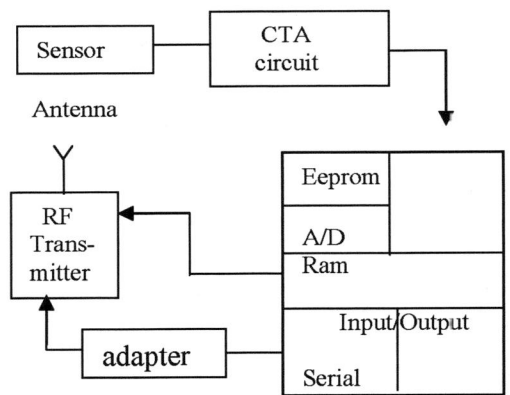

Fig. 9. The micro controller data acquisition

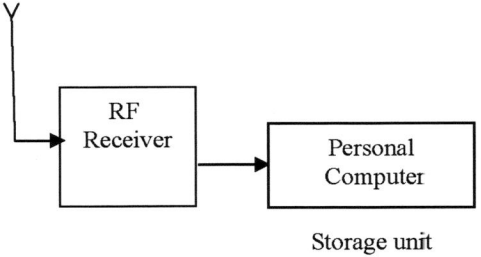

Fig. 10. Data receiver in the laboratory

5. STATIC SPEED CALIBRATION

The sensor was also characterized in a wind tunnel in order to calibrate it for static speed variation.
Experimentally, the hot wire anemometer is mounted at the wall surface of the wind tunnel. A Pitot tube measures the pressure in the centre of the tunnel. This pressure is then transformed in order to know the macroscopic speed.
This calibration was done from 7m/s up to 15m/s.
With this constant temperature configuration, we can find the various parameters α, β and n of the electrical calibration of the sensor described by a law as:

$$E^2 = \alpha.(\overline{u_\tau})^{2n} + \beta \quad (3)$$

where E is the output voltage and $\overline{u_\tau} \approx u_\infty / 22$ the friction speed. In measuring the output voltage variations, we can determine the other coefficients.

According to Levêque solution, n coefficient has to be close to 1/3.
In the figure 11, one can see that the above mentioned law fits perfectly the experimental values when n is taken equal to 1/3.

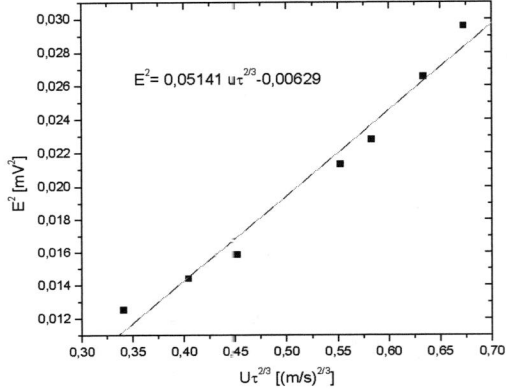

Fig. 11. Static speed calibration in wind tunnel: E^2 in function of $(u_\tau)^{2/3}$.

6. CONCLUSION

We have characterized a wall shear stress sensor with several methods: Temperature Coefficient Resistance measurement, Time response with CCM and CTM circuit measurement and wind tunnel calibration. These measurements have provided coherent results with previous simulation [3,4].
The 25 KHz bandwidth obtained with a CTM circuit (for a $2\times50\times0.5\mu m^3$ hot wire) shows that almost turbulence phenomena will detectable.

REFERENCES

[1] Tardu, S. (1998). *Near wall turbulence control for local periodical blowing*, Experimental thermal and fluid science, Vol 16 , n°1-2, pp.41-53

[2] Meunier. D, D.Tsamados, J.Boussey, S.Tardu; (2002).*Simulation of wall shear stress integrated sensors.* Therminics,02, Madrid

[3] Meunier. D, D.Tsamados, J.Boussey, S.Tardu (2003); *Thermo-fluidic FEM simulation of ultra-miniaturised wall shear sensors.* Eurosime 03, Aix en provence (France), April

[4] Laghrouche . M. , A. Adane, J. Boussey, S. Ameur , D. Meunier, S. Tardu, (2005); *A miniature silicon hot wire sensor for automatic wind speed Measurements,* Renewable energy; Vol. 30; No. 12; pp. 1881-1896.

[5] King, L.V (1914)., *On the convection of heat from small cylinders in a stream of fluid*, Philosophical Transactions of Royal Society, Series A, vol. 214,.

[6] Jiang F., (1998) *Silicon Micromachined Flow Sensors*, Thesis,California Institute of Technology.

Proposal and Study for an Architecture Hardware/software for the Implementation of the Standard H264

K. Messaoudi*   S. Toumi**   E. Bourennane***   M. Boutalbi****

Dpt Electronic / Mentouri University
Ain El-Bey Constantine Cedex, Algeria - (messaoudialg@yahoo.fr)
**Dpt Electronic / Badji Mokhtar University*
Sidi Amar, Annaba Cedex, Algeria - (toumi.salah@hotmail.com)
***LE2I Lab / bourgogne University*
Dijon Cedex, France – (ebourenn@u-bourgogne.fr)
****Dpt Electronic / Badji Mokhtar University*
Sidi Amar, Annaba Cedex, Algeria – (mboutalbi@yahoo.fr)

Abstract: In this study we have presented a proposal architectural hardware/software for the implementation of the H.264 standard, which means a processor-specific for H.264 standard use of it a hard part and a configurable embedded processor, which is LEON2 in our case. The motion estimation and compensation, the integer numbers transformation, the quantification, the algorithm of entropy encoding uses in the coder H.264 are optimized in material form; the other parts over the operating system are implemented by LEON2 processor.

Keywords: H.264coder, hardware/software Architecture, configurable microprocessor, LEON2.

1. INTRODUCTION

Since MPEG-1 in H264, reusable blocks have become standard tools for encoding video compression. But in all standards one finds blocks heavier than the other from the time and complexity of the calculations, which makes mandatory the use of hardware architectures to achieve a schema Joint Hardware/Software; and with the goal of a hardware implementation in real time. Indeed, in a location that can be absorbed on the FPGA the sections of the algorithm that require a important computation time, the processor (the soft part) executes the remaining parts of the algorithm which are simpler from the computing time. In addition, the distribution of functionality between hardware and software can solve some of the problems associated with the design gap. It makes it possible to re-use the same chip in different contexts, thus sharing the fixed costs between several projects. Therefore, the goal of our work is to propose an architecture of reading images from the video source, and a technique for designated blocks of size 8x8 to be used in the different images for the implementation of different blocks of the H.264 standard, and in particular the block of motion estimation.

2. DECOMPOSITION PROPOSED

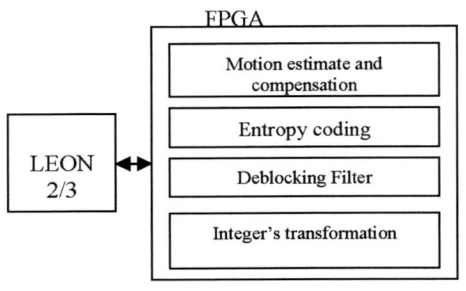

This diagram shows the distribution of work proposed for the implementation of the standard H264, the material location on FPGA and the algorithms to carry out by the processor LEON2. On the FPGA one absorbs the sections of the algorithm that require a number of cycles and important calculation time: The block of motion estimation and compensation, the integer's transformation, the entropy coding and deblocking filter. The processor LEON2 executes the remaining parts on the algorithm which are simpler from computing time.

3. THE VISUAL CODING STANDARD H.264

The standard H.264 has been developed to achieve significant improvements over existing standards for compressing image sequences. The development of international visual coding standards such as MPEG-1, MPEG-2 and H.261 has magnified a diverse range of multimedia applications, including digital video recording, and the system for teleconferences.

3.1 Evolution of visual compression standards for digital:

The following figure shows the evolution of various standards and norms encoding video sequences of the two committees ITU and MPEG:

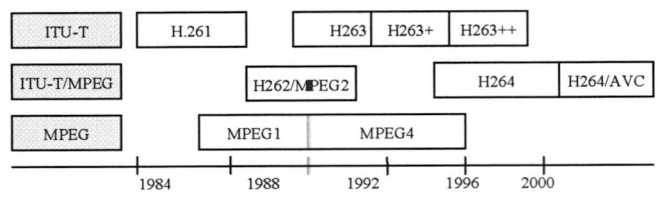

3.2 The H.264 coder structure:

The new H.264 visual standard shares a number of devices with past standards, including H.263 and MPEG-4. Mainly, H.264 is based on a hybrid model for the Adaptive Differential Pulse Code Modulation (ADPCM) and a transformation based on the coding of integers, similar to discrete cosine transform (DCT) used in earlier standards. This complex coding is done to take advantage of the temporal and spatial redundancy occurring in successive visual images. The diagram of the encoder H.264 is shown on the following basis:

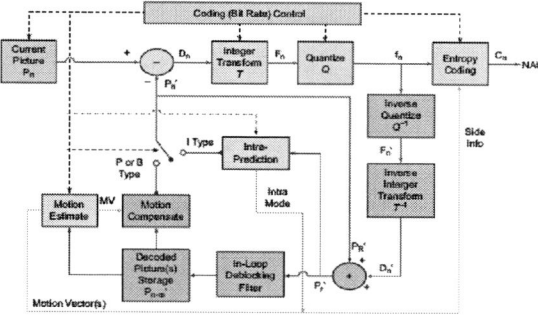

3.3. Type of image in h264 standard:

A video sequence in H.264 is made up of many different types of images. The sequence of images may be such as "I, B, B, B, P, B, B, B '. The image "I" (INTRA) is the image reference encoded independently of the others, it provides access points in the train binary, it is also used to encode images of P and B type, its compression ratio is average. The image "P" (Prédite) is an image composed of macro blocks, this image can be predicted by motion compensation from an image "I" or "P", it serves as the coding of images "B". The compression ratio is higher than that of "I". In the last frame of type "B" (Bidirectional), the macro-blocks can be predicted by simple motion compensation before or back starting from images "P" or "I" or by motion compensation dual before and back.

3.4 Motion estimation and compensation in H.264 :

Compared to the complexity of the standard, and as a first step, we limit ourselves only to the motion estimation block, which remains the most serious part of complexity and quantity of calculations. Indeed, the H.264 standard uses the method of motion compensation per block (block matching) for parameter estimation of movement, this method is one of the most largely used. The starting image is cut into blocks (squares blocks of NxN size), that is applied to the following procedure:

➢ In the reference image, we define a search area (size DxD), in which the block to be estimate in the image 'P' is compared to all blocks of the same size possible in the image 'I'. That giving the "best" comparison is retained. The comparison was made according to a criterion, the most popular being a calculation of energy of residual block to minimize. This process is called 'motion estimation'.

➢ The block selected becomes the predictor of the current block then it is subtracted to form a bloc residual (Motion compensation).

➢ The residual block is coded and transmitted as well as the motion vector between the two blocs.

The decoder use the vector of movement to recreate the bloc predicted and decodes the residual block, and the additions with the predictor and rebuilt a version of the original block.

4. RECONFIGURABLE MICROPROCESSORS

A softcore processor is a processor implemented on a reprogrammable system like as a FPGA. This is called a system on programmable chip (SoPC). With highly flexible architecture from its nature, an implementation softcore can be reconfigured at any time contrary with a processor known as hardcore whose heart has its own chip which can not be changed. A softcore processor therefore adapts to the needs of its developers and hardware constraints (peripheral, performance, consumption…). However, its performance is inferior to those of a hardcore processor. A softcore processor is usually set in a hardware description language such as VHDL or Verilog.

Among the most popular processors softcore are the NIOS of the Altera society, the Microblaze from Xilinx, Openrisc of opencore and LEON from Gaisler Research, which is the object of our work.

5. THE MICROPROCESSOR LEON2

The processor Leon uses an internal architecture of 32-bit, entirely described in VHDL and completely synthetisable. It was developed by the European Space Agency (ESA). Its architecture is a SPARC V8 (Scalable Processor ARChitecture). The schematic diagram of the LEON-2 can be seen in the following figure:

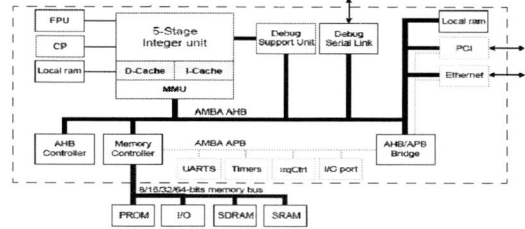

The LEON VHDL model implements a 32-bit processor conforming to the IEEE-1754 (SPARC V8) architecture. It is designed for embedded applications with the following features on-chip: separate instruction and data caches, hardware multiplier and divider, interrupt controller, debug support unit with trace buffer, two 24-bit timers, two UARTs, power-down function, watchdog, 16-bit I/O port, flexible memory controller, ethernet MAC and PCI interface. New modules can easily be added using the on-chip AMBA AHB/APB buses. The VHDL model is fully synthesisable with most synthesis tools and can be implemented on both FPGAs and ASICs. Simulation can be done with all VHDL-87 compliant simulators.

6. THE CODESIGN

The design of systems containing hardware and software components is not a new problem. The classic techniques of designing systems require separate hardware of software of the design cycle. Indeed, the software and hardware are two activities which have always been conducted in relative independence, such kind that the specialists in the software can think that they do not have need to be concerned about the low-level details of the hardware.

The current trend, mainly when dealing with complex real-time systems, is to reflect upon the specifications of the interactions between the two sides (hardware/software). Thus, the separation is done as late as possible in the design cycle to unify the specification so that it includes the hardware and software. This approach is called "hardware/software design", or "codesign."

The system is designed from a single specification describing its architecture and/or his behavior. After the specification phase occurs a phase partitioning system designed to break down the partitions in three parts:

➢ A material part implanted in the form of hardware circuits (FPGAs, ASICs, ...);
➢ A software implanted in the form of an executable program on a general purpose microprocessor;
➢ A communication interface between the two parts.

The partitions obtained should then be checked and validated before moving to phase synthesis and implementation. A feedback can return to the partitioning process by refining different weights associated with constraints obtained if the partitions do not prove satisfactory. However, many problems remain raised by the scientific and industrial communities who are interested in this problem.

The co-design seeks to meet two main requirements which are: determining the best compromise between achieving software and hardware implementation, and increasing productivity by making the design of software and hardware to simultaneously (concurrent design). Otherwise to say, to define tools and methodologies to find the best match between the achievement software and hardware implementation, and accelerating development by making the design of software and hardware in parallel.

6.1. The embedded systems and real time:

Typically, an embedded system must respect strong temporal constraints (Hardware Real Time). It buried an operating system kernel or a Real Time (Real Time Operating System, RTOS). The Real Time is a concept a little vague. This could be defined as: "A system is said Real Time after when information acquisition and processing are still relevant."

This means that in the case of information arriving on a regular basis (in the form of a periodic interruption of the system, such as image sequence TV), the time of acquisition, processing and display must remain below the period refresh this information. This requires that the kernel or the Real Time system is deterministic forever join hands in the next tick down to the task of higher priority ready.

A traditional confusion is to mix the Real Time and the speed of calculation of the system, therefore the power of processor (microprocessor, microcontroller, DSP).

6.2. Codesign: When The Material Rejoint The Software

It now uses description material languages (VHDL, Verilog) to synthesize and test digital circuits. It was a software approach to design equipment. With the increased integration, digital systems have become more complex, while placing on the market should be the fastest possible (taking into account the Time To Market TTM and reuse of things have been achieved Design Reuse).

It has thus emerged the notion of blocks IP (Intellectual Property), which is possible through the use of description languages equipment. It buys blocks of IP as it buys an integrated circuit, for example, the CAN interface, and interface DCT...

With the integration on silicon increasingly important, the challenge now is how to implement the functionality. Example of a video compression algorithm that must be faster by hardware but more expensive, more flexible software but more slowly and which must be able to juggle more with the parameters of cost, speed, strength and packaging.

6.3. Interest on use of a processor:

The use of a processor facilitates the realization of the system in the majority of cases. It is also possible that the processor is oversized compared to the logic functionality to be implemented by software. The use of programmable logic FPGA circuits (or ASIC) is interesting in this case.

The use of a specialized processor as a microcontroller or DSP can also significantly reduce the excessive size and the number of electronic components.

7. PRINCIPLE OF SHIFT AND ROTATION SUGGESTED

The majority of algorithms of image processing, and in particular the blocks of the H264 standard, using blocks of size DxD, and usually blocks of size 8X8. Indeed, in the processing and analysis algorithms of images (digital filters, segmentation, contour detection, motion detection, etc.....) To treat each pixel of the image, and it is necessary to know the other 8 pixels that surround in the same image or in the other preceding images (dynamic image). What gave us idea of use of the registers of shift and rotation for the establishment of some block of the H.264 standard.

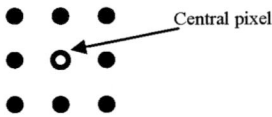

7.1. Reading of images:

Our idea is that in order to build windows 9 pixels in a static image, we must remember at least two lines of the image to be addressed, why offset two registers were reserved in the program in order to use memories internal. These records make it possible to shift a line and two lines every pixel of the image. For example, in a prototype treatment program, we can introduce any filter provided that the memory is sufficient with respect to the size of the image and filter applied.

The following figure illustrates the hardware architecture proposed for reading static images in real time.

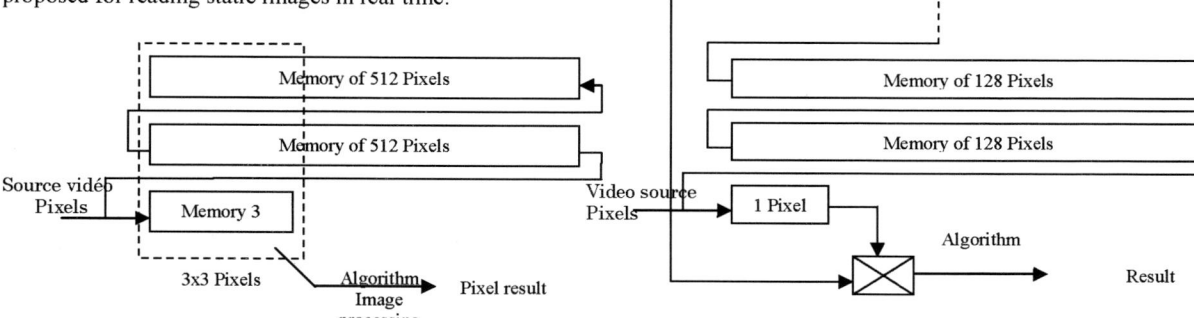

According to the method of reading the matrix corresponding to the input image, 3 pixels are first read on the current line, then the line counter is incremented. At the beginning of each line, it is necessary to read 9 pixels to start calculating. For the following calculation, only 3 pixels are read: the current pixel of the image in pixels, and the two staggered two lines before. To make a gap of two lines (1024 periods clock), it is for each cascade bit 64 blocks of memory synchronous Ram16x1s which contain a bus addresses controlled by a meter 4 bits. Also, a gap of a line is made with the same first 32 blocks of memory ram16x1s used for the offset of two lines.

7.2. Motion detection:

In this part of our work, we started from the idea the simplest, it is to memorize a whole image before reading only one pixel of the following image. This technique enables us to record only one image at the same time and using shift registers. At the output of the block is applied the algorithm desired on two windows of the two successive images (in our case detection it is simply to calculate the difference between the two addresses memories and remember the results). The following figure illustrates the basic principle of the proposed method:

Applying the same principle of rotation and shift of the lines of the images, lead us to more complicated schemes for the establishment of other blocks of H.264 standard, In particular the block of motion estimation.

This technique also allows us to save the successive images of the sequence and implement the desired algorithm in the same memory addresses, which one on good for hardware architectures, and that always with the goal of gaining the calculation time and memory space.

7.3. Blocks localization:

In this proposition, we started from the simplest idea, it is to memorize a whole image (the reference image I) before reading 8 lines of the following image (image P). This technique enables us to record a single image at a time and use the same memory capacity for the reading of the images I and P by simple shift of pixel. Indeed, after each search on a block, one makes a shift of 8 pixels at the video input, and after shift of a line one makes a shift of 7 lines at the same time, allowing us to the end of go to the next block 8x8 on the image P, and to next block DxD on the image I. After scanning total blocks of the image I, the latter will be replaced completely by the image P which is that serves as a reference for the other image P. In addition to scan the block research DxD, simply shift (rotation) of a pixel block itself for a full rotation at the end. This technique has the advantage of manipulating the same memory addresses (at the input program block 'Prg') for the calculation of energy residual block to minimize following the chosen criterion.

The same work could be effective, if we remember the picture completely P, with the disadvantage of total memory space to be used. The advantage in this case is to use a shift of the blocks of size 8x8 instead of shifting 8 pixels. In addition, the motion estimation is preceded by a detection to take decision to launch the inquiry procedure of the most similar block or to move directly to the next block in the image P. This increases the effectiveness of the proposed architecture.

For the program part has the option of using a block material that conducts operations or to assign this task to a processor 'LEON2 in our case' or share calculations block between hardware and processor 'codesign '. The following figure illustrates the principle of rotation and shift to the proposed location and the search for the bloc's used by the block matching method.

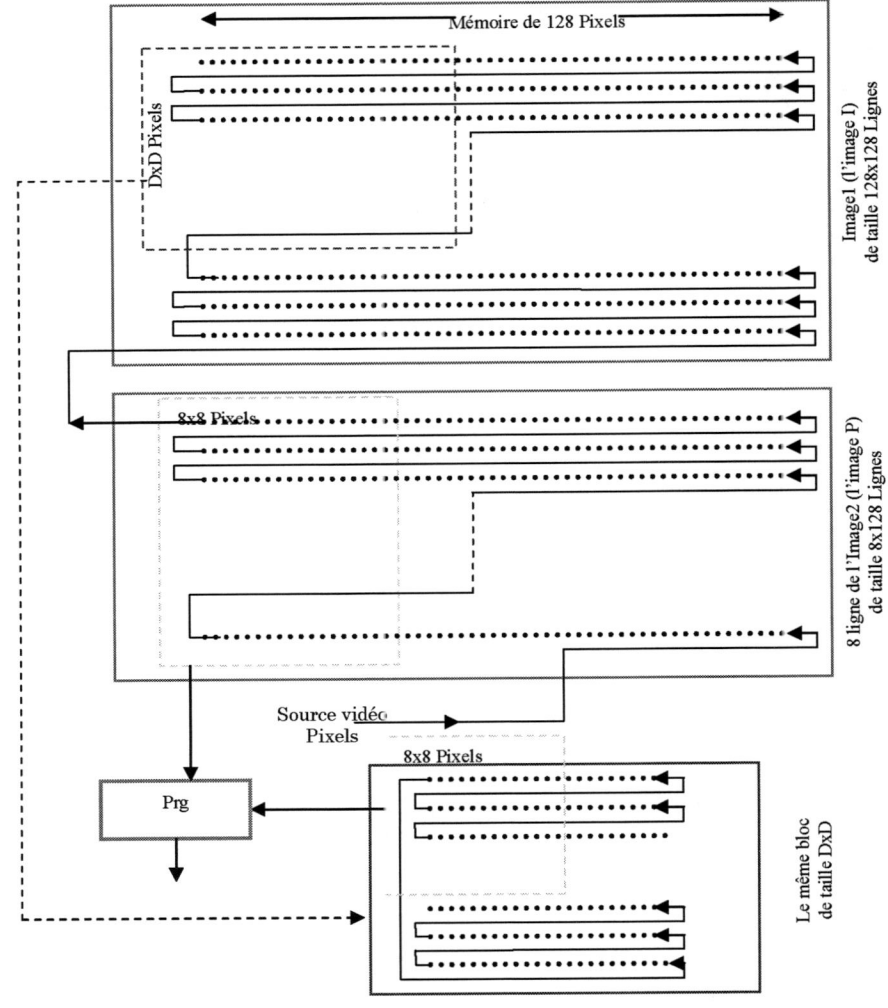

8. CONCLUSION

In this work, we have discussed the aspects of a classical string of acquisition-treatment-restitution of video signals and we presented the case of the H.264 standard. The hardware implementation of this standard involves parallel architectures for real-time processing low-level programmed in VHDL and that with an aim of an establishment of these algorithms in a FPGA. For this we have proposed a hardware architectures for the bloc's most complicated in the standard Video Coding H.264, it is the motion estimation and compensation block. In this work one presents an architecture for motion detection and one for the location of blocks needed for the motion estimation. The calculations and comparisons will be run by a microprocessor. The technique has the advantage of gaining, from the computing time and the memory capacities necessary for the recording of the images and even the advantage of keeping the same memory addresses for the different blocks object of treatment in the various images, which is one of many views physical realization.

REFERENCES

[1] L. E. G. Richardson, '*H.264 and MPEG4 / Video compression*', The Robert Gordon University, Aberdeen, UK, Edition Wiley, 2003.

[2] R. Schäfer, T. Wiegand and H. Schwarz, '*H.264/AVC la norme qui monte*', Heinrich Hertz Institut, Berlin (Allemagne), UER–Revuetechnique –Selection2003.

[3] A. E. Nelson, '*Implementation of image processing algorithms on FPGA hardware*', Faculty of the Graduate School of Vanderbilt University, 2000.

[4] F. Ghozzi, '*Optimisation d'une bibliothèque de modules matériels de traitement d'images*'. Thèse de l'université BordeauxI école doctorale des sciences physiques et de l'ingénieur, 2003.

[5] A. Ouled Zaid, M. Kieffer, '*Reconstruction robuste de vecteurs mouvements appliquée au codeur H.264*', LSS (Laboratoire Signaux et Systèmes)– CNRS SUPELEC–Université Paris-Sud.

[6] Z. L. Wang, '*H.264 Baseline Video Implementation on the CT3400 Multiprocessor DSP*', Cradle Technologies Inc., CA 94041-1525, 2003.

Estimation of Contact Forces of a Four-Wheel Steering Electric Vehicle by Differential Sliding Mode Observer

K. Bouibed[1], A. Aitouche[2], A. Rabhi[3] and M. Bayart[1]

Laboratoire d'Automatique, Génie Informatique et Signal (LAGIS UMR CNRS 8146)
[1]*Polytech-Lille, 59655 Villeneuve d'Ascq, France. E-mail : kamel.bouibed@polytech-lille.fr*
[2]*Hautes Etudes d'Ingénieur (HEI), 13, rue de Toul, 59046 Lille Cedex-France*
[3]*Laboratoire Modélisation Information et Systèmes (MIS), 33 rue Saint Leu, 80039 Amiens Cedex 1, France*

Abstract: This paper deals with the estimation of tire/road contact forces. These forces play a big role on the dynamics of vehicles. Basing on longitudinal, lateral, yaw and wheel dynamics of an electric vehicle ("known as Robucar"), the contact forces can be estimated through second order differential sliding mode observer. A step by step algorithm allows estimating velocities and accelerations and then these forces can be deduced. The results are compared with the empirical model of these forces given by Michelin.

Keywords: Four wheel steering electric vehicle, Robucar, sliding mode observer, tire/road contact forces.

1. INTRODUCTION

In recent years, several research teams and companies have engaged in the field of road safety and driving assistance. An outstanding advantage of electric vehicles over traditional internal combustion engines motor is their ecology as well as the cost and ease of their instrumentation. Those vehicles will be in the middle 21st century more useful.
For this reason, our Laboratory works with such kind of vehicles. The vehicle developed by the company Robosoft is now known as "Robucar". The searchers of our Laboratory develop essentially methods in fault detection and Isolation and fault tolerant control. They suppose that the tire/road contact forces are known. The tire/road contact forces are difficult to measure and their modelling is not easy to obtain. The knowledge of those forces is very important for systems such as ABS (Anti –Lock Brake Systems) or ESP (Electronic Stabilisation Programme). The dynamics of vehicle depends on contact forces which are usually non linear functions. The longitudinal force depends nonlinearly on wheel slip and the lateral force depends nonlinearly on side slip angle.

Tire/road contact forces are usually unknowns. Therefore works are made for their estimations. Observers or formulas given by Michelin [1] can be used; observers are more precise than those formulas and one can obtain also other parameter by estimation.

After the linearization of nonlinear system, unknown input linear observer can be used for their estimations [2, 3]. A simple method [4] based on full and reduced order proportional integral observer for an unknown inputs descriptor system was developed. The observer is solvable by any pole placement algorithm. It achieves *a posteriori* robustness state and unknown input estimation versus to time varying parameters.
Sliding mode observers are largely used in the literature instead to traditionally non linear observer. The advantage of sliding mode observer is easy to construct since the observer converge asymptotically and the disadvantage is the known problem of chattering. The classical non linear observers are difficult to implement in real time situation and the convergence of such observer is not easy to prove.
Extended Kalman Filter [5] is applied to non-linear model in order to estimate parameters of vehicle.

This paper focuses on second order sliding mode observer.
For first order sliding mode observer [6,7] velocity filtering is needed (noise,;..).
This kind of observer can be used to estimate sideslip angle which requires yaw rate sensor and data about vehicle velocities [8].
Then the searchers prefer to use second order observer which have been applied to some practical situations [9,10].

Then, Ouladsine *et al* [11] have estimated longitudinal forces using high order sliding mode observers. For the estimation of the sideslip angle and velocities, an observer based on the hierarchical super twisting algorithm is used.

The paper organisation is as follows: in section 2, the vehicle dynamic's model of Robucar used in our work is presented. Section 3 is devoted for the presentation of the differential sliding mode observer which is used to estimate state and its derivative and tire/road contact forces. Simulation results are presented in section 4 to show the effectiveness of our method. Finally conclusions and perspectives for future works are given.

2. ROBUCAR MODELLING

Robucar is a 4x4 electrical vehicle, with four electromechanical wheel systems. The motor part is actuated by a DC motor delivering a relative important mass torque. This test bench presents built-in some mechanical imperfections as friction and backlash in the transmission system. Front and rear steering angle are each controlled by a DC motor.
One can measure actuators, wheels velocities and rear and front steering angles, from the incremental encoders.
More details on the vehicle can be found in [12].

The following assumptions are given for RobuCar modelling:

- The pitch phenomena is not taken in consideration,
- The road is supposed uniform then the suspension dynamics are not considered,
- Electromechanical systems, longitudinal, lateral and yaw dynamics are modelled.
- The steering angles are supposed small
- The front lateral forces are equal (Fy_1) and the rear lateral forces are equal (Fy_2).
- The vehicle is in single mode (ie $\alpha_r=0$ and $\alpha_f \neq 0$).

Fig.1. Robucar

x and y are the longitudinal and lateral vehicle positions.
F_{xi} and F_{yi} are respectively the longitudinal and lateral efforts.
a and b are the distances between the mass centre and respectively the front and rear wheel axes. d is the distance between the two front (or rear) wheels.
α_f and α_r are respectively the front and rear steering angle.

The Robucar's model is obtained from the different dynamics of the vehicle as: wheels dynamics, longitudinal and lateral dynamics, and yaw dynamic.

- *Wheels dynamics*

$$\begin{cases} \dot{\omega}_{1f} = \frac{1}{J_1}[-f_1\omega_{1f} + R(Fx_1 + \alpha_f Fy_1) + U_1] \\ \dot{\omega}_{2f} = \frac{1}{J_2}[-f_2\omega_{2f} + R(Fx_2 + \alpha_f Fy_1) + U_2] \\ \dot{\omega}_{3r} = \frac{1}{J_3}[-f_3\omega_{3r} + RFx_3 + U_3] \\ \dot{\omega}_{4r} = \frac{1}{J_4}[-f_4\omega_{4r} + RFx_4 + U_4] \end{cases} \quad (1)$$

Where: ω_i is the front or rear velocity of wheel i (r: rear, f: front), $\dot{\omega}_i$ is the front or rear acceleration of wheel i and U_i represents the control input torque applied straightly to wheel i.

- *Longitudinal and lateral dynamics*

The longitudinal and lateral dynamics can be obtained by the fundamental principle of dynamics:

$$\begin{cases} \dot{v}_x = \frac{1}{M}[(Fx_1 + Fx_2 + Fx_3 + Fx_4) + 2\alpha_f Fy_1 - f_x v_x] \\ \dot{v}_y = \frac{1}{M}[2(Fy_1 + Fy_2) + \alpha_f(Fx_1 + Fx_2) - f_y v_y] \end{cases} \quad (2)$$

Where v_x and v_y are respectively the longitudinal and lateral velocities.

- *Yaw dynamic*

The kinetic moment's theorem at the vehicle's centre of gravity allows us to write:

$$\dot{\Gamma} = \frac{a}{Iz}[-\alpha_f(Fx_1+Fx_2)+2Fy_1] - \frac{2b}{Iz}Fy_2 + \frac{d}{2Iz}(Fx_1 - Fx_2 + Fx_3 - Fx_4) - \frac{f_\Gamma}{Iz}\Gamma \quad (3)$$

Where Γ and $\dot{\Gamma}$ are respectively yaw velocity and yaw acceleration.

3. SLIDING MODE OBSERVER

A sliding second order observer based on super-twisting algorithm can be used [13].

For this, firstly, we arrange the equations (1), (2) and (3) and we can regroup these equations in state space form as:

$$\begin{cases} \dot{X} = AX + BU + EF_i \\ Y = CX \end{cases} \quad (4)$$

where $X = [\omega_{1f}\ \omega_{2f}\ \omega_{3r}\ \omega_{4r}\ v_x\ v_y\ \Gamma]^T$ is the state vector, $U = [U_1\ U_2\ U_3\ U_4]^T$ is the input vector, $F_i \in \Re^6$ represents both longitudinal and lateral forces (unknown impact forces) and Y is the output vector.

A is a state matrix, B is the input matrix and E is the unknown input matrix depends on front steering angle α_f which is constant. These matrixes are defined as follows:

$$A = \begin{bmatrix} \frac{-f_1}{J_1} & 0 & 0 & 0 & 0 & 0 & 0 \\ 0 & \frac{-f_2}{J_2} & 0 & 0 & 0 & 0 & 0 \\ 0 & 0 & \frac{-f_3}{J_3} & 0 & 0 & 0 & 0 \\ 0 & 0 & 0 & \frac{-f_4}{J_4} & 0 & 0 & 0 \\ 0 & 0 & 0 & 0 & \frac{-f_x}{M} & 0 & 0 \\ 0 & 0 & 0 & 0 & 0 & \frac{-f_y}{M} & 0 \\ 0 & 0 & 0 & 0 & 0 & 0 & \frac{-f_\Gamma}{Iz} \end{bmatrix}, \quad B = \begin{bmatrix} \frac{1}{J_1} & 0 & 0 & 0 \\ 0 & \frac{1}{J_2} & 0 & 0 \\ 0 & 0 & \frac{1}{J_3} & 0 \\ 0 & 0 & 0 & \frac{1}{J_4} \\ 0 & 0 & 0 & 0 \\ 0 & 0 & 0 & 0 \\ 0 & 0 & 0 & 0 \end{bmatrix}$$

$$C = \begin{bmatrix} 1 & 0 & 0 & 0 & 0 & 0 & 0 \\ 0 & 1 & 0 & 0 & 0 & 0 & 0 \\ 0 & 0 & 1 & 0 & 0 & 0 & 0 \\ 0 & 0 & 0 & 1 & 0 & 0 & 0 \\ 0 & 0 & 0 & 0 & 1 & 0 & 0 \\ 0 & 0 & 0 & 0 & 0 & 1 & 0 \\ 0 & 0 & 0 & 0 & 0 & 0 & 1 \end{bmatrix}$$

$$E = \begin{bmatrix} \frac{R}{J_1} & 0 & 0 & 0 & \frac{R}{J_1}\alpha_f & 0 \\ 0 & \frac{R}{J_2} & 0 & 0 & \frac{R}{J_2}\alpha_f & 0 \\ 0 & 0 & \frac{R}{J_3} & 0 & 0 & 0 \\ 0 & 0 & 0 & \frac{R}{J_4} & 0 & 0 \\ \frac{1}{M} & \frac{1}{M} & \frac{1}{M} & \frac{1}{M} & \frac{2}{M}\alpha_f & 0 \\ \frac{\alpha_f}{M} & \frac{\alpha_f}{M} & 0 & 0 & \frac{2}{M} & \frac{2}{M} \\ \frac{\frac{d}{2}-a\alpha_f}{Iz} & \frac{-\frac{d}{2}+a\alpha_f}{Iz} & \frac{d}{2Iz} & \frac{-d}{2Iz} & \frac{2a}{Iz} & \frac{-2b}{Iz} \end{bmatrix}$$

The objective of our works is to estimate velocities and accelerations in view to obtain an estimation of longitudinal and lateral forces.

The first step is to estimate velocities and accelerations using sliding mode observer. The system is given by the following equation:

$$\dot{X} = AX + BU + EF_i \quad (5)$$

The observer has the following form:

$$\dot{\hat{X}} = A\hat{X} + BU + EF_i + Z \quad (6)$$

with $Z = \lambda |Y - \hat{Y}|^{1/2} sign(Y - \hat{Y})$

where λ is the observer gain matrix. It is determined in the convergence study [13].
The convergence of such observer has been demonstrated for the same kind of system (6) in [13].

To reduce the chattering effects [8], we can replace the function "$sign(e)$" by the function "$atan(q*e)*2/pi$".
The coefficient q is a design parameter to adjust the slope of tan function as shown in Figure 2.
For our simulation, we have taken $q=2$.

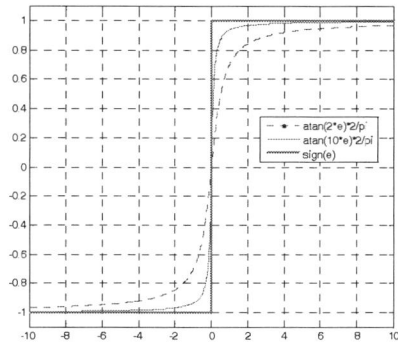

Fig.2. Sign functions for sliding mode observer.

After estimation of different velocities and accelerations, we can deduce the impact forces from the model equations.
The matrix E depends on input steering angle which is known and considered constant in our case. If it's variable, the formulation of eq.(5) is not possible. In this case, the input steering angle must be decoupled from the contact forces.
E is not square, so for calculate the forces from (5), we use the pseudo inverse of E.

$$E^+ = (E^T E)^{-1} E^T \quad (7)$$

This matrix exists if E has full rank. It is the case in our application.

Multiplying the two parts of eq.(7) by E^T and rearranging this equation, finally, we deduce:

$$\hat{F}_i = (E^T E)^{-1} E^T [\dot{\hat{X}} - A\hat{X} - BU] \quad (8)$$

4. SIMULATION RESULTS

In this section, we will present the results obtained by the simulation of the observer developed previously. We set the desired velocities of the wheels, and then by a PID controller, necessary controls are generated for each motor. These motors give the input torques (T_{ij}) applied to each wheel. In this simulation, we suppose that $\alpha_f=\alpha_r=0$. The steering can be provided by making differences between the velocities of left and right wheels. In this case the model becomes:

$$\begin{cases} \dot\omega_{1f} = \frac{1}{J_1}[-f_1\omega_{1f} + RFx_1 + U_1] \\ \dot\omega_{2f} = \frac{1}{J_2}[-f_2\omega_{2f} + RFx_2 + U_2] \\ \dot\omega_{3r} = \frac{1}{J_3}[-f_3\omega_{3r} + RFx_3 + U_3] \\ \dot\omega_{4r} = \frac{1}{J_4}[-f_4\omega_{4r} + RFx_4 + U_4] \\ \dot v_x = \frac{1}{M}[(Fx_1 + Fx_2 + Fx_3 + Fx_4) - f_x v_x] \\ \dot v_y = \frac{1}{M}[2(Fy_1 + Fy_2) - f_y v_y] \\ \dot\Gamma = \frac{a}{Iz}2Fy_1 - \frac{b}{Iz}2Fy_2 + \frac{d}{2Iz}(Fx_1 - Fx_2 + Fx_3 - Fx_4) - \frac{f_\Gamma}{Iz}\Gamma \end{cases} \quad (9)$$

In this case, we can proceed as follows:

- We estimate longitudinal forces from the first four equations of the system (9).

$$\hat{F}x_i = \frac{1}{R}(J_i \dot{\hat\omega}_{ij} + f_i \hat\omega_{ij} - U_i) \quad (10)$$

$i=\{1,2,3,4\}, j=\{f\ (front), r\ (rear)\}$

- We calculate lateral forces from the two last equations, knowing that replaces the longitudinal forces with their estimated values.

$$\begin{bmatrix} \hat{F}y_1 \\ \hat{F}y_2 \end{bmatrix} = \begin{bmatrix} 2 & 2 \\ 2a & 2b \end{bmatrix}^{-1} \begin{bmatrix} M\dot{\hat v}_y + f_y \hat v_y \\ Iz\dot{\hat\Gamma} + f_\Gamma \hat\Gamma - d(\hat{F}x_1 - \hat{F}x_2 + \hat{F}x_3 - \hat{F}x_4) \end{bmatrix} \quad (11)$$

To turn left (for example) we reduce the angular velocity of the left front wheel as shown in figure 3.

In our simulations, the measured values are those given by the equations of the model. They are compared with the results obtained by the observer.

In figure 3, we show the desired angular velocity for each wheel and input torques are shown by figure 4.

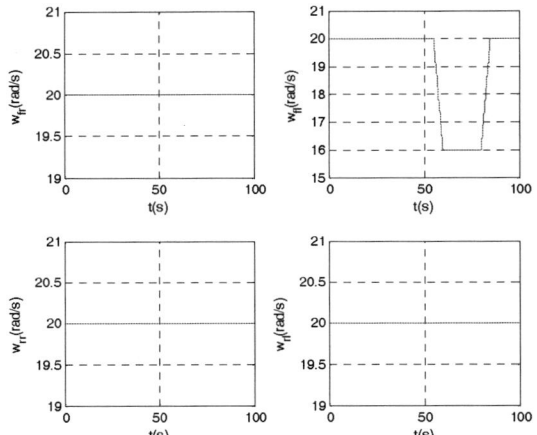

Fig.3. Desired angular velocities of each wheel

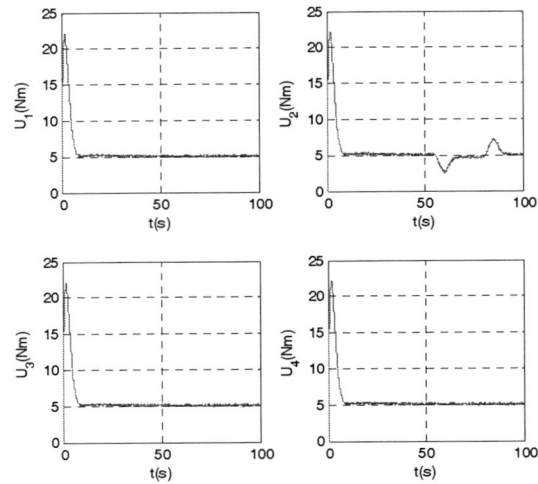

Fig.4. Input Torques applied to each wheel

Between instants 55s and 85s, the angular velocity of the left front wheel is reduced. We notice that the torque applied to this wheel also varies between these two instants.

Angular velocities and accelerations estimated by the sliding mode observer are shown in figure 5.

Figure 6 shows measured and estimated longitudinal, lateral and yaw velocities and estimated accelerations. The vehicle trajectory is shown in figure 7.

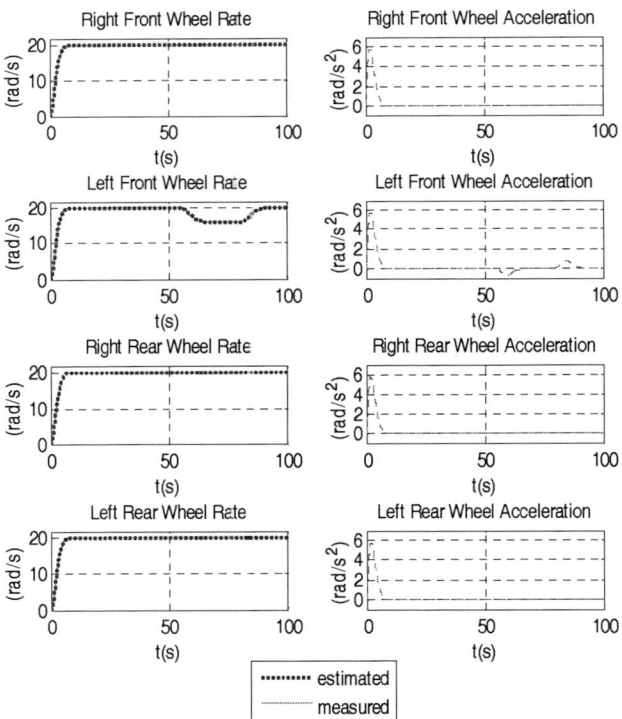

Fig.5. Angular velocities and acceleration of each wheel

In Figure 5, we notice that the velocities estimated by sliding mode observer are close to velocities given by the model. The estimation of accelerations converges asymptotically to wheel's accelerations.

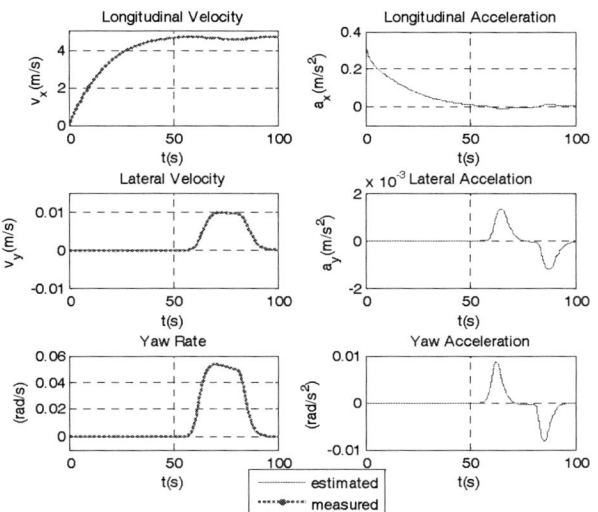

Fig.6. Velocities and accelerations

The left figures show that the estimated velocities, by a sliding mode observer, close to velocities given by the system.

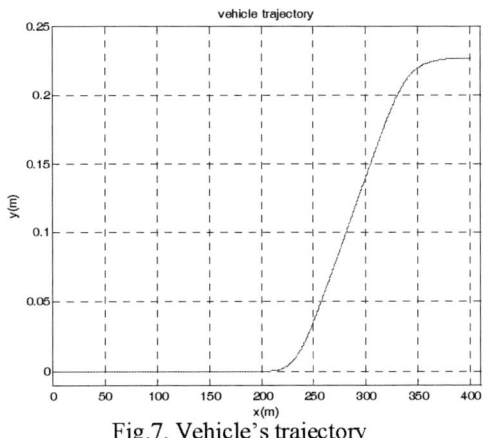

Fig.7. Vehicle's trajectory

The observer allows us a good estimation of different velocities and accelerations. The second step gives us the estimated longitudinal forces F_{xi} and lateral forces F_{yi} which are given in figure 8.

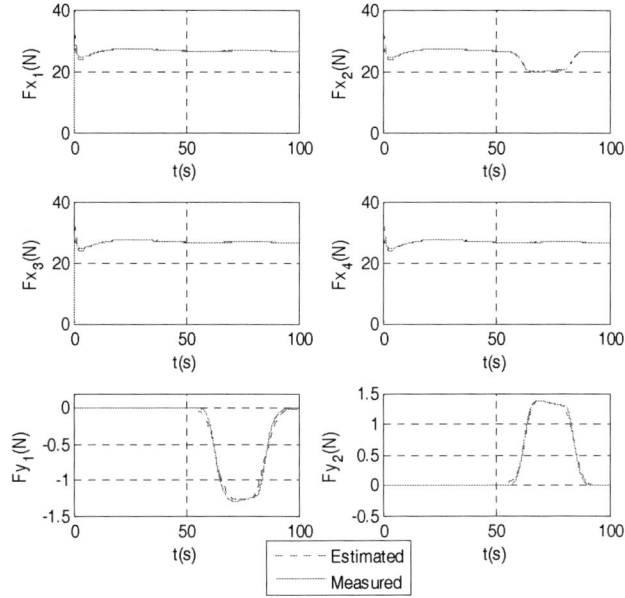

Fig.8. Measured and Estimated impact forces

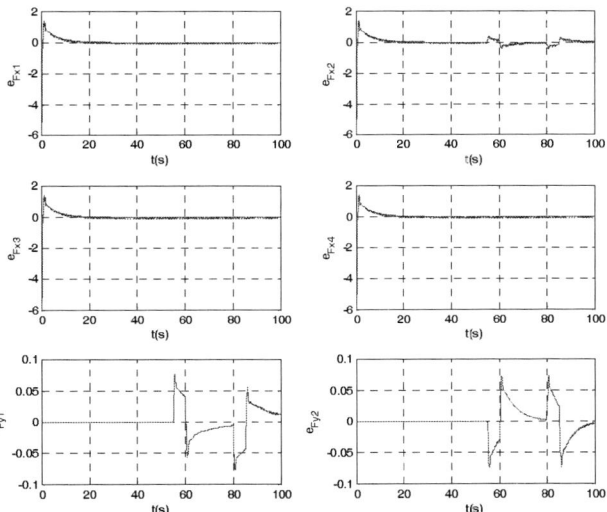

Fig. 9. observation errors

Figure 8 shows that the forces estimated by our observer close to forces measured by the formulas given by Michelin. We notice also that at the instant when we reduce the velocity of one wheel, the lateral forces become different to zero.

Figure 9 shows the observation errors which are the differences between the estimated forces and those obtained from the model equations. We can see that they are almost equal to zero.

5. CONCLUSION

This paper has presented the second order sliding mode observer adapted to the estimation of tire/road contact forces for electric vehicle. It allows the estimation of angular velocities and accelerations (longitudinal, lateral) of wheels and also yaw velocity and acceleration.

Simulation results show the effectiveness in the case where steering angles are null (straight line), and the direction of vehicle is obtained by the adjustment of wheel velocities.

In the future work will take into account the steering angles and more details on the convergence of the observer will be given.

NOMENCLATURE

Notation	signification
$J_i = 3$ kgm^2 $f_i = 0.02$ Nms	Inertia moments and viscous friction coefficients of the electromechanical system.
ω_i $\dot{\omega}_i$	Angular velocity and acceleration of the wheels.
v_x et a_x	Longitudinal velocity and acceleration of the vehicle.
v_y et a_y	Lateral velocity and acceleration of the vehicle.
U_i	Input torques applied for the wheels.
F_{xi} et F_{yi}	Longitudinal and lateral impact forces.
$R = 0.35$m	Wheel ratio.
$D = 0.1$m	Half-width of the tire / road contact area.
$a = 0.4$m $b = 0.8$m	Distance between the centre of gravity and respectively the front and rear axle of the vehicle.
$d = 1$m	Distance between the two front wheels (or both rear wheels)
$M = 350$kg	Total mass of the vehicle
Γ	Yaw rate of the vehicle
$I_z = 82$kgm^2	Moment of inertia of the vehicle around the yaw axis
$f_x = 19$kgms^{-1} $f_y = f_r = 0.01$kgms^{-1}	Tire/road friction coefficients

REFERENCES

[1] Michelin (2001). Le pneu, l'adhéence. *Société de Technlogies Michelin*. Michelin's encyclopedia.

[2] Kudva, P., N. Viswanadham and A. Ramakrishna (1980). Observers for linear systems with unknown inputs. IEEE Transactions on Automatic Control, vol. 25, n° 1, pp. 113-115.

[3] Guan, Y. and M. Saif (1991). A novel approach to the design of unknown input observer. *IEEE Transactions on Automatic Control.*, Vol. 36, N°5, pp. 632-635.

[4] Koenig, D. and S. Mammar (2002). Design of Proportional-Integral Observer for Unknown Input Descriptor Systems. *IEEE Transactions on Automatic Control*, Vol. 47, n°. 12, pp. 2057-2062.

[5] Sentouh, C., S. Glaser, S. Mammar and Y. Bestaoui (2006). Estimation des paramètres d'un modèle dynamique de véhicule par filtrage de Kalman étendu *Conférence Internationale Francophone d'Automatique*, Bordeaux, 30-31 Mai et 1er Juin.

[6] Levant A (1998). Robust exact différentiation via sliding mode technique. *Automatica*, Vol. 34, n°3, pp. 379-384.

[7] Barbot, J.P. M. Djemai and T. Boukhobza (2002). Sliding mode observers; In: Sliding Mode Control in Engineering, ser. Control Engineering, n°11, W. Perruquetti and J.P. Barbot, *Marcel Dekker*, New York.

[8] Stéphant, J., A. Charara and D. Meizel (2007). Evaluation of sliding mode observer for vehicle sideslip angle. *Control Engineering Practice*, Vol.15, pp. 803-812.

[9] Barolini, G., A. Pisano, E. Punta and E. Usai (2003). A survey of applications of second order sliding mode control to mechanical systems. *International Journal of Control*, Vol. 96, pp. 875-892.

[10] Rabhi A., N.K M'Sirdi and A. Elhajjaji (2007). A robust sliding mode observer for vehicle tire side slip angle. *Conference on Systems and Control*, Marrakech, Marocco. 16-18 may

[11] Ouladsine, M et al (2007). Vehicle parameter estimation and stability enhancement using the principles of sliding mode . *IEEE American Control Conference*, New York Cityn, USA, July 11-13, pp. 5224-5229.

[12] Dumont, P.E., A. Aitouche and M. Bayart (2006). Fault tolerant control on electric vehicle. *IEEE International Conference on Industrial Technology* (ICIT), Mumbai, India, 14-17 December.

[13] Davila, J., L. Fridman and A. Levant (2005). Second order sliding mode observer for mechanical systems. *IEEE Transactions on Automatic Control*, vol.50, n° 11, pp. 1785-1789.

Management traceability information system for the food supply chain

S. Bendriss, A. Benabdelhafid, J. Boukachour.

CERENE Laboratory; [Integrated Logistic Information System (I.L.I.S)]
Le Havre University, 25 Rue Philippe Lebon BP 1123, 76063 Le Havre Cedex, France
ben_sabri2003@univ-lehavre.fr; benabdelhafid@univ-lehavre.fr; boukachour@univ-lehavre.fr

Abstract: For a long time, the traceability was applied only for management reasons, but with the advent of new communication and information technologies more and more used in the logistic medium, the notion of the traceability became new extensive to meet the new market needs in term of information by ensuring accessibility the data characteristic or been dependent on the product throughout its life cycle.
On the basis of this postulate, we tried to raise some questions of research, beginning by the presentation of the progress achieved, assumptions and objective relating to the traceability, in the second time we mentioned principal work by showing how evolved the scientific question especially the information systems integrating the traceability were developed very little in the literature.
Based on what was developed in the first part, we present our generic modeling approach of communicating product "smart object", able to take into account the various essential elements for its traceability: the product in its various states, various operations carried out on the product, resources used, its localization, and interactions between the product and its environment carried out on the basis of whole of service. In order to validate our generic modeling, a case of study representing an application in a context of food industry is presented.

Keywords: Traceability, Information Systems, Service Oriented Architecture, food industry.

1. INTRODUCTION

This new paradigm called traceability is defined by ISO: 8402 [7] as "the ability to trace the history, application or location of an entity using the identification recorded" which aims to be able to find at the moment wanted, given advance data on the batch or group of products (also previously given), which extends from one or several IDs [18]. The application of traceability, therefore, involves the development of information systems throughout the entire products life cycle [3]

In product-centered approach for the management and the conduct of processes for traceability, and from a point of view of modeling of the necessary data to cover useful information for management of the products traceability throughout their life cycle [6], the construction of a generic representation facilitates the consistency of the various representations of the product pertaining to each stage of its life cycle. Indeed, thanks to the model of reference the synchronization of the various representations is optimal [4]

So, we propose a modeling approach to transform the product into communicating product "intelligent", with a capacity of communication, perception, action, and management of its own information available locally or through a network [8]. Our approach is founded on the principal researchs which we describe in the next section
In order to represent the product in its environment, interacting with the processes [8], operators and the products of his environment, from its manufacture while passing by its transport, its distribution its storage, and even its recycling, the goal of our modeling is the construction of a generic model of the product which takes into account the various essential elements for the traceability: the in its various states product (raw materials, semi-finished, finished), various operations realized on the product, resources used (human and machines), its localization, and interactions between the product and its environment carried out on the basis of whole of service, in the objective to obtain a functionality (a help, a support, an action and information relating to the product)

Thus, to test, in an industrial context "Food industry", the feasibility and the applicability of our generic model for the specification of all data and information relating to the product, we were based on a audit realized within the framework of a *MAS* Masters of Advanced Studies detailing the chain of supply, manufacture and distribution of the prepared meat products of AUCHAN super market group. The analysis of this work made it possible to know waiting and the requirements in information for each actor present throughout the life cycle for a prepared meat product. Our information system for the management of the traceability is dedicate to a sausage production, where we have subdivide the product lifecycle of the chain into many processes associated each phase of its forming treatment thus into whole of the production line.

The instantiation of our generic model applied to our case of application made it possible to identify the various types of information relating to a given product. The result of this modeling is used in the form of a data base representing the hard core of an information system for the management of the traceability.

2. TRACEABILITY NEW AREA OF RESEARCH

Born in the middle of the 80's, the traceability, declined in multiple forms, will become in the coming years, a tool unavoidable for all the companies. Today, it concerns every branches of industry and either certain sectors such as the agroalimentary one, the pharmaceutical one, aeronautics....it proves to be essential for reasons other than purely logistic: confidence relation towards the consumer, lawful constraints and legal, standardization, recall of defective products, trades electronic, etc. [3].

In this context, new currents research integrating the traceability emergent, in an implicit way in the subjects concerning the intelligent product and more explicitly within the communities centred on the product life-cycle management (PLM).

2.1 The traceability in a vision of intelligent product

Today the traceability of the products can be seen under a new angle centred on the concept of intelligent product [12]. In this part we will seek which relation exists between intelligent product and traceability.

In its work [12], McFarlane mentions, often, the supply chain is disturbed by problems caused by the not-synchronization of the matter flow and the information flow. One of the approaches to mitigate such problems is to establish a direct connection in network between a physical product and its data recording. With through this connection network, a product can interact indirectly with the operations which it undergoes. In this manner, the product becomes "intelligent" or more precisely "communicating", and becomes able to support the information which is necessary to him to the good moment.

Thus, McFarlane states the characteristics of an intelligent product in the following way:

1. To have a single identification;
2. To communicate with its environment;
3. To memorize and manage clean information;
4. To have a language of dialogue and exchange of its information and states;
5. To take part in the decision-making processes during its evolution;
6. To supervise and control its environment.

A basic level of intelligence corresponds to a product which has only the first three characteristics mentioned above, while the level of intelligence highest requires totality of it. In the characteristics mentioned by McFarlane, one notes indeed whom the three first give a base sufficient for saying that an intelligent product is able to develop the basic functions of the traceability: identification, localization, the recording, the historization of the activities and the access to this history (defined by the standard ISO 8402)[7]. The more complex functions of the intelligent products will result in improving each level which composes the logistic chain to reach a complete control of the product life cycle. In short, we can say that the traceability is one of the functions of an intelligent product.

2.2 The traceability in a system centred on the product

Always in the vision of intelligent product, [4] mentions, the rigidity of certain companies structures can handicap them to answer the personalization of the products mass. Integration in the company is made by the information, whereas the element which is central in a production system having to ensure a personalized production on a large scale rather seems to be the product, while placing it in the system main part of the company as preaches it the international initiative of research and development IMS (Intelligent Manufacturing system) particularly by paradigm HMS (Holonic Manufacturing System) [11], [4].

According to Kärkkäinen [8], if it is considered, the intelligence and the decision-making must be distributed to the extreme (working station, transport resources, products...); the product becomes able to control his evolution, to say in which state it is and to collaborate with its environment. This system is described as "system controlled by the product".

2.3 The traceability in the product life cycle

According to Garetti [2], the management paradigm of the product life cycle or PLM is the capacity to manage, to coordinate and to execute all the management and the engineering activities throughout product life cycle until the delivery to the ultimate consumer, with acceptable use and acquisition costs. So the PLM integrates a large variety of disciplines, methods, tools, environments throughout product life cycle: the product development (PD), manufacture systems engineering activities (MSE), tools (CAD, CAPP, CAM, PDM)[1], and the enterprise engineering, activities and management tools (ERP, MRP, CRM, SCM)[2].

If we join the concepts of intelligent product, the PLM and the controlled systems by the product we can say that through an intelligent product and adapted technologies of identification and communication, it would be possible for us to find the history, the use or the localization of the product which represents the principle even traceability (ISO 8402).

We have could understand with what was evoked above that a reliable traceability can not be concretized only by the construction of a coherent model for the representation of the data product and who represents one of the most crucial and complex activities, because if the model is not good, the applications will not answer initial waiting.

3. FIELD STUDIES INCLUDING TRACEABILITY

In what follows we present the literature on the various manners to integrate the traceability in a model or a representation. However, it should be mentioned that if the state of the art on the traceability is rich constantly update, the information systems integrating the traceability were poorly developed in literature.
Moreover, none of the found models shows the way of binding the product and the information nor how the traceability is obtained.

In [5], it is mentioned that the traceability can be seen as a tree structure of genealogical type. Logically its graphical presentation is a tree. Thus, one of the tasks of the traceability is to start to break up the nodes of the tree repeatedly, until the final nodes are found. In the same way, Shinghal [16] indicates, a decomposition of a problem illustrated by a graph AND/OR can be used to represent the problem of the traceability.

One of the first theoretical approaches of the traceability is realized by [9], he proposes an ontology of traceability in project TOVE *(Toronto Virtual Enterprise)*. This ontology introduces two concepts having to be traced: the TRU (Traced Resource Unit) and the primitive activity. The TRU is a single entity i.e., which there is not other identical unit from the view point of traceability. Practically, a TRU

---

[1] CAD: Computer-aided design, CAPP: Computer-aided process planning, CAM: Computer-Aided Manufacturing, PDM: Product Data Management.
[2] ERP: Enterprise Resource Planning, MRP: Materials Requirements Planning, CRM: Customer Relationship Management, SCM: Supply Chain Management

corresponds to a batch of identifiable production. The primitive activity as for it is an activity not consisting of under activities. By their ontology the authors affirm that in an ideal traceability system, the capacity to trace the activities and the products is fundamental. The products and the activities are called the "central entities" and they exist only when they individually are described and considered. The principal contribution of the model of Kim is the explicit assumption on the single identification of each item.

Moe [13] uses the model created by Kim [9] and applies it to food industry. It provides a model more operational than the rhetorical model of Kim. For that, it defines for each central entity a series of essential descriptors which can be included in order to ensure an ideal traceability of the products and activities.

The most recent work on the traceability modeling was made by van Dorp [19] and Terzi [18].

Van Dorp, mentions for its part that the development of the traceability systems of requires a precise representation of the system in question. For that, it is necessary to take into account a representation with the following elements [19]

- The present objects in the system;
- Types of objects to locate or/and to trace;
- The interest information on these objects;
- Relations between the objects;
- The representation and the objects integrity.

Thus, van Dorp shows an approach of design of an information system for the traceability of the goods flow.

He applies a modeling through Gozinto graphs. Those represent a graphic list of the raw materials, parts, sets and subsets, transformed into finished products through a sequence of operations (or process). The representation of the data structure by Gozinto is indicated under acronym BOM *(Bill Of Material)*. Then this graphic list is translated in a reference data model which is the base for the systems design of traceability information.

The composition of certain finished products is represented through the modeling of all the constituting materials and their various intermediate relations. When the sequence supplements assembly to manufacture a finished product is ended, a list with multilevel of batch called BOL (*Bill Of Lots*) is compiled. This list provides necessary information to determine the composition of an articles batch.

In the work of Terzi [18], are illustrated the preliminary results of the holonic approach for the products traceability. The authors propose a model in which the product supports the informational part for the traceability (Holon-product) throughout its life cycle. This model re-uses existing concepts around the company standards like PLCS, MANIDATE, ANSI/ISA-95 and PLM@XML. It formalizes, under UML, the structure of the information system associated with the products traceability data. This model is still conceptual and it must be validated.

Starting from these observations and in order to add a brick in these problematic which attract the interest of many scientific,We present in the following a modeling approach aimed to transform the product into communicating product "intelligent" through a product generic representation of and interactions "product/process", enriched by communication capabilities, information management available locally , or through a network.

4. GENERIC MODEL FOR AGROALIMENTARY INDUSTRY

In what follows, we present our step modeling for the construction of a generic model of the product which takes into account the various essential elements for the traceability: the product in its various states (raw materials, semi-finished, finished), various operations realized on the product, resources used (human and machines), its localization, and interactions between the product and its environment carried out on the basis of whole of service, in the objective to obtain a functionality (a help, a support, an action and product information)

To identify what we must trace, we took inspiration of the Kim work's [9], adopting *the central entities* of their ontology and which are the couple "*TRU and the activity*». As we mentioned previously the *TRU* is a single entity from the traceability, and the elementary activity is an indecomposable primitive activity in under activities. It is thus a basic operation: for example a planning, provisioning, manufacture, delivery.

So, given our generic modeling of the product based on a process approach, in order to show the interactions between the product and its environment. We adopted for the choice of the primitive activities of the TOVE project ontology, the logic of the SCOR model (Supply Chain Operations Reference) [17], which subdivides the supply chain in five great processes: *Plan, Source, Make, Deliver* and *Return*.

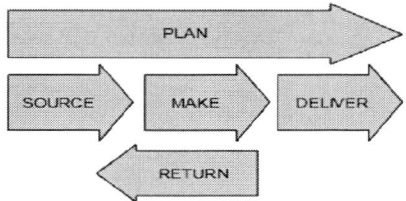

Fig.1. basic processes of model SCOR

Among the five basic processes of SCOR model, we kept only three: *Source* which represents a *reception* operation that we named *"Reception"* in our data model, a *Make* operation named *"Fabrication"*, and a *Deliver* which represents an expedition operation, named in our model *"Expedition"*. *Plan* was isolated of our model. Indeed, the traceability always rests on what was has been achieved without worrying about what was envisaged. It also did not seem necessary to us to create in our generic model a return management operation. Indeed, we consider the returns management as a manufacturing process to whole share where the turned over batch constitutes a new batch in entry .

Now so that the model can take into account the various elements necessary for the traceability, we have supplemented it by allotting to him the human and material resources *"Operateur* and *Ressource"*, the localization *"Site"*, and the interactions between the product *"Produit"* and its environment realized on the basis of whole of service *"Service"* represented by a material resource and/or software *"Systeme_ambiant"* which offers a characteristic functionality locally available or through a network, it can act of a handling or request for information of the product carried out by the process, or of an initiated interaction by the product: car-declaration, request for intervention, localization…

In our approach the product is seen as a material object, characterized and identified by information of intrinsic nature such as size, weight, volume, marks..., formalised in our conceptual model by "info_locale" UML class.

With the physical product can be associated a whole of resources and complementary information such as the specifications of the product: nomenclature composing, manufacturing range, place and storage conditions..., constituting thus an extension of the product " Extension_info", this extension constitutes an artefact of the product represented by an information system; resources distributed between the physical product itself and of the distant resources called ambient systems, the connection of the physical product to its extension, will be realized by automatic identification technologies and systems in network.

To respect our stated objective in the introduction, aiming to propose an modeling approach aiming to transform the product into communicating product "intelligent", equipped with a communication capacity, action perception, and management of its own information locally available or through a network [15].

We have integrated in our model a sensor class *"Capteur"* which represents an apparatus, sensors which gather the significant information on the immediate environment of the object and memorized by the object, and of an actuator *"Actionneur"* which is an apparatus equipping an object and which can generate an action so as to modify the state, the behaviour, the object environment; thus characterizing, perception and action capacities. We added to All that a *Trace Exchange* class *"Trace_Echange"* for the transactions traceability on the product through the Service class by recording the values of each variable state of product for each phase of its life cycle.

For the link between *"produit"* and *"Operation"* class which holds it, let us note that there are two treatment types, an operation can produce or consume a product for example a reception is a consumer operation, producing for expedition and both for manufacture according to whether it is of an assembly or a dismantling from where double association between the *"produit"* and *"Operation"* class in diagram UML (*Unified Modeling Language*) which formalizes the class and the model objects, constituting the skeleton of our generic model for the traceability like continuation:

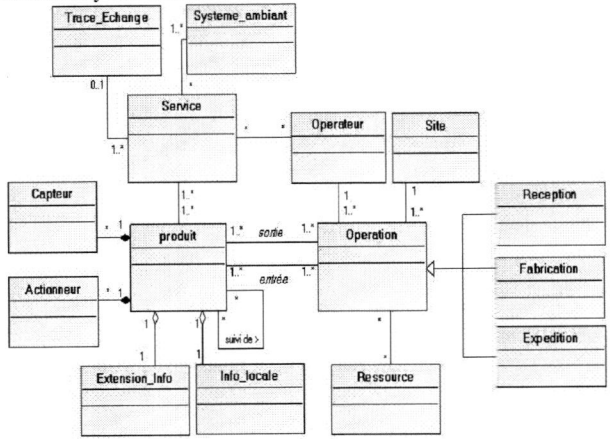

Fig.2. UML class diagram of the generic model

The realization of the services in the mechanisms of interaction requires an exchange and communication infrastructure between the actors. The intrinsic mobility of the products in the environments crossed during their life cycles gets closer the problematic of ambient informatics [10]. Several mechanisms of provision and discovered services are listed in the current activities of research [1]:

Method "centralized pull" Any actor entering a field of devolved work, must record the whole of the services which it can realized near a centralized manager of services. An actor requiring a service carries out a request for this service near the centralized manager. The answer's manager provides him the identification of the provider of service and his conditions of access. The initial actor carries out his request for service near the designated provider of service.

Method "distributed pull" actor requiring service broadcasted in diffusion, "with the round", a request for this service. A qualified provider receiving, available and accessible from this service answers the transmitter then.

Method "distributed push" each actor regularly diffuses with the round a list of its available services. All the actors present in an environment or field of devolved work record these lists.

For our traceability problem, we chose the architecture type *"centralized pull"*, because it has an unquestionable potential of implementation with automatic identification technologies.

The following sequence UML Diagram presents the all interactions between objects during an execution.

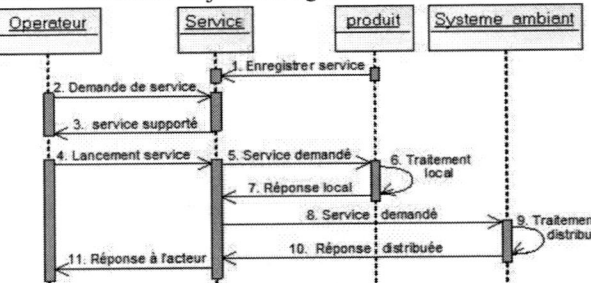

Fig.3. Sequence UML Diagram representing the interactions "operator/product"

5. CASE STUDY

As we announced in the introduction, to validate and test, in an industrial context "agroalimentary industry", the feasibility and the applicability of our product generic model for the specification of the whole of the data and information relating to the product, we were based on a work of audits realized within the framework of a *MAS* Masters of Advanced Studies detailing the chain of supply of manufacture and distribution of the prepared meat products of AUCHAN group.

The problem that we study comes from the manufacturing process of dry sausage of the suppliers of the AUCHAN group whose description of the stages of production is shown in the following figure:

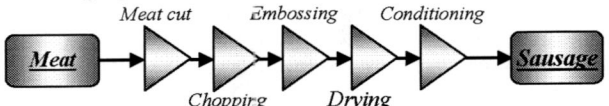

Fig.4. Production stages of dry sausage

To produce dry sausage, the beef meat is cut out in "components" like ham, the shoulder, the hard fats, the bulge... then; these components are mixed and chopped to obtain a paste (or fray). These paste batches are then used for the production of various sausages.

We applied our modeling approach for an industry with productions "by batch"; at each stage of the production process corresponds of the models of the identified batches so there is not other identical batch from the traceability view point. The models of the batches can be linked like bricks to model the complete supply chain. We could identify for our example 11 different batches:

1. *Meat reception batch:* It gathers the meat parts of the same truck, of the same supplier, the same type of articles (Shoulders, ham...) and of the same quality (Certified, French...).
2. *Relegated Hams Batch:* It gathers the relegated ham executives at the same day of production. Indeed, the hams intended for the manufacturing process of raw ham which does not satisfy the quality standards, are displaced to integrate the production chain, via cutting.
3. *Cutting Batch:* it gathers the cutting parts of the same reference exits of the same delivery batch.
4. *Freezing Batch:* It gathers the pallets of articles of the same type, frozen and resulting from by-products cut batches.
5. *Meat reception batch:* It corresponds to a cutting batch or freezing of the same type of article, the same quality and of a truck.
6. *Consumable Batch:* This batch corresponds to a consumables reception: a delivery of a supplier for a given article.
7. *Chopping Batch:* A ton of fray corresponds to 5 carriages of the same paste (mixture of chopped meat and spices) data
8. *Bowels Batch:* This batch corresponds to a bowels reception: a delivery of a supplier for a given article.
9. *Embossing Batch:* This batch corresponds to the whole of the embossed products starting from the same paste and on a given embossing line.
10. *Conditioning Batch:* This batch is consisted of conditioned sausages of a given article code and a day of given embossing.
11. *Expedition Batch:* This batch corresponds to a customer order, which gathers several products types in general.

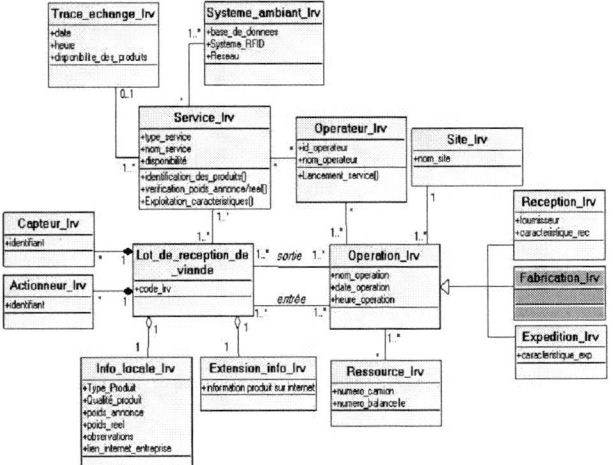

Fig.5. UML class diagram of a meat reception batch *"lrv"*

Above we presented a UML class diagram of the specialization of our generic model in the case of a *Meat reception batch* "lrv".

By the diagram presented, we describe the interaction elements of the product and its environment on the basis of centralized service. The services interface offers to the operator: products identification, exploitation of the products characteristics, the checking of the actual weight compared to the announced weight of the products. In the *info_locale_lrv* class representing the data attached to the product we indicated the intrinsic properties characterizing a meat reception batch (type product, quality, weight, observations as well as bond Internet of the company). The limited capacity of the information attached to the product, is extended by information relating to the batch in accessible subject via an access to a URL reference. Whereas in the Resource_lrv class we find the identifier of the great meat parts of the same truck and the identifier of the truck which are one of the key references for the traceability of a reception batch. In the same way for the *"Operateur_lrv"* class and *"Site_lrv"* which is used to identify the operator handling the product and its localization, we find unique identifiers.

Finally, for the operations on our batch, we could identify two types of operation, because none manufacture task carried out on this batch type.

Then, we have made in the same way for the 10 remaining batches while integrating the characteristics of each one in its specializing model. The link of the batches models the ones to other have allowed to form the whole of the production line of a dry sausage.

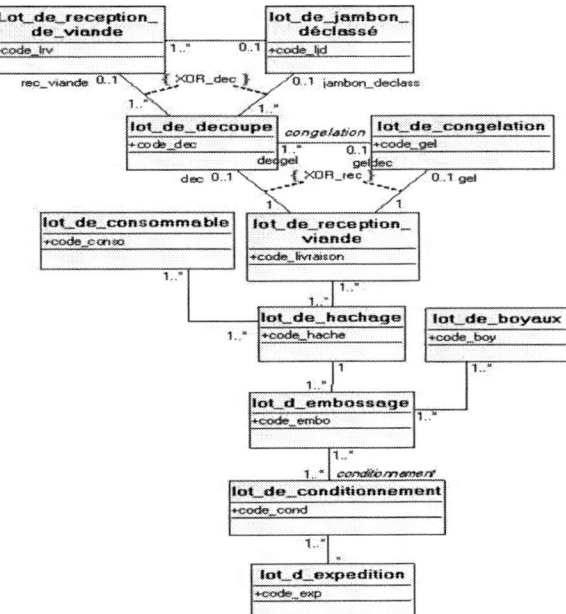

Fig.6. Class diagram of modeling the batches sequence

In this last diagram we established OCL constraints (Object Constraint Language) [14] which is a formal language for the constraints expression and requests applied to UML diagrams. In order to formalize the models semantics and so that the model can take into account the real conditions of manufacture while ensuring the sequence batches models.

For example to link between a Cutting batch *"lot_de_découpe"*, a Meat reception batch *"lot_de_reception_de_viande"* and a Relegated Hams Batch *"lot_de_jambon_déclassé"*, we established a constraint "*XOR_dec* " so that a cutting batch is connected, either with a Meat reception batch only, or with a Relegated Hams Batch only.

Context lot_de_decoupe
inv: rec_viande -> notEmpty() implies jambon_declass -> isEmpty()
         xor
         jambon_declass -> notEmpty() implies rec_viande -> isEmpty()

Likewise for constraint "date_embossage" which verifies that each hash batch with the same date embossed finds himself in the same embossing batch.

Context lot_de_conditionnement
Inv : Lot_embossage.allInstances() ->
       forAll(L1, L2 | date_embossage.L1<>date_embossage.L2
       implies (Lot_de_conditionnement.allInstances() ->
       forAll(C1,C2 | C1<>C2 implies C1.L1 and C2.L2))

Once the built models, we implemented them in the form of a data base constituting the hard core of an information system for the traceability management, for the coherence validation of the system we used a fictitious data set to identify the gaps and the lacks of information of certain paramount information for the correct operation of the system.

6. CONCLUSION AND PROSPECTS

The traceability presented as the miracle solution at the more and more numerous of safety food crises, it is a subject which was never as much topicality. Therefore of this report, the product representation in its environment, interacting with the processes, operators and the products of his entourage could enable us to set up a good traceability solution, according to the product since its creation until its final destination. We thus presented, in this paper our modeling step which led to the construction of a generic model of the product which takes into account the various essential elements for the traceability: the product in its various states (raw materials, semi-finished, finished), various realized operations on the product, resources used (human and machines), its localization, and interactions between the product and its realized environment on the basis of whole of service, in the objective to obtain a functionality (a help, a support, an action and of information relating to the product)

Generic modeling was applied in a context of food industry for the manufacturing process of dry sausage of the AUCHAN group, the validation stage of the model was realized by testing its coherence initially, and then, by making sure seizure and data research followed the built conceptual model well.

In a following phase, the objective is to integrate the information system developed in a real situation (experimental platform), or on a simulation process, this to evaluate the real capacity of the system to be integrated in a real case of company and thus to fix its limits.

REFERENCES

[1] Duda A., 2003 "Ambient Networking". Smart Objects Conference, Grenoble, France.

[2] Garetti M., Macchi M., Van De Berg R., 2003 "Digitally supported engineering of industrial systems in the globally scaled manufacturing", IMS-NoE SIG 1 White Paper, Milano.

[3] GENCOD. 2001 "La traçabilité dans les chaînes d'approvisionnement, de la théorie à la pratique". Issy-les- Moulineaux: GENCOD 2001, pp. 98.

[4] Gouyon D., 2004 "Contrôle par le produit des systèmes d'exécution de la production apport des techniques de synthèse". Thèse de doctorat de l'Université Nancy-I.

[5] Grady J.O., "System Requirements Analysis". McGraw-Hill Inc. 1993.

[6] Helander M.G., Jiao J., 2002 "Research on e-product development (ePD) for mass customization." Technovation 22, pp. 717-724. Pergamon.

[7] ISO. "Norme ISO 8402 : Management de la qualité et assurance de la qualité". 1994, European standard.

[8] Kärkkäinen M., Holmström J., Främling K. and Artto K., 2003 "Intelligent products- A step towards a more effective project delivery chain". Computers in Industry Vol. 50, pp. 41-151.

[9] Kim H.M., Fox M.S., and Gruninger M., 1995 "An ontology of quality for enterprise modelling". 4th Workshop on Enabling Technologies: Infrastructure for Collaborative Enterprises, IEEE Press, pp. 105-116.

[10] KINTZIG C., POULAIN G., PRIVAT G., FAVENNEC P.N., "Objets Communicants. Collection scientifique et technique des télécommunications". Ed Hermès, 2002.

[11] Koestler A., 1967 "The ghost in the machine". ISBN 0-14-019162-5.

[12] McFarlane D., Sarma S., Chirn J.L., Wong C.Y., Ashton K., 2003 "Auto ID System and Intelligent Manufacturing Control". Engineering Applications of Artificial Intelligence, Vol.16, pp. 365-376.

[13] Moe T., 1998 "Perspectives on traceability in food manufacture". Trends in Food Science & Technology, Vol.9, pp. 211-214.

[14] OMG, Inc."Object Constraint Language 1.1 Specification". 1997

[15] Römer K., Schoch T., Mattern F., Dübendorfer T., 2003 "Smart Identification Frameworks for Ubiquitous Computing Applications". IEEE Pervasive Computing and Communications.

[16] Shinghal R., 1992 "Formal Concepts in Artificial Intelligence Fundamentals". London: Chapman & Hall Computing.

[17] Stephens S., 2001 "Supply Chain Operations Reference Model Version 5.0: A New Tool to Improve Supply Chain Efficiency and Achieve Best Practice". Information Systems Frontiers, Vol.3, pp. 471-476.

[18] Terzi S.2005"Elements of Product Lifecycle Management: Definitions, Open Issues and Reference Models". Thèse de doctorat de l'Université Nancy-I.

[19] Van Dorp C.A., 2004 "Reference-data modeling for tracking and tracing". These Wageningen Unieriteit, ISBN 90-8504-005-1.

E-maintenance Scenarios Based on Augmented Reality Software Architecture

S. Benbelkacem *, N. Zenati-Henda *, M. Belhocine*

*Advanced Technologies Development Centre - CDTA,
20 Août 1956, Baba Hassen, 16300 Algiers, Algeria.
(e-mails: sbenbelkacem@cdta.dz, nzenati@cdta.dz, mbelhocine@cdta.dz)*

Abstract: This paper presents architecture of augmented reality for e-maintenance application. In our case, the aim is not to develop a vision system based on augmented reality concept, but to show the relationship between the different actors in the proposed architecture and to facilitate maintenance of the machine. This architecture allows implementing different scenarios which give to the technician possibilities to intervene on a breakdown device with a distant expert help. Each scenario is established according to machine parameters and technician competences. In our case, a hardware platform is designed to carry out e-maintenance scenarios. An example of e-maintenance scenario is then presented.

Keywords: Augmented reality, Global maintenance software architecture, E-maintenance scenarios.

1. INTRODUCTION

The information technology development in the last years led to the emergence of new disciplines. In this case, human beings always interested to enhance its action environment by additional information which can be more reactive and more effective.

Through computer tools, the user aims to combine information from the computer and the physical environments in order to satisfy different applications. The combination of these two environments yields to a new research area called "Augmented Reality".

Augmented Reality (short: AR) is a new way of human-computer-interaction, where virtual objects are added to real scenes provided by a video camera in real time (Mackay, 1996), (Mackay, 1998). They are inserted in the right positions and complement the real image. The digital information merges with the user's environment so that the user can perceive currently important information directly where it is needed.

AR is derived from Virtual Reality (short: VR) in which the user is completely immersed in an artificial world. In virtual reality systems, there is no way for the user to interact with objects in the real world. Using AR technology, users can thus interact with a mixed virtual and real world in a natural way (Mackay, 1998).

Several research fields such as robotics (Drascic, 1993), games (Szalavari, 1998), architecture (Thomas, 2000), maintenance (Da Dalto, 2002), (Schwald, 2003) and medicine (Bockholt, 2003) have a particular interest in the application of this new paradigm and each application adapts the AR concept to its needs and develops its principles design.

Augmented reality application is an excellent domain for maintenance task in industrial environment. Indeed, it allows the user to see computer generated virtual objects superimposed upon the real world through output devices. The technician can thus interact with the virtual world and have additional information which is not directly accessible in the working environment (Kustaborder, 1999). These information can be a set of maintenance task instructions given in form of text, image, video or audio augmentations.

Several projects such as Arvika (Reicher, 2003), Starmate (Da Dalto, 2002), Sear (Goose, 2003) and Amra (Didier, 2005), have been developed in the last decade. Each one is based on its own developed architecture.

So, in this work, we propose a global architecture based on AR paradigm in the context of industrial maintenance applications. The aim is not to establish a vision system of AR system, but to carry out a software platform that shows the role of each operator (technician, expert, AR system, server and machine) in a maintenance operation and the relationship between them.

This paper is organised as follows: section 2 gives our global maintenance architecture based on augmented reality paradigm. Section 3 presents the AR hardware platform corresponding to software architecture. Section 4 presents an overview of the industrial e-maintenance scenarios and describes the different corresponding tasks. An example of scenario is then presented. In the last section a conclusion is given.

2. GLOBAL MAINTENANCE ARCHITECTURE BASED ON AUGMENTED REALITY CONCEPT

The design of Augmented Reality architectures has become a field of investigation in industry and research centres. So, several architectures of Augmented Reality systems have been implemented in various fields (Schwald, 2001), (Da Dalto, 2002), (Didier, 2005). However, in the maintenance domain few works exist and several ones will be developed in the future.

In our laboratory, we develop a global architecture called ARIMA (Augmented Reality and Image processing in Maintenance Application) in order to give solution to technician during maintenance operation.

ARIMA concept is based on the possibility to show the interactions that occur between different actors when technician intervene on the machine. For this reason, we study the relationship between the expert, the technician, the AR system and the machine.

We are focused to establish a set of intervention scenarios from the proposed architecture. The design of the AR vision system is not included in this paper. In our case, we show an application of maintenance scenarios based in AR concept.

The aim of ARIMA project is to use augmented reality to facilitate the technician intervention. The important aspects of this project may be quoted in two points:

1. Providing to maintenance agents assistance by sending relevant information or "augmentations", directly aligned with their workstation. These information are generated from computer and integrated in the view field of the user.

2. Providing help to inexperienced maintainers, thus allowing them to be trained on site.

The aim of our work is not only to achieve a technical device (a set of software animating the hardware platform), but also allow to identify technician activity and his task, the used tools and the recommended scenarios for different types of failures that may occur.

Fig. 1 shows a global structure of augmented reality system aimed to e-maintenance. In this case, the technician operates on the machine through output devices which consist of PDA and/or eye glass and/or Pc tablet.

The e-maintenance operations are given as follows:

First, the supervisor sends to technician the repair order. At the same time, the operator checks his schedule which contains all information about the machine and the intervention history.

The maintenance scenario is then sent in "augmentation" form to the technician through output devices.

When the intervention is achieved, the technician sends the confirmation to supervisor and fills out a maintenance report which indicates the breakdown nature and the intervention type.

Through lack of technician's competence, he collaborates with an expert. The main task of the expert is to guide the technician to perform his intervention.

3. HARDWARE PLATFORM COMPONENTS

This section presents the AR hardware architecture in order to implement e-maintenance scenarios (Fig. 2).

This platform developed in our laboratory consists of two main modules. The *processing unit* module and the *Human Computer Interface (HCI)* module.

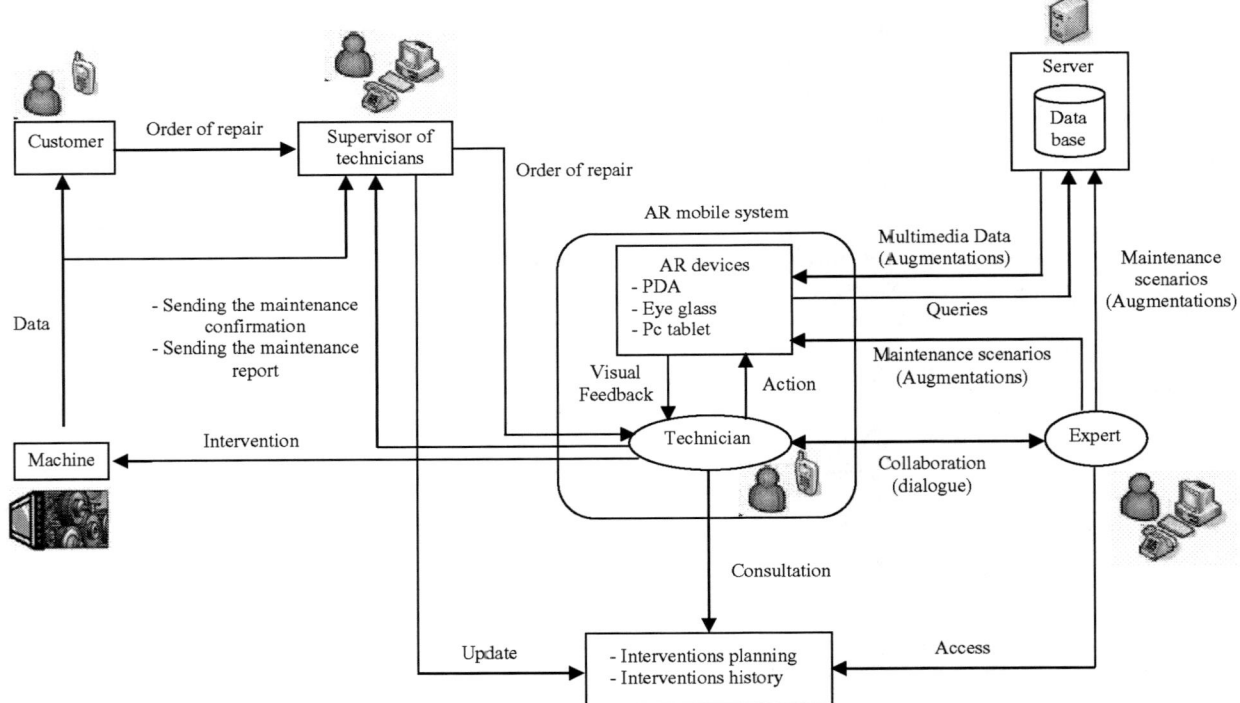

Fig. 1. Global diagram of ARIMA software architecture

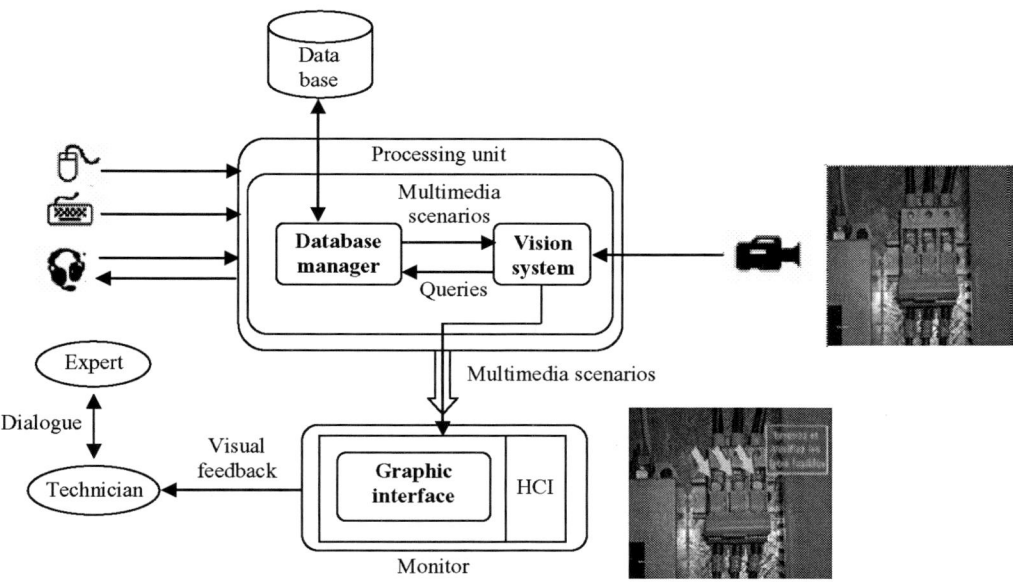

Fig. 2. Overview of AR material platform

The *processing unit* module performs different treatments on the scene image acquired by a camera. This module consists of two packages:

The vision system package: includes various operations of image processing.

First, the vision system receives an image from a camera. Then, it sends queries to the server in order to get multimedia scenarios. These scenarios are inserted in original image to obtain the augmented scene which is sent to the technician's screen.

The database manager package: it manages the database which contains the different augmentations and chooses the adequate one to add it in the real scene image.

The *HCI* module is a GU interface which creates a link between the user and the output device. This module consists to mix the real scene and the virtual object on a monitor.

4. MAINTENANCE SCENARIOS

In a global context, the maintenance scenario is described as follows:

1. The operator connects to the application and reaches his planning where he finds all the machine characteristics, the interventions history and the last interveners.

2. The technician proceeds to maintenance operation. He can use both hands, while consulting the appropriate documentation and the different multimedia augmentations.

3. If the technician has difficulties to repair the machine, he contacts the expert who accesses to the interventions history and guide the operator through "web" by sending graphical, oral and textual indications.

In our case, the technician dialogues with the expert through the following tools: the e-mail (used to send the breakdown machine reports), the on-line dialogue and the transfer files (used to send augmentations).

4. After performing intervention operation, the technician sends a maintenance report to the supervisor who updates the intervention planning.

4.1 Example of scenario

We present in this section an example of scenario done in a mechanical workshop of our laboratory. The aim is to carry out a maintenance operation of a machine in failure.

The device repair is a digital milling machine (Lagun future 1400: Heidenhain) shown in the Fig. 3.

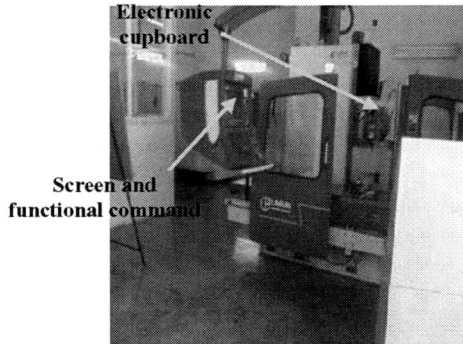

Fig. 3. Digital milling machine.

The e-maintenance scenario is summarized as follows:

Step 1: a web page of e-maintenance network is established to allow customers to register online. During this operation, the system processes and stores all information of this registration (name, company, field, office, level, login and password of customers) in the database.

Step 2: the customers (technician or expert) type their login and password to access to the application. After each connection, each of them can intervene in its own session when expertise or solution is required.

Step 3: the technician performs his maintenance operation when he receives the alert. In this case, the server establishes a link between the expert and the technician. The technician contacts an expert by exchanging computer data which represent the steps of maintenance procedure in the form of multimedia augmentations.

The maintenance operation is done remotely using Internet connection. This connection is initiated by technicians who transmit the data via client – server architecture.

In this case, the expert receives the scene image. He adds the adequate increases and sends the resulting image to the technician.

All increases which are inserted by an automated process were recorded in the database server.

In the case of digital milling machine, a failure was found at the boot of the machine. Thus, an error message is displayed on the screen as shown in Fig. 4.

Fig. 4. Error message of failure

The technician tries to find a solution to this problem. He will access to e-maintenance site (E-maintenance Augmented Reality), and then download and install the application "Emaint-RA v1.0 (Fig. 5).

Fig. 5. Homepage of e-maintenance system

Then, the technician opens its session to access to the application (Fig. 6).

Fig. 6. The session for technician

After completing the failure report and sending it to the expert, they communicate. This communication is done by transferring files in the form of augmented scenarios from the expert to the technician in order to perform the intervention operation.

The dialogue achieved between the expert and the technician to solve the maintenance problem is described as follows:

Technician

Hello, a machine is breakdown and I can't identify the problem.

Expert

Hello, switch off the machine with the selector (Fig. 7).

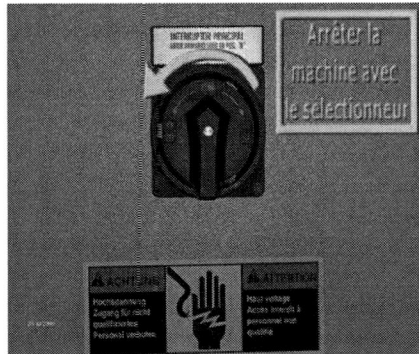

Fig. 7. "Switch off" augmentation

Technician

The machine is stopped successfully.

Expert

Insert the key and unlock the door (Fig. 8 and Fig. 9).

Fig. 8. "Insert the key" augmentation

Fig. 9. "Unlock the door" augmentation

Technician

The door is opened

Expert

Go now to the fuse box and pull the stem of the box, (Fig. 10).

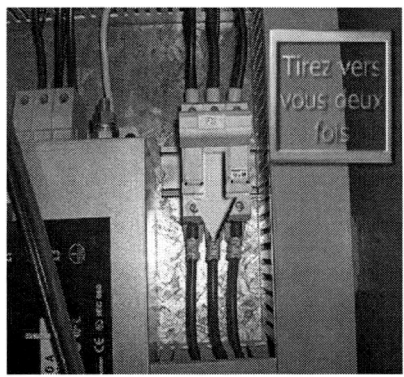

Fig. 10. "Pull the stem" augmentation

Technician

The fuse box is opened.

Expert

Remove and check one by one the three fuses (Fig. 11).

Fig. 11. "Remove and check the three fuses" augmentation

Technician

I found a defective fuse.

Expert

Change this fuse with another having the same characteristics. Close the fuse box and then close the door.

Technician

The machine is ready.

When the machine is repaired, the technician completes a failure report which contains information such as dialogue, transferred files, cause of the failure and summary solution. Then, the report will be sent to the supervisor to be archived in the interventions planning.

5. CONCLUSION

In this paper, a global architecture of e-maintenance application based on Augmented Reality concept is presented. In addition, the corresponding hardware platform is given in order to implement the intervention scenarios.

This approach is applied on a study case which consists in repairing a breakdown machine. The results show the advantages of implementing the AR architecture. Two major advantages are identified:

- The good assistance in maintenance by adding useful information in the augmentations form.
- The easier communication between the expert and the technician using TCP / IP network.

The Augmented Reality concept seems to be an effective tool to improve the efficiency of e-maintenance intervention.

REFERENCES

Bockholt, U., Bisler, A., Becker, M., Muller-Wittig, W., Voss, G. (2003). Augmented Reality for Enhancement of Endoscopic Interventions. *IEEE International Conference Virtual Reality*, pp. 97–101.

Da Dalto, L. (2002). Starmate: using augmented reality for maintenance, training and education. *In Virtual Reality International Conference (VRIC)*, Laval, France.

Didier, J.-Y., Roussel, D., Mallem, M., Otmane, S., Naudet, S., Pham, Q.-C., Bourgeois, S., Mégard, C. Leroux, C., Hocquard, A. (2005). Amra : Augmented reality assistance in train maintenance tasks. *Workshop on Industrial Augmented Reality (ISMAR'05)*, Vienna, Austria.

Drascic, D., Grodski, J.J. (1993). Defence Teleoperation and Stereoscopic Video. *DND workshop on Advanced Technologies in Knowledge Based Systems and Robotics*, Ottawa, Canada.

Goose, S. Sudarsky, S. Xiang Zhang Navab, N. (2003). Speech-enabled augmented reality supporting mobile industrial maintenance. *IEEE Pervasive Computing*, Volume 2, pp 65-70.

Kustaborder, J., Sharma R. (1999). Experimental Evaluation of Augmented Reality for Assembly Training. *2nd International Workshop on Augmented Reality (IWAR'99)*, San Francisco, USA.

Mackay, W.,E. (1996) Réalité Augmentée: le meilleur des deux mondes. *Review « La Recherche »*, n° 285, pp. 80-84.

Mackay, W., E. (1998). Augmented Reality: Linking real and virtual worlds – A new paradigm for interacting with computers. *In Proceedings of AVI'98, ACM Conference on Advanced Visual Interfaces*, l'Aquila, Italy, New York, ACM Press.

Reicher, T., Mac Williams, A., Brugge, B., Klinker, G. (2003). Results of a study on software architectures for augmented reality systems. *The Second IEEE and ACM International Symposium on Mixed and Augmented Reality*, pp 274 – 275.

Schwald, B. and all. (2001). STARMATE: Using Augmented Reality technology for computer guided maintenance of complex mechanical elements. *E-work and E-Commerce*, vol 1, pp17-19, IOS Press, Venice-Italy.

Schwald, B., De Laval, B. (2003). An Augmented Reality System for Training and Assistance to Maintenance in the Industrial Context. *Journal of WSCG'2003*, pp 425-432.

Szalavari, Z., Eckstein, E., Gervautz, M. (1998). Collaborative Gaming in Augmented Reality. *Virtual Reality Software and Technology Symposium*, pp. 195-204.

Thomas, H., Piekarski, W., Gunther, B. (2000). Using augmented reality to visualize architecture designs in an outdoor environment. *International Journal of Design Computing on the Net (DCNet)*, Volume 2.

Numerical approximation of null controllability for $1-D$ linear parabolic-hyperbolic equations

Ali SALEM *

*LIM Laboratory Ecole Polytechnique de Tunisie BP 743, 2078 La Marsa,
Tunisia (e-mail: ali.salem@fsb.rnu.tn).*

Abstract: In this work, we consider the numerical null controllability problem of a linear control parabolic 1-D equation which depends on two parameters, namely the viscosity and the transport coefficient term. We study the dependence, with respect to these parameters and the time of controllability, of the optimal control norm. The aim of these notes is to illustrate, by numerical simulations, an estimation of the controllability time as viscosity tends to 0.

Keywords: Controllability, Parabolic problem, Numerical approximation, Distributed control.

1. INTRODUCTION

Let $(\epsilon, T, M) \in (0, +\infty)^2 \times \mathbb{R}$. We consider the following parabolic linear control system

$$\begin{cases} u_t - \epsilon u_{xx} + M u_x = 0 & \text{in } (0,1) \times (0,T) \\ u(0,t) = v(t), u(1,t) = 0 & \text{on } (0,T) \\ u(x,0) = u^0(x) & \text{in } (0,1) \end{cases} \quad (1)$$

where the state is $u(.,t) \in L^2(0,1)$ and the control is $v(t) \in \mathbb{R}$.
Definition Let $T > 0$ and let's define, for any initial data $u^0 \in L^2(0,1)$, the set of reachable states

$$R(T; u^0) = \{u(T) : u \text{ solution of } (1) \text{ with } v(t) \in L^2(0,T)\} \quad (2)$$

An element of $R(T, u^0)$ is by definition a reachable state of (1) in time T by starting from u^0 with the aid of a convenient control v.
Definition System (1) is null controllable in time T if, for every initial data $u^0 \in L^2(0,1)$, the set of reachable states $R(T; u^0)$ contains the element 0.
It is well known that one of the most important properties of the parabolic equation is its regularizing effect. The solutions of (1) are in $C^\infty(0,1)$. Hence the restriction of the elements of $R(T, u^0)$ to $(0,1)$ are function of C^∞ class. We have to exclude the possibility of exact controllability.
It is already known that system (1) is controllable to the null final state at time $t = T$. This means that, for every $u^0 \in L^2(0,1)$ and for every $(\epsilon, T, M) \in (0, +\infty)^2 \times \mathbb{R}$, there exists $v(t) \in L^2(0,T)$ such that the (weak) solution of (1) satisfies $u(T,) = 0$. This result is due to Fattorini and Russell [1]. See also Fursikov-Imanuvilov [2] and Lebeau-Robbiano [5] for parabolic control systems in dimension larger than 1. We are interested to study the dependance of the cost of the null controllability of system (1) with respect to the three parameters ϵ, T, and M.
Definition For $u^0 \in L^2(0,1)$, we denote by $U(\epsilon, T, M, u^0)$ the set of controls $v \in L^2(0,T)$ such that the corresponding solution of (1) satisfies $u(.,T) = 0$. Next, we define the

quantity which measures the cost of the null controllability for system (1):

$$K(\epsilon, T, L, M) = \sup_{\|u^0\|_{L^2(0,1)} \leq 1} \{min\{\|u\|_{L^2(0,T)} : u \in U(\epsilon, T, M, u^0)\}\} \quad (3)$$

The goal of this work is to give an answer of the following conjecture
Conjecture 1. Is there T^* such that for all $T > T^*$

$$K(\epsilon, T, M) \to 0 \text{ as } \epsilon \to 0^+ \quad (4)$$

Remark 2. For $\epsilon = 0$ the control system (1) is controllable in time T if and only if $T > 1$. This result is due to D. Russel [3].

In [4] J.M. Coron and S. Gerrero have studied the behavior of $K(\epsilon, T, L, M)$ as $\epsilon \to 0$. It is natural to look at the limits of trajectories of the control system (1) as $\epsilon \to 0$. This is done in the following theorem
Theorem 3. [4] Let a be the positive constant

$$a = 4.3 \quad (5)$$

- If $T < 1/M$, one has

$$\lim_{\epsilon \to 0^+} K(\epsilon, T, M) = +\infty \quad (6)$$

- If $T \geq a\frac{1}{M}$

$$\lim_{\epsilon \to 0^+} K(\epsilon, T, M) = 0. \quad (7)$$

Remark 4. In theorem (3) the constant $a = 4.3$ is not optimal. Hence, We are interested on the following question: What is the minimal time for which the system (1) optimal control norm approaches zero when viscosity approaches zero. In the next section, we will prove numerically that this constant can be improved. We will study this problem numerically. The problem (1) is discretized using the difference finite method. A numeric scheme is used as A. Lopez and E.Zuazua [7]. The H.U.M method is implemented to find the control. We find a better controllability time estimation at zero, at witch the system optimal control time approaches zero when viscosity approaches zero.

The paper is organized as follows: In section 2 we describe the numerical algorithm based on HUM for boundary controllability of the parabolic linear control system (1). In section 3

---

* The author would like to acknowledge Prof Azgal Abichou and Prof Jean-Michel Coron for several interesting discussions, helpful comments and for their encouragements.

we describe the numerical scheme that has been implemented. In section 4 we present several numerical simulations and we investigate the relations between the time T needed to control and the control cost as $\epsilon \to 0^+$.

2. THE NUMERICAL ALGORITHM BASED ON HUM

2.1 A short review on controllability

Let $T > 0$ fixed. Consider the linear control system
$$x_t(t) = Ax(t) + Bu(t); \quad (8)$$
where $x(t) \in \mathbb{R}^n$, A is a $(n \times n)$ matrix, B is a $(n \times m)$-matrix, with real coefficients, and $u(.) \in L2(0;T;\mathbb{R}^m)$.
Let $x_0 \in \mathbb{R}^n$. The system (8) is controllable from x_0 in time T if, for every $x_1 \in \mathbb{R}^n$, there exists $u(.) \in L^2(0;T;\mathbb{R}^m)$ so that the corresponding solution $x(.)$ of (8), with $x(0) = x_0$, satisfies $x(T) = x_1$. It is well known that the system (8) is controllable in time T if and only if the matrix
$$\int_0^T e^{(T-t)A} BB^* e^{(T-t)A^*} dt \quad (9)$$
called Gramian of the system, is nonsingular (here, M^* denotes the transpose of the matrix M). In finite dimension, this is equivalent to the existence of $\alpha > 0$ so that
$$\int_0^T \|B^* e^{(T-t)A^*}\psi\|^2 dt \geq \alpha \|\psi\|^2, \quad (10)$$
for every $\psi \in \mathbb{R}^n$ (observability inequality).

2.2 Formulation of the problem

In this section we set $M = 1$, we will study from a numerical viewpoint the following null boundary controllability problem Given $T > 0$, $u^0(x) \in L^2(0,1)$ and $\epsilon > 0$ we search for a control function $v \in L^2(0,T)$ such that the solution u of the boundary initial-value problem
$$\begin{cases} u_t - \epsilon u_{xx} + u_x = 0 & \text{in } (0,1) \times (0,T) \\ u(0,t) = v(t), u(1,t) = 0 & \text{on } (0,T) \\ u(x,0) = u^0(x) & \text{in } (0,1) \end{cases} \quad (11)$$
satisfies
$$u(x,T) = 0. \quad (12)$$
In the present setting, this result is equivalent to an observability inequality for the adjoint equation:
$$\begin{cases} \varphi_t + \varphi_{xx} + \epsilon\varphi_x = 0 & \text{in } (0,1) \times (0,T) \\ \varphi(0,t) = \varphi(1,t) = 0 & \text{on } (0,T) \\ \varphi(x,T) = \varphi^0(x) & \text{in } (0,1) \end{cases} \quad (13)$$
More precisely, it is equivalent to the existence of a positive constant $C > 0$ such that:
$$\|\varphi(0)\|_{L^2(0,1)}^2 \leq C \int_0^T |\varphi_x(0,t)|^2 \quad (14)$$
J.L. Lions presented in [6] a constructive method allowing the calculation of control v, Hilbert Uniqueness Method (H.U.M). It gives the control with minimal norm L^2. In order to apply the HUM algorithm, we look at the following adjoint problem.
$$\begin{cases} \varphi_t + \varphi_{xx} + \epsilon\varphi_x = 0 & \text{in } (0,1) \times (0,T) \\ \varphi(0,t) = \varphi(1,t) = 0 & \text{on } (0,T) \\ \varphi(x,T) = \varphi^0(x) & \text{in } (0,1) \end{cases} \quad (15)$$
We minimize the following fonctionnelle J:
$$J(\varphi^0) = \frac{1}{2}\int_0^T \varphi_x^2(0,t)dt - \int_0^1 \varphi(x,0)u^0(x)dx \quad (16)$$
on the Hilbert space
$$H = \{\varphi^0 \in L^2(0,1) \text{ where } \varphi \text{ is solution of (15) and} $$
$$\varphi_x(0,t) \in L^2(0,T)\} \quad (17)$$
The functional J is continuous and strictly convex and, from the observability Inequality 14, is coercive in H. Thus it has a minimum. We now understand the functioning of HUM:

- Knowing the initial condition of system (11), we minimize J on the Hilbert space H.
- Find φ_0 the minimum of J then solve problem (15).
- The control is then $v(t) = -\varphi_x(0,t)$
- Verify that $u(x,T) = 0$.

2.3 The HUM based algorithm

The previous paragraph proved that the crucial point in characterizing the control v is to minimize the functional (16). By deriving J along ψ, we get:
$$J'(\varphi^0)\psi^0 = \int_0^T \varphi_x(0,t)\psi_x(0,t)dt - \int_0^1 \psi(x,0)u^0(x)dx \quad (18)$$
Where ψ is the solution of the following adjoint problem:
$$\begin{cases} \psi_t + \psi_{xx} + \epsilon\psi_x = 0 & \text{in } (0,1) \times (0,T) \\ \psi(0,t) = \psi(1,t) = 0 & \text{on } (0,T) \\ \psi(x,T) = \psi^0(x) & \text{in } (0,1) \end{cases} \quad (19)$$
We deduce a necessary condition so that (16) has a minimum:
$$\int_0^T \varphi_x(0,t)\psi_x(0,t)dt - \int_0^1 \psi(x,0)u^0(x)dx = 0 \quad (20)$$
We then write problem (20) on the following variational form:
$$\text{Find } \varphi \in H \text{ such that } a(\varphi,\psi) = L(\psi), \forall \psi \in H \quad (21)$$
where
$$a(\varphi,\psi) = \int_0^T \varphi_x(0,t)\psi_x(0,t)dt$$
and
$$L(\psi) = \int_0^1 \psi(x,0)u^0(x)dx$$
It is clear that a is a bilinear continuous symmetric coercive form. We deduce that problem (21) can be solved using the conjugate gradient algorithm. Problem (21) is a particular case of:

- Let V be a real Hilbert space for the scalar product $(.,.)$ and the corresponding norm $\|.\| = (.,.)^{\frac{1}{2}}$;
- Let a be a bilinear, continuous, symmetric and V-elliptic (or coercive) form;
- Let L be a linear and continuous form on V;

Hence, the problem of finding $u \in V$ such that $a(u,v) = L(v)$, $v \in V$ has a unique solution that we can obtain by the conjugate gradient algorithm.

3. PROBLEM DISCRETIZATION

In order to make a discretization of our problem, we are inspired by the method proposed by Lopez and Zuazua [7] for the heat equation. For each $N \in \mathbb{N}$, we consider a partition of $I = (0,L)$, $P = \{x_0 = 0,...,x_j = jh,...,x_N = L\}$ where $h = L/(N-1)$ is the space step. We denote $u_j(t)$ the approximation of the solution to the point x_j. Let δt be the time step and $u_j^n = u(n\delta t, jh)$. Then, using implicit scheme, the problem of semi-discretization is the following system of N ordinary differential equations :

$$\frac{1}{(\delta t)}(u_j^{n+1} - u_j^n) + \frac{1}{h}\left(u_{j+1}^{n+1} - u_j^{n+1}\right)$$
$$-\frac{\epsilon}{h^2}\left(u_{j+1}^{n+1} - 2u_j^{n+1} + u_{j-1}^{n+1}\right) = 0 \quad (22)$$
$$\text{for } j = 1,...N, \quad (23)$$
$$u_0^{n+1} = 0, u_{N+1}^{n+1} = v(n\delta t), \quad (24)$$
$$u_j(0) = u_j^0. \quad (25)$$

Let's denote A and B the matrices defined by,
$$A(i,i) = -2, A(i,i-1) = 1 \text{ et } A(i,i+1) = 1$$
$$B(i,i) = -1 \text{ et } B(i,i+1) = 1$$

Also, let Y^n be a vector from $\mathbb{R}^N$ of components u_j^n and let G^n be a vector from $\mathbb{R}^N$ defined by $G^n = (0,..,0,g(n\delta t))$. The previous scheme becomes:

$$\frac{1}{(\delta t)}(Y^{n+1} - Y^n) = \frac{\epsilon}{h^2}AY^{n+1} - \frac{1}{h}BY^{n+1} + (\frac{\epsilon}{h^2} - \frac{1}{h})G^{n+1}.$$

It is easy to see (with [8]) that this implicit scheme is stable. The discretization of adjoint system is:

$$\frac{1}{(\delta t)}(\varphi_j^{n+1} - \varphi_j^n) + \frac{1}{h}\left(\varphi_{j+1}^{n+1} - \varphi_j^{n+1}\right)$$
$$+\frac{\epsilon}{h^2}\left(\varphi_{j+1}^{n+1} - 2\varphi_j^{n+1} + \varphi_{j-1}^{n+1}\right) = 0 \quad (26)$$
$$\text{for } j = 1,...N, \quad (27)$$
$$\varphi_0^{n+1} = 0, \varphi_{N+1}^{n+1} = 0, \quad (28)$$
$$\varphi_j(0) = \varphi_j^0. \quad (29)$$

The semi discretization of functional (16) is:

$$J(\varphi^0) = \frac{1}{2}\int_0^T \left|\frac{\varphi_N(t)}{h}\right|^2 dt - \int_0^1 \varphi(x,0)u^0(x)\,dx \quad (30)$$

where $\varphi = (\varphi_1, ..., \varphi_N)$ is the solution of problem (26) having the initial condition $\varphi^0 = (\varphi_1^0, ..., \varphi_N^0)$ and $u^0 = (u_1^0, ..., u_N^0)$ is the initial condition of system (22).

4. NUMERICAL STUDY

These numerical experiments have been made with :

- The control in $x = 0$.
- The initial condition is $(sin(\pi x)^2)$.
- The gradient stopping condition is 10^{-12}.
- The space step discretization is $h = 10^{-2}$.
- The time space discretization is $dt = 10^{-2}$.

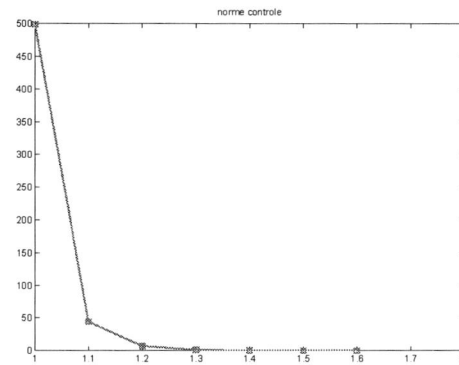

Fig. 2. L^2 norm of the boundary control: $\|g(t)\|_{L^2}$ with $\epsilon = 0.025$.

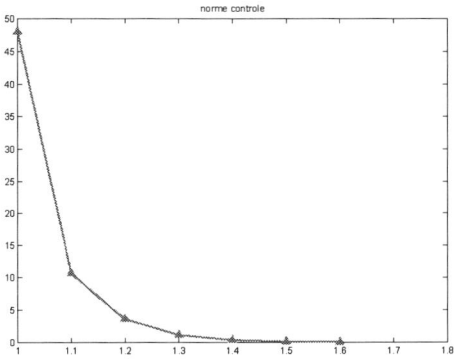

Fig. 3. L^2 norm of the boundary control: $\|g(t)\|_{L^2}$ with $\epsilon = 0.05$.

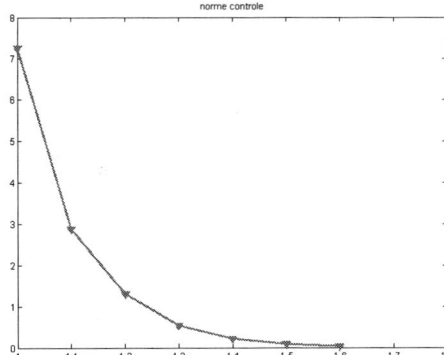

Fig. 1. L^2 norm of the boundary control: $\|g(t)\|_{L^2}$ with $\epsilon = 0.1$.

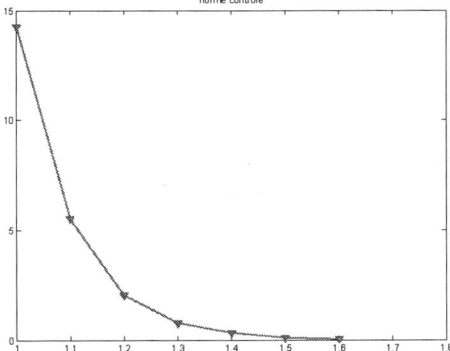

Fig. 4. L^2 norm of the boundary control: $\|g(t)\|_{L^2}$ with $\epsilon = 0.075$.

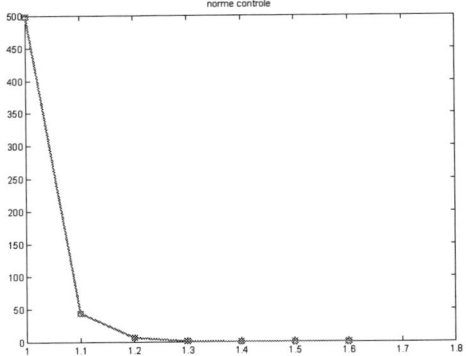

Fig. 5. L^2 norm of the boundary control: $\|g(t)\|_{L^2}$ with $\epsilon = 0.01$.

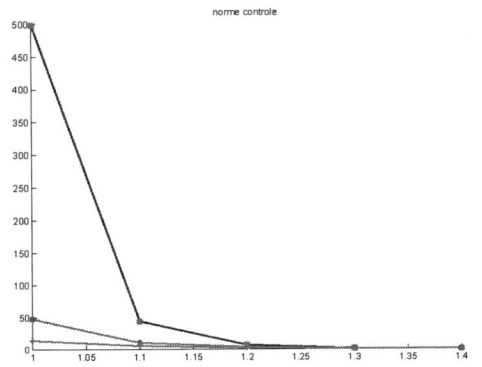

Fig. 6. L^2 norm of the boundary control: $\|g(t)\|_{L^2}$ with $\epsilon = 0.025$, $\epsilon = 0.05$ and $\epsilon = 0.075$.

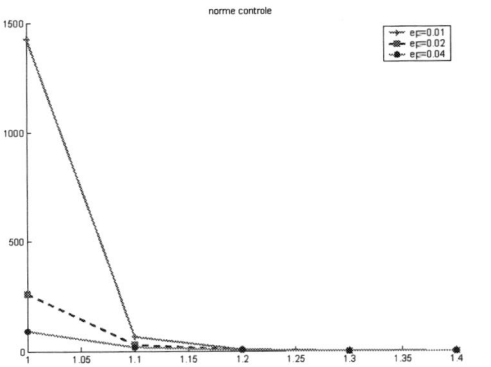

Fig. 7. L^2 norm of the boundary control: $\|g(t)\|_{L^2}$ with $\epsilon = 0.01$, $\epsilon = 0.02$ and $\epsilon = 0.04$.

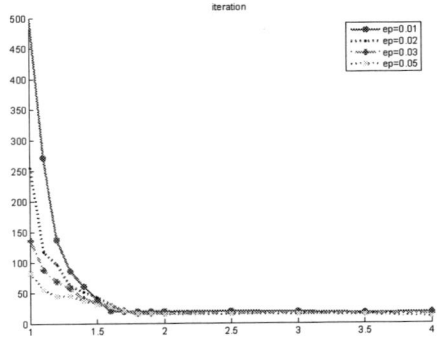

Fig. 8. Number of iteration with $\epsilon = 0.01$, $\epsilon = 0.02$ and $\epsilon = 0.03$ $\epsilon = 0.05$.

In Figures (1)-(7) we illustrate the graphs of $\|g(t)\|_{L^2}$ for different values of ϵ.

5. CONCLUSION

Our numerical results give some quantitative information on a result shown theoretically in [4]. When we consider the control on $x = 0$ we note that the control norm L^2 becomes small for $T = 1.1$ which improves the results obtained by J.M. Coron and S. Guerrero [4]. Obviously the constant $a = 4.3$ is not optimal. The above numerical experiments lead us to formulate the following conjectures: Does the following implication hold?

$$(\text{if } T > 1) \Rightarrow (\lim_{\epsilon \to 0^+} K(\epsilon, T, L, M) = 0) \tag{31}$$

REFERENCES

[1] H. O. Fattorini, D. L. Russell, Exact controllability theorems for linear parabolic equations in one space dimension. *Arch. Rational Mech. Anal.* 43 (1971), 272292.

[2] A. Fursikov, O. Yu. Imanuvilov, Controllability of Evolution Equations, *Lecture Notes 34, Seoul National University,* Korea, 1996.

[3] Russel D.L., Controllability and stabilizability theory for linear partiel differential equations: recent progress and open questions; *SIAM Rev. 20, no.4, 639-739.* (1978) MR MR508380 (80c:93032)

[4] Coron J.M., Guerrero S., Singular optimal control: A linear 1-D parabolic-hyperbolic example, *Asymptot. Anal.* 44 (2005), No 3-4, 237-257.

[5] G. Lebeau, L. Robbiano, Controle exacte de lequation de la chaleur (French), *Comm. Partial Differential Equations,* 20 (1995), p. 335356.

[6] J.L. Lions, Contrôlabilité exacte, Perturbations et Stabilization de Systems distribuites; Tome 1,Contrôlabilité exacte, *Collection de recherch en mathematiques apliquees,* 8, (Masson, Paris), (1988).

[7] A. López, E. Zuazua, Some new results related with the null-controllability of the 1-d heat equation, *Séminaire Equations aux Dérivées Partielles, École Polytechnique,* Paris, 1997-1998.

[8] R. Dautray, J.L Lions, Analyse mathématique et calcul numérique pour les sciences et les techniques; *tome 9 (Masson, Paris)*, (1984).

Interacting with a virtual tool on a real object

Bayart Benjamin * Didier Jean-Yves * Kheddar Abderrahmane **

* IBISC Laboratory, Evry, France (e-mail: bayart,didier@iup.univ-evry.fr).
** Centre National de la Recherche Scientifique (CNRS), Paris, France
(e-mail: kheddar@ieee.org)

Abstract: This paper presents an application enabling the interaction on a real object with a virtual tool. In order to interact within the real world, a real haptic probe is used to interact with the real object so that the user feel the interaction. Furthermore, through the use of a visual diminished reality process and a camera placed on the real scene, the real tool is visually replaced by the virtual one. Since, the real and virtual probes do not match necessarily, a model of the virtual tool is used to modify the haptic feedback, while visually, the virtual tool is modified relatively to the real forces measured. Eventually, proposing a mixed painting application, the painting, applied on the real object, i.e. when the user comes in contact with this latter, is visually displayed such that its form is computed from the virtual tool geometry while its size and intensity from the the real measured forces.

Keywords: Haptics Augmented Reality, Visual Diminished Reality, Interaction, Mixed Painting simulation.

1. INTRODUCTION

To interact within a mix world is not as straightforward as to interact either only in the real world or in the virtual world. The Augmented Reality (AR) domain includes a lot of applications where real and virtual environments are visually merged whereas few works address the haptic merging issue. Bayart *et al.* propose in [1] to interact with a composite object made of real and virtual parts. Our idea is to interact on a real object with a virtual tool. We illustrate our proposal with an application where virtual paint is applied on real objects. In this paper, we address the problem of keeping multi-modal coherence, between visual and haptic feedbacks.

2. A MIXED INTERACTION SYSTEM

We propose to interact in real time on a real object with a virtual tool. Using a tele-operated system, the user has the possibility to interact in real time with an object. At the same time, a camera being at the interplay site, he can visualize the result of the interaction with the chosen virtual tool. For instance, if manipulating a brush, some paint is added onto the object. The main issue is that since the virtual and the real tools do not necessarily correspond (in size, geometry or orientation), the force feedback does not consequently match to the visual one. The user can perceive an inconsistency between the visualization and the haptic feedback. For instance, there is difference if the real probe is rigid while the virtual model is deformable. The proposed solution is to add a coupling, taking into account a model of the virtual tool, in order to modify the visual and haptic information in accordance. Furthermore, a visual Diminished Reality (DR) process is added such that the real haptic probe is visually replaced with the virtual interaction tool.

The figure 1 presents the system, the first possibility (slashed lines), which is to send back directly the force of interaction to the user, and our solution, which adds a coupling, modifying the visual and haptic information.

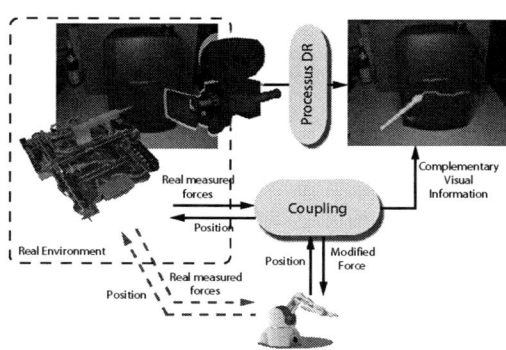

Fig. 1. The master device controls the slave device which interacts with the real object. The camera captures images from the real site. These images are processed accordingly to the diminished reality process and the interaction with the virtual tool is displayed. Real forces are sent back to the user. Adding the coupling, forces and visual information are modified accordingly to the model of the virtual tool to ensure a coherence between the modalities.

In order to ensure the correlation between the haptic and the visual feedback, we propose to add a virtual coupling. This virtual coupling considers a model of the virtual tool and enables:

(1) to modify the visual model of the virtual tool taking in account the real measured forces,
(2) to modify these forces accordingly to the model limitations, i.e. the forces sent back to the user are the result of the real measured forces and the deformation limitations of the visual virtual tool,
(3) to modify the visual result of the interaction accordingly to the model of the virtual tool and the real forces.

3. THE VISUAL DIMINISHED REALITY PROCESS

The term of Diminished Reality (DR) is dedicated to AR applications where real parts appearing in captured video or image sequences are removed or replaced.

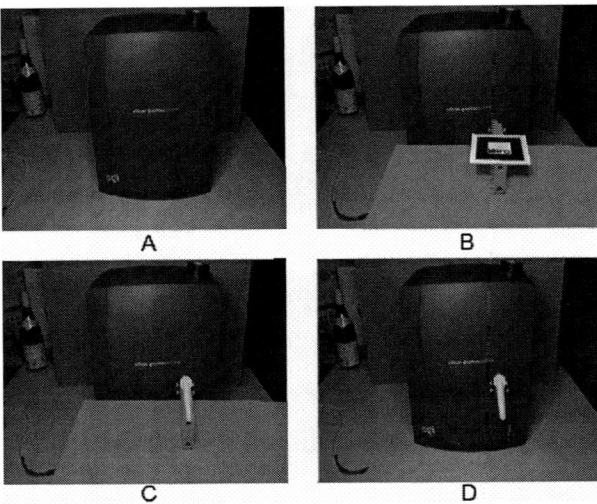

Fig. 2. The DR Process: the real haptic probe is visually replaced by the virtual tool, while the marker is used to find the initial position.

The goal of our DR process aims at removing the real haptic probe and replacing it with the chosen virtual tool. Similarly to [2], we chose to use only one camera and to take a maximum of background information at the start of the process, in order to replaced the removed parts. To improve the robustness and the speed of our technique, we add a chroma keying method as in [3], which enables a fast detection of the parts to be removed. Figure 2 illustrates the process, which works in four steps:

(1) at startup, an image of the background is registered as a reference (*A*)
(2) the haptic probe is placed on the scene (*B*)
(3) the marker is used to place the virtual probe, before being removed (*C*)
(4) the DR process is launch and the real haptic probe is detected (through the search of a specified color on the live image) and removed (i.e. the detected zone is replaced by the similar counterpart from the image of reference - *D*).

4. A MIXED PAINTING APPLICATION

We present an application of mixed painting such that the user can choose different brushes and apply strokes of paint on any kind of object. Three types of brushes are proposed, namely a rigid pencil, a deformable sponge brush and a deformable calligraphic brush. The models taken into account within the coupling are then respectively a rigid and a deformable unilateral beams.
Accordingly to the coupling, the brushes representations are visually modified as the beam model deforms due to the real measured forces.
The boundary of deformation are then used to modify the real forces sent back to the user. The forces sent back to the user correspond to the real ones modified accordingly to a virtual model. Hence, the forces sent back to the user are always equals to a fixed value for the pencil (rigid interaction) while for the two other deformable tools, the maximum forces are sent when the boundary (i.e. limit of deformation of the virtual tool) is reached.
Finally, for the strokes, which are the results of the interaction, their size is proportional to the real forces intensity while their orientation and their form are computed from the virtual tool geometry, as illustrated on figure 3.A. Figures 3.B,C,D show the final result, i.e. some virtual paint added on a real object.

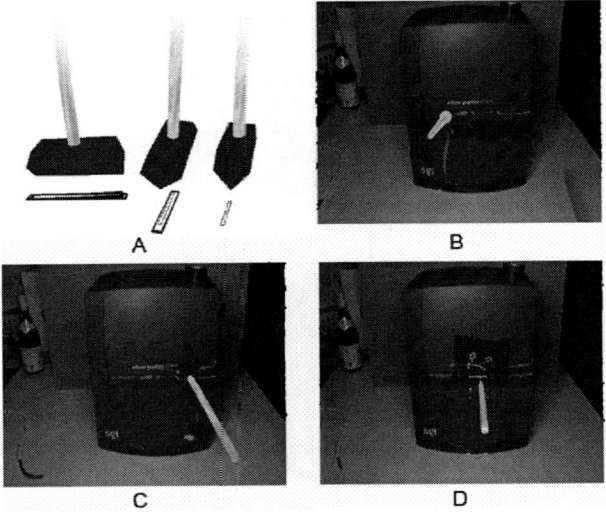

Fig. 3. A: The strokes of paint are computed accordingly to the model of the chosen tool and the intensity of the interaction forces. B,C,D: Applying some paint on a real object with different virtual brushes.

5. CONCLUSION

We present in this paper an application enabling interaction on a real object with different virtual tools. Through the use of a coupling, taking into account a model of the chosen virtual tool, the visual and haptic information are modified in order to ensure a consistency between the two. Furthermore, this solution offers the benefit that since no virtual model of the object is used, the collisions detection process is not required. An application of mixed painting is proposed as a result.

REFERENCES

[1] B. Bayart, A. Drif, A. Kheddar and J.-Y. Didier. Visuo-Haptic Blending Applied to a Tele-Touch-Diagnosis Application. In the proceedings of the 12th International Conference on Human-Computer Interaction 2007 (HCI'07), volume 14, pages 617-626, 2007.

[2] V. Buchmann, T. Nilsen and M. Billinghurst. Interaction with partially transparent hands and objects, In AUIC'05: Proceedings of the Sixth Australasian conference on User interface, pages 17–20, 2005.

[3] Y. Yokokohji, R.L. Hollis and T. Kanade. What you can see is what you can feel. -Development of a visual/haptic interface to virtual environment-. In IEEE Virtual Reality Annual International Symposium, pages 46–53, 1996.

A Robust and Non-Blind Watermarking Scheme for Gray Scale Images Based on the Discrete Wavelet Transform Domain

A. Bakhouche*. N. Doghmane**

*Département d'informatique, université Badji Mokhtar Annaba, 23000, Algérie
bakhouchamara@yahoo.fr
**Département d'électronique, université Badji Mokhtar Annaba, 23000, Algérie
ndoghmane@univ-annaba.org

Abstract: In this paper, a new adaptive watermarking algorithm is proposed for still image based on the wavelet transform. The two major applications for watermarking are protecting copyrights and authenticating photographs. Our robust watermarking [3] [22] is used for copyright protection owners. The main reason for protecting copyrights is to prevent image piracy when the provider distributes the image on the Internet. Embed watermark in low frequency band is most resistant to JPEG compression, blurring, adding Gaussian noise, rescaling, rotation, cropping and sharpening but embedding in high frequency is most resistant to histogram equalization, intensity adjustment and gamma correction. In this paper, we extend the idea to embed the same watermark in two bands (LL and HH bands or LH and HL bands) at the second level of Discrete Wavelet Transform (DWT) decomposition. Our generalization includes all the four bands (LL, HL, LH, and HH) by modifying coefficients of the all four bands in order to compromise between acceptable imperceptibility level and attacks' resistance.

Keywords: Robust Image Watermarking, Discrete Wavelet Transform, copyright, attacks.

1. INTRODUCTION

The rapid growth of the Internet and the great spread of digital media in nowadays, have created a pressing need for copyright protection owners. One area where this problem is most acute is digital image data. As a solution to this problem, digital watermarking has provided a powerful way to claim copyright protection.

In the watermarking the information data "watermark" is hidden in the original image. The digital watermark is an identification code carrying information (we apply an XOR function between logo and pseudo-random sequences) about the copyright owner that is embedded inside an image.

In general, a digital watermark technique must satisfy the following properties: The embedded watermark does should be perceptually invisible, secure and robust to various image attacks [6][18].

The watermarking of images can be grouped into two classes: transform domain methods [4][8][12][13][15][16], which embed the data by modifying the transform domain coefficients and spatial domain techniques [14]. The frequency domain approaches are the most successful for image watermarking. In the present work we use the frequency domain.

Many watermarking schemes for data hiding are developed recently, seeking a compromise robustness-invisibility. Some of them embed the same watermark in two bands (LL and HH) [9][10][19][22]. In this case the watermark is robust against some numbers of attacks (table 1). Other algorithms propose to embed the information data in the other two bands (LH and HL) [1][5][13]. But in this case the watermark is robust against some others numbers of attacks (Table 1.). Other techniques propose to modifying coefficients of all four bands [2].

Table 1. Robustness of the four bands against some attacks.

	Attacks
LL band	JPEG compression, blurring, adding Gaussian noise, rescaling, rotation, cropping, zoom
HH band	histogram equalization, intensity adjustment, and gamma correction
LH,HL bands	Histogram equalization, intensity adjustment, gamma correction, sharpening.
All bands	Collusion

2. PROPOSED WATERMARKING ALGORITHMS

Watermarking in the DWT domain can be split into two procedures: embedding and extraction of the watermark.

2.1 DWT Domain Watermarking

Two-dimensional DWT can be implemented using digital filters and down-samplers. Each level of decomposition produces four bands of data, one corresponding to the (low pass band) (LL), and three other corresponding to horizontal (HL), vertical (LH), and diagonal (HH) (high pass bands) [16].

Several digital filters, such as the Haar filter or the Daubechies filters, can be used to compute the DWT. In our experiments, the use of Haar wavelet is motivated by the following aspects [17] : a simple in conception, fast, can be calculated in place without a temporary array and it is exactly reversible without the edge effects that are a problem with other wavelet transforms.

The decomposed image shows a coarse approximation image in the lowest resolution, low pass band, and three detail images in higher bands. The low pass band can further be decomposed to obtain another level of decomposition.

This process is continued until the desired number of levels determined by the application reached. Fig 1-a- shows two levels of decomposition luminance layer of Lena to be watermarked.

2.2 Watermark Embedding

We use in this work gray scale image as cover image and binary image as watermark.

The diagram of our watermark embedding method is shown in Fig .1. , The watermarked image X' can be obtained by embedding watermark information W into the cover image X.

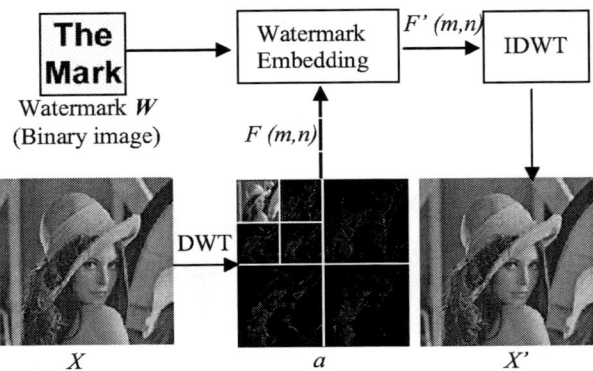

Fig. 1. Embedding diagram

Where:
X – Original image.
X' – Watermarked image.
The F(m,n) and F'(m,n) are respectively the original and modified DWT coefficients for each band (LL, LH, HL, HH).
m, n specify the widths and the heights of each band at the second level of DWT decomposition.
First, the original image is decomposed into several bands using the DWT with the pyramidal structure.
In our approach the decomposition is performed through two levels using the "Haar filter". The watermark, with a size of 128 x 128 pixels, is added in the four bands of the DWT.
The embedding procedure is performed according to the following formula:

$$X^k_{w,ij} = X^k_{ij} + \alpha_k W_{ij}, \quad i,j = 1,...,n/2 \text{ and } k = 1,2,3,4. \quad (1)$$

Where α is a scaling parameter of watermarking $W_{i,j} \in \{0,1\}$, $1 \le i,j \le n/2$.

The watermarked image obtained by applying the inverse of the discrete wavelet transforms (IDWT).

2.3 Watermark Extraction

In the procedure of the mark extraction, the two images, respectively the watermarked image and the original image, are both decomposed in two levels using DWT. It is assumed that the original image is known for mining (extraction not blind).

The extraction procedure is described by the formula:

$$W^*_{ij} = (X^{*k}_{w,ij} - X^*_{ij})/\alpha_k, \quad i,j = 1..n/2, \text{ and } k = 1,2,3,4.$$
$$\text{if } W^*_{ij} > 0.5 \text{ then } W^*_{ij} = 1 \text{ else } W^*_{ij} = 0. \quad (2)$$

Where $X^{*k}_{w,ij}$ are the coefficients of the DWT of the image marked (and can be attacked).
X_{ij} are the DWT coefficients of the original image.
Fig .2 shows the extraction phase of the watermark;

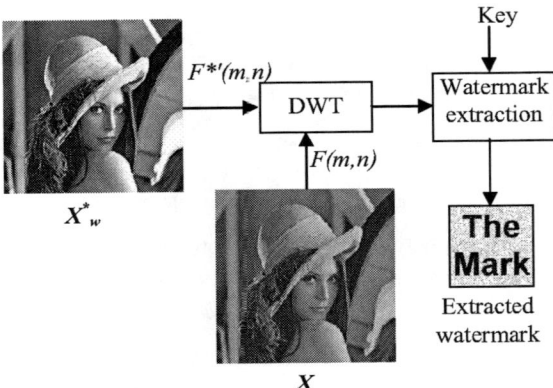

Fig .2. Extraction diagram

X - Original image.
X^*_w - watermarked image (and can be attacked).

3. EXPERIMENTS AND RESULTS

In order to confirm that the proposed image watermarking is effective, some numerical experiments need to be performed on some gray-scale standard images. The description of the experimental results, using the standard image "Lena" (512 x512 pixels, 8 bits/pixel), is shown in fig. 3 (a). The visual watermarks are 128x128 binary patterns. The scaling factor (α) for the LL sub band is 20, and the scaling factor for the other three bands is 3. In this experiment, the visual quality of watermarked and attacked images, is measured by using the Peak-Signal-To-noise Ratio (PSNR).

$$PSNR = 10.\log_{10}\frac{255}{MSE}(db), \quad (3)$$

Where MSE is the Mean Square Error between the original image and the watermarked (distorted) image. The mathematical definition for MSE is:

$$MSE = \frac{1}{MxN}\sum_{i=1}^{M}\sum_{j=1}^{N}(a_{ij} - b_{ij})^2, \quad (4)$$

where a_{ij} and b_{ij} are respectively the gray values of the original and the watermarked images at position (i,j).
(M, N) specify the widths and the heights of the tested image.
The PSNR of the watermarked image (Fig .3 (b)) is 39.7 dB.

a: original image.   **b : watermaked**

Fig .3. (a) Original image, (b) watermarked image

With parameters α = 20 for the LL band and 3 for LH, HL, HH .the PSNR is 39.7 dB.

Fig .4. (b) Illustrates the absolute value of difference between the original image and watermarked one magnified by a factor 20.

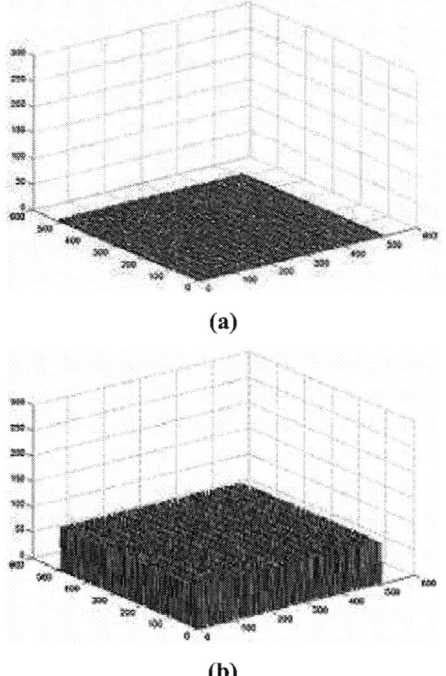

Fig. 4. (a) Absolute value of difference between the original image and watermarked one, (b) absolute value of difference magnified by a factor 20.

3.1 Robustness of the algorithm

In order to demonstrate the influence of different attacks on the extraction of the watermark, The experiments are performed on the single gray scale image Lena, knowing that similar results were obtained on other images with the same size. The performances of the used algorithm are studied in the cases of attacks based on signal processing (filtering, adding noise, gamma correction, intensity adjustment, etc). In this paper, we are more particularly interested, in JPEG compression. Watermarked images are subject to different image processing operations for their robustness test. Fig. 5 shows the result of the algorithm, on the same image ''Lena'', under different alterations including JPEG compression 50%, Gamma correction and Intensity adjustment. The table below summarizes the results of the watermarked image *Lena* under a variety of attacks.

Our algorithm of watermarking offers a good performance according to robustness against certain numbers of attacks. The watermarking remains very resistant against the modifications of luminance and contrast, JPEG compression and adjustments. The method embeds 128x128 bits in each bands of decomposition at the second level of Discrete Wavelet Transform (DWT) decomposition and satisfies the properties of transparency and robustness. It was developed by extending the principles of [7][20][21]. By comparison with the approach of [3][4][8][11][12][15][16][21], the quantity of information data embedded is more significant and algorithm remains robust.

The performances of our algorithm are representing by a graph, comparing the extracted watermark with the desired watermark. We use for this the hamming distance.

Fig. 6. shows the detection response of the real watermarks, and 1000 randomly generated watermarks, the correlation with the real watermark is located at 200 for LL band and in positions 400,600,800 respectively for the other three bands (LH, HL, and HH), it is perfectly detected, and the Hamming distance is zero.

Fig. 7. shows the behaviour of the detector after the three types of attacks JPEG compression of 50% on quality, filtering and zoom of 125%. The Detection is excellent and the watermark is recognised that for LL bands. So the LH, HL and HH, are not robust against these three types of attacks.

Similarly for the three other bands (LH, HL, HH), witch are robust that against certain numbers of attacks (Fig. 8.).

Fig. 9. presents the change in the value of the detector response depending on the type of attack for the four frequency bands. The X-coordinates represent the various attacks according to the order which we gave in Table 2.

The results obtained, under different JPEG compression ratio, are shown in Fig. 10. The fig. 10. (a) represents the change in PSNR between original image and the attacked one depending on the compression ratio. Fig 10 (b) illustrates the variation of detector response according to the compression ratio. If the marked image undergoes a visible degradation after a compression JPEG, it is always possible to extract a good watermark.

Table 2. PSNR and detector response of Watermarked "Lena", under Attacks.

Attacks	PSNR	Hamming distance			
		LL	LH	HL	HH
JPEG 75%	37.41	0.0	0.396	0.360	0.486
JPEG 50%	33.58	0.006	0.349	0.350	0.379
JPEG 25%	32.39	0.138	0.460	0.442	0.466
Filtering	33.84	0.055	0.197	0.415	0.355
Gaussian noise	29.32	0.118	0.423	0.453	0.422
Histogram-Equalization	36.90	0.444	0.138	0.187	0.373
intensity adjustment	25.44	0.494	0.077	0.122	0.056
gamma correction	24.27	0.505	0.100	0.130	0.081
rotate 20%	24.85	0.085	0.243	0.249	0.255
cropping 50%	25.28	0.373	0.371	0.373	0.104
cropping 25%	27.45	0.218	0.462	0.215	0.218
Zoom 90%	36.31	0.139	0.447	0.438	0.435
Zoom 75%	37.24	0.163	0.564	0.564	0.472
Zoom 125%	24.71	0.069	0.462	0.453	0.435

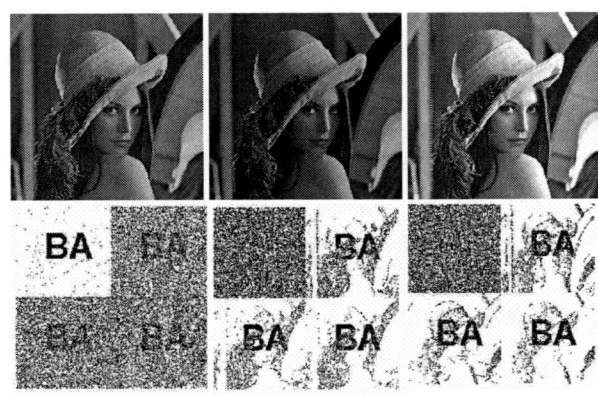

JPEG 50   Gamma correction   Intensity-adj

Fig. 5. Images "Lena" (watermarked and attacked) with the watermarks extracted from the four bands.

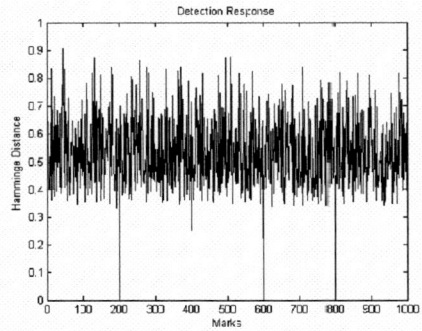

Fig. 6. The detection response.

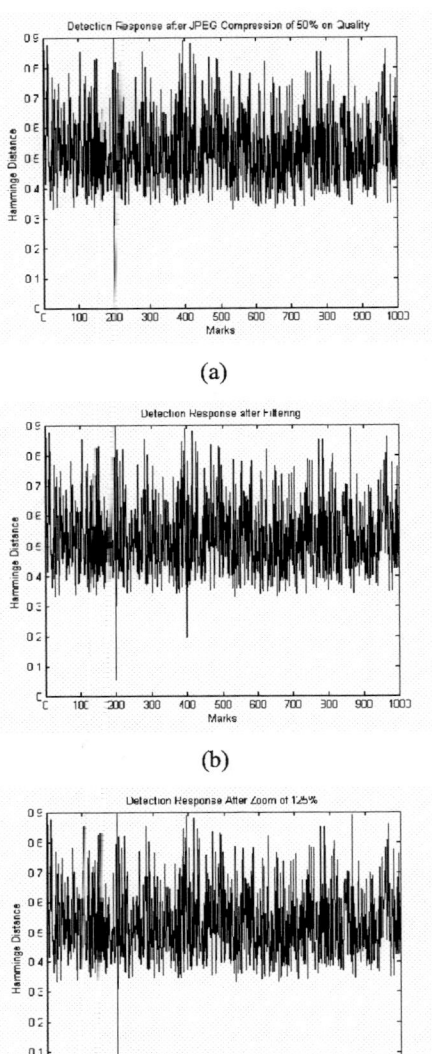

Fig. 7. Detection response after attacks, (a) JPEG compression 50%, (b) filtering, (c) zoom 125%.

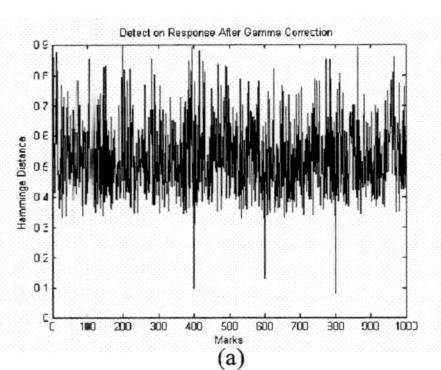

(a)

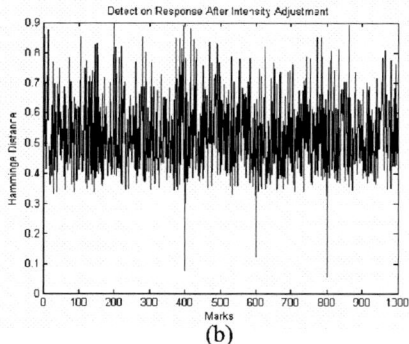

(b)

Fig. 8. Detection response after attacks. (a) Gamma correction, (b) Intensity adjustment.

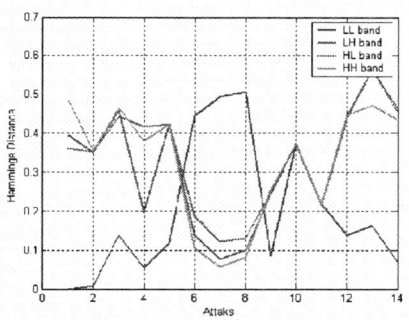

Fig. 9. Detector response depending on the type of attack for the four frequency bands.

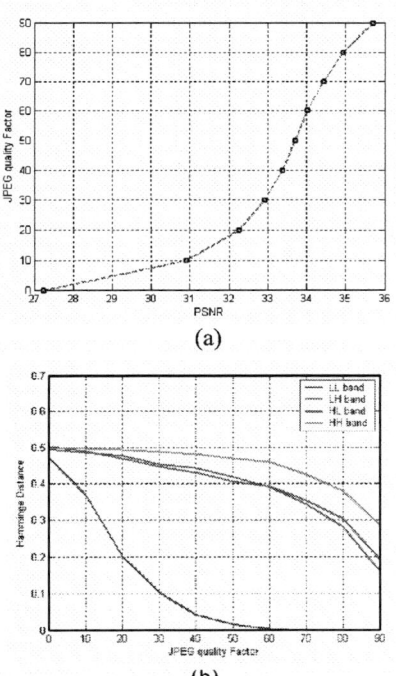

Fig. 10. Robustness under JPEG compression, (a) PSNR between original image and that attacked depending on the compression ratio, (b) variation of detector response according to the compression ratio.

The proposed method is compared with other adaptive and non-blind existing frequency-domain techniques:
- Drira et al [8] used a pyramid of filters for embedding information into significant information of images such as the coefficients of contour.
- Temi et al [22] proposed a method of embedding binary visualized image into the original image by modifying coefficients of wavelet domain in LL bands with appropriate strength factor.

Tab.3 enumerates the different values of PSNR and length of watermark for the different methods.

Table 3. Length of watermark and PSNR obtained by the different tested methods.

	Drira's method	Temi's method	Proposed method
PSNR (dB)	44.25	38	39.7
Length of watermark (bits)	32x32	119	128x128x4

The comparison depends on the correlation computation. A correlation detector is used to calculate the similarity between the original watermark sequence and the extracted one. It is defined by:

$$Corr(X, X') = \frac{X \cdot X'}{\sqrt{X' \cdot X'}} \quad (5)$$

Where X is the original watermark and X' is the extracted watermark.

Fig. 11 represents the variation of correlation obtained by the different tested methods, according to the following iterations: Filtering, Gaussian noise, JPEG 50%, Rotate 20%. Finally, the experimental results presented above, prove the robustness and the imperceptibility of our method of watermarked in the majority of the attacks which can undergo the watermarked image.

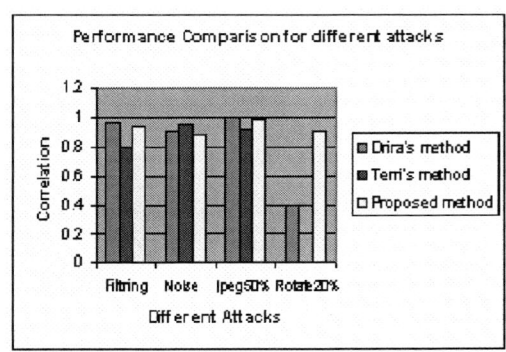

Fig. 11. Performance comparison for different methods.

4. CONCLUSIONS

In this paper an image watermarking (adaptive and not blind) method, based on the coefficients modification of the wavelet transform, is proposed. This approach, robust against several signal processing's, allows the insertion of very large information (16384 bits) and satisfies the properties of transparency and robustness. It was developed by extending the principles of [20] [7] [21]. The mark is introduced in the four bands of DWT to ensure the robustness of the algorithm against several numbers of attacks. Our algorithm has proven its robustness deal with various kinds of attacks, such as JPEG compression and geometric local attacks.

REFERENCES

[1] Chaur-Chin Chen. (2002). Watermarking Experiments Based On Wavelet Transforms. *Electronic Imaging and Multimedia Technology III.* SPIE Vol. 4925, p. 60-68.

[2] Chen, P.Y., H-J. Lin. (2006). A DWT Based Approach for Image Steganography. *International Journal of Applied Science and Engineering,* Vol 4 P 275-290

[3] Chen, V., S. Ruan. (2005). Tatouage d'images robuste appliqué au contrôle de visages photographiques. *CResTIC – LAM. groupe Image IUT de Troyes.* France.

[4] Chouchane, S., W. Puech. (2005). Intégration d'un Nouveau Marqueur dans le Codeur d'Images EZW basé sur les Ondelettes. *CORESA : 10 ème Conférence Nationale de Compression et Représentation des Signaux Audiovisuels.*

[5] Corina, N. (2004). Filigranage dans le domaine des ondelettes Watermarking in the wavelet domain. *Mémoire de diplôme pour obtenir le degré de M.Sc.* L'université "Politehnica" Timisoara Faculté d'Electronique et Télécommunications.

[6] Csurka, G., J.L. Dugelay, C. Mallauran, J.-P. Nguyen and C. Rey. (2001). Attaque malveillante d'images tatouées basée sur l' autosimilarité, In *7èmes Journées d'Études et d'Échanges ``Compression et Représentation des Signaux Audiovisuels" (CORESA)*, Dijon, France.

[7] Dietze, M., S. Jassim. (2002). The choice of filter banks for wavelet-based robust digital watermarking, in *Proc. Multimedia and Security Workshop at ACM Multimedia,* pp. 37-41, French Riviera, France.

[8] Drira, F., F. Denis, A. Baskurt. (2004). Tatouage d'images par techniques multidirectionnelles et multirésolution. *9èmes journée d'études et d'échanges "COmpression et REprésentation des Signaux Audiovisuels -CORESA.*

[9] Elbasi, E., M. Ahmet. (2006). PRN based watermarking scheme for color images. *Istanbul commerce university journal of sciencem.* Vol 2. p 119-131.

[10] Elbasi, E., A. M. Eskicioglu. (2006). A DWT-based Robust Semi-blind Image Watermarking Algorithm Using Two Bands. *IS&T/SPIE's 18th Symposium on Electronic Imaging, Security, Steganography,and Watermarking of Multimedia Contents VIII Conference, San Jose,* CA. Volume 6072, pp. 777-787

[11] Guelvouit, G.L., S.Pateux et J. Delhumeau. (2003). Construction de codes pour tatouage avec prise en compte de l'information adjacente. *in GRETSI Symp. on Image and Signal Processing,* Paris, France.

[12] Guillemot, L. ,J.M, Moureaux. (2003). Tatouage d'images : une nouvelle approche basée sur une méthode de compression. *Proc. CORESA,* Lyon, France.

[13] Hisashi, I., A. Miyazaki, T. Katsura. (1999).An Image Watermarking Method Based on the Wavelet Transform. *Image Processing, ICIP 99. Proceedings. 1999 International Conference on.* Volume 1. p. 296-300.

[14] Kallel, M., J.C Lapayre, M.S Bouhlel. (2007). A multiple Watermarking Scheme for Medical Image in the Spatial Domain. *GVIP Journal,* Volume 7, Issue 1.

[15] Joumaa, H., F. Davoine. (2003).Tatouage substitutif d'images intégrant un masque de pondération visuelle. *In CORESA,* Lyon, France.

[16] Manoury, A. (2001). Tatouage d'images numériques par paquets d'ondelettes. *Thèse De Doctorat, Ecole Centrale de Nantes,* Université de Nantes. France.

[17] Mohamed I. M, Moawad I. M. Dessouky, Salah D, and Fatma H. E. (2007). Comparison between Haar and Daubechies Wavelet Transformions on FPGA Technology. *Proceedings Of World Academy Of Science, Engineering And Technology* – Volume 20 – ISSN 1307-6884.

[18] Rey, C., J-Luc Dugelay. (2001). An overview of watermarking algorithms for image authentication. *Un panorama des méthodes de tatouage permettant d'assurer un service d'intégrité pour les images, Traitement du Signal* – Volume 18 – n° 4 – Spécial.

[19] Santi P. Maity, Malay K. Kundu. (2004). A Blind Cdma Image Watermarking Scheme In Wavelet Domain. *Image Processing,. ICIP 04. International Conference on.* Vol 4. p. 2633- 2636.

[20] Taoa, P. A.M. Eskicioglub. (2004). A robust multiple watermarking scheme in the Discrete Wavelet Transform domain. *International Society for Optical Engineering.* Volume 5601, pp. 133-144. USA.

[21] Terzija, N., M.Repges, K. Luck, W. (2002). Geisselhardt. Digital image watermarking using discrete wavelet transform: Performance comparison of error correction codes. *Visualization, Imaging, and Image Processing.* J.J. Villanueva.

[22] Temi, C., S. Choomchuay, A. Lasakul. (2005). A Robust Image Watermarking Using Multiresolution Analysis of Wavelet. *Proceedings of ISCIT,IEEE.*

PLENARY SESSION PRESENTATIONS
(Abstracts)

Plenary n°:	1
Date:	June 30th 2008
Time:	09h00 - 10h00
Location:	Room 1

1st Mediterranean Conference on Intelligent Systems and Automation (CISA'08)

Title: Higher order sliding modes Observation, identification, and fault detection

Full name: Leonid Fridman

Position: Full Professor, University of Mexico

Web site: http://verona.fi-p.unam.mx/~lfridman/

Chair:

Abstract:

The introduction in higher order sliding modes and higher order sliding mode differentiation is given. Algorithms for observation, identification and fault detection of linear time-invariant strongly observable systems with unknown inputs are developed, based on high order sliding mode differentiation. The possibilities of their extension to strongly detectable linear systems and nonlinear systems are discussed.

The value of the equivalent output injection is used to identify perturbations directly. Continuous time versions of least square methods are proposed to identify unknown time variant and time-invariant parameters.

Some applications of the proposed algorithms to of vehicle estimation parameters, backlash identification are presented actuator and sensors fault detection are discussed.

Biography:

Professor Leonid M. Fridman received his M.S in mathematics from Kuibyshev (Samara) State University, Russia, Ph.D in Applied Mathematics from Institute of Control Science (Moscow), and Dr. of Science degrees in Control Science from Moscow State University of Mathematics and Electronics in 1976,1988 and 1998 correspondingly. In 1976-1999 Dr. Fridman was with the Department of Mathematics at the Samara State Architecture and Civil Engineering Academy, Samara, Russia., 2000 -2002 he served as a Professor of Department of Postgraduate Study and Investigations at the Chihuahua Institute of Technology, Chihuahua, Mexico. In 2002 he joined the Department of Control, Division of Electrical Engineering of Engineering Faculty at National of Autonomous University of Mexico, Mexico.

He is Associate Editor of Conference Editorial Board of IEEE Control Systems Society, Member of TC on Variable Structure Systems and Sliding mode control of IEEE Control Systems Society.

He is an editor of three books and five special issues on sliding mode control and an author of over 170 technical papers.

His research interests include variable structure systems, observation, identification and fault detection.

Plenary n° :
Date :
Time :
Location :

1st Mediterranean Conference on Intelligent Systems and Automation (CISA'08)

Title : Integrated design of mechatronic systems
Full name : Geneviève Dauphin-Tanguy
Position : Professor
Web site : http://lagis.ec-lille.fr

Chair :

Abstract :

The design of controlled systems involving several physical domains and several energy types requires the use of a unified approach allowing specialists of different technical areas to communicate and exchange models and information.

Through different application examples, it will be shown how the bond graph methodology provides the user with a tool with physical insight for modelling, model analysis and simplification, sizing and control designing.

Biography :

Geneviève Dauphin-Tanguy (senior member IEEE), graduated from University of Lille, France (BSc Physics, BSc Mathematics, MSc Mechanics) and from the french Grande Ecole d'Ingénieurs, Ecole Centrale de Lille (State Engineering Diploma), received the Ph.D. and the Doctorate of Sciences both from University of Lille.

She is presently Professor of Control Design at Ecole Centrale de Lille and Head of the research group "Bond graph models" in the Laboratoire d'Automatique Génie Informatique et Signal (LAGIS associated with the CNRS, french National Center for Scientific Research). She is expert for the French Ministry of Education and Research. She is mainly engaged in the research fields of Modelling, Analysis, Control and Monitoring of systems by using the bond-graph tool. Her application domains are mechatronics in car industry and aeronautics, thermofluid processes (fuel cells) and power systems.

Plenary n° :	3
Date :	July 1st 2008
Time :	11h45 - 12h45
Location :	Room 1

1st Mediterranean Conference on Intelligent Systems and Automation (CISA'08)

Title :	Recent results in visual servoing
Full name :	François Chaumette
Position :	Directeur de recherches INRIA
Web site :	http://www.irisa.fr/lagadic

Chair :

Abstract :

Visual servoing techniques consist in using the data provided by a vision sensor in order to control the motions of a dynamic system. Such systems are usually robot arms, mobile robots, aerial robots,... but can also be virtual robots for applications in computer animation, or even a virtual camera for applications in computer vision and augmented reality.

A large variety of positioning tasks, or mobile target tracking, can be implemented by controlling from one to all the degrees of freedom of the system. Whatever the sensor configuration, which can vary from one on-board camera on the robot end-effector to several free-standing cameras, a set of visual features has to be selected at best from the image measurements available, allowing to control the degrees of freedom desired. A control law has also to be designed so that these visual features reach a desired value, defining a correct realization of the task. .

With a vision sensor providing 2D measurements, potential visual features are numerous, since as well 2D data (coordinates of feature points in the image, moments, ...) as 3D data provided by a localization algorithm exploiting the extracted 2D measurements can be considered. It is also possible to combine 2D and 3D visual features to take the advantages of each approach while avoiding their respective drawbacks. From the selected visual features, the behavior of the system will have particular properties as for stability, robustness with respect to noise or to calibration errors, robot 3D trajectory, etc.

The talk will present the main basic aspects of visual servoing, as well as technical advances obtained recently in the field inside the Lagadic group at INRIA/INRISA Rennes. Several application results will be also described.

Biography :

François Chaumette was graduated from Ecole Nationale Supérieure de Mécanique, Nantes, France, in 1987. He received the Ph.D. degree in computer science from the University of Rennes, France, in 1990. Since 1990, he has been with IRISA in Rennes where he is now "Directeur de recherches" INRIA and head of the Lagadic group (http://www.irisa.fr/lagadic).

His research interests include robotics and computer vision, especially visual servoing and active perception.

Dr. Chaumette co-authored more than 150 papers on the topics of robotics and computer vision, and has served in the last five years on the program committees for the main conferences related to robotics and computer vision. He received the AFCET/CNRS Prize for the best French thesis in automatic control in 1991. He also received with Ezio Malis the 2002 King-Sun Fu Memorial Best IEEE Transactions on Robotics and Automation Paper Award. He has been Associate Editor of the IEEE Transactions on Robotics from 2001 to 2005.

Plenary n°:	4
Date:	July 1st 2008
Time:	17h00 - 18h00
Location:	Room 1

1st Mediterranean Conference on Intelligent Systems and Automation (CISA'08)

Title:	Design of Supervision Systems Theory and Practice
Full name:	Belkacem OULD BOUAMAMA
Position:	Professor Ecole Polytechnique de Lille
Web site:	http://sfsd.polytech-lille.net/BelkacemOuldBouamama

Chair:

Abstract:

The term "supervision" means a set of tools and methods used to operate an industrial process in normal situation as well as in the presence of failures or undesired disturbances. The activities concerned with the supervision are the Fault Detection and Isolation (FDI) in the diagnosis level, and the Fault Tolerant Control (FTC) through necessary reconfiguration, whenever possible, in the fault accommodation level. The final goal of a supervision platform is to provide the operator a set of tools that helps to safely run the process and to take appropriate decision in the presence of faults. Different approaches to the design of such decision making tools have been developed in the past twenty years, depending on the kind of knowledge (structural, statistical, fuzzy, expert rules, functional, behavioural ...) used to describe the plant operation.

How to elaborate models for FDI design, how to develop the FDI algorithm, how to avoid false alarms, how to improve the diagnosability of the faults for alarm management design, how to continue to control the process in failure mode, what are the limits of each method,...?. Such are the main purposes concerned by the presented plenary session from an industrial and theoretical point of view.

Biography:

Belkacem OULD BOUAMAMA is full Professor and head of the research at " Ecole Polytechnique Universitaire de Lille, France) " His main research areas, developed at the "Laboratoire d'Automatique Génie Informatique et Signal de Lille " (associated with the CNRS, french Natioanl Center For Scientific Research), concern Integrated Design for Supervision of System Engineering. Their application domains are mainly nuclear, petrochemical, and mechatronic systems. He is the author of several international publications in this domain. He is co-author of three books in bond graph modeling and Fault Detection and Isolation area.

Research and teaching activities can be consulted at: http://sfsd.polytech-lille.net/BelkacemOuldBouamama

Plenary n° :
Date :
Time :
Location :

1st Mediterranean Conference on Intelligent Systems and Automation (CISA'08)

Title :	Applications in Robotics and Controls
Full name :	Kamal Youcef-Toumi
Position :	Professor
Web site :	http://meche.mit.edu/people/faculty/

Chair :

Abstract :

Recent industry trends have set new standards in business dealings and trades. Issues such as time to market, shoter market wondows, product performance, rapid increase of product complexity, costly mistakes, costly late introductions, and customer expectations have changed significantly. These trends have also influenced to a great extend the academic world. Some of these trends will be illustrated through examples which include automated systems, robotics, biotechnollogy, and nanotechnology. The examples will include concepts and prototypes of engineering systems in the macro, micro and nanodomains. The presentation also amphasizes the merging of the traditionally segregated disciplines into one multidisciplinary modeling, design, optimization and control approach.

Biography :

Kamal Youcef-Toumi joined the MIT Mechanical Engineering Department faculty in January 1986 as an Assistant Professor. Prior to his faculty appointment, Professor Youcef-Toumi served as a Research Associate and Lecturer with this Department. He earned his advanced degrees (M.S. 1981 and Sc.D. 1985) in Mechanical Engineering from MIT. His undergraduate degree (B.S. in Mechanical Engineering awarded in 1979) is from the University of Cincinnati.

Professor Youcef-Toumi's research and teaching have focused primarily on design, control theory and its applications to dynamic systems. His research has focused on controller design for systems with unknown dynamics, and in particular, the development of control techniques with fast adaptation, modeling and simulation of dynamic systems. The applications have included manufacturing, robotics, automation, metrology and nano/biotechnology. He teaches courses in the fields of dynamic systems and controls; robotics; and precision machine design and controls.

He has served on several professional committees of The National Science Foundation. Other services include Chairman of the Information Technology program within The Arab Science and Technology Foundation, and Member of the review committee for European Union funded Network of Excellence for Innovative Production Machines and Systems.

Professor Youcef-Toumi is the author of over 100 publications, including a textbook on the theory and practice of direct-drive robots. He is currently completing an additional textbook on Modeling and Controls. Professor Youcef-Toumi has been an invited lecturer at over 80 seminars at companies and universities throughout the world.

1st Mediterranean Conference on Intelligent Systems and Automation (CISA'08)

Plenary n°:	
Date:	
Time:	
Location:	

Title: Some challenging issues in humanoid robotics
Full name: Abderrahmane KHEDDAR
Position: Professor at University of Evry Val d'Essonne
Web site:

Chair:

Abstract:

After a quick review of advanced achievements from the hardware point-of-view and a short overview of trends and possible applications of humanoids, we will address specific problems that we are dealing with at the AIST/CNRS Joint Japanese-French Robotics Laboratory (JRL). We believe autonomy is a common challenge in indoor and outdoor humanoids and several efforts are dedicated to fill the gap between the hardware capabilities and actual achievements (the latter are yet very limited). We are developing efforts in planning of contact support stances that can occur between any part of the humanoid's bodies and any part of the environment; this is to build a richer behavior of acyclic motion. At a lower level, we are also dealing with task description, complex motion optimization and advanced tools for simulation with interactive and haptic feedback. The talk will end by emphasizing on the multidisciplinary link of humanoid research to other fields.

Biography:

Abderrahmane KHEDDAR (*1967) is currently a tenure professor at the university of Evry, France, and was the head of the virtual reality and haptics group since its cration to 2007. He received a DEA (Master of Science by research) and the Ph.D. degree in robotics and computer science, both from the University Paris 6, France; an engineering degree from the Institut National d'Informatique, Algiers, Algeria. He was several months a visiting researcher at the formal Mechanical Engineering Laboratory (Bio-Robotics Division) in Japan, working with Professor Kazuo Tanie. Since 2003, in the frame of his CNRS secondment he took a Director of Research position and is the Codirector of the AIST/CNRS Joint Japanese-French Robotics Laboratory (JRL) in Tsukuba, Japan. His research interests include chronologically teleoperation and telerobotics, haptic interaction with virtual and machined avatars (sensing and display), humanoids (contact planning, haptic, dynamic control), and electro-active polymers for haptic displays. He was a member of the Teleoperation Research Group under the French Nuclear Commissariat (CEA) and the French National Scientific Research Center (CNRS) auspices. He was the coordinator of the CNRS specific action on collision detection and is the main National coordinator of CNRS specific programme on haptics. He Co-chaired with M. Buss the Touch-HapSys workshop on "Touch and Haptics" at IEEE/RSJ IROS'04 in Japan. He was the general chair of EuroHaptics 2006 for its first edition in France. He is the secretary or the EuroHaptics Society settled in Paris in 2006 and member of the EuroHaptics Conference steering committee, he served as referee in many robotics journal and conferences. He is a founding member of the IEEE-RAS chapter on Haptics and a founding member of the IEEE Transactions on Haptics where he serve as an Associate Editor since its creation (Jan. 2008). He is coordinating the ROBOT@CWE FP6 EC project dedicated to collaborative humanoid robotics, CNRS partner in the INTUITION European network of excellence on virtual reality, the FP5 TOUCH-HapSys and FP6 ImmerSence European projects dedicated to haptics, and the PEACH European coordination for presence research.

1st Mediterranean Conference on Intelligent Systems and Automation (CISA'08)

Plenary n°:	7
Date:	July 2nd 2008
Time:	15h30 - 16h30
Location:	Room 1

Title: Computer Assisted Surgery and Current Trends in Orthopaedics Research and Total Joint Replacements

Full name: Farid Amirouche

Position: Professor at University of Illinois, Chicago

Web site: http://www.mie.uic.edu/faculty/amirouche_scholar.htm

Chair:

Abstract:

Musculoskeletal research has brought about revolutionary changes in our ability to perform high precision surgery in joint replacement procedures. Recent advances in computer assisted surgery as well better materials have lead to reduced wear and greatly enhanced the quality of life of patients. The new surgical techniques to reduce the size of the incision and damage to underlying structures have been the primary advance toward this goal. These new techniques are known as MIS or Minimally Invasive Surgery.

Total hip and knee Arthoplasties are at all time high reaching 1.2 million surgeries per year in the USA. Primary joint failures are usually due to osteoarthristis, rheumatoid arthritis, osteocronis and other inflammatory arthritis conditions. The methods for THR and TKA are critical to initial stability and longevity of the prostheses. This research aims at understanding the fundamental mechanics of the joint Arthoplasty and providing an insight into current challenges in patient specific fitting, fixing, and stability. Both experimental and analytical work will be presented.

We will examine Cementless total hip arthroplasty success in the last 10 years and how computer assisted navigation is playing in the follow up studies. Cementless total hip arthroplasty attains permanent fixation by the ingrowth of bone into a porous coated surface. Loosening of an ingrown total hip arthroplasty occurs as a result of osteolysis of the periprosthetic bone and degradation of the bone prosthetic interface. The osteolytic process occurs as a result of polyethylene wear particles produced by the metal polyethylene articulation of the prosthesis. The total hip arthroplasty is a congruent joint and the submicron wear particles produced are phagocytized by macrophages initiating an inflammatory cascade. This cascade produces cytokines ultimately implicated in osteolysis. Resulting bone loss both on the acetabular and femoral sides eventually leads to component instability. As patients are living longer and total hip arthroplasty is performed in younger patients the risks of osteolysis associated with cumulative wear is increased.

Computer-assisted surgery is based on sensing feedback; vision and imaging that help surgeons align the patient's joints during total knee or hip replacement with a degree of accuracy not possible with the naked eye. For the first time, the computer feedback is essential for ligament balancing and longevity of the implants. The computers navigation systems also help surgeons to use smaller incisions instead of the traditional larger openings. Small-incision surgery offers the potential for faster recovery, less bleeding and less pain for patients. The development of SESCAN imaging technique to create a patient based model of a 3D joint will be presented to show the effective solution of complex geometry of joints.

Biography:

Farid Amirouche received his BS in Engineering Computer Science, MS in Aerospace (applied mechanics), and a PhD in Mechanical Engineering from the university of Cincinnati. He joined the university of Illinois at Chicago in 1984 and he is currently a University of Illinois Scholar professor, professor of Mechanical and Bioengineering and a Professor of Orthopedics in the College of Medicine at UIC. He is the author of 5 books and over 200 technical papers. His current research is in biomechanics and medical devices. He is the author of seven patents and currently serves as a CEO and President of a startup Company with emphasis on smart orthopedic implants. He is a recipient of SAE Ralph Teetor Award, a G7 fellowship, DOE fellowship, university of Illinois research award and was indicted into the Palmes Academic of France in 2006. He is an ASME Fellow and serves on 5 editorial board of engineering Journals. He is the director of Orthopedic research at UIC Department of Orthopedics and Director of Biomechanics Research laboratory in the MIE Department.

Author Index

A

Abbassi, H. A., 191, 234
Abdelaziz, M., 369
Abdelkrim, M. N., 221
Abdelmoumene, H., 414
Abed, K., 120
Abichou, A., 397, 498
Achili, B., 228
Achour, K., 3
Achour, N., 324
Adane, A., 531
Ahmed, M. A., 203
Aissani, N., 269
Aitouche, A., 541
Ait Oufroukh, N., 33
Akdag, H., 503
Akli, I., 324
Alla, H., 281
Ameddah, D., 168
Ameur, S., 531
Amimeur, H., 197
Amirouche, F., 579
Arab, Lh., 203
Arfi, F., 375
Arioui, H., 47, 133
Ayad, M., 260
Azizi, H., 308, 503
Azouz, N., 397
Azzaz, M. S., 3
Azzouzi, M., 139

B

Bakhouche, A., 565
Bayart, B., 563
Bayart, M., 541
Bazoula, A., 319
Bechar, H., 260
Behling, S., 53
Beji, L., 498
Beladgham, M., 249
Belala, N., 375
Belbachir, A. H., 297
Beldjilali, B., 269
Belemhedi, A., 203

Belhocine, M., 553
Belhout, K., 211
Belkheiri, M., 115
Belkhiat, S., 420
Benabadji, N., 297
Benabdelhafid, A., 547
Benalla, H., 120
Benamira, A., 375
Benamrane, N., 466
Benarbia, B., 476
Benbelkacem, S., 553
Bendriss, S., 547
Benmahammed, K., 454, 520
Bennaceur, S., 397
Bennia, A., 254
Bensalah, C., 80
Bensalem, S., 487
Bensalem, Y., 221
Benyahia, B., 74
Benyettou, A., 15
Benzaioua, A., 181
Berger, U., 275
Berrached, N. E., 414, 476
Berrahal, S., 85, 168
Bessaid, A., 260
Bestaoui, Y., 313
Bettayeb, M., 127
Bettou, K., 186
Bloch, I., 9
Bonnet, N., 243
Bouamama, B. O., 576
Bouazza, S. E., 194, 234
Boubakir, A., 470
Boubakir, C., 470
Bouchama, Z., 512
Boudjema, F., 115, 470
Boudoin, P., 149
Boudour, R., 439
Boughazi, M., 216, 243, 503, 516
Bouibed, K., 541
Boukachour, J., 547
Boukharouba, A., 254
Boukrouche, A., 408
Bouktir, Y., 353
Boulebtateche, B., 216, 243
Boumous, Z., 420
Bourahala, F., 109

Bourennane, E., 536
Bourquardez, O., 391
Bousbia-Salah, M., 526
Boussey, J., 531
Boutalbi, M., 536
Boyer, F., 346
Bozga, M., 487
Brahmi, Z., 287
Buhrs, S., 53

C

Chaghi, A., 197
Chaibet, A., 27
Charalambous, C. D., 433
Charef, A., 186
Chatila, R., 162
Chaumette, F., 391, 427, 575
Cherki, B., 74, 80
Chettibi, T., 346, 353
Cheviron, T., 67
Chikouche, D., 228
Choukchou-Braham, A., 74, 80
Chriette, A., 67

D

Daachi, B., 228
Daachi, M. E., 228
Dauphin-Tanguy, G., 574
Debakla, M., 15
Devy, M., 3
Didier, J.-Y., 173, 563
Dir, S., 156
Djafri, B., 173
Djaghloul, M., 420
Djennoune, S., 127
Djeridane, B., 333
Djouadi, S. M., 433
Doghmane, N., 565
Doudou, S., 103
Drid, S., 339

E

Eck, L., 391
Eghnam, K. M., 482
El Ani, I., 281

El-Fatah, A., 93
El Kamel, M. A., 498
El Khateb, A., 99
Essabbah, M., 493

F

Fares, T. S., 93
Farfar, D., 439
Farges, J.-L., 162
Ferdi, Y., 143
Fezari, M., 216, 243, 516
Filali, S., 61
Fridman, L., 573
Friha, S., 450

G

Gallien, M., 487
Gammoudi, M. M., 287
Ghedjati, K., 369
Golea, N., 85
Guenard, N., 391
Guitton, J., 162

H

Habani, N., 203
Habert, O., 156
Hacene, I. B., 249
Haddad, M., 346
Hadri, C., 516
Hamad, H., 460
Hamdi, H., 281
Hamed, B., 99
Hamel, T., 391
Hamissi, A., 319
Hamoudi, F., 197
Harmas, M. N., 512
Hegazy, A., 93
Hima, S., 27, 47
Himour, Y., 381
Hussein, M. T., 447

I

Ignat-Coman, A., 383
Ingrand, F. F., 487
Irki, Z., 3
Ishii, C., 154
Isoc, D., 383

J

Jammazi, C., 302
Joldiş, A., 383

K

Kamei, Y., 154
Kermi, A., 9
Khaber, F., 103, 109, 363
Khalil, A. H., 93
Khalil, W., 346
Kheddar, A., 563, 578
Khélif, M., 249
Khensous, G., 466
Kimour, M. T., 439
Klaudel, H., 173
Kretzschmann, R., 275
Krichen, M., 487

L

Laghrouche, M., 531
Laskri, M. T., 9
Lehtihet, H. E., 346
Lezoray, O., 15
Li, Y., 433

M

Madour, F., 520
Mahony, R., 391
Makouf, A., 339
Mallem, M., 149, 493
Mammar, S., 27, 33, 41
Mansouri, N., 450, 507
Mansouri, R., 127
Marey, M., 427

Meftah, B., 15
Mekhnache, C., 143
Mekki, H., 211
Mellit, A., 211
Merabet, A., 181
Merabet, E., 197
Messabih, B., 466
Messaoudène, K., 33
Messaoudi, K., 536
Meunier, D., 531
Mokaddem, S., 363
Mokhnache, S., 308
Mokhtari, M., 85, 168
Moussaoui, A. K., 191, 234

N

Nabti, K., 120
Nait-Said, M.-S., 339
Nehaoua, L., 133
Nouveliere, L., 27, 41

O

Olama, M. M., 433
Otmane, S., 149, 493
Ouhrouche, M., 181

P

Plestan, F., 67
Pruski, A., 156

R

Rabhi, A., 541
Rahal, W. L., 297
Rechid, N., 454
Redjati, A., 526
Rouabhia, C., 20

S

Saïdouni, D. E., 375
Salem, A., 559

Salhi, H., 211
Saoudi, K., 512
Sbita, L., 221
Schulze, H. L., 53
Sentouh, C., 41
Sheta, A. F., 482
Slimi, H., 41

T

Tadjine, M., 339
Tahanout, M., 531
Taleb-Ahmed, A., 143, 249, 260, 454
Tebbikh, H., 20
Toumi, A., 454
Toumi, S., 536
Tripakis, S., 487

V

Vargas, A. V., 275

Y

Youcef-Toumi, K., 577
Yousfi, L., 507

Z

Zehar, K., 512
Zenati, S., 408
Zenati-Henda, N., 553
Ziani, S., 61
Zouaoui, L., 243, 308, 503